ALGEBRA

Lines

Slope of the line through $P_1 = (x_1, y_1)$ and $P_2 = (x_2, y_2)$:

$$m = \frac{y_2 - y_1}{x_2 - x_1}$$

Slope-intercept equation of line with slope m and y-intercept b:

$$y = mx + b$$

Point-slope equation of line through $P_1 = (x_1, y_1)$ with slope m:

$$y - y_1 = m(x - x_1)$$

Point-point equation of line through $P_1 = (x_1, y_1)$ and $P_2 = (x_2, y_2)$:

$$y - y_1 = m(x - x_1) \quad \text{where } m = \frac{y_2 - y_1}{x_2 - x_1}$$

Lines of slope m_1 and m_2 are parallel if and only if $m_1 = m_2$.
Lines of slope m_1 and m_2 are perpendicular if and only if $m_1 = -\frac{1}{m_2}$.

Circles

Equation of the circle with center (a, b) and radius r:

$$(x - a)^2 + (y - b)^2 = r^2$$

Distance and Midpoint Formulas

Distance between $P_1 = (x_1, y_1)$ and $P_2 = (x_2, y_2)$:

$$d = \sqrt{(x_2 - x_1)^2 + (y_2 - y_1)^2}$$

Midpoint of $\overline{P_1 P_2}$: $\left(\frac{x_1 + x_2}{2}, \frac{y_1 + y_2}{2}\right)$

Laws of Exponents

$x^m x^n = x^{m+n}$ $\qquad \frac{x^m}{x^n} = x^{m-n}$ $\qquad (x^m)^n = x^{mn}$

$x^{-n} = \frac{1}{x^n}$ $\qquad (xy)^n = x^n y^n$ $\qquad \left(\frac{x}{y}\right)^n = \frac{x^n}{y^n}$

$x^{1/n} = \sqrt[n]{x}$ $\qquad \sqrt[n]{xy} = \sqrt[n]{x}\,\sqrt[n]{y}$ $\qquad \sqrt[n]{\frac{x}{y}} = \frac{\sqrt[n]{x}}{\sqrt[n]{y}}$

$x^{m/n} = \sqrt[n]{x^m} = \left(\sqrt[n]{x}\right)^m$

Special Factorizations

$x^2 - y^2 = (x + y)(x - y)$
$x^3 + y^3 = (x + y)(x^2 - xy + y^2)$
$x^3 - y^3 = (x - y)(x^2 + xy + y^2)$

Binomial Theorem

$(x + y)^2 = x^2 + 2xy + y^2$
$(x - y)^2 = x^2 - 2xy + y^2$
$(x + y)^3 = x^3 + 3x^2y + 3xy^2 + y^3$
$(x - y)^3 = x^3 - 3x^2y + 3xy^2 - y^3$

$$(x + y)^n = x^n + nx^{n-1}y + \frac{n(n-1)}{2}x^{n-2}y^2 + \cdots + \binom{n}{k}x^{n-k}y^k + \cdots + nxy^{n-1} + y^n$$

where $\binom{n}{k} = \frac{n(n-1)\cdots(n-k+1)}{1 \cdot 2 \cdot 3 \cdot \cdots \cdot k}$

Quadratic Formula

If $ax^2 + bx + c = 0$, then $x = \frac{-b \pm \sqrt{b^2 - 4ac}}{2a}$.

Inequalities and Absolute Value

If $a < b$ and $b < c$, then $a < c$.
If $a < b$, then $a + c < b + c$.
If $a < b$ and $c > 0$, then $ca < cb$.
If $a < b$ and $c < 0$, then $ca > cb$.
$|x| = x \quad$ if $x \geq 0$
$|x| = -x \quad$ if $x \leq 0$

$-a \quad 0 \quad a$
$|x| < a$ means $-a < x < a$.

$c - a \quad c \quad c + a$
$|x - c| < a$ means $c - a < x < c + a$.

GEOMETRY

Formulas for area A, circumference C, and volume V

Triangle
$A = \frac{1}{2}bh$
$= \frac{1}{2}ab\sin\theta$

Circle
$A = \pi r^2$
$C = 2\pi r$

Sector of Circle
$A = \frac{1}{2}r^2\theta$
$s = r\theta$
(θ in radians)

Sphere
$V = \frac{4}{3}\pi r^3$
$A = 4\pi r^2$

Cylinder
$V = \pi r^2 h$

Cone
$V = \frac{1}{3}\pi r^2 h$
$A = \pi r\sqrt{r^2 + h^2}$

Cone with arbitrary base
$V = \frac{1}{3}Ah$
where A is the area of the base

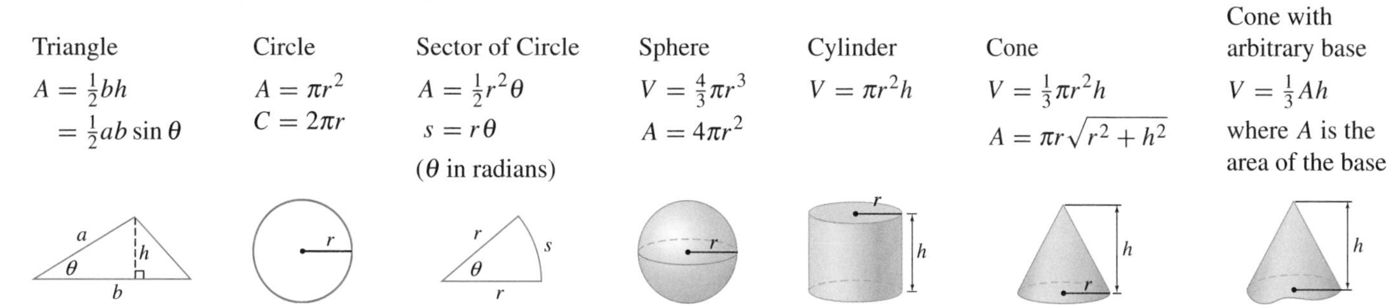

Pythagorean Theorem: For a right triangle with hypotenuse of length c and legs of lengths a and b, $c^2 = a^2 + b^2$.

TRIGONOMETRY

Angle Measurement

π radians $= 180°$

$1° = \dfrac{\pi}{180}$ rad $\qquad 1$ rad $= \dfrac{180°}{\pi}$

$s = r\theta \quad (\theta$ in radians)

Right Triangle Definitions

$\sin\theta = \dfrac{\text{opp}}{\text{hyp}} \qquad \cos\theta = \dfrac{\text{adj}}{\text{hyp}}$

$\tan\theta = \dfrac{\sin\theta}{\cos\theta} = \dfrac{\text{opp}}{\text{adj}} \qquad \cot\theta = \dfrac{\cos\theta}{\sin\theta} = \dfrac{\text{adj}}{\text{opp}}$

$\sec\theta = \dfrac{1}{\cos\theta} = \dfrac{\text{hyp}}{\text{adj}} \qquad \csc\theta = \dfrac{1}{\sin\theta} = \dfrac{\text{hyp}}{\text{opp}}$

Trigonometric Functions

$\sin\theta = \dfrac{y}{r} \qquad \csc\theta = \dfrac{r}{y}$

$\cos\theta = \dfrac{x}{r} \qquad \sec\theta = \dfrac{r}{x}$

$\tan\theta = \dfrac{y}{x} \qquad \cot\theta = \dfrac{x}{y}$

$P = (r\cos\theta, r\sin\theta)$

$\lim\limits_{\theta\to 0} \dfrac{\sin\theta}{\theta} = 1 \qquad \lim\limits_{\theta\to 0} \dfrac{1-\cos\theta}{\theta} = 0$

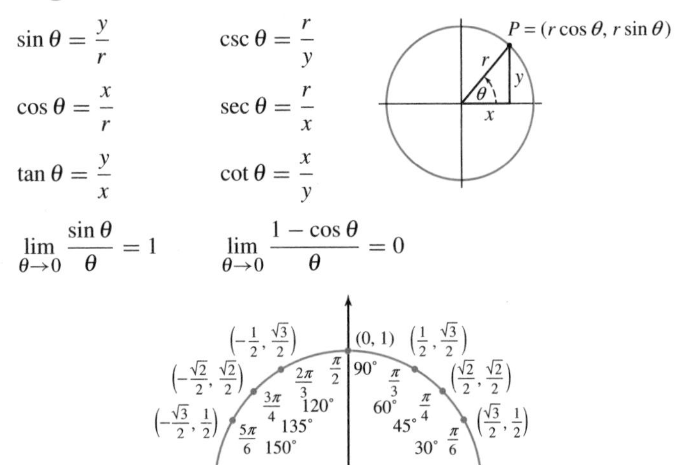

Fundamental Identities

$\sin^2\theta + \cos^2\theta = 1 \qquad \sin(-\theta) = -\sin\theta$

$1 + \tan^2\theta = \sec^2\theta \qquad \cos(-\theta) = \cos\theta$

$1 + \cot^2\theta = \csc^2\theta \qquad \tan(-\theta) = -\tan\theta$

$\sin\left(\dfrac{\pi}{2} - \theta\right) = \cos\theta \qquad \sin(\theta + 2\pi) = \sin\theta$

$\cos\left(\dfrac{\pi}{2} - \theta\right) = \sin\theta \qquad \cos(\theta + 2\pi) = \cos\theta$

$\tan\left(\dfrac{\pi}{2} - \theta\right) = \cot\theta \qquad \tan(\theta + \pi) = \tan\theta$

The Law of Sines

$$\frac{\sin A}{a} = \frac{\sin B}{b} = \frac{\sin C}{c}$$

The Law of Cosines

$$a^2 = b^2 + c^2 - 2bc\cos A$$

Addition and Subtraction Formulas

$\sin(x + y) = \sin x\cos y + \cos x\sin y$

$\sin(x - y) = \sin x\cos y - \cos x\sin y$

$\cos(x + y) = \cos x\cos y - \sin x\sin y$

$\cos(x - y) = \cos x\cos y + \sin x\sin y$

$\tan(x + y) = \dfrac{\tan x + \tan y}{1 - \tan x\tan y}$

$\tan(x - y) = \dfrac{\tan x - \tan y}{1 + \tan x\tan y}$

Double-Angle Formulas

$\sin 2x = 2\sin x\cos x$

$\cos 2x = \cos^2 x - \sin^2 x = 2\cos^2 x - 1 = 1 - 2\sin^2 x$

$\tan 2x = \dfrac{2\tan x}{1 - \tan^2 x}$

$\sin^2 x = \dfrac{1 - \cos 2x}{2} \qquad \cos^2 x = \dfrac{1 + \cos 2x}{2}$

Graphs of Trigonometric Functions

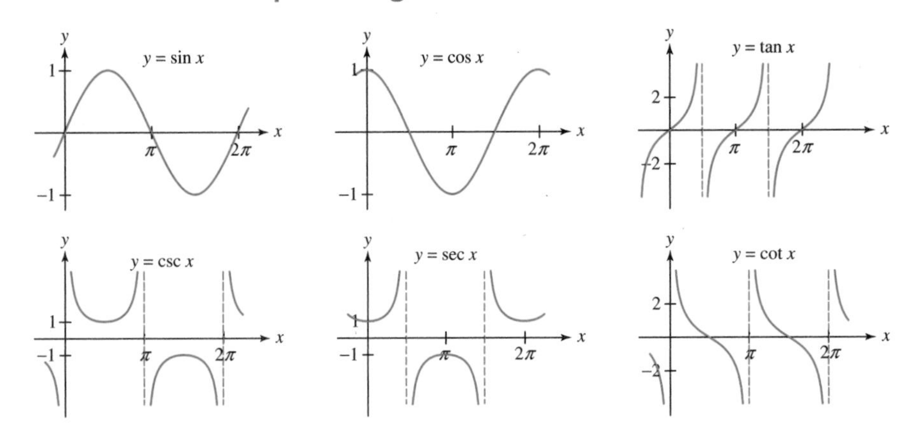

MULTIVARIABLE CALCULUS

Early Transcendentals

MULTIVARIABLE CALCULUS

Early Transcendentals

JON ROGAWSKI

University of California, Los Angeles

W. H. Freeman and Company
New York

Publisher: *Craig Bleyer*
Executive Editor: *Ruth Baruth*
Senior Acquisitions Editor: *Terri Ward*
Development Editor: *Tony Palermino*
Development Editor: *Bruce Kaplan*
Associate Editor: *Brendan Cady*
Market Development: *Steve Rigolosi*
Senior Media Editor: *Roland Cheyney*
Assistant Editor: *Laura Capuano*
Photo Editor: *Ted Szczepanski*
Photo Researcher: *Julie Tesser*
Design Manager: *Blake Logan*
Project Editor: *Vivien Weiss*
Illustrations: *Network Graphics*
Illustration Coordinator: *Susan Timmins*
Production Coordinator: *Paul W. Rohloff*
Composition: *Integre Technical Publishing Co.*
Printing and Binding: *RR Donnelley*

Library of Congress Control Number 2006936453
ISBN-13: 978-0-7167-6070-2
ISBN-10: 0-7167-6070-3

Printed in the United States of America
First printing

W. H. Freeman and Company, 41 Madison Avenue, New York, NY 10010
Houndmills, Basingstoke RG21 6XS, England
www.whfreeman.com

To Julie

CONTENTS | MULTIVARIABLE CALCULUS

Early Transcendentals

ABOUT JON ROGAWSKI

As a successful teacher for more than 25 years, Jon Rogawski has listened to and learned much from his own students. These valuable lessons have made an impact on his thinking, his writing, and his shaping of a calculus text.

Jon Rogawski received his undergraduate degree and simultaneously a master's degree in mathematics from Yale University and a Ph.D. in mathematics from Princeton University, where he studied under Robert Langlands. Before joining the Department of Mathematics at UCLA in 1986, where he is currently a full professor, he held teaching positions at Yale University, University of Chicago, the Hebrew University in Jerusalem, and visiting positions at the Institute for Advanced Study, the University of Bonn, and the University of Paris at Jussieu and at Orsay.

Jon's areas of interest are number theory, automorphic forms, and harmonic analysis on semisimple groups. He has published numerous research articles in leading mathematical journals, including a research monograph titled *Automorphic Representations of Unitary Groups in Three Variables* (Princeton University Press). He is the recipient of a Sloan Fellowship and an editor of the *Pacific Journal of Mathematics*.

Jon and his wife, Julie, a physician in family practice, have four children. They run a busy household and, whenever possible, enjoy family vacations in the mountains of California. Jon is a passionate classical music lover and plays the violin and classical guitar.

PREFACE

ABOUT *CALCULUS* by Jon Rogawski

On Teaching Mathematics

As a young instructor, I enjoyed teaching but didn't appreciate how difficult it is to communicate mathematics effectively. Early in my teaching career, I was confronted with a student rebellion when my efforts to explain epsilon-delta proofs were not greeted with the enthusiasm I anticipated. Experiences of this type taught me two basic principles:

1. We should try to teach students as much as possible, but not more so.
2. As math teachers, how we say it is as important as what we say.

When a concept is wrapped in dry mathematical formalism, the majority of students cannot assimilate it. The formal language of mathematics is intimidating to the uninitiated. By presenting the same concept in everyday language, which may be more casual but no less precise, we open the way for students to understand the underlying idea and integrate it into their way of thinking. Students are then in a better position to appreciate the need for formal definitions and proofs and to grasp their logic.

My most valuable resource in writing this text has been my classroom experience of the past 25 years. I have learned how to teach from my students, and I hope my text reflects a true sensitivity to student needs and capabilities. I also hope it helps students experience the joy of understanding and mastering the beautiful ideas of the subject.

On Writing a New Calculus Text

Calculus has a deservedly central role in higher education. It is not only the key to the full range of scientific and engineering disciplines; it is also a crucial component in a student's intellectual development. I hope my text will help to open up this multifaceted world of ideas for the student. Like a great symphony, the ideas in calculus never grow stale. There is always something new to appreciate and delight in, and I enjoy the challenge of finding the best way to communicate these ideas to students.

My text builds on the tradition of several generations of calculus authors. There is no perfect text, given the choices and inevitable compromises that must be made, but many of the existing textbooks reflect years of careful thought, meticulous craft, and accumulated wisdom, all of which cannot and should not be ignored in the writing of a new text.

I have made a sustained effort to communicate the underlying concepts, ideas, and "reasons why things work" in language that is accessible to students. Throughout the text, student difficulties are anticipated and addressed. Problem-solving skills are systematically developed in the examples and problem sets. I have also included a wide range of applications, both innovative and traditional, which provide additional insight into the mathematics and communicate the important message that calculus plays a vital role in the modern world.

My textbook follows a largely traditional organization, with a few exceptions. One such exception is the placement of Taylor polynomials in Chapter 8.

Placement of Taylor Polynomials

Taylor polynomials appear in Chapter 8, before infinite series in Chapter 10. (For this multivariable volume, see Appendix E.) My goal is to present Taylor polynomials as a natural extension of the linear approximation. When I teach infinite series, the primary focus is on convergence, a topic that many students find challenging. By the time we have covered the basic convergence tests and studied the convergence of power series, students are ready to tackle the issues involved in representing a function by its Taylor series. They can then rely on their previous work with Taylor polynomials and the error bound from Chapter 8. However, the section on Taylor polynomials is written so that you can cover this topic together with the materials on infinite series if this order is preferred. For this reason, the section on Taylor polynomials is included in this multivariable volume as Appendix E.

Careful, Precise Development

W. H. Freeman is committed to high quality and precise textbooks and supplements. From this project's inception and throughout its development and production, quality and precision have been given significant priority. We have in place unparalleled procedures to ensure the accuracy of the text.

These are the steps we took to ensure an accurate first edition for you:

- **Exercises and Examples** Rather than waiting until the book was finished before checking it for accuracy (which is often the practice), we have painstakingly checked all the examples, exercises, and their solutions for accuracy in every draft of the manuscript and each phase of production.
- **Exposition** A team of 12 calculus instructors acted as accuracy reviewers, and made four passes through all exposition, confirming the accuracy and precision of the final manuscript.
- **Figures** Tom Banchoff of Brown University verified the appropriateness and accuracy of all figures throughout the production process.
- **Editing** The author worked with an editor with an advanced degree in mathematics to review each line of text, exercise, and figure.
- **Composition** The compositor used the author's original LaTeX files to prevent the introduction of new errors in the production process.
- **Math Clubs** We engaged math clubs at twenty universities to accuracy check all the exercises and solutions.

Together, these procedures far exceed prior industry standards to safeguard the quality and precision of a calculus textbook.

SUPPLEMENTS

For Instructors

- Instructor Solutions Manual
  Brian Bradie, Christopher Newport University, and Greg Dresden, Washington and Lee University
  Single Variable: 0-7167-9591-4
  Multivariable: 0-7167-9592-2
  Contains worked-out solutions to all exercises in the text.

- Printed Test Bank
  Calculus: 0-7167-9598-1
  Includes multiple-choice and short-answer test items.

- Test Bank CD-ROM
  Calculus: 0-7167-9895-6
  Available online or on a CD-ROM.

- Instructor's Resource Manual
  Vivien Miller, Mississippi State University; Len Miller, Mississippi State University; and Ted Dobson, Mississippi State University
  Calculus: 0-7167-9589-2
  Offers instructors support material for each section of every chapter. Each section includes suggested class time and content emphasis, selected key points, lecture material, discussion topics and class activities, suggested problems, worksheets, and group projects.

- Instructor's Resource CD-ROM
  Calculus: 1-429-20043-X
  Your one-stop resource. Search and export all resources by key term or chapter. Includes text images, Instructor's Solutions Manual, Instructor's Resource Manual, Lecture PowerPoint slides, Test Bank.

- Instructor's Resource Manual for AP Calculus
  Calculus: 0-7167-9590-6
  In conjunction with the text, this manual provides the opportunity for instructors to prepare their students for the AP Exam.

For Students

- Student Solutions Manual
  Brian Bradie, Christopher Newport University, and Greg Dresden, Washington and Lee University
  Single Variable: 0-7167-9594-9
  Multivariable: 0-7167-9880-8
  Offers worked-out solutions to all odd-numbered exercises in the text.

- Companion website at www.whfreeman.com/rogawski

CalcPortal

One click. One place. For all the tools you need.
CalcPortal is the digital gateway to Rogawski's *Calculus*, designed to enrich your course and improve your students' understanding of calculus. For students, CalcPortal integrates review, diagnostic, and tutorial resources right where they are needed in the eBook. For instructors, CalcPortal provides a powerful but easy-to-use course management system complete with a state-of-the-art algorithmic homework and assessment system.

NEW! Next-Generation eBook

CalcPortal is organized around three main teaching and learning tools: (1) The online eBook is a complete version of Rogawski's *Calculus* and includes a Personalized Study Plan that connects students directly to what they need to learn and to the resources on CalcPortal, including important prerequisite content; (2) the homework and assessment center, with an algorithm problem generator; and (3) the CalcResource center with Just-in-Time algebra and precalculus tutorials.

FEATURES

Pedagogical Features in Rogawski's *CALCULUS*

Conceptual Insights
encourage students to develop a conceptual understanding of calculus by explaining important ideas clearly but informally.

CONCEPTUAL INSIGHT The decomposition of Eq. (3) separates out the two ways in which velocity $\mathbf{v}(t)$ can change. The tangential component $a_{\mathbf{T}} = v'(t)$ describes the rate at which the *speed* changes, and the normal component $a_{\mathbf{N}} = \kappa(t)v(t)^2$ describes the rate at which the *direction* changes [this rate depends on both the curvature $\kappa(t)$ and the speed $v(t)$ at which you travel around the curve]. Consider the following scenarios:

- A particle travels in a straight line. Then $\kappa(t) = 0$ and $\mathbf{a}(t) = v'(t)\mathbf{T}$. Thus, $\mathbf{a}(t)$ points in the direction of motion if the particle is speeding up $[v'(t) > 0]$ and opposite to the direction of motion if the particle is slowing down $[v'(t) < 0]$.
- A particle travels with constant speed along a curved path. Then $v'(t) = 0$ and the acceleration $\mathbf{a}(t) = \kappa(t)v(t)^2\mathbf{N}$ is normal to the direction of motion.

General motion is a combination, involving both tangential and normal acceleration.

Ch. 13, p. 774

Graphical Insights
enhance students' visual understanding by making the crucial connection between graphical properties and the underlying concept.

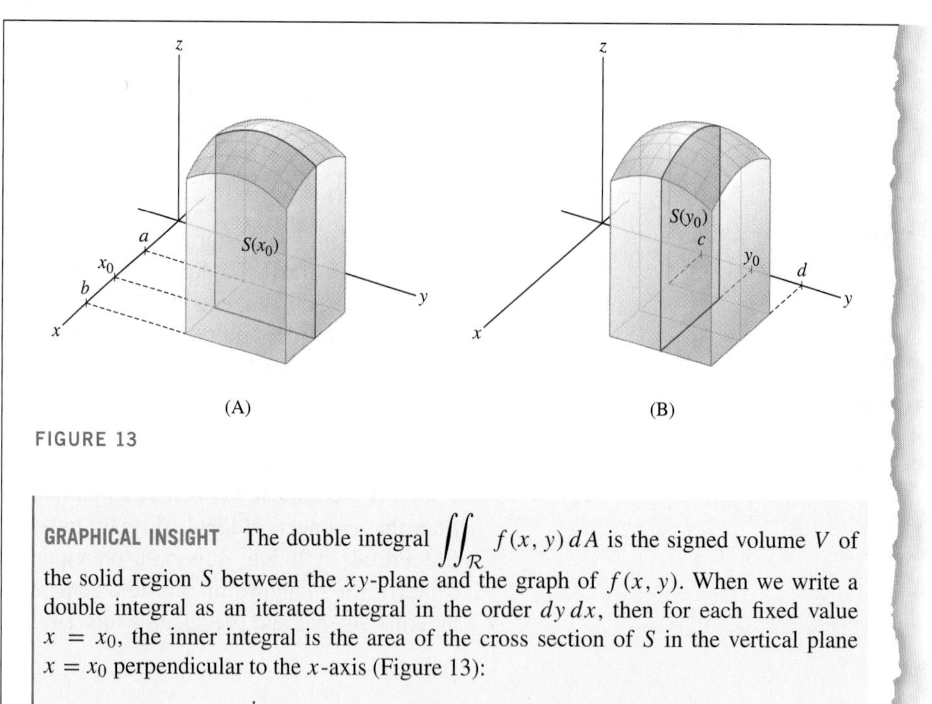

FIGURE 13

GRAPHICAL INSIGHT The double integral $\iint_{\mathcal{R}} f(x, y)\, dA$ is the signed volume V of the solid region S between the xy-plane and the graph of $f(x, y)$. When we write a double integral as an iterated integral in the order $dy\, dx$, then for each fixed value $x = x_0$, the inner integral is the area of the cross section of S in the vertical plane $x = x_0$ perpendicular to the x-axis (Figure 13):

$$S(x_0) = \int_c^d f(x_0, y)\, dy = \text{(area of cross section in vertical plane } x = x_0 \text{ perpendicular to the } x\text{-axis)}$$

Fubini's Theorem thus asserts that the volume V of S may be calculated as the integral

Ch. 15, p. 880

◂·· *REMINDER For any vectors* **u** *and* **v**, *the dot product is*

$$\mathbf{v}\cdot\mathbf{u} = \|\mathbf{v}\|\|\mathbf{u}\|\cos\theta$$

where θ *is the angle between* **v** *and* **u**.

We now discuss two important geometric properties of the gradient. First, suppose that $\nabla f_P \neq 0$ and let **u** be a unit vector (Figure 9). By the properties of the dot product,

$$D_{\mathbf{u}} f(P) = \nabla f_P \cdot \mathbf{u} = \|\nabla f_P\| \cos\theta \qquad \boxed{3}$$

where θ is the angle between ∇f_P and **u**. Note that $\cos\theta$ takes its maximum value when $\theta = 0$. Therefore, $D_{\mathbf{u}} f(P)$ has the largest possible value when $\theta = 0$, that is, when **u** points in the direction of ∇f_P. It follows that the *gradient vector points in the direction*

Ch. 14, p. 832

Reminders are margin notes that link back to important concepts discussed earlier in the text to give students a quick review and make connections with earlier concepts.

CAUTION Do not *use the Quotient Rule to compute the partial derivative in Eq. (1). The denominator does not depend on* y, *so we treat it as a constant when differentiating with respect to* y.

$$g_y(x, y) = \frac{\partial}{\partial y}\frac{y^2}{(1+x^2)^3} = \frac{1}{(1+x^2)^3}\frac{\partial}{\partial y}y^2 = \frac{2y}{(1+x^2)^3} \qquad \boxed{1}$$

$$g_y(1, 3) = \frac{2(3)}{(1+1^2)^3} = \frac{3}{4}$$

Ch. 14, p. 811

Caution Notes

warn students of common pitfalls they can encounter in understanding the material.

Historical Perspectives

are brief vignettes that place key discoveries and conceptual advances in their historical context. They give students a glimpse into past accomplishments of great mathematicians and an appreciation for their significance.

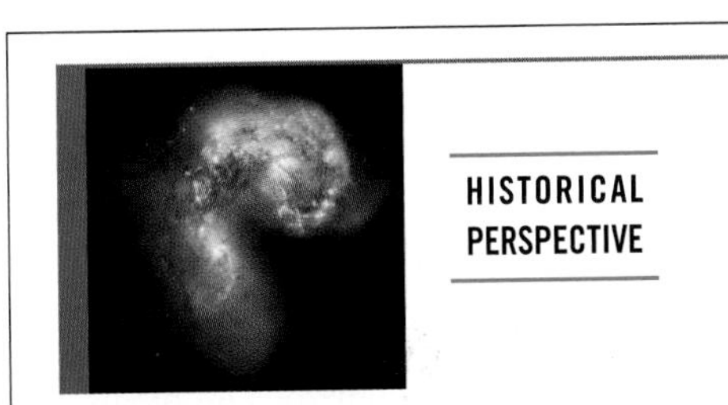

HISTORICAL PERSPECTIVE

The astronomers of the ancient world (Babylon, Egypt, and Greece) mapped out the nighttime sky with impressive accuracy, but their models of planetary motion were based on the erroneous assumption that the planets revolve around the earth. Although the Greek astronomer Aristarchus (310–230 BCE) had suggested that the earth revolves around the sun, this idea was rejected and forgotten for more than fifteen centuries, until the Polish astronomer Nicolaus Copernicus (1473–1543) introduced a revolutionary set of ideas about the solar system, including the hypothesis that the planets revolve around the sun. The ideas of Copernicus, although not entirely correct, paved the way for the next generation, most notably Tycho Brahe (1546–1601), Galileo Galilei (1564–1642), and Johannes Kepler (1571–1630).

The German astronomer Johannes Kepler to work for Tycho Brahe, a Danish astronomer who had revolutionized the art of astronomical measurement and compiled the most complete and accurate data on the planetary orbits then available. When Brahe died in 1601, Kepler succeeded him as "Imperial Mathematician" to the Holy Roman Emperor. Kepler continued to study Brahe's data, and in 1609, he formulated the first two of his laws of planetary motion in a work entitled *Astronomia Nova* (New Astronomy).

In the centuries since Kepler's death, as the observational data improved, astronomers found that planetary orbits are not exactly elliptical. Furthermore, the perihelion (the point on the orbit closest to the sun) shifts slowly over time as shown in Figure 6. Most of these deviations can be explained by the mutual pull of the planets, but the perihelion shift of Mercury is larger than can be accounted for by Newton's Laws. On November 18, 1915, Albert Einstein made a discovery about which he later wrote to a friend, "I was beside myself with ecstasy for days." He had been working for a decade on his famous **General Theory of Relativity**, a theory that would replace Newton's law of gravitation with a new set of much more complicated equations called the Einstein Field Equations.

Ch. 13, p. 784

Assumptions Matter uses short explanations and well-chosen counterexamples to help students appreciate why hypotheses are needed in theorems.

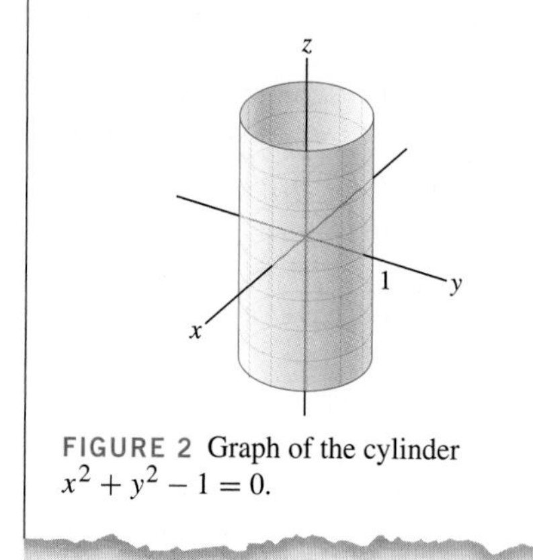

FIGURE 2 Graph of the cylinder $x^2 + y^2 - 1 = 0$.

Assumptions Matter Implicit differentiation is based on the assumption that the equation $F(x, y, z) = 0$ defines z as a function $z = f(x, y)$, at least near a given point $P = (a, b, c)$. According to the Implicit Function Theorem of advanced calculus, such a function $f(x, y)$ exists if F has continuous partial derivatives and if $F_z(P) \neq 0$ [this is also the condition under which the formulas (6) are valid]. Note that $F_z(P) \neq 0$ implies that the tangent plane at P is nonvertical. To see what can go wrong when $F_z(P) = 0$, consider the cylinder (Figure 2)

$$F(x, y, z) = x^2 + y^2 - 1 = 0$$

In this case, $F_z = 0$. The z-coordinate on the cylinder does not depend on x or y, so it is impossible to represent the cylinder as a graph $z = f(x, y)$.

Ch. 14, p. 843

Section Summaries summarize a section's key points in a concise and useful way and emphasize for students what is most important in the section.

Section Exercise Sets offer a comprehensive set of exercises closely coordinated with the text. These exercises vary in levels of difficulty from routine, to moderate, to more challenging. Also included are problems marked with icons that require the student to give a written response or require the use of technology GU CAS:

- ***Preliminary Exercises*** begin each exercise set and need little or no computation. They can be used to check understanding of key concepts of a section before problems from the exercise set are attempted.
- ***Exercises*** offer numerous problems from routine drill problems to moderately challenging problems. These are carefully graded and include many innovative and interesting geometric and real-world applications.
- ***Further Insights and Challenges*** are more challenging problems that require a deeper level of conceptual understanding and sometimes extend a section's material. Many are excellent for use as small-group projects.

Chapter Review Exercises offer a comprehensive set of exercises closely coordinated with the chapter material to provide additional problems for self-study or assignments.

ACKNOWLEDGMENTS

Jon Rogawski and W. H. Freeman and Company are grateful to the many instructors from across the United States and Canada who have offered comments that assisted in the development and refinement of this book. These contributions included class testing, manuscript reviewing, problems reviewing, and participating in surveys about the book and general course needs.

ALABAMA Tammy Potter, *Gadsden State Community College*; David Dempsey, *Jacksonville State University*; Douglas Bailer, *Northeast Alabama Community College*; Michael Hicks, *Shelton State Community College*; Patricia C. Eiland, *Troy University, Montgomery Campus*; James L. Wang, *The University of Alabama*; Stephen Brick, *University of South Alabama*; Joerg Feldvoss, *University of South Alabama* ALASKA Mark A. Fitch, *University of Alaska Anchorage*; Kamal Narang, *University of Alaska Anchorage*; Alexei Rybkin, *University of Alaska Fairbanks*; Martin Getz, *University of Alaska Fairbanks* ARIZONA Stephania Tracogna, *Arizona State University*; Bruno Welfert, *Arizona State University*; Daniel Russow, *Arizona Western College*; Garry Carpenter, *Pima Community College, Northwest Campus*; Katie Louchart, *Northern Arizona University*; Donna M. Krawczyk, *The University of Arizona* ARKANSAS Deborah Parker, *Arkansas Northeastern College*; J. Michael Hall, *Arkansas State University*; Kevin Cornelius, *Ouachita Baptist University*; Hyungkoo Mark Park, *Southern Arkansas University*; Katherine Pinzon, *University of Arkansas at Fort Smith*; Denise LeGrand, *University of Arkansas at Little Rock*; John Annulis, *University of Arkansas at Monticello*; Erin Haller, *University of Arkansas Fayetteville*; Daniel J. Arrigo, *University of Central Arkansas* CALIFORNIA Harvey Greenwald, *California Polytechnic State University, San Luis Obispo*; John M. Alongi, *California Polytechnic State University, San Luis Obispo*; John Hagen, *California Polytechnic State University, San Luis Obispo*; Colleen Margarita Kirk, *California Polytechnic State University, San Luis Obispo*; Lawrence Sze, *California Polytechnic State University, San Luis Obispo*; Raymond Terry, *California Polytechnic State University, San Luis Obispo*; James R. McKinney, *California State Polytechnic University, Pomona*; Charles Lam, *California State University, Bakersfield*; David McKay, *California State University, Long Beach*; Melvin Lax, *California State University, Long Beach*; Wallace A. Etterbeek, *California State University, Sacramento*; Mohamed Allali, *Chapman University*; George Rhys, *College of the Canyons*; Janice Hector, *DeAnza College*; Isabelle Saber, *Glendale Community College*; Peter Stathis, *Glendale Community College*; Kristin Hartford, *Long Beach City College*; Eduardo Arismendi-Pardi, *Orange Coast College*; Mitchell Alves, *Orange Coast College*; Yenkanh Vu, *Orange Coast College*; Yan Tian, *Palomar College*; Donna E. Nordstrom, *Pasadena City College*; Don L. Hancock, *Pepperdine University*; Kevin Iga, *Pepperdine University*; Adolfo J. Rumbos, *Pomona College*; Carlos de la Lama, *San Diego City College*; Matthias Beck, *San Francisco State University*; Arek Goetz, *San Francisco State University*; Nick Bykov, *San Joaquin Delta College*; Eleanor Lang Kendrick, *San Jose City College*; Elizabeth Hodes, *Santa Barbara City College*; William Konya, *Santa Monica College*; John Kennedy, *Santa Monica College*; Peter Lee, *Santa Monica College*; Richard Salome, *Scotts Valley High School*; Norman Feldman, *Sonoma State University*; Elaine McDonald, *Sonoma State University*; Bruno Nachtergaele, *University of California, Davis*; Boumediene Hamzi, *University of California, Davis*; Peter Stevenhagen, *University of California, San Diego*; Jeffrey Stopple, *University of California, Santa Barbara*; Guofang Wei, *University of California, Santa Barbara*; Rick A. Simon, *University of La Verne*; Mohamad A. Alwash, *West Los Angeles College* COLORADO Tony Weathers, *Adams State College*; Erica Johnson, *Arapahoe Community College*; Karen Walters, *Arapahoe Community College*; Joshua D. Laison, *Colorado College*; Gerrald G. Greivel, *Colorado School of Mines*; Jim Thomas, *Colorado State University*; Eleanor Storey, *Front Range Community College*; Larry Johnson, *Metropolitan State College of Denver*; Carol Kuper, *Morgan Community*

College; Larry A. Pontaski, *Pueblo Community College*; Terry Reeves, *Red Rocks Community College*; Debra S. Carney, *University of Denver* **CONNECTICUT** Jeffrey McGowan, *Central Connecticut State University*; Ivan Gotchev, *Central Connecticut State University*; Charles Waiveris, *Central Connecticut State University*; Christopher Hammond, *Connecticut College*; Anthony Y. Aidoo, *Eastern Connecticut State University*; Kim Ward, *Eastern Connecticut State University*; Joan W. Weiss, *Fairfield University*; Theresa M. Sandifer, *Southern Connecticut State University*; Cristian Rios, *Trinity College*; Melanie Stein, *Trinity College* **DELAWARE** Patrick F. Mwerinde, *University of Delaware* **DISTRICT OF COLUMBIA** Jeffrey Hakim, *American University*; Joshua M. Lansky, *American University*; James A. Nickerson, *Gallaudet University* **FLORIDA** Abbas Zadegan, *Florida International University*; Gerardo Aladro, *Florida International University*; Gregory Henderson, *Hillsborough Community College*; Pam Crawford, *Jacksonville University*; Penny Morris, *Polk Community College*; George Schultz, *St. Petersburg College*; Jimmy Chang, *St. Petersburg College*; Carolyn Kistner, *St. Petersburg College*; Aida Kadic-Galeb, *The University of Tampa*; Heath M. Martin, *University of Central Florida*; Constance Schober, *University of Central Florida*; S. Roy Choudhury, *University of Central Florida*; Kurt Overhiser, *Valencia Community College* **GEORGIA** Thomas T. Morley, *Georgia Institute of Technology*; Ralph Wildy, *Georgia Military College*; Shahram Nazari, *Georgia Perimeter College*; Alice Eiko Pierce, *Georgia Perimeter College Clarkson Campus*; Susan Nelson, *Georgia Perimeter College Clarkson Campus*; Shahram Nazari, *Georgia Perimeter College Dunwoody Campus*; Laurene Fausett, *Georgia Southern University*; Scott N. Kersey, *Georgia Southern University*; Jimmy L. Solomon, *Georgia Southern University*; Allen G. Fuller, *Gordon College*; Marwan Zabdawi, *Gordon College*; Carolyn A. Yackel, *Mercer University*; Shahryar Heydari, *Piedmont College*; Dan Kannan, *The University of Georgia* **HAWAII** Shuguang Li, *University of Hawaii at Hilo*; Raina B. Ivanova, *University of Hawaii at Hilo* **IDAHO** Charles Kerr, *Boise State University*; Otis Kenny, *Boise State University*; Alex Feldman, *Boise State University*; Doug Bullock, *Boise State University*; Ed Korntved, *Northwest Nazarene University* **ILLINOIS** Chris Morin, *Blackburn College*; Alberto L. Delgado, *Bradley University*; John Haverhals, *Bradley University*; Herbert E. Kasube, *Bradley University*; Brenda H. Alberico, *College of DuPage*; Marvin Doubet, *Lake Forest College*; Marvin A. Gordon, *Lake Forest Graduate School of Management*; Richard J. Maher, *Loyola University Chicago*; Joseph H. Mayne, *Loyola University Chicago*; Marian Gidea, *Northeastern Illinois University*; Miguel Angel Lerma, *Northwestern University*; Mehmet Dik, *Rockford College*; Tammy Voepel, *Southern Illinois University Edwardsville*; Rahim G. Karimpour, *Southern Illinois University*; Thomas Smith, *University of Chicago* **INDIANA** Julie A. Killingbeck, *Ball State University*; John P. Boardman, *Franklin College*; Robert N. Talbert, *Franklin College*; Robin Symonds, *Indiana University Kokomo*; Henry L. Wyzinski, *Indiana University Northwest*; Melvin Royer, *Indiana Wesleyan University*; Gail P. Greene, *Indiana Wesleyan University*; David L. Finn, *Rose-Hulman Institute of Technology* **IOWA** Nasser Dastrange, *Buena Vista University*; Mark A. Mills, *Central College*; Karen Ernst, *Hawkeye Community College*; Richard Mason, *Indian Hills Community College*; Robert S. Keller, *Loras College*; Eric Robert Westlund, *Luther College* **KANSAS** Timothy W. Flood, *Pittsburg State University*; Sarah Cook, *Washburn University*; Kevin E. Charlwood, *Washburn University* **KENTUCKY** Alex M. McAllister, *Center College*; Sandy Spears, *Jefferson Community & Technical College*; Leanne Faulkner, *Kentucky Wesleyan College*; Donald O. Clayton, *Madisonville Community College*; Thomas Riedel, *University of Louisville*; Manabendra Das, *University of Louisville*; Lee Larson, *University of Louisville*; Jens E. Harlander, *Western Kentucky University*

LOUISIANA William Forrest, *Baton Rouge Community College*; Paul Wayne Britt, *Louisiana State University*; Galen Turner, *Louisiana Tech University*; Randall Wills, *Southeastern Louisiana University*; Kent Neuerburg, *Southeastern Louisiana University* **MAINE** Andrew Knightly, *The University of Maine*; Sergey Lvin, *The University of Maine*; Joel W. Irish, *University of Southern Maine*; Laurie Woodman, *University of Southern Maine* **MARYLAND** Leonid Stern, *Towson University*; Mark E. Williams, *University of Maryland Eastern Shore*; Austin A. Lobo, *Washington College* **MASSACHUSETTS** Sean McGrath, *Algonquin Regional High School*; Norton Starr, *Amherst College*; Renato Mirollo, *Boston College*; Emma Previato, *Boston University*; Richard H. Stout, *Gordon College*; Matthew P. Leingang, *Harvard University*; Suellen Robinson, *North Shore Community College*; Walter Stone, *North Shore Community College*; Barbara Loud, *Regis College*; Andrew B. Perry, *Springfield College*; Tawanda Gwena, *Tufts University*; Gary Simundza, *Wentworth Institute of Technology*; Mikhail Chkhenkeli, *Western New England College*; David Daniels, *Western New England College*; Alan Gorfin, *Western New England College*; Saeed Ghahramani, *Western New England College*; Julian Fleron, *Westfield State College*; Brigitte Servatius, *Worcester Polytechnic Institute* **MICHIGAN** Mark E. Bollman, *Albion College*; Jim Chesla, *Grand Rapids Community College*; Jeanne Wald, *Michigan State University*; Allan A. Struthers, *Michigan Technological University*; Debra Pharo, *Northwestern Michigan College*; Anna Maria Spagnuolo, *Oakland University*; Diana Faoro, *Romeo Senior High School*; Andrew Strowe, *University of Michigan–Dearborn*; Daniel Stephen Drucker, *Wayne State University* **MINNESOTA** Bruce Bordwell, *Anoka-Ramsey Community College*; Robert Dobrow, *Carleton College*; Jessie K. Lenarz, *Concordia College–Moorhead Minnesota*; Bill Tomhave, *Concordia College–Moorhead Minnesota*; David L. Frank, *University of Minnesota*; Steven I. Sperber, *University of Minnesota*; Jeffrey T. McLean, *University of St. Thomas*; Chehrzad Shakiban, *University of St. Thomas*; Melissa Loe, *University of St. Thomas* **MISSISSIPPI** Vivien G. Miller, *Mississippi State University*; Ted Dobson, *Mississippi State University*; Len Miller, *Mississippi State University*; Tristan Denley, *The University of Mississippi* **MISSOURI** Robert Robertson, *Drury University*; Gregory A. Mitchell, *Metropolitan Community College-Penn Valley*; Charles N. Curtis, *Missouri Southern State University*; Vivek Narayanan, *Moberly Area Community College*; Russell Blyth, *Saint Louis University*; Blake Thornton, *Saint Louis University*; Kevin W. Hopkins, *Southwest Baptist University* **MONTANA** Kelly Cline, *Carroll College*; Richard C. Swanson, *Montana State University*; Nikolaus Vonessen, *The University of Montana* **NEBRASKA** Edward G. Reinke Jr., *Concordia University, Nebraska*; Judith Downey, *University of Nebraska at Omaha* **NEVADA** Rohan Dalpatadu, *University of Nevada, Las Vegas*; Paul Aizley, *University of Nevada, Las Vegas* **NEW HAMPSHIRE** Richard Jardine, *Keene State College*; Michael Cullinane, *Keene State College*; Roberta Kieronski, *University of New Hampshire at Manchester* **NEW JERSEY** Paul S. Rossi, *College of Saint Elizabeth*; Mark Galit, *Essex County College*; Katarzyna Potocka, *Ramapo College of New Jersey*; Nora S. Thornber, *Raritan Valley Community College*; Avraham Soffer, *Rutgers The State University of New Jersey*; Chengwen Wang, *Rutgers The State University of New Jersey*; Stephen J. Greenfield, *Rutgers The State University of New Jersey*; John T. Saccoman, *Seton Hall University*; Lawrence E. Levine, *Stevens Institute of Technology* **NEW MEXICO** Kevin Leith, *Central New Mexico Community College*; David Blankenbaker, *Central New Mexico Community College*; Joseph Lakey, *New Mexico State University*; Jurg Bolli, *University of New Mexico*; Kees Onneweer, *University of New Mexico* **NEW YORK** Robert C. Williams, *Alfred University*; Timmy G. Bremer, *Broome Community College State University of New York*; Joaquin O. Carbonara, *Buffalo State College*; Robin Sue

Sanders, *Buffalo State College*; Daniel Cunningham, *Buffalo State College*; Rose Marie Castner, *Canisius College*; Sharon L. Sullivan, *Catawba College*; Camil Muscalu, *Cornell University*; Maria S. Terrell, *Cornell University*; Margaret Mulligan, *Dominican College of Blauvelt*; Robert Andersen, *Farmingdale State University of New York*; Leonard Nissim, *Fordham University*; Jennifer Roche, *Hobart and William Smith Colleges*; James E. Carpenter, *Iona College*; Peter Shenkin, *John Jay College of Criminal Justice/CUNY*; Gordon Crandall, *LaGuardia Community College/CUNY*; Gilbert Traub, *Maritime College, State University of New York*; Paul E. Seeburger, *Monroe Community College Brighton Campus*; Abraham S. Mantell, *Nassau Community College*; Daniel D. Birmajer, *Nazareth College*; Sybil G. Shaver, *Pace University*; Margaret Kiehl, *Rensselaer Polytechnic Institute*; Carl V. Lutzer, *Rochester Institute of Technology*; Michael A. Radin, *Rochester Institute of Technology*; Hossein Shahmohamad, *Rochester Institute of Technology*; Thomas Rousseau, *Siena College*; Jason Hofstein, *Siena College*; Leon E. Gerber, *St. Johns University*; Christopher Bishop, *Stony Brook University*; James Fulton, *Suffolk County Community College*; John G. Michaels, *SUNY Brockport*; Howard J. Skogman, *SUNY Brockport*; Cristina Bacuta, *SUNY Cortland*; Jean Harper, *SUNY Fredonia*; Kelly Black, *Union College*; Thomas W. Cusick, *University at Buffalo/The State University of New York*; Gino Biondini, *University at Buffalo/The State University of New York*; Robert Koehler, *University at Buffalo/The State University of New York* **NORTH CAROLINA** Jeffrey Clark, *Elon University*; William L. Burgin, *Gaston College*; Manouchehr H. Misaghian, *Johnson C. Smith University*; Legunchim L. Emmanwori, *North Carolina A&T State University*; Drew Pasteur, *North Carolina State University*; Demetrio Labate, *North Carolina State University*; Mohammad Kazemi, *The University of North Carolina at Charlotte*; Richard Carmichael, *Wake Forest University*; Gretchen Wilke Whipple, *Warren Wilson College* **NORTH DAKOTA** Anthony J. Bevelacqua, *The University of North Dakota*; Richard P. Millspaugh, *The University of North Dakota* **OHIO** Christopher Butler, *Case Western Reserve University*; Pamela Pierce, *The College of Wooster*; Tzu-Yi Alan Yang, *Columbus State Community College*; Greg S. Goodhart, *Columbus State Community College*; Kelly C. Stady, *Cuyahoga Community College*; Brian T. Van Pelt, *Cuyahoga Community College*; David Robert Ericson, *Miami University*; Frederick S. Gass, *Miami University*; Thomas Stacklin, *Ohio Dominican University*; Vitaly Bergelson, *The Ohio State University*; Darry Andrews, *The Ohio State University*; Robert Knight, *Ohio University*; John R. Pather, *Ohio University, Eastern Campus*; Teresa Contenza, *Otterbein College*; Ali Hajjafar, *The University of Akron*; Jianping Zhu, *The University of Akron*; Ian Clough, *University of Cincinnati Clermont College*; Atif Abueida, *University of Dayton*; Judith McCrory, *The University at Findlay*; Thomas Smotzer, *Youngstown State University*; Angela Spalsbury, *Youngstown State University* **OKLAHOMA** Michael McClendon, *University of Central Oklahoma*; Teri Jo Murphy, *The University of Oklahoma* **OREGON** Lorna TenEyck, *Chemeketa Community College*; Angela Martinek, *Linn-Benton Community College*; Tevian Dray, *Oregon State University* **PENNSYLVANIA** John B. Polhill, *Bloomsburg University of Pennsylvania*; Russell C. Walker, *Carnegie Mellon University*; Jon A. Beal, *Clarion University of Pennsylvania*; Kathleen Kane, *Community College of Allegheny County*; David A. Santos, *Community College of Philadelphia*; David S. Richeson, *Dickinson College*; Christine Marie Cedzo, *Gannon University*; Monica Pierri-Galvao, *Gannon University*; John H. Ellison, *Grove City College*; Gary L. Thompson, *Grove City College*; Dale McIntyre, *Grove City College*; Dennis Benchoff, *Harrisburg Area Community College*; William A. Drumin, *King's College*; Denise Reboli, *King's College*; Chawne Kimber, *Lafeyette College*; David L. Johnson, *Lehigh University*; Zia Uddin, *Lock Haven University of Pennsylvania*; Donna A. Dietz, *Mansfield University of*

publishing experience, Steve Rigolosi for expert market development, and Brendan Cady and Laura Capuano for editorial assistance. My thanks are also due to the superb production team: Blake Logan, Bill Page, Paul Rohloff, Ted Szczepanski, Susan Timmins, and Vivien Weiss, as well as Don DeLand and Leslie Galen of Integre Technical Publishing, for their expert composition.

To my wife, Julie, I owe more than I can say. Thank you for everything. To our wonderful children Rivkah, Dvora, Hannah, and Akiva, thank you for putting up with the calculus book, a demanding extra sibling that dwelled in our home for so many years. And to my mother Elise and late father Alexander Rogawski, MD z"l, thank you for love and support from the beginning.

TO THE STUDENT

Although I have taught calculus for more than 25 years, I still feel excitement when I enter the classroom on the first day of a new semester, as if a great drama is about to unfold. Does the word "drama" seem out of place in a discussion of mathematics?

There is no doubt that calculus is useful—it is used throughout the sciences and engineering, from weather prediction and space flight to nanotechnology and financial modeling. But what is dramatic about it?

For me, one part of the drama lies in the conceptual and logical development of the subject. Starting with just a few basic concepts such as limits and tangent lines, we gradually build the tools for solving innumerable problems of great practical importance. Along the way, there are high points and moments of suspense—computing a derivative using limits for the first time or learning how the Fundamental Theorem of Calculus unifies differential and integral calculus. We also discover that calculus provides the right language for expressing the universal laws of nature, not just Newton's laws of motion for which it was invented, but also the laws of electromagnetism and even the strange laws of quantum mechanics.

Another part of the drama is the learning process itself—the personal voyage of discovery. Certainly, one important part of learning calculus is the acquisition of technical skills. You will learn how to compute derivatives and integrals, solve optimization problems, and so on. These are the skills you need to apply calculus in practical situations. But when you study calculus, you also learn the language of science. You gain access to the thoughts of Newton, Euler, Gauss, and Maxwell, the greatest scientific thinkers, all of whom expressed their insights using calculus. The distinguished mathematician I. M. Gelfand put it this way: "The most important thing a student can get from the study of mathematics is the attainment of a higher intellectual level."

This text is designed to develop both skills and conceptual understanding. In fact, the two go hand-in-hand. As you become proficient in problem solving, you will come to appreciate the underlying ideas. And it is equally true that a solid understanding of the concepts will make you a more effective problem solver. You are likely to devote much of your time to studying examples and working exercises. However, the text also contains numerous down-to-earth explanations, sometimes under the heading "Conceptual Insight" or "Graphical Insight." They are designed to show you how and why calculus works. I urge you to take the time to read these explanations and think about them.

A major challenge for me in writing this textbook was to present calculus in a style that students would find comprehensible and interesting. As I wrote, I continually asked myself: Can it be made simpler? Have I assumed something the student may not be aware of? Can I explain the deeper significance of a concept without confusing the student who is learning the subject for the first time?

I'm afraid that no textbook can make learning calculus entirely painless. According to legend, Alexander the Great once asked the mathematician Menaechmus to show him an easy way to learn geometry. Menaechmus replied, "There is no royal road to geometry." Even kings must work hard to learn geometry, and the same is true of calculus—achieving mastery requires time and effort.

However, I hope my efforts have resulted in a textbook that is "student friendly" and that also encourages you to appreciate the big picture—the beautiful and elegant ideas that hold the entire structure of calculus together. Please let me know if you have comments or suggestions for improving the text. I look forward to hearing from you.

Best wishes and good luck!

Jon Rogawski

MULTIVARIABLE
CALCULUS
Early Transcendentals

10 | INFINITE SERIES

Our knowledge of what stars are made of is based on the study of absorption spectra, the sequences of wavelengths absorbed by gases in the star's atmosphere.

The theory of infinite series is the third branch of calculus, in addition to differential and integral calculus. Infinite series provide us with a new perspective on functions and on many interesting numbers. Two examples are the Gregory–Leibniz series

$$\frac{\pi}{4} = 1 - \frac{1}{3} + \frac{1}{5} - \frac{1}{7} + \frac{1}{9} - \cdots$$

and the infinite series for the exponential function

$$e^x = 1 + x + \frac{x^2}{2!} + \frac{x^3}{3!} + \frac{x^4}{4!} + \cdots$$

The first reveals that π is related to the reciprocals of the odd integers in an unexpected way, whereas the second shows that e^x can be expressed as an "infinite polynomial." Series of this type are widely used in applications, both in computations and in the analysis of functions. To make sense of infinite series, we need to define precisely what it means to add up infinitely many terms. Limits play a key role here, just as they do in differential and integral calculus.

10.1 Sequences

Sequences of numbers appear in diverse situations. If you divide a cake in half, and then divide the remaining half in half, and continue dividing in half indefinitely (Figure 1), then the fraction of cake left at each step forms the sequence

$$1, \quad \frac{1}{2}, \quad \frac{1}{4}, \quad \frac{1}{8}, \quad \ldots$$

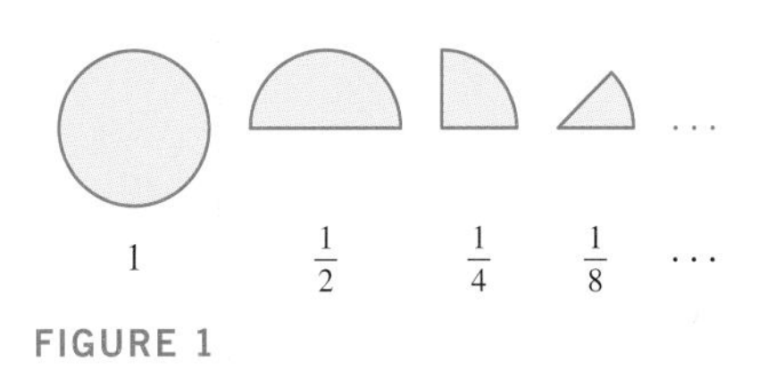

FIGURE 1

This is the sequence of values of $f(n) = \dfrac{1}{2^n}$ for $n = 0, 1, 2, \ldots$.

Formally, a **sequence** is a function $f(n)$ whose domain is a subset of the integers. The values $a_n = f(n)$ are called the **terms** of the sequence and n is called the **index**. We usually think of a sequence informally as a collection of values $\{a_n\}$, or a list of terms:

$$a_1, \quad a_2, \quad a_3, \quad a_4, \quad \ldots$$

When a_n is given by a formula, we refer to a_n as the **general term**.

General Term	Domain	Sequence
$a_n = 1 - \dfrac{1}{n}$	$n \ge 1$	$0, \ \frac{1}{2}, \ \frac{2}{3}, \ \frac{3}{4}, \ \frac{4}{5}, \ \ldots$
$a_n = (-1)^n n$	$n \ge 0$	$0, \ -1, \ 2, \ -3, \ 4, \ \ldots$
$b_n = \dfrac{364.5n^2}{n^2 - 4}$	$n \ge 3$	$656.1, \ 486, \ 433.9, \ 410.1, \ 396.9, \ \ldots$

The sequence $b_n = \dfrac{364.5n^2}{n^2 - 4}$, known as the "Balmer series" in physics and chemistry, plays a key role in spectroscopy. The terms of this sequence are the absorption wavelengths of the hydrogen atom in nanometers.

In the following example, we consider a sequence whose terms are defined *recursively*. The nth term a_n is computed in terms of the preceding term a_{n-1}, but there is no closed formula for the general term.

■ **EXAMPLE 1 Recursively Defined Sequence** Compute a_2, a_3, a_4 for the sequence defined recursively by

$$a_1 = 1, \qquad a_n = \frac{1}{2}\left(a_{n-1} + \frac{2}{a_{n-1}}\right)$$

Solution

$$a_2 = \frac{1}{2}\left(a_1 + \frac{2}{a_1}\right) = \frac{1}{2}\left(1 + \frac{2}{1}\right) = \frac{3}{2} = 1.5$$

$$a_3 = \frac{1}{2}\left(a_2 + \frac{2}{a_2}\right) = \frac{1}{2}\left(\frac{3}{2} + \frac{2}{3/2}\right) = \frac{17}{12} \approx 1.4167$$

$$a_4 = \frac{1}{2}\left(a_3 + \frac{2}{a_3}\right) = \frac{1}{2}\left(\frac{17}{12} + \frac{2}{17/12}\right) = \frac{577}{408} \approx 1.414216$$ ■

You may recognize the sequence in Example 1 as the sequence of approximations to $\sqrt{2} \approx 1.4142136$ produced by Newton's method with starting value $a_1 = 1$. As n tends to infinity, a_n approaches $\sqrt{2}$.

Our main goal is to study convergence of sequences. A sequence $\{a_n\}$ converges to a limit L if the terms a_n get closer and closer to L as $n \to \infty$.

> **DEFINITION Limit of a Sequence** A sequence $\{a_n\}$ **converges** to a limit L, and we write
>
> $$\lim_{n\to\infty} a_n = L \qquad \text{or} \qquad a_n \to L$$
>
> if, for every $\epsilon > 0$, there is a number M such that $|a_n - L| < \epsilon$ for all $n > M$. If no limit exists, we say that $\{a_n\}$ **diverges**.

■ **EXAMPLE 2 Proving Convergence of a Sequence** Let $a_n = \dfrac{n+4}{n+1}$. Prove formally that $\lim_{n\to\infty} a_n = 1$.

Solution The definition requires us to find, for every $\epsilon > 0$, a number M such that

$$|a_n - 1| < \epsilon \qquad \text{for all } n > M \tag{1}$$

We have

$$|a_n - 1| = \left|\frac{n+4}{n+1} - 1\right| = \frac{3}{n+1}$$

Therefore, $|a_n - 1| < \epsilon$ if

$$\frac{3}{n+1} < \epsilon \qquad \text{or} \qquad n > \frac{3}{\epsilon} - 1$$

It follows that (1) is valid with $M = \dfrac{3}{\epsilon} - 1$. For example, if $\epsilon = 0.01$, then we may take $M = \dfrac{3}{0.01} - 1 = 299$. Thus, $|a_n - 1| < 0.01$ for $n = 300, 301, 302, \ldots$. ■

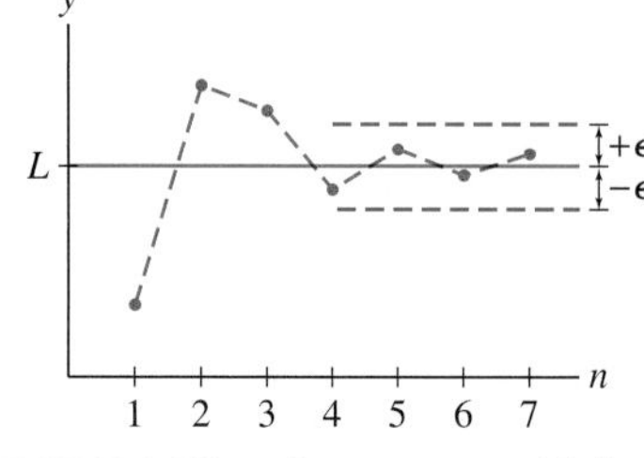

FIGURE 2 Plot of a sequence with limit L. For any ϵ, the dots eventually remain within an ϵ-band around L.

We may visualize a sequence by plotting its "graph," that is, by plotting the points $(1, a_1), (2, a_2), (3, a_3), \ldots$ (Figure 2). The sequence converges to a limit L if, for every $\epsilon > 0$, the plotted points eventually remain within an ϵ-band around the horizontal line $y = L$ (Figure 2). Figure 3 shows the plot of a sequence converging to $L = 1$ and Figure 4 shows the sequence $a_n = \cos n$, which has no limit.

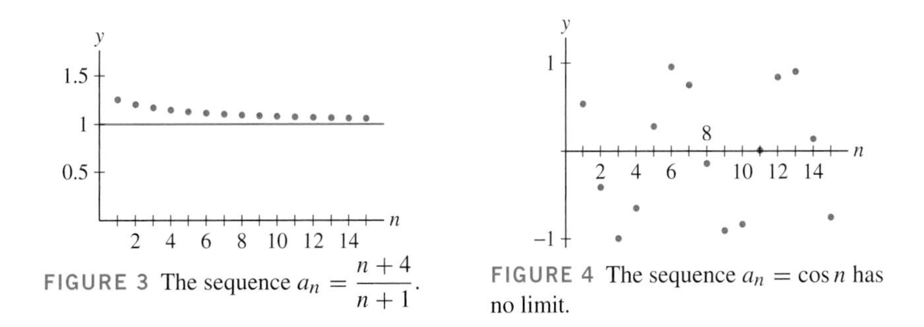

FIGURE 3 The sequence $a_n = \dfrac{n+4}{n+1}$.

FIGURE 4 The sequence $a_n = \cos n$ has no limit.

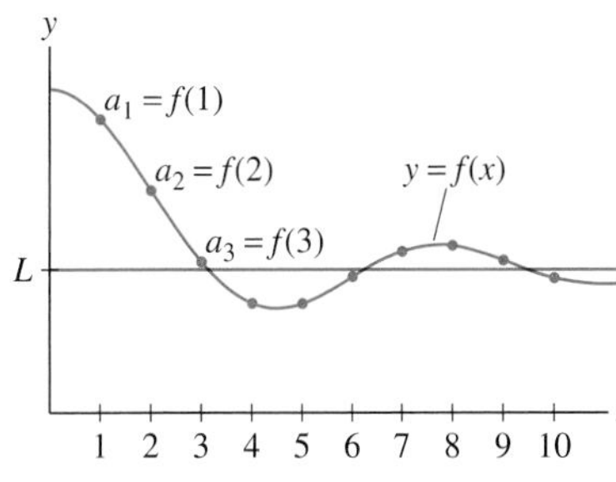

FIGURE 5 If $f(x)$ converges to L, then the sequence $a_n = f(n)$ also converges to L.

We note the following:

- The limit does not change if we modify or drop finitely many terms of the sequence.
- If C is a constant and $a_n = C$ for all n sufficiently large, then $\lim\limits_{n\to\infty} a_n = C$.

Suppose that $f(x)$ is a function and that $f(x)$ approaches a limit L as $x \to \infty$. In this case, the sequence $a_n = f(n)$ approaches the same limit L (Figure 5). Indeed, in this case, for all $\epsilon > 0$, we can find M so that $|f(x) - L| < \epsilon$ for all $x > M$. It follows automatically that $|f(n) - L| < \epsilon$ for all integers $n > M$.

THEOREM 1 Sequence Defined by a Function Let $f(x)$ be a function defined on $[c, \infty)$ for some constant c. If $\lim\limits_{x\to\infty} f(x)$ exists, then the sequence $a_n = f(n)$ converges and

$$\lim_{n\to\infty} a_n = \lim_{x\to\infty} f(x)$$

EXAMPLE 3 Find the limit of the sequence $0, \dfrac{1}{2}, \dfrac{2}{3}, \dfrac{3}{4}, \dfrac{4}{5}, \ldots$.

Solution This is the sequence with general term

$$a_n = \frac{n-1}{n} = 1 - \frac{1}{n}$$

Let $f(x) = 1 - \dfrac{1}{x}$. Then $a_n = f(n)$ and by Theorem 1,

$$\lim_{n\to\infty} a_n = \lim_{x\to\infty} f(x) = \lim_{x\to\infty}\left(1 - \frac{1}{x}\right) = 1 - \lim_{x\to\infty}\frac{1}{x} = 1$$

EXAMPLE 4 Calculate $\lim\limits_{n\to\infty} a_n$, where $a_n = \dfrac{n + \ln n}{n^2}$.

Solution The limit of the sequence is equal to the limit of the function $f(x) = \dfrac{x + \ln x}{x^2}$, which we calculate using L'Hôpital's Rule:

$$\lim_{n\to\infty} a_n = \lim_{x\to\infty} f(x) = \lim_{x\to\infty}\frac{x + \ln x}{x^2} = \lim_{x\to\infty}\frac{1 + (1/x)}{2x} = 0$$

The limit of the Balmer wavelengths b_n defined in the margin on page 553 is of interest in physics and chemistry because it determines the ionization energy of the hydrogen atom. Table 1 suggests that b_n approaches 364.5 as $n \to \infty$. Figure 6 shows the graph of b_n, and in Figure 7, the wavelengths are shown "crowding in" toward their limiting value.

TABLE 1 The Wavelengths in the Balmer Series Approach the Limit $L = 364.5$

n	b_n
3	656.1
4	486
5	433.9
6	410.1
7	396.9
10	379.7
20	368.2
40	365.4
60	364.9
80	364.7
100	364.6

EXAMPLE 5 Limit of Balmer Wavelengths Calculate the limit of the Balmer wavelengths $b_n = \dfrac{364.5n^2}{n^2-4}$.

Solution Observe that $b_n = f(n)$, where $f(x) = \dfrac{364.5x^2}{x^2-4}$. We compute the limit by dividing the numerator and denominator by x^2:

$$\lim_{n\to\infty} b_n = \lim_{x\to\infty} f(x) = 364.5 \lim_{x\to\infty} \frac{x^2}{x^2-4} = 364.5 \lim_{x\to\infty} \frac{1}{1-4/x^2}$$

$$= \frac{364.5}{\lim\limits_{x\to\infty}(1-4/x^2)} = 364.5$$

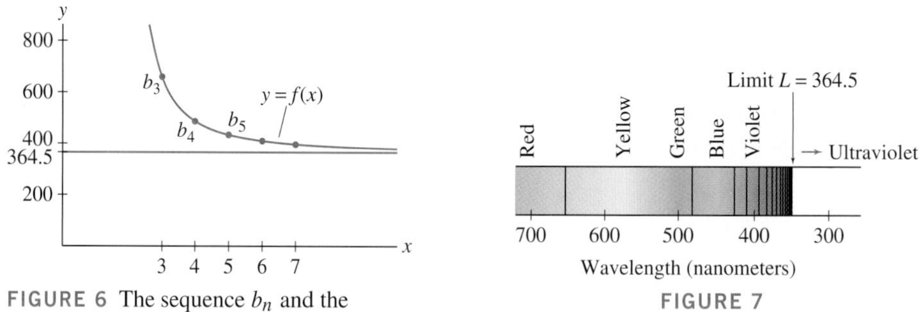

FIGURE 6 The sequence b_n and the function approach the same limit.

FIGURE 7

The geometric sequence $a_n = cr^n$ is the sequence defined by the exponential function $f(x) = cr^x$ whose base r is the common ratio.

A **geometric sequence** is a sequence of the form $a_n = cr^n$, where c and r are nonzero constants. For instance, if $c = 2$ and $r = 3$, we obtain the geometric sequence

$$2, \quad 2\cdot 3, \quad 2\cdot 3^2, \quad 2\cdot 3^3, \quad 2\cdot 3^4, \quad 2\cdot 3^5, \quad \ldots$$

The number r is called the **common ratio**. Each term a_n is r times the previous term a_{n-1}, that is, $a_n/a_{n-1} = r$.

We say that $\{a_n\}$ *diverges to* ∞, and we write $\lim\limits_{n\to\infty} a_n = \infty$, if the terms a_n increase beyond all bounds, that is, if, for every $N > 0$, we have $a_n > N$ for all sufficiently large n (Figure 8).

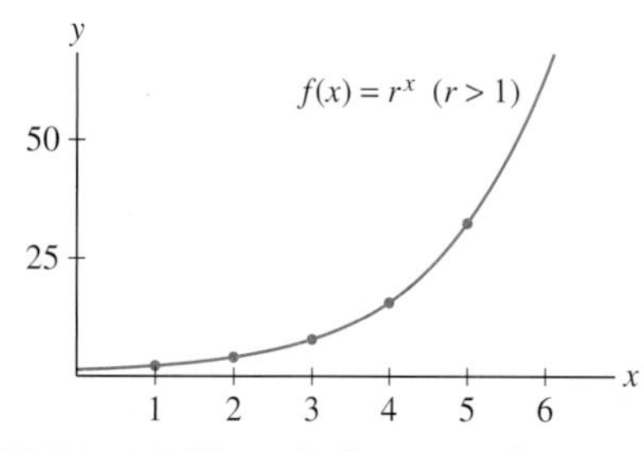

FIGURE 8 If $r > 1$, the geometric sequence $a_n = r^n$ diverges to ∞.

EXAMPLE 6 Limit of a Geometric Sequence Prove that:

$$\lim_{n\to\infty} r^n = \begin{cases} 0 & \text{if} \quad 0 < r < 1 \\ 1 & \text{if} \quad r = 1 \\ \text{diverges to } \infty & \text{if} \quad r > 1 \end{cases}$$

Solution We apply Theorem 1 to the exponential function $f(x) = r^x$. If $0 < r < 1$, then (Figure 9)

$$\lim_{n\to\infty} r^n = \lim_{x\to\infty} r^x = 0$$

Similarly, if $r > 1$, then $f(x)$ diverges to ∞ as $x \to \infty$, so $\{r^n\}$ also diverges to ∞ (Figure 8). If $r = 1$, $r^n = 1$ for all n and the limit is 1.

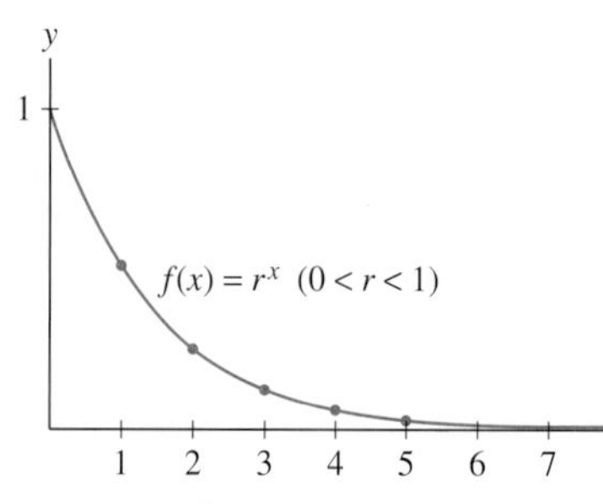

FIGURE 9 If $0 < r < 1$, the geometric sequence $a_n = r^n$ converges to 0.

Most of the limit laws for functions also apply to sequences. The proofs are similar and are omitted.

THEOREM 2 Limit Laws for Sequences Assume that $\{a_n\}$ and $\{b_n\}$ are convergent sequences with

$$\lim_{n\to\infty} a_n = L, \qquad \lim_{n\to\infty} b_n = M$$

Then:

(i) $\lim_{n\to\infty} (a_n \pm b_n) = \lim_{n\to\infty} a_n \pm \lim_{n\to\infty} b_n = L \pm M$

(ii) $\lim_{n\to\infty} a_n b_n = \left(\lim_{n\to\infty} a_n\right)\left(\lim_{n\to\infty} b_n\right) = LM$

(iii) $\lim_{n\to\infty} \dfrac{a_n}{b_n} = \dfrac{\lim_{n\to\infty} a_n}{\lim_{n\to\infty} b_n} = \dfrac{L}{M}$ if $M \neq 0$

(iv) $\lim_{n\to\infty} ca_n = c \lim_{n\to\infty} a_n = cL$ for any constant c

THEOREM 3 Squeeze Theorem for Sequences Let $\{a_n\}$, $\{b_n\}$, $\{c_n\}$ be sequences such that for some number M,

$$b_n \le a_n \le c_n \quad \text{for } n > M \qquad \text{and} \qquad \lim_{n\to\infty} b_n = \lim_{n\to\infty} c_n = L$$

Then $\lim_{n\to\infty} a_n = L$.

EXAMPLE 7 Show that if $\lim_{n\to\infty} |a_n| = 0$, then $\lim_{n\to\infty} a_n = 0$.

Solution We have

$$-|a_n| \le a_n \le |a_n|$$

Since $|a_n|$ tends to zero, $-|a_n|$ also tends to zero and the Squeeze Theorem implies that $\lim_{n\to\infty} a_n = 0$. ■

REMINDER $n!$ *(n-factorial) is the number*

$$n! = n(n-1)(n-2)\cdots 2\cdot 1$$

For example, $4! = 4\cdot 3\cdot 2\cdot 1 = 24$.

As another application of the Squeeze Theorem, consider the sequence

$$a_n = \frac{5^n}{n!}$$

Both the numerator and denominator tend to infinity, so it is not clear in advance whether $\{a_n\}$ converges. Figure 10 and Table 2 suggest that a_n increases initially and then tends to zero. In the next example, we prove that, indeed, $a_n = \dfrac{R^n}{n!}$ converges to zero for all R. We will use this fact in the discussion of Taylor series in Section 10.7.

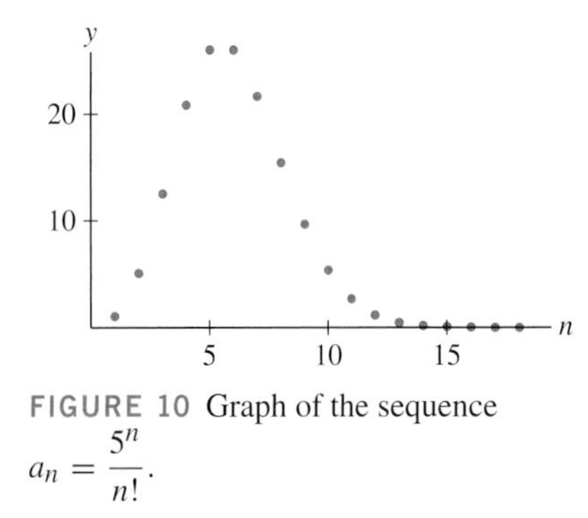

FIGURE 10 Graph of the sequence $a_n = \dfrac{5^n}{n!}$.

EXAMPLE 8 Prove that $\lim_{n\to\infty} \dfrac{R^n}{n!} = 0$ for all R.

Solution We may assume without loss of generality that $R > 0$ by the result of Example 7. Then there is a unique integer $M \ge 0$ such that

$$M \le R < M + 1$$

TABLE 2

n	$a_n = \dfrac{5^n}{n!}$
1	5
2	12.5
3	20.83
4	26.04
10	2.69
15	0.023
20	0.00004

For $n > M$, we write $R^n/n!$ as a product of n factors:

$$\frac{R^n}{n!} = \underbrace{\left(\frac{R}{1}\frac{R}{2}\cdots\frac{R}{M}\right)}_{\text{Call this constant } C}\underbrace{\left(\frac{R}{M+1}\right)\left(\frac{R}{M+2}\right)\cdots\left(\frac{R}{n}\right)}_{\text{Each factor is less than 1}} \le C\left(\frac{R}{n}\right)$$

The first M factors are ≥ 1 and the last $n - M$ factors are < 1. If we lump together the first M factors and call the product C, and drop all the remaining factors except the last factor R/n, we obtain

$$0 \le \frac{R^n}{n!} \le \frac{CR}{n}$$

Since $\dfrac{CR}{n} \to 0$, the Squeeze Theorem implies that $\lim\limits_{n\to\infty} \dfrac{R^n}{n!} = 0$. ■

We can apply a function $f(x)$ to a sequence $\{a_n\}$ to obtain a new sequence $\{f(a_n)\}$. It is useful to know that if $f(x)$ is continuous and $a_n \to L$, then $f(a_n) \to f(L)$. We state this result in the next theorem. See Appendix D for a proof.

THEOREM 4 If $f(x)$ is continuous and the limit $\lim\limits_{n\to\infty} a_n = L$ exists, then

$$\lim_{n\to\infty} f(a_n) = f\left(\lim_{n\to\infty} a_n\right) = f(L)$$

■ **EXAMPLE 9** Calculate $\lim\limits_{n\to\infty} e^{3n/(n+1)}$.

Solution We have $e^{3n/(n+1)} = f(a_n)$, where $a_n = \dfrac{3n}{n+1}$ and $f(x) = e^x$. Furthermore,

$$\lim_{n\to\infty} a_n = \lim_{n\to\infty} \frac{3n}{n+1} = 3$$

By Theorem 4, $\lim\limits_{n\to\infty} f(a_n) = f\left(\lim\limits_{n\to\infty} a_n\right) = f(3)$, that is,

$$\lim_{n\to\infty} e^{3n/(n+1)} = e^{\lim_{n\to\infty} 3n/(n+1)} = e^3$$ ■

We now introduce two concepts that are important for understanding convergence: the concepts of a bounded sequence and a monotonic sequence.

DEFINITION Bounded Sequences A sequence $\{a_n\}$ is:

- **Bounded from above** if there is a number M such that $a_n \le M$ for all n. The number M is called an *upper bound.*
- **Bounded from below** if there exists m such that $a_n \ge m$ for all n. The number m is called a *lower bound.*

If $\{a_n\}$ is bounded from above and below, we say that $\{a_n\}$ is *bounded*. If $\{a_n\}$ is not bounded, we call $\{a_n\}$ an *unbounded sequence*.

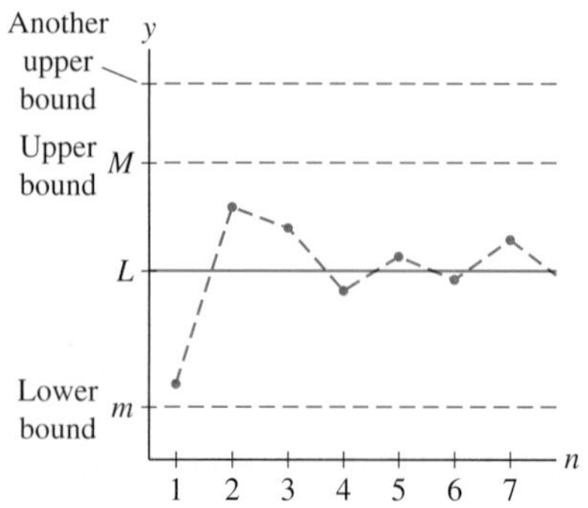

FIGURE 11 A convergent sequence is bounded.

Upper and lower bounds are not unique. If M is an upper bound, then any number larger than M is also an upper bound (Figure 11). Similarly, if m is a lower bound, then any number smaller than m is also a lower bound.

It would seem that a convergent sequence $\{a_n\}$ must be bounded because the terms a_n get closer and closer to the limit (Figure 11). This leads to the next theorem.

> **THEOREM 5 Convergent Sequences Are Bounded** If $\{a_n\}$ converges, then $\{a_n\}$ is bounded.

Proof Let $L = \lim_{n\to\infty} a_n$. Then there exists $N > 0$ such that $|a_n - L| < 1$ for $n > N$. In other words,

$$L - 1 < a_n < L + 1 \qquad \text{for } n > N$$

If M is any number larger than $L + 1$ and also larger than the numbers $a_1, a_2, \dots, a_N$, then $a_n < M$ for all n. Thus, M is an upper bound. Similarly, any number m smaller than $L - 1$ and $a_1, a_2, \dots, a_N$ is a lower bound. ■

There are two ways for a sequence $\{a_n\}$ to be divergent. First, if $\{a_n\}$ is unbounded, then it certainly diverges by Theorem 5. For example, the following sequence diverges:

$$-1, \quad 2, \quad -3, \quad 4, \quad -5, \quad 6, \quad \dots$$

A sequence $\{a_n\}$ is monotonic

- *Increasing if $a_j \le a_{j+1}$ for all j*
- *Decreasing if $a_j \ge a_{j+1}$ for all j.*

On the other hand, a sequence may diverge even if it is bounded if the terms a_n bounce around and never settle down to approach a limit. For example, the sequence $a_n = (-1)^n$ is bounded but does not converge:

$$1, \quad -1, \quad 1, \quad -1, \quad 1, \quad -1, \quad \dots$$

When can we be sure that a sequence converges? One situation is when $\{a_n\}$ is both bounded and **monotonic** increasing or decreasing. The reason, intuitively, is that if $\{a_n\}$ is increasing and bounded above by M, then the terms must eventually bunch up near some limiting value L that is not greater than M (Figure 12). We state this formally in the next theorem, whose proof is given in Appendix B.

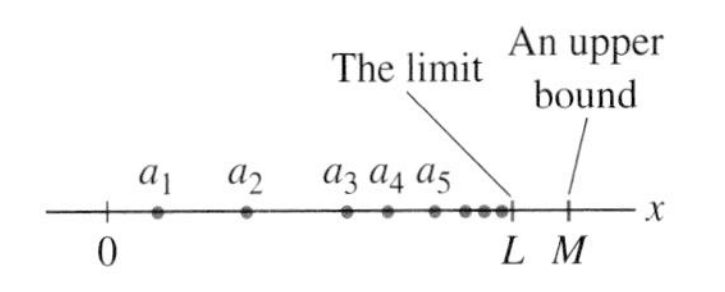

FIGURE 12 An increasing sequence with upper bound M approaches a limit L.

> **THEOREM 6 Bounded Monotonic Sequences Converge**
> - If $\{a_n\}$ is increasing and $a_n \le M$ for all n, then $\{a_n\}$ converges and $\lim_{n\to\infty} a_n \le M$.
> - If $\{a_n\}$ is decreasing and $a_n \ge m$ for all n, then $\{a_n\}$ converges and $\lim_{n\to\infty} a_n \ge m$.

■ **EXAMPLE 10** Verify that $a_n = \sqrt{n+1} - \sqrt{n}$ is decreasing and bounded below. Does $\lim_{n\to\infty} a_n$ exist?

Solution The function $f(x) = \sqrt{x+1} - \sqrt{x}$ is decreasing because its derivative is negative:

$$f'(x) = \frac{1}{2\sqrt{x+1}} - \frac{1}{2\sqrt{x}} < 0 \qquad \text{for } x > 0$$

It follows that $a_n = f(n)$ is also decreasing (see Table 3). The sequence is bounded below by $m = 0$ since $a_n > 0$ for all n. Theorem 6 guarantees that the limit $L = \lim_{n\to\infty} a_n$ exists and $L \ge 0$ (it can be shown that $L = 0$). ■

TABLE 3 **The Sequence $a_n = \sqrt{n+1} - \sqrt{n}$ Is Decreasing**

$a_n = \sqrt{n+1} - \sqrt{n}$
$a_1 \approx 0.414$
$a_2 \approx 0.318$
$a_3 \approx 0.268$
$a_4 \approx 0.213$
$a_5 \approx 0.196$

EXAMPLE 11 Show that the following sequence is bounded and increasing:

$$a_0 = 0, \quad a_1 = \sqrt{2}, \quad a_2 = \sqrt{2+\sqrt{2}}, \quad a_3 = \sqrt{2+\sqrt{2+\sqrt{2}}}, \quad \ldots$$

Prove that $L = \lim_{n\to\infty} a_n$ exists and compute its value.

Solution This sequence is defined recursively by

$$a_0 = 0, \qquad a_{n+1} = \sqrt{2+a_n}$$

It would not be hard to find the limit L if we knew in advance that the limit exists. We could proceed as follows. The sequence $b_n = a_{n+1}$ (the same sequence $\{a_n\}$ but starting with a_1) would converge to the same limit L and we would have, using Theorem 4,

$$L = \lim_{n\to\infty} a_{n+1} = \lim_{n\to\infty} \sqrt{2+a_n} = \sqrt{2+\lim_{n\to\infty} a_n} = \sqrt{2+L}$$

Thus, $L = \sqrt{2+L}$ and hence

$$L^2 = 2 + L \quad \Rightarrow \quad L^2 - L - 2 = (L-2)(L+1) = 0$$

It follows that $L = -1$ or $L = 2$, and since $L \ge 0$, we conclude that $L = 2$ (see Table 4). To justify this conclusion, we must prove that the limit L exists. By Theorem 6, it suffices to show that $\{a_n\}$ is bounded above and increasing.

TABLE 4 **Terms of the Recursive Sequence** $a_{n+1} = \sqrt{2+a_n}$

a_0	0
a_1	1.4142
a_2	1.8478
a_3	1.9616
a_4	1.9904
a_5	1.9976

***Step 1.* Show that $\{a_n\}$ is bounded above by $M = 2$.**
First, we observe that

$$\text{if } a_n < 2, \quad \text{then} \quad a_{n+1} = \sqrt{a_n + 2} < \sqrt{2+2} = 2 \qquad \boxed{2}$$

Now we can prove that $a_n < 2$ for all n. Since $a_0 = 0$, (2) implies that $a_1 < 2$. But then $a_1 < 2$ implies that $a_2 < 2$ by (2), and $a_2 < 2$ implies $a_3 < 2$, etc. for all n (formally speaking, this is a proof by induction).

***Step 2.* Show that $\{a_n\}$ is increasing.**
Since a_n is positive and $a_n < 2$,

$$a_{n+1} = \sqrt{a_n + 2} > \sqrt{a_n + a_n} = \sqrt{2a_n} > \sqrt{a_n \cdot a_n} = a_n$$

Thus, $a_{n+1} > a_n$ for all n and $\{a_n\}$ is increasing. ■

10.1 SUMMARY

- A *sequence* is a function $f(n)$ whose domain is a subset of the integers. We write $a_n = f(n)$ for the nth term and denote the sequence itself by $\{a_n\}$ or simply a_n.
- We say that $\{a_n\}$ *converges* to a limit L, and we write $\lim_{n\to\infty} a_n = L$ or $a_n \to L$ if, for every $\epsilon > 0$, there is a number M such that

$$|a_n - L| < \epsilon \qquad \text{for all } n > M$$

If no limit exists, we say that $\{a_n\}$ *diverges*.
- Let $f(x)$ be a function on $[c, \infty)$ for some number c and let $a_n = f(n)$ for $n \ge c$. If $\lim_{x\to 0} f(x) = L$, then $\lim_{n\to\infty} a_n = L$.

- A *geometric sequence* is a sequence of the form $a_n = cr^n$, where c and r are nonzero.
- The basic Limit Laws and the Squeeze Theorem apply to sequences.
- If $f(x)$ is continuous and $\lim_{n\to\infty} a_n = L$, then $\lim_{n\to\infty} f(a_n) = f(L)$.
- We say that $\{a_n\}$ is *bounded above* by M if $a_n \le M$ for all n and *bounded below* by m if $a_n \ge m$ for all n. If $\{a_n\}$ is bounded above and below, $\{a_n\}$ is called *bounded*.
- A sequence $\{a_n\}$ is *monotonic* if it is increasing ($a_j \le a_{j+1}$) or decreasing ($a_j \ge a_{j+1}$) for all j.
- Theorem 6 states that every increasing sequence that is bounded above and every decreasing sequence that is bounded below converges.

10.1 EXERCISES

Preliminary Questions

1. What is a_4 for the sequence $a_n = n^2 - n$?

2. Which of the following sequences converge to zero?

(a) $\dfrac{n^2}{n^2+1}$ **(b)** 2^n **(c)** $\left(\dfrac{-1}{2}\right)^n$

3. Let a_n be the nth decimal approximation to $\sqrt{2}$. That is, $a_1 = 1$, $a_2 = 1.4$, $a_3 = 1.41$, etc. What is $\lim_{n\to\infty} a_n$?

4. Which sequence is defined recursively?

(a) $a_n = \sqrt{2 + n^{-1}}$ **(b)** $b_n = \sqrt{4 + b_{n-1}}$

5. Theorem 5 says that every convergent sequence is bounded. Which of the following statements follow from Theorem 5 and which are false? If false, give a counterexample.

(a) If $\{a_n\}$ is bounded, then it converges.

(b) If $\{a_n\}$ is not bounded, then it diverges.

(c) If $\{a_n\}$ diverges, then it is not bounded.

Exercises

1. Match the sequence with the general term:

$a_1, a_2, a_3, a_4, \ldots$	General term
(a) $\frac{1}{2}, \frac{2}{3}, \frac{3}{4}, \frac{4}{5}, \ldots$	(i) $\cos \pi n$
(b) $-1, 1, -1, 1, \ldots$	(ii) $\dfrac{n!}{2^n}$
(c) $1, -1, 1, -1, \ldots$	(iii) $(-1)^{n+1}$
(d) $\frac{1}{2}, \frac{2}{4}, \frac{6}{8}, \frac{24}{16} \ldots$	(iv) $\dfrac{n}{n+1}$

2. Let $a_n = \dfrac{1}{2n-1}$ for $n = 1, 2, 3, \ldots$. Write out the first three terms of the following sequences.

(a) $b_n = a_{n+1}$ **(b)** $c_n = a_{n+3}$

(c) $d_n = a_n^2$ **(d)** $e_n = 2a_n - a_{n+1}$

In Exercises 3–10, calculate the first four terms of the following sequences, starting with $n = 1$.

3. $c_n = \dfrac{2^n}{n!}$

4. $b_n = \cos \pi n$

5. $a_1 = 3, \quad a_{n+1} = 1 + a_n^2$

6. $b_n = 2 + (-1)^n$

7. $c_n = 1 + \dfrac{1}{2} + \dfrac{1}{3} + \cdots + \dfrac{1}{n}$

8. $a_n = n + (n+1) + (n+2) + \cdots + (2n)$

9. $b_1 = 2, \quad b_2 = 5, \quad b_n = b_{n-1} + 2b_{n-2}$

10. $c_n = n$th decimal approximation to e^{-1}

11. Find a formula for the nth term of the following sequence:

(a) $\dfrac{1}{1}, \dfrac{-1}{8}, \dfrac{1}{27}, \ldots$ **(b)** $\dfrac{2}{6}, \dfrac{3}{7}, \dfrac{4}{8}, \ldots$

12. Suppose that $\lim_{n\to\infty} a_n = 4$ and $\lim_{n\to\infty} b_n = 7$. Determine:

(a) $\lim_{n\to\infty} (a_n + b_n)$ **(b)** $\lim_{n\to\infty} a_n^3$

(c) $\lim_{n\to\infty} 4b_n$ **(d)** $\lim_{n\to\infty} (a_n^2 - 2a_n b_n)$

In Exercises 13–26, use Theorem 1 to determine the limit of the sequence or state that the sequence diverges.

13. $a_n = 4$

14. $b_n = \dfrac{3n+1}{2n+4}$

15. $a_n = 5 - \dfrac{9}{n^2}$

16. $b_n = (-1)^{2n+1}$

17. $c_n = -2^{-n}$

18. $c_n = 4(2^n)$

19. $z_n = \left(\dfrac{1}{3}\right)^n$

20. $z_n = (0.1)^{-1/n}$

21. $a_n = \dfrac{(-1)^n n^2 + n}{4n^2 + 1}$

22. $a_n = \dfrac{n}{\sqrt{n^2+1}}$

23. $a_n = \dfrac{n}{\sqrt{n^3+1}}$

24. $a_n = \sin \pi n$

25. $a_n = \cos \pi n$

26. $a_n = n((1 + n^{-1})^2 - 1)$

27. Let $a_n = \dfrac{n}{n+1}$. Find a number M such that:

(a) $|a_n - 1| \le 0.001$ for $n \ge M$.

(b) $|a_n - 1| \le 0.00001$ for $n \ge M$.

Then use the limit definition to prove that $\lim_{n\to\infty} a_n = 1$.

28. Let $b_n = (\frac{1}{3})^n$.

(a) Find a value of M such that $|b_n| \le 10^{-5}$ for $n \ge M$.

(b) Use the limit definition to prove that $\lim_{n\to\infty} b_n = 0$.

29. Use the limit definition to prove that $\lim_{n\to\infty} n^{-2} = 0$.

30. Find the limit of $d_n = \sqrt{n+3} - \sqrt{n}$.

31. Find $\lim_{n\to\infty} 2^{1/n}$.

32. Show that $\lim_{n\to\infty} b^{1/n}$ is independent of b for $b > 0$.

33. Find $\lim_{n\to\infty} n^{1/n}$.

34. Find the limit of $a_n = n^2(\sqrt[3]{n^3+1} - n)$. *Hint:* Write $a_n = \dfrac{(1+n^{-3})^{1/3} - 1}{n^{-3}}$ and apply L'Hôpital's Rule.

35. Find $\lim_{n\to\infty} \left(1 + \dfrac{1}{n}\right)^n$.

36. Find $\lim_{n\to\infty} \left(1 + \dfrac{1}{n^2}\right)^n$.

37. Use the Squeeze Theorem to find $\lim_{n\to\infty} a_n$, where $a_n = \dfrac{1}{\sqrt{n^4 + n^8}}$ by proving that

$$\frac{1}{\sqrt{2n^4}} \le a_n \le \frac{1}{\sqrt{2n^2}}$$

38. Evaluate $\lim_{n\to\infty} \dfrac{\cos n}{n}$.

39. Evaluate $\lim_{n\to\infty} n \sin \dfrac{1}{n}$.

40. Evaluate $\lim_{n\to\infty} (2^n + 3^n)^{1/n}$. *Hint:* Show that

$$3 \le a_n \le (2 \cdot 3^n)^{1/n}$$

41. Which statement is equivalent to the assertion $\lim_{n\to\infty} a_n = L$? Explain.

(a) For every $\epsilon > 0$, the interval $(L - \epsilon, L + \epsilon)$ contains at least one element of the sequence $\{a_n\}$.

(b) For every $\epsilon > 0$, the interval $(L - \epsilon, L + \epsilon)$ contains all but at most finitely many elements of the sequence $\{a_n\}$.

42. Which statement is equivalent to the assertion that $\{a_n\}$ is bounded? Explain.

(a) There exists a finite interval $[m, M]$ containing every element of the sequence $\{a_n\}$.

(b) There exists a finite interval $[m, M]$ containing an element of the sequence $\{a_n\}$.

In Exercises 43–63, determine the limit of the sequence or show that the sequence diverges by using the appropriate Limit Laws or theorems.

43. $a_n = \dfrac{3n^2 + n + 2}{2n^2 - 3}$

44. $a_n = \dfrac{\sqrt{n}}{\sqrt{n} + 4}$

45. $a_n = 3 + \left(-\dfrac{1}{2}\right)^n$

46. $a_n = \left(2 + \dfrac{4}{n^2}\right)^{1/3}$

47. $b_n = \tan^{-1}\left(1 - \dfrac{2}{n}\right)$

48. $b_n = e^{n^2 - n}$

49. $c_n = \ln\left(\dfrac{2n+1}{3n+4}\right)$

50. $c_n = \dfrac{n}{n + n^{1/n}}$

51. $y_n = \dfrac{e^n + 3^n}{5^n}$

52. $y_n = \dfrac{e^n}{2^n}$

53. $a_n = \dfrac{e^n}{2^{n^2}}$

54. $a_n = \dfrac{n}{2^n}$

55. $b_n = \dfrac{n^3 + 2e^{-n}}{3n^3 + 4e^{-n}}$

56. $b_n = \dfrac{3 - 4^n}{2 + 7 \cdot 4^n}$

57. $A_n = \dfrac{3 - 4^n}{2 + 7 \cdot 3^n}$

58. $B_n = \dfrac{10^n}{n!}$

59. $A_n = \dfrac{(-4{,}000)^n}{n!}$

60. $B_n = \dfrac{n!}{2^n}$

61. $a_n = \cos \dfrac{\pi}{n}$

62. $a_n = n \sin \dfrac{\pi}{n}$

63. $a_n = \sqrt[n]{n}$

64. Show that $a_n = \dfrac{1}{2n+1}$ is strictly decreasing.

65. Show that $a_n = \dfrac{3n^2}{n^2+2}$ is strictly increasing. Find an upper bound.

66. Show that $a_n = \sqrt[3]{n+1} - n$ is decreasing.

67. Use the limit definition to prove that the limit does not change if a finite number of terms are added or removed from a convergent sequence.

68. Let $b_n = a_{n+1}$. Prove that if $\{a_n\}$ converges, then $\{b_n\}$ also converges and $\lim_{n\to\infty} a_n = \lim_{n\to\infty} b_n$.

69. Let $\{a_n\}$ be a sequence such that $\lim_{n\to\infty} |a_n|$ exists and is nonzero. Show that $\lim_{n\to\infty} a_n$ exists if and only if there exists an integer M such that the sign of a_n does not change for $n > M$.

70. Give an example of a divergent sequence $\{a_n\}$ such that $\lim_{n\to\infty} |a_n|$ converges.

71. Show, by giving an example, that there exist *divergent* sequences $\{a_n\}$ and $\{b_n\}$ such that $\{a_n + b_n\}$ converges.

72. Using the limit definition, prove that if $\{a_n\}$ converges and $\{b_n\}$ diverges, then $\{a_n + b_n\}$ diverges.

73. Use the limit definition to prove that if $\{a_n\}$ is a convergent sequence of integers with limit L, then there exists a number M such that $a_n = L$ for all $n \ge M$.

74. Theorem 1 states that if $\lim_{x\to\infty} f(x) = L$, then the sequence $a_n = f(n)$ converges and $\lim_{n\to\infty} a_n = L$. Show that the *converse* is false. In other words, find a function $f(x)$ such that $a_n = f(n)$ converges but $\lim_{x\to\infty} f(x)$ does not exist.

75. Prove that the following sequence is bounded and increasing. Then find its limit:

$$a_1 = \sqrt{5}, \quad a_2 = \sqrt{5 + \sqrt{5}}, \quad a_3 = \sqrt{5 + \sqrt{5 + \sqrt{5}}}, \dots$$

76. Let $\{a_n\}$ be the sequence

$$\sqrt{2}, \quad \sqrt{2\sqrt{2}}, \quad \sqrt{2\sqrt{2\sqrt{2}}}, \dots$$

Show that $\{a_n\}$ is increasing and $0 \le a_n \le 2$. Then prove that $\{a_n\}$ converges and find the limit.

77. Find the limit of the sequence

$$c_n = \frac{1}{\sqrt{n^2+1}} + \frac{1}{\sqrt{n^2+2}} + \cdots + \frac{1}{\sqrt{n^2+n}}$$

Hint: Show that

$$\frac{n}{\sqrt{n^2+n}} \le c_n \le \frac{n}{\sqrt{n^2+1}}$$

Further Insights and Challenges

78. Show that $a_n = \sqrt[n]{n!}$ diverges. *Hint:* Show that $n! \ge (n/2)^{n/2}$ by observing that half of the factors of $n!$ are greater than or equal to $n/2$.

79. Let $b_n = \dfrac{\sqrt[n]{n!}}{n}$

(a) Show that $\ln b_n = \dfrac{\ln(n!) - n\ln n}{n} = \dfrac{1}{n}\sum_{k=1}^{n} \ln \dfrac{k}{n}$.

(b) Show that $\ln b_n$ converges to $\int_0^1 \ln x\,dx$ and conclude that $b_n \to e^{-1}$.

80. Given positive numbers $a_1 < b_1$, define two sequences recursively by

$$a_{n+1} = \sqrt{a_n b_n}, \qquad b_n = \frac{a_n + b_n}{2}$$

(a) Show that $a_n \le b_n$ for all n (Figure 13).

(b) Show that $\{a_n\}$ is increasing and $\{b_n\}$ is decreasing.

(c) Show that

$$b_{n+1} - a_{n+1} \le \frac{b_n - a_n}{2}$$

Prove that both $\{a_n\}$ and $\{b_n\}$ converge and have the same limit. This limit, denoted AGM(a_1, b_1), is called the **arithmetic-geometric mean** of a_1 and b_1. See Figure 13.

(d) Estimate AGM$(1, \sqrt{2})$ to three decimal places.

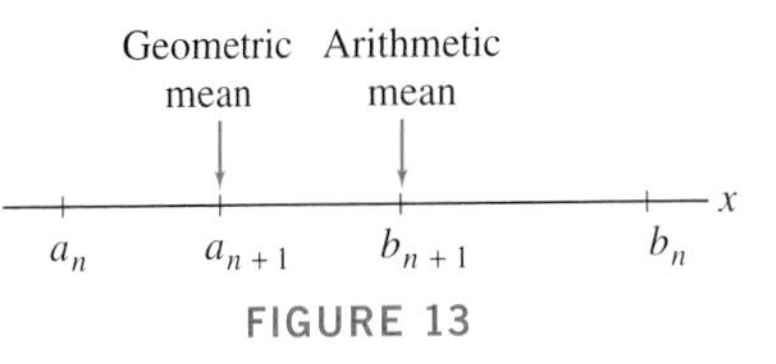

FIGURE 13

81. Let $c_n = \dfrac{1}{n} + \dfrac{1}{n+1} + \dfrac{1}{n+2} + \cdots + \dfrac{1}{2n}$.

(a) Calculate c_1, c_2, c_3, c_4.

(b) Use a comparison of rectangles with the area under $y = x^{-1}$ over the interval $[n, 2n]$ to prove that

$$\int_n^{2n} \frac{dx}{x} + \frac{1}{2n} \le c_n \le \int_n^{2n} \frac{dx}{x} + \frac{1}{n}$$

(c) Use the Squeeze Theorem to determine $\lim_{n\to\infty} c_n$.

82. The nth harmonic number is the number

$$H_n = 1 + \frac{1}{2} + \frac{1}{3} + \cdots + \frac{1}{n}$$

Let $a_n = H_n - \ln n$.

(a) Show that $a_n \ge 0$ for $n \ge 1$. *Hint:* Show that $H_n \ge \int_1^{n+1} \dfrac{dx}{x}$.

(b) Show that $\{a_n\}$ is decreasing by interpreting $a_n - a_{n+1}$ as an area.

(c) Prove that $\lim_{n\to\infty} a_n$ exists. This limit, denoted γ and known as *Euler's Constant*, appears in many areas of mathematics, including analysis and number theory. It has been calculated to more than 100 million decimal places, but it is still not known if γ is an irrational number. The first 10 digits are $\gamma \approx 0.5772156649$.

10.2 Summing an Infinite Series

Quantities that arise in applications often cannot be computed exactly. We cannot write down an exact decimal expression for the number π or for values of the sine function such as sin 1. Sometimes these quantities can be represented as infinite sums. For example,

$$\sin 1 = 1 - \frac{1}{3!} + \frac{1}{5!} - \frac{1}{7!} + \frac{1}{9!} - \frac{1}{11!} + \cdots \qquad \boxed{1}$$

Infinite sums of this type are called **infinite series.**

What precisely does Eq. (1) mean? Although it is impossible to add up infinitely many numbers, we can compute the **partial sums** S_N, defined as the first N terms of the infinite series. Let's compare the first few partial sums of the series above with sin 1:

$$S_1 = 1$$

$$S_2 = 1 - \frac{1}{3!} = 1 - \frac{1}{6} \approx 0.833$$

$$S_3 = 1 - \frac{1}{3!} + \frac{1}{5!} = 1 - \frac{1}{6} + \frac{1}{120} \approx 0.841667$$

$$S_4 = 1 - \frac{1}{6} + \frac{1}{120} - \frac{1}{5{,}040} \approx 0.841468$$

$$S_5 = 1 - \frac{1}{6} + \frac{1}{120} - \frac{1}{5{,}040} + \frac{1}{362{,}880} \approx \mathbf{0.84147098}46$$

$$\sin 1 \approx \mathbf{0.84147098}48079$$

The partial sums appear to converge to sin 1 and, indeed, we will prove that $\lim_{N\to\infty} S_N = \sin 1$ in Section 10.7. This is the precise meaning of Eq. (1).

In general, an infinite series is an expression of the form

$$\sum_{n=1}^{\infty} a_n = a_1 + a_2 + a_3 + \cdots$$

where $\{a_n\}$ is any sequence. For example,

Infinite series may begin with any index. For example,

$$\sum_{n=3}^{\infty} \frac{1}{n} = \frac{1}{3} + \frac{1}{4} + \frac{1}{5} + \cdots$$

When it is not necessary to specify the starting point, we write simply $\sum a_n$. Any letter may be used for the index. Thus, we may write a_m, a_k, a_i, etc.

Sequence	General Term	Infinite Series
$\frac{1}{3}, \frac{1}{9}, \frac{1}{27}, \ldots$	$a_n = \frac{1}{3^n}$	$\sum_{n=1}^{\infty} \frac{1}{3^n} = \frac{1}{3} + \frac{1}{9} + \frac{1}{27} + \frac{1}{81} + \cdots$
$\frac{1}{1}, \frac{1}{4}, \frac{1}{9}, \frac{1}{16}, \ldots$	$a_n = \frac{1}{n^2}$	$\sum_{n=1}^{\infty} \frac{1}{n^2} = \frac{1}{1^2} + \frac{1}{2^2} + \frac{1}{3^2} + \frac{1}{4^2} + \cdots$

The Nth partial sum S_N is the sum of the first N terms of the series:

$$S_N = \sum_{n=1}^{N} a_n = a_1 + a_2 + a_3 + \cdots + a_N$$

The sum of the infinite series $\sum_{n=1}^{\infty} a_n$ is defined as the limit of the partial sums S_N, if this limit exists.

DEFINITION Convergence of an Infinite Series An infinite series $\sum_{n=1}^{\infty} a_n$ *converges* to S if $\lim_{N\to\infty} S_N = S$. The limit S is called the *sum* of the infinite series, and we write $S = \sum_{n=1}^{\infty} a_n$. If the limit does not exist, the infinite series is said to *diverge*.

It is easy to give examples of series that diverge. For example, $\sum_{n=1}^{\infty} 1$ diverges because the partial sums S_N diverge to ∞:

$$S_N = \sum_{n=1}^{N} 1 = \overbrace{1 + 1 + 1 + 1 + \cdots + 1}^{N \text{ times}} = N$$

Similarly, $\sum_{n=1}^{\infty} (-1)^{n-1}$ diverges because the partial sums jump back and forth between 1 and 0:

$$S_1 = 1, \quad S_2 = 1 - 1 = 0, \quad S_3 = 1 - 1 + 1 = 1, \quad S_4 = 1 - 1 + 1 - 1 = 0, \quad \ldots$$

We may investigate series numerically by computing several partial sums S_N. If the partial sums show a trend of convergence to some number S, then we have evidence (but not proof) that the series converges to S. The next example treats a convergent **telescoping series**, where the partial sums are particularly easy to evaluate.

Although there is a simple formula for the partial sums in Example 1, this is the exception rather than the rule. Apart from telescoping series and the geometric series introduced below, there is usually no formula for S_N, and we must develop techniques for studying infinite series that do not rely on formulas.

EXAMPLE 1 Telescoping Series Investigate the following series numerically:

$$S = \sum_{n=1}^{\infty} \frac{1}{n(n+1)} = \frac{1}{1(2)} + \frac{1}{2(3)} + \frac{1}{3(4)} + \frac{1}{4(5)} + \cdots$$

Then compute the sum S using the identity:

$$\frac{1}{n(n+1)} = \frac{1}{n} - \frac{1}{n+1}$$

Solution Table 1 lists some partial sums, computed with the help of a CAS. These numerics suggest convergence to $S = 1$. To evaluate the limit exactly, we use the identity to rewrite the terms of the series. We find that each partial sum collapses down to two terms due to cancellation:

TABLE 1 Partial Sums for $\sum_{n=1}^{\infty} \frac{1}{n(n+1)}$

N	S_N
10	0.90909
50	0.98039
100	0.990099
200	0.995025
300	0.996678

$$S_1 = \frac{1}{1(2)} = \frac{1}{1} - \frac{1}{2}$$

$$S_2 = \frac{1}{1(2)} + \frac{1}{2(3)} = \left(\frac{1}{1} - \frac{1}{2}\right) + \left(\frac{1}{2} - \frac{1}{3}\right) = 1 - \frac{1}{3}$$

$$S_3 = \frac{1}{1(2)} + \frac{1}{2(3)} + \frac{1}{3(4)} = \left(\frac{1}{1} - \frac{1}{2}\right) + \left(\frac{1}{2} - \frac{1}{3}\right) + \left(\frac{1}{3} - \frac{1}{4}\right) = 1 - \frac{1}{4}$$

In general,

$$S_N = \left(\frac{1}{1} - \frac{1}{2}\right) + \left(\frac{1}{2} - \frac{1}{3}\right) + \cdots + \left(\frac{1}{N} - \frac{1}{N+1}\right) = 1 - \frac{1}{N+1}$$

Now we may compute the sum S as the limit of the partial sums:

$$S = \lim_{N\to\infty} S_N = \lim_{N\to\infty}\left(1 - \frac{1}{N+1}\right) = 1$$ ■

It is important to keep in mind the difference between a sequence $\{a_n\}$ and an infinite series $\sum_{n=1}^{\infty} a_n$, which is a sum of terms of the sequence.

■ **EXAMPLE 2** **Difference between a Sequence and a Series** Discuss the difference between $\{a_n\}$ and $\sum_{n=0}^{\infty} a_n$, where $a_n = 3^{-n}$.

Solution The sequence $a_n = 3^{-n}$ converges to zero:

$$1,\quad \frac{1}{3},\quad \frac{1}{3^2},\quad \frac{1}{3^3},\quad \to 0$$

The infinite series defined by this sequence is an infinite sum:

$$\sum_{n=0}^{\infty} 3^{-n} = 1 + \frac{1}{3} + \frac{1}{3^2} + \cdots$$

The value of this sum is nonzero. In fact, the partial sum S_5 gives an approximation to this sum:

$$S_5 = 1 + \frac{1}{3} + \frac{1}{9} + \frac{1}{27} + \frac{1}{81} \approx 1.494$$ ■

One of the most important types of infinite series is the **geometric series**, defined as the sum of terms cr^n, where c and r are fixed numbers:

$$\sum_{n=0}^{\infty} cr^n = c + cr + cr^2 + cr^3 + cr^4 + cr^5 + \cdots$$

The number r is called the **common ratio**.

For $r = \frac{1}{2}$, we can visualize the sum of the geometric series (Figure 1):

$$\sum_{n=1}^{\infty} \frac{1}{2^n} = \frac{1}{2} + \frac{1}{4} + \frac{1}{8} + \frac{1}{16} + \cdots = 1$$

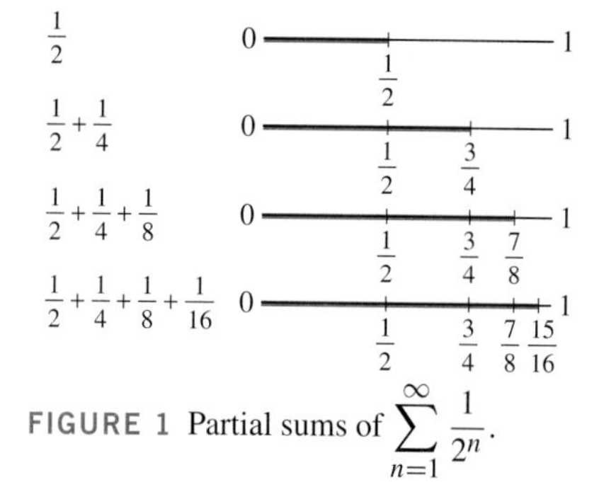

FIGURE 1 Partial sums of $\sum_{n=1}^{\infty} \frac{1}{2^n}$.

Geometric series are important because they

- *Arise often in applications.*
- *Can be evaluated explicitly.*
- *Are used to study other nongeometric series (by comparison).*

The sum is 1 because adding terms in the series corresponds to moving stepwise from 0 to 1, where each step consists of a move to the right by half of the remaining distance.

A simple device exists for computing the partial sums of a geometric series:

$$S_N = c + cr + cr^2 + cr^3 + \cdots + cr^N$$

$$rS_N = cr + cr^2 + cr^3 + \cdots + cr^N + cr^{N+1}$$

$$S_N - rS_N = S_N(1-r) = c - cr^{N+1} = c(1 - r^{N+1})$$

If $r \neq 1$, we may divide by $(1-r)$ to obtain

$$S_N = c + cr + cr^2 + cr^3 + \cdots + cr^N = \frac{c(1-r^{N+1})}{1-r} \qquad \boxed{2}$$

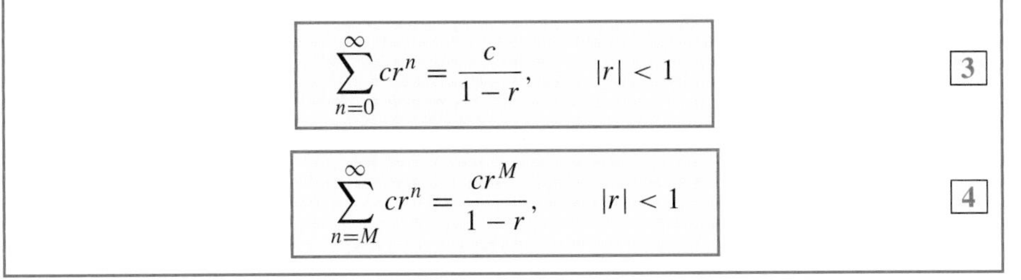

THEOREM 1 Sum of a Geometric Series A geometric series with common ratio r converges if $|r| < 1$ and diverges if $|r| \geq 1$. Furthermore,

$$\sum_{n=0}^{\infty} cr^n = \frac{c}{1-r}, \qquad |r| < 1 \tag{3}$$

$$\sum_{n=M}^{\infty} cr^n = \frac{cr^M}{1-r}, \qquad |r| < 1 \tag{4}$$

Proof If $r \neq 1$, then by Eq. (2),

$$\lim_{N\to\infty} S_N = \lim_{N\to\infty} \frac{c(1-r^{N+1})}{1-r} = \frac{c}{1-r}\left(1 - \lim_{N\to\infty} r^{N+1}\right)$$

If $|r| < 1$, then $\lim\limits_{N\to\infty} r^{N+1} = 0$ and we obtain Eq. (3):

$$\sum_{n=0}^{\infty} cr^n = \lim_{N\to\infty} S_N = \frac{c}{1-r}$$

If $|r| > 1$, $\lim\limits_{N\to\infty} r^{N+1}$ diverges and thus the geometric series diverges. It also diverges in the borderline cases $r = \pm 1$, as we saw in the discussion before Example 1. If the geometric series starts with the term cr^M rather than cr^0, then

$$\sum_{n=M}^{\infty} cr^n = \sum_{n=0}^{\infty} cr^{M+n} = \sum_{n=0}^{\infty} cr^M r^n = r^M \sum_{n=0}^{\infty} cr^n = \frac{cr^M}{1-r} \qquad \blacksquare$$

■ **EXAMPLE 3** Evaluate $\displaystyle\sum_{n=3}^{\infty} 7\left(-\frac{3}{4}\right)^n$.

Solution This is the geometric series with $c = 7$ and $r = -\frac{3}{4}$. The general term is $cr^n = 7\left(-\frac{3}{4}\right)^n$ and the sum starts at $n = 3$. By Eq. (4), the sum is

$$S = \sum_{n=3}^{\infty} 7\left(-\frac{3}{4}\right)^n = \frac{7\left(-\frac{3}{4}\right)^3}{1 - \left(-\frac{3}{4}\right)} = -\frac{27}{16} \qquad \blacksquare$$

A main goal in this chapter is to develop techniques for determining whether a series converges or diverges. Sometimes, it is obvious that an infinite series diverges. For example, $\displaystyle\sum_{k=1}^{\infty} 1$ diverges because its Nth partial sum is $S_N = N$. It is less clear if the following series converges or diverges:

$$\sum_{n=1}^{\infty} \frac{(-1)^{n+1} n}{n+1} = \frac{1}{2} - \frac{2}{3} + \frac{3}{4} - \frac{4}{5} + \frac{5}{6} - \cdots$$

We show that this series diverges in Example 4, using the next theorem.

THEOREM 2 Divergence Test If $\{a_n\}$ does not converge to zero, then $\sum_{n=1}^{\infty} a_n$ diverges.

Proof We use the relation

$$S_n = a_1 + a_2 + \cdots + a_{n-1} + a_n = S_{n-1} + a_n$$

to write $a_n = S_n - S_{n-1}$. If $\sum_{n=1}^{\infty} a_n$ is convergent with sum S, then

$$\lim_{n\to\infty} a_n = \lim_{n\to\infty} (S_n - S_{n-1}) = \lim_{n\to\infty} S_n - \lim_{n\to\infty} S_{n-1} = S - S = 0$$

Therefore, if $\{a_n\}$ does not converge to zero, $\sum_{n=1}^{\infty} a_n$ must diverge. ■

■ **EXAMPLE 4 Using the Divergence Test** Does

$$\sum_{n=1}^{\infty} (-1)^n \frac{n}{n+1} = -\frac{1}{2} + \frac{2}{3} - \frac{3}{4} + \frac{4}{5} - \cdots$$

converge?

Solution The general term $a_n = (-1)^n \frac{n}{n+1}$ does not tend to zero. In fact, $\frac{n}{n+1}$ tends to 1, so the even terms a_{2n} tend to 1 and the odd terms a_{2n+1} to -1. Therefore, the infinite series diverges by Theorem 2. ■

The Divergence Test only tells part of the story. If $\{a_n\}$ does not tend to zero, then $\sum a_n$ certainly diverges. But what if $\{a_n\}$ does tend to zero? In this case, the series may or may not converge. Here is an example of a series that diverges even though its terms tend to zero.

■ **EXAMPLE 5 A Divergent Series Where $\{a_n\}$ Tends to Zero** Show that

$$\sum_{n=1}^{\infty} \frac{1}{\sqrt{n}} = \frac{1}{\sqrt{1}} + \frac{1}{\sqrt{2}} + \frac{1}{\sqrt{3}} + \cdots$$

is divergent.

Solution Each term in the Nth partial sum is greater than or equal to $1/\sqrt{N}$:

$$S_N = \overbrace{\frac{1}{\sqrt{1}} + \frac{1}{\sqrt{2}} + \cdots + \frac{1}{\sqrt{N}}}^{N \text{ terms}} \ge \frac{1}{\sqrt{N}} + \frac{1}{\sqrt{N}} + \cdots + \frac{1}{\sqrt{N}}$$

Therefore,

$$S_N \ge N\left(\frac{1}{\sqrt{N}}\right) = \sqrt{N}$$

Since $S_N \ge \sqrt{N}$, we have $\lim_{N\to\infty} S_N = \infty$ and the series diverges. ■

The next theorem shows that infinite series may be added or subtracted like ordinary sums, *provided that the series are convergent.*

THEOREM 3 Linearity of Infinite Series If $\sum a_n$ and $\sum b_n$ are both convergent, then $\sum (a_n \pm b_n)$ and $\sum c a_n$ are convergent (c any constant), and

$$\sum a_n + \sum b_n = \sum (a_n + b_n)$$

$$\sum a_n - \sum b_n = \sum (a_n - b_n)$$

$$\sum c a_n = c \sum a_n \qquad (c \text{ any constant})$$

Proof These rules follow from the corresponding linearity rules for limits. For the first rule, we have

$$\sum_{n=1}^{\infty} (a_n + b_n) = \lim_{N\to\infty} \sum_{n=1}^{N} (a_n + b_n) = \lim_{N\to\infty} \left(\sum_{n=1}^{N} a_n + \sum_{n=1}^{N} b_n \right)$$

$$= \lim_{N\to\infty} \sum_{n=1}^{N} a_n + \lim_{N\to\infty} \sum_{n=1}^{\infty} b_n = \sum_{n=1}^{\infty} a_n + \sum_{n=1}^{\infty} b_n$$

The remaining statements are proved similarly. ■

■ **EXAMPLE 6** Evaluate $S = \sum_{n=0}^{\infty} \frac{2 + 3^n}{5^n}$.

Solution We write the series as a sum of two geometric series. This is valid by Theorem 3 because both geometric series are convergent:

$$\sum_{n=0}^{\infty} \frac{2 + 3^n}{5^n} = \sum_{n=0}^{\infty} \frac{2}{5^n} + \sum_{n=0}^{\infty} \frac{3^n}{5^n} = \overbrace{2\sum_{n=0}^{\infty} \frac{1}{5^n} + \sum_{n=0}^{\infty} \left(\frac{3}{5}\right)^n}^{\text{Both convergent geometric series}}$$

$$= 2 \cdot \frac{1}{1 - \frac{1}{5}} + \frac{1}{1 - \frac{3}{5}} = 5$$ ■

CONCEPTUAL INSIGHT Sometimes, the following *incorrect argument* is given for summing a geometric series:

$$S = \quad \frac{1}{2} + \frac{1}{4} + \frac{1}{8} + \cdots$$

$$2S = 1 + \frac{1}{2} + \frac{1}{4} + \frac{1}{8} + \cdots = 1 + S$$

Thus, $2S = 1 + S$ or $S = 1$. The answer is correct, so why is the argument wrong? It is wrong because we do not know in advance that the geometric series converges. Observe what happens when this argument is applied to a divergent series:

$$S = 1 - 1 + 1 - 1 + \cdots$$

$$-S = -1 + (1 - 1 + \cdots) = -1 + S$$

This would yield $-S = -1 + S$ or $S = \frac{1}{2}$, which is clearly incorrect since S diverges. Mathematicians developed the formal definition of the sum of an infinite series as the limit of partial sums in order to avoid erroneous conclusions of this type.

Archimedes (287 BC–212 BC), who discovered the law of the lever, said "Give me a place to stand on, and I can move the earth" (quoted by Pappus of Alexandria c. AD 340).

HISTORICAL PERSPECTIVE

Infinite series have been a part of calculus since the beginning of the subject and they have remained an indispensable tool of mathematical analysis ever since. Geometric series were used as early as the third century BC by Archimedes in a brilliant argument for determining the area S of a parabolic segment (shaded region in Figure 2). Archimedes's result is equivalent to our formula for the integral of $f(x) = x^2$, but he discovered it 2,000 years before the invention of calculus. Archimedes expressed his result geometrically rather than in terms of functions (which had not yet been invented). Given any two points A and C on a parabola, we may choose B between A and C so that the tangent at B is parallel to $\overline{AC}$. Let T be the area of triangle $\triangle ABC$. Archimedes proved that if D is chosen in a similar fashion relative to $\overline{AB}$ and E relative to $\overline{BC}$, then

$$\begin{aligned}\frac{1}{4}T &= \frac{1}{4}\text{Area}(\triangle ABC)\\ &= \text{Area}(\triangle ADB) + \text{Area}(\triangle BEC) \qquad \boxed{5}\end{aligned}$$

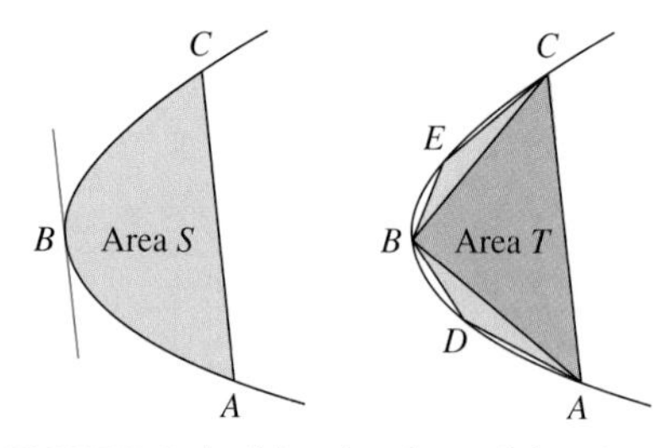

FIGURE 2 Archimedes showed that the area S of the parabolic segment is $\frac{4}{3}T$, where T is the area of $\triangle ABC$.

This construction of triangles can be continued. The next step would be to construct the four triangles on the segments $\overline{AD}$, $\overline{DB}$, $\overline{BE}$, $\overline{EC}$, of total area $(1/4)^2T$, etc. In this way, we obtain infinitely many triangles that completely fill up the parabolic segment. By Eq. (5) and the formula for the sum of a geometric series,

$$S = T + \frac{1}{4}T + \frac{1}{16}T + \cdots = T\sum_{n=0}^{\infty}\frac{1}{4^n} = \frac{4}{3}T$$

For this and many other achievements, Archimedes is ranked as one of the greatest scientists of all time, in the same league as Newton and Gauss.

The modern study of infinite series began in the seventeenth century with Newton, Leibniz, and their contemporaries. The divergence of $\sum_{n=1}^{\infty}\frac{1}{n}$ (called the **harmonic series**) was known to the medieval scholar Nicole d'Oresme (1323–1382), but his proof was lost for centuries and the result was rediscovered on more than one occasion. It was also known that the sum of the reciprocal squares $\sum_{n=1}^{\infty}\frac{1}{n^2}$ converges and, in the 1640s, the Italian Pietro Mengoli put forward the challenge of finding its sum. Despite the efforts of the best mathematicians of the day, including Leibniz and the Bernoulli brothers Jakob and Johann, the problem resisted solution for nearly a century. In 1735, the great master Leonhard Euler astonished his contemporaries by proving that

$$\boxed{\frac{1}{1^2} + \frac{1}{2^2} + \frac{1}{3^2} + \frac{1}{4^2} + \frac{1}{5^2} + \frac{1}{6^2} + \cdots = \frac{\pi^2}{6}} \qquad \boxed{6}$$

This formula is used in a variety of ways in number theory. For example, the probability p that two whole numbers, chosen randomly, have no common factor is $p = 6/\pi^2 \approx 0.6$ (the reciprocal of Euler's result). This application and others like it lie in the realm of "pure mathematics," and for hundreds of years, it seemed that Euler's result had no real-world applications. Surprisingly, there is now evidence that this result and its generalizations may play a role in the area of advanced physics called quantum field theory. History seems to show that even the "purest" branches of mathematics are connected to the real world.

10.2 SUMMARY

- An *infinite series* is an expression

$$\sum_{n=1}^{\infty} a_n = a_1 + a_2 + a_3 + a_4 + \cdots$$

We call a_n the *general term* of the series.

- The Nth *partial sum* is the finite sum:

$$S_N = \sum_{n=1}^{N} a_n = a_1 + a_2 + a_3 + \cdots + a_N$$

If the limit $S = \lim_{N\to\infty} S_N$ exists, we say that the infinite series is *convergent* or *converges* to the sum S. If the limit does not exist, the infinite series is called *divergent*.

- *Divergence Test:* If $\{a_n\}$ does not tend to zero, then $\sum_{n=1}^{\infty} a_n$ diverges. However, a series may diverge, even if its general term $\{a_n\}$ tends to zero.
- A *geometric series* with common ratio r satisfying $|r| < 1$ converges:

$$\sum_{n=M}^{\infty} cr^n = cr^M + cr^{M+1} + cr^{M+2} + \cdots = \frac{cr^M}{1-r}$$

The geometric series diverges for $|r| \geq 1$. There is a formula for the partial sum:

$$c + cr + cr^2 + cr^3 + \cdots + cr^N = \frac{c(1 - r^{N+1})}{1-r}$$

10.2 EXERCISES

Preliminary Questions

1. What role do partial sums play in defining the sum of an infinite series?

2. What is the sum of the following infinite series?

$$\frac{1}{4} + \frac{1}{8} + \frac{1}{16} + \frac{1}{32} + \frac{1}{64} + \cdots$$

3. What happens if you apply the formula for the sum of a geometric series to the following series? Is the formula valid?

$$1 + 3 + 3^2 + 3^3 + 3^4 + \cdots$$

4. Arvind asserts that $\sum_{n=1}^{\infty} \frac{1}{n^2} = 0$ because $\frac{1}{n^2}$ tends to zero. Is this valid reasoning?

5. Colleen claims that $\sum_{n=1}^{\infty} \frac{1}{\sqrt{n}}$ converges because $\lim_{n\to\infty} \frac{1}{\sqrt{n}} = 0$. Is this valid reasoning?

6. Find an N such that $S_N > 25$ for the series $\sum_{n=1}^{\infty} 2$.

7. Does there exist an N such that $S_N > 25$ for the series $\sum_{n=1}^{\infty} 2^{-n}$? Explain.

8. Give an example of a divergent infinite series whose general term tends to zero.

Exercises

1. Find a formula for the general term a_n (not the partial sum) of the infinite series.

(a) $\frac{1}{3} + \frac{1}{9} + \frac{1}{27} + \frac{1}{81} + \cdots$

(b) $\frac{1}{1} + \frac{5}{2} + \frac{25}{4} + \frac{125}{8} + \cdots$

(c) $\frac{1}{1} - \frac{2^2}{2\cdot 1} + \frac{3^3}{3\cdot 2\cdot 1} - \frac{4^4}{4\cdot 3\cdot 2\cdot 1} + \cdots$

(d) $\frac{2}{1^2+1} + \frac{1}{2^2+1} + \frac{2}{3^2+1} + \frac{1}{4^2+1} + \cdots$

2. Write in summation notation:

(a) $1 + \frac{1}{4} + \frac{1}{9} + \frac{1}{16} + \cdots$

(b) $\frac{1}{9} + \frac{1}{16} + \frac{1}{25} + \frac{1}{36} + \cdots$

(c) $1 - \frac{1}{3} + \frac{1}{5} - \frac{1}{7} + \cdots$

(d) $\frac{125}{9} + \frac{625}{16} + \frac{3,125}{25} + \frac{15,625}{36} + \cdots$

In Exercises 3–6, compute the partial sums S_2, S_4, and S_6.

3. $1 + \frac{1}{2^2} + \frac{1}{3^2} + \frac{1}{4^2} + \cdots$

4. $\sum_{k=1}^{\infty} (-1)^k k^{-1}$

5. $\dfrac{1}{1\cdot 2}+\dfrac{1}{2\cdot 3}+\dfrac{1}{3\cdot 4}+\cdots$

6. $\displaystyle\sum_{j=1}^{\infty}\frac{1}{j!}$

7. Compute S_5, S_{10}, and S_{15} for the series

$$S=\frac{1}{2\cdot 3\cdot 4}-\frac{1}{4\cdot 5\cdot 6}+\frac{1}{6\cdot 7\cdot 8}-\frac{1}{8\cdot 9\cdot 10}+\cdots$$

This series S is known to converge to $\dfrac{\pi-3}{4}$. Do your calculations support this conclusion?

8. The series

$$S=\frac{1}{0!}-\frac{1}{1!}+\frac{1}{2!}-\frac{1}{3!}+\cdots$$

is known to converge to e^{-1} (recall that $0!=1$). Find a partial sum that approximates e^{-1} with an error at most 10^{-3}.

9. Calculate S_3, S_4, and S_5 and then find the sum of the telescoping series

$$S=\sum_{n=1}^{\infty}\left(\frac{1}{n+1}-\frac{1}{n+2}\right)$$

10. Calculate S_3, S_4, and S_5 and then find the sum $S=\displaystyle\sum_{n=1}^{\infty}\frac{1}{4n^2-1}$ using the identity

$$\frac{1}{4n^2-1}=\frac{1}{2}\left(\frac{1}{2n-1}-\frac{1}{2n+1}\right)$$

11. Write $\displaystyle\sum_{n=3}^{\infty}\frac{1}{n(n-1)}$ as a telescoping series and find its sum.

12. Find a formula for the partial sums S_N of $\displaystyle\sum_{n=1}^{\infty}(-1)^{n-1}$ and show that the series diverges.

In Exercises 13–16, use Theorem 2 to prove that the following series diverge.

13. $\displaystyle\sum_{n=1}^{\infty}(-1)^n n^2$

14. $\dfrac{0}{1}-\dfrac{1}{2}+\dfrac{2}{3}-\dfrac{3}{4}+\cdots$

15. $\displaystyle\sum_{n=1}^{\infty}\left(\sqrt{n+1}-\sqrt{n}\right)$

16. $\cos\frac{1}{2}+\cos\frac{1}{3}+\cos\frac{1}{4}+\cdots$

17. Which of these series converge?

(a) $\displaystyle\sum_{n=1}^{\infty}\left(\frac{1}{\sqrt{n}}-\frac{1}{\sqrt{n+1}}\right)$ **(b)** $\displaystyle\sum_{n=1}^{\infty}(\ln n-\ln(n+1))$

In Exercises 18–31, use the formula for the sum of a geometric series to find the sum or state that the series diverges.

18. $1+\dfrac{1}{5}+\dfrac{1}{5^2}+\dfrac{1}{5^3}+\cdots$

19. $\dfrac{1}{3^3}+\dfrac{1}{3^4}+\dfrac{1}{3^5}+\cdots$

20. $\displaystyle\sum_{n=0}^{\infty}\frac{3^n}{11^n}$

21. $\displaystyle\sum_{n=3}^{\infty}\frac{3^n}{11^n}$

22. $\displaystyle\sum_{n=0}^{\infty}\frac{7\cdot 3^n}{11^n}$

23. $1+\dfrac{2}{7}+\dfrac{2^2}{7^2}+\dfrac{2^3}{7^3}+\cdots$

24. $\displaystyle\sum_{n=1}^{\infty}e^{-n}$

25. $\displaystyle\sum_{n=2}^{\infty}e^{3-2n}$

26. $\displaystyle\sum_{n=0}^{\infty}\frac{8+2^n}{5^n}$

27. $\displaystyle\sum_{n=0}^{\infty}\frac{93^n+4^{n-2}}{5^n}$

28. $5-\dfrac{5}{4}+\dfrac{5}{4^2}-\dfrac{5}{4^3}+\cdots$

29. $\dfrac{2^3}{7}+\dfrac{2^4}{7^2}+\dfrac{2^5}{7^3}+\dfrac{2^6}{7^4}+\cdots$

30. $\dfrac{7}{8}-\dfrac{49}{64}+\dfrac{343}{512}-\dfrac{2{,}401}{4{,}096}+\cdots$

31. $\dfrac{64}{49}+\dfrac{8}{7}+1+\dfrac{7}{8}+\dfrac{49}{64}+\dfrac{343}{512}+\cdots$

32. Which of the following are *not* geometric series?

(a) $\displaystyle\sum_{n=0}^{\infty}\frac{7^n}{29^n}$ **(b)** $\displaystyle\sum_{n=3}^{\infty}\frac{1}{n^4}$

(c) $\displaystyle\sum_{n=0}^{\infty}\frac{n^2}{2^n}$ **(d)** $\displaystyle\sum_{n=5}^{\infty}\pi^{-n}$

33. Which of the following series are divergent?

(a) $\displaystyle\sum_{n=0}^{\infty}\frac{2^n}{5^n}$ **(b)** $\displaystyle\sum_{n=3}^{\infty}1.5^n$

(c) $\displaystyle\sum_{n=0}^{\infty}\frac{5^n}{2^n}$ **(d)** $\displaystyle\sum_{n=0}^{\infty}(0.4)^n$

34. Explain why each of the following statements is incorrect.

(a) If the general term a_n tends to zero, $\displaystyle\sum_{n=1}^{\infty}a_n=0$.

(b) The Nth partial sum of the infinite series defined by $\{a_n\}$ is a_N.

(c) If a_n tends to zero, then $\displaystyle\sum_{n=1}^{\infty}a_n$ converges.

(d) If a_n tends to L, then $\displaystyle\sum_{n=1}^{\infty}a_n=L$.

35. Let $S=\displaystyle\sum_{n=1}^{\infty}a_n$ be an infinite series such that $S_N=5-\dfrac{2}{N^2}$.

(a) What are the values of $\displaystyle\sum_{n=1}^{10}a_n$ and $\displaystyle\sum_{n=4}^{16}a_n$?

(b) What is the value of a_3?

(c) Find a general formula for a_n.

(d) Find the sum $\sum_{n=1}^{\infty} a_n$.

36. Compute the total area of the (infinitely many) triangles in Figure 3.

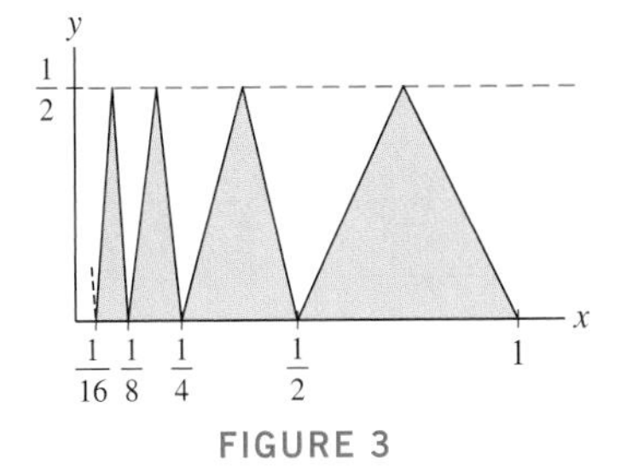

FIGURE 3

37. Use the method of Example 5 to show that $\sum_{k=1}^{\infty} \frac{1}{k^{1/3}}$ diverges.

38. Let S_N be the Nth partial sum of the **harmonic series** $S = \sum_{n=1}^{\infty} \frac{1}{n}$.

(a) Verify the following inequality for $n = 1, 2, 3$. Then prove it for general n.

$$\frac{1}{2^{n-1}+1} + \frac{1}{2^{n-1}+2} + \frac{1}{2^{n-1}+3} + \cdots + \frac{1}{2^n} \geq \frac{1}{2}$$

(b) Prove that S diverges by showing that $S_N \geq 1 + \frac{n}{2}$ for $N = 2^n$. *Hint:* Break up S_N into $n+1$ sums of length $1, 2, 4, 8 \dots$, as in the following:

$$S_{2^3} = 1 + \left(\frac{1}{2}\right) + \left(\frac{1}{3} + \frac{1}{4}\right) + \left(\frac{1}{5} + \frac{1}{6} + \frac{1}{7} + \frac{1}{8}\right)$$

39. A ball dropped from a height of 10 ft begins to bounce. Each time it strikes the ground, it returns to two-thirds of its previous height. What is the total distance traveled by the ball if it bounces infinitely many times?

40. Use partial fractions to rewrite $\sum_{n=1}^{\infty} \frac{1}{n(n+3)}$ as a telescoping series and find its sum.

41. Find the sum of $\frac{1}{1 \cdot 3} + \frac{1}{3 \cdot 5} + \frac{1}{5 \cdot 7} + \cdots$.

42. Let $S = \sum_{n=1}^{\infty} \left(\frac{1}{n} - \frac{1}{n+2}\right)$. Compute S_N for $N = 1, 2, 3, 4$. Find S by showing that

$$S_N = \frac{3}{2} - \frac{1}{N+1} - \frac{1}{N+2}$$

43. Let $\{b_n\}$ be a sequence and let $a_n = b_n - b_{n-1}$. Show that $\sum_{n=1}^{\infty} a_n$ converges if and only if $\lim_{n\to\infty} b_n$ exists.

44. The winner of a lottery receives m dollars at the end of each year for N years. The present value (PV) of this prize in today's dollars is $\text{PV} = \sum_{i=1}^{N} m(1+r)^{-i}$, where r is the interest rate. Calculate PV if $m = \$50{,}000$, $r = 0.06$, and $N = 20$. What is the PV if $N = \infty$?

45. Find the total length of the infinite zigzag path in Figure 4 (each zag occurs at an angle of $\frac{\pi}{4}$).

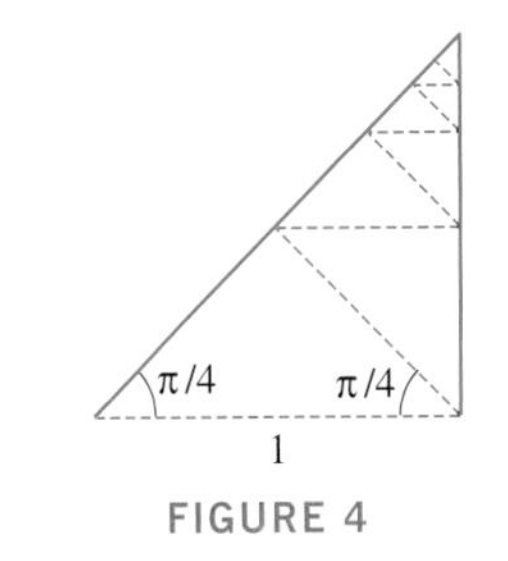

FIGURE 4

46. Evaluate $\sum_{n=1}^{\infty} \frac{1}{n(n+1)(n+2)}$. *Hint:* Find constants A, B, and C such that

$$\frac{1}{n(n+1)(n+2)} = \frac{A}{n} + \frac{B}{(n+1)} + \frac{C}{n+2}$$

47. Show that if a is a positive integer, then

$$\sum_{n=1}^{\infty} \frac{1}{n(n+a)} = \frac{1}{a}\left(1 + \frac{1}{2} + \cdots + \frac{1}{a}\right)$$

48. Assumptions Matter Show, by giving counterexamples, that the assertions of Theorem 3 is not valid if the series $\sum_{n=0}^{\infty} a_n$ and $\sum_{n=0}^{\infty} b_n$ are not convergent.

Further Insights and Challenges

49. Professor George Andrews of Pennsylvania State University observed that geometric sums can be used to calculate the derivative of $f(x) = x^N$ in a new way. By Eq. (2),

$$1 + r + r^2 + \cdots + r^{N-1} = \frac{1 - r^N}{1 - r} \qquad \boxed{7}$$

Assume that $a \neq 0$ and let $x = ra$. Show that

$$f'(a) = \lim_{x \to a} \frac{x^N - a^N}{x - a} = a^{N-1} \lim_{r \to 1} \frac{r^N - 1}{r - 1}$$

Then use Eq. (7) to evaluate the limit on the right.

50. Pierre de Fermat used geometric series to compute the area under the graph of $f(x) = x^N$ over $[0, A]$. For $0 < r < 1$, let $F(r)$ be the sum of the areas of the infinitely many right-endpoint rectangles with endpoints Ar^n, as in Figure 5. As r tends to 1, the rectangles become narrower and $F(r)$ tends to the area under the graph.

(a) Show that $F(r) = A^{N+1} \dfrac{1 - r}{1 - r^{N+1}}$.

(b) Use Eq. (7) to evaluate $\displaystyle\int_0^A x^N \, dx = \lim_{r \to 1} F(r)$.

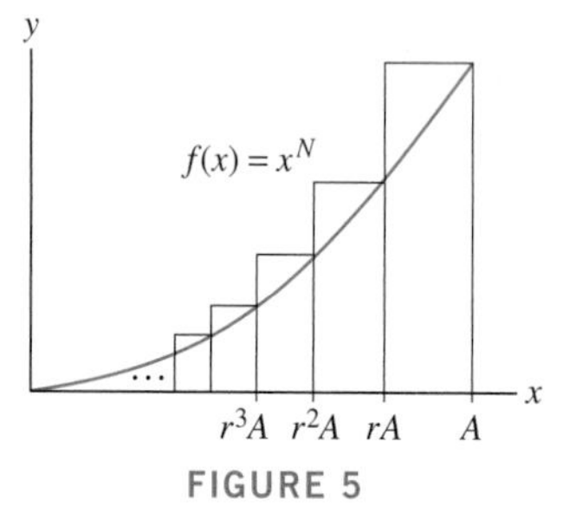

FIGURE 5

51. Cantor's Disappearing Table (following Larry Knop of Hamilton College) Take a table of length L (Figure 6). At stage 1, remove the section of length $L/4$ centered at the midpoint. Two sections remain, each with a length less than $L/2$. At stage 2, remove sections of length $L/4^2$ from each of these two sections (this stage removes $L/8$ of the table). Now four sections remain, each of length less than $L/4$. At stage 3, remove the four central sections of length $L/4^3$, etc.

(a) Show that at the Nth stage, each remaining section has length less than $L/2^N$ and that the total amount of table removed is

$$L\left(\frac{1}{4} + \frac{1}{8} + \frac{1}{16} + \cdots + \frac{1}{2^{N+1}}\right)$$

(b) Show that in the limit as $N \to \infty$, precisely one-half of the table remains.

This result is curious, because there are no nonzero intervals of table left (at each stage, the remaining sections have a length less than $L/2^N$). So the table has "disappeared." However, we can place any object longer than $L/4$ on the table and it will not fall through since it will not fit through any of the removed sections.

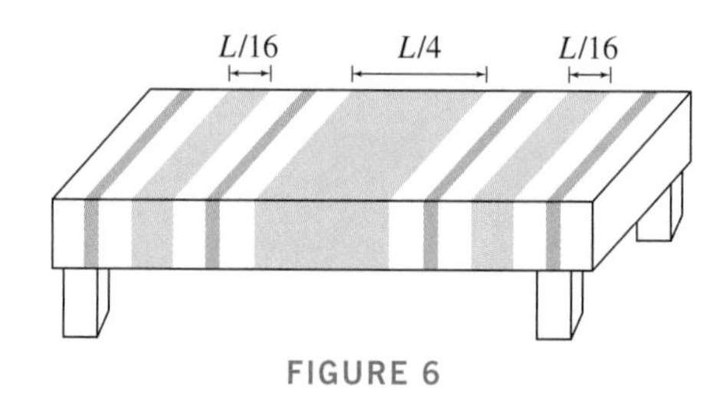

FIGURE 6

52. The **Koch snowflake** (described in 1904 by Swedish mathematician Helge von Koch) is an infinitely jagged "fractal" curve obtained as a limit of polygonal curves (it is continuous, but has no tangent line at any point). Begin with an equilateral triangle (stage 0) and produce stage 1 by replacing each edge with four edges of one-third the length, arranged as in Figure 7. Continue the process: At the nth stage, replace each edge with four edges of one-third the length.

(a) Show that the perimeter P_n of the polygon at the nth stage satisfies $P_n = \frac{4}{3}P_{n-1}$. Prove that $\lim_{n \to \infty} P_n = \infty$. The snowflake has infinite length.

(b) Let A_0 be the area of the original equilateral triangle. Show that $(3)4^{n-1}$ new triangles are added at the nth stage, each with area $A_0/9^n$ (for $n \geq 1$). Show that the total area of the Koch snowflake is $\frac{8}{5}A_0$.

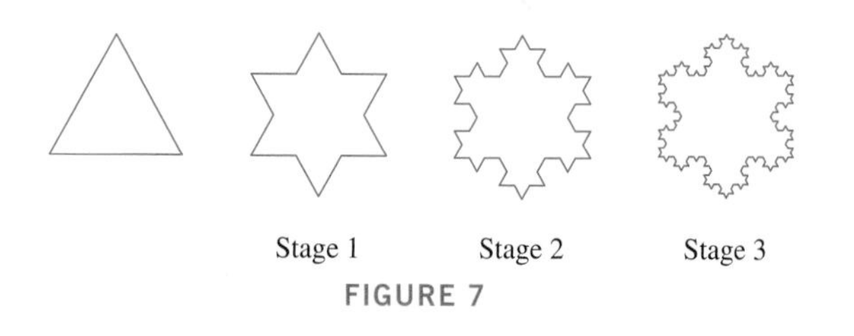

FIGURE 7

10.3 Convergence of Series with Positive Terms

In the next three sections, we focus on the problem of determining whether an infinite series converges or diverges. This is easier than finding the sum of an infinite series, which is possible only in special cases.

There are powerful numerical methods for finding approximations to infinite series. When implemented on a computer, these methods can be used to compute sums to millions of digits (see Exercises 75–77).

In this section, we consider **positive series** $\sum a_n$, that is, series such that $a_n \geq 0$ for all n (thus the terms of the series are nonnegative). The terms of a positive series may be visualized as rectangles of width 1 and height a_n (Figure 1). The partial sum

$$S_N = a_1 + a_2 + \cdots + a_N$$

is equal to the area of the first N rectangles.

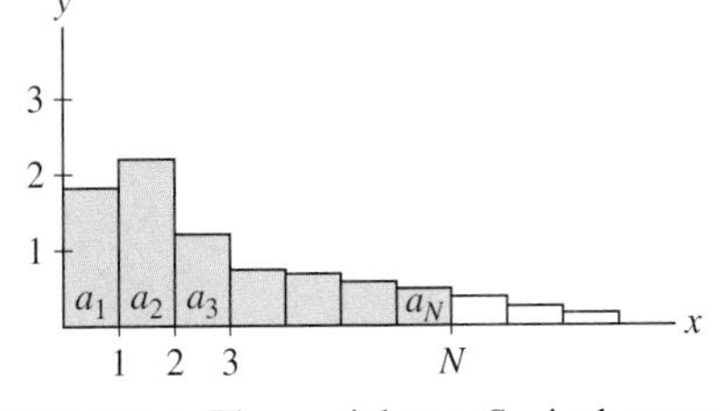

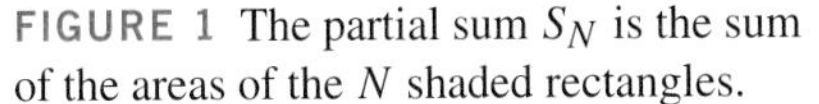
FIGURE 1 The partial sum S_N is the sum of the areas of the N shaded rectangles.

A key property of positive series is that the partial sums S_N form an increasing sequence. Each partial sum is obtained from the previous one by adding a positive number:

$$S_{N+1} = (a_1 + a_2 + \cdots + a_N) + a_{N+1} = S_N + \underbrace{a_{N+1}}_{\text{Nonnegative}}$$

and thus $S_N \le S_{N+1}$. Recall that an increasing sequence converges if it is bounded above and diverges otherwise (Theorem 6, Section 10.1). It follows that there are just two ways a positive series can behave (we refer to this as the "dichotomy").

> **THEOREM 1 Dichotomy Theorem for Positive Series** If $S = \sum_{n=1}^{\infty} a_n$ is a positive series, then there are two possibilities:
>
> **(i)** The partial sums S_N are bounded above. In this case, S converges.
> **(ii)** The partial sums S_N are not bounded above. In this case, S diverges.

Assumptions Matter This dichotomy does not hold in general for nonpositive series. The partial sums of the nonpositive series

$$S = \sum_{n=1}^{\infty} (-1)^n = 1 - 1 + 1 - 1 + 1 - 1 + \cdots$$

are bounded since $S_N = 1$ or 0, but S diverges.

One of the most important applications of Theorem 1 is the following Integral Test, which is useful because integrals are often easier to evaluate than series.

The Integral Test is valid for any series $\sum_{n=k}^{\infty} f(n)$, provided that $f(x)$ is positive, decreasing, and continuous for $x \ge M$ for some M. The convergence of the series is determined by the convergence of

$$\int_M^{\infty} f(x)\, dx$$

> **THEOREM 2 Integral Test** Let $a_n = f(n)$, where $f(x)$ is positive, decreasing, and continuous for $x \ge 1$.
>
> **(i)** If $\int_1^{\infty} f(x)\, dx$ converges, then $\sum_{n=1}^{\infty} a_n$ converges.
>
> **(ii)** If $\int_1^{\infty} f(x)\, dx$ diverges, then $\sum_{n=1}^{\infty} a_n$ diverges.

Proof We compare S_N with the area under the graph of $f(x)$ over the interval $[1, N]$. Since $f(x)$ is decreasing (Figure 2),

$$\underbrace{a_2 + \cdots + a_N}_{\text{Area of shaded rectangles in Figure 2}} \le \int_1^N f(x)\, dx \le \int_1^{\infty} f(x)\, dx$$

FIGURE 2

If the improper integral on the right converges, then the sums $a_2 + \cdots + a_N$ remain bounded. In this case, S_N also remains bounded and the infinite series converges by the Dichotomy Theorem (Theorem 1).

On the other hand (Figure 3),

$$\int_1^N f(x)\, dx \le \underbrace{a_1 + a_2 + \cdots + a_{N-1}}_{\text{Area of shaded rectangles in Figure 3}} \qquad \boxed{1}$$

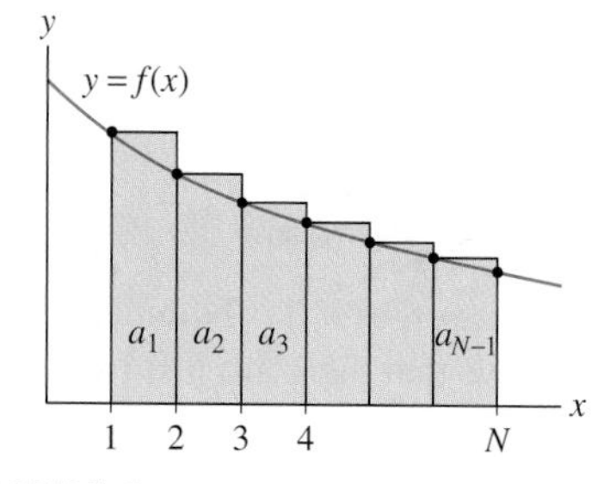

FIGURE 3

The infinite series

$$\sum_{n=1}^{\infty} \frac{1}{n}$$

is called the "harmonic series."

If $\int_1^{\infty} f(x)\,dx$ diverges, then $\int_1^N f(x)\,dx$ tends to ∞ and (1) shows that S_N also tends to ∞. ■

■ **EXAMPLE 1** **Divergence of the Harmonic Series** Show that $\sum_{n=1}^{\infty} \frac{1}{n}$ diverges.

Solution The function $f(x) = \frac{1}{x}$ is positive, decreasing, and continuous for $x \geq 1$, so we may apply the Integral Test:

$$\int_1^{\infty} \frac{dx}{x} = \lim_{R\to\infty} \int_1^R \frac{dx}{x} = \lim_{R\to\infty} \ln R = \infty$$

The integral diverges, and hence the sum $\sum_{n=1}^{\infty} \frac{1}{n}$ also diverges. ■

■ **EXAMPLE 2** Determine whether $\sum_{n=1}^{\infty} \frac{n}{(n^2+1)^2} = \frac{1}{2^2} + \frac{2}{5^2} + \frac{3}{10^2} + \cdots$ converges.

Solution The function $f(x) = \frac{x}{(x^2+1)^2}$ is positive and continuous for $x \geq 1$, and it is decreasing since $f'(x)$ is negative:

$$f'(x) = \frac{1-3x^2}{(x^2+1)^3} < 0 \qquad \text{for } x \geq 1$$

Therefore, we may apply the Integral Test. We use the substitution $u = x^2 + 1$, $du = 2x\,dx$ to evaluate the improper integral:

$$\int_1^{\infty} \frac{x}{(x^2+1)^2}\,dx = \lim_{R\to\infty} \int_1^R \frac{x}{(x^2+1)^2}\,dx = \lim_{R\to\infty} \frac{1}{2} \int_2^R \frac{du}{u^2}$$

$$= \lim_{R\to\infty} \left. \frac{-1}{2u} \right|_2^R = \lim_{R\to\infty} \left(\frac{1}{4} - \frac{1}{2R} \right) = \frac{1}{4}$$

The integral converges and hence $\sum_{n=1}^{\infty} \frac{n}{(n^2+1)^2}$ also converges. ■

The Integral Test applies to the sums of reciprocal powers n^{-p}, called ***p*-series**.

> **THEOREM 3 Convergence of *p*-Series** The infinite series $\sum_{n=1}^{\infty} \frac{1}{n^p}$ converges if $p > 1$ and diverges otherwise.

Proof If $p \neq 1$, we have

$$\int_1^{\infty} \frac{1}{x^p}\,dx = \lim_{R\to\infty} \int_1^R \frac{1}{x^p}\,dx = \lim_{R\to\infty} \left. \frac{x^{1-p}}{1-p} \right|_1^R = \lim_{R\to\infty} \frac{R^{1-p}-1}{1-p}$$

Since R^{1-p} tends to zero if $p > 1$ and to ∞ if $p < 1$, the improper integral converges for $p > 1$ and diverges for $p < 1$. The same is true of the p-series by the Integral Test. For $p = 1$, the series diverges, as shown in Example 1. ■

Here are two examples of p-series:

$$p = \frac{1}{2}: \qquad \sum_{n=1}^{\infty} \frac{1}{\sqrt{n}} = \frac{1}{\sqrt{1}} + \frac{1}{\sqrt{2}} + \frac{1}{\sqrt{3}} + \frac{1}{\sqrt{4}} + \cdots = \infty \quad \text{diverges}$$

$$p = 2: \qquad \sum_{n=1}^{\infty} \frac{1}{n^2} = \frac{1}{1} + \frac{1}{2^2} + \frac{1}{3^2} + \frac{1}{4^2} + \cdots \quad \text{converges}$$

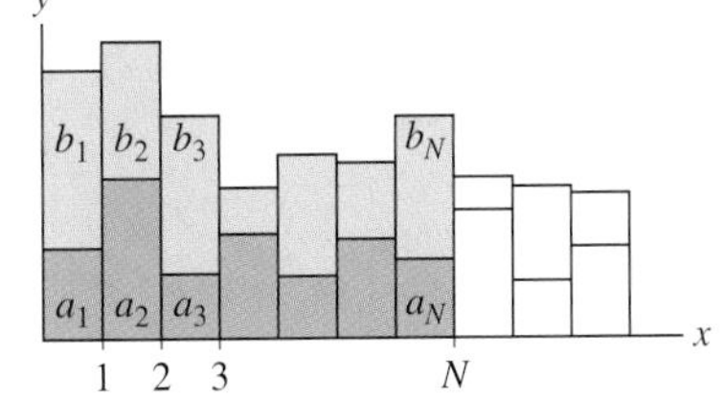

FIGURE 4 The series $\sum a_n$ is dominated by the series $\sum b_n$.

Another powerful method for determining convergence of positive series is comparison. Suppose that $0 \le a_n \le b_n$. Figure 4 suggests that if the larger sum $\sum b_n$ *converges*, then the smaller sum $\sum a_n$ also converges and, similarly, if the smaller sum *diverges*, then the larger sum also diverges.

THEOREM 4 Comparison Test
Assume that there exists $M > 0$ such that $0 \le a_n \le b_n$ for $n \ge M$.

(i) If $\sum_{n=1}^{\infty} b_n$ converges, then $\sum_{n=1}^{\infty} a_n$ also converges.

(ii) If $\sum_{n=1}^{\infty} a_n$ diverges, then $\sum_{n=1}^{\infty} b_n$ also diverges.

The convergence of an infinite series does not depend on where the series begins. Therefore, the Comparison Test remains valid even if the series does not begin with $n = 1$.

Proof Assume without loss of generality that $M = 1$. If $S = \sum_{n=1}^{\infty} b_n$ converges, then

$$a_1 + a_2 + \cdots + a_N \le b_1 + b_2 + \cdots + b_N \le \sum_{n=1}^{\infty} b_n = S \qquad \boxed{2}$$

Thus, the partial sums of $\sum_{n=1}^{\infty} a_n$ are bounded above by S, and $\sum_{n=1}^{\infty} a_n$ converges by the Dichotomy Theorem (Theorem 1). On the other hand, if $\sum_{n=1}^{\infty} a_n$ diverges, then its partial sums increase beyond bound and (2) shows that $\sum_{n=1}^{\infty} b_n$ also diverges. ■

In words, the Comparison Test states that for positive series:

- *Convergence of larger series forces convergence of smaller series.*
- *Divergence of smaller series forces divergence of larger series.*

■ **EXAMPLE 3** Show that $S = \sum_{n=1}^{\infty} 2^{-n!} = 2^{-1} + 2^{-2} + 2^{-6} + 2^{-24} + \cdots$ converges.

Solution We apply the Comparison Test with $a_n = 2^{-n!}$ and $b_n = 2^{-n}$. This is valid because $0 \le 2^{-n!} \le 2^{-n}$ for $n \ge 1$. The geometric series converges:

$$\sum_{n=1}^{\infty} b_n = \sum_{n=1}^{\infty} 2^{-n} = 1 \quad \text{(converges)}$$

Therefore, the smaller series $\sum_{n=1}^{\infty} a_n = \sum_{n=1}^{\infty} 2^{-n!}$ converges. ■

■ **EXAMPLE 4** Show that $\sum_{n=1}^{\infty} \frac{1}{\sqrt{n}\, 3^n}$ converges.

Solution We compare $\sum_{n=1}^{\infty} \frac{1}{\sqrt{n}\,3^n}$ with the geometric series $\sum_{n=1}^{\infty} \frac{1}{3^n}$. For $n \geq 1$,

$$\frac{1}{\sqrt{n}\,3^n} \leq \frac{1}{3^n}$$

The geometric series $\sum_{n=1}^{\infty} \frac{1}{3^n}$ converges, so $\sum_{n=1}^{\infty} \frac{1}{\sqrt{n}\,3^n}$ also converges. ■

In Example 5, the series begins with $n = 2$ because $1/\ln n$ is undefined for $n = 1$.

■ **EXAMPLE 5** Determine whether $\sum_{n=2}^{\infty} \frac{1}{\ln n}$ converges.

Solution We compare $\sum_{n=2}^{\infty} \frac{1}{\ln n}$ with the harmonic series $\sum_{n=2}^{\infty} \frac{1}{n}$ by showing that for $n \geq 2$,

$$\frac{1}{\ln n} \geq \frac{1}{n} \qquad \text{or equivalently,} \qquad n \geq \ln n$$

It suffices to show that $f(x) = x - \ln x$ is positive for $x \geq 2$. However, $f(1) = 1$ and $f'(x) = 1 - x^{-1} > 0$ for $x > 1$. Therefore, $f(x)$ is increasing for $x > 1$ and $f(x) > 0$ as required. Since the harmonic series diverges, the larger series $\sum_{n=2}^{\infty} \frac{1}{\ln n}$ also diverges. ■

■ **EXAMPLE 6** **Using the Comparison Test Correctly** Study the convergence of

$$\sum_{n=2}^{\infty} \frac{1}{n(\ln n)^2}$$

Solution We might be tempted to compare $\sum_{n=2}^{\infty} \frac{1}{n(\ln n)^2}$ to the harmonic series $\sum_{n=2}^{\infty} \frac{1}{n}$ using the inequality (valid for $n \geq 2$)

$$\frac{1}{n(\ln n)^2} \leq \frac{1}{n}$$

However, $\sum_{n=2}^{\infty} \frac{1}{n}$ diverges, so this inequality gives us no information about the *smaller* series $\sum \frac{1}{n(\ln n)^2}$. Fortunately, in this case we can use the Integral Test. The substitution $u = \ln x$ yields

$$\int_2^{\infty} \frac{dx}{x(\ln x)^2} = \int_{\ln 2}^{\infty} \frac{du}{u^2} = \lim_{R\to\infty} \left(\frac{1}{\ln 2} - \frac{1}{\ln R} \right) = \frac{1}{\ln 2} < \infty$$

The Integral Test shows that $\sum_{n=2}^{\infty} \frac{1}{n(\ln n)^2}$ converges. ■

Suppose we wish to study the convergence of

$$S = \sum_{n=2}^{\infty} \frac{n^2}{n^4 - n} \tag{3}$$

For large n, the general term is very close to $\dfrac{1}{n^2}$:

$$\frac{n^2}{n^4 - n} = \frac{1}{n^2 - n^{-1}} \approx \frac{1}{n^2}$$

so we might try to compare S with the convergent series $\sum_{n=2}^{\infty} \dfrac{1}{n^2}$. However, the Comparison Test cannot be used directly because the precise inequality goes in the wrong direction:

$$\frac{n^2}{n^4 - n} > \frac{n^2}{n^4} = \frac{1}{n^2}$$

In this case, we may apply the following variation of the Comparison Test.

CAUTION The Limit Comparison Test may not be applied if the series are not positive. See Exercise 38 in Section 10.4.

THEOREM 5 Limit Comparison Test Let $\{a_n\}$ and $\{b_n\}$ be *positive* sequences. Assume that the following limit exists:

$$L = \lim_{n\to\infty} \frac{a_n}{b_n}$$

- If $L > 0$, then $\sum_{n=1}^{\infty} a_n$ converges if and only if $\sum_{n=1}^{\infty} b_n$ converges.
- If $L = 0$ and $\sum_{n=1}^{\infty} b_n$ converges, then $\sum_{n=1}^{\infty} a_n$ converges.

Proof First we show that if $\sum_{n=1}^{\infty} b_n$ converges, then $\sum_{n=1}^{\infty} a_n$ converges for $L > 0$ or $L = 0$. Choose a positive number $R > L$. Since the sequences are positive and a_n/b_n approaches L, we have $0 \le a_n/b_n \le R$ and thus $a_n \le Rb_n$ for all n sufficiently large. Since $\sum_{n=1}^{\infty} Rb_n$ also converges, $\sum_{n=1}^{\infty} a_n$ converges by the Comparison Test.

Now suppose that $L > 0$ and $\sum_{n=1}^{\infty} a_n$ converges. We may choose r such that $0 < r < L$. Since a_n/b_n approaches L, we have $a_n/b_n \ge r$ and thus $a_n \ge rb_n$ for all n sufficiently large. Therefore, $\sum_{n=1}^{\infty} rb_n$ converges by the Comparison Test and hence $\sum_{n=1}^{\infty} b_n$ converges. ■

■ **EXAMPLE 7** Show that $\sum_{n=2}^{\infty} \dfrac{n^2}{n^4 - n}$ converges.

Solution Let $a_n = \dfrac{n^2}{n^4 - n}$. We observed above that $a_n \approx n^{-2}$ for large n, so it makes sense to apply the Limit Comparison Test with $b_n = n^{-2}$:

$$\frac{a_n}{b_n} = \left(\frac{n^2}{n^4 - n}\right)n^2 = \frac{n^4}{n^4 - n} = \frac{1}{1 - n^{-3}}$$

$$L = \lim_{n\to\infty} \frac{a_n}{b_n} = \lim_{n\to\infty} \frac{1}{1 - n^{-3}} = 1$$

Since L exists and $\sum_{n=2}^{\infty} b_n = \sum_{n=2}^{\infty} n^{-2}$ converges, $\sum_{n=2}^{\infty} a_n$ also converges. ■

EXAMPLE 8 Determine whether $\sum_{n=3}^{\infty} \frac{1}{\sqrt{n^2-4}}$ converges.

Solution Let $a_n = \frac{1}{\sqrt{n^2-4}}$ and $b_n = \frac{1}{n}$. Then

$$L = \lim_{n\to\infty} \frac{a_n}{b_n} = \lim_{n\to\infty} \frac{n}{\sqrt{n^2-4}} = \lim_{n\to\infty} \frac{1}{\sqrt{1-4/n^2}} = 1$$

Since $\sum_{n=3}^{\infty} \frac{1}{n}$ diverges and $L > 0$, the series $\sum_{n=3}^{\infty} \frac{1}{\sqrt{n^2-4}}$ also diverges. ■

10.3 SUMMARY

- The partial sums S_N of a positive series $S = \sum a_n$ form an increasing sequence.
- *Dichotomy Theorem:* A positive series S converges if its partial sums S_N remain bounded. Otherwise, it diverges.
- *Integral Test:* If f is positive, decreasing, and continuous, then $S = \sum f(n)$ converges (or diverges) if $\int_M^{\infty} f(x)\,dx$ converges (or diverges) for some $M > 0$.
- *p-Series:* The series $\sum_{n=1}^{\infty} \frac{1}{n^p}$ converges if $p > 1$ and diverges if $p \le 1$.
- *Comparison Test:* Assume that there exists $M > 0$ such that $0 \le a_n \le b_n$ for $n \ge M$. If $\sum_{n=1}^{\infty} b_n$ converges, then $\sum_{n=1}^{\infty} a_n$ converges; if $\sum_{n=1}^{\infty} a_n$ diverges, then $\sum_{n=1}^{\infty} b_n$ diverges.
- *Limit Comparison Test:* Let $\{a_n\}$ and $\{b_n\}$ be positive sequences and assume that the following limit exists:

$$L = \lim_{n\to\infty} \frac{a_n}{b_n}$$

  – If $L > 0$, then $\sum_{n=1}^{\infty} a_n$ converges if and only if $\sum_{n=1}^{\infty} b_n$ converges.

  – If $L = 0$ and $\sum_{n=1}^{\infty} b_n$ converges, then $\sum_{n=1}^{\infty} a_n$ converges.

10.3 EXERCISES

Preliminary Questions

1. Let $S = \sum_{n=1}^{\infty} a_n$. If the partial sums S_N are increasing, then (choose correct conclusion)

(a) $\{a_n\}$ is an increasing sequence.
(b) $\{a_n\}$ is a positive sequence.

2. What are the hypotheses of the Integral Test?

3. Which test would you use to determine whether $\sum_{n=1}^{\infty} n^{-3.2}$ converges?

4. Which test would you use to determine whether $\sum_{n=1}^{\infty} \frac{1}{2^n + \sqrt{n}}$ converges?

5. Ralph hopes to investigate the convergence of $\sum_{n=1}^{\infty} \frac{e^{-n}}{n}$ by comparing it with $\sum_{n=1}^{\infty} \frac{1}{n}$. Is Ralph on the right track?

Exercises

In Exercises 1–14, use the Integral Test to determine whether the infinite series is convergent.

1. $\sum_{n=1}^{\infty} \frac{1}{n^4}$

2. $\sum_{n=1}^{\infty} \frac{1}{n+3}$

3. $\sum_{n=1}^{\infty} n^{-1/3}$

4. $\sum_{n=5}^{\infty} \frac{1}{\sqrt{n-4}}$

5. $\sum_{n=25}^{\infty} \frac{n^2}{(n^3+9)^{5/2}}$

6. $\sum_{n=1}^{\infty} \frac{n}{\sqrt{n^2+1}}$

7. $\sum_{n=1}^{\infty} \frac{1}{n^2+1}$

8. $\sum_{n=1}^{\infty} \frac{n}{(n^2+1)^2}$

9. $\sum_{n=1}^{\infty} ne^{-n^2}$

10. $\sum_{n=2}^{\infty} \frac{1}{n(\ln n)^2}$

11. $\sum_{n=1}^{\infty} \frac{1}{2^{\ln n}}$

12. $\sum_{n=4}^{\infty} \frac{1}{n^2-1}$

13. $\sum_{n=1}^{\infty} \frac{\ln n}{n^2}$

14. $\sum_{n=1}^{\infty} \frac{n}{2^n}$

15. Use the Comparison Test to show that $\sum_{n=1}^{\infty} \frac{1}{n^3+8n}$ converges.

Hint: Compare with $\sum_{n=1}^{\infty} n^{-3}$.

16. Show that $\sum_{n=2}^{\infty} \frac{1}{\sqrt{n^2-3}}$ diverges by comparing with $\sum_{n=2}^{\infty} n^{-1}$.

17. Let $S = \sum_{n=1}^{\infty} \frac{1}{n+\sqrt{n}}$. Verify that for $n \geq 1$

$$\frac{1}{n+\sqrt{n}} \leq \frac{1}{n}, \qquad \frac{1}{n+\sqrt{n}} \leq \frac{1}{\sqrt{n}}$$

Can either inequality be used to show that S diverges? Show that $\frac{1}{n+\sqrt{n}} \geq \frac{1}{2n}$ and conclude that S diverges.

18. Which of the following inequalities can be used to study the convergence of $\sum_{n=2}^{\infty} \frac{1}{n^2+\sqrt{n}}$? Explain.

$$\frac{1}{n^2+\sqrt{n}} \leq \frac{1}{\sqrt{n}}, \qquad \frac{1}{n^2+\sqrt{n}} \leq \frac{1}{n^2}$$

In Exercises 19–31, use the Comparison Test to determine whether the infinite series is convergent.

19. $\sum_{n=1}^{\infty} \frac{1}{n2^n}$

20. $\sum_{n=1}^{\infty} \frac{1}{\sqrt{n}+2^n}$

21. $\sum_{k=1}^{\infty} \frac{k^{1/3}}{k^2+k}$

22. $\sum_{n=4}^{\infty} \frac{\sqrt{n}}{n-3}$

23. $\sum_{n=1}^{\infty} \frac{1}{\sqrt{n^3+1}}$

24. $\sum_{n=1}^{\infty} \frac{n^3}{n^5+4n+1}$

25. $\sum_{k=1}^{\infty} \frac{\sin^2 k}{k^2}$

26. $\sum_{n=2}^{\infty} \frac{1}{(\ln n)2^n}$

27. $\sum_{m=1}^{\infty} \frac{4}{m!+4^m}$

28. $\sum_{n=1}^{\infty} \frac{2}{3^n+3^{-n}}$

29. $\sum_{k=1}^{\infty} 2^{-k^2}$

30. $\sum_{n=1}^{\infty} \frac{\ln n}{n^3}$

31. $\sum_{n=1}^{\infty} \frac{\ln n}{n^3+3\ln n}$

32. Show that $\sum_{n=1}^{\infty} \sin\frac{1}{n^2}$ is a positive, convergent series. *Hint:* Use the inequality $\sin x \leq x$ for $x \geq 0$.

33. Does $\sum_{n=1}^{\infty} \frac{n}{\sqrt{n^2+c}}$ converge for any c?

In Exercises 34–42, use the Limit Comparison Test to prove convergence or divergence of the infinite series.

34. $\sum_{n=2}^{\infty} \frac{n^2}{n^4-1}$

35. $\sum_{n=2}^{\infty} \frac{1}{n^2-\sqrt{n}}$

36. $\sum_{n=2}^{\infty} \frac{n}{\sqrt{n^3-1}}$

37. $\sum_{n=3}^{\infty} \frac{n^3}{\sqrt{n^4-2n^2+1}}$

38. $\sum_{n=3}^{\infty} \frac{3n+5}{n(n-1)(n-2)}$

39. $\sum_{n=1}^{\infty} \frac{e^n+n}{e^{2n}-n^2}$

40. $\sum_{n=1}^{\infty} \frac{\ln n}{n^2}$. *Hint:* Use L'Hôpital's Rule to compare with $\sum_{n=1}^{\infty} \frac{1}{n^{3/2}}$.

41. $\sum_{n=1}^{\infty} \left(1 - \cos\frac{1}{n}\right)$ *Hint:* Compare with $\sum_{n=1}^{\infty} n^{-2}$.

42. $\sum_{n=1}^{\infty} (1 - 2^{-1/n})$ *Hint:* Compare with the harmonic series.

43. Show that if $a_n \geq 0$ and $\lim_{n\to\infty} n^2 a_n$ exists, then $\sum_{n=1}^{\infty} a_n$ converges.

Hint: Show that if M is larger than $\lim_{n\to\infty} n^2 a_n$, then $a_n \leq M/n^2$ for n sufficiently large.

44. Show that $\sum_{n=2}^{\infty} \frac{1}{n^{\ln n}}$ converges. *Hint:* Show that $n^{\ln n} \geq n^2$ for $n > e^2$.

45. Show that $\sum_{n=2}^{\infty} (\ln n)^{-2}$ diverges. *Hint:* Show that for x sufficiently large, $\ln x < x^{1/2}$.

46. For which a does $\sum_{n=2}^{\infty} \frac{1}{n(\ln n)^a}$ converge?

47. For which a does $\sum_{n=2}^{\infty} \frac{1}{n^a \ln n}$ converge?

48. Use the Integral Test to show that $\sum_{n=2}^{\infty} \frac{(\ln n)^k}{n^2}$ converges for all exponents k. You may use that $\int_1^{\infty} u^k e^{-u}\, du$ converges for all k.

In Exercises 49–74, determine convergence or divergence using any method covered so far.

49. $\sum_{n=4}^{\infty} \frac{1}{n^2 - 9}$

50. $\sum_{n=1}^{\infty} \frac{\cos^2 n}{n^2}$

51. $\sum_{n=1}^{\infty} \frac{\sqrt{n}}{4n + 9}$

52. $\sum_{n=1}^{\infty} e^{-n}$

53. $\sum_{n=1}^{\infty} \frac{1}{3^{n^2}}$

54. $\sum_{n=1}^{\infty} \frac{1}{n^2 + \sin n}$

55. $\sum_{n=2}^{\infty} \frac{1}{n^{3/2} \ln n}$

56. $\sum_{k=1}^{\infty} 2^{1/k}$

57. $\sum_{n=2}^{\infty} \frac{1}{n^{1/2} \ln n}$

58. $\sum_{n=2}^{\infty} \frac{n}{e^{n^2}}$

59. $\sum_{n=2}^{\infty} \frac{n}{e^{n^2}}$

60. $\sum_{n=1}^{\infty} \frac{n - \sin n}{n^3}$

61. $\sum_{n=1}^{\infty} \frac{2^n}{3^n - n}$

62. $\sum_{n=2}^{\infty} \frac{1}{n \ln n - n}$

63. $\sum_{n=1}^{\infty} \frac{\tan^{-1} n}{n^2}$

64. $\sum_{n=1}^{\infty} \frac{1}{n^n}$

65. $\sum_{n=1}^{\infty} \frac{\ln n}{n^3}$

66. $\sum_{n=1}^{\infty} \frac{2 + (-1)^n}{n}$

67. $\sum_{n=1}^{\infty} \frac{2 + (-1)^n}{n^{3/2}}$

68. $\sum_{n=1}^{\infty} \sin \frac{1}{n}$

69. $\sum_{n=1}^{\infty} \frac{2n + 1}{4^n}$

70. $\sum_{n=3}^{\infty} \frac{1}{n^4 - \sqrt{n}}$

71. $\sum_{n=1}^{\infty} \frac{n^2 - n}{n^5 + n}$

72. $\sum_{n=2}^{\infty} \frac{n^2 + n}{n^5 - n}$

73. $\sum_{n=1}^{\infty} \frac{1}{n^{1.2} \ln n}$

74. $\sum_{n=1}^{\infty} \frac{\ln n}{n^{1.2}}$

Approximating Infinite Sums *In Exercises 75–77, let $a_n = f(n)$, where $f(x)$ is a continuous, decreasing function such that*

$$\int_1^{\infty} f(x)\, dx$$

converges.

75. Show that

$$\int_1^{\infty} f(x)\, dx \le \sum_{n=1}^{\infty} a_n \le a_1 + \int_1^{\infty} f(x)\, dx \qquad \boxed{4}$$

76. CAS Using Eq. (4), show that

$$5 \le \sum_{n=1}^{\infty} \frac{1}{n^{1.2}} \le 6$$

This series converges slowly. Use a computer algebra system to verify that $S_N < 5$ for $N \le 43{,}128$ and $S_{43{,}129} \approx 5.00000021$.

77. Let $S = \sum_{n=1}^{\infty} a_n$. Arguing as in Exercise 75, show that

$$\sum_{n=1}^{M} a_n + \int_{M+1}^{\infty} f(x)\, dx \le S \le \sum_{n=1}^{M+1} a_n + \int_{M+1}^{\infty} f(x)\, dx \qquad \boxed{5}$$

Conclude that

$$0 \le S - \left(\sum_{n=1}^{M} a_n + \int_{M+1}^{\infty} f(x)\, dx \right) \le a_{M+1} \qquad \boxed{6}$$

This yields a method for approximating S with an error of at most a_{M+1}.

78. CAS Use Eq. (5) with $M = 43{,}129$ to prove that

$$5.5915810 \le \sum_{n=1}^{\infty} \frac{1}{n^{1.2}} \le 5.5915839$$

79. CAS Apply Eq. (5) with $M = 40{,}000$ to show that

$$1.644934066 \le \sum_{n=1}^{\infty} \frac{1}{n^2} \le 1.644934068$$

Is this consistent with Euler's result, according to which this infinite series has sum $\pi^2/6$?

80. CAS Using a CAS and Eq. (6), determine the value of $\sum_{n=1}^{\infty} n^{-4}$ to within an error less than 10^{-4}. Check that your result is consistent with that of Euler, who proved that the sum is equal to $\pi^4/90$.

81. CAS Using a CAS and Eq. (6), determine the value of $\sum_{n=1}^{\infty} n^{-5}$ to within an error less than 10^{-4}.

82. The harmonic series diverges, but it does so very slowly. Show that the partial sum S_N satisfies

$$\ln N \le 1 + \frac{1}{2} + \frac{1}{3} + \cdots + \frac{1}{N} \le 1 + \ln N$$

Verify that $S_N \le 10$ for $N \le 8{,}200$. Find an N such that $S_N \ge 100$.

83. Let p_n denote the nth prime number ($p_1 = 2$, $p_2 = 3$, etc.). It is known that there is a constant C such that $p_n \le Cn \ln n$. Prove the divergence of

$$\sum_{n=1}^{\infty} \frac{1}{p_n} = \frac{1}{2} + \frac{1}{3} + \frac{1}{5} + \frac{1}{7} + \frac{1}{11} + \cdots$$

84. How far can a stack of identical books (each of unit length) extend without tipping over? The stack will not tip over if the $(n+1)$st book is placed at the bottom of the stack with its left edge located at the center of mass of the first n books (Figure 5). Let c_n be the center of mass of the first n books, measured along the x-axis.

(a) Prove that $c_{n+1} = c_n + \frac{1}{2(n+1)}$. Recall that if objects of mass $m_1, \dots, m_n$ are placed along the x-axis with their centers of mass at $x_1, \dots, x_n$, then the center of mass of the system is located at

$$\frac{m_1 x_1 + \cdots + m_n x_n}{x_1 + \cdots + x_n}$$

(b) Prove that $\lim_{n\to\infty} c_n = \infty$. Thus, by using enough books, the stack can be extended as far as desired without tipping over.

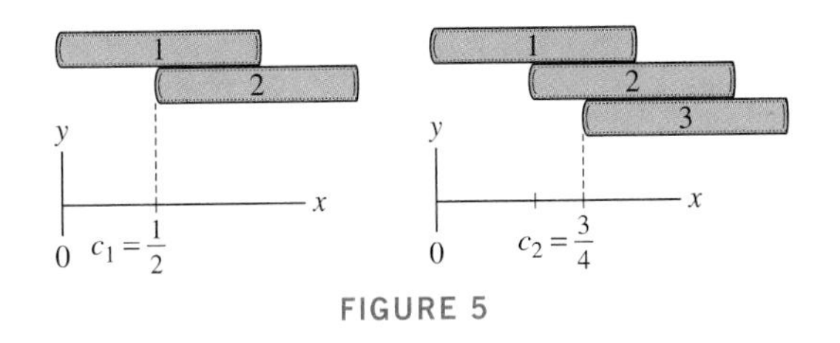

FIGURE 5

Further Insights and Challenges

85. Use the Integral Test to prove again that the geometric series $\sum_{n=1}^{\infty} r^n$ converges if $0 < r < 1$ and diverges if $r > 1$.

86. Let $S = \sum_{n=2}^{\infty} a_n$, where $a_n = (\ln(\ln n))^{-\ln n}$.

(a) Show, by taking logarithms, that $a_n = n^{-\ln(\ln(\ln n))}$.

(b) Show that $\ln(\ln(\ln n)) \ge 2$ if $n > C$, where $C = e^{e^{e^2}}$.

(c) Show that S converges.

87. Kummer's Acceleration Method Suppose we wish to approximate $S = \sum_{n=1}^{\infty} \frac{1}{n^2}$. There is a similar telescoping series whose value can be computed exactly (see Example 1 in Section 10.2):

$$\sum_{n=1}^{\infty} \frac{1}{n(n+1)} = 1$$

(a) Verify that

$$S = \sum_{n=1}^{\infty} \frac{1}{n(n+1)} + \sum_{n=1}^{\infty} \left(\frac{1}{n^2} - \frac{1}{n(n+1)} \right)$$

Thus for M large,

$$S \approx 1 + \sum_{n=1}^{M} \frac{1}{n^2(n+1)} \qquad \boxed{7}$$

(b) Explain what has been gained. Why is (7) a better approximation to S than $\sum_{n=1}^{M} \frac{1}{n^2}$?

(c) CAS Compute

$$\sum_{n=1}^{1{,}000} \frac{1}{n^2}, \qquad 1 + \sum_{n=1}^{100} \frac{1}{n^2(n+1)}$$

Which is a better approximation to S, whose exact value is $\pi^2/6$?

88. CAS The series $S = \sum_{k=1}^{\infty} k^{-3}$ has been computed to more than 100 million digits. The first 30 digits are

$$S = 1.202056903159594285399738161511$$

Approximate S using the Acceleration Method of Exercise 87 with $M = 100$ and auxiliary series $R = \sum_{n=1}^{\infty} (n(n+1)(n+2))^{-3}$. According to Exercise 46 in Section 10.2, R is a telescoping series with the sum $R = \frac{1}{4}$.

10.4 Absolute and Conditional Convergence

In the previous section, we studied the convergence of positive series, but we still lack the tools to analyze series with both positive and negative terms such as

$$\frac{1}{1^2} - \frac{1}{2^2} + \frac{1}{3^2} - \frac{1}{4^2} + \cdots \qquad \boxed{1}$$

One of the keys to understanding these more general series is the concept of absolute convergence.

DEFINITION Absolute Convergence $\sum a_n$ is called **absolutely convergent** if $\sum |a_n|$ converges.

As an example, consider the series (1). It is absolutely convergent because the sequence of absolute values is a convergent p-series:

$$\frac{1}{1^2} + \frac{1}{2^2} + \frac{1}{3^2} + \frac{1}{4^2} + \cdots$$

The following theorem states that if the series of absolute values converges, then the original series also converges.

THEOREM 1 Absolute Convergence Implies Convergence If $\sum a_n$ is absolutely convergent, then $\sum a_n$ converges.

Proof We have

$$0 \le a_n + |a_n| \le 2|a_n|$$

and thus $\sum (a_n + |a_n|)$ is a positive series. If $\sum |a_n|$ converges, then $\sum 2|a_n|$ also converges and hence $\sum (a_n + |a_n|)$ converges by the Comparison Test. Our original series is the difference of two convergent series and hence it converges:

$$\sum a_n = \sum (a_n + |a_n|) - \sum |a_n| \qquad \blacksquare$$

■ **EXAMPLE 1** Verify the convergence of $S = \frac{1}{1^2} - \frac{1}{2^2} + \frac{1}{3^2} - \frac{1}{4^2} + \cdots$.

Solution The positive series $\sum_{n=1}^{\infty} \frac{1}{n^2}$ converges because it is a p-series with $p = 2$. Thus, S converges absolutely. By Theorem 1, S converges. ■

■ **EXAMPLE 2** Does $S = \sum_{n=2}^{\infty} \frac{(-1)^n}{n \ln n}$ converge absolutely?

Solution We apply the Integral Test to the positive series $\sum_{n=2}^{\infty} \frac{1}{n \ln n}$. Using the substitution $u = \ln x$, $du = x^{-1}\,dx$, we find that the improper integral diverges:

$$\int_2^{\infty} \frac{dx}{x \ln x} = \int_{\ln 2}^{\infty} \frac{du}{u} = \lim_{R\to\infty} \int_{\ln 2}^{R} \frac{du}{u} = \lim_{R\to\infty} \big(\ln R - \ln(\ln 2)\big) = \infty$$

Therefore, $\sum_{n=2}^{\infty} \frac{1}{n \ln n}$ diverges and S does not converge absolutely. ■

In the previous example, we showed that the series $\sum_{n=2}^{\infty} \frac{(-1)^n}{n \ln n}$ does not converge *absolutely*, but we still do not know whether or not it converges. A series $\sum a_n$ may converge without converging absolutely. In this case, we say that $\sum a_n$ is conditionally convergent.

> **DEFINITION Conditional Convergence** An infinite series $\sum a_n$ is called **conditionally convergent** if it converges but $\sum |a_n|$ diverges.

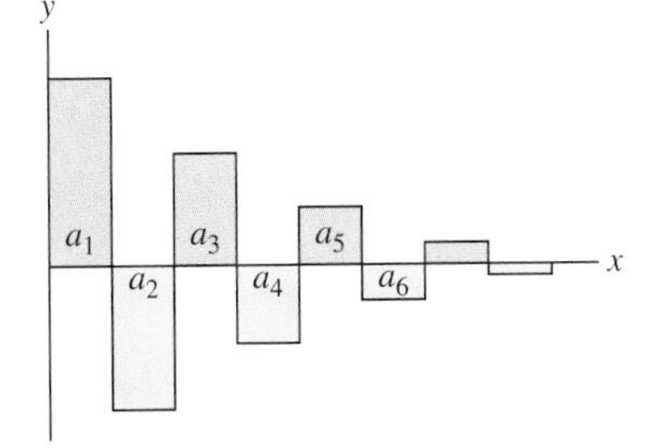

FIGURE 1 An alternating series with decreasing terms. The sum of the signed areas is positive and at most a_1.

Assumptions Matter *The Leibniz Test is not valid if we drop the assumption that a_n is decreasing (see Exercise 35).*

If we encounter a series that is not absolutely convergent, how can we determine whether it is conditionally convergent? The Integral Test and Comparison Test cannot be used because they apply only to positive series. Our next test applies to **alternating series**, that is, series in which the terms alternate in sign. Such a series has the following form, where $\{a_n\}$ is a *positive* sequence (Figure 1):

$$\sum_{n=1}^{\infty}(-1)^{n-1}a_n = a_1 - a_2 + a_3 - a_4 + a_5 + \cdots$$

> **THEOREM 2 Leibniz Test for Alternating Series** Let $\{a_n\}$ be a decreasing positive sequence that converges to 0:
>
> $$a_1 \geq a_2 \geq a_3 \geq a_4 \geq \cdots \geq 0, \qquad \lim_{n\to\infty} a_n = 0$$
>
> Then the following alternating series converges:
>
> $$S = \sum_{n=1}^{\infty}(-1)^{n-1}a_n = a_1 - a_2 + a_3 - a_4 + \cdots$$
>
> Furthermore, $0 \leq S \leq a_1$ and $S_{2N} \leq S \leq S_{2N+1}$ for all N.

Proof We analyze the even and odd partial sums separately. First, we observe that the even partial sums form an increasing sequence:

$$S_{2N+2} = \underbrace{(a_1 - a_2) + \cdots + (a_{2N-1} - a_{2N})}_{S_{2N}} + (a_{2N+1} - a_{2N+2})$$
$$= S_{2N} + (a_{2N+1} - a_{2N+2})$$

Since $\{a_n\}$ is decreasing, the quantity $a_{2N+1} - a_{2N+2}$ is positive and thus S_{2N+2} is obtained by adding the positive quantity $(a_{2N+1} - a_{2N+2})$ to S_{2N}. On the other hand, we may regroup the terms:

$$S_{2N} = a_1 - (a_2 - a_3) - (a_4 - a_5) - \cdots - (a_{2N-2} - a_{2N-1}) - a_{2N}$$

We conclude that $S_{2N} \leq a_1$ because each term $(a_{2j} - a_{2j+1})$ is nonnegative. This shows that $\{S_{2N}\}$ is a positive increasing sequence with upper bound a_1. Consequently, the limit $S = \lim S_{2N}$ exists and $0 \leq S_{2N} \leq S \leq a_1$ by Theorem 6 in Section 10.1.

The odd partial sums form a decreasing sequence:

$$S_{2N+1} = S_{2N-1} - (a_{2N} - a_{2N+1}) \leq S_{2N-1}$$

The Leibniz Test is the only test for conditional convergence developed in this text. More sophisticated tests such as Abel's Criterion are treated in analysis textbooks.

Furthermore, the odd partial sums converge to the same limit:

$$\lim_{N\to\infty} S_{2N+1} = \lim_{N\to\infty} (S_{2N} + a_{2N+1}) = \lim_{N\to\infty} S_{2N} + \lim_{N\to\infty} a_{2N+1} = S + 0 = S$$

This proves that the alternating series converges to S and that $S_{2N+1} \geq S$ for all N. ■

■ **EXAMPLE 3** Show that $S = \sum_{n=1}^{\infty} \frac{(-1)^{n-1}}{\sqrt{n}}$ is conditionally convergent and that $0 \leq S \leq 1$.

(A) Partial sums of $S = \sum_{n=1}^{\infty} (-1)^{n-1} \frac{1}{\sqrt{n}}$

(B) Partial sums of $\sum_{n=1}^{\infty} \frac{1}{\sqrt{n}}$

FIGURE 2 The series $S = \sum_{n=1}^{\infty} \frac{(-1)^{n-1}}{\sqrt{n}}$ converges conditionally but not absolutely.

Solution The terms $a_n = \frac{1}{\sqrt{n}}$ form a decreasing sequence that tends to zero. The Leibniz Test implies that $S = \sum_{n=1}^{\infty} \frac{(-1)^{n-1}}{\sqrt{n}}$ converges and that $0 \leq S \leq a_1 = 1$ [Figure 2(A)]. However, S is only conditionally convergent because the positive series $\sum_{n=1}^{\infty} \frac{1}{\sqrt{n}}$ is a divergent p-series [Figure 2(B)]. ■

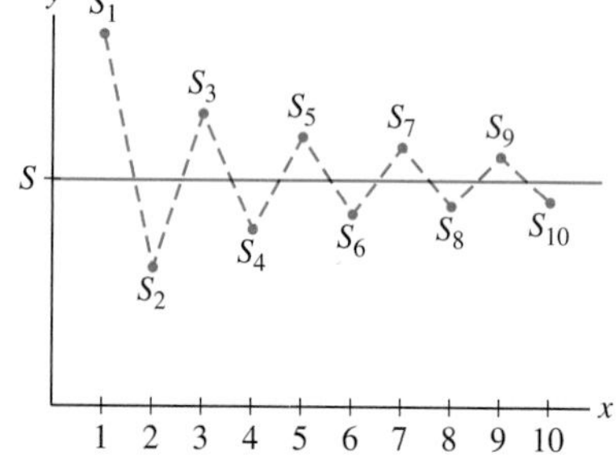

FIGURE 3 The partial sums of an alternating series zigzag, above and below the limit.

According to Theorem 2, if $\{a_n\}$ is decreasing and positive, then the partial sums S_N of the alternating series satisfy

$$S_{2N} \leq S \leq S_{2N+1}$$

Therefore, the partial sums zigzag above and below the limit (Figure 3). It follows that the error $|S_N - S|$ is not greater than $|S_N - S_{N+1}| = a_{N+1}$:

$$\boxed{|S_N - S| \leq a_{N+1}} \qquad \boxed{2}$$

In other words, *the error committed when we approximate S by S_N is at most the size of the first omitted term a_{N+1}*.

Leibniz proved that

$$\sum_{n=1}^{\infty} \frac{(-1)^{n+1}}{n} = \ln 2 \approx 0.6931$$

We will prove this in Exercise 82 in Section 10.7. The result of Example 4 is consistent with this fact.

■ **EXAMPLE 4 The Alternating Harmonic Series** Show that $S = \sum_{n=1}^{\infty} \frac{(-1)^{n+1}}{n}$ converges conditionally. Then

(a) Show that $|S_6 - S| \leq \frac{1}{7}$.

(b) Find an N such that S_N approximates S with an error less than 10^{-3}.

Solution The harmonic series $\sum_{n=1}^{\infty} n^{-1}$ diverges, but S converges conditionally by the Leibniz Test, since $a_n = n^{-1}$ is a positive, decreasing sequence.

(a) By Eq. (2), $|S_6 - S| \le a_7 = \frac{1}{7}$.

(b) The error in the Nth partial sum is at most a_{N+1}:

$$\left| \sum_{n=1}^{N} \frac{(-1)^{n+1}}{n} - S \right| \le a_{N+1} = \frac{1}{N+1}$$

To make the error less than 10^{-3}, we choose N so that $\frac{1}{N+1} < 10^{-3}$ or $N + 1 > 10^3$. This gives $N > 999$. Using a computer algebra system, we find that $S_{1,000} \approx 0.6926$. ■

CONCEPTUAL INSIGHT The convergence of an infinite series $\sum a_n$ depends on two factors: (1) how quickly a_n tends to zero, and (2) how much cancellation takes place among the terms. Consider the series

Harmonic series (diverges):	$1 + \frac{1}{2} + \frac{1}{3} + \frac{1}{4} + \frac{1}{5} + \cdots$
p-Series with $p = 2$ (converges):	$1 + \frac{1}{2^2} + \frac{1}{3^2} + \frac{1}{4^2} + \frac{1}{5^2} + \cdots$
Alternating harmonic series (converges):	$1 - \frac{1}{2} + \frac{1}{3} - \frac{1}{4} + \frac{1}{5} - \cdots$

The harmonic series diverges because reciprocals $\frac{1}{n}$ do not tend to zero quickly enough. By contrast, the reciprocal squares $\frac{1}{n^2}$ tend to zero quickly enough for the series $\sum_{n=1}^{\infty} \frac{1}{n^2}$ to converge. The alternating harmonic series converges as a result of cancellation among the terms.

10.4 SUMMARY

- An infinite series $\sum a_n$ is called *absolutely convergent* if the positive series $\sum |a_n|$ converges.
- Theorem: Absolute convergence implies convergence. Namely, if $\sum |a_n|$ converges, then $\sum a_n$ also converges.
- An infinite series $\sum a_n$ is *conditionally convergent* if it converges but $\sum |a_n|$ diverges.
- *Leibniz Test:* If $\{a_n\}$ is a positive decreasing sequence such that $\lim_{n\to\infty} a_n = 0$, then the following alternating series converges:

$$S = \sum_{n=1}^{\infty} (-1)^{n-1} a_n = a_1 - a_2 + a_3 - a_4 + a_5 - \cdots$$

Furthermore, $|S - S_N| \le a_{N+1}$.

- We have developed two ways to handle nonpositive series: Either show absolute convergence or use the Leibniz Test, if applicable.

10.4 EXERCISES

Preliminary Questions

1. Suppose that $S = \sum_{n=0}^{\infty} a_n$ is conditionally convergent. Which of the following statements are correct?

(a) $\sum_{n=0}^{\infty} |a_n|$ may or may not converge.

(b) S may or may not converge.

(c) $\sum_{n=0}^{\infty} |a_n|$ diverges.

2. Which of the following statements is equivalent to Theorem 1?

(a) If $\sum_{n=0}^{\infty} |a_n|$ diverges, then $\sum_{n=0}^{\infty} a_n$ also diverges.

(b) If $\sum_{n=0}^{\infty} a_n$ diverges, then $\sum_{n=0}^{\infty} |a_n|$ also diverges.

(c) If $\sum_{n=0}^{\infty} a_n$ converges, then $\sum_{n=0}^{\infty} |a_n|$ also converges.

3. Lathika argues that $\sum_{n=1}^{\infty} (-1)^n \sqrt{n}$ is an alternating series and therefore converges. Is Lathika right?

4. Give an example of a series such that $\sum a_n$ converges but $\sum |a_n|$ diverges.

Exercises

1. Show that $\sum_{n=0}^{\infty} \frac{(-1)^n}{2^n}$ converges absolutely.

2. Show that the following series converges conditionally:

$$\sum_{n=1}^{\infty} (-1)^{n-1} \frac{1}{n^{2/3}} = \frac{1}{1^{2/3}} - \frac{1}{2^{2/3}} + \frac{1}{3^{2/3}} - \frac{1}{4^{2/3}} + \cdots$$

In Exercises 3–12, determine whether the series converges absolutely, conditionally, or not at all.

3. $\sum_{n=1}^{\infty} \frac{(-1)^n}{\sqrt{n}}$

4. $\sum_{n=1}^{\infty} \frac{(-1)^n n^4}{n^3+1}$

5. $\sum_{n=1}^{\infty} \frac{(-1)^{n-1}}{(1.1)^n}$

6. $\sum_{n=1}^{\infty} \frac{\sin n}{n^2}$

7. $\sum_{n=2}^{\infty} \frac{(-1)^{n+1}}{n \ln n}$

8. $\sum_{n=1}^{\infty} \frac{(-1)^n}{1+\frac{1}{n}}$

9. $\sum_{n=1}^{\infty} \frac{\sin n\pi}{\sqrt{n}}$

10. $\sum_{n=4}^{\infty} (-1)^n \tan \frac{1}{n}$

11. $\sum_{n=1}^{\infty} \frac{\cos \frac{1}{n}}{n^2}$

12. $\sum_{n=1}^{\infty} \frac{\cos \frac{1}{n}}{n}$

13. Let $S = \sum_{n=1}^{\infty} (-1)^{n+1} \frac{1}{n^3}$.

(a) Calculate S_n for $1 \le n \le 10$.

(b) Use Eq. (2) to show that $0.9 \le S \le 0.902$.

14. Use Eq. (2) to approximate $\sum_{n=1}^{\infty} \frac{(-1)^{n+1}}{n!}$ to four decimal places.

15. Approximate $\sum_{n=1}^{\infty} \frac{(-1)^{n+1}}{n^4}$ to three decimal places.

16. CAS Let $S = \sum_{n=1}^{\infty} (-1)^{n-1} \frac{n}{n^2+1}$. Use a computer algebra system to calculate and plot the partial sums S_n for $1 \le n \le 100$. Observe that the partial sums zigzag above and below the limit.

In Exercises 17–18, use Eq. (2) to approximate the value of the series to within an error of at most 10^{-5}.

17. $\sum_{n=1}^{\infty} \frac{(-1)^{n+1}}{n(n+2)(n+3)}$

18. $\sum_{n=1}^{\infty} \frac{(-1)^{n+1} \ln n}{n!}$

In Exercises 19–26, determine convergence or divergence by any method.

19. $\sum_{n=1}^{\infty} \frac{1}{3^n + 5^n}$

20. $\sum_{n=2}^{\infty} \frac{n}{n^2 - n}$

21. $\sum_{n=1}^{\infty} \frac{(-1)^n}{\sqrt{n^2+1}}$

22. $\sum_{n=1}^{\infty} \frac{1}{\sqrt{n^2+1}}$

23. $\sum_{n=1}^{\infty} \frac{3^n + (-1)^n 2^n}{5^n}$

24. $\sum_{n=1}^{\infty} \frac{(-1)^{n+1}}{(2n+1)!}$

25. $\sum_{n=1}^{\infty} (-1)^n n e^{-n}$

26. $\sum_{n=1}^{\infty} (-1)^n n^4 2^{-n}$

27. Show that

$$S = \frac{1}{2} - \frac{1}{2} + \frac{1}{3} - \frac{1}{3} + \frac{1}{4} - \frac{1}{4}$$

converges by computing the partial sums. Does it converge absolutely?

28. The Leibniz Test cannot be applied to

$$\frac{1}{2}-\frac{1}{3}+\frac{1}{2^2}-\frac{1}{3^2}+\frac{1}{2^3}-\frac{1}{3^3}+\cdots$$

Why not? Show that it converges by another method.

29. Determine whether the following series converges conditionally:

$$1-\frac{1}{3}+\frac{1}{2}-\frac{1}{5}+\frac{1}{3}-\frac{1}{7}+\frac{1}{4}-\frac{1}{9}+\frac{1}{5}-\frac{1}{11}+\cdots$$

30. Prove that if $\sum a_n$ converges absolutely, then $\sum a_n^2$ also converges. Then show by giving a counterexample that $\sum a_n^2$ need not converge if $\sum a_n$ is only conditionally convergent.

Further Insights and Challenges

31. Prove the following variant of the Leibniz Test: If $\{a_n\}$ is a positive, decreasing sequence with $\lim_{n\to\infty} a_n = 0$, then the series

$$a_1 + a_2 - 2a_3 + a_4 + a_5 - 2a_6 + \cdots$$

converges. *Hint:* Show that S_{3N} is increasing and bounded by $a_1 + a_2$, and continue as in the proof of the Leibniz Test.

32. Use Exercise 31 to show that the following series converges:

$$S = \frac{1}{\ln 2}+\frac{1}{\ln 3}-\frac{2}{\ln 4}+\frac{1}{\ln 5}+\frac{1}{\ln 6}-\frac{2}{\ln 7}+\cdots$$

33. Prove the conditional convergence of

$$R = 1+\frac{1}{2}+\frac{1}{3}-\frac{3}{4}+\frac{1}{5}+\frac{1}{6}+\frac{1}{7}-\frac{3}{8}+\cdots$$

34. Show that the following series diverges:

$$S = 1+\frac{1}{2}+\frac{1}{3}-\frac{2}{4}+\frac{1}{5}+\frac{1}{6}+\frac{1}{7}-\frac{2}{8}+\cdots$$

Hint: Use the result of Exercise 33 to write S as the sum of a convergent and a divergent series.

35. **Assumptions Matter** Show by counterexample that the Leibniz Test does not remain true if $\{a_n\}$ tends to zero but we drop the assumption that the sequence a_n is decreasing. *Hint:* Consider

$$R = \frac{1}{2}-\frac{1}{4}+\frac{1}{3}-\frac{1}{8}+\frac{1}{4}-\frac{1}{16}+\cdots+\left(\frac{1}{n}-\frac{1}{2^n}\right)+\cdots$$

36. Prove that $\sum_{n=1}^{\infty}(-1)^{n+1}\frac{(\ln n)^a}{n}$ converges for all exponents a.

Hint: Show that $f(x) = \frac{(\ln x)^a}{x}$ is decreasing for x sufficiently large.

37. We say that $\{b_n\}$ is a rearrangement of $\{a_n\}$ if $\{b_n\}$ has the same terms as $\{a_n\}$ but occurring in a different order. Show that if $\{b_n\}$ is a rearrangement of $\{a_n\}$ and $S = \sum_{n=1}^{\infty} a_n$ converges absolutely, then $T = \sum_{n=1}^{\infty} b_n$ also converges absolutely. (This result does not hold if S is only conditionally convergent.) *Hint:* Prove that the partial sums $\sum_{n=1}^{N} |b_n|$ are bounded. It can be shown further that $S = T$.

38. **Assumptions Matter** In 1829, Lejeune Dirichlet pointed out that the great French mathematician Augustin Louis Cauchy made a mistake in a published paper by improperly assuming the Limit Comparison Test to be valid for nonpositive series. Here are Dirichlet's two series:

$$\sum_{n=1}^{\infty}\frac{(-1)^n}{\sqrt{n}}, \qquad \sum_{n=1}^{\infty}\frac{(-1)^n}{\sqrt{n}}\left(1+\frac{(-1)^n}{\sqrt{n}}\right)$$

Explain how they provide a counterexample to the Limit Comparison Test when the series are not assumed to be positive.

10.5 The Ratio and Root Tests

As we will show in Section 10.7, the number e has a well-known expression as an infinite series:

$$e = 1+\frac{1}{1!}+\frac{1}{2!}+\frac{1}{3!}+\cdots$$

However, the convergence tests developed so far cannot be easily applied to this series. This points to the need for the following test, which is also of key importance in the study of power series (Section 10.6).

The symbol ρ, pronounced "rho," is the seventeenth letter of the Greek alphabet.

THEOREM 1 Ratio Test Let $\{a_n\}$ be a sequence and assume that the following limit exists:

$$\rho = \lim_{n\to\infty} \left| \frac{a_{n+1}}{a_n} \right|$$

(i) If $\rho < 1$, then $\sum a_n$ converges absolutely.
(ii) If $\rho > 1$, then $\sum a_n$ diverges.
(iii) If $\rho = 1$, the Ratio Test is inconclusive (the series may converge or diverge).

Proof If $\rho < 1$, we may choose a number r such that $\rho < r < 1$. Since $\left| \frac{a_{n+1}}{a_n} \right|$ converges to ρ, there exists a number M such that $\left| \frac{a_{n+1}}{a_n} \right| < r$ for $n \ge M$. Therefore,

$$|a_{M+1}| < r|a_M|$$

$$|a_{M+2}| < r|a_{M+1}| < r^2|a_M|$$

$$|a_{M+3}| < r|a_{M+2}| < r^3|a_M|$$

In general, $|a_{M+n}| < r^n|a_M|$ and therefore

$$\sum_{n=M}^{\infty} |a_n| = \sum_{n=0}^{\infty} |a_{M+n}| \le \sum_{n=0}^{\infty} |a_M| r^n = |a_M| \sum_{n=0}^{\infty} r^n$$

The geometric series on the right converges because $0 < r < 1$, so $\sum_{n=M}^{\infty} |a_n|$ converges by the Comparison Test. Thus, $\sum_{n=1}^{\infty} a_n$ converges absolutely.

If $\rho > 1$, choose a number r such that $1 < r < \rho$. Since $\left| \frac{a_{n+1}}{a_n} \right|$ converges to ρ, there exists a number M such that $\left| \frac{a_{n+1}}{a_n} \right| > r$ for $n \ge M$. Arguing as before with the inequalities reversed, we find that $|a_{M+n}| \ge r^n|a_M|$. Since r^n tends to ∞, we see that the terms a_{M+n} do not tend to zero and, consequently, $\sum_{n=1}^{\infty} a_n$ diverges. Finally, Example 4 below shows that when $\rho = 1$, both convergence and divergence are possible, so the test is inconclusive. ■

■ **EXAMPLE 1** Prove that $S = \sum_{n=1}^{\infty} \frac{1}{n!}$ converges.

Solution We compute the limit ρ. Let $a_n = \frac{1}{n!}$. Then

$$\frac{a_{n+1}}{a_n} = \frac{1}{(n+1)!} \frac{n!}{1} = \frac{n!}{(n+1)!} = \frac{1}{n+1}$$

$$\rho = \lim_{n\to\infty} \left| \frac{a_{n+1}}{a_n} \right| = \lim_{n\to\infty} \frac{1}{n+1} = 0$$

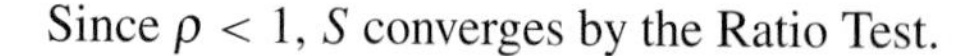

Since $\rho < 1$, S converges by the Ratio Test. ■

EXAMPLE 2 Apply the Ratio Test to determine if $\sum_{n=1}^{\infty} \frac{n^2}{2^n}$ converges.

Solution Let $a_n = \frac{n^2}{2^n}$. We have

$$\left|\frac{a_{n+1}}{a_n}\right| = \frac{(n+1)^2}{2^{n+1}}\frac{2^n}{n^2} = \frac{1}{2}\left(\frac{n^2+2n+1}{n^2}\right) = \frac{1}{2}\left(1+\frac{2}{n}+\frac{1}{n^2}\right)$$

$$\rho = \lim_{n\to\infty}\left|\frac{a_{n+1}}{a_n}\right| = \frac{1}{2}\lim_{n\to\infty}\left(1+\frac{2}{n}+\frac{1}{n^2}\right) = \frac{1}{2}$$

Since $\rho < 1$, the series converges by the Ratio Test. ■

EXAMPLE 3 Determine whether $\sum_{n=0}^{\infty}(-1)^n \frac{n!}{100^n}$ converges.

Solution Let $a_n = (-1)^n \frac{n!}{100^n}$. Then

$$\left|\frac{a_{n+1}}{a_n}\right| = \frac{(n+1)!}{100^{n+1}}\frac{100^n}{n!} = \frac{n+1}{100}$$

We see that the ratio of the coefficients tends to infinity:

$$\rho = \lim_{n\to\infty}\left|\frac{a_{n+1}}{a_n}\right| = \lim_{n\to\infty}\frac{n+1}{100} = \infty$$

Since $\rho > 1$, $\sum_{n=0}^{\infty}(-1)^n \frac{n!}{100^n}$ diverges by the Ratio Test. ■

EXAMPLE 4 Ratio Test Inconclusive Show that $\rho = 1$ for both $\sum_{n=1}^{\infty} n^2$ and $\sum_{n=1}^{\infty} n^{-2}$. Conclude that the Ratio Test is inconclusive when $\rho = 1$.

Solution For $a_n = n^2$, we have

$$\rho = \lim_{n\to\infty}\left|\frac{a_{n+1}}{a_n}\right| = \lim_{n\to\infty}\frac{(n+1)^2}{n^2} = \lim_{n\to\infty}\frac{n^2+2n+1}{n^2} = \lim_{n\to\infty}\left(1+\frac{2}{n}+\frac{1}{n^2}\right) = 1$$

On the other hand, for $b_n = n^{-2}$,

$$\rho = \lim_{n\to\infty}\left|\frac{b_{n+1}}{b_n}\right| = \lim_{n\to\infty}\left|\frac{a_n}{a_{n+1}}\right| = \frac{1}{\lim_{n\to\infty}\frac{a_{n+1}}{a_n}} = 1$$

Thus, $\rho = 1$ in both cases, but $\sum_{n=1}^{\infty} n^2$ diverges and $\sum_{n=1}^{\infty} n^{-2}$ converges (a p-series with $p = 2$). This shows that both convergence and divergence are possible when $\rho = 1$. ■

For some series, it is more convenient to use the following Root Test, based on the limit of the nth roots $\sqrt[n]{a_n}$ rather than the ratios a_{n+1}/a_n. The proof of the Root Test, like that of the Ratio Test, is based on a comparison with geometric series (see Exercise 53).

THEOREM 2 Root Test Let $\{a_n\}$ be a sequence and assume that the following limit exists:

$$L = \lim_{n\to\infty} \sqrt[n]{|a_n|}$$

(i) If $L < 1$, then $\sum a_n$ converges absolutely.
(ii) If $L > 1$, then $\sum a_n$ diverges.
(iii) If $L = 1$, the Root Test is inconclusive: The series may converge or diverge.

■ **EXAMPLE 5** Determine whether $\sum_{n=1}^{\infty} \left(\frac{n}{2n+3}\right)^n$ converges.

Solution Let $a_n = \left(\frac{n}{2n+3}\right)^n$. Then

$$L = \lim_{n\to\infty} \sqrt[n]{a_n} = \lim_{n\to\infty} \frac{n}{2n+3} = \frac{1}{2}$$

Since $L < \frac{1}{2}$, the series $\sum_{n=1}^{\infty} \left(\frac{n}{2n+3}\right)^n$ converges. ■

10.5 SUMMARY

- *Ratio Test:* Assume that the following limit exists:

$$\rho = \lim_{n\to\infty} \left|\frac{a_{n+1}}{a_n}\right|$$

Then $\sum a_n$ converges absolutely if $\rho < 1$ and it diverges if $\rho > 1$. The test is inconclusive if $\rho = 1$.
- *Root Test:* Assume that the following limit exists: $L = \lim_{n\to\infty} \sqrt[n]{|a_n|}$. Then $\sum a_n$ converges if $L < 1$ and it diverges if $L > 1$. The test is inconclusive if $L = 1$.

10.5 EXERCISES

Preliminary Questions

1. In the Ratio Test, is ρ equal to $\lim_{n\to\infty} \left|\frac{a_{n+1}}{a_n}\right|$ or $\lim_{n\to\infty} \left|\frac{a_n}{a_{n+1}}\right|$?

2. Is the Ratio Test conclusive for $\sum_{n=1}^{\infty} \frac{1}{2^n}$? Is it conclusive for $\sum_{n=1}^{\infty} \frac{1}{n}$?

3. Can the Ratio Test be used to show convergence if the series is only conditionally convergent?

Exercises

In Exercises 1–18, apply the Ratio Test to determine convergence or divergence, or state that the Ratio Test is inconclusive.

1. $\sum_{n=1}^{\infty} \frac{1}{5^n}$

2. $\sum_{n=1}^{\infty} \frac{(-1)^{n-1}n}{5^n}$

3. $\sum_{n=1}^{\infty} \frac{(-1)^{n-1}}{n^n}$

4. $\sum_{n=0}^{\infty} \frac{3n+2}{5n^3+1}$

5. $\sum_{n=1}^{\infty} \frac{n}{n^2+1}$

6. $\sum_{n=1}^{\infty} \frac{2^n}{n}$

7. $\sum_{n=1}^{\infty} \frac{2^n}{n^{100}}$

8. $\sum_{n=1}^{\infty} \frac{n^3}{3^{n^2}}$

9. $\sum_{n=1}^{\infty} \frac{10^n}{2^{n^2}}$

10. $\sum_{n=1}^{\infty} \frac{e^n}{n!}$

11. $\sum_{n=1}^{\infty} \frac{e^n}{n^n}$

12. $\sum_{n=1}^{\infty} \frac{n^{50}}{n!}$

13. $\sum_{n=0}^{\infty} (-1)^n \frac{n!}{4^n}$

14. $\sum_{n=1}^{\infty} \frac{n!}{n^4}$

15. $\sum_{n=2}^{\infty} \frac{1}{n \ln n}$

16. $\sum_{n=1}^{\infty} \frac{1}{(2n)!}$

17. $\sum_{n=1}^{\infty} \frac{n^2}{(2n+1)!}$

18. $\sum_{n=1}^{\infty} \frac{(n!)^2}{(2n)!}$

19. Show that $\sum_{n=1}^{\infty} n^k 3^{-n}$ converges for all exponents k.

20. Show that $\sum_{n=1}^{\infty} n^2 x^n$ converges if $|x| < 1$.

21. Show that $\sum_{n=1}^{\infty} 2^n x^n$ converges if $|x| < \frac{1}{2}$.

22. Show that $\sum_{n=1}^{\infty} \frac{r^n}{n!}$ converges for all r.

23. Show that $\sum_{n=1}^{\infty} \frac{r^n}{n}$ converges if $|r| < 1$.

24. Is there any value of k such that $\sum_{n=1}^{\infty} \frac{2^n}{n^k}$ converges?

25. Show that $\sum_{n=1}^{\infty} \frac{n!}{n^n}$ converges. *Hint:* Use $\lim_{n\to\infty} \left(1 + \frac{1}{n}\right)^n = e$.

In Exercises 26–31, assume that $|a_{n+1}/a_n|$ converges to $\rho = \frac{1}{3}$. What can you say about the convergence of the given series?

26. $\sum_{n=1}^{\infty} n a_n$

27. $\sum_{n=1}^{\infty} n^3 a_n$

28. $\sum_{n=1}^{\infty} 2^n a_n$

29. $\sum_{n=1}^{\infty} 3^n a_n$

30. $\sum_{n=1}^{\infty} 4^n a_n$

31. $\sum_{n=1}^{\infty} a_n^2$

32. Assume that $\left|\frac{a_{n+1}}{a_n}\right|$ converges to $\rho = 4$. Does $\sum_{n=1}^{\infty} a_n^{-1}$ converge (assume that $a_n \neq 0$ for all n)?

33. Is the Ratio Test conclusive for the p-series $\sum_{n=1}^{\infty} \frac{1}{n^p}$?

In Exercises 34–39, use the Root Test to determine convergence or divergence (or state that the test is inconclusive).

34. $\sum_{n=0}^{\infty} \frac{1}{10^n}$

35. $\sum_{n=1}^{\infty} \frac{1}{n^n}$

36. $\sum_{k=0}^{\infty} \left(\frac{k}{k+10}\right)^k$

37. $\sum_{k=0}^{\infty} \left(\frac{k}{3k+1}\right)^k$

38. $\sum_{n=1}^{\infty} \left(1 + \frac{1}{n}\right)^{-n}$

39. $\sum_{n=4}^{\infty} \left(1 + \frac{1}{n}\right)^{-n^2}$

40. Prove that $\sum_{n=1}^{\infty} \frac{2^{n^2}}{n!}$ diverges. *Hint:* Use $2^{n^2} = (2^n)^n$ and $n! \le n^n$.

In Exercises 41–52, determine convergence or divergence using any method covered in the text so far.

41. $\sum_{n=1}^{\infty} \frac{2^n + 4^n}{7^n}$

42. $\sum_{n=1}^{\infty} \frac{n^3}{n!}$

43. $\sum_{n=1}^{\infty} \frac{n^3}{5^n}$

44. $\sum_{n=2}^{\infty} \frac{1}{n(\ln n)^3}$

45. $\sum_{n=2}^{\infty} \frac{1}{\sqrt{n^3 - n^2}}$

46. $\sum_{k=1}^{\infty} 4^{-2k+1}$

47. $\sum_{n=1}^{\infty} \frac{n^2 + 4n}{3n^4 + 9}$

48. $\sum_{n=1}^{\infty} (-1)^n \cos \frac{1}{n}$

49. $\sum_{n=1}^{\infty} \sin \frac{1}{n^2}$

50. $\sum_{n=1}^{\infty} \frac{(-1)^{n-1}}{\sqrt{n}}$

51. $\sum_{n=1}^{\infty} \left(\frac{n}{n+12}\right)^n$

52. $\sum_{n=1}^{\infty} \frac{(-2)^n}{\sqrt{n}}$

Further Insights and Challenges

53. Proof of the Root Test Let $S = \sum_{n=0}^{\infty} a_n$ be a positive series and assume that $L = \lim_{n\to\infty} \sqrt[n]{a_n}$ exists.

(a) Show that S converges if $L < 1$. *Hint:* Choose R with $\rho < R < 1$ and show that $a_n \le R^n$ for n sufficiently large. Then compare with the geometric series $\sum R^n$.

(b) Show that S converges if $L > 1$.

54. Show that the Ratio Test is inconclusive but the Root Test indicates convergence for the series

$$\frac{1}{2} + \frac{1}{3^2} + \frac{1}{2^3} + \frac{1}{3^4} + \frac{1}{2^5} + \cdots$$

55. Let $S = \sum_{n=1}^{\infty} \frac{c^n n!}{n^n}$, where c is a constant.

(a) Prove that S converges absolutely if $|c| < e$ and diverges if $|c| > e$.

(b) It is known that $\lim_{n\to\infty} \frac{e^n n!}{n^{n+1/2}} = \sqrt{2\pi}$. Verify this numerically.

(c) Use the Limit Comparison Test to prove that S diverges for $c = e$.

10.6 Power Series

Most functions that arise in applications can be represented as power series. This includes not only the familiar trigonometric, exponential, logarithm, and root functions, but also the host of more advanced "special functions" of physics and engineering such as Bessel functions and elliptic functions.

In the introduction to this chapter, we mentioned that e^x can be expressed as an "infinite polynomial" called a power series:

$$e^x = 1 + \frac{x}{1!} + \frac{x^2}{2!} + \frac{x^3}{3!} + \frac{x^4}{4!} + \cdots$$

In this section, we develop the basic properties of power series, especially the key concept of *radius of convergence*.

A **power series** centered at the point $x = c$ is an infinite series of the form

$$F(x) = \sum_{n=0}^{\infty} a_n(x-c)^n = a_0 + a_1(x-c) + a_2(x-c)^2 + a_3(x-c)^3 + \cdots$$

To make use of a power series, we must determine the values of x for which the series converges. It certainly converges at its center $x = c$:

$$F(c) = a_0 + a_1(c-c) + a_2(c-c)^2 + a_3(c-c)^3 + \cdots = a_0$$

Where else does it converge? The following basic theorem states that every power series converges absolutely on an interval that is symmetric around the center $x = c$ (the interval may be infinite or possibly reduced to the single point c).

THEOREM 1 Radius of Convergence Let $F(x) = \sum_{n=0}^{\infty} a_n(x-c)^n$. There are three possibilities:

(i) $F(x)$ converges only for $x = c$, or

(ii) $F(x)$ converges for all x, or

(iii) There is a number $R > 0$ such that $F(x)$ converges absolutely if $|x - c| < R$ and diverges if $|x - c| > R$. It may or may not converge at the endpoints $|x - c| = R$.

In Case (i), set $R = 0$, and in Case (ii), set $R = \infty$. We call R the **radius of convergence** of $F(x)$.

Proof We assume that $c = 0$ to simplify the notation. The key observation is that if $F(x)$ converges for some nonzero value $x = B$, then it converges absolutely for all $|x| < |B|$. To prove this, note that if $F(B) = \sum_{n=0}^{\infty} a_n B^n$ converges, then the general term $a_n B^n$ must tend to zero. In particular, there exists $M > 0$ such that $|a_n B^n| < M$ for all n, and therefore,

$$\sum_{n=0}^{\infty} |a_n x^n| = \sum_{n=0}^{\infty} |a_n B^n| \left|\frac{x}{B}\right|^n < M \sum_{n=0}^{\infty} \left|\frac{x}{B}\right|^n$$

If $|x| < |B|$, then $|x/B| < 1$ and the series on the right is a convergent geometric series. By the Comparison Test, the series on the left also converges and thus $F(x)$ converges absolutely if $|x| < |B|$.

Least Upper Bound Property: If S is a set of real numbers with an upper bound M (that is, $x \le M$ for all $x \in S$), then S has a least upper bound L. See Appendix B.

Let S be the set of numbers x such that $F(x)$ converges. Then S contains 0. If $S = \{0\}$, then $F(x)$ converges only for $x = 0$ and Case (i) holds. Otherwise, S contains a number $B \neq 0$. In this case, S contains the open interval $(-|B|, |B|)$ by the previous paragraph. If S is bounded, then S has a least upper bound $L > 0$ (see marginal note). Since there exist numbers $B \in S$ smaller than but arbitrarily close to L, S contains $(-B, B)$ for all $0 < B < L$. It follows that S contains the open interval $(-L, L)$. S cannot contain any number x with $|x| > L$, but S may contain one or both of the endpoints $x = \pm L$. This is Case (iii). If S is not bounded, then S contains intervals $(-B, B)$ for B arbitrarily large. Thus $S = \mathbf{R}$ and we are in Case (ii). ■

According to Theorem 1, if the radius of convergence R is nonzero and finite, then $F(x)$ converges absolutely on an interval around c of radius R (Figure 1). In some cases, the Ratio Test can be used to find the radius of convergence.

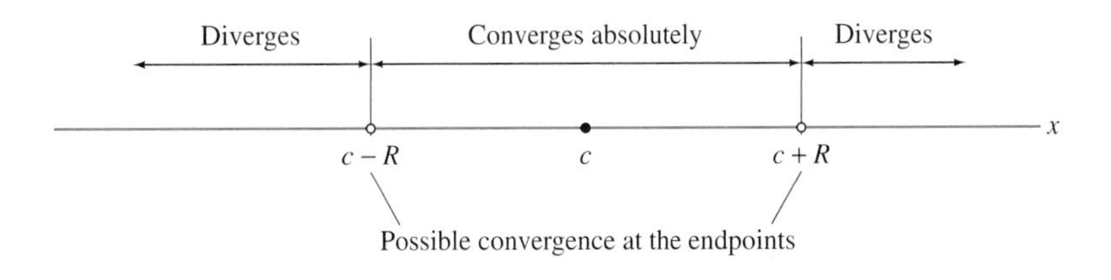

FIGURE 1 Interval of convergence of a power series.

■ **EXAMPLE 1** **Using the Ratio Test** For which values of x does $F(x) = \sum_{n=0}^{\infty} \frac{x^n}{2^n}$ converge?

Solution Let $b_n = \frac{x^n}{2^n}$ and let us compute the ratio ρ of the Ratio Test:

$$\rho = \lim_{n\to\infty} \left|\frac{b_{n+1}}{b_n}\right| = \lim_{n\to\infty} \left|\frac{x^{n+1}}{2^{n+1}}\right| \cdot \left|\frac{2^n}{x^n}\right| = \lim_{n\to\infty} \left|\frac{2^{-(n+1)}x^{n+1}}{2^{-n}x^n}\right| = \lim_{n\to\infty} \frac{1}{2}|x| = \frac{1}{2}|x|$$

By the Ratio Test, $F(x)$ converges if $\rho = \frac{1}{2}|x| < 1$, that is, for $|x| < 2$. Similarly, $F(x)$ diverges if $\rho = \frac{1}{2}|x| > 1$ or $|x| > 2$. Therefore, the radius of convergence is $R = 2$.

What about the endpoints? The Ratio Test is inconclusive for $x = \pm 2$, so we must check these cases directly. Both series diverge:

$$F(2) = \sum_{n=0}^{\infty} \frac{2^n}{2^n} = 1 + 1 + 1 + 1 + 1 + 1 \cdots$$

$$F(-2) = \sum_{n=0}^{\infty} \frac{(-2)^n}{2^n} = 1 - 1 + 1 - 1 + 1 - 1 \cdots$$

Therefore, $F(x)$ converges only for $|x| < 2$ (Figure 2). ■

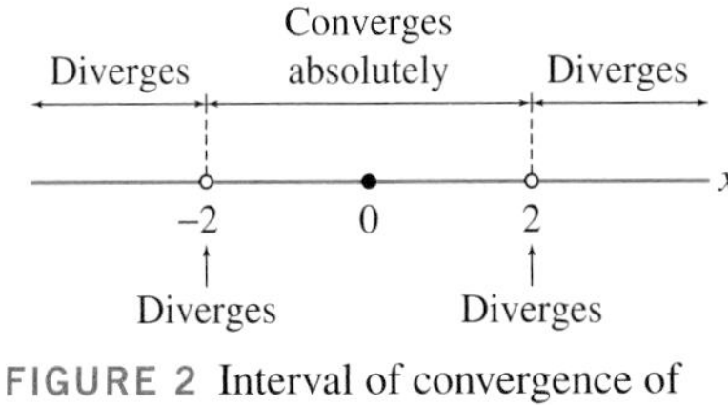

FIGURE 2 Interval of convergence of $\sum_{n=0}^{\infty} \frac{x^n}{2^n}$.

The method of the previous example may be applied more generally to any power series $F(x) = \sum b_n(x-c)^n$ for which the following limit exists:

$$r = \lim_{n\to\infty} \left| \frac{b_{n+1}}{b_n} \right|$$

We compute the ratio ρ of the Ratio Test applied to $F(x)$:

$$\rho = \lim_{n\to\infty} \left| \frac{b_{n+1}(x-c)^{n+1}}{b_n(x-c)^n} \right| = |x-c| \left(\lim_{n\to\infty} \left| \frac{b_{n+1}}{b_n} \right| \right) = r|x-c|$$

By the Ratio Test, $F(x)$ converges if $\rho = r|x-c| < 1$ and diverges if $\rho = r|x-c| > 1$. Thus, if r is finite and nonzero, then $F(x)$ converges if $|x-c| < r^{-1}$ and the radius of convergence is $R = r^{-1}$. If $r = 0$, then $F(x)$ converges for all x and the radius of convergence is $R = \infty$. If $r = \infty$, then $F(x)$ diverges for all $x \neq c$ and $R = 0$.

THEOREM 2 Finding the Radius of Convergence Let $F(x) = \sum b_n(x-c)^n$ and assume that the following limit exists:

$$r = \lim_{n\to\infty} \left| \frac{b_{n+1}}{b_n} \right|$$

Then $F(x)$ has radius of convergence $R = r^{-1}$ (where we set $R = \infty$ if $r = 0$ and $R = 0$ if $r = \infty$).

EXAMPLE 2 Determine the convergence of $F(x) = \sum_{n=1}^{\infty} \frac{(-1)^n}{n}(x-5)^n$.

Solution Let $b_n = \frac{(-1)^n}{n}$. Then

$$r = \lim_{n\to\infty} \left| \frac{b_{n+1}}{b_n} \right| = \lim_{n\to\infty} \left| \frac{1}{n+1} \frac{n}{1} \right| = \lim_{n\to\infty} \left| \frac{n}{n+1} \right| = 1$$

The radius of convergence is $R = r^{-1} = 1$. Therefore, the power series converges absolutely if $|x-5| < 1$ and diverges if $|x-5| > 1$. In other words, $F(x)$ converges absolutely on the open interval $(4, 6)$. At the endpoints we have

$$x = 6: \quad \sum_{n=1}^{\infty} \frac{(-1)^n}{n}(6-5)^n = \sum_{n=1}^{\infty} \frac{(-1)^n}{n} \qquad \text{convergent by the Leibniz Test}$$

$$x = 4: \quad \sum_{n=1}^{\infty} \frac{(-1)^n}{n}(4-5)^n = \sum_{n=1}^{\infty} \frac{1}{n} \qquad \text{divergent (harmonic series)}$$

Therefore, the power series converges on the half-open interval $(4, 6]$. ■

EXAMPLE 3 Infinite Radius of Convergence Show that $\sum_{n=0}^{\infty} \frac{x^n}{n!}$ converges for all x.

Solution Let $b_n = \frac{1}{n!}$. Then

$$r = \lim_{n\to\infty} \left| \frac{b_{n+1}}{b_n} \right| = \lim_{n\to\infty} \frac{1}{(n+1)!} \frac{n!}{1} = \lim_{n\to\infty} \frac{1}{n+1} = 0$$

Thus, $R = r^{-1} = \infty$ and the series converges for all x by Theorem 2. ■

When a function $f(x)$ is represented by a power series on an interval I, we refer to the power series as the "power series expansion" of $f(x)$ on I. In the next section, we show that a function has at most one power series expansion with a given center c on an interval.

An important example of a power series is provided by the geometric series. Recall that $\sum_{n=0}^{\infty} r^n = \frac{1}{1-r}$ for $|r| < 1$. Writing x instead of r, we view this formula as a power series expansion:

$$\frac{1}{1-x} = \sum_{n=0}^{\infty} x^n \qquad \text{for } |x| < 1 \tag{1}$$

The next two examples show how this formula may be adapted to find the power series representations of other functions.

■ **EXAMPLE 4 Using the Formula for Geometric Series** Prove that $\frac{1}{1-2x} = \sum_{n=0}^{\infty} 2^n x^n$ for $|x| < \frac{1}{2}$.

Solution Substitute $2x$ for x in Eq. (1):

$$\frac{1}{1-2x} = \sum_{n=0}^{\infty} (2x)^n = \sum_{n=0}^{\infty} 2^n x^n \tag{2}$$

Expansion (1) is valid for $|x| < 1$ and thus expansion (2) is valid for $|2x| < 1$ or $|x| < \frac{1}{2}$. ■

■ **EXAMPLE 5** Prove that $\frac{1}{2+x^2} = \sum_{n=0}^{\infty} \frac{(-1)^n x^{2n}}{2^{n+1}}$. For which x is this formula valid?

Solution We first rewrite $\frac{1}{2+x^2}$ in the form $\frac{1}{1-u}$ so we can use Eq. (1):

$$\frac{1}{2+x^2} = \frac{1}{2} \frac{1}{1+\frac{1}{2}x^2} = \frac{1}{2} \frac{1}{1-(-\frac{1}{2}x^2)}$$

We may substitute $-\frac{1}{2}x^2$ for x in Eq. (1), provided that $\left|\frac{1}{2}x^2\right| < 1$, to obtain

$$\frac{1}{2+x^2} = \frac{1}{2} \sum_{n=0}^{\infty} \left(-\frac{x^2}{2}\right)^n = \sum_{n=0}^{\infty} \frac{(-1)^n x^{2n}}{2^{n+1}}$$

This expansion is valid if $\left|-x^2/2\right| < 1$ or $|x| < \sqrt{2}$. ■

Our next theorem states, in essence, that power series are well-behaved functions in the following sense: A power series $F(x)$ is differentiable within its interval of convergence and we may differentiate and integrate $F(x)$ as if it were a polynomial.

THEOREM 3 Term-by-Term Differentiation and Integration Suppose that

$$F(x) = \sum_{n=0}^{\infty} a_n (x-c)^n$$

has radius of convergence $R > 0$. Then $F(x)$ is differentiable on $(c - R, c + R)$ and its derivative and antiderivative may be computed term by term. More precisely, for $x \in (c - R, c + R)$ we have

$$F'(x) = \sum_{n=1}^{\infty} n a_n (x-c)^{n-1}$$

$$\int F(x)\,dx = A + \sum_{n=0}^{\infty} \frac{a_n}{n+1}(x-c)^{n+1} \qquad (A \text{ any constant})$$

These series have the same radius of convergence R.

See Exercise 58 for a proof that $F(x)$ is continuous. The proofs of the remaining statements are omitted.

EXAMPLE 6 Differentiating a Power Series Prove that

$$\frac{1}{(1-x)^2} = 1 + 2x + 3x^2 + 4x^3 + 5x^4 + \cdots$$

for $-1 < x < 1$.

Solution Noting that

$$\frac{d}{dx}\frac{1}{1-x} = \frac{1}{(1-x)^2}$$

we obtain the result by differentiating the geometric series term by term for $|x| < 1$:

$$\frac{d}{dx}\frac{1}{1-x} = \frac{d}{dx}\left(1 + x + x^2 + x^3 + x^4 + \cdots\right)$$

$$\frac{1}{(1-x)^2} = 1 + 2x + 3x^2 + 4x^3 + 5x^4 + \cdots \qquad \boxed{3}$$

Expansion (3) is valid for $|x| < 1$ because the geometric series has radius of convergence $R = 1$. ■

EXAMPLE 7 The Power Series for $f(x) = \tan^{-1} x$ via Integration Prove that for $-1 < x < 1$,

$$\tan^{-1} x = \sum_{n=0}^{\infty} \frac{(-1)^n x^{2n+1}}{2n+1} = x - \frac{x^3}{3} + \frac{x^5}{5} - \frac{x^7}{7} + \cdots \qquad \boxed{4}$$

Solution First, substitute $-x^2$ for x in (1) to obtain

$$\frac{1}{1+x^2} = 1 - x^2 + x^4 - x^6 + \cdots$$

Since the geometric series has radius of convergence $R = 1$, this expansion is valid for $|x^2| < 1$, that is, $|x| < 1$. Now apply Theorem 3 to integrate this series term by term, recalling that $\tan^{-1} x$ is an antiderivative $(1+x^2)^{-1}$:

$$\tan^{-1} x = \int \frac{dx}{1+x^2} = \int \left(1 - x^2 + x^4 - x^6 + \cdots\right) dx = A + x - \frac{x^3}{3} + \frac{x^5}{5} - \frac{x^7}{7} + \cdots$$

To determine the constant A, set $x = 0$. We obtain $\tan^{-1} 0 = 0 = A$ and therefore $A = 0$. This proves Eq. (4) for $-1 < x < 1$. ■

GRAPHICAL INSIGHT Let's examine the expansion of the previous example graphically. The partial sums of the power series for $f(x) = \tan^{-1} x$ are

$$S_N(x) = \sum_{n=0}^{N} (-1)^n \frac{x^{2n+1}}{2n+1} = x - \frac{x^3}{3} + \frac{x^5}{5} - \cdots + (-1)^N \frac{x^{2N+1}}{2N+1}$$

We can expect $S_N(x)$ to provide a good approximation to $f(x) = \tan^{-1} x$ on the interval $(-1, 1)$, where the power series expansion is valid. Figure 3 confirms this: The graphs of the partial sums $S_{50}(x)$ and $S_{51}(x)$ are nearly indistinguishable from the graph of $\tan^{-1} x$ on $(-1, 1)$. Thus we may use the partial sums to approximate values of the arctangent. For example, an approximation to $\tan^{-1}(0.3)$ is given by

$$S_4(0.3) = 0.3 - \frac{(0.3)^3}{3} + \frac{(0.3)^5}{5} - \frac{(0.3)^7}{7} + \frac{(0.3)^9}{9} \approx 0.2914569$$

Since the power series is an alternating series, the error is not greater than the first omitted term:

$$|\tan^{-1}(0.3) - S_4(0.3)| \le \frac{(0.3)^{11}}{11} \approx 1.61 \times 10^{-7}$$

The situation changes drastically in the region $|x| > 1$, where the power series diverges. The partial sums $S_N(x)$ deviate sharply from $\tan^{-1} x$ outside $(-1, 1)$.

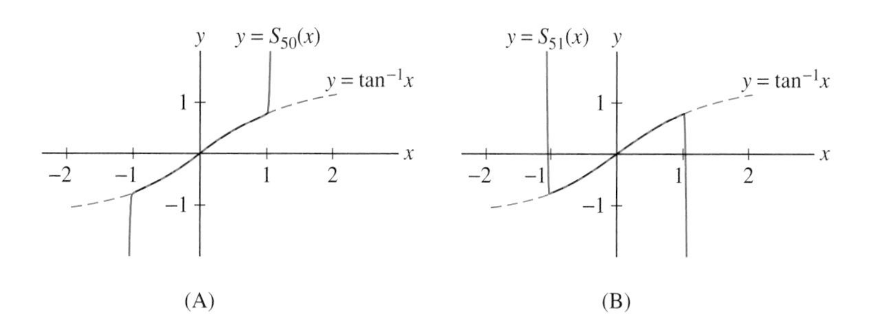

FIGURE 3 $S_{50}(x)$ and $S_{51}(x)$ are nearly indistinguishable from $\tan^{-1} x$ on $(-1, 1)$.

Power Series Solutions of Differential Equations

In the next section, we use the theory of Taylor series to prove that the exponential function $f(x) = e^x$ is represented by a power series. However, we can already show this with the tools at our disposal by making use of the differential equation satisfied by $f(x) = e^x$. Recall that by Theorem 1 in Section 5.8, $y = e^x$ is the unique function satisfying the differential equation $y' = y$ with initial condition $y(0) = 1$. Let's try to find a power series $P(x) = \sum_{n=0}^{\infty} a_n x^n$ that also satisfies $P'(x) = P(x)$ and $P(0) = 1$.

We have

$$P(x) = \sum_{n=0}^{\infty} a_n x^n = a_0 + a_1 x + a_2 x^2 + a_3 x^3 + \cdots$$

$$P'(x) = \sum_{n=0}^{\infty} n a_n x^{n-1} = a_1 + 2a_2 x + 3a_3 x^2 + 4a_4 x^3 + \cdots$$

We see that $P(x)$ satisfies $P'(x) = P(x)$ if

$$a_0 = a_1, \quad a_1 = 2a_2, \quad a_2 = 3a_3, \quad a_3 = 4a_4, \quad \ldots$$

In general, $a_{n-1} = na_n$, or

$$\boxed{a_n = \frac{a_{n-1}}{n}}$$

This equation is called a *recursion relation*. It allows us to successively determine all of the coefficients a_n from the first coefficient a_0, which may be chosen arbitrarily. For example, the recursion relation yields

$$n = 1: \quad a_1 = \frac{a_0}{1}$$

$$n = 2: \quad a_2 = \frac{a_1}{2} = \frac{a_0}{2 \cdot 1} = \frac{a_0}{2!}$$

$$n = 3: \quad a_3 = \frac{a_2}{3} = \frac{a_1}{3 \cdot 2} = \frac{a_0}{3 \cdot 2 \cdot 1} = \frac{a_0}{3!}$$

To obtain a general formula for a_n, apply the recursion relation n times:

$$a_n = \frac{a_{n-1}}{n} = \frac{a_{n-2}}{n(n-1)} = \frac{a_{n-3}}{n(n-1)(n-2)} = \cdots = \frac{a_0}{n!}$$

We conclude that $P(x) = a_0 \sum_{n=0}^{\infty} \frac{x^n}{n!}$. As we showed in Example 3, this power series has an infinite radius of convergence and thus $P(x)$ is a solution of $y' = y$ for all x.

Now observe that $P(0) = a_0$, so we set $a_0 = 1$ to obtain a solution satisfying the initial condition $y(0) = 1$. Now, since $f(x) = e^x$ and $P(x)$ both satisfy the differential condition with initial condition, they are equal. Thus we have proven that for all x,

$$e^x = \sum_{n=0}^{\infty} \frac{x^n}{n!} = 1 + x + \frac{x^2}{2!} + \frac{x^3}{3!} + \frac{x^4}{4!} + \cdots \qquad \boxed{5}$$

The method just employed is a powerful tool in the study of differential equations. We knew in advance that $y = e^x$ is a solution of $y' = y$, but suppose we are given a differential equation whose solution is unknown. We may try to find a solution in the form of a power series $P(x) = \sum_{n=0}^{\infty} a_n x^n$. In favorable cases, the differential equation leads to a recursion relation that enables us to determine the coefficients a_n.

The solution of Eq. (6) satisfying $y'(0) = 1$ is called the "Bessel function of order one." The Bessel function of order n is a solution of

$$x^2 y'' + xy' + (x^2 - n^2)y = 0$$

Bessel functions appear in many areas of physics and engineering.

■ **EXAMPLE 8** Find a power series solution to the differential equation

$$x^2 y'' + xy' + (x^2 - 1)y = 0 \qquad \boxed{6}$$

and the initial condition $y'(0) = 1$.

Solution Assume that Eq. (6) has a power series solution $P(x) = \sum_{n=0}^{\infty} a_n x^n$. Then

$$y' = P'(x) = \sum_{n=0}^{\infty} n a_n x^{n-1}$$

$$y'' = P''(x) = \sum_{n=0}^{\infty} n(n-1) a_n x^{n-2}$$

Now substitute the series for y, y', and y'' into the differential equation (6) to determine the recursion relation satisfied by the coefficients a_n:

$$\begin{aligned} & x^2 y'' + xy' + (x^2 - 1)y \\ &\quad = x^2 \sum_{n=0}^{\infty} n(n-1) a_n x^{n-2} + x \sum_{n=0}^{\infty} n a_n x^{n-1} + (x^2 - 1) \sum_{n=0}^{\infty} a_n x^n \\ &\quad = \sum_{n=0}^{\infty} n(n-1) a_n x^n + \sum_{n=0}^{\infty} n a_n x^n - \sum_{n=0}^{\infty} a_n x^n + \sum_{n=0}^{\infty} a_n x^{n+2} \\ &\quad = \sum_{n=0}^{\infty} (n^2 - 1) a_n x^n + \sum_{n=2}^{\infty} a_{n-2} x^n \end{aligned}$$

We see that the equation $x^2 y'' + xy' + (x^2 - 1)y = 0$ is satisfied if

$$\sum_{n=0}^{\infty} (n^2 - 1) a_n x^n = -\sum_{n=2}^{\infty} a_{n-2} x^n \qquad \boxed{7}$$

The first few terms on each side of this equation are

$$-a_0 + 0 \cdot x + 3a_2 x^2 + 8a_3 x^3 + 15a_4 x^4 + \cdots = 0 + 0 \cdot x + -a_0 x^2 - a_1 x^3 - a_2 x^4 - \cdots$$

Matching up the coefficients of x^n, we find that

$$-a_0 = 0, \qquad 3a_2 = -a_0, \qquad 8a_3 = -a_1, \qquad 15a_4 = -a_2 \qquad \boxed{8}$$

In general, $(n^2 - 1)a_n = -a_{n-2}$, and this yields the recursion relation

$$\boxed{a_n = -\frac{a_{n-2}}{n^2 - 1} \quad \text{for } n \geq 2} \qquad \boxed{9}$$

Note that $a_0 = 0$ by (8). The recursion relation implies that all of the even coefficients $a_2, a_4, a_6, \ldots$ are zero:

$$a_2 = \frac{a_0}{2^2 - 1} \text{ so } a_2 = 0, \qquad \text{and then} \qquad a_4 = \frac{a_2}{4^2 - 1} = 0 \text{ so } a_4 = 0, \qquad \text{etc.}$$

As for the odd coefficients, note that we may choose a_1 arbitrarily. However, $P'(0) = a_1$. Thus, we set $a_1 = 1$ so that $P(x)$ satisfies the initial condition $y'(0) = 1$. Now apply

Eq. (9):

$$n = 3 \qquad a_3 = -\frac{a_1}{3^2 - 1} = -\frac{1}{3^2 - 1}$$

$$n = 5: \qquad a_5 = -\frac{a_3}{5^2 - 1} = \frac{1}{(5^2 - 1)(3^2 - 1)}$$

$$n = 7: \qquad a_7 = -\frac{a_5}{7^2 - 1} = -\frac{1}{(7^2 - 1)(3^2 - 1)(5^2 - 1)}$$

This shows the general pattern of coefficients. To express the coefficients in a compact form, let $n = 2k + 1$. Then

$$n^2 - 1 = (2k + 1)^2 - 1 = 4k^2 + 4k = 4k(k + 1)$$

and the recursion relation may be written

$$a_{2k+1} = -\frac{a_{2k-1}}{4k(k + 1)}$$

Applying this recursion relation k times, we obtain the closed formula

$$a_{2k+1} = (-1)^k \left(\frac{1}{4k(k + 1)}\right)\left(\frac{1}{4(k - 1)k}\right) \cdots \left(\frac{a_1}{4(1)(2)}\right) = \frac{(-1)^k}{4^k\, k!\, (k + 1)!}$$

Thus we obtain a power series representation of our solution:

$$P(x) = \sum_{k=0}^{\infty} \frac{(-1)^k}{4^k k!(k + 1)!} x^{2k+1}$$

A straightforward application of the Ratio Test shows that $P(x)$ has an infinite radius of convergence. Therefore $P(x)$ is a solution of the initial value problem for all x. ■

10.6 SUMMARY

- A *power series* is an infinite series of the form

$$F(x) = \sum_{n=0}^{\infty} a_n (x - c)^n$$

The constant c is called the *center* of $F(x)$.

- A power series behaves in one of three ways:

(i) $F(x)$ converges only for $x = c$, or

(ii) $F(x)$ converges for all x, or

(iii) There exists $R > 0$ such that $F(x)$ converges absolutely for $|x - c| < R$ and diverges for $|x - c| > R$.

The number R is called the *radius of convergence* of $F(x)$. Convergence at the endpoints $x = c \pm R$ must be checked separately. We set $R = 0$ in Case (i), and $R = \infty$ in Case (ii).

- If $r = \lim_{n\to\infty} \left|\frac{a_{n+1}}{a_n}\right|$ exists, then $F(x)$ has radius of convergence $R = r^{-1}$ (where $R = 0$ if $r = \infty$ and $R = \infty$ if $r = 0$).

- If $R > 0$, then $F(x)$ is differentiable on $(c - R, c + R)$ and may be differentiated and integrated term by term:

$$F'(x) = \sum_{n=1}^{\infty} n a_n (x - c)^{n-1}, \qquad \int F(x)\,dx = A + \sum_{n=0}^{\infty} \frac{a_n}{n+1}(x - c)^{n+1}$$

(A is any constant). The power series for $F'(x)$ and $\int F(x)\,dx$ have the same radius of convergence R.

- The power series expansion $\frac{1}{1-x} = \sum_{n=0}^{\infty} x^n$ is valid for $|x| < 1$. It may be used to derive expansions of other related functions by substitution, integration, or differentiation.

10.6 EXERCISES

Preliminary Questions

1. Suppose that $\sum a_n x^n$ converges for $x = 5$. Must it also converge for $x = 4$? What about $x = -3$?

2. Suppose that $\sum a_n (x - 6)^n$ converges for $x = 10$. At which of the points (a)–(d) must it also converge?

(a) $x = 8$ **(b)** $x = 12$ **(c)** $x = 2$ **(d)** $x = 0$

3. Suppose that $F(x)$ is a power series with radius of convergence $R = 12$. What is the radius of convergence of $F(3x)$?

4. The power series $F(x) = \sum_{n=1}^{\infty} n x^n$ has radius of convergence $R = 1$. What is the power series expansion of $F'(x)$ and what is its radius of convergence?

Exercises

1. Use the Ratio Test to determine the radius of convergence of $\sum_{n=0}^{\infty} \frac{x^n}{2^n}$.

2. Use the Ratio Test to show that $\sum_{n=1}^{\infty} \frac{x^n}{\sqrt{n}2^n}$ has radius of convergence $R = 2$. Then determine whether it converges absolutely or conditionally at the endpoints $R = \pm 2$.

3. Show that the following three power series have the same radius of convergence. Then show that (a) diverges at both endpoints, (b) converges at one endpoint but diverges at the other, and (c) converges at both endpoints.

(a) $\sum_{n=1}^{\infty} \frac{x^n}{3^n}$ **(b)** $\sum_{n=1}^{\infty} \frac{x^n}{n3^n}$ **(c)** $\sum_{n=1}^{\infty} \frac{x^n}{n^2 3^n}$

4. Repeat Exercise 3 for the following series:

(a) $\sum_{n=1}^{\infty} \frac{(x-5)^n}{9^n}$ **(b)** $\sum_{n=1}^{\infty} \frac{(x-5)^n}{n9^n}$ **(c)** $\sum_{n=1}^{\infty} \frac{(x-5)^n}{n^2 9^n}$

5. Show that $\sum_{n=0}^{\infty} n^n x^n$ diverges for all $x \neq 0$.

6. **(a)** Find the radius of convergence of $\sum_{n=1}^{\infty} \frac{x^n}{n^2}$.

(b) Determine whether the series converges at the endpoints of the interval of convergence.

In Exercises 7–26, find the values of x for which the following power series converge.

7. $\sum_{n=1}^{\infty} n x^n$

8. $\sum_{n=1}^{\infty} n(x-3)^n$

9. $\sum_{n=1}^{\infty} \frac{2^n x^n}{n}$

10. $\sum_{n=1}^{\infty} \frac{(-5)^n (x-3)^n}{n^2}$

11. $\sum_{n=2}^{\infty} \frac{x^n}{\ln n}$

12. $\sum_{n=1}^{\infty} \frac{x^n}{n2^n}$

13. $\sum_{n=1}^{\infty} \frac{x^n}{(n!)^2}$

14. $\sum_{n=4}^{\infty} \frac{x^n}{n^5}$

15. $\sum_{n=1}^{\infty} (-1)^n n^4 (x+4)^n$

16. $\sum_{n=0}^{\infty} \frac{(-1)^n x^n}{\sqrt{n^2+1}}$

17. $\sum_{n=0}^{\infty} \frac{n}{2^n} x^n$

18. $\sum_{n=0}^{\infty} 4^n x^n$

19. $\sum_{n=1}^{\infty} \frac{(x-4)^n}{n^4}$

20. $\sum_{n=1}^{\infty} \frac{2^n}{3n}(x+3)^n$

21. $\sum_{n=10}^{\infty} n!\,(x+5)^n$

22. $\sum_{n=15}^{\infty} \frac{x^{2n+1}}{3n+1}$

23. $\sum_{n=12}^{\infty} e^n (x-2)^n$

24. $\sum_{n=0}^{\infty} \frac{x^n}{n^4+2}$

25. $\sum_{n=2}^{\infty} \frac{x^n}{\ln n}$

26. $\sum_{n=2}^{\infty} \frac{(x-2)^n}{(n \ln n)^2}$

In Exercises 27–34, use Eq. (1) to expand the function in a power series with center $c = 0$ and determine the set of x for which the expansion is valid.

27. $f(x) = \frac{1}{1-3x}$

28. $f(x) = \frac{1}{1+3x}$

29. $f(x) = \frac{1}{3-x}$

30. $f(x) = \frac{1}{4+3x}$

31. $f(x) = \frac{1}{1+x^9}$

32. $f(x) = \frac{1}{5-x^2}$

33. $f(x) = \frac{1}{1+3x^7}$

34. $f(x) = \frac{1}{4-2x^3}$

35. Use the equalities

$$\frac{1}{1-x} = \frac{1}{-3-(x-4)} = \frac{-\frac{1}{3}}{1+\left(\frac{x-4}{3}\right)}$$

to show that for $|x-4| < 3$

$$\frac{1}{1-x} = \sum_{n=0}^{\infty} (-1)^{n+1} \frac{(x-4)^n}{3^{n+1}}$$

36. Use the method of Exercise 35 to expand $\frac{1}{1-x}$ in power series with centers $c = 2$ and $c = -2$. Determine the set of x for which the expansions are valid.

37. Use the method of Exercise 35 to expand $\frac{1}{4-x}$ in a power series with center $c = 5$. Determine the set of x for which the expansion is valid.

38. Evaluate $\sum_{n=1}^{\infty} \frac{n}{2^n}$. *Hint:* Show that

$$(1-x)^{-2} = \sum_{n=1}^{\infty} nx^{n-1}$$

39. Give an example of a power series that converges for x in $[2, 6)$.

40. Prove that for $-1 < x < 1$

$$\frac{1}{1+x} = \sum_{n=0}^{\infty} (-1)^n x^n = 1 - x + x^2 - x^3 + \cdots$$

$$\ln(1+x) = \sum_{n=1}^{\infty} \frac{(-1)^{n-1} x^n}{n} = x - \frac{x^2}{2} + \frac{x^3}{3} - \frac{x^4}{4} + \cdots$$

41. Use Exercise 40 to prove that

$$\ln \frac{3}{2} = \frac{1}{2} - \frac{1}{2 \cdot 2^2} + \frac{1}{3 \cdot 2^3} - \frac{1}{4 \cdot 2^4} + \cdots$$

Use your knowledge of alternating series to find an N such that the partial sum S_N approximates $\ln \frac{3}{2}$ to within an error of at most 10^{-3}. Confirm this using a calculator to compute both S_N and $\ln \frac{3}{2}$.

42. Show that the following series converges absolutely for $|x| < 1$ and compute its sum:

$$F(x) = 1 - x - x^2 + x^3 - x^4 - x^5 + x^6 - x^7 - x^8 + \cdots$$

Hint: Write $F(x)$ as a sum of three geometric series with common ratio x^3.

43. Show that for $|x| < 1$

$$\frac{1+2x}{1+x+x^2} = 1 + x - 2x^2 + x^3 + x^4 - 2x^5 + x^6 + x^7 - 2x^8 + \cdots$$

Hint: Use the hint from Exercise 42.

44. Find a power series $P(x) = \sum_{n=0}^{\infty} a_n x^n$ satisfying the differential equation $y' = -y$ with initial condition $y(0) = 1$. Then use Theorem 1 of Section 5.8 to conclude that $P(x) = e^{-x}$.

45. Use the power series for $y = e^x$ to show that

$$\frac{1}{e} = \frac{1}{2!} - \frac{1}{3!} + \frac{1}{4!} - \cdots$$

Use your knowledge of alternating series to find an N such that the partial sum S_N approximates e^{-1} to within an error of at most 10^{-3}. Confirm this using a calculator to compute both S_N and e^{-1}.

46. Let $P(x) = \sum_{n=0} a_n x^n$ be a power series solution to $y' = 2xy$ with initial condition $y(0) = 1$.

(a) Show that the odd coefficients a_{2k+1} are all zero.

(b) Prove that $a_{2k} = \frac{a_{2k-2}}{k}$ and use this result to determine the coefficients a_{2k}.

47. Find a power series $P(x)$ satifying the differential equation:

$$y'' - xy' + y = 0 \qquad \boxed{10}$$

with initial condition $y(0) = 1$, $y'(0) = 0$. What is the radius of convergence of the power series?

48. Find a power series satisfying Eq. (10) with initial condition $y(0) = 0$, $y'(0) = 1$.

49. Prove that $J_2(x) = \sum_{k=0}^{\infty} \frac{(-1)^k}{2^{2k+2}\, k!\,(k+3)!} x^{2k+2}$ is a solution of the Bessel differential equation of order two:

$$x^2 y'' + xy' + (x^2 - 4)y = 0$$

50. Use Eq. (4) to approximate $\tan^{-1}(0.5)$ to three decimal places.

51. Let $C(x) = 1 - \frac{x^2}{2!} + \frac{x^4}{4!} - \frac{x^6}{6!} + \cdots$.

(a) Show that $C(x)$ has an infinite radius of convergence.

(b) Prove that $C(x)$ and $f(x) = \cos x$ are both solutions of $y'' = -y$ with initial conditions $y(0) = 1$, $y'(0) = 0$. This initial value problem has a unique solution, so it follows that $C(x) = \cos x$ for all x.

52. Find all values of x such that $\sum_{n=1}^{\infty} \frac{x^{n^2}}{n!}$ converges.

53. Find all values of x such that the following series converges:

$$F(x) = 1 + 3x + x^2 + 27x^3 + x^4 + 243x^5 + \cdots$$

54. Explain why Theorem 2 cannot be applied directly to find the radius of convergence of $\sum_{n=1}^{\infty} \frac{x^{3n}}{5^n}$. What is the radius of convergence of this series?

55. Why is it impossible to expand $f(x) = |x|$ as a power series that converges in an interval around $x = 0$? Explain this using Theorem 3.

Further Insights and Challenges

56. Prove that for $b \neq 0$

$$\sum_{n=1}^{\infty} \frac{1}{n(n+b)} = \frac{1}{b}\int_0^1 \frac{1-x^b}{1-x}\,dx$$

Conclude that the sum on the left has the value

$$\frac{1}{b}\left(1 + \frac{1}{2} + \frac{1}{3} + \cdots + \frac{1}{b}\right)$$

57. Suppose that the coefficients of $F(x) = \sum_{n=0}^{\infty} a_n x^n$ are *periodic*, that is, for some whole number $M > 0$, we have $a_{M+n} = a_n$. Prove that $F(x)$ converges absolutely for $|x| < 1$ and that

$$F(x) = \frac{a_0 + a_1 x + \cdots + a_{M-1}x^{M-1}}{1 - x^M}$$

Hint: Use the hint for Exercise 42.

58. Continuity of Power Series Let $F(x) = \sum_{n=0}^{\infty} a_n x^n$ be a power series with radius of convergence $R > 0$.

(a) Prove the inequality

$$|x^n - y^n| \le n|x - y|(|x|^{n-1} + |y|^{n-1}) \qquad \boxed{11}$$

Hint: $x^n - y^n = (x - y)(x^{n-1} + x^{n-2}y + \cdots + y^{n-1})$.

(b) Choose R_1 with $0 < R_1 < R$. Use the Ratio Test to show that the infinite series $M = \sum_{n=0}^{\infty} 2n|a_n|R_1^n$ converges.

(c) Use Eq. (11) to show that if $|x| < R_1$ and $|y| < R_1$, then $|F(x) - F(y)| \le M|x - y|$.

(d) Prove that if $|x| < R$, then $F(x)$ is continuous at x. *Hint:* Choose R_1 such that $|x| < R_1 < R$. Show that if $\epsilon > 0$ is given, then $|F(x) - F(y)| \le \epsilon$ for all y such that $|x - y| < \delta$, where δ is any positive number that is less than ϵ/M and $R_1 - |x|$ (see Figure 4).

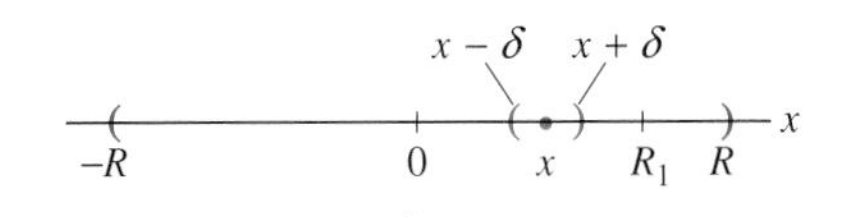

FIGURE 4 If $x > 0$, choose $\delta > 0$ less than ϵ/M and $R_1 - x$.

10.7 Taylor Series

We saw in the previous section that functions such as $f(x) = e^x$ and $f(x) = \tan^{-1} x$ can be represented as power series. These power series give us a certain tangible insight into the function represented and they allow us to approximate the values of $f(x)$ to any desired degree of accuracy. Thus, it is desirable to develop general methods for finding power series representations.

Suppose that $f(x)$ has a power series expansion centered at $x = c$ that is valid for all x in an interval $(c - R, c + R)$ with $R > 0$:

$$f(x) = \sum_{n=0}^{\infty} a_n(x - c)^n = a_0 + a_1(x - c) + a_2(x - c)^2 + \cdots$$

Then we may differentiate the series term by term (Theorem 3 in Section 10.6) to obtain

$$f(x) = a_0 + a_1(x-c) + a_2(x-c)^2 + a_3(x-c)^3 + \cdots$$
$$f'(x) = a_1 + 2a_2(x-c) + 3a_3(x-c)^2 + 4a_4(x-c)^3 + \cdots$$
$$f''(x) = 2a_2 + 2\cdot 3a_3(x-c) + 3\cdot 4a_4(x-c)^2 + \cdots$$
$$\vdots$$
$$f^{(k)}(x) = k!a_k + \big(2\cdot 3\cdots(k+1)\big)a_{k+1}(x-c) + \cdots$$

Setting $x = c$ in each of these series, we find that

$$f(c) = a_0, \quad f'(c) = a_1, \quad f''(c) = 2a_2, \quad \ldots, \quad f^{(k)}(c) = k!a_k$$

This shows that the coefficients are given by the formula (and proves Theorem 1 below):

$$a_k = \frac{f^{(k)}(c)}{k!} \qquad \boxed{1}$$

Recall that these are the coefficients of the Taylor polynomials. In summary:

$$f(x) = \sum_{n=0}^{\infty} \frac{f^{(n)}(c)}{n!}(x-c)^n$$
$$= f(c) + f'(c)(x-c) + \frac{f''(c)}{2!}(x-c)^2 + \frac{f'''(c)}{3!}(x-c)^3 + \cdots$$

The power series on the right is called the **Taylor series** of $f(x)$ centered at $x = c$. In the special case $c = 0$, the Taylor series is also called the **Maclaurin series**:

$$f(x) = f(0) + f'(0)x + \frac{f''(0)}{2!}x^2 + \frac{f'''(0)}{3!}x^3 + \frac{f^{(4)}(0)}{4!}x^4 + \cdots$$

THEOREM 1 Uniqueness of the Power Series Expansion If $f(x)$ is represented by a power series $F(x)$ centered at c on an interval $(c - R, c + R)$ with $R > 0$, then $F(x)$ is the Taylor series of $f(x)$ centered at $x = c$.

EXAMPLE 1 Find the Maclaurin series for $f(x) = e^x$.

Solution The nth derivative $f^{(n)}(x)$ is $f^{(n)}(x) = e^x$ for all n and thus

$$f(0) = f'(0) = f''(0) = \cdots = e^0 = 1$$

Therefore, the coefficients of the Maclaurin series are $a_k = \dfrac{f^{(k)}(0)}{k!} = \dfrac{1}{k!}$ and the Maclaurin series is

$$1 + \frac{x}{1} + \frac{x^2}{2!} + \frac{x^3}{3!} + \cdots$$ ■

Theorem 1 tells us that if we want to represent $f(x)$ by a power series centered at c, the only candidate for the job is the Taylor series:

$$T(x) = \sum_{n=0}^{\infty} \frac{f^{(n)}(c)}{n!}(x-c)^n$$

However, there is no guarantee that $T(x)$ converges to $f(x)$. To study convergence, we consider the kth partial sum, which is the Taylor polynomial of degree k:

$$T_k(x) = f(c) + f'(c)(x-c) + \frac{f''(c)}{2!}(x-c)^2 + \cdots + \frac{f^{(k)}(c)}{k!}(x-c)^k$$

Recall that the remainder is defined by

$$R_k(x) = f(x) - T_k(x)$$

Since $T(x)$ is the limit of the partial sums $T_k(x)$, we see that

The Taylor series converges to $f(x)$ if and only if $\lim_{k\to\infty} R_k(x) = 0$

Although there is no general method for determining whether $R_k(x)$ tends to zero, the following theorem can often be applied.

← REMINDER *$f(x)$ is called "infinitely differentiable" if $f^{(n)}(x)$ exists for all n.*

THEOREM 2 Let $f(x)$ be an infinitely differentiable function on the open interval $I = (c-R, c+R)$ with $R > 0$. Assume there exists $K \ge 0$ such that for all $k \ge 0$,

$$|f^{(k)}(x)| \le K \qquad \text{for all} \quad x \in I$$

Then $f(x)$ is represented by its Taylor series in I:

$$f(x) = \sum_{n=0}^{\infty} \frac{f^{(n)}(c)}{n!}(x-c)^n \qquad \text{for all} \quad x \in I$$

Proof We apply the Error Bound for Taylor polynomials:

$$|R_k(x)| = |f(x) - T_k(x)| \le K \frac{|x-c|^{k+1}}{(k+1)!}$$

If $x \in I$, then $|x - c| < R$ and

$$|R_k(x)| \le \underbrace{K \frac{R^{k+1}}{(k+1)!}}_{\text{This tends to zero as } k \to \infty}$$

As shown in Example 8 of Section 10.1, the quantity $R^k/k!$ tends to zero as $k \to \infty$ for every number R. We conclude that $\lim_{k\to\infty} R_k(x) = 0$ for all $x \in (c-R, c+R)$ and Theorem 2 follows. ■

Taylor expansions were studied throughout the seventeenth and eighteenth centuries by Euler, Gregory, Leibniz, Maclaurin, Newton, Taylor, and others. These developments in Europe and England were anticipated by the great Indian mathematician Madhava (c. 1340–1425), who discovered the expansions of sine and cosine and many other results two centuries earlier.

■ **EXAMPLE 2 Maclaurin Expansions of Sine and Cosine** Show that the following Taylor expansions are valid for all x:

$$\sin x = \sum_{n=0}^{\infty} (-1)^n \frac{x^{2n+1}}{(2n+1)!} = x - \frac{x^3}{3!} + \frac{x^5}{5!} - \frac{x^7}{7!} + \cdots$$

$$\cos x = \sum_{n=0}^{\infty} (-1)^n \frac{x^{2n}}{(2n)!} = 1 - \frac{x^2}{2!} + \frac{x^4}{4!} - \frac{x^6}{6!} + \cdots$$

← *REMINDER The derivatives of $f(x) = \sin x$ form a repeating pattern of length 4:*

$f(x)$	$f'(x)$	$f''(x)$	$f'''(x)$	...
$\sin x$	$\cos x$	$-\sin x$	$-\cos x$	...

Solution For $f(x) = \sin x$, we have

$$f^{(2n)}(x) = (-1)^n \sin x, \qquad f^{(2n+1)}(x) = (-1)^n \cos x$$

Therefore, $f^{(2n)}(0) = 0$ and $f^{(2n+1)}(0) = (-1)^n$. The nonzero Taylor coefficients for $\sin x$ are $a_{2n+1} = \dfrac{(-1)^n}{(2n+1)!}$. Similarly, for $f(x) = \cos x$,

$$f^{(2n)}(x) = (-1)^n \cos x, \qquad f^{(2n+1)}(x) = (-1)^{n+1} \sin x$$

Therefore, $f^{(2n)}(0) = (-1)^n$ and $f^{(2n+1)}(0) = 0$. The nonzero Taylor coefficients for $\cos x$ are $a_{2n} = \dfrac{(-1)^n}{(2n)!}$.

In both cases, $|f^{(n)}(x)| \le 1$ for all x and n. Thus, we may apply Theorem 2 with $M = 1$ and any R to conclude that the Taylor series converges to $f(x)$ for $|x| < R$. Since R is arbitrary, the Taylor expansions hold for all x. ■

■ **EXAMPLE 3** **Taylor Expansion $f(x) = e^x$ at $x = c$** Find the Taylor series $T(x)$ of $f(x) = e^x$ at $x = c$.

Solution We have $f^{(n)}(c) = e^c$ for all x and thus

$$T(x) = \sum_{n=0}^{\infty} \frac{e^c}{n!}(x-c)^n$$

To prove convergence, we note that e^x is increasing and therefore, for any R, $|f^{(k)}(x)| \le e^{c+R}$ for $x \in (c-R, c+R)$. Applying Theorem 2 with $M = e^{c+R}$, we conclude that $T(x)$ converges to $f(x)$ for all $x \in (c-R, c+R)$. Since R is arbitrary, the Taylor expansion holds for all x. For $c = 0$ we obtain the standard Taylor series

$$e^x = 1 + x + \frac{x^2}{2!} + \frac{x^3}{3!} + \cdots$$ ■

Shortcuts to Finding Taylor Series

Since a Taylor series is a power series, we may differentiate and integrate a Taylor series term by term within its interval of convergence by Theorem 3 in Section 10.6. We may also multiply two Taylor series or substitute one Taylor series into another (we omit the proofs of these facts). This leads to shortcuts for generating new Taylor series from known ones.

■ **EXAMPLE 4** Find the Maclaurin series for $f(x) = x^2e^x$.

Solution We obtain the Maclaurin series of $f(x)$ by multiplying the known Maclaurin series for x^2 and e^x:

$$x^2e^x = x^2\left(1 + x + \frac{x^2}{2!} + \frac{x^3}{3!} + \frac{x^4}{4!} + \frac{x^5}{5!} + \cdots\right)$$

$$= x^2 + x^3 + \frac{x^4}{2!} + \frac{x^5}{3!} + \frac{x^6}{4!} + \frac{x^7}{5!} + \cdots = \sum_{n=2}^{\infty} \frac{x^n}{(n-2)!}$$ ■

In some cases, there is no convenient formula for the Taylor coefficients of a product, but we can compute as many coefficients as desired numerically.

EXAMPLE 5 Multiplying Taylor Series Write out the first five terms in the Maclaurin series for $f(x) = e^x \cos x$.

Solution We multiply the fifth-order Taylor polynomials of e^x and $\cos x$ together, dropping the terms of degree greater than 5:

$$\left(1 + x + \frac{x^2}{2} + \frac{x^3}{6} + \frac{x^4}{24} + \frac{x^5}{120}\right)\left(1 - \frac{x^2}{2} + \frac{x^4}{24}\right)$$

Distributing the term on the left (and ignoring terms of degree greater than 5), we obtain

$$\left(1 + x + \frac{x^2}{2} + \frac{x^3}{6} + \frac{x^4}{24} + \frac{x^5}{120}\right) - \left(\frac{x^2}{2}\right)\left(1 + x + \frac{x^2}{2} + \frac{x^3}{6}\right) + \left(\frac{x^4}{24}\right)(1 + x)$$

$$= \underbrace{1 + x - \frac{x^3}{3} - \frac{x^4}{6} - \frac{x^5}{30}}_{\text{Retain terms of degree} \le 5} + \text{higher-order terms}$$

We conclude that the first five terms of the Taylor series for $f(x) = e^x \cos x$ are

$$T_5(x) = 1 + x - \frac{x^3}{3} - \frac{x^4}{6} - \frac{x^5}{30}$$

EXAMPLE 6 Substitution Use substitution to determine the Maclaurin series for e^{-x^2}.

Solution The Maclaurin series for e^{-x^2} is obtained by substituting $-x^2$ in the Maclaurin series for e^x:

$$e^{-x^2} = \sum_{n=0}^{\infty} \frac{(-1)^n x^{2n}}{n!} = 1 - x^2 + \frac{x^4}{2!} - \frac{x^6}{3!} + \frac{x^8}{4!} - \cdots \qquad \boxed{2}$$

Since the Taylor expansion of e^x is valid for all x, this expansion is also valid for all x.

EXAMPLE 7 Integrating a Taylor Series Find the Maclaurin series for $f(x) = \ln(1 + x)$.

Solution In our study of Taylor polynomials, we computed the Maclaurin polynomials of $\ln(1 + x)$ directly. Here we obtain the same result by integrating the geometric series with common ratio $-x$:

$$\frac{1}{1 + x} = 1 - x + x^2 - x^3 + \cdots \qquad \boxed{3}$$

$$\ln(1 + x) = \int \frac{dx}{1 + x} = x - \frac{x^2}{2} + \frac{x^3}{3} - \frac{x^4}{4} + \cdots \qquad \boxed{4}$$

In principle, we might need a constant of integration on the right-hand side of Eq. (4). However, the constant of integration is zero because both $\ln(1 + x)$ and the power series take the value 0 at $x = 0$.

Furthermore, (3) is valid for $|x| < 1$, so the expansion of $\ln(1 + x)$ is also valid for $|x| < 1$. It also holds for $x = 1$ (see Exercise 82).

Taylor series may be used to express definite integrals as infinite series. This is useful when the integrand does not have an explicit antiderivative and the FTC cannot be applied. To justify this use of Taylor series, we appeal to Theorem 3 in Section 10.6, which implies the following: If a power series $P(x)$ centered at c has radius of convergence R, then the definite integral $\int_a^b P(x)\,dx$ over an interval $[a, b]$ contained in $(c - R, c + R)$ may be evaluated term by term.

EXAMPLE 8 Let $J = \int_0^1 \sin(x^2)\,dx$.

(a) Express J as an infinite series.

(b) Determine J to within an error less than 10^{-4}.

Solution

(a) The Maclaurin expansion for $\sin x$ is valid for all x, so we have

$$\sin x = \sum_{n=0}^{\infty} \frac{(-1)^n}{(2n+1)!} x^{2n+1} \quad \Rightarrow \quad \sin(x^2) = \sum_{n=0}^{\infty} \frac{(-1)^n}{(2n+1)!} x^{4n+2}$$

We obtain an infinite series for J by integration:

$$J = \int_0^1 \sin(x^2)\,dx = \sum_{n=0}^{\infty} \frac{(-1)^n}{(2n+1)!} \int_0^1 x^{4n+2} dx$$

$$= \sum_{n=0}^{\infty} \frac{(-1)^n}{(2n+1)!} \left(\frac{1}{4n+3} \right)$$

$$= \frac{1}{3} - \frac{1}{42} + \frac{1}{1{,}320} - \frac{1}{75{,}600} + \cdots \qquad \boxed{5}$$

(b) The infinite series for J is an alternating series with decreasing terms, so the sum of the first N terms is accurate to within an error not greater than the $(N + 1)$st term. In other words,

$$\left| J - \sum_{n=0}^{N} \frac{(-1)^n}{(4n+3)(2n+1)!} \right| \le \frac{1}{(4N+7)(2N+3)!}$$

For $N = 2$, we obtain

$$J \approx \frac{1}{3} - \frac{1}{42} + \frac{1}{1{,}320} \approx 0.31028$$

with an error

$$\left| J - \left(\frac{1}{3} - \frac{1}{42} + \frac{1}{1{,}320} \right) \right| \le \frac{1}{(4(2)+7)(2(2)+3)!} = \frac{1}{75{,}600} \approx 1.3 \times 10^{-5}$$

We see that three terms of the series suffice to compute the integral with an error less than 10^{-4}. ■

Binomial Series

Taylor series yield a generalization of the Binomial Theorem that was first discovered by Isaac Newton around 1665. For any number a (integer or not) and integer $n \ge 0$, we define the **binomial coefficient**:

$$\binom{a}{n} = \frac{a(a-1)(a-2)\cdots(a-n+1)}{n!}, \qquad \binom{a}{0} = 1$$

For example,

$$\binom{6}{3} = \frac{6\cdot 5\cdot 4}{3\cdot 2\cdot 1} = 20, \qquad \binom{\frac{4}{3}}{3} = \frac{\frac{4}{3}\cdot\frac{1}{3}\cdot(-\frac{2}{3})}{3\cdot 2\cdot 1} = -\frac{4}{81}$$

The **Binomial Theorem** of algebra (see Appendix C) states that for any whole number a,

$$(r+s)^a = r^a + \binom{a}{1}r^{a-1}s + \binom{a}{2}r^{a-2}s^2 + \cdots + \binom{a}{a-1}rs^{a-1} + s^a$$

Setting $r = 1$ and $s = x$, we obtain an expansion of $f(x) = (1+x)^a$:

$$f(x) = (1+x)^a = 1 + \binom{a}{1}x + \binom{a}{2}x^2 + \cdots + \binom{a}{a-1}x^{a-1} + x^a$$

To derive Newton's generalization, we compute the Taylor series of $f(x) = (1+x)^a$ without assuming that a is a whole number. Observe that the derivatives of $f^{(k)}(0)$ follow a pattern:

$$\begin{aligned}
f(x) &= (1+x)^a & f(0) &= 1\\
f'(x) &= a(1+x)^{a-1} & f'(0) &= a\\
f''(x) &= a(a-1)(1+x)^{a-2} & f''(0) &= a(a-1)\\
f'''(x) &= a(a-1)(a-2)(1+x)^{a-3} & f'''(0) &= a(a-1)(a-2)
\end{aligned}$$

In general, $f^{(n)}(0) = a(a-1)(a-2)\cdots(a-n+1)$ and

$$\frac{f^{(n)}(0)}{n!} = \frac{a(a-1)(a-2)\cdots(a-n+1)}{n!} = \binom{a}{n}$$

Hence the Taylor series for $(1+x)^a$ is the binomial series

When a is a whole number, $\binom{a}{n}$ is zero for $n > a$, and in this case, the binomial series breaks off at degree n. The binomial series is infinite when a is not a whole number.

$$\sum_{n=0}^{\infty}\binom{a}{n}x^n = 1 + ax + \frac{a(a-1)}{2!}x^2 + \frac{a(a-1)(a-2)}{3!}x^3 + \cdots + \binom{a}{n}x^n + \cdots$$

The Ratio Test shows that this series has radius of convergence $R = 1$ (see Exercise 83). Furthermore, the binomial series converges to $(1+x)^a$ for $|x| < 1$ (see Exercise 84).

THEOREM 3 The Binomial Series For any exponent a, the following expansion is valid for $|x| < 1$:

$$(1+x)^a = 1 + ax + \frac{a(a-1)}{2!}x^2 + \frac{a(a-1)(a-2)}{3!}x^3 + \cdots + \binom{a}{n}x^n + \cdots \qquad (6)$$

EXAMPLE 9 Find the first five terms in the Maclaurin expansion of

$$f(x) = (1+x)^{1/3}$$

Solution The binomial coefficients $\binom{a}{n}$ for $a = \frac{1}{3}$ are

$$1, \quad \frac{1}{3}, \quad \frac{\frac{1}{3}(-\frac{2}{3})}{2!} = -\frac{1}{9}, \quad \frac{\frac{1}{3}(-\frac{2}{3})(-\frac{5}{3})}{3!} = \frac{5}{81}, \quad \frac{\frac{1}{3}(-\frac{2}{3})(-\frac{5}{3})(-\frac{8}{3})}{4!} = -\frac{10}{243}$$

Therefore, $(1+x)^{1/3} \approx 1 + \frac{1}{3}x - \frac{1}{9}x^2 + \frac{5}{81}x^3 - \frac{10}{243}x^4 + \cdots$. ■

■ **EXAMPLE 10** Find the Maclaurin series for $f(x) = \dfrac{1}{\sqrt{1-x^2}}$.

Solution First, we compute the coefficients $\binom{-\frac{1}{2}}{n}$ in the binomial series for $(1+x)^{-1/2}$. The coefficients for $n = 0, 1, 2, 3$ are

$$1, \quad \frac{(-\frac{1}{2})}{1} = -\frac{1}{2}, \quad \frac{(-\frac{1}{2})(-\frac{3}{2})}{2!} = \frac{1 \cdot 3}{2 \cdot 4}, \quad \frac{(-\frac{1}{2})(-\frac{3}{2})(-\frac{5}{2})}{3!} = -\frac{1 \cdot 3 \cdot 5}{2 \cdot 4 \cdot 6}$$

The general pattern is

$$\binom{-\frac{1}{2}}{n} = (-1)^n \frac{1 \cdot 3 \cdot 5 \cdots (2n-1)}{2^n\, n!} = (-1)^n \frac{1 \cdot 3 \cdot 5 \cdots (2n-1)}{2 \cdot 4 \cdot 6 \cdots 2n}$$

Thus, the following binomial expansion is valid for $|x| < 1$:

$$\frac{1}{\sqrt{1+x}} = 1 + \sum_{n=1}^{\infty} (-1)^n \frac{1 \cdot 3 \cdot 5 \cdots (2n-1)}{2 \cdot 4 \cdot 6 \cdots (2n)} x^n = 1 - \frac{1}{2}x + \frac{1 \cdot 3}{2 \cdot 4}x^2 - \cdots$$

Now substitute $-x^2$ for x. We obtain for $|x^2| < 1$ or $|x| < 1$,

$$\frac{1}{\sqrt{1-x^2}} = 1 + \sum_{n=1}^{\infty} \frac{1 \cdot 3 \cdot 5 \cdots (2n-1)}{2 \cdot 4 \cdot 6 \cdots 2n} x^{2n} = 1 + \frac{1}{2}x^2 + \frac{1 \cdot 3}{2 \cdot 4}x^4 + \cdots \qquad \boxed{7}$$

■

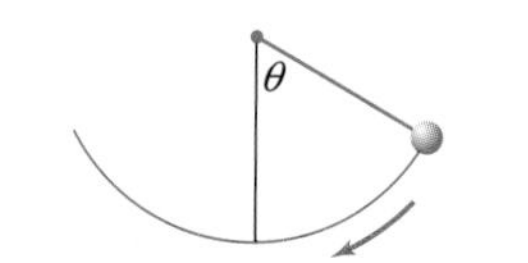

FIGURE 1 Pendulum released at an angle θ.

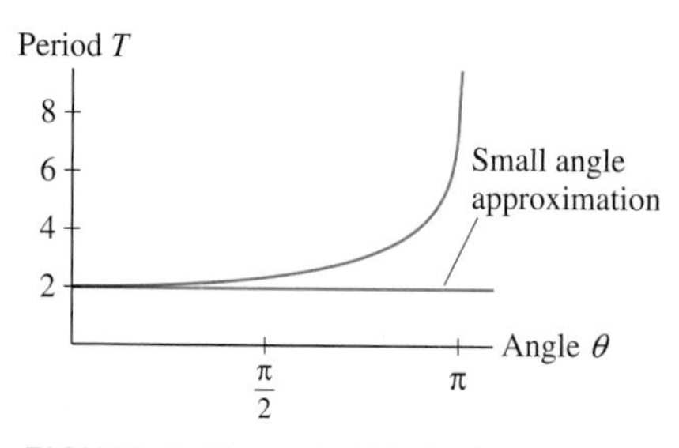

FIGURE 2 The period T of a 1-m pendulum as a function of the angle θ at which it is released.

Taylor series are used to study the transcendental functions occurring in physics and engineering, such as Bessel functions and hypergeometric functions. Though less familiar than the trigonometric and exponential functions, these so-called *special functions* appear in a wide range of applications. An example is the following **elliptic function of the first kind**:

$$E(k) = \int_0^{\pi/2} \frac{dt}{\sqrt{1 - k^2 \sin^2 t}}$$

In physics, one shows that the period T of pendulum of length L released from an angle θ is equal to $T = 4\sqrt{L/g}\,E(k)$, where $k = \sin \frac{1}{2}\theta$ (Figure 1). For small θ, we have the "small angle approximation" $T \approx 2\pi\sqrt{L/g}$, but this approximation breaks down for large angles (as we see in Figure 2).

■ **EXAMPLE 11** **Elliptic Functions** Find the Maclaurin series for $E(k)$ and estimate $E(k)$ for $k = \sin \dfrac{\pi}{6}$.

Solution Substitute $x = k \sin t$ in the Taylor expansion (7):

$$\frac{1}{\sqrt{1-k^2\sin^2 t}} = 1 + \frac{1}{2}k^2\sin^2 t + \frac{1\cdot 3}{2\cdot 4}k^4\sin^4 t + \frac{1\cdot 3\cdot 5}{2\cdot 4\cdot 6}k^6\sin^6 t + \cdots$$

Then $E(k)$ is equal to

$$\int_0^{\pi/2} \frac{dt}{\sqrt{1-k^2\sin^2 t}} = \int_0^{\pi/2} dt + \sum_{n=1}^{\infty} \frac{1\cdot 3\cdots(2n-1)}{2\cdot 4\cdot(2n)} \left(\int_0^{\pi/2} \sin^{2n} t\, dt\right) k^{2n}$$

According to Exercise 75 in Section 7.3,

$$\int_0^{\pi/2} \sin^{2n} t\, dt = \left(\frac{1\cdot 3\cdots(2n-1)}{2\cdot 4\cdot(2n)}\right)\frac{\pi}{2}$$

This yields

$$E(k) = \frac{\pi}{2} + \frac{\pi}{2}\sum_{n=1}^{\infty}\left(\frac{1\cdot 3\cdots(2n-1)^2}{2\cdot 4\cdots(2n)}\right)^2 k^{2n}$$

We approximate $E(k)$ for $k = \sin(\frac{\pi}{6}) = \frac{1}{2}$ using the first five terms:

$$F\left(\frac{1}{2}\right) \approx \frac{\pi}{2}\left(1 + \left(\frac{1}{2}\right)^2\left(\frac{1}{2}\right)^2 + \left(\frac{1\cdot 3}{2\cdot 4}\right)^2\left(\frac{1}{2}\right)^4\right.$$
$$\left. + \left(\frac{1\cdot 3\cdot 5}{2\cdot 4\cdot 6}\right)^2\left(\frac{1}{2}\right)^6 + \left(\frac{1\cdot 3\cdot 5\cdot 7}{2\cdot 4\cdot 6\cdot 8}\right)^2\left(\frac{1}{2}\right)^8\right)$$
$$\approx 1.68517$$

The value given by a computer algebra system to seven places is $F(\frac{1}{2}) \approx 1.6856325$. ■

10.7 SUMMARY

- The *Taylor series* of $f(x)$ centered at $x = c$ is

$$T(x) = \sum_{n=0}^{\infty} \frac{f^{(n)}(c)}{n!}(x-c)^n$$

The partial sum $T_k(x)$ of $T(x)$ is the kth Taylor polynomial.

- If $f(x)$ is represented as a power series $\sum_{n=0}^{\infty} a_n(x-c)^n$ on an interval $(c-R, c+R)$ with $R > 0$, then this power series is the Taylor series centered at $x = c$.
- When $c = 0$, $T(x)$ is called the *Maclaurin series* of $f(x)$.
- The equality $f(x) = T(x)$ holds if and only if the remainder, defined as $R_k(x) = f(x) - T_k(x)$, tends to zero as $k \to \infty$.
- Suppose that $f(x)$ is infinitely differentiable on an interval $I = (c-R, c+R)$ with $R > 0$ and assume there exists $K > 0$ such that $|f^{(k)}(x)| < K$ for $x \in I$. Then $f(x)$ is represented by its Taylor series on I, that is, $f(x) = T(x)$ for $x \in I$.

TABLE 1

Function $f(x)$	Maclaurin Series	Converges to $f(x)$ for
e^x	$\sum_{n=0}^{\infty} \frac{x^n}{n!} = 1 + x + \frac{x^2}{2!} + \frac{x^3}{3!} + \frac{x^4}{4!} + \cdots$	All x
$\sin x$	$\sum_{n=0}^{\infty} \frac{(-1)^n x^{2n+1}}{(2n+1)!} = x - \frac{x^3}{3!} + \frac{x^5}{5!} - \frac{x^7}{7!} + \cdots$	All x
$\cos x$	$\sum_{n=0}^{\infty} \frac{(-1)^n x^{2n}}{(2n)!} = 1 - \frac{x^2}{2!} + \frac{x^4}{4!} - \frac{x^6}{6!} + \cdots$	All x
$\frac{1}{1-x}$	$\sum_{n=0}^{\infty} x^n = 1 + x + x^2 + x^3 + x^4 + \cdots$	$\lvert x\rvert < 1$
$\frac{1}{1+x}$	$\sum_{n=0}^{\infty} (-1)^n x^n = 1 - x + x^2 - x^3 + x^4 - \cdots$	$\lvert x\rvert < 1$
$\ln(1+x)$	$\sum_{n=1}^{\infty} \frac{(-1)^{n-1} x^n}{n} = x - \frac{x^2}{2} + \frac{x^3}{3} - \frac{x^4}{4} + \cdots$	$\lvert x\rvert < 1$ and $x = 1$
$\tan^{-1} x$	$\sum_{n=0}^{\infty} \frac{(-1)^n x^{2n+1}}{2n+1} = x - \frac{x^3}{3} + \frac{x^5}{5} - \frac{x^7}{7} + \cdots$	$\lvert x\rvert \le 1$
$(1+x)^a$	$\sum_{n=0}^{\infty} \binom{a}{n} x^n = 1 + ax + \frac{a(a-1)}{2!}x^2 + \frac{a(a-1)(a-2)}{3!}x^3 + \cdots$	$\lvert x\rvert < 1$

- A good way to find the Taylor series of a function is to start with known Taylor expansions and apply one of the following operations: multiplication, substitution, differentiation, or integration.

10.7 EXERCISES

Preliminary Questions

1. Determine $f(0)$ and $f'''(0)$ for a function $f(x)$ with Maclaurin series

$$T(x) = 3 + 2x + 12x^2 + 5x^3 + \cdots$$

2. Determine $f(-2)$ and $f^{(4)}(-2)$ for a function with Taylor series

$$T(x) = 3(x+2) + (x+2)^2 - 4(x+2)^3 + 2(x+2)^4 + \cdots$$

3. What is the easiest way to find the Maclaurin series for the function $f(x) = \sin(x^2)$?

4. What is the Taylor series for $f(x)$ centered at $c = 3$ if $f(3) = 4$ and $f'(x)$ has a Taylor expansion

$$f'(x) = \sum_{n=1}^{\infty} \frac{(x-3)^n}{n}$$

5. Let $T(x)$ be the Maclaurin series of $f(x)$. Which of the following guarantees that $f(2) = T(2)$?

(a) $T(x)$ converges for $x = 2$.

(b) The remainder $R_k(2)$ approaches a limit as $k \to \infty$.

(c) The remainder $R_k(2)$ approaches zero as $k \to \infty$.

Exercises

1. Write out the first four terms of the Maclaurin of $f(x)$ if

$$f(0) = 2, \quad f'(0) = 3, \quad f''(0) = 4, \quad f'''(0) = 12$$

2. Write out the first four terms of the Taylor series of $f(x)$ centered at $c = 3$ if

$$f(3) = 1, \quad f'(3) = 2, \quad f''(3) = 12, \quad f'''(3) = 3$$

In Exercises 3–20, find the Maclaurin series.

3. $f(x) = \dfrac{1}{1-2x}$

4. $f(x) = \dfrac{x}{1-x^4}$

5. $f(x) = \cos 3x$

6. $f(x) = \sin(2x)$

7. $f(x) = \sin(x^2)$

8. $f(x) = e^{4x}$

9. $f(x) = \ln(1-x^2)$

10. $f(x) = (1-x)^{-1/2}$

11. $f(x) = \tan^{-1}(x^2)$

12. $f(x) = x^2 e^{x^2}$

13. $f(x) = e^{x-2}$

14. $f(x) = \cos\sqrt{x}$

15. $f(x) = \ln(1-5x)$

16. $f(x) = (x^2+2x)e^x$

17. $f(x) = \sinh x$

18. $f(x) = \cosh x$

19. $f(x) = \dfrac{1-\cos(x^2)}{x}$

20. $f(x) = \dfrac{e^x - \cos x}{x}$

21. Use multiplication to find the first four terms in the Maclaurin series for $f(x) = e^x \sin x$.

22. Find the first five terms of the Maclaurin series for $f(x) = \dfrac{\sin x}{1-x}$.

23. Find the first four terms of the Maclaurin series for $f(x) = e^x \ln(1-x)$.

24. Write out the first five terms of the binomial series for $f(x) = (1+x)^{1/3}$.

25. Write out the first five terms of the binomial series for $f(x) = (1+x)^{-3/2}$.

26. Differentiate the Maclaurin series for $\dfrac{1}{1-x}$ twice to find the Maclaurin series of $\dfrac{1}{(1-x)^3}$.

27. Find the first four terms of the Maclaurin for $f(x) = e^{(e^x)}$.

28. Find the first three terms of the Maclaurin series for $f(x) = \dfrac{1}{1+\sin x}$. *Hint:* First expand $f(x)$ as a geometric series.

29. Find the Taylor series for $\sin x$ at $c = \dfrac{\pi}{2}$.

30. What is the Maclaurin series for $f(x) = x^4 - 2x^2 + 3$? What is the Taylor series centered at $c = 2$?

In Exercises 31–40, find the Taylor series centered at c.

31. $f(x) = \dfrac{1}{x}$, $c = 1$

32. $f(x) = \sqrt{x}$, $c = 4$

33. $f(x) = \dfrac{1}{1-x}$, $c = 5$

34. $f(x) = x^4 + 3x - 1$, $c = 0$

35. $f(x) = x^4 + 3x - 1$, $c = 2$

36. $f(x) = \dfrac{1}{x^2}$, $c = 4$

37. $f(x) = e^{3x}$, $c = -1$

38. $f(x) = \dfrac{1}{1-4x}$, $c = -2$

39. $f(x) = \dfrac{1}{1-x^2}$, $c = 3$

40. $f(x) = \dfrac{1}{3x-2}$, $c = -1$

41. Find the Maclaurin series for $f(x) = \dfrac{1}{\sqrt{1-9x^2}}$ (see Example 10).

42. Show, by integrating the Maclaurin series for $f(x) = \dfrac{1}{\sqrt{1-x^2}}$, that for $|x| < 1$,

$$\sin^{-1} x = x + \sum_{n=1}^{\infty} \frac{1 \cdot 3 \cdot 5 \cdots (2n-1)}{2 \cdot 4 \cdot 6 \cdots (2n)} \frac{x^{2n+1}}{2n+1}$$

43. Use the first five terms of the Maclaurin series in Exercise 42 to approximate $\sin^{-1} \frac{1}{2}$. Compare the result with the calculator value.

44. Show that for $|x| < 1$

$$\tanh^{-1} x = x + \frac{x^3}{3} + \frac{x^5}{5} + \cdots$$

Hint: Recall that $\dfrac{d}{dx} \tanh^{-1} x = \dfrac{1}{1-x^2}$.

45. Use the Maclaurin series for $\ln(1+x)$ and $\ln(1-x)$ to show that

$$\frac{1}{2} \ln\left(\frac{1+x}{1-x}\right) = x + \frac{x^3}{3} + \frac{x^5}{5} + \cdots$$

What can you conclude by comparing this result with that of Exercise 44?

46. Use the Taylor series for $\cos x$ to compute cos 1 to within an error of at most 10^{-6}. Use the fact that $\cos x$ is an alternating series with decreasing terms to estimate the error.

47. Use the Maclaurin expansion for e^{-t^2} to express $\int_0^x e^{-t^2}\,dt$ as an alternating power series in t.

(a) How many terms of the infinite series are needed to approximate the integral for $x = 1$ to within an error of at most 0.001?

(b) CAS Carry out the computation and check your answer using a computer algebra system.

48. Let $F(x) = \displaystyle\int_0^x \frac{\sin t\,dt}{t}$. Show that

$$F(x) = x - \frac{x^3}{3 \cdot 3!} + \frac{x^5}{5 \cdot 5!} - \frac{x^7}{7 \cdot 7!} + \cdots$$

Evaluate $F(1)$ to three decimal places.

In Exercises 49–52, express the definite integral as an infinite series and find its value to within an error of at most 10^{-4}.

49. $\int_0^1 \cos(x^2)\,dx$

50. $\int_0^1 \tan^{-1}(x^2)\,dx$

51. $\int_0^2 e^{-x^3}\,dx$

52. $\int_0^1 \frac{dx}{\sqrt{x^4+1}}$

In Exercises 53–56, express the integral as an infinite series.

53. $\int_0^x \frac{1-\cos(t)}{t}\,dt$, for all x

54. $\int_0^x \frac{t-\sin t}{t}\,dt$, for all x

55. $\int_0^x \ln(1+t^2)\,dt$, for $|x|<1$

56. $\int_0^x \frac{dt}{\sqrt{1-t^4}}$, for $|x|<1$

57. Which function has Maclaurin series $\sum_{n=0}^{\infty}(-1)^n 2^n x^n$?

58. Which function has Maclaurin series

$$\sum_{k=0}^{\infty}\frac{(-1)^k}{3^{k+1}}(x-3)^k?$$

For which values of x is the expansion valid?

In Exercises 59–62, find the first four terms of the Taylor series.

59. $f(x)=\sin(x^2)\cos(x^2)$

60. $f(x)=e^x\tan^{-1}x$

61. $f(x)=e^{\sin x}$

62. $f(x)=\sin(x^3+2x)$

In Exercises 63–66, find the functions with the following Maclaurin series (refer to Table 1).

63. $1+x^3+\frac{x^6}{2!}+\frac{x^9}{3!}+\frac{x^{12}}{4!}+\cdots$

64. $1-4x+4^2x^2-4^3x^3+4^4x^5-4^5x^5+\cdots$

65. $1-\frac{5^3x^3}{3!}+\frac{5^5x^5}{5!}-\frac{5^7x^7}{7!}+\cdots$

66. $x^4-\frac{x^{12}}{3}+\frac{x^{20}}{5}-\frac{x^{28}}{7}+\cdots$

67. When a voltage V is applied to a series circuit consisting of a resistor R and an inductor L, the current at time t is

$$I(t)=\left(\frac{V}{R}\right)\left(1-e^{-Rt/L}\right)$$

Expand $I(t)$ in a Maclaurin series. Show that $I(t)\approx Vt/L$ if R is small.

68. Use substitution to write out the first three terms of the Maclaurin series for $f(x)=e^{x^{20}}$. Explain how the result implies that $f^{(k)}(0)=0$ for $1\le k\le 19$.

69. Find the Maclaurin series for $f(x)=\cos(\sqrt{x})$ and use it to determine $f^{(5)}(0)$.

70. Find $f^{(7)}(0)$ and $f^{(8)}(0)$ for $f(x)=\tan^{-1}x$.

71. Use the binomial series to find $f^{(8)}(0)$ for $f(x)=\sqrt{1-x^2}$.

72. Show that $\pi-\frac{\pi^3}{3!}+\frac{\pi^5}{5!}-\frac{\pi^7}{7!}+\cdots$ converges to zero. How many terms must be computed to get within 0.01 of zero?

73. Does the Taylor series for $f(x)=(1+x)^{3/4}$ converge to $f(x)$ at $x=2$? Give numerical evidence to support your answer.

74. Explain the steps required to verify that the Maclaurin series for $f(x)=\sin x$ converges to $f(x)$ at $x=1$.

75. Explain the steps required to verify that the Maclaurin series for $f(x)=\tan^{-1}x$ converges to $f(x)$ at $x=0.5$.

76. GU Let $f(x)=\sqrt{1+x}$.

(a) Use a graphing calculator to compare the graph of f with the graphs of the first five Taylor polynomials for f. What do they suggest about the interval of convergence of the Taylor series?

(b) Investigate numerically whether or not the Taylor expansion for f is valid for $x=1$ and $x=-1$.

77. How many terms of the Maclaurin series of $f(x)=\ln(1+x)$ are needed to compute $\ln 1.2$ to within an error of at most 0.0001? Make the computation and compare the result with the calculator value.

In Exercises 78–79, let

$$f(x)=\frac{1}{(1-x)(1-2x)}$$

78. Find the Maclaurin series of $f(x)$ using the identity

$$f(x)=\frac{2}{1-2x}-\frac{1}{1-x}$$

79. Find the Taylor series for $f(x)$ at $c=2$. *Hint:* Rewrite the identity of Exercise 78 as

$$f(x)=\frac{2}{-3-2(x-2)}-\frac{1}{-1-(x-2)}$$

80. Use the first five terms of the Maclaurin series for the elliptic function $E(k)$ to estimate the period T of a 1-m pendulum released at an angle $\theta=\frac{\pi}{4}$ (see Example 11).

81. Use Example 11 and the approximation $\sin x\approx x$ to show that the period T of a pendulum released at an angle θ has the following second-order approximation:

$$T\approx 2\pi\sqrt{\frac{L}{g}}\left(1+\frac{\theta^2}{16}\right)$$

Further Insights and Challenges

82. In this exercise we show that the Maclaurin expansion of the function $f(x) = \ln(1+x)$ is valid for $x = 1$.

(a) Show that for all $x \neq -1$,

$$\frac{1}{1+x} = \sum_{n=0}^{N} (-1)^n x^n + \frac{(-1)^{N+1} x^{N+1}}{1+x}$$

(b) Integrate from 0 to 1 to obtain

$$\ln 2 = \sum_{n=1}^{N} \frac{(-1)^{n-1}}{n} + (-1)^{N+1} \int_0^1 \frac{x^{N+1}\, dx}{1+x}$$

(c) Verify that the integral on the right tends to zero as $N \to \infty$ by showing that it is smaller than $\int_0^1 x^{N+1}\, dx$.

(d) Prove Leibniz's formula

$$\ln 2 = 1 - \frac{1}{2} + \frac{1}{3} - \frac{1}{4} + \cdots$$

In Exercises 83–84, we investigate the convergence of the binomial series

$$T_a(x) = \sum_{n=0}^{\infty} \binom{a}{n} x^n$$

83. Prove that $T_a(x)$ has radius of convergence $R = 1$ if a is not a whole number. What is the radius of convergence if a is a whole number?

84. By Exercise 83, $T_a(x)$ converges for $|x| < 1$, but we do not yet know whether $T_a(x) = (1+x)^a$.

(a) Verify the identity

$$a\binom{a}{n} = n\binom{a}{n} + (n+1)\binom{a}{n+1}$$

(b) Use (a) to show that $y = T_a(x)$ satisfies the differential equation $(1+x)y' = ay$ with initial condition $y(0) = 1$.

(c) Prove that $T_a(x) = (1+x)^a$ for $|x| < 1$ by showing that the derivative of the ratio $\dfrac{T_a(x)}{(1+x)^a}$ is constant.

85. The function $G(k) = \int_0^{\pi/2} \sqrt{1 - k^2 \sin^2 t}\, dt$ is called an **elliptic function of the second kind**. Prove that for $|k| < 1$

$$G(k) = \frac{\pi}{2} - \frac{\pi}{2} \sum_{n=1}^{\infty} \left(\frac{1 \cdot 3 \cdots (2n-1)}{2 \cdots 4 \cdot (2n)} \right)^2 \frac{k^{2n}}{2n-1}$$

86. Assume that $a < b$ and let L be the arc length (circumference) of the ellipse $\left(\frac{x}{a}\right)^2 + \left(\frac{y}{b}\right)^2 = 1$ shown in Figure 3. There is no explicit formula for L, but it is known that $L = 4bG(k)$, where $k = \sqrt{1 - a^2/b^2}$. Use the first three terms of the expansion of Exercise 85 to estimate the arc length when $a = 4$ and $b = 5$.

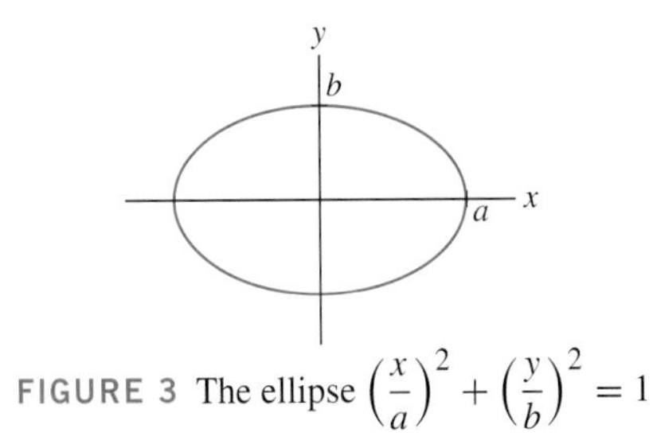

FIGURE 3 The ellipse $\left(\frac{x}{a}\right)^2 + \left(\frac{y}{b}\right)^2 = 1$

87. Use Exercise 85 to prove that if $a < b$ and a/b is near 1 (a nearly circular ellipse), then

$$L \approx \frac{\pi}{2}\left(3b + \frac{a^2}{b}\right)$$

Hint: Use the first two terms of the series for $G(k)$.

88. Irrationality of *e* Prove that e is an irrational number using the following argument by contradiction. Suppose that $e = M/N$, where M, N are nonzero integers.

(a) Show that $M!\,e^{-1}$ is a whole number.

(b) Use the power series for e^x at $x = -1$ to show that there is an integer B such that $M!\,e^{-1}$ equals

$$B + (-1)^{M+1}\left(\frac{1}{M+1} - \frac{1}{(M+1)(M+2)} + \cdots\right)$$

(c) Use your knowledge of alternating series with decreasing terms to conclude that $0 < |M!\,e^{-1} - B| < 1$. However, this contradicts (a). Hence, e is not equal to M/N.

CHAPTER REVIEW EXERCISES

1. Let $a_n = \dfrac{n-3}{n!}$ and $b_n = a_{n+3}$. Calculate the first three terms in the sequence:

(a) a_n^2

(b) b_n

(c) $a_n b_n$

(d) $2a_{n+1} - 3a_n$

2. Prove that $\lim_{n\to\infty} \dfrac{2n-1}{3n+2} = 1$ using the limit definition.

In Exercises 3–8, compute the limit (or state that it does not exist) assuming that $\lim_{n\to\infty} a_n = 2$.

3. $\lim_{n\to\infty} (5a_n - 2a_n^2)$

4. $\lim_{n\to\infty} \frac{1}{a_n}$

5. $\lim_{n\to\infty} e^{a_n}$

6. $\lim_{n\to\infty} \cos(\pi a_n)$

7. $\lim_{n\to\infty} (-1)^n a_n$

8. $\lim_{n\to\infty} \frac{a_n + n}{a_n + n^2}$

In Exercises 9–22, determine the limit of the sequence or show that the sequence diverges.

9. $a_n = \sqrt{n+5} - \sqrt{n+2}$

10. $a_n = \frac{3n^3 - n}{1 - 2n^3}$

11. $a_n = 2^{1/n^2}$

12. $a_n = \frac{10^n}{n!}$

13. $b_m = 1 + (-1)^m$

14. $b_m = \frac{1 + (-1)^m}{m}$

15. $b_n = \tan^{-1}\left(\frac{n+2}{n+5}\right)$

16. $a_n = \ln(n+1) - \ln n$

17. $b_n = \sqrt{n^2+n} - \sqrt{n^2+1}$

18. $c_n = \sqrt{n^2+n} - \sqrt{n^2-n}$

19. $a_n = \frac{100^n}{n!} - \frac{3+\pi^n}{5^n}$

20. $b_m = \left(1 + \frac{1}{m}\right)^m$

21. $c_n = \left(1 + \frac{3}{n}\right)^n$

22. $b_n = n(\ln(n+1) - \ln n)$

23. Use the Squeeze Theorem to show that $\lim_{n\to\infty} \frac{\arctan(n^2)}{\sqrt{n}} = 0$.

24. Give an example of a divergent sequence $\{a_n\}$ such that $\{\sin a_n\}$ is convergent.

25. Given $a_n = \frac{1}{2}3^n - \frac{1}{3}2^n$,

(a) Calculate $\lim_{n\to\infty} a_n$.

(b) Calculate $\lim_{n\to\infty} \frac{a_{n+1}}{a_n}$.

26. Define $a_{n+1} = \sqrt{a_n + 6}$ with $a_1 = 2$.

(a) Compute a_n for $n = 2, 3, 4, 5$.

(b) Show that $\{a_n\}$ is increasing and bounded by 3.

(c) Prove that $\lim_{n\to\infty} a_n$ exists and find its value.

27. Calculate the partial sums S_4 and S_7 of the series $\sum_{n=1}^{\infty} \frac{n-2}{n^2+2n}$.

28. Find the sum $1 - \frac{1}{4} + \frac{1}{4^2} - \frac{1}{4^3} + \cdots$.

29. Find the sum $\frac{4}{9} + \frac{8}{27} + \frac{16}{81} + \frac{32}{243} + \cdots$.

30. Find the sum $\sum_{n=2}^{\infty} \left(\frac{2}{e}\right)^n$.

31. Find the sum $\sum_{n=0}^{\infty} \frac{2^{n+1}}{3^n}$.

32. Show that $\sum_{n=1}^{\infty} \left(b - \tan^{-1}\frac{1}{n^2}\right)$ diverges if $b \neq \frac{\pi}{2}$.

33. Give an example of divergent series $\sum_{n=1}^{\infty} a_n$, $\sum_{n=1}^{\infty} b_n$ such that $\sum_{n=1}^{\infty} (a_n + b_n) = 1$.

34. Find the total area of the infinitely many circles on the interval [0, 1] in Figure 1.

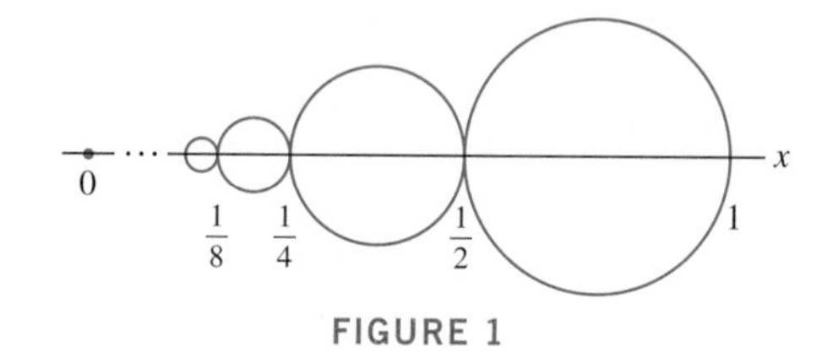

FIGURE 1

In Exercises 35–38, use the Integral Test to determine if the infinite series converges.

35. $\sum_{n=1}^{\infty} \frac{n^2}{n^3+1}$

36. $\sum_{n=1}^{\infty} \frac{n^2}{(n^3+1)^{1.01}}$

37. $\sum_{n=1}^{\infty} \frac{n^3}{e^{n^4}}$

38. $\sum_{n=1}^{\infty} \frac{1}{(2n+1)(\ln(2n+1))^2}$

In Exercises 39–46, use the Comparison or Limit Comparison Test to determine whether the infinite series converges.

39. $\sum_{n=1}^{\infty} \frac{1}{(n+1)^2}$

40. $\sum_{n=1}^{\infty} \frac{1}{\sqrt{n}+n}$

41. $\sum_{n=2}^{\infty} \frac{n^2+1}{n^{3.5}-2}$

42. $\sum_{n=1}^{\infty} \frac{1}{n-\ln n}$

43. $\sum_{n=2}^{\infty} \frac{\ln n}{1.5^n}$

44. $\sum_{n=2}^{\infty} \frac{n}{\sqrt{n^5+5}}$

45. $\sum_{n=1}^{\infty} \frac{1}{3^n-2^n}$

46. $\sum_{n=1}^{\infty} \frac{n^{10}+10^n}{n^{11}+11^n}$

47. Show that $\sum_{n=2}^{\infty} \left(1-\sqrt{1-\frac{1}{n}}\right)$ diverges. *Hint:* Show that

$$1-\sqrt{1-\frac{1}{n}} \geq \frac{1}{2n}$$

48. Determine whether $\sum_{n=2}^{\infty} \left(1-\sqrt{1-\frac{1}{n^2}}\right)$ converges.

49. Let $S = \sum_{n=1}^{\infty} \frac{n}{(n^2+1)^2}$.

(a) Show that S converges.

(b) CAS Use Eq. (5) in Exercise 77 of Section 10.3 with $M = 99$ to approximate S. What is the maximum size of the error?

In Exercises 50–53, determine whether the series converges absolutely. If not, determine whether it converges conditionally.

50. $\sum_{n=1}^{\infty} \frac{(-1)^n}{\sqrt[3]{n}+2n}$

51. $\sum_{n=1}^{\infty} \frac{(-1)^n}{n^{1.1}\ln(n+1)}$

52. $\sum_{n=1}^{\infty} \frac{\cos\left(\frac{\pi}{4}+\pi n\right)}{\sqrt{n}}$

53. $\sum_{n=1}^{\infty} \frac{\cos\left(\frac{\pi}{4}+2\pi n\right)}{\sqrt{n}}$

54. CAS Use a computer algebra system to approximate $\sum_{n=1}^{\infty} \frac{(-1)^n}{n^3+\sqrt{n}}$ to within an error of at most 10^{-5}.

55. How many terms of the series are needed to calculate Catalan's constant $K = \sum_{k=0}^{\infty} \frac{(-1)^k}{(2k+1)^2}$ to three decimal places? Carry out the calculation.

56. Give an example of conditionally convergent series $\sum_{n=1}^{\infty} a_n$, $\sum_{n=1}^{\infty} b_n$ such that $\sum_{n=1}^{\infty} (a_n+b_n)$ converges absolutely.

57. Let $\sum_{n=1}^{\infty} a_n$ be an absolutely convergent series. Determine whether the following series are convergent or divergent:

(a) $\sum_{n=1}^{\infty} \left(a_n+\frac{1}{n^2}\right)$

(b) $\sum_{n=1}^{\infty} (-1)^n a_n$

(c) $\sum_{n=1}^{\infty} \frac{1}{1+a_n^2}$

(d) $\sum_{n=1}^{\infty} \frac{|a_n|}{n}$

In Exercises 58–65, apply the Ratio Test to determine convergence or divergence, or state that the Ratio Test is inconclusive.

58. $\sum_{n=1}^{\infty} \frac{n^5}{5^n}$

59. $\sum_{n=1}^{\infty} \frac{\sqrt{n+1}}{n^8}$

60. $\sum_{n=1}^{\infty} \frac{1}{n2^n+n^3}$

61. $\sum_{n=1}^{\infty} \frac{n^4}{n!}$

62. $\sum_{n=1}^{\infty} \frac{2^{n^2}}{n!}$

63. $\sum_{n=4}^{\infty} \frac{\ln n}{n^{3/2}}$

64. $\sum_{n=1}^{\infty} \left(\frac{n}{2}\right)^n \frac{1}{n!}$

65. $\sum_{n=1}^{\infty} \left(\frac{n}{4}\right)^n \frac{1}{n!}$

In Exercises 66–69, apply the Root Test to determine convergence or divergence, or state that the Root Test is inconclusive.

66. $\sum_{n=1}^{\infty} \frac{1}{4^n}$

67. $\sum_{n=1}^{\infty} \left(\frac{2}{n}\right)^n$

68. $\sum_{n=1}^{\infty} \left(\frac{3}{4n}\right)^n$

69. $\sum_{n=1}^{\infty} \left(\cos\frac{1}{n}\right)^{n^3}$

70. Let $\{a_n\}$ be a positive sequence such that $\lim_{n\to\infty} \sqrt[n]{a_n} = \frac{1}{2}$. Determine whether the following series converge or diverge:

(a) $\sum_{n=1}^{\infty} 2a_n$

(b) $\sum_{n=1}^{\infty} 3^n a_n$

(c) $\sum_{n=1}^{\infty} \sqrt{a_n}$

In Exercises 71–84, determine convergence or divergence using any method covered in the text.

71. $\sum_{n=1}^{\infty} \left(\frac{2}{3}\right)^n$

72. $\sum_{n=1}^{\infty} \frac{n+2^n}{3^n-1}$

73. $\sum_{n=1}^{\infty} e^{-0.02n}$

74. $\sum_{n=1}^{\infty} ne^{-0.02n}$

75. $\sum_{n=1}^{\infty} \frac{(-1)^{n-1}}{\sqrt{n}+\sqrt{n+1}}$

76. $\sum_{n=10}^{\infty} \frac{1}{n(\log n)^{3/2}}$

77. $\sum_{n=10}^{\infty} \frac{(-1)^n}{\log n}$

78. $\sum_{n=1}^{\infty} \frac{1}{n\sqrt{n}+\ln n}$

79. $\sum_{n=1}^{\infty} \frac{e^n}{n!}$

80. $\sum_{n=1}^{\infty} \frac{1}{\sqrt[3]{n}(1+\sqrt{n})}$

81. $\displaystyle\sum_{n=1}^{\infty} \frac{1}{n - 100.1}$

82. $\displaystyle\sum_{n=2}^{\infty} \frac{\cos(\pi n)}{n^{2/3}}$

83. $\displaystyle\sum_{n=1}^{\infty} \sin^2 \frac{\pi}{n}$

84. $\displaystyle\sum_{n=0}^{\infty} \frac{2^{2n}}{n!}$

In Exercises 85–90, find the values of x for which the power series converges.

85. $\displaystyle\sum_{n=0}^{\infty} \frac{2^n x^n}{n!}$

86. $\displaystyle\sum_{n=0}^{\infty} \frac{x^n}{n+1}$

87. $\displaystyle\sum_{n=0}^{\infty} \frac{n^6(x-3)^n}{n^8+1}$

88. $\displaystyle\sum_{n=0}^{\infty} nx^n$

89. $\displaystyle\sum_{n=0}^{\infty} (nx)^n$

90. $\displaystyle\sum_{n=0}^{\infty} \frac{(2x-3)^n}{n \ln n}$

91. Expand the function $f(x) = \dfrac{2}{4-3x}$ as a power series centered at $c = 0$. Determine the values of x for which the series converges.

92. Prove that $\displaystyle\sum_{n=0}^{\infty} ne^{-nx} = \frac{e^{-x}}{(1-e^{-x})^2}$. *Hint:* Express the left-hand side as the derivative of a geometric series.

93. Let $F(x) = \displaystyle\sum_{k=0}^{\infty} \frac{x^{2k}}{2^k \cdot k!}$.

(a) Show that $F(x)$ has infinite radius of convergence.

(b) Show that $y = F(x)$ is a solution to the differential equation

$$y'' = xy' + y$$

satisfying $y(0) = 1$, $y'(0) = 0$.

(c) CAS Plot the partial sums S_N for $N = 1, 3, 5, 7$ on the same set of axes.

94. Find a power series solution $P(x) = \displaystyle\sum_{n=0}^{\infty} a_n x^n$ to the Laguerre differential equation

$$xy'' + (1-x)y' - y = 0$$

satisfying $P(0) = 1$.

95. Use the Maclaurin series for $f(x) = \cos x$ to calculate the limit $\displaystyle\lim_{x\to 0} \frac{1 - \frac{x^2}{2} - \cos x}{x^4}$.

In Exercises 96–103, find the Taylor series centered at c.

96. $f(x) = e^{4x}$, $c = 0$

97. $f(x) = e^{2x}$, $c = -1$

98. $f(x) = x \sin x$, $c = \pi$

99. $f(x) = \ln \dfrac{x}{2}$, $c = 2$

100. $f(x) = x \ln\left(1 + \dfrac{x}{2}\right)$, $c = 0$

101. $f(x) = \sqrt{x} \arctan \sqrt{x}$, $c = 0$

102. $f(x) = \dfrac{1}{1-2x}$, $c = -2$

103. $f(x) = e^{x-1}$, $c = -1$

104. Find the Maclaurin series of $f(x) = x^5 - 4x^2 + 10$. Determine its Taylor series centered at $c = 2$.

105. Use the Maclaurin series of $\sin x$ and $\sqrt{1+x}$ to calculate $f^{(4)}(0)$, where $f(x) = (\sin x)\sqrt{1+x}$.

106. Calculate $\dfrac{\pi}{2} - \dfrac{\pi^3}{2^3 3!} + \dfrac{\pi^5}{2^5 5!} - \dfrac{\pi^7}{2^7 7!} + \cdots$.

107. Find the Maclaurin series of the function $F(x) = \displaystyle\int_0^x \frac{e^t - 1}{t}\,dt$.

11 PARAMETRIC EQUATIONS, POLAR COORDINATES, AND CONIC SECTIONS

The beautiful shell of a chambered nautilus grows in the shape of an equiangular spiral, a curve described in polar coordinates by an equation $r = e^{a\theta}$.

This chapter develops several new tools of calculus. First, we introduce parametric equations, which are used to analyze motion and to describe curves that cannot be represented as graphs of functions or equations. We then discuss polar coordinates, which are an alternative to rectangular coordinates and are useful in many applications. The chapter closes with a discussion of conic sections (ellipses, hyperbolas, and parabolas). The conic sections were first defined and studied by the ancient Greek mathematicians Menaechmus (c. 380–320 BC) and Apollonius (c. 262–190 BC) and they have played an important role in the mathematics ever since.

11.1 Parametric Equations

In previous chapters, we have studied curves that are graphs of functions or equations. In this section, we introduce a new and important way of describing a curve, via parametric equations. Imagine a particle P moving along a path $\mathcal{C}$ as a function of time t (Figure 1). Then the coordinates of P are functions of t:

$$x = f(t), \qquad y = g(t) \tag{1}$$

Parametric equations are useful in multivariable calculus, especially in three dimensions, where it is no longer possible to describe a curve as a graph of a function (the graph of a function of two variables is a surface in $\mathbf{R}^3$ rather than a curve).

The equations (1) are called **parametric equations** of the path $\mathcal{C}$ with **parameter** t. We also write

$$c(t) = (f(t), g(t))$$

As t varies in a domain such as an interval or the real line, $c(t)$ represents a point moving along the path $\mathcal{C}$. We refer to $\mathcal{C}$ as a **parametrized** or **parametric** curve. The direction of motion is often indicated by an arrow next to the plot of $c(t)$ as in Figure 1. Finally, since x and y are functions of t, we often write $c(t) = (x(t), y(t))$ instead of $(f(t), g(t))$.

In physical problems, t often represents time, but we are free to use other variables such as s or θ for the parameter.

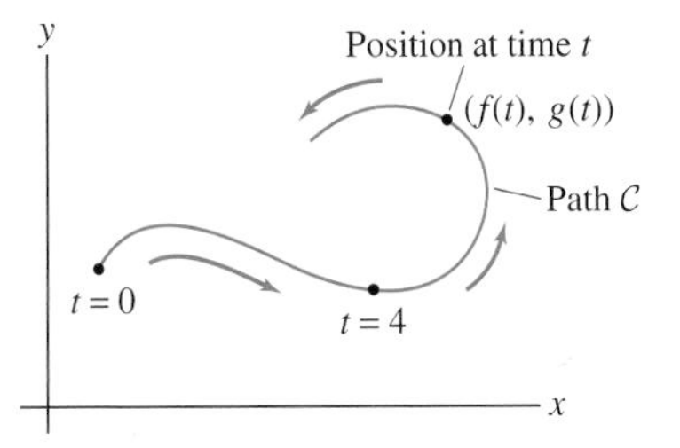

FIGURE 1 Path of a particle moving along a path $\mathcal{C}$ in the plane.

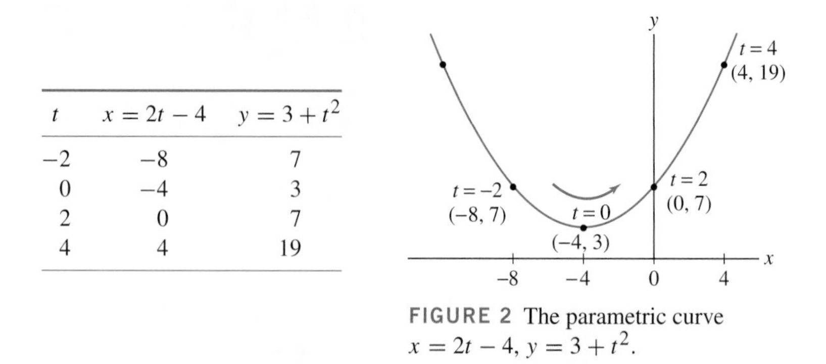

t	$x = 2t - 4$	$y = 3 + t^2$
−2	−8	7
0	−4	3
2	0	7
4	4	19

FIGURE 2 The parametric curve $x = 2t - 4$, $y = 3 + t^2$.

Consider the parametric equations

$$x = 2t - 4, \qquad y = 3 + t^2 \tag{2}$$

The table above gives the x and y coordinates for several values of t. In Figure 2, we plot these points and join them by a smooth curve, indicating the direction of motion with an arrow.

In some cases, a parametric curve can be expressed as the graph of a function.

EXAMPLE 1 Eliminating the Parameter Describe the parametric curve

$$c(t) = (2t - 4, 3 + t^2)$$

in the form $y = f(x)$.

Solution We "eliminate the parameter" by solving for y as a function of x. First, express t in terms of x: Since $x = 2t - 4$, we have $t = \frac{1}{2}x + 2$. Then substitute

$$y = 3 + t^2 = 3 + \left(\frac{1}{2}x + 2\right)^2 = 7 + 2x + \frac{1}{4}x^2$$

Thus, $c(t)$ traces out the parabola $y = 7 + 2x + \frac{1}{4}x^2$ (Figure 2). ■

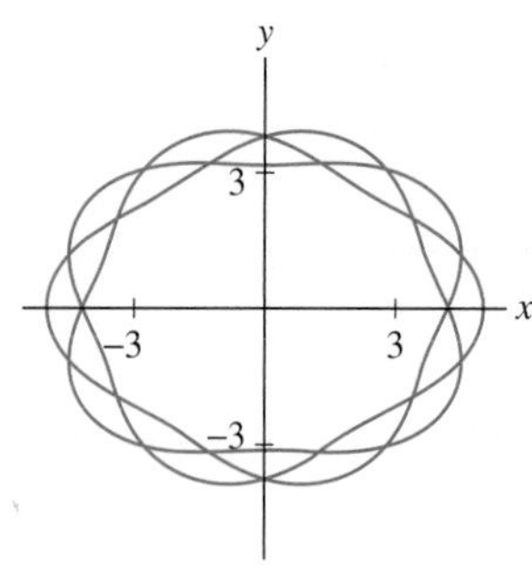

FIGURE 3 The parametric curve:
$x = 5\cos(3t)\cos\left(\frac{2}{3}\sin(5t)\right)$
$y = 4\sin(3t)\cos\left(\frac{2}{3}\sin(5t)\right)$.

CONCEPTUAL INSIGHT The graph of a function $y = f(x)$ can always be parametrized as $c(t) = (t, f(t))$. For example, the parabola $y = x^2$ is parametrized by $c(t) = (t, t^2)$ and $y = e^t$ by $c(t) = (t, e^t)$. An advantage of parametric equations is that they allow us to describe curves that are not graphs of functions, such as the curve in Figure 3.

EXAMPLE 2 A bullet follows the trajectory (t in seconds, distance in feet):

$$x(t) = 200t, \qquad y(t) = 400t - 16t^2 \qquad 0 \le t \le 25$$

Find the bullet's height at $t = 5$ and its maximum height.

Solution The bullet's height is $y(t) = 400t - 16t^2$. At $t = 5$, the height is

$$y(5) = 400(5) - 16(5^2) = 1{,}600 \text{ ft}$$

The function $y(t)$ is quadratic and takes on a maximum value where $y'(t) = 0$:

$$y'(t) = \frac{d}{dt}(400t - 16t^2) = 400 - 32t = 0 \qquad \Rightarrow \qquad t = \frac{400}{32} = 12.5$$

Thus, the maximum height is $y(12.5) = 400(12.5) - 16(12.5)^2 = 2{,}500$ ft (Figure 4). ■

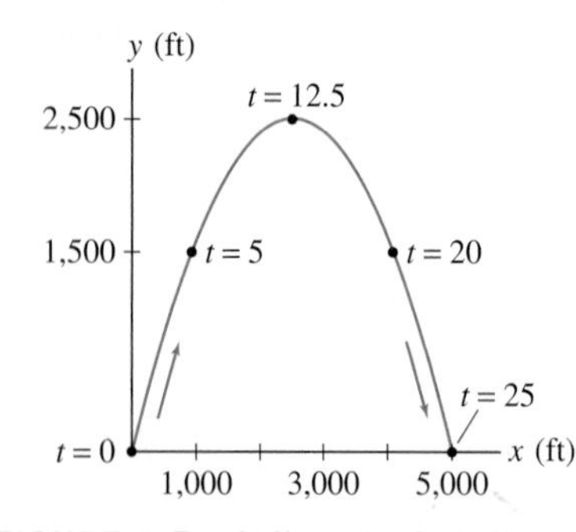

FIGURE 4 Parabolic path of a bullet.

We now discuss parametrizations of straight lines and circles. These are used frequently in multivariable calculus.

EXAMPLE 3 Parametric Representation of a Line Show that the line through the point $P = (a, b)$ with slope m has parametrization

$$c(t) = (a + t, b + mt) \qquad -\infty < t < \infty \tag{3}$$

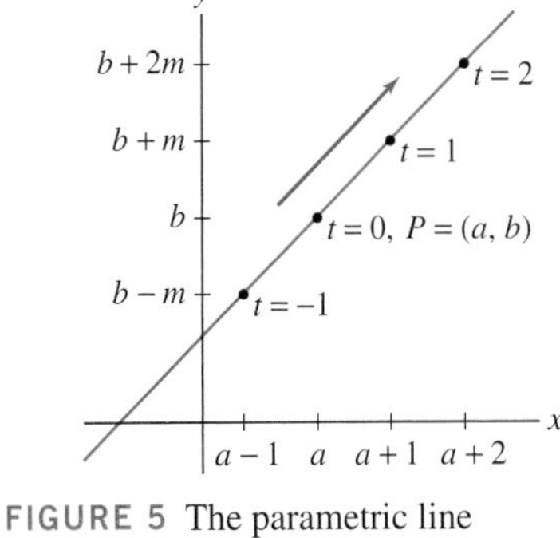

FIGURE 5 The parametric line $c(t) = (a + t, b + mt)$.

Find parametric equations for the line through $P = (3, -1)$ of slope $m = 4$.

Solution We eliminate the parameter $x = a + t$. Thus, $t = x - a$ and

$$y = b + mt = b + m(x - a) \qquad \text{or} \qquad y - b = m(x - a)$$

This is the equation of the line through $P = (a, b)$ of slope m as claimed (Figure 5).

The line through $P = (3, -1)$ of slope $m = 4$ has parametrization

$$c(t) = (3 + t, -1 + 4t)$$

A circle of radius R centered at the origin has the parametrization

$$c(\theta) = (R\cos\theta, R\sin\theta)$$

The parameter θ represents the angle corresponding to the point (x, y) on the circle (Figure 6). The circle is traversed once in the counterclockwise direction as θ varies from 0 to 2π.

To move or "translate" a parametric curve a units horizontally and b units vertically, we replace $c(t) = (x(t), y(t))$ by $c(t) = (a + x(t), b + y(t))$. In particular, the circle of radius R with center (a, b) has parametrization (Figure 6)

$$x = a + R\cos\theta, \qquad y = b + R\sin\theta \tag{4}$$

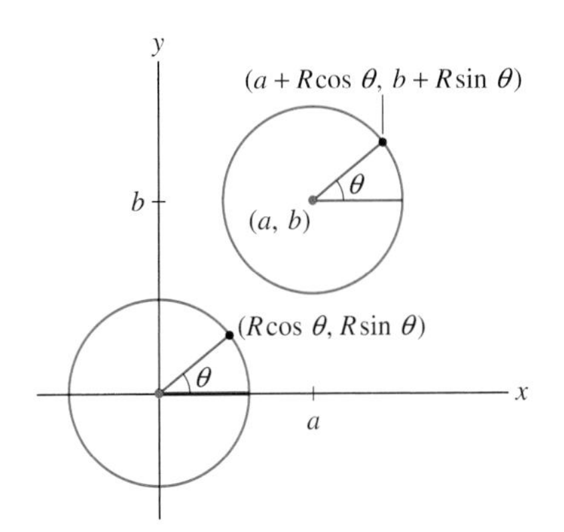

FIGURE 6 Parametrization of a circle of radius R with center (a, b).

As a check, let's verify that the point (x, y) satisfies the equation of the circle of radius R centered at (a, b):

$$\begin{aligned}(x - a)^2 + (y - b)^2 &= (a + R\cos\theta - a)^2 + (b + R\sin\theta - b)^2 \\ &= R^2\cos^2\theta + R^2\sin^2\theta = R^2\end{aligned}$$

A modification of these equations may be used to parametrize ellipses.

EXAMPLE 4 Parametrization of an Ellipse Show that the ellipse $\left(\frac{x}{a}\right)^2 + \left(\frac{y}{b}\right)^2 = 1$ is parametrized by

$$c(t) = (a\cos t, b\sin t) \qquad 0 \le t < 2\pi$$

Graph the case $a = 4$, $b = 2$, and indicate the points corresponding to $t = 0, \frac{\pi}{6}, \frac{\pi}{3}, \frac{\pi}{2}$.

Solution We check that $x = a\cos t$, $y = b\sin t$ satisfies the equation of the ellipse:

$$\left(\frac{x}{a}\right)^2 + \left(\frac{y}{b}\right)^2 = \left(\frac{a\cos t}{a}\right)^2 + \left(\frac{b\sin t}{b}\right)^2 = \cos^2 t + \sin^2 t = 1$$

As t varies from 0 to π, $c(t)$ traces the top half of the ellipse because x decreases from a to $-a$ and y varies from 0 to b and back again to 0. Similarly, $c(t)$ traces the bottom half as t varies from π to 2π. For $a = 4$, $b = 2$, we obtain $c(t) = (4\cos t, 2\sin t)$, which parametrizes the ellipse $\left(\frac{x}{4}\right)^2 + \left(\frac{y}{2}\right)^2 = 1$. Figure 7 shows the ellipse and points indicated. ■

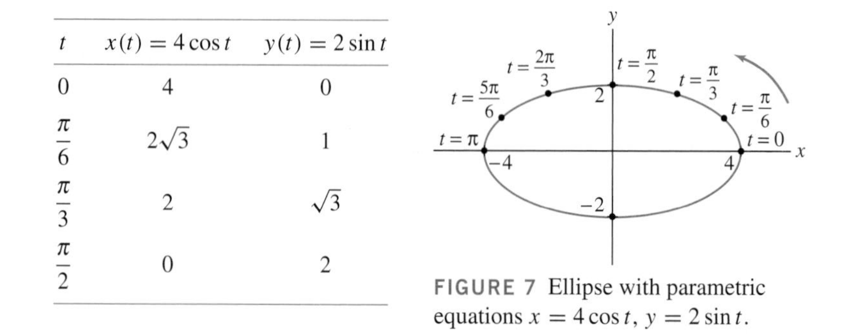

t	$x(t) = 4\cos t$	$y(t) = 2\sin t$
0	4	0
$\frac{\pi}{6}$	$2\sqrt{3}$	1
$\frac{\pi}{3}$	2	$\sqrt{3}$
$\frac{\pi}{2}$	0	2

FIGURE 7 Ellipse with parametric equations $x = 4\cos t$, $y = 2\sin t$.

Before continuing, we make two important remarks:

- There is a difference between a path $c(t)$—such as the orbit of a moon around a planet—and the underlying curve $\mathcal{C}$. The curve $\mathcal{C}$ is the set of points in the plane. The path $c(t)$ (the orbit) describes not just the curve, but also the location of a point along the curve as a function of the parameter. The path may traverse all or part of $\mathcal{C}$ several times. For example, $c(t) = (\cos t, \sin t)$ describes a *path* that goes around the unit circle twice as t varies from 0 to 4π.
- Parametrizations are not unique, and in fact, every curve may be parametrized in infinitely many different ways. For instance, the parabola $y = x^2$ is parametrized by (t, t^2), but also (t^3, t^6), or (t^5, t^{10}), etc.

■ **EXAMPLE 5** **Parametrizations and Paths Versus Curves** Describe the motion of a particle moving along the paths (for $-\infty \le t < \infty$):

(a) $c_1(t) = (t^3, t^6)$ **(b)** $c_2(t) = (t^2, t^4)$ **(c)** $c_3(t) = (\cos t, \cos^2 t)$

Solution The relation $y = x^2$ holds for each of these parametrizations, so all three parametrize portions of the parabola $y = x^2$.

(a) As t varies from $-\infty$ to ∞, the function t^3 also varies from $-\infty$ to ∞. Therefore, a particle following the path $c_1(t) = (t^3, t^6)$ traces the entire parabola $y = x^2$, moving from left to right and passing through each point once [Figure 8(A)].

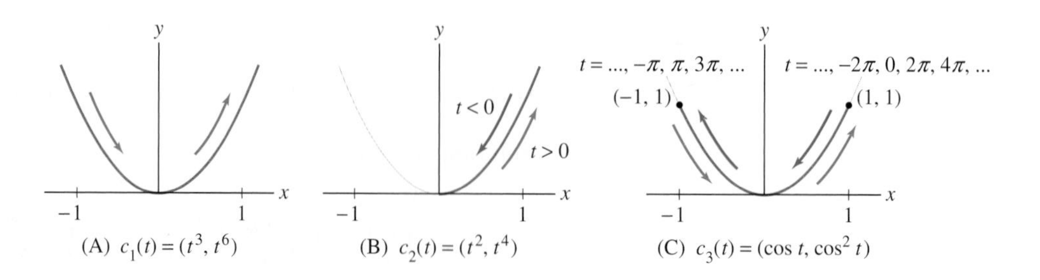

FIGURE 8 Three parametrizations of portions of the parabola.

(b) In this parametrization, $x = t^2 > 0$, so the path $c_2(t) = (t^2, t^4)$ only traces the right half of the parabola. The particle comes in toward the origin as t varies from $-\infty$ to 0, and goes back out to the right as t varies from 0 to ∞ [Figure 8(B)].

(c) The function $\cos t$ oscillates between 1 to -1. Therefore, a particle following the path $c_3(t) = (\cos t, \cos^2 t)$ oscillates back and forth between the points $(1, 1)$ and $(-1, 1)$ on the parabola [Figure 8(C)]. The motion is repeated as t varies on intervals of length 2π. ■

■ **EXAMPLE 6 Using Symmetry to Sketch a Loop** Sketch the curve parametrized by

$$c(t) = (t^2 + 1, t^3 - 4t)$$

Label the points corresponding to $t = 0, \pm 1, \pm 2, \pm 2.5$.

Solution We note that $x(t) = t^2 + 1$ is an even function and $y(t) = t^3 - 4t$ is an odd function:

$$x(-t) = x(t), \qquad y(-t) = -y(t)$$

It follows that if a point $P = c(t)$ lies on the curve, then its reflection across the x-axis also lies on the curve and corresponds to "time" $-t$:

$$c(-t) = (x(t), -y(t)) = \text{reflection of } c(t) \text{ across the } x\text{-axis}$$

In other words, the curve is symmetric with respect to the x-axis. Therefore, it suffices to sketch the curve for $t \geq 0$ and then use symmetry.

Next, we observe that the coordinate $x(t) = t^2 + 1$ tends to ∞ as $t \to \infty$. To analyze the y-coordinate, we graph $y(t) = t^3 - 4t$ as a function of t (*not* as a function of x). Figure 9(A) shows that

$$\begin{aligned} y(t) &< 0 \quad &&\text{for} \quad 0 < t < 2, \\ y(t) &> 0 \quad &&\text{for} \quad t > 2, \\ y(t) &\to \infty \quad &&\text{as} \quad t \to \infty \end{aligned}$$

Therefore, starting at $c(0) = (1, 0)$, the curve first dips below the x-axis. Then it turns up and tends to ∞. The points $c(0)$, $c(1)$, $c(2)$, and $c(2.5)$ are tabulated in Table 1 and plotted in Figure 9(B). The part of the path for $t \leq 0$ is obtained by reflecting across the x-axis as shown in Figure 9(C). ■

TABLE 1

t	$x = t^2 + 1$	$y = t^3 - 4t$
0	1	0
1	2	-3
2	5	0
2.5	7.25	5.625

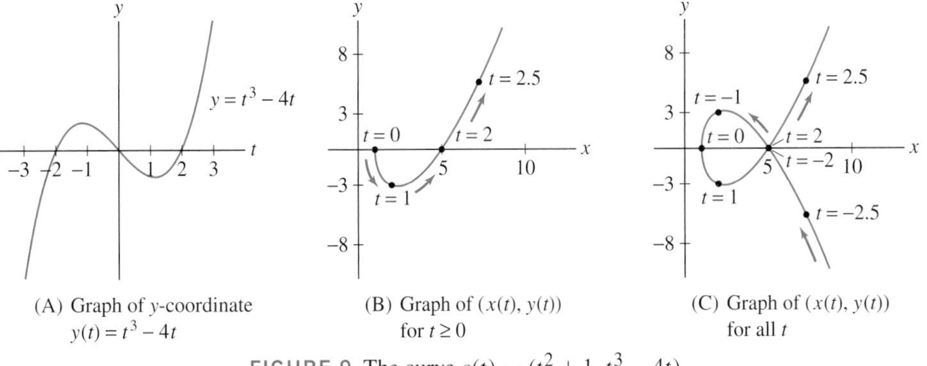

(A) Graph of y-coordinate $y(t) = t^3 - 4t$

(B) Graph of $(x(t), y(t))$ for $t \geq 0$

(C) Graph of $(x(t), y(t))$ for all t

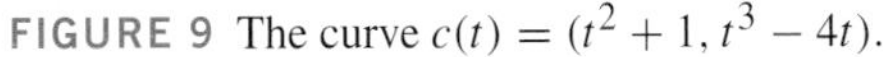

FIGURE 9 The curve $c(t) = (t^2 + 1, t^3 - 4t)$.

The **cycloid** is the curve traced by a point on the circumference of a rolling wheel (Figure 10).

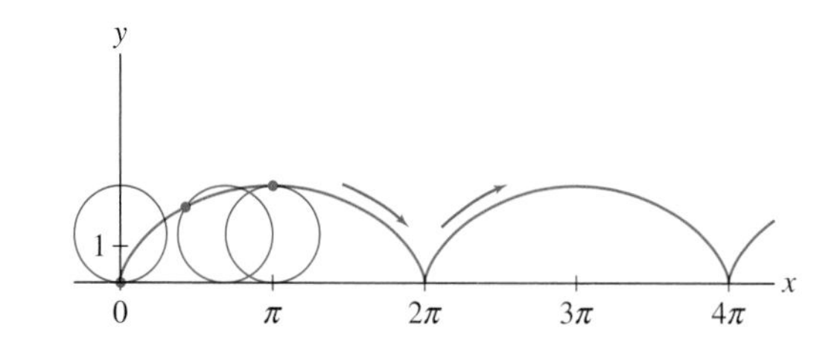

FIGURE 10 A cycloid.

The cycloid was studied intensively in the sixteenth and seventeenth centuries by a stellar cast of mathematicians (including Galileo, Pascal, Newton, Leibniz, Huygens, and Bernoulli) who discovered many of its remarkable properties. For example, to build a slide with the property that an object sliding down (without friction) reaches the bottom in the least amount of time, the slide must have the shape of an inverted cycloid. This is called the "brachistochrone property" from the Greek brachistos, *"the shortest," and* chronos, *"time."*

EXAMPLE 7 Parametrizing the Cycloid Find parametric equations for the cycloid generated by a point P on the unit circle.

Solution We refer to Figure 11. Let P be the point on the rolling unit circle located at the origin at $t = 0$. Now let the circle roll a distance t along the x-axis. Then the length of the circular arc $\overset{\frown}{QP}$ is also t and the angle $\angle PCQ$ has radian measure t.

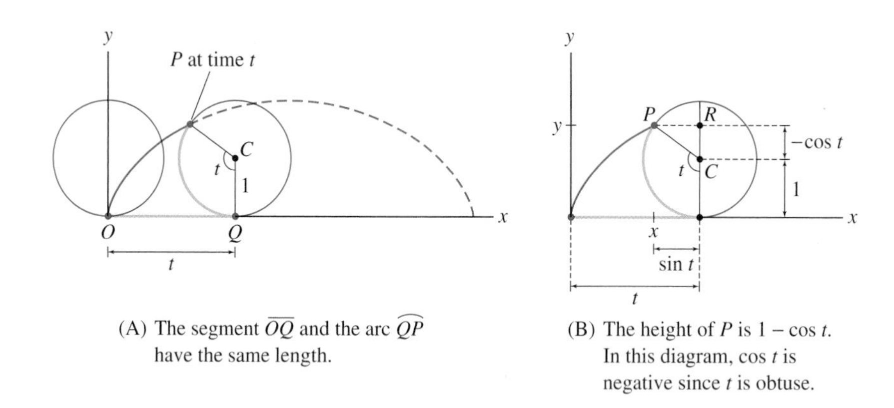

FIGURE 11 Parametric representation of the cycloid.

(A) The segment $\overline{OQ}$ and the arc $\overset{\frown}{QP}$ have the same length.

(B) The height of P is $1 - \cos t$. In this diagram, $\cos t$ is negative since t is obtuse.

To calculate the coordinates of P, observe in Figure 11(B) that the height y of P is $y = 1 - \cos t$. Note that in the figure, $\cos t$ is negative since t is obtuse and the segment $\overline{CR}$ has length $-\cos t$. The x-coordinate of P is $x = t - PR = t - \sin t$, so we obtain

$$x(t) = t - \sin t, \qquad y(t) = 1 - \cos t \tag{5}$$

■

The argument in Example 7 may be modified to show that the cycloid generated by a circle of radius R has parametric equations

$$x = Rt - R\sin t, \qquad y = R - R\cos t \tag{6}$$

Next, we address the problem of finding the tangent line for a curve in parametric form. For any curve in the xy-plane, the slope of the tangent line is the derivative dy/dx (whether described parametrically or not), but for a parametric curve, we must use the Chain Rule to compute dy/dx because y is not given explicitly as a function of x. If $x = f(t)$, $y = g(t)$, then

$$g'(t) = \frac{dy}{dt} = \frac{dy}{dx}\frac{dx}{dt} = \frac{dy}{dx}f'(t)$$

NOTATION In this section, we write $f'(t), x'(t), y'(t)$, etc. to denote the derivative with respect to t.

If $f'(t) \neq 0$, we may divide by $f'(t)$ to obtain

$$\frac{dy}{dx} = \frac{g'(t)}{f'(t)}$$

This calculation is valid on an interval I where $f(t)$ and $g(t)$ are differentiable, $f'(t)$ is continuous, and $f'(t) \neq 0$. In this case, $t = f^{-1}(x)$ exists and the composite $y = g(f^{-1}(x))$ is a differentiable function of x.

CAUTION Do not confuse dy/dx (the slope of the tangent line) with the derivatives dx/dt and dy/dt, which are derivatives with respect to the parameter t.

THEOREM 1 Slope of the Tangent Line Let $c(t) = (x(t), y(t))$, where $x(t)$ and $y(t)$ are differentiable. Assume that $x'(t)$ is continuous and $x'(t) \neq 0$. Then

$$\frac{dy}{dx} = \frac{dy/dt}{dx/dt} = \frac{y'(t)}{x'(t)} \tag{7}$$

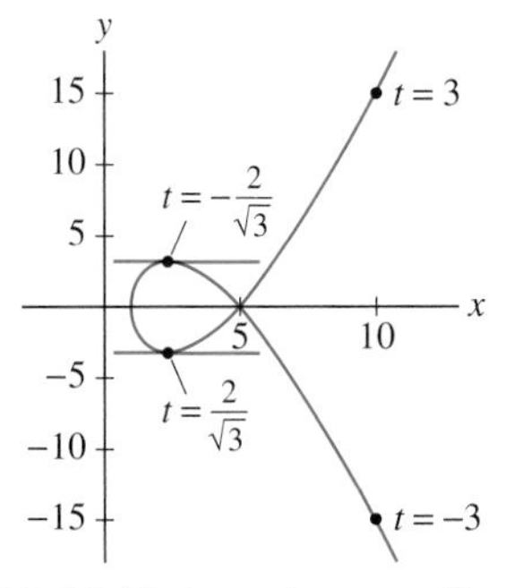

FIGURE 12 Horizontal tangent lines on $c(t) = (t^2 + 1, t^3 - 4t)$.

■ **EXAMPLE 8** Let $c(t) = (t^2 + 1, t^3 - 4t)$. Find the equation of the tangent line at $t = 3$ and find the points where the tangent is horizontal.

Solution We have

$$\frac{dy}{dx} = \frac{y'(t)}{x'(t)} = \frac{(t^3 - 4t)'}{(t^2 + 1)'} = \frac{3t^2 - 4}{2t}$$

The slope at $t = 3$ is

$$\frac{dy}{dx} = \left.\frac{3t^2 - 4}{2t}\right|_{t=3} = \frac{3(3)^2 - 4}{2(3)} = \frac{23}{6}$$

Since $c(3) = (10, 15)$, the equation of the tangent line in point-slope form is

$$y - 15 = \frac{23}{6}(x - 10)$$

The slope $\frac{dy}{dx}$ is zero if $3t^2 - 4 = 0$ and $2t \neq 0$. This gives $t = \pm 2/\sqrt{3}$ (Figure 12). Therefore, the tangent line is horizontal at the points

$$c\left(-\frac{2}{\sqrt{3}}\right) = \left(\frac{7}{3}, \frac{16}{3\sqrt{3}}\right), \qquad c\left(\frac{2}{\sqrt{3}}\right) = \left(\frac{7}{3}, -\frac{16}{3\sqrt{3}}\right)$$ ■

Bézier curves were invented in the 1960s by Pierre Bézier (1910–1999), a French engineer who worked for the Renault car company. They are based on the properties of Bernstein polynomials, which were introduced 50 years earlier (in 1911) by the Russian mathematician Sergei Bernstein to study the approximation of continuous functions by polynomials. Today, Bézier curves are used in standard graphics programs such as Adobe Illustrator™ and Corel Draw™, and in the construction and storage of computer fonts such as TrueType™ and PostScript™ fonts.

Parametric curves are widely used in the field of computer graphics. A particularly important class of curves are **Bézier curves**, which we discuss here briefly in the cubic case. Given four "control points" (Figure 13):

$$P_0 = (a_0, b_0), \qquad P_1 = (a_1, b_1), \qquad P_2 = (a_2, b_2), \qquad P_3 = (a_3, b_3)$$

we define the Bézier curve $c(t) = (x(t), y(t))$ (for $0 \leq t \leq 1$), where

$$x(t) = a_0(1-t)^3 + 3a_1t(1-t)^2 + 3a_2t^2(1-t) + a_3t^3 \tag{8}$$

$$y(t) = b_0(1-t)^3 + 3b_1t(1-t)^2 + 3b_2t^2(1-t) + b_3t^3 \tag{9}$$

Note that $c(0) = (a_0, b_0)$ and $c(1) = (a_3, b_3)$, so the Bézier curve begins at P_0 and ends at P_3 (Figure 13). It can also be shown that the Bézier curve is contained within the quadrilateral (shown in yellow) with vertices P_0, P_1, P_2, P_3. However, $c(t)$ does not pass

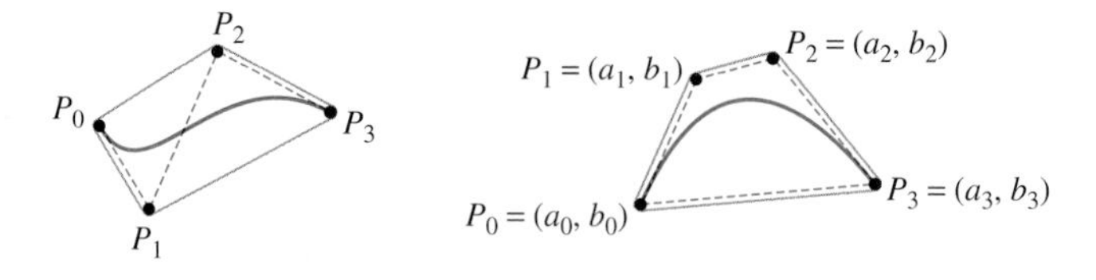

FIGURE 13 Cubic Bézier curves specifying four control points.

through P_1 and P_2. Instead, these intermediate control points determine the slopes of the tangent lines at P_0 and P_3, as we show in the next example (also, see Exercises 58–60).

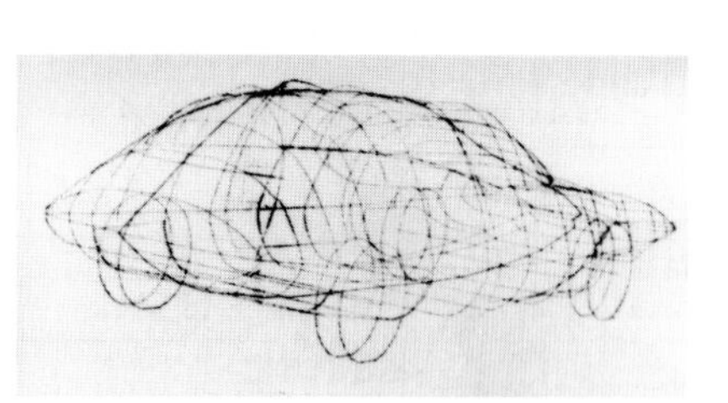

Hand sketch by Pierre Bézier for the French automobile manufacturer Renault, made in 1964.

■ **EXAMPLE 9** Show that the Bézier curve is tangent to the segment $\overline{P_0P_1}$ at P_0.

Solution The Bézier curve passes through P_0 at $t = 0$, so we must show that the slope of the tangent line at $c(0) = P_0 = (a_0, b_0)$ is equal to the slope of $\overline{P_0P_1}$. To find the slope, we compute the derivatives:

$$x'(t) = -3a_0(1-t)^2 + 3a_1(1-4t+3t^2) + a_2(2t-3t^2) + 3a_3t^2$$

$$y'(t) = -3b_0(1-t)^2 + 3b_1(1-4t+3t^2) + b_2(2t-3t^2) + 3b_3t^2$$

Evaluating at $t = 0$, we obtain $x'(0) = 3(a_1 - a_0)$, $y'(0) = 3(b_1 - b_0)$, and

$$\left.\frac{dy}{dx}\right|_{t=0} = \frac{y'(0)}{x'(0)} = \frac{3(b_1 - b_0)}{3(a_1 - a_0)} = \frac{b_1 - b_0}{a_1 - a_0}$$

This is equal to the slope of the line segment through $P_0 = (a_0, b_0)$ and $P_1 = (a_1, b_1)$ as claimed (provided that $a_1 \neq a_0$). ■

11.1 SUMMARY

- A path traced by a point $P = (x, y)$, where x and y are functions of a parameter t, is called a *parametric* or *parametrized* path or curve. We write $c(t) = (f(t), g(t))$ or $c(t) = (x(t), y(t))$.
- Keep in mind that the path $c(t) = (x(t), y(t))$ and the curve that it traces are different. The path $c(t)$ describes the location of a point as a function of t, whereas the underlying curve is a set of points in the plane. For example, the path $(\cos t, \sin t)$ moves around the unit circle infinitely many times as t varies from 0 to ∞.
- Parametrizations are not unique: Every path may be parametrized in infinitely many ways.
- Let $c(t) = (x(t), y(t))$, where $x(t)$ and $y(t)$ are differentiable and $x'(t)$ is continuous. Then the slope of the tangent line at $c(t)$ is the derivative

$$\frac{dy}{dx} = \frac{dy/dt}{dx/dt} = \frac{y'(t)}{x'(t)} \quad [\text{valid if } x'(t) \neq 0]$$

- Do not confuse the derivatives with respect to t, dy/dt, and dx/dt, with the derivative dy/dx (the slope of the tangent line).

11.1 EXERCISES

Preliminary Questions

1. Describe the shape of the curve $x = 3\cos t$, $y = 3\sin t$.

2. How does $x = 4 + 3\cos t$, $y = 5 + 3\sin t$ differ from the curve in the previous question?

3. What is the maximum height of a particle whose path has parametric equations $x = t^9$, $y = 4 - t^2$?

4. Can the parametric curve $(t, \sin t)$ be represented as a graph $y = f(x)$? What about $(\sin t, t)$?

5. Match the derivatives with a verbal description:

(a) $\dfrac{dx}{dt}$ **(b)** $\dfrac{dy}{dt}$ **(c)** $\dfrac{dy}{dx}$

(i) Slope of the tangent line to the curve

(ii) Vertical rate of change with respect to time

(iii) Horizontal rate of change with respect to time

Exercises

1. Find the coordinates at times $t = 0, 2, 4$ of a particle following the path $x = 1 + t^3$, $y = 9 - 3t^2$.

2. Find the coordinates at $t = 0, \frac{\pi}{4}, \pi$ of a particle moving along the path $c(t) = (\cos 2t, \sin^2 t)$.

3. Show that the path traced by the bullet in Example 2 is a parabola by eliminating the parameter.

4. Use the table of values to sketch the parametric curve $(x(t), y(t))$, indicating the direction of motion.

t	-3	-2	-1	0	1	2	3
x	-15	0	3	0	-3	0	15
y	5	0	-3	-4	-3	0	5

5. Graph the parametric curves. Include arrows indicating the direction of motion.

(a) (t, t), $-\infty < t < \infty$ **(b)** $(\sin t, \sin t)$, $0 \le t \le 2\pi$

(c) (e^t, e^t), $-\infty < t < \infty$ **(d)** (t^3, t^3), $-1 \le t \le 1$

6. Give two different parametrizations of the line through $(4, 1)$ with slope 2.

In Exercises 7–14, express in the form $y = f(x)$ by eliminating the parameter.

7. $x = t + 3$, $y = 4t$ **8.** $x = t^{-1}$, $y = t^{-2}$

9. $x = t$, $y = \tan^{-1}(t^3 + e^t)$ **10.** $x = t^2$, $y = t^3 + 1$

11. $x = e^{-2t}$, $y = 6e^{4t}$ **12.** $x = 1 + t^{-1}$, $y = t^2$

13. $x = \ln t$, $y = 2 - t$ **14.** $x = \cos t$, $y = \tan t$

In Exercises 15–18, graph the curve and draw an arrow specifying the direction corresponding to motion.

15. $x = \frac{1}{2}t$, $y = 2t^2$ **16.** $x = 2 + 4t$, $y = 3 + 2t$

17. $x = \pi t$, $y = \sin t$ **18.** $x = t^2$, $y = t^3$

19. Match the parametrizations (a)–(d) below with their plots in Figure 14 and draw an arrow indicating the direction of motion.

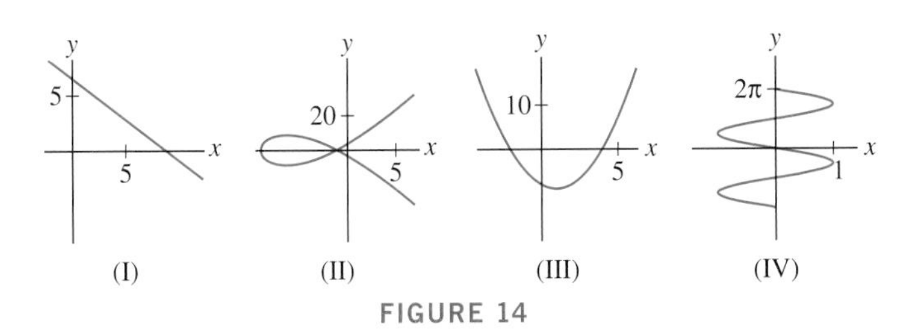

FIGURE 14

(a) $c(t) = (\sin t, -t)$ **(b)** $c(t) = (t^2 - 9, -t^3 - 8)$

(c) $c(t) = (1 - t, t^2 - 9)$ **(d)** $c(t) = (4t + 2, 5 - 3t)$

20. The graphs of $x(t)$ and $y(t)$ as functions of t are shown in Figure 15(A). Which of (I)–(III) is the plot of $c(t) = (x(t), y(t))$? Explain.

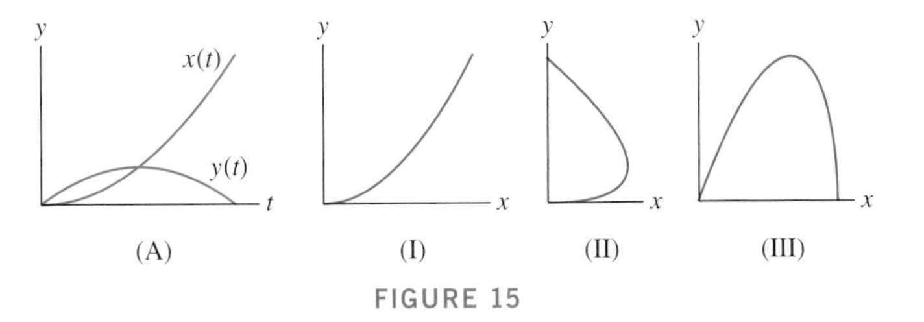

FIGURE 15

21. Find an interval of t-values such that $c(t) = (\cos t, \sin t)$ traces the lower half of the unit circle.

22. Find an interval of t-values such that $c(t) = (2t + 1, 4t - 5)$ parametrizes the segment from $(0, -7)$ to $(7, 7)$.

In Exercises 23–34, find parametric equations for the given curve.

23. $y = 9 - 4x$ **24.** $y = 8x^2 - 3x$

25. $4x - y^2 = 5$ **26.** $x^2 + y^2 = 49$

27. $(x + 9)^2 + (y - 4)^2 = 49$

28. Line of slope 8 through $(-4, 9)$

29. Line through $(2, 5)$ perpendicular to $y = 3x$

30. Circle of radius 4 with center $(3, 9)$

31. $\left(\frac{x}{4}\right)^2 + \left(\frac{y}{9}\right)^2 = 1$

32. Ellipse of Exercise 31, with its center translated to $(2, 11)$

33. The parabola $y = x^2$ translated so that its minimum occurs at $(2, 3)$

34. The curve $y = \cos x$ translated so that a maximum occurs at $(3, 5)$

35. Describe the parametrized curve $c(t) = (\sin^2 t, \cos^2 t)$ for $0 \le t \le \pi$.

36. Find the graph $y = f(x)$ traced by the path $x = \sec t$, $y = \tan t$. Which intervals of t-values trace the graph exactly once?

37. Find a parametrization $c(t)$ of the line $y = 3x - 4$ such that $c(0) = (2, 2)$.

38. Find a parametrization $c(t)$ of $y = x^2$ such that $c(0) = (3, 9)$.

39. Show that $x = \cosh t$, $y = \sinh t$ parametrizes the hyperbola $x^2 - y^2 = 1$. Calculate $\frac{dy}{dx}$ as a function of t. Generalize to obtain a parametrization of $\left(\frac{x}{a}\right)^2 - \left(\frac{y}{b}\right)^2 = 1$.

40. A particle moves along the path $x(t) = \frac{1}{4}t^3 + 2t$, $y(t) = 20t - t^2$ (in centimeters).

(a) What is the maximum height attained by the object?

(b) At what time does the object hit the ground?

(c) How far is the object from the origin when it hits the ground?

41. Which of (I) or (II) is the graph of $x(t)$ for the parametric curve in Figure 16(A)? Which represents $y(t)$?

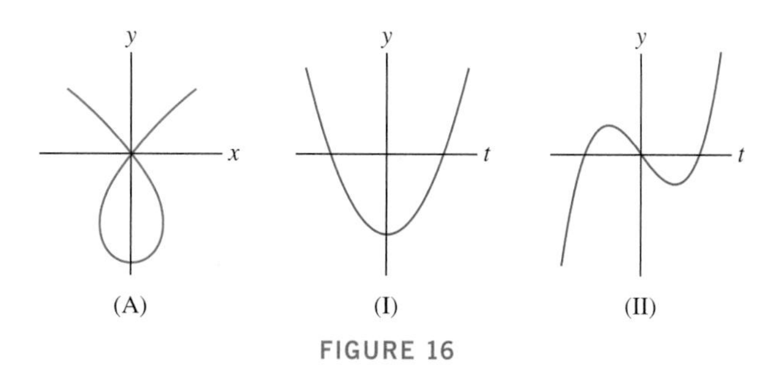

FIGURE 16

42. Sketch the graph of $c(t) = (t^3 - 4t, t^2)$ following the steps in Example 6.

43. Sketch $c(t) = (t^2, \sin t)$ for $-2\pi \le t \le 2\pi$.

44. Sketch $c(t) = (t^2 - 4t, 9 - t^2)$ for $-4 \le t \le 10$.

In Exercises 45–48, use Eq. (7) to find $\frac{dy}{dx}$ *at the given point.*

45. $(t^3, t^2 - 1)$, $t = -4$

46. $(2t + 9, 7t - 9)$, $t = 1$

47. $(s^{-1} - 3s, s^3)$, $s = -1$

48. $(\sin 2\theta, \cos 3\theta)$, $\theta = \frac{\pi}{4}$

In Exercises 49–52, find an equation $y = f(x)$ for the parametric curve and compute $\frac{dy}{dx}$ *in two ways: using Eq. (7) and by differentiating $f(x)$.*

49. $c(t) = (2t + 1, 1 - 9t)$

50. $c(t) = \left(\frac{1}{2}t, \frac{1}{4}t^2 - t\right)$

51. $x = s^3$, $y = s^6 + s^{-3}$

52. $x = \cos\theta$, $y = \cos\theta + \sin^2\theta$

In Exercises 53–56, let $c(t) = (t^2 - 9, t^2 - 8t)$ (see Figure 17).

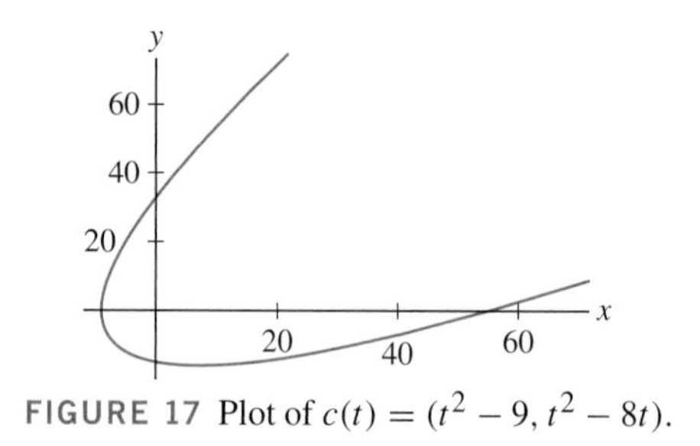

FIGURE 17 Plot of $c(t) = (t^2 - 9, t^2 - 8t)$.

53. Draw an arrow indicating the direction of motion and determine the interval of t values corresponding to the portion of the curve in each of the four quadrants.

54. Find the equation of the tangent line at $t = 4$.

55. Find the points where the tangent has slope $\frac{1}{2}$.

56. Find the points where the tangent is horizontal or vertical.

57. Find the equation of the ellipse represented parametrically by $x = 4\cos t$, $y = 7\sin t$. Calculate the slope of the tangent line at the point $(2\sqrt{2}, 7\sqrt{2}/2)$.

In Exercises 58–60, refer to the Bézier curve defined by Eqs. (8) and (9).

58. Show that the Bézier curve with control points

$$P_0 = (1, 4), \quad P_1 = (3, 12), \quad P_2 = (6, 15), \quad P_3 = (7, 4)$$

has parametrization

$$c(t) = (1 + 6t + 3t^2 - 3t^3, 4 + 24t - 15t^2 - 9t^3)$$

Verify that the slope at $t = 0$ is equal to the slope of the segment $\overline{P_0P_1}$.

59. Find and plot the Bézier curve $c(t)$ passing through the control points

$$P_0 = (3, 2), \quad P_1 = (0, 2), \quad P_2 = (5, 4), \quad P_3 = (2, 4)$$

60. Show that a cubic Bézier curve is tangent to the segment $\overline{P_2P_3}$ at P_3.

61. A bullet fired from a gun follows the trajectory

$$x = at, \quad y = bt - 16t^2 \quad (a, b > 0)$$

Show that the bullet leaves the gun at an angle $\theta = \tan^{-1}\left(\frac{b}{a}\right)$ and lands at a distance $\frac{ab}{16}$ from the origin.

62. CAS Plot $c(t) = (t^3 - 4t, t^4 - 12t^2 + 48)$ for $-3 \le t \le 3$. Find the points where the tangent line is horizontal and vertical.

63. CAS Plot the astroid $x = \cos^3\theta$, $y = \sin^3\theta$ and find the equation of the tangent line at $\theta = \frac{\pi}{3}$.

64. Find the equation of the tangent line at $t = \frac{\pi}{4}$ to the cycloid generated by the unit circle with parametric equation (5).

65. Find the points with horizontal tangent line on the cycloid with parametric equation (5).

66. Property of the Cycloid Prove that the tangent line at a point P on the cycloid always passes through the top point on the rolling circle as indicated in Figure 18. Use the parametrization (6).

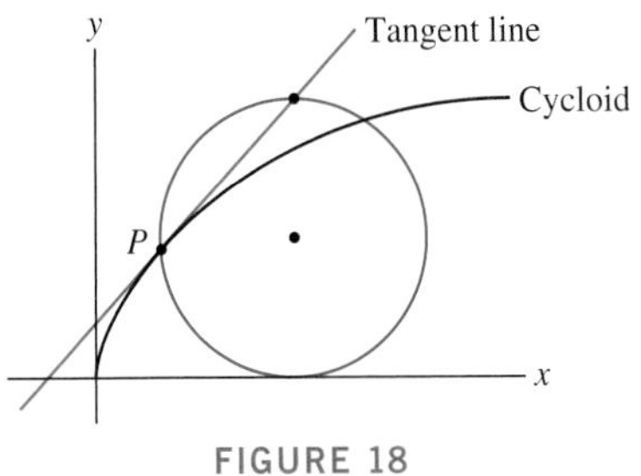

FIGURE 18

67. A *curtate cycloid* (Figure 19) is the curve traced by a point at a distance h from the center of a circle of radius R rolling along the x-axis where $h < R$. Show that this curve has parametric equations $x = Rt - h\sin t$, $y = R - h\cos t$.

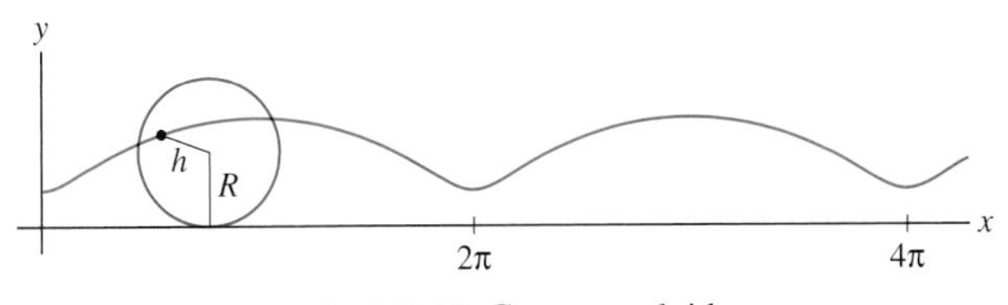

FIGURE 19 Curtate cycloid.

68. CAS Use a computer algebra system to explore what happens when $h > R$ in the parametric equations of Exercise 67. Describe the result.

69. Show that the line of slope t through $(-1, 0)$ intersects the unit circle in the point with coordinates

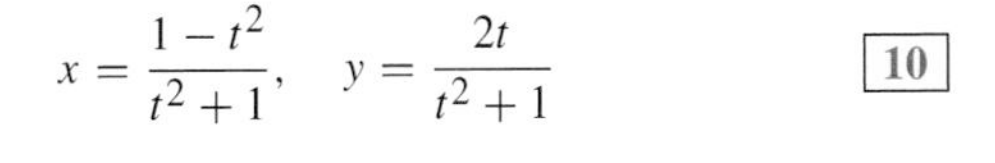

$$x = \frac{1 - t^2}{t^2 + 1}, \quad y = \frac{2t}{t^2 + 1} \qquad \boxed{10}$$

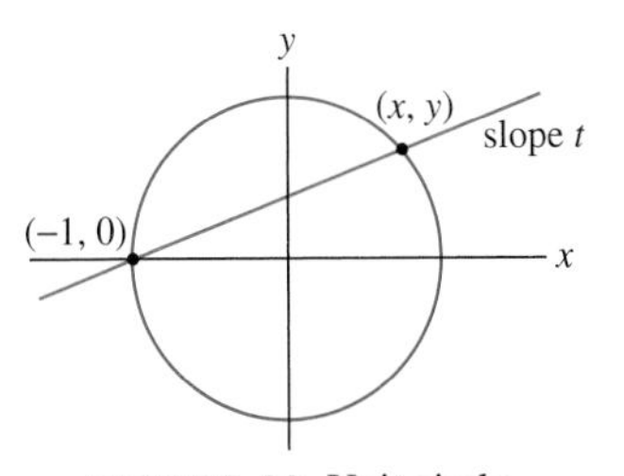

FIGURE 20 Unit circle.

Conclude that these equations parametrize the unit circle with the point $(-1, 0)$ excluded (Figure 20). Show further that $t = \dfrac{y}{x + 1}$.

70. The **folium of Descartes** is the curve with equation $x^3 + y^3 = 3axy$, where $a \neq 0$ is a constant (Figure 21).

(a) Show that for $t \neq -1, 0$, the line $y = tx$ intersects the folium at the origin and at one other point P. Express the coordinates of P in terms of t to obtain a parametrization of the folium. Indicate the direction of the parametrization on the graph.

(b) Describe the interval of t values parametrizing the parts of the curve in quadrants I, II, and IV. Note that $t = -1$ is a point of discontinuity of the parametrization.

(c) Calculate dy/dx as a function of t and find the points with horizontal or vertical tangent.

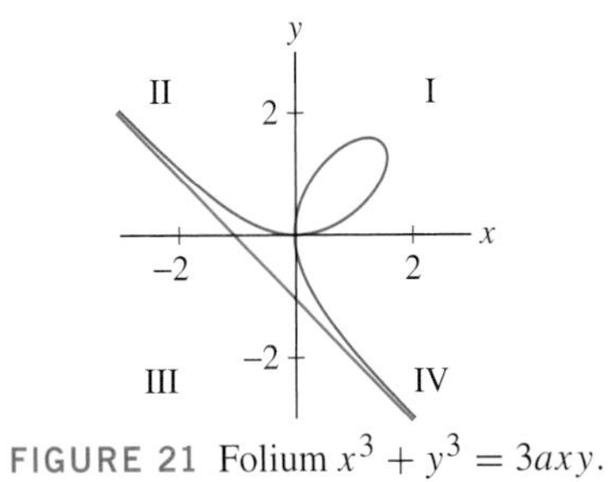

FIGURE 21 Folium $x^3 + y^3 = 3axy$.

71. Use the results of Exercise 70 to show that the asymptote of the folium is the line $x + y = -a$. Hint: show that $\lim\limits_{t \to -1} (x + y) = -a$.

72. Find a parametrization of $x^{2n+1} + y^{2n+1} = ax^n y^n$, where a and n are constants.

73. Verify that the **tractrix** curve ($\ell > 0$)

$$c(t) = \left(t - \ell \tanh\frac{t}{\ell}, \ \ell \operatorname{sech}\frac{t}{\ell}\right)$$

has the following property: For all t, the segment from $c(t)$ to $(0, t)$ is tangent to the curve and has length ℓ (Figure 22).

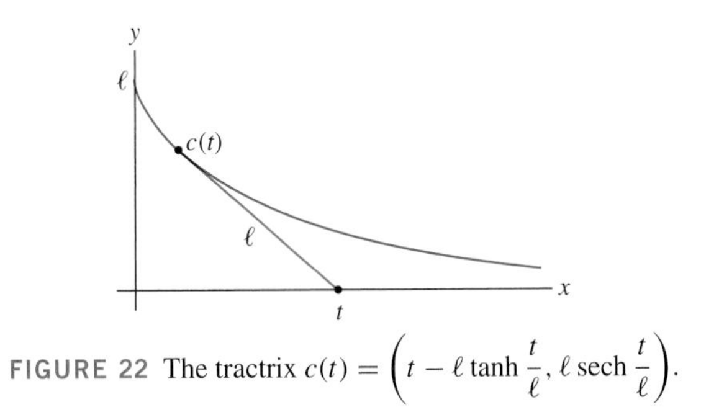

FIGURE 22 The tractrix $c(t) = \left(t - \ell \tanh\frac{t}{\ell}, \ \ell \operatorname{sech}\frac{t}{\ell}\right)$.

74. In Exercise 54 of Section 9.1, we described the tractrix by the differential equation

$$\frac{dy}{dx} = -\frac{y}{\sqrt{\ell^2 - y^2}}$$

Show that the curve $c(t)$ identified as the tractrix in Exercise 73 satisfies this differential equation. Note that the derivative on the left is taken with respect to x, not t.

75. Let A and B be the points where the ray of angle θ intersects the two concentric circles of radii $r < R$ centered at the origin (Figure 23). Let P be the point of intersection of the horizontal line through A and the vertical line through B. Express the coordinates of P as a function of θ and describe the curve traced by P for $0 \le \theta \le 2\pi$.

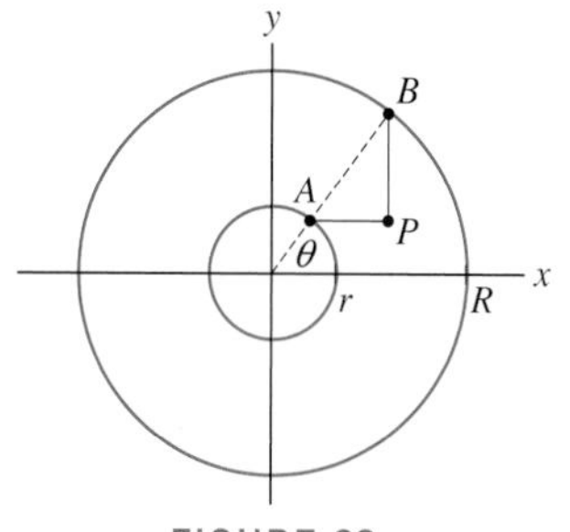

FIGURE 23

76. A 10-ft ladder slides down a wall as its bottom B is pulled away from the wall (Figure 24). Using the angle θ as parameter, find the parametric equations for the path followed by (a) the top of the ladder A, (b) the bottom of the ladder B, and (c) the point P located 4 ft from the top of the ladder. Show that P describes an ellipse.

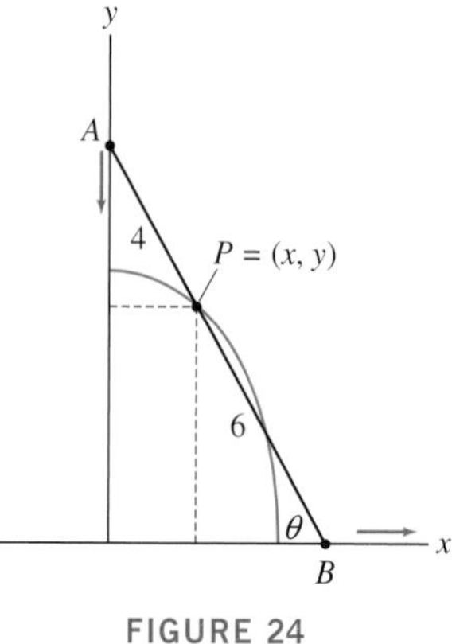

FIGURE 24

Further Insights and Challenges

77. Derive the formula for the slope of the tangent line to a parametric curve $c(t) = (x(t), y(t))$ using a different method than that presented in the text. Assume that $x'(t_0)$ and $y'(t_0)$ exist and that $x'(t_0) \neq 0$. Show that

$$\lim_{h \to 0} \frac{y(t_0 + h) - y(t_0)}{x(t_0 + h) - x(t_0)} = \frac{y'(t_0)}{x'(t_0)}$$

Then explain why this limit is equal to the slope dy/dx. Draw a diagram showing that the ratio in the limit is the slope of a secant line.

78. Second Derivative for a Parametrized Curve Given a parametrized curve $c(t) = (x(t), y(t))$, show that

$$\frac{d}{dt}\left(\frac{dy}{dx}\right) = \frac{x'(t)y''(t) - y'(t)x''(t)}{x'(t)^2}$$

Use this to prove the formula

$$\frac{d^2y}{dx^2} = \frac{x'(t)y''(t) - y'(t)x''(t)}{x'(t)^3} \qquad \boxed{11}$$

In Exercises 79–82, use Eq. (11) to find $\dfrac{d^2y}{dx^2}$.

79. $x = t^3 + t^2, \quad y = 7t^2 - 4, \quad t = 2$

80. $x = s^{-1} + s, \quad y = 4 - s^{-2}, \quad s = 1$

81. $x = 8t + 9, \quad y = 1 - 4t, \quad t = -3$

82. $x = \cos\theta, \quad y = \sin\theta, \quad \theta = \frac{\pi}{4}$

83. Use Eq. (11) to find the t-intervals on which $c(t) = (t^2, t^3 - 4t)$ is concave up.

84. Use Eq. (11) to find the t-intervals on which $c(t) = (t^2, t^4 - 4t)$ is concave up.

85. Area under a Parametrized Curve Let $c(t) = (x(t), y(t))$ be a parametrized curve such that $x'(t) > 0$ and $y(t) > 0$ (Figure 25). Show that the area A under $c(t)$ for $t_0 \le t \le t_1$ is

$$A = \int_{t_0}^{t_1} y(t)x'(t)\,dt \qquad \boxed{12}$$

Hint: $x(t)$ is increasing and therefore has an inverse, say, $t = g(x)$. Observe that $c(t)$ is the graph of the function $y(g(x))$ and apply the Change of Variables formula to $A = \displaystyle\int_{x(t_0)}^{x(t_1)} y(g(x))\,dx$.

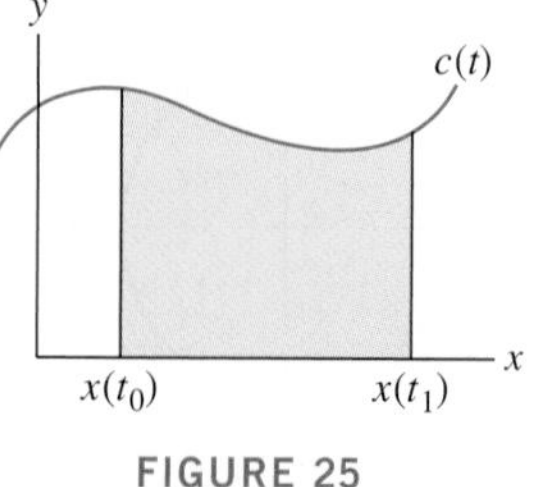

FIGURE 25

86. Calculate the area under $y = x^2$ over $[0, 1]$ using Eq. (12) with the parametrizations (t^3, t^6) and (t^2, t^4).

87. What does Eq. (12) say if $c(t) = (t, f(t))$?

88. Sketch the graph of $c(t) = (\ln t, 2 - t)$ for $1 \le t \le 2$ and compute the area under the graph using Eq. (12).

89. Use Eq. (12) to show that the area under one arch of the cycloid $c(t)$ (Figure 26) generated by a circle of radius R is equal to three times the area of the circle. Recall that

$$c(t) = (Rt - R\sin t,\ R - R\cos t)$$

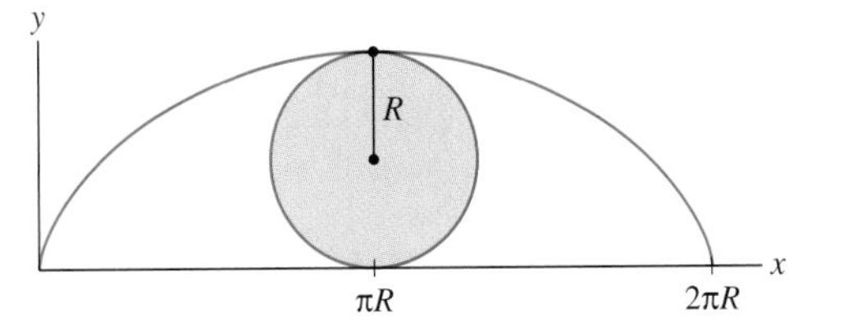

FIGURE 26 The area of the generating circle is one-third the area of one arch of the cycloid.

In Exercises 90–91, refer to Figure 27.

90. The parameter t in the standard parametrization of the ellipse $c(t) = (a\cos t, b\sin t)$ is *not* an angular parameter (unless $a = b$), but it can be interpreted as an area parameter. Show that if $c(t) = (x, y)$, then $t = 2A/ab$, where A is the area of the shaded region in Figure 27. *Hint:* Use Eq. (12).

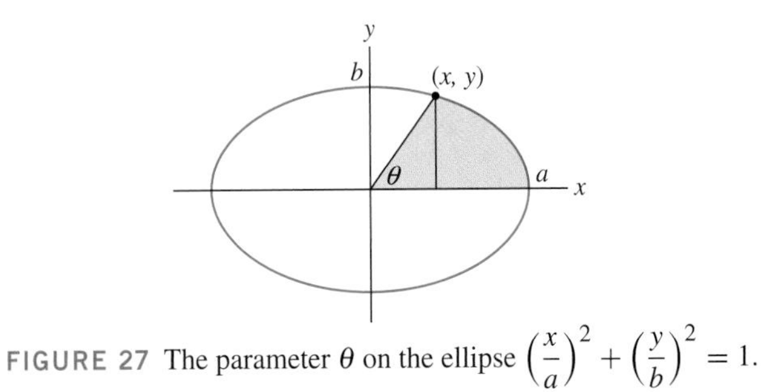

FIGURE 27 The parameter θ on the ellipse $\left(\frac{x}{a}\right)^2 + \left(\frac{y}{b}\right)^2 = 1$.

91. Show that the parametrization of the ellipse by the angle θ is

$$x = \frac{ab\cos\theta}{\sqrt{a^2\sin^2\theta + b^2\cos^2\theta}}$$

$$y = \frac{ab\sin\theta}{\sqrt{a^2\sin^2\theta + b^2\cos^2\theta}}$$

11.2 Arc Length and Speed

In Section 8.1, the length (or "arc length") s of a curve was defined as the limit of the lengths of polygonal approximations. We showed that if $f(x)$ has a continuous derivative, then the arc length of the graph of $y = f(x)$ over $[a, b]$ is equal to

$$s = \text{arc length} = \int_a^b \sqrt{1 + f'(x)^2}\,dx \qquad \boxed{1}$$

There is a more general formula for the arc length of a path $c(t) = (x(t), y(t))$ for $a \le t \le b$. To derive this formula, choose a partition P of $[a, b]$ by an increasing sequence of parameter values:

$$t_0 = a < t_1 < t_2 < \cdots < t_N = b$$

and let L be the polygonal path obtained by joining the points

$$P_0 = c(t_0), \quad P_1 = c(t_1), \ldots, P_N = c(t_N)$$

by segments as in Figure 1. We define the length s of the path to be the limit of the lengths $|L|$ of the polygonal approximations L as the norm $\|P\|$ tends to zero (recall that $\|P\|$ is the maximum of the widths $\Delta t_i = t_i - t_{i-1}$). Note that N tends to ∞ as $\|P\| \to 0$.

By the distance formula, the length of the ith segment in a polygonal approximation L is

$$\overline{P_{i-1}P_i} = \sqrt{(x(t_i) - x(t_{i-1}))^2 + (y(t_i) - y(t_{i-1}))^2} \qquad \boxed{2}$$

We rewrite this expression by applying the Mean Value Theorem (MVT) to both $x(t)$ and $y(t)$. Assuming that $x(t)$ and $y(t)$ are differentiable, the MVT states that there exist

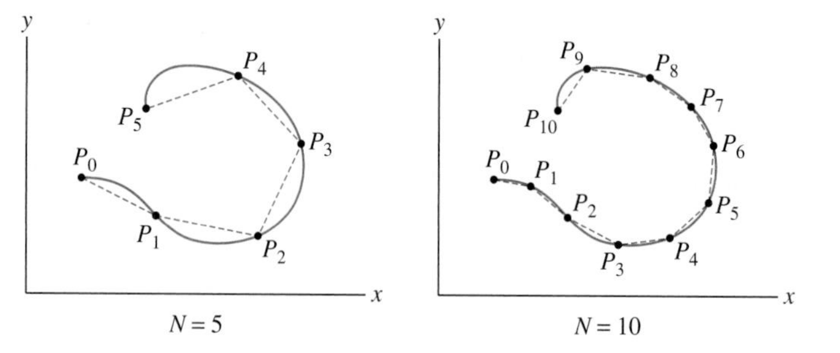

FIGURE 1 Polygonal approximations for $N = 5$ and $N = 10$.

intermediate values t_i^* and t_i^{**} in the interval $[t_{i-1}, t_i]$ such that

$$x(t_i) - x(t_{i-1}) = x'(t_i^*)\Delta t_i, \qquad y(t_i) - y(t_{i-1}) = y'(t_i^{**})\Delta t_i$$

where $\Delta t_i = t_i - t_{i-1}$. Therefore,

$$P_{i-1}P_i = \sqrt{x'(t_i^*)^2 \Delta t_i^2 + y'(t_i^{**})^2 \Delta t_i^2} = \sqrt{x'(t_i^*)^2 + y'(t_i^{**})^2}\,\Delta t_i$$

The length $|L|$ of L is the sum of the lengths of the segments:

$$|L| = \sum_{i=1}^{N} P_{i-1}P_i = \sum_{i=1}^{N} \sqrt{x'(t_i^*)^2 + y'(t_i^{**})^2}\,\Delta t_i \tag{3}$$

This is *nearly* a Riemann sum for the function $\sqrt{x'(t)^2 + y'(t)^2}$. It would be a true Riemann sum if the two intermediate values t_i^* and t_i^{**} were equal. Although they need not be equal, it can be shown (and we will take it for granted) that if $x'(t)$ and $y'(t)$ are continuous, then the sum in Eq. (3) still approaches the integral as $\|P\| \to 0$. Thus,

$$\lim_{\|P\| \to 0} \sum_{i=1}^{N} P_{i-1}P_i = \int_a^b \sqrt{x'(t)^2 + y'(t)^2}\,dt$$

> **THEOREM 1 Arc Length** Let $c(t) = (x(t), y(t))$ and assume that $x'(t)$ and $y'(t)$ exist and are continuous. Then the length s of $c(t)$ for $a \le t \le b$ is equal to
>
> $$s = \int_a^b \sqrt{x'(t)^2 + y'(t)^2}\,dt \tag{4}$$

It is often impossible to evaluate the arc length integral (4) explicitly, but we can always approximate it numerically.

Keep in mind that the length of a path $c(t)$ is the distance traveled by a particle following the path. This is greater than the length of the underlying curve if parts of the curve are traversed more than once.

■ **EXAMPLE 1 Length of a Circular Arc** Calculate the length s of the arc $0 \le \theta \le \theta_0$ of a circle of radius R.

Solution We use the parametrization $x = R\cos\theta$, $y = R\sin\theta$:

$$x'(\theta)^2 + y'(\theta)^2 = (-R\sin\theta)^2 + (R\cos\theta)^2 = R^2(\sin^2\theta + \cos^2\theta) = R^2$$

$$s = \int_0^{\theta_0} \sqrt{x'(\theta)^2 + y'(\theta)^2}\,d\theta = \int_0^{\theta_0} R\,d\theta = R\theta_0$$

This agrees with the standard formula for the length of a circular arc (Figure 2). ■

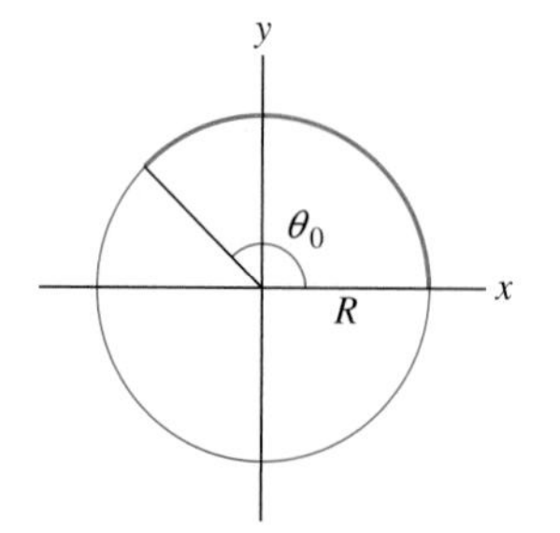

FIGURE 2 An arc of angle θ_0 on a circle of radius R has length $R\theta_0$.

It turns out that the cycloid is a curve whose arc length can be computed explicitly.

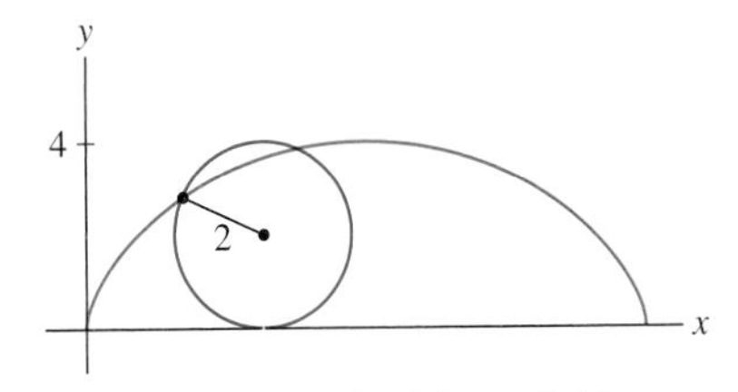

FIGURE 3 One arch of the cycloid generated by a circle of radius 2.

EXAMPLE 2 Length of the Cycloid Calculate the length s of one arch of the cycloid generated by a circle of radius $R = 2$ (Figure 3).

Solution By Eq. (6) in Section 1, the cycloid generated by a circle of radius $R = 2$ has parametric equations

$$x = 2(t - \sin t), \qquad y = 2(1 - \cos t)$$

Using the identity $\dfrac{1 - \cos t}{2} = \sin^2 \dfrac{t}{2}$, we find

$$\begin{aligned} x'(t)^2 + y'(t)^2 &= 2^2(1 - \cos t)^2 + 2^2 \sin^2 t \\ &= 4 - 8\cos t + 4\cos^2 t + 4\sin^2 t \\ &= 8 - 8\cos t = 16\left(\frac{1 - \cos t}{2}\right) = 16 \sin^2 \frac{t}{2} \end{aligned}$$

and thus,

$$\sqrt{x'(t)^2 + y'(t)^2} = 4\left|\sin \frac{t}{2}\right|$$

One arch of the cycloid is traced as t varies from 0 to 2π, so

$$s = \int_0^{2\pi} \sqrt{x'(t)^2 + y'(t)^2}\, dt = 4\int_0^{2\pi} \sin \frac{t}{2}\, dt = -8\cos \frac{t}{2}\bigg|_0^{2\pi} = -8(-1) + 8 = 16$$

Note that we may drop the absolute value in the integral over $[0, 2\pi]$ because $\sin \frac{t}{2} \geq 0$ for $0 \leq t \leq 2\pi$. ■

The arc length integral leads to an expression for the **speed** of a particle moving along a path $c(t) = (x(t), y(t))$. By definition, speed is the rate of change of distance traveled with respect to time. The distance traveled over the time interval $[t_0, t]$ is given by the arc length integral:

In Chapter 13, we will discuss not just the speed but also the velocity of a particle moving along a curved path. Velocity is "speed plus direction" and is represented by a "vector."

$$s(t) = \text{distance traveled} = \int_{t_0}^{t} \sqrt{x'(u)^2 + y'(u)^2}\, du$$

Therefore, by the Fundamental Theorem of Calculus,

$$\text{Speed} = \frac{ds}{dt} = \frac{d}{dt}\int_0^t \sqrt{x'(u)^2 + y'(u)^2}\, du = \sqrt{x'(t)^2 + y'(t)^2}$$

THEOREM 2 Speed along a Parametrized Path The speed at time t of a particle with trajectory $c(t) = (x(t), y(t))$ is the derivative of the arc length integral

$$s(t) = \int_{t_0}^{t} \sqrt{x'(u)^2 + y'(u)^2}\, du$$

Namely,

$$\text{Speed} = \frac{ds}{dt} = \sqrt{x'(t)^2 + y'(t)^2}$$

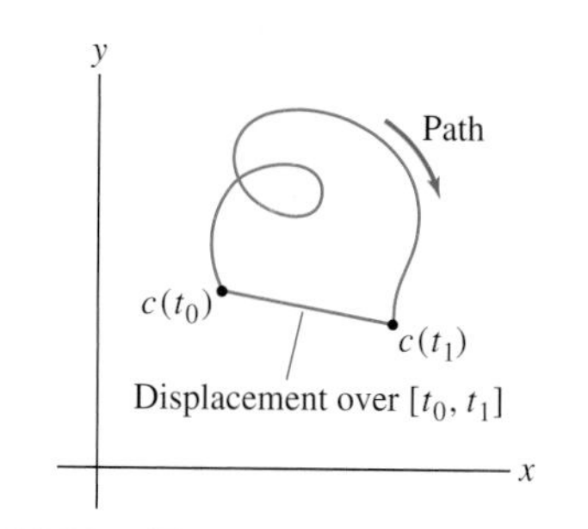

FIGURE 4 The distance traveled is greater than or equal to the displacement.

Before continuing, let us emphasize the distinction between distance traveled and **displacement** (also called net change in position). The displacement over a time interval $[t_0, t_1]$ is the distance between the initial point $c(t_0)$ and the endpoint $c(t_1)$. The distance traveled is greater than the displacement unless the particle happens to move in a straight line (Figure 4).

EXAMPLE 3 A particle travels along the path $c(t) = (2t, 1 + t^{3/2})$ (t in minutes, distance in feet).

(a) Find the speed at $t = 1$.

(b) Compute distance traveled s and displacement d during the first 4 min.

Solution We have $x'(t) = 2$, $y'(t) = \frac{3}{2}t^{1/2}$, so the particle's speed at time t is

$$s'(t) = \sqrt{x'(t)^2 + y'(t)^2} = \sqrt{4 + \frac{9}{4}t} \quad \text{ft/min}$$

The speed at $t = 1$ is $s'(1) = \sqrt{4 + \frac{9}{4}} = 2.5$ ft/min. The distance traveled during the first 4 min is equal to

$$s = \int_0^4 \sqrt{4 + \frac{9}{4}t}\, dt = \frac{8}{27}\left(4 + \frac{9}{4}t\right)^{3/2}\Bigg|_0^4 = \frac{8}{27}(13^{3/2} - 8) \approx 11.52 \text{ ft}$$

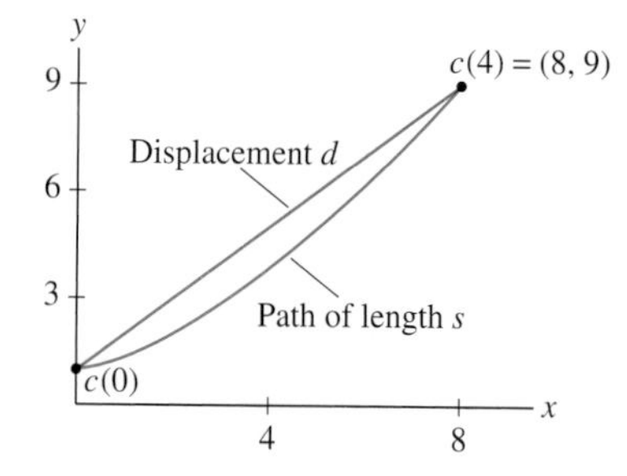

FIGURE 5 The path $c(t) = (2t, 1 + t^{3/2})$ for $0 \le t \le 4$.

The displacement d is the distance from the initial point $c(0) = (0, 1)$ to the endpoint $c(4) = (8, 1 + 4^{3/2}) = (8, 9)$ (see Figure 5):

$$d = \sqrt{(8 - 0)^2 + (9 - 1)^2} = 8\sqrt{2} \approx 11.31 \text{ ft}$$

EXAMPLE 4 Angular Velocity Let $c(t) = (R\cos\omega t, R\sin\omega t)$ be a circular path, where ω is a constant (called the **angular velocity**, in units of radians per unit time). Show that $c(t)$ is a path of constant speed. Find the angular velocity ω of a counterclockwise path if $R = 3$ m and the speed is 12 m/s.

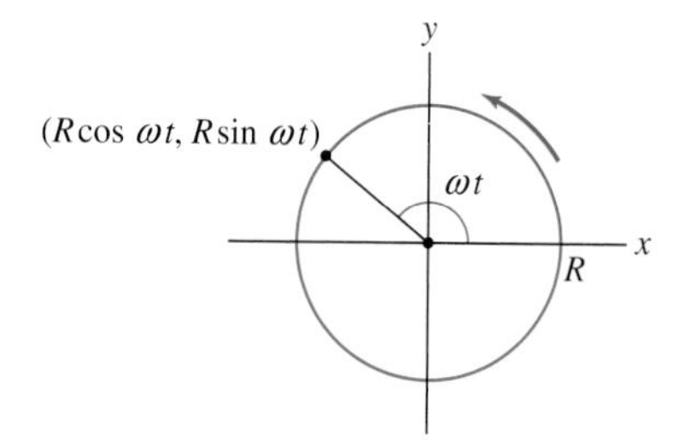

FIGURE 6 A particle rotating on a circular path with angular velocity ω has speed $|\omega R|$.

Solution We have $x = R\cos\omega t$ and $y = R\sin\omega t$, and

$$x'(t) = -\omega R\sin\omega t, \qquad y'(t) = \omega R\cos\omega t$$

We see that the speed is a constant independent of t:

$$\frac{ds}{dt} = \sqrt{x'(t)^2 + y'(t)^2} = \sqrt{(-\omega R\sin\omega t)^2 + (\omega R\cos\omega t)^2}$$

$$= \sqrt{\omega^2 R^2(\sin^2\omega t + \cos^2\omega t)} = |\omega|R$$

If $R = 3$ and the speed is 12 m/s, then $|\omega|R = 3|\omega| = 12$. Therefore $|\omega| = 4$ and $\omega = 4$ since the path is counterclockwise (Figure 6).

The surface area S of a surface of revolution obtained by rotating a parametric curve $c(t) = (x(t), y(t))$ about the x-axis may be computed by an integral that is closely related to the arc length integral. The formula, stated in the next theorem (without proof), may be justified using polygonal approximations as in the case of the arc length formula. Note that we assume $y(t) \ge 0$ so that the curve $c(t)$ lies above the x-axis.

THEOREM 3 Surface Area Let $c(t) = (x(t), y(t))$ and assume that $x'(t)$ and $y'(t)$ exist and are continuous. Assume further that $y(t) \geq 0$. Then the surface area S of the surface obtained by rotating $c(t)$ about the x-axis for $a \leq t \leq b$ is equal to

$$S = 2\pi \int_a^b y(t)\sqrt{x'(t)^2 + y'(t)^2}\, dt \qquad \boxed{5}$$

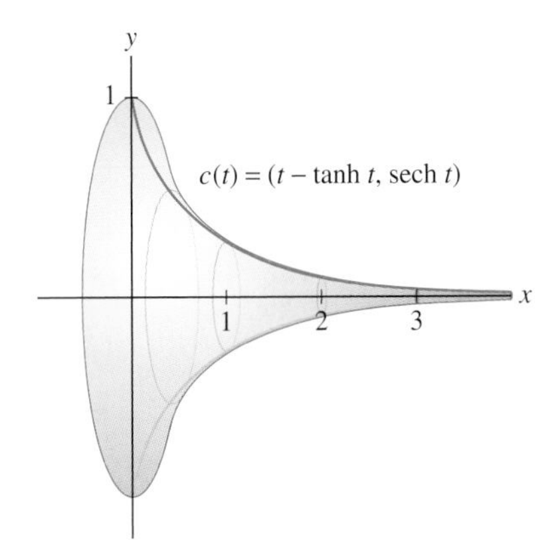

FIGURE 7 Surface generated by revolving the tractrix about the x-axis.

EXAMPLE 5 The curve $c(t) = (t - \tanh t, \operatorname{sech} t)$ is called a *tractrix*. Calculate the surface area of the infinite surface generated by revolving the tractrix about the x-axis for $0 \leq t < \infty$ (Figure 7).

Solution We have

$$x'(t) = \frac{d}{dt}(t - \tanh t) = 1 - \operatorname{sech}^2 t, \qquad y'(t) = \frac{d}{dt}\operatorname{sech} t = -\operatorname{sech} t \tanh t$$

Using the identities $1 - \operatorname{sech}^2 t = \tanh^2 t$ and $\operatorname{sech}^2 t = 1 - \tanh^2 t$, we obtain

$$\begin{aligned} x'(t)^2 + y'(t)^2 &= (1 - \operatorname{sech}^2 t)^2 + (-\operatorname{sech} t \tanh t)^2 \\ &= (\tanh^2 t)^2 + (1 - \tanh^2 t)\tanh^2 t = \tanh^2 t \end{aligned}$$

REMINDER

$$\operatorname{sech} t = \frac{1}{\cosh t} = \frac{2}{e^t + e^{-t}}$$

$$1 - \operatorname{sech}^2 t = \tanh^2 t$$

$$\frac{d}{dt}\tanh t = \operatorname{sech}^2 t$$

$$\frac{d}{dt}\operatorname{sech} t = -\operatorname{sech} t \tanh t$$

$$\int \operatorname{sech} t \tanh t\, dt = -\operatorname{sech} t + C$$

The surface area is given by an improper integral, which we evaluate using the integral formula recalled in the margin:

$$\begin{aligned} S &= 2\pi \int_0^\infty \operatorname{sech} t \sqrt{\tanh^2 t}\, dt = 2\pi \int_0^\infty \operatorname{sech} t \tanh t\, dt = 2\pi \lim_{R\to\infty} \int_0^R \operatorname{sech} t \tanh t\, dt \\ &= 2\pi \lim_{R\to\infty} (-\operatorname{sech} t)\Big|_0^R = 2\pi \lim_{R\to\infty} (\operatorname{sech} 0 - \operatorname{sech} R) = 2\pi \operatorname{sech} 0 = 2\pi \end{aligned}$$

Here we use that $\operatorname{sech} t = \dfrac{2}{e^t + e^{-t}}$ tends to 0 as $t \to \infty$ because the term e^t in the denominator tends to ∞. ■

11.2 SUMMARY

- The *arc length* s of a path $c(t) = (x(t), y(t))$ for $a \leq t \leq b$ is

$$s = \text{arc length} = \int_a^b \sqrt{x'(t)^2 + y'(t)^2}\, dt$$

- The *arc length integral* $s(t) = \int_{t_0}^t \sqrt{x'(u)^2 + y'(u)^2}\, du$ is equal to the *distance traveled* over the time interval $[t_0, t]$. The *displacement* is the distance between the starting point $c(t_0)$ and endpoint $c(t)$.
- The speed at time t of a particle with trajectory $c(t) = (x(t), y(t))$ is

$$\frac{ds}{dt} = \sqrt{x'(t)^2 + y'(t)^2}$$

11.2 EXERCISES

Preliminary Questions

1. What is the definition of arc length?

2. What is the interpretation of $\sqrt{x'(t)^2 + y'(t)^2}$ for a particle following the trajectory $(x(t), y(t))$?

3. A particle travels along a path from $(0, 0)$ to $(3, 4)$. What is the displacement? Can the distance traveled be determined from the information given?

4. A particle traverses the parabola $y = x^2$ with constant speed 3 cm/s. What is the distance traveled during the first minute? *Hint:* No computation is necessary.

Exercises

1. Use Eq. (4) to calculate the length of the semicircle

$$x = 3\sin t, \quad y = 3\cos t, \quad 0 \le t \le \pi$$

2. Find the speed at $t = 4$ s of a particle whose position at time t seconds is $c(t) = (4 - t^2, t^3)$.

In Exercises 3–12, use Eq. (4) to find the length of the path over the given interval.

3. $(3t + 1, 9 - 4t), \quad 0 \le t \le 2$

4. $(1 + 2t, 2 + 4t), \quad 1 \le t \le 4$

5. $(2t^2, 3t^2 - 1), \quad 0 \le t \le 4$

6. $(3t, 4t^{3/2}), \quad 0 \le t \le 1$

7. $(3t^2, 4t^3), \quad 1 \le t \le 4$

8. $(t^3 + 1, t^2 - 3), \quad 0 \le t \le 1$

9. $(\sin 3t, \cos 3t), \quad 0 \le t \le \pi$

10. $(2\cos t - \cos 2t, 2\sin t - \sin 2t), \quad 0 \le t \le \frac{\pi}{2}$

11. $(\sin\theta - \theta\cos\theta, \cos\theta + \theta\sin\theta), \quad 0 \le \theta \le 2$

12. $(5(\theta - \sin\theta), 5(1 - \cos\theta)), \quad 0 \le \theta \le 2\pi$

13. Show that one arch of a cycloid generated by a circle of radius R has length $8R$.

14. Find the length of the tractrix (Figure 22 in Section 11.1) with parametrization

$$c(t) = (t - \tanh(t), \operatorname{sech}(t)), \quad 0 \le t \le A$$

15. Find the length of the spiral $c(t) = (t\cos t, t\sin t)$ for $0 \le t \le 2\pi$ to three decimal places (Figure 8). *Hint:* Use the formula

$$\int \sqrt{1 + t^2}\, dt = \frac{1}{2}t\sqrt{1 + t^2} + \frac{1}{2}\ln\left(t + \sqrt{1 + t^2}\right)$$

In Exercises 16–19, determine the speed $s(t)$ of a particle with a given trajectory at time t_0 (in units of meters and seconds).

16. $(t^3, t^2), \quad t = 2$

17. $(3\sin 5t, 8\cos 5t), \quad t = \frac{\pi}{4}$

18. $(5t + 1, 4t - 3), \quad t = 9$

19. $(\ln(t^2 + 1), t^3), \quad t = 1$

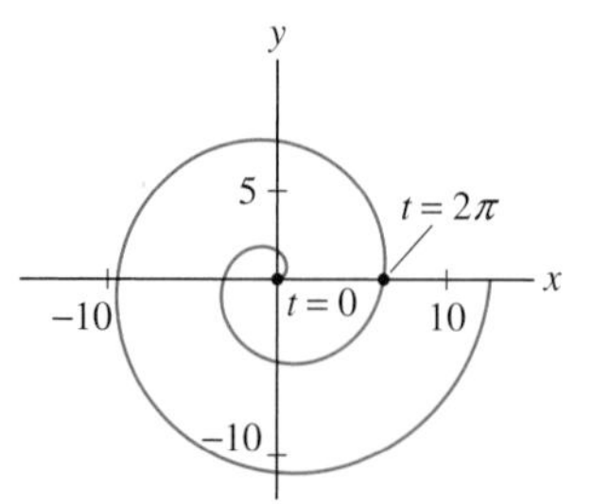

FIGURE 8 The spiral $c(t) = (t\cos t, t\sin t)$.

20. Find the speed of a particle whose path around a circle of radius r is described by the parametric curve $c(t) = (r\cos\omega t, r\sin\omega t)$.

21. Find the minimum speed of a particle with trajectory $c(t) = (t^3 - 4t, t^2 + 1)$ for $t \ge 0$. *Hint:* It is easier to find the minimum of the square of the speed.

22. Find the minimum speed of a particle with trajectory $c(t) = (t^3, t^{-2})$ for $t \ge 0.5$.

23. Find the speed of the cycloid $c(t) = (4t - 4\sin t, 4 - 4\cos t)$ at points where the tangent line is horizontal.

24. If you unwind thread from a stationary circular spool, keeping the thread taut at all times, then the endpoint traces a curve C called the **involute** of the circle (Figure 9). Observe that $\overline{PQ}$ has length $R\theta$. Show that C is parametrized by

$$c(\theta) = \big(R(\cos\theta + \theta\sin\theta), R(\sin\theta - \theta\cos\theta)\big)$$

Then find the length of the involute for $0 \le \theta \le 2\pi$.

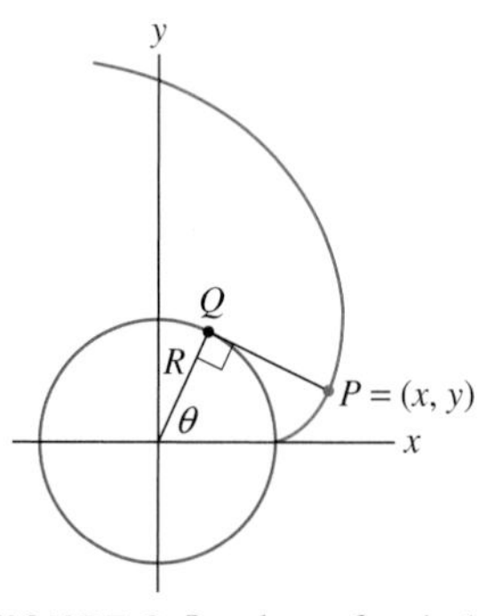

FIGURE 9 Involute of a circle.

CAS *In Exercises 25–28, plot the curve and use the Midpoint Rule with N = 10, 20, 30, and 50 to approximate its length.*

25. $c(t) = (\cos t, e^{\sin t})$ for $0 \le t \le 2\pi$

26. $c(t) = (t - \sin 2t, 1 - \cos 2t)$ for $0 \le t \le 2\pi$

27. The ellipse $\left(\frac{x}{5}\right)^2 + \left(\frac{y}{3}\right)^2 = 1$

28. $x = \sin 2t, \quad y = \sin 3t$ for $0 \le t \le 2\pi$

29. Let $a > b$ and set $k = \sqrt{1 - \frac{b^2}{a^2}}$. Use a parametric representation to show that the ellipse $\left(\frac{x}{a}\right)^2 + \left(\frac{y}{b}\right)^2 = 1$ has length $L = 4aG\left(\frac{\pi}{2}, k\right)$, where

$$G(\theta, k) = \int_0^\theta \sqrt{1 - k^2 \sin^2 t}\, dt$$

is the *elliptic integral of the second kind.*

In Exercises 30–33, use Eq. (5) to compute the surface area of the given surface.

30. The cone generated by revolving $c(t) = (t, mt)$ about the x-axis for $0 \le t \le A$

31. The surface generated by revolving the astroid with parametrization $c(t) = (\cos^3 t, \sin^3 t)$ about the x-axis for $0 \le t \le \frac{\pi}{2}$

32. A sphere of radius R

33. The surface generated by revolving one arch of the cycloid $c(t) = (t - \sin t, 1 - \cos t)$ about the x-axis

Further Insights and Challenges

34. CAS Let $b(t)$ be the "Butterfly Curve":

$$x(t) = \sin t \left(e^{\cos t} - 2\cos 4t - \sin\left(\frac{t}{12}\right)^5\right)$$

$$y(t) = \cos t \left(e^{\cos t} - 2\cos 4t - \sin\left(\frac{t}{12}\right)^5\right)$$

(a) Use a computer algebra system to plot $b(t)$ and the speed $s(t)$ for $0 \le t \le 12\pi$.

(b) Approximate the length $b(t)$ for $0 \le t \le 10\pi$.

35. CAS Let $a \ge b > 0$ and set $k = \frac{2\sqrt{ab}}{a - b}$. Show that the **trochoid**

$$x = at - b\sin t, \qquad y = a - b\cos t, \qquad 0 \le t \le T$$

has length $2(a - b)G\left(\frac{T}{2}, k\right)$ with $G(\theta, k)$ as in Exercise 29.

36. The path of a satellite orbiting at a distance R from the center of the earth is parametrized by $x = R\cos \omega t$, $y = R\sin \omega t$.

(a) Show that the period T (the time of one revolution) is $T = 2\pi/\omega$.

(b) According to Newton's laws of motion and gravity,

$$x''(t) = -Gm_e \frac{x}{R^3}, \qquad y''(t) = -Gm_e \frac{y}{R^3}$$

where G is the universal gravitational constant and m_e is the mass of the earth. Prove that $\frac{R^3}{T^2} = \frac{Gm_e}{4\pi^2}$. Thus, $\frac{R^3}{T^2}$ has the same value for all orbits (a special case of Kepler's Third Law).

37. The acceleration due to gravity on the surface of the earth is $g = \frac{Gm_e}{R_e^2} = 9.8 \text{ m/s}^2$, where $R_e = 6{,}378$ km. Use Exercise 36(b) to show that a satellite orbiting at the earth's surface would have period $T_e = 2\pi\sqrt{R_e/g} \approx 84.5$ min. Then estimate the distance R_m from the moon to the center of the earth. Assume that the period of the moon (sidereal month) is $T_m \approx 27.43$ days.

11.3 Polar Coordinates

Polar coordinates are an alternative system of labeling points P in the plane. Instead of specifying the x- and y-coordinates, we specify coordinates (r, θ), where r is the distance from P to the origin O and θ is the angle between $\overline{OP}$ and the positive x-axis (Figure 1). By convention, an angle is positive if the corresponding rotation is counterclockwise. We call r the **radial coordinate** and θ the **angular coordinate**.

Polar coordinates are useful for problems in which the distance from the origin or angle plays a role, especially if the problem involves rotational symmetry. They are a natural choice for describing the gravitational force exerted by an object such as the sun, since the magnitude of the force depends only on distance from the sun.

For example, the point P in Figure 2 has polar coordinates $(4, \frac{2\pi}{3})$. It is located at a distance 4 from the origin (so it lies on the circle of radius 4) and it lies on the ray of angle $\frac{2\pi}{3}$.

To relate polar and rectangular coordinates, we refer to Figure 1. We see that if P has polar coordinates (r, θ), then its rectangular coordinates are $x = r\cos\theta$ and $y = r\sin\theta$. On the other hand, $r^2 = x^2 + y^2$ by the distance formula and $\tan\theta = \frac{y}{x}$ if $x \ne 0$.

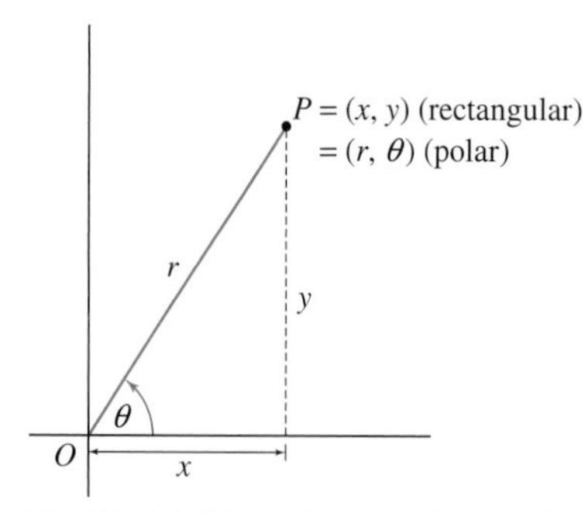

FIGURE 1 The polar coordinates of P satisfy $r = \sqrt{x^2 + y^2}$, $\tan\theta = y/x$.

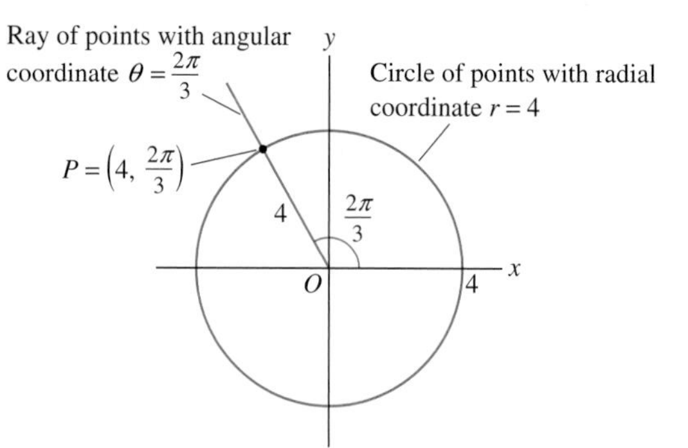

FIGURE 2 The point P lies at the intersection of the circle $r = 4$ and the ray $\theta = \frac{2\pi}{3}$.

Polar to Rectangular	Rectangular to Polar
$x = r\cos\theta$	$r = \sqrt{x^2 + y^2}$
$y = r\sin\theta$	$\tan\theta = \dfrac{y}{x}$

A few remarks are in order before proceeding.

- The angular coordinate θ is not uniquely determined because the polar coordinates (r, θ) and $(r, \theta + 2\pi n)$ *label the same point* for any integer n. For instance, the point $(x, y) = (0, 2)$ has polar coordinates $\left(2, \frac{\pi}{2}\right)$ and $\left(2, \frac{5\pi}{2}\right)$ and $\left(2, \frac{9\pi}{2}\right)$, etc. (see Figure 3).
- The origin O does not have a well-defined angular coordinate, so we assign to O the polar coordinates $(0, \theta)$ for any angle θ.
- By convention, we also allow *negative* radial coordinates. For $r > 0$, the point $(-r, \theta)$ is defined to be the reflection of (r, θ) through the origin (see Figure 4). With this convention, $(-r, \theta)$ and $(r, \theta + \pi)$ represent the same point.
- We may specify unique polar coordinates for points other than the origin by placing restrictions on r and θ. We commonly choose $r > 0$ and $0 \le \theta < 2\pi$.

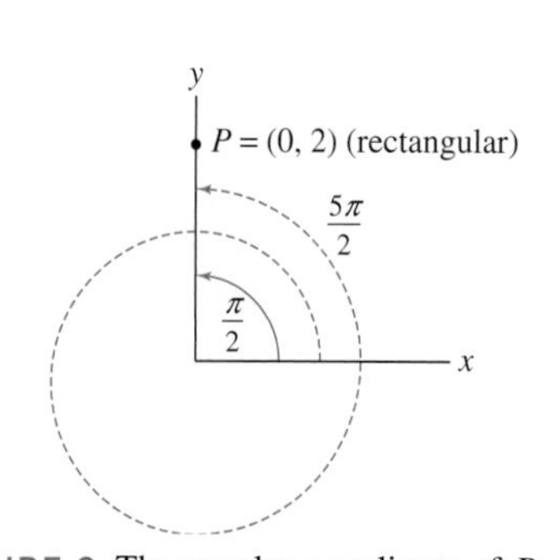

FIGURE 3 The angular coordinate of P can take on the value $\frac{\pi}{2} + 2\pi n$ for any integer n.

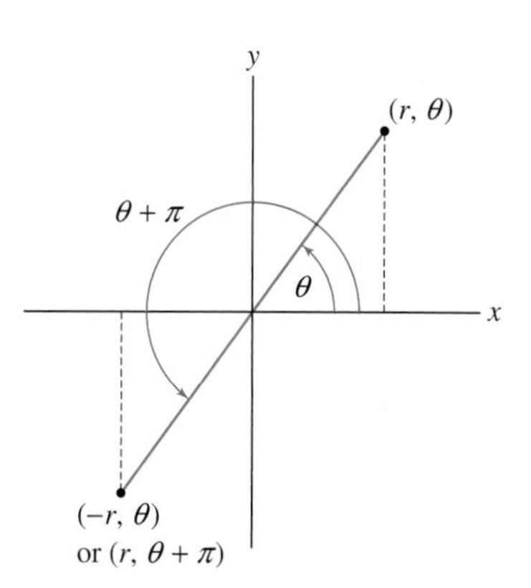

FIGURE 4 Relation between (r, θ) and $(-r, \theta)$.

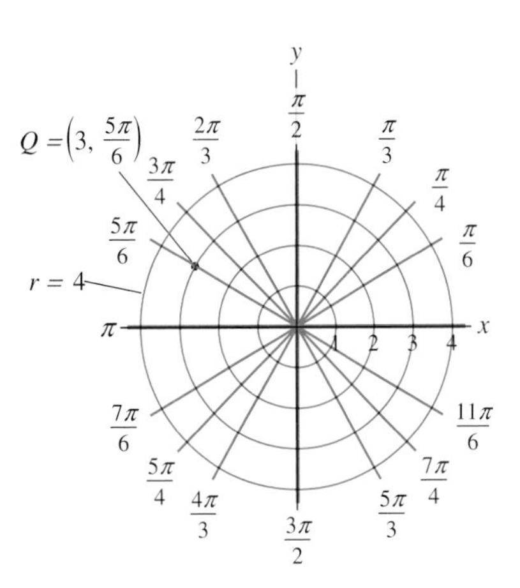

FIGURE 5 Grid lines in polar coordinates.

Figure 5 shows the two families of **grid lines** in polar coordinates:

$$\text{Circle centered at } O \quad \longleftrightarrow \quad r = \text{constant}$$

$$\text{Ray starting at } O \quad \longleftrightarrow \quad \theta = \text{constant}$$

Every point in the plane other than the origin lies at the intersection of two grid lines that determine its polar coordinates. For example, the point Q in Figure 5 lies on the circle $r = 3$ and the ray $\theta = \frac{5\pi}{6}$ so $Q = (3, \frac{5\pi}{6})$ in polar coordinates.

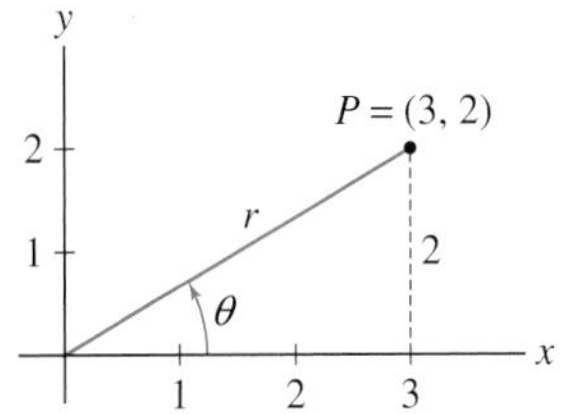

FIGURE 6 The polar coordinates of P satisfy $r = \sqrt{3^2 + 2^2}$ and $\tan\theta = \frac{2}{3}$.

■ **EXAMPLE 1 From Polar to Rectangular Coordinates** Find the rectangular coordinates of the point Q with polar coordinates $(r, \theta) = (3, \frac{5\pi}{6})$.

Solution The point $Q = (r, \theta) = (3, \frac{5\pi}{6})$ has rectangular coordinates:

$$x = r\cos\theta = 3\cos\left(\frac{5\pi}{6}\right) = 3\left(-\frac{\sqrt{3}}{2}\right) = -\frac{3\sqrt{3}}{2}$$

$$y = r\sin\theta = 3\sin\left(\frac{5\pi}{6}\right) = 3\left(\frac{1}{2}\right) = \frac{3}{2}$$ ■

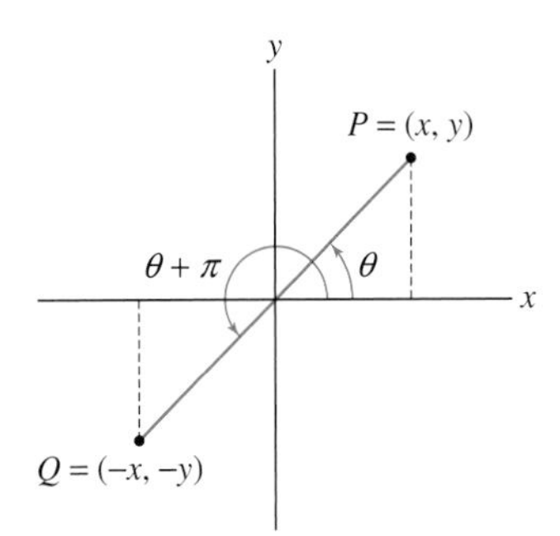

FIGURE 7

■ **EXAMPLE 2 From Rectangular to Polar Coordinates** Find polar coordinates of the point P with rectangular coordinates $(x, y) = (3, 2)$.

Solution Since $P = (x, y) = (3, 2)$,

$$r = \sqrt{x^2 + y^2} = \sqrt{3^2 + 2^2} = \sqrt{13} \approx 3.6$$

$$\tan\theta = \frac{y}{x} = \frac{2}{3}$$

$$\theta = \tan^{-1}\frac{y}{x} = \tan^{-1}\frac{2}{3} \approx 0.588$$

Thus, P has polar coordinates $(r, \theta) \approx (3.6, 0.588)$ (see Figure 6). ■

By definition, the arctangent satisfies

$$-\frac{\pi}{2} < \tan^{-1} x < \frac{\pi}{2}$$

An angular coordinate θ of $P = (x, y)$ is

$$\theta = \begin{cases} \tan^{-1}\frac{y}{x} & \text{if } x > 0 \\ \tan^{-1}\frac{y}{x} + \pi & \text{if } x < 0 \\ \pm\frac{\pi}{2} & \text{if } x = 0 \end{cases}$$

The next example calls attention to a subtlety in determining the angular coordinate θ of a point $P = (x, y)$. There are two angles between 0 and 2π satisfying $\tan\theta = \frac{y}{x}$ because $\tan(\theta + \pi) = \tan\theta$. We must choose θ so that (r, θ) lies in the quadrant containing P and not its reflection through the origin Q (Figure 7).

FIGURE 8

EXAMPLE 3 Choosing θ Correctly Find two polar representations of $P = (-1, 1)$, one using a positive radial coordinate and one using a negative radial coordinate.

Solution The positive radial coordinate of $P = (x, y) = (-1, 1)$ is

$$r = \sqrt{(-1)^2 + 1^2} = \sqrt{2}$$

To find the angular coordinate, we solve

$$\tan\theta = \frac{y}{x} = -1 \quad \Rightarrow \quad \theta = \frac{3\pi}{4}, \quad \frac{3\pi}{4} + \pi = \frac{7\pi}{4}$$

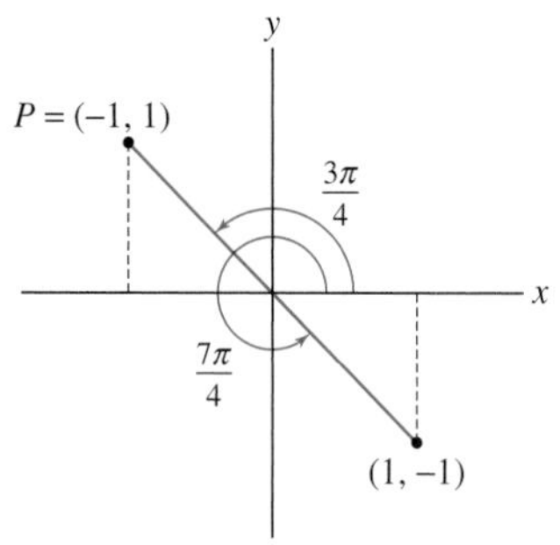

Since P lies in the second quadrant, the appropriate angular coordinate of P is $\theta = \frac{3\pi}{4}$ (Figure 8). Alternatively, P may be represented with the negative radial coordinate $r = -\sqrt{2}$ and angle $\theta = \frac{7\pi}{4}$. Thus,

$$P = \left(\sqrt{2}, \frac{3\pi}{4}\right) \quad \text{or} \quad \left(-\sqrt{2}, \frac{7\pi}{4}\right)$$

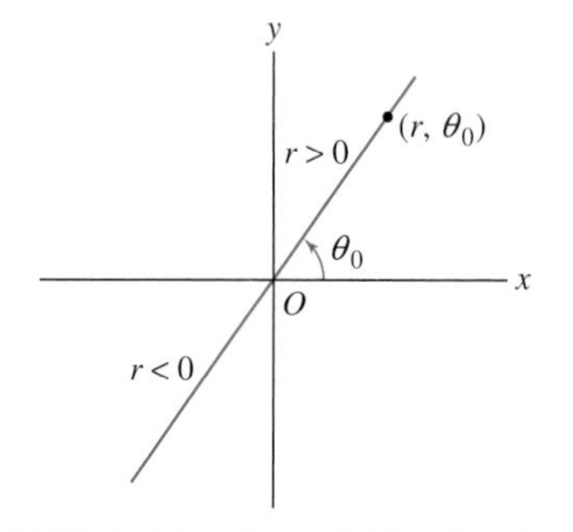

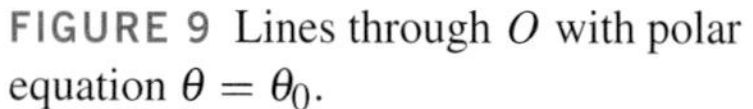

FIGURE 9 Lines through O with polar equation $\theta = \theta_0$.

Curves are described in polar coordinates by equations relating r and θ. We refer to an equation in polar coordinates as a **polar equation** to avoid confusion with the usual equation in rectangular coordinates. By convention, we allow solutions with negative r when considering equations in polar coordinates.

The polar equation of a line through the origin O has the simple form $\theta = \theta_0$, where θ_0 is the angle between the line and the x-axis (Figure 9). Indeed, the solutions of $\theta = \theta_0$ are (r, θ_0), where r is arbitrary (positive, negative, or zero).

EXAMPLE 4 Line Through the Origin Find the polar equation of the line through the origin of slope $\frac{3}{2}$.

Solution A line of slope m makes an angle θ_0 with the x-axis, where $m = \tan\theta_0$. In our case, Figure 10 shows that $\tan\theta_0 = \frac{3}{2}$ and so $\theta_0 = \tan^{-1}\frac{3}{2} \approx 0.98$. The equation of the line is $\theta = \tan^{-1}\frac{3}{2}$.

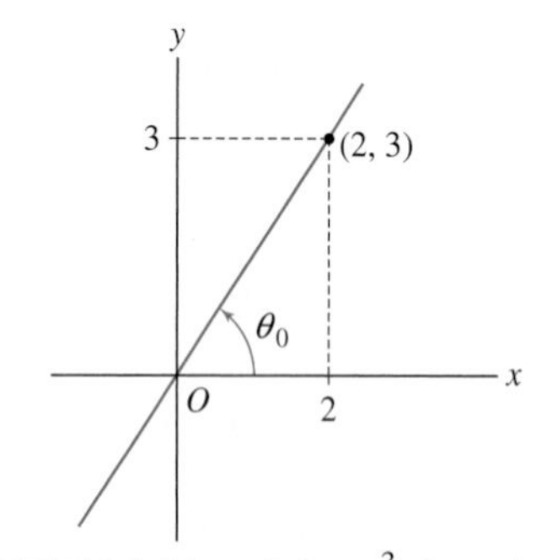

FIGURE 10 Line of slope $\frac{3}{2}$ through the origin.

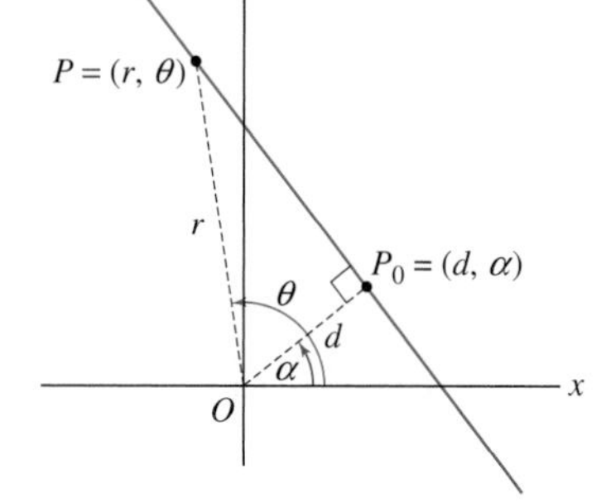

FIGURE 11 Line not passing through the origin.

To describe a line not passing through the origin in polar coordinates, the key is to consider the point P_0 on the line that is closest to the origin (Figure 11).

■ **EXAMPLE 5 Line Not Passing Through the Origin** Let $P_0 = (d, \alpha)$ be polar coordinates of the point on a line $\mathcal{L}$ closest to the origin. Show that $\mathcal{L}$ has polar equation

$$\boxed{r = d\sec(\theta - \alpha)} \qquad \boxed{1}$$

Solution The point P_0 on $\mathcal{L}$ closest to the origin is obtained by dropping a perpendicular from the origin to $\mathcal{L}$ (Figure 11). Therefore, if $P = (r, \theta)$ is any point on $\mathcal{L}$ other than P_0, then $\triangle OPP_0$ is a right triangle and as we see in Figure 11, $\dfrac{d}{r} = \cos(\theta - \alpha)$. We obtain the equation $r = d\sec(\theta - \alpha)$ as claimed. ■

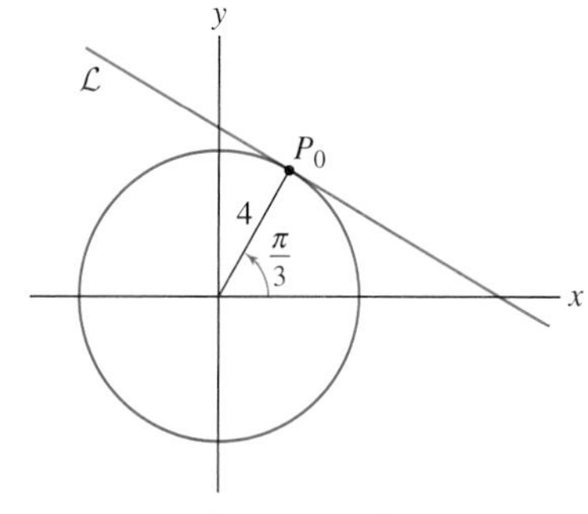

FIGURE 12 The tangent line has equation $r = 4\sec\left(\theta - \dfrac{\pi}{3}\right)$.

■ **EXAMPLE 6** Find the polar equation of the line $\mathcal{L}$ tangent to the circle $r = 4$ at the point with polar coordinates $P_0 = (4, \frac{\pi}{3})$.

Solution Since $\mathcal{L}$ is tangent to the circle at P_0, all points $P \neq P_0$ on $\mathcal{L}$ lie outside the circle. Therefore, P_0 is the point on $\mathcal{L}$ closest to the origin (Figure 12). We take $(d, \alpha) = (4, \frac{\pi}{3})$ in Eq. (1) to obtain the equation $r = 4\sec(\theta - \frac{\pi}{3})$. ■

It is often difficult to guess the shape of a graph of a polar equation. One way of sketching the graph of an equation $r = f(\theta)$ is to rewrite the equation in rectangular coordinates.

■ **EXAMPLE 7 Converting to Rectangular Coordinates** Identify the curve with polar equation $r = 2a\cos\theta$ (a a constant).

Solution Substitute $r = \sqrt{x^2 + y^2}$ and $\cos\theta = \dfrac{x}{r} = \dfrac{x}{\sqrt{x^2 + y^2}}$ in the polar equation:

$$r = 2a\cos\theta \quad \Rightarrow \quad \sqrt{x^2 + y^2} = 2a\frac{x}{\sqrt{x^2 + y^2}} \quad \Rightarrow \quad x^2 + y^2 = 2ax$$

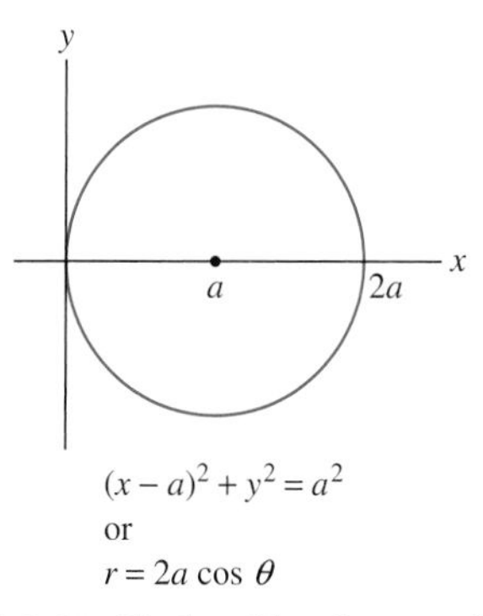

FIGURE 13 Circle with polar equation $r = 2a\cos\theta$.

Now complete the square:

$$x^2 - 2ax + y^2 = 0$$

$$(x - a)^2 - a^2 + y^2 = 0 \qquad \text{or} \qquad \boxed{(x - a)^2 + y^2 = a^2}$$

Thus, $r = 2a\cos\theta$ is the circle of radius a and center $(a, 0)$ (Figure 13). ■

A similar calculation shows that $x^2 + (y - a)^2 = a^2$ has polar equation $r = 2a\sin\theta$. In the next example, we make use of symmetry. Note that the points (r, θ) and $(r, -\theta)$ are symmetric about the x-axis (Figure 14).

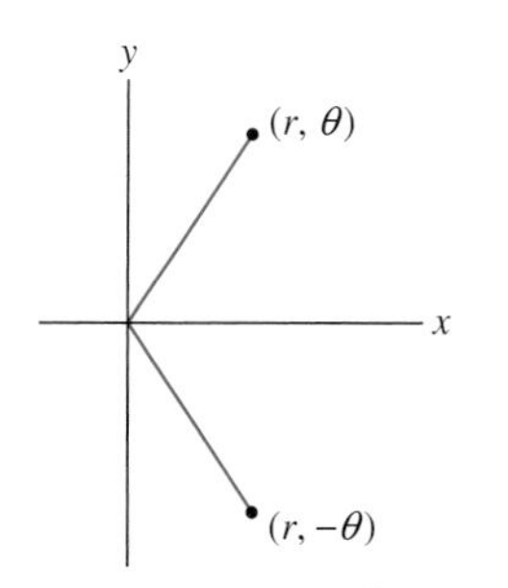

FIGURE 14 The points (r, θ) and $(r, -\theta)$ are symmetric with respect to the x-axis.

■ **EXAMPLE 8 The *Limaçon*: Symmetry About the x-Axis** Sketch the graph of $r = 2\cos\theta - 1$.

Solution To get started, we plot the points A–G on a grid and join them by a smooth curve (Figure 15). However, for a better understanding, it is helpful to graph r as a function of θ in rectangular coordinates. Figure 16(A) shows that

As θ varies from 0 to $\frac{\pi}{3}$, $\quad r$ varies from 1 to 0

As θ varies from $\frac{\pi}{3}$ to π, $\quad r$ is *negative* and varies from 0 to -3

	A	B	C	D	E	F	G
θ	0	$\pi/6$	$\pi/3$	$\pi/2$	$2\pi/3$	$5\pi/6$	π
r	1	0.73	0	-1	-2	-2.73	-3

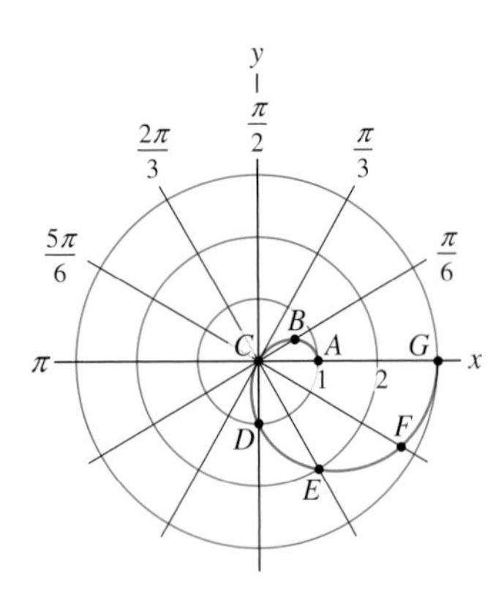

FIGURE 15 Plotting $r = 2\cos\theta - 1$ using a grid.

Since $r(\theta) = 2\cos\theta - 1$ is periodic, it suffices to plot points for $-\pi \le \theta \le \pi$.

- The graph begins at point A with polar coordinates $(r, \theta) = (1, 0)$ and moves in toward the origin as θ varies from 0 to $\pi/3$ [Figure 16(B)].
- For $\pi/3 \le \theta \le \pi$, r is negative. Therefore the curve continues into the third and fourth quadrants (rather than into the first and second quadrants), moving toward the point G with polar coordinates $(r, \theta) = (-3, \pi)$ as in Figure 16(C).
- Since $r(\theta) = r(-\theta)$, the curve is symmetric with respect to the x-axis. As θ varies from 0 to $-\pi$, we obtain the reflection through the x-axis of the first half of the curve, where $0 \le \theta \le \pi$, as in Figure 16(D). ■

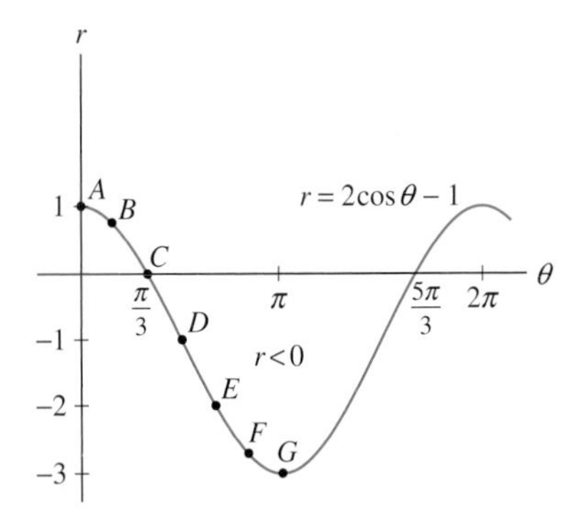

(A) Variation of r as a function of θ.

(B) As θ varies from 0 to $\pi/3$, r varies from 1 to 0.

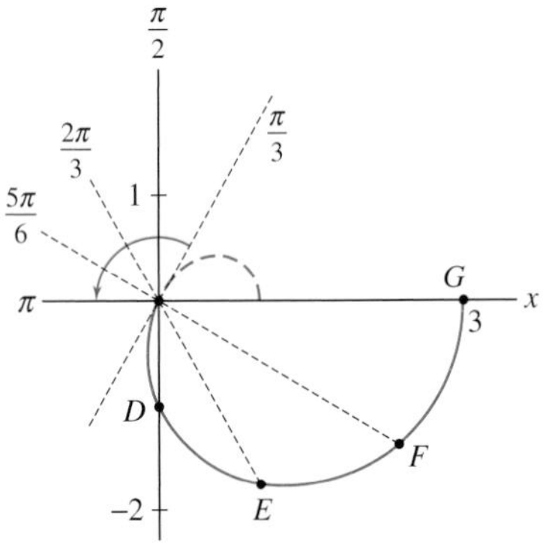

(C) As θ varies from $\pi/3$ to π, r is negative and varies from 0 to -3.

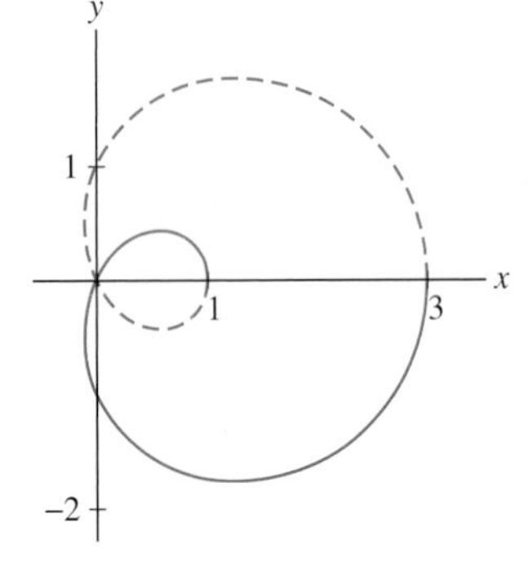

(D) The entire limaçon.

FIGURE 16 The curve $r = 2\cos\theta - 1$ is called the *limaçon*, from the Latin word for "snail." It was first described in 1525 by the German artist Albrecht Dürer.

11.3 SUMMARY

- A point P in the plane has polar coordinates (r, θ), where r is the distance from P to the origin and θ is the angle between the positive x-axis and the segment $\overline{OP}$, measured in the counterclockwise direction. The rectangular coordinates of P are

$$x = r\cos\theta, \qquad y = r\sin\theta$$

- If P has rectangular coordinates (x, y), then

$$r = \sqrt{x^2 + y^2}, \qquad \tan\theta = \frac{y}{x}$$

The angular coordinate θ must be chosen so that (r, θ) lies in the proper quadrant. We have

$$\theta = \begin{cases} \tan^{-1}\dfrac{y}{x} & \text{if } x > 0 \\ \tan^{-1}\dfrac{y}{x} + \pi & \text{if } x < 0 \\ \pm\dfrac{\pi}{2} & \text{if } x = 0 \end{cases}$$

- Nonuniqueness: (r, θ) and $(r, \theta + 2n\pi)$ represent the same point for all integers n. The origin O has polar coordinates $(0, \theta)$ for any θ.
- Negative radial coordinates: $(-r, \theta)$ and $(r, \theta + \pi)$ represent the same point.
- Polar equations:

Curve	**Polar Equation**
Circle of radius R, center at the origin	$r = R$
Line through origin of slope $m = \tan\theta_0$	$\theta = \theta_0$
Line on which $P_0 = (d, \alpha)$ is the point closest to the origin	$r = d\sec(\theta - \alpha)$
Circle of radius a, center at $(a, 0)$ $(x-a)^2 + y^2 = a^2$	$r = 2a\cos\theta$
Circle of radius a, center at $(0, a)$ $x^2 + (y-a)^2 = a^2$	$r = 2a\sin\theta$

11.3 EXERCISES

Preliminary Questions

1. If P and Q have the same radial coordinate, then (choose the correct answer):

(a) P and Q lie on the same circle with the center at the origin.
(b) P and Q lie on the same ray based at the origin.

2. Give two polar coordinate representations for the point $(x, y) = (0, 1)$, one with negative r and one with positive r.

3. Does a point (r, θ) have more than one representation in rectangular coordinates?

4. Describe the curves with polar equations

(a) $r = 2$ **(b)** $r^2 = 2$ **(c)** $r\cos\theta = 2$

5. If $f(-\theta) = f(\theta)$, then the curve $r = f(\theta)$ is symmetric with respect to the (choose the correct answer):

(a) x-axis **(b)** y-axis **(c)** origin

Exercises

1. Find polar coordinates for each of the seven points plotted in Figure 17.

2. Plot the points with polar coordinates

(a) $(2, \frac{\pi}{6})$ **(b)** $(4, \frac{3\pi}{4})$ **(c)** $(3, -\frac{\pi}{2})$ **(d)** $(0, \frac{\pi}{6})$

3. Convert from rectangular to polar coordinates:

(a) $(1, 0)$ **(b)** $(3, \sqrt{3})$
(c) $(-2, 2)$ **(d)** $(-1, \sqrt{3})$

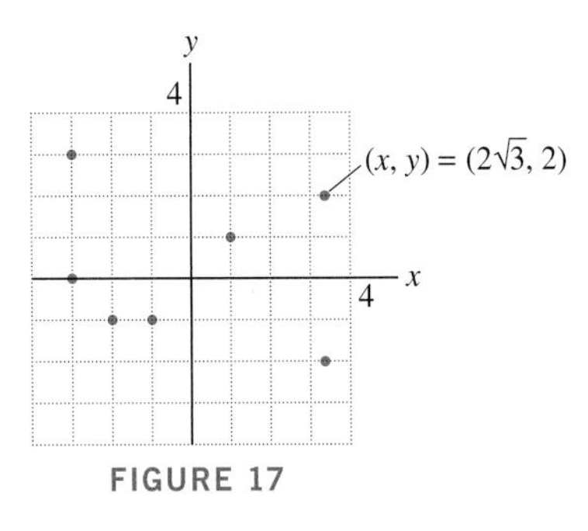

FIGURE 17

4. Use a calculator to convert from rectangular to polar coordinates (make sure your choice of θ gives the correct quadrant):

(a) $(2, 3)$ **(b)** $(4, -7)$
(c) $(-3, -8)$ **(d)** $(-5, 2)$

5. Convert from polar to rectangular coordinates:

(a) $(3, \frac{\pi}{6})$ **(b)** $(6, \frac{3\pi}{4})$ **(c)** $(5, -\frac{\pi}{2})$

6. Convert from polar to rectangular coordinates:

(a) $(0, 0)$ **(b)** $(-4, \frac{\pi}{3})$ **(c)** $(0, \frac{\pi}{6})$

7. Which of the following are possible polar coordinates for the point P with rectangular coordinates $(0, -2)$?

(a) $\left(2, \dfrac{\pi}{2}\right)$ **(b)** $\left(2, \dfrac{7\pi}{2}\right)$

(c) $\left(-2, -\dfrac{3\pi}{2}\right)$ **(d)** $\left(-2, \dfrac{7\pi}{2}\right)$

(e) $\left(-2, -\dfrac{\pi}{2}\right)$ **(f)** $\left(2, -\dfrac{7\pi}{2}\right)$

8. Describe each shaded sector in Figure 18 by inequalities in r and θ.

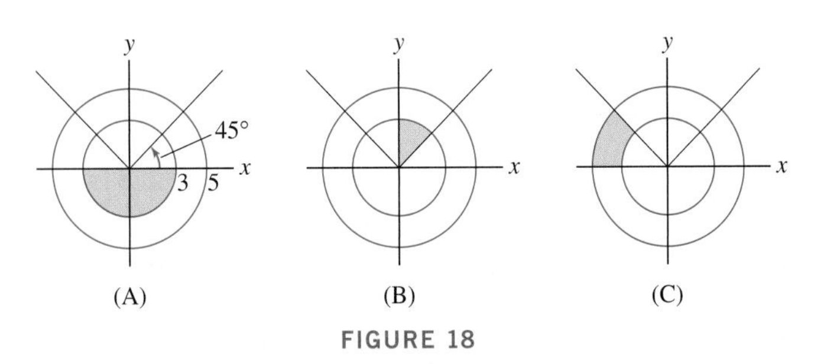

FIGURE 18

9. Find the equation in polar coordinates of the line through the origin with slope $\frac{1}{2}$.

10. What is the slope of the line $\theta = \frac{3\pi}{5}$?

11. Which of the two equations, $r = 2\sec\theta$ and $r = 2\csc\theta$, defines a horizontal line?

In Exercises 12–17, convert to an equation in rectangular coordinates.

12. $r = 7$ **13.** $r = \sin\theta$

14. $r = 2\sin\theta$ **15.** $r = 2\csc\theta$

16. $r = \dfrac{1}{\cos\theta - \sin\theta}$ **17.** $r = \dfrac{1}{2 - \cos\theta}$

In Exercises 18–21, convert to an equation in polar coordinates.

18. $x^2 + y^2 = 5$ **19.** $x = 5$

20. $y = x^2$ **21.** $xy = 1$

22. Match the equation with its description:

(a) $r = 2$ **(i)** Vertical line
(b) $\theta = 2$ **(ii)** Horizontal line
(c) $r = 2\sec\theta$ **(iii)** Circle
(d) $r = 2\csc\theta$ **(iv)** Line through origin

23. Find the values of θ in the plot of $r = 4\cos\theta$ corresponding to points A, B, C, D in Figure 19. Then indicate the portion of the graph traced out as θ varies in the following intervals:

(a) $0 \le \theta \le \frac{\pi}{2}$ **(b)** $\frac{\pi}{2} \le \theta \le \pi$ **(c)** $\pi \le \theta \le \frac{3\pi}{2}$

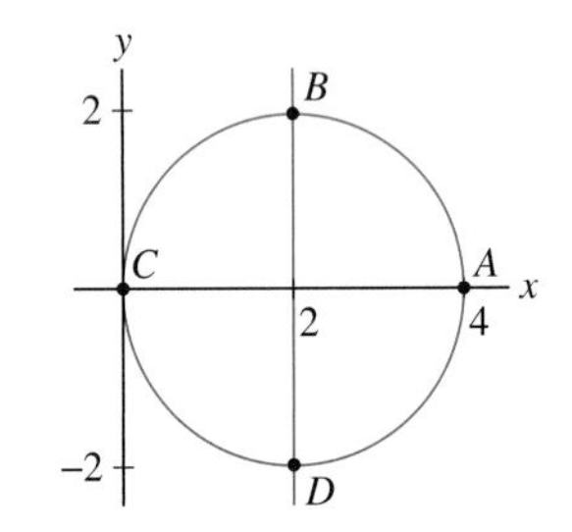

FIGURE 19 Plot of $r = 4\cos\theta$.

24. Suppose that (x, y) has polar coordinates (r, θ). Find the polar coordinates of the following:

(a) $(x, -y)$ **(b)** $(-x, -y)$ **(c)** $(-x, y)$ **(d)** (y, x)

25. Match each equation in rectangular coordinates with its equation in polar coordinates.

(a) $x^2 + y^2 = 2$ **(i)** $r^2(1 + 2\sin^2\theta) = 4$
(b) $x^2 + (y-1)^2 = 1$ **(ii)** $r(\cos\theta + \sin\theta) = 4$
(c) $x^2 - y^2 = 4$ **(iii)** $r = 2\sin\theta$
(d) $x + y = 4$ **(iv)** $r = 2$

26. What are the polar equations of the lines parallel to the line $r\cos(\theta - \frac{\pi}{3}) = 1$?

27. Show that $r = \sin\theta + \cos\theta$ is the equation of the circle of radius $1/\sqrt{2}$ whose center in rectangular coordinates is $(\frac{1}{2}, \frac{1}{2})$. Then find the values of θ between 0 and π such that $(\theta, r(\theta))$ yields the points A, B, C, and D in Figure 20.

28. Sketch the curve $r = \frac{1}{2}\theta$ (the spiral of Archimedes) for θ between 0 and 2π by plotting the points for $\theta = 0, \frac{\pi}{4}, \frac{\pi}{2}, \ldots, 2\pi$.

29. Sketch the graph of $r = 3\cos\theta - 1$ (see Example 8).

30. Sketch the graph of $r = \cos\theta - 1$.

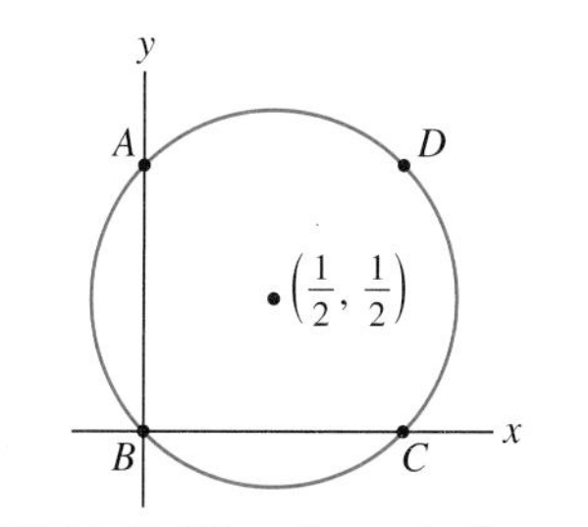

FIGURE 20 Plot of $r = \sin\theta + \cos\theta$.

31. Figure 21 displays the graphs of $r = \sin 2\theta$ in rectangular coordinates and in polar coordinates, where it is a "rose with four petals." Identify (a) the points in (B) corresponding to the points labeled A–I in (A), and (b) the parts of the curve in (B) corresponding to the angle intervals $\left[0, \frac{\pi}{2}\right]$, $\left[\frac{\pi}{2}, \pi\right]$, $\left[\pi, \frac{3\pi}{2}\right]$, and $\left[\frac{3\pi}{2}, 2\pi\right]$.

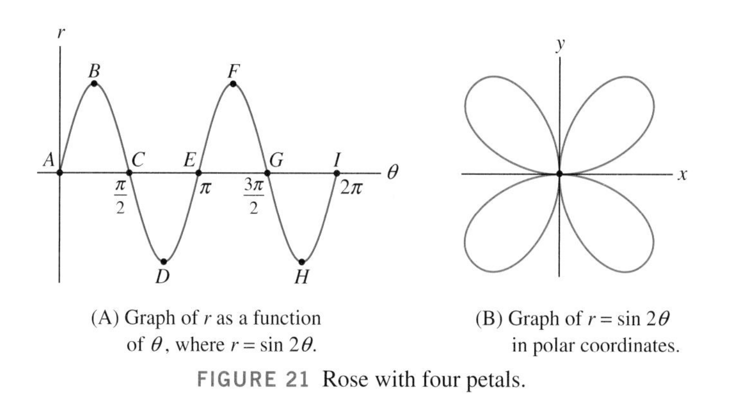

FIGURE 21 Rose with four petals.

32. Sketch the curve $r = \sin 3\theta$. First fill in the table of r-values below and plot the corresponding points of the curve. Notice that the three petals of the curve correspond to the angle intervals $\left[0, \frac{\pi}{3}\right]$, $\left[\frac{\pi}{3}, \frac{2\pi}{3}\right]$, and $\left[\frac{2\pi}{3}, \pi\right]$. Then plot $r = \sin 3\theta$ in rectangular coordinates and label the points on this graph corresponding to (r, θ) in the table.

θ	0	$\frac{\pi}{12}$	$\frac{\pi}{6}$	$\frac{\pi}{4}$	$\frac{\pi}{3}$	$\frac{5\pi}{12}$	$\cdots$	$\frac{11\pi}{12}$	π
r									

33. CAS Plot the **cissoid** $r = 2\sin\theta\tan\theta$ and show that its equation in rectangular coordinates is $y^2 = \dfrac{x^3}{2 - x}$.

34. Prove that $r = 2a\cos\theta$ is the equation of the circle in Figure 22 using only the fact that a triangle inscribed in a circle with one side a diameter is a right triangle.

35. Show that $r = a\cos\theta + b\sin\theta$ is the equation of a circle passing through the origin. Express the radius and center (in rectangular coordinates) in terms of a and b.

36. Use the previous exercise to write the equation of the circle of radius 5 and center (3, 4) in the form $r = a\cos\theta + b\sin\theta$.

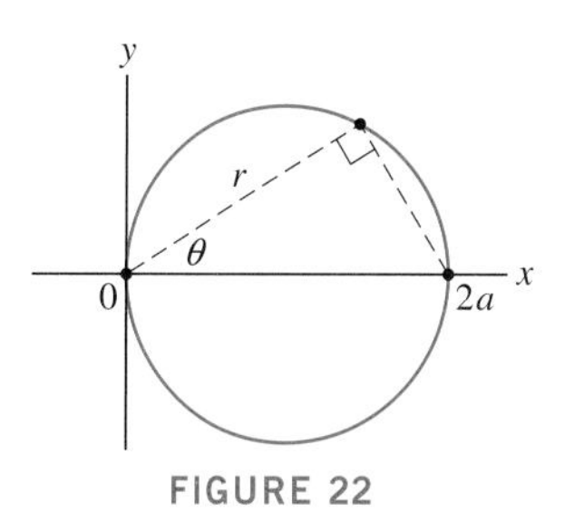

FIGURE 22

37. Use the identity $\cos 2\theta = \cos^2\theta - \sin^2\theta$ to find a polar equation of the hyperbola $x^2 - y^2 = 1$.

38. Find an equation in rectangular coordinates for the curve $r^2 = \cos 2\theta$.

39. Show that $\cos 3\theta = \cos^3\theta - 3\cos\theta\sin^2\theta$ and use this identity to find an equation in rectangular coordinates for the curve $r = \cos 3\theta$.

40. Use the addition formula for the cosine to show that the line $\mathcal{L}$ with polar equation $r\cos(\theta - \alpha) = d$ has the equation in rectangular coordinates $(\cos\alpha)x + (\sin\alpha)y = d$. Show that $\mathcal{L}$ has slope $m = -\cot\alpha$ and y-intercept $\dfrac{d}{\sin\alpha}$.

In Exercises 41–45, find an equation in polar coordinates of the line $\mathcal{L}$ with given description.

41. The point on $\mathcal{L}$ closest to the origin has polar coordinates $\left(2, \frac{\pi}{9}\right)$.

42. The point on $\mathcal{L}$ closest to the origin has rectangular coordinates $(-2, 2)$.

43. $\mathcal{L}$ is tangent to the circle $r = 2\sqrt{10}$ at the point with rectangular coordinates $(-2, -6)$.

44. $\mathcal{L}$ has slope 3 and is tangent to the unit circle in the fourth quadrant.

45. $y = 4x - 9$.

46. Show that the polar equation of the line $y = ax + b$ can be written in the form

$$r = \frac{b}{\sin\theta - a\cos\theta}$$

47. Distance Formula Use the Law of Cosines (Figure 23) to show that the distance d between two points with polar coordinates (r, θ) and (r_0, θ_0) is

$$d^2 = r^2 + r_0^2 - 2rr_0\cos(\theta - \theta_0) \qquad \boxed{2}$$

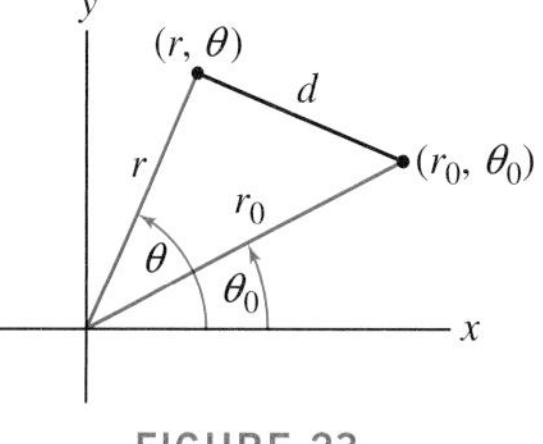

FIGURE 23

48. Use the distance formula (2) to show that the circle of radius 9 whose center has polar coordinates $(5, \frac{\pi}{4})$ has equation:

$$r^2 - 10r\cos\left(\theta - \frac{\pi}{4}\right) = 56$$

49. Show that the cardiod $r = 1 + \sin\theta$ has equation

$$x^2 + y^2 = (x^2 + y^2 - x)^2$$

50. For $a > 0$, a **lemniscate curve** is the set of points P such that the product of the distances from P to $(a, 0)$ and $(-a, 0)$ is a^2. Show that the equation of the lemniscate is:

$$(x^2 + y^2)^2 = 2a^2(x^2 - y^2)$$

Then find the equation in polar coordinates. To obtain the simplest form of the equation, use the identity $\cos 2\theta = \cos^2\theta - \sin^2\theta$. Plot the lemniscate for $a = 2$ if you have a computer algebra system.

51. The Derivative in Polar Coordinates A polar curve $r = f(\theta)$ has parametric equations (since $x = r\cos\theta$ and $y = r\sin\theta$):

$$x = f(\theta)\cos\theta, \quad y = f(\theta)\sin\theta$$

Apply Theorem 1 of Section 11.1 to prove the formula

$$\frac{dy}{dx} = \frac{f(\theta)\cos\theta + f'(\theta)\sin\theta}{-f(\theta)\sin\theta + f'(\theta)\cos\theta} \qquad \boxed{3}$$

where $f'(\theta) = df/d\theta$.

52. Use Eq. (3) to find the slope of the tangent line to $r = \theta$ at $\theta = \frac{\pi}{2}$ and $\theta = \pi$.

53. Find the equation in rectangular coordinates of the tangent line to $r = 4\cos 3\theta$ at $\theta = \frac{\pi}{6}$.

54. Show that for the circle $r = \sin\theta + \cos\theta$,

$$\frac{dy}{dx} = \frac{\cos 2\theta + \sin 2\theta}{\cos 2\theta - \sin 2\theta}$$

Calculate the slopes of the tangent lines at points A, B, C in Figure 20 and find the polar coordinates of the points at which the tangent line is horizontal.

Further Insights and Challenges

55. Let c be a fixed constant. Explain the relationship between the graphs of:

(a) $y = f(x + c)$ and $y = f(x)$ (rectangular)

(b) $r = f(\theta + c)$ and $r = f(\theta)$ (polar)

(c) $y = f(x) + c$ and $y = f(x)$ (rectangular)

(d) $r = f(\theta) + c$ and $r = f(\theta)$ (polar)

56. Let $f(x)$ be a periodic function of period 2π, that is, $f(x) = f(x + 2\pi)$. Explain how this periodicity is reflected in the graph of:

(a) $y = f(x)$ in rectangular coordinates

(b) $r = f(\theta)$ in polar coordinates

57. GU Use a graphing utility to convince yourself that graphs of the polar equations $r = f_1(\theta) = 2\cos\theta - 1$ and $r = f_2(\theta) = 2\cos\theta + 1$ have the same graph. Then explain why. *Hint:* Show that the points $(f_1(\theta + \pi), \theta + \pi)$ and $(f_2(\theta), \theta)$ coincide.

58. CAS Plot the limaçon curves $r = a + \cos\theta$ for several values of a.

(a) Describe how the shape changes as a increases. What happens for $a < 0$?

(b) A closed curve is called convex if, for any two points P and Q in the interior of the curve, the segment $\overline{PQ}$ is also contained in the interior. Figure 24 shows that the limaçon is convex if a is large and is not convex if a is small. Experiment with a computer algebra system to estimate the smallest value a for which it is convex.

(c) Use Eq. (3) of Exercise 51 to show that if the tangent line is vertical, then either $\theta = 0, \pi$ or $\cos\theta = -a/2$.

(d) Show that the limaçon has three vertical tangent lines to the left of the y-axis if $1 < a < 2$ and a unique vertical tangent line to the left of the y-axis there if $a \geq 2$. What does this suggest about the convexity of the limaçon for $a > 2$?

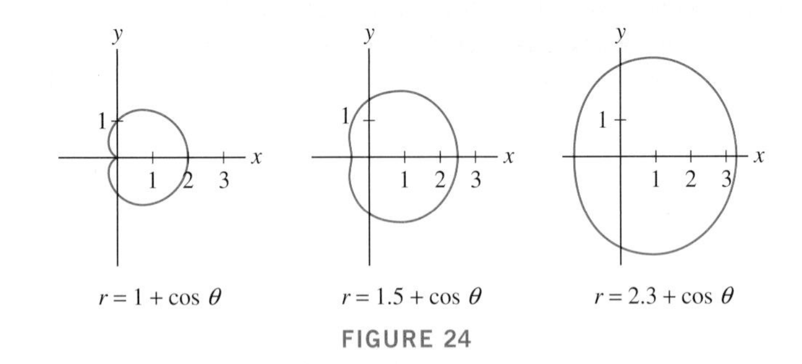

FIGURE 24

11.4 Area and Arc Length in Polar Coordinates

In this section, we derive the formulas for area and arc length in polar coordinates. First, we compute the area of a sector bounded by a curve $r = f(\theta)$ and two rays $\theta = \alpha$ and $\theta = \beta$ [shaded region in Figure 1(A)]. Divide the region in N narrow sectors of angle

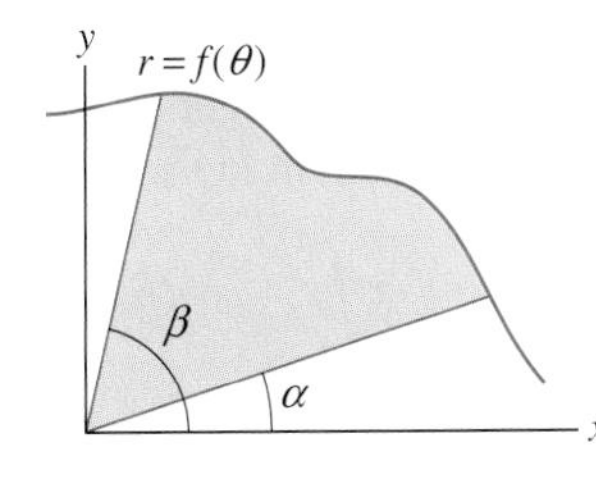

(A) Region defined by $\alpha \le \theta \le \beta$

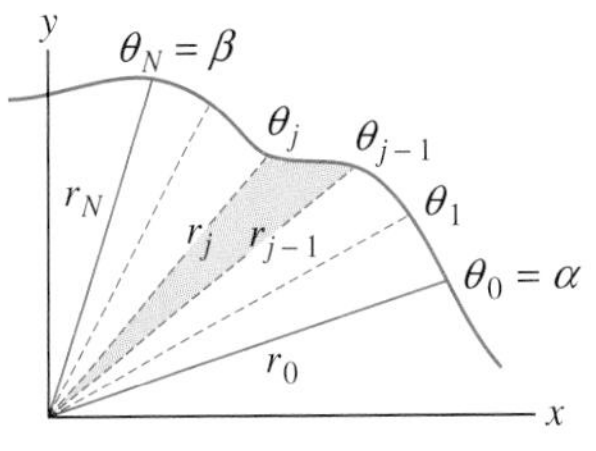

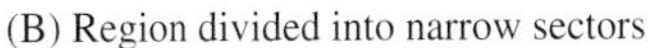
(B) Region divided into narrow sectors

FIGURE 1 Area bounded by the curve $r = f(\theta)$ and the two rays $\theta = \alpha$ and $\theta = \beta$.

Rectangular and polar coordinates are suited for different kinds of area calculations. Rectangular coordinates should be used to compute the signed area under the graph $y = f(x)$ over an interval $[a, b]$. Polar coordinates should be used to calculate the area bounded by a curve in an angular sector $\alpha \le \theta \le \beta$.

$\Delta\theta = \frac{\beta-\alpha}{N}$ corresponding to a partition of the interval $[\alpha, \beta]$:

$$\theta_0 = \alpha \le \theta_1 \le \theta_2 \cdots \le \theta_N = \beta$$

Recall that a circular sector of width $\Delta\theta$ and radius r has area $\frac{1}{2}r^2\,\Delta\theta$ (Figure 2). Each narrow sector in our region is nearly circular, of radius $r_j = f(\theta_j)$, and therefore has area approximately equal to $\frac{1}{2}r_j^2\,\Delta\theta$ (Figure 3). The total area is approximated by the sum:

$$\text{Area of region} \approx \sum_{j=1}^{N} \frac{1}{2}r_j^2\,\Delta\theta = \sum_{j=1}^{N} \frac{1}{2}f(\theta_j)^2\,\Delta\theta \qquad \boxed{1}$$

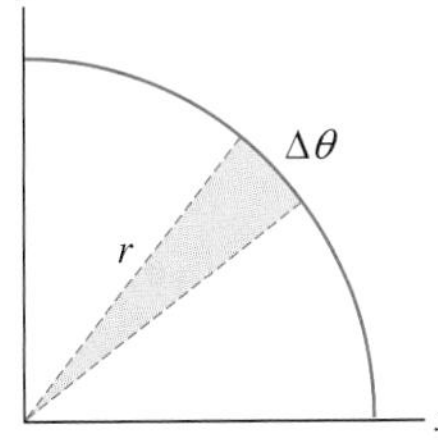

FIGURE 2 The area of a circular sector is *exactly* $\frac{1}{2}r^2\,\Delta\theta$.

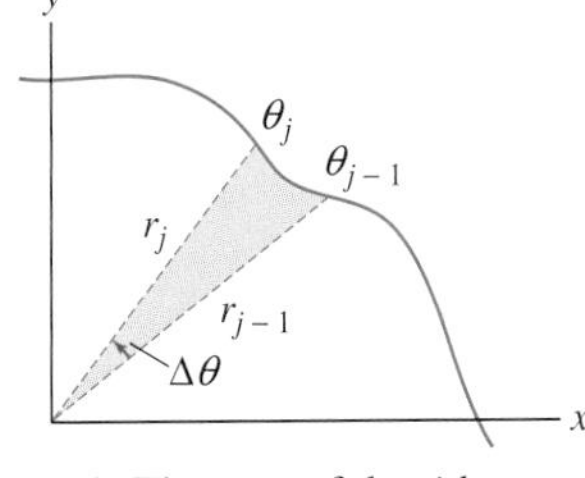

FIGURE 3 The area of the jth sector is *approximately* $\frac{1}{2}r_j^2\,\Delta\theta$.

This is a Riemann sum for the integral $\int_\alpha^\beta \frac{1}{2}f(\theta)^2\,d\theta$. If $f(\theta)$ is continuous, the sum approaches the integral as $N \to \infty$, and we obtain the following formula.

CAUTION In rectangular coordinates, the integral computes signed *area, but in polar coordinates, formula (2) gives the actual (positive) area, not signed area. The formula remains valid if $f(\theta)$ takes on negative values.*

THEOREM 1 Area in Polar Coordinates If $f(\theta)$ is continuous function, then the area bounded by a curve in polar form $r = f(\theta)$ and the rays $\theta = \alpha$ and $\theta = \beta$ is equal to

$$\frac{1}{2}\int_\alpha^\beta r^2\,d\theta = \frac{1}{2}\int_\alpha^\beta f(\theta)^2\,d\theta \qquad \boxed{2}$$

As a check on this formula, note that the graph of $f(r) = R$ is a circle of radius R. According to Eq. (2), the area of the circle is $\frac{1}{2}\int_0^{2\pi} R^2\,d\theta = \frac{1}{2}R^2(2\pi) = \pi R^2$, as expected.

■ **EXAMPLE 1** Find the area of the right semicircle with equation $r = 4\sin\theta$.

REMINDER In Eq. (4), we use the identity

$$\sin^2\theta = \frac{1}{2}(1 - \cos 2\theta) \qquad \boxed{3}$$

to integrate $y = \sin^2\theta$.

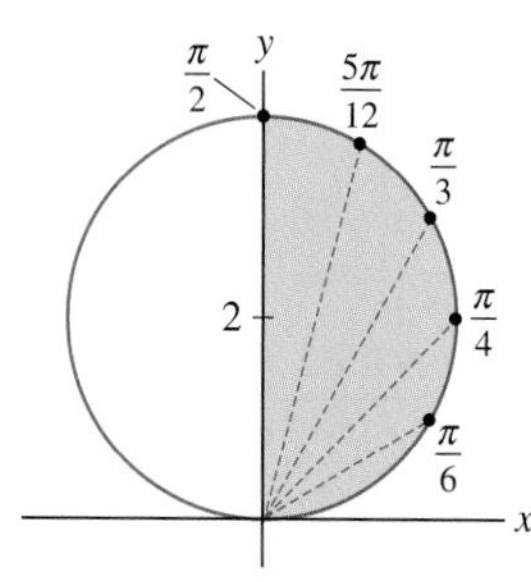

FIGURE 4 The circle with equation $r = 4\sin\theta$.

Solution The equation $r = 4\sin\theta$ defines a circle of radius 2 tangent to the x-axis at the origin (Figure 4). The right semicircle is "swept out" as θ varies from 0 to $\frac{\pi}{2}$. By Eq. (2), the area of the semicircle is

$$\frac{1}{2}\int_0^{\pi/2} r^2\,d\theta = \frac{1}{2}\int_0^{\pi/2} (4\sin\theta)^2\,d\theta = 8\int_0^{2\pi} \sin^2\theta\,d\theta \qquad \boxed{4}$$

$$= 8\int_0^{\pi/2} \frac{1}{2}(1 - \cos 2\theta)\,d\theta$$

$$= (4\theta - 2\sin 2\theta)\Big|_0^{\pi/2} = 4\left(\frac{\pi}{2}\right) - 0 = 2\pi$$

This is the expected result since the total area of a circle of radius $R = 2$ is 4π. ■

CONCEPTUAL INSIGHT Keep in mind that the integral $\frac{1}{2}\int_\alpha^\beta r^2\,d\theta$ does not compute the area *under* a curve, but rather the area "swept out" by a radial segment as θ varies from α to β. For example, integral $\frac{1}{2}\int_0^{\pi/3} (4\sin\theta)^2\,d\theta$ is equal to the area of the shaded region in Figure 5(A), not the area under the curve in Figure 5(B).

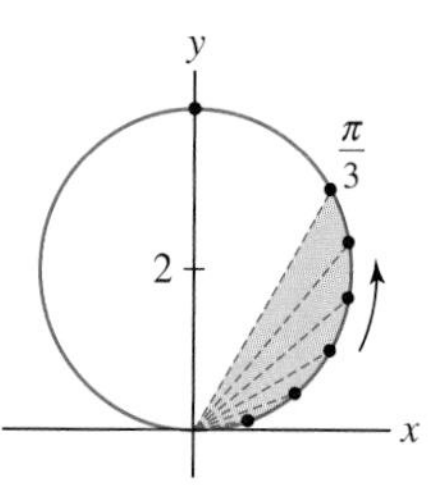

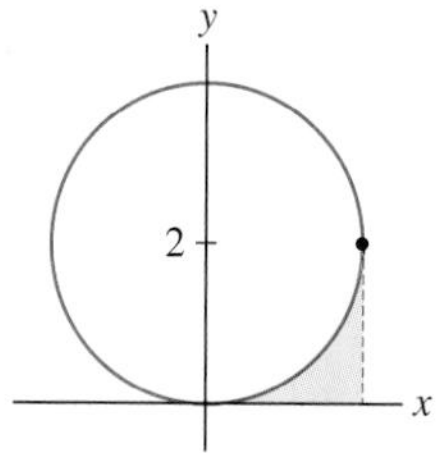

(A) Region swept out by radial segment has area

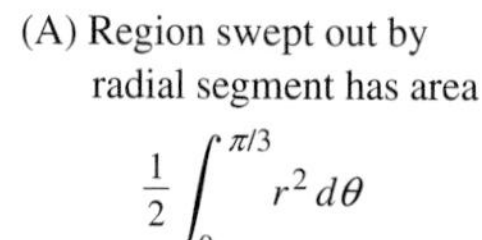

$$\frac{1}{2}\int_0^{\pi/3} r^2\,d\theta$$

(B) Region under the curve.

FIGURE 5

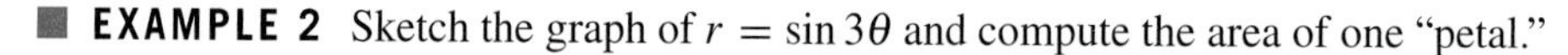

■ **EXAMPLE 2** Sketch the graph of $r = \sin 3\theta$ and compute the area of one "petal."

Solution To sketch the curve, we first graph $r = \sin 3\theta$ in rectangular coordinates. Figure 6 shows that as θ varies from 0 to $\frac{\pi}{3}$, the radius r varies from 0 to 1, and then back to 0. This gives petal A (Figure 7). Petal B is traced as θ varies from $\frac{\pi}{3}$ to $\frac{2\pi}{3}$, and petal C for $\frac{2\pi}{3} \le \theta \le \pi$. We compute the area of petal A using identity (3):

$$\frac{1}{2}\int_0^{\pi/3} (\sin 3\theta)^2\,d\theta = \frac{1}{4}\int_0^{\pi/3} (1 - \cos 6\theta)\,d\theta = \left(\frac{1}{4}\theta - \frac{1}{24}\sin 6\theta\right)\Big|_0^{\pi/3} = \frac{\pi}{12} \quad ■$$

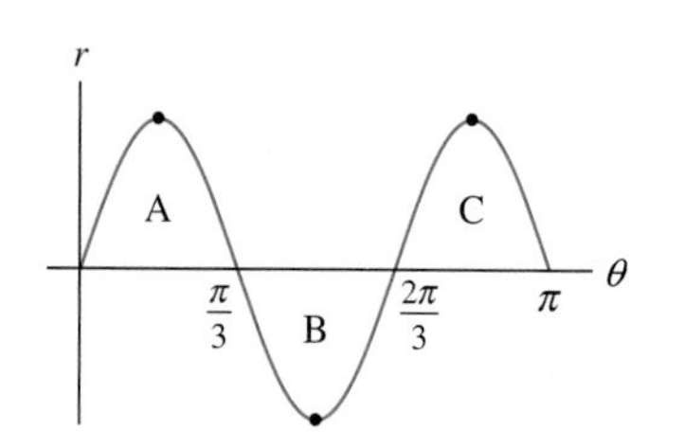

FIGURE 6 Graph of $r = \sin 3\theta$ as a function of θ.

If $r = f_1(\theta)$ and $r = f_2(\theta)$ are two polar curves with $f_2(\theta) \ge f_1(\theta)$, then we may compute the area of the region between the two curves within the sector $\alpha \le \theta \le \beta$ by the integral (Figure 8):

$$\text{Area between two curves} = \frac{1}{2}\int_\alpha^\beta (f_2(\theta)^2 - f_1(\theta)^2)\,d\theta \qquad \boxed{5}$$

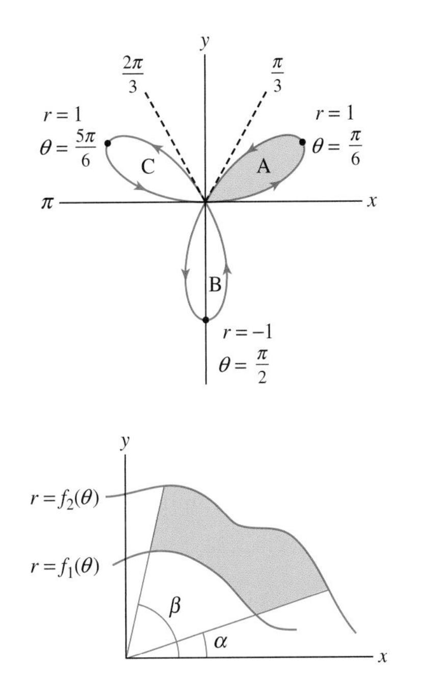

FIGURE 7 The "rose with three petals" $r = \sin 3\theta$.

FIGURE 8 Area between two polar graphs in a sector.

■ **EXAMPLE 3 Area Between Two Curves** Find the area of the region inside the circle $r = 2\cos\theta$ but outside the circle $r = 1$.

Solution Recall that $r = 2a\cos\theta$ is the polar equation of the circle of radius a with center $(a, 0)$ (Section 11.3, Example 7). Thus $r = 2\cos\theta$ is a circle of radius 1 with center $(1, 0)$. We wish to compute the area of the region within this circle and outside the circle $r = 1$ of radius 1 centered at the origin. This is the region labeled (I) in Figure 9(A). The two circles intersect at the points where $r = 2\cos\theta = 1$ or $\cos\theta = \frac{1}{2}$. Thus the intersection points occur at $\theta = \pm\frac{\pi}{3}$. The area of region (I) is equal to

$$\text{Area of (I)} = \text{area of (II)} - \text{area of (III)}$$

$$= \frac{1}{2}\int_{-\pi/3}^{\pi/3} (2\cos\theta)^2\, d\theta - \frac{1}{2}\int_{-\pi/3}^{\pi/3} (1)^2\, d\theta$$

$$= \frac{1}{2}\int_{-\pi/3}^{\pi/3} (4\cos^2\theta - 1)\, d\theta = \frac{1}{2}\int_{-\pi/3}^{\pi/3} (2\cos 2\theta + 1)\, d\theta \qquad \boxed{6}$$

$$= \frac{1}{2}(\sin 2\theta + \theta)\Big|_{-\pi/3}^{\pi/3} = \frac{\sqrt{3}}{2} + \frac{\pi}{3} \approx 1.91$$ ■

← REMINDER In (6), we use the identity

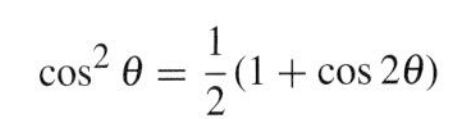

$$\cos^2\theta = \frac{1}{2}(1 + \cos 2\theta)$$

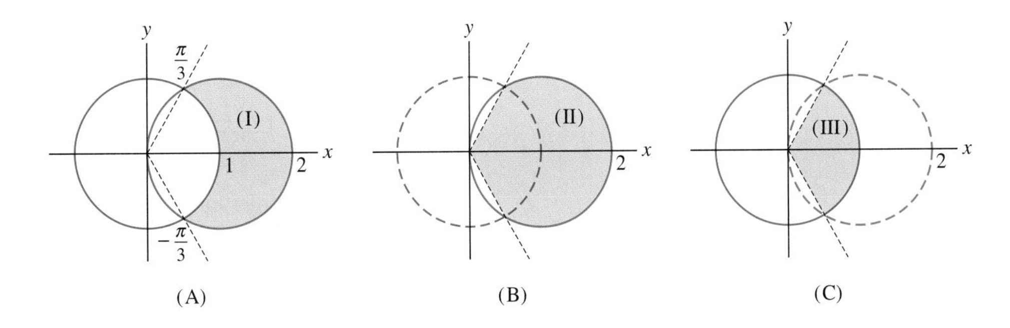

FIGURE 9 Region I is the difference of regions II and III.

To derive a formula for arc length in polar coordinates, we observe that a polar curve $r = f(\theta)$ has a natural parametrization with θ as a parameter:

$$x = r\cos\theta = f(\theta)\cos\theta, \qquad y = r\sin\theta = f(\theta)\sin\theta$$

Using a prime to denote the derivative with respect to θ, we have

$$x'(\theta) = \frac{dx}{d\theta} = -f(\theta)\sin\theta + f'(\theta)\cos\theta$$

$$y'(\theta) = \frac{dy}{d\theta} = f(\theta)\cos\theta + f'(\theta)\sin\theta$$

We check using algebra that $x'(\theta)^2 + y'(\theta)^2 = f(\theta)^2 + f'(\theta)^2$. Now recall from Section 11.2 that the length s of the arc traced out for $\alpha \le \theta \le \beta$ is $\int_\alpha^\beta \sqrt{x'(\theta)^2 + y'(\theta)^2}\,d\theta$. It follows that

$$\text{Arc length } s = \int_\alpha^\beta \sqrt{f(\theta)^2 + f'(\theta)^2}\,d\theta \qquad \boxed{7}$$

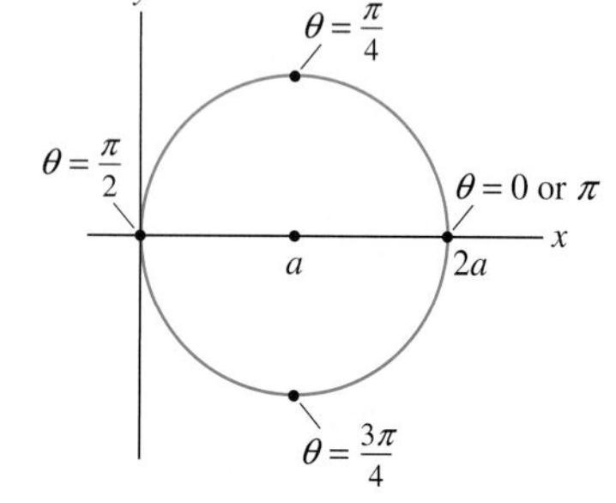

FIGURE 10 Graph of $r = 2a\cos\theta$.

■ **EXAMPLE 4** Find the total length of the circle with equation $r = 2a\cos\theta$ for $a > 0$.

Solution In this case, $f(\theta) = 2a\cos\theta$ and

$$f(\theta)^2 + f'(\theta)^2 = 4a^2\cos^2\theta + 4a^2\sin^2\theta = 4a^2$$

The total length of this circle of radius a has the expected value:

$$\int_0^\pi \sqrt{f(\theta)^2 + f'(\theta)^2}\,d\theta = \int_0^\pi (2a)\,d\theta = 2\pi a$$

Note that the upper limit of integration is π rather than 2π because the entire circle is traced out as θ varies from 0 to π (see Figure 10). ■

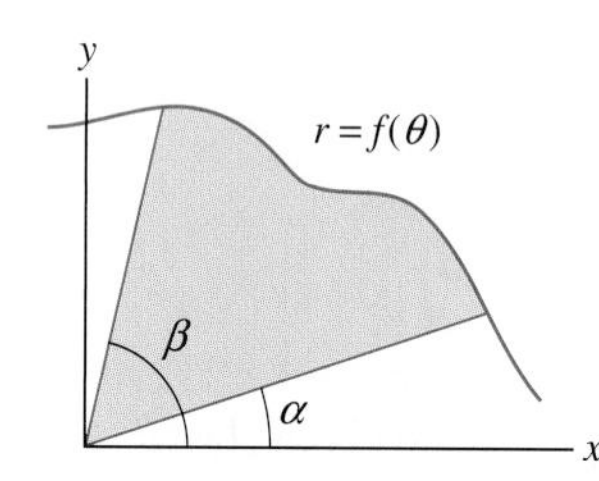

FIGURE 11 Region bounded by the polar curve $r = f(\theta)$ and the rays $\theta = \alpha$, $\theta = \beta$.

11.4 SUMMARY

- Polar coordinates are suited to calculating areas bounded by a polar curve $r = f(\theta)$ and two rays $\theta = \alpha$ and $\theta = \beta$ (Figure 11):

$$\text{Area} = \frac{1}{2}\int_\alpha^\beta f(\theta)^2\,d\theta$$

- The area between the polar curves $r = f_1(\theta)$ and $r = f_2(\theta)$, where $f_2(\theta) \ge f_1(\theta)$, is equal to (Figure 12)

$$\text{Area} = \frac{1}{2}\int_\alpha^\beta \left(f_2(\theta)^2 - f_1(\theta)^2\right)d\theta$$

- The arc length of the polar curve $r = f(\theta)$ for $\alpha \le \theta \le \beta$ is

$$\text{Arc length} = \int_\alpha^\beta \sqrt{f(\theta)^2 + f'(\theta)^2}\,d\theta$$

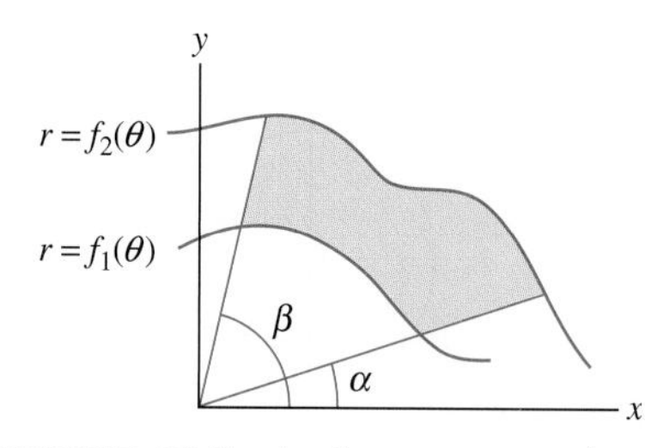

FIGURE 12 Region between two polar curves.

11.4 EXERCISES

Preliminary Questions

1. True or False: The area under the curve with polar equation $r = f(\theta)$ is equal to the integral of $f(\theta)$.

2. Polar coordinates are best suited to finding the area (choose one):

(a) Under a curve between $x = a$ and $x = b$.

(b) Bounded by a curve and two rays through the origin.

3. True or False: The formula for area in polar coordinates is valid only if $f(\theta) \geq 0$.

4. The horizontal line $y = 1$ has polar equation $r = \csc\theta$. Which area is represented by the integral $\frac{1}{2}\int_{\pi/6}^{\pi/2} \csc^2\theta\, d\theta$ (Figure 13)?

(a) $\square ABCD$ **(b)** $\triangle ABC$ **(c)** $\triangle ACD$

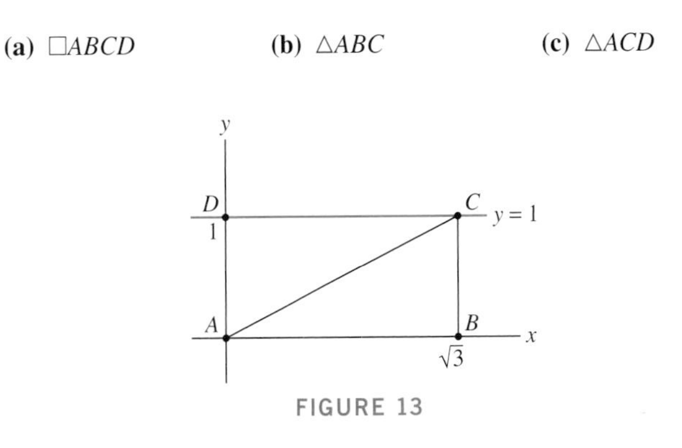

FIGURE 13

Exercises

1. Sketch the area bounded by the circle $r = 5$ and the rays $\theta = \frac{\pi}{2}$ and $\theta = \pi$, and compute its area as an integral in polar coordinates.

2. Sketch the region bounded by the line $r = \sec\theta$ and the rays $\theta = 0$ and $\theta = \frac{\pi}{3}$. Compute its area in two ways: as an integral in polar coordinates and using geometry.

3. Calculate the area of the circle $r = 4\sin\theta$ as an integral in polar coordinates (see Figure 4). Be careful to choose the correct limits of integration.

4. Compute the area of the shaded region in Figure 14 as an integral in polar coordinates.

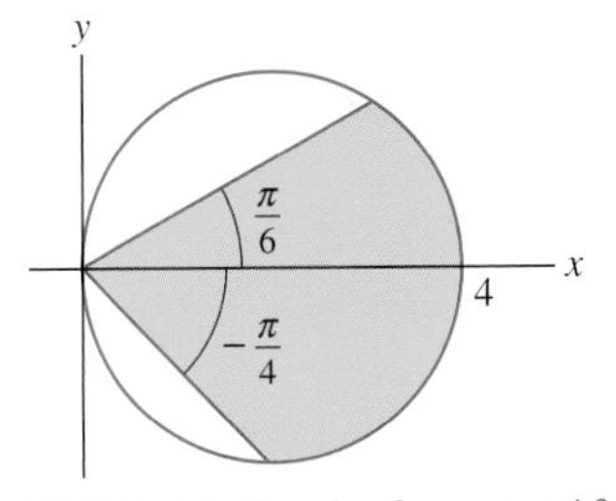

FIGURE 14 Graph of $r = \cos 4\theta$.

5. Find the total area enclosed by the cardioid $r = 1 - \cos\theta$ (Figure 15).

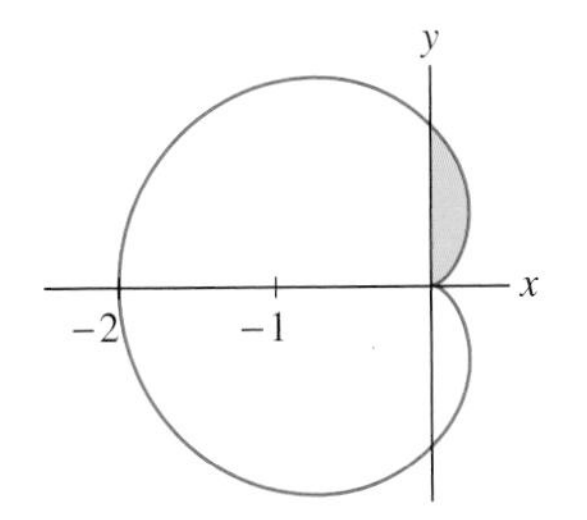

FIGURE 15 The cardioid $r = 1 - \cos\theta$.

6. Find the area of the shaded region in Figure 15.

7. Find the area of one leaf of the "four-petaled rose" $r = \sin 2\theta$ (Figure 16).

8. Prove that the total area of the four-petaled rose $r = \sin 2\theta$ is equal to one-half the area of the circumscribed circle (Figure 16).

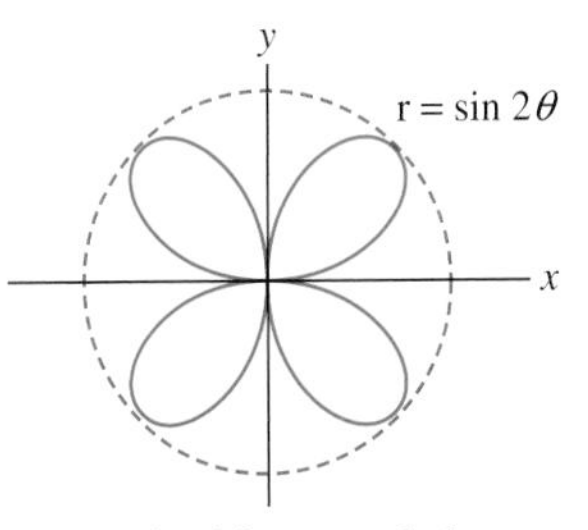

FIGURE 16 Graph of four-petaled rose $r = \sin 2\theta$.

9. Find the area enclosed by one loop of the lemniscate with equation $r^2 = \cos 2\theta$ (Figure 17). Choose your limits of integration carefully.

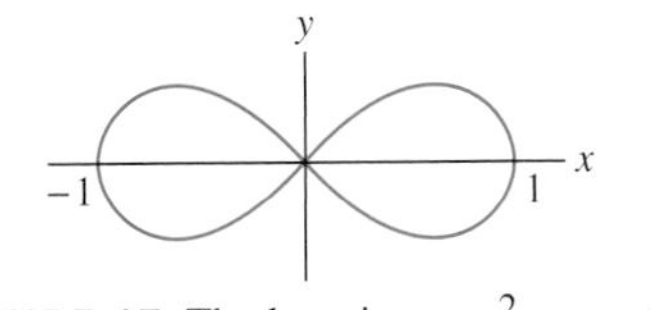

FIGURE 17 The lemniscate $r^2 = \cos 2\theta$.

10. Sketch the spiral $r = \theta$ for $0 \leq \theta \leq 2\pi$ and find the area bounded by the curve and the first quadrant.

11. Find the area enclosed by the cardioid $r = a(1 + \cos\theta)$, where $a > 0$.

12. Find the area of the intersection of the circles $r = \sin\theta$ and $r = \cos\theta$.

13. Find the area of region A in Figure 18.

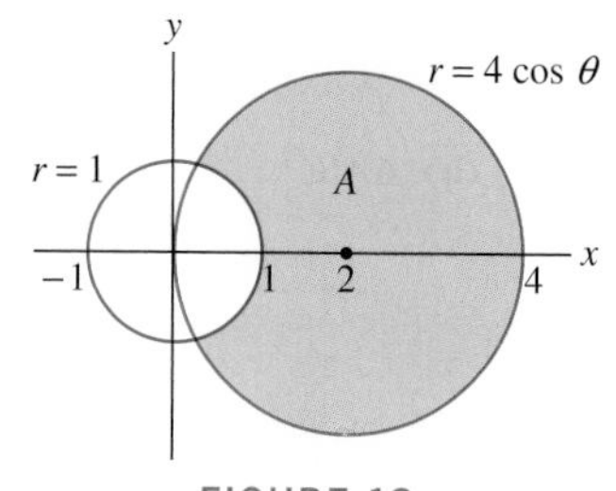

FIGURE 18

14. Find the area of the shaded region in Figure 19, enclosed by the circle $r = \frac{1}{2}$ and a petal of the curve $r = \cos 3\theta$. *Hint:* Compute the area of both the petal and the region inside the petal and outside the circle.

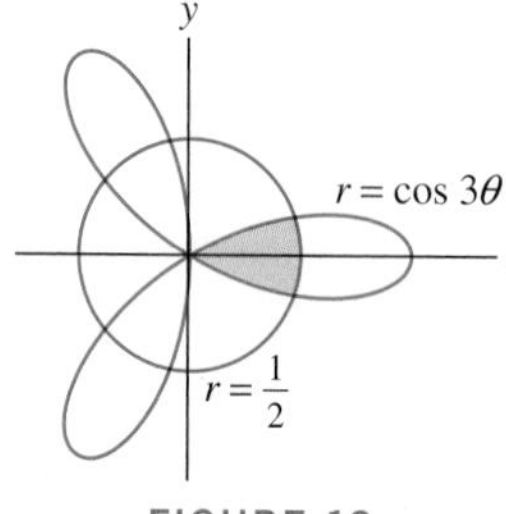

FIGURE 19

15. Find the area of the inner loop of the limaçon with polar equation $r = 2\cos\theta - 1$ (Figure 20).

16. Find the area of the region between the inner and outer loop of the limaçon $r = 2\cos\theta - 1$.

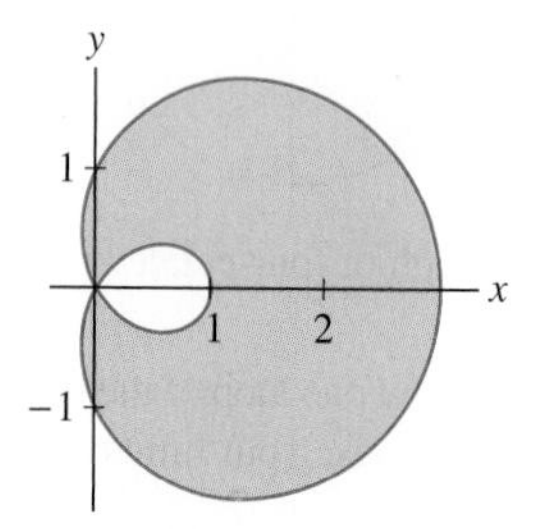

FIGURE 20 The limaçon $r = 2\cos\theta - 1$.

17. Find the area of the part of the circle $r = \sin\theta + \cos\theta$ in the fourth quadrant (see Exercise 27 in Section 11.3).

18. Compute the area of the shaded region in Figure 21.

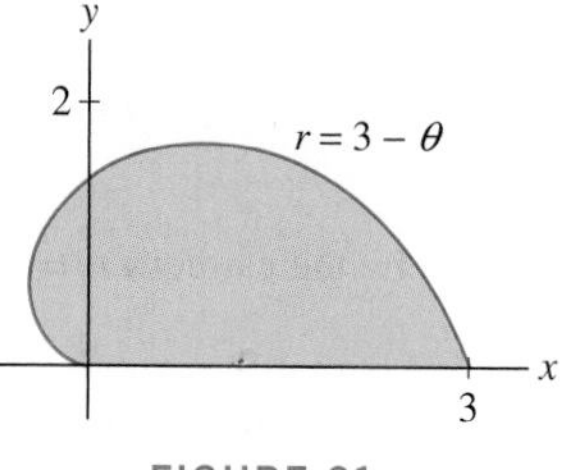

FIGURE 21

19. Find the area between the two curves in Figure 22(A).

20. Find the area between the two curves in Figure 22(B).

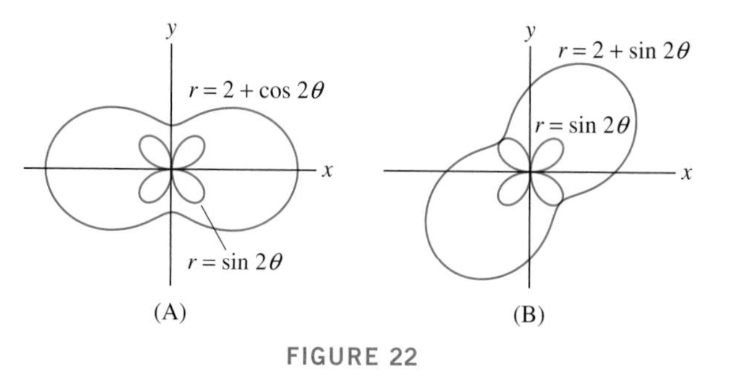

FIGURE 22

21. Find the area inside both curves in Figure 23.

22. Find the area of the region that lies inside one but not both of the curves in Figure 23.

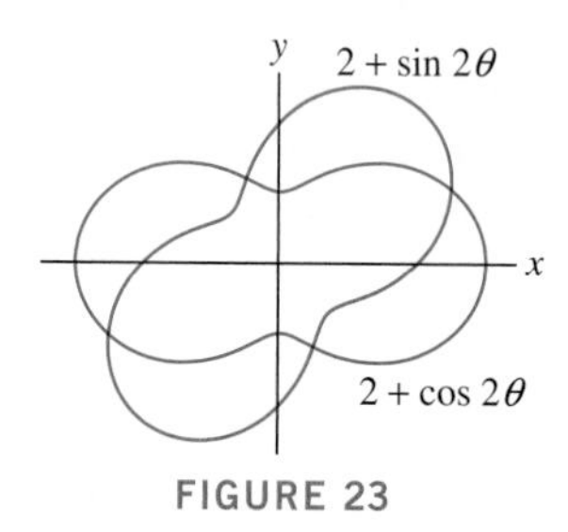

FIGURE 23

23. Figure 24 suggests that the circle $r = \sin\theta$ lies inside the spiral $r = \theta$. Which inequality from Chapter 2 assures us that this is the case? Find the area between the curves $r = \theta$ and $r = \sin\theta$ in the first quadrant.

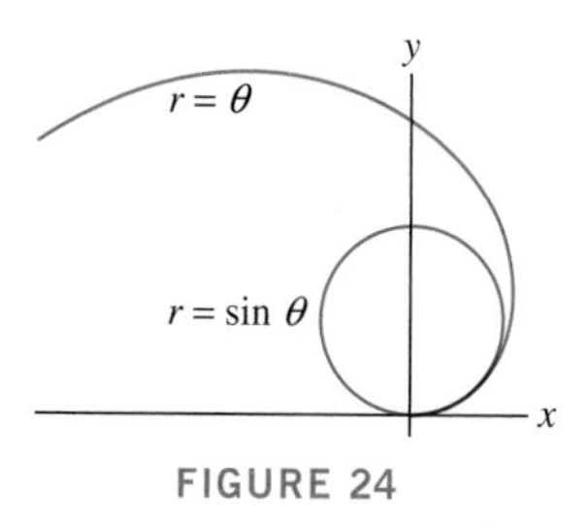

FIGURE 24

24. Calculate the total length of the circle $r = 4\sin\theta$ as an integral in polar coordinates.

25. Find the length of the spiral $r = \theta$ for $0 \le \theta \le A$.

26. Find the length of $r = \theta^2$ for $0 \le \theta \le \pi$.

27. Sketch the segment $r = \sec\theta$ for $0 \le \theta \le A$. Then compute its length in two ways: as an integral in polar coordinates and using trigonometry.

28. Sketch the circle $r = \sin\theta + \cos\theta$. Then compute its length in two ways: as an integral in polar coordinates and using trigonometry.

29. Find the length of the cardioid $r = 1 + \cos\theta$. *Hint:* Use the identity $1 + \cos\theta = 2\cos^2\left(\frac{\theta}{2}\right)$ to evaluate the arc length integral.

30. Find the length of the cardioid with equation $r = 1 + \cos\theta$ located in the first quadrant.

31. Find the length of the *equiangular spiral* $r = e^{\theta}$ for $0 \le \theta \le 2\pi$.

32. Find the length of the curve $r = \cos^2\theta$.

In Exercises 33–36, express the length of the curve as an integral but do not evaluate it.

33. $r = e^{a\theta}, \quad 0 \le \theta \le \pi$

34. $r = \sin 2\theta, \quad 0 \le \theta \le \pi$

35. $r = (2 - \cos\theta)^{-1}, \quad 0 \le \theta \le 2\pi$

36. The inner loop of $r = 2\cos\theta - 1$ (see Exercise 15)

Further Insights and Challenges

37. Suppose that the polar coordinates of a moving particle at time t are $(r(t), \theta(t))$. Prove that the particle's speed is equal to $\sqrt{(dr/dt)^2 + r^2(d\theta/dt)^2}$.

38. Compute the speed at time $t = 1$ of a particle whose polar coordinates at time t are $r = t$, $\theta = t$ (use Exercise 37). What would the speed be if the particle's rectangular coordinates are $x = t$, $y = t$? Why is the speed increasing in one case and constant in the other?

11.5 Conic Sections

The three familiar families of curves—ellipses, hyperbolas, and parabolas—appear throughout mathematics and its applications. They were first studied by the ancient Greek mathematicians, who recognized that they are obtained by intersecting a plane with a cone (Figure 1) and hence referred to them as "conic sections." Our main goal in this section is to derive the equations of the conic sections from their geometric definitions as curves in the plane.

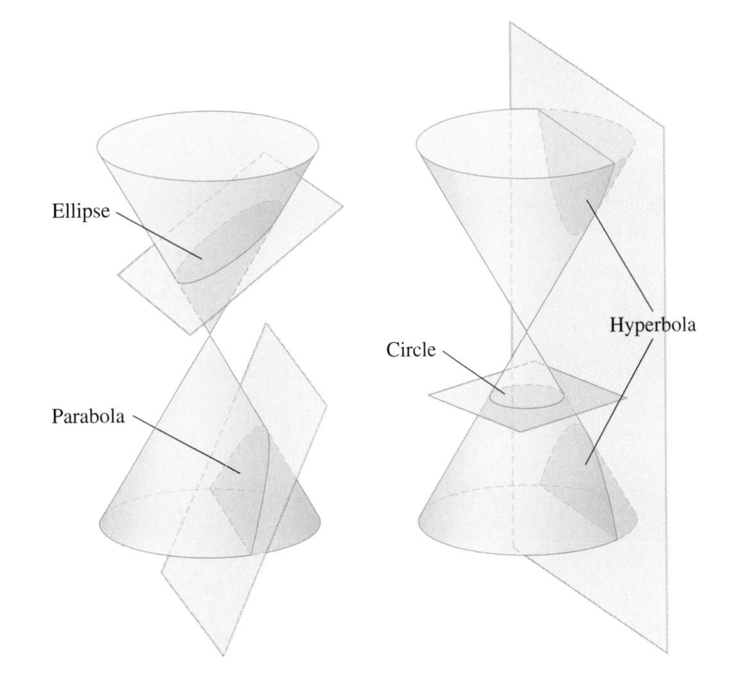

FIGURE 1 The conic sections are obtained by intersecting a plane and a cone.

A circle is an ellipse whose foci coincide. If $F_1 = F_2$, Eq. (1) reduces to $PF_1 = K/2$, which defines a circle of radius $K/2$ and center F_1.

An ellipse is the set of all points P such that the sum of the distances to two fixed points F_1 and F_2 is a constant $K > 0$:

$$PF_1 + PF_2 = K \tag{1}$$

The points F_1 and F_2 are called the **foci** (plural of "focus") of the ellipse. We assume that K is greater than the distance F_1F_2 between the foci (if $K = F_1F_2$, the ellipse reduces to the line segment $\overline{F_1F_2}$; if $K < F_1F_2$, the ellipse has no points). Figure 2(A) shows an ellipse with foci at F_1 and F_2.

We use the following terminology:

- The midpoint of $\overline{F_1F_2}$ is the **center** of the ellipse.
- The line through the foci is the **focal axis**.
- The line through the center perpendicular to the focal axis is the **conjugate axis**.

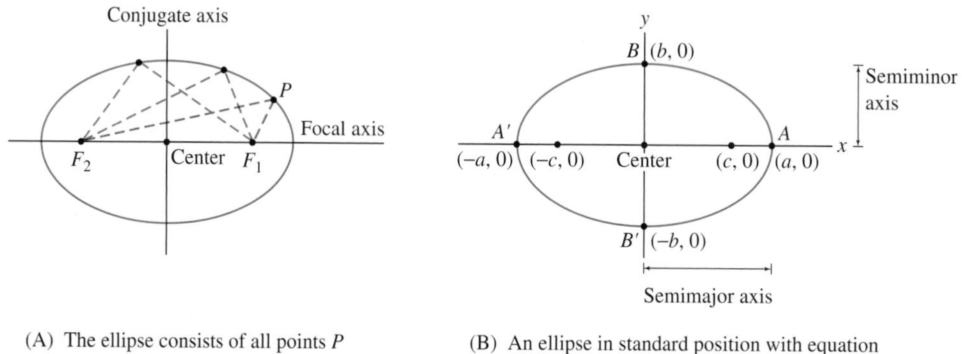

(A) The ellipse consists of all points P such that $PF_1 + PF_2 =$ constant.

(B) An ellipse in standard position with equation $\left(\frac{x}{a}\right)^2 + \left(\frac{y}{b}\right)^2 = 1$.

FIGURE 2

The ellipse is said to be in **standard position** if the focal and conjugate axes are the x- and y-axes, as shown in Figure 2(B). In this case, the foci have coordinates $F_1 = (c, 0)$ and $F_2 = (-c, 0)$ for some $c > 0$. Let us prove that the equation of this ellipse has the particularly simple form

$$\left(\frac{x}{a}\right)^2 + \left(\frac{x}{b}\right)^2 = 1 \tag{2}$$

where $a = K/2$ and $b = \sqrt{a^2 - c^2}$. By definition, a point $P = (x, y)$ lies on the ellipse if $PF_1 + PF_2 = 2a$ and therefore, by the distance formula,

$$\sqrt{(x+c)^2 + y^2} + \sqrt{(x-c)^2 + y^2} = 2a \tag{3}$$

Move the second term on the left over to the right and square both sides:

$$(x+c)^2 + y^2 = 4a^2 - 4a\sqrt{(x-c)^2 + y^2} + (x-c)^2 + y^2$$

$$4a\sqrt{(x-c)^2 + y^2} = 4a^2 + (x-c)^2 - (x+c)^2 = 4a^2 - 4cx$$

Now divide by 4, square, and simplify:

Strictly speaking, we must also show that every point (x, y) satisfying (4) also satisfies (3). When we begin with (4) and reverse the algebraic steps, the process of taking square roots leads to the relation

$$\sqrt{(x-c)^2+y^2} \pm \sqrt{(x+c)^2+y^2} = \pm 2a$$

However, since $a > c$, this equation cannot hold unless both signs are positive.

$$a^2(x^2 - 2cx + c^2 + y^2) = a^4 - 2a^2cx + c^2x^2$$

$$(a^2 - c^2)x^2 + a^2y^2 = a^4 - a^2c^2 = a^2(a^2 - c^2)$$

$$\frac{x^2}{a^2} + \frac{y^2}{a^2 - c^2} = 1 \qquad \boxed{4}$$

This is Eq. (2) with $b^2 = a^2 - c^2$ as claimed.

The axes intersect the ellipse in four points called **vertices** [labeled A, A', B, and B' in Figure 2(B)]. The segments $\overline{AA'}$ and $\overline{BB'}$ are called the major and minor axes of the ellipse. Following common usage, the numbers a and b are referred to as the semimajor axis and semiminor axis (even though they are numbers rather than axes).

THEOREM 1 Equation of an Ellipse in Standard Position Let $a, b > 0$ be constants with $a > b > 0$, and set $c = \sqrt{a^2 - b^2}$. Then

$$\left(\frac{x}{a}\right)^2 + \left(\frac{y}{b}\right)^2 = 1 \qquad \boxed{5}$$

is an equation of the ellipse

$$PF_1 + PF_2 = 2a$$

with foci $F_1 = (c, 0)$ and $F_2 = (-c, 0)$. Furthermore, the ellipse has:

- Semimajor axis a, semiminor axis b.
- Vertices $(\pm a, 0)$, $(0, \pm b)$.

If $b > a > 0$, then (5) defines an ellipse with foci $(0, \pm c)$, where $c = \sqrt{b^2 - a^2}$.

EXAMPLE 1 Find the equation of the ellipse with foci $(\pm\sqrt{11}, 0)$ and semimajor axis $a = 6$. Then find the semiminor axis and sketch the graph.

Solution The foci are $(\pm c, 0)$ with $c = \sqrt{11}$ and the semimajor axis is $a = 6$, so we may use the relation $c = \sqrt{a^2 - b^2}$ to find b:

$$b^2 = a^2 - c^2 = 6^2 - (\sqrt{11})^2 = 25$$

Thus, the semiminor axis is $b = 5$ and the ellipse has equation $\left(\frac{x}{6}\right)^2 + \left(\frac{y}{5}\right)^2 = 1$. To sketch an ellipse, plot the vertices $(\pm 6, 0)$ and $(0, \pm 5)$ and connect them as in Figure 3. ■

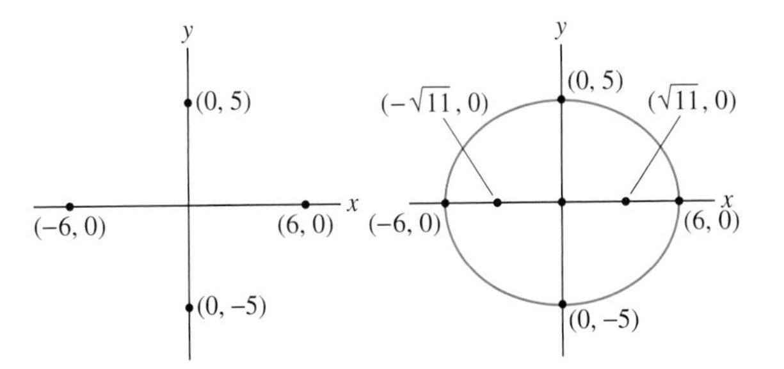

FIGURE 3 To sketch an ellipse, first draw the vertices.

We can easily write down the equation of an ellipse with axes parallel to the x- and y-axes and center translated to the point $C = (x_0, y_0)$ (Figure 4). The equation of the

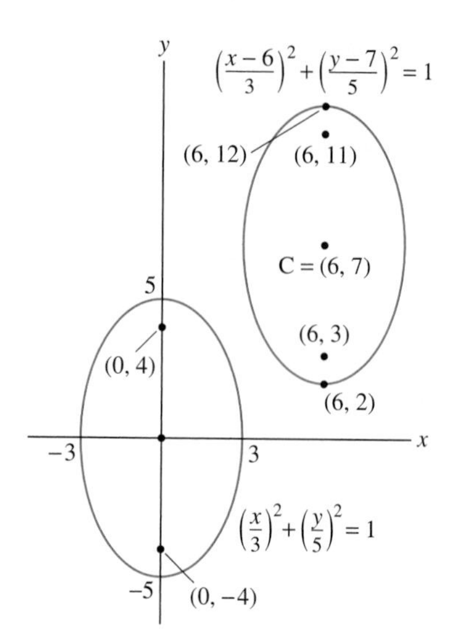

FIGURE 4 An ellipse with vertical major axis and its translate with center $C = (6, 7)$.

translated ellipse is

$$\left(\frac{x - x_0}{a}\right)^2 + \left(\frac{y - y_0}{b}\right)^2 = 1$$

EXAMPLE 2 Translating the Center of an Ellipse Find the equation of the ellipse with vertical focal axis, center at $C = (6, 7)$, semimajor axis 5, and semiminor axis 3. Where are the foci located?

Solution First, we find the equation of the ellipse centered at the origin. Since the focal axis is vertical, we use Eq. (5) with $a < b$. We set $a = 3$ and $b = 5$ to obtain $\left(\frac{x}{3}\right)^2 + \left(\frac{y}{5}\right)^2 = 1$. Furthermore, $c = \sqrt{b^2 - a^2} = \sqrt{5^2 - 3^2} = 4$, so the foci are $F_1 = (0, 4)$ and $F_2 = (0, -4)$ on the y-axis. When we translate the ellipse so that its center is $(6, 7)$, the equation becomes

$$\left(\frac{x - 6}{3}\right)^2 + \left(\frac{y - 7}{5}\right)^2 = 1$$

The foci are still ± 4 vertical units from the center, so the foci of the translated ellipse are $F_1 = (6, 11)$ and $F_2 = (6, 3)$ (Figure 4). ■

A hyperbola is the set of all points P such that the difference of the distances to two fixed points F_1 and F_2 is $\pm K$, where $K > 0$ is a constant:

$$PF_1 - PF_2 = \pm K \qquad \boxed{6}$$

The points F_1 and F_2 are called the foci of the hyperbola. We assume that K is less than the distance F_1F_2 between the foci. Note that a hyperbola consists of two branches corresponding to the choices of sign $\pm$ (Figure 5).

As before, the midpoint of $\overline{F_1F_2}$ is the center of the hyperbola, the line through F_1 and F_2 is the focal axis, and the line through the center perpendicular to the focal axis is the conjugate axis. The vertices are the points where the focal axis intersects the hyperbola, labeled A and A' in Figure 5. The hyperbola is said to be in standard position

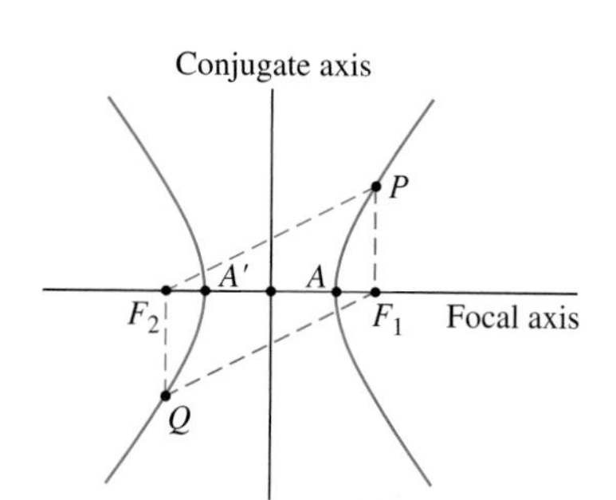

FIGURE 5 A hyperbola with center (0, 0).

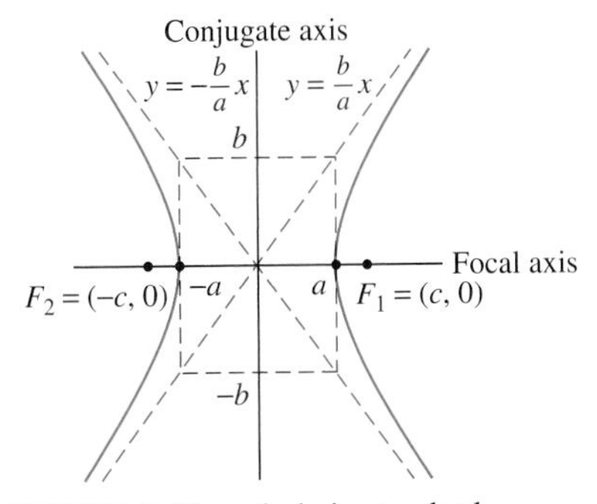

FIGURE 6 Hyperbola in standard position.

when the focal and conjugate axes are the x- and y-axes as in Figure 6. In this case, the foci are $(\pm c, 0)$ for some $c > 0$ and the equation of the hyperbola has the simple form (7) below, with $a = K/2$ and $b = \sqrt{c^2 - a^2}$. The vertices are $A = (a, 0)$ and $A' = (-a, 0)$ (see Exercise 64).

THEOREM 2 Equation of a Hyperbola in Standard Position Let a, b be positive constants and set $c = \sqrt{a^2 + b^2}$. Then

$$\left(\frac{x}{a}\right)^2 - \left(\frac{y}{b}\right)^2 = 1 \qquad \boxed{7}$$

is the equation of the hyperbola

$$PF_1 - PF_2 = \pm 2a$$

with foci $F_1 = (c, 0)$ and $F_2 = (-c, 0)$.

A hyperbola has two **asymptotes** that may be determined by drawing the rectangle whose sides pass through $(\pm a, 0)$ and $(0, \pm b)$ as in Figure 6. Let us prove that the asymptotes are the diagonals of this rectangle, that is, the lines with equations $y = \frac{b}{a}x$. Consider a point (x, y) on the hyperbola in the first quadrant. Thus we assume that $x > 0$ and $y > 0$. Equation (7) yields

$$y = \sqrt{\frac{b^2}{a^2}x^2 - b^2} = \frac{b}{a}\sqrt{x^2 - a^2}$$

To show that the point (x, y) on the hyperbola approaches the line $y = \frac{b}{a}x$ as $x \to \infty$, we verify that the following limit is zero:

$$\begin{aligned}
\lim_{x\to\infty}\left(y - \frac{b}{a}x\right) &= \frac{b}{a}\lim_{x\to\infty}\left(\sqrt{x^2 - a^2} - x\right) \\
&= \frac{b}{a}\lim_{x\to\infty}\left(\sqrt{x^2 - a^2} - x\right)\left(\frac{\sqrt{x^2 - a^2} + x}{\sqrt{x^2 - a^2} + x}\right) \\
&= \frac{b}{a}\lim_{x\to\infty}\left(\frac{-a^2}{\sqrt{x^2 - a^2} + x}\right) = 0
\end{aligned}$$

The asymptotic behavior in the remaining quadrants may be verified by a similar computation.

EXAMPLE 3 Find the foci of the hyperbola $9x^2 - 4y^2 = 36$. Sketch its graph and asymptotes.

Solution First divide by 36 to write the equation in standard form:

$$\frac{x^2}{4} - \frac{y^2}{9} = 1 \qquad \text{or} \qquad \left(\frac{x}{2}\right)^2 - \left(\frac{y}{3}\right)^2 = 1$$

Thus, $a = 2$, $b = 3$, and $c = \sqrt{a^2 + b^2} = \sqrt{4 + 9} = \sqrt{13}$. The foci are $F_1 = (\sqrt{13}, 0)$ and $F_2 = (-\sqrt{13}, 0)$. To sketch the graph, we draw the rectangle through the points

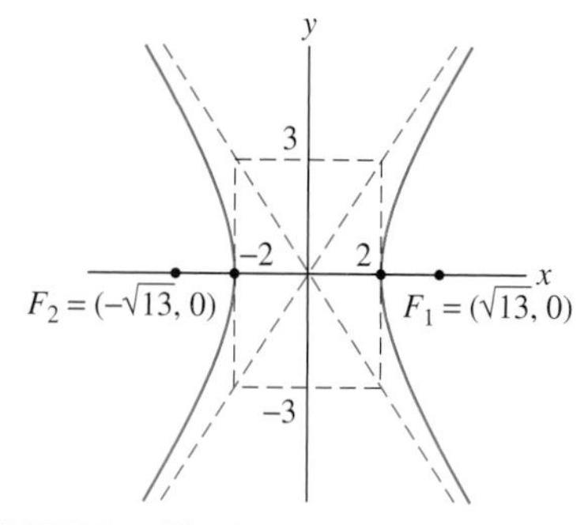

FIGURE 7 The hyperbola $9x^2 - 4y^2 = 36$.

$(\pm 2, 0)$ and $(0, \pm 3)$ as in Figure 7. The diagonals of the rectangle are the asymptotes $y = \pm \frac{3}{2}x$. We sketch the hyperbola so that it passes through the vertices $(\pm 2, 0)$ and approaches the asymptotes. ■

Finally, we consider the parabola. Unlike the ellipse and hyperbola, which are defined in terms of two foci, a parabola is the set of points P equidistant from a focus F and a line $\mathcal{D}$ called the **directrix**:

$$PF = P\mathcal{D} \tag{8}$$

Here, when we speak of the *distance* from a point P to a line $\mathcal{D}$, we mean the distance from P to the point Q on $\mathcal{D}$ closest to P, obtained by dropping a perpendicular from P to $\mathcal{D}$ (Figure 8). We denote this distance by $P\mathcal{D}$.

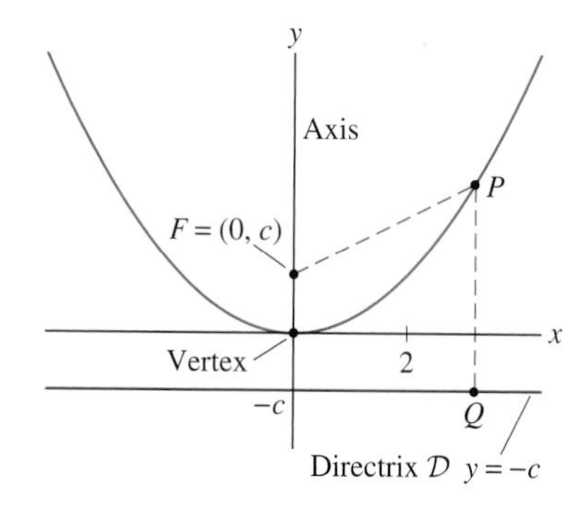

FIGURE 8 Parabola with focus $(0, c)$ and directrix $y = -c$.

The line through the focus F perpendicular to $\mathcal{D}$ is an axis of symmetry and is called the axis of the parabola. It intersects the parabola at a unique point called the vertex. We say that the parabola is in standard position if, for some c, the focus is $F = (0, c)$ and the directrix is $y = -c$, as shown in Figure 8. As we verify in Exercise 71, the vertex is then located at the origin and the equation of the parabola is $y = \dfrac{x^2}{4c}$. If $c < 0$, then the parabola opens downward.

> **THEOREM 3 Equation of a Parabola in Standard Position** Let $c \neq 0$. The parabola with focus $F = (0, c)$ and directrix $y = -c$ has equation
>
> $$y = \frac{1}{4c}x^2 \tag{9}$$
>
> The vertex is located at the origin. If $c < 0$, then the parabola opens downward.

■ **EXAMPLE 4** Find the equation of the standard parabola with directrix $y = -2$. Where are its focus and vertex located? What is the equation if this parabola is translated so that its vertex is located at $(2, 4)$, and what are the directrix and focus in this case?

Solution We apply Eq. (9) with $c = 2$. The standard parabola with directrix $y = -2$ has equation $y = \dfrac{x^2}{8}$ (Figure 9). The focus is $(0, c) = (0, 2)$ and the vertex is the origin $(0, 0)$. Note that the focus is two units above the vertex.

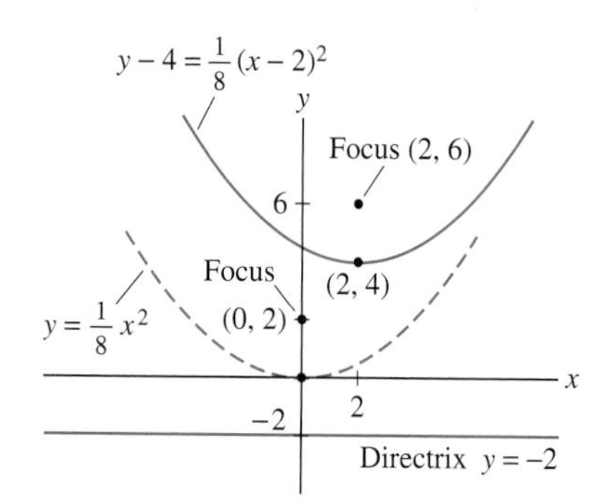

FIGURE 9 A parabola and its translate.

When the vertex is translated to (2, 4), the equation becomes

$$y - 4 = \frac{(x-2)^2}{8} \qquad \text{or} \qquad y = \frac{1}{8}x^2 - \frac{1}{2}x + \frac{9}{2}$$

The focus F is still two units above the vertex, so now $F = (2, 6)$ (see Figure 9). The directrix is the horizontal line located two units below the vertex, namely $y = 2$. ■

Eccentricity

Some ellipses are flatter than others, just as some hyperbolas are steeper. The "shape" of a conic section is measured by a quantity e called the **eccentricity**. For an ellipse or hyperbola,

$$e = \frac{\text{distance betweeen foci}}{\text{distance between vertices on focal axis}}$$

A parabola is defined to have eccentricity $e = 1$.

The eccentricity of an ellipse or hyperbola in standard position is given by the formula $e = c/a$ in the notation used above. To check this, recall that for a standard ellipse with equation $\left(\frac{x}{a}\right)^2 + \left(\frac{y}{b}\right)^2 = 1$, where $a > b > 0$, we set $c = \sqrt{a^2 - b^2}$. The foci are located at $(\pm c, 0)$ and the vertices on the focal axis at $(\pm a, 0)$. Therefore,

$$e = \frac{\text{distance between foci}}{\text{distance between vertices on focal axis}} = \frac{2c}{2a} = \frac{c}{a}$$

Similarly, the standard hyperbola $\left(\frac{x}{a}\right)^2 - \left(\frac{y}{b}\right)^2 = 1$ has foci at $(\pm c, 0)$, where $c = \sqrt{a^2 + b^2}$, and the vertices on the focal axis at $(\pm a, 0)$. Again, we obtain $e = c/a$. In summary:

$$e = \frac{c}{a} \qquad \text{where} \quad c = \begin{cases} \sqrt{a^2 - b^2} & \text{for an ellipse} \\ \sqrt{a^2 + b^2} & \text{for a hyperbola} \end{cases} \qquad \boxed{10}$$

The eccentricity of an ellipse satisfies $0 \le e < 1$ since $c = \sqrt{a^2 - b^2} < a$. If $e = 0$, the ellipse is a circle because in this case, $a = b$. Circles are the "roundest" possible ellipses and, as we see in Figure 10(A), the larger the eccentricity, the flatter the ellipse. In fact, it is straightforward to check that

$$\frac{b}{a} = \sqrt{1 - e^2}$$

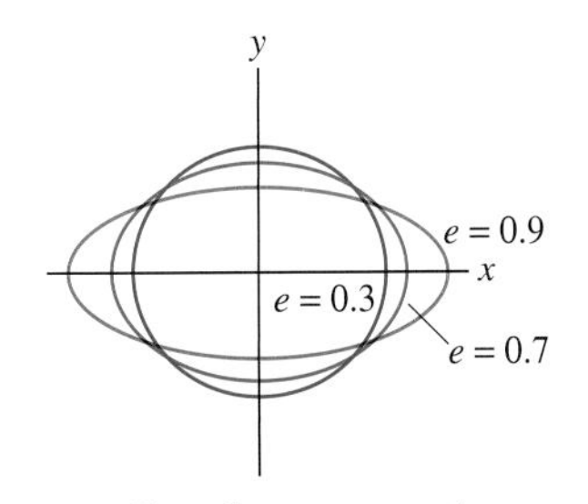

(A) Ellipse flattens as $e \to 1$.

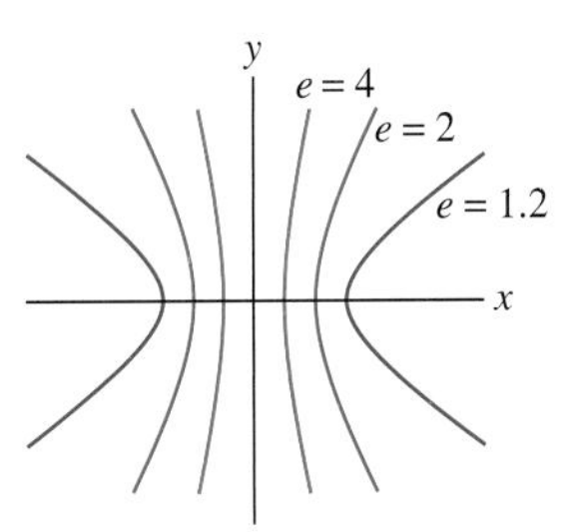

(B) Asymptotes of the hyperbola become more vertical as $e \to \infty$.

FIGURE 10 Effect of changing eccentricity.

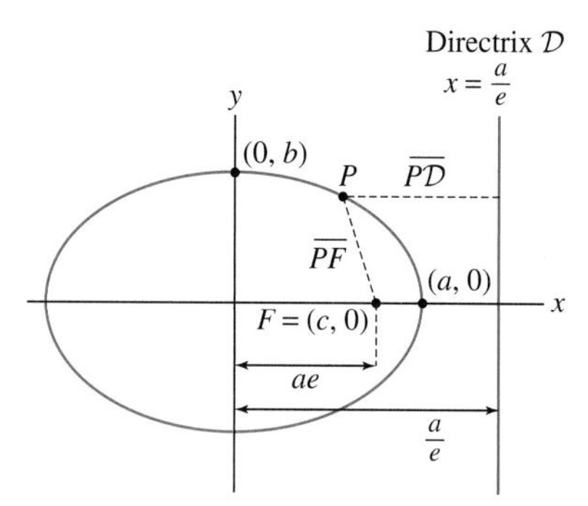

FIGURE 11 The ellipse consists of points P such that $PF = ePD$.

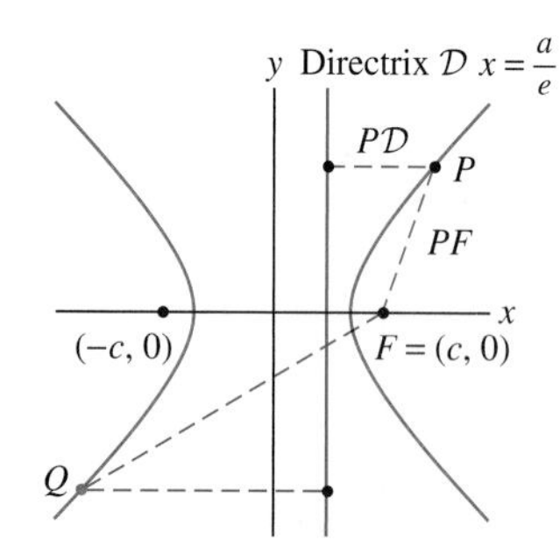

FIGURE 12 The hyperbola consists of points P such that $PF = eP\mathcal{D}$.

As $e \to 1$, the ratio b/a tends to zero. But b/a is the ratio of the semiminor axis to the semimajor axis, and thus a smaller value of b/a results in a flatter ellipse.

A hyperbola has eccentricity $e > 1$ since $c = \sqrt{a^2 + b^2} < a$. As e increases, the asymptotes of the hyperbola become steeper. We check this by observing that $b/a = \sqrt{1 + e^2}$. The asymptotes of the hyperbola have slope $\pm b/a$ and thus the slopes of asymptotes approach infinity as $e \to \infty$.

The notion of eccentricity may be used to give a unified focus-directrix definition of all three types of conics. Given a point F (the focus), a line $\mathcal{D}$ (the directrix), and a number $e > 0$, we consider the set of all points P such that

$$PF = eP\mathcal{D} \qquad (11)$$

For $e = 1$, this is precisely our definition of a parabola. According to the next theorem, (11) defines a conic section of eccentricity e for all $e > 0$ (Figures 11 and 12). Note, however, that there is no focus-directrix definition for circles ($e = 0$).

THEOREM 4 Focus-Directrix Definition Let F be a point, $\mathcal{D}$ a line, and let $e > 0$. Then the set of points satisfying (11) is a conic section of eccentricity e. Furthermore,

- **Ellipse:** Let $a > b > 0$ and $c = \sqrt{a^2 - b^2}$. The ellipse

$$\left(\frac{x}{a}\right)^2 + \left(\frac{y}{b}\right)^2 = 1$$

satisfies the focus-directrix definition (11) with focus $F = (c, 0)$, eccentricity $e = \frac{c}{a}$, and vertical directrix $x = \frac{a}{e}$.

- **Hyperbola:** Let $a, b > 0$ and $c = \sqrt{a^2 + b^2}$. The hyperbola

$$\left(\frac{x}{a}\right)^2 - \left(\frac{y}{b}\right)^2 = 1$$

satisfies the focus-directrix definition (11) with focus $F = (c, 0)$, eccentricity $e = \frac{c}{a}$, and vertical directrix $x = \frac{a}{e}$.

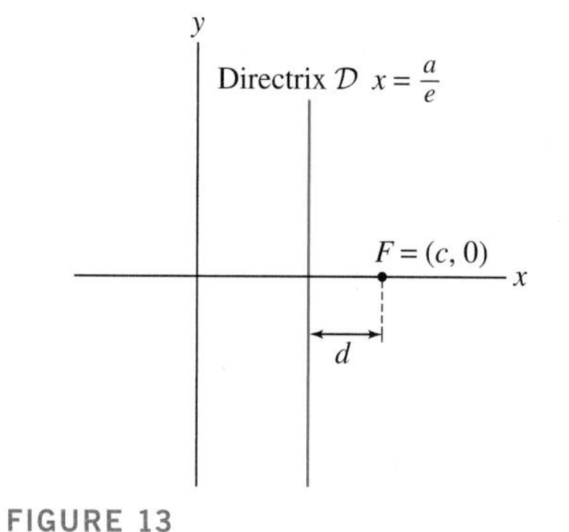

FIGURE 13

Proof We assume that $e > 1$ and prove that $PF = eP\mathcal{D}$ defines a hyperbola (the case $e < 1$ is similar, see Exercise 65). We may choose our coordinate axes so that the focus F lies on the x-axis and the directrix is vertical, lying to the left of F as in Figure 13. Anticipating the final result, we let d be the distance from the focus F to the directrix $\mathcal{D}$ and set

$$c = \frac{d}{1 - e^{-2}}, \quad a = \frac{c}{e}, \quad b = \sqrt{c^2 - a^2}$$

Since we are free to shift the y-axis, let us choose the y-axis so that the focus has coordinates $F = (c, 0)$. Then the directrix is the line

$$x = c - d = c - c(1 - e^{-2}) = c\,e^{-2} = \frac{a}{e}$$

Now, the equation $PF = eP\mathcal{D}$ for a point $P = (x, y)$ may be written

$$\underbrace{\sqrt{(x - c)^2 + y^2}}_{PF} = e\underbrace{\sqrt{(x - (a/e))^2}}_{P\mathcal{D}}$$

Algebraic manipulation yields

$$(x-c)^2 + y^2 = e^2(x-(a/e))^2 \qquad \text{(square)}$$

$$x^2 - 2cx + c^2 + y^2 = e^2x^2 - 2aex + a^2 \qquad \text{(note: } c = ae\text{)}$$

$$(e^2-1)x^2 - y^2 = c^2 - a^2 = a^2(e^2-1) \qquad \text{(rearrange)}$$

$$\frac{x^2}{a^2} - \frac{y^2}{a^2(e^2-1)} = 1 \qquad \text{(divide)}$$

Because $a^2(e^2-1) = c^2 - a^2 = b^2$, we obtain the equation of the hyperbola

$$\left(\frac{x}{a}\right)^2 - \left(\frac{y}{b}\right)^2 = 1$$

as claimed. ■

■ **EXAMPLE 5** Find the equation of the standard ellipse with eccentricity $e = 0.8$ and vertices $(\pm 10, 0)$. What are the foci and directrix?

Solution The vertices are $(\pm a, 0)$ with $a = 10$, so the ellipse has equation

$$\left(\frac{x}{10}\right)^2 + \left(\frac{y}{b}\right)^2 = 1$$

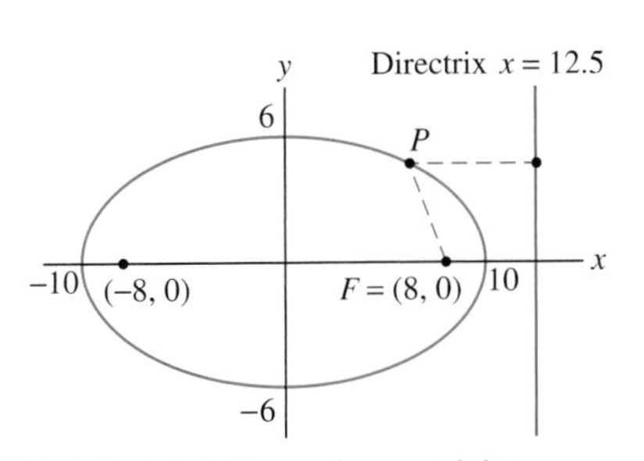

FIGURE 14 Ellipse of eccentricity $e = 0.8$ with focus at $(8, 0)$.

(Figure 14). By Theorem 4, $c = ae = 10(0.8) = 8$, and since $c = \sqrt{a^2 - b^2}$, we obtain $8 = \sqrt{10^2 - b^2}$ or $b = 6$. Thus, our ellipse has equation

$$\left(\frac{x}{10}\right)^2 + \left(\frac{y}{6}\right)^2 = 1$$

The foci are $(\pm c, 0) = (\pm 8, 0)$ and the directrix is $x = \dfrac{a}{e} = \dfrac{10}{0.8} = 12.5$. ■

CONCEPTUAL INSIGHT We can prove that if two conic sections C_1 and C_2 have the same eccentricity e, then they have the same shape in the following precise sense: It is possible to scale C_1 so that C_1 and C_2 become congruent. By scaling, we mean changing the units along the x- and y-axes by a common positive factor. A curve scaled by a factor of 10 would have the same shape but would be ten times as large. This would correspond, for example, to changing units from centimeters to millimeters (smaller units make for a larger figure). On the other hand, by "congruent" we mean that after scaling, it is possible to move C_1 by a rigid motion (that is, without stretching or bending) so that it lies directly on top of C_2.

All circles ($e = 0$) have the same shape because scaling by a factor $r > 0$ transforms a circle of radius R into a circle of radius rR. Similarly, any two parabolas ($e = 1$) become congruent after suitable scaling. However, an ellipse of eccentricity $e = 0.5$ cannot be made congruent to an ellipse of eccentricity $e = 0.8$ by scaling (see Exercise 72).

In Section 13.6, we discuss the famous law of Johannes Kepler stating that the orbit of a planet around the sun is an ellipse with one focus at the sun. In this discussion, we will need to write the equation of an ellipse in polar coordinates. To derive the polar equations of the conic sections, it is convenient to use the focus-directrix definition with focus F at the origin O, and vertical line $x = d$ as directrix $\mathcal{D}$. By Eq. (11), the equation

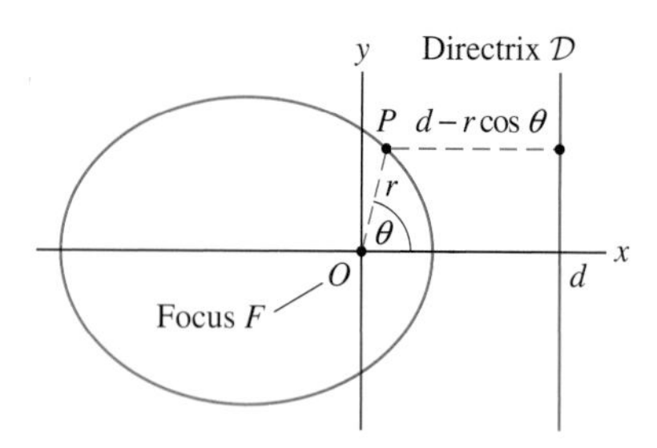

FIGURE 15 Focus-directrix definition of the ellipse in polar coordinates.

of the ellipse is $PF = eP\mathcal{D}$. Referring to Figure 15, we see that if $P = (r, \theta)$, then

$$PF = r, \qquad P\mathcal{D} = d - r\cos\theta$$

Thus, the equation of the ellipse $PF = eP\mathcal{D}$ becomes $r = e(d - r\cos\theta)$ or $r(1 + e\cos\theta) = ed$. This proves the following result, which is also valid for the hyperbola and parabola (see Exercise 66).

Polar Equation of a Conic Section The conic section of eccentricity $e > 0$ with focus at the origin and directrix at $x = d$ has polar equation

$$r = \frac{ed}{1 + e\cos\theta} \tag{12}$$

Reflective Properties of Conic Sections

The conic sections have numerous geometric properties that are of importance in applications, especially in optics and communications (for example, in antenna and telescope design; Figure 16). Here, we limit ourselves to mentioning (without proof) the *reflective properties* of each family of conics. See Exercises 67–69 for a proof of the reflective property of an ellipse.

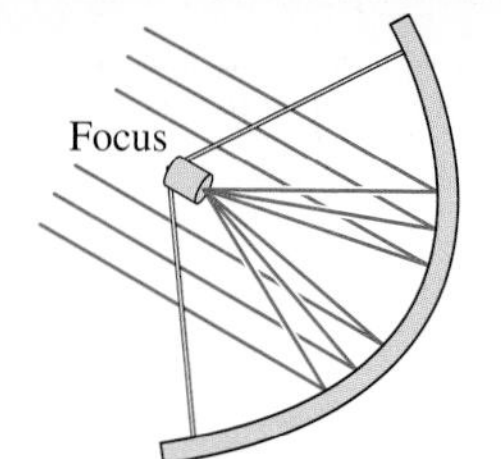

FIGURE 16 The paraboloid shape of this radio telescope directs the incoming signal to the focus.

- **Ellipse:** The segments F_1P and F_2P make equal angles with the tangent line at a point P on the ellipse. Therefore, a beam of light originating at focus F_1 is reflected off the ellipse toward the second focus F_2 [Figure 17(A)].

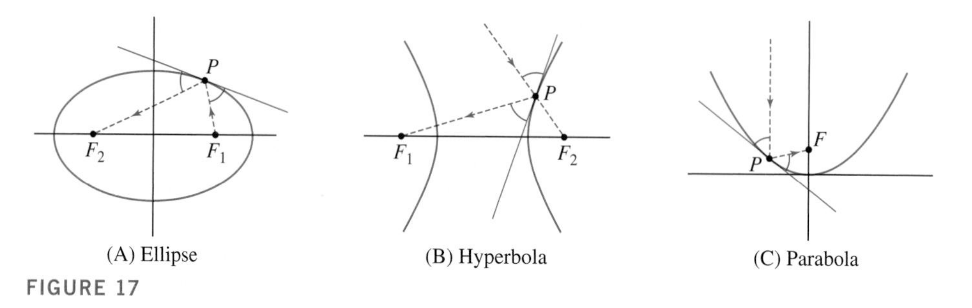

FIGURE 17

- **Hyperbola:** The tangent line at a point P on the hyperbola bisects the angle formed by the segments F_1P and F_2P. Therefore, a beam of light directed toward F_2 is reflected off the hyperbola toward the second focus F_1 [Figure 17(B)].
- **Parabola:** The segment FP and the line through P perpendicular to the directrix make equal angles with the tangent line at a point P on the parabola. Therefore, a beam of light approaching P from above, in the direction perpendicular to the directrix, is reflected off the parabola toward the focus F [Figure 17(C)].

General Equations of Degree Two

We now consider briefly the general equation of degree two in x and y:

$$Ax^2 + Bxy + Cy^2 + Dx + Ey + F = 0 \tag{13}$$

where A, B, C, D, E, F are constants with A, B, C not all zero. The equations of the conic sections in standard position are special cases of this general quadratic equation. For example, the equation of a standard ellipse may be written in the form

$$Ax^2 + Cy^2 - 1 = 0$$

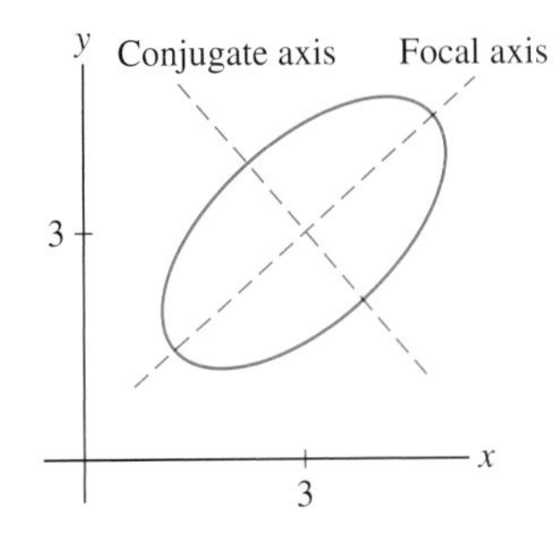

FIGURE 18 The ellipse with equation $6x^2 - 8xy + 8y^2 - 12x - 24y + 38 = 0$.

It turns out that the general quadratic equation does not give rise to any new types of curves. Apart from certain "degenerate cases," Eq. (13) defines a conic section with arbitrary center whose focal and conjugate axes may be rotated relative to the coordinate axes. For example, the equation

$$6x^2 - 8xy + 8y^2 - 12x - 24y + 38 = 0$$

defines an ellipse with center at (3, 3) whose axes are rotated (Figure 18).

We say that Eq. (13) is degenerate if the set of solutions is a pair of intersecting lines, a pair of parallel lines, a single line, a point, or the empty set. For example:

- $x^2 - y^2 = 0$ defines a pair of intersecting lines $y = x$ and $y = -x$.
- $x^2 - x = 0$ defines a pair of parallel lines $x = 0$ and $x = 1$.
- $x^2 = 0$ defines a single line (the y-axis).
- $x^2 + y^2 = 0$ has just one solution $(0, 0)$.
- $x^2 + y^2 = -1$ has no solutions.

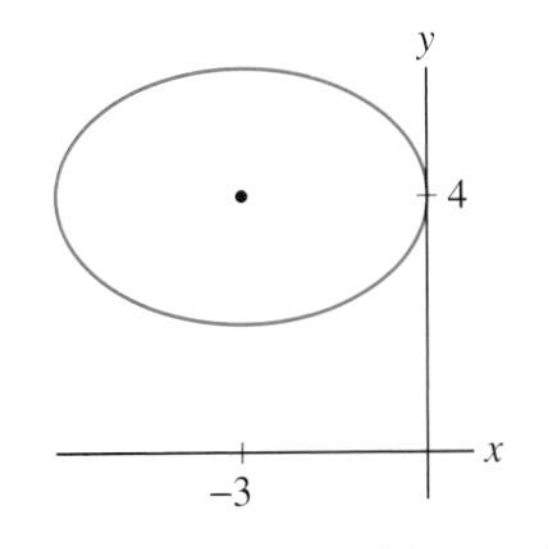

FIGURE 19 The ellipse with equation $4x^2 + 9y^2 + 24x - 72y + 144 = 0$.

Now assume that Eq. (13) is nondegenerate. The term Bxy is called the *cross term*. When the cross term is zero (i.e., $B = 0$), we can "complete the square" to show that Eq. (13) defines a translate of conic in standard position. In other words, the axes of the conic are parallel to the coordinate axes. This is illustrated in the next example.

■ **EXAMPLE 6 Completing the Square** Show that $4x^2 + 9y^2 + 24x - 72y + 144 = 0$ defines a translate of a conic section in standard position (Figure 19).

Solution Since there is no cross term, we may complete the square of the terms involving x and y terms separately:

$$4x^2 + 9y^2 + 24x - 72y + 144 = 4(x^2 + 6x) + 9(y^2 - 8y) + 144 = 0$$

$$4(x+3)^2 - 36 + 9(y-4)^2 - 144 + 144 = 0$$

$$4(x+3)^2 + 9(y-4)^2 = 36$$

Therefore, this quadratic equation can be rewritten in the form

$$\left(\frac{x+3}{3}\right)^2 + \left(\frac{y-4}{2}\right)^2 = 1$$ ■

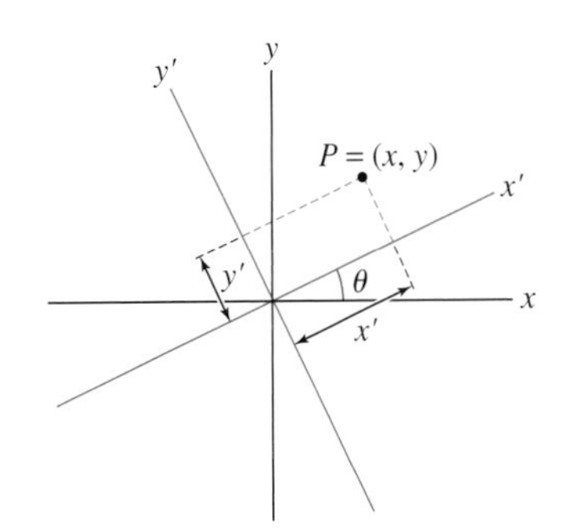

FIGURE 20

More generally, if (x', y') are coordinates relative to axes rotated by an angle θ as in Figure 20, then

$$x = x' \cos\theta - y' \sin\theta \qquad \boxed{14}$$

$$y = x' \sin\theta + y' \cos\theta \qquad \boxed{15}$$

In Exercise 73, we show that the cross term disappears when Eq. (13) is rewritten in terms of x' and y' for the angle

$$\theta = \frac{1}{2}\cot^{-1}\frac{A-C}{B} \qquad \boxed{16}$$

When the cross-term Bxy is nonzero, Eq. (13) defines a conic whose axes are rotated relative to the coordinate axes. The marginal note describes how this may be verified in general. We illustrate with the following example.

■ **EXAMPLE 7** Show that $2xy = 1$ defines a conic section whose focal and conjugate axes are rotated relative to the coordinate axes.

Solution Figure 21(A) shows axes labeled x' and y' that are rotated by 45° relative to the standard coordinate axes. A point P with coordinates (x, y) may also be described by coordinates (x', y') relative to these rotated axes. Applying (14) and (15) with $\theta = \frac{\pi}{4}$, we find that (x, y) and (x', y') are related by the formulas

$$x = \frac{x' - y'}{\sqrt{2}}, \qquad y = \frac{x' + y'}{\sqrt{2}}$$

Therefore, if $P = (x, y)$ lies on the hyperbola, that is, if $2xy = 1$, then

$$2xy = 2\left(\frac{x' - y'}{\sqrt{2}}\right)\left(\frac{x' + y'}{\sqrt{2}}\right) = x'^2 - y'^2 = 1$$

Thus, the coordinates (x', y') satisfy the equation of the standard hyperbola $x'^2 - y'^2 = 1$ whose focal and conjugate axes are the x'- and y'-axes, respectively. ■

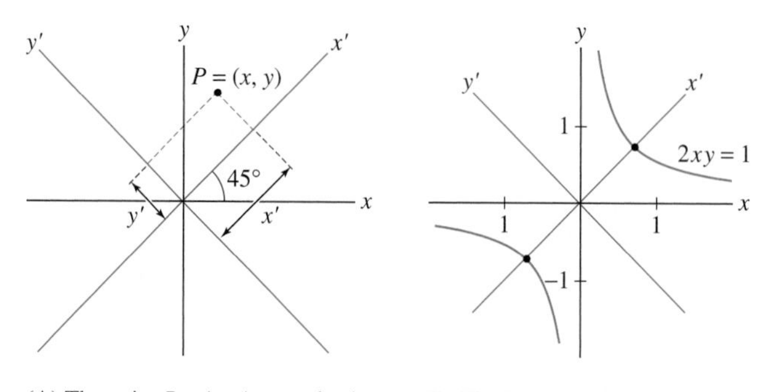

FIGURE 21 The x'- and y'-axes are rotated at a 45° angle relative to the x- and y-axes.

(A) The point $P = (x, y)$ may also be described by coordinates (x', y') relative to the rotated axis.

(B) The hyperbola $2xy = 1$ is in standard position relative to the x', y' axes.

We conclude our discussion of conics by stating the Discriminant Test. Suppose that the equation

$$Ax^2 + Bxy + Cy^2 + Dx + Ey + F = 0$$

is nondegenerate and thus defines a conic section. The type of conic is determined by a quantity D called the **discriminant** of the equation:

$$D = B^2 - 4AC$$

According to the Discriminant Test (which we state without proof), the curve is

- an ellipse if $D < 0$.
- a hyperbola if $D > 0$.
- a parabola if $D = 0$.

For example, the discriminant of the equation $2xy = 1$ is

$$D = B^2 - 4AC = 2^2 - 0 = 4 > 0$$

According to the Discriminant Test, $2xy = 1$ defines a hyperbola. This agrees with our conclusion in Example 7.

11.5 SUMMARY

- An *ellipse* with foci F_1 and F_2 is the set of points P such that $PF_1 + PF_2 = K$, where $K > 0$ is a constant. The equation of an ellipse in standard position is

$$\left(\frac{x}{a}\right)^2 + \left(\frac{y}{b}\right)^2 = 1$$

	Focal Axis	Foci	Vertices
$a > b$	x-axis	$(\pm c, 0)$ with $c = \sqrt{a^2 - b^2}$	$(\pm a, 0)$, $(0, \pm b)$
$a < b$	y-axis	$(0, \pm c)$ with $c = \sqrt{b^2 - a^2}$	$(\pm a, 0)$, $(0, \pm b)$

- A *hyperbola* with foci F_1 and F_2 is the set of points P such that

$$PF_1 - PF_2 = \pm K$$

where $K > 0$ is a constant. The equation of a hyperbola in standard position is

$$\left(\frac{x}{a}\right)^2 - \left(\frac{y}{b}\right)^2 = 1$$

The focal axis is the x-axis and the foci are $(\pm c, 0)$ with $c = \sqrt{a^2 + b^2}$. The vertices are $(\pm a, 0)$ and the asymptotes are the lines $y = \pm \frac{b}{a} x$.

- A *parabola* with focus F and directrix $\mathcal{D}$ is the set of points P such that $PF = P\mathcal{D}$. A parabola in standard position, with focus $F = (0, c)$ and directrix $y = -c$ with $c \neq 0$, has equation $y = \frac{1}{4c}x^2$. The vertex of a parabola in standard position is the origin $(0, 0)$.
- To translate a conic so that its center is located at (x_0, y_0) (while keeping its axes parallel to the x- and y-axes), replace x and y by $x - x_0$ and $y - y_0$, respectively, in the equation of the conic.
- The *eccentricity* e of an ellipse or a hyperbola is the quantity

$$e = \frac{\text{distance betweeen foci}}{\text{distance between vertices on focal axis}}$$

In standard position, $e = \frac{c}{a}$. For an ellipse, $0 \le e < 1$; for a hyperbola, $e > 1$. By definition, a parabola has eccentricity $e = 1$.

- *Focus-directrix definition of conic sections:* Given a point P (a focus), a line $\mathcal{D}$ (a directrix), and a number $e > 0$, the set of points P such that $PF = eP\mathcal{D}$ is a conic section of eccentricity e.
- The conic section of eccentricity $e > 0$ with focus at the origin and directrix at $x = d$ has polar equation

$$r = \frac{ed}{1 + e\cos\theta}$$

11.5 EXERCISES

Preliminary Questions

1. Which of the following equations defines an ellipse? Which does not define a conic section?

(a) $4x^2 - 9y^2 = 12$ **(b)** $-4x + 9y^2 = 0$

(c) $4y^2 + 9x^2 = 12$ **(d)** $4x^3 + 9y^3 = 12$

2. For which conic sections do the vertices lie between the foci?

3. What are the foci of $\left(\frac{x}{a}\right)^2 + \left(\frac{y}{b}\right)^2 = 1$ if $a < b$?

4. For a hyperbola in standard position, the set of points equidistant from the foci is the y-axis. Use the definition $PF_1 - PF_2 = \pm K$ to explain why the hyperbola does not intersect the y-axis.

5. What is the geometric interpretation of the quantity $\frac{b}{a}$ in the equation of a hyperbola in standard position?

Exercises

In Exercises 1–8, find the vertices and foci of the conic section.

1. $\left(\frac{x}{9}\right)^2 + \left(\frac{y}{4}\right)^2 = 1$

2. $\left(\frac{x}{4}\right)^2 + \left(\frac{y}{9}\right)^2 = 1$

3. $\frac{x^2}{9} + \frac{y^2}{4} = 1$

4. $\left(\frac{x}{9}\right)^2 - \left(\frac{y}{4}\right)^2 = 1$

5. $\left(\frac{x}{4}\right)^2 - \left(\frac{y}{9}\right)^2 = 1$

6. $\frac{x^2}{9} - \frac{y^2}{4} = 1$

7. $\left(\frac{x-3}{7}\right)^2 - \left(\frac{y+1}{4}\right)^2 = 1$

8. $\left(\frac{x-3}{4}\right)^2 + \left(\frac{y+1}{7}\right)^2 = 1$

In Exercises 9–12, consider the ellipse

$$\left(\frac{x-12}{5}\right)^2 + \left(\frac{y-9}{7}\right)^2 = 1$$

Find the equation of the translated ellipse.

9. Translated so that its center is at the origin

10. Translated four units to the right

11. Translated three units down

12. Translated so its center is $(-4, -9)$

In Exercises 13–16, find the equation of the ellipse with the given properties.

13. Vertices at $(\pm 9, 0)$ and $(0, \pm 16)$

14. Foci $(\pm 6, 0)$ and two vertices at $(\pm 10, 0)$

15. Foci $(0, \pm 6)$ and two vertices at $(\pm 4, 0)$

16. Foci $(0, \pm 3)$ and eccentricity $\frac{3}{4}$

In Exercises 17–22, find the equation of the hyperbola with the given properties.

17. Vertices $(\pm 3, 0)$ and foci at $(\pm 5, 0)$

18. Vertices $(0, \pm 5)$ and foci $(0, \pm 8)$

19. Vertices $(\pm 4, 0)$ and asymptotes $y = \pm 3x$

20. Vertices $(\pm 3, 0)$ and asymptotes $y = \pm\frac{1}{2}x$

21. Vertices $(0, -5)$, $(0, 4)$ and foci $(0, -8)$, $(0, 7)$

22. Foci $(0, \pm 3)$ and eccentricity $\frac{3}{4}$

In Exercises 23–30, find the equation of the parabola with the given properties.

23. Vertex $(0, 0)$, focus $(2, 0)$

24. Vertex $(0, 0)$, focus $(0, \frac{1}{2})$

25. Vertex $(0, 0)$, directrix $y = -5$

26. Vertex $(0, 0)$, directrix $y = -\frac{1}{8}$

27. Focus $(0, 4)$, directrix $y = -4$

28. Focus $(0, -4)$, directrix $y = 4$

29. Focus $(2, 0)$, directrix $x = -2$

30. Focus $(-2, 0)$, directrix $x = 2$

In Exercises 31–40, find the vertices, foci, axes, center (if an ellipse or a hyperbola) and asymptotes (if a hyperbola) of the conic section.

31. $x^2 + 4y^2 = 16$

32. $\left(\frac{x-3}{16}\right)^2 - \left(\frac{y+5}{49}\right)^2 = 1$

33. $4x^2 + y^2 = 16$

34. $3x^2 - 27y^2 = 12$

35. $4x^2 - 3y^2 + 8x + 30y = 215$

36. $y = 4x^2$

37. $y = 4(x-4)^2$

38. $8y^2 + 6x^2 - 36x - 64y + 134 = 0$

39. $4x^2 + 25y^2 - 8x - 10y = 20$

40. $y^2 - 2x^2 + 8x - 9y - 12 = 0$

In Exercises 41–44, use the Discriminant Test to determine the type of the conic section defined by the equation. You may assume that the equation is nondegenerate. Plot the curve if you have a computer algebra system.

41. $4x^2 + 5xy + 7y^2 = 24$

42. $2x^2 - 8xy + 3y^2 - 4 = 0$

43. $2x^2 - 8xy - 3y^2 - 4 = 0$

44. $x^2 - 2xy + y^2 + 24x - 8 = 0$

45. Show that $\frac{b}{a} = \sqrt{1 - e^2}$ for a standard ellipse of eccentricity e.

46. Show that the eccentricity of a hyperbola in standard position is $e = \sqrt{1 + m^2}$, where $\pm m$ are the slopes of the asymptotes.

47. Explain why the dots in Figure 22 lie on a parabola. Where are the focus and directrix located?

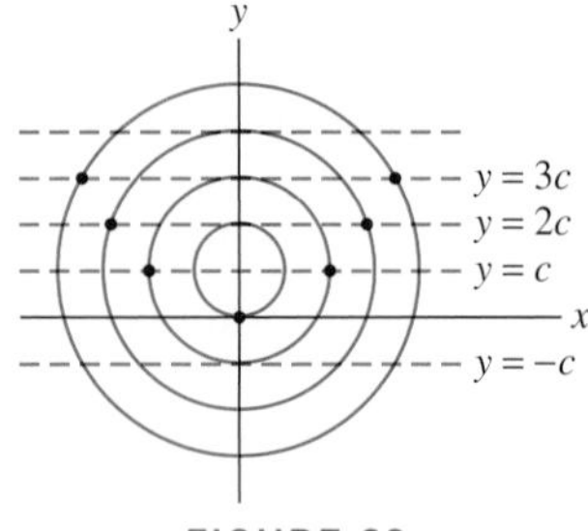

FIGURE 22

48. Show that the equation of the tangent line to the hyperbola $\left(\frac{x}{a}\right)^2 - \left(\frac{y}{b}\right)^2 = 1$ at a point (x_0, y_0) is

$$Ax - By = 1$$

where $A = \frac{x_0}{a^2}$ and $B = \frac{y_0}{b^2}$.

49. Kepler's First Law states that the orbits of the planets around the sun are ellipses with the sun at one focus. The orbit of Pluto has an eccentricity of approximately $e = 0.25$ and the **perihelion** (closest distance to the sun) of Pluto's orbit is approximately 2.7 billion miles. Find the **aphelion** (farthest distance from the sun).

50. Kepler's Third Law states that the ratio $T/a^{3/2}$ is equal to a constant C for all planetary orbits around the sun, where T is the period (time for a complete orbit) and a is the semimajor axis.

(a) Compute C in units of days and kilometers, given that the semimajor axis of the earth's orbit is 150×10^6 km.

(b) Compute the period of Saturn's orbit, given that its semimajor axis is approximately 1.43×10^9 km.

(c) Saturn's orbit has eccentricity $e = 0.056$. Find the perihelion and aphelion of Saturn.

In Exercises 51–54, find the polar equation of the conic with given eccentricity and directrix.

51. $e = \frac{1}{2}, \quad x = 3$

52. $e = \frac{1}{2}, \quad x = -3$

53. $e = 1, \quad x = 4$

54. $e = \frac{3}{2}, \quad x = -4$

In Exercises 55–58, identify the type of conic, the eccentricity, and the equation of the directrix.

55. $r = \dfrac{8}{1 + 4\cos\theta}$

56. $r = \dfrac{8}{4 + \cos\theta}$

57. $r = \dfrac{8}{4 + 3\cos\theta}$

58. $r = \dfrac{12}{4 + 3\cos\theta}$

59. Show that $r = f_1(\theta)$ and $r = f_2(\theta)$ define the same curves in polar coordinates if $f_1(\theta) = -f_2(\theta + \pi)$, and use this to show that the following define the same conic section:

$$r = \frac{de}{1 - e\cos\theta}, \qquad r = \frac{-de}{1 + e\cos\theta}$$

60. Find a polar equation for the hyperbola with focus at the origin, directrix $x = -2$, and eccentricity $e = 1.2$.

61. Find the equation of the ellipse $r = \dfrac{4}{2 + \cos\theta}$ in rectangular coordinates.

62. Let C be the ellipse $r = \dfrac{de}{1 + e\cos\theta}$, where $e < 1$. Express the x-coordinates of the vertices A, A', the center C, and the second focus F_2 in terms of d and e (Figure 23).

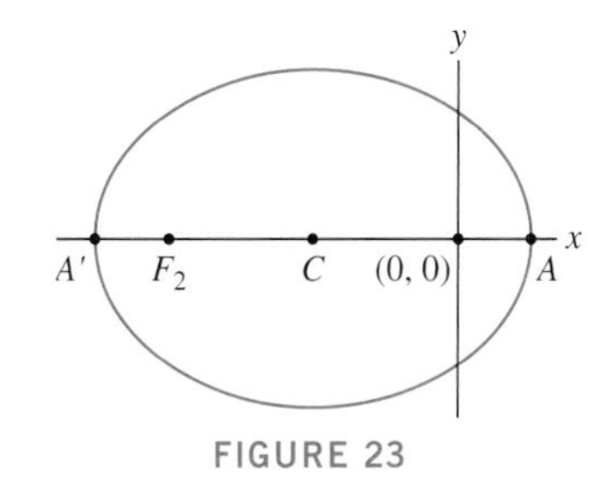

FIGURE 23

63. Let $e > 1$. Show that the vertices of the hyperbola $r = \dfrac{de}{1 + e\cos\theta}$ have x-coordinates $\dfrac{ed}{e+1}$ and $\dfrac{ed}{e-1}$.

Further Insights and Challenges

64. Verify Theorem 2.

65. Verify Theorem 4 in the case $0 < e < 1$. *Hint:* Repeat the proof of Theorem 4, but set $c = d/(e^{-2} - 1)$.

66. Verify that if $e > 1$, then Eq. (12) defines a hyperbola of eccentricity e, with its focus at the origin and directrix at $x = d$.

Reflective Property of the Ellipse *In Exercises 67–69, we prove that the focal radii at a point on an ellipse make equal angles with the tangent line. Let $P = (x_0, y_0)$ be a point on the ellipse $\left(\frac{x}{a}\right)^2 + \left(\frac{y}{b}\right)^2 = 1$ $(a > b)$ with foci $F_1 = (-c, 0)$ and $F_2 = (c, 0)$, and eccentricity e (Figure 24).*

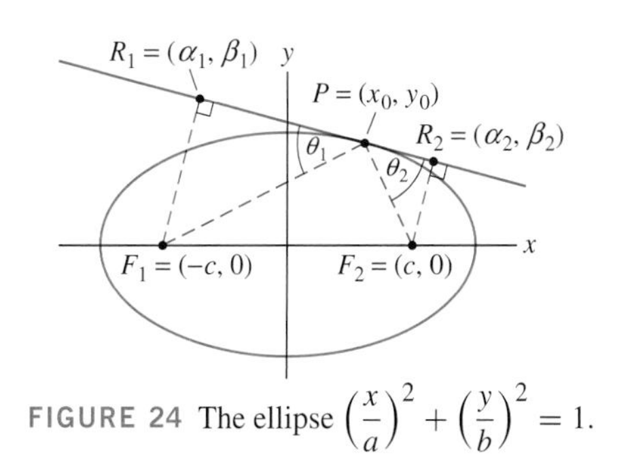

FIGURE 24 The ellipse $\left(\frac{x}{a}\right)^2 + \left(\frac{y}{b}\right)^2 = 1$.

67. Show that $PF_1 = a + x_0 e$ and $PF_2 = a - x_0 e$. *Hints:*

(a) Show that $PF_1^{\,2} - PF_2^{\,2} = 4x_0 c$.

(b) Divide the previous relation by $PF_1 + PF_2 = 2a$, and conclude that $PF_1 - PF_2 = 2x_0 e$.

68. Show that the equation of the tangent line at P is $Ax + By = 1$, where $A = \frac{x_0}{a^2}$ and $B = \frac{y_0}{b^2}$.

69. Define R_1 and R_2 as in the figure, so that $\overline{F_1R_1}$ and $\overline{F_2R_2}$ are perpendicular to the tangent line.

(a) Show that $\dfrac{\alpha_1 + c}{\beta_1} = \dfrac{\alpha_2 - c}{\beta_2} = \dfrac{A}{B}$.

(b) Use (a) and the distance formula to show that

$$\frac{F_1 R_1}{F_2 R_2} = \frac{\beta_1}{\beta_2}$$

(c) Solve for β_1 and β_2:

$$\beta_1 = \frac{B(1 + Ac)}{A^2 + B^2}, \qquad \beta_2 = \frac{B(1 - Ac)}{A^2 + B^2}$$

(d) Show that $\dfrac{F_1 R_1}{F_2 R_2} = \dfrac{PF_1}{PF_2}$. Conclude that $\theta_1 = \theta_2$.

70. Show that the length QR in Figure 25 is independent of the point P.

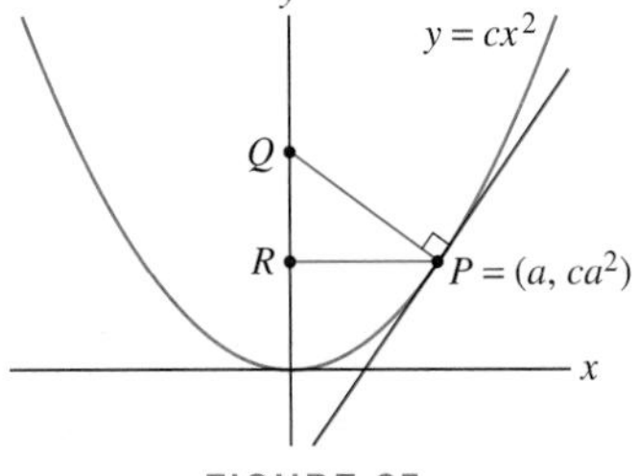

FIGURE 25

71. Show that $y = \dfrac{x^2}{4c}$ is the equation of a parabola with directrix $y = -c$, focus $(0, c)$, and the vertex at the origin, as stated in Theorem 3.

72. Consider two ellipses in standard position:

$$E_1: \quad \left(\frac{x}{a_1}\right)^2 + \left(\frac{y}{b_1}\right)^2 = 1$$

$$E_2: \quad \left(\frac{x}{a_2}\right)^2 + \left(\frac{y}{b_2}\right)^2 = 1$$

We say that E_1 is similar to E_2 under scaling if there exists a factor $r > 0$ such that for all (x, y) on E_1, the point (rx, ry) lies on E_2. Show that E_1 and E_2 are similar under scaling if and only if they have the same eccentricity. Show that any two circles are similar under scaling.

73. If we rewrite the general equation of degree two (13) in terms of variables x' and y' that are related to x and y by equations (14) and (15), we obtain a new equation of degree two in x' and y' of the same form but with different coefficients:

$$A'x^2 + B'xy + C'y^2 + D'x + E'y + F' = 0$$

(a) Show that $B' = B\cos 2\theta + (C - A)\sin 2\theta$.

(b) Show that if $B \neq 0$, then we obtain $B' = 0$ for

$$\theta = \frac{1}{2}\cot^{-1}\frac{A - C}{B}$$

This proves that it is always possible to eliminate the cross term Bxy by rotating the axes through a suitable angle.

CHAPTER REVIEW EXERCISES

1. Which of the following curves pass through the point $(1, 4)$?

(a) $c(t) = (t^2, t + 3)$ **(b)** $c(t) = (t^2, t - 3)$

(c) $c(t) = (t^2, 3 - t)$ **(d)** $c(t) = (t - 3, t^2)$

2. Find parametric equations for the line through $P = (2, 5)$ perpendicular to the line $y = 4x - 3$.

3. Find parametric equations for the circle of radius 2 with center $(1, 1)$. Use the equations to find the points of intersection of the circle with the x- and the y-axes.

4. Find a parametrization $c(t)$ of the line $y = 5 - 2x$ such that $c(0) = (2, 1)$.

5. Find a parametrization $c(\theta)$ of the unit circle such that $c(0) = (-1, 0)$.

6. Find a path $c(t)$ that traces the parabolic arc $y = x^2$ from $(0, 0)$ to $(3, 9)$ for $0 \le t \le 1$.

7. Find a path $c(t)$ that traces the line $y = 2x + 1$ from $(1, 3)$ to $(3, 7)$ for $0 \le t \le 1$.

8. Sketch the graph $c(t) = (1 + \cos t, \sin 2t)$ for $0 \le t \le 2\pi$ and draw arrows specifying the direction of motion.

In Exercises 9–12, express the parametric curve in the form $y = f(x)$.

9. $c(t) = (4t - 3, 10 - t)$ **10.** $c(t) = (t^3 + 1, t^2 - 4)$

11. $c(t) = \left(3 - \dfrac{2}{t}, t^3 + \dfrac{1}{t}\right)$ **12.** $x = \tan t, \quad y = \sec t$

13. Find all points visited twice by the path $c(t) = (t^2, \sin t)$. Plot $c(t)$ with a graphing utility.

In Exercises 14–17, calculate $\dfrac{dy}{dx}$ at the point indicated.

14. $c(t) = (t^3 + t, t^2 - 1), \quad t = 3$

15. $c(\theta) = (\tan^2\theta, \cos\theta), \quad \theta = \frac{\pi}{4}$

16. $c(t) = (e^t - 1, \sin t), \quad t = 20$

17. $c(t) = (\ln t, 3t^2 - t), \quad P = (0, 2)$

18. CAS Find the point on the cycloid $c(t) = (t - \sin t, 1 - \cos t)$ where the tangent line has slope $\frac{1}{2}$.

19. Find the points on $(t + \sin t, t - 2\sin t)$ where the tangent is vertical or horizontal.

20. Find the equation of the Bézier curve with control points

$$P_0 = (-1, -1), \quad P_1 = (-1, 1), \quad P_2 = (1, 1), \quad P_3(1, -1)$$

21. Find the speed at $t = \frac{\pi}{4}$ of a particle whose position at time t seconds is $c(t) = (\sin 4t, \cos 3t)$.

22. Find the speed (as a function of t) of a particle whose position at time t seconds is $c(t) = (\sin t + t, \cos t + t)$. What is the particle's maximal speed?

23. Find the length of $(3e^t - 3, 4e^t + 7)$ for $0 \le t \le 1$.

In Exercises 24–25, let $c(t) = (e^{-t}\cos t, e^{-t}\sin t)$.

24. Show that $c(t)$ for $0 \le t < \infty$ has finite length and calculate its value.

25. Find the first positive value of t_0 such that the tangent line to $c(t_0)$ is vertical and calculate the speed at $t = t_0$.

26. CAS Plot $c(t) = (\sin 2t, 2\cos t)$ for $0 \le t \le \pi$. Express the length of the curve as a definite integral and approximate it using a computer algebra system.

27. Convert the points $(x, y) = (1, -3), (3, -1)$ from rectangular to polar coordinates.

28. Convert the points $(r, \theta) = \left(1, \frac{\pi}{6}\right), \left(3, \frac{5\pi}{4}\right)$ from polar to rectangular coordinates.

29. Write $(x + y)^2 = xy + 6$ as an equation in polar coordinates.

30. Write $r = \dfrac{2\cos\theta}{\cos\theta - \sin\theta}$ as an equation in rectangular coordinates.

31. Show that $r = \dfrac{4}{7\cos\theta - \sin\theta}$ is the polar equation of a line.

32. GU Convert the equation

$$9(x^2 + y^2) = (x^2 + y^2 - 2y)^2$$

to polar coordinates and plot with a graphing utility.

33. Calculate the area of the circle $r = 3\sin\theta$ bounded by the rays $\theta = \frac{\pi}{3}$ and $\theta = \frac{2\pi}{3}$.

34. GU Plot the graph of $r = \sin 4\theta$ and calculate the area of one petal.

35. The equation $r = \sin(n\theta)$, where $n \ge 2$ is even, is a "rose" of $2n$ petals (Figure 1). Compute the total area of the flower and show that it does not depend on n.

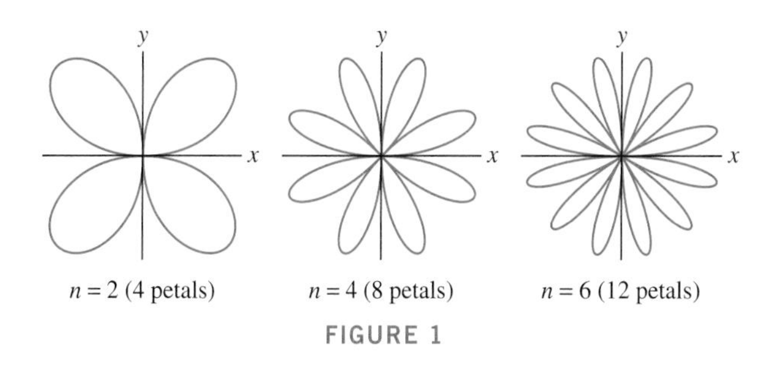

FIGURE 1

36. Calculate the total area enclosed by the curve $r^2 = \cos\theta e^{\sin\theta}$ (Figure 2).

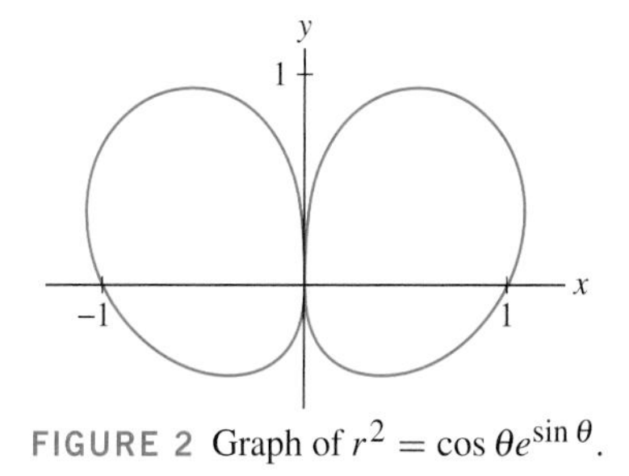

FIGURE 2 Graph of $r^2 = \cos\theta e^{\sin\theta}$.

37. Find the shaded area in Figure 3.

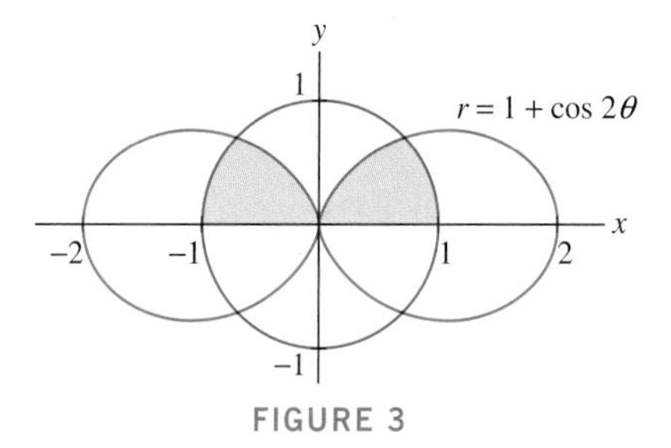

FIGURE 3

38. Calculate the length of the curve with polar equation $r = \theta$ in Figure 4.

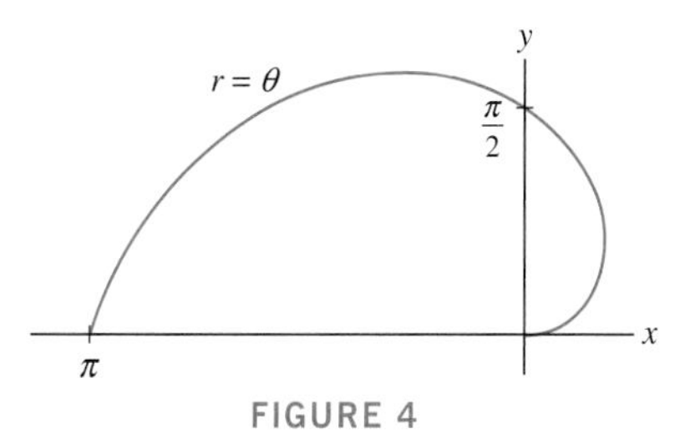

FIGURE 4

39. CAS Figure 5 shows the graph of $r = e^{0.5\theta} \sin\theta$ for $0 \le \theta \le 2\pi$. Use a CAS to approximate the difference in length between the outer and inner loops.

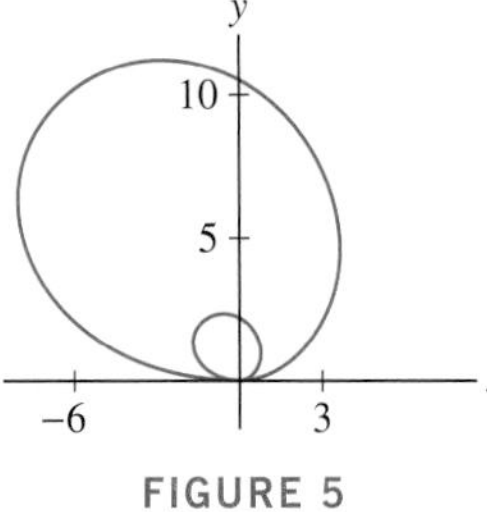

FIGURE 5

In Exercises 40–43, identify the conic section. Find the vertices and foci.

40. $\left(\frac{x}{3}\right)^2 + \left(\frac{y}{2}\right)^2 = 1$

41. $x^2 - 2y^2 = 4$

42. $(2x + \frac{1}{2}y)^2 = 4 - (x - y)^2$

43. $(y - 3)^2 = 2x^2 - 1$

44. Find the equation of a standard ellipse with two vertices at $(\pm 8, 0)$ and foci $(\pm\sqrt{3}, 0)$.

45. Find the equation of a standard hyperbola with vertices at $(\pm 8, 0)$ and asymptotes $y = \pm\frac{3}{4}x$.

46. Find the equation of a standard parabola with focus $(8, 0)$ and directrix $x = -8$.

47. Find the equation of a standard ellipse with foci at $(\pm 8, 0)$ and eccentricity $\frac{1}{8}$.

48. Find the asymptotes of the hyperbola $3x^2 + 6x - y^2 - 10y = 1$.

49. Show that the "conic section" with equation $x^2 - 4x + y^2 + 5 = 0$ has no points.

50. Show that the relation $\frac{dy}{dx} = (e^2 - 1)\frac{x}{y}$ holds on a standard ellipse or hyperbola of eccentricity e.

51. The orbit of Jupiter is an ellipse with the sun at a focus. Find the eccentricity of the orbit if the perihelion (closest distance to the sun) equals 740×10^6 km and the aphelion (farthest distance to the sun) equals 816×10^6 km.

52. Refer to Figure 24 in Section 11.5. Prove that the product of the perpendicular distances F_1R_1 and F_2R_2 from the foci to a tangent line of an ellipse is equal to the square b^2 of the semiminor axes.

12 | VECTOR GEOMETRY

The tension in the cables of the Tri-City Bridge over the Columbia River in Washington State is described using vectors.

Vectors play a role in nearly all areas of mathematics and its applications. In physics and engineering, vectors are used to represent quantities that have direction as well as magnitude, such as velocity or force. More advanced physical applications of vectors include aerodynamics, electromagnetic theory, quantum theory, and more recent fields such as computer graphics, image processing, and robotics. This chapter develops the basic geometric and algebraic properties of vectors. Although this chapter does not use any calculus, the concepts developed will be used throughout the remainder of the text.

12.1 Vectors in the Plane

A two-dimensional **vector v** is determined by two points in the plane: an initial point P (also called the "tail" or basepoint) and a terminal point Q (also called the "head"). We write $\mathbf{v} = \overrightarrow{PQ}$ and we draw $\mathbf{v}$ as an arrow pointing from P to Q [Figure 1(A)]. The vector $\mathbf{v} = \overrightarrow{PQ}$ is said to be based at P. The **position vector** of a point R is the vector $\mathbf{v} = \overrightarrow{OR}$, based at the origin with endpoint R [Figure 1(B)].

The term "vector" was first used in 1844 by the Irish mathematician William Rowan Hamilton (1805–1865) who made fundamental contributions to both mathematics and physics. In his work Lectures on Quaternions, *he wrote*

> *But in the new mode of speaking which it is here proposed to introduce, ... the* vector *of the sun has (itself)* direction, *as well as* length. *It is, therefore not sufficiently characterized by any single number, but requires, for its complete numerical expression, a system of three numbers.*

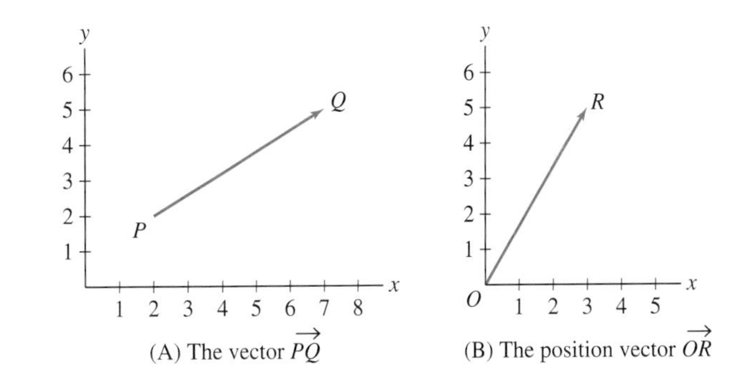

(A) The vector $\overrightarrow{PQ}$ (B) The position vector $\overrightarrow{OR}$

FIGURE 1

NOTATION *In this text, vectors are represented by boldface lowercase letters such as* **v**, **w**, **a**, **b**, *etc.*

When referring to vectors, we use the terms "length" and "magnitude" interchangeably. The term "norm" is also commonly used.

The **length** or **magnitude** of $\mathbf{v} = \overrightarrow{PQ}$, denoted $\|\mathbf{v}\|$, is defined as the distance from P to Q. We denote the distance from P to Q by $|P - Q|$. If $P = (a_1, b_1)$ and $Q = (a_2, b_2)$, then by the distance formula,

$$\|\mathbf{v}\| = \|\overrightarrow{PQ}\| = |P - Q| = \sqrt{(a_2 - a_1)^2 + (b_2 - b_1)^2}$$

■ **EXAMPLE 1** Find the length of the vector $\mathbf{v} = \overrightarrow{PQ}$ in Figure 1(A).

Solution Observe in Figure 1(A) that the initial point is $P = (2, 2)$ and the terminal point is $Q = (7, 5)$. By the distance formula,

$$\|\mathbf{v}\| = \text{Distance from } P \text{ to } Q = \sqrt{(7-2)^2 + (5-2)^2} = \sqrt{5^2 + 3^2} = \sqrt{34} \qquad ■$$

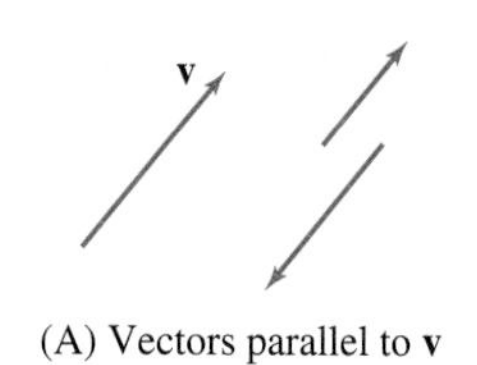

(A) Vectors parallel to **v**

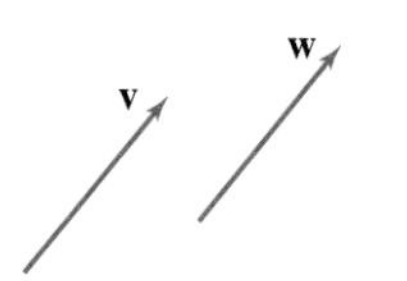

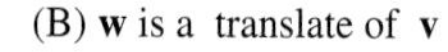

(B) **w** is a translate of **v**

FIGURE 2

Two vectors **v** and **w** of nonzero length are called **parallel** if the lines through **v** and **w** are parallel. Parallel vectors point either in the same or in opposite directions [Figure 2(A)]. A vector **v** is said to undergo a **translation** when it is moved parallel to itself without changing in length or direction. The resulting vector **w** is called a translate of **v** [Figure 2(B)]. Translates have the same length and direction but different basepoints.

As our study of vectors progresses, we will see that it is convenient to treat two vectors with the same length and direction as equivalent, even if they have different basepoints. With this in mind, we call vectors **v** and **w** **equivalent** if **w** is a translate of **v** [Figure 3(A)]. Every vector can be translated so that its tail is at the origin [Figure 3(C)]. Therefore,

Every vector **v** *is equivalent to a unique vector* $\mathbf{v}_0$ *based at the origin.*

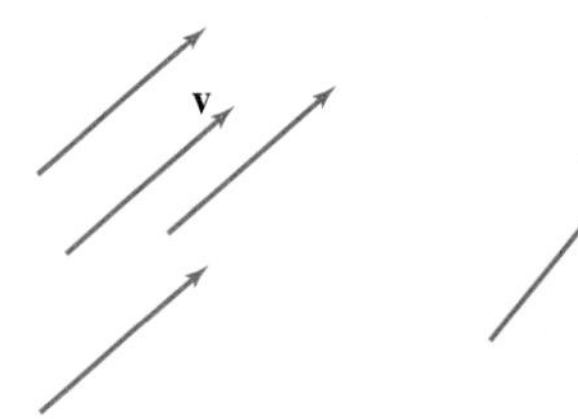

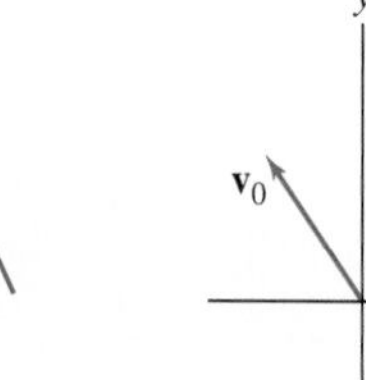

(A) Vectors equivalent to **v** (translates of **v**)

(B) Inequivalent vectors

(C) $\mathbf{v}_0$ is the unique vector based at the origin equivalent to **v**

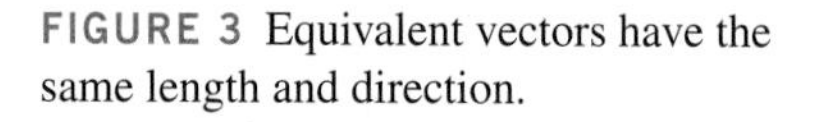

FIGURE 3 Equivalent vectors have the same length and direction.

To work with vectors algebraically, we define the components of a vector.

DEFINITION Components of a Vector Let $\mathbf{v} = \overrightarrow{PQ}$, where $P = (a_1, b_1)$ and $Q = (a_2, b_2)$. The **components** of **v** are the quantities

$$a = a_2 - a_1 \quad \text{(the } x\text{-component)}, \qquad b = b_2 - b_1 \quad \text{(the } y\text{-component)}$$

The pair of components is denoted $\langle a, b \rangle$.

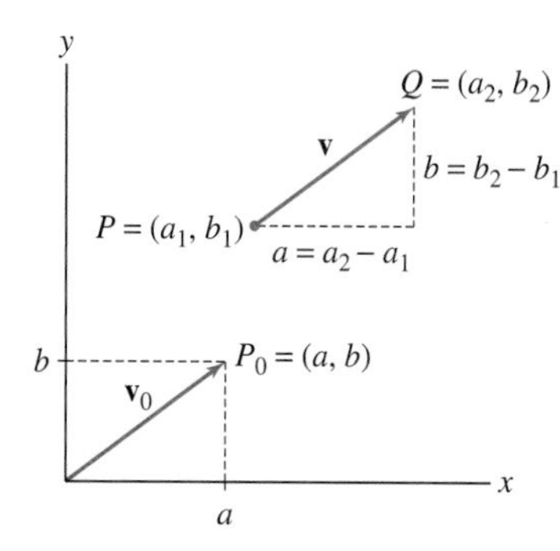

FIGURE 4 The vectors **v** and $\mathbf{v}_0$ have components $\langle a, b \rangle$.

NOTATION In this text, vectors in component form are written using "angle brackets." This allows us to clearly distinguish between the vector $\mathbf{v} = \langle a, b \rangle$ *and the point* $P = (a, b)$. *Some textbooks denote both* **v** *and P by the same symbol* (a, b).

As we see in Figure 4, the components $\langle a, b \rangle$ of **v** are the horizontal and vertical displacements from P to Q. By the Pythagorean Theorem,

$$\|\mathbf{v}\| = \sqrt{a^2 + b^2}$$

Note that the components $\langle a, b \rangle$ determine the length and direction of **v**, but not its basepoint. Therefore, two vectors are equivalent if and only if they have the same components. Furthermore, **v** is equivalent to the position vector $\mathbf{v}_0 = \overrightarrow{OP_0}$ with terminal point $P_0 = (a, b)$ (Figure 4).

We write $\mathbf{v} = \langle a, b \rangle$ if **v** has components a and b. This notation is ambiguous because it does not indicate the basepoint of **v**. However, in practice this rarely causes confusion. To further avoid confusion, the following convention will be in force for the remainder of the text:

Convention: *We assume all vectors are based at the origin unless otherwise stated.*

The **zero vector** (whose head and tail coincide) is the vector $\mathbf{0} = \langle 0, 0 \rangle$ of length zero.

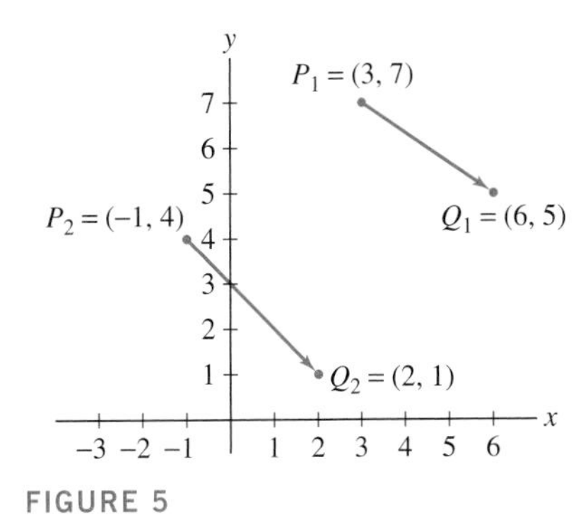

FIGURE 5

■ **EXAMPLE 2** Find the components of $\mathbf{v}_1 = \overrightarrow{P_1Q_1}$ and $\mathbf{v}_2 = \overrightarrow{P_2Q_2}$, where

$$P_1 = (3, 7),\ Q_1 = (6, 5) \qquad \text{and} \qquad P_2 = (-1, 4),\ Q_2 = (2, 1)$$

Are $\mathbf{v}_1$ and $\mathbf{v}_2$ equivalent?

Solution We compute the components of $\mathbf{v}_1$ and $\mathbf{v}_2$:

$$\mathbf{v}_1 = \langle 6 - 3, 5 - 7 \rangle = \langle 3, -2 \rangle, \qquad \mathbf{v}_2 = \langle 2 - (-1), 1 - 4 \rangle = \langle 3, -3 \rangle$$

The vectors $\mathbf{v}_1$ and $\mathbf{v}_2$ do not have the same components and hence are not equivalent (Figure 5). ■

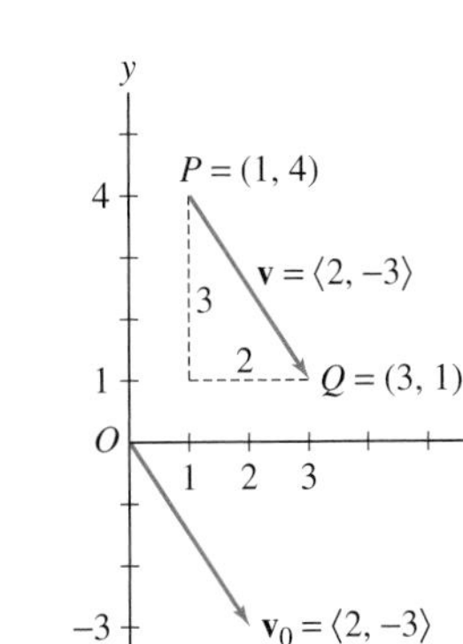

FIGURE 6 The vectors $\mathbf{v}$ and $\mathbf{v}_0$ have the same components $\langle 2, -3 \rangle$ but different basepoints.

■ **EXAMPLE 3** Sketch the vector $\mathbf{v} = \langle 2, -3 \rangle$ based at $P = (1, 4)$ and the vector $\mathbf{v}_0$ equivalent to $\mathbf{v}$ based at the origin.

Solution The vector $\mathbf{v} = \langle 2, -3 \rangle$ based at P has terminal point Q located two units to the right and three units down from P as shown in Figure 6. Therefore, $Q = (1 + 2, 4 - 3) = (3, 1)$. The vector $\mathbf{v}_0$ equivalent to $\mathbf{v}$ based at O has terminal point $(2, -3)$. ■

Vector Algebra

We now define two basic vector operations: vector addition and scalar multiplication. The term **scalar** is another word for "real number," and we often speak of scalar versus vector quantities. Thus, the number 8 is a scalar, while $\langle 8, 2 \rangle$ is a vector.

NOTATION λ *is the eleventh letter in the Greek alphabet. We often (but not exclusively) use the symbol* λ *to denote a scalar.*

If λ is a scalar and $\mathbf{v}$ is a nonzero vector, the scalar multiple $\lambda\mathbf{v}$ is defined as the vector of length $|\lambda|\ \|\mathbf{v}\|$ pointing in the same direction as $\mathbf{v}$ if $\lambda > 0$ and in the opposite direction if $\lambda < 0$ (Figure 7). For $\lambda = 0$, we set $0\mathbf{v} = \mathbf{0}$. In all cases,

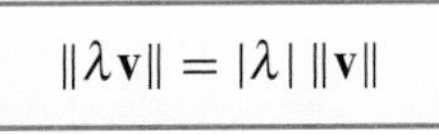

$$\|\lambda\mathbf{v}\| = |\lambda|\ \|\mathbf{v}\|$$

In particular, $-\mathbf{v}$ has the same length as $\mathbf{v}$ but points in the opposite direction. A vector $\mathbf{w}$ is parallel to $\mathbf{v}$ if and only if $\mathbf{w} = \lambda\mathbf{v}$ for some scalar λ.

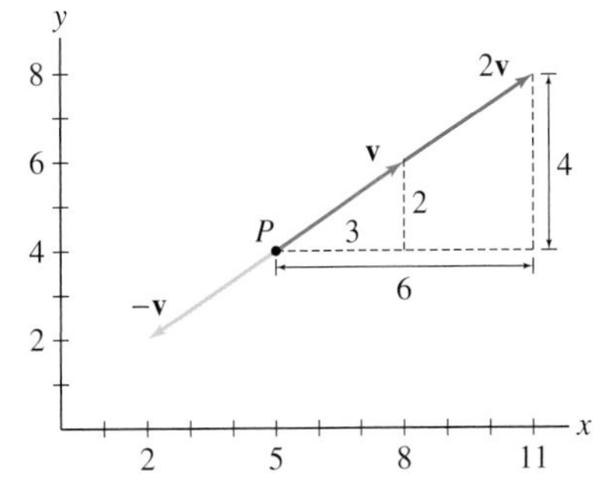

FIGURE 7 The vector $\mathbf{v} = \langle 3, 2 \rangle$ based at $P = (5, 4)$. The vector $2\mathbf{v}$ is also based at P and is twice as long as $\mathbf{v}$.

The vector sum $\mathbf{v} + \mathbf{w}$ is defined when $\mathbf{v}$ and $\mathbf{w}$ have the same basepoint. The sum can be described in two ways. First, translate $\mathbf{w}$ to the equivalent vector $\mathbf{w}'$ whose tail coincides with the head of $\mathbf{v}$. Then $\mathbf{v} + \mathbf{w}$ is the vector pointing from the tail of $\mathbf{v}$ to the head of $\mathbf{w}'$ [Figure 8(A)]. Alternatively, we may use the **Parallelogram Law**: $\mathbf{v} + \mathbf{w}$ is the vector pointing from the basepoint to the opposite vertex of the parallelogram formed by $\mathbf{v}$ and $\mathbf{w}$ [Figure 8(B)].

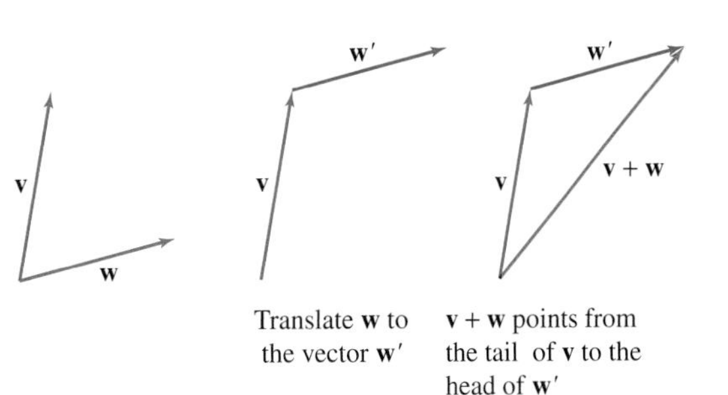

(A) The vector sum $\mathbf{v} + \mathbf{w}$.

(B) $\mathbf{v} + \mathbf{w}$ can also be formed using the Parallelogram Law.

FIGURE 8

To add several vectors $\mathbf{v}_1, \mathbf{v}_2, \ldots, \mathbf{v}_n$, translate the vectors to $\mathbf{v}_1 = \mathbf{v}'_1, \mathbf{v}'_2, \ldots, \mathbf{v}'_n$ so that they lie head to tail as in Figure 9. The vector sum $\mathbf{v} = \mathbf{v}_1 + \mathbf{v}_2 + \cdots + \mathbf{v}_n$ is the vector whose terminal point is the terminal point of $\mathbf{v}'_n$.

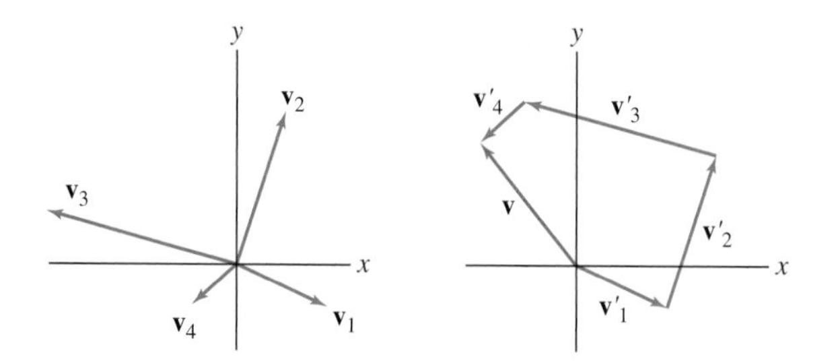

FIGURE 9 The sum $\mathbf{v} = \mathbf{v}_1 + \mathbf{v}_2 + \mathbf{v}_3 + \mathbf{v}_4$.

Vector subtraction $\mathbf{v} - \mathbf{w}$ is carried out by adding $-\mathbf{w}$ to $\mathbf{v}$ as in Figure 10(A). More simply, draw the vector pointing from $\mathbf{w}$ to $\mathbf{v}$ as in Figure 10(B), and translate it back to the basepoint to obtain $\mathbf{v} - \mathbf{w}$.

Remember that the vector $\mathbf{v} - \mathbf{w}$ *points in the direction* from $\mathbf{w}$ to $\mathbf{v}$ *(not from* $\mathbf{v}$ *to* $\mathbf{w}$*).*

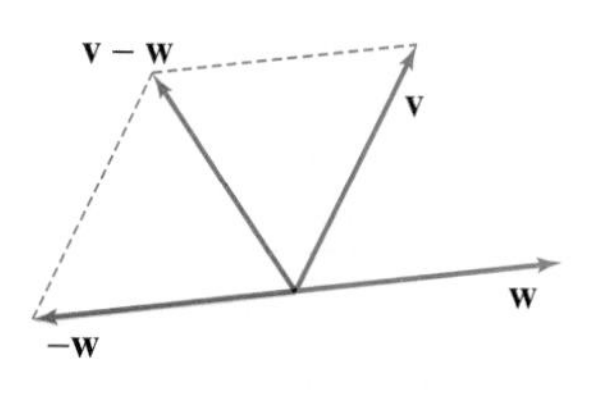

(A) The difference $\mathbf{v} - \mathbf{w}$ is obtained by adding $\mathbf{v}$ and $-\mathbf{w}$.

(B) More simply, to form $\mathbf{v} - \mathbf{w}$, draw the vector pointing from $\mathbf{w}$ to $\mathbf{v}$ and translate it back to the basepoint.

FIGURE 10 Vector subtraction.

The vector operations are easily performed using components. To add (or subtract) two vectors, we add (or subtract) their components. This follows from the parallelogram law as indicated in Figure 11(A). Similarly, if $\mathbf{v} = \langle a, b\rangle$, then $\lambda\mathbf{v} = \langle \lambda a, \lambda b\rangle$ because $\langle \lambda a, \lambda b\rangle$ is a vector of length $|\lambda|\,\|\mathbf{v}\|$ that, if λ and $\mathbf{v}$ are nonzero, points in the same or opposite direction as $\langle a, b\rangle$ depending on whether $\lambda > 0$ or $\lambda < 0$ [Figures 11(B) and (C)]. Therefore,

Vector Operations Using Components If $\mathbf{v} = \langle a, b\rangle$ and $\mathbf{w} = \langle c, d\rangle$, then:

(i) $\mathbf{v} + \mathbf{w} = \langle a + c, b + d\rangle$

(ii) $\mathbf{v} - \mathbf{w} = \langle a - c, b - d\rangle$

(iii) $\lambda\mathbf{v} = \langle \lambda a, \lambda b\rangle$

(iv) $\mathbf{v} + \mathbf{0} = \mathbf{0} + \mathbf{v} = \mathbf{v}$ and $\mathbf{v} - \mathbf{v} = \mathbf{0}$

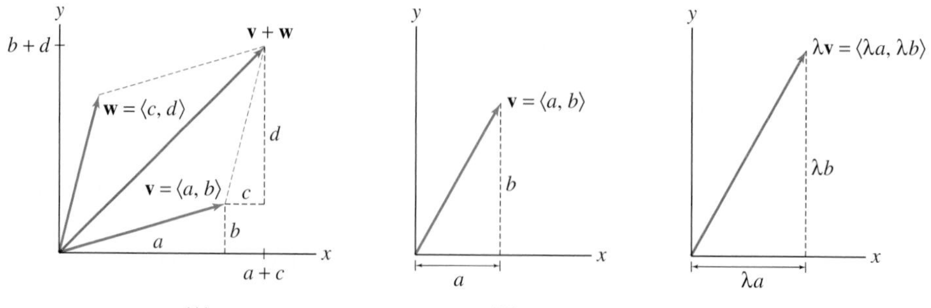

(A) (B) (C)

FIGURE 11 Vector operations using components.

We also note that if $P = (a_1, b_1)$ and $Q = (a_2, b_2)$, then components of the vector $\mathbf{v} = \overrightarrow{PQ}$ are conveniently computed as the difference:

$$\overrightarrow{PQ} = \overrightarrow{OQ} - \overrightarrow{OP} = \langle a_2, b_2 \rangle - \langle a_1, b_1 \rangle = \langle a_2 - a_1, b_2 - b_1 \rangle$$

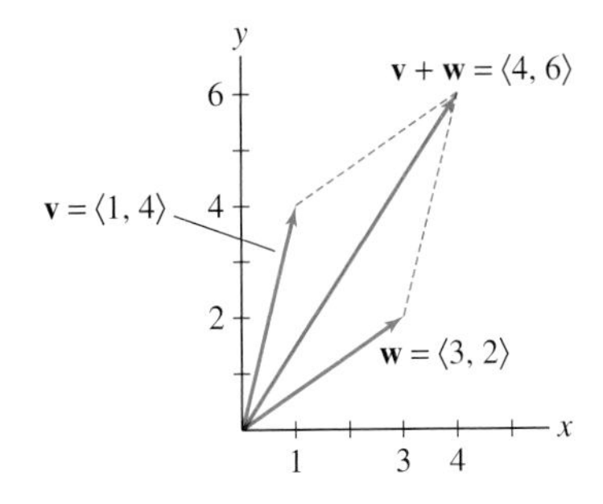

FIGURE 12

EXAMPLE 4 Let $\mathbf{v} = \langle 1, 4 \rangle$ and $\mathbf{w} = \langle 3, 2 \rangle$. Calculate **(a)** $\mathbf{v} + \mathbf{w}$ and **(b)** $5\mathbf{v}$.

Solution

$$\mathbf{v} + \mathbf{w} = \langle 1, 4 \rangle + \langle 3, 2 \rangle = \langle 1 + 3, 4 + 2 \rangle = \langle 4, 6 \rangle$$

$$5\mathbf{v} = 5\,\langle 1, 4 \rangle = \langle 5, 20 \rangle$$

The vector sum is illustrated in Figure 12. ■

The vector operations obey the usual laws of algebra.

THEOREM 1 Basic Properties of Vector Algebra For all vectors $\mathbf{u}$, $\mathbf{v}$, $\mathbf{w}$ and scalars λ,

Commutative Law:	$\mathbf{v} + \mathbf{w} = \mathbf{w} + \mathbf{v}$
Associative Law:	$\mathbf{u} + (\mathbf{v} + \mathbf{w}) = (\mathbf{u} + \mathbf{v}) + \mathbf{w}$
Distributive Law for Scalars:	$\lambda(\mathbf{v} + \mathbf{w}) = \lambda\mathbf{v} + \lambda\mathbf{w}$

These properties are easily verified using components. For example, we may check that vector addition is commutative as follows:

$$\langle a, b \rangle + \langle c, d \rangle = \underbrace{\langle a + c, b + d \rangle = \langle c + a, d + b \rangle}_{\text{Commutativity of ordinary addition}} = \langle c, d \rangle + \langle a, b \rangle$$

A **linear combination** of vectors $\mathbf{v}$ and $\mathbf{w}$ is a vector of the form $r\mathbf{v} + s\mathbf{w}$, where r and s are scalars. If $\mathbf{v}$ and $\mathbf{w}$ are not parallel, then every vector $\mathbf{u}$ can be expressed as a linear combination $\mathbf{u} = r\mathbf{v} + s\mathbf{w}$ [Figure 13(A)]. We refer to the parallelogram $\mathcal{P}$ with vertices at the origin and the terminal points of $\mathbf{v}$, $\mathbf{w}$ and $\mathbf{v} + \mathbf{w}$ as the parallelogram **spanned** by $\mathbf{v}$ and $\mathbf{w}$ [Figure 13(B)]. Observe that $\mathcal{P}$ consists of the linear combinations $r\mathbf{v} + s\mathbf{w}$ with $0 \le r, s \le 1$.

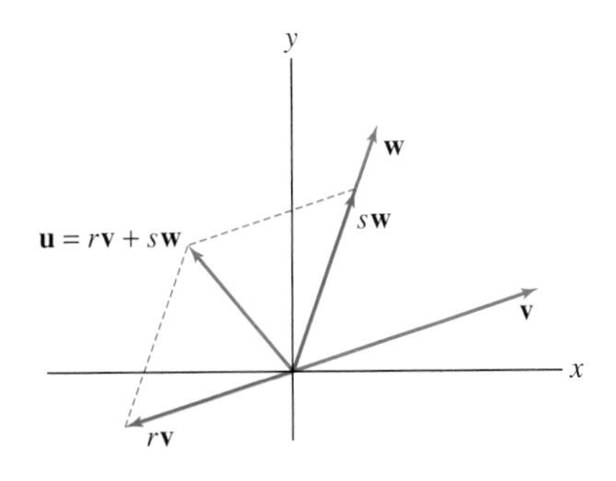

(A) The vector **u** can be expressed as a linear combination $\mathbf{u} = r\mathbf{v} + s\mathbf{w}$. In this figure, $r < 0$.

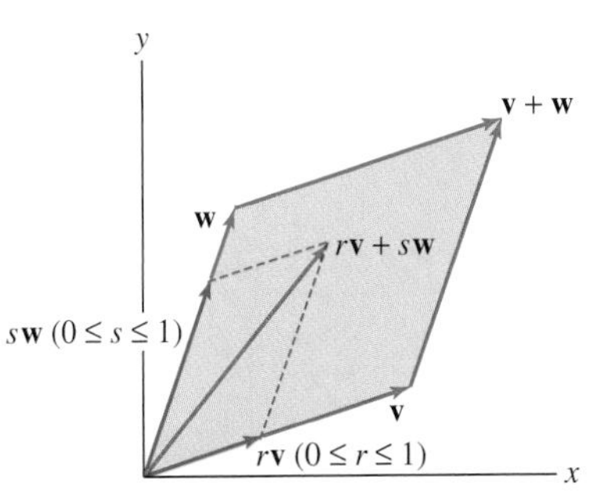

(B) The parallelogram $\mathcal{P}$ spanned by **v** and **w** consists of all linear combinations $r\mathbf{v} + s\mathbf{w}$ with $0 \le r, s \le 1$.

FIGURE 13

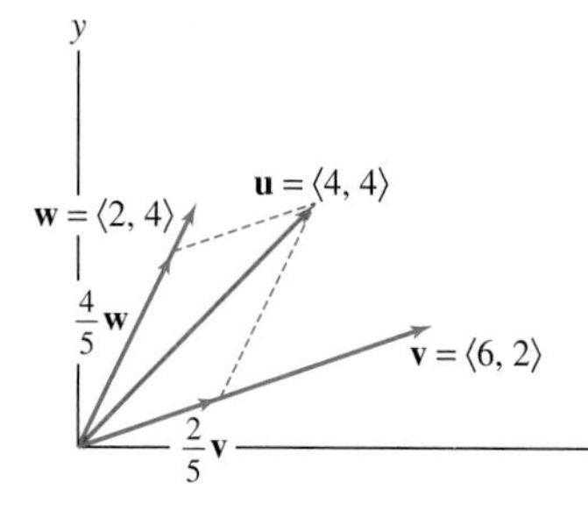

FIGURE 14

■ **EXAMPLE 5 Linear Combinations** Express $\mathbf{u} = \langle 4, 4\rangle$ as a linear combination of $\mathbf{v} = \langle 6, 2\rangle$ and $\mathbf{w} = \langle 2, 4\rangle$.

Solution We must find r and s such that $r\mathbf{v} + s\mathbf{w} = \langle 4, 4\rangle$, or

$$r\langle 6, 2\rangle + s\langle 2, 4\rangle = \langle 6r + 2s, 2r + 4s\rangle = \langle 4, 4\rangle$$

This gives $6r + 2s = 4$ and $2r + 4s = 4$. Subtracting the equations, we obtain $4r - 2s = 0$ or $s = 2r$. Substitution in the first equation yields $6r + 4r = 4$ or $r = \frac{2}{5}$ and $s = \frac{4}{5}$. Thus (Figure 14),

$$\mathbf{u} = \langle 4, 4\rangle = \frac{2}{5}\langle 6, 2\rangle + \frac{4}{5}\langle 2, 4\rangle$$ ■

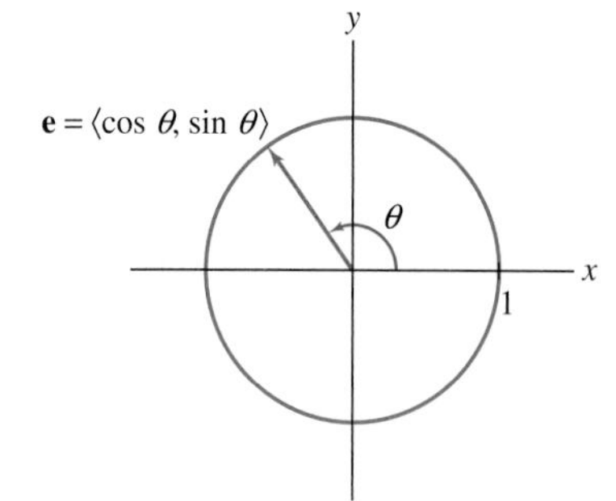

FIGURE 15 The head of a unit vector lies on the unit circle.

A vector of length one is called a **unit vector**. Unit vectors are often used to indicate direction, when it is not necessary to specify length. The head of a unit vector $\mathbf{e}$ based at the origin lies on the unit circle and has components $\mathbf{e} = \langle \cos\theta, \sin\theta\rangle$, where θ is the angle between the x-axis and $\mathbf{e}$, measured counterclockwise (Figure 15).

If $\mathbf{v} = \langle a, b\rangle$ is a nonzero vector, then the vector

$$\mathbf{e_v} = \frac{1}{\|\mathbf{v}\|}\mathbf{v}$$

is the unit vector pointing in the direction of $\mathbf{v}$ (Figure 16). Indeed,

$$\|\mathbf{e_v}\| = \left\|\frac{1}{\|\mathbf{v}\|}\mathbf{v}\right\| = \frac{1}{\|\mathbf{v}\|}\|\mathbf{v}\| = 1$$

and $\mathbf{e_v}$ is a positive scalar multiple of $\mathbf{v}$, so $\mathbf{e_v}$ and $\mathbf{v}$ point in the same direction. If $\mathbf{v}$ makes an angle θ with the positive x-axis, then (Figure 16)

$$\mathbf{e_v} = \langle \cos\theta, \sin\theta\rangle$$

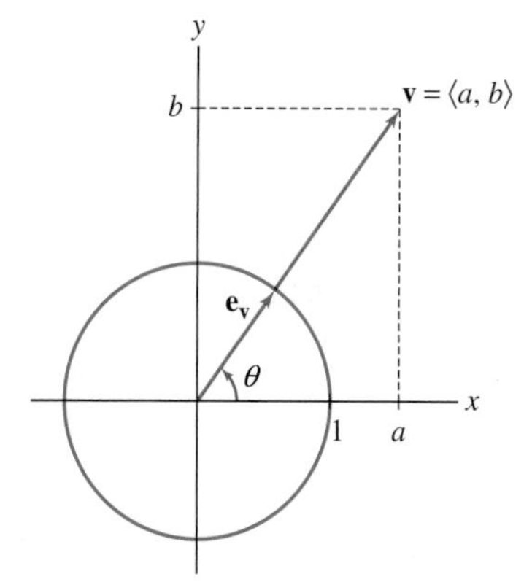

FIGURE 16 Unit vector in the direction of **v**.

Furthermore, the components of $\mathbf{v}$ are $a = \|\mathbf{v}\|\cos\theta$, $b = \|\mathbf{v}\|\sin\theta$. Thus

$$\mathbf{v} = \langle a, b\rangle = \|\mathbf{v}\|\mathbf{e_v} = \|\mathbf{v}\|\langle \cos\theta, \sin\theta\rangle \qquad \boxed{1}$$

■ **EXAMPLE 6** Find the unit vector in the direction of $\mathbf{v} = \langle 3, 5\rangle$.

Solution $\|\mathbf{v}\| = \sqrt{3^2 + 5^2} = \sqrt{34}$ and $\mathbf{e_v} = \frac{1}{\sqrt{34}}\mathbf{v} = \left\langle \frac{3}{\sqrt{34}}, \frac{5}{\sqrt{34}}\right\rangle$. ■

It is customary to introduce a special notation for the unit vectors in the direction of the positive x- and y-axes (Figure 17):

$$\mathbf{i} = \langle 1, 0\rangle, \qquad \mathbf{j} = \langle 0, 1\rangle$$

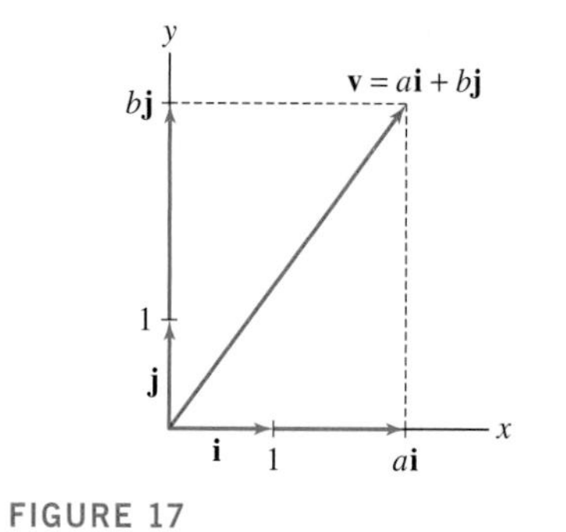

FIGURE 17

The vectors $\mathbf{i}$ and $\mathbf{j}$ are called the **standard basis vectors**. Every vector can be written as a linear combination of these basis vectors (Figure 17):

$$\langle a, b\rangle = a\mathbf{i} + b\mathbf{j}$$

For example, $\langle 4, -2\rangle = 4\mathbf{i} - 2\mathbf{j}$. Vector addition is performed by adding the $\mathbf{i}$ and $\mathbf{j}$ coefficients. For example,

$$(4\mathbf{i} - 2\mathbf{j}) + (5\mathbf{i} + 7\mathbf{j}) = (4 + 5)\mathbf{i} + (-2 + 7)\mathbf{j} = 9\mathbf{i} + 5\mathbf{j}$$

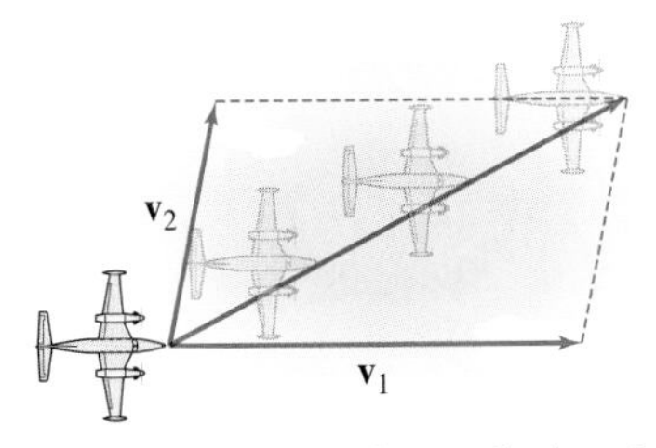

FIGURE 18 The resultant velocity of an airplane traveling with velocity $\mathbf{v}_1$ when it encounters a wind of velocity $\mathbf{v}_2$ is the vector sum $\mathbf{v}_1 + \mathbf{v}_2$.

CONCEPTUAL INSIGHT We commonly say that quantities such as force or velocity are vectors because they have both magnitude and direction, but there is more to this statement than meets the eye. A vector quantity must obey the law of vector addition, so when we say that force is a vector quantity, we are implicitly claiming that forces add according to the Parallelogram Law (Figure 18). In other words, if two forces $\mathbf{F}_1$ and $\mathbf{F}_2$ act on an object, then the resultant force is the **vector sum** $\mathbf{F}_1 + \mathbf{F}_2$. This is a physical fact that must be verified experimentally. It was well known to physicists in the time of Galileo and Newton, who used the Parallelogram Law long before the formal introduction of vectors by Hamilton and others in the 1800s.

EXAMPLE 7 Calculate the magnitudes of the forces on cables 1 and 2 in Figure 19(A).

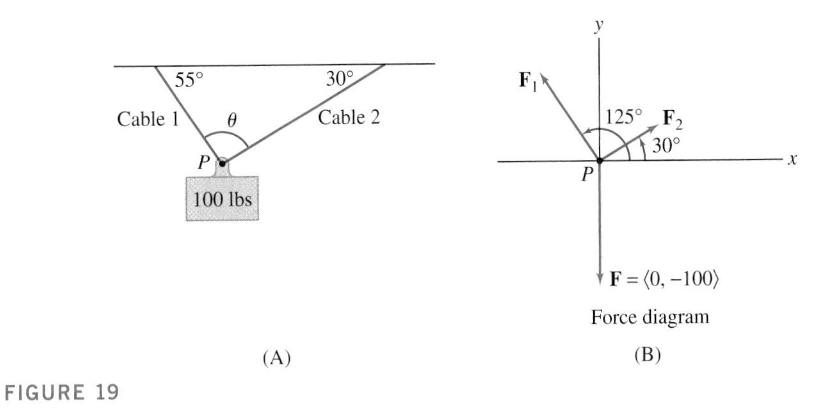

FIGURE 19

Solution Three forces act on the point P in Figure 19(A). A 100-lb force $\mathbf{F}$ due to gravity acts vertically downward and two unknown forces $\mathbf{F}_1$ and $\mathbf{F}_2$ act through cables 1 and 2. The point P is not in motion, so the net force acting on P must be zero, that is,

$$\mathbf{F}_1 + \mathbf{F}_2 + \mathbf{F} = \mathbf{0} \qquad \boxed{2}$$

The three forces are indicated in the force diagram in Figure 19(B). Notice that the angle between $\mathbf{F}_1$ and $\mathbf{F}_2$ is $\theta = 180 - 55 - 30 = 95°$. The vector $\mathbf{F}_2$ makes an angle of 30° with the horizontal and hence $\mathbf{F}_1$ makes an angle of $95 + 30 = 125°$ with the positive x-axis. Let $f_1 = \|\mathbf{F}_1\|$ and $f_2 = \|\mathbf{F}_2\|$ be the magnitudes of $\mathbf{F}_1$ and $\mathbf{F}_2$. We use Eq. (1) and the table in the margin to write the force vectors in component form:

θ	$\cos\theta$	$\sin\theta$
125°	−0.573	0.819
30°	0.866	0.5

$$\mathbf{F}_1 = f_1 \left\langle \cos 125°, \sin 125° \right\rangle \approx f_1 \langle -0.573, 0.819 \rangle$$

$$\mathbf{F}_2 = f_2 \left\langle \cos 30°, \sin 30° \right\rangle \approx f_2 \langle 0.866, 0.5 \rangle$$

$$\mathbf{F} = \langle 0, -100 \rangle$$

Therefore, Eq. (2) becomes

$$\mathbf{F}_1 + \mathbf{F}_2 + \mathbf{F} = f_1 \langle -0.573, 0.819 \rangle + f_2 \langle 0.866, 0.5 \rangle + \langle 0, -100 \rangle = \langle 0, 0 \rangle$$

This gives us two equations:

$$-0.573 f_1 + 0.866 f_2 = 0, \qquad 0.819 f_1 + 0.5 f_2 - 100 = 0$$

By the first equation, $f_2 = \left(\frac{0.573}{0.866}\right) f_1$. The second equation then gives

$$0.819 f_1 + 0.5 f_2 - 100 = 0.819 f_1 + 0.5\left(\frac{0.573}{0.866}\right) f_1 - 100 \approx 1.15 f_1 - 100 = 0$$

Therefore,

$$f_1 \approx \frac{100}{1.15} \approx 87 \text{ lb} \qquad \text{and} \qquad f_2 \approx \left(\frac{0.573}{0.866}\right) 87 \approx 58 \text{ lb}$$

■

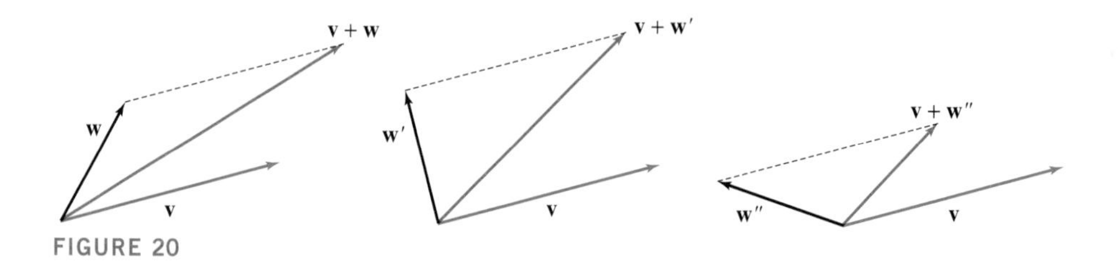

FIGURE 20

We close this section by stating the Triangle Inequality. As we see in Figure 20, the magnitude of a sum $\|\mathbf{v} + \mathbf{w}\|$ depends on the angle between $\mathbf{v}$ and $\mathbf{w}$. So in general, $\|\mathbf{v} + \mathbf{w}\|$ is not equal to the sum $\|\mathbf{v}\| + \|\mathbf{w}\|$. What we can say is that $\|\mathbf{v} + \mathbf{w}\|$ is *at most* equal to the sum $\|\mathbf{v}\| + \|\mathbf{w}\|$. This corresponds to the fact that the length of one side of a triangle is at most the sum of the lengths of the other two sides. A formal proof may be given using the dot product (see Exercise 81 in Section 12.3).

THEOREM 2 Triangle Inequality For any two vectors $\mathbf{v}$ and $\mathbf{w}$,

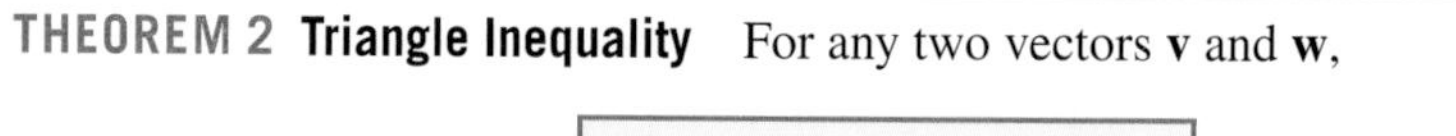

$$\|\mathbf{v} + \mathbf{w}\| \le \|\mathbf{v}\| + \|\mathbf{w}\|$$

Equality holds if and only if $\mathbf{w} = \lambda \mathbf{v}$, where $\lambda \ge 0$ or $\mathbf{v} = \mathbf{0}$.

12.1 SUMMARY

- A *vector* $\mathbf{v} = \overrightarrow{PQ}$ is determined by a basepoint P (the "tail") and a terminal point Q (the "head"). The *length* of $\mathbf{v} = \overrightarrow{PQ}$, denoted $\|\mathbf{v}\|$, is the distance from P to Q.
- The vector $\mathbf{v} = \overrightarrow{OP_0}$ based at the origin $O = (0, 0)$, with head $P_0 = (a, b)$ is called the *position vector* of P_0.
- Two vectors $\mathbf{v}$ and $\mathbf{w}$ are *equivalent* if they are translates of each other, that is, if they have the same magnitude and direction, but possibly different basepoints.
- The vector $\mathbf{v} = \overrightarrow{PQ}$ with $P = (a_1, b_1)$ and $Q = (a_2, b_2)$ has *components* $a = a_2 - a_1$, $b = b_2 - b_1$.
- Two vectors are equivalent if and only if they have the same components. If $\mathbf{v}$ has components a and b, we write $\mathbf{v} = \langle a, b\rangle$. Although this notation is ambiguous (because equivalent vectors have the same components), it rarely causes confusion. Our convention is that all vectors are assumed to be based at the origin unless otherwise indicated.
- If $\mathbf{v} = \langle a, b\rangle$, then $\|\mathbf{v}\| = \sqrt{a^2 + b^2}$. The *zero vector* is the vector $\mathbf{0} = \langle 0, 0\rangle$ of length 0.

- *Vector addition* is defined geometrically by the *Parallelogram Law*. If λ is a scalar, then the *scalar multiple* $\lambda\mathbf{v}$ is the vector of length $|\lambda|\,\|\mathbf{v}\|$ in the same direction as $\mathbf{v}$ if $\lambda > 0$ and in the opposite direction if $\lambda < 0$. For $\lambda = 0$, we set $0\mathbf{v} = \mathbf{0}$. In components,

$$\langle a_1, b_1\rangle + \langle a_2, b_2\rangle = \langle a_1 + a_2, b_1 + b_2\rangle$$
$$\lambda\,\langle a, b\rangle = \langle \lambda a, \lambda b\rangle$$

- Nonzero vectors $\mathbf{v}$ and $\mathbf{w}$ are *parallel* if $\mathbf{w} = \lambda\mathbf{v}$ for some scalar λ (they point in the same direction if $\lambda > 0$).
- A *unit vector* is a vector of length one. If $\mathbf{v} \neq \mathbf{0}$, the unit vector pointing in the direction $\mathbf{v}$ is $\mathbf{e_v} = \dfrac{1}{\|\mathbf{v}\|}\mathbf{v}$.
- Every vector can be expressed as a linear combination of the *standard basis vectors* $\mathbf{i} = \langle 1, 0\rangle$ and $\mathbf{j} = \langle 0, 1\rangle$. If $\mathbf{v} = \langle a, b\rangle$, then $\mathbf{v} = a\mathbf{i} + b\mathbf{j}$.
- If the vector $\mathbf{v} = \langle a, b\rangle$ is nonzero and makes an angle θ with the positive x-axis, then $a = \|\mathbf{v}\|\cos\theta$ and $b = \|\mathbf{v}\|\sin\theta$, and $\mathbf{e_v} = \langle \cos\theta, \sin\theta\rangle$.
- The *Triangle Inequality:* For any two vectors $\mathbf{v}$ and $\mathbf{w}$,

$$\|\mathbf{v} + \mathbf{w}\| \le \|\mathbf{v}\| + \|\mathbf{w}\|$$

Equality holds if and only if $\mathbf{w} = \lambda\mathbf{v}$, where $\lambda \ge 0$ or $\mathbf{v} = \mathbf{0}$.

12.1 EXERCISES

Preliminary Questions

1. Answer true or false. Every nonzero vector is:

(a) Equivalent to a vector based at the origin.
(b) Equivalent to a unit vector based at the origin.
(c) Parallel to a vector based at the origin.
(d) Parallel to a unit vector based at the origin.

2. What is the length of $-3\mathbf{a}$ if $\|\mathbf{a}\| = 5$?

3. Suppose that $\mathbf{v}$ has components $\langle 3, 1\rangle$. How, if at all, do the components change if you translate $\mathbf{v}$ horizontally two units to the left?

4. What are the components of the zero vector based at $P = (3, 5)$?

5. Are the following true or false?

(a) The vectors $\mathbf{v}$ and $-2\mathbf{v}$ are parallel.

(b) The vectors $\mathbf{v}$ and $-2\mathbf{v}$ point in the same direction.

6. Explain the commutativity of vector addition in terms of the Parallelogram Law.

Exercises

1. Sketch the vectors $\mathbf{v}_1, \mathbf{v}_2, \mathbf{v}_3, \mathbf{v}_4$ with tail P and head Q, and compute their lengths. Are any two of these vectors equivalent?

	$\mathbf{v}_1$	$\mathbf{v}_2$	$\mathbf{v}_3$	$\mathbf{v}_4$
P	$(2, 4)$	$(-1, 3)$	$(-1, 3)$	$(4, 1)$
Q	$(4, 4)$	$(1, 3)$	$(2, 4)$	$(6, 3)$

2. Sketch the vector $\mathbf{b} = \langle 3, 4\rangle$ based at $P = (-2, -1)$.

3. What is the terminal point of the vector $\mathbf{a} = \langle 1, 3\rangle$ based at $P = (2, 2)$? Sketch $\mathbf{a}$ and the vector $\mathbf{a}_0$ based at the origin equivalent to $\mathbf{a}$.

4. Let $\mathbf{v} = \overrightarrow{PQ}$, where $P = (1, 1)$ and $Q = (2, 2)$. What is the head of the vector $\mathbf{v}'$ equivalent to $\mathbf{v}$ based at $(2, 4)$? What is the head of the vector $\mathbf{v}_0$ equivalent to $\mathbf{v}$ based at the origin? Sketch $\mathbf{v}$, $\mathbf{v}_0$, and $\mathbf{v}'$.

In Exercises 5–8, find the components of $\overrightarrow{PQ}$.

5. $P = (3, 2), \quad Q = (2, 7)$

6. $P = (1, -4), \quad Q = (3, 5)$

7. $P = (3, 5), \quad Q = (1, -4)$

8. $P = (0, 2), \quad Q = (5, 0)$

In Exercises 9–14, calculate.

9. $\langle 2, 1\rangle + \langle 3, 4\rangle$

10. $\langle -4, 6\rangle - \langle 3, -2\rangle$

11. $5\,\langle 6, 2\rangle$

12. $4(\langle 1, 1\rangle + \langle 3, 2\rangle)$

13. $\left\langle -\frac{1}{2}, \frac{5}{3}\right\rangle + \left\langle 3, \frac{10}{3}\right\rangle$

14. $\langle \ln 2, e\rangle + \langle \ln 3, \pi\rangle$

15. Which of the vectors (A)–(C) in Figure 21 is equivalent to $\mathbf{v} - \mathbf{w}$?

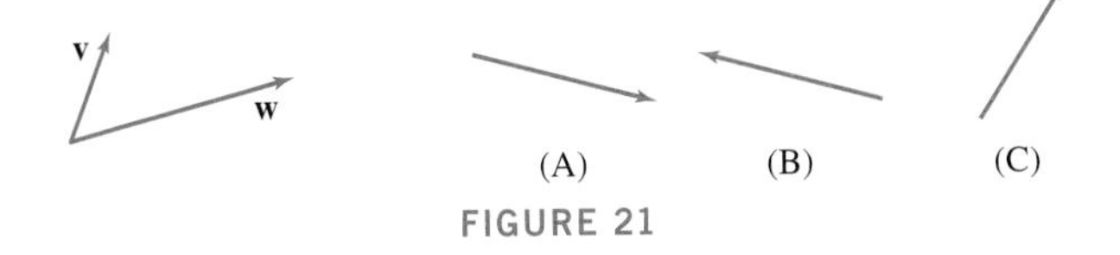

FIGURE 21

16. Sketch $\mathbf{v}+\mathbf{w}$ and $\mathbf{v}-\mathbf{w}$ for the vectors in Figure 22.

FIGURE 22

17. Sketch $2\mathbf{v}$, $-\mathbf{w}$, $\mathbf{v}+\mathbf{w}$, and $2\mathbf{v}-\mathbf{w}$ for the vectors in Figure 23.

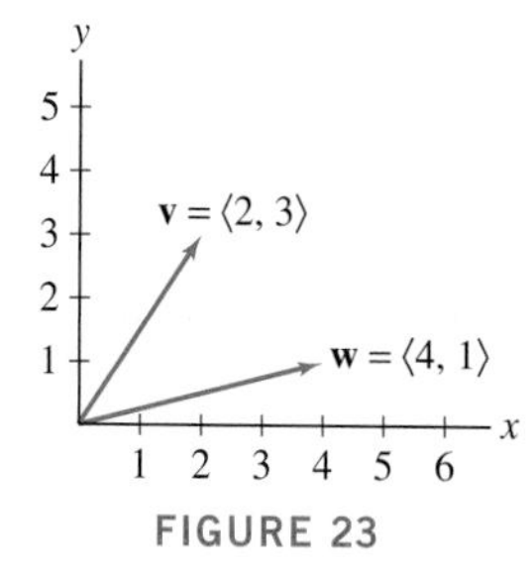

FIGURE 23

18. Sketch $\mathbf{v}=\langle 1,3\rangle$, $\mathbf{w}=\langle 2,-2\rangle$, $\mathbf{v}+\mathbf{w}$, $\mathbf{v}-\mathbf{w}$.

19. Sketch $\mathbf{v}=\langle 0,2\rangle$, $\mathbf{w}=\langle -2,4\rangle$, $3\mathbf{v}+\mathbf{w}$, $2\mathbf{v}-2\mathbf{w}$.

20. Sketch $\mathbf{v}=\langle -2,1\rangle$, $\mathbf{w}=\langle 2,2\rangle$, $\mathbf{v}+2\mathbf{w}$, $\mathbf{v}-2\mathbf{w}$.

21. Sketch the vector $\mathbf{v}$ such that $\mathbf{v}+\mathbf{v}_1+\mathbf{v}_2=\mathbf{0}$ for $\mathbf{v}_1$ and $\mathbf{v}_2$ in Figure 24(A).

22. Sketch the vector sum $\mathbf{v}=\mathbf{v}_1+\mathbf{v}_2+\mathbf{v}_3+\mathbf{v}_4$ in Figure 24(B).

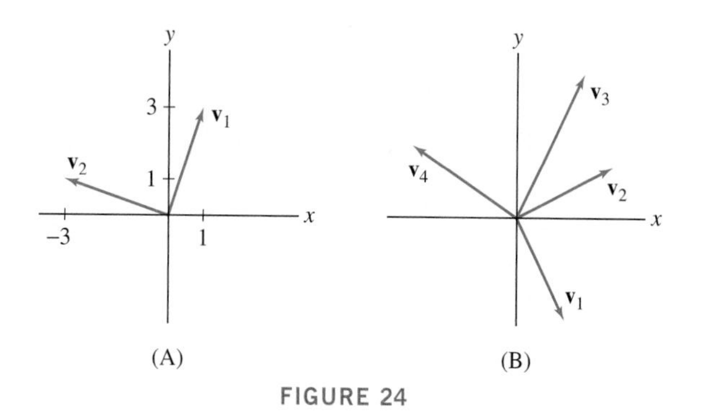

FIGURE 24

23. Let $\mathbf{v}=\overrightarrow{PQ}$, where $P=(-2,5)$, $Q=(1,-2)$. Which of the vectors with the following given tails and heads are equivalent to $\mathbf{v}$?

(a) $(-3,3)$, $(0,4)$
(b) $(0,0)$, $(3,-7)$
(c) $(-1,2)$, $(2,-5)$
(d) $(4,-5)$, $(1,4)$

24. Let $\mathbf{v}=\langle 6,9\rangle$. Which of the following vectors are parallel to $\mathbf{v}$ and which point in the same direction?

(a) $\langle 12,18\rangle$
(b) $\langle 3,2\rangle$
(c) $\langle 2,3\rangle$
(d) $\langle -6,-9\rangle$
(e) $\langle -24,-27\rangle$
(f) $\langle -24,-36\rangle$

In Exercises 25–28, sketch the vectors $\overrightarrow{AB}$ and $\overrightarrow{PQ}$, and determine whether they are equivalent.

25. $A=(1,1)$, $B=(3,7)$, $P=(4,-1)$, $Q=(6,5)$

26. $A=(1,4)$, $B=(-6,3)$, $P=(1,4)$, $Q=(6,3)$

27. $A=(-3,2)$, $B=(0,0)$, $P=(0,0)$, $Q=(3,-2)$

28. $A=(5,8)$, $B=(1,8)$, $P=(1,8)$, $Q=(-3,8)$

In Exercises 29–32, are $\overrightarrow{AB}$ and $\overrightarrow{PQ}$ parallel (and if so, do they point in the same direction)?

29. $A=(1,1)$, $B=(3,4)$, $P=(1,1)$, $Q=(7,10)$

30. $A=(-3,2)$, $B=(0,0)$, $P=(0,0)$, $Q=(3,2)$

31. $A=(2,2)$, $B=(-6,3)$, $P=(9,5)$, $Q=(17,4)$

32. $A=(5,8)$, $B=(2,2)$, $P=(2,2)$, $Q=(-3,8)$

In Exercises 33–36, let $R=(-2,7)$. Calculate the following.

33. The length of $\overrightarrow{OR}$

34. The components of $\mathbf{u}=\overrightarrow{PR}$, where $P=(1,2)$

35. The point P such that $\overrightarrow{PR}$ has components $\langle -2,7\rangle$

36. The point Q such that $\overrightarrow{RQ}$ has components $\langle 8,-3\rangle$

In Exercises 37–44, find the given vector.

37. Unit vector $\mathbf{e_v}$ where $\mathbf{v}=\langle 3,4\rangle$

38. Unit vector $\mathbf{e_w}$ where $\mathbf{w}=\langle 21,-20\rangle$

39. Unit vector in the direction of $\mathbf{u}=\langle -1,-1\rangle$

40. Vector of length 4 in the direction of $\mathbf{u}=\langle -1,-1\rangle$

41. Unit vector in the direction opposite to $\mathbf{v}=\langle -2,4\rangle$

42. Unit vector $\mathbf{e}$ making an angle of $\frac{4\pi}{7}$ with the x-axis

43. Vector of length 2 making an angle of 30° with the x-axis

44. Vector of length 4 making an angle of $\frac{7\pi}{3}$ with the y-axis

45. Find all scalars λ such that $\lambda\langle 2,3\rangle$ has length one.

46. What are the coordinates of the point P in the parallelogram in Figure 25(A)?

47. What are the coordinates a and b in the parallelogram in Figure 25(B)?

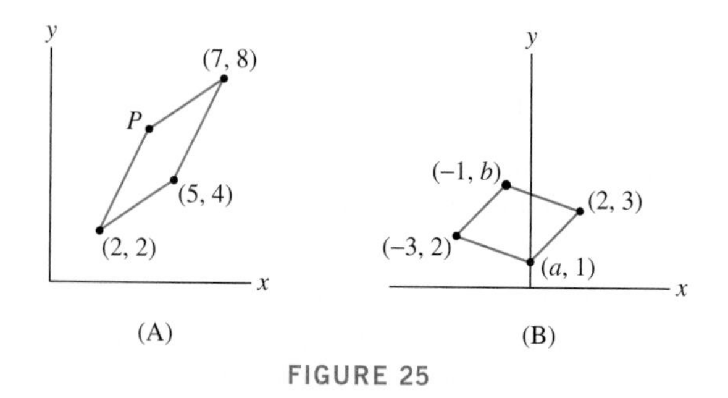

FIGURE 25

48. Let $\mathbf{v} = \overrightarrow{AB}$ and $\mathbf{w} = \overrightarrow{AC}$, where A, B, C are three distinct points in the plane. Match (a)–(d) with (i)–(iv). (*Hint:* Draw a picture.)

(a) $-\mathbf{w}$ **(b)** $-\mathbf{v}$ **(c)** $\mathbf{w} - \mathbf{v}$ **(d)** $\mathbf{v} - \mathbf{w}$

(i) $\overrightarrow{CB}$ **(ii)** $\overrightarrow{CA}$ **(iii)** $\overrightarrow{BC}$ **(iv)** $\overrightarrow{BA}$

49. Find the components and length of the following vectors:

(a) $4\mathbf{i} + 3\mathbf{j}$ **(b)** $2\mathbf{i} - 3\mathbf{j}$ **(c)** $\mathbf{i} + \mathbf{j}$ **(d)** $\mathbf{i} - 3\mathbf{j}$

In Exercises 50–53, calculate the linear combination.

50. $(-2\mathbf{i} + 9\mathbf{j}) + (3\mathbf{i} - 4\mathbf{j})$

51. $6(\mathbf{i} + 9\mathbf{j}) + 2(\mathbf{i} - 4\mathbf{j})$

52. $(\mathbf{i} + \mathbf{j}) + (\mathbf{i} - 2\mathbf{j})$

53. $3(\mathbf{i} + \mathbf{j}) + 7(3\mathbf{i} - \mathbf{j})$

54. For each of the position vectors $\mathbf{u}$ with endpoints A, B, and C in Figure 26, indicate with a diagram the multiples $r\mathbf{v}$ and $s\mathbf{w}$ such that $\mathbf{u} = r\mathbf{v} + s\mathbf{w}$. A sample is shown for $\mathbf{u} = \overrightarrow{OQ}$.

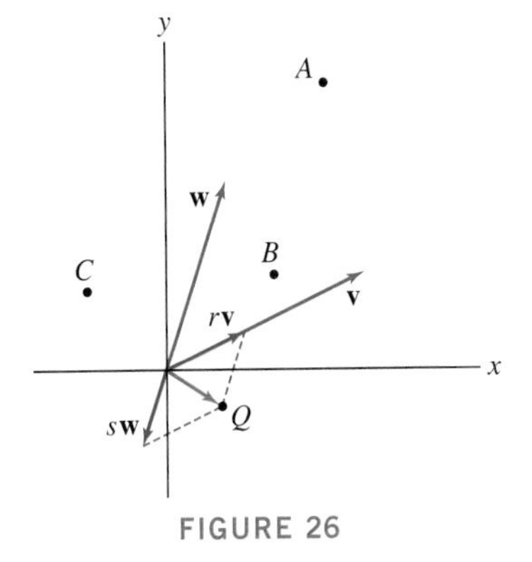

FIGURE 26

55. Express $\mathbf{u} = \langle 3, -1 \rangle$ as a linear combination $\mathbf{u} = r\mathbf{v} + s\mathbf{w}$, where $\mathbf{v} = \langle 2, 1 \rangle$ and $\mathbf{w} = \langle 1, 3 \rangle$. Sketch the vectors $\mathbf{u}$, $\mathbf{v}$, $\mathbf{w}$ and the parallelogram formed by $r\mathbf{v}$ and $s\mathbf{w}$.

56. Express $\mathbf{u} = \langle 6, -2 \rangle$ as a linear combination $\mathbf{u} = r\mathbf{v} + s\mathbf{w}$, where $\mathbf{v} = \langle 1, 1 \rangle$ and $\mathbf{w} = \langle 1, -1 \rangle$. Sketch the vectors $\mathbf{u}$, $\mathbf{v}$, $\mathbf{w}$ and the parallelogram formed by $r\mathbf{v}$ and $s\mathbf{w}$.

57. Sketch the parallelogram spanned by $\mathbf{v} = \langle 1, 4 \rangle$ and $\mathbf{w} = \langle 5, 2 \rangle$. Add the vector $\mathbf{u} = \langle 2, 3 \rangle$ to the sketch and express $\mathbf{u}$ as a linear combination of $\mathbf{v}$ and $\mathbf{w}$.

58. Calculate the magnitude of the force on cables 1 and 2 in Figure 27.

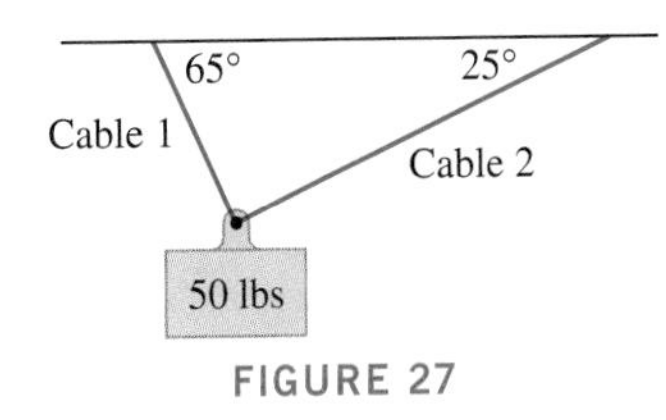

FIGURE 27

59. Determine the magnitude of the forces $\mathbf{F}_1$ and $\mathbf{F}_2$ in Figure 28, assuming that there is no net force on the object.

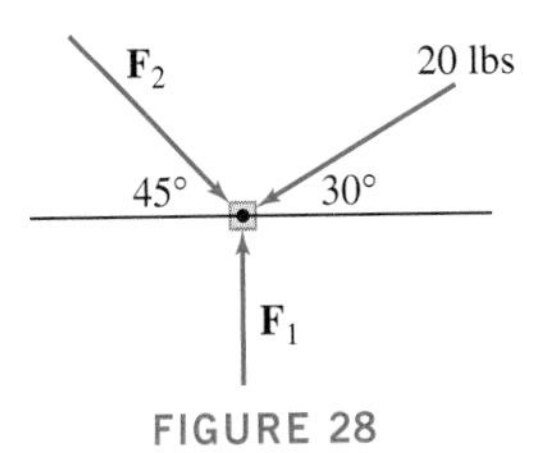

FIGURE 28

60. A plane flying due east at 200 km/hour encounters a 40 km/hour north-easterly wind. The resultant velocity of the plane is the vector sum $\mathbf{v} = \mathbf{v}_1 + \mathbf{v}_2$, where $\mathbf{v}_1$ is the velocity vector of the plane and $\mathbf{v}_2$ is the velocity vector of the wind (Figure 29). The angle between $\mathbf{v}_1$ and $\mathbf{v}_2$ is $\frac{\pi}{4}$. Determine the resultant *speed* of the plane (the length of the $\mathbf{v}$).

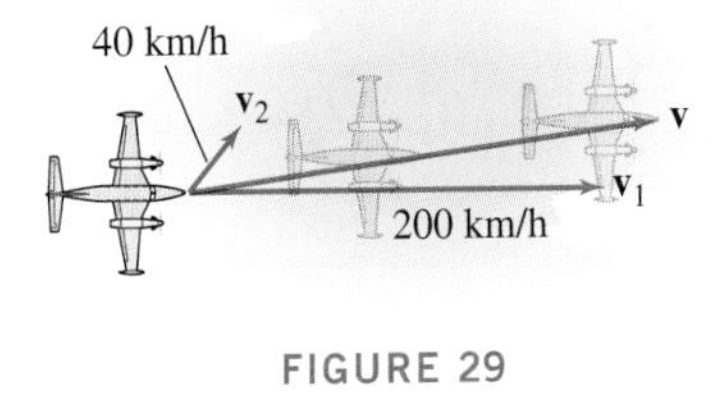

FIGURE 29

Further Insights and Challenges

In Exercises 61–63, refer to Figure 30, which shows a robotic arm consisting of two segments of lengths L_1 and L_2.

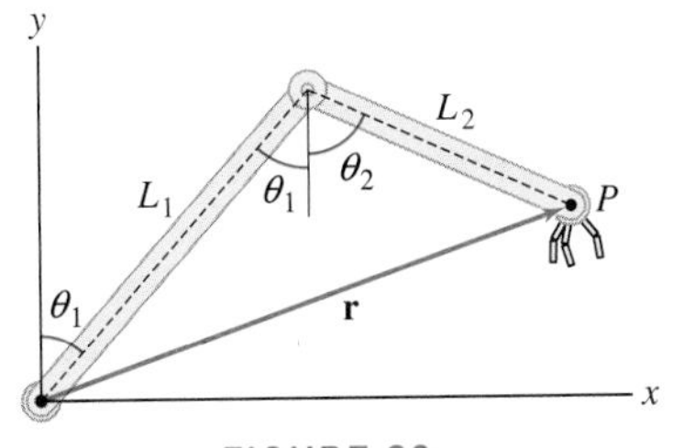

FIGURE 30

61. Find the components of the vector $\mathbf{r} = \overrightarrow{OP}$ in terms of θ_1 and θ_2.

62. Let $L_1 = 5$ and $L_2 = 3$. Find $\mathbf{r}$ for $\theta_1 = \frac{\pi}{3}$, $\theta_2 = \frac{\pi}{4}$.

63. Let $L_1 = 5$ and $L_2 = 3$. Show that the set of points reachable by the robotic arm with $\theta_1 = \theta_2$ is an ellipse.

64. Use vectors to prove that the diagonals $\overline{AC}$ and $\overline{BD}$ of a parallelogram bisect each other (Figure 31). *Hint:* Observe that the midpoint of $\overline{BD}$ is the terminal point of $\mathbf{w} + \frac{1}{2}(\mathbf{v} - \mathbf{w})$.

65. Use vectors to prove that the segments joining the midpoints of opposite sides of a quadrilateral bisect each other (Figure 32). *Hint:*

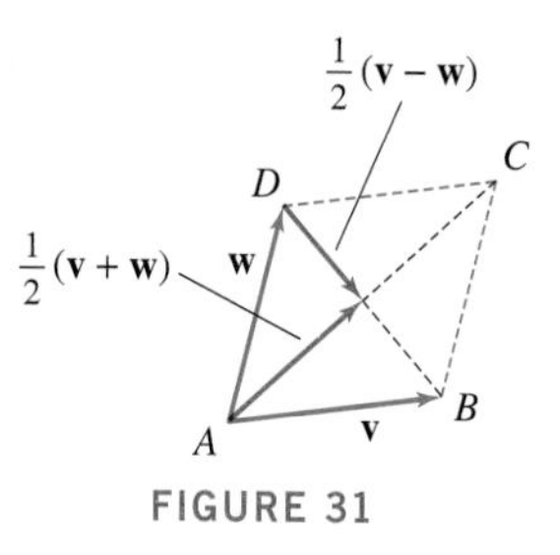

FIGURE 31

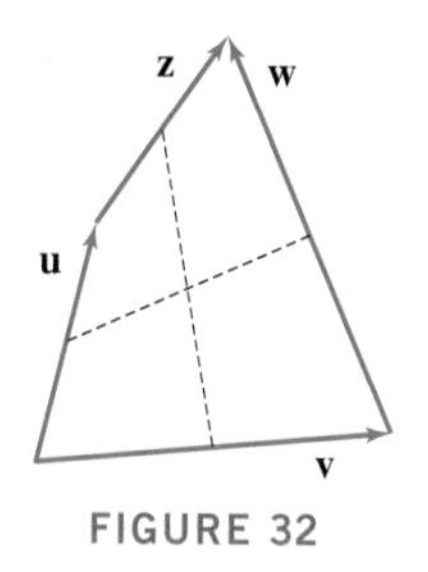

FIGURE 32

Show that the midpoints of these segments are the terminal points of $\frac{1}{4}(2\mathbf{u}+\mathbf{v}+\mathbf{z})$ and $\frac{1}{4}(2\mathbf{v}+\mathbf{w}+\mathbf{u})$.

66. Prove that two vectors $\mathbf{v} = \langle a, b\rangle$ and $\mathbf{w} = \langle c, d\rangle$ are perpendicular if and only if $ac + bd = 0$.

12.2 Vectors in Three Dimensions

This section extends the vector concepts introduced in the previous section to three-dimensional space. We begin with some introductory remarks about the three-dimensional coordinate system.

By convention, we label the x-, y-, and z-axes as in Figure 1(A). This labeling satisfies the **right-hand rule**, which means that when you position your right hand so that your fingers curl from the positive x-axis toward the positive y-axis, your thumb points in the positive z-direction. The axes in Figure 1(B) are not labeled according to the right-hand rule.

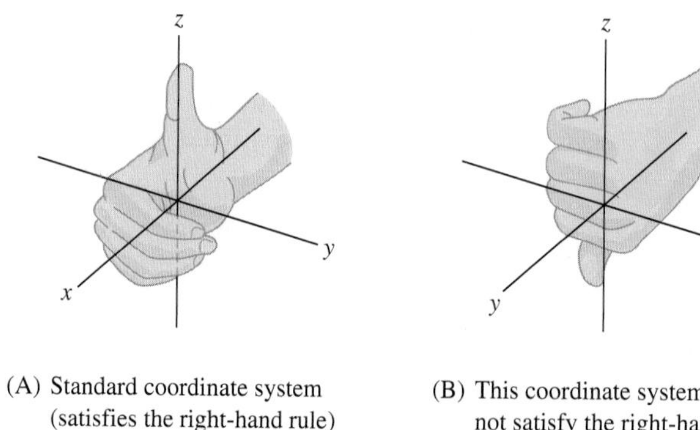

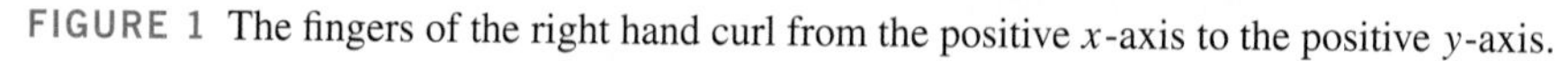

(A) Standard coordinate system (satisfies the right-hand rule)

(B) This coordinate system does not satisfy the right-hand rule because the thumb points in the negative z-direction.

FIGURE 1 The fingers of the right hand curl from the positive x-axis to the positive y-axis.

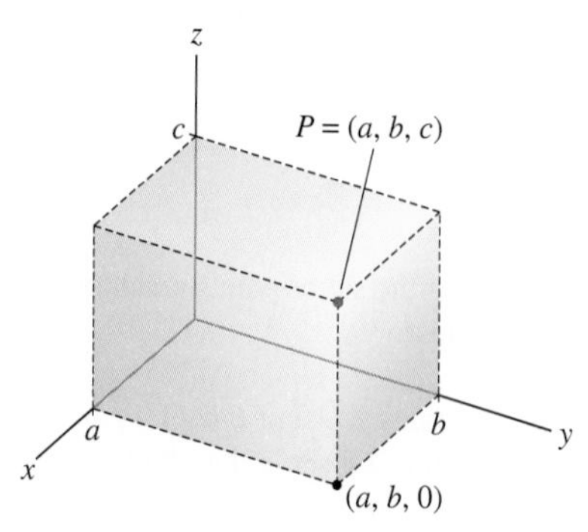

FIGURE 2

Each point in space has unique coordinates (a, b, c) relative to the axes (Figure 2). We denote the set of all triples (a, b, c) by $\mathbf{R}^3$. The **coordinate planes** in $\mathbf{R}^3$ are defined by setting one of the coordinates equal to zero (Figure 3). The xy-plane consists of the points $(a, b, 0)$ and is defined by the equation $z = 0$. Similarly, $x = 0$ defines the yz-plane consisting of the points $(0, b, c)$ and $y = 0$ defines the xz-plane consisting of the points $(a, 0, c)$. The coordinate planes divide $\mathbf{R}^3$ into eight **octants** (analogous to the four quadrants in the plane). Each octant corresponds to a possible combination of signs of the coordinates. The set of points (a, b, c) with $a, b, c > 0$ is called the **first octant**.

As in two dimensions, we derive the distance formula in $\mathbf{R}^3$ from the Pythagorean Theorem.

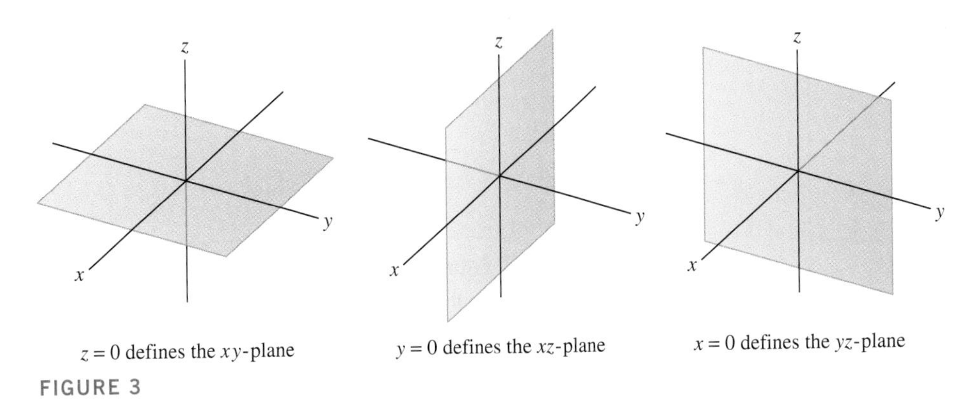

FIGURE 3

THEOREM 1 Distance Formula in $\mathbf{R}^3$ The distance $|P - Q|$ between the points $P = (a_1, b_1, c_1)$ and $Q = (a_2, b_2, c_2)$ is

$$|P - Q| = \sqrt{(a_2 - a_1)^2 + (b_2 - b_1)^2 + (c_2 - c_1)^2} \tag{1}$$

Proof First apply the distance formula in the plane to the points P and R (Figure 4):

$$|P - R|^2 = (a_2 - a_1)^2 + (b_2 - b_1)^2$$

Then observe that $\triangle PRQ$ is a right triangle [Figure 4(B)] and use the Pythagorean Theorem:

$$|P - Q|^2 = |P - R|^2 + |R - Q|^2 = (a_2 - a_1)^2 + (b_2 - b_1)^2 + (c_2 - c_1)^2 \qquad \blacksquare$$

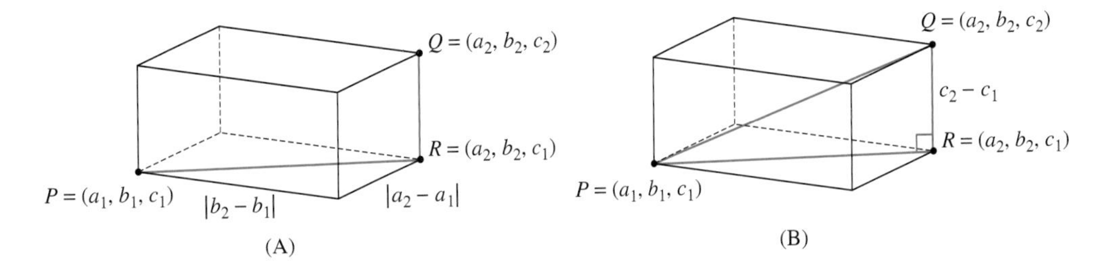

FIGURE 4 Compute $|P - Q|$ using the right triangle $\triangle PQR$.

The sphere of radius R with center $Q = (a, b, c)$ consists of all points $P = (x, y, z)$ located a distance R from Q (Figure 5). By the distance formula, the coordinates of $P = (x, y, z)$ must satisfy

$$\sqrt{(x - a)^2 + (y - b)^2 + (z - c)^2} = R$$

On squaring both sides, we obtain the standard equation of the sphere [Eq. (3) below].

Now consider the equation

$$(x - a)^2 + (y - b)^2 = R^2 \tag{2}$$

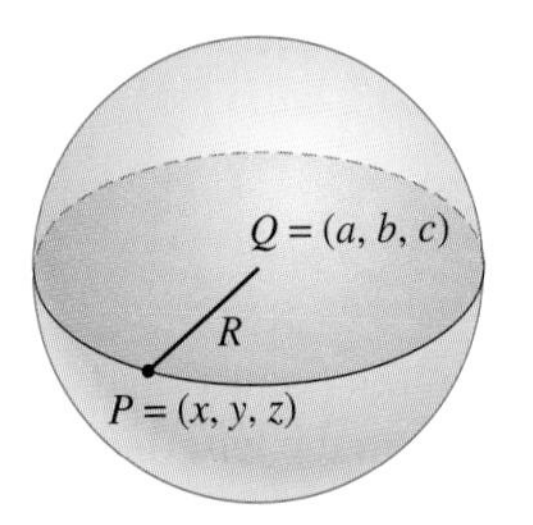

FIGURE 5 Sphere of radius R centered at (a, b, c).

In the xy-plane, Eq. (2) defines the circle of radius R with center (a, b). However, as an equation in $\mathbf{R}^3$, it defines the cylinder of radius R whose central axis is the vertical line through $(a, b, 0)$ (Figure 6). Indeed, a point (x, y, z) satisfies Eq. (2) for any value of z

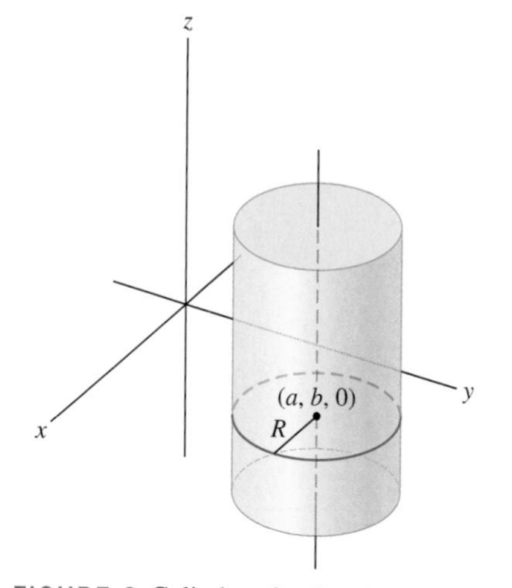

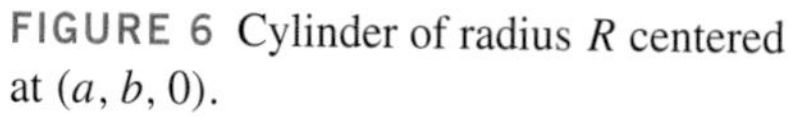
FIGURE 6 Cylinder of radius R centered at $(a, b, 0)$.

if (x, y) lies on the circle. It is usually clear from the context which of the following is intended:

$$\text{Circle} = \{(x, y) : (x - a)^2 + (y - b)^2 = R^2\}$$

$$\text{Cylinder} = \{(x, y, z) : (x - a)^2 + (y - b)^2 = R^2\}$$

Equations of Spheres and Cylinders An equation of the sphere in $\mathbf{R}^3$ of radius R centered at $Q = (a, b, c)$ is

$$(x - a)^2 + (y - b)^2 + (z - c)^2 = R^2 \qquad \boxed{3}$$

An equation of the cylinder in $\mathbf{R}^3$ of radius R whose central axis is the vertical line through $(a, b, 0)$ is

$$(x - a)^2 + (y - b)^2 = R^2 \qquad \boxed{4}$$

EXAMPLE 1 Describe the sets of points defined by the following conditions:

(a) $x^2 + y^2 + z^2 = 4, \quad y \geq 0$ **(b)** $(x - 3)^2 + (y - 2)^2 = 1, \quad z \geq -1$

Solution

(a) The equation $x^2 + y^2 + z^2 = 4$ defines a sphere of radius 2 centered at the origin. The inequality $y \geq 0$ holds for points lying on the positive side of the xz-plane. We obtain the right hemisphere of radius 2 illustrated in Figure 7(A).

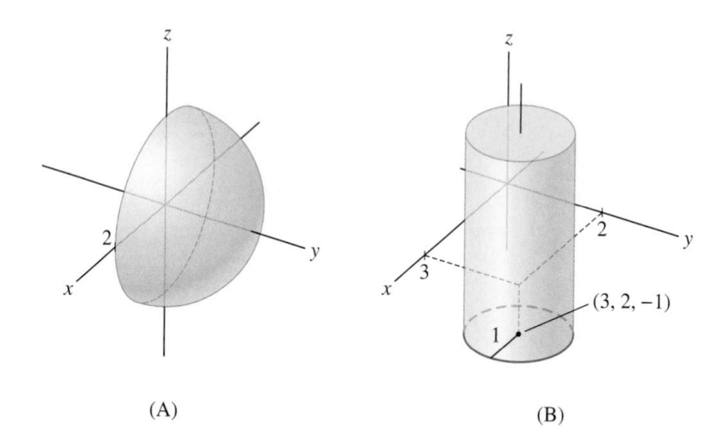

FIGURE 7 Hemisphere and upper cylinder.

(b) The equation $(x - 3)^2 + (y - 2)^2 = 1$ defines a cylinder of radius 1 whose central axis is the vertical line through $(3, 2, 0)$. The part of the cylinder where $z \geq -1$ is illustrated in Figure 7(B). ■

Vector Concepts

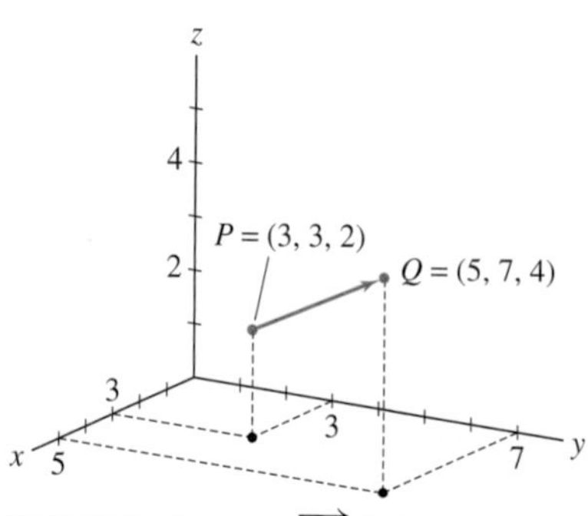

FIGURE 8 A vector $\overrightarrow{PQ}$ in 3-space.

As in the plane, a vector $\mathbf{v} = \overrightarrow{PQ}$ in $\mathbf{R}^3$ is determined by an initial point P and a terminal point Q (Figure 8). If $P = (a_1, b_1, c_1)$ and $Q = (a_2, b_2, c_2)$, then the **length** of $\mathbf{v} = \overrightarrow{PQ}$, denoted $\|\mathbf{v}\|$, is the distance from P to Q:

$$\|\mathbf{v}\| = \|\overrightarrow{PQ}\| = \sqrt{(a_2 - a_1)^2 + (b_2 - b_1)^2 + (c_2 - c_1)^2}$$

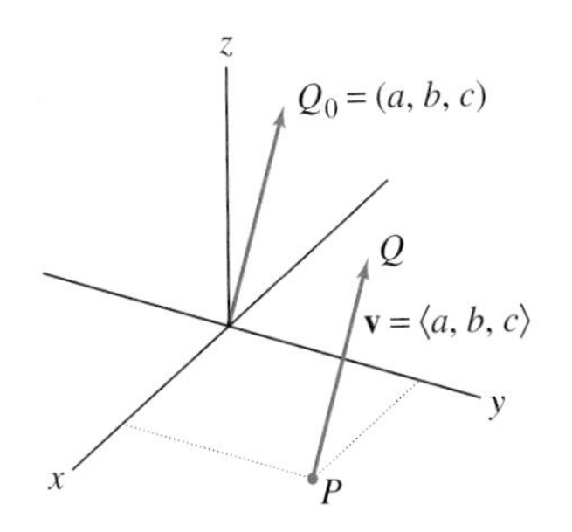

FIGURE 9 A vector **v** and its translate based at the origin.

Our convention about basepoints remains in force: All vectors are assumed to be based at the origin unless otherwise indicated.

The terminology of the previous section carries to $\mathbf{R}^3$ with little change.

- A vector is said to undergo a **translation** **v** if it is moved without changing direction or magnitude.
- Two vectors **v** and **w** are called **equivalent** if **w** is a translate of **v** or, equivalently, if **v** and **w** have the same length and direction.
- The **position vector** of a point Q_0 is the vector $\mathbf{v} = \overrightarrow{OQ_0}$ based at the origin (Figure 9).
- If $P = (a_1, b_1, c_1)$ and $Q = (a_2, b_2, c_2)$, then the **components** of $\mathbf{v} = \overrightarrow{PQ}$ are the differences $a = a_2 - a_1$, $b = b_2 - b_1$, $c = c_2 - c_1$, that is,

$$\mathbf{v} = \overrightarrow{PQ} = \overrightarrow{OQ} - \overrightarrow{OP} = \langle a_2, b_2, c_2 \rangle - \langle a_1, b_1, c_1 \rangle$$

For example, if $P = (3, -4, -4)$ and $Q = (2, 5, -1)$, then

$$\mathbf{v} = \overrightarrow{PQ} = \langle 2, 5, -1 \rangle - \langle 3, -4, -4 \rangle = \langle -1, 9, 3 \rangle$$

- A vector $\mathbf{v} = \overrightarrow{PQ}$ with components $\langle a, b, c \rangle$ is equivalent to the vector $\mathbf{v}_0 = \overrightarrow{OQ_0}$ based at the origin with $Q_0 = (a, b, c)$ (Figure 9).
- The operations of vector addition and scalar multiplication are defined as in the two-dimensional case. Vector addition is commutative and associative, and satisfies the distributive property with respect to scalar multiplication (Theorem 1 in Section 12.1).
- Vector addition is visualized geometrically using the Parallelogram Law (Figure 10). In terms of components, if $\mathbf{v} = \langle a_1, b_1, c_1 \rangle$ and $\mathbf{w} = \langle a_2, b_2, c_2 \rangle$, then

$$\lambda \mathbf{v} = \lambda \langle a_1, b_1, c_1 \rangle = \langle \lambda a_1, \lambda b_1, \lambda c_1 \rangle$$
$$\mathbf{v} + \mathbf{w} = \langle a_1, b_1, c_1 \rangle + \langle a_2, b_2, c_2 \rangle = \langle a_1 + a_2, b_1 + b_2, c_1 + c_2 \rangle$$

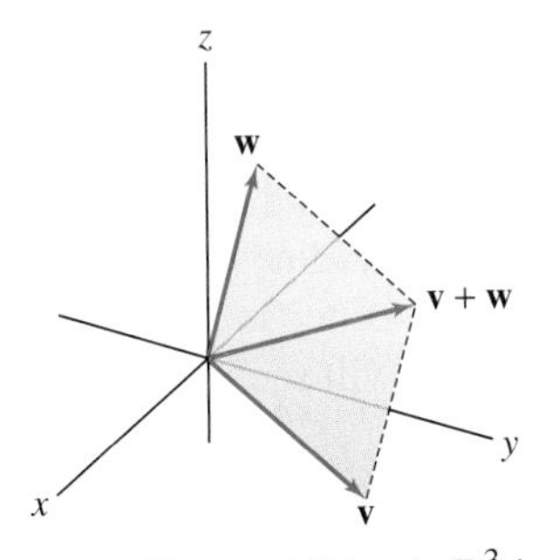

FIGURE 10 Vector addition in $\mathbf{R}^3$ is defined using the Parallelogram Law as in $\mathbf{R}^2$.

EXAMPLE 2 Vector Calculations Calculate $\|\mathbf{v}\|$ and $6\mathbf{v} - \frac{1}{2}\mathbf{w}$, where $\mathbf{v} = \langle 3, -1, 2 \rangle$ and $\mathbf{w} = \langle 4, 6, -8 \rangle$.

Solution

$$\|\mathbf{v}\| = \sqrt{3^2 + (-1)^2 + 2^2} = \sqrt{14}$$

$$6\mathbf{v} - \frac{1}{2}\mathbf{w} = 6\langle 3, -1, 2 \rangle - \frac{1}{2}\langle 4, 6, -8 \rangle$$
$$= \langle 18, -6, 12 \rangle - \langle 2, 3, -4 \rangle = \langle 16, -9, 16 \rangle$$

The **standard basis vectors** in $\mathbf{R}^3$ are

$$\mathbf{i} = \langle 1, 0, 0 \rangle, \qquad \mathbf{j} = \langle 0, 1, 0 \rangle, \qquad \mathbf{k} = \langle 0, 0, 1 \rangle$$

Every vector can be written as a **linear combination** of the standard basis vectors (Figure 11):

$$\langle a, b, c \rangle = a\langle 1, 0, 0 \rangle + b\langle 0, 1, 0 \rangle + c\langle 0, 0, 1 \rangle = a\mathbf{i} + b\mathbf{j} + c\mathbf{k}$$

For example, $\langle -9, -4, 17 \rangle = -9\mathbf{i} - 4\mathbf{j} + 17\mathbf{k}$.

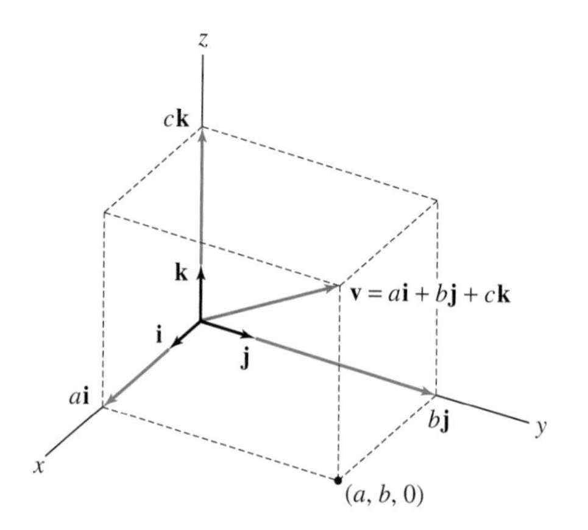

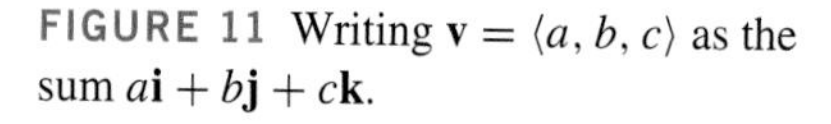

FIGURE 11 Writing $\mathbf{v} = \langle a, b, c \rangle$ as the sum $a\mathbf{i} + b\mathbf{j} + c\mathbf{k}$.

■ **EXAMPLE 3** Find the unit vector $\mathbf{e_v}$ in the direction of $\mathbf{v} = 3\mathbf{i} + 2\mathbf{j} - 4\mathbf{k}$.

Solution Since $\|\mathbf{v}\| = \sqrt{3^2 + 2^2 + (-4)^2} = \sqrt{29}$,

$$\mathbf{e_v} = \frac{1}{\|\mathbf{v}\|}\mathbf{v} = \frac{1}{\sqrt{29}}(3\mathbf{i} + 2\mathbf{j} - 4\mathbf{k}) = \left\langle \frac{3}{\sqrt{29}}, \frac{2}{\sqrt{29}}, \frac{-4}{\sqrt{29}} \right\rangle$$ ■

Parametric Equations of a Line

Although the basic vector concepts are essentially the same in two and three dimensions, there is an important difference in the way lines are described. A line in $\mathbf{R}^2$ can be described by a single linear equation such as $y = mx + b$. This is not so in $\mathbf{R}^3$ because a linear equation in $\mathbf{R}^3$ defines a plane rather than a line (as we will see in Section 12.5). Therefore, we describe lines in $\mathbf{R}^3$ parametrically.

We may specify a line $\mathcal{L}$ by stating that $\mathcal{L}$ passes through a given point $P_0 = (x_0, y_0, z_0)$ in the direction of a vector $\mathbf{v} = \langle a, b, c \rangle$ (Figure 12). The vector $\mathbf{v}$ is called a **direction vector** for $\mathcal{L}$. Figure 12 shows that a point on this line is the terminal point of the vector obtained by adding a multiple $t\mathbf{v}$ to the position vector $\overrightarrow{OP_0}$. Thus, $\mathcal{L}$ has the following parametrization:

$$\mathbf{r}(t) = \overrightarrow{OP_0} + t\mathbf{v} = \langle x_0, y_0, z_0 \rangle + t\,\langle a, b, c \rangle$$

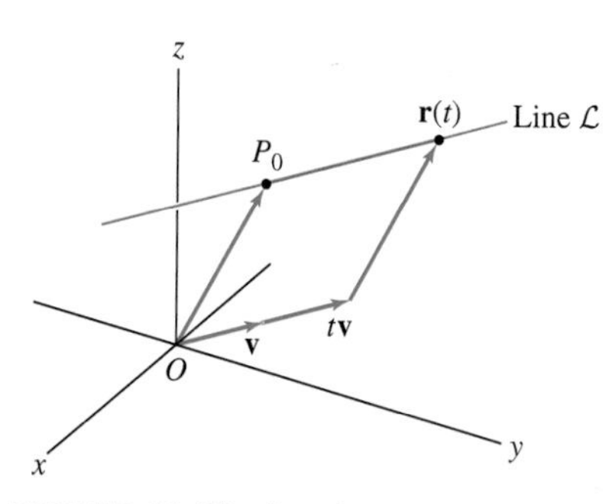

FIGURE 12 The line through P_0 in the direction $\mathbf{v}$.

In this vector parametrization, the terminal point of $\mathbf{r}(t)$ traces the line as t varies from $-\infty$ to ∞. The components of $\mathbf{r}(t)$ are given by the parametric equations

$$x = x_0 + at, \quad y = y_0 + bt, \quad z = z_0 + ct$$

Equation of a Line (Point-Direction Form) The line $\mathcal{L}$ through $P_0 = (x_0, y_0, z_0)$ in the direction $\mathbf{v} = \langle a, b, c \rangle$ is described in vector or parametric form by

Vector parametrization: $\mathbf{r}(t) = \overrightarrow{OP_0} + t\mathbf{v} = \langle x_0, y_0, z_0 \rangle + t\,\langle a, b, c \rangle$ [5]

Parametric equations: $x = x_0 + at, \quad y = y_0 + bt, \quad z = z_0 + ct$ [6]

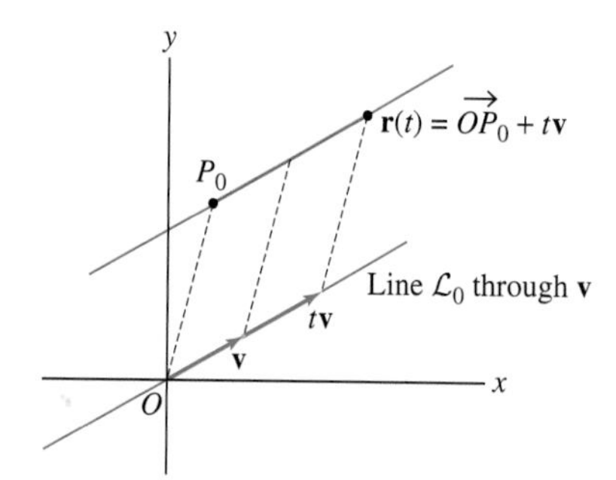

FIGURE 13 The line $\mathbf{r}(t)$ is obtained by translating the line $\mathcal{L}_0$ through $\mathbf{v}$ so that it passes through P_0.

A direction vector $\mathbf{v}$ for a line is a substitute for the slope (which only makes sense in the plane). However, the direction vector is not unique—any nonzero scalar multiple of $\mathbf{v}$ is also a direction vector for $\mathcal{L}$.

GRAPHICAL INSIGHT For any nonzero vector $\mathbf{v}$, the line $\mathcal{L}_0$ through $\mathbf{v}$ consists of the multiples $\mathbf{r}_0(t) = t\mathbf{v}$. Observe in Figure 13 (illustrating the case of a line in the plane) that the line through P_0 in the direction of $\mathbf{v}$ is obtained by translating $\mathcal{L}_0$ in a parallel fashion so that it passes through P_0. This is the graphical interpretation of the parameterization $\mathbf{r}(t) = \overrightarrow{OP_0} + t\mathbf{v}$.

■ **EXAMPLE 4** Find a parametrization of the line through $P_0 = (3, -1, 4)$ with direction vector $\mathbf{v} = \langle 2, 1, 7\rangle$.

Vector parametrizations are not unique. Every line $\mathcal{L}$ can be parametrized in infinitely many ways because we are free to choose any point P_0 on $\mathcal{L}$ and we may replace the direction vector by any nonzero scalar multiple.

Solution By Eq. (5),

$$\mathbf{r}(t) = \underbrace{\langle 3, -1, 4\rangle}_{\text{Coordinates of } P_0} + t\underbrace{\langle 2, 1, 7\rangle}_{\text{Direction vector}} = \langle 3+2t, -1+t, 4+7t\rangle$$

The parametric equations are $x = 3 + 2t, \quad y = -1 + t, \quad z = 4 + 7t$. ■

■ **EXAMPLE 5 Intersection of Two Lines** Determine if the following two lines intersect:

$$\mathbf{r}_1(t) = \langle 1, 0, 1\rangle + t\langle 3, 3, 5\rangle, \qquad \mathbf{r}_2(t) = \langle 3, 6, 1\rangle + t\langle 4, -2, 7\rangle$$

CAUTION We cannot assume in Eq. (7) that the parameter values t_1 and t_2 are equal. The point of intersection may correspond to different parameter values on the two lines.

Solution The two lines intersect if there exist parameter values t_1 and t_2 such that $\mathbf{r}_1(t_1) = \mathbf{r}_2(t_2)$, that is,

$$\langle 1, 0, 1\rangle + t_1\langle 3, 3, 5\rangle = \langle 3, 6, 1\rangle + t_2\langle 4, -2, 7\rangle \qquad \boxed{7}$$

This is equivalent to the three equations for the components:

$$x = 1 + 3t_1 = 3 + 4t_2, \qquad y = 3t_1 = 6 - 2t_2, \qquad z = 1 + 5t_1 = 1 + 7t_2 \qquad \boxed{8}$$

Let's solve the first two equations for t_1 and t_2. Subtracting the second equation from the first, we get $1 = 6t_2 - 3$ or $t_2 = \frac{2}{3}$. Using this value in the second equation, we get $t_1 = 2 - \frac{2}{3}t_2 = \frac{14}{9}$. The values $t_1 = \frac{14}{9}$ and $t_2 = \frac{2}{3}$ satisfy the first two equations and thus $\mathbf{r}_1(t_1)$ and $\mathbf{r}_2(t_2)$ have the same x- and y-coordinates (Figure 14). However, t_1 and t_2 do not satisfy the third equation in (8):

$$1 + 5\left(\frac{14}{9}\right) \neq 1 + 7\left(\frac{2}{3}\right)$$

Therefore, Eq. (7) has no solution and the lines do not intersect. ■

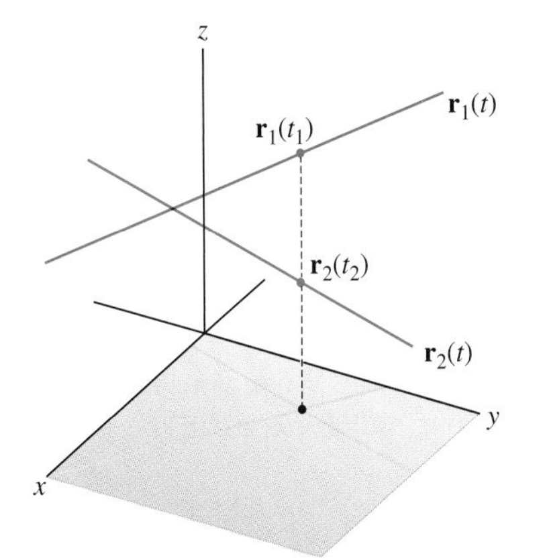

FIGURE 14 The lines $\mathbf{r}_1(t)$ and $\mathbf{r}_2(t)$ do not intersect, but $\mathbf{r}_1(t_1)$ and $\mathbf{r}_2(t_2)$ have the same x- and y-coordinates.

Suppose we wish to parametrize the line $\mathcal{L}$ passing through two points:

$$P = (a_1, b_1, c_1), \qquad Q = (a_2, b_2, c_2)$$

The vector

$$\mathbf{v} = \overrightarrow{PQ} = \langle a_2 - a_1, b_2 - b_1, c_2 - c_1\rangle$$

points from P to Q, so it is a direction vector for $\mathcal{L}$ (Figure 15). Thus, a possible vector parametrization is

$$\mathbf{r}(t) = \overrightarrow{OP} + t\mathbf{v} = \langle a_1, b_1, c_1\rangle + t\langle a_2 - a_1, b_2 - b_1, c_2 - c_1\rangle$$
$$= (1 - t)\langle a_1, b_1, c_1\rangle + t\langle a_2, b_2, c_2\rangle$$

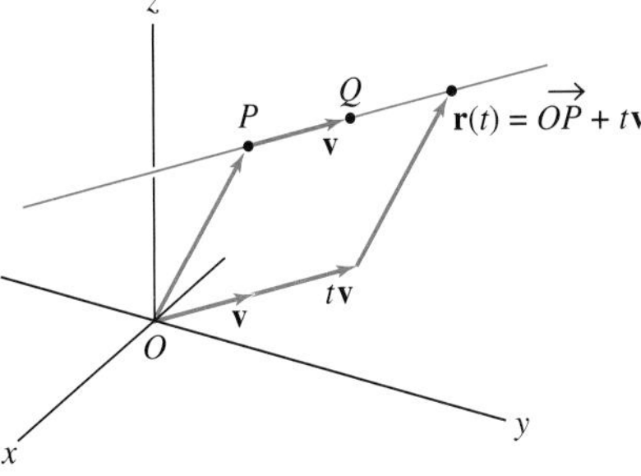

FIGURE 15 Line through two points P and Q.

In this parametrization,

$$\mathbf{r}(0) = \langle a_1, b_1, c_1\rangle = \overrightarrow{OP}, \qquad \mathbf{r}(1) = \langle a_2, b_2, c_2\rangle = \overrightarrow{OQ}$$

Thus, $\mathbf{r}(t)$ begins at P and traces the segment $\overline{PQ}$ joining P and Q for $0 \le t \le 1$. For $t = \frac{1}{2}$, we obtain the **midpoint**:

$$\text{Midpoint of } \overline{PQ} = \left\langle \frac{a_1 + a_2}{2}, \frac{b_1 + b_2}{2}, \frac{c_1 + c_2}{2} \right\rangle$$

Equation of Line Through Two Points The line through the points $P = (a_1, b_1, c_1)$ and $Q = (a_2, b_2, c_2)$ has **vector parametrization**

$$\mathbf{r}(t) = (1-t)\overrightarrow{OP} + t\overrightarrow{OQ} = (1-t)\langle a_1, b_1, c_1\rangle + t\langle a_2, b_2, c_2\rangle \qquad \boxed{9}$$

and **parametric equations:**

$$x = a_1 + (a_2 - a_1)t, \qquad y = b_1 + (b_2 - b_1)t, \qquad z = c_1 + (c_2 - c_1)t \qquad \boxed{10}$$

In this parametrization, $\mathbf{r}(0) = \overrightarrow{OP}$, $\mathbf{r}(1) = \overrightarrow{OQ}$, and $\mathbf{r}(t)$ traces the segment $\overline{PQ}$ from P to Q for $0 \le t \le 1$.

EXAMPLE 6 Parametrize the line through $P = (1, 0, 4)$ and $Q = (3, 2, 1)$, and find the midpoint of $\overline{PQ}$.

Solution The line through $P = (1, 0, 4)$ and $Q = (3, 2, 1)$ has the parametrization

$$\mathbf{r}(t) = (1-t)\langle 1, 0, 4\rangle + t\langle 3, 2, 1\rangle = \langle 1 + 2t, 2t, 4 - 3t\rangle$$

The midpoint of $\overline{PQ}$ is the endpoint of $\mathbf{r}(\frac{1}{2})$:

$$\mathbf{r}\left(\frac{1}{2}\right) = \frac{1}{2}\langle 1, 0, 4\rangle + \frac{1}{2}\langle 3, 2, 1\rangle = \left\langle 2, 1, \frac{5}{2}\right\rangle$$

12.2 SUMMARY

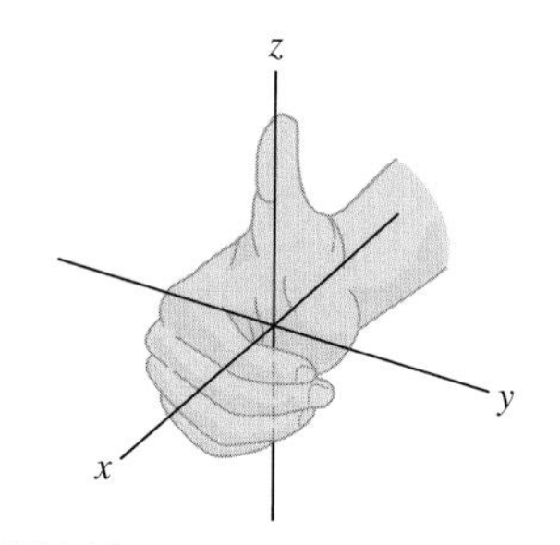

FIGURE 16

- The axes in $\mathbf{R}^3$ are labeled so as to satisfy the *right-hand rule*: When your fingers curl from the positive x-axis toward the positive y-axis, your thumb points in the positive z-direction (Figure 16).
-

Sphere of radius R and center (a, b, c)	$(x-a)^2 + (y-b)^2 + (z-c)^2 = R^2$
Cylinder of radius R with vertical axis through $(a, b, 0)$	$(x-a)^2 + (y-b)^2 = R^2$

- The notation and terminology for vectors in the plane carry over to vectors in $\mathbf{R}^3$.
- If $P = (a_1, b_1, c_1)$ and $Q = (a_2, b_2, c_2)$, then the length of $\mathbf{v} = \overrightarrow{PQ}$ is

$$\|\mathbf{v}\| = \|\overrightarrow{PQ}\| = \sqrt{(a_2 - a_1)^2 + (b_2 - b_1)^2 + (c_2 - c_1)^2}$$

- Equations for the line through $P_0 = (x_0, y_0, z_0)$ with direction vector $\mathbf{v} = \langle a, b, c\rangle$:

Vector parametrization: $\mathbf{r}(t) = P_0 + t\mathbf{v} = \langle x_0, y_0, z_0\rangle + t\langle a, b, c\rangle$

Parametric equations: $x = x_0 + at, \quad y = y_0 + bt, \quad z = z_0 + ct$

- Equation of the line through $P = (a_1, b_1, c_1)$ and $Q = (a_2, b_2, c_2)$:

Vector parametrization: $\mathbf{r}(t) = (1-t)\langle a_1, b_1, c_1\rangle + t\langle a_2, b_2, c_2\rangle$

Parametric equations: $x = a_1 + (a_2 - a_1)t, \quad y = b_1 + (b_2 - b_1)t,$

$z = c_1 + (c_2 - c_1)t$

The segment $\overline{PQ}$ is parametrized by $\mathbf{r}(t)$ for $0 \le t \le 1$ and the midpoint of $\overline{PQ}$ is the terminal point of $\mathbf{r}(\frac{1}{2})$.

12.2 EXERCISES

Preliminary Questions

1. What is the terminal point of the vector $\mathbf{v} = \langle 3, 2, 1 \rangle$ based at the point $P = (1, 1, 1)$?

2. What are the components of the vector $\mathbf{v} = \langle 3, 2, 1 \rangle$ based at the point $P = (1, 1, 1)$?

3. If $\mathbf{v} = -3\mathbf{w}$, then (choose the correct answer):

(a) $\mathbf{v}$ and $\mathbf{w}$ are parallel.

(b) $\mathbf{v}$ and $\mathbf{w}$ point in the same direction.

4. Which of the following is a direction vector for the line through $P = (3, 2, 1)$ and $Q = (1, 1, 1)$?

(a) $\langle 3, 2, 1 \rangle$ **(b)** $\langle 1, 1, 1 \rangle$ **(c)** $\langle 2, 1, 0 \rangle$

5. How many different direction vectors does a line have?

6. True or false: If $\mathbf{v}$ is a direction vector for a line $\mathcal{L}$, then $-\mathbf{v}$ is also a direction vector for $\mathcal{L}$.

Exercises

1. Sketch the vector $\mathbf{v} = \langle 1, 3, 2 \rangle$ and compute its length.

2. Let $\mathbf{v} = \overrightarrow{P_0Q_0}$, where $P_0 = (1, -2, 5)$ and $Q_0 = (0, 1, -4)$. Which of the following vectors (with tail P and head Q) are equivalent to $\mathbf{v}$?

	$\mathbf{v}_1$	$\mathbf{v}_2$	$\mathbf{v}_3$	$\mathbf{v}_4$
P	$(1, 2, 4)$	$(1, 5, 4)$	$(0, 0, 0)$	$(2, 4, 5)$
Q	$(0, 5, -5)$	$(0, -8, 13)$	$(-1, 3, -9)$	$(1, 7, 4)$

3. Sketch the vector $\mathbf{v} = \langle 1, 1, 0 \rangle$ based at $P = (0, 1, 1)$. Describe this vector in the form $\overrightarrow{PQ}$ for some point Q and sketch the vector $\mathbf{v}_0$ based at the origin equivalent to $\mathbf{v}$.

4. Determine if the coordinate systems (A)–(C) in Figure 17 satisfy the right-hand rule.

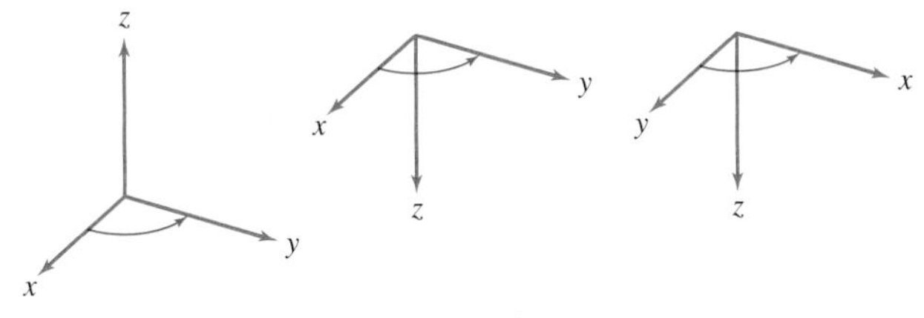

FIGURE 17

In Exercises 5–8, find the components of the vector $\overrightarrow{PQ}$.

5. $P = (1, 0, 1)$, $Q = (2, 1, 0)$

6. $P = (-3, -4, 2)$, $Q = (1, -4, 3)$

7. $P = (4, 6, 0)$, $Q = (-\frac{1}{2}, \frac{9}{2}, 1)$

8. $P = (-\frac{1}{2}, \frac{9}{2}, 1)$, $Q = (4, 6, 0)$

In Exercises 9–12, let $R = (1, 4, 3)$.

9. Calculate the length of $\overrightarrow{OR}$.

10. Find the point Q such that $\mathbf{v} = \overrightarrow{RQ}$ has components $\langle 4, 1, 1 \rangle$ and sketch $\mathbf{v}$.

11. Find the point P such that $\mathbf{w} = \overrightarrow{PR}$ has components $\langle 3, -2, 3 \rangle$ and sketch $\mathbf{w}$.

12. Find the components of $\mathbf{u} = \overrightarrow{PR}$, where $P = (1, 2, 2)$.

13. Let $\mathbf{v} = \langle 4, 8, 12 \rangle$. Which of the following vectors is parallel to $\mathbf{v}$? Which point in the same direction?

(a) $\langle 2, 4, 6 \rangle$ **(b)** $\langle -1, -2, 3 \rangle$

(c) $\langle -7, -14, -21 \rangle$ **(d)** $\langle 6, 10, 14 \rangle$

In Exercises 14–17, determine whether $\overrightarrow{AB}$ is equivalent to $\overrightarrow{PQ}$.

14. $A = (1, 1, 1)$, $B = (3, 3, 3)$, $P = (1, 4, 5)$, $Q = (3, 6, 7)$

15. $A = (1, 4, 1)$, $B = (-2, 2, 0)$, $P = (2, 5, 7)$, $Q = (-3, 2, 1)$

16. $A = (0, 0, 0)$, $B = (-4, 2, 3)$, $P = (4, -2, -3)$, $Q = (0, 0, 0)$

17. $A = (1, 1, 0)$, $B = (3, 3, 5)$, $P = (2, -9, 7)$, $Q = (4, -7, 13)$

In Exercises 18–21, calculate the linear combinations.

18. $5\langle 2, 2, -3 \rangle + 3\langle 1, 7, 2 \rangle$

19. $-2\langle 8, 11, 3 \rangle + 4\langle 2, 1, 1 \rangle$

20. $6\langle 2, 0, -1 \rangle - 3\langle 8, 6, 9 \rangle$

21. $-\langle 4, 3, 8 \rangle + \langle 8, 3, 3 \rangle$

In Exercises 22–25, find the given vector.

22. $\mathbf{e}_{\mathbf{v}}$, where $\mathbf{v} = \langle 1, 1, 2 \rangle$

23. $\mathbf{e_w}$, where $\mathbf{w} = \langle 4, -2, -1 \rangle$

24. Unit vector in the direction of $\mathbf{u} = \langle 1, 0, 7 \rangle$

25. Unit vector in the direction opposite to $\mathbf{v} = \langle -4, 4, 2 \rangle$

26. Sketch the following vectors, and find their components and lengths.

(a) $4\mathbf{i} + 3\mathbf{j} - 2\mathbf{k}$ **(b)** $\mathbf{i} + \mathbf{j} + \mathbf{k}$

(c) $4\mathbf{j} + 3\mathbf{k}$ **(d)** $12\mathbf{i} + 8\mathbf{j} - \mathbf{k}$

In Exercises 27–34, find a vector parametrization for the line with the given description.

27. Passes through $P = (1, 2, -8)$, direction vector $\mathbf{v} = \langle 2, 1, 3 \rangle$

28. Passes through $P = (4, 0, 8)$, direction vector $\mathbf{v} = \langle 1, 0, 1 \rangle$

29. Passes through $P = (4, 0, 8)$, direction vector $\mathbf{v} = 7\mathbf{i} + 4\mathbf{k}$

30. Passes through O, direction vector $\mathbf{v} = \langle 4, 2, -1 \rangle$

31. Passes through $(1, 1, 1)$ and $(3, -5, 2)$

32. Passes through $(-2, 0, -2)$ and $(4, 3, 7)$

33. Passes through O and $(4, 1, 1)$

34. Passes through $(1, 1, 1)$ parallel to the line through $(2, 0, -1)$ and $(4, 1, 3)$

In Exercises 35–40, find parametric equations of the line with the given description.

35. Perpendicular to the xy-plane passing through the origin

36. Perpendicular to the xz-plane passing through the point $(1, -1, 2)$

37. Perpendicular to the yz-plane passing through the point $(0, 0, 2)$

38. Passes through the origin perpendicular to the yz-plane

39. Passes through $(2, 3, 1)$ in the direction $\mathbf{v} = \langle 2, 1, -2 \rangle$

40. Passes through the origin with direction vector $\mathbf{v} = \langle 3, 4, -2 \rangle$

In Exercises 41–44, let $P = (2, 1, -1)$ and $Q = (4, 7, 7)$. Find the coordinates of each of the following.

41. The midpoint of $\overline{PQ}$

42. The point on $\overline{PQ}$ lying two-thirds of the way from P to Q

43. The point R such that Q is the midpoint of $\overline{PR}$

44. All points on the line through $\overline{PQ}$ whose distance from P is twice its distance from Q

45. Show that $\mathbf{r}_1(t)$ and $\mathbf{r}_2(t)$ define the same line, where

$$\mathbf{r}_1(t) = \langle 3, -1, 4 \rangle + t\,\langle 8, 12, -6 \rangle$$

$$\mathbf{r}_2(t) = \langle 11, 11, -2 \rangle + t\,\langle 4, 6, -3 \rangle$$

Hint: Show that $\mathbf{r}_2$ passes through $(3, -1, 4)$ and that the direction vectors for $\mathbf{r}_1$ and $\mathbf{r}_2$ are parallel.

46. Show that $\mathbf{r}_1(t)$ and $\mathbf{r}_2(t)$ define the same line, where

$$\mathbf{r}_1(t) = t\,\langle 2, 1, 3 \rangle, \qquad \mathbf{r}_2(t) = \langle -6, -3, -9 \rangle + t\,\langle 8, 4, 12 \rangle$$

47. Find two different vector parametrizations of the line through $P = (5, 5, 2)$ with direction vector $\mathbf{v} = \langle 0, -2, 1 \rangle$.

48. Find the point of intersection of the lines $\mathbf{r}(t) = \langle 1, 0, 0 \rangle + t\,\langle -3, 1, 0 \rangle$ and $\langle 0, 1, 1 \rangle + t\,\langle 2, 0, 1 \rangle$.

49. Show that the lines $\mathbf{r}_1(t) = \langle -1, 2, 2 \rangle + t\,\langle 4, -2, 1 \rangle$ and $\mathbf{r}_2(t) = \langle 0, 1, 1 \rangle + t\,\langle 2, 0, 1 \rangle$ do not intersect.

50. Determine if the lines $\mathbf{r}_1(t) = \langle 2, 1, 1 \rangle + t\,\langle -4, 0, 1 \rangle$ and $\mathbf{r}_2(t) = \langle -4, 1, 5 \rangle + s\,\langle 2, 1, -2 \rangle$ intersect and, if so, find the point of intersection.

51. Determine if the lines $\mathbf{r}_1(t) = \langle 0, 1, 1 \rangle + t\,\langle 1, 1, 2 \rangle$ and $\mathbf{r}_2(t) = \langle 2, 0, 3 \rangle + t\,\langle 1, 4, 4 \rangle$ intersect and, if so, find the point of intersection.

52. Find the intersection of the lines $\mathbf{r}_1(t) = \langle -1, 1 \rangle + t\,\langle 2, 4 \rangle$ and $\mathbf{r}_2(t) = \langle 2, 1 \rangle + t\,\langle -1, 6 \rangle$ in $\mathbf{R}^2$.

53. Find the components of the vector $\mathbf{v}$ whose tail and head are the midpoints of segments $\overline{AC}$ and $\overline{BC}$ in Figure 18.

54. Find the components of the vector $\mathbf{w}$ whose tail is C and head is the midpoint of $\overline{AB}$ in Figure 18.

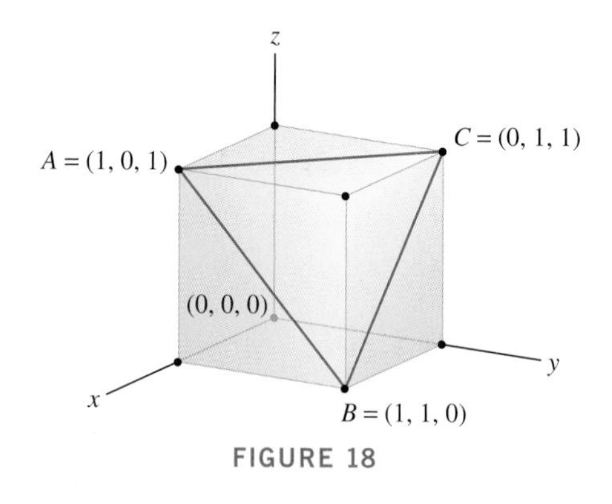

FIGURE 18

Further Insights and Challenges

In Exercises 55–59, we consider the equations of a line in ***symmetric form.***

$$\frac{x - x_0}{a} = \frac{y - y_0}{b} = \frac{z - z_0}{c} \qquad \boxed{11}$$

55. Let $\mathcal{L}$ be the line through $P_0 = (x_0, y_0, c_0)$ with direction vector $\mathbf{v} = \langle a, b, c \rangle$. Show that $\mathcal{L}$ is defined by the symmetric equations (11). *Hint:* Use the vector parametrization to show that every point on $\mathcal{L}$ satisfies (11).

56. Find the symmetric equations of the line through $P_0 = (-2, 3, 3)$ with direction vector $\mathbf{v} = \langle 2, 4, 3 \rangle$.

57. Find a vector parametrization for the line with symmetric equations

$$\frac{x-5}{9} = \frac{y+3}{7} = z - 10$$

58. Find the symmetric equations of the line through $P = (1, 1, 2)$ and $Q = (-2, 4, 0)$.

59. Show that the line in the plane through (x_0, y_0) of slope m has symmetric equations

$$\frac{x - x_0}{1} = \frac{y - y_0}{m}$$

60. A median of a triangle is a segment joining a vertex to the midpoint of the opposite side. It is known that the three medians intersect at a point P called the *centroid*. Prove this for a triangle whose vertices are the terminal points of vectors $\mathbf{u}$, $\mathbf{v}$, and $\mathbf{w}$ as in Figure 19(A). More precisely, show that the centroid is the terminal point of $\frac{1}{3}(\mathbf{u}+\mathbf{v}+\mathbf{w})$.

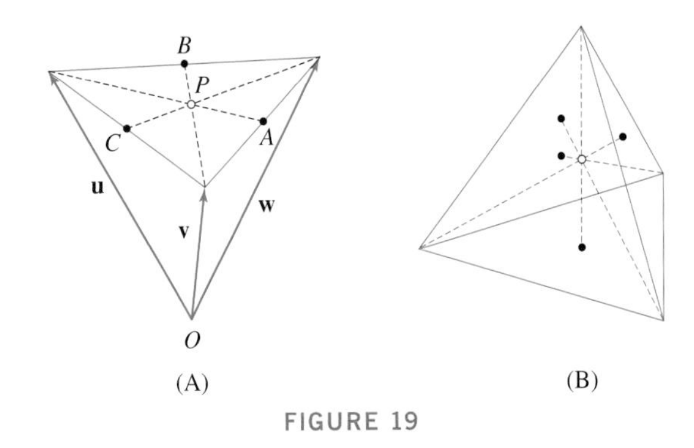

FIGURE 19

61. A median of a tetrahedron is a segment joining a vertex to the centroid of the opposite face. Show that the medians of the tetrahedron with vertices at the terminal points of vectors $\mathbf{u}$, $\mathbf{v}$, $\mathbf{w}$, $\mathbf{z}$ intersect at the terminal point of $\frac{1}{4}(\mathbf{u}+\mathbf{v}+\mathbf{w}+\mathbf{z})$ [Figure 19(B)].

12.3 Dot Product and the Angle Between Two Vectors

We now introduce one of the most important vector operations, the dot product. It plays a fundamental role in nearly all aspects of multivariable calculus.

Important concepts in mathematics often have multiple names or notations either for historical reasons or because they arise in more than one context. The dot product *is also called the "scalar product" or "inner product" and in many texts,* $\mathbf{v}\cdot\mathbf{w}$ *is denoted* $(\mathbf{v}, \mathbf{w})$ *or* $\langle \mathbf{v}, \mathbf{w}\rangle$.

DEFINITION Dot Product The **dot product** $\mathbf{v}\cdot\mathbf{w}$ of two vectors

$$\mathbf{v} = \langle a_1, b_1, c_1 \rangle, \qquad \mathbf{w} = \langle a_2, b_2, c_2 \rangle$$

is defined by

$$\mathbf{v}\cdot\mathbf{w} = a_1a_2 + b_1b_2 + c_1c_2$$

The dot product of vectors $\mathbf{v} = \langle a_1, b_1\rangle$ *and* $\mathbf{w} = \langle a_2, b_2\rangle$ *in* $\mathbf{R}^2$ *is defined similarly:*

$$\mathbf{v}\cdot\mathbf{w} = a_1a_2 + b_1b_2$$

In words, to compute the dot product, *multiply the corresponding components and add*. For example,

$$\langle 2, 3, 1\rangle \cdot \langle -4, 2, 5\rangle = 2(-4) + 3(2) + 1(5) = -8 + 6 + 5 = 3$$

As we will see in a moment, the dot product is closely related to the angle between $\mathbf{v}$ and $\mathbf{w}$. Before getting to this, we describe some elementary properties of dot products.

First, the dot product is *commutative*: $\mathbf{v}\cdot\mathbf{w} = \mathbf{w}\cdot\mathbf{v}$. This follows because we may multiply the components in either order. Second, the dot product of a vector with itself is the square of the length: If $\mathbf{v} = \langle a, b, c\rangle$, then

$$\mathbf{v}\cdot\mathbf{v} = a\cdot a + b\cdot b + c\cdot c = a^2 + b^2 + c^2 = \|\mathbf{v}\|^2$$

The dot product also satisfies a Distributive Law and a scalar property as summarized in the next theorem (see Exercises 77–78).

THEOREM 1 Properties of the Dot Product

(i) $\mathbf{0} \cdot \mathbf{v} = \mathbf{v} \cdot \mathbf{0} = 0$

(ii) **Commutativity:** $\mathbf{v} \cdot \mathbf{w} = \mathbf{w} \cdot \mathbf{v}$

(iii) **Pulling out scalars:** $(\lambda \mathbf{v}) \cdot \mathbf{w} = \mathbf{v} \cdot (\lambda \mathbf{w}) = \lambda(\mathbf{v} \cdot \mathbf{w})$

(iv) **Distributive Law:** $\mathbf{u} \cdot (\mathbf{v} + \mathbf{w}) = \mathbf{u} \cdot \mathbf{v} + \mathbf{u} \cdot \mathbf{w}$

$(\mathbf{v} + \mathbf{w}) \cdot \mathbf{u} = \mathbf{v} \cdot \mathbf{u} + \mathbf{w} \cdot \mathbf{u}$

(v) **Relation with length:** $\mathbf{v} \cdot \mathbf{v} = \|\mathbf{v}\|^2$

EXAMPLE 1 Verify the Distributive Law $\mathbf{u} \cdot (\mathbf{v} + \mathbf{w}) = \mathbf{u} \cdot \mathbf{v} + \mathbf{u} \cdot \mathbf{w}$ for the vectors

$$\mathbf{u} = \langle 4, 3, 3 \rangle, \quad \mathbf{v} = \langle 1, 2, 2 \rangle, \quad \mathbf{w} = \langle 3, -2, 5 \rangle$$

Solution We compute both sides and check that they are equal:

$$\begin{aligned} \mathbf{u} \cdot (\mathbf{v} + \mathbf{w}) &= \langle 4, 3, 3 \rangle \cdot (\langle 1, 2, 2 \rangle + \langle 3, -2, 5 \rangle) \\ &= \langle 4, 3, 3 \rangle \cdot \langle 4, 0, 7 \rangle = 4(4) + 3(0) + 3(7) = 37 \\ \mathbf{u} \cdot \mathbf{v} + \mathbf{u} \cdot \mathbf{w} &= \langle 4, 3, 3 \rangle \cdot \langle 1, 2, 2 \rangle + \langle 4, 3, 3 \rangle \langle 3, -2, 5 \rangle \\ &= \big(4(1) + 3(2) + 3(2)\big) + \big(4(3) + 3(-2) + 3(5)\big) \\ &= 16 + 21 = 37 \end{aligned}$$

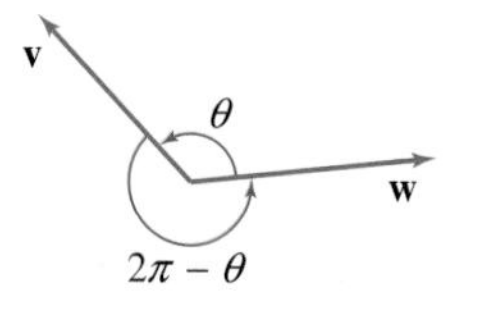

FIGURE 1 By convention, the angle θ between two vectors is chosen so that $0 \le \theta \le \pi$.

As mentioned above, the dot product $\mathbf{v} \cdot \mathbf{w}$ is related to the angle θ between $\mathbf{v}$ and $\mathbf{w}$. This angle θ is not uniquely defined because, as we see in Figure 1, both θ and $2\pi - \theta$ can serve as an angle between $\mathbf{v}$ and $\mathbf{w}$. Furthermore, any multiple of 2π may be added to θ. We adopt the convention:

The angle between two vectors is chosen to satisfy $0 \le \theta \le \pi$.

The next theorem establishes the fundamental relation between the dot product and the angle between two vectors.

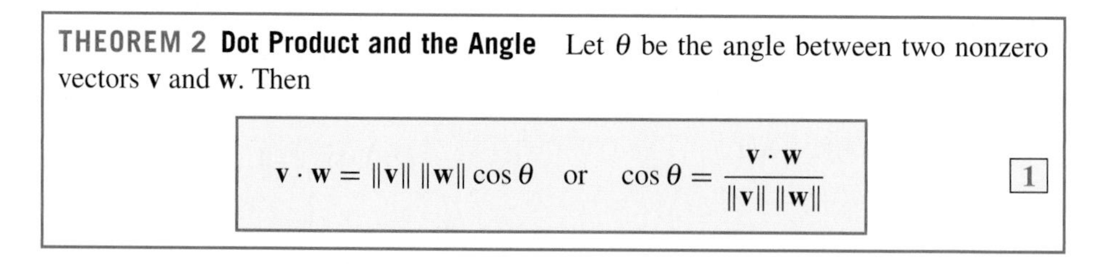

THEOREM 2 Dot Product and the Angle Let θ be the angle between two nonzero vectors $\mathbf{v}$ and $\mathbf{w}$. Then

$$\mathbf{v} \cdot \mathbf{w} = \|\mathbf{v}\| \, \|\mathbf{w}\| \cos \theta \quad \text{or} \quad \cos \theta = \frac{\mathbf{v} \cdot \mathbf{w}}{\|\mathbf{v}\| \, \|\mathbf{w}\|} \qquad \boxed{1}$$

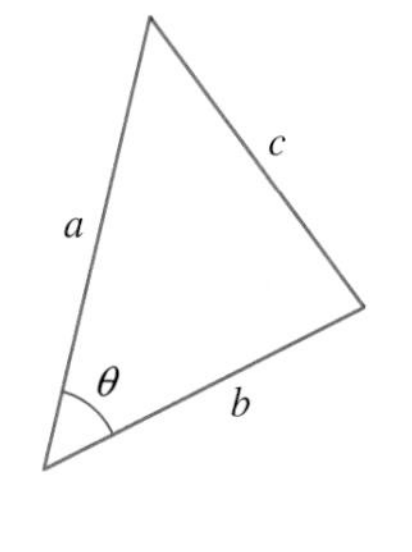

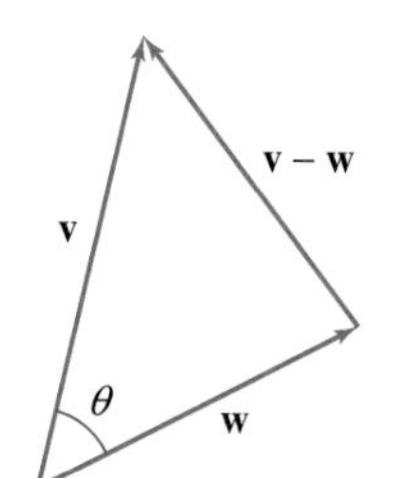

FIGURE 2

Proof According to the Law of Cosines, the three sides of a triangle satisfy (Figure 2)

$$c^2 = a^2 + b^2 - 2ab \cos \theta$$

If two sides of the triangle are $\mathbf{v}$ and $\mathbf{w}$, then the third side is $\mathbf{v} - \mathbf{w}$, as in the figure, and the Law of Cosines gives

$$\|\mathbf{v} - \mathbf{w}\|^2 = \|\mathbf{v}\|^2 + \|\mathbf{w}\|^2 - 2 \cos \theta \|\mathbf{v}\| \, \|\mathbf{w}\| \qquad \boxed{2}$$

Now, by property (v) of Theorem 1 and the Distributive Law,

$$\|\mathbf{v}-\mathbf{w}\|^2 = (\mathbf{v}-\mathbf{w})\cdot(\mathbf{v}-\mathbf{w}) = \mathbf{v}\cdot\mathbf{v} - 2\mathbf{v}\cdot\mathbf{w} + \mathbf{w}\cdot\mathbf{w}$$

$$= \|\mathbf{v}\|^2 + \|\mathbf{w}\|^2 - 2\mathbf{v}\cdot\mathbf{w} \qquad \boxed{3}$$

Comparing (2) and (3), we obtain $-2\cos\theta\|\mathbf{v}\|\,\|\mathbf{w}\| = -2\mathbf{v}\cdot\mathbf{w}$, and Eq. (1) follows. ■

By definition of the arccosine, the angle $\theta = \cos^{-1} x$ is the angle in the interval $[0, \pi]$ satisfying $\cos\theta = x$. Thus, for nonzero vectors **v** and **w**, we have

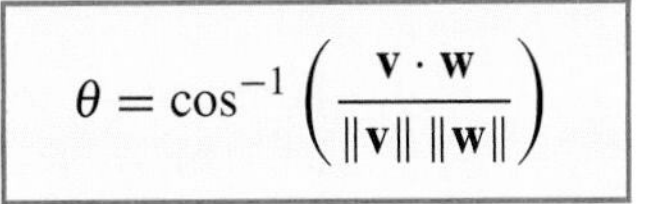

$$\theta = \cos^{-1}\left(\frac{\mathbf{v}\cdot\mathbf{w}}{\|\mathbf{v}\|\,\|\mathbf{w}\|}\right)$$

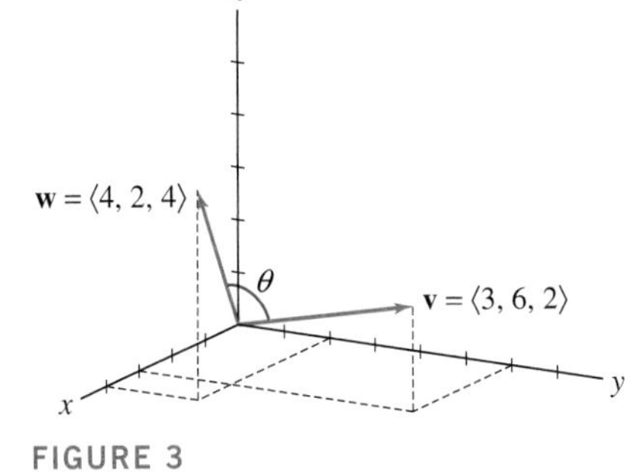

FIGURE 3

■ **EXAMPLE 2** Find the angle θ between $\mathbf{v} = \langle 3, 6, 2\rangle$ and $\mathbf{w} = \langle 4, 2, 4\rangle$.

Solution Compute $\cos\theta$ using the dot product:

$$\|\mathbf{v}\| = \sqrt{3^2+6^2+2^2} = \sqrt{49} = 7, \qquad \|\mathbf{w}\| = \sqrt{4^2+2^2+4^2} = \sqrt{36} = 6$$

$$\cos\theta = \frac{\mathbf{v}\cdot\mathbf{w}}{\|\mathbf{v}\|\,\|\mathbf{w}\|} = \frac{\langle 3,6,2\rangle\cdot\langle 4,2,4\rangle}{7\cdot 6} = \frac{3\cdot 4 + 6\cdot 2 + 2\cdot 4}{42} = \frac{32}{42} = \frac{16}{21}$$

The angle itself is $\theta = \cos^{-1}(\frac{16}{21}) \approx 0.705$ rad (Figure 3). ■

The terms "orthogonal" and "perpendicular" are synonymous and are used interchangeably, although orthogonal *is commonly used when dealing with vectors.*

Two nonzero vectors **v** and **w** are called **perpendicular** or **orthogonal** if the angle between them is $\frac{\pi}{2}$. (The zero vector is considered orthogonal to every vector.) The dot product gives us a simple way to test whether **v**, **w** are orthogonal. An angle between 0 and π satisfies $\cos\theta = 0$ if and only if $\theta = \frac{\pi}{2}$, and therefore,

$$\mathbf{v}\cdot\mathbf{w} = \|\mathbf{v}\|\,\|\mathbf{w}\|\cos\theta = 0 \quad\Leftrightarrow\quad \theta = \frac{\pi}{2}$$

We write $\mathbf{v} \perp \mathbf{w}$ if **v** and **w** are orthogonal. Thus, we have

$$\boxed{\mathbf{v} \perp \mathbf{w} \text{ if and only if } \mathbf{v}\cdot\mathbf{w} = 0}$$

■ **EXAMPLE 3** **Testing for Orthogonality** Determine if $\mathbf{v} = \langle 2, 6, 1\rangle$ is orthogonal to $\mathbf{u} = \langle 2, -1, 1\rangle$ or $\mathbf{w} = \langle -4, 1, 2\rangle$.

Solution We test for orthogonality by computing the dot products (Figure 4):

$$\mathbf{v}\cdot\mathbf{u} = \langle 2,6,1\rangle\cdot\langle 2,-1,1\rangle = 2(2)+6(-1)+1(1) = -1 \quad \text{(not orthogonal)}$$

$$\mathbf{v}\cdot\mathbf{w} = \langle 2,6,1\rangle\cdot\langle -4,1,2\rangle = 2(-4)+6(1)+1(2) = 0 \quad \text{(orthogonal)}$$ ■

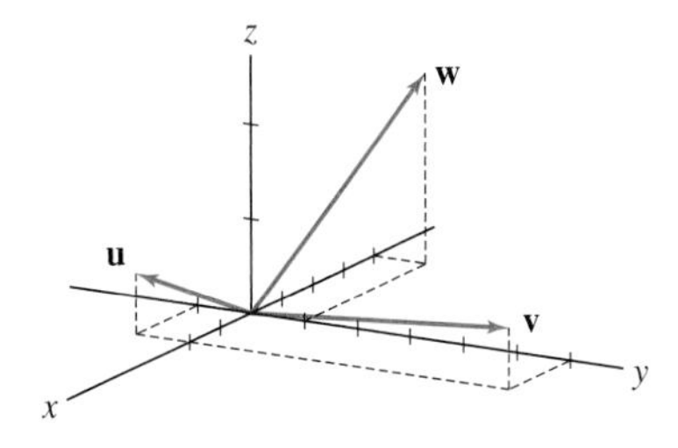

FIGURE 4

■ **EXAMPLE 4** **Testing for Obtuseness** Determine if the angles between $\mathbf{v} = \langle 3, 1, -2\rangle$ and the vectors $\mathbf{u} = \left\langle \frac{1}{2}, \frac{1}{2}, 5\right\rangle$ and $\mathbf{w} = \langle 4, -3, 0\rangle$ are obtuse.

Solution The angle θ between two vectors is obtuse if $\frac{\pi}{2} < \theta \le \pi$, and this is the case if $\cos\theta < 0$. Since $\mathbf{v}\cdot\mathbf{u} = \|\mathbf{v}\|\,\|\mathbf{u}\|\cos\theta$ and the lengths $\|\mathbf{v}\|$ and $\|\mathbf{u}\|$ are positive, we see that

$$\boxed{\text{The angle between } \mathbf{v} \text{ and } \mathbf{w} \text{ is obtuse if } \mathbf{v}\cdot\mathbf{u} < 0.}$$

We have

$$\mathbf{v}\cdot\mathbf{u} = \langle 3, 1, -2\rangle \cdot \left\langle \frac{1}{2}, \frac{1}{2}, 5\right\rangle = \frac{3}{2} + \frac{1}{2} - 10 = -8 < 0 \quad \text{(angle is obtuse)}$$

$$\mathbf{v}\cdot\mathbf{w} = \langle 3, 1, -2\rangle \cdot \langle 4, -3, 0\rangle = 12 - 3 + 0 = 9 > 0 \quad \text{(angle is acute)}$$ ■

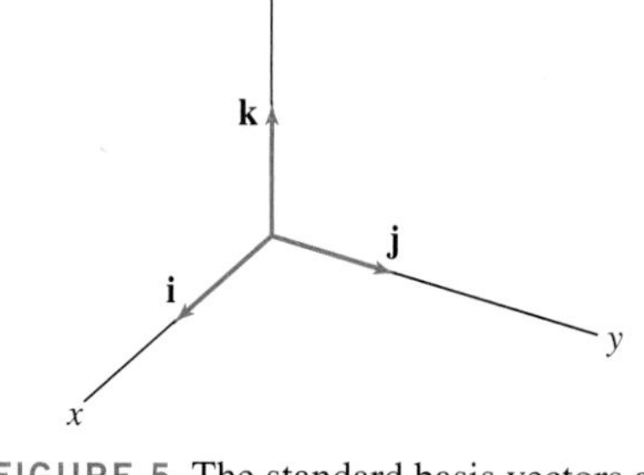

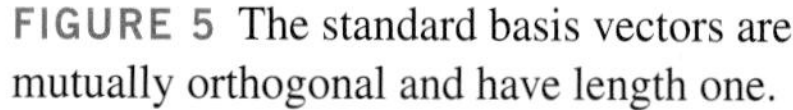
FIGURE 5 The standard basis vectors are mutually orthogonal and have length one.

The standard basis vectors are mutually orthogonal and have length one (Figure 5). In terms of dot products,

$$\mathbf{i}\cdot\mathbf{j} = \mathbf{i}\cdot\mathbf{k} = \mathbf{j}\cdot\mathbf{k} = 0, \qquad \mathbf{i}\cdot\mathbf{i} = \mathbf{j}\cdot\mathbf{j} = \mathbf{k}\cdot\mathbf{k} = 1$$

Furthermore, the components of a vector $\mathbf{v} = \langle a, b, c\rangle$ are simply the dot products with the standard basis vectors:

$$\mathbf{v}\cdot\mathbf{i} = \langle a, b, c\rangle \cdot \langle 1, 0, 0\rangle = a$$

$$\mathbf{v}\cdot\mathbf{j} = \langle a, b, c\rangle \cdot \langle 0, 1, 0\rangle = b$$

$$\mathbf{v}\cdot\mathbf{k} = \langle a, b, c\rangle \cdot \langle 0, 0, 1\rangle = c$$

■ **EXAMPLE 5 Using the Distributive Law** Calculate $\mathbf{v}\cdot\mathbf{w}$, where $\mathbf{v} = 4\mathbf{i} - 3\mathbf{j}$ and $\mathbf{w} = \mathbf{i} + 2\mathbf{j} + \mathbf{k}$.

Solution Using the Distributive Law and the orthogonality of **i**, **j**, and **k**, we obtain

$$\begin{aligned}\mathbf{v}\cdot\mathbf{w} &= (4\mathbf{i} - 3\mathbf{j})\cdot(\mathbf{i} + 2\mathbf{j} + \mathbf{k})\\ &= 4\mathbf{i}\cdot(\mathbf{i} + 2\mathbf{j} + \mathbf{k}) - 3\mathbf{j}\cdot(\mathbf{i} + 2\mathbf{j} + \mathbf{k})\\ &= 4\mathbf{i}\cdot\mathbf{i} - 3\mathbf{j}\cdot(2\mathbf{j}) = 4 - 6 = -2\end{aligned}$$ ■

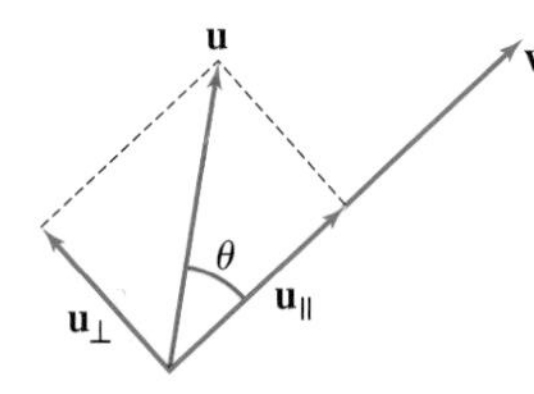

FIGURE 6 Resolution of **u** into parallel and orthogonal components with respect to **v**.

Given vectors **u** and **v** with $\mathbf{v} \neq \mathbf{0}$, it is possible to decompose **u** as a sum:

$$\mathbf{u} = \mathbf{u}_{||} + \mathbf{u}_{\perp}$$

where $\mathbf{u}_{||}$ is parallel to **v** and $\mathbf{u}_{\perp}$ is orthogonal to **v** (Figure 6). The vector $\mathbf{u}_{||}$ is called the **projection** of **u** along **v** and is also denoted $\text{proj}_{\mathbf{v}}(\mathbf{u})$.

To determine the projection $\mathbf{u}_{||}$, we note that $\mathbf{u}_{||}$ is parallel to **v**, and therefore, $\mathbf{u}_{||}$ is a multiple of the unit vector $\mathbf{e}_{\mathbf{v}} = \dfrac{\mathbf{v}}{\|\mathbf{v}\|}$. Let $\mathbf{u}_{||} = c\mathbf{e}_{\mathbf{v}}$ for some constant c. Since $\mathbf{u}_{\perp}\cdot\mathbf{e}_{\mathbf{v}} = 0$,

$$\mathbf{u}\cdot\mathbf{e}_{\mathbf{v}} = (\mathbf{u}_{||} + \mathbf{u}_{\perp})\cdot\mathbf{e}_{\mathbf{v}} = \mathbf{u}_{||}\cdot\mathbf{e}_{\mathbf{v}} + \mathbf{u}_{\perp}\cdot\mathbf{e}_{\mathbf{v}} = \mathbf{u}_{||}\cdot\mathbf{e}_{\mathbf{v}} = c\mathbf{e}_{\mathbf{v}}\cdot\mathbf{e}_{\mathbf{v}} = c$$

Thus, $c = \mathbf{u}\cdot\mathbf{e}_{\mathbf{v}}$ and $\mathbf{u}_{||} = (\mathbf{u}\cdot\mathbf{e}_{\mathbf{v}})\mathbf{e}_{\mathbf{v}}$. We can also write $\mathbf{u}_{||}$ directly in terms of **v**:

$$\mathbf{u}_{||} = (\mathbf{u}\cdot\mathbf{e}_{\mathbf{v}})\mathbf{e}_{\mathbf{v}} = \left(\mathbf{u}\cdot\frac{\mathbf{v}}{\|\mathbf{v}\|}\right)\frac{\mathbf{v}}{\|\mathbf{v}\|} = \left(\frac{\mathbf{u}\cdot\mathbf{v}}{\|\mathbf{v}\|^2}\right)\mathbf{v} = \left(\frac{\mathbf{u}\cdot\mathbf{v}}{\mathbf{v}\cdot\mathbf{v}}\right)\mathbf{v}$$

In summary,

$$\mathbf{u}_{||} = \text{proj}_{\mathbf{v}}(\mathbf{u}) = (\mathbf{u}\cdot\mathbf{e}_{\mathbf{v}})\mathbf{e}_{\mathbf{v}} = \left(\frac{\mathbf{u}\cdot\mathbf{v}}{\mathbf{v}\cdot\mathbf{v}}\right)\mathbf{v} \qquad \boxed{4}$$

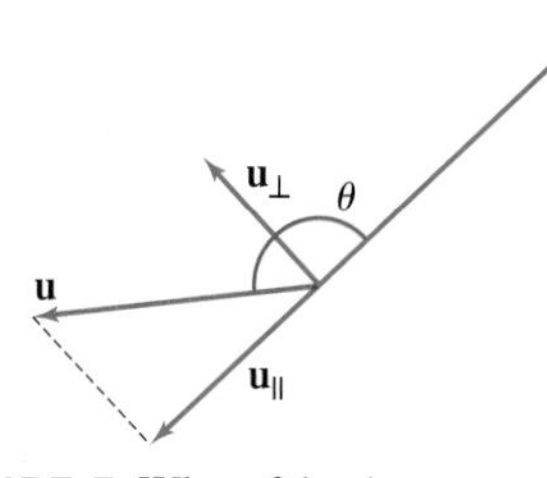

FIGURE 7 When θ is obtuse, $\mathbf{u}_{||}$ and **v** point in opposite directions.

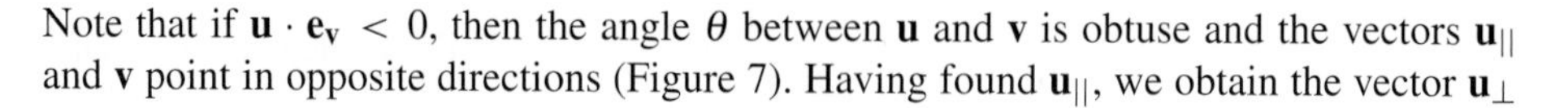

Note that if $\mathbf{u}\cdot\mathbf{e}_{\mathbf{v}} < 0$, then the angle θ between **u** and **v** is obtuse and the vectors $\mathbf{u}_{||}$ and **v** point in opposite directions (Figure 7). Having found $\mathbf{u}_{||}$, we obtain the vector $\mathbf{u}_{\perp}$

as the difference

$$\mathbf{u}_{\perp} = \mathbf{u} - \mathbf{u}_{||} \qquad (5)$$

The vector $\mathbf{u}_{\perp}$ is orthogonal to $\mathbf{v}$ since

$$\mathbf{u}_{\perp} \cdot \mathbf{v} = \left(\mathbf{u} - \left(\frac{\mathbf{u}\cdot\mathbf{v}}{\mathbf{v}\cdot\mathbf{v}}\right)\mathbf{v}\right)\cdot\mathbf{v} = \mathbf{u}\cdot\mathbf{v} - \left(\frac{\mathbf{u}\cdot\mathbf{v}}{\mathbf{v}\cdot\mathbf{v}}\right)\mathbf{v}\cdot\mathbf{v} = 0$$

EXAMPLE 6 Find the decomposition $\mathbf{u} = \mathbf{u}_{||} + \mathbf{u}_{\perp}$ of $\mathbf{u} = \langle 5, 1, -3\rangle$ with respect to $\mathbf{v} = \langle 4, 4, 2\rangle$.

Solution We compute $\mathbf{u}_{||}$ in two steps:

***Step 1.* Compute u · v and v · v.**

$$\mathbf{u}\cdot\mathbf{v} = \langle 5, 1, -3\rangle \cdot \langle 4, 4, 2\rangle = 20 + 4 - 6 = 18, \qquad \mathbf{v}\cdot\mathbf{v} = 4^2 + 4^2 + 2^2 = 36$$

***Step 2.* Use formulas (4) and (5).**

$$\mathbf{u}_{||} = \mathrm{proj}_{\mathbf{v}}(\mathbf{u}) = \left(\frac{\mathbf{u}\cdot\mathbf{v}}{\mathbf{v}\cdot\mathbf{v}}\right)\mathbf{v} = \left(\frac{18}{36}\right)\langle 4, 4, 2\rangle = \langle 2, 2, 1\rangle$$

$$\mathbf{u}_{\perp} = \mathbf{u} - \mathbf{u}_{||} = \langle 5, 1, -3\rangle - \langle 2, 2, 1\rangle = \langle 3, -1, -4\rangle$$

We obtain the decomposition

$$\mathbf{u} = \langle 5, 1, -3\rangle = \mathbf{u}_{||} + \mathbf{u}_{\perp} = \underbrace{\langle 2, 2, 1\rangle}_{\text{Projection along } \mathbf{v}} + \underbrace{\langle 3, -1, -4\rangle}_{\text{Orthogonal to } \mathbf{v}}$$

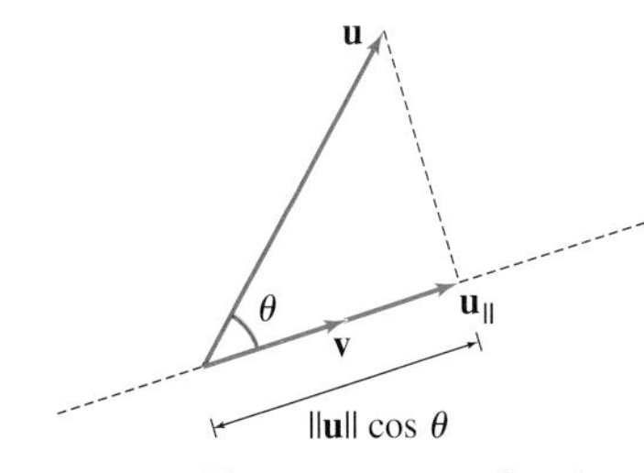

FIGURE 8 The component of $\mathbf{u}$ along $\mathbf{v}$ is $\mathbf{u}\cdot\mathbf{e_v} = \|\mathbf{u}\|\cos\theta$.

The coefficient of $\mathbf{e_v}$ in Eq. (4) is called the **component** of $\mathbf{u}$ along $\mathbf{v}$:

$$\text{Component of } \mathbf{u} \text{ along } \mathbf{v} = \mathbf{u}\cdot\mathbf{e_v} = \frac{\mathbf{u}\cdot\mathbf{v}}{\|\mathbf{v}\|} = \|\mathbf{u}\|\cos\theta \qquad (6)$$

The component $\mathbf{u}\cdot\mathbf{e_v}$ tells us "how much" of $\mathbf{u}$ points in the direction $\mathbf{v}$.

where θ is the angle between $\mathbf{u}$ and $\mathbf{v}$ (Figure 8). Notice that by Eq. (4), the absolute value $|\mathbf{u}\cdot\mathbf{e_v}|$ is the length of $\mathrm{proj}_{\mathbf{v}}(\mathbf{u})$.

EXAMPLE 7 What is the minimum force you must apply to pull a 35-lb wagon up a frictionless ramp, inclined at an angle $\theta = 15°$?

Solution Referring to Figure 9, we resolve the force due to gravity $\mathbf{F}_g$ into a sum $\mathbf{F}_g = \mathbf{F}_{||} + \mathbf{F}_{\perp}$, where $\mathbf{F}_{||}$ is the force along the ramp and $\mathbf{F}_{\perp}$ is the force orthogonal to the ramp (the **normal force**). The force $\mathbf{F}_{\perp}$ is canceled by the force of the ramp pushing against the wagon in the normal direction. To move the wagon, you need only pull against the projection $\mathbf{F}_{||}$ of the force along the ramp. The angle between $\mathbf{F}_g$ and the ramp is $90° - \theta = 75°$, so by Eq. (6),

$$\text{Component along ramp} = \|\mathbf{F}_{||}\| = \|\mathbf{F}_g\|\cos 75° \approx 35(0.26) = 9.1 \text{ lb}$$

Therefore, a minimum force of 9.1 lb is required to pull the wagon up the ramp.

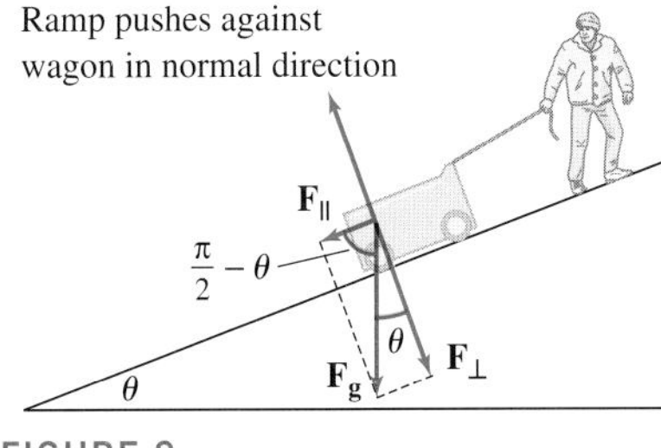

FIGURE 9

12.3 SUMMARY

- The *dot product* of $\mathbf{v} = \langle a_1, b_1, c_1 \rangle$ and $\mathbf{w} = \langle a_2, b_2, c_2 \rangle$ is

$$\mathbf{v} \cdot \mathbf{w} = a_1a_2 + b_1b_2 + c_1c_2$$

- By convention, the angle θ between two vectors is chosen to satisfy $0 \le \theta \le \pi$.
- The dot product and the angle θ between nonzero vectors $\mathbf{v}$ and $\mathbf{w}$ are related by

$$\mathbf{v} \cdot \mathbf{w} = \|\mathbf{v}\| \, \|\mathbf{w}\| \cos\theta, \qquad \cos\theta = \frac{\mathbf{v} \cdot \mathbf{w}}{\|\mathbf{v}\| \, \|\mathbf{w}\|}, \qquad \theta = \cos^{-1}\left(\frac{\mathbf{v} \cdot \mathbf{w}}{\|\mathbf{v}\| \, \|\mathbf{w}\|}\right)$$

- Test for orthogonality: $\mathbf{v} \perp \mathbf{w}$ if and only if $\mathbf{v} \cdot \mathbf{w} = 0$.
- The angle between $\mathbf{v}$ and $\mathbf{w}$ is acute if $\mathbf{v} \cdot \mathbf{w} > 0$ and obtuse if $\mathbf{v} \cdot \mathbf{w} < 0$.
- Assume $\mathbf{v} \neq \mathbf{0}$. Then every vector $\mathbf{u}$ has a decomposition $\mathbf{u} = \mathbf{u}_{||} + \mathbf{u}_{\perp}$, where $\mathbf{u}_{||}$ is parallel to $\mathbf{v}$ and $\mathbf{u}_{\perp}$ is orthogonal to $\mathbf{v}$. The vector $\mathbf{u}_{||}$ is called the *projection of* $\mathbf{u}$ along $\mathbf{v}$ and is also denoted $\text{proj}_{\mathbf{v}}(\mathbf{u})$.
- Let $\mathbf{e}_{\mathbf{v}} = \dfrac{\mathbf{v}}{\|\mathbf{v}\|}$ for some nonzero vector $\mathbf{v}$. Then

$$\mathbf{u}_{||} = \text{proj}_{\mathbf{v}}(\mathbf{u}) = (\mathbf{u} \cdot \mathbf{e}_{\mathbf{v}})\mathbf{e}_{\mathbf{v}} = \left(\frac{\mathbf{u} \cdot \mathbf{v}}{\mathbf{v} \cdot \mathbf{v}}\right)\mathbf{v}, \qquad \mathbf{u}_{\perp} = \mathbf{u} - \mathbf{u}_{||}$$

- The *component* of $\mathbf{u}$ along $\mathbf{v}$ is the quantity

$$\text{Component of } \mathbf{u} \text{ along } \mathbf{v} = \mathbf{u} \cdot \mathbf{e}_{\mathbf{v}} = \frac{\mathbf{u} \cdot \mathbf{v}}{\|\mathbf{v}\|} = \|\mathbf{u}\| \cos\theta$$

12.3 EXERCISES

Preliminary Questions

1. Is the dot product of two vectors a scalar or a vector?

2. What can you say about the angle between $\mathbf{a}$ and $\mathbf{b}$ if $\mathbf{a} \cdot \mathbf{b} < 0$?

3. Suppose that $\mathbf{v}$ is orthogonal to both $\mathbf{u}$ and $\mathbf{w}$. Which property of dot products allows us to conclude that $\mathbf{v}$ is orthogonal to $\mathbf{u} + \mathbf{w}$?

4. What is $\text{proj}_{\mathbf{v}}(\mathbf{v})$?

5. What is the difference, if any, between the projection of $\mathbf{u}$ along $\mathbf{v}$ and the projection along the unit vector $\mathbf{e}_{\mathbf{v}}$?

6. Suppose that $\text{proj}_{\mathbf{v}}(\mathbf{u}) = \mathbf{v}$. Determine:

(a) $\text{proj}_{2\mathbf{v}}(\mathbf{u})$ **(b)** $\text{proj}_{\mathbf{v}}(2\mathbf{u})$

7. Let $\mathbf{u}_{||}$ be the projection of $\mathbf{u}$ along $\mathbf{v}$. Which of the following is the projection $\mathbf{u}$ along the vector $2\mathbf{v}$?

(a) $\frac{1}{2}\mathbf{u}_{||}$ **(b)** $\mathbf{u}_{||}$ **(c)** $2\mathbf{u}_{||}$

8. Let θ be the angle between $\mathbf{u}$ and $\mathbf{v}$. Which of the following is equal to the $\cos\theta$?

(a) $\mathbf{u} \cdot \mathbf{v}$ **(b)** $\mathbf{u} \cdot \mathbf{e}_{\mathbf{v}}$ **(c)** $\mathbf{e}_{\mathbf{u}} \cdot \mathbf{e}_{\mathbf{v}}$

Exercises

In Exercises 1–12, compute the dot product.

1. $\langle 1, 2, 1 \rangle \cdot \langle 4, 3, 5 \rangle$

2. $\langle 3, -2, 2 \rangle \cdot \langle 1, 0, 1 \rangle$

3. $\langle 0, 1, 0 \rangle \cdot \langle 7, 41, -3 \rangle$

4. $\langle 1, 1, 1 \rangle \cdot \langle 6, 4, 2 \rangle$

5. $\langle 3, 1 \rangle \cdot \langle 4, -7 \rangle$

6. $\left\langle \frac{1}{6}, \frac{1}{2} \right\rangle \cdot \left\langle 3, \frac{1}{2} \right\rangle$

7. $\mathbf{i} \cdot \mathbf{k}$

8. $\mathbf{j} \cdot \mathbf{j}$

9. $(\mathbf{i} + \mathbf{j}) \cdot (\mathbf{i} + \mathbf{k})$

10. $(3\mathbf{j} + 2\mathbf{k}) \cdot (\mathbf{i} - 4\mathbf{k})$

11. $(\mathbf{i}+\mathbf{j}+\mathbf{k})\cdot(3\mathbf{i}+2\mathbf{j}-5\mathbf{k})$

12. $(-\mathbf{k})\cdot(\mathbf{i}-2\mathbf{j}+7\mathbf{k})$

In Exercises 13–18, determine whether the two vectors are orthogonal and if not, whether the angle between them is acute or obtuse.

13. $\langle 1,1,1\rangle,\quad \langle 1,-2,-2\rangle$

14. $\langle 1,1,1\rangle,\quad \langle 3,-2,-1\rangle$

15. $\langle 0,2,4\rangle,\quad \langle 3,1,0\rangle$

16. $\langle 0,2,4\rangle,\quad \langle -5,0,0\rangle$

17. $\langle 4,3\rangle,\quad \langle 2,-4\rangle$

18. $\langle 12,6\rangle,\quad \langle 2,-4\rangle$

In Exercises 19–22, find the cosine of the angle between the vectors.

19. $\langle 0,3,1\rangle,\quad \langle 4,0,0\rangle$

20. $\langle 1,1,1\rangle,\quad \langle 2,-1,2\rangle$

21. $\mathbf{i}+\mathbf{j},\quad \mathbf{j}+2\mathbf{k}$

22. $3\mathbf{i}+\mathbf{k},\quad \mathbf{i}+\mathbf{j}+\mathbf{k}$

In Exercises 23–28, find the angle between the vectors.

23. $\langle 1,2\rangle,\quad \langle 3,5\rangle$

24. $\langle 0,-4\rangle,\quad \langle 1,3\rangle$

25. $\langle 1,1,1\rangle,\quad \langle 1,-1,1\rangle$

26. $\langle 2,4,1\rangle,\quad \langle 1,-3,5\rangle$

27. $\langle \pi,e,3\rangle,\quad \left\langle \cos 1,\tan 1,e^2\right\rangle$

28. $\langle 1,1,5\rangle,\quad \langle 1,-1,5\rangle$

29. Find all values of b for which the vectors are orthogonal.
(a) $\langle b,3,2\rangle,\quad \langle 1,b,1\rangle$
(b) $\langle 4,-2,7\rangle,\quad \left\langle b^2,b,0\right\rangle$

30. Find a vector that is orthogonal to $\langle -1,2,2\rangle$.

31. Find two vectors (which are not multiples of each other) that are both orthogonal to $\langle 2,0,-3\rangle$.

32. The angle between two unit vectors $\mathbf{e}$ and $\mathbf{f}$ is 50°. Find $\mathbf{e}\cdot\mathbf{f}$.

In Exercises 33–36, assume that $\mathbf{u}\cdot\mathbf{v}=2$, $\|\mathbf{u}\|=1$, *and* $\|\mathbf{v}\|=3$ *and evaluate the expression.*

33. $\mathbf{u}\cdot(4\mathbf{v})$

34. $(\mathbf{u}+\mathbf{v})\cdot\mathbf{v}$

35. $2\mathbf{u}\cdot(3\mathbf{u}-\mathbf{v})$

36. $(\mathbf{u}+\mathbf{v})\cdot(\mathbf{u}-\mathbf{v})$

In Exercises 37–40, simplify the expression.

37. $(\mathbf{v}-\mathbf{w})\cdot\mathbf{v}+\mathbf{v}\cdot\mathbf{w}$

38. $(\mathbf{v}+\mathbf{w})\cdot(\mathbf{v}+\mathbf{w})-2\mathbf{v}\cdot\mathbf{w}$

39. $(\mathbf{v}+\mathbf{w})\cdot\mathbf{v}-(\mathbf{v}+\mathbf{w})\cdot\mathbf{w}$

40. $(\mathbf{v}+\mathbf{w})\cdot\mathbf{v}-(\mathbf{v}-\mathbf{w})\cdot\mathbf{w}$

In Exercises 41–48, find the projection $\mathrm{proj}_{\mathbf{v}}(\mathbf{u})$.

41. $\mathbf{u}=\langle -1,2,0\rangle,\quad \mathbf{v}=\langle 2,0,1\rangle$

42. $\mathbf{u}=\langle 2,0\rangle,\quad \mathbf{v}=\langle 4,3\rangle$

43. $\mathbf{u}=\langle 1,1,1\rangle,\quad \mathbf{v}=\langle 1,1,0\rangle$

44. $\mathbf{u}=\langle 3,2,1\rangle,\quad \mathbf{v}=\langle 6,4,2\rangle$

45. $\mathbf{u}=5\mathbf{i}+7\mathbf{j}-4\mathbf{k},\quad \mathbf{v}=\mathbf{k}$

46. $\mathbf{u}=\mathbf{i}+29\mathbf{k},\quad \mathbf{v}=\mathbf{j}$

47. $\mathbf{u}=\langle a,b,c\rangle,\quad \mathbf{v}=\mathbf{i}$

48. $\mathbf{u}=\langle a,a,b\rangle,\quad \mathbf{v}=\mathbf{i}-\mathbf{j}$

In Exercises 49–54, find the decomposition $\mathbf{a}=\mathbf{a}_{\|}+\mathbf{a}_{\perp}$ *with respect to* $\mathbf{b}$.

49. $\mathbf{a}=\langle -2,1,1\rangle,\quad \mathbf{b}=\langle 1,0,0\rangle$

50. $\mathbf{a}=\langle 4,-1,0\rangle,\quad \mathbf{b}=\langle 0,1,1\rangle$

51. $\mathbf{a}=\langle 1,0\rangle,\quad \mathbf{b}=\langle 1,1\rangle$

52. $\mathbf{a}=\langle 4,4,-5\rangle,\quad \mathbf{b}=\langle 2,-1,1\rangle$

53. $\mathbf{a}=\langle 4,2\rangle,\quad \mathbf{b}=\langle -1,2\rangle$

54. $\mathbf{a}=\langle 3,5,5\rangle,\quad \mathbf{b}=\langle 5,2,2\rangle$

55. Let $\mathbf{e}_\theta=\langle\cos\theta,\sin\theta\rangle$. Show that $\mathbf{e}_\theta$ is a unit vector making an angle θ with the x-axis. Show that $\mathbf{e}_\theta\cdot\mathbf{e}_\psi=\cos(\theta-\psi)$ for any two angles θ and ψ.

56. Let $\mathbf{v}$ and $\mathbf{w}$ be vectors in the plane.
(a) Use Theorem 2 to explain why the dot product $\mathbf{v}\cdot\mathbf{w}$ does not change if both $\mathbf{v}$ and $\mathbf{w}$ are rotated by the same angle θ.
(b) Sketch the vectors $\mathbf{e}_1=\langle 1,0\rangle$ and $\mathbf{e}_2=\left\langle \frac{\sqrt{2}}{2},\frac{\sqrt{2}}{2}\right\rangle$, and determine the vectors $\mathbf{e}_1',\mathbf{e}_2'$ obtained by rotating $\mathbf{e}_1,\mathbf{e}_2$ through an angle $\frac{\pi}{4}$. Verify that $\mathbf{e}_1\cdot\mathbf{e}_2=\mathbf{e}_1'\cdot\mathbf{e}_2'$.

57. Determine $\|\mathbf{v}+\mathbf{w}\|$ if $\mathbf{v}$ and $\mathbf{w}$ are unit vectors separated by an angle of 30°.

58. What is the angle between $\mathbf{v}$ and $\mathbf{w}$ if:
(a) $\mathbf{v}\cdot\mathbf{w}=-\|\mathbf{v}\|\,\|\mathbf{w}\|$
(b) $\mathbf{v}\cdot\mathbf{w}=\frac{1}{2}\|\mathbf{v}\|\,\|\mathbf{w}\|$

59. Suppose that $\|\mathbf{v}\|=2$ and $\|\mathbf{w}\|=3$, and the angle between $\mathbf{v}$ and $\mathbf{w}$ is 120°. Determine:
(a) $\mathbf{v}\cdot\mathbf{w}$
(b) $\|2\mathbf{v}+\mathbf{w}\|$
(c) $\|2\mathbf{v}-3\mathbf{w}\|$

60. Find the angle θ in the triangle in Figure 10.

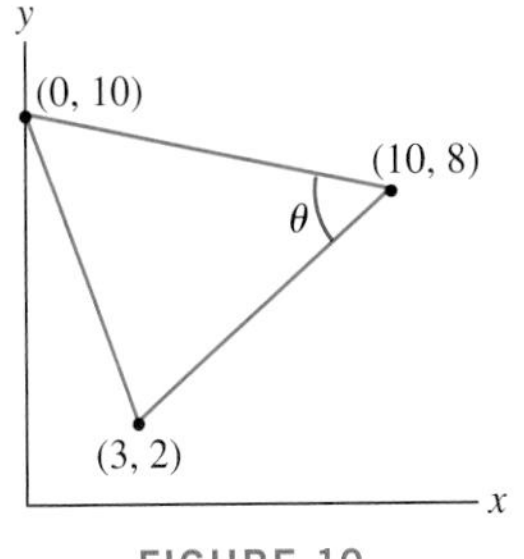

FIGURE 10

61. Find all three angles in the triangle in Figure 11.

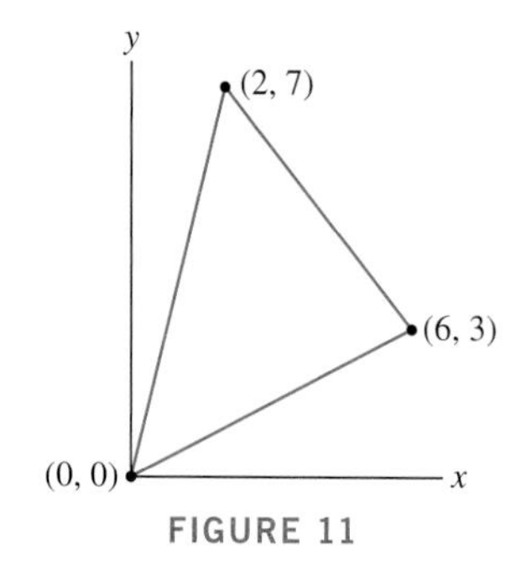

FIGURE 11

In Exercises 62–65, refer to Figure 12.

62. Find the angle between $\overline{AB}$ and $\overline{AC}$.

63. Find the angle between $\overline{AB}$ and $\overline{AD}$.

64. Calculate the projection of $\overrightarrow{AC}$ along $\overrightarrow{AD}$.

65. Calculate the projection of $\overrightarrow{AD}$ along $\overrightarrow{AB}$.

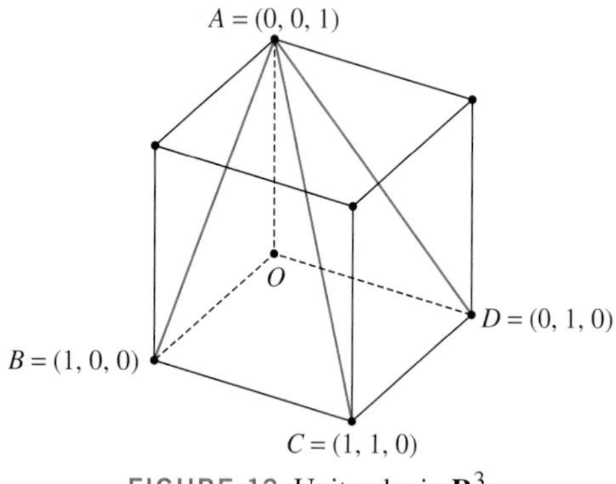

FIGURE 12 Unit cube in $\mathbf{R}^3$.

66. Let $\mathbf{v}$ and $\mathbf{w}$ be nonzero vectors and set $\mathbf{u} = \mathbf{e_v} + \mathbf{e_w}$. Use the dot product to show that the angle between $\mathbf{u}$ and $\mathbf{v}$ is equal to the angle between $\mathbf{u}$ and $\mathbf{w}$. Explain this result geometrically with a diagram.

67. Let $\mathbf{v}$, $\mathbf{w}$, and $\mathbf{a}$ be nonzero vectors such that $\mathbf{v} \cdot \mathbf{a} = \mathbf{w} \cdot \mathbf{a}$. Is it true that $\mathbf{v} = \mathbf{w}$? Either prove this or give a counterexample.

68. Prove that if $\mathbf{v}$ and $\mathbf{w}$ are nonzero and orthogonal, and if $c_1\mathbf{v} + c_2\mathbf{w} = \mathbf{0}$, then $c_1 = c_2 = 0$. *Hint:* Take the dot product of the equation with $\mathbf{v}$ and $\mathbf{w}$.

69. Calculate the force (in Newtons) required to push a 40-kg wagon up a 10° incline (Figure 13). One N is equal to 1 kg-m/s^2, and the force due to gravity on the wagon is $\mathbf{F} = 40g$ N, where $g = 9.8$.

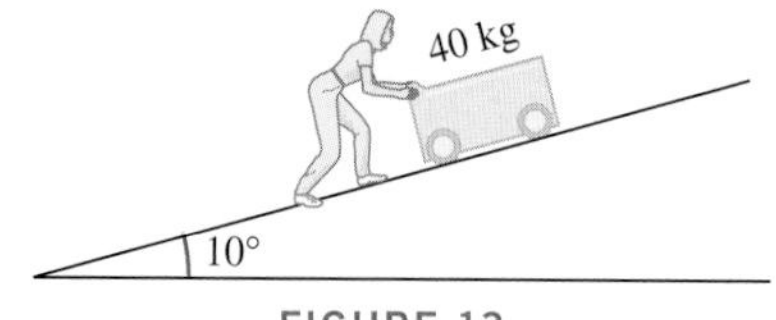

FIGURE 13

70. A 75-lb wagon is pulled by a rope (of negligible weight) making an angle of 35° with the ground (Figure 14). What is the maximum magnitude of force that can be applied along the rope without lifting the wagon off the ground?

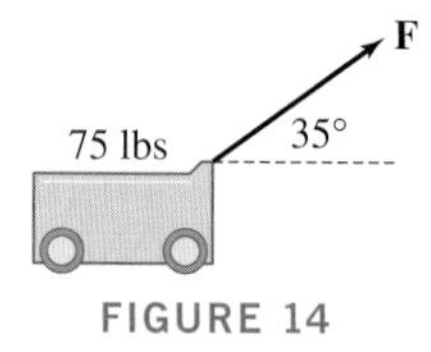

FIGURE 14

71. A 10-kg mass hangs from two ropes (of negligible weight) as in Figure 15. Rope 1 exerts a force of magnitude $\mathbf{F}_1$ acting in the direction $\overrightarrow{PA}$ and rope 2 exerts a force $\mathbf{F}_2$ in the direction $\overrightarrow{PB}$. The sum of these forces balances the force of gravity $\mathbf{F}_g = 10g$ N acting downward at P. Determine $\mathbf{F}_1$ and $\mathbf{F}_2$.

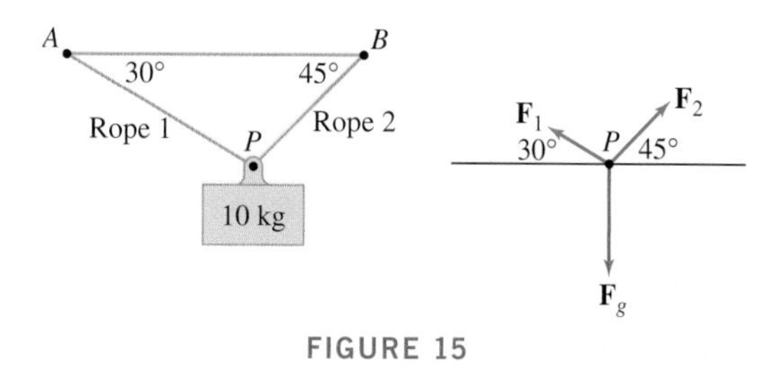

FIGURE 15

72. A light beam travels along the ray determined by a unit vector $\mathbf{L}$, strikes a flat surface at point P, and is reflected along the ray determined by a unit vector $\mathbf{R}$ where $\theta_1 = \theta_2$ (Figure 16). Show that if $\mathbf{N}$ is the unit vector orthogonal to the surface, then $\mathbf{R} = 2(\mathbf{L} \cdot \mathbf{N})\mathbf{N} - \mathbf{L}$.

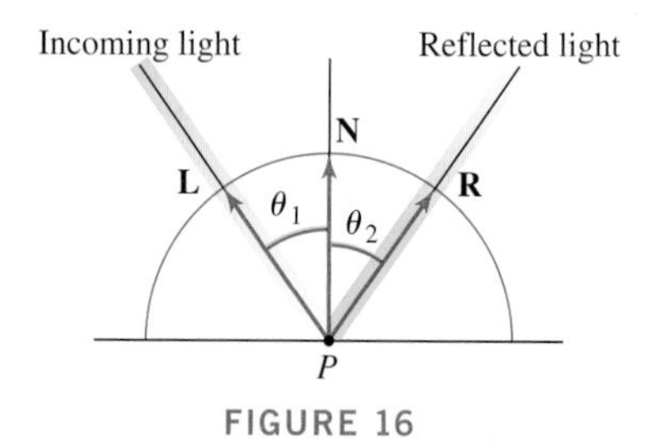

FIGURE 16

73. Let P and Q be antipodal (opposite) points on a sphere of radius r centered at the origin and let R be a third point on the sphere (Figure 17). Prove that $\overline{PR}$ and $\overline{QR}$ are orthogonal.

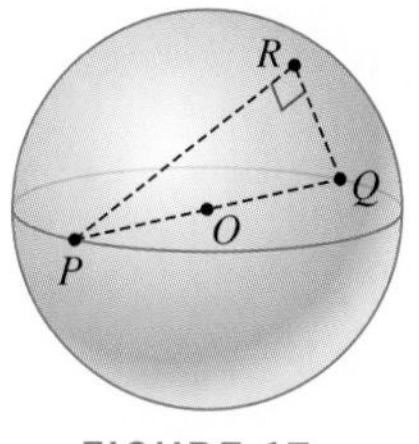

FIGURE 17

74. Prove that $\|\mathbf{v}+\mathbf{w}\|^2 - \|\mathbf{v}-\mathbf{w}\|^2 = 4\mathbf{v}\cdot\mathbf{w}$. *Hint:* Expand the dot products $(\mathbf{v}\pm\mathbf{w})\cdot(\mathbf{v}\pm\mathbf{w})$.

75. Use Exercise 74 to show that $\mathbf{v}$ and $\mathbf{w}$ are orthogonal if and only if $\|\mathbf{v}-\mathbf{w}\| = \|\mathbf{v}+\mathbf{w}\|$.

76. Show that the two diagonals of a parallelogram are perpendicular if and only if its sides have equal length. *Hint:* Use Exercise 75 to show that $\mathbf{v}-\mathbf{w}$ and $\mathbf{v}+\mathbf{w}$ are orthogonal if and only if $\|\mathbf{v}\| = \|\mathbf{w}\|$.

77. Verify the Distributive Law:

$$\mathbf{u}\cdot(\mathbf{v}+\mathbf{w}) = \mathbf{u}\cdot\mathbf{v} + \mathbf{u}\cdot\mathbf{w}$$

78. Verify that $(\lambda\mathbf{v})\cdot\mathbf{w} = \lambda(\mathbf{v}\cdot\mathbf{w})$ for any scalar λ.

Further Insights and Challenges

79. Prove the Law of Cosines: $c^2 = a^2 + b^2 - 2ab\cos\theta$ by referring to Figure 18. *Hint:* Consider the right triangle $\triangle PQR$.

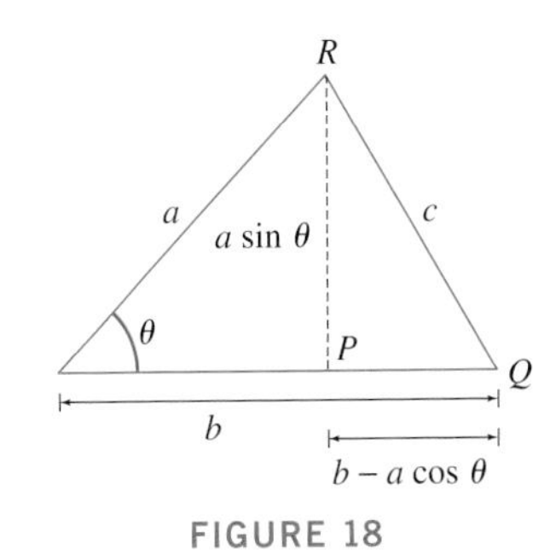

FIGURE 18

80. In this exercise, we prove the Cauchy–Schwarz inequality: If $\mathbf{v}$ and $\mathbf{w}$ are any two vectors, then

$$|\mathbf{v}\cdot\mathbf{w}| \le \|\mathbf{v}\|\,\|\mathbf{w}\| \qquad \boxed{7}$$

(a) Let $f(x) = \|x\mathbf{v}+\mathbf{w}\|^2$ for x a scalar. Show that $f(x) = ax^2 + bx + c$, where $a = \|\mathbf{v}\|^2$, $b = 2\mathbf{v}\cdot\mathbf{w}$, and $c = \|\mathbf{w}\|^2$.
(b) Conclude that $b^2 - 4ac \le 0$. *Hint:* Observe that $f(x) \ge 0$ for all x.

81. Use (7) to prove the Triangle Inequality

$$\|\mathbf{v}+\mathbf{w}\| \le \|\mathbf{v}\| + \|\mathbf{w}\|$$

Hint: First use the Triangle Inequality for numbers to prove

$$|(\mathbf{v}+\mathbf{w})\cdot(\mathbf{v}+\mathbf{w})| \le |(\mathbf{v}+\mathbf{w})\cdot\mathbf{v}| + |(\mathbf{v}+\mathbf{w})\cdot\mathbf{w}|$$

82. This exercise gives another proof of the relation between the dot product and the angle θ between two vectors $\mathbf{v} = \langle a_1, b_1\rangle$ and $\mathbf{w} = \langle a_2, b_2\rangle$ in the plane. Let θ_1 and θ_2 be the angles between $\mathbf{v}$ and $\mathbf{w}$ and the positive x-axis as in Figure 19. Use the addition formula $\cos(\theta_1 - \theta_2) = \cos\theta_1\cos\theta_2 + \sin\theta_1\sin\theta_2$ to prove that $\mathbf{v}\cdot\mathbf{w} = \|\mathbf{v}\|\,\|\mathbf{w}\|\cos\theta$. *Hint:* $\mathbf{v} = \|\mathbf{v}\|\langle\cos\theta_1, \sin\theta_1\rangle$ and $\mathbf{w} = \|\mathbf{w}\|\langle\cos\theta_2, \sin\theta_2\rangle$.

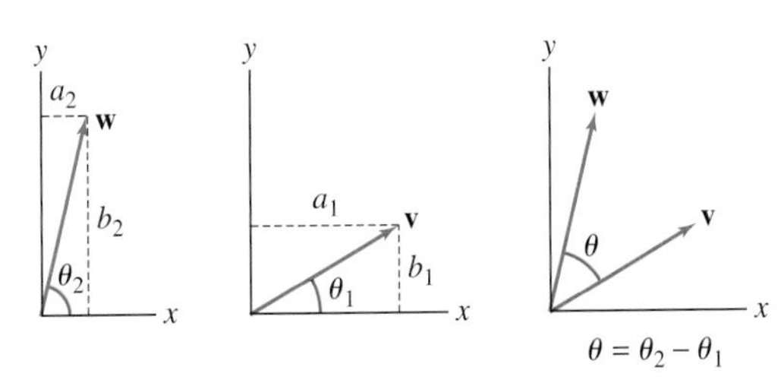

FIGURE 19

83. Let $\mathbf{v} = \langle x, y\rangle$ and

$$\mathbf{v}_\theta = \langle x\cos\theta + y\sin\theta, -x\sin\theta + y\cos\theta\rangle$$

Prove that the angle between $\mathbf{v}$ and $\mathbf{v}_\theta$ is θ.

84. Let $\mathbf{v}$ be a nonzero vector. The angles α, β, γ between $\mathbf{v}$ and the unit vectors $\mathbf{i}, \mathbf{j}, \mathbf{k}$ are called the direction angles of $\mathbf{v}$ (Figure 20). The cosines of these angles are called the **direction cosines** of $\mathbf{v}$. Prove that $\cos^2\alpha + \cos^2\beta + \cos^2\gamma = 1$.

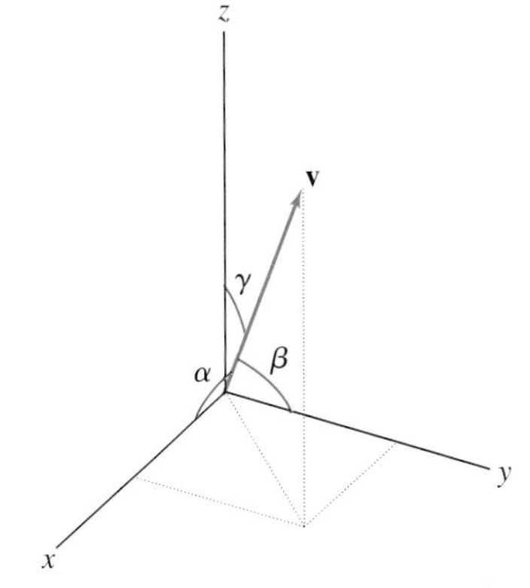

FIGURE 20 Direction angles of $\mathbf{v}$.

85. Find the direction cosines of $\mathbf{v} = \langle 3, 6, -2\rangle$.

86. The set of all points $X = (x, y, z)$ equidistant from two points P, Q in $\mathbf{R}^3$ is a plane (Figure 21). Show that X lies on this plane if

$$\overrightarrow{PQ}\cdot\overrightarrow{OX} = \frac{1}{2}\left(\|\overrightarrow{OQ}\|^2 - \|\overrightarrow{OP}\|^2\right) \qquad \boxed{8}$$

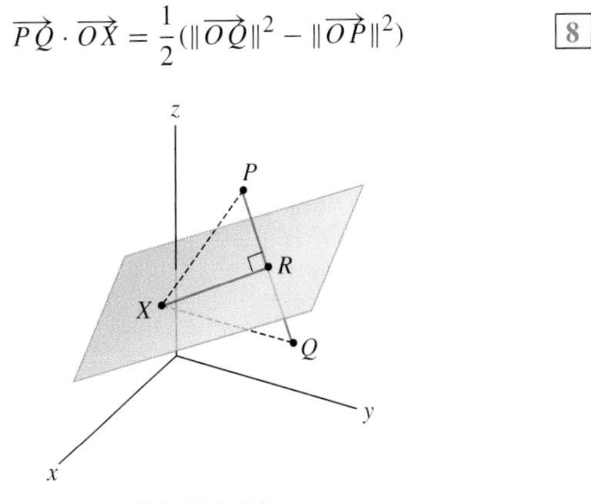

FIGURE 21

Hint: If R is the midpoint of $\overline{PQ}$, then X is equidistant from P and Q if and only if $\overrightarrow{XR}$ is orthogonal to $\overrightarrow{PQ}$.

87. Sketch the plane consisting of all points $X = (x, y, z)$ equidistant from the points $P = (0, 1, 0)$ and $Q = (0, 0, 1)$. Use Eq. (8) to show that X lies on this plane if and only if $y = z$.

88. Use Eq. (8) to find the equation of the plane consisting of all points $X = (x, y, z)$ equidistant from $P = (2, 1, 1)$ and $Q = (1, 0, 2)$.

12.4 The Cross Product

FIGURE 1 The spiral paths of charged particles in a bubble chamber in the presence of a magnetic field is described using cross products.

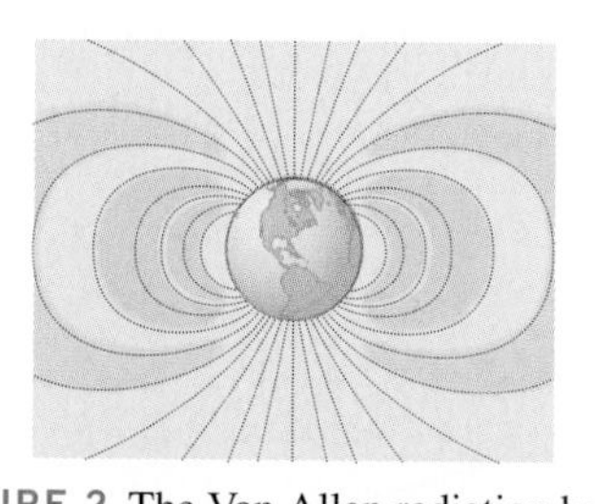

FIGURE 2 The Van Allen radiation belts, located approximately 4,000 miles above the earth's surface, are made up of streams of protons and electrons that oscillate back and forth in helical paths between two "magnetic mirrors" set up by the earth's magnetic field. This helical motion is explained by the "cross-product" nature of magnetic forces.

The theory of matrices and determinants is part of linear algebra, *a subject of great importance throughout mathematics. In this section, we discuss just a few basic definitions and facts needed for our treatment of multivariable calculus.*

This section introduces the **cross product v × w** of two vectors **v** and **w**. The cross product (sometimes called the **vector product**) is used in physics and engineering to describe quantities involving rotation such as torque and angular momentum. In electromagnetic theory, magnetic forces are described using cross products (Figures 1 and 2).

Unlike the dot product $\mathbf{v} \cdot \mathbf{w}$ (which is a scalar), the cross product $\mathbf{v} \times \mathbf{w}$ is again a vector. It is defined using determinants, which we now define in the 2×2 and 3×3 cases. A 2×2 determinant is a number formed from an array A of numbers with two rows and two columns (called a **matrix**) according to the formula

$$\det(A) = \begin{vmatrix} a_{11} & a_{12} \\ a_{21} & a_{22} \end{vmatrix} = a_{11}a_{22} - a_{12}a_{21} \tag{1}$$

For example,

$$\begin{vmatrix} 3 & 2 \\ \frac{1}{2} & 4 \end{vmatrix} = 3 \cdot 4 - 2 \cdot \frac{1}{2} = 11$$

The determinant of a 3×3 matrix is defined by the formula

$$\begin{vmatrix} a_{11} & a_{12} & a_{13} \\ a_{21} & a_{22} & a_{23} \\ a_{31} & a_{32} & a_{33} \end{vmatrix} = a_{11}\underbrace{\begin{vmatrix} a_{22} & a_{23} \\ a_{32} & a_{33} \end{vmatrix}}_{(1,1)\text{-minor}} - a_{12}\underbrace{\begin{vmatrix} a_{21} & a_{23} \\ a_{31} & a_{33} \end{vmatrix}}_{(1,2)\text{-minor}} + a_{13}\underbrace{\begin{vmatrix} a_{21} & a_{22} \\ a_{31} & a_{32} \end{vmatrix}}_{(1,3)\text{-minor}} \tag{2}$$

This formula expresses the 3×3 determinant in terms of 2×2 determinants called **minors**. The minors are obtained by crossing out the first row and one of the three columns of the 3×3 matrix. For example, the minor labeled $(1, 2)$ above is obtained as follows:

$$\underbrace{\begin{vmatrix} \not{a_{11}} & \textcircled{a_{12}} & \not{a_{13}} \\ a_{21} & \not{a_{22}} & a_{23} \\ a_{31} & \not{a_{32}} & a_{33} \end{vmatrix}}_{\text{Cross out row 1 and column 2}} \quad \text{to obtain the } (1, 2)\text{-minor} \quad \underbrace{\begin{vmatrix} a_{21} & a_{23} \\ a_{31} & a_{33} \end{vmatrix}}_{(1,2)\text{-minor}}$$

■ **EXAMPLE 1 A 3 × 3 Determinant** Calculate $\begin{vmatrix} 2 & 4 & 3 \\ 0 & 1 & -7 \\ -1 & 5 & 3 \end{vmatrix}$.

Solution

$$\begin{vmatrix} \textcircled{2} & \textcircled{4} & \textcircled{3} \\ 0 & 1 & -7 \\ -1 & 5 & 3 \end{vmatrix} = \textcircled{2}\begin{vmatrix} 1 & -7 \\ 5 & 3 \end{vmatrix} - \textcircled{4}\begin{vmatrix} 0 & -7 \\ -1 & 3 \end{vmatrix} + \textcircled{3}\begin{vmatrix} 0 & 1 \\ -1 & 5 \end{vmatrix}$$

$$= 2(38) - 4(-7) + 3(1) = 107$$

■

We will see below that determinants are related to area and volume. Our immediate goal is to introduce the cross product, which is defined as a "symbolic" determinant whose first row has the vector entries **i**, **j**, **k**.

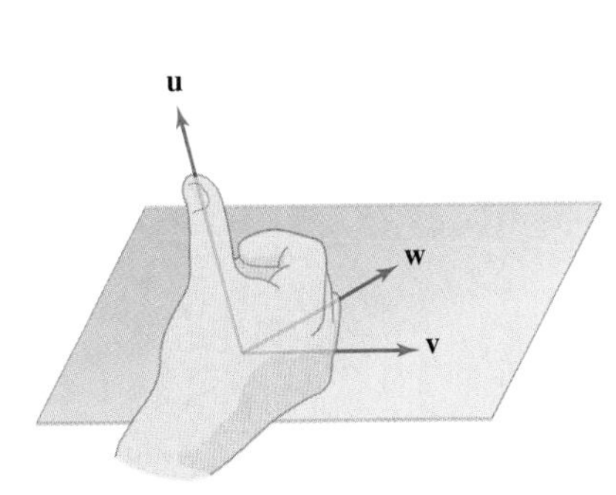

FIGURE 3 {**v**, **w**, **u**} forms a right-handed system.

DEFINITION The Cross Product The cross product of $\mathbf{v} = \langle a_1, b_1, c_1 \rangle$ and $\mathbf{w} = \langle a_2, b_2, c_2 \rangle$ is the vector

$$\mathbf{v} \times \mathbf{w} = \begin{vmatrix} \mathbf{i} & \mathbf{j} & \mathbf{k} \\ a_1 & b_1 & c_1 \\ a_2 & b_2 & c_2 \end{vmatrix} = \begin{vmatrix} b_1 & c_1 \\ b_2 & c_2 \end{vmatrix} \mathbf{i} - \begin{vmatrix} a_1 & c_1 \\ a_2 & c_2 \end{vmatrix} \mathbf{j} + \begin{vmatrix} a_1 & b_1 \\ a_2 & b_2 \end{vmatrix} \mathbf{k} \qquad \boxed{3}$$

■ **EXAMPLE 2** Calculate $\mathbf{v} \times \mathbf{w}$, where $\mathbf{v} = \langle -2, 1, 4 \rangle$ and $\mathbf{w} = \langle 3, 2, 5 \rangle$.

Solution

$$\mathbf{v} \times \mathbf{w} = \begin{vmatrix} \mathbf{i} & \mathbf{j} & \mathbf{k} \\ -2 & 1 & 4 \\ 3 & 2 & 5 \end{vmatrix} = \begin{vmatrix} 1 & 4 \\ 2 & 5 \end{vmatrix} \mathbf{i} - \begin{vmatrix} -2 & 4 \\ 3 & 5 \end{vmatrix} \mathbf{j} + \begin{vmatrix} -2 & 1 \\ 3 & 2 \end{vmatrix} \mathbf{k}$$

$$= (-3)\mathbf{i} - (-22)\mathbf{j} + (-7)\mathbf{k} = \langle -3, 22, -7 \rangle \qquad ■$$

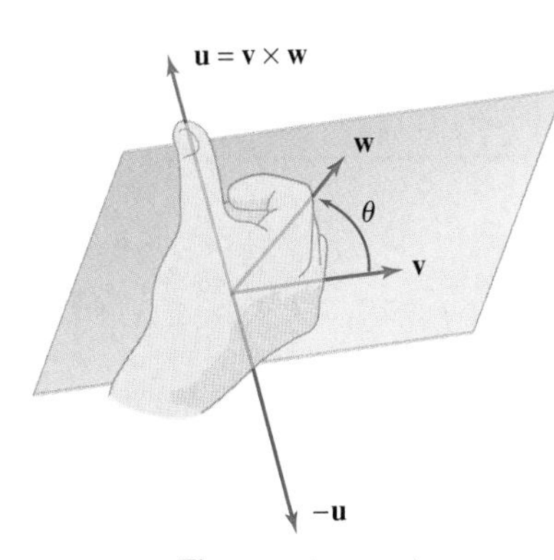

FIGURE 4 There are two vectors orthogonal to **v** and **w** with length $\|\mathbf{v}\|\,\|\mathbf{w}\| \sin\theta$. The right-hand rule determines which is $\mathbf{v} \times \mathbf{w}$.

Formula (3) gives no hint of the geometric meaning of the cross product. However, there is a simple way to visualize the vector $\mathbf{v} \times \mathbf{w}$ using the **right-hand rule**. Suppose that **v**, **w**, and **u** are nonzero vectors that do not all lie in a plane. We say that {**v**, **w**, **u**} forms a **right-handed system** if the direction of **u** is determined by the right-hand rule: *When the fingers of your right hand curl from* **v** *to* **w**, *your thumb points to the same side of the plane spanned by* **v** *and* **w** *as* **u** (Figure 3). The following theorem is proved at the end of this section.

THEOREM 1 Geometric Description of the Cross Product The cross product $\mathbf{v} \times \mathbf{w}$ is the unique vector with the following three properties:

(i) $\mathbf{v} \times \mathbf{w}$ is orthogonal to **v** and **w**.

(ii) $\mathbf{v} \times \mathbf{w}$ has length $\|\mathbf{v}\|\,\|\mathbf{w}\|\,|\sin\theta|$ (θ = angle between **v** and **w**).

(iii) $\{\mathbf{v}, \mathbf{w}, \mathbf{v} \times \mathbf{w}\}$ forms a right-handed system.

How do the three properties in Theorem 1 determine $\mathbf{v} \times \mathbf{w}$? By property (i), $\mathbf{v} \times \mathbf{w}$ lies on the line orthogonal to **v** and **w**. By property (ii), $\mathbf{v} \times \mathbf{w}$ is one of the two vectors on this line of length $\|\mathbf{v}\|\,\|\mathbf{w}\|\,|\sin\theta|$. Finally, property (iii) tells us which of these two vectors is $\mathbf{v} \times \mathbf{w}$, namely, the vector for which {**v**, **w**, **u**} is right-handed (Figure 4).

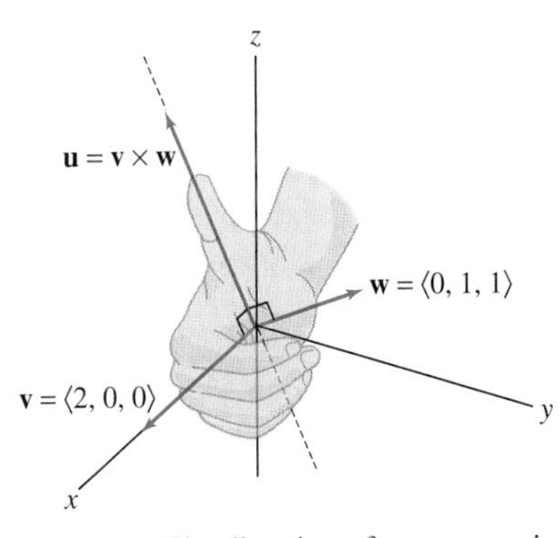

FIGURE 5 The direction of $\mathbf{u} = \mathbf{v} \times \mathbf{w}$ is determined by the right-hand rule. Thus, **u** has a positive z-component.

■ **EXAMPLE 3** Let $\mathbf{v} = \langle 2, 0, 0 \rangle$ and $\mathbf{w} = \langle 0, 1, 1 \rangle$. Determine $\mathbf{u} = \mathbf{v} \times \mathbf{w}$ using the geometric properties of the cross product rather than the formula.

Solution We use Theorem 1. By (i), $\mathbf{u} = \mathbf{v} \times \mathbf{w}$ is orthogonal to **v** and **w**. Since **v** lies along the x-axis, **u** must lie in the yz-plane (Figure 5). In other words, $\mathbf{u} = \langle 0, b, c \rangle$. But **u** is also orthogonal to $\mathbf{w} = \langle 0, 1, 1 \rangle$, so $\mathbf{u} \cdot \mathbf{w} = b + c = 0$ and thus $\mathbf{u} = \langle 0, b, -b \rangle$.

Direct computation shows that $\|\mathbf{v}\| = 2$ and $\|\mathbf{w}\| = \sqrt{2}$. Furthermore, the angle between **v** and **w** is $\theta = \frac{\pi}{2}$ since $\mathbf{v} \cdot \mathbf{w} = 0$. By property (ii),

$$\|\mathbf{u}\| = \sqrt{b^2 + (-b)^2} = |b|\sqrt{2} \qquad \text{is equal to} \qquad \|\mathbf{v}\|\,\|\mathbf{w}\| \sin\frac{\pi}{2} = 2\sqrt{2}$$

Therefore, $|b| = 2$ and $b = \pm 2$. Finally, property (iii) tells us that $\mathbf{u}$ points in the positive z-direction (Figure 5). Thus, $b = -2$ and $\mathbf{u} = \langle 0, -2, 2 \rangle$. You can verify that the formula for the cross product yields the same answer. ■

One of the most striking properties of the cross product is that it is *anticommutative*. Reversing the order changes the sign:

$$\boxed{\mathbf{w} \times \mathbf{v} = -\mathbf{v} \times \mathbf{w}} \qquad \boxed{4}$$

This follows from Eq. (3) and the fact that a 2×2 determinant changes sign when the rows are interchanged:

$$\begin{vmatrix} a_1 & b_1 \\ a_2 & b_2 \end{vmatrix} = a_1b_2 - b_1a_2 = -(b_1a_2 - a_1b_2) = -\begin{vmatrix} a_2 & b_2 \\ a_1 & b_1 \end{vmatrix}$$

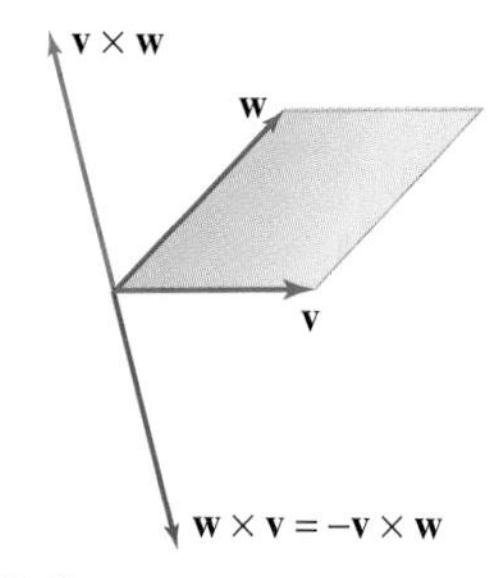

FIGURE 6

Anticommutativity also follows from the geometric description of the cross product. By properties (i) and (ii) in Theorem 1, $\mathbf{v} \times \mathbf{w}$ and $\mathbf{w} \times \mathbf{v}$ are both orthogonal to $\mathbf{v}$ and $\mathbf{w}$ and have the same length. However, $\mathbf{v} \times \mathbf{w}$ and $\mathbf{w} \times \mathbf{v}$ point in opposite directions by the right-hand rule and thus $\mathbf{v} \times \mathbf{w} = -\mathbf{w} \times \mathbf{v}$ (Figure 6). In particular, $\mathbf{v} \times \mathbf{v} = -\mathbf{v} \times \mathbf{v}$ and hence $\mathbf{v} \times \mathbf{v} = \mathbf{0}$.

The next theorem lists some further properties of cross products (the proofs are given as Exercises 49–51).

Note an important distinction between the dot and cross products of a vector with itself:

$$\mathbf{v} \times \mathbf{v} = \mathbf{0}$$

$$\mathbf{v} \cdot \mathbf{v} = \|\mathbf{v}\|^2$$

THEOREM 2 Basic Properties of the Cross Product

(i) $\mathbf{w} \times \mathbf{v} = -\mathbf{v} \times \mathbf{w}$

(ii) $\mathbf{v} \times \mathbf{v} = \mathbf{0}$

(iii) $\mathbf{v} \times \mathbf{w} = \mathbf{0}$ if and only if $\mathbf{w} = \lambda \mathbf{v}$ for some scalar λ or $\mathbf{v} = \mathbf{0}$.

(iv) $(\lambda \mathbf{v}) \times \mathbf{w} = \mathbf{v} \times (\lambda \mathbf{w}) = \lambda(\mathbf{v} \times \mathbf{w})$

(v) $(\mathbf{u} + \mathbf{v}) \times \mathbf{w} = \mathbf{u} \times \mathbf{w} + \mathbf{v} \times \mathbf{w}$

$\mathbf{v} \times (\mathbf{u} + \mathbf{w}) = \mathbf{v} \times \mathbf{u} + \mathbf{v} \times \mathbf{w}$

The cross product of any two of the standard basis vectors $\mathbf{i}$, $\mathbf{j}$, and $\mathbf{k}$ is equal to the third, possibly with a minus sign. More precisely (see Exercise 52),

$$\mathbf{i} \times \mathbf{j} = \mathbf{k}, \qquad \mathbf{j} \times \mathbf{k} = \mathbf{i}, \qquad \mathbf{k} \times \mathbf{i} = \mathbf{j} \qquad \boxed{5}$$

$$\mathbf{i} \times \mathbf{i} = \mathbf{j} \times \mathbf{j} = \mathbf{k} \times \mathbf{k} = \mathbf{0}$$

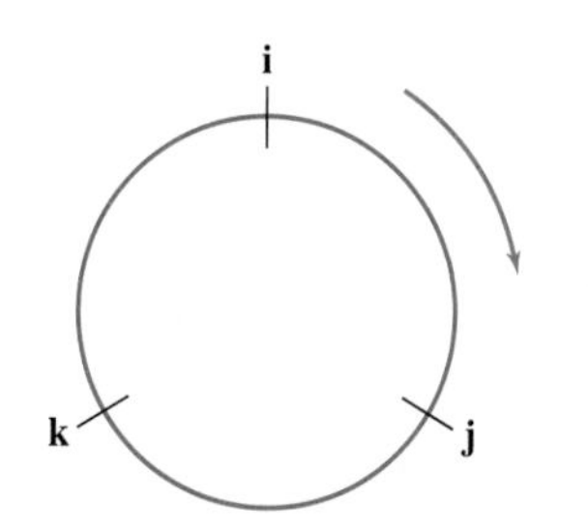

FIGURE 7 Circle for computing the cross products of the basic vectors.

Since the cross product is anticommutative, minus signs occur when the cross products are taken in the opposite order. An easy way to remember these relations is to draw $\mathbf{i}$, $\mathbf{j}$, and $\mathbf{k}$ in a circle as in Figure 7. Go around the circle in the clockwise direction (starting at any point) and you obtain one of the relations (5). For example, starting at $\mathbf{i}$ and moving clockwise yields $\mathbf{i} \times \mathbf{j} = \mathbf{k}$. If you go around in the counterclockwise direction, you obtain the relations with a minus sign. Thus, starting at $\mathbf{k}$ and going counterclockwise gives the relation $\mathbf{k} \times \mathbf{j} = -\mathbf{i}$.

■ **EXAMPLE 4 Using the ijk Relations** Compute $(2\mathbf{i} + \mathbf{k}) \times (3\mathbf{j} + 5\mathbf{k})$.

Solution We use the Distributive Law for cross products:

$$
\begin{aligned}
(2\mathbf{i}+\mathbf{k})\times(3\mathbf{j}+5\mathbf{k}) &= (2\mathbf{i})\times(3\mathbf{j})+(2\mathbf{i})\times(5\mathbf{k})+\mathbf{k}\times(3\mathbf{j})+\mathbf{k}\times(5\mathbf{k})\\
&= 6(\mathbf{i}\times\mathbf{j})+10(\mathbf{i}\times\mathbf{k})+3(\mathbf{k}\times\mathbf{j})+5(\mathbf{k}\times\mathbf{k})\\
&= 6\mathbf{k}-10\mathbf{j}-3\mathbf{i}+5(\mathbf{0}) = -3\mathbf{i}-10\mathbf{j}+6\mathbf{k}
\end{aligned}
$$

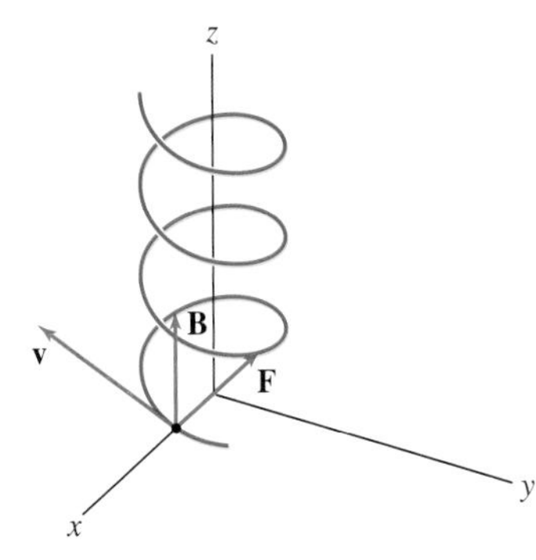

FIGURE 8 A proton in a uniform magnetic field travels in a helical path.

EXAMPLE 5 Velocity in a Magnetic Field The force $\mathbf{F}$ on a proton moving at velocity $\mathbf{v}$ m/s in a uniform magnetic field $\mathbf{B}$ (in teslas) is $\mathbf{F} = q(\mathbf{v}\times\mathbf{B})$ in newtons, where $q = 1.6\times10^{-19}$ C (Figure 8). Calculate $\mathbf{F}$ if $\mathbf{B} = 0.0004\mathbf{k}$ T and $\mathbf{v}$ has magnitude 10^6 m/s in the direction $-\mathbf{j}+\mathbf{k}$.

Solution The vector $-\mathbf{j}+\mathbf{k}$ has length $\sqrt{2}$, and since $\mathbf{v}$ has magnitude 10^6,

$$
\mathbf{v} = 10^6\left(\frac{-\mathbf{j}+\mathbf{k}}{\sqrt{2}}\right)
$$

Therefore, the force (in newtons) is

$$
\begin{aligned}
\mathbf{F} = q(\mathbf{v}\times\mathbf{B}) &= 10^6 q\left(\frac{-\mathbf{j}+\mathbf{k}}{\sqrt{2}}\right)\times(0.0004\mathbf{k}) = \frac{400q}{\sqrt{2}}\left((-\mathbf{j}+\mathbf{k})\times\mathbf{k}\right)\\
&= -\frac{400q}{\sqrt{2}}\mathbf{i} = \frac{-400(1.6\times10^{-19})}{\sqrt{2}}\mathbf{i} \approx -(4.5\times10^{-17})\mathbf{i}
\end{aligned}
$$

Relation Between the Cross Product, Area, and Volume

The cross product and determinants are closely related to area and volume. Consider the parallelogram $\mathcal{P}$ spanned by nonzero vectors $\mathbf{v}$ and $\mathbf{w}$ with a common basepoint [Figure 9(A)]. Recall that $\mathcal{P}$ is the parallelogram whose vertices are the basepoint and the terminal points of $\mathbf{v}$, $\mathbf{w}$, and $\mathbf{v}+\mathbf{w}$. This parallelogram has base $b = \|\mathbf{v}\|$ and height $h = \|\mathbf{w}\|\sin\theta$, where θ is the angle between $\mathbf{v}$ and $\mathbf{w}$. Therefore, the area of $\mathcal{P}$ is equal to $A = bh = \|\mathbf{v}\|\,\|\mathbf{w}\|\sin\theta = \|\mathbf{v}\times\mathbf{w}\|$.

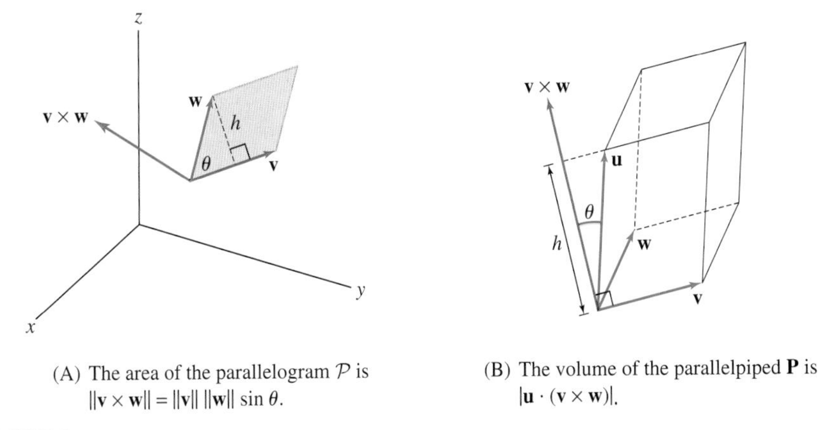

(A) The area of the parallelogram $\mathcal{P}$ is $\|\mathbf{v}\times\mathbf{w}\| = \|\mathbf{v}\|\,\|\mathbf{w}\|\sin\theta$.

(B) The volume of the parallelpiped $\mathbf{P}$ is $|\mathbf{u}\cdot(\mathbf{v}\times\mathbf{w})|$.

FIGURE 9

A "parallelepiped" is the solid spanned by three vectors. Each face is a parallelogram.

Next, consider the **parallelepiped P** spanned by three nonzero vectors $\mathbf{u}$, $\mathbf{v}$, $\mathbf{w}$ in $\mathbf{R}^3$ [the three-dimensional prism in Figure 9(B)]. The base of $\mathbf{P}$ is the parallelogram spanned by $\mathbf{v}$ and $\mathbf{w}$, so the area of the base is $\|\mathbf{v}\times\mathbf{w}\|$. The height of $\mathbf{P}$ is $h = \|\mathbf{u}\|\cdot|\cos\theta|$, where θ is the angle between $\mathbf{u}$ and $\mathbf{v}\times\mathbf{w}$. Therefore,

$$
\text{Volume of } \mathbf{P} = (\text{area of base})(\text{height}) = \|\mathbf{v}\times\mathbf{w}\|\cdot\|\mathbf{u}\|\cdot|\cos\theta| = |\mathbf{u}\cdot(\mathbf{v}\times\mathbf{w})|
$$

We use the following notation for the determinant of the matrix whose rows are the vectors $\mathbf{v}, \mathbf{w}, \mathbf{u}$:

$$\det\begin{pmatrix}\mathbf{u}\\ \mathbf{v}\\ \mathbf{w}\end{pmatrix} = \begin{vmatrix} a_1 & b_1 & c_1\\ a_2 & b_2 & c_2\\ a_3 & b_3 & c_3\end{vmatrix}$$

It is awkward to write the absolute value of a determinant in the notation on the right, but we may denote it

$$\left|\det\begin{pmatrix}\mathbf{u}\\ \mathbf{v}\\ \mathbf{w}\end{pmatrix}\right|$$

The quantity $\mathbf{u}\cdot(\mathbf{v}\times\mathbf{w})$, which is sometimes called the **vector triple product**, can be expressed as a determinant. Let

$$\mathbf{u} = \langle a_1, b_1, c_1\rangle, \qquad \mathbf{v} = \langle a_2, b_2, c_2\rangle, \qquad \mathbf{w} = \langle a_3, b_3, c_3\rangle$$

Then

$$\begin{aligned}\mathbf{u}\cdot(\mathbf{v}\times\mathbf{w}) &= \mathbf{u}\cdot\left(\begin{vmatrix} b_2 & c_2\\ b_3 & c_3\end{vmatrix}\mathbf{i} - \begin{vmatrix} a_2 & c_2\\ a_3 & c_3\end{vmatrix}\mathbf{j} + \begin{vmatrix} a_2 & b_2\\ a_3 & b_3\end{vmatrix}\mathbf{k}\right)\\ &= a_1\begin{vmatrix} b_2 & c_2\\ b_3 & c_3\end{vmatrix} - b_1\begin{vmatrix} a_2 & c_2\\ a_3 & c_3\end{vmatrix} + c_1\begin{vmatrix} a_2 & b_2\\ a_3 & b_3\end{vmatrix}\\ &= \begin{vmatrix} a_1 & b_1 & c_1\\ a_2 & b_2 & c_2\\ a_3 & b_3 & c_3\end{vmatrix} = \det\begin{pmatrix}\mathbf{u}\\ \mathbf{v}\\ \mathbf{w}\end{pmatrix}\end{aligned} \qquad \boxed{6}$$

We obtain the following formulas for area and volume.

THEOREM 3 Area and Volume via Cross Products and Determinants Let $\mathbf{u}, \mathbf{v}, \mathbf{w}$ be nonzero vectors in $\mathbf{R}^3$. Then

(i) The parallelogram $\mathcal{P}$ spanned by $\mathbf{v}$ and $\mathbf{w}$ has area $A = \|\mathbf{v}\times\mathbf{w}\|$.

(ii) The parallelepiped $\mathbf{P}$ spanned by $\mathbf{u}$, $\mathbf{v}$, and $\mathbf{w}$ has volume

$$V = |\mathbf{u}\cdot(\mathbf{v}\times\mathbf{w})| = \left|\det\begin{pmatrix}\mathbf{u}\\ \mathbf{v}\\ \mathbf{w}\end{pmatrix}\right| \qquad \boxed{7}$$

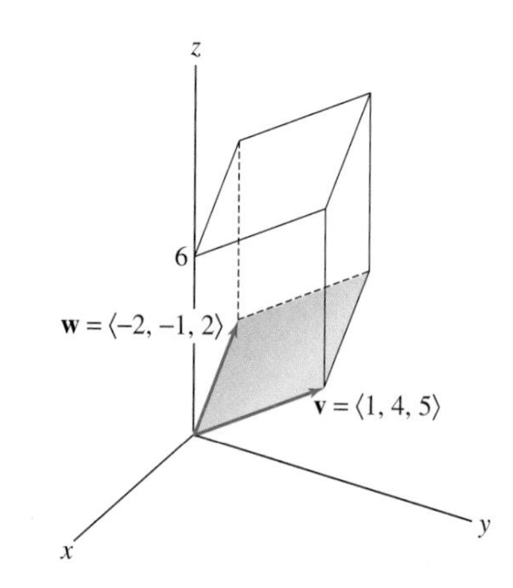

FIGURE 10

EXAMPLE 6 Let $\mathbf{v} = \langle 1, 4, 5\rangle$ and $\mathbf{w} = \langle -2, -1, 2\rangle$. Calculate:

(a) The area A of the parallelogram spanned by $\mathbf{v}$ and $\mathbf{w}$

(b) The volume V of the parallelepiped in Figure 10

Solution We compute the cross product and apply Theorem 3:

$$\mathbf{v}\times\mathbf{w} = \begin{vmatrix} 4 & 5\\ -1 & 2\end{vmatrix}\mathbf{i} - \begin{vmatrix} 1 & 5\\ -2 & 2\end{vmatrix}\mathbf{j} + \begin{vmatrix} 1 & 4\\ -2 & -1\end{vmatrix}\mathbf{k} = \langle 13, -12, 7\rangle$$

(a) The area of the parallelogram spanned by $\mathbf{v}$ and $\mathbf{w}$ is

$$A = \|\mathbf{v}\times\mathbf{w}\| = \sqrt{13^2 + (-12)^2 + 7^2} = \sqrt{362} \approx 19$$

(b) The vertical leg of the parallelepiped is the vector $6\mathbf{k}$, so by Eq. (7),

$$V = |(6\mathbf{k})\cdot(\mathbf{v}\times\mathbf{w})| = |\langle 0, 0, 6\rangle\cdot\langle 13, -12, 7\rangle| = 6(7) = 42$$ ■

If $\mathbf{v} = \langle a, b\rangle$ and $\mathbf{w} = \langle c, d\rangle$ are vectors in the xy-plane, we may compute the area A of the parallelogram spanned by $\mathbf{v}$ and $\mathbf{w}$ as follows. We regard $\mathbf{v}$ and $\mathbf{w}$ as vectors in $\mathbf{R}^3$ with zero component in the z-direction (Figure 11). Thus, we write $\mathbf{v} = \langle a, b, 0\rangle$ and $\mathbf{w} = \langle c, d, 0\rangle$. The cross product $\mathbf{v}\times\mathbf{w}$ is a vector pointing in the z-direction:

$$\mathbf{v}\times\mathbf{w} = \begin{vmatrix} \mathbf{i} & \mathbf{j} & \mathbf{k}\\ a & b & 0\\ c & d & 0\end{vmatrix} = \begin{vmatrix} b & 0\\ d & 0\end{vmatrix}\mathbf{i} - \begin{vmatrix} a & 0\\ b & 0\end{vmatrix}\mathbf{j} + \begin{vmatrix} a & b\\ c & d\end{vmatrix}\mathbf{k} = \begin{vmatrix} a & b\\ c & d\end{vmatrix}\mathbf{k}$$

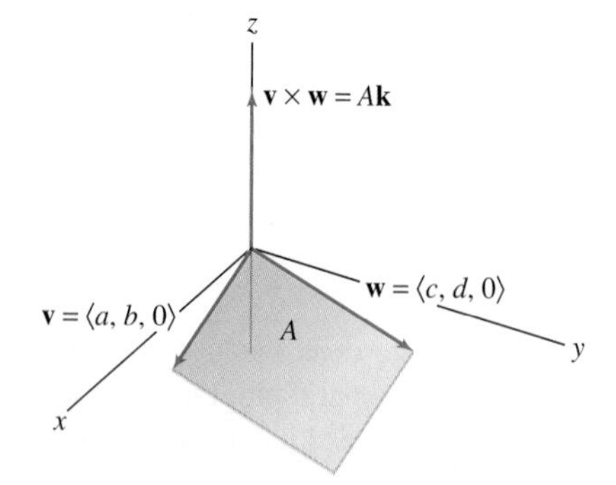

FIGURE 11 Parallelogram spanned by $\mathbf{v}$ and $\mathbf{w}$ in the xy-plane.

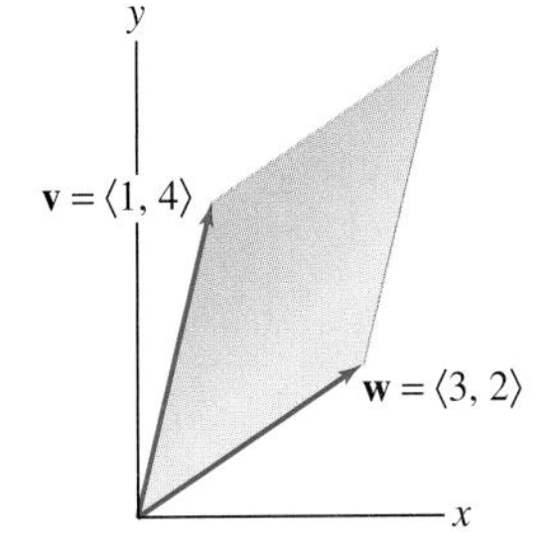

FIGURE 12

By Theorem 3, the parallelogram spanned by **v** and **w** has area $A = \|\mathbf{v} \times \mathbf{w}\|$, and thus,

$$A = \left|\det\begin{pmatrix} \mathbf{v} \\ \mathbf{w} \end{pmatrix}\right| = \left|\det\begin{pmatrix} a & b \\ c & d \end{pmatrix}\right| \qquad \boxed{8}$$

■ **EXAMPLE 7** Compute the area A of the parallelogram spanned by $\mathbf{v} = \langle 1, 4\rangle$ and $\mathbf{w} = \langle 3, 2\rangle$.

Solution We have $\begin{vmatrix} \mathbf{v} \\ \mathbf{w} \end{vmatrix} = \begin{vmatrix} 1 & 4 \\ 3 & 2 \end{vmatrix} = 1 \cdot 2 - 3 \cdot 4 = -10$. The area is the absolute value $A = |-10| = 10$ (Figure 12). ■

Proofs of Cross-Product Properties

We now derive the properties of the cross product listed in Theorem 1. Let

$$\mathbf{v} = \langle a_1, b_1, c_1\rangle, \qquad \mathbf{w} = \langle a_2, b_2, c_2\rangle$$

To show that $\mathbf{v} \times \mathbf{w}$ is orthogonal to **v**, it suffices to show that $\mathbf{v} \cdot (\mathbf{v} \times \mathbf{w}) = 0$. By (6), we have

$$\mathbf{v} \cdot (\mathbf{v} \times \mathbf{w}) = \det\begin{pmatrix} \mathbf{v} \\ \mathbf{v} \\ \mathbf{w} \end{pmatrix} = a_1\begin{vmatrix} b_1 & c_1 \\ b_2 & c_2 \end{vmatrix} - b_1\begin{vmatrix} a_1 & c_1 \\ a_2 & c_2 \end{vmatrix} + c_1\begin{vmatrix} a_1 & b_1 \\ a_2 & b_2 \end{vmatrix} \qquad \boxed{9}$$

Straightforward algebra (left to the reader) shows that (9) is equal to zero. Similarly, we have $\mathbf{w} \cdot (\mathbf{v} \times \mathbf{w}) = -\mathbf{w} \cdot (\mathbf{w} \times \mathbf{v}) = 0$. Thus, **w** is also orthogonal to $\mathbf{v} \times \mathbf{w}$. This proves part (i) of Theorem 1. To prove (ii), we use the following identity:

$$\|\mathbf{v} \times \mathbf{w}\|^2 = \|\mathbf{v}\|^2\|\mathbf{w}\|^2 - (\mathbf{v} \cdot \mathbf{w})^2 \qquad \boxed{10}$$

To verify this identity, we compute $\|\mathbf{v} \times \mathbf{w}\|^2$ as the sum of the squares of the components of $\mathbf{v} \times \mathbf{w}$:

$$\begin{aligned} \|\mathbf{v} \times \mathbf{w}\|^2 &= \begin{vmatrix} b_1 & c_1 \\ b_2 & c_2 \end{vmatrix}^2 + \begin{vmatrix} a_1 & c_1 \\ a_2 & c_2 \end{vmatrix}^2 + \begin{vmatrix} a_1 & b_1 \\ a_2 & b_2 \end{vmatrix}^2 \\ &= (b_1c_2 - c_1b_2)^2 + (a_1c_2 - c_1a_2)^2 + (a_1b_2 - b_1a_2)^2 \end{aligned} \qquad \boxed{11}$$

On the other hand, by definition,

$$\|\mathbf{v}\|^2\|\mathbf{w}\|^2 - (\mathbf{v} \cdot \mathbf{w})^2 = (a_1^2 + b_1^2 + c_1^2)^2(a_2^2 + b_2^2 + c_2^2)^2 - (a_1a_2 + b_1b_2 + c_1c_2)^2 \qquad \boxed{12}$$

Again, algebra (left to the reader) shows that (11) is equal to (12).

Now let θ be the angle between **v** and **w**. By (10),

$$\begin{aligned} \|\mathbf{v} \times \mathbf{w}\|^2 &= \|\mathbf{v}\|^2\|\mathbf{w}\|^2 - (\mathbf{v} \cdot \mathbf{w})^2 = \|\mathbf{v}\|^2\|\mathbf{w}\|^2 - \|\mathbf{v}\|^2\|\mathbf{w}\|^2\cos^2\theta \\ &= \|\mathbf{v}\|^2\|\mathbf{w}\|^2(1 - \cos^2\theta) = \|\mathbf{v}\|^2\|\mathbf{w}\|^2\sin^2\theta \end{aligned}$$

Therefore, $\|\mathbf{v} \times \mathbf{w}\| = \|\mathbf{v}\|\|\mathbf{w}\|\sin\theta$. Note that $\sin\theta \geq 0$ since, by convention, θ lies between 0 and π. This proves (ii).

Part (iii) of Theorem 1 asserts that $\{\mathbf{v}, \mathbf{w}, \mathbf{v} \times \mathbf{w}\}$ is a right-handed system. This is a more subtle property that cannot be verified by algebra alone. We must rely on the

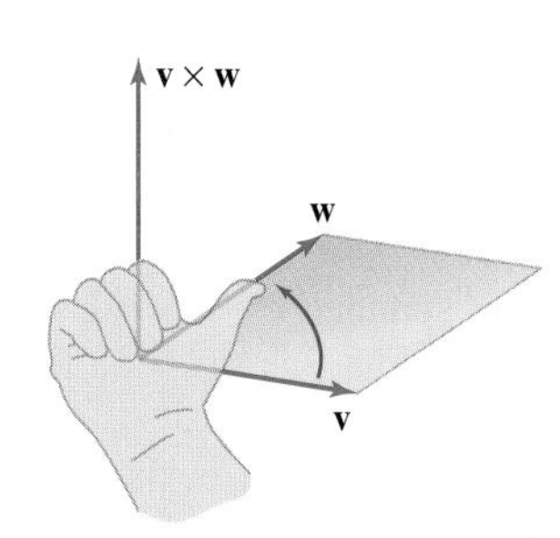

FIGURE 13 Both $\{\mathbf{v}\times\mathbf{w}, \mathbf{v}, \mathbf{w}\}$ and $\{\mathbf{v}, \mathbf{w}, \mathbf{v}\times\mathbf{w}\}$ are right-handed.

following relation between right-handedness and the sign of the determinant which can only be established using continuity:

$$\det\begin{pmatrix}\mathbf{u}\\ \mathbf{v}\\ \mathbf{w}\end{pmatrix} > 0 \quad \text{if and only if } \{\mathbf{u}, \mathbf{v}, \mathbf{w}\} \text{ is a right-handed system}$$

Granting this, we see that $\{\mathbf{v}\times\mathbf{w}, \mathbf{v}, \mathbf{w}\}$ is right-handed:

$$\det\begin{pmatrix}\mathbf{v}\times\mathbf{w}\\ \mathbf{v}\\ \mathbf{w}\end{pmatrix} = (\mathbf{v}\times\mathbf{w})\cdot(\mathbf{v}\times\mathbf{w}) = \|\mathbf{v}\times\mathbf{w}\|^2 > 0$$

It follows that $\{\mathbf{v}, \mathbf{w}, \mathbf{v}\times\mathbf{w}\}$ is also right-handed (Figure 13).

12.4 SUMMARY

- The determinants of 2×2 and 3×3 matrices are defined by the formulas

$$\begin{vmatrix} a_{11} & a_{12}\\ a_{21} & a_{22}\end{vmatrix} = a_{11}a_{22} - a_{12}a_{21}$$

$$\begin{vmatrix} a_{11} & a_{12} & a_{13}\\ a_{21} & a_{22} & a_{23}\\ a_{31} & a_{32} & a_{33}\end{vmatrix} = a_{11}\begin{vmatrix} a_{22} & a_{23}\\ a_{32} & a_{33}\end{vmatrix} - a_{12}\begin{vmatrix} a_{21} & a_{23}\\ a_{31} & a_{33}\end{vmatrix} + a_{13}\begin{vmatrix} a_{21} & a_{22}\\ a_{31} & a_{32}\end{vmatrix}$$

- The *cross product* of $\mathbf{v} = \langle a_1, b_1, c_1\rangle$ and $\mathbf{w} = \langle a_2, b_2, c_2\rangle$ is the symbolic determinant

$$\mathbf{v}\times\mathbf{w} = \begin{vmatrix} \mathbf{i} & \mathbf{j} & \mathbf{k}\\ a_1 & b_1 & c_1\\ a_2 & b_2 & c_2\end{vmatrix} = \begin{vmatrix} b_1 & c_1\\ b_2 & c_2\end{vmatrix}\mathbf{i} - \begin{vmatrix} a_1 & c_1\\ a_2 & c_2\end{vmatrix}\mathbf{j} + \begin{vmatrix} a_1 & b_1\\ a_2 & b_2\end{vmatrix}\mathbf{k}$$

- Properties of the cross product:

  (i) $\mathbf{w}\times\mathbf{v} = -\mathbf{v}\times\mathbf{w}$
  (ii) $\mathbf{v}\times\mathbf{w} = 0$ if and only if $\mathbf{w} = \lambda\mathbf{v}$ for some scalar or $\mathbf{v} = \mathbf{0}$
  (iii) $(\lambda\mathbf{v})\times\mathbf{w} = \mathbf{v}\times(\lambda\mathbf{w}) = \lambda(\mathbf{v}\times\mathbf{w})$
  (iv) $(\mathbf{u}+\mathbf{v})\times\mathbf{w} = \mathbf{u}\times\mathbf{w} + \mathbf{v}\times\mathbf{w}$

  $\mathbf{v}\times(\mathbf{u}+\mathbf{w}) = \mathbf{v}\times\mathbf{u} + \mathbf{v}\times\mathbf{w}$

- The cross product $\mathbf{v}\times\mathbf{w}$ is the unique vector with the following three properties:

  (i) $\mathbf{v}\times\mathbf{w}$ is orthogonal to $\mathbf{v}$ and $\mathbf{w}$.
  (ii) $\mathbf{v}\times\mathbf{w}$ has length $\|\mathbf{v}\|\,\|\mathbf{w}\|\,|\sin\theta|$ (θ is the angle between $\mathbf{v}$ and $\mathbf{w}$).
  (iii) $\{\mathbf{v}, \mathbf{w}, \mathbf{v}\times\mathbf{w}\}$ is a right-handed system.

- The *vector triple product* is defined by $\mathbf{u}\cdot(\mathbf{v}\times\mathbf{w})$. We have

$$\mathbf{u}\cdot(\mathbf{v}\times\mathbf{w}) = \det\begin{pmatrix}\mathbf{u}\\ \mathbf{v}\\ \mathbf{w}\end{pmatrix}$$

The volume of the parallelepiped spanned by $\mathbf{u}$, $\mathbf{v}$, and $\mathbf{w}$ is equal to the absolute value $|\mathbf{u}\cdot(\mathbf{v}\times\mathbf{w})|$.

- $\|\mathbf{v}\times\mathbf{w}\|$ is equal to the area of the parallelogram spanned by $\mathbf{v}$ and $\mathbf{w}$.
- Cross-product identity: $\|\mathbf{v}\times\mathbf{w}\|^2 = \|\mathbf{v}\|^2\|\mathbf{w}\|^2 - (\mathbf{v}\cdot\mathbf{w})^2$.

12.4 EXERCISES

Preliminary Questions

1. What is the (1, 3) minor of the matrix $\begin{vmatrix} 3 & 4 & 2 \\ -5 & -1 & 1 \\ 4 & 0 & 3 \end{vmatrix}$?

2. The angle between two unit vectors **e** and **f** is $\frac{\pi}{6}$. What is the length of $\mathbf{e} \times \mathbf{f}$?

3. What is $\mathbf{u} \times \mathbf{w}$, assuming that $\mathbf{w} \times \mathbf{u} = \langle 2, 2, 1 \rangle$?

4. Find the cross product without using the formula:

(a) $\langle 4, 8, 2 \rangle \times \langle 4, 8, 2 \rangle$ **(b)** $\langle 4, 8, 2 \rangle \times \langle 2, 4, 1 \rangle$

5. What are $\mathbf{i} \times \mathbf{j}$ and $\mathbf{i} \times \mathbf{k}$?

6. When is the cross product $\mathbf{v} \times \mathbf{w}$ equal to zero?

Exercises

In Exercises 1–4, calculate the 2×2*-determinant.*

1. $\begin{vmatrix} 1 & 2 \\ 4 & 3 \end{vmatrix}$

2. $\begin{vmatrix} \frac{2}{3} & \frac{1}{6} \\ -5 & 2 \end{vmatrix}$

3. $\begin{vmatrix} -6 & 9 \\ 1 & 1 \end{vmatrix}$

4. $\begin{vmatrix} 9 & 25 \\ 5 & 14 \end{vmatrix}$

In Exercises 5–10, calculate the 3×3*-determinant.*

5. $\begin{vmatrix} 1 & 2 & 1 \\ 4 & -3 & 0 \\ 1 & 0 & 1 \end{vmatrix}$

6. $\begin{vmatrix} 1 & 0 & 1 \\ -2 & 0 & 3 \\ 1 & 3 & -1 \end{vmatrix}$

7. $\begin{vmatrix} 1 & 0 & 1 \\ 1 & 3 & 1 \\ -2 & 0 & 3 \end{vmatrix}$

8. $\begin{vmatrix} 0 & 1 & 0 \\ 5 & 93 & 6 \\ 4 & 78 & 5 \end{vmatrix}$

9. $\begin{vmatrix} 1 & 0 & 0 \\ 0 & 1 & 0 \\ 0 & 0 & 1 \end{vmatrix}$

10. $\begin{vmatrix} 1 & 2 & 3 \\ 2 & 4 & 6 \\ -3 & -4 & 2 \end{vmatrix}$

In Exercises 11–16, calculate $\mathbf{v} \times \mathbf{w}$.

11. $\mathbf{v} = \langle 1, 2, 1 \rangle, \quad \mathbf{w} = \langle 3, 1, 1 \rangle$

12. $\mathbf{v} = \langle 2, 0, 0 \rangle, \quad \mathbf{w} = \langle -1, 0, 1 \rangle$

13. $\mathbf{v} = \left\langle \frac{2}{3}, 1, \frac{1}{2} \right\rangle, \quad \mathbf{w} = \langle 4, -6, 3 \rangle$

14. $\mathbf{v} = \langle 0, 1, -1 \rangle, \quad \mathbf{w} = \langle 1, -1, 0 \rangle$

15. $\mathbf{v} = \left\langle \frac{1}{3}, 1, \frac{1}{3} \right\rangle, \quad \mathbf{w} = \langle -1, -1, 2 \rangle$

16. $\mathbf{v} = \langle 1, 1, 0 \rangle, \quad \mathbf{w} = \langle 0, 1, 1 \rangle$

In Exercises 17–20, calculate the cross product.

17. $(\mathbf{i} + \mathbf{j}) \times \mathbf{k}$

18. $(\mathbf{j} - \mathbf{k}) \times (\mathbf{j} + \mathbf{k})$

19. $(\mathbf{i} + 2\mathbf{k}) \times (\mathbf{j} - \mathbf{k})$

20. $(2\mathbf{i} - 3\mathbf{j} + 4\mathbf{k}) \times (\mathbf{i} + \mathbf{j} - 7\mathbf{k})$

In Exercises 21–26, calculate the cross product assuming that $\mathbf{u} \times \mathbf{v} = \langle 1, 1, 0 \rangle$, $\mathbf{u} \times \mathbf{w} = \langle 0, 3, 1 \rangle$, *and* $\mathbf{v} \times \mathbf{w} = \langle 2, -1, 1 \rangle$.

21. $\mathbf{v} \times \mathbf{u}$

22. $\mathbf{w} \times (4\mathbf{w})$

23. $(3\mathbf{u} + 4\mathbf{w}) \times \mathbf{w}$

24. $(\mathbf{u} - 2\mathbf{v}) \times (\mathbf{u} + 2\mathbf{v})$

25. $\mathbf{w} \times (\mathbf{u} + \mathbf{v})$

26. $(\mathbf{v} + \mathbf{w}) \times (3\mathbf{u} + 2\mathbf{v})$

27. Let $\mathbf{v} = \langle a, b, c \rangle$. Calculate $\mathbf{v} \times \mathbf{i}$, $\mathbf{v} \times \mathbf{j}$, and $\mathbf{v} \times \mathbf{k}$.

28. Find $\mathbf{v} \times \mathbf{w}$, where **v** and **w** are vectors of length 3 in the xz-plane, oriented as in Figure 14, and the angle between **v** and **w** is $\theta = \frac{\pi}{6}$.

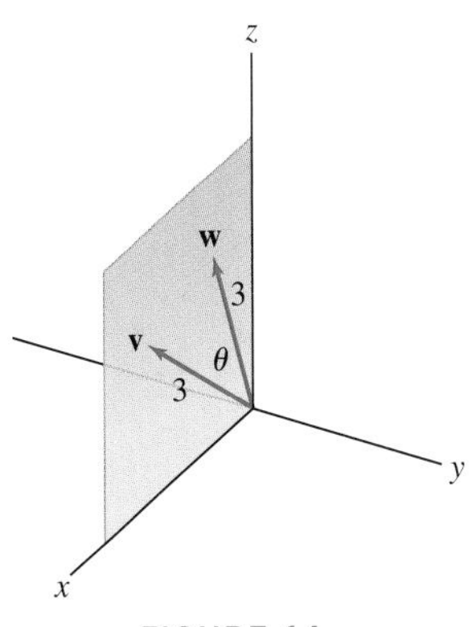

FIGURE 14

In Exercises 29–30, refer to Figure 15.

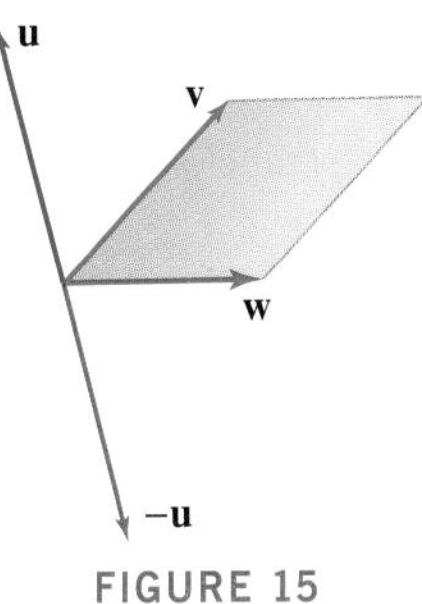

FIGURE 15

29. Which of $\mathbf{u}$ and $-\mathbf{u}$ is equal to $\mathbf{v} \times \mathbf{w}$?

30. Which of the following form a right-handed system?

(a) $\{\mathbf{v}, \mathbf{w}, \mathbf{u}\}$ **(b)** $\{\mathbf{w}, \mathbf{v}, \mathbf{u}\}$ **(c)** $\{\mathbf{v}, \mathbf{u}, \mathbf{w}\}$

(d) $\{\mathbf{u}, \mathbf{v}, \mathbf{w}\}$ **(e)** $\{\mathbf{w}, \mathbf{v}, -\mathbf{u}\}$ **(f)** $\{\mathbf{v}, -\mathbf{u}, \mathbf{w}\}$

31. Let $\mathbf{v} = \langle 3, 0, 0 \rangle$ and $\mathbf{w} = \langle 0, 1, -1 \rangle$. Determine $\mathbf{u} = \mathbf{v} \times \mathbf{w}$ using the geometric properties of the cross product rather than the formula.

32. What are the possible angles θ between two unit vectors $\mathbf{e}$ and $\mathbf{f}$ if $\|\mathbf{e} \times \mathbf{f}\| = \frac{1}{2}$?

33. Show that if $\mathbf{v}$ and $\mathbf{w}$ lie in the yz-plane, then $\mathbf{v} \times \mathbf{w}$ is a multiple of $\mathbf{i}$.

34. Find the two unit vectors orthogonal to both $\mathbf{a} = \langle 3, 1, 1 \rangle$ and $\mathbf{b} = \langle -1, 2, 1 \rangle$.

35. Let $\mathbf{e}$ and $\mathbf{e}'$ be unit vectors in $\mathbf{R}^3$ such that $\mathbf{e} \perp \mathbf{e}'$. Reason geometrically to find the cross product $\mathbf{e} \times (\mathbf{e}' \times \mathbf{e})$.

36. Calculate the force $\mathbf{F}$ on an electron (charge $q = -1.6 \times 10^{-19}$C) moving with velocity 10^5 m/s in the direction $\mathbf{i}$ in a uniform magnetic field $\mathbf{B}$, where $\mathbf{B} = 0.0004\mathbf{i} + 0.0001\mathbf{j}$ (units of teslas) (see Example 5).

37. An electron moving with velocity $\mathbf{v}$ in the plane experiences a force $\mathbf{F} = q(\mathbf{v} \times \mathbf{B})$, where q is the charge on the electron and $\mathbf{B}$ is a uniform magnetic field pointing directly out of the page. Which of the two vectors $\mathbf{F}_1$ or $\mathbf{F}_2$ in Figure 16 represents the force on the electron? Remember that q is negative.

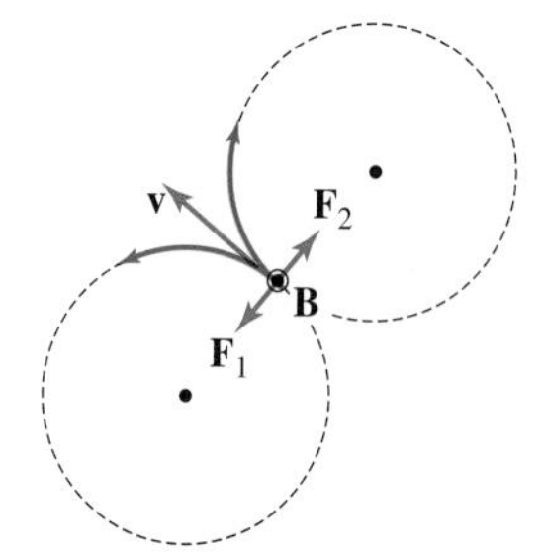

FIGURE 16 The magnetic field vector **B** points directly out of the page.

38. Calculate the scalar triple product $\mathbf{u} \cdot (\mathbf{v} \times \mathbf{w})$, where $\mathbf{u} = \langle 1, 1, 0 \rangle$, $\mathbf{v} = \langle 3, -2, 2 \rangle$, and $\mathbf{w} = \langle 4, -1, 2 \rangle$.

39. Verify identity (10) for vectors $\mathbf{v} = \langle 3, -2, 2 \rangle$ and $\mathbf{w} = \langle 4, -1, 2 \rangle$.

40. Find the volume of the parallelepiped spanned by $\mathbf{u}$, $\mathbf{v}$, and $\mathbf{w}$ in Figure 17.

41. Find the area of the parallelogram spanned by $\mathbf{v}$ and $\mathbf{w}$ in Figure 17.

42. Calculate the volume of the parallelepiped spanned by

$$\mathbf{u} = \langle 2, 2, 1 \rangle, \qquad \mathbf{v} = \langle 1, 0, 3 \rangle, \qquad \mathbf{w} = \langle 0, -4, 0 \rangle$$

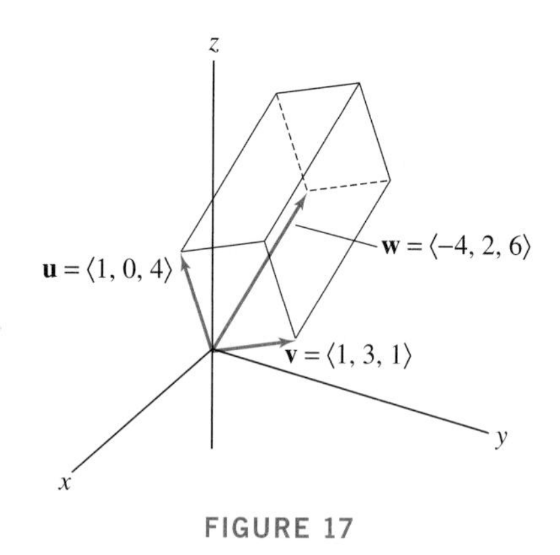

FIGURE 17

43. Sketch and compute the volume of the parallelepiped spanned by

$$\mathbf{u} = \langle 1, 0, 0 \rangle, \qquad \mathbf{v} = \langle 0, 2, 0 \rangle, \qquad \mathbf{w} = \langle 1, 1, 2 \rangle$$

44. Sketch the parallelogram spanned by $\mathbf{u} = \langle 1, 1, 1 \rangle$ and $\mathbf{v} = \langle 0, 0, 4 \rangle$, and compute its area.

45. Calculate the area of the parallelogram spanned by $\mathbf{u} = \langle 1, 0, 3 \rangle$ and $\mathbf{v} = \langle 2, 1, 1 \rangle$.

46. Find the area of the parallelogram determined by the vectors $\langle a, 0, 0 \rangle$ and $\langle 0, b, c \rangle$.

47. Sketch the triangle with vertices O, $P = (3, 3, 0)$, and $Q = (0, 3, 3)$, and compute its area using cross products.

48. Use the cross product to find the area of the triangle with vertices $P = (1, 1, 5)$, $Q = (3, 4, 3)$, and $R = (1, 5, 7)$ (Figure 18).

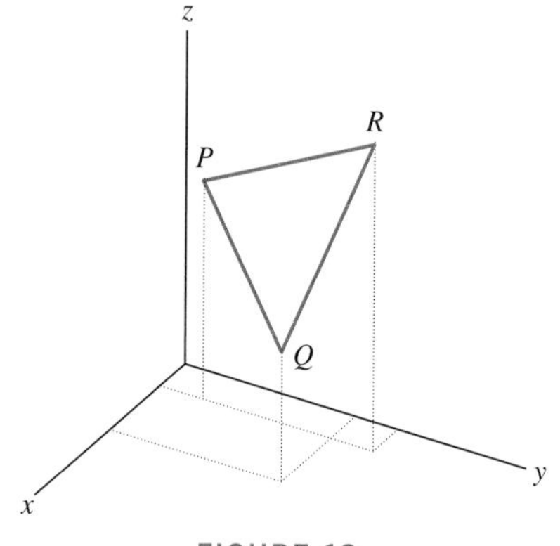

FIGURE 18

In Exercises 49–51, verify the identity using the formula for the cross product.

49. $\mathbf{v} \times \mathbf{w} = -\mathbf{w} \times \mathbf{v}$

50. $(\lambda \mathbf{v}) \times \mathbf{w} = \lambda(\mathbf{v} \times \mathbf{w})$ (λ a scalar)

51. $(\mathbf{u} + \mathbf{v}) \times \mathbf{w} = \mathbf{u} \times \mathbf{w} + \mathbf{v} \times \mathbf{w}$

52. Verify the relations (5).

53. Show that $(\mathbf{i} \times \mathbf{j}) \times \mathbf{j} \neq \mathbf{i} \times (\mathbf{j} \times \mathbf{j})$. Conclude that the Associative Law does not hold for cross products.

54. Use Eq. (9) to verify that $\mathbf{v} \cdot (\mathbf{v} \times \mathbf{w}) = 0$ for all $\mathbf{v}$ and $\mathbf{w}$.

55. Use Theorem 1(ii) to show that $\mathbf{v} \times \mathbf{w} = \mathbf{0}$ if and only if $\mathbf{w} = \lambda \mathbf{v}$ for some scalar λ or $\mathbf{v} = \mathbf{0}$.

56. The components of the cross product have a geometric interpretation. Let $\mathbf{v}_0$, $\mathbf{w}_0$ be the projections onto the xy-plane of vectors $\mathbf{v}$, $\mathbf{w}$ as in Figure 19. Show that the area of the parallelogram spanned by $\mathbf{v}_0$ and $\mathbf{w}_0$ is equal to the absolute value of the $\mathbf{k}$-component of $\mathbf{v} \times \mathbf{w}$.

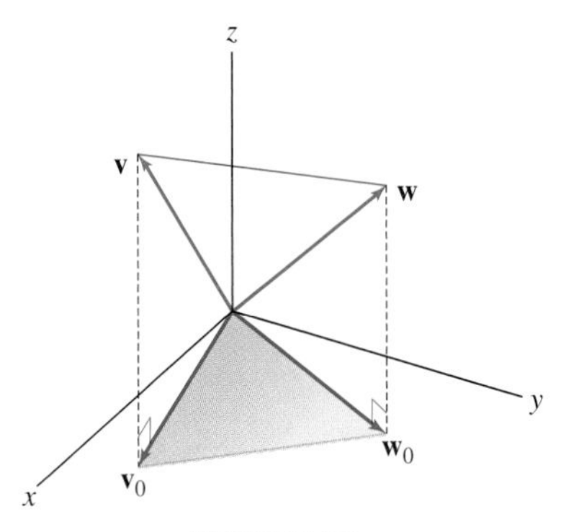

FIGURE 19

57. Formulate and prove analogs of the result in Exercise 56 for the $\mathbf{i}$ and $\mathbf{j}$ components of $\mathbf{v} \times \mathbf{w}$.

58. Determine the angle θ between two vectors $\mathbf{v}$ and $\mathbf{w}$ if $\mathbf{v} \cdot \mathbf{w} = \|\mathbf{v} \times \mathbf{w}\|$.

59. Show that three points P, Q, R are collinear (lie on a line) if and only if $\overrightarrow{PQ} \times \overrightarrow{PR} = \mathbf{0}$.

60. Use the result of Exercise 59 to determine whether the points P, Q, and R are collinear and, if not, find a vector normal to the plane containing them.

(a) $P = (2, 1, 0)$, $Q = (1, 5, 2)$, $R = (-1, 13, 6)$

(b) $P = (2, 1, 0)$, $Q = (-3, 21, 10)$, $R = (5, -2, 9)$

(c) $P = (1, 1, 0)$, $Q = (1, -2, -1)$, $R = (3, 2, -4)$

61. Find a vector $\mathbf{X}$ such that $\langle 1, 1, 1 \rangle \times \mathbf{X} = \langle 1, -1, 0 \rangle$.

62. Explain geometrically why the equation $\langle 1, 1, 1 \rangle \times \mathbf{X} = \langle 1, 0, 0 \rangle$ has no solution.

63. Suppose that vectors $\mathbf{u}$, $\mathbf{v}$, and $\mathbf{w}$ are mutually orthogonal, that is, $\mathbf{u} \perp \mathbf{v}$, $\mathbf{u} \perp \mathbf{w}$, and $\mathbf{v} \perp \mathbf{w}$. Prove that $(\mathbf{u} \times \mathbf{v}) \times \mathbf{w} = \mathbf{0}$ and $\mathbf{u} \times (\mathbf{v} \times \mathbf{w}) = \mathbf{0}$.

Exercises 64–67 deal with torque: When a force $\mathbf{F}$ *acts on an object with position vector* $\mathbf{r}$*, the* ***torque*** *about the origin* O *is the vector quantity* $\tau = \mathbf{r} \times \mathbf{F}$. *If several forces* $\mathbf{F}_j$ *act at positions* $\mathbf{r}_j$*, then the net torque is the sum* $\tau = \sum \mathbf{r}_j \times \mathbf{F}_j$. *By Newton's Laws,* τ *is the rate of change of angular momentum. Torque is a measure of how much the force causes the object to rotate about* O.

64. Calculate the torque τ about O acting at the point P on the mechanical arm in Figure 20(A), assuming that a 25-N force acts as indicated. Ignore the weight of the arm itself.

65. Calculate the net torque about O at P, assuming that a 30-kg mass is attached at P [Figure 20(B)]. The force $\mathbf{F}_g$ due to gravity on a mass m has magnitude $9.8m$ m/s^2 in the downward direction.

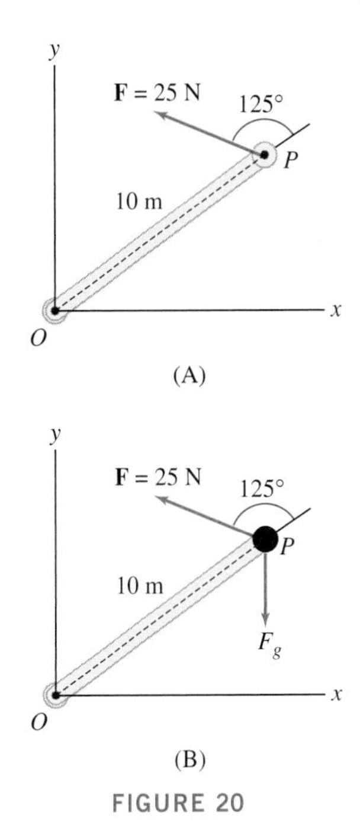

FIGURE 20

66. Let τ be the net torque about O acting on the robotic arm of Figure 21. Assume that the two segments of the arms have weight m_1 and m_2 (in pounds) and that a weight of m_3 pounds is located at the endpoint P. In calculating the torque, we may assume that the entire mass of each arm segment lies at the midpoint of the arm (its center of mass). Show that the position vectors of the masses m_1, m_2, and m_3 are

$$\mathbf{r}_1 = \frac{1}{2}L_1(\sin\theta_1\mathbf{i} + \cos\theta_1\mathbf{j})$$

$$\mathbf{r}_2 = L_1(\sin\theta_1\mathbf{i} + \cos\theta_1\mathbf{j}) + \frac{1}{2}L_2(\sin\theta_2\mathbf{i} - \cos\theta_2\mathbf{j})$$

$$\mathbf{r}_3 = L_1(\sin\theta_1\mathbf{i} + \cos\theta_1\mathbf{j}) + L_2(\sin\theta_2\mathbf{i} - \cos\theta_2\mathbf{j})$$

Then show that

$$\tau = -\left(L_1\left(\frac{1}{2}m_1 + m_2 + m_3\right)\sin\theta_1 + L_2\left(\frac{1}{2}m_2 + m_3\right)\sin\theta_2\right)\mathbf{k}$$

To simplify the computation, note that all three gravitational forces act in the $-\mathbf{j}$ direction, so the $\mathbf{j}$ components of the position vectors $\mathbf{r}_i$ do not contribute to the torque.

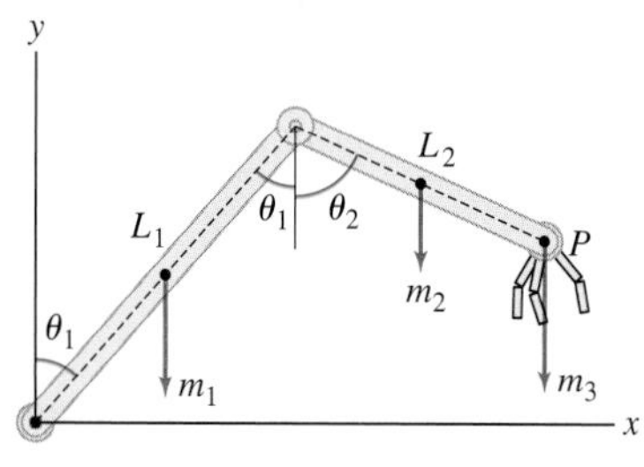

FIGURE 21

67. Continuing with Exercise 66, suppose that $L_1 = 5$ ft, $L_2 = 3$ ft, and $m_1 = 30$ lb, $m_2 = 20$ lb, $m_3 = 50$ lb. If the angles θ_1, θ_2 are equal (say, to θ), what is the maximum allowable value of θ if we assume that the robotic arm can sustain a maximum torque of 400 ft-lb?

Further Insights and Challenges

68. Show that 3×3 determinants can be computed using the **diagonal rule**: Repeat the first two columns of the matrix and form the products of the numbers along the six diagonals indicated. Then add the products for the diagonals that slant from left to right and subtract the products for the diagonals that slant from right to left.

$$\det(A) = \left| \begin{array}{ccc|cc} a_{11} & a_{12} & a_{13} & a_{11} & a_{12} \\ a_{21} & a_{22} & a_{23} & a_{21} & a_{22} \\ a_{31} & a_{32} & a_{33} & a_{31} & a_{32} \\ - & - & - & + & + \quad + \end{array} \right.$$

$$= a_{11}a_{22}a_{33} + a_{12}a_{23}a_{31} + a_{13}a_{21}a_{32}$$

$$- a_{13}a_{22}a_{31} - a_{11}a_{23}a_{32} - a_{12}a_{21}a_{33}$$

69. Use the diagonal rule to calculate $\begin{vmatrix} 2 & 4 & 3 \\ 0 & 1 & -7 \\ -1 & 5 & 3 \end{vmatrix}$.

70. Prove that $\mathbf{v} \times \mathbf{w} = \mathbf{v} \times \mathbf{u}$ if and only if $\mathbf{u} = \mathbf{w} + \lambda\mathbf{v}$ for some scalar λ. Assume that $\mathbf{v} \neq \mathbf{0}$.

71. Use Eq. (10) to prove the Cauchy–Schwarz inequality:

$$|\mathbf{v} \cdot \mathbf{w}| \leq \|\mathbf{v}\| \, \|\mathbf{w}\|$$

Show that equality holds if and only if $\mathbf{w}$ is a multiple of $\mathbf{v}$ or at least one of $\mathbf{v}$ and $\mathbf{w}$ is zero.

72. Show that if $\mathbf{u}$, $\mathbf{v}$, and $\mathbf{w}$ are nonzero vectors and $(\mathbf{u} \times \mathbf{v}) \times \mathbf{w} = \mathbf{0}$, then either: (i) $\mathbf{u}$ and $\mathbf{v}$ are parallel, or (ii) $\mathbf{w}$ is orthogonal to $\mathbf{u}$ and $\mathbf{v}$.

73. Suppose that $\mathbf{u}$, $\mathbf{v}$, $\mathbf{w}$ are nonzero and

$$(\mathbf{u} \times \mathbf{v}) \times \mathbf{w} = \mathbf{u} \times (\mathbf{v} \times \mathbf{w}) = \mathbf{0}$$

Show that $\mathbf{u}$, $\mathbf{v}$, and $\mathbf{w}$ are either mutually parallel or mutually perpendicular. *Hint:* Use Exercise 72.

74. Let $\mathbf{a}$, $\mathbf{b}$, $\mathbf{c}$ be nonzero vectors and set

$$\mathbf{v} = \mathbf{a} \times (\mathbf{b} \times \mathbf{c}), \qquad \mathbf{w} = (\mathbf{a} \cdot \mathbf{c})\mathbf{b} - (\mathbf{a} \cdot \mathbf{b})\mathbf{c}$$

(a) Prove that

(i) $\mathbf{v}$ lies in the plane spanned by $\mathbf{b}$ and $\mathbf{c}$.

(ii) $\mathbf{v}$ is orthogonal to $\mathbf{a}$.

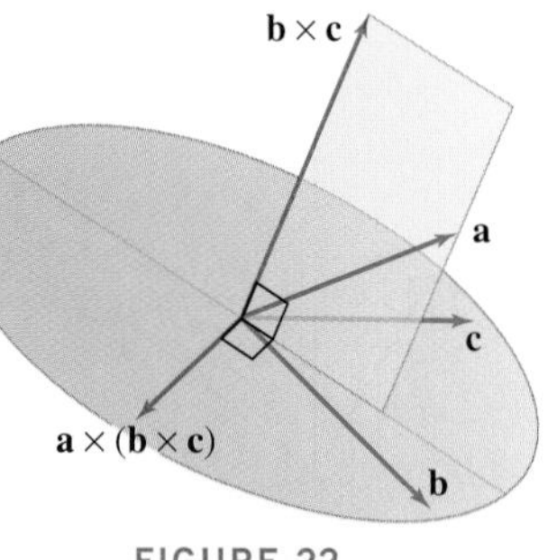

FIGURE 22

(b) Prove that $\mathbf{w}$ also satisfies (i) and (ii). Conclude that $\mathbf{v}$ and $\mathbf{w}$ are parallel.

(c) Show algebraically that $\mathbf{v} = \mathbf{w}$ (Figure 22).

75. Use Exercise 74 to prove the identity

$$(\mathbf{a} \times \mathbf{b}) \times \mathbf{c} - \mathbf{a} \times (\mathbf{b} \times \mathbf{c}) = (\mathbf{a} \cdot \mathbf{b})\mathbf{c} - (\mathbf{b} \cdot \mathbf{c})\mathbf{a}$$

76. Assume that $\mathbf{v}$ and $\mathbf{w}$ are in the first quadrant in $\mathbf{R}^2$ as in Figure 23. Use geometry to prove that the area of the parallelogram is equal to $\det\begin{pmatrix} \mathbf{v} \\ \mathbf{w} \end{pmatrix}$.

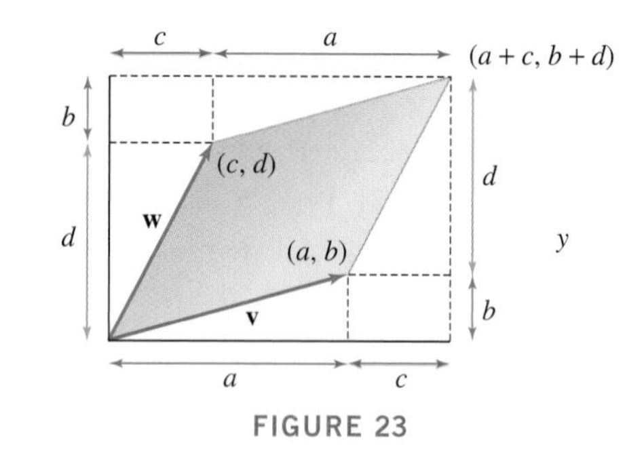

FIGURE 23

77. Show that if $\mathbf{a}$, $\mathbf{b}$ are nonzero vectors such that $\mathbf{a} \perp \mathbf{b}$, then there exists a vector $\mathbf{x}$ such that

$$\mathbf{x} \times \mathbf{a} = \mathbf{b} \qquad \boxed{13}$$

Hint: Show that if $\mathbf{x}$ is orthogonal to $\mathbf{b}$ and is not a multiple of $\mathbf{a}$, then $\mathbf{x} \times \mathbf{a}$ is a multiple of $\mathbf{b}$.

78. Show that if $\mathbf{a}, \mathbf{b}$ are nonzero vectors such that $\mathbf{a} \perp \mathbf{b}$, then the set of all solutions of Eq. (13) is a line with $\mathbf{a}$ as direction vector. *Hint:* Let $\mathbf{x}_0$ be any solution (which exists by Exercise 77) and show that every other solution is of the form $\mathbf{x}_0 + \lambda\mathbf{a}$ for some scalar λ.

79. Consider the tetrahedron spanned by vectors $\mathbf{a}$, $\mathbf{b}$, and $\mathbf{c}$ as in Figure 24(A). Let A, B, C be the faces containing the origin O and let D be the fourth face opposite O. For each face F, let $\mathbf{v}_F$ be the vector normal to the face, pointing outside the tetrahedron, of magnitude equal to twice the area of F. Prove the relations

$$\mathbf{v}_A + \mathbf{v}_B + \mathbf{v}_C = \mathbf{a} \times \mathbf{b} + \mathbf{b} \times \mathbf{c} + \mathbf{c} \times \mathbf{a}$$

$$\mathbf{v}_A + \mathbf{v}_B + \mathbf{v}_C + \mathbf{v}_D = 0$$

Hint: Show that $\mathbf{v}_D = (\mathbf{c} - \mathbf{b}) \times (\mathbf{b} - \mathbf{a})$.

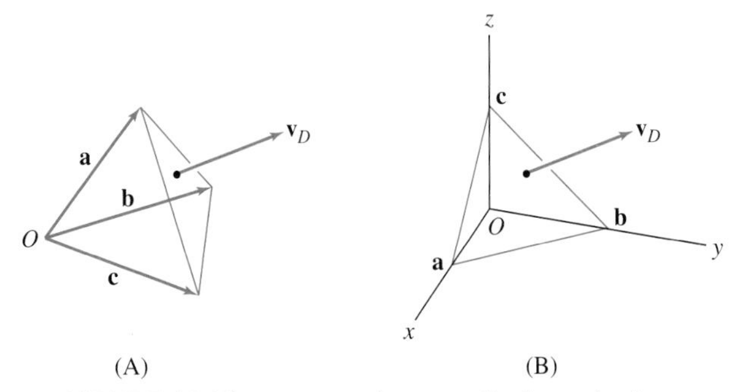

FIGURE 24 The vector $\mathbf{v}_D$ is perpendicular to the face.

80. In the notation of Exercise 79, suppose that $\mathbf{a}, \mathbf{b}, \mathbf{c}$ are mutually perpendicular as in Figure 24(B). Let S_F be the area of face F. Prove the following three-dimensional version of the Pythagorean Theorem:

$$S_A^2 + S_B^2 + S_C^2 = S_D^2$$

12.5 Planes in Three-Space

As we know, a line $\mathbf{R}^2$ is defined by a linear equation $ax + by = c$. In this section, we discuss planes in $\mathbf{R}^3$. Our first goal is to show that a plane is defined by a linear equation $ax + by + cz = d$ in three variables.

← *REMINDER the term "normal" is another word for "perpendicular."*

Consider a plane $\mathcal{P}$ passing through a point $P_0 = (x_0, y_0, z_0)$ as in Figure 1. Although there are infinitely many planes through P_0, we can identify the particular plane $\mathcal{P}$ by specifying a vector $\mathbf{n} = \langle a, b, c \rangle$ normal to $\mathcal{P}$. Only one plane through P_0 is perpendicular to $\mathbf{n}$. Furthermore, we see in Figure 2 that a point $P = (x, y, z)$ lies on $\mathcal{P}$ if and only if $\overrightarrow{P_0P}$ is perpendicular to $\mathbf{n}$. Thus, the dot product is zero:

$$\mathbf{n} \cdot \overrightarrow{P_0P} = 0 \quad \boxed{1}$$

To write this out explicitly, let $\mathbf{r}_0 = \langle x_0, y_0, z_0 \rangle$ and $\mathbf{r} = \langle x, y, z \rangle$ be the position vectors of P_0 and P. Then

$$\overrightarrow{P_0P} = \mathbf{r} - \mathbf{r}_0 = \langle x - x_0, y - y_0, z - z_0 \rangle$$

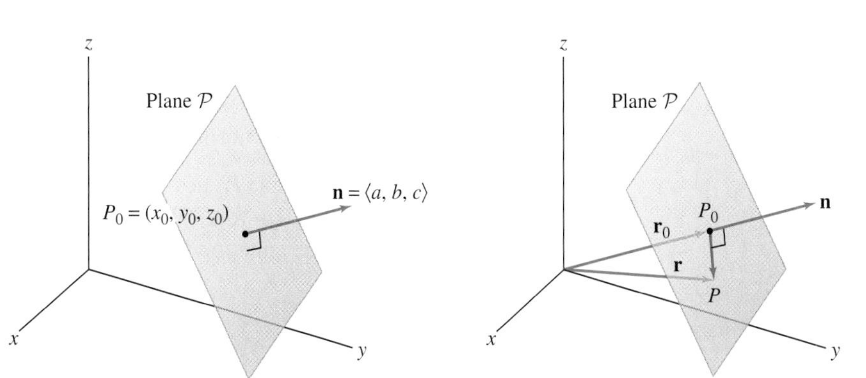

FIGURE 1 A plane through P_0 is uniquely determined by the normal vector $\mathbf{n}$.

FIGURE 2 A point P lies on the plane $\mathcal{P}$ if $\overrightarrow{P_0P}$ is orthogonal to the normal vector $\mathbf{n}$.

Equation (1) becomes

$$\mathbf{n} \cdot (\mathbf{r} - \mathbf{r}_0) = 0 \quad \text{or} \quad \langle a, b, c \rangle \cdot \langle x - x_0, y - y_0, z - z_0 \rangle = 0$$

This yields the equation

$$a(x - x_0) + b(y - y_0) + c(z - z_0) = 0$$

Alternatively, we may write $\mathbf{n} \cdot (\mathbf{r} - \mathbf{r}_0) = 0$ as

$$\mathbf{n} \cdot \mathbf{r} = \mathbf{n} \cdot \mathbf{r}_0 \quad \text{or} \quad \langle a, b, c \rangle \cdot \langle x, y, z \rangle = \langle a, b, c \rangle \cdot \langle x_0, y_0, z_0 \rangle$$

Keep in mind that the equation of a plane is not unique since we are free to multiply the equation by any nonzero scalar. This amounts to multiplying $\mathbf{n}$ *by a nonzero scalar.*

This yields the equivalent equation

$$ax + by + cz = \underbrace{ax_0 + by_0 + cz_0}_{\text{Call this } d}$$

THEOREM 1 Equation of a Plane in Point-Normal Form An equation of the plane through $P_0 = (x_0, y_0, z_0)$ with (nonzero) normal vector $\mathbf{n} = \langle a, b, c \rangle$ can be written in the following three ways:

Vector form: $$\mathbf{n} \cdot \langle x, y, z \rangle = d \qquad (2)$$

Scalar forms: $$a(x - x_0) + b(y - y_0) + c(z - z_0) = 0 \qquad (3)$$

$$ax + by + cz = d \qquad (4)$$

where $d = \mathbf{n} \cdot \langle x_0, y_0, z_0 \rangle = ax_0 + by_0 + cz_0$.

EXAMPLE 1 Find an equation of the plane through $P_0 = (3, 1, 0)$ with normal vector $\mathbf{n} = \langle 3, 2, -5 \rangle$.

Solution By Eq. (2), the vector equation is

$$\underbrace{\langle 3, 2, -5 \rangle \cdot \langle x, y, z \rangle}_{\mathbf{n} \cdot \langle x, y, z \rangle} = \underbrace{\langle 3, 2, -5 \rangle \cdot \langle 3, 1, 0 \rangle}_{d} = 11$$

This yields the scalar equation $3x + 2y - 5z = 11$. We can also write the equation in the form (3)

$$3(x - 3) + 2(y - 1) - 5z = 0$$

■

EXAMPLE 2 An Equation for the xy-Plane Find an equation of the plane $\mathcal{P}$ through $P_0 = (1, 2, 0)$ with normal vector $\mathbf{n} = \langle 0, 0, 3 \rangle$.

Solution Note that P_0 lies in the xy-plane and $\mathbf{n}$ points along the z-axis, orthogonal to the xy-plane. Therefore, $\mathcal{P}$ must be the xy-plane itself (Figure 3). We confirm this by determining the equation in vector form:

$$\mathbf{n} \cdot \langle x, y, z \rangle = d = \mathbf{n} \cdot \langle 1, 2, 0 \rangle$$

$$\langle 0, 0, 3 \rangle \cdot \langle x, y, z \rangle = \langle 0, 0, 3 \rangle \cdot \langle 1, 2, 0 \rangle$$

$$3z = 0 \quad \text{or} \quad z = 0$$

As expected, $z = 0$ is the equation of the xy-plane. ■

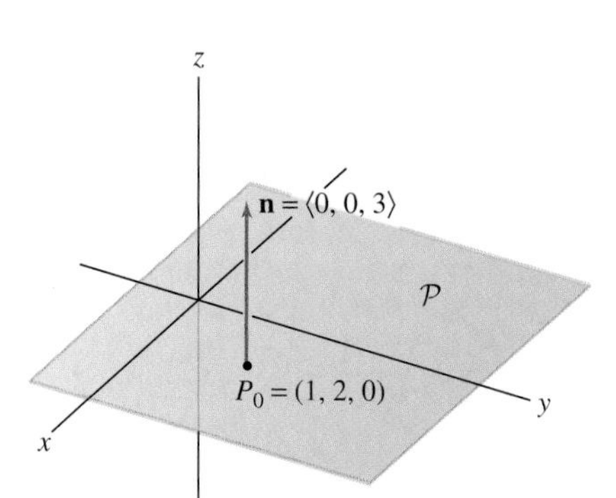

FIGURE 3 The xy-plane can be described as the plane through $P_0 = (1, 2, 0)$ with normal vector $\mathbf{n} = \langle 0, 0, 3 \rangle$.

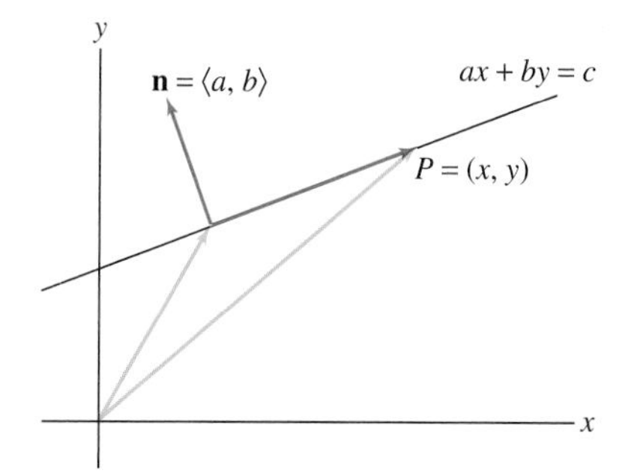

FIGURE 4 The line with normal vector **n**.

CONCEPTUAL INSIGHT The direction of a line in the plane is specified by the slope m, which is a number. By contrast, it takes a vector rather than a single number to specify the orientation of a plane in $\mathbf{R}^3$. Thus, the normal vector serves as a substitute for the slope. Also, note that a line in $\mathbf{R}^2$ may be described via normal vectors: The line with equation $ax + by = c$ has normal vector $\mathbf{n} = \langle a, b \rangle$ (Figure 4).

Suppose we are given a linear equation

$$ax + by + cz = d$$

This equation defines a plane $\mathcal{P}$ by Theorem 1. Furthermore, the vector of coefficients $\mathbf{n} = \langle a, b, c \rangle$ is a normal vector to $\mathcal{P}$. For example, the plane $2x - 5y + z = 2$ has normal vector $\mathbf{n} = \langle 2, -5, 1 \rangle$. Now observe that two planes are *parallel* if and only if they have the same normal vector. Therefore, the planes parallel to $\mathcal{P}$ are precisely the planes with equation $ax + by + cz = d'$ for some d'. The unique plane parallel to $\mathcal{P}$ passing through the origin has equation $ax + by + cz = 0$.

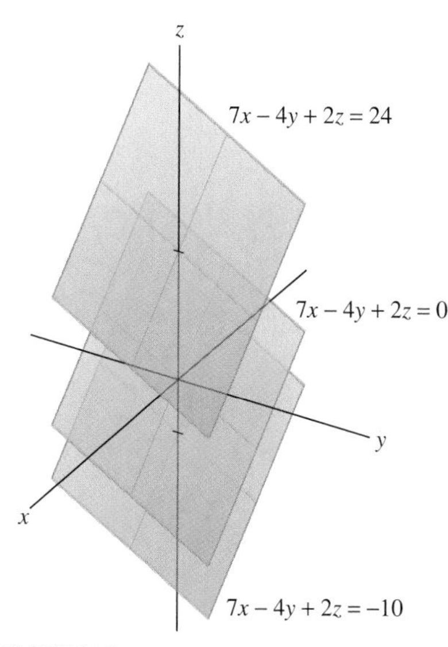

FIGURE 5

EXAMPLE 3 Planes Parallel to a Given Plane Let $\mathcal{P}$ be the plane with equation $7x - 4y + 2z = -10$. Find an equation of the plane parallel to $\mathcal{P}$ and passing through:

(a) The origin. **(b)** The point $Q = (2, -1, 3)$.

Solution As noted in the previous paragraph, the planes parallel to $\mathcal{P}$ are the planes with an equation of the form

$$7x - 4y + 2z = d' \qquad \boxed{5}$$

(a) The plane parallel to $\mathcal{P}$ passing through the origin has equation $7x - 4y + 2z = 0$.

(b) The point $Q = (2, -1, 3)$ satisfies Eq. (5) with

$$d' = 7(2) - 4(-1) + 2(3) = 24$$

Therefore, the plane parallel to $\mathcal{P}$ passing through Q has equation $7x - 4y + 2z = 24$ (Figure 5). ■

Another way of specifying a plane is to choose three points P, Q, and R that do not lie on a line. In this case, there is a unique plane passing through P, Q, and R (Figure 6). The next example shows how to find an equation of this plane.

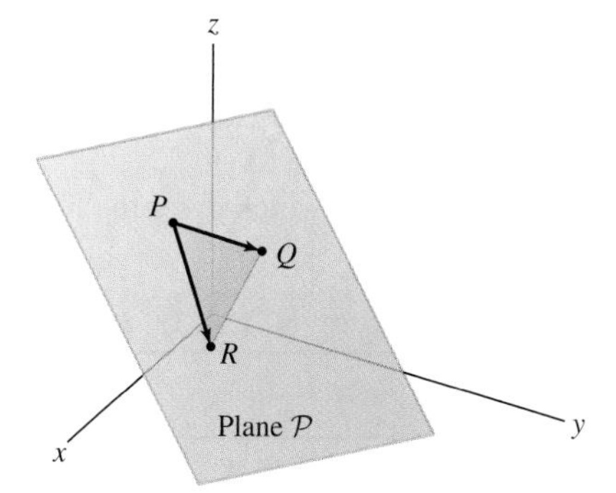

FIGURE 6 Three points P, Q, and R determine a plane (assuming they do not lie in a straight line).

In Example 4, we could just as well have used the vectors $\mathbf{a} = \overrightarrow{QP}$ *and* $\mathbf{b} = \overrightarrow{QR}$ *(or* $\overrightarrow{RP}$ *and* $\overrightarrow{RQ}$*) to find a normal vector* $\mathbf{n}$.

EXAMPLE 4 Equation of a Plane Determined by Three Points Find an equation of the plane $\mathcal{P}$ determined by the points

$$P = (1, 0, -1), \quad Q = (2, 2, 1), \quad R = (4, 1, 2)$$

Solution

***Step 1.* Find a normal vector.**

The vectors $\mathbf{a} = \overrightarrow{PQ}$ and $\mathbf{b} = \overrightarrow{PR}$ lie in $\mathcal{P}$ (Figure 7), so the cross product $\mathbf{n} = \mathbf{a} \times \mathbf{b}$ is normal to $\mathcal{P}$:

$$\mathbf{a} = \overrightarrow{PQ} = \langle 2, 2, 1 \rangle - \langle 1, 0, -1 \rangle = \langle 1, 2, 2 \rangle$$

$$\mathbf{b} = \overrightarrow{PR} = \langle 4, 1, 2 \rangle - \langle 1, 0, -1 \rangle = \langle 3, 1, 3 \rangle$$

$$\mathbf{n} = \mathbf{a} \times \mathbf{b} = \begin{vmatrix} \mathbf{i} & \mathbf{j} & \mathbf{k} \\ 1 & 2 & 2 \\ 3 & 1 & 3 \end{vmatrix} = 4\mathbf{i} + 3\mathbf{j} - 5\mathbf{k} = \langle 4, 3, -5 \rangle$$

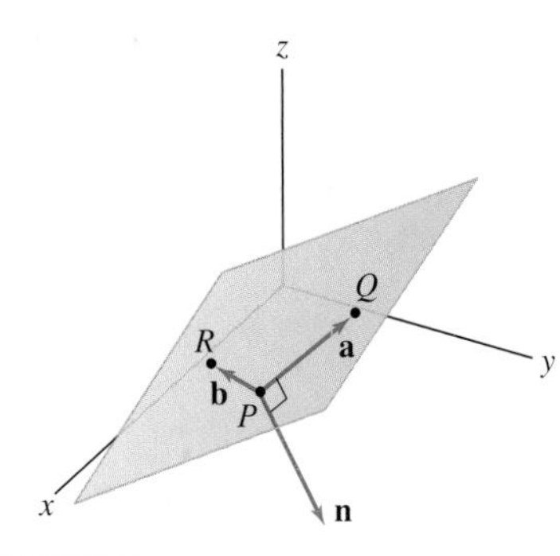

FIGURE 7

CAUTION When you find a normal vector to the plane containing points P, Q, R, be sure to compute a cross product such as $\overrightarrow{PQ} \times \overrightarrow{PR}$. A common mistake is to use a cross product such as $\overrightarrow{OP} \times \overrightarrow{OQ}$ or $\overrightarrow{OP} \times \overrightarrow{OR}$, which need not be normal to the plane.

***Step 2.* Choose a point on the plane and compute d.**
Now choose any one of the three points on the plane, say, $P = (1, 0, -1)$. By Eq. (4), the vector equation of $\mathcal{P}$ is $4x + 3y - 5z = d$, where

$$d = \mathbf{n} \cdot \overrightarrow{OP} = \langle 4, 3, -5 \rangle \cdot \langle 1, 0, -1 \rangle = 9$$

Therefore, $\mathcal{P}$ has equation $4x + 3y - 5z = 9$. ■

■ EXAMPLE 5 Intersection of a Plane and a Line Find the intersection of the plane $3x - 9y + 2z = 7$ and the line $\mathbf{r}(t) = \langle 1, 2, 1 \rangle + t \langle -2, 0, 1 \rangle$.

Solution The line has parametric equations

$$x = 1 - 2t, \qquad y = 2, \qquad z = 1 + t$$

We want to find a value of t for which $\mathbf{r}(t)$ lies on the plane, so we substitute the parametric equations in the equation of the plane and solve for t:

$$\begin{aligned} 3x - 9y + 2z &= 7 \\ 3(1 - 2t) - 9(2) + 2(1 + t) &= 7 \\ -4t - 13 &= 7 \qquad \text{(simplify)} \end{aligned}$$

The last equation yields $-4t = 20$ or $t = -5$. Therefore, the point P of intersection has coordinates

$$x = 1 - 2(-5) = 11, \qquad y = 2, \qquad z = 1 + (-5) = -4$$

and thus, $P = (11, 2, -4)$. ■

The intersection of a plane $\mathcal{P}$ with a coordinate plane or a plane parallel to a coordinate plane is called a **trace**. The trace is a line unless $\mathcal{P}$ is parallel to the coordinate plane (in which case the trace is empty or $\mathcal{P}$ itself).

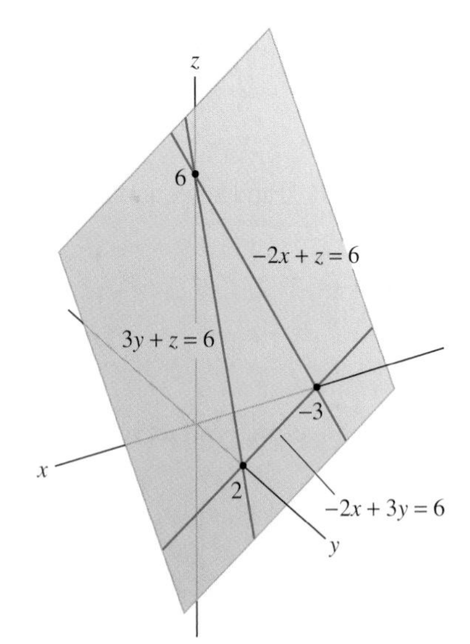

FIGURE 8 Traces of plane $-2x + 3y + z = 6$ in the coordinate planes.

■ EXAMPLE 6 Traces of the Plane Find the traces of the plane $-2x + 3y + z = 6$ in the coordinate planes.

Solution We obtain the trace in the xy-plane by setting $z = 0$ in the equation of the plane. Thus, the trace is the line $-2x + 3y = 6$ in the xy-plane (Figure 8).

Similarly, the trace in the xz-plane is obtained by setting $y = 0$, which gives the line $-2x + z = 6$ in the xz-plane. Finally, the trace in the yz-plane is $3y + z = 6$. ■

12.5 SUMMARY

- An equation of the plane through the point $P_0 = (x_0, y_0, z_0)$ with normal vector $\mathbf{n} = \langle a, b, c \rangle$ can be written in the following three ways:

Vector form:	$\mathbf{n} \cdot \langle x, y, z \rangle = d$
Scalar forms:	$a(x - x_0) + b(y - y_0) + c(z - z_0) = 0$
	$ax + by + cz = d$

where $d = \mathbf{n} \cdot \langle x_0, y_0, z_0 \rangle = ax_0 + by_0 + cz_0$.

- A linear equation $ax + by + cz = d$ defines a plane $\mathcal{P}$ with normal vector $\mathbf{n} = \langle a, b, c \rangle$. The planes parallel to $\mathcal{P}$ are the planes with equation $ax + by + cz = d'$ for some d'.
- To find an equation of the plane through three points P, Q, R not on a line, first compute a normal vector as a cross product $\mathbf{n} = \overrightarrow{PQ} \times \overrightarrow{PR}$. Then, if $P = (x_0, y_0, z_0)$, an equation is $\mathbf{n} \cdot \langle x, y, z \rangle = d$, where $d = \mathbf{n} \cdot \langle x_0, y_0, z_0 \rangle$.
- The intersection of a plane $\mathcal{P}$ with a coordinate plane or a plane parallel to a coordinate plane is called a *trace*. The trace in the yz-plane is obtained by setting $x = 0$ in the equation of the plane (and similarly for the xz and xy traces).

12.5 EXERCISES

Preliminary Questions

1. What is the equation of the plane parallel to $3x + 4 - z = 5$ passing through the origin?

2. The vector $\mathbf{k}$ is normal to which of the following planes?

(a) $x = 1$ **(b)** $y = 1$ **(c)** $z = 1$

3. Which of the following planes are not parallel to the plane $x + y + z = 1$?

(a) $2x + 2y + 2z = 1$ **(b)** $x + y + z = 3$

(c) $x - y + z = 0$

4. To which coordinate plane is the plane $y = 1$ parallel?

5. Which of the following planes contains the z-axis?

(a) $z = 1$ **(b)** $x + y = 1$ **(c)** $x + y = 0$

6. Suppose that a plane $\mathcal{P}$ with normal vector $\mathbf{n}$ and a line $\mathcal{L}$ with direction vector $\mathbf{v}$ both pass through the origin and that $\mathbf{n} \cdot \mathbf{v} = 0$. Which of the following statements is correct?

(a) $\mathcal{L}$ is contained in the $\mathcal{P}$.

(b) $\mathcal{L}$ is orthogonal to $\mathcal{P}$.

Exercises

In Exercises 1–8, write an equation of the plane with normal vector $\mathbf{n}$ *passing through the given point in vector form and the scalar forms (3) and (4).*

1. $\mathbf{n} = \langle 1, 3, 2 \rangle$, $(4, -1, 1)$

2. $\mathbf{n} = \langle -1, 2, 1 \rangle$, $(3, 1, 9)$

3. $\mathbf{n} = \langle -1, 2, 1 \rangle$, $(4, 1, 5)$

4. $\mathbf{n} = \langle 2, -4, 1 \rangle$, $\left(\frac{1}{3}, \frac{2}{3}, 1\right)$

5. $\mathbf{n} = \mathbf{i}$, $(3, 1, -9)$ **6.** $\mathbf{n} = \mathbf{j}$, $\left(-5, \frac{1}{2}, \frac{1}{2}\right)$

7. $\mathbf{n} = \mathbf{k}$, $(6, 7, 2)$ **8.** $\mathbf{n} = \mathbf{i} - \mathbf{k}$, $(4, 2, -8)$

9. Find an equation of any plane with normal $\mathbf{n} = \langle 2, 1, 1 \rangle$ *not* passing through the point $P = (0, 1, 1)$.

10. Find an equation of any plane passing through the origin but not passing through $P = (2, 1, 1)$.

In Exercises 11–14, find an equation of the plane passing through the three points given.

11. $P = (2, -1, 4)$, $Q = (1, 1, 1)$, $R = (3, 1, -2)$

12. $P = (5, 1, 1)$, $Q = (1, 1, 2)$, $R = (2, 1, 1)$

13. $P = (1, 0, 0)$, $Q = (0, 1, 1)$, $R = (2, 0, 1)$

14. $P = (2, 0, 0)$, $Q = (0, 4, 0)$, $R = (0, 0, 2)$

In Exercises 15–20, find a vector normal to the plane with the given equation.

15. $9x - 4y - 11z = 2$ **16.** $3x + 25y + 9z = 7$

17. $3x + 25y + 9z = 45$ **18.** $4y - 8z = -1$

19. $x - z = 0$ **20.** $x = 1$

In Exercises 21–26, find the equation of the plane with the given description.

21. Passes through O and is parallel to $4x - 9y + z = 3$

22. Passes through $(4, 1, 9)$ and is parallel to $x + y + z = 3$

23. Passes through $(4, 1, 9)$ and is parallel to $x = 3$

24. Passes through $(-2, -3, 5)$ and has normal vector $\mathbf{i} + \mathbf{k}$

25. Contains the lines $\mathbf{r}_1(t) = \langle t, 2t, 3t \rangle$ and $\mathbf{r}_2(t) = \langle 3t, t, 8t \rangle$

26. Contains $P = (-1, 0, 1)$ and $\mathbf{r}(t) = \langle t + 1, 2t, 3t - 1 \rangle$

27. Are the planes $\frac{1}{2}x + 2x - y = 5$ and $3x + 12x - 6y = 1$ parallel?

28. Let a, b, c be constants. Which two of the following equations define the plane passing through $(a, 0, 0)$, $(0, b, 0)$, $(0, 0, c)$?

(a) $ax + by + cz = 1$

(b) $bcx + acy + abz = abc$

(c) $bx + cy + az = 1$

(d) $\frac{x}{a} + \frac{y}{b} + \frac{z}{c} = 1$

29. Find an equation of the plane $\mathcal{P}$ in Figure 9.

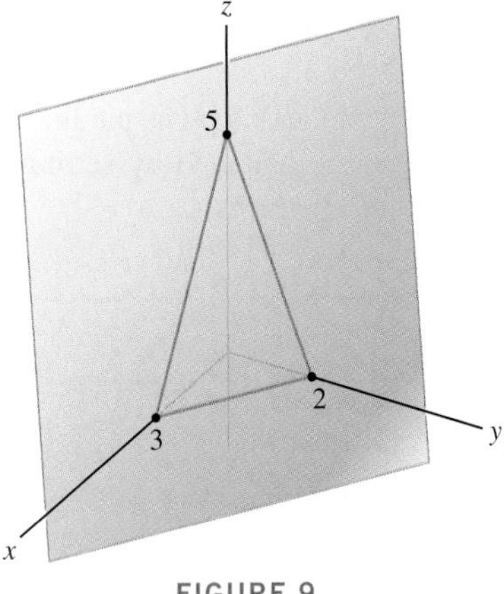

FIGURE 9

In Exercises 30–33, find the intersection of the line and plane.

30. $x + y + z = 14, \quad \mathbf{r}(t) = \langle 1, 1, 0\rangle + t\,\langle 0, 2, 4\rangle$

31. $2x + y = 3, \quad \mathbf{r}(t) = \langle 2, -1, -1\rangle + t\,\langle 1, 2, -4\rangle$

32. $z = 12, \quad \mathbf{r}(t) = t\,\langle -6, 9, 36\rangle$

33. $x - z = 6, \quad \mathbf{r}(t) = \langle 1, 0, -1\rangle + t\,\langle 4, 9, 2\rangle$

34. Verify that the plane $x - y + 5z = 10$ and the line $\mathbf{r}(t) = \langle 1, 0, 1\rangle + t\,\langle -2, 1, 1\rangle$ intersect at $P = (-3, 2, 3)$.

35. The plane $3x - 4y + 2z = 8$ intersects the xy-plane in a line. What is the equation of this line?

In Exercises 36–41, find the trace of the plane in the given coordinate plane.

36. $3x - 9y + 4z = 5, \quad yz$

37. $3x - 9y + 4z = 5, \quad xz$

38. $3x + 4z = -2, \quad xy$

39. $3x + 4z = -2, \quad xz$

40. $-x + y = 4, \quad xz$

41. $-x + y = 4, \quad yz$

42. Does the plane $x = 5$ have a trace in the yz-plane? Explain.

43. Give equations for two distinct planes whose trace in the xy-plane has equation $4x + 3y = 8$.

44. Find parametric equations for the line through $P_0 = (3, -1, 1)$ perpendicular to the plane $3x + 5y - 7z = 29$.

45. Find all planes in $\mathbf{R}^3$ whose intersection with the xy-plane is the line $\mathbf{r}(t) = t\,\langle 2, 1, 0\rangle$.

46. Find all planes in $\mathbf{R}^3$ whose intersection with the xz-plane is the line with equation $3x + 2z = 5$.

47. Let $\mathcal{P}$ be the plane $\mathbf{n} \cdot \langle x, y, z\rangle = d$, where $\mathbf{n} \neq \mathbf{0}$, and let $\mathcal{P}_1$ be the parallel plane $\mathbf{n} \cdot \langle x, y, z\rangle = d_1$ (Figure 10).

(a) Show that the line through $\mathbf{n}$ intersects $\mathcal{P}$ at the terminal point of the vector $\left(\dfrac{d}{\|\mathbf{n}\|}\right)\mathbf{e_n}$.

(b) Show that the distance between $\mathcal{P}$ and $\mathcal{P}_1$ is $\dfrac{|d - d_1|}{\|\mathbf{n}\|}$.

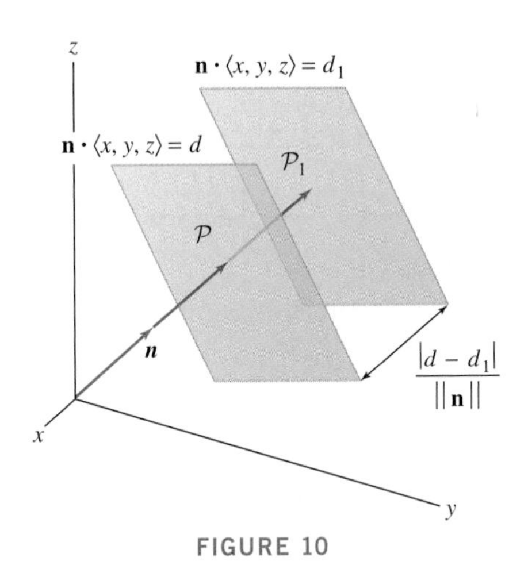

FIGURE 10

48. Use Exercise 47 to compute the distance between the parallel planes $x + 2y + 3z = 5$ and $x + 2y + 3z = 9$.

In Exercises 49–54, compute the angle between the two planes, defined as the angle θ (between 0 and π) between their normal vectors (Figure 11).

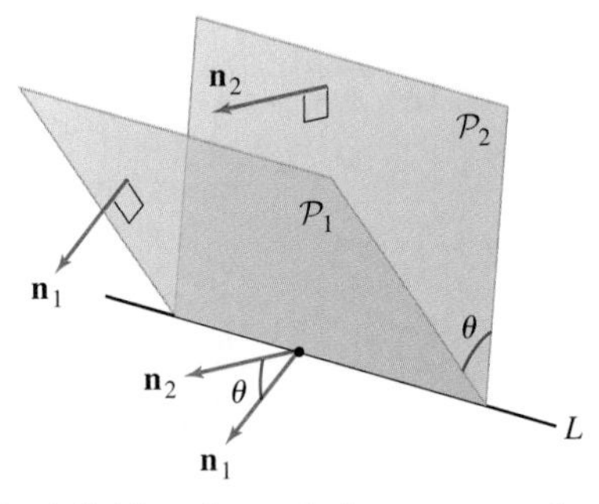

FIGURE 11 By definition, the angle between two planes is the angle between their normal vectors.

49. Planes with normals $\mathbf{n}_1 = \langle 1, 0, 1\rangle$, $\mathbf{n}_2 = \langle -1, 1, 1\rangle$

50. Planes with normals $\mathbf{n}_1 = \langle 1, 2, 1\rangle$, $\mathbf{n}_2 = \langle 4, 1, 3\rangle$

51. Planes with normals $\mathbf{n}_1 = \langle 1, 0, 0\rangle$, $\mathbf{n}_2 = \langle 0, 1, 0\rangle$

52. $2x + 3y + 7z = 2$ and $4x - 2y + 2z = 4$

53. $x - 3y + z = 3$ and $2x - 3z = 4$

54. The plane through $(1, 0, 0)$, $(0, 1, 0)$, and $(0, 0, 1)$ and the yz-plane

55. Find an equation of a plane making an angle of $\frac{\pi}{2}$ with the plane $3x + y - 4z = 2$.

56. Let $\mathcal{P}_1$ and $\mathcal{P}_2$ be planes with normal vectors $\mathbf{n}_1$ and $\mathbf{n}_2$. Assume that the planes are not parallel and let $\mathcal{L}$ be their intersection (a line). Show that $\mathbf{n}_1 \times \mathbf{n}_2$ is a direction vector for $\mathcal{L}$.

57. Find a plane that is perpendicular to the two planes $x + y = 3$ and $x + 2y - z = 4$.

58. Let $\mathcal{L}$ be the intersection of the planes $x + y + z = 1$ and $x + 2y + 3z = 1$. Use Exercise 56 to find a direction vector for $\mathcal{L}$. Then find a point P on $\mathcal{L}$ by *inspection* and write down the parametric equations for $\mathcal{L}$.

59. Let $\mathcal{L}$ be the intersection of the planes $x - y - z = 1$ and $2x + 3y + z = 2$. Find parametric equations for the line $\mathcal{L}$. *Hint:* To find a point $\mathcal{L}$, substitute an arbitrary value for z (say, $z = 2$) and then solve the resulting pair of equations for x and y.

60. Find parametric equations for the intersection of the planes $2x + y - 3z = 0$ and $x + y = 1$.

61. Find parametric equations for the line that is perpendicular to the plane $3x + 5y - 7z = 29$ and passes through the point $P_0 = (3, -1, 1)$.

Further Insights and Challenges

In Exercises 62–72, if $\mathcal{P}$ is a plane and Q is a point not lying on $\mathcal{P}$, then the nearest point to Q on $\mathcal{P}$ is the unique point P on $\mathcal{P}$ such that $\overline{PQ}$ is orthogonal to $\mathcal{P}$ (Figure 12).

62. Let $\mathcal{P}$ be the plane $ax + by + cz = d$ with normal $\mathbf{n} = \langle a, b, c \rangle$ and let Q be a point not on $\mathcal{P}$. Show that the point P on $\mathcal{P}$ closest to Q is determined by the equation

$$\overrightarrow{OP} = \overrightarrow{OQ} + \left(\frac{d - \overrightarrow{OQ} \cdot \mathbf{n}}{\mathbf{n} \cdot \mathbf{n}} \right) \mathbf{n} \qquad \boxed{6}$$

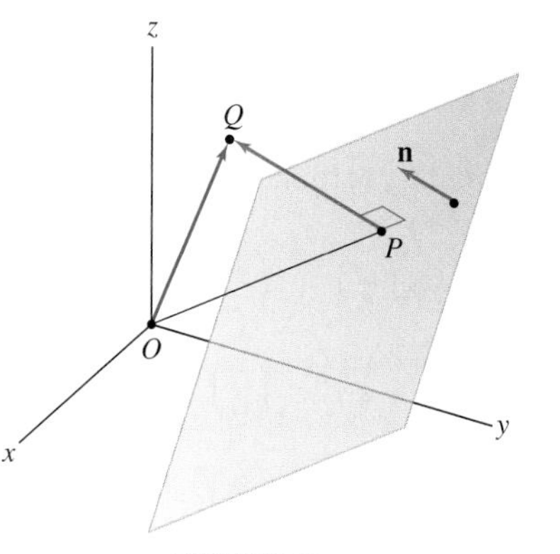

FIGURE 12

63. Use Eq. (6) to find the nearest point to $Q = (2, 1, 2)$ on the plane $x + y + z = 1$.

64. Find the nearest point to $Q = (-1, 3, -1)$ on the plane

$$x - 4z = 2$$

65. In the notation of Exercise 62, the *distance* from Q to $\mathcal{P}$ is the length $\|\overrightarrow{PQ}\|$. Show that if $Q = (x_1, y_1, z_1)$, then

$$\text{Distance from } Q \text{ to } \mathcal{P} = \frac{|ax_1 + by_1 + cz_1 - d|}{\|\mathbf{n}\|} \qquad \boxed{7}$$

66. Use Eq. (7) to find the distance from $Q = (1, 1, 1)$ to the plane $2x + y + 5z = 2$.

67. Find the distance from $Q = (1, 2, 2)$ to the plane $\mathbf{n} \cdot \langle x, y, z \rangle = 3$, where $\mathbf{n} = \left\langle \frac{3}{5}, \frac{4}{5}, 0 \right\rangle$.

68. What is the distance from $P = (a, b, c)$ to the plane $x = 0$? Visualize your answer geometrically and explain without computation. Then verify that Eq. (7) yields the same answer.

69. Show that the point Q on the plane $\mathbf{n} \cdot \langle x, y, z \rangle = d$ nearest the origin lies on the ray through $\mathbf{n}$. More precisely, show that Q is the terminal point of $(d/\|\mathbf{n}\|^2)\mathbf{n}$.

70. The equation of a plane $\mathbf{n} \cdot \langle x, y, z \rangle = d$ is said to be in **normal form** if $\mathbf{n}$ is a unit vector. Show that in this case, $|d|$ is the distance from the plane to the origin. Write the equation of the plane $4x - 2y + 4z = 24$ in normal form.

71. Let $\mathbf{n} = \overrightarrow{OP}$, where $P = (x_0, y_0, z_0)$ is a point on the sphere $x^2 + y^2 + z^2 = r^2$, and let $\mathcal{P}$ be the plane with equation $\mathbf{n} \cdot \langle x, y, z \rangle = r^2$. Show that the point on $\mathcal{P}$ nearest the origin is P itself and conclude that $\mathcal{P}$ is tangent to the sphere at P (Figure 13).

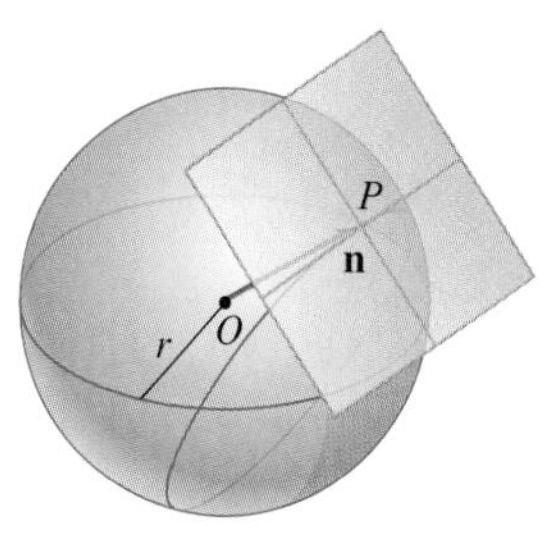

FIGURE 13 The terminal point of **n** lies on the sphere of radius r.

72. Use Exercise 71 to find the equation of the plane tangent to the unit sphere at $P = \left(\frac{1}{\sqrt{3}}, \frac{1}{\sqrt{3}}, \frac{1}{\sqrt{3}} \right)$.

12.6 A Survey of Quadric Surfaces

Quadric surfaces are the surface analogs of conic sections. Recall that a conic section is a curve in the $\mathbf{R}^2$ defined by a quadratic equation in two variables. A quadric surface is defined by a quadratic equation in *three* variables:

To insure that Eq. (1) is genuinely quadratic, we assume that the degree-two coefficients A, B, C, D, E, F are not all zero.

$$Ax^2 + By^2 + Cz^2 + Dxy + Eyz + Fzx + ax + by + cz + d = 0 \tag{1}$$

Like conic sections, quadric surfaces are classified into a small number of types. When the coordinate axes are chosen to coincide with the axes of the quadric, the equation of the quadric has a simple form. The quadric is then said to be in **standard position**. In this short survey of quadric surfaces, we restrict our attention to quadrics in standard position.

The surface analogs of ellipses are the egg-shaped **ellipsoids** (Figure 1). In standard form, an ellipsoid has the equation

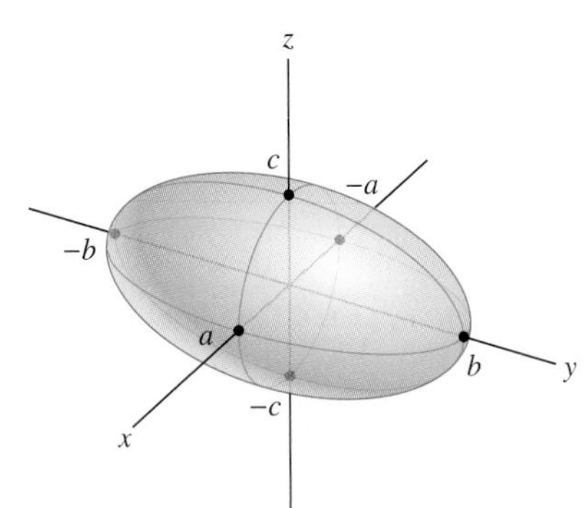

FIGURE 1 Ellipsoid with equation $\left(\frac{x}{a}\right)^2 + \left(\frac{y}{b}\right)^2 + \left(\frac{z}{c}\right)^2 = 1$.

$$\left(\frac{x}{a}\right)^2 + \left(\frac{y}{b}\right)^2 + \left(\frac{z}{c}\right)^2 = 1$$

For $a = b = c$, this equation is equivalent to $x^2 + y^2 + z^2 = a^2$ and the ellipsoid is a sphere of radius a.

Surfaces are often represented graphically by a mesh of curves called **traces**, obtained by intersecting the surface with planes parallel to one of the coordinate planes. Algebraically, this corresponds to **freezing** one of the three variables (holding it constant). For example, the intersection of the horizontal plane $z = z_0$ with the surface is a horizontal trace curve (Figure 2).

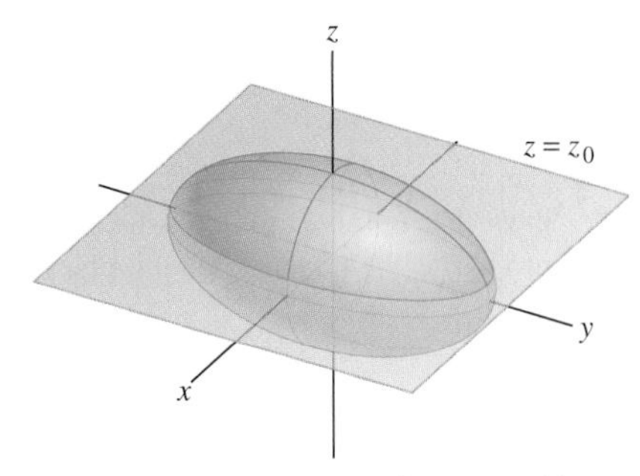

FIGURE 2 Intersection of $z = z_0$ with an ellipsoid.

■ **EXAMPLE 1 The Traces of an Ellipsoid** Describe the traces of the ellipsoid

$$\left(\frac{x}{5}\right)^2 + \left(\frac{y}{7}\right)^2 + \left(\frac{z}{9}\right)^2 = 1$$

Solution First we observe that the traces in the coordinate planes are ellipses:

$$xy\text{-trace (set } z = 0\text{):} \qquad \left(\frac{x}{5}\right)^2 + \left(\frac{y}{7}\right)^2 = 1$$

$$yz\text{-trace (set } x = 0\text{):} \qquad \left(\frac{y}{7}\right)^2 + \left(\frac{z}{9}\right)^2 = 1$$

$$xz\text{-trace (set } y = 0\text{):} \qquad \left(\frac{x}{5}\right)^2 + \left(\frac{z}{9}\right)^2 = 1$$

In fact, all the traces of an ellipsoid are ellipses. For example, the horizontal trace defined by setting $z = z_0$ is the ellipse [Figure 3(B)]

$$\text{Trace at height } z_0 = \left(\frac{x}{5}\right)^2 + \left(\frac{y}{7}\right)^2 + \left(\frac{z_0}{9}\right)^2 = 1 \quad \text{or} \quad \frac{x^2}{25} + \frac{y^2}{49} = \underbrace{1 - \frac{z_0^2}{81}}_{\text{A constant}}$$

For $z_0 = 9$, we obtain $\frac{x^2}{25} + \frac{y^2}{49} = 0$, and the trace reduces to a single point $(0, 0, 9)$. Similarly, for $z_0 = -9$, the trace is the point $(0, 0, -9)$. If $|z_0| > 9$, the constant $1 - z_0^2/81$ is negative and the trace has no points. The traces in the vertical planes $x = x_0$ and $y = y_0$ have a similar description [Figure 3(C)]. ■

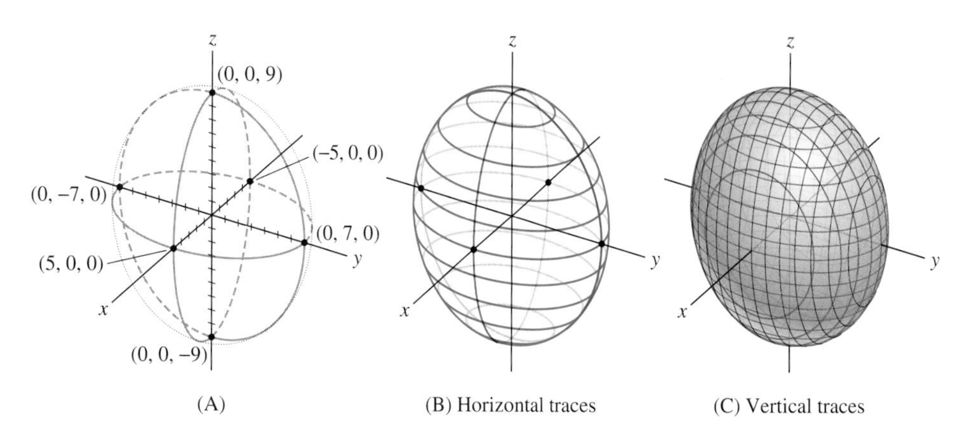

FIGURE 3 The ellipsoid $\left(\frac{x}{5}\right)^2 + \left(\frac{y}{7}\right)^2 + \left(\frac{z}{9}\right)^2 = 1$.

The analogs of the hyperbolas are the **hyperboloids**, which come in two types, depending on whether the surface has one or two components. We refer to these types as hyperboloids of one or two sheets (Figure 4). Their equations in standard position are

$$\underbrace{\left(\frac{x}{a}\right)^2 + \left(\frac{y}{b}\right)^2 = \left(\frac{z}{c}\right)^2 + 1}_{\textbf{Hyperboloid of one sheet}}, \qquad \underbrace{\left(\frac{x}{a}\right)^2 + \left(\frac{y}{b}\right)^2 = \left(\frac{z}{c}\right)^2 - 1}_{\textbf{Hyperboloid of two sheets}} \tag{2}$$

Notice that a hyperboloid of two sheets does not contain any points whose z-coordinate satisfies $-c < z < c$ because the right-hand side $\left(\frac{z}{c}\right)^2 - 1$ is then negative, but the left-hand side of the equation is greater than or equal to zero.

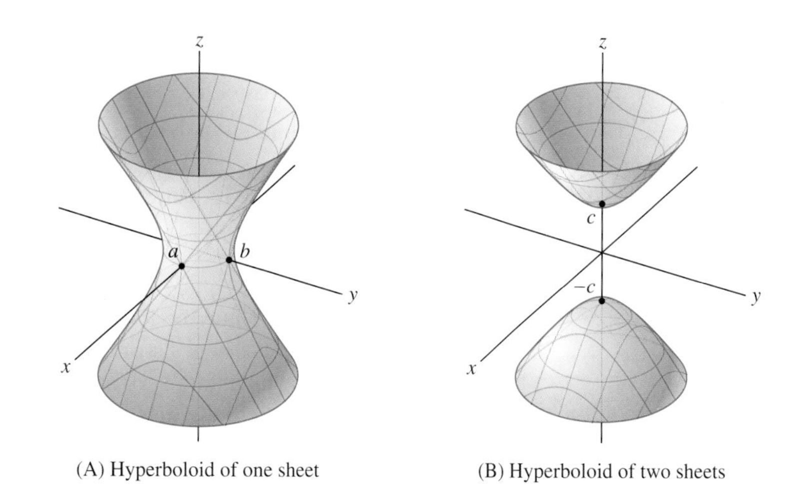

FIGURE 4 Hyperboloids of one and two sheets.

EXAMPLE 2 The Traces of a Hyperboloid of One Sheet Determine the traces of the hyperboloid $\left(\frac{x}{2}\right)^2 + \left(\frac{y}{3}\right)^2 = \left(\frac{z}{4}\right)^2 + 1$.

Solution Unlike the ellipsoid, the traces of a hyperboloid are of two types. The horizontal traces are ellipses and the two families of vertical traces (parallel to the yz- and xz-planes) are hyperbolas (Figure 5):

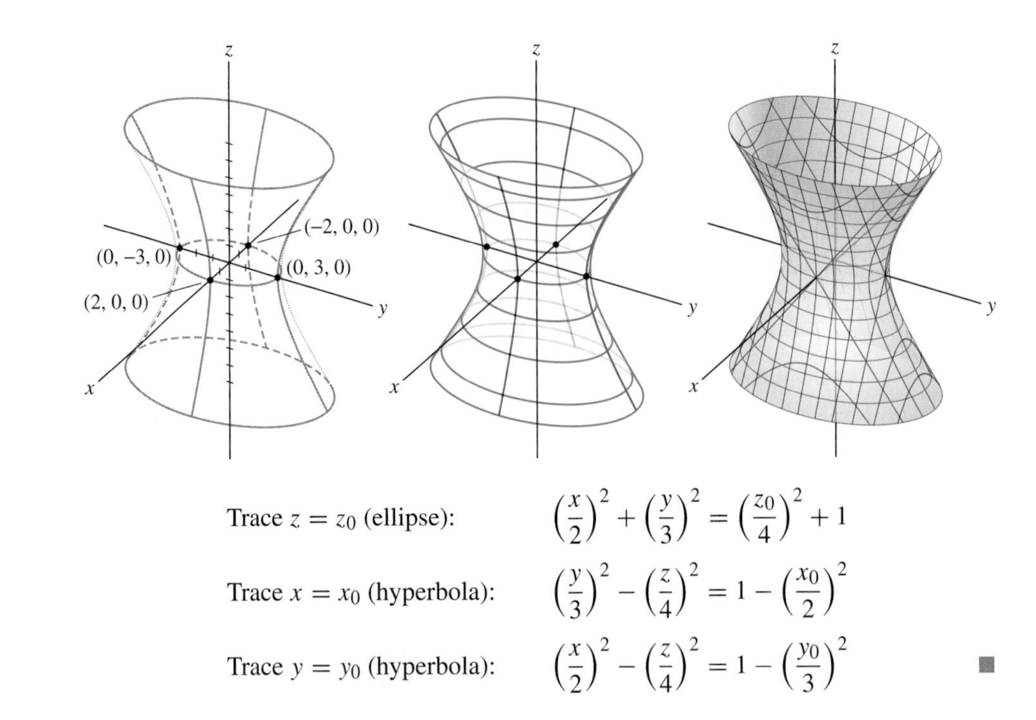

FIGURE 5 The hyperboloid $\left(\frac{x}{2}\right)^2 + \left(\frac{y}{3}\right)^2 = \left(\frac{z}{4}\right)^2 + 1$.

Trace $z = z_0$ (ellipse): $\quad \left(\frac{x}{2}\right)^2 + \left(\frac{y}{3}\right)^2 = \left(\frac{z_0}{4}\right)^2 + 1$

Trace $x = x_0$ (hyperbola): $\quad \left(\frac{y}{3}\right)^2 - \left(\frac{z}{4}\right)^2 = 1 - \left(\frac{x_0}{2}\right)^2$

Trace $y = y_0$ (hyperbola): $\quad \left(\frac{x}{2}\right)^2 - \left(\frac{z}{4}\right)^2 = 1 - \left(\frac{y_0}{3}\right)^2$ ■

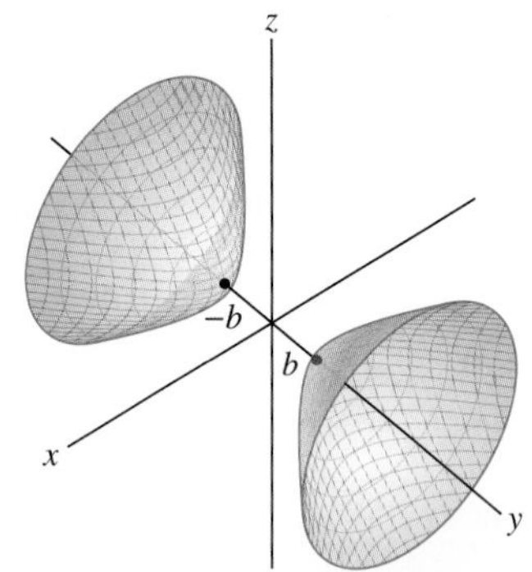

Hyperboloid of two sheets

FIGURE 6 The two-sheeted hyperboloid $\left(\frac{x}{a}\right)^2 + \left(\frac{z}{c}\right)^2 = \left(\frac{y}{b}\right)^2 - 1$.

■ **EXAMPLE 3 Hyperboloid of Two Sheets Symmetric About the y-axis** Show that $\left(\frac{x}{a}\right)^2 + \left(\frac{z}{c}\right)^2 = \left(\frac{y}{b}\right)^2 - 1$ has no points for $-b < y < b$.

Solution This equation does not have the same form as Eq. (2) because the variables y and z have been interchanged. This hyperboloid is symmetrical about the y-axis rather than the z-axis (Figure 6). The left-hand side of the equation is always ≥ 0. Thus, there are no solutions with $|y| < b$ because the right-hand side is $\left(\frac{y}{b}\right)^2 - 1 < 0$. Therefore, the hyperboloid has two sheets, corresponding to $y \geq b$ and $y \leq -b$. ■

The following equation defines an **elliptic cone** (Figure 7):

$$\left(\frac{x}{a}\right)^2 + \left(\frac{y}{b}\right)^2 = \left(\frac{z}{c}\right)^2$$

An elliptic cone may be thought of as a limiting case of a hyperboloid of one sheet in which we "pinch the waist" down to a point.

The third main family of quadric surfaces are the **paraboloids**. There are two types—elliptic and hyperbolic. In standard position, their equations are

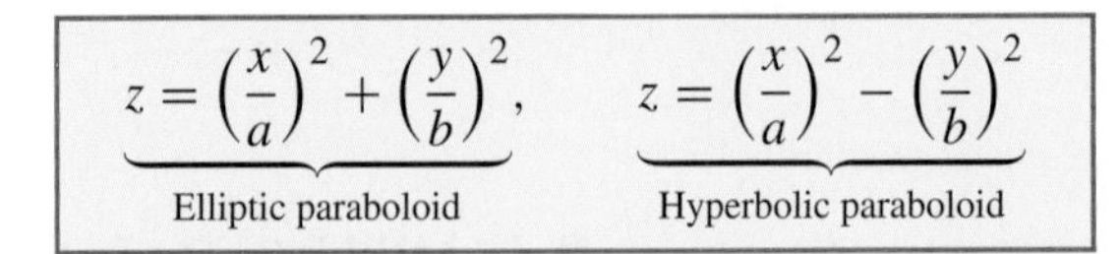

In both cases, the vertical trace curves are parabolas. For example, the traces defined by $x = x_0$ are the parabolas

$$\underbrace{z = \left(\frac{y}{b}\right)^2 + C,}_{\text{Trace } x = x_0 \text{ of elliptic paraboloid}} \qquad \underbrace{z = -\left(\frac{y}{b}\right)^2 + C}_{\text{Trace } x = x_0 \text{ of hyperbolic paraboloid}}$$

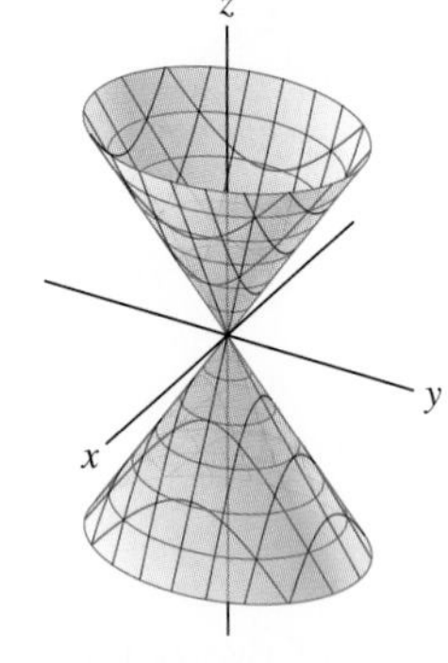

FIGURE 7 Elliptic cone $\left(\frac{x}{a}\right)^2 + \left(\frac{y}{b}\right)^2 = \left(\frac{z}{c}\right)^2$.

where $C = (x_0/a)^2$. The two types of paraboloids differ in their horizontal traces. As the names suggest, the horizontal traces of an elliptic paraboloid are ellipses and the horizontal traces of a hyperbolic paraboloid are hyperbolas (Figure 8).

Paraboloids play an important role in the optimization of functions of two variables. The elliptic paraboloid in Figure 8 has a local minimum at the origin. The hyperbolic paraboloid is a "saddle shape" at the origin, which is an analog for surfaces of a point of inflection.

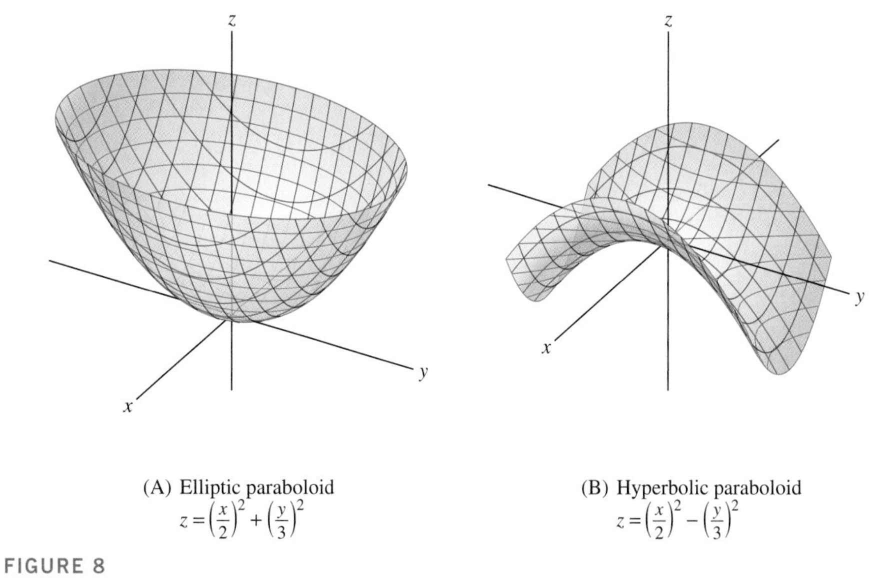

(A) Elliptic paraboloid $z = \left(\frac{x}{2}\right)^2 + \left(\frac{y}{3}\right)^2$

(B) Hyperbolic paraboloid $z = \left(\frac{x}{2}\right)^2 - \left(\frac{y}{3}\right)^2$

FIGURE 8

■ **EXAMPLE 4 Alternate Form of a Hyperbolic Paraboloid** Show that $z = 4xy$ is a hyperbolic paraboloid by writing the equation in terms of the variables $u = x + y$ and $v = x - y$.

Solution Note that $u + v = 2x$ and $u - v = 2y$. Therefore,

$$4xy = (u + v)(u - v) = u^2 - v^2$$

and thus the equation takes the form $z = u^2 - v^2$ in the coordinates $\{u, v, z\}$. The coordinates $\{u, v, z\}$ are obtained by rotating the coordinates $\{x, y, z\}$ by 45° about the z-axis (Figure 9). ■

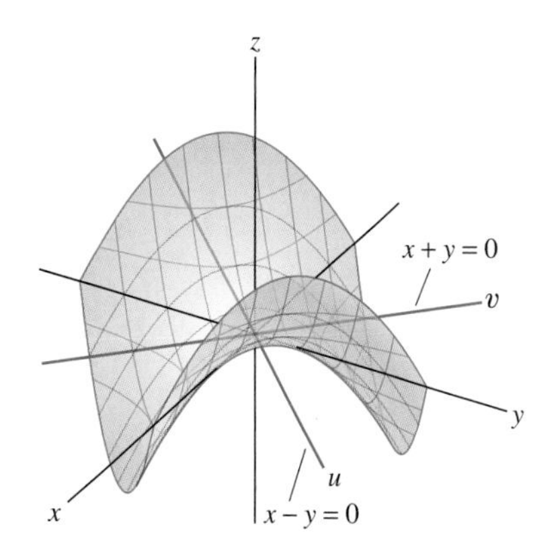

FIGURE 9 The hyperbolic paraboloid is defined by $z = 4xy$ or $z = u^2 - v^2$.

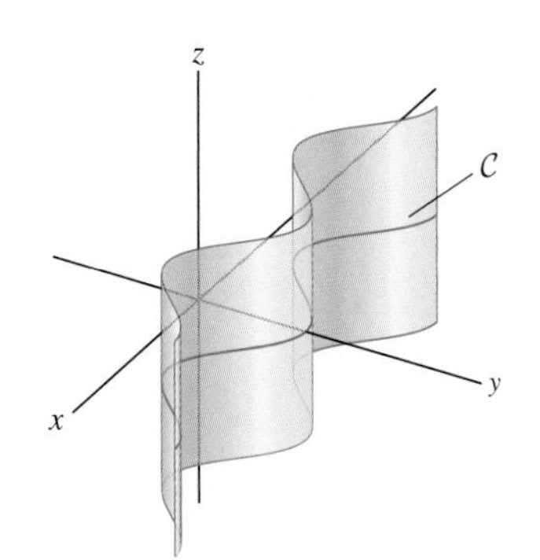

FIGURE 10 The cylinder with base $\mathcal{C}$.

Further examples of quadric surfaces are the **quadratic cylinders**. We use the term *cylinder* in the following general sense: Given a curve $\mathcal{C}$ in the xy-plane, the cylinder with base $\mathcal{C}$ is the surface consisting of all vertical lines passing through $\mathcal{C}$ (see Figure 10). The equation $x^2 + y^2 = r^2$ defines a standard cylinder of radius r with the z-axis as central axis. Figure 11 shows a standard cylinder and three other types of quadratic cylinders.

The ellipsoids, hyperboloids, paraboloids, and quadratic cylinders are called **nondegenerate** quadric surfaces. There are also a certain number of "degenerate" quadric surfaces. For example, $x^2 + y^2 + z^2 = 0$ is a quadric that reduces to a single point $(0, 0, 0)$, and $(x + y + z)^2 = 1$ reduces to the union of the two planes $x + y + z = \pm 1$.

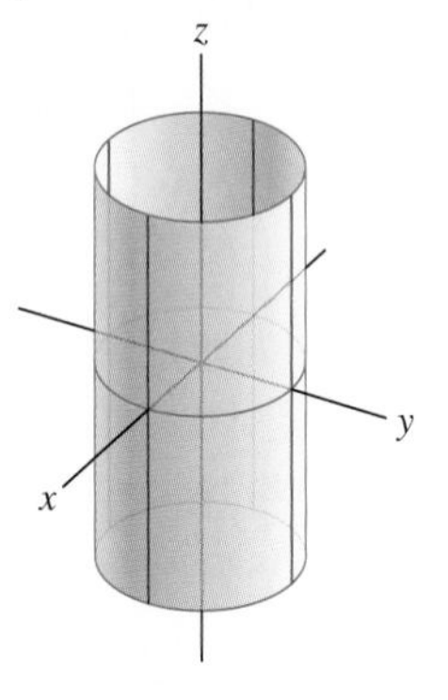

$x^2 + y^2 = r^2$
Standard cylinder of radius r

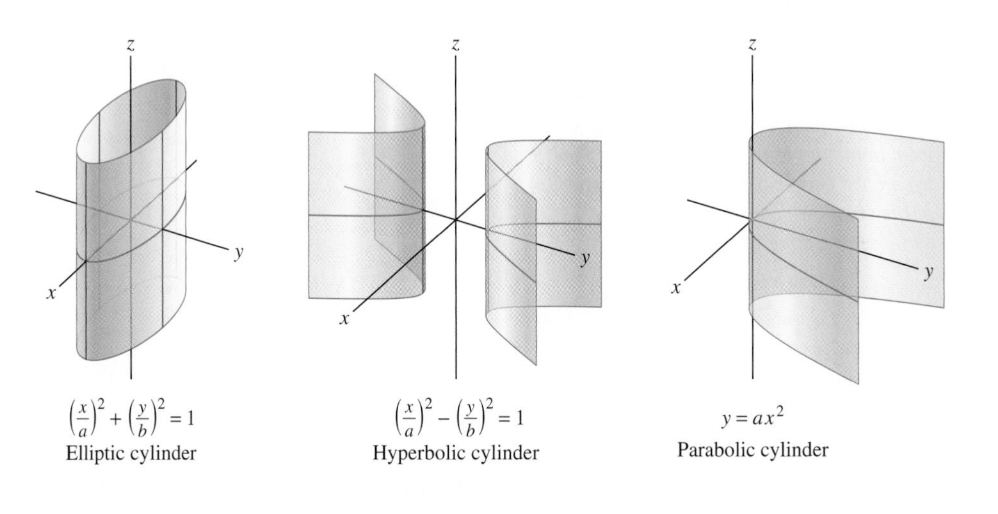

FIGURE 11

12.6 SUMMARY

- A *quadric surface* is a surface defined by a quadratic equation in three variables in which the coefficients A–F are not all zero:

$$Ax^2 + By^2 + Cz^2 + Dxy + Eyz + Fzx + ax + by + cz + d = 0$$

- Quadric surfaces in standard position:

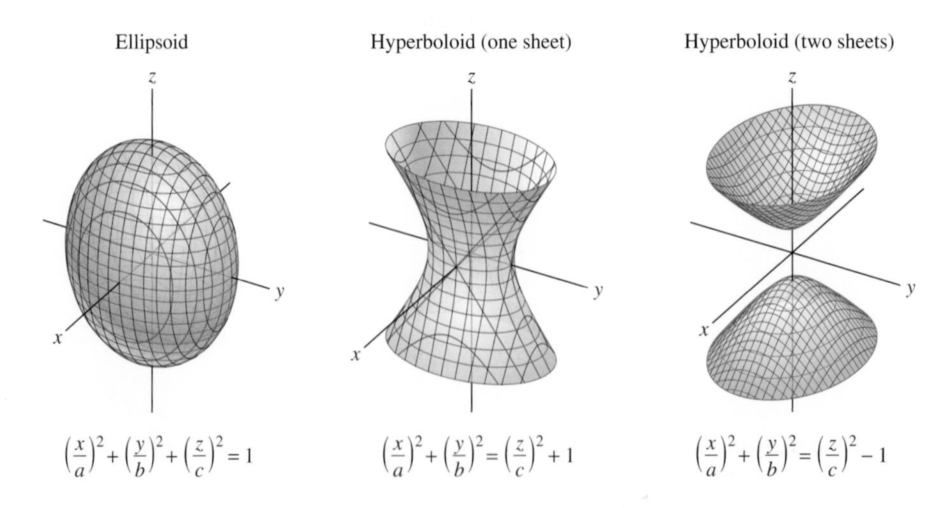

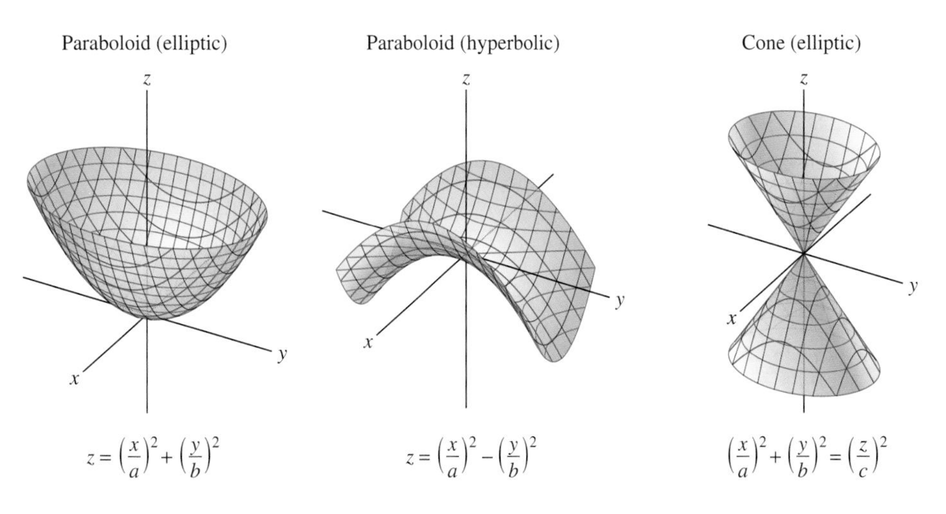

- A (vertical) cylinder is a surface consisting of all vertical lines passing through a curve (called the base) in the xy-plane. A quadratic cylinder is a cylinder whose base is a conic section. There are three types:

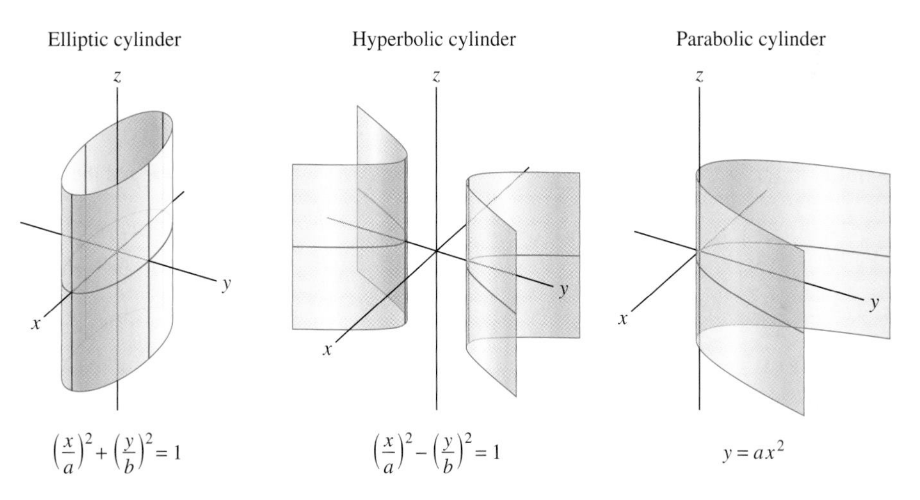

12.6 EXERCISES

Preliminary Questions

1. True or false: All traces of an ellipsoid are ellipses.

2. True or false: All traces of a hyperboloid are hyperbolas.

3. Which quadric surfaces have both hyperbolas and parabolas as traces?

4. Is there any quadric surface whose traces are all parabolas?

5. A surface is called **bounded** if there exists $M > 0$ such that every point on the surfaces lies at a distance of at most M from the origin. Which of the quadric surfaces are bounded?

6. Give an equation for a quadric surface that consists of two separate components.

7. What is the definition of a parabolic cylinder?

Exercises

In Exercises 1–6, state whether the given equation defines an ellipsoid or hyperboloid, and if a hyperboloid, whether it is of one or two sheets.

1. $\left(\frac{x}{2}\right)^2 + \left(\frac{y}{3}\right)^2 + \left(\frac{z}{5}\right)^2 = 1$

2. $\left(\frac{x}{5}\right)^2 + \left(\frac{y}{5}\right)^2 - \left(\frac{z}{7}\right)^2 = 1$

3. $x^2 + 3y^2 + 9z^2 = 1$

4. $-\left(\frac{x}{2}\right)^2 - \left(\frac{y}{3}\right)^2 + \left(\frac{z}{5}\right)^2 = 1$

5. $x^2 - 3y^2 + 9z^2 = 1$

6. $x^2 - 3y^2 - 9z^2 = 1$

In Exercises 7–12, state whether the given equation defines an elliptic paraboloid, hyperbolic paraboloid, or elliptic cone.

7. $z = \left(\frac{x}{4}\right)^2 + \left(\frac{y}{3}\right)^2$

8. $z^2 = \left(\frac{x}{4}\right)^2 + \left(\frac{y}{3}\right)^2$

9. $z = \left(\frac{x}{9}\right)^2 - \left(\frac{y}{12}\right)^2$

10. $4z = 9x^2 + 5y^2$

11. $3x^2 - 7y^2 = z$

12. $3x^2 + 7y^2 = 14z^2$

In Exercises 13–20, state the type of the quadric surface and describe the trace obtained by intersecting with the given plane.

13. $x^2 + \left(\frac{y}{4}\right)^2 + z^2 = 1, \quad y = 0$

14. $x^2 + \left(\frac{y}{4}\right)^2 + z^2 = 1, \quad y = 5$

15. $x^2 + \left(\frac{y}{4}\right)^2 + z^2 = 1, \quad z = \frac{1}{4}$

16. $\left(\frac{x}{2}\right)^2 + \left(\frac{y}{5}\right)^2 - 5z^2 = 1, \quad x = 0$

17. $\left(\frac{x}{3}\right)^2 + \left(\frac{y}{5}\right)^2 - 5z^2 = 1, \quad y = 1$

18. $4x^2 - \left(\frac{y}{3}\right)^2 - 2z^2 = 1, \quad z = 1$

19. $y = 3x^2, \quad z = 27$

20. $y = 3x^2, \quad y = 27$

21. Match the ellipsoids in Figure 12 with the equation:
(a) $x^2 + 4y^2 + 4z^2 = 16$
(b) $4x^2 + y^2 + 4z^2 = 16$
(c) $4x^2 + 4y^2 + z^2 = 16$

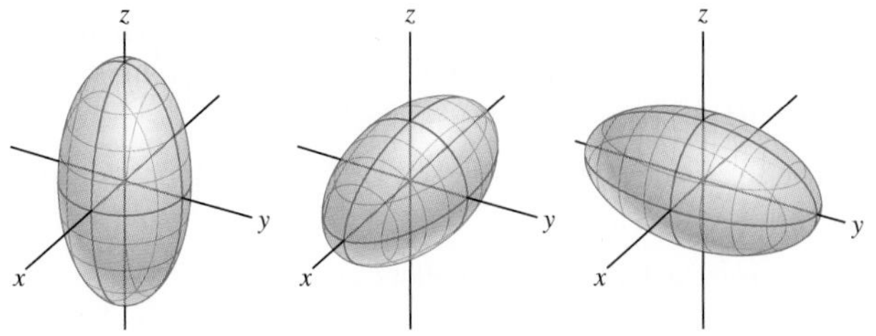

FIGURE 12

22. Describe the surface obtained if, in the equation $\pm 8x^2 \pm 3y^2 \pm z^2 = 1$, we choose (a) all plus signs, (b) one minus sign, and (c) two minus signs.

23. What is the equation of the surface obtained when the elliptic paraboloid $z = \left(\frac{x}{2}\right)^2 + \left(\frac{y}{4}\right)^2$ is rotated about the x-axis by 90°? Refer to Figure 13.

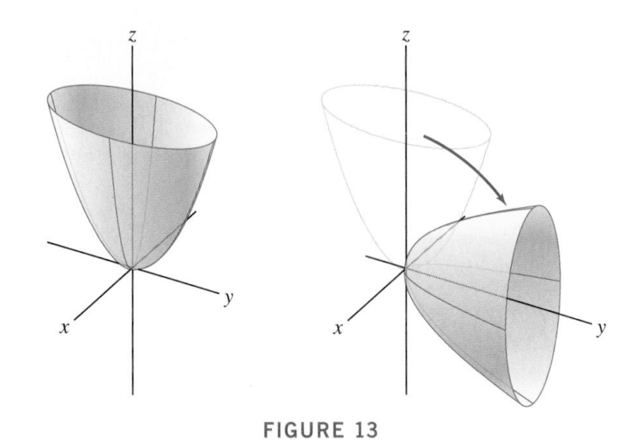

FIGURE 13

24. Describe the intersection of the horizontal plane $z = h$ and the hyperboloid $-x^2 - 4y^2 + 4z^2 = 1$. For which values of h is the intersection empty?

25. For which values of h is the intersection of the horizontal plane $x = h$ and the hyperboloid $\left(\frac{x}{2}\right)^2 - \left(\frac{y}{4}\right)^2 - \left(\frac{z}{9}\right)^2 = 1$ empty?

26. For which values of h is the intersection of the horizontal plane $x = h$ and the elliptic paraboloid $\left(\frac{y}{2}\right)^2 + \left(\frac{z}{4}\right)^2 = \frac{x}{9}$ empty?

In Exercises 27–32, sketch the given surface.

27. $x^2 + y^2 - z^2 = 1$

28. $\left(\frac{x}{4}\right)^2 + \left(\frac{y}{8}\right)^2 + \left(\frac{z}{12}\right)^2 = 1$

29. $z = \left(\frac{x}{4}\right)^2 + \left(\frac{y}{8}\right)^2$

30. $z = \left(\frac{x}{4}\right)^2 - \left(\frac{y}{8}\right)^2$

31. $z^2 = \left(\frac{x}{4}\right)^2 + \left(\frac{y}{8}\right)^2$

32. $z = -x^2$

33. Find the equation of the ellipsoid passing through the points marked in Figure 14(A).

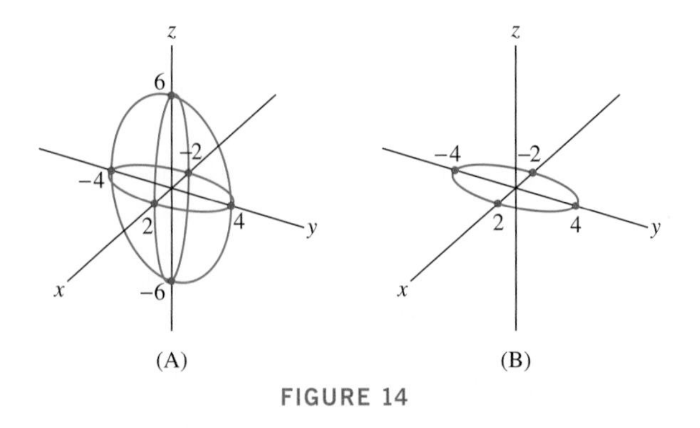

FIGURE 14

34. Find the equation of the elliptic cylinder passing through the points marked in Figure 14(B).

35. Find the equation of the hyperboloid shown in Figure 15(A).

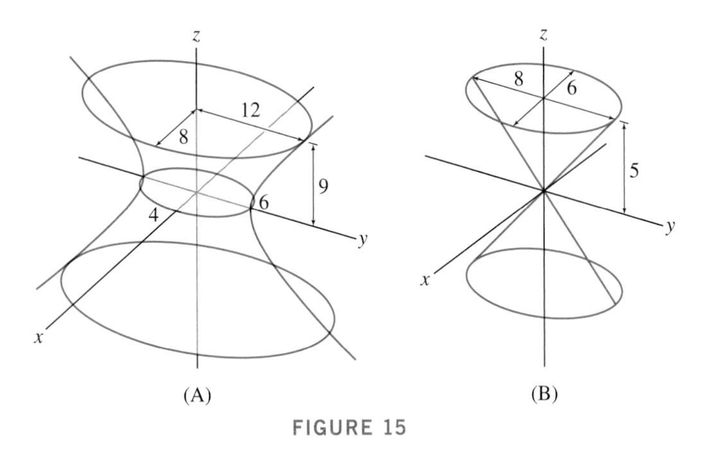

FIGURE 15

36. Find the equation of the quadric surface shown in Figure 15(B).

37. Determine the vertical traces of elliptic and parabolic cylinders in standard form.

38. What is the equation of a hyperboloid of one or two sheets in standard form if every horizontal trace is a circle?

39. Let $\mathcal{C}$ be an ellipse in a horizonal plane lying above the xy-plane. Which type of quadric surface is made up of all lines passing through the origin and a point on $\mathcal{C}$?

40. Show that the horizontal traces of the ellipsoid

$$\left(\frac{x}{a}\right)^2 + \left(\frac{y}{b}\right)^2 + \left(\frac{z}{c}\right)^2 = 1$$

are ellipses of the same eccentricity (apart from the traces at height $h = \pm c$, which reduce to a single point). Find the eccentricity.

Further Insights and Challenges

41. Let $Q = (m, n, r)$ be a point on the ellipsoid with equation $\left(\frac{x}{a}\right)^2 + \left(\frac{y}{b}\right)^2 + \left(\frac{z}{c}\right)^2 = 1$ (Figure 16). Let $\mathcal{P}$ be the plane with equation

$$\frac{mx}{a^2} + \frac{ny}{b^2} + \frac{rz}{c^2} = 1$$

Show that Q lies on $\mathcal{P}$ and that the tangent lines to the trace curves through P are each contained in $\mathcal{P}$. This shows that $\mathcal{P}$ is the tangent plane at P, as defined in Section 14.4.

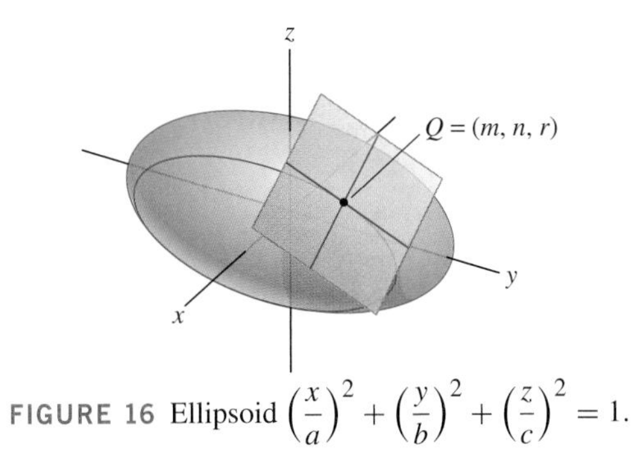

FIGURE 16 Ellipsoid $\left(\frac{x}{a}\right)^2 + \left(\frac{y}{b}\right)^2 + \left(\frac{z}{c}\right)^2 = 1$.

42. Let $\mathcal{S}$ be the hyperboloid $x^2 + y^2 = z^2 + 1$ and let $P = (\alpha, \beta, 0)$ be a point on $\mathcal{S}$ in the (x, y)-plane. Show that there are precisely two lines through P entirely contained in $\mathcal{S}$ (Figure 17). *Hint:* Consider the line $\mathbf{r}(t) = \langle \alpha + at, \beta + bt, t\rangle$ through P. Show that $\mathbf{r}(t)$ is contained in $\mathcal{S}$ if (a, b) is one of the two points on the unit circle obtained by rotating (α, β) through $\pm\frac{\pi}{2}$. This proves that a hyperboloid of one sheet is a **doubly ruled surface**, which means that it can be swept out by moving a line in space in two different ways.

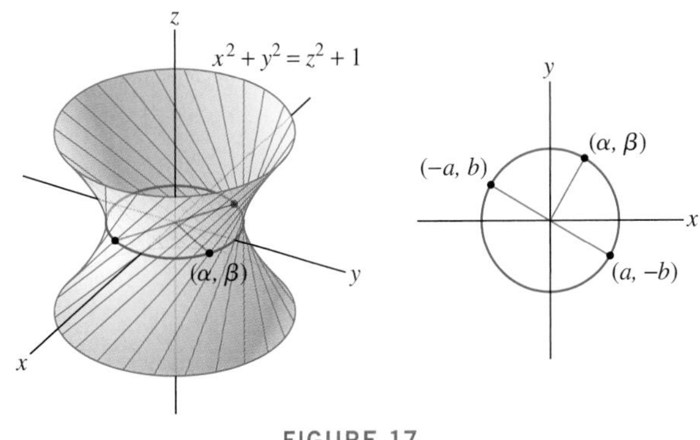

FIGURE 17

In Exercises 43–44, let $\mathcal{C}$ be a curve in $\mathbf{R}^3$ not passing through the origin. The cone on $\mathcal{C}$ is the surface consisting of all lines passing through the origin and a point on $\mathcal{C}$.

43. Show that the elliptic cone $\left(\frac{z}{c}\right)^2 = \left(\frac{x}{a}\right)^2 + \left(\frac{y}{b}\right)^2$ is, in fact, a cone on the ellipse $\mathcal{C}$ consisting of all points (x, y, c) such that $\left(\frac{x}{a}\right)^2 + \left(\frac{y}{b}\right)^2 = 1$.

44. In this exercise, we show that if $\mathcal{C}$ is a parabola, then the cone on $\mathcal{C}$ is also an elliptic cone. Let S be the cone on the curve $\mathcal{C} = \{(x, y, c) : y = ax^2\}$, where a, c are constants.

(a) Show that S has equation $yz = acx^2$.

(b) Show that under the change of variables $y = u + v$ and $z = u - v$, this equation becomes $acx^2 = u^2 - v^2$ or $u^2 = acx^2 + v^2$ (the equation of an elliptic cone in the variables x, v, u).

12.7 Cylindrical and Spherical Coordinates

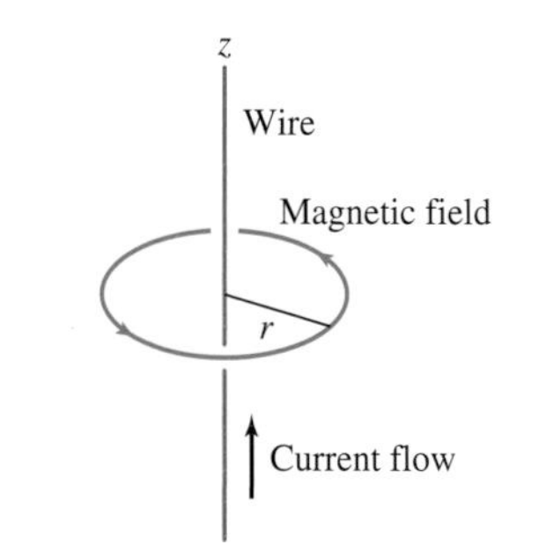

FIGURE 1

This section introduces two generalizations of polar coordinates to $\mathbf{R}^3$: cylindrical and spherical coordinates. These coordinate systems are commonly used in problems having rotational symmetry or symmetry about an axis. For example, the magnetic field generated by a current flowing in a long straight wire is conveniently expressed in cylindrical coordinates (Figure 1). We will also see the benefits of cylindrical and spherical coordinates when we study change of variables for multiple integrals.

Cylindrical Coordinates

In cylindrical coordinates, we replace the x- and y-coordinates of a point $P = (x, y, z)$ by polar coordinates. Thus, the **cylindrical coordinates** of P are (r, θ, z), where (r, θ) are polar coordinates of (x, y). Note that (r, θ) are polar coordinates of the projection Q of P onto the xy-plane (Figure 2). We convert between rectangular and cylindrical coordinates using the rectangular-polar formulas of Section 11.3.

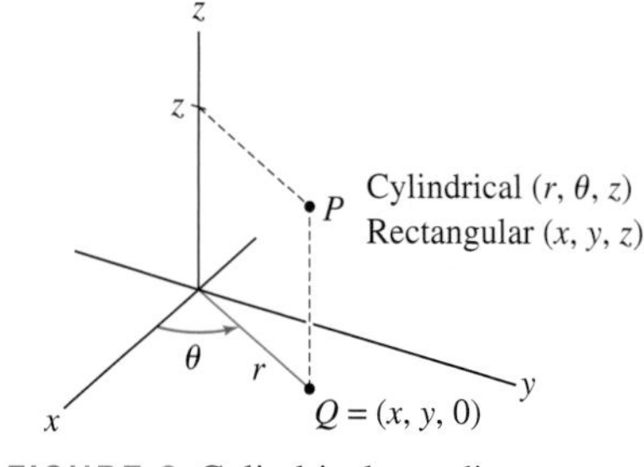

FIGURE 2 Cylindrical coordinates.

Cylindrical to Rectangular	Rectangular to Cylindrical
$x = r\cos\theta$	$r = \sqrt{x^2 + y^2}$
$y = r\sin\theta$	$\tan\theta = \dfrac{y}{x}$
$z = z$	$z = z$

■ **EXAMPLE 1 Converting from Cylindrical to Rectangular Coordinates** Convert $(r, \theta, z) = (2, \frac{3\pi}{4}, 5)$ to rectangular coordinates.

Solution Conversion to rectangular coordinates is straightforward (Figure 3):

$$x = r\cos\theta = 2\cos\frac{3\pi}{4} = 2\left(-\frac{\sqrt{2}}{2}\right) = -\sqrt{2}$$

$$y = r\sin\theta = 2\sin\frac{3\pi}{4} = 2\left(\frac{\sqrt{2}}{2}\right) = \sqrt{2}$$

The z-coordinate is unchanged, so $(x, y, z) = (-\sqrt{2}, \sqrt{2}, 5)$. ■

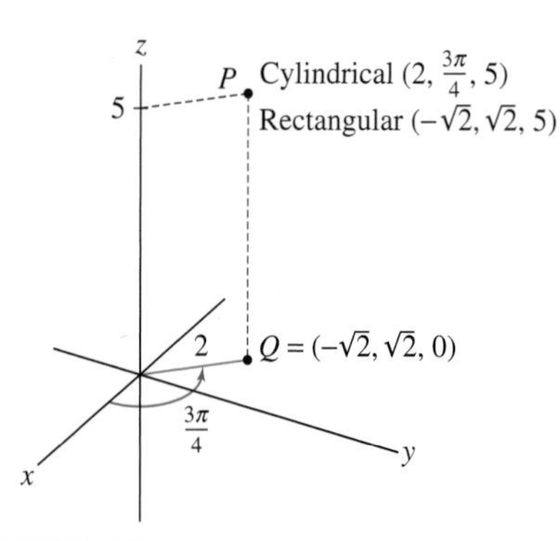

FIGURE 3

■ **EXAMPLE 2 Converting from Rectangular to Cylindrical Coordinates** Convert the rectangular coordinates $(x, y, z) = (-3\sqrt{3}, -3, 5)$ to cylindrical coordinates.

Solution We have $r = \sqrt{x^2 + y^2} = \sqrt{(-3\sqrt{3})^2 + (-3)^2} = 6$. The angle θ is determined by the location of the projection $Q = (-3\sqrt{3}, -3, 0)$. Although

$$\tan^{-1}\left(\frac{y}{x}\right) = \tan^{-1}\frac{1}{\sqrt{3}} = \frac{\pi}{6}$$

the angle θ is not equal to $\frac{\pi}{6}$ because Q lies in the third quadrant (Figure 4). The correct angle is $\theta = \frac{\pi}{6} + \pi = \frac{7\pi}{6}$ and the cylindrical coordinates are $(r, \theta, z) = (6, \frac{7\pi}{6}, 5)$. ■

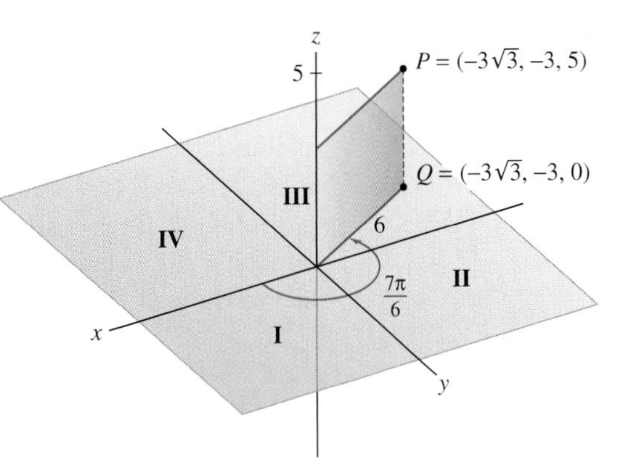

FIGURE 4 The projection Q lies in the third quadrant. Therefore, $\theta = \tan^{-1}\frac{1}{\sqrt{3}} + \pi = \frac{7\pi}{6}$.

The **level surfaces** of a coordinate system are the surfaces obtained by setting one of the coordinates equal to a constant. In rectangular coordinates, the level surfaces are the planes $x = x_0$, $y = y_0$, and $z = z_0$. In cylindrical coordinates, the level surfaces come in three types (Figure 5). The surface $r = R$ is the cylinder of radius R consisting of all points located a distance R from the z-axis. The equation $\theta = \theta_0$ defines the half-plane of all points that project onto the ray $\theta = \theta_0$ in the (x, y)-plane. Finally, $z = c$ is the horizontal plane at height c.

Cylindrical coordinates *are named after the surfaces $r = R$, which are cylinders.*

Level Surfaces in Cylindrical Coordinates

$r = R$	*Cylinder of radius R with the z-axis as axis of symmetry*
$\theta = \theta_0$	*Half-plane through the z-axis making an angle θ_0 with the xz-plane*
$z = c$	*Horizontal plane at height c*

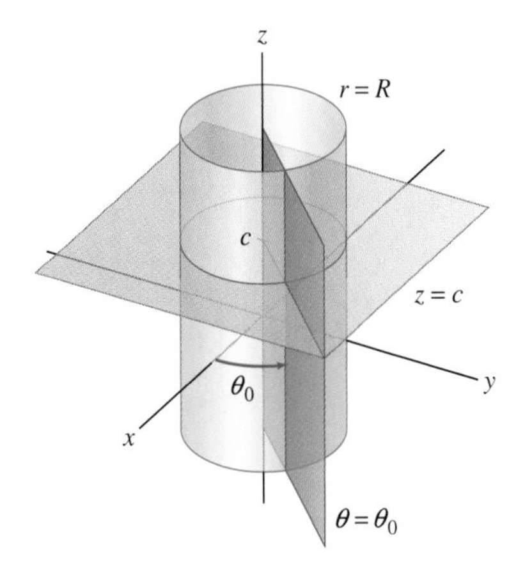

FIGURE 5 Level surfaces in spherical coordinates.

EXAMPLE 3 Equations in Cylindrical Coordinates Find an equation of the form $z = f(r, \theta)$ for the surfaces

(a) $x^2 + y^2 + z^2 = 9$ **(b)** $x + y + z = 1$

Solution We use the formulas

$$x^2 + y^2 = r^2, \qquad x = r\cos\theta, \qquad y = r\sin\theta$$

(a) The equation $x^2 + y^2 + z^2 = 9$ becomes $r^2 + z^2 = 9$ or $z = \pm\sqrt{9 - r^2}$. This is a sphere of radius 3.

(b) The plane $x + y + z = 1$ becomes

$$z = 1 - x - y = 1 - r\cos\theta - r\sin\theta \qquad \text{or} \qquad z = 1 - r(\cos\theta + \sin\theta)$$

Spherical Coordinates

The symbol ϕ (usually pronounced "fee," but sometimes pronounced "fie") is the twenty-first letter of the Greek alphabet.

A point $P = (x, y, z)$ in 3-space is described in spherical coordinates by a triple (ρ, θ, ϕ) as illustrated in Figure 6. The radial coordinate ρ is the distance from P to the origin: $\rho = \sqrt{x^2 + y^2 + z^2}$. The other two coordinates are angular: θ is the polar angle of the projection $Q = (x, y, 0)$ and ϕ is the angle between the z-axis and the ray from the origin through P. The angle ϕ is called the **angle of declination** because it measures how much the ray through P declines from the vertical.

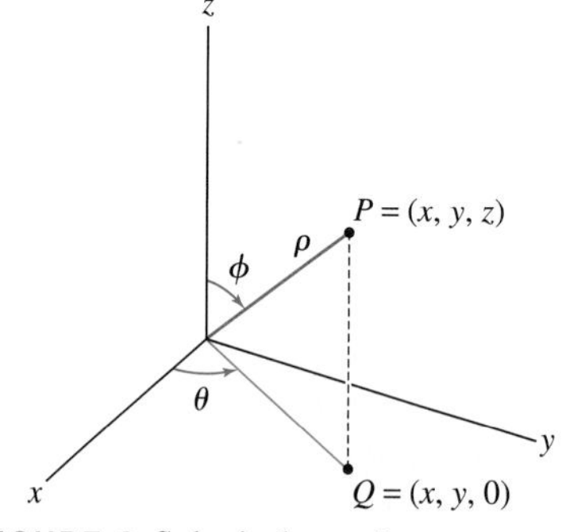

FIGURE 6 Spherical coordinates (ρ, θ, ϕ).

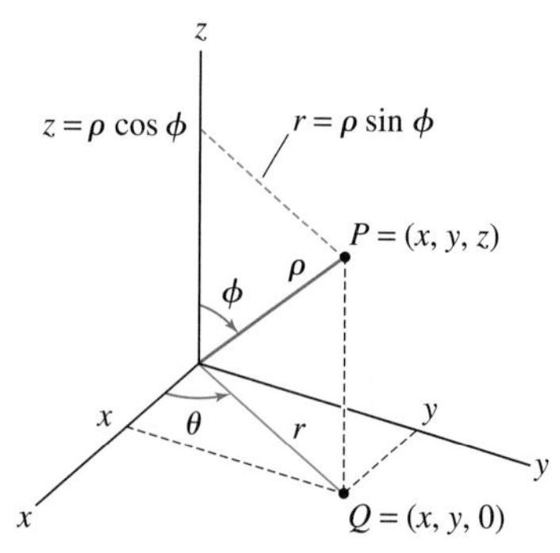

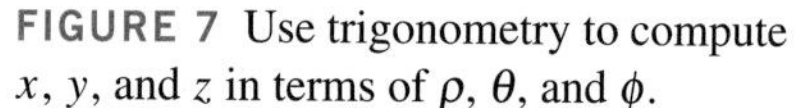

FIGURE 7 Use trigonometry to compute x, y, and z in terms of ρ, θ, and ϕ.

We use trigonometry to convert between spherical and rectangular coordinates. From Figure 7, we see that

$$\rho = \sqrt{x^2 + y^2 + z^2}, \qquad \tan\theta = \frac{y}{x}, \qquad \cos\phi = \frac{z}{\rho} \qquad \boxed{1}$$

Spherical Coordinates

ρ = distance from origin
θ = polar angle in the xy-plane
ϕ = angular declination from the vertical

In some textbooks, θ is referred to as the azimuthal angle *and ϕ as the* polar angle.

The radial coordinate r of $Q = (x, y, 0)$ is $r = \rho \sin\phi$, and therefore,

$$x = r\cos\theta = \rho\cos\theta\sin\phi, \qquad y = r\cos\theta = \rho\sin\theta\sin\phi, \qquad z = \rho\cos\phi$$

Spherical to Rectangular	Rectangular to Spherical
$x = \rho\cos\theta\sin\phi$	$\rho = \sqrt{x^2 + y^2 + z^2}$
$y = \rho\sin\theta\sin\phi$	$\tan\theta = \dfrac{y}{x}$
$z = \rho\cos\phi$	$\cos\phi = \dfrac{z}{\rho}$

■ **EXAMPLE 4 From Spherical to Rectangular Coordinates** Find the rectangular coordinates of $P = (\rho, \theta, \phi) = (3, \frac{\pi}{3}, \frac{\pi}{4})$ and find the radial coordinate r of its projection Q onto the xy-plane.

Solution By the formulas above,

$$x = \rho\cos\theta\sin\phi = 3\cos\frac{\pi}{3}\sin\frac{\pi}{4} = 3\left(\frac{1}{2}\right)\frac{\sqrt{2}}{2} = \frac{3\sqrt{2}}{4}$$

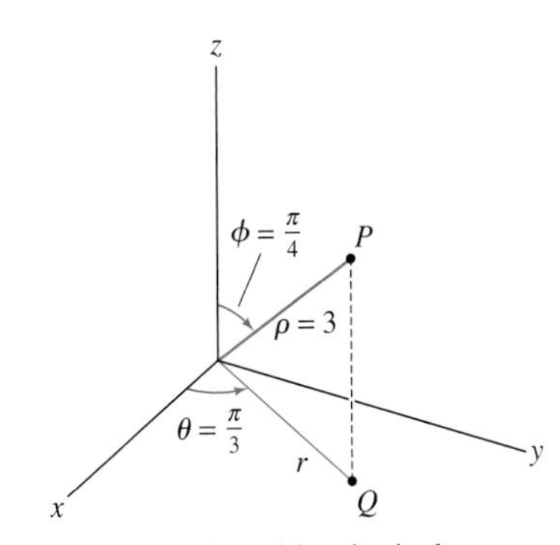

FIGURE 8 Point with spherical coordinates $(3, \frac{\pi}{3}, \frac{\pi}{4})$.

$$y = \rho \sin\theta \sin\phi = 3 \sin\frac{\pi}{3} \sin\frac{\pi}{4} = 3\left(\frac{\sqrt{3}}{2}\right)\frac{\sqrt{2}}{2} = \frac{3\sqrt{6}}{4}$$

$$z = \rho\cos\phi = 3\cos\frac{\pi}{4} = 3\frac{\sqrt{2}}{2} = \frac{3\sqrt{2}}{2}$$

Now consider the projection $Q = (x, y, 0) = \left(\frac{3\sqrt{2}}{4}, \frac{3\sqrt{6}}{4}, 0\right)$ (Figure 8). The radial coordinate r of Q satisfies

$$r^2 = x^2 + y^2 = \left(\frac{3\sqrt{2}}{4}\right)^2 + \left(\frac{3\sqrt{6}}{4}\right)^2 = \frac{9}{2}$$

Therefore, $r = \frac{3}{\sqrt{2}}$. ■

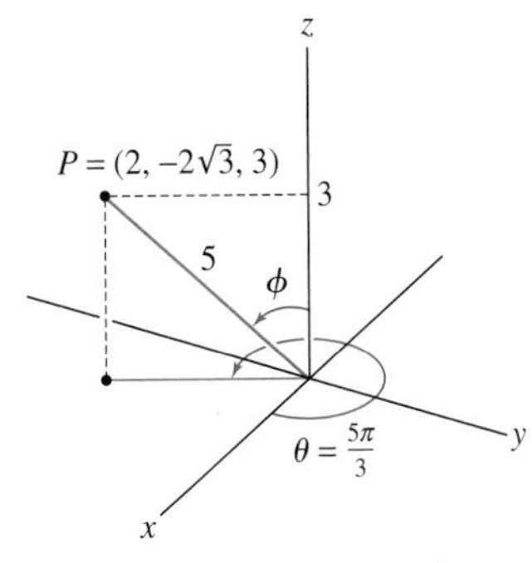

FIGURE 9 Point with rectangular coordinates $(2, -2\sqrt{3}, 3)$.

■ **EXAMPLE 5 From Rectangular to Spherical Coordinates** Find the spherical coordinates of the point $P = (x, y, z) = (2, -2\sqrt{3}, 3)$.

Solution The radial coordinate is $\rho = \sqrt{2^2 + (-2\sqrt{3})^2 + 3^2} = \sqrt{25} = 5$. The angular coordinate θ satisfies

$$\tan\theta = \frac{y}{x} = \frac{-2\sqrt{3}}{2} = -\sqrt{3} \quad\Rightarrow\quad \theta = \frac{2\pi}{3} \text{ or } \frac{5\pi}{3}$$

Since the point $(x, y) = (2, -2\sqrt{3})$ lies in the fourth quadrant, the correct choice is $\theta = \frac{5\pi}{3}$ (Figure 9). Finally, $\cos\phi = \frac{z}{\rho} = \frac{3}{5}$ and so $\phi = \cos^{-1}\frac{3}{5} \approx 0.93$. Therefore, P has spherical coordinates $(5, \frac{5\pi}{3}, 0.93)$. ■

Level Surfaces in Spherical Coordinates

$\rho = R$	*Sphere of radius R with its center at the origin*
$\theta = \theta_0$	*Half-plane through the z-axis making an angle θ_0 with the xz-plane*
$\phi = \phi_0$	*Right-circular cone with the following three exceptions:*
$\phi = 0$	*Positive z-axis*
$\phi = \frac{\pi}{2}$	*xy-plane*
$\phi = \pi$	*Negative z-axis*

Figure 10 shows the three types of level surfaces in spherical coordinates. Notice that if $\phi \neq \frac{\pi}{2}$ or π, then the level surface $\phi = \phi_0$ is the right-circular cone consisting of points P such that $\overline{OP}$ makes an angle ϕ_0 with the z-axis.

For fixed radius $\rho = R$, the angular coordinates (θ, ϕ) provide useful coordinates on the sphere of radius R. They are closely related to the longitude-latitude system used to identify points on the surface of the earth (Figure 11). In this system, $\theta = \theta_0$ defines a

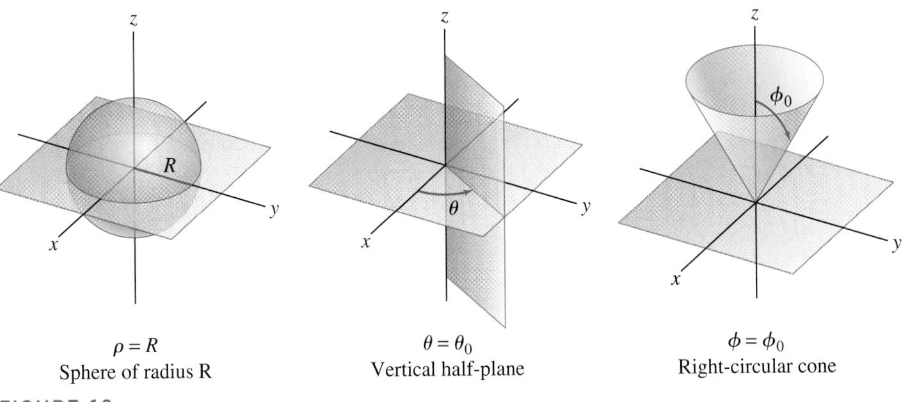

FIGURE 10

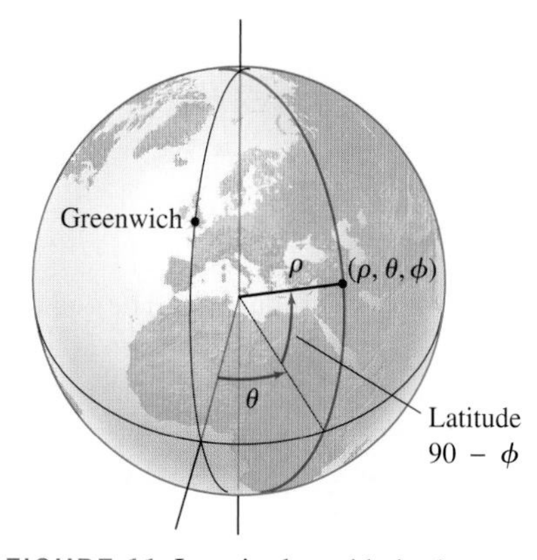

FIGURE 11 Longitude and latitude provide spherical coordinates on the surface of the earth.

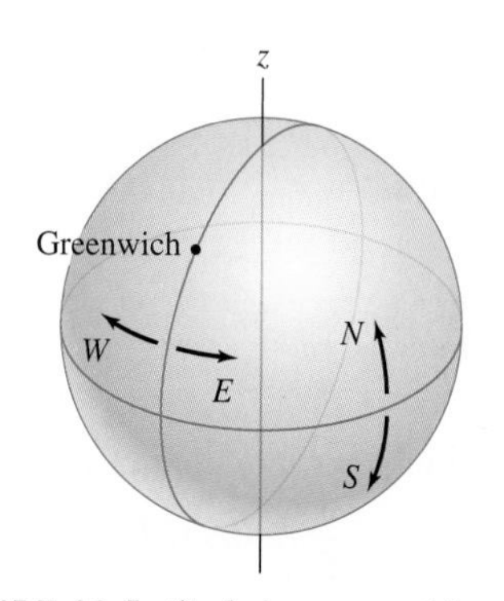

FIGURE 12 Latitude is measured from the equator and is labeled N (north) in the upper hemisphere, and S (south) in the lower hemisphere.

longitude, that is, a half-circle stretching from the North to the South Pole (Figure 12). By convention, the axes are chosen so that $\theta = 0$ passes through Greenwich, England (this longitude is called the *prime meridian*). The longitude of a point is given as an angle between 0 and 180° and is designated E or W according to whether the point lies to the east or west of the prime meridian. The set of points on the sphere satisfying $\phi = \phi_0$ is a horizontal circle called a **latitude**. By convention, on a globe we measure latitudes from the equator and use the labels N or S to specify the Northern or Southern Hemisphere. Thus, the latitude at $\phi = \phi_0$ is $90° - \phi_0$ if $\theta \le \phi_0 \le 90°$, and it is $\phi_0 - 90°$ if $90° \le \phi_0 \le 180°$.

EXAMPLE 6 Spherical Coordinates via Longitude and Latitude Find the angles (θ, ϕ) for Nairobi (1.17° S, 36.48° E) and Ottawa (45.27° N, 75.42° W).

Solution For Nairobi, $\theta = 36.48°$ since the longitude lies to the east of Greenwich. Nairobi's latitude is south of the equator, so $\phi = 90 + 1.17 = 91.17°$.

For Ottawa, we have $\theta = 360 - 75.42 = 284.58°$ since 75.42° W refers to 75.42 degrees in the negative θ direction. Since the latitude of Ottawa is north of the equator, $\phi = 90 - 45.27 = 44.73°$. ■

EXAMPLE 7 Finding a Polar Equation Find an equation of the form $\rho = f(\theta, \phi)$ for the surfaces:

(a) $x^2 + y^2 + z^2 = 9$ **(b)** $z = x^2 - y^2$

Solution

(a) The equation $x^2 + y^2 + z^2 = 9$ defines the sphere of radius 3 centered at the origin. Since $\rho^2 = x^2 + y^2 + z^2$, the equation in spherical coordinates is $\rho = 3$.

(b) To convert $z = x^2 - y^2$ to spherical coordinates, we substitute the formulas for x, y, and z in terms of ρ, θ, and ϕ:

$$\overbrace{\rho\cos\phi}^{z} = \overbrace{(\rho\cos\theta\sin\phi)^2}^{x^2} - \overbrace{(\rho\sin\theta\sin\phi)^2}^{y^2}$$

$$\cos\phi = \rho\sin^2\phi(\cos^2\theta - \sin^2\theta) \qquad \text{(divide by } \rho \text{ and factor)}$$

$$\cos\phi = \rho\sin^2\phi\cos 2\theta \qquad \text{(since } \cos^2\theta - \sin^2\theta = \cos 2\theta\text{)}$$

Solving for ρ, we obtain $\rho = \dfrac{\cos\phi}{\sin^2\phi\cos 2\theta}$. ■

12.7 SUMMARY

- Conversion from rectangular to cylindrical and spherical coordinates:

Cylindrical	Spherical
$r = \sqrt{x^2 + y^2}$	$\rho = \sqrt{x^2 + y^2 + z^2}$
$\tan\theta = \dfrac{y}{x}$	$\tan\theta = \dfrac{y}{x}$
$z = z$	$\cos\phi = \dfrac{z}{\rho}$

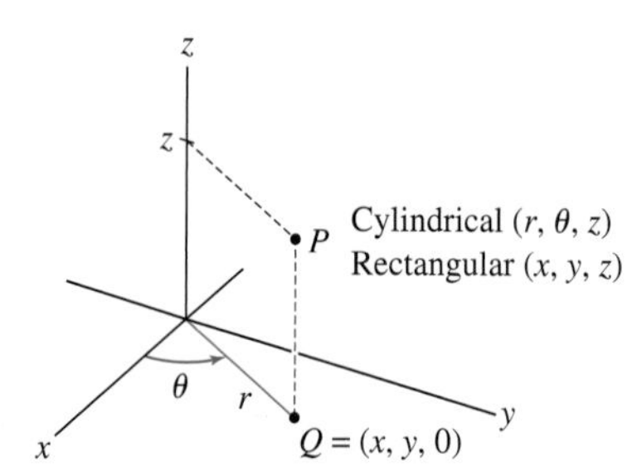

FIGURE 13 Cylindrical coordinates (r, θ, z).

The angles are chosen so that

$$0 \le \theta < 2\pi \quad \text{(cylindrical or spherical)}, \qquad 0 \le \phi \le \pi \quad \text{(spherical)}$$

- Conversion to rectangular coordinates:

Cylindrical (r, θ, z)	Spherical (ρ, θ, ϕ)
$x = r\cos\theta$	$x = \rho\cos\theta\sin\phi$
$y = r\sin\theta$	$y = \rho\sin\theta\sin\phi$
$z = z$	$z = \rho\cos\phi$

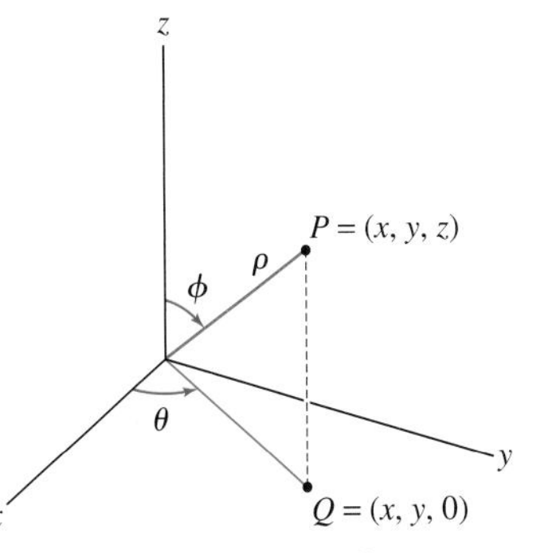

FIGURE 14 Spherical coordinates (ρ, θ, ϕ).

- Level surfaces:

Cylindrical		Spherical	
$r = R$:	Cylinder of radius R	$\rho = R$:	Sphere of radius R
$\theta = \theta_0$:	Vertical half-plane	$\theta = \theta_0$:	Vertical half-plane
$z = c$:	Horizontal plane	$\phi = \phi_0$:	Right-circular cone

12.7 EXERCISES

Preliminary Questions

1. Describe the surfaces $r = R$ in cylindrical coordinates and $\rho = R$ in spherical coordinates.

2. Which statement about the cylindrical coordinates is correct?
(a) If $\theta = 0$, then P lies on the z-axis.
(b) If $\theta = 0$, then P lies in the xz-plane.

3. Which statement about spherical coordinates is correct?
(a) If $\phi = 0$, then P lies on the z-axis.
(b) If $\phi = 0$, then P lies in the xy-plane.

4. The level surface $\phi = \phi_0$ in spherical coordinates, usually a cone, reduces to a half-line for two values of ϕ_0. Which two values?

5. For which value of ϕ_0 is $\phi = \phi_0$ a plane? Which plane?

Exercises

In Exercises 1–4, convert from cylindrical to rectangular coordinates.

1. $(4, \pi, 4)$
2. $\left(2, \frac{\pi}{3}, -8\right)$
3. $\left(0, \frac{\pi}{5}, \frac{1}{2}\right)$
4. $\left(1, \frac{\pi}{2}, -2\right)$

In Exercises 5–10, convert from rectangular to cylindrical coordinates.

5. $(1, -1, 1)$
6. $(2, 2, 1)$
7. $(1, \sqrt{3}, 7)$
8. $\left(\frac{3}{2}, \frac{3\sqrt{3}}{2}, 9\right)$
9. $\left(\frac{5}{\sqrt{2}}, \frac{5}{\sqrt{2}}, 2\right)$
10. $(3, 3\sqrt{3}, 2)$

In Exercises 11–16, describe the set in cylindrical coordinates.

11. $x^2 + y^2 \le 1$
12. $y^2 + z^2 \le 4, \quad x = 0$
13. $x^2 + y^2 + z^2 = 4, \quad x \ge 0, \quad y \ge 0, \quad z \ge 0$
14. $x^2 + y^2 = 9, \quad x \ge 0, \quad y \ge 0, \quad x = y$
15. $y^2 + z^2 \le 9, \quad x \ge y$
16. $x^2 + y^2 + z^2 \le 1$

In Exercises 17–19, sketch the level surface.

17. $r = 4$
18. $\theta = \frac{\pi}{3}$
19. $z = -2$

In Exercises 20–23, sketch the set (described in cylindrical coordinates).

20. $1 \le r \le 3, \quad 0 \le z \le 4$
21. $z^2 + r^2 \le 4$
22. $1 \le r \le 3, \quad 0 \le \theta \le \frac{\pi}{2}, \quad 0 \le z \le 4$
23. $r \le 3, \quad \pi \le \theta \le \frac{3\pi}{2}, \quad z = 4$

In Exercises 24–29, find an equation of the form $r = f(\theta, z)$ in cylindrical coordinates for the following surfaces.

24. $z = x + y$
25. $x^2 + y^2 + z^2 = 4$
26. $\frac{x^2}{yz} = 1$
27. $x^2 - y^2 = 4$
28. $x^2 + y^2 = 4$
29. $z = 3xy$

In Exercises 30–35, convert from spherical to rectangular coordinates.

30. $\left(3, 0, \frac{\pi}{2}\right)$ **31.** $\left(2, \frac{\pi}{4}, \frac{\pi}{3}\right)$ **32.** $(3, \pi, 0)$

33. $\left(5, \frac{3\pi}{4}, \frac{\pi}{4}\right)$ **34.** $\left(6, \frac{\pi}{6}, \frac{5\pi}{6}\right)$ **35.** $(0.5, 3.7, 2)$

In Exercises 36–41, convert from rectangular to spherical coordinates.

36. $(\sqrt{3}, 0, 1)$ **37.** $\left(\frac{\sqrt{3}}{2}, \frac{3}{2}, 1\right)$

38. $(1, 1, 1)$ **39.** $(1, -1, 1)$

40. $\left(\frac{1}{2}, \frac{\sqrt{3}}{2}, \sqrt{3}\right)$ **41.** $\left(\frac{\sqrt{2}}{2}, \frac{\sqrt{2}}{2}, \sqrt{3}\right)$

In Exercises 42–47, describe the given set in spherical coordinates.

42. $x^2 + y^2 + z^2 \leq 1$

43. $x^2 + y^2 + z^2 = 1, \quad z \geq 0$

44. $x^2 + y^2 + z^2 = 1, \quad x \geq 0, \quad y \geq 0, \quad z \geq 0$

45. $x^2 + y^2 + z^2 \leq 1, \quad x = y, \quad x \geq 0, \quad y \geq 0$

46. $y^2 + z^2 \leq 4, \quad x = 0$

47. $x^2 + y^2 = 3z^2$

In Exercises 48–50, sketch the level surface.

48. $\rho = 4$ **49.** $\theta = \frac{\pi}{3}$ **50.** $\phi = \frac{\pi}{4}$

In Exercises 51–54, sketch the set of points.

51. $\rho = 2, \quad 0 \leq \phi \leq \frac{\pi}{2}$ **52.** $\theta = \frac{\pi}{2}, \quad \phi = \frac{\pi}{4}, \quad \rho \geq 1$

53. $\rho \leq 2, \quad 0 \leq \theta \leq \frac{\pi}{2}, \quad \frac{\pi}{2} \leq \phi \leq \pi$

54. $\rho = 1, \quad \frac{\pi}{3} \leq \phi \leq \frac{2\pi}{3}$

In Exercises 55–60, find an equation of the form $\rho = f(\theta, \phi)$ in spherical coordinates for the following surfaces.

55. $z = 2$ **56.** $z^2 = 3(x^2 + y^2)$ **57.** $x = z^2$

58. $z = x^2 + y^2$ **59.** $x^2 - y^2 = 4$ **60.** $xy = z$

61. Which of (a)–(c) is the equation of the cylinder of radius R in spherical coordinates? Refer to Figure 15.

(a) $R\rho = \sin\phi$ **(b)** $\rho \sin\phi = R$ **(c)** $\rho = R\sin\phi$

62. Let $P_1 = (1, -\sqrt{3}, 5)$ and $P_2 = (-1, \sqrt{3}, 5)$ in rectangular coordinates. In which quadrants do the projections of P_1 and P_2 onto the xy-plane lie? Find the polar angle θ of each point.

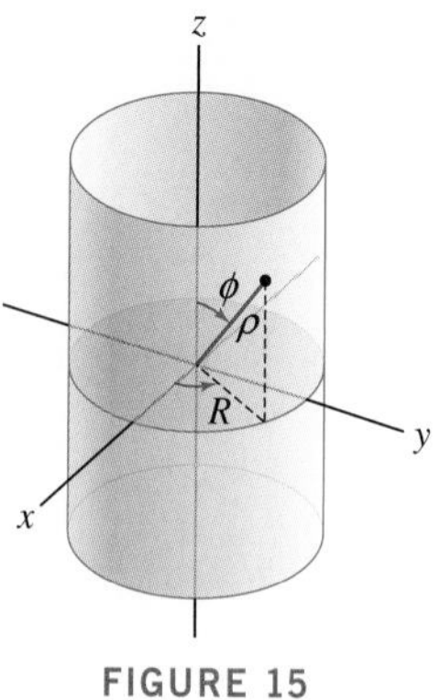

FIGURE 15

63. Consider a rectangular coordinate system with origin at the center of the earth, z-axis through the North Pole, and x-axis through the prime meridian. Find the rectangular coordinates of Sydney, Australia (34° S, 151° E), and Bogota, Colombia (4° 32′ N, 74° 15′ W). A minute is $1/60°$. Assume that the earth is a sphere of radius $R = 6{,}367$ km.

64. Find the equation in rectangular coordinates of the quadric surface consisting of the two cones $\phi = \frac{\pi}{4}$ and $\phi = \frac{3\pi}{4}$.

65. Find an equation of the form $z = f(r, \theta)$ in cylindrical coordinates for $z^2 = x^2 - y^2$.

66. Show that $\rho = 2\cos\phi$ is the equation of a sphere with its center on the z-axis. Find its radius and center.

67. Explain the following statement: If the equation of a surface in cylindrical or spherical coordinates does not involve the coordinate θ, then the surface is rotationally symmetric with respect to the z-axis.

68. CAS Plot the surface $\rho = 1 - \cos\phi$. Then plot the trace of S in the xz-plane and explain why S is obtained by rotating this trace.

69. Find equations $r = g(\theta, z)$ (cylindrical) and $\rho = f(\theta, \phi)$ (spherical) for the hyperboloid $x^2 + y^2 = z^2 + 1$ (Figure 16). Do there exist points on the hyperboloid with $\phi = 0$ or π? Which values of ϕ occur for points on the hyperboloid?

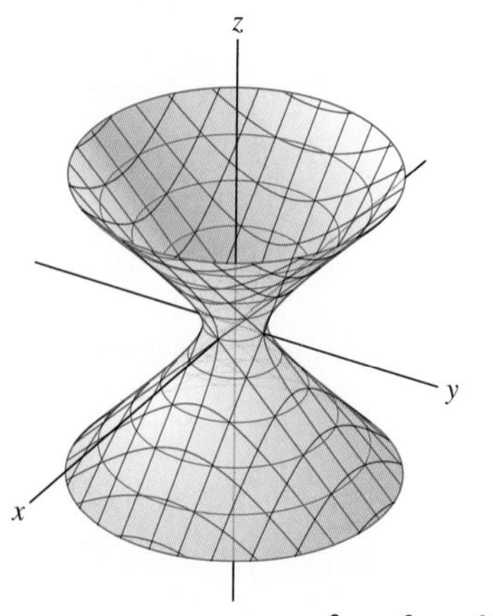

FIGURE 16 The hyperboloid $x^2 + y^2 = z^2 + 1$.

Further Insights and Challenges

*In Exercises 70–74, a **great circle** on a sphere S with center O is a circle obtained by intersecting S with a plane that passes through O (Figure 17). If P and Q are not antipodal (on opposite sides), there is a unique great circle through P and Q on S (intersect S with the plane through O, P, and Q). The geodesic distance from P to Q is defined as the length of the smaller of the two circular arcs of this great circle.*

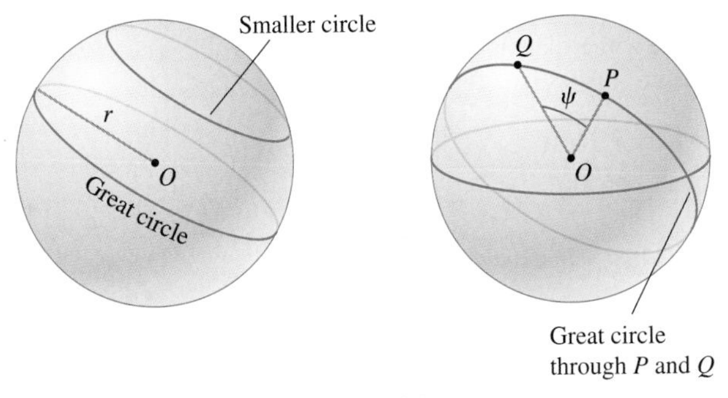

FIGURE 17

70. Show that the geodesic distance from P to Q is equal to $R\psi$, where ψ is the *central angle* between P and Q (the angle between the vectors $\mathbf{v} = \overrightarrow{OP}$ and $\mathbf{u} = \overrightarrow{OQ}$).

71. Show that the geodesic distance from $Q = (a, b, c)$ to the North Pole $P = (0, 0, R)$ is equal to $R\cos^{-1}\left(\frac{c}{R}\right)$.

72. The coordinates of Los Angeles are 34° N and 118° W. Find the geodesic distance from the North Pole to Los Angeles, assuming that the earth is a sphere of radius 3,960 miles.

73. Show that the central angle ψ between points P and Q on a sphere (of any radius) with angular coordinates (θ, ϕ) and (θ', ϕ') is equal to

$$\psi = \cos^{-1}\left(\sin\phi \sin\phi' \cos(\theta - \theta') + \cos\phi\cos\phi'\right)$$

Hint: Compute the dot product of $\overrightarrow{OP}$ and $\overrightarrow{OQ}$. Check this formula by computing the geodesic distance between the North and South Poles.

74. Use Exercise 73 to find the geodesic distance between Los Angeles (34° N, 118° W) and Bombay (19° N, 72.8° E).

CHAPTER REVIEW EXERCISES

In Exercises 1–6, let $\mathbf{v} = \langle -2, 5\rangle$, *and* $\mathbf{w} = \langle 3, -2\rangle$.

1. Calculate $5\mathbf{w} - 3\mathbf{v}$ and $5\mathbf{v} - 3\mathbf{w}$.

2. Sketch $\mathbf{v}$, $\mathbf{w}$, and $2\mathbf{v} - 3\mathbf{w}$.

3. Find the unit vector in the direction of $\mathbf{v}$.

4. Find the length of $\mathbf{v} + \mathbf{w}$.

5. Express $\mathbf{i}$ as a linear combination $r\mathbf{v} + s\mathbf{w}$.

6. Find a scalar α such that $\|\mathbf{v} + \alpha\mathbf{w}\| = 6$.

7. If $P = (1, 4)$ and $Q = (-3, 5)$, what are the components of $\overrightarrow{PQ}$? What is the length of $\overrightarrow{PQ}$?

8. Let $A = (2, -1)$, $B = (1, 4)$, and $P = (2, 3)$. Find the point Q such that $\overrightarrow{PQ}$ is equivalent to $\overrightarrow{AB}$. Sketch $\overrightarrow{PQ}$ and $\overrightarrow{AB}$.

9. Find the vector with length 3 making an angle of $\frac{7\pi}{4}$ with the positive x-axis.

10. Calculate $3(\mathbf{i} - 2\mathbf{j}) - 6(\mathbf{i} + 6\mathbf{j})$.

11. Find the value of β for which $\mathbf{w} = \langle -2, \beta\rangle$ is parallel to $\mathbf{v} = \langle 4, -3\rangle$.

12. Let $P = (1, 4, -3)$.

(a) Find the point Q such that $\overrightarrow{PQ}$ is equivalent to $\langle 3, -1, 5\rangle$.

(b) Find a unit vector $\mathbf{e}$ equivalent to $\overrightarrow{PQ}$.

13. Let $\mathbf{w} = \langle 2, -2, 1\rangle$ and $\mathbf{v} = \langle 4, 5, -4\rangle$. Solve for $\mathbf{u}$ if $\mathbf{v} + 5\mathbf{u} = 3\mathbf{w} - \mathbf{u}$.

14. Let $\mathbf{v} = 3\mathbf{i} - \mathbf{j} + 4\mathbf{k}$. Find the length of $\mathbf{v}$ and the vector $2\mathbf{v} + 3(4\mathbf{i} - \mathbf{k})$.

15. Find a parametrization $\mathbf{r}_1(t)$ of the line passing through $(1, 4, 5)$ and $(-2, 3, -1)$. Then find a parametrization $\mathbf{r}_2(t)$ of the line parallel to $\mathbf{r}_1$ passing through $(1, 0, 0)$.

16. Let $\mathbf{r}_1(t) = \mathbf{v}_1 + t\mathbf{w}_1$ and $\mathbf{r}_2(t) = \mathbf{v}_2 + t\mathbf{w}_2$ be the parametrizations of lines $\mathcal{L}_1$ and $\mathcal{L}_2$. For each statement (a)–(e), either prove it or provide a counterexample showing that it is false.

(a) If $\mathcal{L}_1 = \mathcal{L}_2$, then $\mathbf{v}_1 = \mathbf{v}_2$ and $\mathbf{w}_1 = \mathbf{w}_2$.

(b) If $\mathcal{L}_1 = \mathcal{L}_2$ and $\mathbf{v}_1 = \mathbf{v}_2$, then $\mathbf{w}_1 = \mathbf{w}_2$.

(c) If $\mathcal{L}_1 = \mathcal{L}_2$ and $\mathbf{w}_1 = \mathbf{w}_2$, then $\mathbf{v}_1 = \mathbf{v}_2$.

(d) If $\mathcal{L}_1$ is parallel to $\mathcal{L}_2$, then $\mathbf{w}_1 = \mathbf{w}_2$.

(e) If $\mathcal{L}_1$ is parallel to $\mathcal{L}_2$, then $\mathbf{w}_1 = \lambda\mathbf{w}_2$ for some scalar λ.

17. Find a and b such that the lines $\mathbf{r}_1 = \langle 1, 2, 1\rangle + t\langle 1, -1, 1\rangle$ and $\mathbf{r}_2 = \langle 3, -1, 1\rangle + t\langle a, b, -2\rangle$ are parallel.

18. Find a such that the lines $\mathbf{r}_1 = \langle 1, 2, 1\rangle + t\langle 1, -1, 1\rangle$ and $\mathbf{r}_2 = \langle 3, -1, 1\rangle + t\langle a, 4, -2\rangle$ intersect.

19. Sketch the vector sum $\mathbf{v} = \mathbf{v}_1 - \mathbf{v}_2 + \mathbf{v}_3$ for the vectors in Figure 1(A).

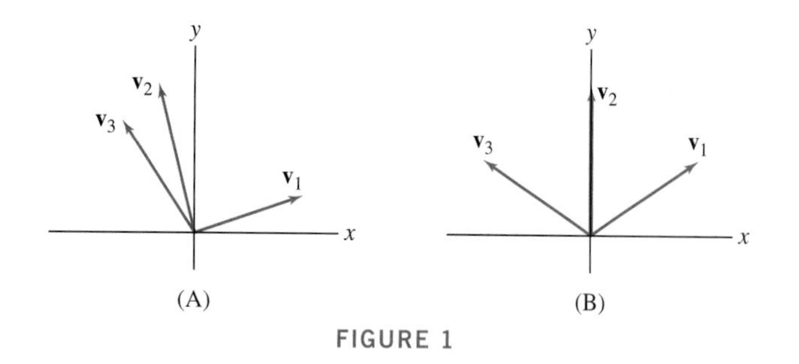

FIGURE 1

20. Sketch the sums $\mathbf{v}_1 + \mathbf{v}_2 + \mathbf{v}_3$, $\mathbf{v}_1 + 2\mathbf{v}_2$, and $\mathbf{v}_2 - \mathbf{v}_3$ for the vectors in Figure 1(B).

21. Use vectors to prove that the line connecting the midpoints of two sides of a triangle is parallel to the third side.

In Exercises 22–27, let $\mathbf{v} = \langle 1, 3, -2\rangle$ *and* $\mathbf{w} = \langle 2, -1, 4\rangle$.

22. Compute $\mathbf{v} \cdot \mathbf{w}$.

23. Compute the angle between $\mathbf{v}$ and $\mathbf{w}$.

24. Compute $\mathbf{v} \times \mathbf{w}$.

25. Find the area of the parallelogram spanned by $\mathbf{v}$ and $\mathbf{w}$.

26. Find the volume of the parallelopiped spanned by $\mathbf{v}$, $\mathbf{w}$, and $\mathbf{u} = \langle 1, 2, 6\rangle$.

27. Find all the vectors orthogonal to both $\mathbf{v}$ and $\mathbf{w}$.

28. Let $\mathbf{v} = \langle 1, -1, 3\rangle$ and $\mathbf{w} = \langle 4, -2, 1\rangle$.
(a) Find the decomposition $\mathbf{v} = \mathbf{v}_{\parallel} + \mathbf{v}_{\perp}$ with respect to $\mathbf{w}$.
(b) Find the decomposition $\mathbf{w} = \mathbf{w}_{\parallel} + \mathbf{w}_{\perp}$ with respect to $\mathbf{v}$.

29. A 50-kg wagon is pulled to the right by a force $\mathbf{F}_1$ making an angle of 30° with the ground. At the same time the wagon is pulled to the left by a horizontal force $\mathbf{F}_2$.
(a) Find the magnitude of $\mathbf{F}_1$ in terms of the magnitude of $\mathbf{F}_2$ if the wagon does not move.
(b) What is the maximal magnitude of $\mathbf{F}_1$ that can be applied to the wagon without lifting it?

30. Let $\mathbf{v}$, $\mathbf{w}$, and $\mathbf{u}$ be the vectors in $\mathbf{R}^3$. Which of the following is a scalar?
(a) $\mathbf{v} \times (\mathbf{u} + \mathbf{w})$
(b) $(\mathbf{u} + \mathbf{w}) \cdot (\mathbf{v} \times \mathbf{w})$
(c) $(\mathbf{u} \times \mathbf{w}) + (\mathbf{w} - \mathbf{v})$

In Exercises 31–34, let $\mathbf{v} = \langle 1, 2, 4\rangle$, $\mathbf{u} = \langle 6, -1, 2\rangle$, *and* $\mathbf{w} = \langle 1, 0, -3\rangle$. *Calculate the given quantity.*

31. $\mathbf{v} \times \mathbf{w}$

32. $\mathbf{w} \times \mathbf{u}$

33. $\det \begin{pmatrix} \mathbf{u} \\ \mathbf{v} \\ \mathbf{w} \end{pmatrix}$

34. $\mathbf{v} \cdot (\mathbf{u} \times \mathbf{w})$

35. Use the cross product to find the area of the triangle whose vertices are $(1, 3, -1)$, $(2, -1, 3)$, and $(4, 1, 1)$.

36. Calculate $\|\mathbf{v} \times \mathbf{w}\|$ if $\|\mathbf{v}\| = 2$, $\mathbf{v} \cdot \mathbf{w} = 3$, and the angle between $\mathbf{v}$ and $\mathbf{w}$ is $\frac{\pi}{6}$.

37. Show that if the vectors $\mathbf{v}$, $\mathbf{w}$ are orthogonal, then $\|\mathbf{v} + \mathbf{w}\|^2 = \|\mathbf{v}\|^2 + \|\mathbf{w}\|^2$.

38. Find the angle between $\mathbf{v}$, $\mathbf{w}$ if $\|\mathbf{v} + \mathbf{w}\| = \|\mathbf{v}\| = \|\mathbf{w}\|$.

39. Show that the equation $\langle 1, 2, 3\rangle \times \mathbf{v} = \langle -1, 2, a\rangle$ has no solution for $a \neq -1$.

40. Prove with a diagram the following: If $\mathbf{e}$ is a unit vector orthogonal to $\mathbf{v}$, then $\mathbf{e} \times (\mathbf{v} \times \mathbf{e}) = (\mathbf{e} \times \mathbf{v}) \times \mathbf{e} = \mathbf{v}$.

41. Use the identity

$$\mathbf{u} \times (\mathbf{v} \times \mathbf{w}) = (\mathbf{u} \cdot \mathbf{w})\,\mathbf{v} - (\mathbf{u} \cdot \mathbf{v})\,\mathbf{w}$$

to prove that

$$\mathbf{u} \times (\mathbf{v} \times \mathbf{w}) + \mathbf{v} \times (\mathbf{w} \times \mathbf{u}) + \mathbf{w} \times (\mathbf{u} \times \mathbf{v}) = \mathbf{0}$$

42. Find an equation of the plane through $(1, -3, 5)$ with normal vector $\mathbf{n} = \langle 2, 1, -4\rangle$.

43. Write the equation of the plane $\mathcal{P}$ with vector equation

$$\langle 1, 4, -3\rangle \cdot \langle x, y, z\rangle = 7$$

in the form

$$a\,(x - x_0) + b\,(y - y_0) + c\,(z - z_0) = 0$$

Hint: You must find a point $P = (x_0, y_0, z_0)$ on $\mathcal{P}$.

44. Find all the planes parallel to the plane passing through the points $(1, 2, 3)$, $(1, 2, 7)$, and $(1, 1, -3)$.

45. Find the plane through $P = (4, -1, 9)$ containing the line $\mathbf{r}(t) = \langle 1, 4, -3\rangle + t\langle 2, 1, 1\rangle$.

46. Find the intersection of the line $\mathbf{r}(t) = \langle 3t + 2, 1, -7t\rangle$ and the plane $2x - 3y + z = 5$.

47. Find the trace of the plane $3x - 2y + 5z = 4$ in the xy-plane.

48. Find the intersection of the planes $x + y + z = 1$ and $3x - 2y + z = 5$.

In Exercises 49–54, determine the type of the quadric surface.

49. $\left(\frac{x}{3}\right)^2 + \left(\frac{y}{4}\right)^2 + 2z^2 = 1$

50. $\left(\frac{x}{3}\right)^2 - \left(\frac{y}{4}\right)^2 + 2z^2 = 1$

51. $\left(\frac{x}{3}\right)^2 + \left(\frac{y}{4}\right)^2 - 2z = 0$

52. $\left(\frac{x}{3}\right)^2 - \left(\frac{y}{4}\right)^2 - 2z = 0$

53. $\left(\frac{x}{3}\right)^2 - \left(\frac{y}{4}\right)^2 - 2z^2 = 0$

54. $\left(\frac{x}{3}\right)^2 - \left(\frac{y}{4}\right)^2 - 2z^2 = 1$

55. Determine the type of the quadric surface $ax^2 + by^2 - z^2 = 1$ if:

(a) $a < 0, \quad b < 0$

(b) $a > 0, \quad b > 0$

(c) $a > 0, \quad b < 0$

56. Describe the traces of the surface

$$\left(\frac{x}{2}\right)^2 - y^2 + \left(\frac{z}{2}\right)^2 = 1$$

in the three coordinate planes.

57. Convert $(x, y, z) = (3, 4, -1)$ from rectangular to cylindrical and spherical coordinates.

58. Convert $(r, \theta, z) = \left(3, \frac{\pi}{6}, 4\right)$ from cylindrical to spherical coordinates.

59. Convert the point $(\rho, \theta, \phi) = \left(3, \frac{\pi}{6}, \frac{\pi}{3}\right)$ from spherical to cylindrical coordinates.

60. Describe the set of all points $P = (x, y, z)$ satisfying $x^2 + y^2 \le 4$ in both cylindrical and spherical coordinates.

61. Sketch the graph of the cylindrical equation $z = 2r\cos\theta$ and write the equation in rectangular coordinates.

62. Write the surface $x^2 + y^2 - z^2 = 2(x + y)$ as an equation $r = f(\theta, z)$ in cylindrical coordinates.

63. Show that the cylindrical equation

$$r^2(1 - 2\sin^2\theta) + z^2 = 1$$

is a hyperboloid of one sheet.

64. Sketch the graph of the spherical equation $\rho = 2\cos\theta\sin\phi$ and write the equation in rectangular coordinates.

65. Describe how the surface with spherical equation

$$\rho^2(1 + A\cos^2\phi) = 1$$

depends on the constant A.

66. Show that the spherical equation $\cot\phi = 2\cos\theta + \sin\theta$ defines a plane through the origin (with the origin excluded). Find a normal vector to this plane.

67. Let c be a scalar, let $\mathbf{a}$ and $\mathbf{b}$ be vectors, and let $\mathbf{X} = \langle x, y, z \rangle$. Show that the equation $(\mathbf{X} - \mathbf{a})(\mathbf{X} - \mathbf{b}) = c^2$ defines a sphere with center $\mathbf{m} = \frac{1}{2}(\mathbf{a} + \mathbf{b})$ and radius R, where $R^2 = c^2 + \|\frac{1}{2}(\mathbf{a} - \mathbf{b})\|^2$.

13 CALCULUS OF VECTOR-VALUED FUNCTIONS

The mathematics of space flight is based on vector calculus. In this photo, Astronauts Piers Sellers and Michael Fossum work on the space shuttle's Remote Manipulator System.

In this chapter, we study vector-valued functions and their derivatives, and use them to analyze curves and motion in three-space. Although many techniques from single-variable calculus carry over to the vector setting, there is an important new aspect to the derivative. A real-valued function $f(x)$ can change in just one of two ways: it can increase or decrease. By contrast, a vector-valued function can change, not just in magnitude but also in direction, and the rate of change is not a single number but is itself a vector. To develop these new concepts, we begin with an introduction to vector-valued functions.

13.1 Vector-Valued Functions

Consider a particle moving in $\mathbf{R}^3$ and suppose that its coordinates at time t are $(x(t), y(t), z(t))$. It is convenient to represent the particle's path by the **vector-valued function**

Functions $f(x)$ (with real number values) are often called scalar-valued, *to distinguish them from vector-valued functions.*

$$\mathbf{r}(t) = \langle x(t), y(t), z(t)\rangle = x(t)\mathbf{i} + y(t)\mathbf{j} + z(t)\mathbf{k} \qquad \boxed{1}$$

The vector $\mathbf{r}(t)$ points from the origin to the position of the particle at time t (Figure 1).

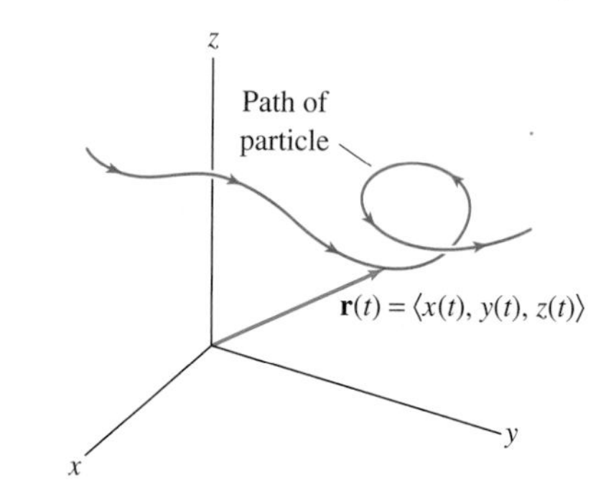

FIGURE 1

The letter t is a convenient choice for the parameter (which often represents time), but we are free to use any other variable such as s or θ. It is best to avoid writing $\mathbf{r}(x)$ or $\mathbf{r}(y)$ to prevent confusion with the x- and y-components of $\mathbf{r}$.

More generally, a vector-valued function is any function $\mathbf{r}(t)$ of the form (1) whose domain $\mathcal{D}$ is a set of real numbers and whose range is a set of position vectors. The variable t is called a **parameter**, and the functions $x(t)$, $y(t)$, $z(t)$ are called the **components** or **coordinate functions**. We usually take as domain the set of all values of t for which $\mathbf{r}(t)$ is defined, that is, all values of t that belong to the domains of all three coordinate functions $x(t)$, $y(t)$, $z(t)$. For example,

$$\mathbf{r}(t) = \langle t^2, e^t, 4-7t\rangle \qquad \text{domain } \mathcal{D} = \mathbf{R}$$

$$\mathbf{r}(s) = \langle \sqrt{s}, e^s, s^{-1}\rangle \qquad \text{domain } \mathcal{D} = \{s \in \mathbf{R} : s > 0\}$$

If the components of $\mathbf{r}(t)$ are continuous functions, then the terminal point of $\mathbf{r}(t)$ traces a path in $\mathbf{R}^3$ as t varies, and we refer to $\mathbf{r}(t)$ as a **vector parametrization** of this path. The set of all points $(x(t), y(t), z(t))$ for t in the domain of $\mathbf{r}$ is called a **space curve** $\mathcal{C}$.

The description of a path by a vector-valued function $\mathbf{r}(t) = \langle x(t), y(t)\rangle$ is equivalent to the description as a parametrized curve $c(t) = (x(t), y(t))$ that we considered in Chapter 11. The difference lies only in whether we visualize the path as traced by a "moving vector" $\mathbf{r}(t)$ or a "moving point" $c(t)$. The vector form $\mathbf{r}(t)$ is more natural in the context of vector-valued derivatives, and we will use it rather than $c(t)$ in this chapter.

We treated a special case of vector-valued functions in Chapter 12 when we studied vector parametrizations of lines. Recall that the line through $P_0 = (x_0, y_0, z_0)$ with direction vector $\mathbf{v} = \langle a, b, c\rangle$ has the vector parametrization

$$\mathbf{r}(t) = \langle x_0, y_0, z_0\rangle + t\mathbf{v} = \langle x_0 + ta, y_0 + tb, z_0 + tc\rangle$$

In some cases, it is helpful to consider the projections onto the coordinate planes. The projection of $\mathbf{r}(t) = \langle x(t), y(t), z(t)\rangle$ onto the xy-plane is the path $\mathbf{p}(t) = \langle x(t), y(t), 0\rangle$ (Figure 1). Similarly, the projections onto the yz- and xz-planes are the paths $\langle 0, y(t), z(t)\rangle$ and $\langle x(t), 0, z(t)\rangle$, respectively.

EXAMPLE 1 Helix The curve traced by $\mathbf{r}(t) = \langle -\sin t, \cos t, t\rangle$ is a helix. Describe this curve and its projections onto the coordinate planes.

Solution The projection onto the xy-plane is the path $\mathbf{p}(t) = \langle -\sin t, \cos t, 0\rangle$, which describes a point moving counterclockwise around the unit circle. The function $\mathbf{r}(t)$ itself describes a point whose projection traces out the circle while the height $z = t$ increases linearly with time, resulting in the helix of Figure 2. The projection onto the xz-plane is

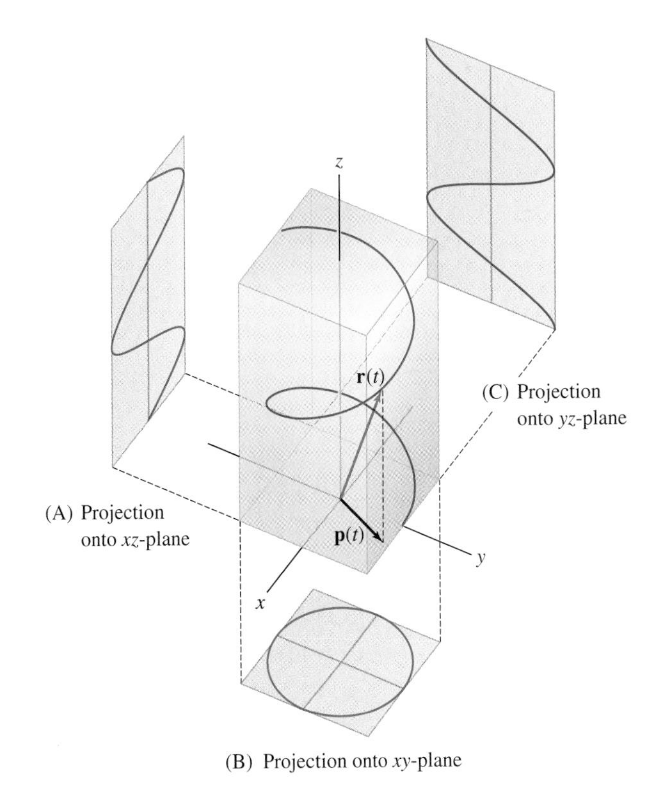

FIGURE 2 Projections of the helix $\mathbf{r}(t) = \langle \cos t, \sin t, t\rangle$.

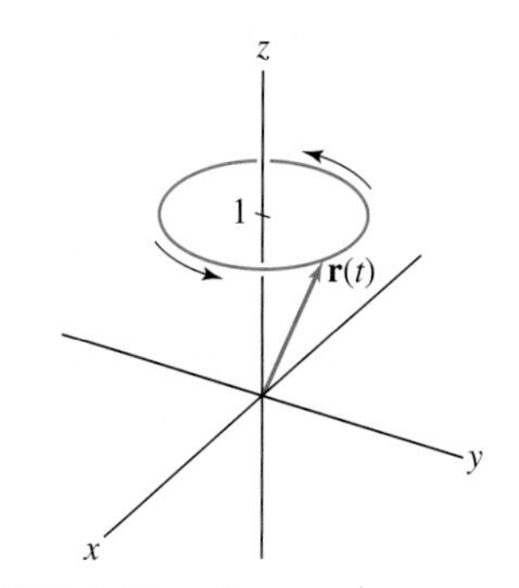

FIGURE 3 Plot of $\mathbf{r}(t) = \langle \cos t, \sin t, 1 \rangle$.

the path $\langle -\sin t, 0, t \rangle$, which is a wave moving in the positive z-direction. Similarly, the projection onto the yz-plane is the wave $\langle 0, \cos t, t \rangle$. ■

It is important to distinguish between the path $\mathbf{r}(t)$ and the underlying space curve $\mathcal{C}$. The path is a particular way of traversing the curve; it may traverse the curve several times, reverse direction, or move back and forth, etc. For example, the path $\mathbf{r}(t) = \langle \cos t, \sin t, 1 \rangle$ moves around a unit circle at height $z = 1$ infinitely many times as t varies from $-\infty$ to ∞ (Figure 3).

In general, space curves may be quite complicated and difficult to sketch by hand. Fortunately, computers produce good graphics from different perspectives (Figure 4). As an aid to visualization, it is helpful to plot a "thickened" curve as in Figures 4 and 5, but keep in mind that space curves are one-dimensional and have no thickness.

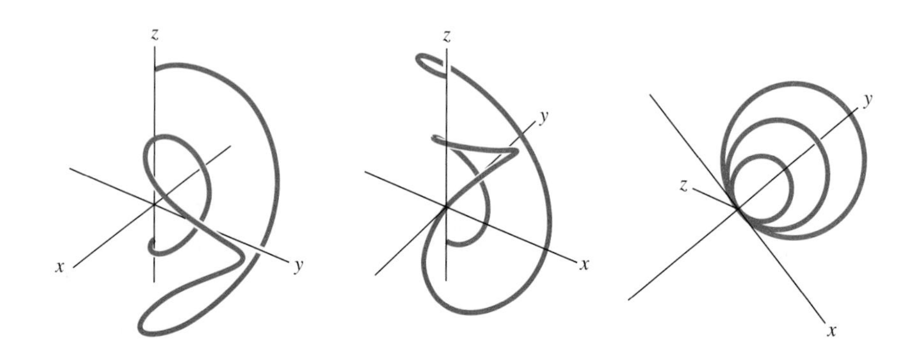

FIGURE 4 The curve $\mathbf{r}(t) = \langle t \sin t \cos t, t \sin^2 t, t \cos t \rangle$ for $\pi \le t \le 4\pi$.

Keep in mind that every curve can be parametrized in infinitely many ways. The next example describes two very different parametrizations of the same curve.

■ **EXAMPLE 2 Parametrizing the Intersection of Surfaces** Parametrize the intersection of the surfaces $x^2 - y^2 = z - 1$ and $x^2 + y^2 = 4$ (Figure 5).

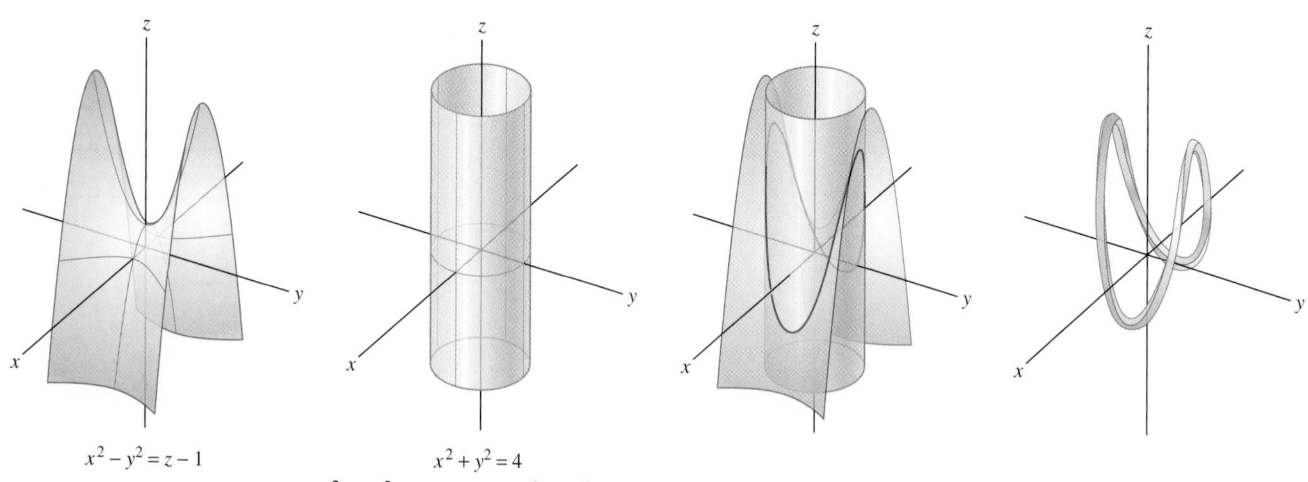

FIGURE 5 Intersection of surfaces $x^2 - y^2 = z - 1$ and $x^2 + y^2 = 4$.

Solution Our goal is to express the coordinates of a point on the curve as functions of a parameter t. We solve this problem in two ways. First, let's solve for y and z in terms of x. The two equations may be rewritten $y^2 = 4 - x^2$ and $z = x^2 - y^2 + 1$. Thus,

$$y = \pm\sqrt{4 - x^2}, \qquad z = x^2 - y^2 + 1 = x^2 - (4 - x^2) + 1 = 2x^2 - 3$$

Taking $t = x$ as parameter, we have $y = \pm\sqrt{4 - t^2}$, $z = 2t^2 - 3$. The two signs of the square root correspond to the two halves of the curve where $y > 0$ and $y < 0$ (Figure 6). Therefore, we need two vector-valued functions to parametrize the entire curve:

$$\mathbf{r}_1(t) = \left\langle t, \sqrt{4 - t^2}, 2t^2 - 3 \right\rangle, \quad \mathbf{r}_2(t) = \left\langle t, -\sqrt{4 - t^2}, 2t^2 - 3 \right\rangle, \qquad -2 \le t \le 2$$

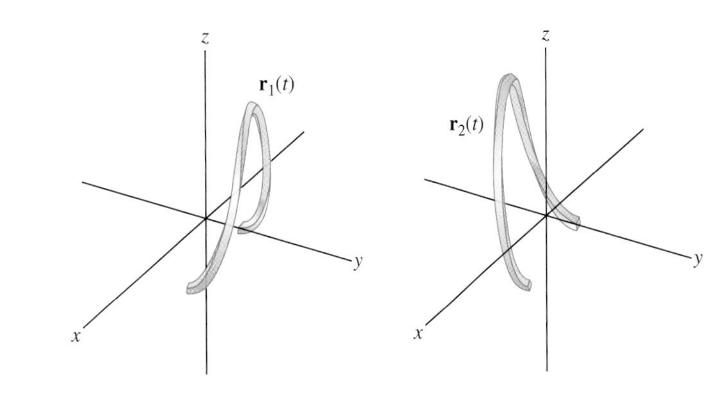

FIGURE 6 Two halves of the curve of intersection in Example 2.

A second and perhaps better parametrization is obtained by noting that the equation $x^2 + y^2 = 4$ has a trigonometric parametrization: $x = 2\cos t$, $y = 2\sin t$. The second equation yields

$$z = x^2 - y^2 + 1 = 4\cos^2 t - 4\sin^2 t + 1 = 4\cos 2t + 1$$

Thus, we may parametrize the entire curve by a single vector-valued function:

$$\mathbf{r}(t) = \langle 2\cos t, 2\sin t, 4\cos 2t + 1 \rangle, \qquad 0 \le t < 2\pi$$

■

■ **EXAMPLE 3** Parametrize the circle of radius 3 with center $P = (2, 6, 8)$ located in a plane:

(a) Parallel to the xy-plane **(b)** Parallel to the xz-plane

Solution A circle of radius R in the xy-plane with center at the origin has parametrization $\langle R\cos t, R\sin t \rangle$. To place the circle in a three-dimensional coordinate system, we use the parametrization $\langle R\cos t, R\sin t, 0 \rangle$.

(a) The circle of radius 3 centered at the origin in the xy-plane has parametrization $\langle 3\cos t, 3\sin t, 0 \rangle$. To move the circle in a parallel fashion so that its center lies at $P = (2, 6, 8)$, we translate by the vector $\langle 2, 6, 8 \rangle$:

$$\mathbf{r}_1(t) = \langle 2, 6, 8 \rangle + \langle 3\cos t, 3\sin t, 0 \rangle = \langle 2 + 3\cos t, 6 + 3\sin t, 8 \rangle$$

(b) The parametrization $\langle 3\cos t, 0, 3\sin t \rangle$ gives us a circle of radius 3 centered at the origin in the xz-plane. To move the circle in a parallel fashion so that its center lies at $(2, 6, 8)$, we translate by the vector $\langle 2, 6, 8 \rangle$:

$$\mathbf{r}_2(t) = \langle 2, 6, 8 \rangle + \langle 3\cos t, 0, 3\sin t \rangle = \langle 2 + 3\cos t, 6, 8 + 3\sin t \rangle$$

These two circles are shown in Figure 7. ■

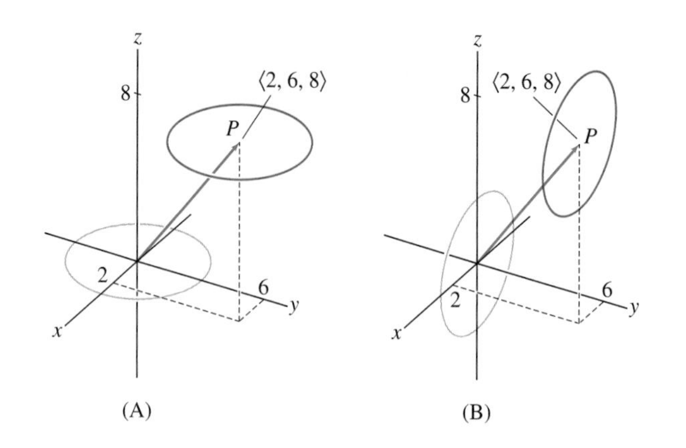

FIGURE 7 Horizontal and vertical circles of radius 3 and center $P = (2, 6, 8)$ obtained by translation.

13.1 SUMMARY

- A *vector-valued function* is a function of the form

$$\mathbf{r}(t) = \langle x(t), y(t), z(t)\rangle = x(t)\mathbf{i} + y(t)\mathbf{j} + z(t)\mathbf{k}$$

We often think of the parameter t as time and $\mathbf{r}(t)$ as a "moving vector" whose terminal point traces out a path as a function of time. We call $\mathbf{r}(t)$ a *vector parametrization* of the path.

- If the components $x(t)$, $y(t)$, and $z(t)$ are continuous, the set of points $(x(t), y(t), z(t))$ in $\mathbf{R}^3$ for t in the domain of $\mathbf{r}(t)$ is called a *space curve*.
- Every space curve can be parametrized in infinitely many ways.
- The projection of $\mathbf{r}(t)$ onto the xy-plane is the curve traced by $\langle x(t), y(t), 0\rangle$. The projection onto the xz-plane is $\langle x(t), 0, z(t)\rangle$ and the projection onto the yz-plane is $\langle 0, y(t), z(t)\rangle$.

13.1 EXERCISES

Preliminary Questions

1. Which one of the following does *not* parametrize a line?

(a) $\mathbf{r}_1(t) = \langle 8 - t, 2t, 3t\rangle$

(b) $\mathbf{r}_2(t) = t^3\mathbf{i} - 7t^3\mathbf{j} + t^3\mathbf{k}$

(c) $\mathbf{r}_3(t) = \langle 8 - 4t^3, 2 + 5t^2, 9t^3\rangle$

2. What is the projection of $\mathbf{r}(t) = t\mathbf{i} + t^4\mathbf{j} + e^t\mathbf{k}$ onto the xz-plane?

3. Which projection of $\langle \cos t, \cos 2t, \sin t\rangle$ is a circle?

4. What is the center of the circle with parametrization

$$\mathbf{r}(t) = (-2 + \cos t)\mathbf{i} + 2\mathbf{j} + (3 - \sin t)\mathbf{k}?$$

5. How do the paths $\mathbf{r}_1(t) = \langle \cos t, \sin t\rangle$ and $\mathbf{r}_2(t) = \langle \sin t, \cos t\rangle$ around the unit circle differ?

6. Which three of the following vector-valued functions parametrize the same space curve?

(a) $(-2 + \cos t)\mathbf{i} + 9\mathbf{j} + (3 - \sin t)\mathbf{k}$

(b) $(2 + \cos t)\mathbf{i} - 9\mathbf{j} + (-3 - \sin t)\mathbf{k}$

(c) $(-2 + \cos 3t)\mathbf{i} + 9\mathbf{j} + (3 - \sin 3t)\mathbf{k}$

(d) $(-2 - \cos t)\mathbf{i} + 9\mathbf{j} + (3 + \sin t)\mathbf{k}$

(e) $(2 + \cos t)\mathbf{i} + 9\mathbf{j} + (3 + \sin t)\mathbf{k}$

Exercises

1. What is the domain of $\mathbf{r}(t) = e^t\mathbf{i} + \frac{1}{t}\mathbf{j} + (t + 1)^{-3}\mathbf{k}$?

2. What is the domain of $\mathbf{r}(s) = e^s\mathbf{i} + \sqrt{s}\mathbf{j} + \cos s\mathbf{k}$?

3. Find a vector parametrization of the line through $P = (3, -5, 7)$ in the direction $\mathbf{v} = \langle 3, 0, 1\rangle$.

4. Find a direction vector for the line with parametrization $\mathbf{r}(t) = (4-t)\mathbf{i} + (2+5t)\mathbf{j} + \frac{1}{2}t\mathbf{k}$.

5. Match the space curves in Figure 8 with their projections onto the xy-plane in Figure 9.

6. Match the space curves in Figure 8 with the following vector-valued functions:

(a) $\mathbf{r}_1(t) = \langle \cos 2t, \cos t, \sin t \rangle$
(b) $\mathbf{r}_2(t) = \langle t, \cos 2t, \sin 2t \rangle$
(c) $\mathbf{r}_3(t) = \langle 1, t, t \rangle$

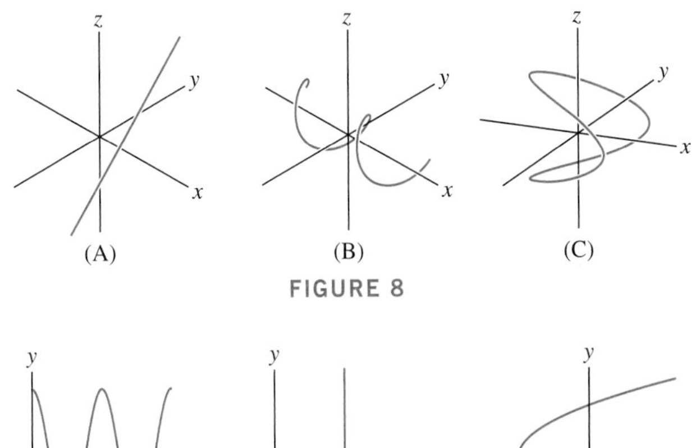

FIGURE 8

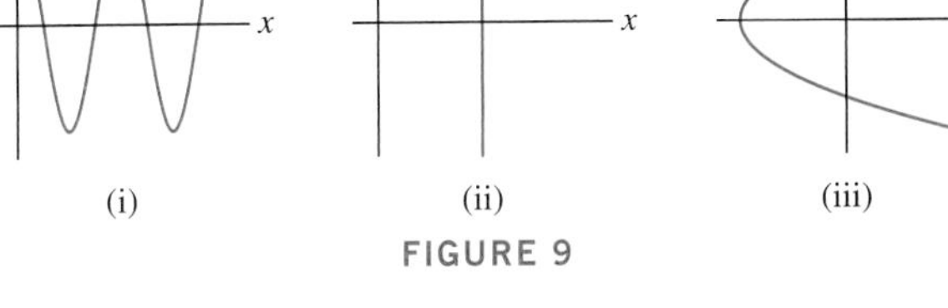

FIGURE 9

7. Match the vector-valued functions (a)–(f) with the space curves (i)–(vi) in Figure 10.

(a) $\mathbf{r}(t) = \left\langle t+15, e^{0.08t}\cos t, e^{0.08t}\sin t \right\rangle$
(b) $\mathbf{r}(t) = \left\langle \cos t, \sin t, \sin 12t \right\rangle$

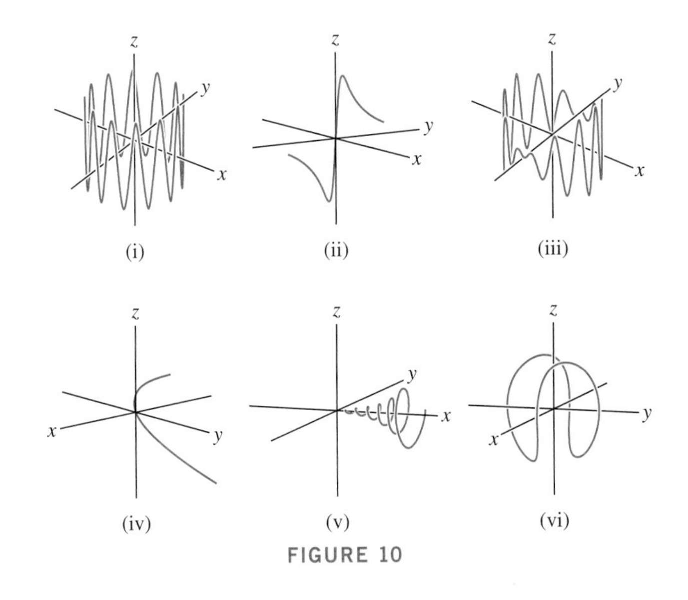

FIGURE 10

(c) $\mathbf{r}(t) = \left\langle t, t, \dfrac{25t}{1+t^2} \right\rangle$
(d) $\mathbf{r}(t) = \left\langle \cos^3 t, \sin^3 t, \sin 2t \right\rangle$
(e) $\mathbf{r}(t) = \left\langle t, t^2, 2t \right\rangle$
(f) $\mathbf{r}(t) = \left\langle \cos t, \sin t, \cos t \sin 12t \right\rangle$

8. Which of the following curves have the same projection onto the xy-plane?

(a) $\mathbf{r}_1(t) = \left\langle t, t^2, e^t \right\rangle$
(b) $\mathbf{r}_2(t) = \left\langle e^t, t^2, t \right\rangle$
(c) $\mathbf{r}_3(t) = \left\langle t, t^2, \cos t \right\rangle$

9. Match the space curves (A)–(C) in Figure 11 with their projections (i)–(iii) onto the xy-plane.

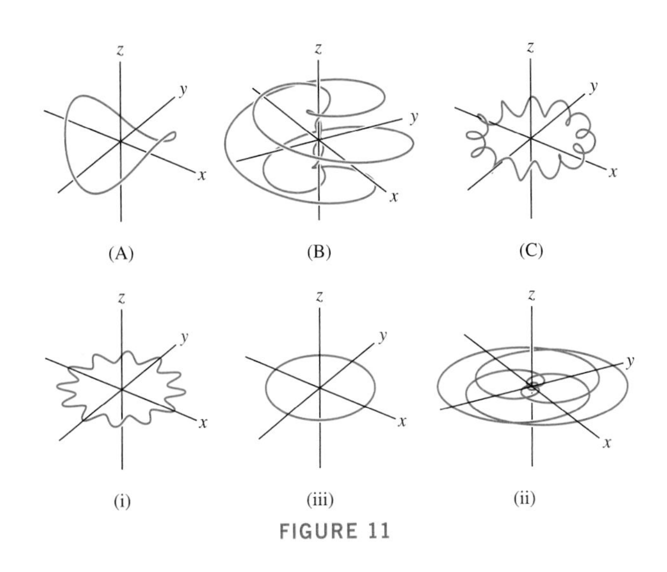

FIGURE 11

In Exercises 10–13, the function $\mathbf{r}(t)$ *traces a circle. Determine the radius, center, and plane containing the circle.*

10. $\mathbf{r}(t) = (9\cos t)\mathbf{i} + (9\sin t)\mathbf{j}$

11. $\mathbf{r}(t) = 7\mathbf{i} + (12\cos t)\mathbf{j} + (12\sin t)\mathbf{k}$

12. $\mathbf{r}(t) = \langle \sin t, 0, 4+\cos t \rangle$

13. $\mathbf{r}(t) = \langle 6+3\sin t, 9, 4+3\cos t \rangle$

14. Describe the projections of the circle $\mathbf{r}(t) = \langle \sin t, 0, 4+\cos t \rangle$ onto the coordinate planes.

15. Do either of $P = (4, 11, 20)$ or $Q = (-1, 6, 16)$ lie on the curve $\mathbf{r}(t) = \left\langle 1+t, 2+t^2, t^4 \right\rangle$?

16. **(a)** Describe the curve $\mathbf{r}(t) = \langle t\cos t, t\sin t, t \rangle$ and its projections onto the xy- and xz-planes.

(b) CAS Plot $\mathbf{r}(t)$ with a computer algebra system if you have one.

17. Find the points where the path $\mathbf{r}(t) = \langle \sin t, \cos t, \sin t \cos 2t \rangle$ intersects the xy-plane.

18. Parametrize the intersection of the surfaces

$$y^2 - z^2 = x - 2, \qquad y^2 + z^2 = 9$$

using $t = y$ as the parameter (two vector functions are needed as in Example 2).

19. Find a parametrization of the curve in Exercise 18 using trigonometric functions.

20. Viviani's Curve $\mathcal{C}$ is the intersection of the surfaces $x^2 + y^2 = z^2$, $y = z^2$ (Figure 12).

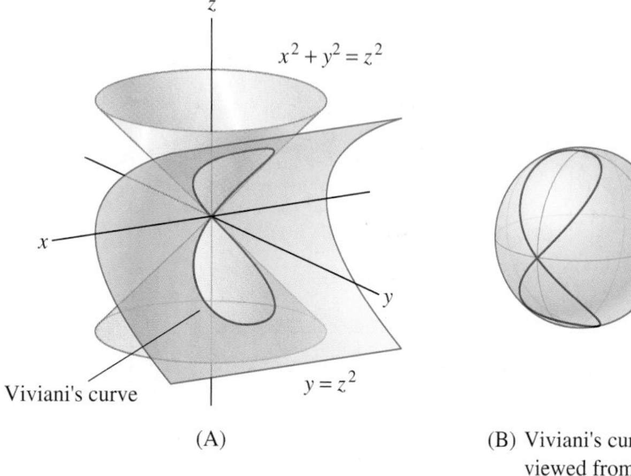

FIGURE 12 Viviani's curve is the intersection of the surfaces $x^2 + y^2 = z^2$ and $y = z^2$.

(a) Parametrize each of the two parts of $\mathcal{C}$ corresponding to $y \geq 0$ and $y \leq 0$ using $t = z$ as parameter.

(b) Describe the projection of $\mathcal{C}$ onto the xy-plane.

(c) Show that $\mathcal{C}$ lies on the sphere of radius 1 with center $(0, 1, 0)$. This curve looks like a figure eight lying on a sphere [Figure 12(B)].

21. Show that any point on $x^2 + y^2 = z^2$ can be written in the form $(z \cos \theta, z \sin \theta, z)$ for some θ. Use this to find a parametrization of Viviani's curve (Exercise 20) with θ as parameter.

22. Use sine and cosine to parametrize the intersection of the cylinders $x^2 + y^2 = 1$ and $x^2 + z^2 = 1$ (use two vector-valued functions). Then describe the projections of this curve on the three coordinate planes.

23. CAS Use sine and cosine to parametrize the intersection of the surfaces $x^2 + y^2 = 1$ and $z = 4x^2$, and plot this curve using a CAS (Figure 13).

24. Use hyperbolic functions to parametrize the intersection of the surfaces $x^2 - y^2 = 4$, $z = xy$.

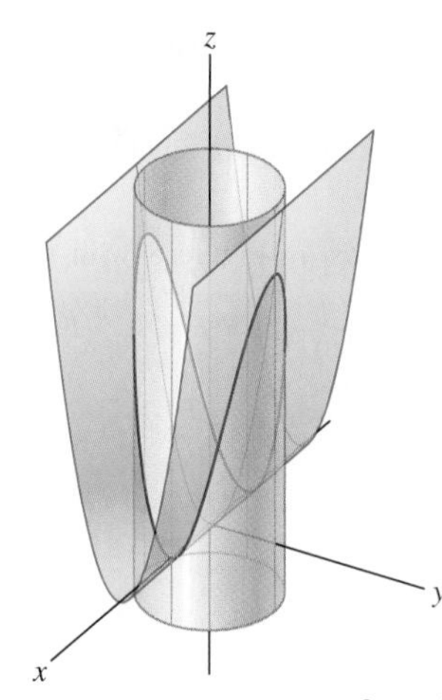

FIGURE 13 Intersection of the surfaces $x^2 + y^2 = 1$ and $z = 4x^2$.

In Exercises 25–34, find a parametrization of the curve.

25. The vertical line passing through the point $(3, 2, 0)$

26. The line passing through $(1, 0, 4)$ and $(4, 1, 2)$

27. The line through the origin whose projection on the xy-plane is a line of slope 3 and on the yz-plane is a line of slope 5 (i.e., $\Delta y / \Delta z = 5$)

28. The horizontal circle of radius 1 with center $(2, -1, 4)$

29. The circle of radius 2 with center $(1, 2, 5)$ in a plane parallel to the yz-plane

30. The ellipse $\left(\frac{x}{2}\right)^2 + \left(\frac{y}{3}\right)^2 = 1$ in the xy-plane, translated to have center $(9, -4, 0)$

31. The intersection of the plane $y = \frac{1}{2}$ with the sphere $x^2 + y^2 + z^2 = 1$

32. The intersection of the surfaces

$$z = x^2 - y^2 \qquad \text{and} \qquad z = x^2 + xy - 1$$

33. The ellipse $\left(\frac{x}{2}\right)^2 + \left(\frac{z}{3}\right)^2 = 1$ in the xz-plane, translated to have center $(3, 1, 5)$ [Figure 14(A)]

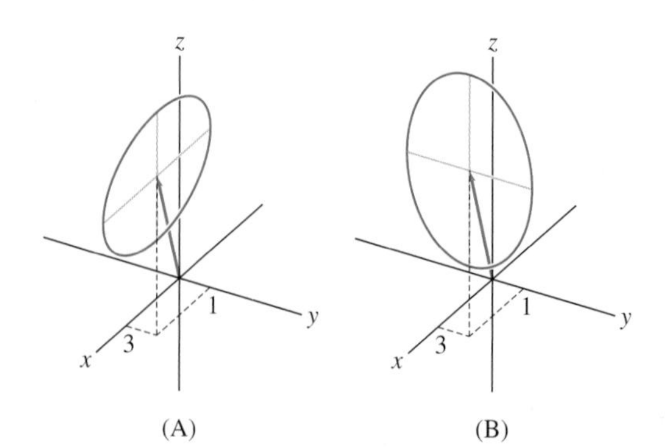

FIGURE 14 The ellipses described in Exercise 33 and 34.

34. The ellipse $\left(\frac{y}{2}\right)^2 + \left(\frac{z}{3}\right)^2 = 1$, translated to have center $(3, 1, 5)$ [Figure 14(B)]

In Exercises 35–37, assume that two paths $\mathbf{r}_1(t)$ *and* $\mathbf{r}_2(t)$ *intersect if there is a point P lying on both curves. We say that* $\mathbf{r}_1(t)$ *and* $\mathbf{r}_2(t)$ *collide if* $\mathbf{r}_1(t_0) = \mathbf{r}_2(t_0)$ *at some time* t_0.

35. Which of the following are true?

(a) If $\mathbf{r}_1$ and $\mathbf{r}_2$ intersect, then they collide.
(b) If $\mathbf{r}_1$ and $\mathbf{r}_2$ collide, then they intersect.
(c) Intersection depends only on the underlying curves traced by $\mathbf{r}_1$ and $\mathbf{r}_2$ but collision depends on the actual parametrizations.

36. Determine whether $\mathbf{r}_1$ and $\mathbf{r}_2$ collide or intersect:

$$\mathbf{r}_1(t) = \langle t^2 + 3, t + 1, 6t^{-1} \rangle$$
$$\mathbf{r}_2(t) = \langle 4t, 2t - 2, t^2 - 7 \rangle$$

37. Determine whether $\mathbf{r}_1$ and $\mathbf{r}_2$ collide or intersect:

$$\mathbf{r}_1(t) = \langle t, t^2, t^3 \rangle, \qquad r_2(t) = \langle 4t + 6, 4t^2, 7 - t \rangle$$

Further Insights and Challenges

38. Sketch the curve parametrized by $\mathbf{r}(t) = \langle |t| + t, |t| - t \rangle$.

39. Find the maximum height above the xy-plane of a point on $\mathbf{r}(t) = \langle e^t, \sin t, t(4 - t) \rangle$.

40. Let C be the curve obtained by intersecting a cylinder of radius r and a plane. Insert two spheres of radius r into the cylinder above and below the plane, and let F_1 and F_2 be the points where the plane is tangent to the sphere [Figure 15(A)]. Let K be the vertical distance between the equators of the two spheres. Rediscover Archimedes's proof that C is an ellipse by showing that every point P on C satisfies

$$PF_1 + PF_2 = K \qquad \boxed{2}$$

Hint: If two lines through a point P are tangent to a sphere and intersect the sphere at Q_1 and Q_2 as in Figure 15(B), then the segments $\overline{PQ_1}$ and $\overline{PQ_2}$ have equal length. Use this to show that $PF_1 = PR_1$ and $PF_2 = PR_2$.

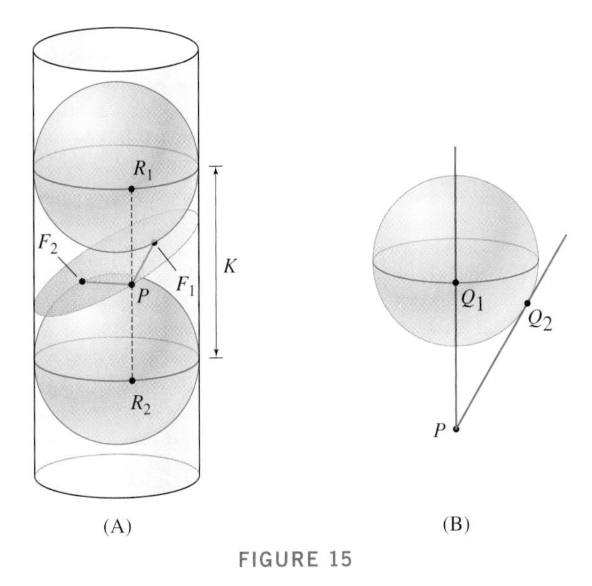

FIGURE 15

41. CAS Now reprove the result of Exercise 40 using vector geometry. Assume that the cylinder has equation $x^2 + y^2 = r^2$ and the plane has equation $z = ax + by$.

(a) Show that the upper and lower spheres in Figure 15 have centers

$$C_1 = \left(0, 0, r\sqrt{a^2 + b^2 + 1}\right)$$
$$C_2 = \left(0, 0, -r\sqrt{a^2 + b^2 + 1}\right)$$

(b) Show that the points where the plane is tangent to the sphere are

$$F_1 = \frac{r}{\sqrt{a^2 + b^2 + 1}} \langle a, b, a^2 + b^2 \rangle$$
$$F_2 = \frac{-r}{\sqrt{a^2 + b^2 + 1}} \langle a, b, a^2 + b^2 \rangle$$

Hint: Show that $\overline{C_1F_1}$ and $\overline{C_2F_2}$ have length r and are orthogonal to the plane.

(c) Verify, with the aid of a computer algebra system, that Eq. (2) holds with $K = 2r\sqrt{a^2 + b^2 + 1}$. To simplify the algebra, observe that since a and b are arbitrary, it suffices to verify Eq. (2) for the point $P = (r, 0, ar)$.

13.2 Calculus of Vector-Valued Functions

In this section, we extend differentiation and integration to vector-valued functions and we discuss the geometric interpretation of the derivative as a tangent vector. The first step is to define limits of vector-valued functions.

DEFINITION Limit of a Vector-Valued Function A vector-valued function $\mathbf{r}(t)$ approaches the limit $\mathbf{v}$ (a vector) as t approaches t_0 if $\lim_{t\to t_0} \|\mathbf{r}(t) - \mathbf{v}\| = 0$. In this case, we write

$$\lim_{t\to t_0} \mathbf{r}(t) = \mathbf{v}$$

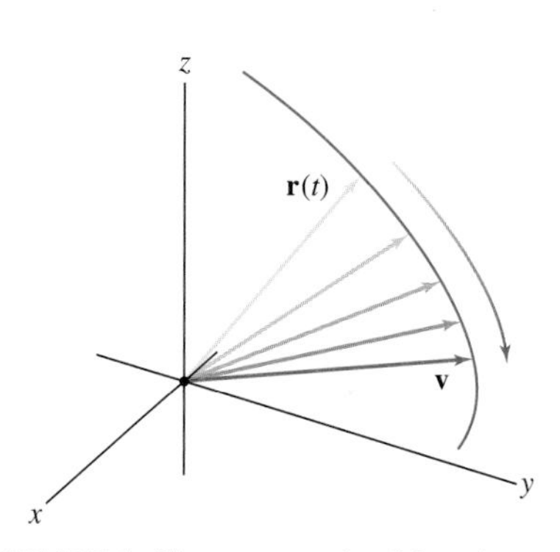

FIGURE 1 The vector-valued function $\mathbf{r}(t)$ approaches $\mathbf{v}$ as $t \to t_0$.

We may visualize the limit of a vector-valued function as a vector $\mathbf{r}(t)$ "moving" toward the limit $\mathbf{v}$ (Figure 1). According to the following theorem, vector limits may be computed componentwise.

THEOREM 1 Vector-Valued Limits Are Computed Componentwise A vector-valued function $\mathbf{r}(t) = \langle x(t), y(t), z(t)\rangle$ approaches a limit as $t \to t_0$ if and only if each component approaches a limit, and in this case,

$$\lim_{t\to t_0} \mathbf{r}(t) = \left\langle \lim_{t\to t_0} x(t), \lim_{t\to t_0} y(t), \lim_{t\to t_0} z(t)\right\rangle \quad \boxed{1}$$

The Limit Laws of scalar functions remain valid in the vector-valued case. They are verified by applying the Limit Laws to the components.

Proof Suppose that $\lim_{t\to t_0} \mathbf{r}(t) = \mathbf{v}$, where $\mathbf{v} = \langle a, b, c\rangle$. Then the length $\|\mathbf{r}(t) - \mathbf{v}\|$ tends to zero and its square

$$\|\mathbf{r}(t) - \mathbf{v}\|^2 = (x(t) - a)^2 + (y(t) - b)^2 + (z(t) - c)^2 \quad \boxed{2}$$

also tends to zero as $t \to t_0$. Since the terms on the right-hand side of Eq. (2) are nonnegative, each term individually tends to zero. Thus,

$$\left\langle \lim_{t\to t_0} x(t), \lim_{t\to t_0} y(t), \lim_{t\to t_0} z(t)\right\rangle = \langle a, b, c\rangle = \mathbf{v} \quad \boxed{3}$$

Conversely, if (3) holds, then $\|\mathbf{r}(t) - \mathbf{v}\|^2$ tends to zero by Eq. (2) and we have

$$\lim_{t\to t_0} \mathbf{r}(t) = \mathbf{v}$$

■

■ **EXAMPLE 1** Calculate $\lim_{t\to 3} \mathbf{r}(t)$, where $\mathbf{r}(t) = \langle t^2, 1 - t, t^{-1}\rangle$.

Solution By Theorem 1,

$$\lim_{t\to 3} \mathbf{r}(t) = \lim_{t\to 3} \langle t^2, 1 - t, t^{-1}\rangle = \left\langle \lim_{t\to 3} t^2, \lim_{t\to 3} (1 - t), \lim_{t\to 3} t^{-1}\right\rangle = \left\langle 9, -2, \frac{1}{3}\right\rangle$$

■

Continuity and the derivative for vector-valued functions are defined in much the same way as in the scalar case. A vector-valued function $\mathbf{r}(t) = \langle x(t), y(t), z(t)\rangle$ is **continuous** at t_0 if

$$\lim_{t\to t_0} \mathbf{r}(t) = \mathbf{r}(t_0)$$

By Theorem 1, $\mathbf{r}(t)$ is continuous at t_0 if and only if the components $x(t)$, $y(t)$, $z(t)$ are continuous at t_0. The derivative of $\mathbf{r}(t)$ is the limit of the difference quotient:

$$\mathbf{r}'(t) = \frac{d}{dt}\mathbf{r}(t) = \lim_{h\to 0} \frac{\mathbf{r}(t + h) - \mathbf{r}(t)}{h} \quad \boxed{4}$$

The terminology and notation of differential calculus carry over to the vector-valued setting. We say that $\mathbf{r}(t)$ is *differentiable* at t if the limit in Eq. (4) exists. In Leibniz notation, the derivative is written $\frac{d\mathbf{r}}{dt}$. Higher-order derivatives are defined by repeated differentiation:

$$\mathbf{r}''(t) = \frac{d}{dt}\mathbf{r}'(t), \qquad \mathbf{r}'''(t) = \frac{d}{dt}\mathbf{r}''(t), \ \ldots$$

The components of the difference quotient are difference quotients:

$$\lim_{h\to 0} \frac{\mathbf{r}(t+h)-\mathbf{r}(t)}{h} = \lim_{h\to 0} \left\langle \frac{x(t+h)-x(t)}{h}, \frac{y(t+h)-y(t)}{h}, \frac{z(t+h)-z(t)}{h} \right\rangle$$

It follows from Theorem 1 that $\mathbf{r}(t)$ is differentiable if and only if the components are differentiable, and in this case, $\mathbf{r}'(t)$ is equal to the vector of derivatives $\langle x'(t), y'(t), z'(t)\rangle$.

Theorems 1 and 2 show that vector-valued limits and derivatives may be computed "componentwise" and thus are not more difficult to compute than ordinary limits and derivatives.

THEOREM 2 Vector-Valued Derivatives Are Computed Componentwise A vector-valued function $\mathbf{r}(t) = \langle x(t), y(t), z(t)\rangle$ is differentiable if and only if each component is differentiable. In this case,

$$\mathbf{r}'(t) = \frac{d}{dt}\mathbf{r}(t) = \langle x'(t), y'(t), z'(t)\rangle$$

Here are some vector-valued derivatives, computed componentwise:

$$\frac{d}{dt}\langle t^2, t^3, \sin t\rangle = \langle 2t, 3t^2, \cos t\rangle, \qquad \frac{d}{dt}\langle \cos t, -1, e^{2t}\rangle = \langle -\sin t, 0, 2e^{2t}\rangle$$

EXAMPLE 2 Calculate $\mathbf{r}''(3)$, where $\mathbf{r}(t) = \langle \ln t, t, t^2\rangle$.

Solution We perform the differentiation componentwise:

$$\mathbf{r}'(t) = \frac{d}{dt}\langle \ln t, t, t^2\rangle = \langle t^{-1}, 1, 2t\rangle$$

$$\mathbf{r}''(t) = \frac{d}{dt}\langle t^{-1}, 1, 2t\rangle = \langle -t^{-2}, 0, 2\rangle$$

Therefore, $\mathbf{r}''(3) = \left\langle -\frac{1}{9}, 0, 2\right\rangle$. ■

Since vector differentiation is performed componentwise, the differentiation rules of single-variable calculus carry over to the vector setting.

Differentiation Rules Assume that $\mathbf{r}(t)$, $\mathbf{r}_1(t)$, and $\mathbf{r}_2(t)$ are differentiable. Then

- **Sum Rule:** $(\mathbf{r}_1(t) + \mathbf{r}_2(t))' = \mathbf{r}_1'(t) + \mathbf{r}_2'(t)$
- **Constant Multiple Rule:** For any constant c, $(c\,\mathbf{r}(t))' = c\,\mathbf{r}'(t)$.
- **Product Rule:** For any differentiable scalar-valued function $f(t)$,

$$\frac{d}{dt}\big(f(t)\mathbf{r}(t)\big) = f(t)\mathbf{r}'(t) + f'(t)\mathbf{r}(t)$$

- **Chain Rule:** For any differentiable scalar-valued function $g(t)$,

$$\frac{d}{dt}\mathbf{r}(g(t)) = g'(t)\mathbf{r}'(g(t))$$

We do not state a Quotient Rule because $\mathbf{r}(t)/g(t)$ equals $g(t)^{-1}\mathbf{r}(t)$ and may be differentiated using the Product Rule.

Proof Each rule is proved by applying the differentiation rules to the components. For example, to prove the Product Rule (we consider vector-valued functions in the plane, to keep the notation simple), we write

$$f(t)\mathbf{r}(t) = f(t)\langle x(t), y(t)\rangle = \langle f(t)x(t), f(t)y(t)\rangle$$

Now apply the Product Rule to each component:

$$\begin{aligned}\frac{d}{dt}f(t)\mathbf{r}(t) &= \left\langle \frac{d}{dt}f(t)x(t), \frac{d}{dt}f(t)y(t)\right\rangle \\ &= \langle f'(t)x(t) + f(t)x'(t), f'(t)y(t) + f(t)y'(t)\rangle \\ &= \langle f'(t)x(t), f'(t)y(t)\rangle + \langle f(t)x'(t), f(t)y'(t)\rangle \\ &= f'(t)\langle x(t), y(t)\rangle + f(t)\langle x'(t), y'(t)\rangle = f'(t)\mathbf{r}(t) + f(t)\mathbf{r}'(t)\end{aligned}$$

The remaining proofs are left as exercises (Exercises 64–65). ■

■ **EXAMPLE 3** Let $\mathbf{r}(t) = \langle t^2, 5t, 1\rangle$ and $f(t) = e^t$. Calculate:

(a) $\frac{d}{dt}f(t)\mathbf{r}(t)$ **(b)** $\frac{d}{dt}\mathbf{r}(f(t))$

Solution We have $\mathbf{r}'(t) = \langle 2t, 5, 0\rangle$ and $f'(t) = e^t$.

(a) By the Product Rule,

$$\begin{aligned}\frac{d}{dt}f(t)\mathbf{r}(t) &= f(t)\mathbf{r}'(t) + f'(t)\mathbf{r}(t) = e^t\langle 2t, 5, 0\rangle + e^t\langle t^2, 5t, 1\rangle \\ &= \langle (t^2 + 2t)e^t, (5t + 5)e^t, e^t\rangle\end{aligned}$$

(b) By the Chain Rule,

$$\frac{d}{dt}\mathbf{r}(f(t)) = f'(t)\mathbf{r}'(f(t)) = e^t\mathbf{r}'(e^t) = e^t\langle 2e^t, 5, 0\rangle = \langle 2e^{2t}, 5e^t, 0\rangle$$ ■

There are three different Product Rules for vector-valued functions. We have already stated the rule for the product of a scalar function $f(t)$ and a vector-valued function $\mathbf{r}(t)$ above. The Product Rules for the dot and cross products of two vector-valued functions are even more important in applications.

CAUTION Order is important in the Product Rule for cross products. The first term is $\mathbf{r}_1(t) \times \mathbf{r}_2'(t)$ [not $\mathbf{r}_2'(t) \times \mathbf{r}_1(t)$] and the second term is $\mathbf{r}_1'(t) \times \mathbf{r}_2(t)$. Why is order not a concern for dot products?

THEOREM 3 Product Rule for Dot and Cross Products Assume that $\mathbf{r}_1(t)$ and $\mathbf{r}_2(t)$ are differentiable. Then

Dot Products: $$\frac{d}{dt}(\mathbf{r}_1(t) \cdot \mathbf{r}_2(t)) = \mathbf{r}_1(t) \cdot \mathbf{r}_2'(t) + \mathbf{r}_1'(t) \cdot \mathbf{r}_2(t) \qquad (5)$$

Cross Products: $$\frac{d}{dt}(\mathbf{r}_1(t) \times \mathbf{r}_2(t)) = [\mathbf{r}_1(t) \times \mathbf{r}_2'(t)] + [\mathbf{r}_1'(t) \times \mathbf{r}_2(t)] \qquad (6)$$

Proof We verify (5) for vector-valued functions in the plane for simplicity of notation. If $\mathbf{r}_1(t) = \langle x_1(t), y_1(t)\rangle$ and $\mathbf{r}_2(t) = \langle x_2(t), y_2(t)\rangle$, then

$$\begin{aligned}\frac{d}{dt}\big(\mathbf{r}_1(t)\cdot\mathbf{r}_2(t)\big) &= \frac{d}{dt}\big(x_1(t)x_2(t)+y_1(t)y_2(t)\big)\\ &= x_1(t)x_2'(t)+x_1'(t)x_2(t)+y_1(t)y_2'(t)+y_1'(t)y_2(t)\\ &= \big(x_1(t)x_2'(t)+y_1(t)y_2'(t)\big)+\big(x_1'(t)x_2(t)+y_1'(t)y_2(t)\big)\\ &= \mathbf{r}_1(t)\cdot\mathbf{r}_2'(t)+\mathbf{r}_1'(t)\cdot\mathbf{r}_2(t)\end{aligned}$$

The proof of (6) is left as an exercise (Exercise 66). ■

In the next example and throughout this chapter, *all vector-valued functions are assumed differentiable, unless otherwise stated.*

■ **EXAMPLE 4** Prove the formula $\frac{d}{dt}\big(\mathbf{r}(t)\times\mathbf{r}'(t)\big)=\mathbf{r}(t)\times\mathbf{r}''(t)$.

Solution By the Product Formula for cross products,

$$\frac{d}{dt}\big(\mathbf{r}(t)\times\mathbf{r}'(t)\big)=\mathbf{r}(t)\times\mathbf{r}''(t)+\underbrace{\mathbf{r}'(t)\times\mathbf{r}'(t)}_{\text{Equals }\mathbf{0}}=\mathbf{r}(t)\times\mathbf{r}''(t)$$

Here, $\mathbf{r}'\times\mathbf{r}'=\mathbf{0}$ because the cross product of a vector with itself is the zero vector. ■

The Derivative as a Tangent Vector

The vector $\mathbf{r}'(t_0)$ has an important geometric interpretation: It points in the direction tangent to the path traced by $\mathbf{r}(t)$ at $t=t_0$. To understand why, consider the difference quotient where we write $\Delta\mathbf{r}=\mathbf{r}(t_0+h)-\mathbf{r}(t_0)$ and $\Delta t=h$ with $h\neq 0$:

$$\frac{\Delta\mathbf{r}}{\Delta t}=\frac{\mathbf{r}(t_0+h)-\mathbf{r}(t_0)}{h} \qquad \boxed{7}$$

The vector $\Delta\mathbf{r}$ points from the terminal point of $\mathbf{r}(t)$ to the terminal point of $\mathbf{r}(t+h)$ as in Figure 2(A). The difference quotient $\Delta\mathbf{r}/\Delta t$ is a scalar multiple of $\Delta\mathbf{r}$ and therefore points in the same direction, along the secant line, in the direction from $\mathbf{r}(t_0)$ to $\mathbf{r}(t_0+h)$ [Figure 2(B)].

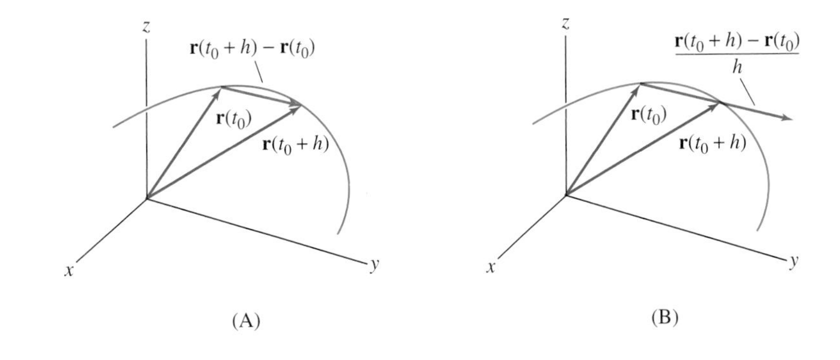

FIGURE 2 The difference quotient points in the direction of $\Delta\mathbf{r}=\mathbf{r}(t_0+h)-\mathbf{r}(t_0)$.

As $h\to 0$, $\Delta\mathbf{r}$ itself tends to zero but the quotient $\Delta\mathbf{r}/\Delta t$ approaches a vector $\mathbf{r}'(t_0)$ which, if nonzero, points in the direction tangent to the curve (Figure 3). For this reason, $\mathbf{r}'(t_0)$ is called the **tangent vector** at $\mathbf{r}(t_0)$. We also refer to $\mathbf{r}'(t_0)$ as the **velocity vector**.

Although it has been our convention so far to regard all vectors as based at the origin, we usually view the tangent vector $\mathbf{r}'(t)$ as a vector based at the terminal point of $\mathbf{r}(t)$. This is more natural because $\mathbf{r}'(t)$ then appears as a vector tangent to the curve (Figure 3).

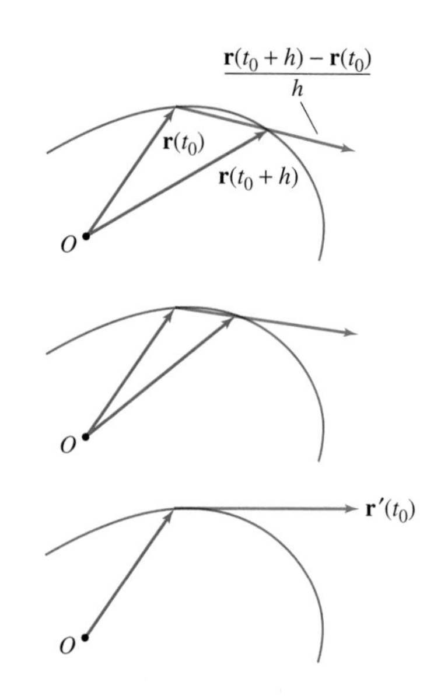

FIGURE 3 The difference quotient converges to a vector $\mathbf{r}'(t)$, tangent to the curve $\mathbf{r}(t)$.

If $\mathbf{r}'(t_0) \neq \mathbf{0}$, then the tangent line itself has vector parametrization:

$$\text{Tangent line at } \mathbf{r}(t_0): \qquad \mathbf{L}(t) = \mathbf{r}(t_0) + t\mathbf{r}'(t_0) \tag{8}$$

Thus, $\mathbf{r}'(t_0)$ is a direction vector for the tangent line.

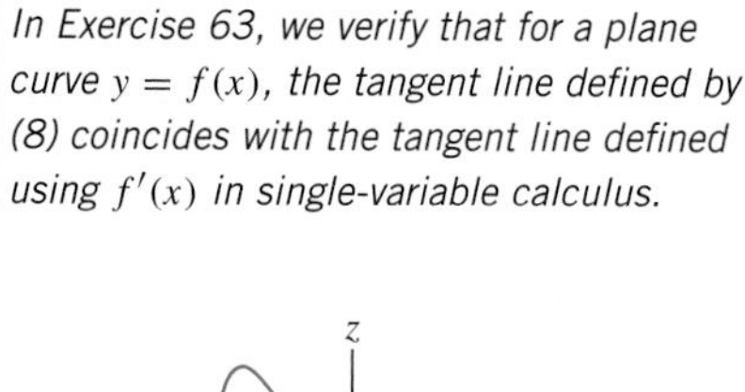

In Exercise 63, we verify that for a plane curve $y = f(x)$, the tangent line defined by (8) coincides with the tangent line defined using $f'(x)$ in single-variable calculus.

EXAMPLE 5 Plotting Tangent Vectors *CAS* Plot $\mathbf{r}(t) = \langle \cos t, \sin t, 4\cos^2 t \rangle$ together with its tangent vectors at $t = \frac{\pi}{4}, \frac{3\pi}{2}$. Find a parametrization of the tangent line at $t = \frac{\pi}{4}$.

Solution Figure 4 shows the plot of $\mathbf{r}(t)$ produced using the parametric plot function on a computer algebra system. Observe that

$$\mathbf{r}'(t) = \langle -\sin t, \cos t, -8\cos t \sin t \rangle$$

and

$$\mathbf{r}\left(\frac{\pi}{4}\right) = \left\langle \frac{\sqrt{2}}{2}, \frac{\sqrt{2}}{2}, 2 \right\rangle, \qquad \mathbf{r}'\left(\frac{\pi}{4}\right) = \left\langle -\frac{\sqrt{2}}{2}, \frac{\sqrt{2}}{2}, -4 \right\rangle$$

$$\mathbf{r}\left(\frac{3\pi}{2}\right) = \langle 0, -1, 0 \rangle, \qquad \mathbf{r}'\left(\frac{3\pi}{2}\right) = \langle 1, 0, 0 \rangle$$

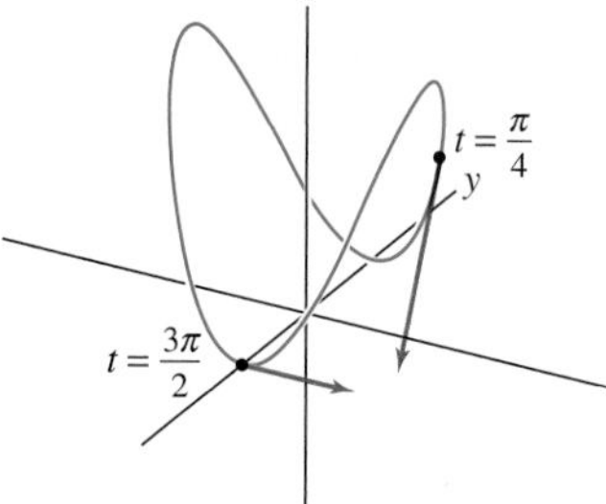

FIGURE 4 Tangent vectors to $\mathbf{r}(t) = \langle \cos t, \sin t, 4\cos^2 t \rangle$ at $t = \frac{\pi}{4}$ and $t = \frac{3\pi}{2}$.

We plot the vector $\mathbf{r}'(\frac{\pi}{4})$ based at $\mathbf{r}(\frac{\pi}{4})$ and the vector $\mathbf{r}'(\frac{3\pi}{2})$ based at $\mathbf{r}(\frac{3\pi}{2})$ as in Figure 4. The tangent line at $t = \frac{\pi}{4}$ has parametrization

$$\mathbf{L}(t) = \mathbf{r}\left(\frac{\pi}{4}\right) + t\mathbf{r}'\left(\frac{\pi}{4}\right) = \left\langle \frac{\sqrt{2}}{2}, \frac{\sqrt{2}}{2}, 2 \right\rangle + t\left\langle -\frac{\sqrt{2}}{2}, \frac{\sqrt{2}}{2}, -4 \right\rangle \qquad \blacksquare$$

The next example establishes an important property of vector-valued functions that will be used in Sections 13.4–13.6.

EXAMPLE 6 Orthogonality of r and r′ When r Has Constant Length Prove that $\mathbf{r}(t)$ has constant length if and only if $\mathbf{r}(t)$ is orthogonal to $\mathbf{r}'(t)$.

Solution We may assume that $\mathbf{r}(t)$ is nonzero. By the Product Rule,

$$\frac{d}{dt}\|\mathbf{r}(t)\|^2 = \frac{d}{dt}\big(\mathbf{r}(t)\cdot\mathbf{r}(t)\big) = \mathbf{r}(t)\cdot\mathbf{r}'(t) + \mathbf{r}'(t)\cdot\mathbf{r}(t) = 2\mathbf{r}(t)\cdot\mathbf{r}'(t)$$

The derivative on the left is zero if and only if $\mathbf{r}(t)$ has constant length and the dot product on the right is zero if and only if $\mathbf{r}'(t)$ is orthogonal to $\mathbf{r}(t)$ [or $\mathbf{r}'(t) = \mathbf{0}$]. The result follows. ■

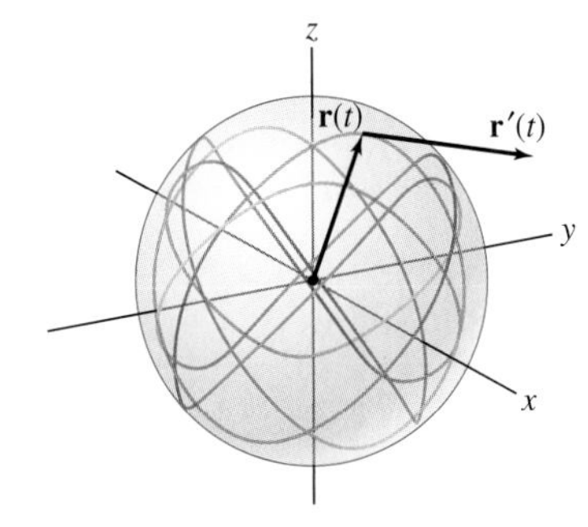

FIGURE 5

GRAPHICAL INSIGHT The result of Example 6 has a geometric explanation. If $\mathbf{r}(t)$ has constant length R, then it traces a curve on the surface of a sphere of radius R with center at the origin (Figure 5). The vector $\mathbf{r}'(t)$ is then tangent to this sphere. However, any segment tangent to a sphere at a point P is orthogonal to the radial vector through P and thus $\mathbf{r}'(t)$ is orthogonal to $\mathbf{r}(t)$.

To highlight the difference between vector- and scalar-valued derivatives, consider the problem of finding points on a plane curve where the tangent line is horizontal. The tangent line to the curve $y = f(x)$ is horizontal at x_0 if $f'(x_0) = 0$. However, the tangent vector $\mathbf{r}'(t_0) = \big\langle x'(t_0), y'(t_0)\big\rangle$ is horizontal and nonzero if $y'(t_0) = 0$ but $x'(t_0) \neq 0$.

■ **EXAMPLE 7** **Horizontal Tangent Vector on the Cycloid** The function

$$\mathbf{r}(t) = \langle t - \sin t, 1 - \cos t\rangle$$

traces a cycloid. Find the points where:

(a) $\mathbf{r}'(t)$ is horizontal and nonzero. **(b)** $\mathbf{r}'(t)$ is the zero vector.

Solution The y-component of the tangent vector $\mathbf{r}'(t) = \langle 1 - \cos t, \sin t\rangle$ is zero if $\sin t = 0$, that is, if $t = 0, \pi, 2\pi, \ldots$. We have

$$\mathbf{r}(0) = \langle 0, 0\rangle, \quad \mathbf{r}'(0) = \langle 1 - \cos 0, \sin 0\rangle = \langle 0, 0\rangle \quad \text{(zero vector)}$$

$$\mathbf{r}(\pi) = \langle \pi, 2\rangle, \quad \mathbf{r}'(\pi) = \langle 1 - \cos \pi, \sin \pi\rangle = \langle 2, 0\rangle \quad \text{(horizontal)}$$

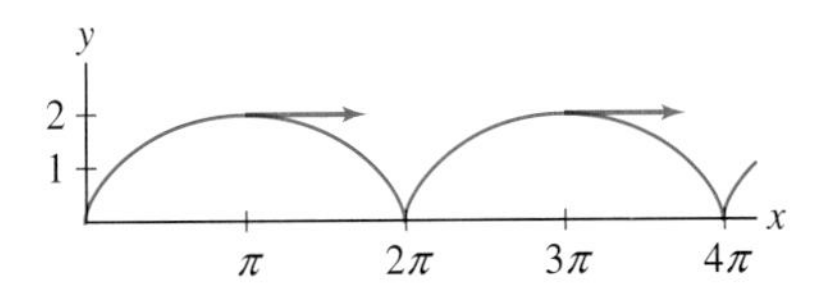

FIGURE 6 Points on the cycloid where the tangent vector is horizontal.

By periodicity, we conclude that $\mathbf{r}'(t)$ is nonzero and horizontal for $t = \pi, 3\pi, 5\pi, \ldots$ and $\mathbf{r}'(t) = \mathbf{0}$ for $t = 0, 2\pi, 4\pi, \ldots$ (Figure 6). ■

CONCEPTUAL INSIGHT The cycloid in Figure 6 shows another difference between vector and scalar derivatives. If we describe the cycloid as the graph of a function $y = f(x)$, then $f'(x)$ does not exist at the points $x = 0, 2\pi, 4\pi, \ldots$, where the curve has a sharp point called a **cusp**. However, the vector parametrization of the cycloid $\mathbf{r}(t) = \langle t - \sin t, 1 - \cos t\rangle$ is differentiable for all t and $\mathbf{r}'(t)$ is the zero vector at the cusps.

Vector-Valued Integration

The integral of a vector-valued function may be defined in terms of Riemann sums as in Chapter 5. We will define it more simply via componentwise integration (the two definitions are equivalent). In other words, we define

$$\int_a^b \mathbf{r}(t)\,dt = \left\langle \int_a^b x(t)\,dt, \int_a^b y(t)\,dt, \int_a^b z(t)\,dt \right\rangle$$

The integral exists if each of the components $x(t)$, $y(t)$, $z(t)$ is integrable. For example,

$$\int_0^{\pi} \langle 1, t, \sin t \rangle \, dt = \left\langle \int_0^{\pi} 1 \, dt, \int_0^{\pi} t \, dt, \int_0^{\pi} \sin t \, dt \right\rangle = \left\langle \pi, \frac{1}{2}\pi^2, 2 \right\rangle$$

Vector-valued integrals obey the linearity rules as scalar-valued integrals (see Exercise 68).

An **antiderivative** of $\mathbf{r}(t)$ is a vector-valued function $\mathbf{R}(t)$ such that $\mathbf{R}'(t) = \mathbf{r}(t)$. In the single-variable case, two functions $f_1(x)$ and $f_2(x)$ with the same derivative differ by a constant. Similarly, two vector-valued functions with the same derivative differ by a *constant vector* (i.e., a vector that does not depend on t). This is proved by applying the scalar result to each component of $\mathbf{r}(t)$.

THEOREM 4 Let $\mathbf{R}_1(t)$ and $\mathbf{R}_2(t)$ be differentiable vector-valued functions such that $\mathbf{R}_1'(t) = \mathbf{R}_2'(t)$. Then $\mathbf{R}_1(t) = \mathbf{R}_2(t) + \mathbf{c}$ for some constant vector $\mathbf{c}$.

The general antiderivative of $\mathbf{r}(t)$ is written

$$\int \mathbf{r}(t) \, dt = \mathbf{R}(t) + \mathbf{c}$$

where $\mathbf{c} = \langle c_1, c_2, c_3 \rangle$ is an arbitrary constant vector. For example,

$$\int \langle 1, t, \sin t \rangle \, dt = \left\langle t, \frac{1}{2}t^2, -\cos t \right\rangle + \mathbf{c} = \left\langle t + c_1, \frac{1}{2}t^2 + c_2, -\cos t + c_3 \right\rangle$$

Fundamental Theorem of Calculus for Vector-Valued Functions Let $\mathbf{r}(t)$ be a continuous vector-valued function on the interval $[a, b]$, and let $\mathbf{R}(t)$ be an antiderivative of $\mathbf{r}(t)$. Then

$$\int_a^b \mathbf{r}(t) \, dt = \mathbf{R}(b) - \mathbf{R}(a)$$

EXAMPLE 8 Finding Position via Vector-Valued Differential Equations The path of a particle satisfies $\frac{d\mathbf{r}}{dt} = \langle 8t, 5 - 8\sin 2t \rangle$. Where is the particle located at $t = 3$ if $\mathbf{r}(0) = \langle 0, 10 \rangle$?

Note that in Example 8, $d\mathbf{r}/dt$ is the particle's velocity function, and just as in the single-variable case, the integral of the velocity yields the position $\mathbf{r}(t)$.

Solution The general solution is obtained by integration:

$$\mathbf{r}(t) = \int \langle 8t, 5 - 8\sin 2t \rangle \, dt = \langle 4t^2, 5t + 4\cos 2t \rangle + \mathbf{c}$$

We have

$$\mathbf{r}(0) = \langle 4(0)^2, 5(0) + 4\cos 0 \rangle + \mathbf{c} = \langle 0, 4 \rangle + \mathbf{c}$$

Using the initial condition $\mathbf{r}(0) = \langle 0, 4 \rangle + \mathbf{c} = \langle 0, 10 \rangle$, we obtain $\mathbf{c} = \langle 0, 6 \rangle$ and

$$\mathbf{r}(t) = \langle 4t^2, 5t + 4\cos 2t \rangle + \langle 0, 6 \rangle = \langle 4t^2, 5t + 4\cos 2t + 6 \rangle$$

The particle's position at $t = 3$ is

$$\mathbf{r}(3) = \langle 4(3)^2, 5(3) + 4\cos 6 + 6 \rangle \approx \langle 36, 24.84 \rangle$$

■

13.2 SUMMARY

- Limits, differentiation, and integration of vector-valued functions are performed componentwise.
- The differentiation rules apply in the vector-valued setting, but there are three Product Rules for vector-valued functions. If $f(t)$, $\mathbf{r}(t)$, $\mathbf{r}_1(t)$, and $\mathbf{r}_2(t)$ are differentiable, then

Scalar times vector: $$\frac{d}{dt}\big(f(t)\mathbf{r}(t)\big) = f(t)\mathbf{r}'(t) + f'(t)\mathbf{r}(t)$$

Dot product: $$\frac{d}{dt}\big(\mathbf{r}_1(t)\cdot\mathbf{r}_2(t)\big) = \mathbf{r}_1(t)\cdot\mathbf{r}_2'(t) + \mathbf{r}_1'(t)\cdot\mathbf{r}_2(t)$$

Cross product: $$\frac{d}{dt}\big(\mathbf{r}_1(t)\times\mathbf{r}_2(t)\big) = \big[\mathbf{r}_1(t)\times\mathbf{r}_2'(t)\big] + \big[\mathbf{r}_1'(t)\times\mathbf{r}_2(t)\big]$$

- The derivative $\mathbf{r}'(t_0)$ is referred to as the *tangent vector* or *velocity vector* at $\mathbf{r}(t_0)$. If $\mathbf{r}'(t_0)$ is nonzero, then it points in the direction tangent to the curve at $\mathbf{r}(t_0)$.
- If $\mathbf{r}'(t_0) \neq \mathbf{0}$, then the tangent line at t_0 has vector parametrization

$$\mathbf{L}(t) = \mathbf{r}(t_0) + t\mathbf{r}'(t_0)$$

- If $\mathbf{R}_1(t)$ and $\mathbf{R}_2(t)$ are differentiable, vector-valued functions such that $\mathbf{R}_1'(t) = \mathbf{R}_2'(t)$, then $\mathbf{R}_2(t) = \mathbf{R}_1(t) + \mathbf{c}$ for some constant vector $\mathbf{c}$.
- The Fundamental Theorem for vector-valued functions: If $\mathbf{r}(t)$ is continuous on the interval $[a, b]$ and if $\mathbf{R}(t)$ is an antiderivative of $\mathbf{r}(t)$, then

$$\int_a^b \mathbf{r}(t)\,dt = \mathbf{R}(b) - \mathbf{R}(a)$$

13.2 EXERCISES

Preliminary Questions

1. State the three forms of the Product Rule for vector-valued functions.

In Questions 2–6, indicate whether true or false and if false, provide a correct statement.

2. The derivative of a vector-valued function is defined as the limit of the difference quotient, just as in the scalar-valued case.

3. There are two Chain Rules for vector-valued functions, one for the composite of two vector-valued functions and one for the composite of a vector-valued and scalar-valued function.

4. The terms "velocity vector" and "tangent vector" for a path $\mathbf{r}(t)$ mean one and the same thing.

5. The derivative of a vector-valued function is the slope of the tangent line, just as in the scalar case.

6. The derivative of the cross product is the cross product of the derivatives.

7. State whether the following derivatives of vector-valued functions $\mathbf{r}_1(t)$ and $\mathbf{r}_2(t)$ are scalars or vectors:

(a) $\frac{d}{dt}\mathbf{r}_1(t)$

(b) $\frac{d}{dt}\big(\mathbf{r}_1(t)\cdot\mathbf{r}_2(t)\big)$

(c) $\frac{d}{dt}\big(\mathbf{r}_1(t)\times\mathbf{r}_2(t)\big)$

Exercises

In Exercises 1–4, evaluate the limit.

1. $\lim\limits_{t\to 3}\left\langle t^2, 4t, \frac{1}{t}\right\rangle$

2. $\lim\limits_{t\to \pi} \sin 2t\mathbf{i} + \cos t\mathbf{j} + \tan 4t\mathbf{k}$

3. $\lim\limits_{t\to 0} e^{2t}\mathbf{i} + \ln(t+1)\mathbf{j} + 4\mathbf{k}$

4. $\lim\limits_{t\to 0}\left\langle \frac{1}{t+1}, \frac{e^t-1}{t}, 4t\right\rangle$

5. Evaluate $\lim\limits_{h\to 0}\frac{\mathbf{r}(t+h)-\mathbf{r}(t)}{h}$ for $\mathbf{r}(t) = \left\langle t^{-1}, \sin t, 4\right\rangle$.

6. Evaluate $\lim_{t\to 0} \frac{\mathbf{r}(t)}{t}$ for $\mathbf{r}(t) = \langle \sin t, 1 - \cos t, -2t \rangle$.

In Exercises 7–14, compute the derivative.

7. $\mathbf{r}(t) = \langle t, t^2, t^3 \rangle$

8. $\mathbf{v}(t) = \langle \sin 3t, \cos 3t \rangle$

9. $\mathbf{w}(s) = \langle e^s, e^{-2s} \rangle$

10. $\mathbf{r}(\theta) = \langle \tan\theta, 4\theta - 2, \sin\theta \rangle$

11. $\mathbf{r}(t) = \langle t - t^{-1}, 4t^2, 8 \rangle$

12. $\mathbf{c}(t) = t^{-1}\mathbf{i} - e^{2t}\mathbf{k}$

13. $\mathbf{a}(\theta) = (\cos 2\theta)\mathbf{i} + (\sin 2\theta)\mathbf{k} + (\sin 4\theta)\mathbf{k}$

14. $\mathbf{b}(t) = \langle e^{4t-3}, \sin(t^2), (4t+3)^{-1} \rangle$

15. Calculate $\mathbf{r}'(t)$ and $\mathbf{r}''(t)$ for $\mathbf{r}(t) = \langle t, t^2, t^3 \rangle$.

16. Sketch the curve $\mathbf{r}(t) = \langle 1 - t^2, t \rangle$ for $-1 \le t \le 1$. Compute the tangent vector at $t = 1$ and add it to the sketch.

17. Sketch the curve $\mathbf{r}_1(t) = \langle t, t^2 \rangle$ together with its tangent vector at $t = 1$. Then do the same for $\mathbf{r}_2(t) = \langle t^3, t^6 \rangle$.

18. Sketch the cycloid $\mathbf{r}(t) = \langle t - \sin t, 1 - \cos t \rangle$ together with its tangent vectors at $t = \frac{\pi}{3}$ and $\frac{3\pi}{4}$.

In Exercises 19–22, use the appropriate Product Rule to evaluate the derivative, where

$$\mathbf{r}_1(t) = \langle 8t, 4, -t^3 \rangle, \qquad \mathbf{r}_2(t) = \langle 0, e^t, -6 \rangle$$

19. $\frac{d}{dt}(\mathbf{r}_1(t) \cdot \mathbf{r}_2(t))$

20. $\frac{d}{dt}(t^4\mathbf{r}_1(t))$

21. $\frac{d}{dt}(\mathbf{r}_1(t) \times \mathbf{r}_2(t))$

22. $\frac{d}{dt}(\mathbf{r}_1(t) \cdot \mathbf{r}_3(t))\Big|_{t=5}$, assuming that $\mathbf{r}_3(5) = \langle 3, 1, 2 \rangle$ and $\mathbf{r}_3'(5) = \langle -1, 2, 7 \rangle$.

In Exercises 23–25, let

$$\mathbf{r}_1(t) = \langle t^2, t^3, 4t \rangle, \qquad \mathbf{r}_2(t) = \langle t^{-1}, 1 + t, 2 \rangle$$

23. Let $F(t) = \mathbf{r}_1(t) \cdot \mathbf{r}_2(t)$.

(a) Calculate $F'(t)$ using the Product Rule.

(b) Expand the product $\mathbf{r}_1(t) \cdot \mathbf{r}_2(t)$ and differentiate. Compare with part (a).

24. Let $G(t) = \mathbf{r}_1(t) \times \mathbf{r}_2(t)$.

(a) Calculate $G'(t)$ using the Product Rule.

(b) Expand the cross product $\mathbf{r}_1(t) \times \mathbf{r}_2(t)$ and differentiate. Compare with part (a).

25. Find the rate of change of the angle θ between $\mathbf{r}_1(t)$ and $\mathbf{r}_2(t)$ at $t = 0$, assuming that t is measured in seconds.

In Exercises 26–29, evaluate $\frac{d}{dt}\mathbf{r}(g(t))$ using the Chain Rule.

26. $\mathbf{r}(t) = \langle t^2, 2t, 4 \rangle$, $\quad g(t) = e^t$

27. $\mathbf{r}(t) = \langle e^t, e^{2t} 4 \rangle$, $\quad g(t) = 4t + 9$

28. $\mathbf{r}(t) = \langle 4\sin 2t, 6\cos 2t \rangle$, $\quad g(t) = t^2$

29. $\mathbf{r}(t) = \langle 3^t, \tan^{-1} t \rangle$, $\quad g(t) = \sin t$

30. Let $\mathbf{v}(s) = s^2\mathbf{i} + 2s\mathbf{j} + 9s^{-2}\mathbf{k}$. Evaluate $\frac{d}{ds}\mathbf{v}(g(s))$ at $s = 4$, assuming that $g(4) = 3$ and $g'(4) = -9$.

31. Let $\mathbf{r}(t) = \langle t^2, 1 - t, 4t \rangle$. Calculate the derivative of $\mathbf{r}(t) \cdot \mathbf{a}(t)$ at $t = 2$, assuming that $\mathbf{a}(2) = \langle 1, 3, 3 \rangle$ and $\mathbf{a}'(2) = \langle -1, 4, 1 \rangle$.

32. Let $\mathbf{r}(t) = \langle t^2, t^3, e^t \rangle$. Use Example 4 to calculate $\frac{d}{dt}(\mathbf{r} \times \mathbf{r}')$.

In Exercises 33–37, find a parametrization of the tangent line at the point indicated.

33. $\mathbf{r}(t) = \langle 1 - t^2, 5t, 2t^3 \rangle$, $\quad t = 2$

34. $\mathbf{r}(t) = \langle \cos 2t, \sin 3t, \sin 4t \rangle$, $\quad t = \frac{\pi}{4}$

35. $\mathbf{r}(s) = 4s^{-1}\mathbf{i} - 8s^{-3}\mathbf{k}$, $\quad s = 2$

36. $\mathbf{r}(t) = \langle t^2, t^4 \rangle$, $\quad t = 1$

37. $\mathbf{r}(s) = \ln s\,\mathbf{i} + s^{-1}\mathbf{j} + 9s\mathbf{k}$, $\quad s = 1$

38. Let $\mathbf{r}(t) = \langle \sin 2t \cos t, \sin 2t \sin t, \cos 2t \rangle$. Show that $\|\mathbf{r}(t)\|$ is constant and conclude using Example 6 that $\mathbf{r}(t)$ and $\mathbf{r}'(t)$ are orthogonal. Then compute $\mathbf{r}'(t)$ and verify directly that it is orthogonal to $\mathbf{r}(t)$.

39. Show, by finding a counterexample, that in general $\|\mathbf{r}'(t)\|$ need not equal $\|\mathbf{r}(t)\|'$.

In Exercises 40–45, evaluate the integrals.

40. $\int_0^1 \langle 2t, 4t, -\cos 3t \rangle\, dt$

41. $\int_1^4 (t^{-1}\mathbf{i} + 4\sqrt{t}\,\mathbf{j} - 8t^{3/2}\mathbf{k})\, dt$

42. $\int_0^1 \langle te^{-t^2}, t\ln(t^2 + 1) \rangle\, dt$

43. $\int_{-2}^2 (u^3\mathbf{i} + u^5\mathbf{j} + u^7\mathbf{k})\, du$

44. $\int_0^1 \left\langle \frac{1}{1+s^2}, \frac{s}{1+s^2} \right\rangle ds$

45. $\int_0^t (3s\mathbf{i} + 6s^2\mathbf{j} + 9\mathbf{k})\, ds$

In Exercises 46–53, find the general solution $\mathbf{r}(t)$ of the differential equation and the solution with the given initial condition.

46. $\dfrac{d\mathbf{r}}{dt} = \langle 1 - 2t, 4t \rangle, \quad \mathbf{r}(0) = \langle 3, 1 \rangle$

47. $\mathbf{r}'(t) = \mathbf{i} - \mathbf{j}, \quad \mathbf{r}(0) = 2\mathbf{i} + 3\mathbf{k}$

48. $\mathbf{r}'(t) = t^2\mathbf{i} + 5t\mathbf{j} + \mathbf{k}, \quad \mathbf{r}(0) = \mathbf{j} + 2\mathbf{k}$

49. $\mathbf{r}'(t) = \langle \sin 3t, \sin 3t, t \rangle, \quad \mathbf{r}(0) = \langle 0, 1, 8 \rangle$

50. $\dfrac{d\mathbf{r}}{dt} = \langle e^{2t}, e^t, e^{-2t} \rangle, \quad \mathbf{r}(0) = \langle 4, -2, 3 \rangle$

51. $\mathbf{r}''(t) = 16\mathbf{k}, \quad \mathbf{r}(0) = \langle 1, 0, 0 \rangle, \quad \mathbf{r}'(0) = \langle 0, 1, 0 \rangle$

52. $\mathbf{r}''(t) = \langle 0, 0, 1 \rangle, \quad \mathbf{r}(0) = \langle 2, 1, 1 \rangle, \quad \mathbf{r}'(0) = \langle 3, 1, 1 \rangle$

53. $\mathbf{r}''(t) = \langle e^t, \sin t, \cos t \rangle, \quad \mathbf{r}(0) = \langle 1, 0, 1 \rangle, \quad \mathbf{r}'(0) = \langle 0, 2, 2 \rangle$

54. Show that $\mathbf{w}(t) = \langle \sin(3t + 4), \sin(3t - 2), \cos 3t \rangle$ satisfies the differential equation $\mathbf{w}''(t) = -9\mathbf{w}(t)$.

55. The path $\mathbf{r}(t)$ of a particle satisfies $\dfrac{d\mathbf{r}}{dt} = \langle 8, 5 - 3t, 4t^2 \rangle$. Where is the particle located at $t = 4$ if $\mathbf{r}(0) = \langle 1, 6, 0 \rangle$?

56. A fighter plane, which can only shoot bullets straight ahead, travels along the path $\mathbf{r}(t) = \langle 5 - t, 21 - t^2, 3 - t^3/27 \rangle$. Show that there is precisely one time t at which the pilot can hit a target located at the origin.

57. Find all solutions to $\mathbf{r}'(t) = \mathbf{v}$, where $\mathbf{v}$ is a constant vector in $\mathbf{R}^3$.

58. Let $\mathbf{u}$ be a constant vector in $\mathbf{R}^3$. Find the solution of $\mathbf{r}'(t) = (\sin t)\mathbf{u}$ satisfying $\mathbf{r}'(0) = 0$.

59. Find all solutions to $\mathbf{r}'(t) = 2\mathbf{r}(t)$ where $\mathbf{r}(t)$ is a vector-valued function in three-space.

60. Show that $\dfrac{d}{dt}(\mathbf{a} \times \mathbf{r}) = \mathbf{a} \times \mathbf{r}'$ for any constant vector $\mathbf{a}$.

61. Prove that $\mathbf{r}(t_0)$ and $\mathbf{r}'(t_0)$ are orthogonal at values $t = t_0$ where $\|\mathbf{r}(t)\|$ takes on a local minimum or maximum value. Explain how this result is related to Figure 7. *Hint:* In the figure, $\|\mathbf{r}(t_0)\|$ is a minimum and the path $\mathbf{r}(t)$ intersects the sphere of radius $\|\mathbf{r}(t_0)\|$ in a single point (and hence is tangent at that point).

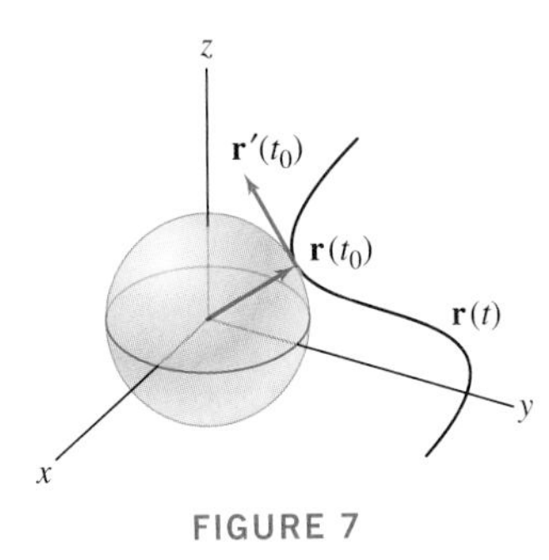

FIGURE 7

62. Newton's Second Law of Motion in vector form states that $\mathbf{F} = \dfrac{d\mathbf{p}}{dt}$ where $\mathbf{F}$ is the force acting on an object of mass m and $\mathbf{p} = m\mathbf{r}'(t)$ is the object's momentum. The analogs of force and momentum for rotational motion are the **torque** $\tau = \mathbf{r} \times \mathbf{F}$ and **angular momentum** $\mathbf{J} = m\mathbf{r}(t) \times \mathbf{r}'(t)$. Use the Second Law to prove that $\tau = \dfrac{d\mathbf{J}}{dt}$. This is a rotational version of the Second Law.

Further Insights and Challenges

63. In this exercise, we verify that the definition of the tangent line using vector-valued functions agrees with the usual definition in terms of the scalar derivative in the case of a plane curve. Suppose that $\mathbf{r}(t) = \langle x(t), y(t) \rangle$ traces a plane curve $\mathcal{C}$.

(a) Show that $\dfrac{dy}{dx} = \dfrac{y'(t)}{x'(t)}$ at any point such that $x'(t) \neq 0$. *Hint:* By the Chain Rule, $\dfrac{dy}{dt} = \dfrac{dy}{dx}\dfrac{dx}{dt}$.

(b) Show that if $x'(t_0) \neq 0$, then the line $\mathbf{L}(t) = \mathbf{r}(t_0) + t\mathbf{r}'(t_0)$ passes through $\mathbf{r}(t_0)$ and has slope $\left.\dfrac{dy}{dx}\right|_{t=t_0}$.

64. Verify the Sum and Product Rules for derivatives of vector-valued functions.

65. Verify the Chain Rule for vector-valued functions.

66. Verify the Product Rule for cross products [Eq. (6)].

67. Prove that $\dfrac{d}{dt}(\mathbf{r} \cdot (\mathbf{r}' \times \mathbf{r}'')) = \mathbf{r} \cdot (\mathbf{r}' \times \mathbf{r}''')$

Exercises 68–71 establish additional properties of vector-valued integrals. Assume that all functions are integrable.

68. Prove the linearity properties

$$\int c\mathbf{r}(t)\,dt = c\int \mathbf{r}(t)\,dt \qquad (c \text{ any constant})$$

$$\int (\mathbf{r}_1(t) + \mathbf{r}_2(t))\,dt = \int \mathbf{r}_1(t)\,dt + \int \mathbf{r}_2(t)\,dt$$

69. Prove the Substitution Rule [where $g(t)$ is a differentiable scalar function]:

$$\int_a^b \mathbf{r}(t)\,dt = \int_{g(a)}^{g(b)} \mathbf{r}(g(t))g'(t)\,dt$$

70. Formulate and verify a version of Integration by Parts for vector-valued integrals.

71. Show that if $\|\mathbf{r}(t)\| \le K$ for $t \in [a, b]$, then

$$\left\| \int_a^b \mathbf{r}(t)\,dt \right\| \le K(b - a)$$

13.3 Arc Length and Speed

← REMINDER The length of a path or curve is referred to as the arc length.

In Section 11.2, we derived a formula for the arc length of a plane curve. That discussion applies with little change to paths in three-space. Recall that arc length is defined as the limit of polygonal approximations. A polygonal approximation to a path

$$\mathbf{r}(t) = \langle x(t), y(t), z(t) \rangle, \qquad a \le t \le b$$

is obtained by choosing a partition $a = t_0 < t_1 < t_2 < \cdots < t_N = b$ and joining the terminal points of the vectors $\mathbf{r}(t_j)$ by segments, as in Figure 1. Arguing as in Section 11.2, we find that if $\mathbf{r}'(t)$ exists and is continuous on $[a, b]$, then the lengths of the polygonal approximations approach a limit L as the maximum of the widths $|t_j - t_{j-1}|$ tends to zero. This limit is the arc length and is given by the formula

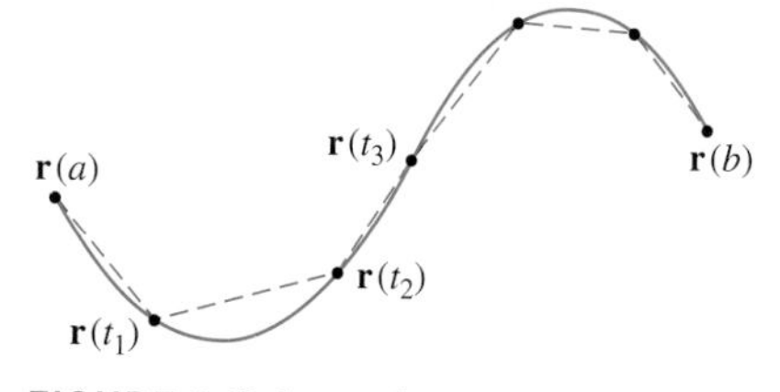

FIGURE 1 Polygonal approximation to the arc $\mathbf{r}(t)$ for $a \le t \le b$.

$$L = \int_a^b \sqrt{x'(t)^2 + y'(t)^2 + z'(t)^2}\, dt \tag{1}$$

The quantity L in Eq. (2) is the distance traveled by a particle following the trajectory $\mathbf{r}(t)$. However, $\mathbf{r}(t)$ may traverse all or part of a curve more than once and it may reverse direction. Thus, L is equal to the length of the underlying curve only if the curve is traversed once without reversal of direction for $a \le t \le b$.

Arc Length of a Path Assume that $\mathbf{r}(t)$ is differentiable and that $\mathbf{r}'(t)$ is continuous on $[a, b]$. Then the length L of the path $\mathbf{r}(t)$ for $a \le t \le b$ is equal to

$$L = \int_a^b \|\mathbf{r}'(t)\|\, dt = \int_a^b \sqrt{x'(t)^2 + y'(t)^2 + z'(t)^2}\, dt \tag{2}$$

■ **EXAMPLE 1** Find the arc length L of $\mathbf{r}(t) = \langle \cos 3t, \sin 3t, 3t \rangle$ for $0 \le t \le 2\pi$.

Solution The derivative is $\mathbf{r}'(t) = \langle -3\sin 3t, 3\cos 3t, 3 \rangle$, and

$$\|\mathbf{r}'(t)\|^2 = 9\sin^2 3t + 9\cos^2 3t + 9 = 9(\sin^2 3t + \cos^2 3t) + 9 = 18$$

Therefore, $L = \displaystyle\int_0^{2\pi} \|\mathbf{r}'(t)\|\, dt = \int_0^{2\pi} \sqrt{18}\, dt = 6\sqrt{2}\pi.$ ■

As we noted in the previous section, the derivative $\mathbf{r}'(t)$ is referred to as both the *tangent vector* and the *velocity vector* of the path $\mathbf{r}(t)$. The term "tangent vector" seems justified because $\mathbf{r}'(t)$ is indeed tangent to the path when it is nonzero. To justify the term "velocity vector," assume that t is a time parameter and let us relate $\mathbf{r}'(t)$ to the speed of the path. By definition, the speed of a path, denoted $v(t)$, is the rate of change of distance traveled with respect to t. To calculate the speed, we define the arc length function:

$$s(t) = \int_a^t \|\mathbf{r}'(u)\|\, du$$

Then $s(t)$ is the distance traveled during the time interval $[a, t]$ and the derivative $v(t) = \dfrac{ds}{dt}$ is the speed. By the Fundamental Theorem of Calculus,

$$v(t) = \text{speed at time } t = \frac{ds}{dt} = \|\mathbf{r}'(t)\|$$

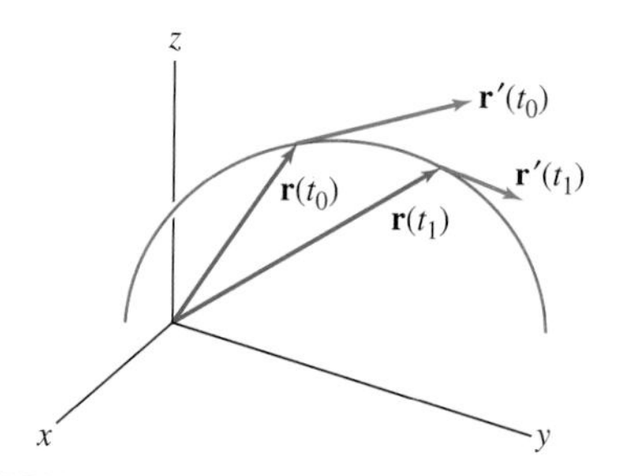

FIGURE 2 The particle is moving faster at t_0 than t_1 since the velocity vector is longer at t_1.

Thus, speed is the length of the velocity vector (Figure 2). In summary, we have the following interpretation of the velocity vector:

- $\mathbf{r}'(t)$ points in the instantaneous direction of motion (when it is nonzero).
- $v(t) = \|\mathbf{r}'(t)\|$ is the speed at time t.

■ **EXAMPLE 2** Find the speed of a particle at time $t = 2$ s if the position vector of the particle is $\mathbf{r}(t) = t^3\mathbf{i} - e^t\mathbf{j} + 4t\mathbf{k}$ (in units of feet).

Solution The velocity vector is $\mathbf{r}'(t) = 3t^2\mathbf{i} - e^t\mathbf{j} + 4\mathbf{k}$, and at $t = 2$,

$$\mathbf{r}'(2) = 12\mathbf{i} - e^2\mathbf{j} + 4\mathbf{k}$$

The particle's speed is $\|\mathbf{r}'(2)\| = \sqrt{12^2 + (-e^2)^2 + 4^2} \approx 14.65$ ft/s. ■

Arc Length Parametrization

Keep in mind that a parametrization $\mathbf{r}(t)$ does more than just describe a curve. It also tells you how a particle traverses the curve, possibly speeding up, slowing down, or reversing direction along the way. Changing the parametrization amounts to describing a different way of traversing the same underlying curve.

We have seen that parametrizations are not unique. For example, $\mathbf{r}_1(t) = \langle t, t^2\rangle$ and $\mathbf{r}_2(s) = \langle s^3, s^6\rangle$ both parametrize the parabola $y = x^2$. Notice that $\mathbf{r}_2(s)$ is obtained by substituting $t = s^3$ in $\mathbf{r}_1(t)$.

In general, we obtain a new parametrization by making a substitution $t = \varphi(s)$, that is, by replacing $\mathbf{r}(t)$ with $\mathbf{r}_1(s) = \mathbf{r}(\varphi(s))$. If $\mathbf{r}(t)$ and $\varphi(s)$ are differentiable, then $\mathbf{r}_1(s)$ is also differentiable. Furthermore, if $t = \varphi(s)$ increases from a to b as s varies from c to d, then path $\mathbf{r}(t)$ for $a \le t \le b$ is also parametrized by $\mathbf{r}_1(s)$ for $c \le s \le d$.

■ **EXAMPLE 3** Parametrize the path $\mathbf{r}(t) = (t^2, \sin t, t)$ for $3 \le t \le 9$ in terms of the parameter s, where $t = \varphi(s) = e^s$.

Solution Substituting $t = e^s$ in $\mathbf{r}(t)$, we obtain the parametrization

$$\mathbf{r}_1(s) = \mathbf{r}(\varphi(s)) = \langle e^{2s}, \sin e^s, e^s\rangle$$

Furthermore, $s = \ln t$, so t varies from 3 to 9 as s varies from $\ln 3$ to $\ln 9$. Therefore, the path is parametrized by $\mathbf{r}_1(s)$ for $\ln 3 \le s \le \ln 9$. ■

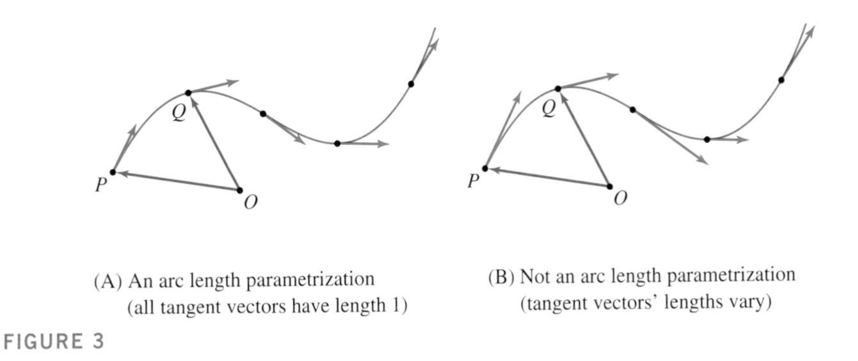

(A) An arc length parametrization (all tangent vectors have length 1)

(B) Not an arc length parametrization (tangent vectors' lengths vary)

FIGURE 3

A special type of parametrization, called an **arc length parametrization**, plays an important role in the study of curves. It is defined by the condition $\|\mathbf{r}'(t)\| = 1$ for all t [Figure 3(A)]. An arc length parametrization is also called the **unit speed parametrization** because the speed $\|\mathbf{r}'(t)\|$ is equal to 1 for all t. Informally speaking, *the arc length parametrization corresponds to walking along the curve at* "1 m/s," and the distance

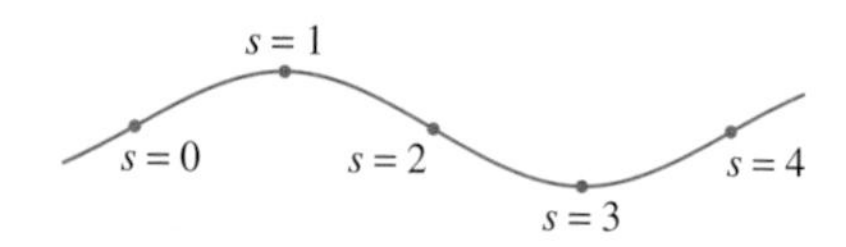

FIGURE 4 In an arc length parametrization, the particle is located s units past $\mathbf{r}(0)$ at time s.

traveled over any interval is equal to the length of the interval (Figure 4):

$$\underbrace{\int_a^b \|\mathbf{r}'(u)\|\,du}_{\text{Distance traveled during interval }[a,b]} = \int_a^b du = b - a$$

How do we find an arc length parametrization? Let $\mathbf{r}(t)$ be any parametrization and consider the arc length function:

$$s(t) = \int_a^t \|\mathbf{r}'(u)\|\,du$$

Then $\varphi'(s) = \|\mathbf{r}'(t)\| \geq 0$, and if we assume that $\mathbf{r}'(t) \neq \mathbf{0}$ for all t, then $s(t)$ is an increasing function of t. In this case, $s(t)$ has an inverse $t = \varphi(s)$. We claim that $\mathbf{r}_1(s) = \mathbf{r}(\varphi(s))$ is an arc length parametrization. To check this, we must verify that $\|\mathbf{r}_1'(s)\| = 1$. By the formula for the derivative of an inverse,

$$\varphi'(s) = \frac{dt}{ds} = \frac{1}{s'(t)} = \frac{1}{s'(\varphi(s))} = \frac{1}{\|\mathbf{r}'(\varphi(s))\|}$$

It is an important fact that arc length parametrizations exist (we will use them to define curvature in Section 13.4). However, it is often difficult or impossible to find an arc length parametrization explicitly. This requires finding the inverse of the arc length function $s(t)$ and can only be done in special cases.

Hence, by the Chain Rule,

$$\|\mathbf{r}_1'(s)\| = \|\mathbf{r}'(\varphi(s))\varphi'(s)\| = \|\mathbf{r}'(\varphi(s))\|\varphi'(s) = \|\mathbf{r}'(\varphi(s))\| \cdot \frac{1}{\|\mathbf{r}'(\varphi(s))\|} = 1$$

The letter s is often used as the parameter in an arc length parametrization. Thus, $\mathbf{r}_1(s)$ is the position vector of the point located at a distance s along the path from $\mathbf{r}_1(0)$.

EXAMPLE 4 Finding an Arc Length Parametrization Find the arc length parametrization of the line $\mathbf{r}(t) = \langle 2t, 1 - 2t, t\rangle$.

Solution

***Step 1.* Find the inverse of the arc length function.**
We have

$$\|\mathbf{r}'(t)\| = \|\langle 2, -2, 1\rangle\| = \sqrt{2^2 + (-2)^2 + 1^2} = \sqrt{9} = 3$$

Thus, the arc length function is

$$s(t) = \int_0^t \|\mathbf{r}'(t)\|\,dt = \int_0^t 3\,dt = 3t$$

The inverse of $s(t) = 3t$ is $t = \varphi(s) = \dfrac{s}{3}$.

***Step 2.* Reparametrize the curve.**
As explained above, an arc length parametrization is

$$\mathbf{r}_1(s) = \mathbf{r}(\varphi(s)) = \mathbf{r}\left(\frac{s}{3}\right) = \left\langle \frac{2s}{3}, 1 - \frac{2s}{3}, \frac{s}{3}\right\rangle$$

As a check, we verify that $\mathbf{r}_1(s)$ has unit speed: $\mathbf{r}_1'(s) = \left\langle \frac{2}{3}, -\frac{2}{3}, \frac{1}{3}\right\rangle$ and

$$\|\mathbf{r}_1'(s)\| = \left\|\left\langle \frac{2}{3}, -\frac{2}{3}, \frac{1}{3}\right\rangle\right\| = \sqrt{\frac{4}{9} + \frac{4}{9} + \frac{1}{9}} = 1$$ ■

CONCEPTUAL INSIGHT Although the arc length L of a curve is defined as the limit of polygonal approximations, we never compute this limit directly. Instead, we choose a parametrization $\mathbf{r}(t)$ that traverses the curve exactly once for $a \le t \le b$ and determine L by evaluating the arc length integral. It follows that the value of the arc length integral does not depend on the choice of parametrization (the value is L for all choices of parametrization). In Exercise 26, we use the Change of Variables formula to give another proof that the arc length integral is independent of the parametrization.

13.3 SUMMARY

- The length L of a path $\mathbf{r}(t) = \langle x(t), y(t), z(t)\rangle$ for $a \le t \le b$ is

$$L = \int_a^b \|\mathbf{r}'(t)\|\, dt = \int_a^b \sqrt{x'(t)^2 + y'(t)^2 + z'(t)^2}\, dt$$

- The distance traveled by a particle following the trajectory $\mathbf{r}(t)$ over the time interval $[a, t]$ is

$$s(t) = \int_a^t \|\mathbf{r}'(u)\|\, du$$

The particle's speed $v(t)$ is the rate of change of distance traveled with respect to time:

$$v(t) = \text{speed at time } t = \frac{ds}{dt} = \|\mathbf{r}'(u)\|$$

The velocity vector $\mathbf{r}'(t)$ points in the direction of motion [provided that $\mathbf{r}'(t) \neq \mathbf{0}$] and its magnitude $\|\mathbf{r}'(t)\|$ is the particle's speed.

- We say that $\mathbf{r}(s)$ is an *arc length parametrization* or *unit speed parametrization* if $\|\mathbf{r}'(s)\| = 1$ for all s. In this case, the length of the path for $a \le s \le b$ is $b - a$.
- If $\mathbf{r}(t)$ is a parametrization such that $\mathbf{r}'(t) \neq \mathbf{0}$ for all t, then $\mathbf{r}_1(s) = \mathbf{r}(\varphi(s))$ is an arc length parametrization, where $t = \varphi(s)$ is the inverse of the arc length function

$$s(t) = \int_a^t \|\mathbf{r}'(u)\|\, du$$

13.3 EXERCISES

Preliminary Questions

1. At a given instant, a car on a roller coaster has velocity vector $\mathbf{r}' = \langle 25, -35, 10\rangle$ (in miles per hour). What would the velocity vector be if the speed were doubled? What would it be if the car's direction were reversed but its speed remained unchanged?

2. Two cars travel in the same direction along the same roller coaster (at different times). Which of the following statements about their velocity vectors at a given point P on the roller coaster are true?

(a) The velocity vectors are identical.

(b) The velocity vectors point in the same direction but may have different lengths.

(c) The velocity vectors may point in opposite directions.

3. A mosquito flies along a parabola with speed $v(t) = t^2$. Let $L(t)$ be the total distance traveled at time t.

(a) How fast is $L(t)$ changing at $t = 2$?

(b) Is $L(t)$ equal to the mosquito's distance from the origin?

4. What is the length of the path traced by $\mathbf{r}(t)$ for $4 \le t \le 10$ if $\mathbf{r}(t)$ is an arc length parametrization?

Exercises

In Exercises 1–6, compute the length of the curve over the given interval.

1. $\mathbf{r}(t) = \langle 3t, 4t-3, 6t+1\rangle, \quad 0 \le t \le 3$

2. $\mathbf{r}(t) = 2t\mathbf{i} - 3t\mathbf{k}, \quad 11 \le t \le 15$

3. $\mathbf{r}(t) = \langle 2t, \ln t, t^2\rangle, \quad 1 \le t \le 4$

4. $\mathbf{r}(t) = \langle 2t^2+1, 2t^2-1, t^3\rangle, \quad 0 \le t \le 4$

5. $\mathbf{r}(t) = t\mathbf{i} + 2t\mathbf{j} + (t^2-3)\mathbf{k}, \quad 0 \le t \le 2$. *Hint:*

$$\int \sqrt{t^2+a^2} = \frac{1}{2}t\sqrt{t^2+a^2} + \frac{1}{2}a^2 \ln\left(t + \sqrt{t^2+a^2}\right)$$

6. $\mathbf{r}(t) = \langle t\cos t, t\sin t, 3t\rangle, \quad 0 \le t \le 2\pi$

7. Compute $s(t) = \int_0^t \|\mathbf{r}'(u)\|\,du$ for $\mathbf{r}(t) = \langle t^2, 2t^2, t^3\rangle$.

In Exercises 8–11, find the speed at the given value of t.

8. $\mathbf{r}(t) = \langle 2t+3, 4t-3, 5-t\rangle, \quad t = 4$

9. $\mathbf{r}(t) = \langle e^{t-3}, 12, 3t^{-1}\rangle, \quad t = 3$

10. $\mathbf{r}(t) = \langle \sin 3t, \cos 4t, \cos 5t\rangle, \quad t = \frac{\pi}{2}$

11. $\mathbf{r}(t) = \langle \cosh t, \sinh t, t\rangle, \quad t = 0$

12. What is the velocity vector of a particle traveling to the right along the curve $y = t^{-1}$ with constant speed 5 cm/s when the particle's location is $(2, \frac{1}{2})$?

13. A bee with velocity vector $\mathbf{r}'(t)$ starts out at the origin at $t = 0$ and flies around for T seconds. Where is the bee located at time T if $\int_0^T \mathbf{r}'(u)\,du = \mathbf{0}$? What does the quantity $\int_0^T \|\mathbf{r}'(u)\|\,du$ represent?

14. Which of the following is an arc length parametrization of a circle of radius 4 centered at the origin?

(a) $\mathbf{r}_1(t) = \langle 4\sin t, 4\cos t\rangle$

(b) $\mathbf{r}_2(t) = \langle 4\sin 4t, 4\cos 4t\rangle$

(c) $\mathbf{r}_3(t) = \langle 4\sin \frac{t}{4}, 4\cos \frac{t}{4}\rangle$

15. Let $\mathbf{r}(t) = \langle 3t+1, 4t-5, 2t\rangle$.

(a) Calculate $s(t) = \int_0^t \|\mathbf{r}'(u)\|\,du$ as a function of t.

(b) Find the inverse $\varphi(s) = t(s)$ and show that $\mathbf{r}_1(s) = \mathbf{r}(\varphi(s))$ is an arc length parametrization.

16. Find an arc length parametrization of the circle in the plane $z = 9$ with radius 4 and center $(1, 4, 10)$.

17. Find a path that traces the circle in the plane $y = 10$ with radius 4 and center $(2, 10, -3)$ with constant speed 8.

18. Show that one arch of the cycloid $\mathbf{r}(t) = \langle t - \sin t, 1 - \cos t\rangle$ has length 8. Find the value of t in $[0, 2\pi]$ where the speed is at a maximum.

19. Find an arc length parametrization of $\mathbf{r}(t) = \langle e^t \sin t, e^t \cos t, e^t\rangle$.

20. Find an arc length parametrization of $\mathbf{r}(t) = \langle t^2, t^3\rangle$.

21. Express the arc length L of $y = x^3$ for $0 \le x \le 8$ as an integral in two ways, using the parametrizations $\mathbf{r}_1(t) = \langle t, t^3\rangle$ and $\mathbf{r}_2(t) = \langle t^3, t^9\rangle$. Do not evaluate the integrals, but use substitution to show that they yield the same result.

22. Show that a helix of radius R and height h making N complete turns has the parametrization

$$\left\langle R\cos\left(\frac{2\pi N t}{h}\right), R\sin\left(\frac{2\pi N t}{h}\right), t\right\rangle, \qquad 0 \le t \le h$$

23. Consider the two springs in Figure 5. One has radius 5 cm, height 4 cm, and makes three complete turns. The other has height 3 cm, radius 4 cm, and makes five complete turns.

(a) Take a guess as to which spring uses more wire.

(b) Compute the lengths of the two springs (use Exercise 22) and compare.

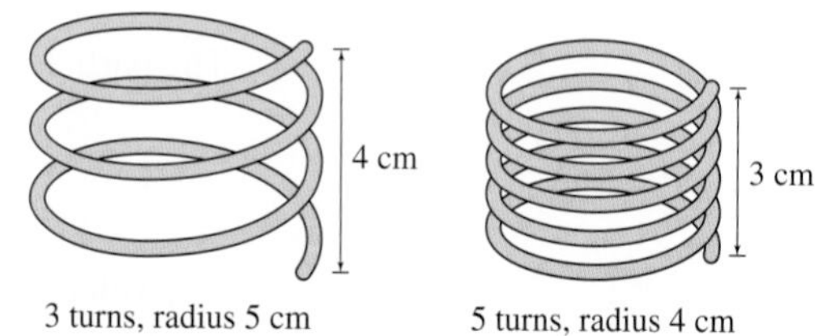

FIGURE 5 Which spring uses more wire?

24. Use Exercise 22 to find a general formula for the length of a helix of radius R and height h that makes N complete turns.

25. Evaluate $s(t) = \int_{-\infty}^t \|\mathbf{r}'(u)\|\,du$ for the **Bernoulli spiral** $\mathbf{r}(t) = \langle e^t \cos 4t, e^t \sin 4t\rangle$ (Figure 6). It is convenient to take $-\infty$ as the lower limit since $s(-\infty) = 0$. Then:

(a) Use s to obtain an arc length parametrization of $\mathbf{r}(t)$.

(b) Prove that the angle between the position vector and the tangent vector is constant.

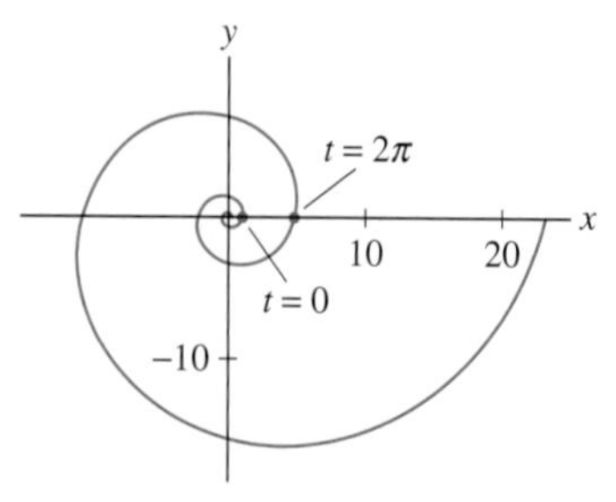

FIGURE 6 Bernoulli spiral.

Further Insights and Challenges

26. Prove that the length of a curve as computed using the arc length integral does not depend on its parametrization. More precisely, let $\mathcal{C}$ be the curve traced by $\mathbf{r}(t)$ for $a \le t \le b$. Let $\varphi(s)$ be a differentiable function such that $\varphi'(s) > 0$ and that $\varphi(c) = a$ and $\varphi(d) = b$. Then $\mathbf{r}_1(s) = \mathbf{r}(\varphi(s))$ parametrizes $\mathcal{C}$ for $c \le s \le d$. Verify that

$$\int_a^b \|\mathbf{r}'(t)\|\,dt = \int_c^d \|\mathbf{r}_1'(s)\|\,ds$$

27. Show that path $\mathbf{r}(t) = \left\langle \frac{1-t^2}{1+t^2}, \frac{2t}{1+t^2} \right\rangle$ parametrizes the unit circle with the point $(0, -1)$ excluded for $-\infty < t < \infty$. Use this parametrization to compute the length of the unit circle as an improper integral. *Hint:* The expression for $\|\mathbf{r}'(t)\|$ simplifies.

28. The involute of a circle is the curve traced by a point at the end of a thread unwinding from a circular spool of radius R. Parametrize the involute by finding the position vector $\mathbf{r}(t)$ of the point P in Figure 7 as a function of t. Determine the arc length function and the arc length parametrization.

29. The curve $\mathbf{r}(t) = \langle t - \tanh t, \operatorname{sech} t \rangle$ is called a **tractrix**.

(a) Show that the arc length function $s(t) = \int_0^t \|\mathbf{r}'(u)\|\,du$ is equal to $s(t) = \ln(\cosh t)$.

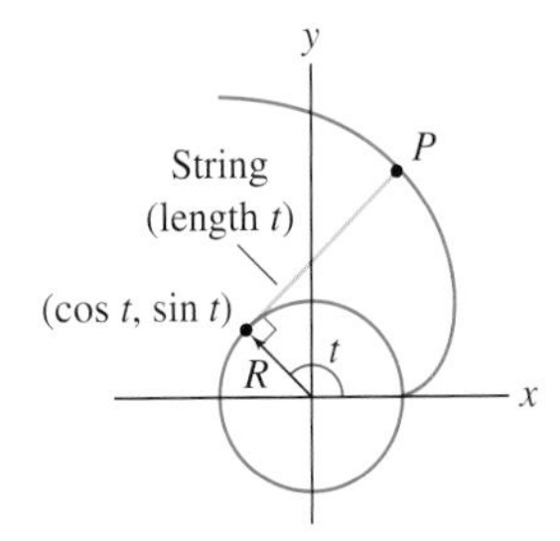

FIGURE 7 The involute of a circle.

(b) Show that $t = \varphi(s) = \ln(e^s + \sqrt{e^{2s} - 1})$ is an inverse of $s(t)$ and verify that

$$\mathbf{r}_1(s) = \left\langle \tanh^{-1}\left(\sqrt{1 - e^{-2s}}\right) - \sqrt{1 - e^{-2s}}, e^{-s} \right\rangle$$

is an arc length parametrization of the tractrix.

(c) Plot the tractrix if you have a computer algebra system.

13.4 Curvature

FIGURE 1 Curvature is a key ingredient in roller coaster design.

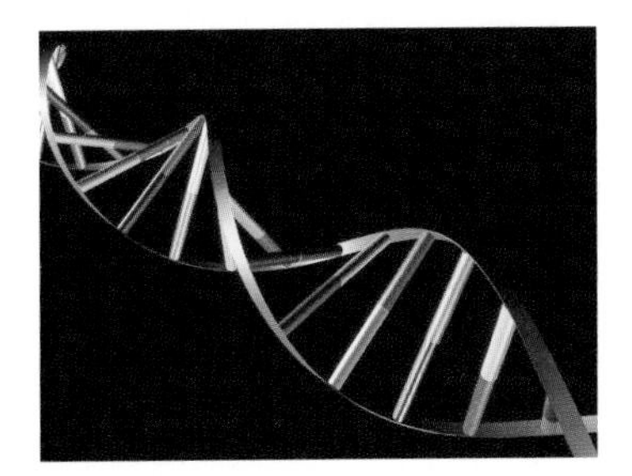

FIGURE 2 Biochemists study the effect of the curvature of DNA strands on biological processes.

In this section, we apply the vector tools we have developed to study curvature which, as the name suggests, is a measure of how much a curve bends. Curvature is used to study geometric properties of curves and motion along curves. It has applications in diverse areas such as highway and roller coaster design (Figure 1), computer-aided design, optics, and even eye surgery (see Exercise 58). Biochemists have found that the curvature of DNA strands plays a role in biological processes in ways that are not yet fully understood (Figure 2).

In Chapter 4, we used the second derivative $f''(x)$ to measure the bending or "concavity" of a graph $y = f(x)$. It might seem natural, therefore, to take $f''(x)$ as our definition of curvature. However, there are two reasons why this will not work. First, $f''(x)$ only makes sense for a graph $y = f(x)$ in the plane, and our goal is to define curvature for curves in three-space. A second more serious problem is that $f''(x)$ does not truly capture the curvature of a curve, as we see by considering the unit circle. The circle is symmetrical, so the curvature ought to be the same at every point (Figure 3). But note that the upper semicircle is the graph of $f(x) = (1 - x^2)^{1/2}$. The second derivative $f''(x) = -(1 - x^2)^{-3/2}$ does not have the same value at every point and thus is not a good measure of curvature.

To arrive at the proper definition of curvature, let us consider a path in vector form $\mathbf{r}(t) = \langle x(t), y(t), z(t) \rangle$. The parametrization is called **regular** if $\mathbf{r}'(t) \ne \mathbf{0}$ for all t in the domain of $\mathbf{r}(t)$. We assume that $\mathbf{r}(t)$ is regular and define the **unit tangent vector** in the direction $\mathbf{r}'(t)$, denoted $\mathbf{T}(t)$:

$$\text{Unit tangent vector} = \mathbf{T}(t) = \frac{\mathbf{r}'(t)}{\|\mathbf{r}'(t)\|}$$

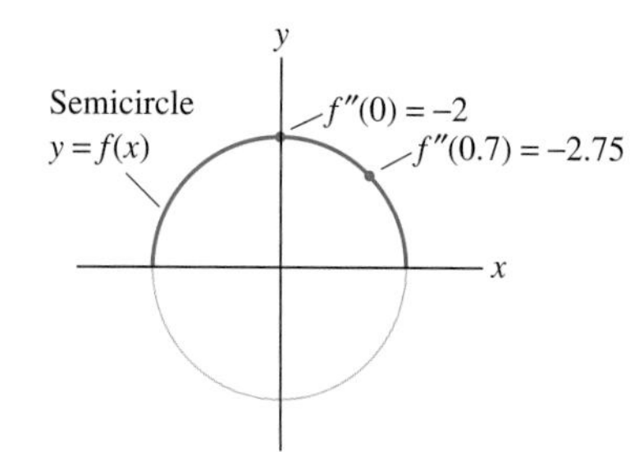

FIGURE 3 The second derivative of $f(x) = \sqrt{1 - x^2}$ does not capture the curvature of the circle, which by symmetry should be the same at all points.

Now imagine walking along a path and observing how the unit tangent vector $\mathbf{T}(t)$ changes direction (Figure 4). A change in $\mathbf{T}(t)$ indicates that the path is bending, and the more rapidly $\mathbf{T}(t)$ changes, the more the path is bent. Thus, it might seem reasonable to define curvature as the magnitude $\|\mathbf{T}'(t)\|$ of the derivative of $\mathbf{T}(t)$. This is almost correct, but $\|\mathbf{T}'(t)\|$ also depends on how fast you walk. To eliminate the effect of your walking speed and capture the "intrinsic" bending rate of the curve itself, we define curvature as the magnitude $\kappa(t) = \|\mathbf{T}'(t)\|$ for an *arc length parametrization* (i.e., we assume that you walk with unit speed). Recall that $\mathbf{r}(t)$ is an arc length parametrization if $\|\mathbf{r}'(t)\| = 1$.

DEFINITION Curvature Let $\mathbf{r}(s)$ be an arc length parametrization and $\mathbf{T}$ the unit tangent vector. The **curvature** at $\mathbf{r}(s)$ is the quantity

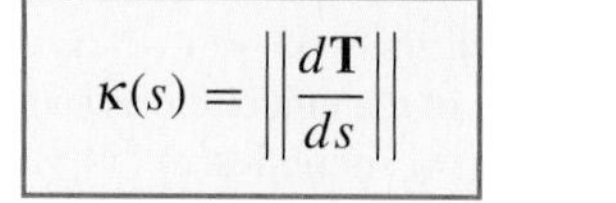

$$\kappa(s) = \left\| \frac{d\mathbf{T}}{ds} \right\| \qquad \boxed{1}$$

Let us compute the curvature in the two basic cases of a line and a circle.

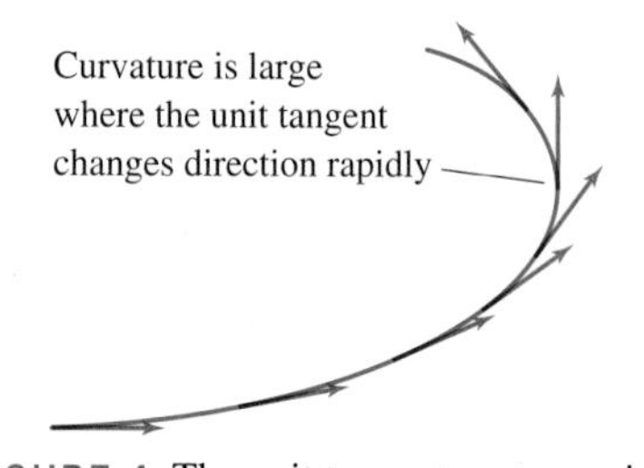

FIGURE 4 The unit tangent vector varies in direction but not length.

EXAMPLE 1 A Line Has Zero Curvature Let $\mathcal{L}$ be the line $\mathbf{r}(s) = \langle x_0, y_0, z_0 \rangle + s\mathbf{u}$, where $\|\mathbf{u}\| = 1$. Compute the unit tangent vector and curvature at each point.

Solution First, we note that $\mathbf{r}(s)$ is an arc length parametrization because $\mathbf{r}'(s) = \mathbf{u}$ and $\|\mathbf{r}'(s)\| = \|\mathbf{u}\| = 1$. In an arc length parametrization, $\mathbf{T}(s) = \mathbf{r}'(s)$. Therefore, $\mathbf{T}(s) = \mathbf{u}$ is constant and the curvature is zero at all points:

$$\kappa(s) = \left\| \frac{d\mathbf{T}}{ds} \right\| = \left\| \frac{d}{ds}\mathbf{u} \right\| = 0$$

■

EXAMPLE 2 Curvature of a Circle of Radius R Is $1/R$ Compute the curvature of a circle of radius R.

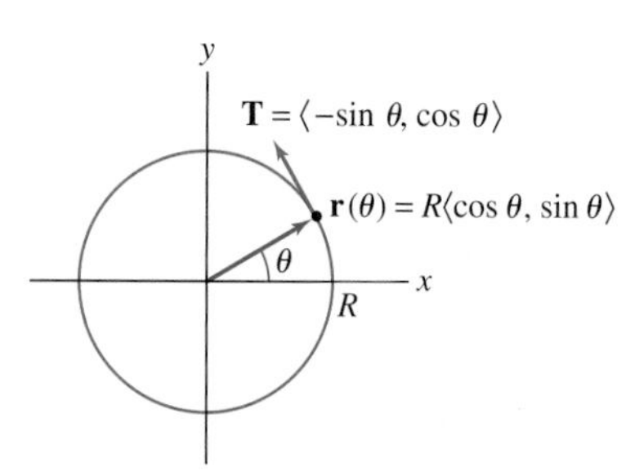

FIGURE 5 The unit tangent vector at a point on a circle of radius R.

Since a circle of radius R has curvature $1/R$, a large circle has small curvature. This makes sense if you consider an arc of fixed length 1 on a circle of radius R. As R increases, this arc gets straighter.

Solution Assume the circle is centered at the origin and consider the parametrization $\mathbf{r}(\theta) = \langle R\cos\theta, R\sin\theta \rangle$ (Figure 5). The tangent vector $\mathbf{r}'(\theta) = \langle -R\sin\theta, R\cos\theta \rangle$ has length R, so $\mathbf{r}(\theta)$ is not an arc length parametrization unless $R = 1$. Recall from Section 13.3 that to find an arc length parametrization, we compute the arc length function:

$$s(\theta) = \int_0^\theta \|\mathbf{r}'(t)\|\, dt = \int_0^\theta R\, dt = R\theta$$

and form the inverse of $s = R\theta$, which is $\theta = \varphi(s) = \frac{s}{R}$. The arc length parametrization is

$$\mathbf{r}_1(s) = \mathbf{r}(\varphi(s)) = \mathbf{r}\left(\frac{s}{R}\right) = \left\langle R\cos\frac{s}{R}, R\sin\frac{s}{R} \right\rangle$$

This is an arc length parametrization, and we have

$$\mathbf{T}(s) = \mathbf{r}_1'(s) = \left\langle -\sin\frac{s}{R}, \cos\frac{s}{R} \right\rangle$$

$$\frac{d\mathbf{T}}{ds} = -\frac{1}{R}\left\langle \cos\frac{s}{R}, \sin\frac{s}{R} \right\rangle$$

$$\kappa(s) = \left\| \frac{d\mathbf{T}}{ds} \right\| = \frac{1}{R}$$

This shows that the curvature is $1/R$ at all points on the circle. ■

Formula (1) for curvature is often hard to use because it requires an arc length parametrization. Fortunately, there are formulas for the curvature $\kappa(t)$ in terms of any regular parametrization $\mathbf{r}(t)$. Let $s(t) = \int_a^t \|\mathbf{r}'(u)\|\,du$ be the arc length function (with any lower limit a) and let $t = \varphi(s)$ be the inverse of $s(t)$. As shown in Section 13.3, $\mathbf{r}(\varphi(s))$ is an arc length parametrization. Note that $\mathbf{T}(\varphi(s))$ is the unit tangent vector along $\mathbf{r}(\varphi(s))$ and thus $\kappa(t) = \left\|\frac{d\mathbf{T}}{ds}\right\|$. By the Chain Rule,

REMINDER $\mathbf{T}(t)$ *is the unit tangent vector*

$$\mathbf{T}(t) = \frac{\mathbf{r}'(t)}{\|\mathbf{r}'(t)\|}$$

$$\mathbf{T}'(t) = \frac{d\mathbf{T}}{ds}\frac{ds}{dt} = \frac{d\mathbf{T}}{ds}\|\mathbf{r}'(t)\|$$

$$\|\mathbf{T}'(t)\| = \left\|\frac{d\mathbf{T}}{ds}\right\| \|\mathbf{r}'(t)\| = \kappa(t)\,\|\mathbf{r}'(t)\|$$

This yields the formula

$$\kappa(t) = \frac{\|\mathbf{T}'(t)\|}{\|\mathbf{r}'(t)\|} \qquad \boxed{2}$$

The following theorem provides another useful formula for curvature directly in terms of the path $\mathbf{r}(t)$.

THEOREM 1 Formula for Curvature If $\mathbf{r}(t)$ is a regular parametrization, then the curvature at $\mathbf{r}(t)$ is

$$\kappa(t) = \frac{\|\mathbf{r}'(t) \times \mathbf{r}''(t)\|}{\|\mathbf{r}'(t)\|^3} \qquad \boxed{3}$$

To apply Eq. (3) to plane curves, replace $\mathbf{r}(t) = \langle x(t), y(t)\rangle$ *by* $\mathbf{r}(t) = \langle x(t), y(t), 0\rangle$ *and compute the cross product.*

Proof The vectors $\mathbf{T}(t)$ and $\mathbf{T}'(t)$ are orthogonal. Indeed, $\mathbf{T}(t)$ is a unit vector and, as we showed in Example 6 of Section 13.2, any vector-valued function of constant length is orthogonal to its derivative (see the marginal note). Set $v(t) = \|\mathbf{r}'(t)\|$. Then $\|\mathbf{T}'(t)\| = \kappa(t)v(t)$ by Eq. (2), so we have

$$\|\mathbf{T}(t) \times \mathbf{T}'(t)\| = \|\mathbf{T}(t)\|\,\|\mathbf{T}'(t)\| = \|\mathbf{T}'(t)\| = \kappa(t)\|\mathbf{r}'(t)\| = \kappa(t)v(t) \qquad \boxed{4}$$

REMINDER To prove that $\mathbf{T}(t)$ *and* $\mathbf{T}'(t)$ *are orthogonal, differentiate the equation* $\mathbf{T}(t)\cdot\mathbf{T}(t) = 1$:

$$\frac{d}{dt}\mathbf{T}(t)\cdot\mathbf{T}(t) = 2\mathbf{T}(t)\cdot\mathbf{T}'(t) = 0$$

Thus, $\mathbf{T}(t)\cdot\mathbf{T}'(t) = 0$. *In Eq. (4), we use the formula*

$$\|\mathbf{v}\times\mathbf{w}\| = \|\mathbf{v}\|\,\|\mathbf{w}\|\sin\theta$$

to conclude that $\|\mathbf{v}\times\mathbf{w}\| = \|\mathbf{v}\|\,\|\mathbf{w}\|$ *if* $\mathbf{v}$ *and* $\mathbf{w}$ *are orthogonal.*

The next step is to express this cross product in terms of $\mathbf{r}(t)$ and $\mathbf{r}'(t)$. By definition, $\mathbf{r}'(t) = v(t)\mathbf{T}(t)$, and by the Product Rule,

$$\mathbf{r}''(t) = \frac{d}{dt}\mathbf{r}'(t) = v'(t)\mathbf{T}(t) + v(t)\mathbf{T}'(t)$$

Using $\mathbf{T}(t) \times \mathbf{T}(t) = \mathbf{0}$, we find that

$$\mathbf{r}'(t) \times \mathbf{r}''(t) = v(t)\mathbf{T}(t) \times \big(v'(t)\mathbf{T}(t) + v(t)\mathbf{T}'(t)\big) = v(t)^2(\mathbf{T}(t) \times \mathbf{T}'(t))$$

To finish the proof, compute the length and apply Eq. (4):

$$\|\mathbf{r}'(t) \times \mathbf{r}''(t)\| = v(t)^2\|\mathbf{T}(t) \times \mathbf{T}'(t)\| = \kappa(t)v(t)^3$$

We obtain $\kappa(t) = \dfrac{\|\mathbf{r}'(t) \times \mathbf{r}''(t)\|}{v(t)^3}$ as claimed. ■

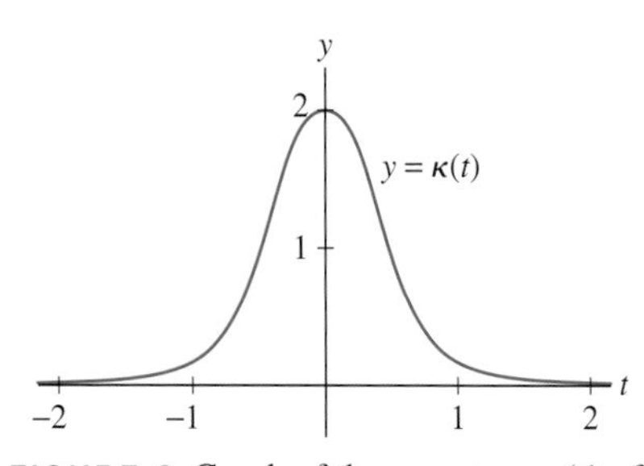

FIGURE 6 Graph of the curvature $\kappa(t)$ of the twisted cubic $\mathbf{r}(t) = \langle t, t^2, t^3 \rangle$.

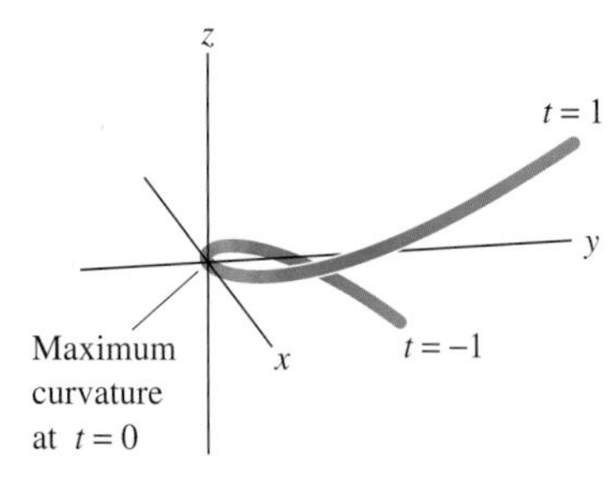

FIGURE 7 Graph of twisted cubic $\mathbf{r}(t) = \langle t, t^2, t^3 \rangle$ colored by curvature.

■ **EXAMPLE 3 Twisted Cubic Curve** CAS Calculate the curvature $\kappa(t)$ of the twisted cubic $\mathbf{r}(t) = \langle t, t^2, t^3 \rangle$. Then plot the graph of $\kappa(t)$ and determine where the curvature is largest.

Solution We use Eq. (3):

$$\mathbf{r}'(t) = \langle 1, 2t, 3t^2 \rangle, \qquad \mathbf{r}''(t) = \langle 0, 2, 6t \rangle$$

$$\mathbf{r}'(t) \times \mathbf{r}''(t) = \begin{vmatrix} \mathbf{i} & \mathbf{j} & \mathbf{k} \\ 1 & 2t & 3t^2 \\ 0 & 2 & 6t \end{vmatrix} = 6t^2\mathbf{i} - 6t\mathbf{j} + 2\mathbf{k}$$

$$\kappa(t) = \frac{\|\mathbf{r}'(t) \times \mathbf{r}''(t)\|}{\|\mathbf{r}'(t)\|^3} = \frac{\sqrt{36t^4 + 36t^2 + 4}}{(1 + 4t^2 + 9t^4)^{3/2}}$$

The graph of $\kappa(t)$ in Figure 6 shows that the curvature is largest at $t = 0$. The curve $\mathbf{r}(t)$ is illustrated in Figure 7. The plot is colored by curvature, with large curvature represented in blue, small curvature in green. ■

In the second paragraph of this section, we pointed out that the curvature of a graph $y = f(x)$ must involve more than just the second derivative $f''(x)$. We now show that the curvature can be expressed in terms of both $f''(x)$ and $f'(x)$.

THEOREM 2 Curvature of a Graph in the Plane The curvature at the point $(x, f(x))$ on the graph of $y = f(x)$ is equal to

$$\kappa(x) = \frac{|f''(x)|}{(1 + f'(x)^2)^{3/2}} \qquad \boxed{5}$$

Proof The curve $y = f(x)$ is parametrized by $\mathbf{r}(x) = \langle x, f(x) \rangle$. Therefore, $\mathbf{r}'(x) = \langle 1, f'(x) \rangle$ and $\mathbf{r}''(x) = \langle 0, f''(x) \rangle$. To compute the cross product, we treat $\mathbf{r}'(x)$ and $\mathbf{r}''(x)$ as vectors in $\mathbf{R}^3$ with z-component equal to zero:

$$\mathbf{r}'(x) \times \mathbf{r}''(x) = \begin{vmatrix} \mathbf{i} & \mathbf{j} & \mathbf{k} \\ 1 & f'(x) & 0 \\ 0 & f''(x) & 0 \end{vmatrix} = f''(x)\mathbf{k}$$

Since $\|\mathbf{r}'(t)\| = \|\langle 1, f'(x) \rangle\| = (1 + f'(x)^2)^{1/2}$, Eq. (3) yields

$$\kappa(x) = \frac{\|\mathbf{r}'(x) \times \mathbf{r}''(x)\|}{\|\mathbf{r}'(x)\|^3} = \frac{|f''(x)|}{(1 + f'(x)^2)^{3/2}}$$ ■

■ **EXAMPLE 4** Compute the curvature of $f(x) = x^3 - 3x^2 + 4$ at $x = 0, 1, 2, 3$.

Solution We apply Eq. (5):

$$f'(x) = 3x^2 - 6x = 3x(x - 2), \qquad f''(x) = 6x - 6$$

$$\kappa(x) = \frac{|f''(x)|}{(1 + f'(x)^2)^{3/2}} = \frac{|6x - 6|}{(1 + 9x^2(x - 2)^2)^{3/2}}$$

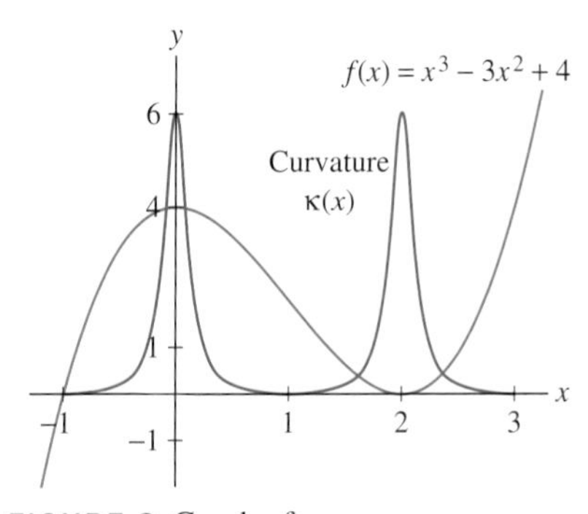

FIGURE 8 Graph of $f(x) = x^3 - 3x^2 + 4$ and the curvature $\kappa(x)$.

We obtain the following values:

$$\kappa(0) = \frac{6}{(1+0)^{3/2}} = 6, \quad \kappa(1) = \frac{0}{(1+9)^{3/2}} = 0$$

$$\kappa(2) = \frac{6}{(1+0)^{3/2}} = 6, \quad \kappa(3) = \frac{12}{82^{3/2}} \approx 0.016$$

Figure 8 shows that the graph bends more where the curvature is large. ■

Unit Normal Vector

The unit tangent vector $\mathbf{T}(t)$ is orthogonal to its derivative $\mathbf{T}'(t)$, as we observed in the proof of Theorem 1. Assuming that $\mathbf{T}'(t)$ is nonzero, the unit vector in the direction $\mathbf{T}'(t)$ is called the **unit normal vector**; it is denoted $\mathbf{N}(t)$ or simply $\mathbf{N}$:

$$\text{Unit normal vector} = \mathbf{N}(t) = \frac{\mathbf{T}'(t)}{\|\mathbf{T}'(t)\|} \tag{6}$$

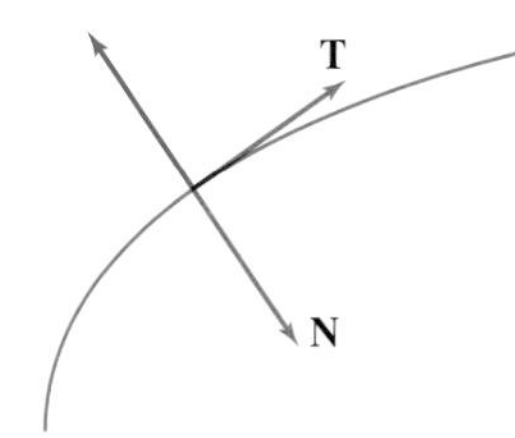

FIGURE 9 For a plane curve, the unit normal points in the direction of bending.

Intuitively speaking, the unit normal $\mathbf{N}$ indicates the direction in which the curve is turning (Figure 10). This is particularly clear for a plane curve. In this case, there are two unit vectors orthogonal to $\mathbf{T}$ (Figure 9), and of these two, $\mathbf{N}$ is the vector that points to the "inside" of the curve.

■ **EXAMPLE 5 Unit Normal to a Helix** Find the unit normal vector to the helix $\mathbf{r}(t) = \langle \cos t, \sin t, t \rangle$ at $t = \frac{\pi}{4}$.

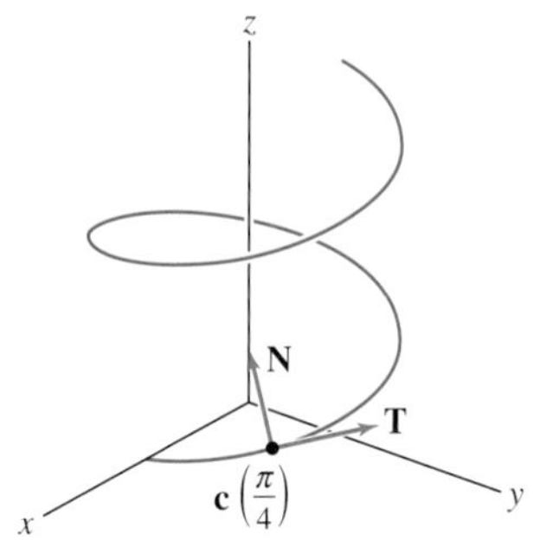

FIGURE 10 Unit tangent and normal vectors at $t = \frac{\pi}{4}$ on the helix in Example 5.

Solution The tangent vector $\mathbf{r}'(t) = \langle -\sin t, \cos t, 1 \rangle$ has length $\sqrt{2}$, so

$$\mathbf{T}(t) = \frac{\mathbf{r}'(t)}{\|\mathbf{r}'(t)\|} = \frac{1}{\sqrt{2}} \langle -\sin t, \cos t, 1 \rangle$$

$$\mathbf{T}'(t) = \frac{1}{\sqrt{2}} \langle -\cos t, -\sin t, 0 \rangle$$

$$\mathbf{N}(t) = \frac{\mathbf{T}'(t)}{\|\mathbf{T}'(t)\|} = \langle -\cos t, -\sin t, 0 \rangle$$

Hence, $\mathbf{N}\left(\frac{\pi}{4}\right) = \left\langle -\frac{\sqrt{2}}{2}, -\frac{\sqrt{2}}{2}, 0 \right\rangle$ (Figure 10). ■

We conclude this section by mentioning an important geometric interpretation of curvature, in terms of the "best fitting circle." Let P be a point on a plane curve $\mathcal{C}$ with parametrization $\mathbf{r}(t)$ and let κ_P be the curvature at P. If $\kappa_P \neq 0$, then there is a unique circle through P, called the **osculating circle** and denoted Osc_P, such that

(i) Osc_P and $\mathcal{C}$ have the same tangent line and unit normal vectors at P.

(ii) Both Osc_P and $\mathcal{C}$ have curvature κ_P at P.

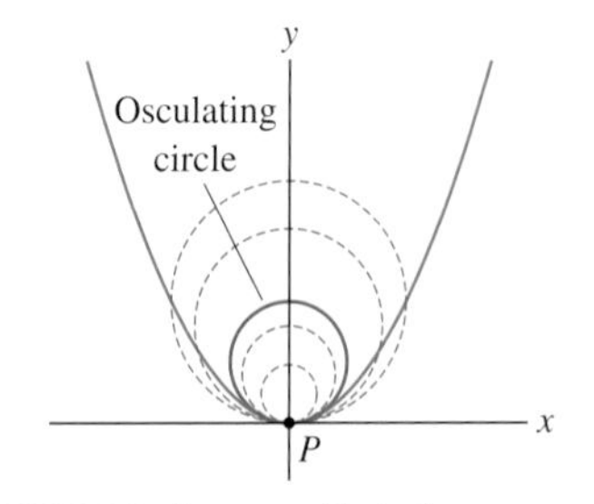

FIGURE 11 Among all circles tangent to the $\mathcal{C}$ at P, the osculating circle is the "best fit" to the curve.

Since the curvature of a circle is the reciprocal of its radius (by Example 2), the osculating circle has radius $R = 1/\kappa_P$. We refer to $R = 1/\kappa_P$ as the **radius of curvature** at P. The center of Osc_P is called the **center of curvature** at P. Among all circles tangent to $\mathcal{C}$ at P, Osc_P is the circle that "best fits" the curve (Figure 11; see also Exercise 70).

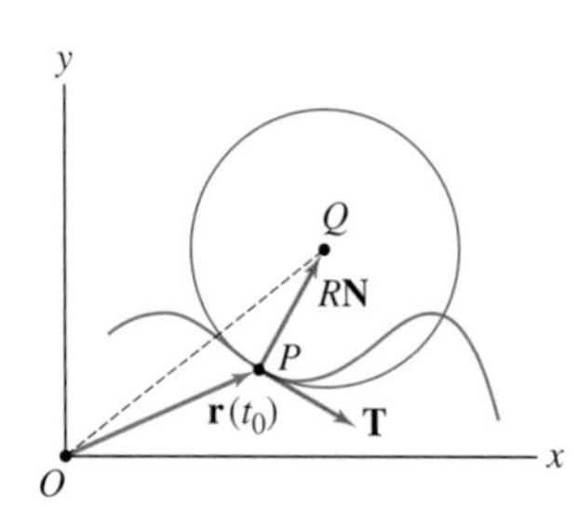

FIGURE 12 The center Q of the osculating circle at P lies at a distance $R = \kappa_P^{-1}$ from P in the normal direction.

Let Q be the center of Osc_P. We can determine Q by observing that it is located at a distance $R = 1/\kappa_P$ from P in the normal direction $\mathbf{N}$ (Figure 12). Therefore, if P is the terminal point of $\mathbf{r}(t)$ at $t = t_0$, then

$$\overrightarrow{OQ} = \mathbf{r}(t_0) + \kappa_P^{-1}\mathbf{N} \qquad \boxed{7}$$

■ **EXAMPLE 6** Parametrize the osculating circle to the parabola $y = x^2$ at $x = \frac{1}{2}$.

Solution Let $f(x) = x^2$. We use the parametrization

$$\mathbf{r}(x) = \langle x, f(x)\rangle = \left\langle x, x^2\right\rangle$$

and proceed by the following steps.

Step 1. **Find κ and N.**
We apply Eq. (5) to $f(x) = x^2$ to compute the curvature:

$$\kappa(x) = \frac{|f''(x)|}{(1 + f'(x)^2)^{3/2}} = \frac{2}{(1 + 4x^2)^{3/2}}, \qquad \kappa\left(\frac{1}{2}\right) = \frac{2}{2^{3/2}} = \frac{1}{\sqrt{2}}$$

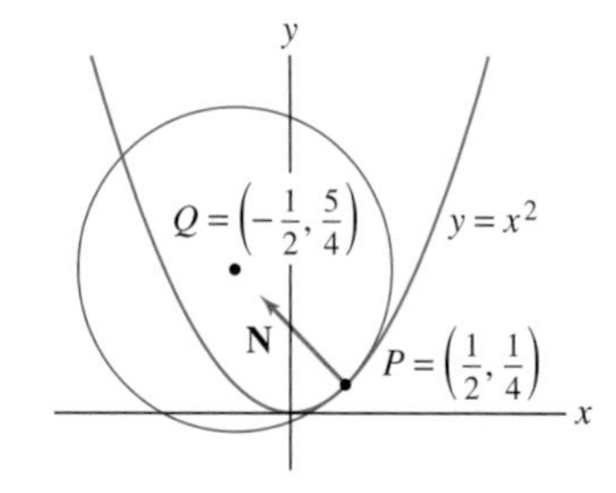

FIGURE 13 Osculating circle to $y = x^2$ at $x = 1/2$.

There is an easy way to find $\mathbf{N}$ for a plane curve that saves us the trouble of computing $\mathbf{T}'$. The tangent vector is $\mathbf{r}'(x) = \langle 1, 2x\rangle$, and we simply observe that $\langle 2x, -1\rangle$ is orthogonal to $\mathbf{r}'(t) = \langle 1, 2x\rangle$ (since their dot product is zero). Therefore, $\mathbf{N}(x)$ is the unit vector in one of the two directions $\pm\langle 2x, -1\rangle$. Figure 13 shows that the unit normal vector points in the positive y-direction (the direction of bending), so we conclude that

$$\mathbf{N}(x) = \frac{\langle -2x, 1\rangle}{\|\langle -2x, 1\rangle\|} = \frac{\langle -2x, 1\rangle}{\sqrt{1 + 4x^2}}, \qquad \mathbf{N}\left(\frac{1}{2}\right) = \frac{1}{\sqrt{2}}\langle -1, 1\rangle$$

Step 2. **Find the center of the osculating circle.**
To find the center Q of the osculating circle at $\mathbf{r}(\frac{1}{2}) = \langle \frac{1}{2}, \frac{1}{4}\rangle$, we use Eq. (7). The position vector of Q is

$$\overrightarrow{OQ} = \mathbf{r}\left(\frac{1}{2}\right) + \kappa\left(\frac{1}{2}\right)^{-1}\mathbf{N}\left(\frac{1}{2}\right) = \left\langle \frac{1}{2}, \frac{1}{4}\right\rangle + \sqrt{2}\left(\frac{\langle -1, 1\rangle}{\sqrt{2}}\right) = \left\langle -\frac{1}{2}, \frac{5}{4}\right\rangle$$

Step 3. **Parametrize the osculating circle.**
The osculating circle has radius $R = 1/\kappa = \sqrt{2}$ and parametrization

$$\mathbf{c}(t) = \underbrace{\left\langle -\frac{1}{2}, \frac{5}{4}\right\rangle}_{\text{Center}} + \sqrt{2}\,\langle \cos t, \sin t\rangle \qquad ■$$

If a curve $\mathcal{C}$ lies in a plane, then this plane is the osculating plane. For a general curve in three-space, the osculating plane varies from point to point.

To define the osculating circle for space curve $\mathcal{C}$, we must first specify the plane in which the circle lies. The **osculating plane** at a point P on $\mathcal{C}$ is the plane through P determined by the unit tangent $\mathbf{T}$ and the unit normal $\mathbf{N}$ at P (we assume that $\mathbf{T}' \neq 0$ so $\mathbf{N}$ is defined at P). Intuitively, the osculating plane is the plane that "most nearly" contains the curve $\mathcal{C}$ near P. The osculating circle is then defined as the circle of radius $R = 1/\kappa_P$ through P in the osculating plane having $\mathbf{T}$ and $\mathbf{N}$ as its unit tangent and normal vectors. Equation (7) remains valid for space curves. Figure 14 shows the osculating circle at two points on the curve $\mathbf{r}(t) = \langle \cos t, \sin t, \sin 2t\rangle$. The osculating plane at each of these points is the plane containing the circle.

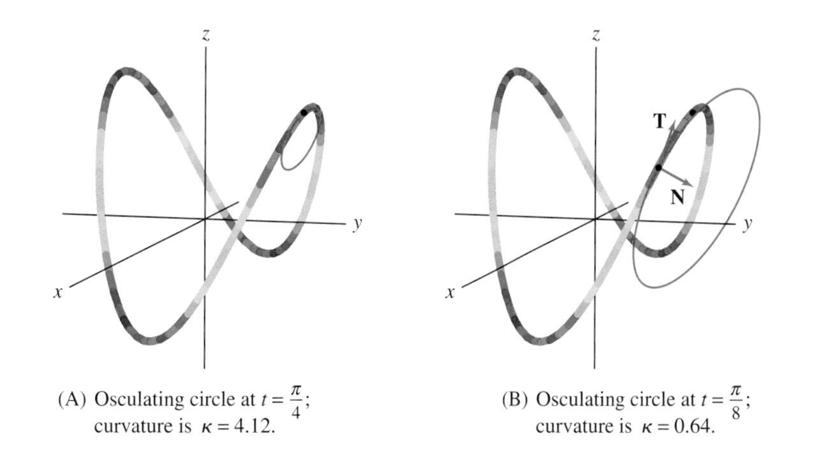

FIGURE 14 Osculating circles to $\mathbf{r}(t) = \langle \cos t, \sin t, \sin 2t \rangle$.

(A) Osculating circle at $t = \frac{\pi}{4}$; curvature is $\kappa = 4.12$.

(B) Osculating circle at $t = \frac{\pi}{8}$; curvature is $\kappa = 0.64$.

13.4 SUMMARY

- A parametrization $\mathbf{r}(t)$ is called *regular* if $\mathbf{r}'(t) \neq \mathbf{0}$ for all t. If $\mathbf{r}(t)$ is regular, we define the *unit tangent vector* $\mathbf{T}(t) = \dfrac{\mathbf{r}'(t)}{\|\mathbf{r}'(t)\|}$.
- *Curvature* is defined by $\kappa(s) = \left\| \dfrac{d\mathbf{T}}{ds} \right\|$, where $\mathbf{r}(s)$ is an arc length parametrization. In practice, we compute curvature using the following formulas, valid for arbitrary regular parametrizations:

$$\kappa(t) = \frac{\|\mathbf{T}'(t)\|}{\|\mathbf{r}'(t)\|} = \frac{\|\mathbf{r}'(t) \times \mathbf{r}''(t)\|}{\|\mathbf{r}'(t)\|^3}$$

- The curvature at a point on a graph $y = f(x)$ in the plane is

$$\kappa(x) = \frac{|f''(x)|}{(1 + f'(x)^2)^{3/2}}$$

- If $\|\mathbf{T}'(t)\| \neq 0$, we define the *unit normal vector* $\mathbf{N}(t) = \dfrac{\mathbf{T}'(t)}{\|\mathbf{T}'(t)\|}$.
- The *osculating plane* at a point P on a curve $\mathcal{C}$ is the plane through P determined by the vectors $\mathbf{T}$ and $\mathbf{N}$. It is defined only if the curvature κ_P at P is nonzero.
- The *osculating circle at* P is the circle Osc_P of radius $R = 1/\kappa_P$ in the osculating plane having $\mathbf{T}$ and $\mathbf{N}$ as its unit tangent and normal vectors. The center of Osc_P is called the *center of curvature* and R is called the *radius of curvature*.
- If $\overrightarrow{OP} = \mathbf{r}(t_0)$, the center Q of Osc_P satisfies $\overrightarrow{OQ} = \mathbf{r}(t_0) + \kappa_P^{-1}\mathbf{N}$.

13.4 EXERCISES

Preliminary Questions

1. What is the unit tangent vector of a line with direction vector $\mathbf{v} = \langle 2, 1, -2 \rangle$?

2. What is the curvature of a circle of radius 4?

3. Which has larger curvature, a circle of radius 2 or a circle of radius 4?

4. What is the curvature of $\mathbf{r}(t) = \langle 2 + 3t, 7t, 5 - t \rangle$?

5. What is the curvature at a point where $\mathbf{T}'(s) = \langle 1, 2, 3\rangle$ in an arc length parametrization $\mathbf{r}(s)$?

6. What is the radius of curvature of a circle of radius 4?

7. What is the radius of curvature at P if $\kappa_P = 9$?

Exercises

In Exercises 1–6, calculate $\mathbf{r}'(t)$ *and* $\mathbf{T}(t)$, *and evaluate* $\mathbf{T}(1)$.

1. $\mathbf{r}(t) = \langle 12t^3, 18t^2, 9t^4\rangle$

2. $\mathbf{r}(t) = \langle \cos \pi t, \sin \pi t, t\rangle$

3. $\mathbf{r}(t) = \langle 3 + 4t, 3 - 5t, 9t\rangle$

4. $\mathbf{r}(t) = \langle 1 + 2t, t^2, 3 - t^2\rangle$

5. $\mathbf{r}(t) = \langle 4t^2, 9t\rangle$

6. $\mathbf{r}(t) = \langle e^t, t^2\rangle$

In Exercises 7–12, use Eq. (3) to calculate $\kappa(t)$.

7. $\mathbf{r}(t) = \langle 1, e^t, t\rangle$

8. $\mathbf{r}(t) = \langle \cos t, t \sin t, t\rangle$

9. $\mathbf{r}(t) = \langle 4 \cos t, t, 4 \sin t\rangle$

10. $\mathbf{r}(t) = \langle 4t + 1, 4t - 3, 2t\rangle$

11. $\mathbf{r}(t) = \langle t^{-1}, 1, t\rangle$

12. $\mathbf{r}(t) = \langle \cosh t, \sinh t, t\rangle$

In Exercises 13–16, find the curvature of the plane curve at the point indicated.

13. $y = e^t, \quad t = 3$

14. $y = \cos x, \quad x = 0$

15. $y = t^4, \quad t = 2$

16. $y = t^n, \quad t = 1$

17. Find the curvature of $\mathbf{r}(t) = \langle 2 \sin t, \cos 3t, t\rangle$ at $t = \frac{\pi}{3}$ and $t = \frac{\pi}{2}$ (Figure 15).

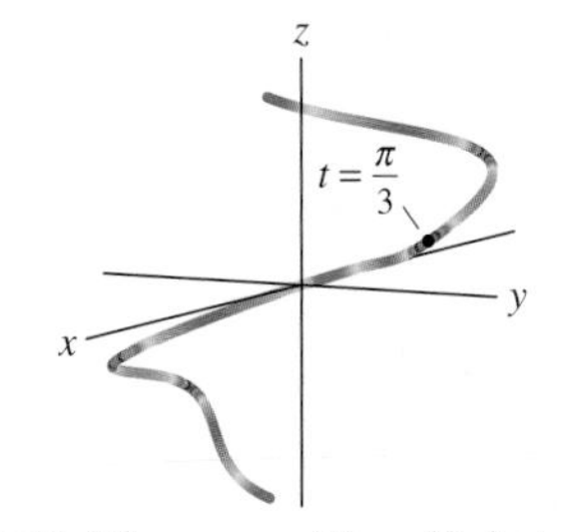

FIGURE 15 The curve $\mathbf{r}(t) = \langle 2 \sin t, \cos 3t, t\rangle$.

18. CAS Find the curvature function $\kappa(x)$ for $y = \sin x$. Use a computer algebra system to plot $\kappa(x)$ for $0 \le x \le 2\pi$. Prove that the curvature takes its maximum at $x = \frac{\pi}{2}$ and $\frac{3\pi}{2}$. *Hint:* To simplify the calculation, find the maximum of $\kappa(x)^2$.

19. Show that curvature at an inflection point of a plane curve $y = f(x)$ is zero.

20. Show that the tractrix $\mathbf{r}(t) = \langle t - \tanh t, \operatorname{sech} t\rangle$ has the curvature function $\kappa(t) = \operatorname{sech} t$.

21. Find the value of α such that the curvature of $y = e^{\alpha x}$ at $x = 0$ is as large as possible.

22. Find the point of maximum curvature on $y = e^x$.

23. Show that the curvature function of the parametrization $\mathbf{r}(t) = \langle a \cos t, b \sin t\rangle$ of the ellipse $\left(\frac{x}{a}\right)^2 + \left(\frac{y}{b}\right)^2 = 1$ is

$$\kappa(t) = \frac{ab}{(b^2 \cos^2 t + a^2 \sin^2 t)^{3/2}} \qquad \boxed{8}$$

24. Use a sketch to predict where the points of minimal and maximal curvature occur on an ellipse. Then use Eq. (8) to confirm or refute your prediction.

25. In the notation of Exercise 23, assume that $a \ge b$. Show that $b/a^2 \le \kappa(t) \le a/b^2$ for all t.

26. Use Eq. (3) to prove that for a plane curve $\mathbf{r}(t) = \langle x(t), y(t)\rangle$,

$$\kappa(t) = \frac{|x'(t)y''(t) - x''(t)y'(t)|}{(x'(t)^2 + y'(t)^2)^{3/2}} \qquad \boxed{9}$$

In Exercises 27–30, use Eq. (9) to compute the curvature at the given point.

27. $\langle t^2, t^3\rangle, \quad t = 2$

28. $\langle \cosh s, s\rangle, \quad s = 0$

29. $\langle t \cos t, \sin t\rangle, \quad t = \pi$

30. $\langle \sin 3s, 2 \sin 4s\rangle, \quad s = \frac{\pi}{2}$

31. Let $s(t) = \int_{-\infty}^{t} \|\mathbf{r}'(u)\|\, du$ for the Bernoulli spiral $\mathbf{r}(t) = \langle e^t \cos 4t, e^t \sin 4t\rangle$ (see Exercise 25 in Section 13.3). Show that the radius of curvature is proportional to $s(t)$.

32. The **Cornu spiral** is the plane curve $\mathbf{r}(t) = \langle x(t), y(t)\rangle$, where

$$x(t) = \int_0^t \sin \frac{u^2}{2}\, du, \qquad y(t) = \int_0^t \cos \frac{u^2}{2}\, du$$

Verify that $\kappa(t) = |t|$. Since the curvature increases linearly, the Cornu spiral is used in highway design to create transitions between straight and curved road segments (Figure 16).

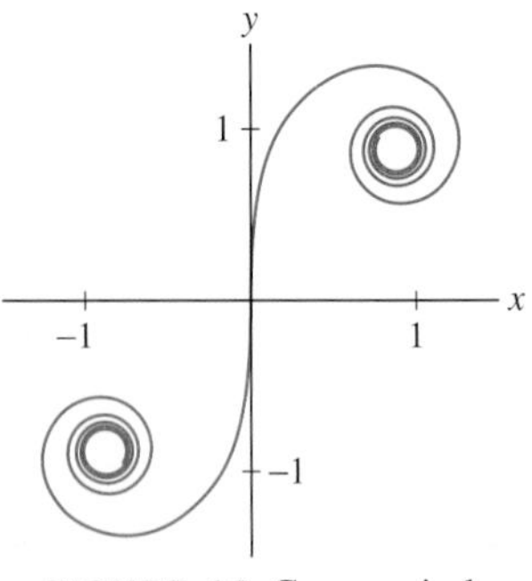

FIGURE 16 Cornu spiral.

33. CAS Plot and compute the curvature $\kappa(t)$ of the **clothoid** $\mathbf{r}(t) = \langle x(t), y(t)\rangle$, where

$$x(t) = \int_0^t \sin \frac{u^3}{3}\, du, \qquad y(t) = \int_0^t \cos \frac{u^2}{3}\, du$$

34. Find the unit normal vector $\mathbf{N}(\theta)$ to $\mathbf{r}(\theta) = R\langle\cos\theta, \sin\theta\rangle$, the circle of radius R. Does $\mathbf{N}(\theta)$ point inside or outside the circle? Draw $\mathbf{N}(\theta)$ at $\theta = \frac{\pi}{4}$ with $R = 4$.

35. Find the unit normal vector $\mathbf{N}(t)$ to $\mathbf{r}(t) = \langle 4, \sin 2t, \cos 2t\rangle$.

36. Sketch the graph of $\mathbf{r}(t) = \langle t, t^3\rangle$. Since $\mathbf{r}'(t) = \langle 1, 3t^2\rangle$, the unit normal $\mathbf{N}(t)$ points in one of the two directions $\pm\langle -3t^2, 1\rangle$. Which sign is correct at $t = 1$? Which is correct at $t = -1$?

37. Find the normal vectors to $\mathbf{r}(t) = \langle t, \cos t\rangle$ at $t = \frac{\pi}{4}$ and $t = \frac{3\pi}{4}$.

38. Find the unit normal to the Cornu spiral (Exercise 32) at $t = \sqrt{\pi}$.

39. Find the unit normal to the clothoid (Exercise 33) at $t = \pi^{1/3}$.

40. Method for Computing N Let $v(t) = \|\mathbf{r}'(t)\|$. Show that

$$\mathbf{N}(t) = \frac{\mathbf{r}''(t) - v'(t)\mathbf{T}(t)}{\|\mathbf{r}''(t) - v'(t)\mathbf{T}(t)\|} \qquad \boxed{10}$$

Hint: Differentiate $\mathbf{r}'(t) = v(t)\mathbf{T}(t)$ and note that $\mathbf{N}$ is the unit vector in the direction $\mathbf{T}'$.

In Exercises 41–46, use Eq. (10) to find $\mathbf{N}$ *at the point indicated.*

41. $\langle 1 + t^2, 2t, t^3\rangle, \quad t = 1$

42. $\langle t, e^t, t\rangle, \quad t = 2$

43. $\langle t - \sin t, 1 - \cos t\rangle, \quad t = \pi$

44. $\langle t^2, t^3\rangle, \quad t = 1$

45. $\langle t^{-1}, t, t^2\rangle, \quad t = -1$

46. $\langle \cosh t, \sinh t, t\rangle, \quad t = 0$

47. Let $f(x) = x^2$. Show that the center of the osculating circle at (x_0, x_0^2) is $\left(-4x_0^3, \frac{1}{2} + 3x_0^2\right)$.

48. Find a parametrization of the osculating circle to $y = x^2$ at $x = 1$.

49. Find a parametrization of the osculating circle to $y = \sin x$ at $x = \frac{\pi}{2}$.

50. Use Eq. (7) to find the center of curvature to $\mathbf{r}(t) = \langle t^2, t^3\rangle$ at $t = 1$.

In Exercises 51–55, find a parametrization of the osculating circle at the point indicated.

51. $\langle \cos t, \sin t\rangle, \quad t = \frac{\pi}{4}$

52. $\langle \sin t, \cos t\rangle, \quad t = 0$

53. $\langle t - \sin t, 1 - \cos t\rangle, \quad t = \pi$ (use Exercise 43)

54. $\langle 1 + t^2, 2t, t^3\rangle, \quad t = 1$ (use Exercise 41)

55. $\mathbf{r}(t) = \langle \cosh t, \sinh t, t\rangle$ (use Exercise 12)

56. Find the curvature and unit normal vector to the helix $\mathbf{r}(t) = \langle \cos t, \sin t, t\rangle$ at $t = 0$. Then use Eq. (7) to find the center of curvature and a parametrization of the osculating circle at $t = 0$.

57. Figure 17 shows the graph of the half-ellipse $y = \pm\sqrt{2rx - px^2}$, where r and p are positive constants. Show that the radius of curvature at the origin is equal to r. *Hint:* One way of proceeding is to write the ellipse in the form of Exercise 23 and apply Eq. (8).

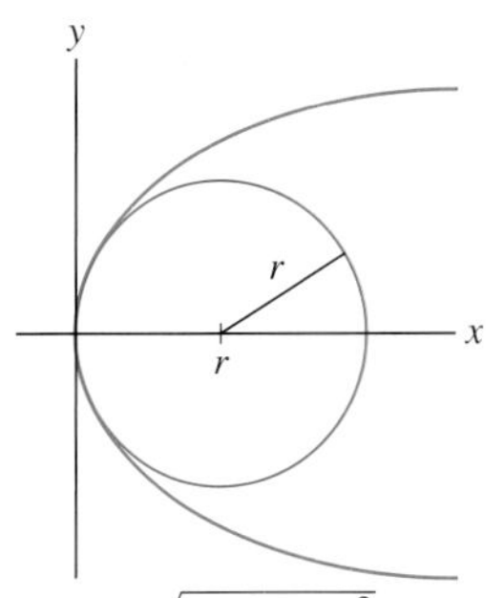

FIGURE 17 The curve $y = \sqrt{2rx - px^2}$ and the osculating circle at the origin.

58. In a recent study of laser eye surgery by Gatinel, Hoang-Xuan, and Azar, a vertical cross section of the cornea is modeled by the half-ellipse of Exercise 57. Show that the half-ellipse can be written in the form $x = f(y)$, where $f(y) = p^{-1}\left(r - \sqrt{r^2 - py^2}\right)$. During surgery, tissue is removed to a depth $t(y)$ at height y for $-S \le y \le S$, where $t(y)$ is given by Munnerlyn's equation (for some $R > r$):

$$t(y) = \sqrt{R^2 - S^2} - \sqrt{R^2 - y^2} - \sqrt{r^2 - S^2} + \sqrt{r^2 - y^2}$$

After surgery, the cross section of the cornea has the shape $x = f(y) + t(y)$ (Figure 18). Show that after surgery, the radius of curvature at the point P (where $y = 0$) is R.

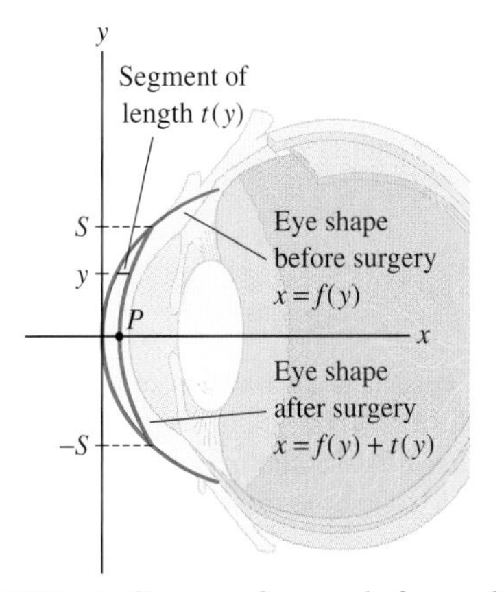

FIGURE 18 Contour of cornea before and after surgery.

59. The **angle of inclination** of a plane curve with parametrization $\mathbf{r}(t)$ is defined as the angle $\theta(t)$ between the unit tangent vector $\mathbf{T}(t)$ and the x-axis (Figure 19). Show that $\|\mathbf{T}'(t)\| = |\theta'(t)|$ and conclude that if $\mathbf{r}(s)$ is a parametrization by arc length, then

$$\kappa(s) = \left|\frac{d\theta}{ds}\right| \qquad \boxed{11}$$

Hint: Observe that $\mathbf{T}(t) = \langle \cos\theta(t), \sin\theta(t)\rangle$.

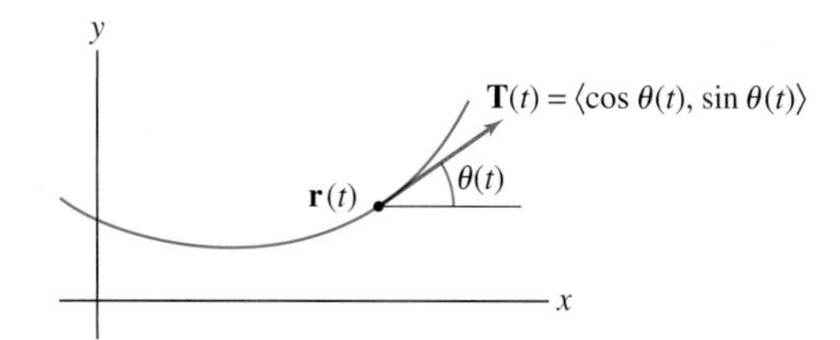

FIGURE 19 The curvature is the rate of change of $\theta(t)$.

60. A particle moves along the path $y = x^3$ with unit speed. How fast is the tangent turning (i.e., how fast is the angle of inclination changing) when the particle passes through the point (2, 8)?

61. Verify Eq. (11) for a circle of radius R. Suppose that a particle traverses the circle at unit speed. Show that the change in the angle during an interval of length Δt is $\Delta\theta = \Delta t/R$ and conclude that $\theta'(s) = 1/R$.

62. Let $\theta(x)$ be the angle of inclination at a point on the graph $y = f(x)$ (see Exercise 59).

(a) Use the relation $f'(x) = \tan\theta$ to prove that $\dfrac{d\theta}{dx} = \dfrac{f''(x)}{(1 + f'(x)^2)}$.

(b) Use the arc length integral to show that $\dfrac{ds}{dx} = \sqrt{1 + f'(x)^2}$.

(c) Now give a proof of Eq. (5) using Eq. (11).

63. Use the parametrization $\mathbf{r}(\theta) = \langle f(\theta)\cos\theta, f(\theta)\sin\theta\rangle$ to show that a curve $r = f(\theta)$ in polar coordinates has curvature

$$\kappa(\theta) = \frac{|f(\theta)^2 + 2f'(\theta)^2 - 2f(\theta)f''(\theta)|}{(f(\theta)^2 + f'(\theta)^2)^{3/2}} \qquad \boxed{12}$$

In Exercises 64–66, use Eq. (12) to find the curvature of the curve given in polar form.

64. $f(\theta) = 2\cos\theta$ **65.** $f(\theta) = \theta$ **66.** $f(\theta) = e^\theta$

67. Use Eq. (12) to find the curvature of the general Bernoulli spiral $r = ae^{b\theta}$ in polar form (a and b are constants).

68. Show that both $\mathbf{r}'(t)$ and $\mathbf{r}''(t)$ lie in the osculating plane for a vector function $\mathbf{r}(t)$. *Hint:* Differentiate $\mathbf{r}'(t) = v(t)\mathbf{T}(t)$.

69. Show that

$$\gamma(s) = \frac{1}{\kappa}\mathbf{N} + \frac{1}{\kappa}\big((\sin\kappa s)\mathbf{T} - (\cos\kappa s)\mathbf{N}\big)$$

is an arc length parametrization of the osculating circle.

70. Two vector-valued functions $\mathbf{r}_1(s)$ and $\mathbf{r}_2(s)$ are said to *agree to order 2* at s_0 if

$$\mathbf{r}_1(s_0) = \mathbf{r}_2(s_0), \quad \mathbf{r}_1'(s_0) = \mathbf{r}_2'(s_0), \quad \mathbf{r}_1''(s_0) = \mathbf{r}_2''(s_0)$$

Let $\mathbf{r}(s)$ be an arc length parametrization of a path C and let P be the terminal point of $\mathbf{r}(0)$. Let $\gamma(s)$ be the arc length parametrization of the osculating circle given in Exercise 69. Show that $\mathbf{r}(s)$ and $\gamma(s)$ agree to order 2 at $s = 0$ (in fact, the osculating circle is the unique circle that approximates C to order 2 at P).

71. Let $\mathbf{r}(t) = \langle x(t), y(t), z(t)\rangle$ be a path with curvature $\kappa(t)$ and define the scaled path $\mathbf{r}_1(t) = \langle\lambda x(t), \lambda y(t), \lambda z(t)\rangle$, where $\lambda \neq 0$ is a constant. Prove that curvature varies inversely with the scale factor, that is, the curvature $\kappa_1(t)$ of $\mathbf{r}_1(t)$ is $\kappa_1(t) = \lambda^{-1}\kappa(t)$. This explains why the curvature of a circle of radius R is proportional to $1/R$ (in fact, it is equal to $1/R$). *Hint:* Use Eq. (3).

Further Insights and Challenges

72. Show that the curvature of Viviani's curve, given by $\mathbf{r}(t) = \langle 1 + \cos t, \sin t, 2\sin(t/2)\rangle$, is

$$\kappa(t) = \frac{\sqrt{13 + 3\cos t}}{(3 + \cos t)^{3/2}}$$

73. Let $\mathbf{r}(s)$ be an arc length parametrization of a closed curve C of length L. We call C an **oval** if $d\theta/ds > 0$ (see Exercise 59). Observe that $-\mathbf{N}$ points to the *outside* of C. For $k > 0$, the curve C_1 defined by $\mathbf{r}_1(s) = \mathbf{r}(s) - k\mathbf{N}$ is called the expansion of $c(s)$ in the normal direction.

(a) Show that $\|\mathbf{r}_1'(s)\| = \|\mathbf{r}'(s)\| + k\kappa(s)$.

(b) As P moves around the oval counterclockwise, θ increases by 2π [Figure 20(A)]. Use this and a change of variables to prove that $\displaystyle\int_0^L \kappa(s)\,ds = 2\pi$.

(c) Show that C_1 has length $L + 2\pi k$.

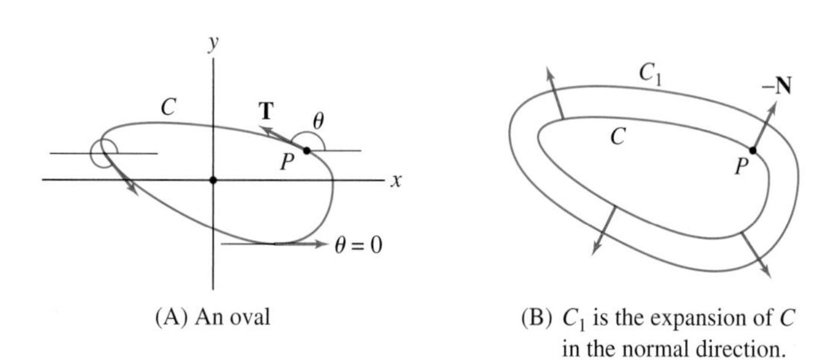

FIGURE 20 As P moves around the oval, θ increases by 2π.

In Exercises 74–81, let **B** *denote the* binormal vector *at a point on a space curve C, defined by* $\mathbf{B} = \mathbf{T} \times \mathbf{N}$.

74. Show that **B** is a unit vector.

75. Follow steps (a)–(c) to prove that there is a number τ (lowercase Greek tau) called the **torsion** such that

$$\frac{d\mathbf{B}}{ds} = -\tau\mathbf{N} \qquad \boxed{13}$$

(a) Show that $\dfrac{d\mathbf{B}}{ds} = \mathbf{T} \times \dfrac{d\mathbf{N}}{ds}$ and conclude that $d\mathbf{B}/ds$ is orthogonal to **T**.

(b) Differentiate $\mathbf{B} \cdot \mathbf{B} = 1$ with respect to s to show that $d\mathbf{B}/ds$ is orthogonal to $\mathbf{B}$.

(c) Conclude that $d\mathbf{B}/ds$ is a multiple of $\mathbf{N}$.

76. Show that if $\mathcal{C}$ is contained in a plane $\mathcal{P}$, then $\mathbf{B}$ is a unit vector normal to $\mathcal{P}$. Conclude that $\tau = 0$ for a plane curve.

77. Torsion means "twisting." Is this an appropriate name for τ? Explain by interpreting τ geometrically.

78. Use the identity

$$\mathbf{a} \times (\mathbf{b} \times \mathbf{c}) = (\mathbf{a} \cdot \mathbf{c})\mathbf{b} - (\mathbf{a} \cdot \mathbf{b})\mathbf{c}$$

to prove

$$\mathbf{N} \times \mathbf{B} = \mathbf{T}, \qquad \mathbf{B} \times \mathbf{T} = \mathbf{N} \tag{14}$$

79. Follow steps (a)–(b) to prove

$$\frac{d\mathbf{N}}{ds} = -\kappa\mathbf{T} + \tau\mathbf{B} \tag{15}$$

(a) Show that $d\mathbf{N}/ds$ is orthogonal to $\mathbf{N}$. Conclude that $d\mathbf{N}/ds$ lies in the plane spanned by $\mathbf{T}$ and $\mathbf{B}$, and hence, $d\mathbf{N}/ds = a\mathbf{T} + b\mathbf{B}$ for some scalars a, b.

(b) Use $\mathbf{N} \cdot \mathbf{T} = 0$ to show that $\mathbf{T} \cdot \frac{d\mathbf{N}}{ds} = -\mathbf{N} \cdot \frac{d\mathbf{T}}{ds}$ and compute a. Compute b similarly.

Equations (13) and (15) together with $d\mathbf{T}/dt = \kappa\mathbf{N}$ are called the **Frenet formulas** and were discovered by the French geometer Jean Frenet (1816–1900).

80. Show that $\mathbf{r}' \times \mathbf{r}''$ is a multiple of $\mathbf{B}$. Conclude that

$$\mathbf{B} = \frac{\mathbf{r}' \times \mathbf{r}''}{\|\mathbf{r}' \times \mathbf{r}''\|} \tag{16}$$

81. The vector $\mathbf{N}$ can be computed using $\mathbf{N} = \mathbf{B} \times \mathbf{T}$ [Eq. (14)] with $\mathbf{B}$, as in Eq. (16). Use this method to find $\mathbf{N}$ in the following cases:

(a) $\mathbf{r}(t) = \langle \cos t, t, t^2 \rangle$ at $t = 0$

(b) $\mathbf{r}(t) = \langle t^2, t^{-1}, t \rangle$ at $t = 1$

13.5 Motion in Three-Space

FIGURE 1 The flight of the space shuttle is analyzed using vector calculus.

In this section, we study the motion of a particle traveling along a path $\mathbf{r}(t)$. Recall that the velocity vector, which we now denote $\mathbf{v}(t)$, is the derivative

$$\mathbf{v}(t) = \mathbf{r}'(t) = \lim_{h \to 0} \frac{\mathbf{r}(t+h) - \mathbf{r}(t)}{h}$$

As we have seen, the velocity vector points in the direction of motion (if it is nonzero) and its magnitude $v(t) = \|\mathbf{v}(t)\|$ is the particle's speed. The **acceleration vector** is the second derivative $\mathbf{r}''(t)$, which we shall denote $\mathbf{a}(t)$. In summary,

$$\mathbf{v}(t) = \mathbf{r}'(t), \qquad v(t) = \|\mathbf{v}(t)\|, \qquad \mathbf{a}(t) = \mathbf{r}''(t)$$

■ **EXAMPLE 1** Find the acceleration vector of $\mathbf{r}(t) = t\mathbf{i} - t^3\mathbf{j} + \ln t\,\mathbf{k}$ at $t = 1$.

Solution

$$\mathbf{v}(t) = \mathbf{r}'(t) = \mathbf{i} - 3t^2\mathbf{j} + t^{-1}\mathbf{k}$$

$$\mathbf{a}(t) = \mathbf{r}''(t) = -6t\mathbf{j} - t^{-2}\mathbf{k}$$

Therefore, $\mathbf{a}(1) = -6\mathbf{j} - \mathbf{k}$. ■

In linear motion, acceleration is the scalar derivative of velocity. The acceleration is the rate at which the object is speeding up or slowing down, so it is zero if the object travels with constant speed. One of the surprising things about acceleration in two or three dimensions is that an object may have nonzero acceleration even when its speed is constant. This happens when $v(t) = \|\mathbf{v}(t)\|$ is constant but the *direction* of the velocity vector $\mathbf{v}(t)$ is changing. The simplest example is uniform circular motion, in which an object travels in a circular path at constant speed. In this case, $\mathbf{v}(t)$ has constant length

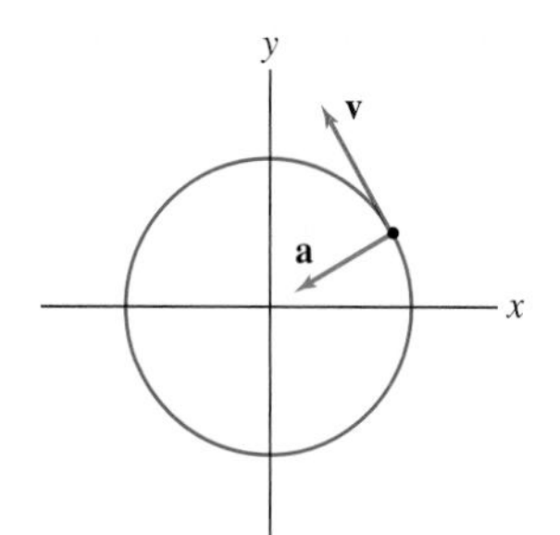

FIGURE 2 In uniform circular motion, the acceleration vector points toward the center of motion.

but turns continuously as the object moves around the circle. This change in direction is measured by the acceleration vector $\mathbf{a}(t)$, which points in the normal direction, toward the center of the circle (Figure 2). We will see that general motion combines the two types of acceleration: changing speed and changing direction.

EXAMPLE 2 **Acceleration Vector for Uniform Circular Motion** Find $\mathbf{a}(t)$ and $\|\mathbf{a}(t)\|$ for motion around a circle of radius R with constant speed v.

Solution The term **uniform circular motion** refers to motion around a circle at constant speed (Figure 2). If the circle is centered at the origin and the particle moves counterclockwise starting at position $(R, 0)$ at time $t = 0$, then for some constant ω (lowercase Greek omega) we have

$$\mathbf{r}(t) = R\langle \cos \omega t, \sin \omega t\rangle, \qquad \mathbf{v}(t) = R\omega\langle -\sin \omega t, \cos \omega t\rangle$$

The constant ω is called the angular speed *because the particle's angle along the circle changes at a rate of ω radians per unit time.*

To determine ω, we observe that $v = \|\mathbf{v}(t)\| = R\omega$. Thus, $\omega = v/R$, and accordingly,

$$\mathbf{a}(t) = \mathbf{v}'(t) = -R\omega^2\langle \cos \omega t, \sin \omega t\rangle, \qquad \|\mathbf{a}(t)\| = R\omega^2 = R\left(\frac{v}{R}\right)^2 = \frac{v^2}{R}$$

We see that $\mathbf{a}(t)$ *has length* $\dfrac{v^2}{R}$ *and points in toward the origin* [since $\mathbf{a}(t)$ is a negative multiple of $\mathbf{r}(t)$] as in Figure 2. ■

According to Newton's Law of Motion $\mathbf{F} = m\mathbf{a}$, a force is required to make an object change direction. In uniform circular motion, the force $m\mathbf{a}(t)$, which points in the normal direction by Example 2, is called the "centripetal force."

If the acceleration $\mathbf{a}(t) = \mathbf{r}''(t)$ of an object is given, we may solve for $\mathbf{v}(t)$ and $\mathbf{r}(t)$ by integrating:

$$\mathbf{v}(t) = \int_0^t \mathbf{a}(u)\,du + \mathbf{v}_0 \qquad (\mathbf{v}_0 = \text{initial velocity vector})$$

$$\mathbf{r}(t) = \int_0^t \mathbf{v}(u)\,du + \mathbf{v}_0 t + \mathbf{r}_0 \qquad (\mathbf{r}_0 = \text{initial position})$$

EXAMPLE 3 Find $\mathbf{r}(t)$ if $\mathbf{a}(t) = 2\mathbf{i} + 12t\mathbf{j}$, $\mathbf{v}(0) = 7\mathbf{i}$, and $\mathbf{r}(0) = 2\mathbf{i} + 9\mathbf{k}$.

Solution We have

$$\mathbf{v}(t) = \int_0^t \mathbf{a}(u)\,du = 2t\mathbf{i} + 6t^2\mathbf{j} + \mathbf{v}_0$$

$$\mathbf{r}(t) = \int_0^t \mathbf{v}'(u)\,du = t^2\mathbf{i} + 2t^3\mathbf{j} + \mathbf{v}_0 t + \mathbf{r}_0$$

The initial conditions give $\mathbf{v}(0) = \mathbf{v}_0 = 7\mathbf{i}$ and $\mathbf{r}(0) = \mathbf{r}_0 = 2\mathbf{i} + 9\mathbf{k}$, so

$$\mathbf{r}(t) = (t^2\mathbf{i} + 2t^3\mathbf{j}) + 7t\mathbf{i} + (2\mathbf{i} + 9\mathbf{k}) = (t^2 + 7t + 2)\mathbf{i} + 2t^3\mathbf{j} + 9\mathbf{k}$$ ■

Newton's Second Law of Motion is often stated in the scalar form $F = ma$, but a more general statement is the vector law $\mathbf{F} = m\mathbf{a}$, where $\mathbf{F}$ is the net force vector acting on the object and $\mathbf{a}$ is the acceleration vector. When the force varies from position to position, we write $\mathbf{F}(\mathbf{r}(t))$ for the force acting on a particle with position vector $\mathbf{r}(t)$ at time t. Then Newton's Second Law reads

$$\mathbf{F}(\mathbf{r}(t)) = m\mathbf{a}(t) \qquad \text{or} \qquad \mathbf{F}(\mathbf{r}(t)) = m\mathbf{r}''(t) \tag{1}$$

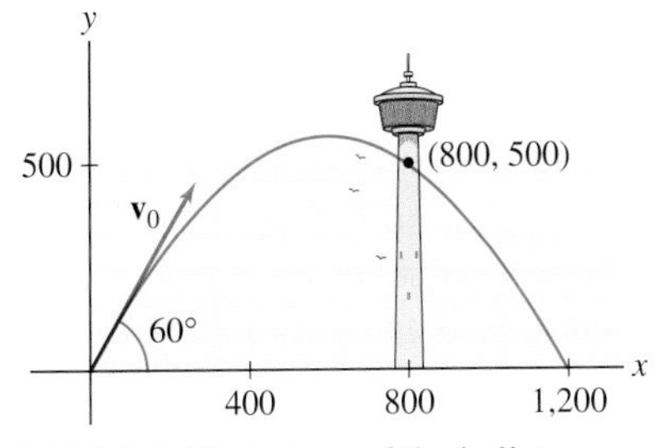

FIGURE 3 Trajectory of the bullet.

EXAMPLE 4 A bullet is fired from the ground at an angle of 60° above the horizontal. What initial speed v_0 must the bullet have in order to hit a point 500 ft high on a tower located 800 ft away (ignoring air resistance)?

Solution Place the gun at the origin, and let $\mathbf{r}(t)$ be the position vector of the bullet (Figure 3).

***Step 1.* Use Newton's Law.**

Gravity exerts a downward force of magnitude mg, where m is the mass of the bullet and $g = 32\ \text{ft/s}^2$. In vector form,

$$\mathbf{F} = \langle 0, -gm \rangle = m\langle 0, -g \rangle$$

In this case, Newton's Second Law $\mathbf{F} = m\mathbf{r}''(t)$ reduces to $\mathbf{r}''(t) = \langle 0, -g \rangle$. We determine $\mathbf{r}(t)$ by integrating twice:

$$\mathbf{r}'(t) = \int_0^t \mathbf{r}''(u)\, du = \int_0^t \langle 0, -32 \rangle\, du = \langle 0, -32t \rangle + \mathbf{v}_0$$

$$\mathbf{r}(t) = \int_0^t \mathbf{r}'(u)\, du = \int_0^t \left(\langle 0, -32u \rangle + \mathbf{v}_0\right) du = \langle 0, -16t^2 \rangle + t\mathbf{v}_0 + \mathbf{r}_0$$

Here, $\mathbf{r}_0$ is the initial position and $\mathbf{v}_0$ is the initial velocity. By our choice of coordinates, $\mathbf{r}_0 = \mathbf{0}$. The initial velocity $\mathbf{v}_0$ has unknown length v_0, but we know that it points in the direction of the unit vector $\langle \cos 60°, \sin 60° \rangle$. Therefore,

$$\mathbf{v}_0 = v_0\langle \cos 60°, \sin 60° \rangle = v_0 \left\langle \frac{1}{2}, \frac{\sqrt{3}}{2} \right\rangle$$

$$\mathbf{r}(t) = -\langle 0, -16t^2 \rangle + tv_0 \left\langle \frac{1}{2}, \frac{\sqrt{3}}{2} \right\rangle$$

***Step 2.* Solve for $\mathbf{v}_0$.**

The position vector of the point on the tower is $\langle 800, 500 \rangle$, so the bullet will hit a point on the tower 500 ft high if there exists a time t such that

$$\mathbf{r}(t) = -\langle 0, -16t^2 \rangle + tv_0 \left\langle \frac{1}{2}, \frac{\sqrt{3}}{2} \right\rangle = \langle 800, 500 \rangle$$

Equating components, we obtain the equations

$$\frac{1}{2}tv_0 = 800, \qquad -16t^2 + \frac{\sqrt{3}}{2}tv_0 = 500$$

The first equation yields $t = \dfrac{1{,}600}{v_0}$. Now substitute in the second equation and solve:

$$-16\left(\frac{1{,}600}{v_0}\right)^2 + \frac{\sqrt{3}}{2}\left(\frac{1{,}600}{v_0}\right)v_0 = 500$$

$$\left(\frac{1{,}600}{v_0}\right)^2 = \frac{800\sqrt{3} - 500}{16}$$

$$v_0^2 = \frac{16(1{,}600)^2}{800\sqrt{3} - 500}$$

$$v_0 = \frac{6{,}400}{\sqrt{800\sqrt{3} - 500}} \approx 215\ \text{ft/s}$$

Understanding the Acceleration Vector

The velocity vector $\mathbf{v}(t)$ has a clear interpretation: It points in the direction of motion and its length is the speed. What about the acceleration vector $\mathbf{a}(t)$? Acceleration is the rate of change of velocity $\mathbf{v}(t)$, but to understand $\mathbf{a}(t)$, we must keep in mind that $\mathbf{v}(t)$ can change in two ways: in length and in direction. The acceleration vector "encodes" both types of change, but to extract this information, we must decompose $\mathbf{a}(t)$ into a sum of tangential and normal components.

Recall the definition of unit tangent and unit normal vectors:

$$\mathbf{T}(t) = \frac{\mathbf{v}(t)}{\|\mathbf{v}(t)\|}, \qquad \mathbf{N}(t) = \frac{\mathbf{T}'(t)}{\|\mathbf{T}'(t)\|}$$

Thus, $\mathbf{v}(t) = v(t)\mathbf{T}(t)$, where $v(t) = \|\mathbf{v}(t)\|$, and by the Product Rule,

$$\mathbf{a}(t) = \frac{d\mathbf{v}}{dt} = \frac{d}{dt}v(t)\mathbf{T}(t) = v'(t)\mathbf{T}(t) + v(t)\mathbf{T}'(t) \qquad \boxed{2}$$

On the other hand, $\|\mathbf{T}'(t)\| = \kappa(t)v(t)$, where $\kappa(t)$ is the curvature, by Eq. (2) of Section 13.4. Therefore,

$$v(t)\mathbf{T}'(t) = v(t)\|\mathbf{T}'(t)\|\left(\frac{\mathbf{T}'(t)}{\|\mathbf{T}'(t)\|}\right) = \kappa(t)v(t)^2\mathbf{N}(t)$$

We may rewrite Eq. (2) in the form

$$\boxed{\mathbf{a}(t) = v'(t)\mathbf{T}(t) + \kappa(t)v(t)^2\,\mathbf{N}(t)} \qquad \boxed{3}$$

We call $a_{\mathbf{T}}(t) = v'(t)$ the **tangential component** and $a_{\mathbf{N}}(t) = \kappa(t)v(t)^2$ the **normal component** of acceleration (Figure 4). We often write Eq. (3) more simply as

$$\boxed{\mathbf{a} = a_{\mathbf{T}}\mathbf{T} + a_{\mathbf{N}}\,\mathbf{N}}$$

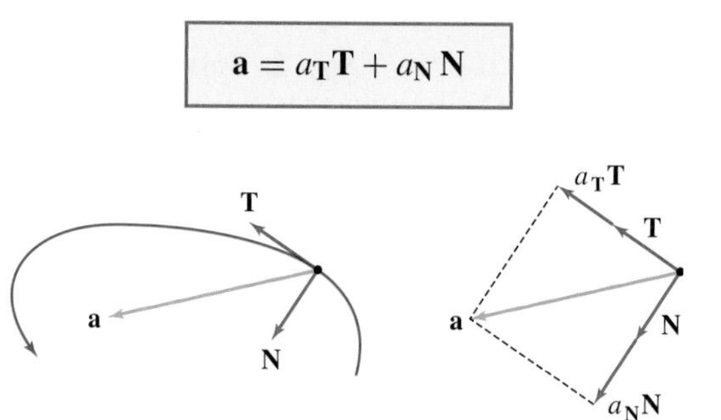

FIGURE 4 Decomposition of **a** into tangential and normal components.

For a good example of normal acceleration, imagine making a left turn in an automobile. If you drive at constant speed, your tangential acceleration is zero and you will not be pushed back against your seat. But the car seat (via friction) pushes you to the left against the car door, causing you to accelerate in the normal direction (due to inertia, you feel as if you are being pushed to the right toward the passenger's seat). This force is proportional to κv^2, so a sharp turn (large κ) or high speed (large v) produces a strong normal force.

CONCEPTUAL INSIGHT The decomposition of Eq. (3) separates out the two ways in which velocity $\mathbf{v}(t)$ can change. The tangential component $a_{\mathbf{T}} = v'(t)$ describes the rate at which the *speed* changes, and the normal component $a_{\mathbf{N}} = \kappa(t)v(t)^2$ describes the rate at which the *direction* changes [this rate depends on both the curvature $\kappa(t)$ and the speed $v(t)$ at which you travel around the curve]. Consider the following scenarios:

- A particle travels in a straight line. Then $\kappa(t) = 0$ and $\mathbf{a}(t) = v'(t)\mathbf{T}$. Thus, $\mathbf{a}(t)$ points in the direction of motion if the particle is speeding up [$v'(t) > 0$] and opposite to the direction of motion if the particle is slowing down [$v'(t) < 0$].
- A particle travels with constant speed along a curved path. Then $v'(t) = 0$ and the acceleration $\mathbf{a}(t) = \kappa(t)v(t)^2\mathbf{N}$ is normal to the direction of motion.

General motion is a combination, involving both tangential and normal acceleration.

FIGURE 5 The Giant Ferris Wheel in Vienna, Austria, erected in 1897 to celebrate the 50th anniversary of the coronation of Emperor Franz Joseph I.

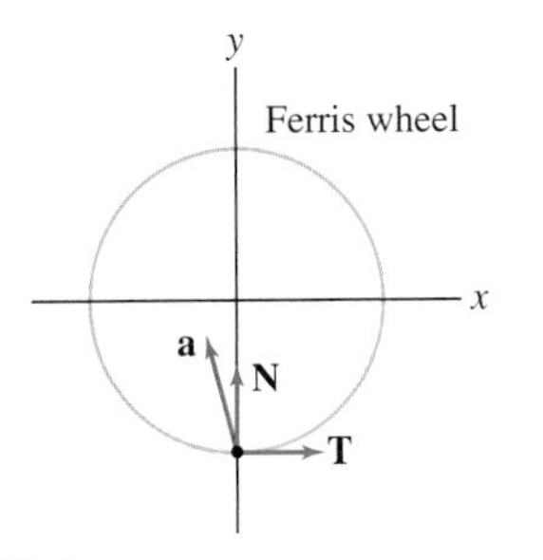

FIGURE 6

EXAMPLE 5 The Giant Ferris Wheel in Vienna has radius $R = 30$ m (Figure 5). Assume that at time $t = t_0$, the wheel rotates counterclockwise with a speed of 40 m/min, and is slowing at a rate of 15 m/min^2. Find the acceleration vector $\mathbf{a}$ for a person seated in a car at the lowest point of the wheel (Figure 6).

Solution At the bottom of the wheel, $\mathbf{T} = \langle 1, 0 \rangle$ and $\mathbf{N} = \langle 0, 1 \rangle$. We are told that $a_{\mathbf{T}} = v' = -15$ at time t_0. The curvature of the wheel is $\kappa = 1/R = 1/30$, so the normal component is $a_{\mathbf{N}} = \kappa v^2 = v^2/R = (40)^2/30 \approx 53.3$. Therefore,

$$\mathbf{a} \approx -15\mathbf{T} + 53.3\mathbf{N} = \langle -15, 53.3 \rangle \ \text{m/min}^2$$

The following theorem provides some useful formulas for the tangential and normal components of the acceleration vector.

THEOREM 1 Tangential and Normal Components of Acceleration Let $\mathbf{a} = \mathbf{r}''(t)$ and $\mathbf{v} = \mathbf{r}'(t)$ be the acceleration and velocity vectors of a path $\mathbf{r}(t)$ at time t. Let $v(t) = \|\mathbf{v}(t)\|$ be the speed at time t. Then the coefficients $a_{\mathbf{T}} = v'(t)$ and $a_{\mathbf{N}} = \kappa(t)v(t)^2$ in the decomposition

$$\mathbf{a} = a_{\mathbf{T}}\mathbf{T} + a_{\mathbf{N}}\mathbf{N}$$

are given by the formulas

$$a_{\mathbf{T}} = \frac{\mathbf{a} \cdot \mathbf{v}}{\|\mathbf{v}\|} \qquad \boxed{4}$$

$$a_{\mathbf{N}} = \frac{\|\mathbf{a} \times \mathbf{v}\|}{\|\mathbf{v}\|} = \sqrt{\|\mathbf{a}\|^2 - |a_{\mathbf{T}}|^2} \qquad \boxed{5}$$

Proof Recall that $\mathbf{T} = \dfrac{\mathbf{v}}{\|\mathbf{v}\|}$. Since $\mathbf{T} \cdot \mathbf{T} = 1$ and $\mathbf{N} \cdot \mathbf{T} = 0$, we obtain (4) as follows:

$$\frac{\mathbf{a} \cdot \mathbf{v}}{\|\mathbf{v}\|} = \mathbf{a} \cdot \mathbf{T} = (a_{\mathbf{T}}\mathbf{T} + a_{\mathbf{N}}\mathbf{N}) \cdot \mathbf{T} = a_{\mathbf{T}}\mathbf{T} \cdot \mathbf{T} + a_{\mathbf{N}}\mathbf{N} \cdot \mathbf{T} = a_{\mathbf{T}}$$

On the other hand, $\mathbf{T} \times \mathbf{T} = \mathbf{0}$, so

$$\frac{\mathbf{a} \times \mathbf{v}}{\|\mathbf{v}\|} = \mathbf{a} \times \mathbf{T} = (a_{\mathbf{T}}\mathbf{T} + a_{\mathbf{N}}\mathbf{N}) \times \mathbf{T} = a_{\mathbf{T}}\mathbf{T} \times \mathbf{T} + a_{\mathbf{N}}\mathbf{N} \times \mathbf{T} = a_{\mathbf{N}}(\mathbf{N} \times \mathbf{T})$$

REMINDER *The cross product* $\mathbf{v} \times \mathbf{w}$ *has length*

$$\|\mathbf{v}\| \, \|\mathbf{w}\| \sin \theta$$

where θ *is the angle between* $\mathbf{v}$ *and* $\mathbf{w}$. *If* $\mathbf{v}$ *and* $\mathbf{w}$ *are unit vectors and are orthogonal to each other* ($\theta = \frac{\pi}{2}$), *then* $\mathbf{v} \times \mathbf{w}$ *has length 1.*

Note that $\mathbf{N} \times \mathbf{T}$ is a unit vector since $\mathbf{T}$ and $\mathbf{N}$ are orthogonal (see the marginal note). Thus

$$\frac{\|\mathbf{a} \times \mathbf{v}\|}{\|\mathbf{v}\|} = a_{\mathbf{N}}\|\mathbf{N} \times \mathbf{T}\| = a_{\mathbf{N}}$$

This proves the first equality in (5). Finally, since $\mathbf{T}$ and $\mathbf{N}$ are orthogonal unit vectors, $\mathbf{a}$ is the hypotenuse of a right triangle with sides of length $a_{\mathbf{T}}$ and $a_{\mathbf{N}}$ (see Figure 4), and the Pythagorean Theorem gives

$$\|\mathbf{a}\|^2 = |a_{\mathbf{T}}|^2 + |a_{\mathbf{N}}|^2$$

This yields the second equality in (5), namely $a_{\mathbf{N}} = \sqrt{\|\mathbf{a}\|^2 - |a_{\mathbf{T}}|^2}$.

Equation (5) is convenient if you only need to compute $a_{\mathbf{N}}$. To compute both $a_{\mathbf{N}}$ and the vector $\mathbf{N}$, we observe that

$$a_{\mathbf{N}}\mathbf{N} = \mathbf{a} - a_{\mathbf{T}}\mathbf{T} = \mathbf{a} - \left(\frac{\mathbf{a}\cdot\mathbf{v}}{\|\mathbf{v}\|}\right)\frac{\mathbf{v}}{\|\mathbf{v}\|} = \mathbf{a} - \left(\frac{\mathbf{a}\cdot\mathbf{v}}{\mathbf{v}\cdot\mathbf{v}}\right)\mathbf{v} \qquad \boxed{6}$$

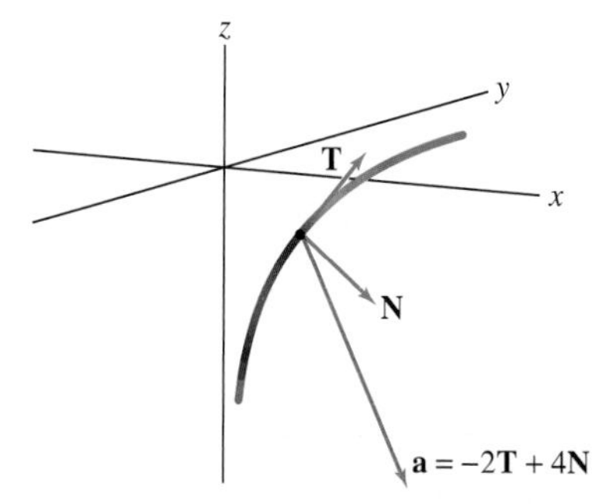

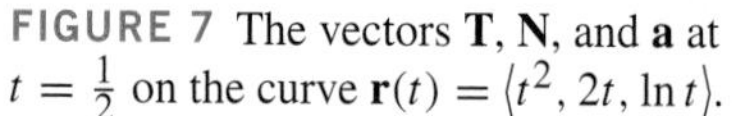

FIGURE 7 The vectors $\mathbf{T}$, $\mathbf{N}$, and $\mathbf{a}$ at $t = \frac{1}{2}$ on the curve $\mathbf{r}(t) = \langle t^2, 2t, \ln t\rangle$.

EXAMPLE 6 Decompose the acceleration vector $\mathbf{a}$ of $\mathbf{r}(t) = \langle t^2, 2t, \ln t\rangle$ into tangential and normal components at $t = \frac{1}{2}$. Refer to Figure 7.

Solution

***Step 1.* Compute T and $a_{\mathbf{T}}$.**
We have $\mathbf{v}(t) = \mathbf{r}'(t) = \langle 2t, 2, t^{-1}\rangle$ and $\mathbf{a}(t) = \mathbf{r}''(t) = \langle 2, 0, -t^{-2}\rangle$. Thus at $t = \frac{1}{2}$,

$$\mathbf{v} = \mathbf{r}'\left(\frac{1}{2}\right) = \left\langle 2\left(\frac{1}{2}\right), 2, \left(\frac{1}{2}\right)^{-1}\right\rangle = \langle 1, 2, 2\rangle$$

$$\mathbf{a} = \mathbf{r}''\left(\frac{1}{2}\right) = \left\langle 2, 0, -\left(\frac{1}{2}\right)^{-2}\right\rangle = \langle 2, 0, -4\rangle$$

$$\mathbf{T} = \frac{\mathbf{v}}{\|\mathbf{v}\|} = \frac{\langle 1, 2, 2\rangle}{\sqrt{1^2+2^2+2^2}} = \left\langle \frac{1}{3}, \frac{2}{3}, \frac{2}{3}\right\rangle$$

$$a_{\mathbf{T}} = \mathbf{a}\cdot\mathbf{T} = \langle 2, 0, -4\rangle \cdot \left\langle \frac{1}{3}, \frac{2}{3}, \frac{2}{3}\right\rangle = \frac{-6}{3} = -2 \quad \text{[by Eq. (4)]}$$

Here is a summary of the steps in Example 6:

$$a_{\mathbf{T}} = \mathbf{a}\cdot\mathbf{T}$$
$$a_{\mathbf{N}}\mathbf{N} = \mathbf{a} - a_{\mathbf{T}}\mathbf{T}$$
$$a_{\mathbf{N}} = \|a_{\mathbf{N}}\mathbf{N}\| = \|\mathbf{a} - a_{\mathbf{T}}\mathbf{T}\|$$
$$\mathbf{N} = \frac{\mathbf{a} - a_{\mathbf{T}}\mathbf{T}}{a_{\mathbf{N}}}$$

***Step 2.* Compute $a_{\mathbf{N}}$ and N.**
By Eq. (6),

$$a_{\mathbf{N}}\mathbf{N} = \mathbf{a} - a_{\mathbf{T}}\mathbf{T} = \langle 2, 0, -4\rangle - (-2)\left\langle \frac{1}{3}, \frac{2}{3}, \frac{2}{3}\right\rangle = \left\langle \frac{8}{3}, \frac{4}{3}, -\frac{8}{3}\right\rangle$$

Since $\mathbf{N}$ is a unit vector and $a_{\mathbf{N}} \ge 0$,

$$a_{\mathbf{N}} = \|a_{\mathbf{N}}\mathbf{N}\| = \left\|\left\langle \frac{8}{3}, \frac{4}{3}, -\frac{8}{3}\right\rangle\right\| = \sqrt{\frac{64}{9}+\frac{16}{9}+\frac{64}{9}} = \sqrt{\frac{144}{9}} = 4$$

$$\mathbf{N} = \frac{a_{\mathbf{N}}\mathbf{N}}{a_{\mathbf{N}}} = \frac{\left\langle \frac{8}{3}, \frac{4}{3}, -\frac{8}{3}\right\rangle}{4} = \left\langle \frac{2}{3}, \frac{1}{3}, -\frac{2}{3}\right\rangle$$

***Step 3.* Write the decomposition.**
Since $\mathbf{a} = \langle 2, 0, -4\rangle$, $a_{\mathbf{T}} = -2$, and $a_{\mathbf{N}} = 4$, the decomposition of $\mathbf{a}$ into tangential and normal components is

$$\mathbf{a} = \langle 2, 0, -4\rangle = -2\mathbf{T} + 4\mathbf{N}$$

where $\mathbf{T} = \left\langle \frac{1}{3}, \frac{2}{3}, \frac{2}{3}\right\rangle$ and $\mathbf{N} = \left\langle \frac{2}{3}, \frac{1}{3}, -\frac{2}{3}\right\rangle$. ■

EXAMPLE 7 Nonuniform Circular Motion Figure 8 shows the acceleration vectors of three particles moving *counterclockwise* around a circle. In each case, state whether the particle's speed v is increasing, decreasing, or momentarily constant.

Solution In (A), the acceleration vector has a negative component in the tangential direction $\mathbf{T}$, so the particle's speed is decreasing ($v' < 0$). In (B), the acceleration vector

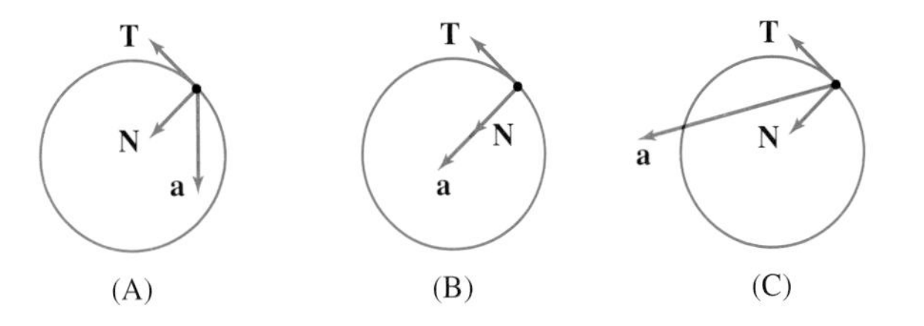

FIGURE 8 Acceleration vectors of particles moving counterclockwise (in the direction of **T**) around a circle.

points in the normal direction, so the particle's speed is constant at this moment, that is, $v' = 0$. In (C), the acceleration vector has positive tangential component in the tangential direction **T**, so the particle's speed is increasing ($v' > 0$).

13.5 SUMMARY

- The path of an object in 3-space is described by a vector-valued function $\mathbf{r}(t)$.
- The *velocity vector* $\mathbf{v}(t) = \mathbf{r}'(t)$ points in the direction of motion and its length $v(t) = \|\mathbf{v}(t)\|$ is the object's speed.
- The *acceleration vector* is the second derivative $\mathbf{a}(t) = \mathbf{r}''(t)$.
- The acceleration vector **a** is the sum of a tangential component (reflecting change in speed) and a normal component (reflecting change in direction).

Unit tangent vector	$\mathbf{T}(t) = \dfrac{\mathbf{v}(t)}{\|\mathbf{v}(t)\|}$
Unit normal vector	$\mathbf{N}(t) = \dfrac{\mathbf{T}'(t)}{\|\mathbf{T}'(t)\|}$
Decomposition of acceleration	$\mathbf{a}(t) = a_{\mathbf{T}}(t)\mathbf{T}(t) + a_{\mathbf{N}}(t)\mathbf{N}(t)$
Tangential component	$a_{\mathbf{T}} = v'(t) = \dfrac{\mathbf{a}\cdot\mathbf{v}}{\|\mathbf{v}\|}$
Normal component	$a_{\mathbf{N}} = \kappa(t)v(t)^2 = \dfrac{\|\mathbf{a}\times\mathbf{v}\|}{\|\mathbf{v}\|}$
	$a_{\mathbf{N}}\mathbf{N} = \mathbf{a} - a_{\mathbf{T}}\mathbf{T} = \mathbf{a} - \left(\dfrac{\mathbf{a}\cdot\mathbf{v}}{\mathbf{v}\cdot\mathbf{v}}\right)\mathbf{v}$

13.5 EXERCISES

Preliminary Questions

1. If a particle travels with constant speed, must its acceleration vector be zero? Explain.

2. For a particle in uniform circular motion around a circle, which of the vectors $\mathbf{v}(t)$ or $\mathbf{a}(t)$ always points toward the center of the circle?

3. Two objects travel to the right along the parabola $y = x^2$ with nonzero speed. Which of the following must be true?

(a) Their velocity vectors point in the same direction.
(b) Their velocity vectors have the same length.
(c) Their acceleration vectors point in the same direction.

4. Use the decomposition of acceleration into tangential and normal components to explain the following statement: If the speed is constant, then the acceleration and velocity vectors are orthogonal.

5. If a particle travels along a straight line, then the acceleration and velocity vectors are (choose the correct statement):

(a) Orthogonal **(b)** Parallel

6. What is the length of the acceleration vector of a particle traveling around a circle of radius 2 cm with constant velocity 4 cm/s?

7. Two cars are racing around a circular track. If, at a certain moment, both of their speedometers read 110 mph. then the two cars have the same (choose one):

(a) $a_{\mathbf{T}}$ **(b)** $a_{\mathbf{N}}$

Exercises

1. Use the table here to calculate the difference quotients $\dfrac{\mathbf{r}(1+h)-\mathbf{r}(1)}{h}$ for $h=-0.2, -0.1, 0.1, 0.2$. Then estimate the velocity and speed at $t=1$.

$\mathbf{r}(0.8)$	$\langle 1.557, 2.459, -1.970\rangle$
$\mathbf{r}(0.9)$	$\langle 1.559, 2.634, -1.740\rangle$
$\mathbf{r}(1)$	$\langle 1.540, 2.841, -1.443\rangle$
$\mathbf{r}(1.1)$	$\langle 1.499, 3.078, -1.035\rangle$
$\mathbf{r}(1.2)$	$\langle 1.435, 3.342, -0.428\rangle$

2. Draw the vectors $\mathbf{r}(2+h)-\mathbf{r}(2)$ and $\dfrac{\mathbf{r}(2+h)-\mathbf{r}(2)}{h}$ for $h=0.5$ for the path in Figure 9. Draw $\mathbf{v}(2)$ (using a rough estimate for its length).

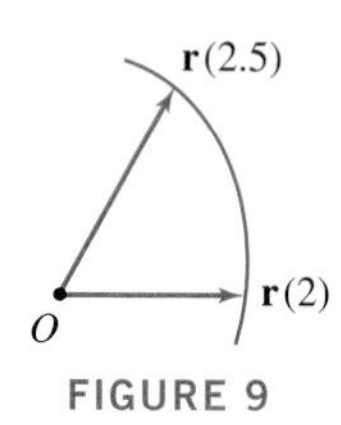

FIGURE 9

In Exercises 3–6, calculate the velocity and acceleration vectors and the speed at the time indicated.

3. $\mathbf{r}(t)=\langle t^3, 1-t, 4t^2\rangle,\quad t=1$

4. $\mathbf{r}(t)=e^t\mathbf{j}-\cos(2t)\mathbf{k},\quad t=0$

5. $\mathbf{r}(\theta)=\langle\sin\theta,\cos\theta,\cos 3\theta\rangle,\quad \theta=\frac{\pi}{3}$

6. $\mathbf{r}(s)=\left\langle\dfrac{1}{1+s^2},\dfrac{s}{1+s^2}\right\rangle,\quad s=2$

7. Find $\mathbf{a}(t)$ for a particle moving around a circle of radius 8 cm at a constant speed of $v=4$ cm/s (see Example 2). Draw the path and acceleration vector at $t=\frac{\pi}{4}$.

8. Sketch the path $\mathbf{r}(t)=\langle 1-t^2, 1-t\rangle$ for $-2\le t\le 2$, indicating the direction of motion. Draw the velocity and acceleration vectors at $t=0$ and $t=1$.

9. Sketch the path $\mathbf{r}(t)=\langle t^2, t^3\rangle$ together with the velocity and acceleration vectors at $t=1$.

10. The paths $\mathbf{r}(t)=\langle t^2, t^3\rangle$ and $\mathbf{r}_1(t)=\langle t^4, t^6\rangle$ trace the same curve and $\mathbf{r}_1(1)=\mathbf{r}(1)$. Do you expect either the velocity or acceleration vectors of these paths at $t=1$ to point in the same direction? Compute these vectors and draw them on a single plot of the path.

In Exercises 11–14, find $\mathbf{v}(t)$ given $\mathbf{a}(t)$ and the initial velocity.

11. $\mathbf{a}(t)=\langle t, 4\rangle,\quad \mathbf{v}(0)=\langle\frac{1}{3}, -2\rangle$

12. $\mathbf{a}(t)=\langle e^t, 0, t+1\rangle,\quad \mathbf{v}(0)=\langle 1, -3, \sqrt{2}\rangle$

13. $\mathbf{a}(t)=\mathbf{k},\quad \mathbf{v}(0)=\mathbf{i}$

14. $\mathbf{a}(t)=t^2\mathbf{k},\quad \mathbf{v}(0)=\mathbf{i}-\mathbf{j}$

In Exercises 15–18, find $\mathbf{r}(t)$ and $\mathbf{v}(t)$ given $\mathbf{a}(t)$ and the initial velocity and position.

15. $\mathbf{a}(t)=\langle t, 4\rangle,\quad \mathbf{v}(0)=\langle 3, -2\rangle,\quad \mathbf{r}(0)=\langle 0, 0\rangle$

16. $\mathbf{a}(t)=\langle e^t, 2t, t+1\rangle,\quad \mathbf{v}(0)=\langle 1, 0, 1\rangle,\quad \mathbf{r}(0)=\langle 2, 1, 1\rangle$

17. $\mathbf{a}(t)=t\mathbf{k},\quad \mathbf{v}(0)=\mathbf{i},\ \mathbf{r}(0)=\mathbf{j}$

18. $\mathbf{a}(t)=\cos t\mathbf{k},\quad \mathbf{v}(0)=\mathbf{i}-\mathbf{j},\quad \mathbf{r}(0)=\mathbf{i}$

19. A bullet is fired from the ground at an angle of 45°. What initial speed must the bullet have in order to hit the top of a 400-ft tower located 600 ft away?

20. A bullet is fired from the ground at an angle of 60° with initial speed $v_0=30$ ft/s. How far does the bullet travel? What is the bullet's velocity vector and speed when it hits the ground?

21. A projectile fired at an angle of 60° lands 1,200 ft away. What was its initial speed?

22. Show that a projectile fired at an angle θ with initial speed v_0 travels a total distance $(v_0^2/g)\sin 2\theta$ before hitting the ground. Conclude that the maximum distance (for a given v_0) is attained for $\theta=45°$.

23. A baseball is thrown to another player standing 80 ft away with initial speed 60 ft/s. Use the result of Exercise 22 to find two angles θ at which the ball can be released. Which angle gets the ball there faster?

24. Show that a bullet fired at an angle θ will hit the top of an h-ft tower located d ft away if its initial speed is

$$v_0=\frac{4d\sec\theta}{\sqrt{d\tan\theta-h}}$$

25. At a certain moment, a moving particle has velocity $\mathbf{v}=\langle 2, 2, -1\rangle$ and $\mathbf{a}=\langle 0, 4, 3\rangle$. Find $\mathbf{T}$, $\mathbf{N}$, and the decomposition of $\mathbf{a}$ into tangential and normal components.

26. At a certain moment, a moving particle has velocity $\mathbf{v}=\langle 12, 20, 20\rangle$ and acceleration $\mathbf{a}=\langle 2, 1, -3\rangle$. Is the particle speeding up or slowing down?

27. A particle follows a path $\mathbf{r}(t)$ for $0\le t\le T$, beginning at the origin O. The vector $\overline{\mathbf{v}}=\dfrac{1}{T}\displaystyle\int_0^T\mathbf{r}'(t)\,dt$ is called the **average velocity**

vector. Suppose that $\overline{\mathbf{v}} = \mathbf{0}$. Answer and explain the following:

(a) Where is the particle located at time T if $\overline{\mathbf{v}} = \mathbf{0}$?

(b) Is the particle's average speed necessarily equal to zero?

28. One consequence of Kepler's Laws of Planetary Motion is that a planet moves faster when it is closer to the sun. Which of Figure 10(A) or 10(B) represents a planetary orbit (with the velocity vectors as shown)?

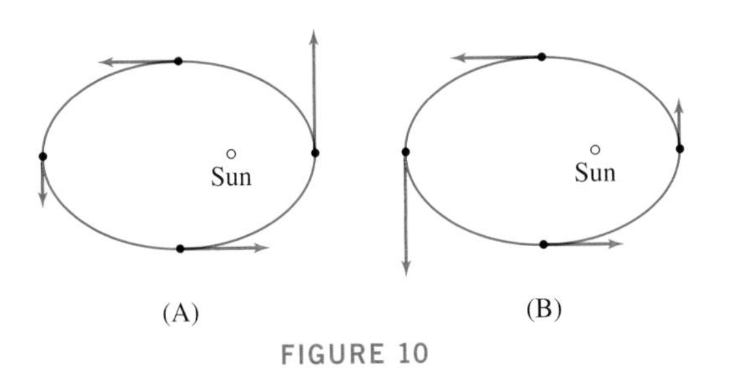

FIGURE 10

29. A space shuttle orbits the earth at an altitude 200 miles above the earth's surface, with constant speed $v = 17{,}000$ mph. Find the magnitude of the shuttle's acceleration (in ft/s^2), assuming that the radius of the earth is 4,000 miles (Figure 11).

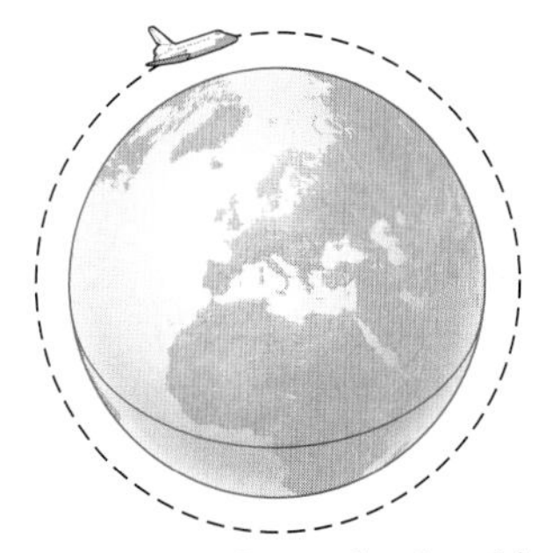

FIGURE 11 Space shuttle orbit.

In Exercises 30–33, use (4) and (5) to find $a_{\mathbf{T}}$ and $a_{\mathbf{N}}$ as a function of t or at the point indicated.

30. $\mathbf{r}(t) = \langle t^2, t^3 \rangle$

31. $\mathbf{r}(t) = \langle t, \cos t, \sin t \rangle$

32. $\mathbf{r}(t) = \langle t^{-1}, \ln t, t^2 \rangle, \quad t = 1$

33. $\mathbf{r}(t) = \langle e^t, t, e^{-t} \rangle, \quad t = 0$

In Exercise 34–41, use (4) and (6) to find the decomposition of $\mathbf{a}(t)$ into tangential and normal components at the point indicated, as in Example 6.

34. $\mathbf{r}(t) = \langle 4 - t, t + 1, t^2 \rangle, \quad t = 1$

35. $\mathbf{r}(t) = \langle t, e^t, te^t \rangle, \quad t = 0$

36. $\mathbf{r}(t) = \left\langle t, \frac{1}{2}t^2, \frac{1}{6}t^3 \right\rangle, \quad t = 1$

37. $\mathbf{r}(t) = \left\langle t, \frac{1}{2}t^2, \frac{1}{6}t^3 \right\rangle, \quad t = 4$

38. $\mathbf{r}(t) = \langle e^t, 1 - t \rangle, \quad t = 0$

39. $\mathbf{r}(\theta) = \langle \cos\theta, \sin\theta, \theta \rangle, \quad \theta = 0$

40. $\mathbf{r}(t) = \left\langle \frac{1}{3}t^3, 1 - 3t \right\rangle, \quad t = -1$

41. $\mathbf{r}(t) = \langle t, \cos t, t \sin t \rangle, \quad t = \frac{\pi}{2}$

42. Let $\mathbf{r}(t) = \langle t^2, 4t - 3 \rangle$. Find $\mathbf{T}(t)$ and $\mathbf{N}(t)$, and show that the decomposition of $\mathbf{a}(t)$ into tangential and normal components is

$$\mathbf{a}(t) = \left(\frac{2t}{\sqrt{t^2 + 4}} \right) \mathbf{T} + \left(\frac{4}{t^2 + 4} \right) \mathbf{N}$$

43. Find the components $a_{\mathbf{T}}$ and $a_{\mathbf{N}}$ of the acceleration vector of a particle moving along a circular path of radius $R = 100$ cm with constant velocity $v_0 = 5$ cm/s.

44. At time t_0, a moving particle has velocity vector $\mathbf{v} = 2\mathbf{i}$ and acceleration vector $\mathbf{a} = 3\mathbf{i} + 18\mathbf{k}$. Determine the curvature $\kappa(t_0)$ of the particle's path at time t_0.

45. A car proceeds along a circular path of radius $R = 1{,}000$ ft centered at the origin. Starting at rest, its speed increases at a rate of t ft/s^2. Find the acceleration vector $\mathbf{a}$ at time $t = 3$ s and determine its decomposition into normal and tangential components.

46. In the notation of Example 5, find the acceleration vector for a person seated in a car at (a) the highest point of the ferris wheel and (b) the two points level with the center of the wheel.

47. Suppose that $\mathbf{r} = \mathbf{r}(t)$ lies on a sphere of radius R for all t. Let $\mathbf{J} = \mathbf{r} \times \mathbf{r}'$. Show that $\mathbf{r}' = (\mathbf{J} \times \mathbf{r})/\|\mathbf{r}\|^2$. *Hint:* Observe that $\mathbf{r}$ and $\mathbf{r}'$ are perpendicular.

48. A particle moves counterclockwise around a circle. Which of the vectors in Figure 12 is not a possible acceleration vector? Explain. For the two possible acceleration vectors, state whether the particle is speeding up or slowing down.

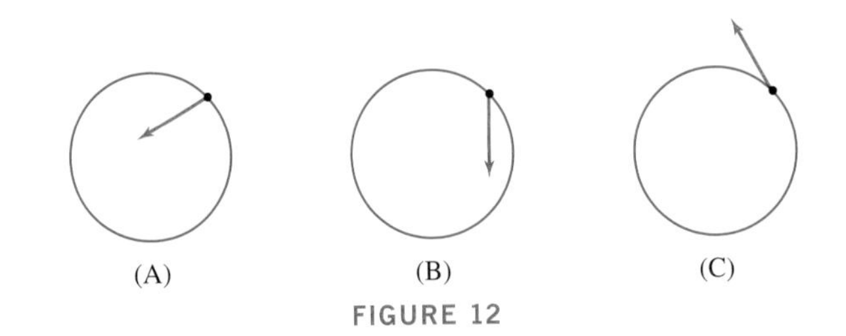

FIGURE 12

Further Insights and Challenges

49. The orbit of a planet is an ellipse with the sun at one focus. The sun's gravitational force acts along the radial line from the planet to the sun (the dashed lines in Figure 13), and by Newton's Second Law, the acceleration vector points in the same direction. Explain in words why the planet must slow down in the upper half of the orbit (as it moves away from the sun) and speed up in the lower half. Kepler's Second Law, discussed in the next section, gives a more precise version of this qualitative conclusion. *Hint:* Consider the decomposition of **a** into normal and tangential components.

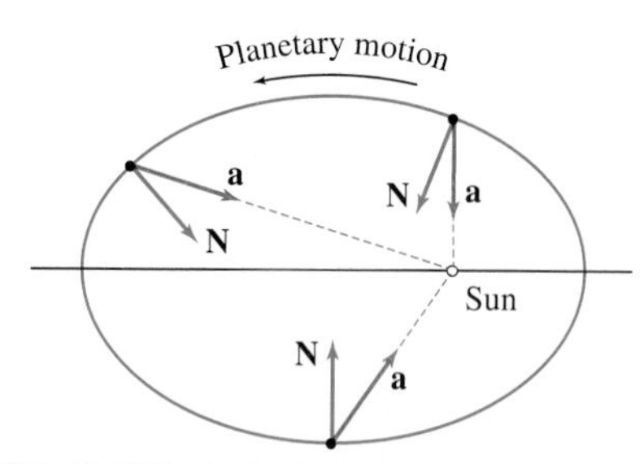

FIGURE 13 Elliptical orbit of a planet around the sun.

In Exercises 50–54, consider a car of mass m traveling along a curved but level road. To avoid skidding, the road must supply a frictional force $\mathbf{F} = m\mathbf{a}$, *where* **a** *is the car's acceleration vector. The maximum magnitude of the frictional force is* $\mu m g$, *where* μ *is the coefficient of friction and* $g = 32$ ft/s^2. *Let* v *be the car's speed in feet per second.*

50. Show that the car will not skid if the curvature κ of the road is such that (with $R = 1/\kappa$)

$$(v')^2 + \left(\frac{v^2}{R}\right)^2 < (\mu g)^2 \qquad \boxed{7}$$

Note that braking ($v' < 0$) and speeding up ($v' > 0$) contribute equally to skidding.

51. Suppose that the maximum radius of curvature along a curved highway is $R = 600$ ft. How fast can a car travel (at constant speed) along the highway without skidding if the coefficient of friction is $\mu = 0.5$?

52. Beginning at rest, a car drives around a circular track of radius $R = 1{,}000$ ft, accelerating at a rate of 1 ft/s^2. After how many seconds will the car begin to skid if the coefficient of friction is $\mu = 0.6$?

53. You want to reverse your direction in the shortest possible time by driving around a semicircular bend (Figure 14). If you travel at the maximum possible *constant speed* v that will not cause skidding, is it faster to hug the inside curve (radius r) or the outside curb (radius R)? *Hint:* Use Eq. (7) to show that at maximum speed, the time required to drive around the semicircle is proportional to the square root of the radius.

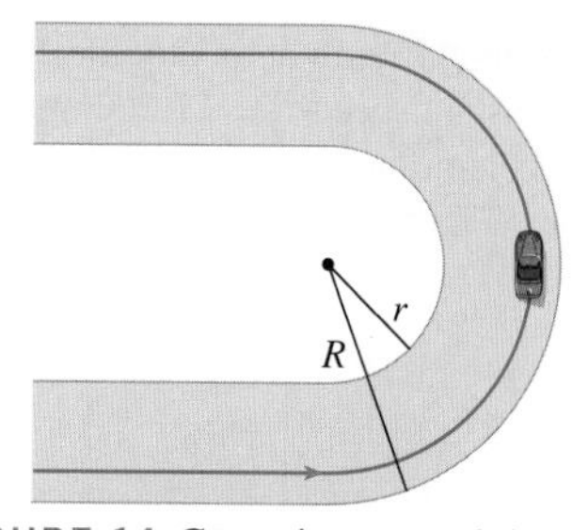

FIGURE 14 Car going around the bend.

54. What is the smallest radius R about which a car can turn without skidding at 60 mph if $\mu = 0.75$ (a typical value)?

13.6 Planetary Motion According to Kepler and Newton

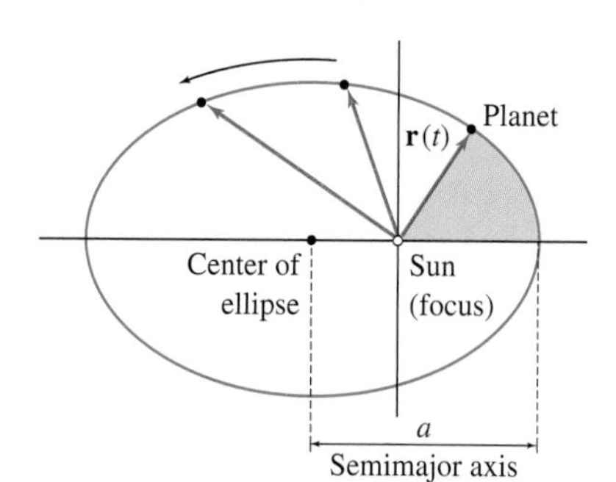

FIGURE 1 The planet travels along an ellipse with the sun at one focus. As it travels, the radial vector $\mathbf{r}(t)$ sweeps out area at a constant rate.

No events are more emblematic of the scientific revolution of the seventeenth century than the discovery of the laws of planetary motion by the German astronomer Johannes Kepler around 1609 and Isaac Newton's publication in 1687 of a mathematical derivation of these laws from the universal laws of motion and gravitation. This dramatic success of mathematics in rendering the natural world comprehensible made a great impression on succeeding generations of scientists and encouraged them to search for mathematical laws governing other phenomena such as electricity and magnetism, aerodynamics, and atomic processes.

To formulate Kepler's Laws, we recall that an ellipse is the set of all points P such that $F_1P + F_2P = K$, where K is a constant and the points F_1 and F_2 are the **foci** of the ellipse. As we will state formally in a moment, the planetary orbits are ellipses with the sun at one focus. Furthermore, if we imagine a radial vector $\mathbf{r}(t)$ pointing from the sun to the planet as in Figure 1, then this radial vector sweeps out area at a constant rate or, as Kepler stated in his Second Law, it sweeps out equal areas in equal times (Figure

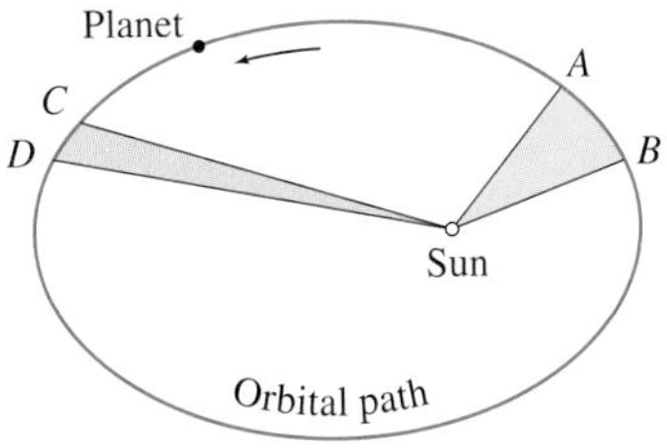

Planet
C
A
D
B
Sun
Orbital path

FIGURE 2 The two shaded regions have equal areas, and by Kepler's Second Law, the planet sweeps them out in equal times. To do so, the planet must travel faster along the path from A to B than from C to D.

2). The Second Law governs how speed changes along the orbit: To sweep out area at a constant rate, the planet must move faster when it is near the sun and slower when it is farther away. Kepler's Third Law concerns the **period** T of the orbit, defined as the time required to complete one full revolution.

Kepler's Three Laws

(i) Law of Ellipses: The orbit of a planet is an ellipse with the sun at one focus.

(ii) Law of Equal Area in Equal Time: The position vector pointing from the sun to the planet sweeps out equal areas in equal times.

(iii) Law of the Period of Motion: $T^2 = \left(\dfrac{4\pi^2}{GM}\right)a^3$, where a is the length of the semimajor axis of the ellipse (see Figure 1).

In the Third Law:

- G is Newton's universal gravitational constant: $6.673 \times 10^{-11}\ \mathrm{m^3\,kg^{-1}\,s^{-2}}$.
- M is the mass of the sun, approximately 1.989×10^{30} kg.

Kepler's original version of the Third Law stated only that T^2 is proportional to a^3. It was Newton who discovered that the constant of proportionality is $4\pi^2/GM$. Newton further realized that if you can measure T and a through observation, then you can use the Third Law to solve for the mass M. This method is used by astronomers to find the masses of the planets (by measuring T and a for moons revolving around the planet) as well as the masses of binary stars and galaxies. See Exercises 4–6, 13.

Our treatment of Kepler's Laws makes several simplifying assumptions. We treat the sun and planet as point masses and ignore the gravitation attraction of the planets on each other. Furthermore, both the sun and the planet revolve around their mutual center of mass. However, the sun is much more massive than the planet, and it is justifiable to ignore the sun's motion and assume that the planet revolves around the center of the sun.

We place the sun at the origin of the coordinate system and let $\mathbf{r} = \mathbf{r}(t)$ denote the vector from the origin to a planet of mass m, as in Figure 1. Newton's Universal Law of Gravitation tells us that the sun attracts the planet with a gravitational force of magnitude $GMm/\|\mathbf{r}(t)\|^2$ in the direction of the vector $-\mathbf{r}(t)$ pointing from the planet to the sun. Therefore,

$$\mathbf{F}(\mathbf{r}(t)) = -\left(\frac{GMm}{\|\mathbf{r}(t)\|^2}\right)\underbrace{\frac{\mathbf{r}(t)}{\|\mathbf{r}(t)\|}}_{\text{Unit radial vector}} = -\frac{GMm}{\|\mathbf{r}(t)\|^3}\mathbf{r}(t)$$

Combining this with Newton's Second Law of Motion, $\mathbf{F}(\mathbf{r}(t)) = m\mathbf{r}''(t)$, we obtain

$$m\mathbf{r}''(t) = -\frac{GMm}{\|\mathbf{r}(t)\|^3}\mathbf{r}(t) \qquad \boxed{1}$$

To simplify the notation, we set $k = GM$. Then (1) yields

$$\boxed{\mathbf{r}''(t) = -\frac{k}{\|\mathbf{r}(t)\|^3}\mathbf{r}(t)} \qquad \boxed{2}$$

The constant k has the value

$$k = GM \approx 1.327 \times 10^{20}\ \mathrm{m^3\,s^{-2}}$$

Each of Kepler's Laws is a consequence of this *differential equation* for the vector-valued function $\mathbf{r}(t)$.

Proof of Kepler's Second Law

Let $A(t)$ be the area swept out by the radial vector in the time interval $[0, t]$ (Figure 3). Kepler's Second Law states that the rate of change $\dfrac{dA(t)}{dt}$ is constant. We prove this by expressing the rate of change as a cross product.

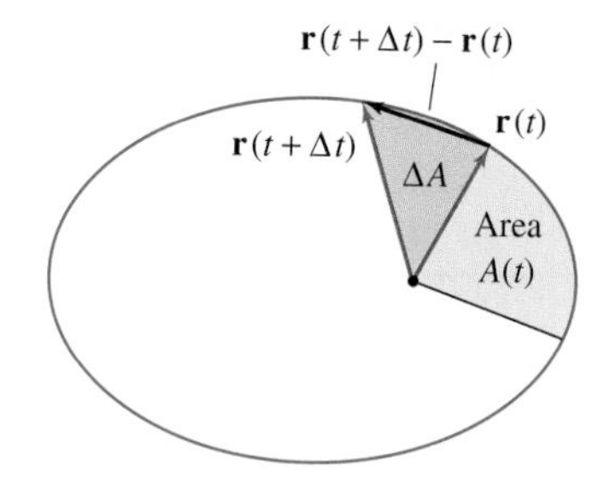

FIGURE 3 Area swept out during time interval Δt is approximately equal to the area of the triangle.

The area swept out in a small time interval $[t, t + \Delta t]$ is

$$\Delta A = A(t + \Delta t) - A(t)$$

Figure 3 shows that ΔA is approximately equal to the area of the triangle whose sides are the vectors $\mathbf{r}(t)$, $\mathbf{r}(t) + \Delta t$, and $\mathbf{r}(t + \Delta t) - \mathbf{r}(t)$. By Theorem 3 of Section 12.4, this area is equal to one-half the length of the cross product:

$$\Delta A \approx \frac{1}{2} \|\mathbf{r}(t) \times (\mathbf{r}(t + \Delta t) - \mathbf{r}(t))\|$$

Thus, we have

$$\begin{aligned} \frac{dA}{dt} &= \lim_{\Delta t \to 0} \frac{\Delta A}{\Delta t} = \frac{1}{2} \lim_{\Delta t \to 0} \left\| \mathbf{r}(t) \times \left(\frac{\mathbf{r}(t + \Delta t) - \mathbf{r}(t)}{\Delta t} \right) \right\| \\ &= \frac{1}{2} \|\mathbf{r}(t) \times \mathbf{r}'(t)\| \end{aligned}$$

Define

See Exercise 24 for a more rigorous derivation of Eq. (3).

$$\boxed{\mathbf{J} = \mathbf{r}(t) \times \mathbf{r}'(t)}$$

We have shown that

$$\frac{dA}{dt} = \frac{1}{2} \|\mathbf{J}\| \qquad \boxed{3}$$

To finish the proof, it is necessary to show that $\|\mathbf{J}\|$ is constant. In fact, we prove that $\mathbf{J}$ itself is constant by showing that $\mathbf{J}'(t) = 0$ for all t. By the Product Rule,

$$\frac{d\mathbf{J}}{dt} = \frac{d}{dt}\big(\mathbf{r}(t) \times \mathbf{r}'(t)\big) = \mathbf{r}(t) \times \mathbf{r}''(t) + \mathbf{r}'(t) \times \mathbf{r}'(t) \qquad \boxed{4}$$

← REMINDER Recall that $\mathbf{a} \times \mathbf{b} = \mathbf{0}$ if $\mathbf{a}$ and $\mathbf{b}$ are proportional.

Equation (2) tells us that $\mathbf{r}''(t)$ and $\mathbf{r}(t)$ are proportional, so $\mathbf{r}(t) \times \mathbf{r}''(t) = \mathbf{0}$. Similarly, $\mathbf{r}'(t) \times \mathbf{r}'(t) = \mathbf{0}$, and thus we obtain the desired result

In physics, $m\mathbf{J}$ is called the angular momentum vector, and when $m\mathbf{J}$ is constant, we say that angular momentum is conserved. *Our proof actually shows that angular momentum is conserved whenever the force on an object acts in the radial direction (whether or not its magnitude follows an inverse square law).*

$$\boxed{\frac{d\mathbf{J}}{dt} = 0} \qquad \boxed{5}$$

Equation (5) has another important consequence. The vectors $\mathbf{r}(t)$ and $\mathbf{r}'(t)$ are orthogonal to $\mathbf{J} = \mathbf{r}(t) \times \mathbf{r}'(t)$ for all t. Since $\mathbf{J}$ is constant, we may conclude that both $\mathbf{r}(t)$ and $\mathbf{r}'(t)$ are confined to the plane orthogonal to $\mathbf{J}$ (Figure 4). In other words, the *motion of a planet around the sun takes place in a plane.*

Proof of the Law of Ellipses

In the centuries since Newton's original work on Kepler's Laws, several different proofs of the the law of ellipses have been devised. One the most elegant methods, presented here, is based on the **Lenz vector**:

$$\mathbf{L} = \left(\frac{1}{k} \mathbf{r}' \times \mathbf{J} \right) - \frac{\mathbf{r}}{\|\mathbf{r}\|}$$

FIGURE 4 The orbit is contained in the plane orthogonal to $\mathbf{J}$.

The first step is to show that $\mathbf{L}$ is constant (thus $\mathbf{L}$, like $\mathbf{J}$, is conserved). Although it is far from obvious at this stage, we will eventually show that $\mathbf{L}$ points in the direction of the semimajor axis of the orbit.

THEOREM 1 The Lenz Vector Is Conserved

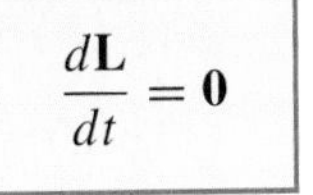

$$\frac{d\mathbf{L}}{dt} = \mathbf{0}$$

Equation (6) can be verified by direct calculation (see Exercise 74 in Section 12.4). Equation (7) may be verified using Eq. (6) in Section 12.4.

Proof We use two cross product identities in this proof and the discussion below:

$$\mathbf{u} \times (\mathbf{v} \times \mathbf{w}) = (\mathbf{u} \cdot \mathbf{w})\mathbf{v} - (\mathbf{u} \cdot \mathbf{v})\mathbf{w} \tag{6}$$

$$\mathbf{u} \cdot (\mathbf{v} \times \mathbf{w}) = \mathbf{v} \cdot (\mathbf{w} \times \mathbf{u}) = \mathbf{w} \cdot (\mathbf{u} \times \mathbf{v}) \tag{7}$$

By the definition of $\mathbf{L}$,

$$\frac{d\mathbf{L}}{dt} = \overbrace{\frac{1}{k}\frac{d}{dt}(\mathbf{r}' \times \mathbf{J})}^{\text{Call this } \mathbf{A}.} - \overbrace{\frac{d}{dt}\left(\frac{\mathbf{r}}{\|\mathbf{r}\|}\right)}^{\text{Call this } \mathbf{B}.} = \mathbf{A} - \mathbf{B}$$

We compute $\mathbf{A}$ and $\mathbf{B}$ separately. First, we have

$$\mathbf{A} = \frac{d}{dt}\left(\frac{1}{k}\mathbf{r}' \times \mathbf{J}\right) = \left(\frac{1}{k}\mathbf{r}'' \times \mathbf{J}\right) + \left(\frac{1}{k}\mathbf{r}' \times \mathbf{J}'\right) = \frac{1}{k}\mathbf{r}'' \times \mathbf{J} \qquad (\text{since } \mathbf{J}' = \mathbf{0})$$

$$= \frac{1}{k}\mathbf{r}'' \times (\mathbf{r} \times \mathbf{r}') \qquad (\text{since } \mathbf{J} = \mathbf{r} \times \mathbf{r}')$$

$$= -\frac{1}{\|\mathbf{r}\|^3}\mathbf{r} \times (\mathbf{r} \times \mathbf{r}') \qquad (\text{since } \mathbf{r}'' = -k\mathbf{r}/\|\mathbf{r}\|^3)$$

$$= -\frac{1}{\|\mathbf{r}\|^3}\left((\mathbf{r} \cdot \mathbf{r}')\mathbf{r} - (\mathbf{r} \cdot \mathbf{r})\mathbf{r}'\right) \qquad [\text{by Eq. (6)}]$$

$$= \frac{\mathbf{r}'}{\|\mathbf{r}\|} - \frac{(\mathbf{r} \cdot \mathbf{r}')\mathbf{r}}{\|\mathbf{r}\|^3} \tag{8}$$

Newton's demonstration of the law of ellipses is a watershed that separates the ancient world from the modern world—the culmination of the Scientific Revolution. It is one of the crowning achievements of the human mind, comparable to Beethoven's symphonies or Shakespeare's plays, or Michelangelo's Sistine Chapel.

—From Feynmann's Lost Lecture, *David Goodstein and Judith Goodstein, W. W. Norton, New York, 1996, p. 19.*

Before continuing, we compute the following derivative:

$$\frac{d}{dt}\|\mathbf{r}\| = \frac{d}{dt}(\mathbf{r} \cdot \mathbf{r})^{1/2} = \frac{1}{2}(\mathbf{r} \cdot \mathbf{r})^{-1/2}\frac{d}{dt}\mathbf{r} \cdot \mathbf{r} = \frac{1}{2}(\mathbf{r} \cdot \mathbf{r})^{-1/2}(\mathbf{r} \cdot \mathbf{r}' + \mathbf{r}' \cdot \mathbf{r}) = \frac{\mathbf{r} \cdot \mathbf{r}'}{\|\mathbf{r}\|} \tag{9}$$

Using Eq. (9) and the Quotient Rule,

$$\mathbf{B} = \frac{d}{dt}\left(\frac{\mathbf{r}}{\|\mathbf{r}\|}\right) = \frac{\|\mathbf{r}\|\mathbf{r}' - (d\|\mathbf{r}\|/dt)\mathbf{r}}{\|\mathbf{r}\|^2} = \frac{\|\mathbf{r}\|\mathbf{r}' - \|\mathbf{r}\|^{-1}(\mathbf{r} \cdot \mathbf{r}')\mathbf{r}}{\|\mathbf{r}\|^2}$$

$$= \frac{\mathbf{r}'}{\|\mathbf{r}\|} - \frac{(\mathbf{r} \cdot \mathbf{r}')\mathbf{r}}{\|\mathbf{r}\|^3} \tag{10}$$

Comparing Eq. (8) and Eq. (10), we see that $\mathbf{A} = \mathbf{B}$ and therefore $\frac{d}{dt}\mathbf{L} = \mathbf{0}$ as claimed. ■

To derive an equation of the planet's orbit, assume first that $\mathbf{L} \neq \mathbf{0}$. Choose coordinates so that $\mathbf{L}$ points in the direction of the positive x-axis and let (r, θ) be the polar coordinates of the planet (Figure 5). Define

$$e = \|\mathbf{L}\|, \qquad p = \frac{\|\mathbf{J}\|^2}{k}, \qquad r = \|\mathbf{r}\|$$

Then the dot product $\mathbf{r} \cdot \mathbf{L}$ is equal to

$$\mathbf{r} \cdot \mathbf{L} = r\, e \cos\theta$$

On the other hand, by the definition of $\mathbf{L}$ and identity (7),

$$\mathbf{r} \cdot \mathbf{L} = \frac{1}{k}\mathbf{r} \cdot (\mathbf{r}' \times \mathbf{J}) - \frac{\mathbf{r} \cdot \mathbf{r}}{\|\mathbf{r}\|} = \frac{1}{k}\mathbf{J} \cdot (\mathbf{r} \times \mathbf{r}') - \|\mathbf{r}\| = \frac{1}{k}\mathbf{J} \cdot \mathbf{J} - \|\mathbf{r}\| = p - r$$

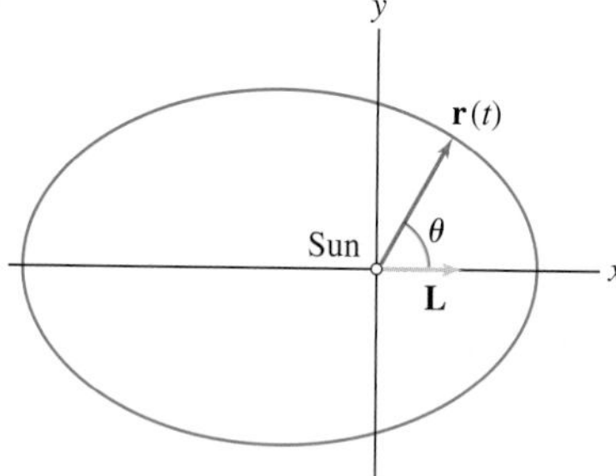

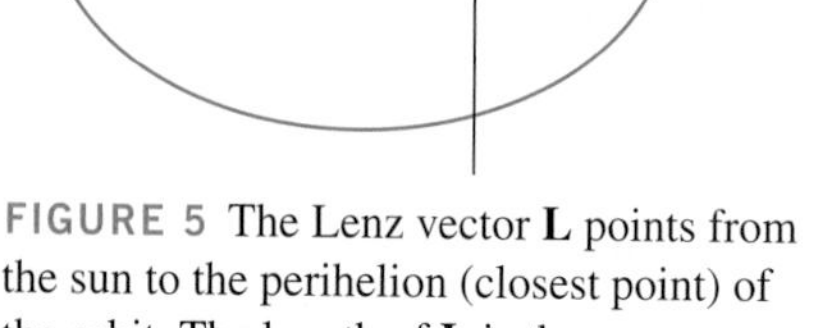

FIGURE 5 The Lenz vector $\mathbf{L}$ points from the sun to the perihelion (closest point) of the orbit. The length of $\mathbf{L}$ is the eccentricity of the orbit.

This equation shows that if $\mathbf{L} = \mathbf{0}$, then $r = p$ and the orbit is a circle of radius p. Assuming $\mathbf{L} \neq \mathbf{0}$, we equate the two expressions for $\mathbf{r} \cdot \mathbf{L}$ to obtain $re\cos\theta = p - r$ or

$$r = \frac{p}{1 + e\cos\theta}$$

This is the polar equation of a conic section (ellipse, parabola, or hyperbola) with eccentricity e and focus at the origin [see Eq. (12) in Section 11.5]. If we assume that the

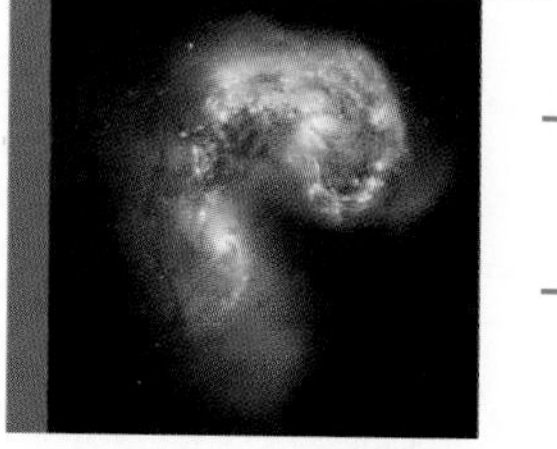

The Hubble Space Telescope produced this image of the Antenna galaxies, a pair of spiral galaxies that began to collide hundreds of millions of years ago.

HISTORICAL PERSPECTIVE

The astronomers of the ancient world (Babylon, Egypt, and Greece) mapped out the nighttime sky with impressive accuracy, but their models of planetary motion were based on the erroneous assumption that the planets revolve around the earth. Although the Greek astronomer Aristarchus (310–230 BCE) had suggested that the earth revolves around the sun, this idea was rejected and forgotten for more than fifteen centuries, until the Polish astronomer Nicolaus Copernicus (1473–1543) introduced a revolutionary set of ideas about the solar system, including the hypothesis that the planets revolve around the sun. The ideas of Copernicus, although not entirely correct, paved the way for the next generation, most notably Tycho Brahe (1546–1601), Galileo Galilei (1564–1642), and Johannes Kepler (1571–1630).

The German astronomer Johannes Kepler was the son of a mercenary soldier who died when Johannes was 5. He was raised by his mother in his grandfather's inn. His mathematical brilliance won him a scholarship to the University of Tübingen. At the age of 29 he went to work for Tycho Brahe, a Danish astronomer who had revolutionized the art of astronomical measurement and compiled the most complete and accurate data on the planetary orbits then available. When Brahe died in 1601, Kepler succeeded him as "Imperial Mathematician" to the Holy Roman Emperor. Kepler continued to study Brahe's data, and in 1609, he formulated the first two of his laws of planetary motion in a work entitled *Astronomia Nova* (New Astronomy).

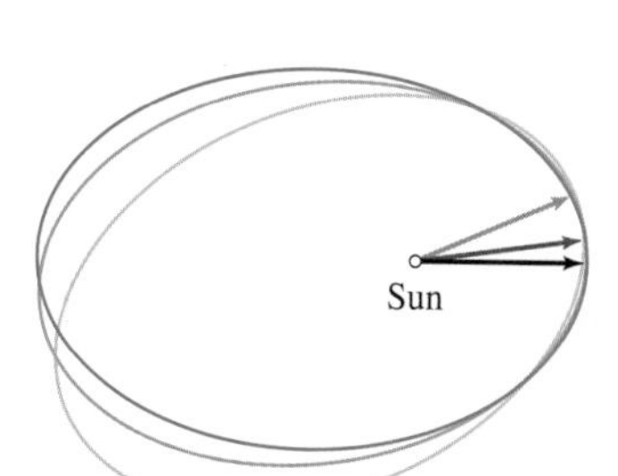

FIGURE 6 The perihelion of an orbit shifts slowly over time. For Mercury, the semimajor axis makes a full rotation approximately once every 24,000 years.

In the centuries since Kepler's death, as the observational data improved, astronomers found that planetary orbits are not exactly elliptical. Furthermore, the perihelion (the point on the orbit closest to the sun) shifts slowly over time as shown in Figure 6. Most of these deviations can be explained by the mutual pull of the planets, but the perihelion shift of Mercury is larger than can be accounted for by Newton's Laws. On November 18, 1915, Albert Einstein made a discovery about which he later wrote to a friend, "I was beside myself with ecstasy for days." He had been working for a decade on his famous **General Theory of Relativity**, a theory that would replace Newton's law of gravitation with a new set of much more complicated equations called the Einstein Field Equations. On that 18th of November, Einstein showed that Mercury's perihelion shift was accurately explained by his new theory. At the time, this was the only substantial piece of evidence that the General Theory of Relativity was correct.

planet travels around the sun in a bounded orbit, this orbit must be an ellipse. We also see that there are "open orbits," parabolic and hyperbolic, that could describe a comet which passes by the sun and then continues into space never to return. If $\mathbf{J} = 0$, then the orbit is a straight line and the planet falls directly into the sun.

CONCEPTUAL INSIGHT You may not have noticed, but we managed to derive Kepler's Laws without ever finding a formula for the position vector $\mathbf{r}(t)$ of the planet. In fact, $\mathbf{r}(t)$ cannot be expressed in terms of elementary functions. This illustrates an important principle: We may be able to understand how the solutions to a differential equation behave even if it is hard or impossible to determine the solutions explicitly.

13.6 SUMMARY

- Kepler's three laws of planetary motion are the Law of Ellipses, the Law of Equal Area in Equal Time, and the Law of the Period $T^2 = \left(\frac{4\pi^2}{GM}\right) a^3$.
- Let $k = GM$. According to Newton's Laws of Motion and Gravitation, the position vector of a planet satisfies the differential equation

$$\mathbf{r}''(t) = -\frac{k}{\|\mathbf{r}\|^3}\mathbf{r}(t)$$

- We define the vector $\mathbf{J} = \mathbf{r}(t) \times \mathbf{r}'(t)$. The planet's position vector $\mathbf{r}(t)$ sweeps out area at the rate $\frac{dA}{dt} = \frac{1}{2}\|\mathbf{J}\|$. Using vector differentiation and the differential equation, we show that $\frac{d\mathbf{J}}{dt} = \mathbf{0}$, which implies Kepler's Second Law.
- The Lenz vector $\mathbf{L} = \frac{1}{k}\mathbf{r}' \times \mathbf{J} - \frac{\mathbf{r}}{\|\mathbf{r}\|}$ is also a constant of planetary motion. It points along the semimajor axis of the elliptical orbit and $\|\mathbf{L}\|$ is the orbit eccentricity.

13.6 EXERCISES

Preliminary Questions

1. Describe the relation between the vector $\mathbf{J} = \mathbf{r} \times \mathbf{r}'$ and the rate at which the radial vector sweeps out area.

2. Equation (1) shows that $\mathbf{r}''$ is proportional to $\mathbf{r}$. Explain how this fact is used to prove Kepler's Second Law.

3. How is the period T affected if the semimajor axis a is increased four-fold?

Exercises

1. Kepler's Third Law states that T^2/a^3 has the same value for each planetary orbit. Do the data in the following table support this conclusion? Estimate the length of Jupiter's period, assuming that $a = 77.8 \times 10^{10}$ m.

Planet	Mercury	Venus	Earth	Mars
a (10^{10} m)	5.79	10.8	15.0	22.8
T (years)	0.241	0.615	1.00	1.88

2. A satellite has initial position $\mathbf{r} = \langle 1{,}000, 2{,}000, 0\rangle$ and initial velocity $\mathbf{r}' = \langle 1, 2, 2\rangle$ (units of kilometrs and seconds). Find the equation of the plane containing the satellite's orbit. *Hint:* This plane is orthogonal to $\mathbf{J}$.

3. The earth's orbit is nearly circular with radius $R = 93 \times 10^6$ miles (the eccentricity is $e = 0.017$). Find the rate at which the earth's radial vector sweeps out area in units of ft^2/s. What is the magnitude of the vector $\mathbf{J} = \mathbf{r} \times \mathbf{r}'$ for the earth (in units of squared feet per second)?

4. Finding the Mass of a Star Using Kepler's Third Law, show that if a planet revolves around a star with period T and semimajor axis a, then the mass of the star is $M = \left(\frac{4\pi^2}{G}\right)\left(\frac{a^3}{T^2}\right)$.

5. Ganymede, one of Jupiter's moons discovered by Galileo, has an orbital period of 7.154 days and a semimajor axis of 1.07×10^9 m. Use Exercise 4 to estimate the mass of Jupiter.

6. An astronomer observes a planet orbiting a star with a period of 9.5 years and a semimajor axis of 3×10^8 km. Find the mass of the star.

7. Use the fact that $\mathbf{J}$ is constant to show that a planet in a circular orbit travels at constant speed.

8. Prove that if a planetary orbit is circular, then $vT = 2\pi R$, where v is the planet's speed (constant by Exercise 7) and T is the period. Then use Kepler's Third Law to prove that $v = \sqrt{\frac{GM}{R}}$.

9. Show directly that the circular orbit

$$\mathbf{r}(t) = \langle R\cos\omega t, R\sin\omega t\rangle$$

satisfies the differential equation, Eq. (2), provided that $\omega^2 = kR^{-3}$. Then deduce Kepler's Third Law $T^2 = \left(\frac{4\pi^2}{k}\right)R^3$ for this orbit.

10. The orbit of a satellite orbiting above the equator of the earth is called **geosynchronous** if the period is $T = 24$ hours (in this case, the satellite stays over a fixed point on the equator). Use Kepler's Third Law to find the altitude h above the earth's surface of a geosynchronous orbit. The earth has mass $M \approx 5.974 \times 10^{24}$ kg and radius $R \approx 6{,}371$ km.

11. Use the results of Exercises 8 and 10 to find the velocity of a satellite in geosynchronous orbit.

12. Show that if a planet revolves around a star of mass M in a circular orbit of radius R with speed v, then $M = \frac{Rv^2}{G}$.

13. Mass of the Milky Way The sun revolves around the center of mass of the Milky Way galaxy in an orbit that is approximately circular, of radius $a \approx 2.8 \times 10^{17}$ km and velocity $v \approx 250$ km/s. Use the result of Exercise 12 to estimate the mass of the portion of the Milky Way inside the sun's orbit (place all of this mass at the center of the orbit).

14. Conservation of Energy The total mechanical energy (kinetic plus potential) of a planet of mass m orbiting a sun of mass M with position $\mathbf{r}$ and speed $v = \|\mathbf{r}'\|$ is

$$E = \frac{1}{2}mv^2 - \frac{GMm}{\|\mathbf{r}\|} \qquad \boxed{11}$$

Use (2) and (9) to show that E is conserved, that is, $\frac{dE}{dt} = 0$.

15. Show that the total energy (11) of a planet in a circular orbit of radius R is $E = -GMm/(2R)$. *Hint:* Use Exercise 8.

In Exercises 16–20, we consider a planetary orbit with orbital parameters p and e. The perihelion and aphelion of the orbits are the points on the orbit closest to and farthest from the sun (Figure 7). Denote the distances from the sun at the perihelion and aphelion by r_{per} and r_{ap} and the speeds of the planet at the perihelion and aphelion by v_{per} and v_{ap}.

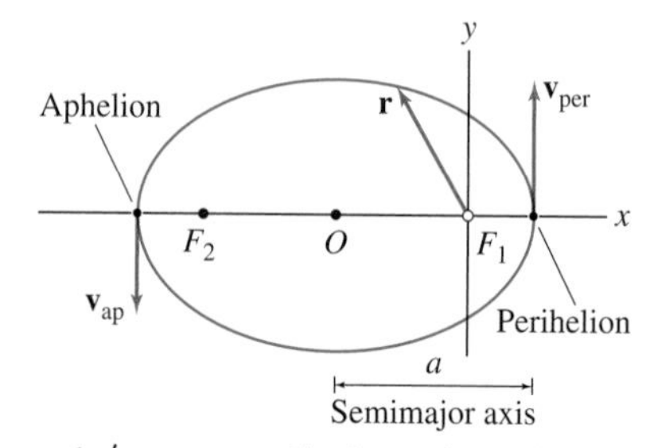

FIGURE 7 $\mathbf{r}$ and $\mathbf{r}'$ are perpendicular at the perihelion and aphelion.

16. Use the polar equation

$$r = \frac{p}{1 + e\cos\theta}$$

to show that $r_{per} = a(1-e)$ and $r_{ap} = a(1+e)$.

17. Compute r_{per} and r_{ap} for the orbit of Mercury, which has eccentricity $e = 0.244$ (see the table in Exercise 1 for the semimajor axis).

18. Prove the formulas

$$e = \frac{r_{ap} - r_{per}}{r_{ap} + r_{per}}, \qquad p = \frac{2r_{ap}r_{per}}{r_{ap} + r_{per}}$$

19. Prove that $v_{per}(1-e) = v_{ap}(1+e)$. *Hint:* $\mathbf{r} \times \mathbf{r}'$ is constant by Eq. (5). Compute this cross product at the perihelion and aphelion, noting that $\mathbf{r}$ is perpendicular to $\mathbf{r}'$ at these two points.

20. Prove that $v_{per} = \sqrt{\left(\frac{GM}{a}\right)\frac{1+e}{1-e}}$ as follows:

(a) Use Conservation of Energy (see Exercise 14) to show that

$$v_{per}^2 - v_{ap}^2 = 2GM(r_{per}^{-1} - r_{ap}^{-1})$$

(b) Show $r_{per}^{-1} - r_{ap}^{-1} = \frac{2e}{a(1-e^2)}$ using Exercise 16.

(c) Show that $v_{per}^2 - v_{ap}^2 = 4\frac{e}{(1+e)^2}v_{per}^2$ using Exercise 19. Then solve for v_{per} using (a) and (b).

21. Show that the total mechanical energy E of a planet in an elliptical orbit with semimajor axis a is $E = -\frac{GMm}{2a}$. *Hint:* Use Exercise 20 to compute the total energy at the perihelion.

22. Prove that $v^2 = GM\left(\frac{2}{r} - \frac{1}{a}\right)$ at any point on an elliptical orbit with semimajor axis a, where $r = \|\mathbf{r}\|$.

23. Two space shuttles A and B orbit the earth along the solid trajectory in Figure 8. Hoping to catch up to B, the pilot of A applies a forward thrust to increase her shuttle's kinetic energy. Use Exercise 21 to show that shuttle A will move off into a larger orbit as shown in the figure. Then use Kepler's Third Law to show that A's orbital period T will increase (and she will fall farther and farther behind B)!

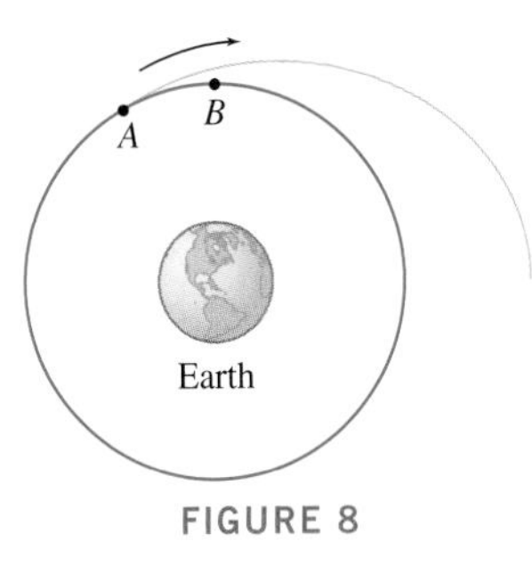

FIGURE 8

Further Insights and Challenges

24. In this exercise, we prove Eq. (3) in a rigorous fashion. Let $r(t) = \|\mathbf{r}(t)\|$ and let $\theta(t)$ be the angle between $\mathbf{r}(t)$ and the x-axis. Then $\mathbf{r} = r(t)\langle\cos\theta(t), \sin\theta(t)\rangle$.

(a) Prove that $\dfrac{dA}{dt} = \dfrac{1}{2}r(t)^2\theta'(t)$ by applying the Fundamental Theorem of Calculus to the formula for area in polar coordinates:

$$A(\theta) = \frac{1}{2}\int_0^\theta r(u)^2\,du$$

(b) Show that $\|\mathbf{J}\| = \|\mathbf{r} \times \mathbf{r}'\| = r(t)^2\theta'(t)$.

(c) Conclude that $\dfrac{dA}{dt} = \dfrac{1}{2}\|\mathbf{J}\|$.

In Exercises 25–26, we prove Kepler's Third Law. Figure 9 shows an elliptical orbit with polar equation

$$r = \frac{p}{1 + e\cos\theta}$$

where $p = \|\mathbf{J}\|^2/k$. Let a and b be the semimajor and semiminor axes, respectively. The origin is located at F_1.

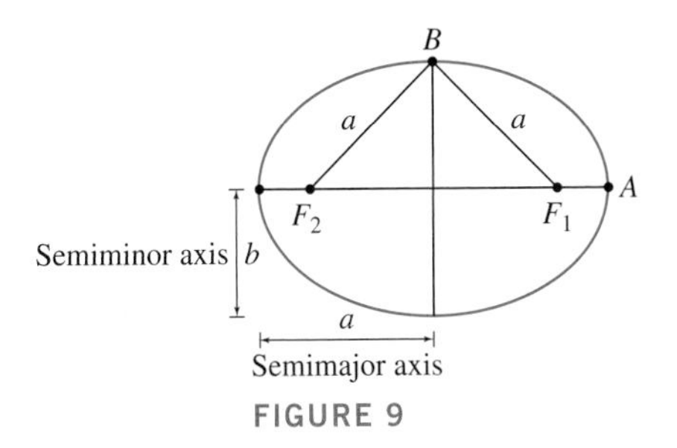

FIGURE 9

25. The goal of this exercise is to show that $b = \sqrt{pa}$.

(a) Show that $F_1A + F_2A = 2a$. Conclude that $F_1B + F_2B = 2a$ and hence $F_1B = F_2B = a$.

(b) Show that $F_1A = \dfrac{p}{1+e}$ and $F_2A = \dfrac{p}{1-e}$, and conclude that $a = \dfrac{p}{1-e^2}$.

(c) Use the Pythagorean Theorem to prove that

$$b = \frac{p}{\sqrt{1-e^2}} = \sqrt{pa}$$

26. We prove Kepler's Third Law by computing the area A of the ellipse in two ways.

(a) Use Exercise 25 to show that $A = (\pi\sqrt{p})a^{3/2}$.

(b) Use Kepler's First Law to show that $A = \frac{1}{2}\|\mathbf{J}\|T$.

(c) Deduce that $T^2 = \dfrac{4\pi^2}{GM}a^3$.

27. Let $\mathbf{e_r} = \langle\cos\theta, \sin\theta\rangle$ and $\mathbf{e}_\theta = \langle-\sin\theta, \cos\theta\rangle$. Write the position vector of a planet as $\mathbf{r} = r\mathbf{e_r}$, where $r = \|\mathbf{r}\|$.

(a) Show that $\dfrac{d\mathbf{e}_\theta}{d\theta} = -\mathbf{e_r}$.

(b) Write Eq. (2) in the form $\dfrac{d\mathbf{v}}{dt} = -\dfrac{k}{r^2}\mathbf{e_r}$ and use the Chain Rule to show that

$$\frac{d\mathbf{v}}{d\theta}\frac{d\theta}{dt} = \frac{k}{r^2}\frac{d\mathbf{e}_\theta}{d\theta}$$

(c) Show that $\dfrac{d\mathbf{v}}{d\theta} = \dfrac{k}{\|\mathbf{J}\|}\dfrac{d\mathbf{e}_\theta}{d\theta}$. *Hint:* Use Exercise 24 to show that $\dfrac{d\theta}{dt} = \|\mathbf{J}\|/r^2$.

(d) Conclude that there is a constant vector $\mathbf{w}$ such that

$$\mathbf{v}(\theta) = \frac{k}{\|\mathbf{J}\|}\mathbf{e}_\theta + \mathbf{w}$$

This shows that as θ varies from 0 to 2π, the velocity vector $\mathbf{v}$ traces out a circle of radius $k/\|\mathbf{J}\|$ with center at the terminal point of $\mathbf{w}$ (Figure 10).

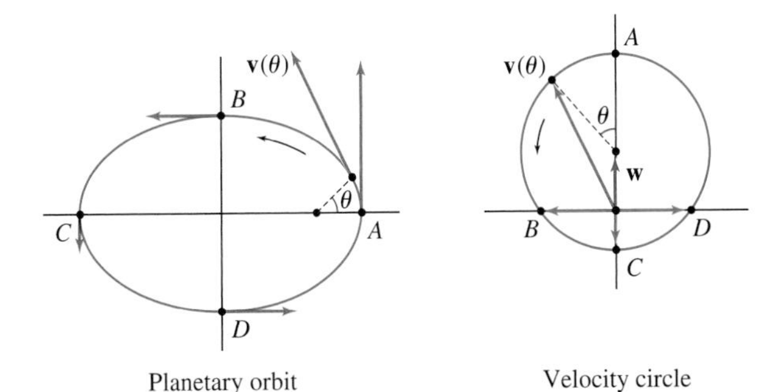

FIGURE 10 The terminal point of the velocity vector traces out the circle as the planet travels along its orbit.

CHAPTER REVIEW EXERCISES

1. Determine the domains of the vector-valued functions.

(a) $\mathbf{r}_1(t) = \langle t^{-1}, (t+1)^{-1}, \sin^{-1} t\rangle$

(b) $\mathbf{r}_2(t) = \langle \sqrt{8-t^3}, \ln t, e^{\sqrt{t}}\rangle$

2. Sketch the paths $\mathbf{r}_1(\theta) = \langle \theta, \cos\theta\rangle$ and $\mathbf{r}_2(\theta) = \langle \cos\theta, \theta\rangle$ in the xy-plane.

3. Find a vector parametrization of the intersection of the surfaces $x^2 + y^4 + 2z^3 = 6$ and $x = y^2$ in $\mathbf{R}^3$.

4. Find a vector parametrization using trigonometric functions of the intersection of the plane $x + y + z = 1$ and the elliptical cylinder $\left(\frac{y}{3}\right)^2 + \left(\frac{z}{8}\right)^2 = 1$ in $\mathbf{R}^3$.

In Exercises 5–10, calculate the derivative indicated.

5. $\mathbf{r}'(t)$, where $\mathbf{r}(t) = \langle 1-t, t^{-2}, \ln t\rangle$

6. $\mathbf{r}'''(1)$, where $\mathbf{r}(t) = \langle t^3, 4t^2, 7t\rangle$

7. $\mathbf{r}'(0)$, where $\mathbf{r}(t) = \langle e^{2t}, e^{-4t^2}, e^{6t}\rangle$

8. $\mathbf{r}''(2)$, where $\mathbf{r}(t) = \langle e^{2t}, e^{-4t^2}, e^{6t}\rangle$

9. $\dfrac{d}{dt} e^t \langle 1, t, t^2\rangle$

10. $\dfrac{d}{d\theta}\mathbf{r}(\cos\theta)$, where $\mathbf{r}(s) = \langle s, 2s, s^2\rangle$

In Exercises 11–14, calculate the derivative at $t = 3$ assuming that

$$\mathbf{r}_1(3) = \langle 1, 1, 0\rangle, \quad \mathbf{r}_2(3) = \langle 1, 1, 0\rangle$$

$$\mathbf{r}_1'(3) = \langle 0, 0, 1\rangle, \quad \mathbf{r}_2'(3) = \langle 0, 2, 4\rangle$$

11. $\dfrac{d}{dt}(6\mathbf{r}_1(t) - 4 \cdot \mathbf{r}_2(t))$

12. $\dfrac{d}{dt}\left(e^t \mathbf{r}_2(t)\right)$

13. $\dfrac{d}{dt}\left(\mathbf{r}_1(t) \cdot \mathbf{r}_2(t)\right)$

14. $\dfrac{d}{dt}\left(\mathbf{r}_1(t) \times \mathbf{r}_2(t)\right)$

15. Calculate $\displaystyle\int_0^3 \langle 4t+3, t^2, -4t^3\rangle\, dt$.

16. Calculate $\displaystyle\int_0^\pi \langle \sin\theta, \theta, \cos 2\theta\rangle\, d\theta$.

17. Find the unit tangent vector to $\mathbf{r}(t) = \langle \sin t, t, \cos t\rangle$ at $t = \pi$.

18. Find the unit tangent vector to $\mathbf{r}(t) = \langle t^2, \tan^{-1} t, t\rangle$ at $t = 1$.

19. A particle located at $(1, 1, 0)$ at time $t = 0$ follows a path whose velocity vector is $\mathbf{v}(t) = \langle 1, t, 2t^2\rangle$. Find the particle's location at $t = 2$.

20. Find the vector-valued function $\mathbf{r}(t) = \langle x(t), y(t)\rangle$ in $\mathbf{R}^2$ satisfying $\mathbf{r}'(t) = -\mathbf{r}(t)$ with initial conditions $\mathbf{r}(0) = \langle 1, 2\rangle$.

21. Compute the length of the path $\mathbf{r}(t) = \langle \sin 2t, \cos 2t, 3t - 1\rangle$ for $1 \le t \le 3$.

22. CAS Express the length of the path $\mathbf{r}(t) = \langle \ln t, t, e^t\rangle$ for $1 \le t \le 2$ as a definite integral and use a computer algebra system to find its value to two decimal places.

23. A string in the shape of a helix has a height of 20 cm and makes four full rotations over a circle of radius 5 cm. Find a parametrization $\mathbf{r}(t)$ of the string and compute its length.

24. Find the minimum speed of a particle with trajectory $\mathbf{r}(t) = \langle t, e^{t-3}, e^{4-t}\rangle$.

25. Calculate the curvature $\kappa(t)$ for $\mathbf{r}(t) = \langle t^{-1}, \ln t, t\rangle$ and find the unit tangent and normal vectors at $t = 1$.

26. A specially trained mouse runs counterclockwise in a circle of radius 2 ft on the floor of an elevator with speed 1 ft/s while the elevator ascends from ground level (along the z-axis) at a speed of 40 ft/s. Find the mouse's acceleration vector as a function of time. Assume that the circle is centered at the origin of the xy-plane and the mouse is at $(2, 0, 0)$ at $t = 0$.

In Exercises 27–30, let $\mathbf{r}(t) = \langle t, e^{-t^2}\rangle$.

27. Compute the curvature function $\kappa(t)$.

28. GU Plot $\mathbf{r}(t)$ and $\kappa(t)$ on the same set of axes and estimate the values of t where the curvature has a maximum value.

29. Find the unit tangent and normal vectors at $t = 0$ and $t = 1$.

30. Write the acceleration vector at $t = 1$ as a sum of tangential and normal components.

31. Find the curvature $\kappa(t)$ and unit normal vector $\mathbf{N}(t)$ for $\mathbf{r}(t) = \langle \sin t, \sin t, \cos t\rangle$.

32. Find the curvature $\kappa(t)$ and unit normal vector $\mathbf{N}(t)$ for $\mathbf{r}(t) = \langle \ln t, t^{-1}, t^{-2}\rangle$.

In Exercises 33–34, write the acceleration vector **a** *at the point indicated as a sum of tangential and normal components.*

33. $\mathbf{r}(\theta) = \langle \cos\theta, \sin 2\theta\rangle, \quad \theta = \frac{\pi}{4}$

34. $\mathbf{r}(t) = \langle t^2, t^3\rangle, \quad t = 2$

35. Find the osculating circle to the curve $y = e^{-x^2}$ at $x = 0$.

36. Find the osculating circle to the curve $y = \ln x$ at $x = 1$.

37. If a planet is in "orbit" around a sun whose mass is zero, Newton's Laws imply that the position vector of the plane satisfies $\mathbf{r}''(t) = \mathbf{0}$. Show that in this case, the orbit is the straight line with parametrization $\mathbf{r}(t) = \mathbf{r} + t\mathbf{v}$, where $\mathbf{r} = \mathbf{r}(0)$ and $\mathbf{v} = \mathbf{r}'(0)$ (Figure 1).

38. Continuing with Exercise 37, show that the area A swept out by the radial vector $\mathbf{r}(t)$ over the time interval $[0, t]$ is equal to $A = \frac{1}{2}\|\mathbf{r}(0) \times \mathbf{v}\|t$. Thus, Kepler's Second Law continues to hold since area is swept out at the rate $\frac{1}{2}\|\mathbf{r}(0) \times \mathbf{v}\|$.

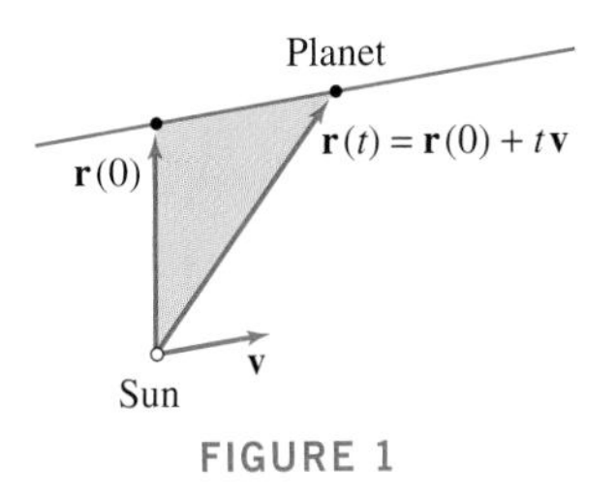

FIGURE 1

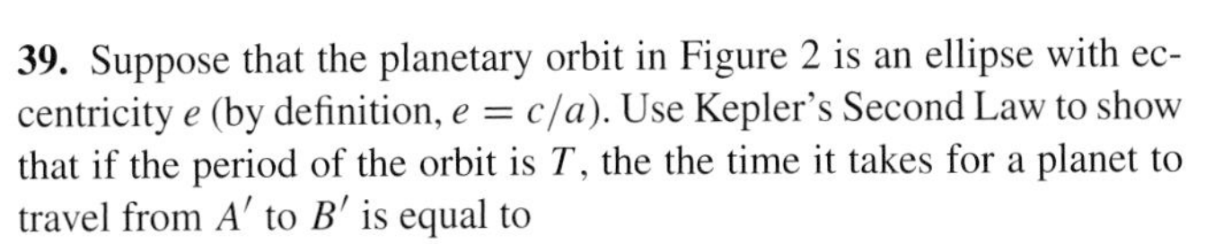

39. Suppose that the planetary orbit in Figure 2 is an ellipse with eccentricity e (by definition, $e = c/a$). Use Kepler's Second Law to show that if the period of the orbit is T, the the time it takes for a planet to travel from A' to B' is equal to

$$\left(\frac{1}{4} + \frac{e}{2\pi}\right) T$$

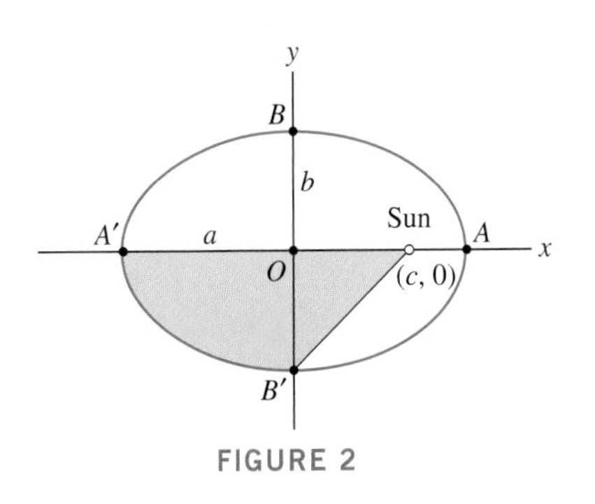

FIGURE 2

40. The period of Mercury is approximately 88 days and its orbit has eccentricity 0.205. How much longer does it take Mercury to travel from A' to B' than from B' to A (Figure 2)?

14 DIFFERENTIATION IN SEVERAL VARIABLES

Yosemite National Park. The steepness at a point in a mountain range is measured by the gradient vector, a concept defined in this chapter.

The goal of this chapter is to extend the concepts and techniques of differential calculus to functions of several variables. As we will see, a function f that depends on two or more variables has not just one derivative but rather a set of *partial derivatives*, one for each variable. The partial derivatives make up the components of the gradient vector, which provides important geometric insight into the behavior of the function. In the last two sections of the chapter, we apply the insights gained to optimization in several variables.

14.1 Functions of Two or More Variables

Functions of several variables are a basic tool in the study of natural phenomena. For example, scientists study variations in the density ρ of seawater (mass per unit volume). These variations are responsible for deep ocean currents that distribute heat throughout the earth and influence the global climate (Figure 1). However, ρ depends on several variables, of which the most important are salinity S and temperature T. We write $\rho = f(S, T)$ to indicate that ρ is a function of S and T. There is no simple formula for $f(S, T)$, but scientists have measured $f(S, T)$ experimentally (Figure 2). According to Table 1, the density at $S = 32$ (parts per thousand or ppt) and $T = 10°\text{C}$ is

$$\rho = f(32, 10) = 1.0246 \text{ kg/m}^3$$

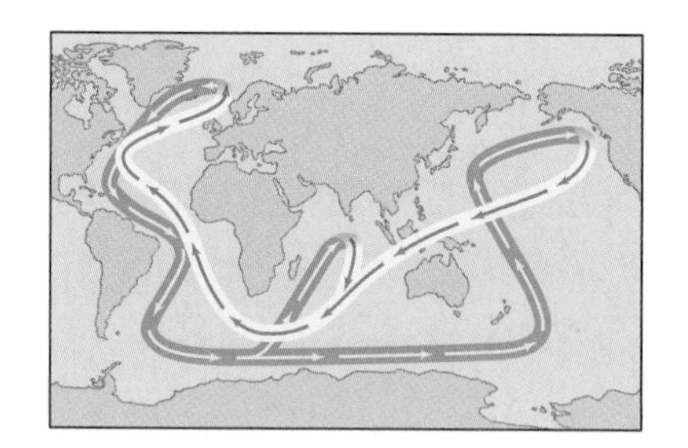

FIGURE 1 Ocean "conveyer belt," the system of deep currents driven by variations in seawater density.

TABLE 1 Seawater Density ρ (kg/m^3) as a Function of Temperature and Salinity.

	Salinity (ppt)		
°C	32	32.5	33
5	1.0253	1.0257	1.0261
10	1.0246	1.0250	1.0254
15	1.0237	1.0240	1.0244
20	1.0224	1.0229	1.0232

A function of n-variables is a real-valued function $f(x_1, \dots, x_n)$ whose domain $\mathcal{D}$ is a set of n-tuples $(x_1, \dots, x_n)$ in $\mathbf{R}^n$. When f is defined by a formula, we usually take as domain the set of all n-tuples for which $f(x_1, \dots, x_n)$ is defined. The range of f is the set of all values $f(x_1, \dots, x_n)$ for $(x_1, \dots, x_n)$ in the domain. For instance, $f(x_1, x_2, x_3) = x_1 + x_2 e^{x_3}$ has domain $\mathcal{D} = \mathbf{R}^3$ since f is defined for all (x_1, x_2, x_3). The range is $\mathbf{R}$, as we may show by observing that $f(x_1, 0, 0) = x_1$. Thus, f takes on every value. We will often focus our attention on functions of two and three variables.

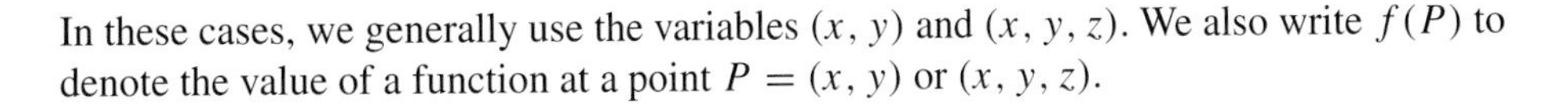

FIGURE 2 A researcher lowers a Conductivity-Temperature-Depth (CDT) instrument into the ocean to measure seawater variables such as density, temperature, pressure, and salinity.

In these cases, we generally use the variables (x, y) and (x, y, z). We also write $f(P)$ to denote the value of a function at a point $P = (x, y)$ or (x, y, z).

EXAMPLE 1 Sketch the domains of:

(a) $f(x, y) = \sqrt{9 - x^2 - y^2}$ **(b)** $g(x, y, z) = x\sqrt{y} + \ln(z - 1)$

Solution

(a) $f(x, y) = \sqrt{9 - x^2 - y^2}$ is defined when $9 - x^2 - y^2 \geq 0$, so the domain of f is $\mathcal{D} = \{(x, y) : x^2 + y^2 \leq 9\}$, a disk of radius 3 [Figure 3(A)].

(b) The function $g(x, y, z)$ is defined if both $\sqrt{y}$ and $\ln(z - 1)$ are defined. Thus we require that $y \geq 0$ and $z > 1$, and the domain is $\{(x, y, z) : y \geq 0, z > 1\}$ [Figure 3(B)]. ■

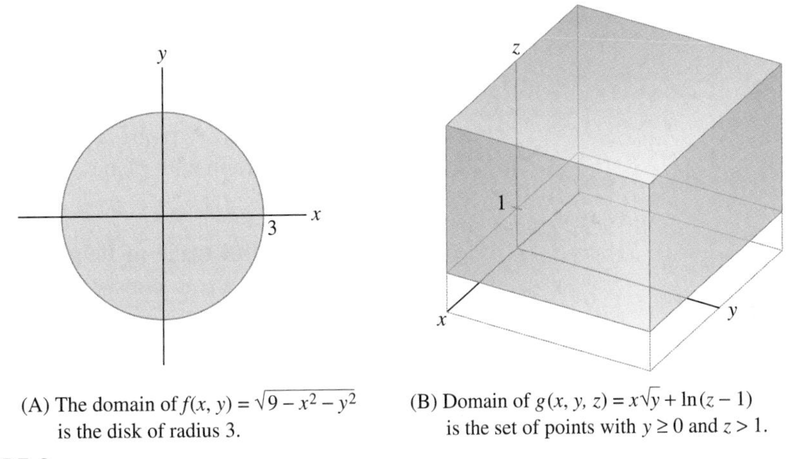

(A) The domain of $f(x, y) = \sqrt{9 - x^2 - y^2}$ is the disk of radius 3.

(B) Domain of $g(x, y, z) = x\sqrt{y} + \ln(z - 1)$ is the set of points with $y \geq 0$ and $z > 1$.

FIGURE 3

Graphing Functions of Two Variables

In our study of single-variable calculus, we have seen that a graph displays the important features of a function at a glance. Graphs are also useful for visualizing functions of two variables. The graph of $f(x, y)$ is the set of all points $(a, b, f(a, b))$ in $\mathbf{R}^3$ for (a, b) in the domain $\mathcal{D}$ of f. If we assume that f is continuous (as defined in the next section), then the graph is a surface whose *height* above or below the xy-plane at (a, b) is the function value $f(a, b)$ (Figure 4). Note that the graph of $f(x, y)$ is also the graph of the equation $z - f(x, y) = 0$.

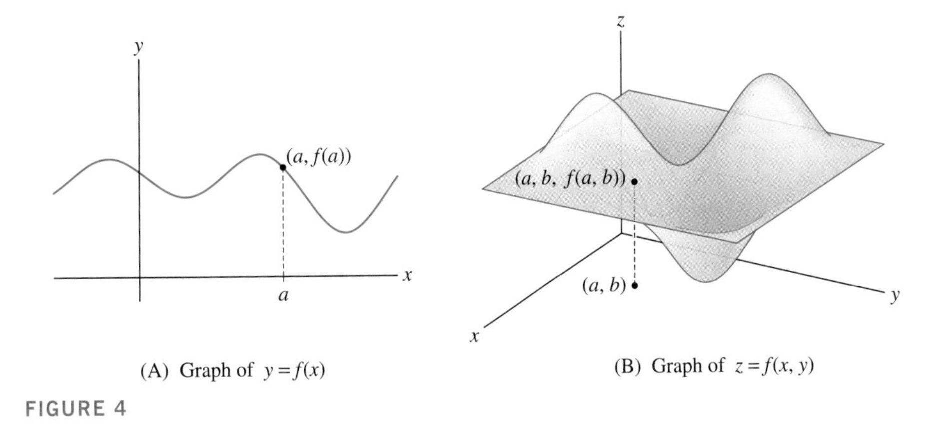

(A) Graph of $y = f(x)$

(B) Graph of $z = f(x, y)$

FIGURE 4

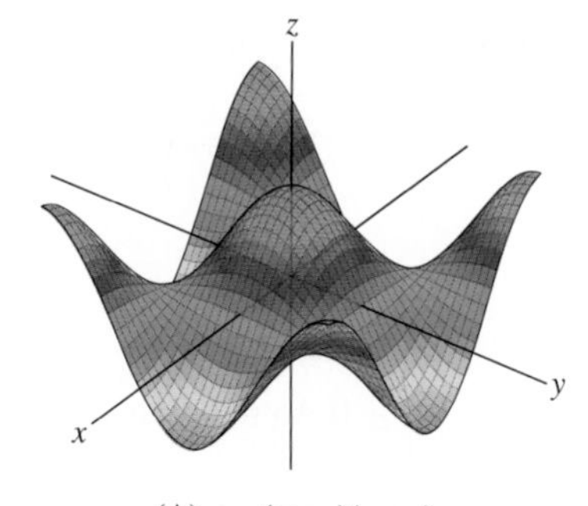

(A) $z = (\cos x)(\cos y)$

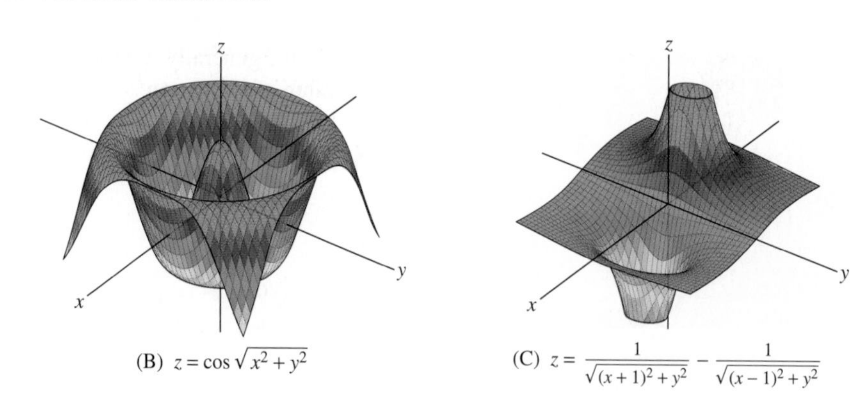

(B) $z = \cos\sqrt{x^2 + y^2}$

(C) $z = \dfrac{1}{\sqrt{(x+1)^2 + y^2}} - \dfrac{1}{\sqrt{(x-1)^2 + y^2}}$

FIGURE 5 Gallery of graphs.

Plotting graphs of functions of two variables by hand is often a difficult task. Fortunately, computer algebra systems eliminate much of the labor and greatly enhance our ability to explore functions graphically from different perspectives (Figure 5).

To help analyze graphs, we define the **traces**, which are the curves obtained by intersecting the graph with planes parallel to a coordinate plane (Figure 6). There are three types of traces:

- **Horizontal trace at height c:** The intersection of the graph with the horizontal plane $z = c$, consisting of points (x, y, c) such that $f(x, y) = c$. This trace projects down to the curve $f(x, y) = c$ in the xy-plane.
- **Vertical trace in the plane $x = a$:** The intersection of the graph with the vertical plane $x = a$, consisting of points $(a, y, f(a, y))$.
- **Vertical trace in the plane $y = b$:** The intersection of the graph with the vertical plane $y = b$, consisting of points $(x, b, f(x, b))$.

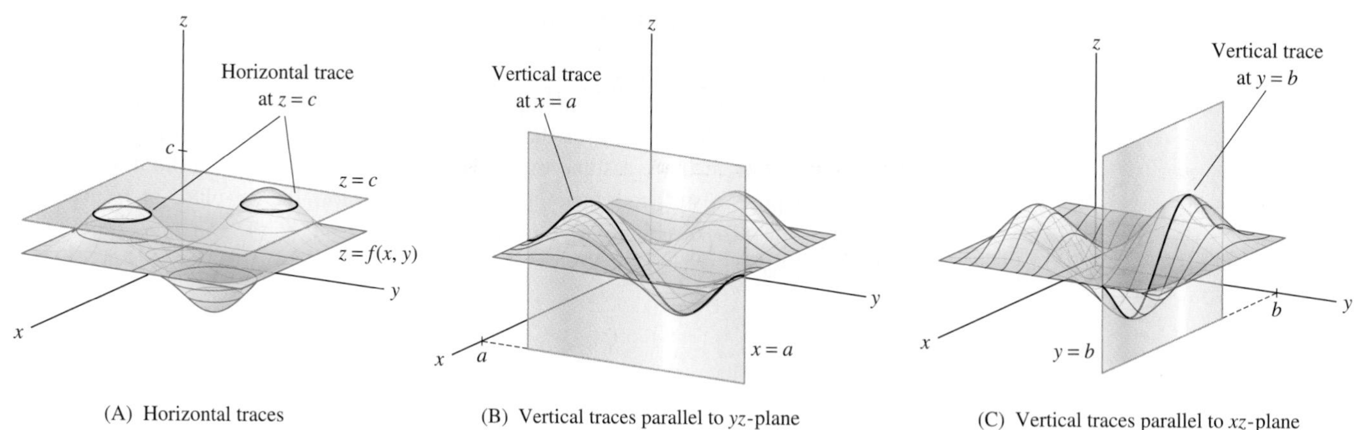

(A) Horizontal traces

(B) Vertical traces parallel to yz-plane

(C) Vertical traces parallel to xz-plane

FIGURE 6

■ **EXAMPLE 2 Traces of a Saddle-Shaped Surface** Describe the horizontal and vertical traces of $f(x, y) = x^2 - y^2$.

Solution The horizontal trace at height c lies above the hyperbola $f(x, y) = x^2 - y^2 = c$ in the xy-plane. In other words, the trace is the hyperbola raised to height c [Figure 7(A)]. The vertical traces are parabolas. The trace in the plane $x = a$ is the parabola

$z = a^2 - y^2$ moved to the plane $x = a$. The trace in the plane $y = b$ is the parabola $z = x^2 - b^2$ moved to the plane $y = b$ [Figure 7(B)]. ■

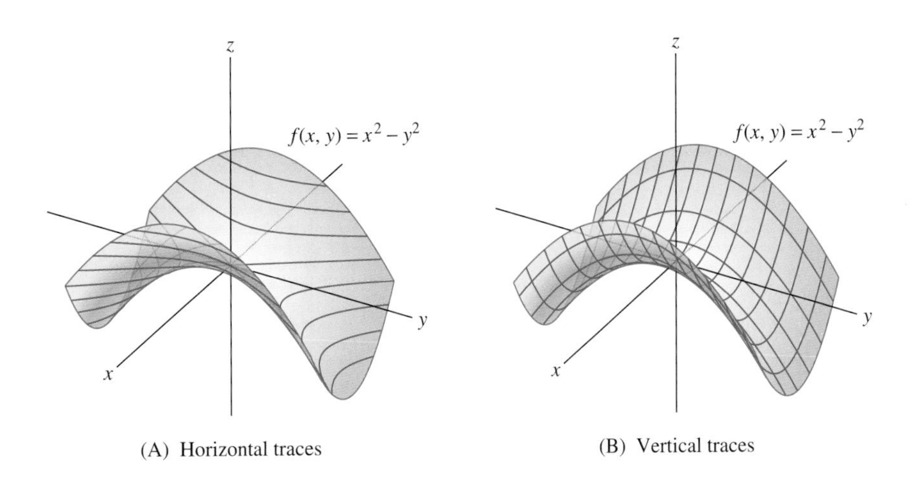

FIGURE 7 The graph of $f(x, y) = x^2 - y^2$ has a "saddle" shape.

REMINDER A linear function of two variables is a function of the form $f(x, y) = mx + ny + r$, where m, n, and r are constants. The graph of a linear function is a plane.

■ **EXAMPLE 3 The Traces of a Linear Function Are Lines** Describe the traces of the linear function $f(x, y) = 4 - 2x - y$.

Solution The graph of f is the plane with equation $z = 4 - 2x - y$. The horizontal trace at height c projects to the line $f(x, y) = c$ or $4 - 2x - y = c$. The trace itself is obtained by raising this line to height c [Figure 8(A)]. There are two families of vertical traces:

- Trace in $x = a$: The line $z = 4 - 2a - y$ in the plane $x = a$ [Figure 8(B)]
- Trace in $y = b$: The line $z = 4 - 2x - b$ in the plane $y = b$ [Figure 8(C)] ■

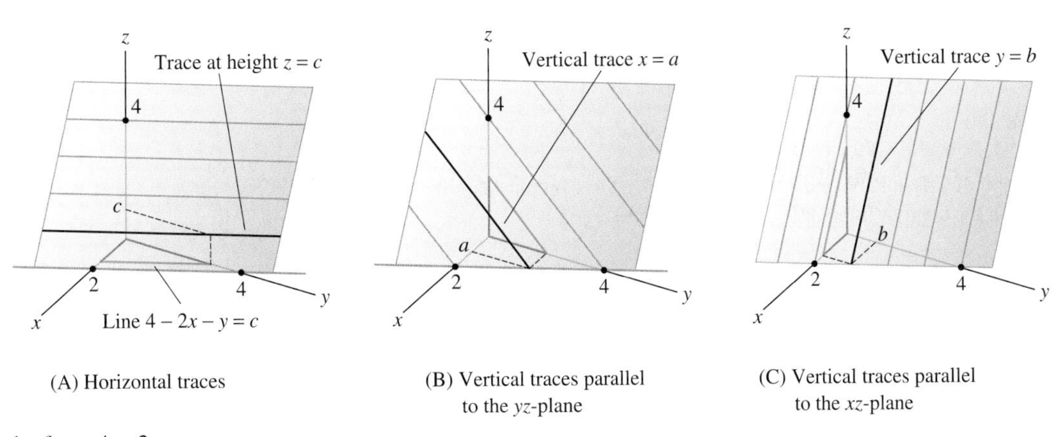

FIGURE 8 Graph of $z = 4 - 2x - y$.

Contour Maps

Another useful way of displaying the main features of a function is to draw a contour map showing level curves. The **level curves** of a function $f(x, y)$ are the curves in the xy-plane with equation $f(x, y) = c$. Thus, the level curve $f(x, y) = c$ is the projection of the horizontal trace at height c onto the xy-plane.

A **contour map** shows the level curves of $f(x, y)$ for equally spaced values of c. The spacing m is called the **contour interval**. Figure 9 is a contour map showing the level curves with contour interval $m = 10$.

EXAMPLE 4 Draw a contour map for $f(x, y) = x^2y$ with contour interval $m = 2$.

Solution The level curves are the curves with equation $x^2y = c$ or $y = cx^{-2}$. In Figure 10, we plot $y = cx^{-2}$ for $c = 0, \pm2, \pm4, \pm6, \pm8, \pm10$ (contour interval $m = 2$). ■

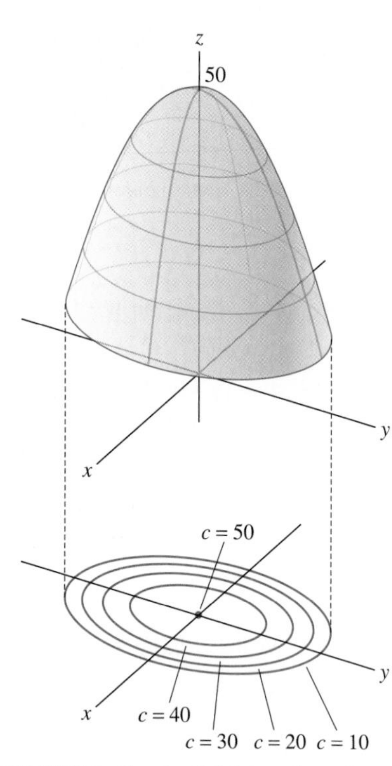

FIGURE 9 Graph of f lying above its contour map with the contour interval $m = 10$.

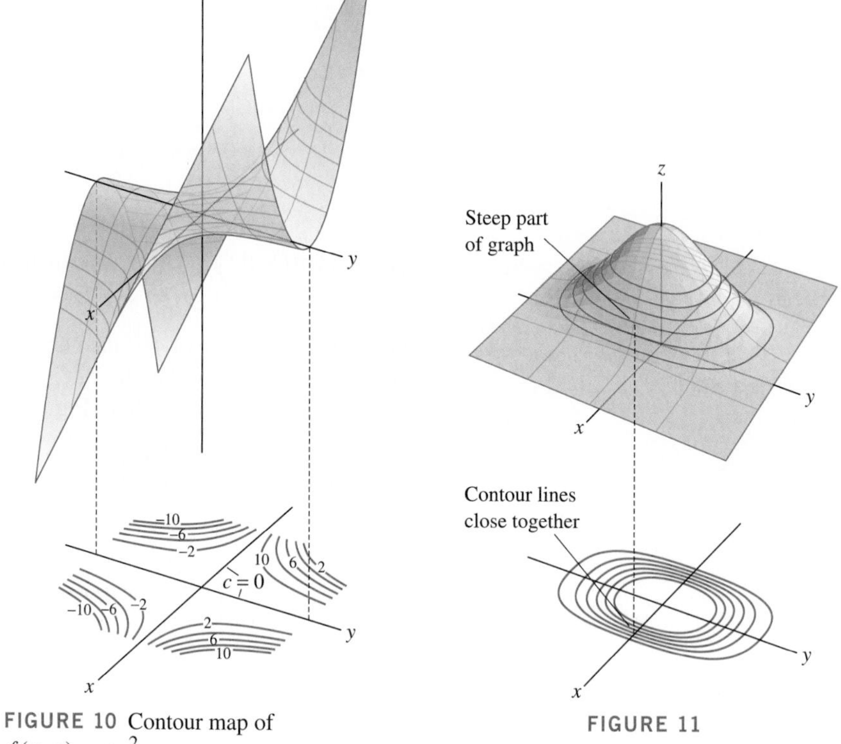

FIGURE 10 Contour map of $f(x, y) = xy^2$.

FIGURE 11

The level curve $f(x, y) = c$ shows us where the function takes the value c. When you move from one level curve to another in a contour map, the value of $f(x, y)$ changes by $\pm m$, where m is the contour interval. Therefore, *the distance between the level curves in a contour map indicates the steepness of the graph* (Figure 11). The level curves are close together where f changes quickly, and thus, where the graph is steep.

EXAMPLE 5 Spacing of Level Curves CAS Draw a contour map for the function $f(x, y) = x^2 + 3y^2$ with contour interval $m = 10$. Discuss the spacing of the level curves.

Solution The graph of $z = f(x, y)$ is a paraboloid and the level curves are the ellipses $f(x, y) = x^2 + 3y^2 = c$ (Figure 12). The level curve is empty if $c < 0$, and for $c = 0$, the level curve reduces to the point $(0, 0)$. Since the contour interval is $m = 10$, the function changes by 10 when we go from one level curve to the next. As we move away from the origin, $f(x, y)$ increases more rapidly and the level curves get closer together. ■

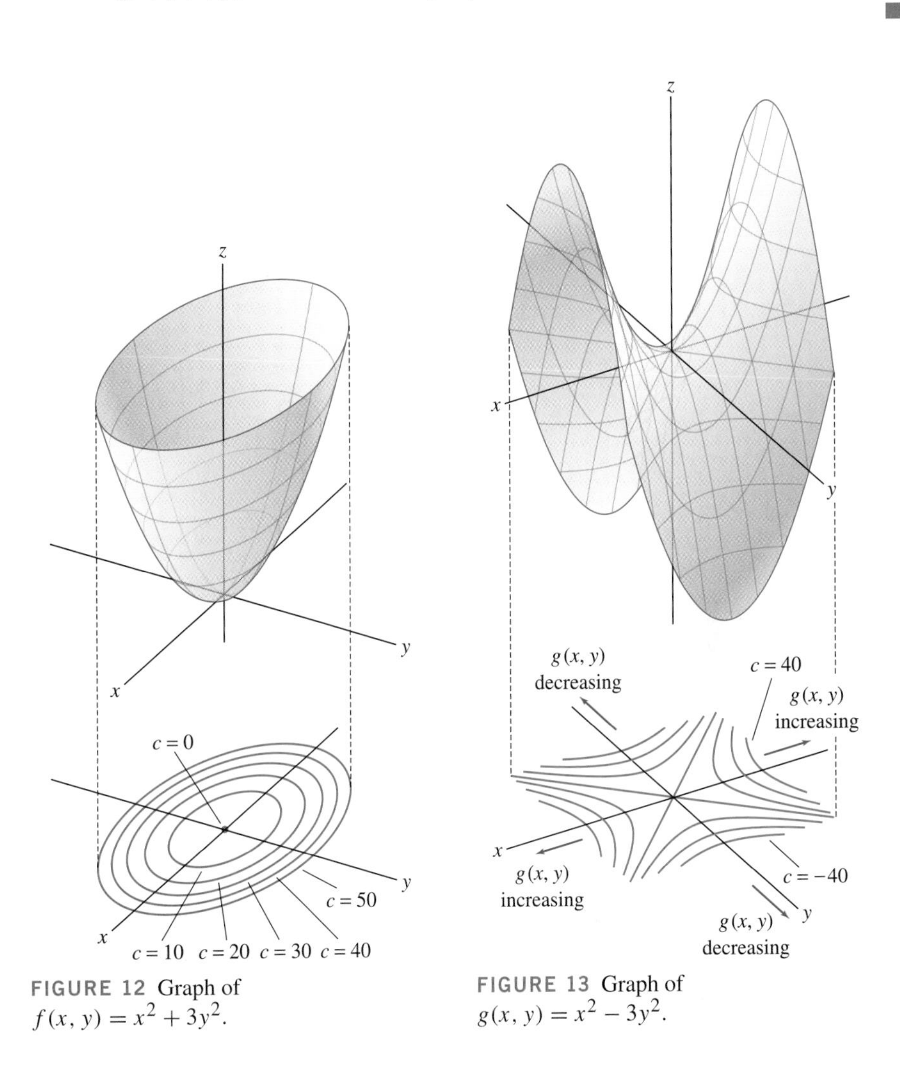

FIGURE 12 Graph of $f(x, y) = x^2 + 3y^2$.

FIGURE 13 Graph of $g(x, y) = x^2 - 3y^2$.

■ **EXAMPLE 6 Contour Map of a Saddle** CAS Draw a contour map for the function $g(x, y) = x^2 - 3y^2$ with contour interval $m = 10$.

Solution The surface $z = g(x, y)$ is a hyperbolic paraboloid and for $c \neq 0$, the level curve is the hyperbola $x^2 - 3y^2 = c$ (Figure 13). The level curve for $c = 0$ consists of the two lines $x = \pm\sqrt{3}y$ because the equation $g(x, y) = 0$ factors:

$$x^2 - 3y^2 = 0 = (x - \sqrt{3}y)(x + \sqrt{3}y) = 0$$

Notice that when you stand at the origin, $g(x, y)$ is increasing in either direction along the x-axis and decreasing in either direction along the y-axis. The resulting graph is called a *saddle* because of its saddle-like shape. ■

■ **EXAMPLE 7 Contour Map of a Linear Function** Draw a contour map for the function $f(x, y) = 12 - 2x - 3y$ with contour interval $m = 4$.

Solution The level curves of a linear function $f(x, y) = mx + ny + r$ are the straight lines with equation $mx + ny + r = c$. Therefore, *the contour map of a linear function consists of equally spaced parallel lines.* In our case, the level curves are the lines $12 - 2x - 3y = c$ or $2x + 3y = 12 - c$ (Figure 14). ■

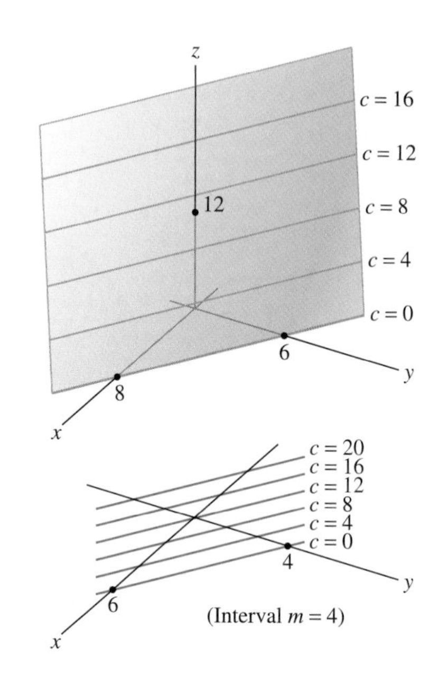

FIGURE 14

Two things to keep in mind when reading a contour map:

- *Your altitude does not change if you hike along a level curve.*
- *Your altitude increases or decreases by m (the contour interval) when you hike from one level curve to the next.*

To examine some further properties of contour maps, let's imagine that the surface $z = f(x, y)$ depicts a mountain range. We place the xy-plane at sea level, so that $f(a, b)$ is the height (also called altitude or elevation) of the mountain above sea level at the point above (a, b) in the plane. The contour map in Figure 15 shows the elevation of the Mount Whitney range in California.

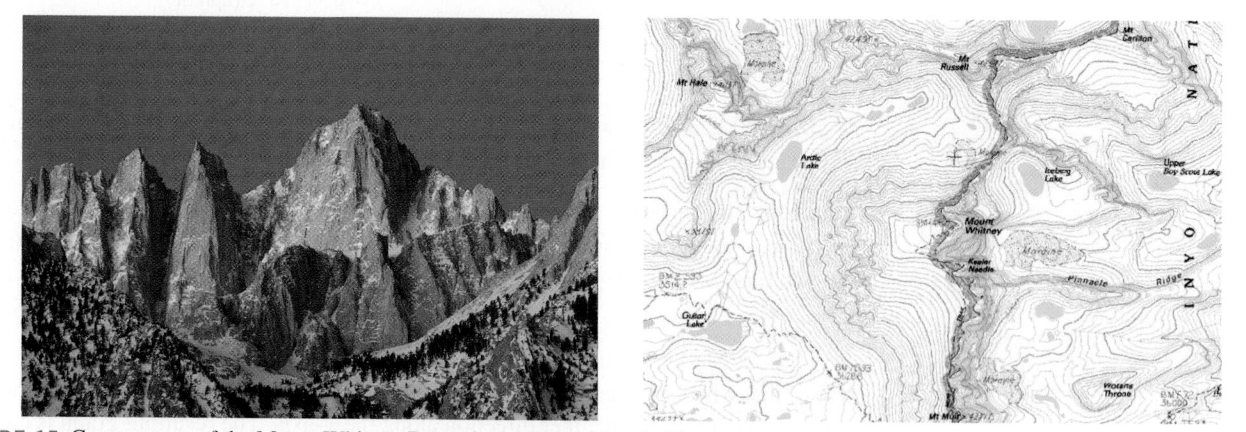

FIGURE 15 Contour map of the Mount Whitney Range in the Sierra Nevada Mountains.

Consider the "mountain" in Figure 16. To measure steepness quantitatively, we define the **average rate of change (ROC)** between points P and Q:

$$\text{Average ROC from } P \text{ to } Q = \frac{\Delta \text{ altitude}}{\Delta \text{ horizontal distance}}$$

Keep in mind that P and Q are points in the domain of f in the xy-plane and the change in altitude is the difference in heights of the points P' and Q' on the mountain (graph) that lie above P and Q.

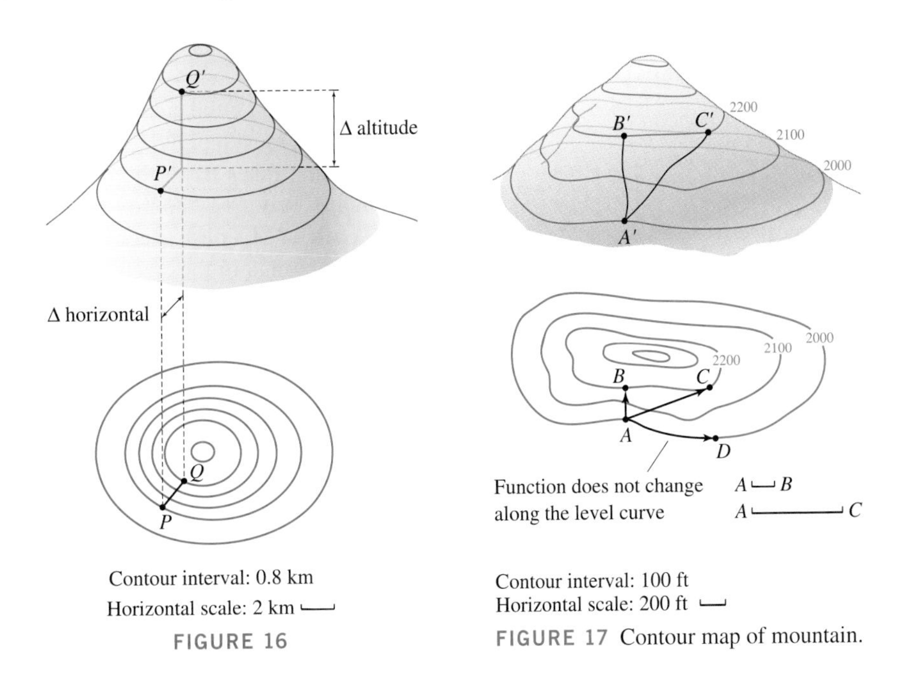

Contour interval: 0.8 km
Horizontal scale: 2 km

FIGURE 16

Contour interval: 100 ft
Horizontal scale: 200 ft

FIGURE 17 Contour map of mountain.

EXAMPLE 8 Calculating Average ROC Calculate the average ROC of $f(x, y)$ from P to Q for the function whose contour map is shown in Figure 16.

Solution The segment $\overline{PQ}$ spans three level curves and the contour interval is $m = 0.8$, so the change in altitude from P' to Q' is $3(0.8) = 2.4$ km. From the horizontal scale of the contour map, we see that the horizontal distance from P to Q is 2 km, so

$$\text{Average ROC from } P \text{ to } Q = \frac{\Delta \text{ altitude}}{\Delta \text{ horizontal distance}} = \frac{3(0.8)}{2} = 1.2$$

CONCEPTUAL INSIGHT In single-variable calculus, we measure the ROC by the derivative $f'(a)$. In the multivariable case, there is no single ROC because the change in $f(x, y)$ depends on the direction. For example, $f(x, y)$ is constant along level curves so its ROC along a level curve is zero. On the other hand, the ROC is nonzero in a direction pointing from one level curve to the next (Figure 17). We will discuss this again when we come to directional derivatives in Section 14.5.

EXAMPLE 9 Average ROC Depends on Direction Compute the average ROC from A to the points B, C, and D in Figure 17.

Solution Both segments $\overline{AB}$ and $\overline{AC}$ span two level curves, so the change in altitude is 200 ft in both cases. The horizontal scale shows that $\overline{AB}$ is two units long, corresponding to a horizontal change of 400 ft, and $\overline{AC}$ is four units long, corresponding to a horizontal change of 800 ft. On the other hand, there is no change in altitude from A to D. Therefore:

$$\text{Average ROC from } A \text{ to } B = \frac{\Delta \text{ altitude}}{\Delta \text{ horizontal distance}} \approx \frac{200}{400} = 0.5$$

$$\text{Average ROC from } A \text{ to } C = \frac{\Delta \text{ altitude}}{\Delta \text{ horizontal distance}} \approx \frac{200}{800} = 0.25$$

$$\text{Average ROC from } A \text{ to } D = \frac{\Delta \text{ altitude}}{\Delta \text{ horizontal distance}} = 0$$

■

When we stand on the side of a mountain, there is a *steepest* direction in which the altitude increases most rapidly. On a contour map, the steepest direction is approximately the direction that takes us to the closest point on the next highest level curve [Figure 18(A)]. This is only true "approximately" because the terrain may vary between level curves. A **path of steepest ascent** is a path that begins at a point P and, everywhere along the way, points in the steepest direction. Again, we may approximate the path of steepest ascent by drawing a sequence of segments that move as directly as possible from one level curve to the next. Figure 18(B) shows two paths from P to Q. The solid path is a path of steepest ascent but the dashed path is not, because it does not move from one level curve to the next along the shortest possible segment.

We obtain a path of steepest descent *by following a path of steepest ascent in the opposite direction. Water flowing down the side of a mountain follows a path of steepest descent.*

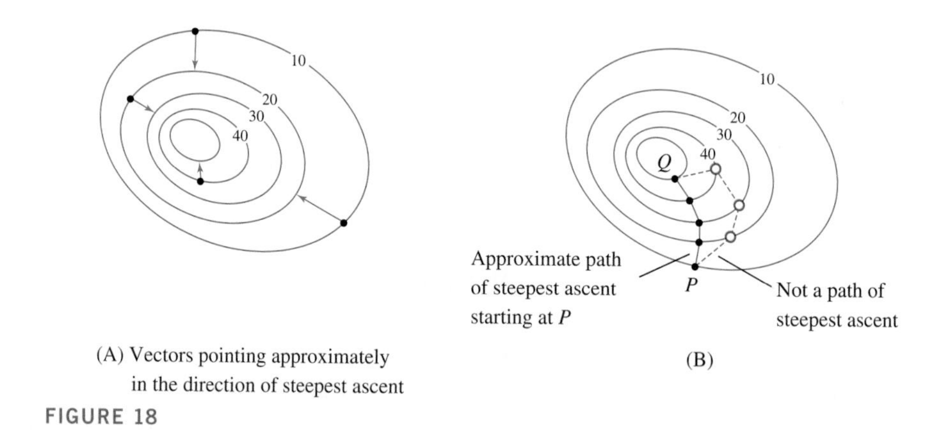

(A) Vectors pointing approximately in the direction of steepest ascent

(B)

FIGURE 18

More Than Two Variables

It is not possible to draw the graph of a function of more than two variables. The graph of a function $f(x, y, z)$ would consist of the set of points $(x, y, z, f(x, y, z))$ in four-dimensional space $\mathbf{R}^4$. However, it is possible to draw the **level surfaces** of a function of three variables $f(x, y, z)$. These are the surfaces with equation $f(x, y, z) = c$. For example, the level surfaces of

$$f(x, y, z) = x^2 + y^2 + z^2$$

are the spheres with equation $x^2 + y^2 + z^2 = c$ (Figure 19). For functions of four or more variables, we can no longer visualize the graph or the level surfaces. We must rely on intuition developed through the study of functions of two and three variables.

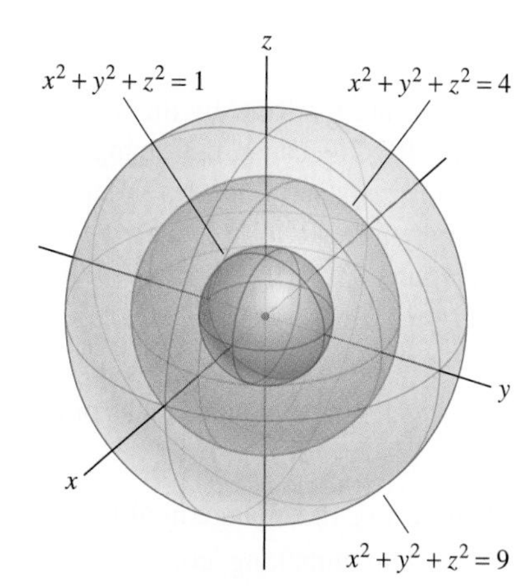

FIGURE 19 The level surfaces of $f(x, y, z) = x^2 + y^2 + z^2$ are spheres.

■ **EXAMPLE 10** Describe the level surfaces of $g(x, y, z) = x^2 + y^2 - z^2$.

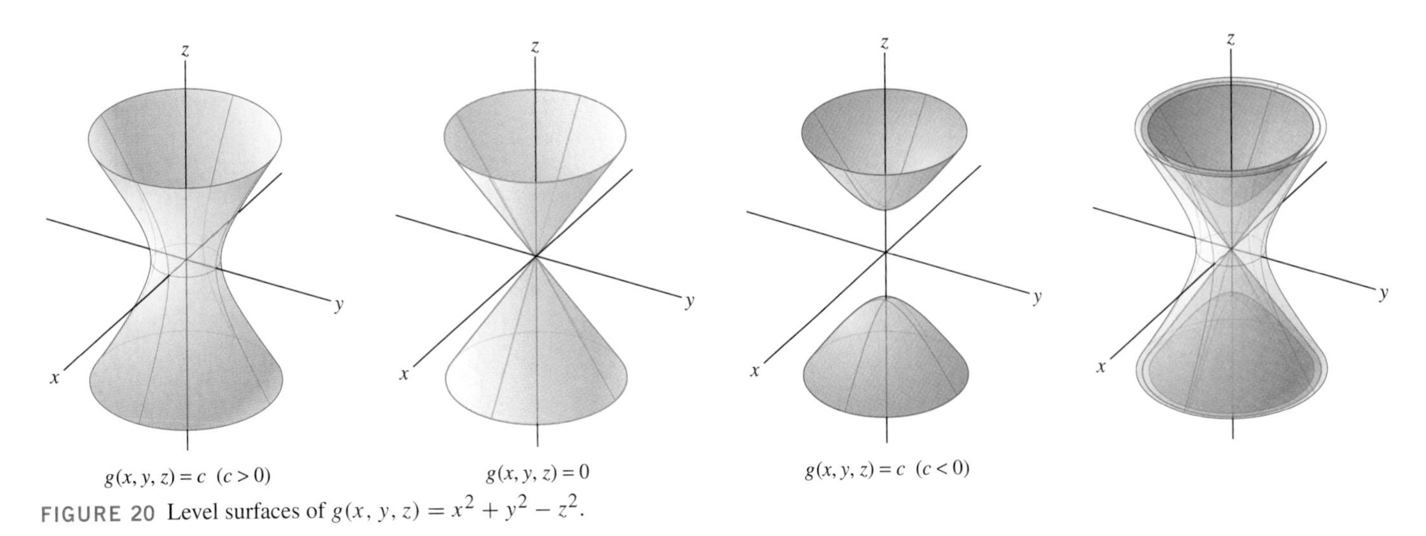

FIGURE 20 Level surfaces of $g(x, y, z) = x^2 + y^2 - z^2$.

Solution The level surface for $c = 0$ is the cone $x^2 + y^2 - z^2 = 0$. For $c \neq 0$, the level surfaces are the hyperboloids $x^2 + y^2 - z^2 = c$. The hyperboloid has one sheet if $c > 0$ and two sheets if $c < 0$ (Figure 20). ■

14.1 SUMMARY

- The domain $\mathcal{D}$ of a function $f(x_1, \dots, x_n)$ of n variables is the set of n-tuples $(a_1, \dots, a_n)$ in $\mathbf{R}^n$ for which $f(a_1, \dots, a_n)$ is defined. The range of f is the set of values taken by f.
- The graph of a continuous real-valued function $f(x, y)$ is the surface in $\mathbf{R}^3$ consisting of the points $(a, b, f(a, b))$ for (a, b) in the domain $\mathcal{D}$ of f.
- A *trace* is a curve obtained by intersecting the graph with a horizontal plane $z = c$ or a vertical plane $x = a$ or $y = b$.
- A *level curve* is a curve in the xy-plane defined by an equation $f(x, y) = c$. The level curve $f(x, y) = c$ is the projection onto the xy-plane of the trace curve at height $z = c$.
- A *contour map* shows the level curves $f(x, y) = c$ for equally spaced values of c. The spacing m is called the *contour interval*. The level curves are closer together where the graph is steeper.
- A direction of steepest ascent at a point P is a direction along which $f(x, y)$ changes most rapidly. The steepest direction is obtained (approximately) by drawing the vector from P to the nearest point on the next level curve.

14.1 EXERCISES

Preliminary Questions

1. What is the difference between a horizontal trace and a level curve? How are they related?

2. Describe the trace of $f(x, y) = x^2 - \sin(x^3 y)$ in the xz-plane.

3. Is it possible for two different level curves of a function to intersect? Explain.

4. Describe the contour map of $f(x, y) = x$ with contour interval 1.

5. How will the contour maps of $f(x, y) = x$ and $g(x, y) = 2x$ with contour interval 1 look different?

Exercises

In Exercises 1–4, evaluate the function at the specified points.

1. $f(x, y) = x + yx^3$, $(2, 2)$, $(-1, 4)$, $(6, \frac{1}{2})$

2. $g(x, y) = \dfrac{y}{x^2 + y^2}$, $(1, 3)$, $(3, -2)$

3. $h(x, y, z) = xyz^{-2}$, $(3, 8, 2)$, $(3, -2, -6)$

4. $Q(y, z) = y^2 + y \sin z$, $(y, z) = (2, \frac{\pi}{2})$, $(-2, \frac{\pi}{6})$

In Exercises 5–16, sketch the domain of the function.

5. $f(x, y) = 4x - 7y$

6. $f(x, y) = \sqrt{9 - x^2}$

7. $f(x, y) = \ln(y - 2x)$

8. $h(x, t) = \dfrac{1}{x + t}$

9. $G(x, t) = e^{1/(x+t)}$

10. $g(y, z) = \dfrac{1}{z + y^2}$

11. $f(x, y) = \sin \dfrac{y}{x}$

12. $H(r, s) = 3r^2 s^4$

13. $F(I, R) = \sqrt{IR}$

14. $f(x, y) = \cos^{-1}(x + y)$

15. $g(r, t) = \dfrac{1}{r^2 - t}$

16. $f(x, y) = \ln(4x^2 - y)$

In Exercises 17–19, describe the domain and range of the function.

17. $f(x, y, z) = xz + e^y$

18. $P(r, s) = e^{r/s}$

19. $f(x, y, z) = \sqrt{9 - x^2 - y^2 - z^2}$

20. Match the functions (a)–(f) with their graphs (A)–(F) in Figure 21.

(a) $f(x, y) = |x| + |y|$

(b) $f(x, y) = \cos(x - y)$

(c) $f(x, y) = \dfrac{-1}{1 + 9x^2 + y^2}$

(d) $f(x, y) = \cos(x^2)e^{-0.1(x^2+y^2)}$

(e) $f(x, y) = \dfrac{-1}{1 + 9x^2 + 9y^2}$

(f) $f(x, y) = \cos(x^2 + y^2)e^{-0.1(x^2+y^2)}$

In Exercises 21–30, sketch the graph and describe the vertical and horizontal traces.

21. $f(x, y) = 12 - 3x - 4y$

22. $f(x, y) = x^2 + y$

23. $f(x, y) = x^2 + 4y^2$

24. $f(x, y) = y^2$

25. $f(x, y) = \dfrac{1}{x^2 + y^2 + 1}$

26. $f(x, y) = 7$

27. $f(x, y) = x + |y|$

28. $f(x, y) = 9 - x^2 - y^2$

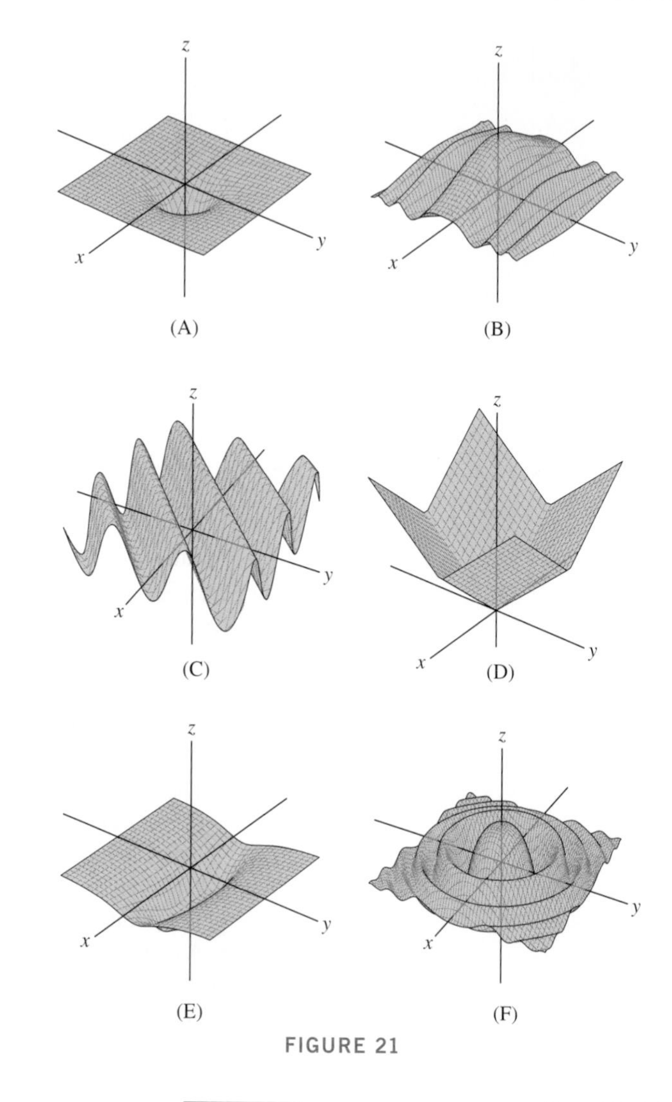

FIGURE 21

29. $f(x, y) = \sqrt{1 - x^2 - y^2}$

30. $f(x, y) = \sqrt{1 + x^2 + y^2}$

31. Sketch the contour maps of $f(x, y) = x + y$ with contour intervals $m = 1$ and 2.

32. Sketch the contour map of $f(x, y) = x^2 + y^2$ with level curves $c = 0, 4, 8, 12, 16$.

33. The function $f(x, t) = t^{-1/2}e^{-x^2/t}$, whose graph is shown in Figure 22, models the temperature along a metal bar after an intense burst of heat is applied at its center point.

(a) Sketch the graphs of the vertical traces at times $t = 0.5, 1, 1.5, 2$. What do these traces tell us about the way heat diffuses through the bar?

(b) Sketch the vertical trace at $x = 0$. Describe how temperature varies in time at the center point.

(c) Sketch the vertical trace $x = c$ for $c = \pm 0.2, \pm 0.4$. Describe how temperature varies in time at points near the center.

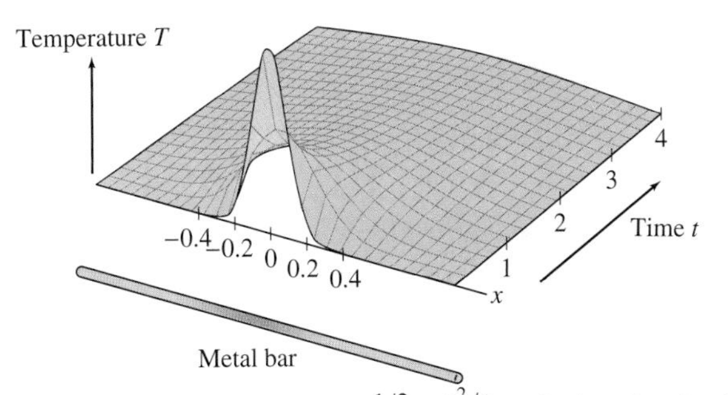

FIGURE 22 Graph of $f(x, t) = t^{-1/2}e^{-x^2/t}$ beginning shortly after $t = 0$.

In Exercises 34–43, draw a contour map of $f(x, y)$ with an appropriate contour interval, showing at least six level curves.

34. $f(x, y) = \dfrac{y}{x}$

35. $f(x, y) = e^{y/x}$

36. $f(x, y) = \dfrac{y}{x^2}$

37. $f(x, y) = xy$

38. $f(x, y) = x - y$

39. $f(x, y) = x + 2y - 1$

40. $f(x, y) = x^2 - y$

41. $f(x, y) = x^2$

42. $f(x, y) = 3x^2 - y^2$

43. $f(x, y) = \dfrac{10}{1 + x^2 + y^2}$

44. Match the contour maps (A) and (B) in Figure 23 with the two functions $f(x, y) = x - 2y$ and $g(x, y) = 2x - y$.

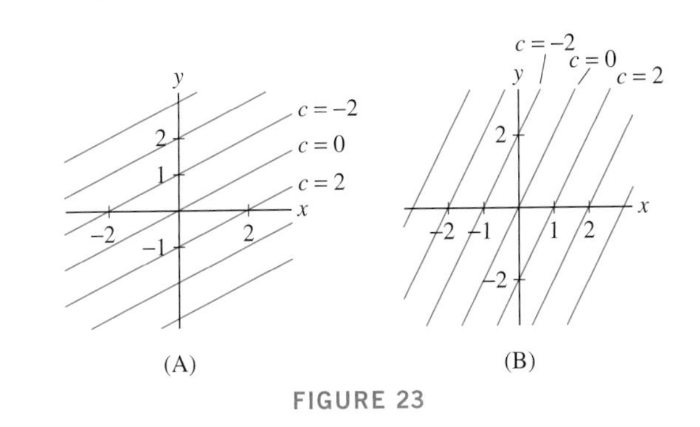

FIGURE 23

45. Which linear function has the contour map shown in Figure 24 (with level curve $c = 0$ as indicated), assuming that the contour interval is $m = 6$? What if $m = 3$?

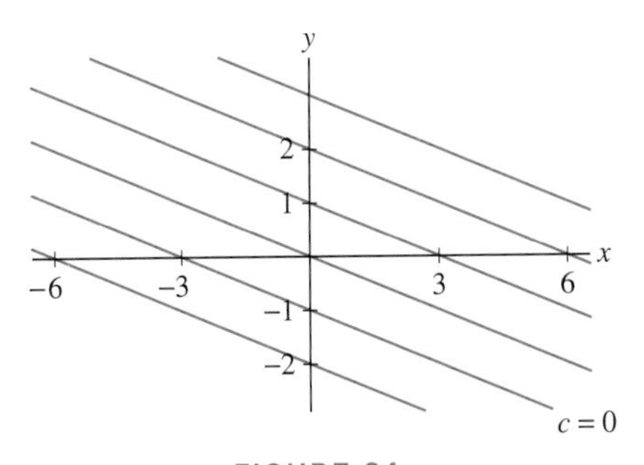

FIGURE 24

In Exercises 46–49, $f(S, T)$ denotes the density of seawater at salinity level S (parts per thousand) and temperature T (degrees Celsius). Refer to the contour map of $f(S, T)$ in Figure 25.

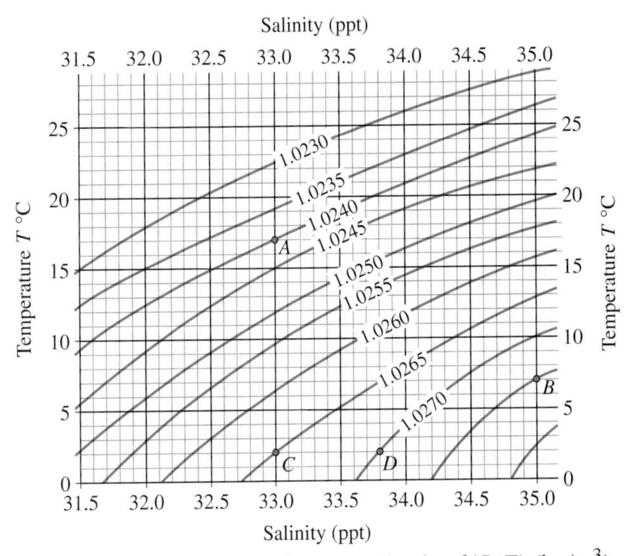

FIGURE 25 Contour map of seawater density $f(S, T)$ (kg/m^3).

46. Calculate the average ROC of density with respect to temperature from C to A.

47. Calculate the average ROC of density with respect to salinity from C to D.

48. At a fixed level of salinity, is seawater density an increasing or decreasing function of temperature?

49. Does water density appear to be more sensitive to a change in temperature at point A or point B?

In Exercises 50–53, refer to Figure 26.

50. Find the change in elevation from A and B.

51. Estimate the average ROC from A and B and from A to C.

52. Estimate the average ROC from P to points i, ii, iii, and iv.

53. Sketch the paths of steepest ascent beginning at D and E.

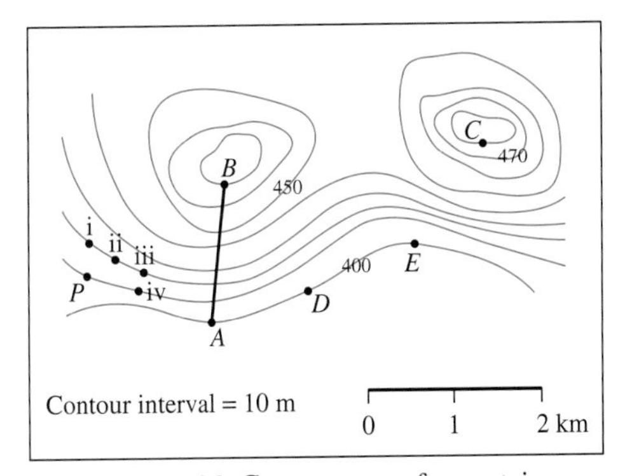

FIGURE 26 Contour map of mountain.

54. Refer to Figure 27 to answer the following questions.

(a) At which of (A)–(E) is temperature increasing in the easterly direction?

(b) At which of (A)–(E) is temperature decreasing most rapidly in the northern direction?

(c) In which direction at (B) is temperature increasing most rapidly?

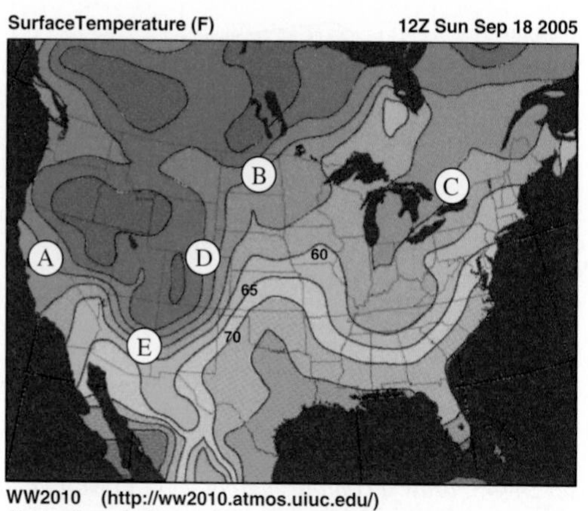

FIGURE 27

Further Insights and Challenges

55. Let $f(x, y) = \dfrac{x}{\sqrt{x^2 + y^2}}$ for $(x, y) \neq 0$. Write f as a function $f(r, \theta)$ in polar coordinates and use this to find the level curves of f.

56. CAS Use a computer algebra system to draw the graph of $f(x, y) = \dfrac{xy}{x^2 + y^2 + 1}$ and its contour map.

(a) Show that the level curve $f(x, y) = c$ has equation

$$\left(\frac{1}{4c^2} - 1\right) y^2 - \left(x - \frac{1}{2c} y\right)^2 = 1$$

(b) Show that the level curve is empty if $|c| > \frac{1}{2}$ and is a hyperbola if $|c| < \frac{1}{2}$. What is the level curve for $c = \frac{1}{2}$?

14.2 Limits and Continuity in Several Variables

In this section, we extend the basic concepts of limit and continuity to the multivariable setting, focusing on functions of two variables (the cases of three or more variables are similar).

To formulate the limit definition, we define the **open disk** $D(P, r)$ and the **punctured disk** $D^*(P, r)$ of radius r and center $P = (a, b)$ (Figure 1):

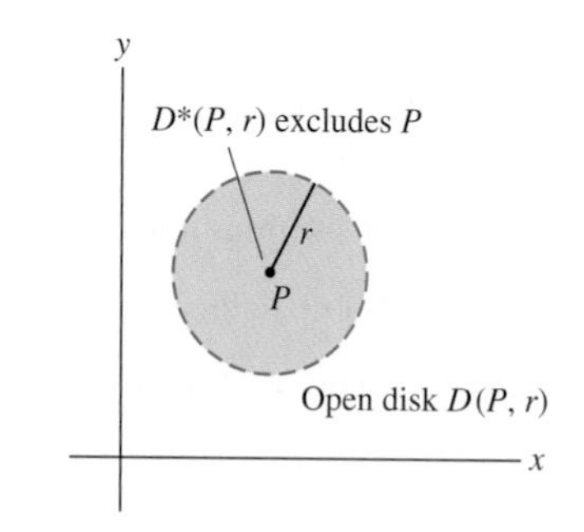

FIGURE 1 The open disk $D(P, r)$ does not include the boundary circle.

$$D(P, r) = \{(x, y) \in \mathbf{R}^2 : (x - a)^2 + (y - b)^2 < r^2\}$$

$$D^*(P, r) = \{(x, y) \in \mathbf{R}^2 : 0 < (x - a)^2 + (y - b)^2 < r^2\}$$

Observe that $D^*(P, r)$ consists of all points in $D(P, r)$ other than P itself.

Recall that in one variable, $\lim_{x \to a} f(x) = L$ if $|f(x) - L|$ is as small as desired for x sufficiently close (but not equal) to a. In two variables, we say that a function $f(x, y)$ with domain $\mathcal{D}$ approaches the limit L as (x, y) approaches $P = (a, b)$ if $|f(x, y) - L|$ is as small as desired for all (x, y) in $\mathcal{D}$ lying in a sufficiently small punctured disk centered at P [Figure 2(C)]. We write

$$\lim_{(x,y) \to P} f(x, y) = \lim_{(x,y) \to (a,b)} f(x, y) = L$$

Here is the formal ϵ-δ definition.

DEFINITION Limit Let $P = (a, b)$ be a point in the domain $\mathcal{D}$ of $f(x, y)$. Then

$$\lim_{(x,y)\to(a,b)} f(x, y) = L$$

if, for any $\epsilon > 0$, there exists $\delta > 0$ such that

$$|f(x, y) - L| < \epsilon \quad \text{for all} \quad (x, y) \in \mathcal{D} \text{ lying in } D^*(P, \delta)$$

Observe that in a one-variable limit, $f(x)$ approaches L as x approaches a from the left or right as in Figure 2(A). In a multivariable limit, $f(x, y)$ must approach L no matter how (x, y) approaches (a, b) [Figure 2(B)].

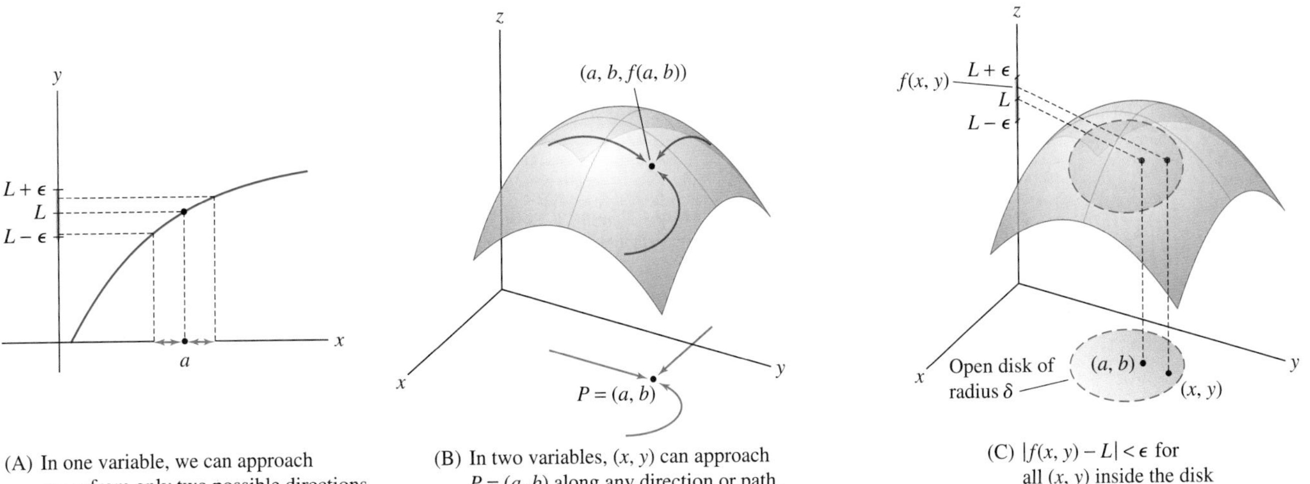

(A) In one variable, we can approach $x = a$ from only two possible directions.

(B) In two variables, (x, y) can approach $P = (a, b)$ along any direction or path.

(C) $|f(x, y) - L| < \epsilon$ for all (x, y) inside the disk

FIGURE 2

EXAMPLE 1 Show that for all (a, b), **(a)** $\lim_{(x,y)\to(a,b)} x = a$ and **(b)** $\lim_{(x,y)\to(a,b)} y = b$.

Solution To verify (a), let $f(x, y) = x$ and $L = a$. We must show that for any $\epsilon > 0$, we can find $\delta > 0$ such that

$$|f(x, y) - L| = |x - a| < \epsilon \quad \text{for all} \quad (x, y) \in D^*(P, \delta) \qquad \boxed{1}$$

But if $(x, y) \in D^*(P, \delta)$, then

$$(x - a)^2 + (y - b)^2 < \delta^2 \quad \Rightarrow \quad (x - a)^2 < \delta^2 \quad \Rightarrow \quad |x - a| < \delta$$

Therefore, (1) holds with $\delta = \epsilon$:

$$|x - a| < \epsilon \quad \text{for all} \quad (x, y) \in D^*(P, \epsilon)$$

This proves (a). The limit (b) is similar. ■

The following theorem lists the basic laws for working with limits. The proofs are similar to the proofs in the single-variable Limit Laws and are omitted.

THEOREM 1 Limit Laws Assume that $\lim_{(x,y)\to(a,b)} f(x, y)$ and $\lim_{(x,y)\to(a,b)} g(x, y)$ exist. Then:

(i) Sum Law:

$$\lim_{(x,y)\to(a,b)} (f(x, y) + g(x, y)) = \lim_{(x,y)\to(a,b)} f(x, y) + \lim_{(x,y)\to(a,b)} g(x, y)$$

(ii) Constant Multiple Law: For any number k,

$$\lim_{(x,y)\to(a,b)} kf(x, y) = k \lim_{(x,y)\to(a,b)} f(x, y)$$

(iii) Product Law:

$$\lim_{(x,y)\to(a,b)} f(x, y)\, g(x, y) = \left(\lim_{(x,y)\to(a,b)} f(x, y)\right)\left(\lim_{(x,y)\to(a,b)} g(x, y)\right)$$

(iv) Quotient Law: If $\lim_{(x,y)\to(a,b)} g(x, y) \neq 0$, then

$$\lim_{(x,y)\to(a,b)} \frac{f(x, y)}{g(x, y)} = \frac{\lim_{(x,y)\to(a,b)} f(x, y)}{\lim_{(x,y)\to(a,b)} g(x, y)}$$

Continuity for functions of two variables is defined by imitating the single-variable case: f is continuous at (a, b) if $f(x, y)$ approaches the function value $f(a, b)$ as $(x, y) \to (a, b)$.

DEFINITION Continuity A function $f(x, y)$ is **continuous** at a point $P = (a, b)$ in its domain if

$$\lim_{(x,y)\to(a,b)} f(x, y) = f(a, b)$$

We say that f is continuous on a set $\mathcal{D}$ if f is continuous at each point of $\mathcal{D}$ (or simply continuous if $\mathcal{D}$ is the domain of f).

It follows from the Limit Laws that arbitrary sums, multiples, and products of continuous functions are continuous. For instance, $f(x, y) = x$ and $g(x, y) = y$ are continuous by Example 1, so we may conclude that the power functions $f(x, y) = x^m y^n$ are continuous for all whole numbers m, n. We may also conclude that polynomials and rational functions [quotients $P(x, y)/Q(x, y)$, where P and Q are polynomials] are continuous at all points (a, b) where $Q(a, b) \neq 0$. As in one variable, we may evaluate limits of continuous functions using substitution.

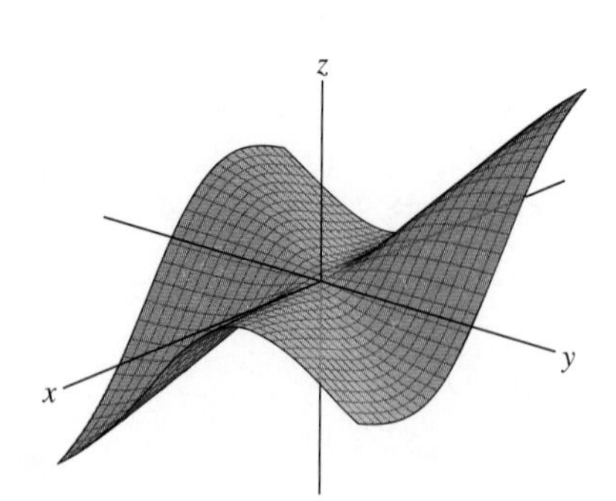

FIGURE 3 Graph of $f(x, y) = \dfrac{2x^2y - 3xy^2}{x^2 + y^2 + 1}$.

■ **EXAMPLE 2 Evaluating Limits by Substitution** Show that $f(x, y) = \dfrac{2x^2y - 3xy^2}{x^2 + y^2 + 1}$ is continuous. Evaluate $\lim_{(x,y)\to(1,2)} f(x, y)$.

Solution The function $f(x, y)$ is continuous everywhere because it is a rational function whose denominator $Q(x, y) = x^2 + y^2 + 1$ is never zero (Figure 3). We evaluate the limit using substitution:

$$\lim_{(x,y)\to(1,2)} \frac{2x^2y - 3xy^2}{x^2 + y^2 + 1} = \frac{2(1^2)(2) - 3(1)(2^2)}{1^2 + 2^2 + 1} = -\frac{4}{3}$$ ■

Consider the special case that $f(x, y)$ is a product $f(x, y) = h(x)g(y)$, where $h(x)$ and $g(y)$ are continuous. By the Product Law, the limit is equal to a product:

$$\lim_{(x,y)\to(a,b)} f(x, y) = \lim_{(x,y)\to(a,b)} h(x)g(y) = \left(\lim_{x\to a} h(x)\right)\left(\lim_{y\to b} g(y)\right)$$

■ **EXAMPLE 3** **Product of Functions of One Variable** Evaluate $\displaystyle\lim_{(x,y)\to(-1,1)} e^x \tan^{-1} y$.

Solution We evaluate the limit as a product of limits:

$$\lim_{(x,y)\to(-1,1)} e^x \tan^{-1} y = \left(\lim_{x\to -1} e^x\right)\left(\lim_{y\to 1} \tan^{-1} y\right) = e^{-1}(\tan^{-1} 1) = \frac{e^{-1}\pi}{4}$$ ■

Composition is another important way to build functions. If $f(x, y)$ is a function of two variables and $G(u)$ a function of one variable, then the composite $G \circ f$ is the function $G(f(x, y))$. According to the next theorem, a composite of continuous functions is again continuous (the proof is omitted).

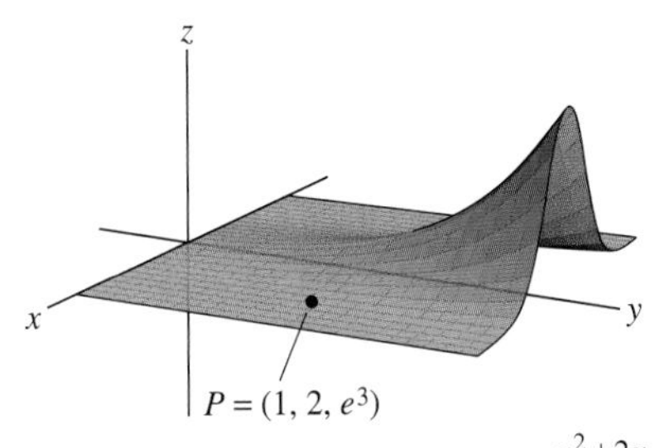

FIGURE 4 Graph of $H(x, y) = e^{-x^2+2y}$.

> **THEOREM 2 A Composite of Continuous Functions Is Continuous** If $f(x, y)$ is continuous at (a, b) and $G(u)$ is continuous at $c = f(a, b)$, then the composite function $G(f(x, y))$ is continuous at (a, b).

■ **EXAMPLE 4** Write $H(x, y) = e^{-x^2+2y}$ as a composite function and evaluate

$$\lim_{(x,y)\to(1,2)} H(x, y)$$

Solution We have $H(x, y) = G \circ f$, where $G(u) = e^u$ and $f(x, y) = -x^2 + 2y$ (Figure 4). Since G and f (and hence H) are continuous, we may evaluate the limit using substitution:

$$\lim_{(x,y)\to(1,2)} H(x, y) = \lim_{(x,y)\to(1,2)} e^{-x^2+2y} = e^{-(1)^2+2(2)} = e^3$$ ■

We have emphasized that if $\displaystyle\lim_{(x,y)\to(a,b)} f(x, y) = L$, then $f(x, y)$ must tend to L as (x, y) approaches (a, b) along any path. In the next example, we prove that a limit *does not exist* by showing that $f(x, y)$ approaches *different limits* along lines through the origin (Figure 5).

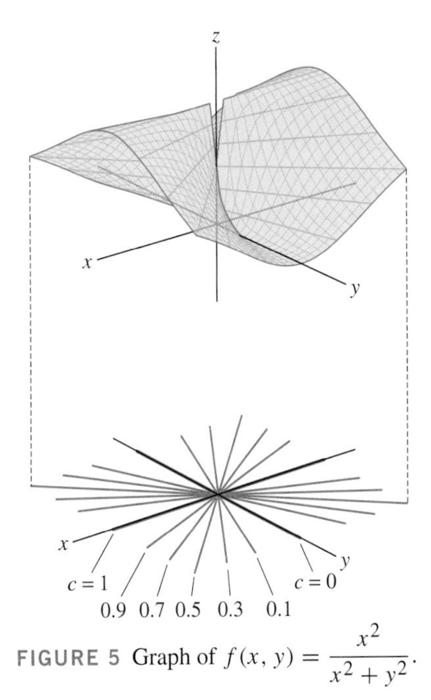

FIGURE 5 Graph of $f(x, y) = \dfrac{x^2}{x^2 + y^2}$.

■ **EXAMPLE 5** **Showing a Limit Does Not Exist** Examine $\displaystyle\lim_{(x,y)\to(0,0)} \frac{x^2}{x^2 + y^2}$ numerically. Prove that the limit does not exist.

Solution First, let's examine a table of values of $f(x, y) = \dfrac{x^2}{x^2 + y^2}$. Table 1 suggests that $f(x, y)$ takes on all values between 0 and 1, no matter how close (x, y) gets to $(0, 0)$.

TABLE 1 Values of $f(x,y) = \dfrac{x^2}{x^2+y^2}$

y \ x	−0.5	−0.4	−0.3	−0.2	−0.1	0	0.1	0.2	0.3	0.4	0.5
0.5	**0.5**	0.39	0.265	0.138	0.038	**0**	0.038	0.138	0.265	0.39	**0.5**
0.4	0.61	**0.5**	0.36	0.2	0.059	**0**	0.059	0.2	0.36	**0.5**	0.61
0.3	0.735	0.64	**0.5**	0.308	0.1	**0**	0.1	0.308	**0.5**	0.64	0.735
0.2	0.862	0.8	0.692	**0.5**	0.2	**0**	0.2	**0.5**	0.692	0.8	0.862
0.1	0.962	0.941	0.9	0.8	**0.5**	**0**	**0.5**	0.8	0.9	0.941	0.962
0	**1**	**1**	**1**	**1**	**1**		**1**	**1**	**1**	**1**	**1**
−0.1	0.962	0.941	0.9	0.8	**0.5**	**0**	**0.5**	0.8	0.9	0.941	0.962
−0.2	0.862	0.8	0.692	**0.5**	0.2	**0**	0.2	**0.5**	0.692	0.8	0.862
−0.3	0.735	0.640	**0.5**	0.308	0.1	**0**	0.1	0.308	**0.5**	0.640	0.735
−0.4	0.610	**0.5**	0.360	0.2	0.059	**0**	0.059	0.2	0.36	**0.5**	0.61
−0.5	**0.5**	0.39	0.265	0.138	0.038	**0**	0.038	0.138	0.265	0.390	**0.5**

For example,

$$f(0.1, 0) = 1, \qquad f(0.1, 0.1) = 0.5, \qquad f(0, 0.1) = 0$$

Thus, $f(x, y)$ does not seem to approach any fixed value L as $(x, y) \to (0, 0)$.

Now let's prove that the limit does not exist by showing that $f(x, y)$ approaches different limits along the x- and y-axes (Figure 5). Since $f(x, 0) = x^2/x^2 = 1$ for $x \neq 0$ and $f(0, y) = 0/y^2 = 0$ for $y \neq 0$,

$$\text{Limit along } x\text{-axis} = \lim_{x\to 0} f(x, 0) = \lim_{x\to 0} 1 = 1$$

$$\text{Limit along } y\text{-axis} = \lim_{y\to 0} f(0, y) = \lim_{y\to 0} 0 = 0$$

These two limits are different and hence $\displaystyle\lim_{(x,y)\to(0,0)} f(x, y)$ does not exist. ■

GRAPHICAL INSIGHT To better understand the behavior of $f(x, y) = \dfrac{x^2}{x^2+y^2}$ near $(0, 0)$, observe that f is constant on each line $y = mx$:

$$f(x, mx) = \frac{x^2}{x^2 + (mx)^2} = \frac{1}{m^2+1} \qquad (\text{for } x \neq 0)$$

In other words, the lines through the origin $y = mx$ are the level curves of f (Figure 6). As m varies, $(m^2+1)^{-1}$ varies between 0 and 1, and thus, $f(x, y)$ takes on all values between 0 and 1 in every disk around the origin $(0, 0)$, no matter how small. This shows in greater detail that there is no number L such that $|f(x, y) - L|$ is arbitrarily small for all (x, y) close to $(0, 0)$.

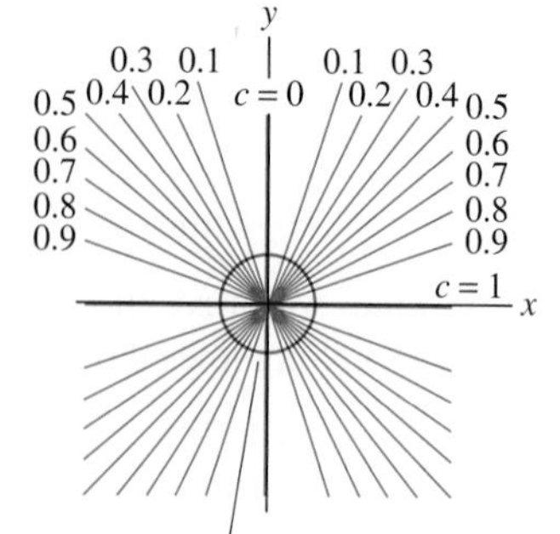

FIGURE 6 Contour map of $f(x, y) = \dfrac{x^2}{x^2+y^2}$.

As we know from single-variable calculus, there is no general method for computing limits. The next example shows that polar coordinates are useful in some cases.

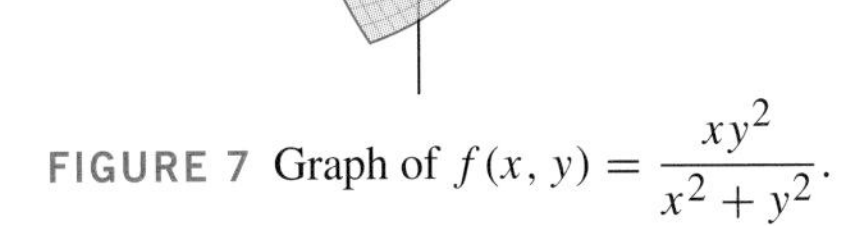

FIGURE 7 Graph of $f(x, y) = \dfrac{xy^2}{x^2 + y^2}$.

EXAMPLE 6 Two Methods for Verifying a Limit Prove that the following function is continuous:

$$f(x, y) = \begin{cases} \dfrac{xy^2}{x^2 + y^2} & \text{for } (x, y) \neq (0, 0) \\ 0 & \text{for } (x, y) = (0, 0) \end{cases}$$

Solution The denominator of f is nonzero for $(x, y) \neq (0, 0)$, so f is continuous on the domain $\{(x, y) \neq (0, 0)\}$ (Figure 7). To prove that $f(x, y)$ is also continuous at $(0, 0)$, we must verify that

$$\lim_{(x,y)\to(0,0)} \frac{xy^2}{x^2 + y^2} = f(0, 0) = 0$$

First Method We use the inequality

$$|xy^2| = |x|y^2 \leq |x|x^2 + |x|y^2 \leq |x|(x^2 + y^2)$$

The inequality holds because $|x|x^2 \geq 0$. For $(x, y) \neq (0, 0)$, division by $(x^2 + y^2)$ yields

$$0 \leq \left| \frac{xy^2}{x^2 + y^2} \right| \leq |x|$$

Now use the Squeeze Theorem (which is valid for limits in several variables):

$$\lim_{(x,y)\to(0,0)} 0 \leq \lim_{(x,y)\to(0,0)} \left| \frac{xy^2}{x^2 + y^2} \right| \leq \lim_{(x,y)\to(0,0)} |x| = 0$$

We conclude that $f(x, y)$ approaches $f(0, 0) = 0$ as $(x, y) \to (0, 0)$.

Second Method Use polar coordinates: $x = r\cos\theta$, $y = r\sin\theta$. Then $x^2 + y^2 = r^2$ and for $r \neq 0$,

$$\left| \frac{xy^2}{x^2 + y^2} \right| = \left| \frac{(r\cos\theta)(r^2\sin^2\theta)}{r^2} \right| = r|\cos\theta \sin^2\theta| \leq r$$

As (x, y) approaches $(0, 0)$, the variable r also approaches 0, so again, the Squeeze Theorem yields

$$0 \leq \lim_{(x,y)\to(0,0)} \left| \frac{xy^2}{x^2 + y^2} \right| \leq \lim_{r\to 0} r = 0$$ ■

14.2 SUMMARY

- The *open disk* of radius r centered at $P = (a, b)$ is defined by

$$D(P, r) = \{Q = (x, y) \in \mathbf{R}^2 : (x - a)^2 + (y - b)^2 < r^2\}$$

The *punctured disk* $D^*(P, r)$ is $D(P, r)$ with the point P removed.

- Suppose $P = (a, b)$ belongs to the domain $\mathcal{D}$ of $f(x, y)$. Then

$$\lim_{(x,y)\to(a,b)} f(x, y) = L$$

if, for any $\epsilon > 0$, there exists $\delta > 0$ such that

$$|f(x, y) - L| < \epsilon \quad \text{for all} \quad (x, y) \in \mathcal{D} \text{ lying in } D^*(P, \delta)$$

- A function $f(x, y)$ is *continuous* at $P = (a, b)$ if

$$\lim_{(x,y)\to(a,b)} f(x, y) = f(a, b)$$

14.2 EXERCISES

Preliminary Questions

1. What is the difference between $D(P, r)$ and $D^*(P, r)$?

2. Suppose that $f(x, y)$ is continuous at (2, 3) and that $f(2, y) = y^3$ for $y \neq 3$. What is the value $f(2, 3)$?

3. Suppose that $Q(x, y)$ is a function such that $\frac{1}{Q(x, y)}$ is continuous for all (x, y). Which of the following statements are true?

(a) $Q(x, y)$ is continuous at all (x, y).

(b) $Q(x, y)$ is continuous for $(x, y) \neq (0, 0)$.

(c) $Q(x, y) \neq 0$ for all (x, y).

4. Suppose that $f(x, 0) = 3$ for all $x \neq 0$ and $f(0, y) = 5$ for all $y \neq 0$. What can you conclude about $\lim_{(x,y)\to(0,0)} f(x, y)$?

Exercises

In Exercises 1–10, use continuity to evaluate the limit.

1. $\lim_{(x,y)\to(1,2)} (x^2 + y)$

2. $\lim_{(x,y)\to(\frac{4}{9},\frac{2}{9})} \frac{x}{y}$

3. $\lim_{(x,y)\to(3,-1)} (3x^2y - 2xy^3)$

4. $\lim_{(x,y)\to(-2,4)} \frac{x^2 - 3y^2}{4x + y}$

5. $\lim_{(x,y)\to(1,2)} \frac{x^2 + y}{x + y^2}$

6. $\lim_{(x,y)\to(\frac{\pi}{4},0)} \tan x \cos y$

7. $\lim_{(x,y)\to(4,4)} \arctan \frac{y}{x}$

8. $\lim_{(x,y)\to(0,0)} e^{x^2-y^2}$

9. $\lim_{(x,y)\to(1,1)} \frac{e^{x^2} - e^{-y^2}}{x + y}$

10. $\lim_{(x,y)\to(1,0)} \ln(x - y)$

In Exercises 11–18, evaluate the limit or determine that it does not exist.

11. $\lim_{(x,y)\to(0,1)} \frac{x}{y}$

12. $\lim_{(x,y)\to(1,0)} \frac{x}{y}$

13. $\lim_{(x,y)\to(1,1)} e^{xy} \ln(1 + xy)$

14. $\lim_{(x,y)\to(0,0)} \frac{x^2 + y^2}{1 + y^2}$

15. $\lim_{(x,y)\to(-1,-2)} x^2|y|^3$

16. $\lim_{(x,y)\to(-1,-2)} \frac{xy^2}{|x|}$

17. $\lim_{(x,y)\to(4,2)} \frac{y - 2}{\sqrt{x^2 - 4}}$

18. $\lim_{(x,y)\to(\pi,0)} \frac{\sin x}{\sin y}$

In Exercises 19–22, assume that

$$\lim_{(x,y)\to(2,5)} f(x, y) = 3, \qquad \lim_{(x,y)\to(2,5)} g(x, y) = 7$$

19. $\lim_{(x,y)\to(2,5)} \big(f(x, y) + 4g(x, y)\big)$

20. $\lim_{(x,y)\to(2,5)} f(x, y)g(x, y)^2$

21. $\lim_{(x,y)\to(2,5)} e^{f(x,y)}$

22. $\lim_{(x,y)\to(2,5)} \ln\big(g(x, y) - 2f(x, y)\big)$

In Exercises 23–32, evaluate the limit or determine that the limit does not exist. You may evaluate the limit of a product function as a product of limits as in Example 3.

23. $\lim_{(x,y)\to(0,0)} \frac{(\sin x)(\sin y)}{xy}$

24. $\lim_{(x,y)\to(2,1)} e^{x^2-y^2}$

25. $\lim_{(z,w)\to(-1,2)} (z^2w - 9z)$

26. $\lim_{(x,y)\to(0,0)} \sin x \cos \frac{1}{y}$

27. $\lim_{(h,k)\to(2,0)} h^4 \frac{(2 + k)^2 - 4}{k}$

28. $\lim_{(h,k)\to(0,0)} \frac{h(k^2 + 4)}{k}$

29. $\lim_{(x,y)\to(0,0)} e^{1/x} \tan^{-1} \frac{1}{y}$

30. $\lim_{(x,k)\to(1,0)} x^2 \left(\frac{e^k - 1}{k}\right)$

31. $\lim_{(x,y)\to(0,0)} x \ln y$

32. $\lim_{(x,y)\to(0,0)} \left(\frac{1}{2xy} - \frac{1}{xy(x + 2)}\right)$

33. Let $f(x, y) = \frac{x^3 + y^3}{x^2 + y^2}$.

(a) Show that

$$|x^3| \le |x|(x^2+y^2), \quad |y^3| \le |y|(x^2+y^2)$$

(b) Show that $|f(x,y)| \le |x|+|y|$.
(c) Use (b) and the formal definition of the limit to prove that $\lim_{(x,y)\to(0,0)} f(x,y) = 0$.
(d) Verify the conclusion of (c) again using polar coordinates as in Example 6.

34. Show that $\displaystyle\lim_{(x,y)\to(0,0)} \frac{y^2}{x^2+y^2}$ does not exist.

35. Show that $\displaystyle\lim_{(x,y)\to(0,0)} \frac{x}{x+y}$ does not exist. *Hint:* Consider the limits along the lines $y = mx$.

36. Show that $\displaystyle\lim_{(x,y)\to(0,0)} \frac{x}{\sqrt{x^2+y^2}}$ does not exist. *Hint:* Use polar coordinates.

37. Let $a, b \ge 0$. Show that $\displaystyle\lim_{(x,y)\to(0,0)} \frac{x^a y^b}{x^2+y^2} = 0$ if $a+b > 2$ and the limit does not exist if $a + b \le 2$.

38. Evaluate $\displaystyle\lim_{(x,y)\to(0,0)} \frac{x^2+y^2}{\sqrt{x^2+y^2+1}-1}$.

39. Figure 8 shows the contour maps of two functions with contour interval $m = 2$. Explain why the limit $\lim_{(x,y)\to P} f(x,y)$ does not exist. Does $\lim_{(x,y)\to P} g(x,y)$ appear to exist in (B)? If so, what is its limit?

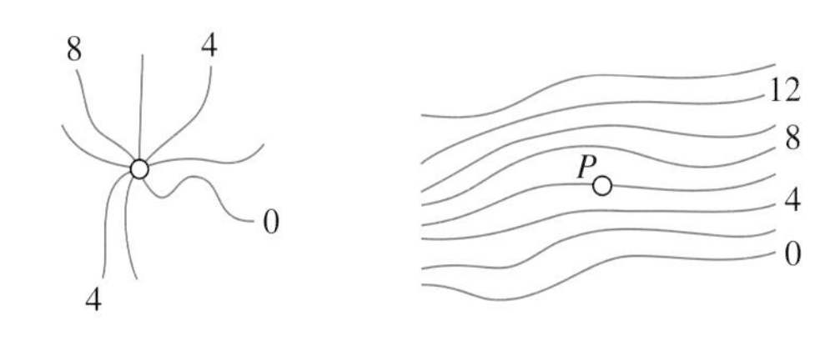

(A) Contour map of $f(x, y)$ (B) Contour map of $g(x, y)$

FIGURE 8

Further Insights and Challenges

40. Evaluate $\displaystyle\lim_{(x,y)\to(0,2)} (1+x)^{y/x}$.

41. CAS The function $f(x,y) = \dfrac{\sin(xy)}{xy}$ is defined for $xy \ne 0$.
(a) Is it possible to extend the domain of $f(x, y)$ to all of $\mathbf{R}^2$ so that the result is a continuous function?
(b) Use a computer algebra system to plot $f(x, y)$. Does the result support your conclusion in (a)?

42. CAS Repeat Exercise 41 for the functions $f(x,y) = \dfrac{xy}{x+y}$ and $g(x,y) = \dfrac{xy}{x^2+y^2}$.

43. The function $f(x,y) = \dfrac{x^2y}{x^4+y^2}$ provides an interesting example where the limit as $(x, y) \to (0, 0)$ does not exist, even though the limit along every line $y = mx$ exists and is zero (Figure 9).
(a) Show that the limit along any line $y = mx$ exists and is equal to 0.
(b) Calculate $f(x, y)$ at the points $(10^{-1}, 10^{-2})$, $(10^{-5}, 10^{-10})$, $(10^{-20}, 10^{-40})$. Do not use a calculator.
(c) Show that $\lim_{(x,y)\to(0,0)} f(x,y)$ does not exist. *Hint:* Compute the limit along the parabola $y = x^2$.

44. Is the following function continuous?

$$f(x,y) = \begin{cases} x^2+y^2 & \text{if } x^2+y^2 < 1 \\ 1 & \text{if } x^2+y^2 \ge 1 \end{cases}$$

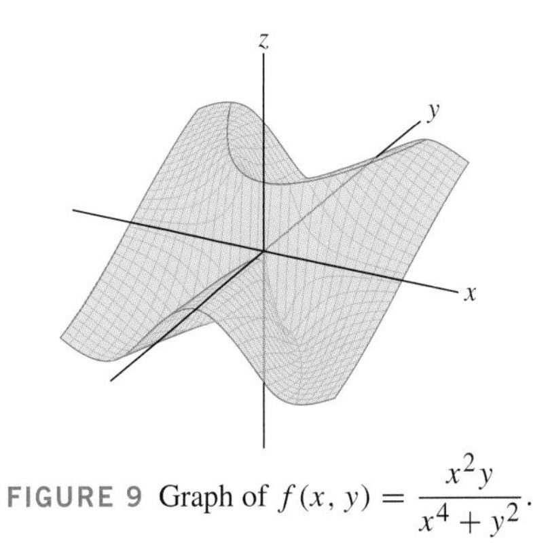

FIGURE 9 Graph of $f(x,y) = \dfrac{x^2y}{x^4+y^2}$.

14.3 Partial Derivatives

In calculus of one variable, the derivative $f'(a)$ is the rate of change of $f(x)$ at $x = a$. By contrast, a function f of two or more variables does not have a unique rate of change because each variable may affect f in different ways. For example, the current I in a circuit is a function of both voltage V and resistance R as given by Ohm's Law:

$$I(V, R) = \frac{V}{R}$$

The current I is *increasing* as function of V but *decreasing* as a function of R.

The **partial derivatives** are the rates of change with respect to each variable separately. A function $f(x, y)$ of two variables has two partial derivatives, denoted f_x and f_y, if each of the following limits exist:

$$f_x(a, b) = \lim_{h \to 0} \frac{f(a+h, b) - f(a, b)}{h}, \qquad f_y(a, b) = \lim_{k \to 0} \frac{f(a, b+k) - f(a, b)}{k}$$

The partial derivative symbol ∂ is a rounded "d." The symbols $\frac{\partial f}{\partial x}$ and $\frac{\partial f}{\partial y}$ are read as follows: "dee-eff dee-ex" and "dee-eff dee-why."

Thus, f_x is the derivative at $f(x, b)$ as a function of x alone and f_y is the derivative at $f(a, y)$ as a function of y alone. The Leibniz notation for partial derivatives is

$$\frac{\partial f}{\partial x} = f_x, \quad \frac{\partial f}{\partial y} = f_y \qquad \text{and} \qquad \left.\frac{\partial f}{\partial x}\right|_{(a,b)} = f_x(a, b), \quad \left.\frac{\partial f}{\partial y}\right|_{(a,b)} = f_y(a, b)$$

If $z = f(x, y)$, then we also write $\frac{\partial z}{\partial x}$ and $\frac{\partial z}{\partial y}$.

Partial derivatives are computed just like ordinary derivatives in one variable with this difference: To compute f_x, treat y as a constant, and to compute f_y, treat x as a constant.

EXAMPLE 1 Compute the partial derivatives of $f(x, y) = x^2y^5$.

Solution

$$\frac{\partial f}{\partial x} = \underbrace{\frac{\partial}{\partial x}(x^2y^5) = y^5\frac{\partial}{\partial x}(x^2)}_{\text{Treat } y^5 \text{ as a constant}} = y^5(2x) = 2xy^5$$

$$\frac{\partial f}{\partial y} = \underbrace{\frac{\partial}{\partial y}(x^2y^5) = x^2\frac{\partial}{\partial x}(y^5)}_{\text{Treat } x^2 \text{ as a constant}} = x^2(5y^4) = 5x^2y^4$$

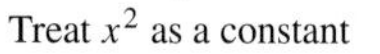

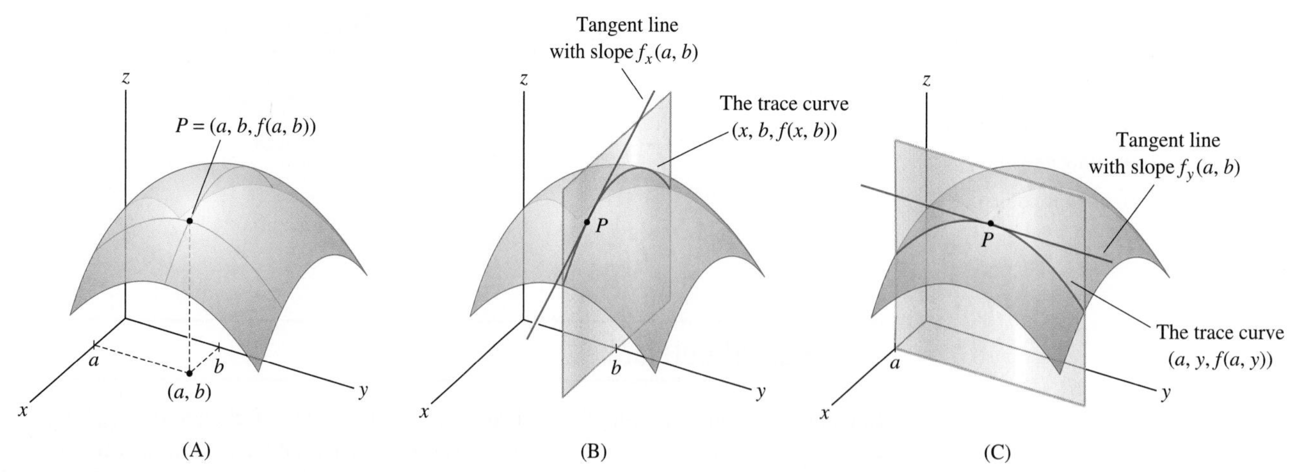

FIGURE 1 Graph of $z = f(x, y)$ and its vertical trace curves.

GRAPHICAL INSIGHT The partial derivatives at $P = (a, b)$ are the slopes of the tangent lines to the vertical trace curves passing through the point $(a, b, f(a, b))$ on the graph of $z = f(x, y)$ [Figure 1(A)]. The trace curve in the vertical plane $y = b$ is the set of points $(x, b, f(x, b))$, and $f_x(a, b)$ is the slope of the tangent line to this curve at $x = a$ [Figure 1(B)]. Similarly, $f_y(a, b)$ is the slope of the tangent line at $y = b$ to the trace curve obtained by intersecting with the vertical plane $x = a$ [Figure 1(C)].

All of the differentiation rules from calculus of one variable are valid for partial derivatives.

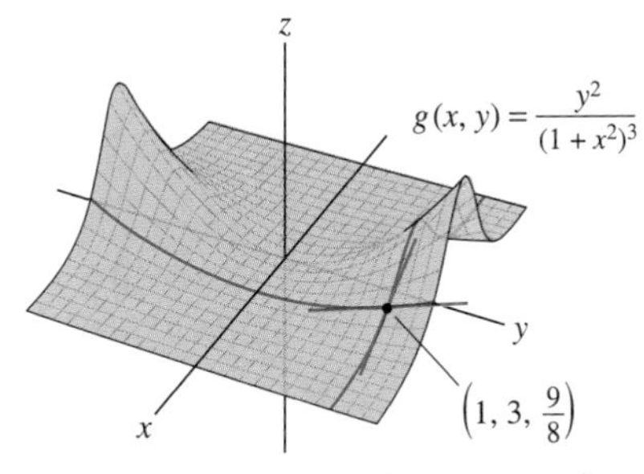

FIGURE 2 The slopes of the tangent lines to the trace curves are $g_x(1, 3)$ and $g_y(1, 3)$.

■ **EXAMPLE 2** Calculate $g_x(1, 3)$ and $g_y(1, 3)$, where $g(x, y) = \dfrac{y^2}{(1 + x^2)^3}$.

Solution To calculate $\dfrac{\partial g}{\partial x}$, treat y (and therefore y^2) as a constant:

$$g_x(x, y) = \frac{\partial}{\partial x} y^2(1 + x^2)^{-3} = y^2 \frac{\partial}{\partial x}(1 + x^2)^{-3} = \frac{-6xy^2}{(1 + x^2)^4}$$

$$g_x(1, 3) = \frac{-6(1)3^2}{(1 + 1^2)^4} = -\frac{27}{8}$$

To differentiate with respect to y, treat x (and therefore $1 + x^2$) as a constant:

$$g_y(x, y) = \frac{\partial}{\partial y} \frac{y^2}{(1 + x^2)^3} = \frac{1}{(1 + x^2)^3} \frac{\partial}{\partial y} y^2 = \frac{2y}{(1 + x^2)^3} \qquad \boxed{1}$$

$$g_y(1, 3) = \frac{2(3)}{(1 + 1^2)^3} = \frac{3}{4}$$

CAUTION Do not *use the Quotient Rule to compute the partial derivative in Eq. (1). The denominator does not depend on y, so we treat it as a constant when differentiating with respect to y.*

See Figure 2 for the geometric interpretation of these partial derivatives. ■

We use the Chain Rule in the usual way to compute the partial derivative of a composite function $f(x, y) = F(g(x, y))$, where $F(u)$ is a function of one variable and $u = g(x, y)$:

$$\frac{\partial f}{\partial x} = \frac{dF}{du}\frac{\partial u}{\partial x}, \qquad \frac{\partial f}{\partial y} = \frac{dF}{du}\frac{\partial u}{\partial y}$$

■ **EXAMPLE 3 The Chain Rule in Partial Derivatives** Compute $\dfrac{\partial}{\partial x}\sin(x^2y^3)$.

Solution We have $\sin(x^2y^3) = F(u)$, where $F(u) = \sin u$ and $u = x^2y^3$, so

$$\underbrace{\frac{\partial}{\partial x}\sin(x^2y^3) = \frac{dF}{du}\frac{\partial u}{\partial x} = \cos(x^2y^3)\frac{\partial}{\partial x}x^2y^3}_{\text{Chain Rule}} = 2xy^3\cos(x^2y^3)$$ ■

Partial derivatives are defined for functions of any number of variables. We compute the partial derivative with respect to any one of the variables by holding the remaining variables constant.

■ **EXAMPLE 4** **More Than Two Variables** Calculate $f_z(0, 1, 0, 1)$, where

$$f(x, y, z, w) = \frac{e^{xz+yw}}{z+w}$$

In Example 4, note that the calculation

$$\frac{\partial}{\partial z} e^{xz+yw} = x\, e^{xz+yw}$$

follows from the Chain Rule, just like

$$\frac{d}{dz} e^{az+b} = a\, e^{az+b}$$

Solution Use the Quotient Rule, treating x, y, and w as constants:

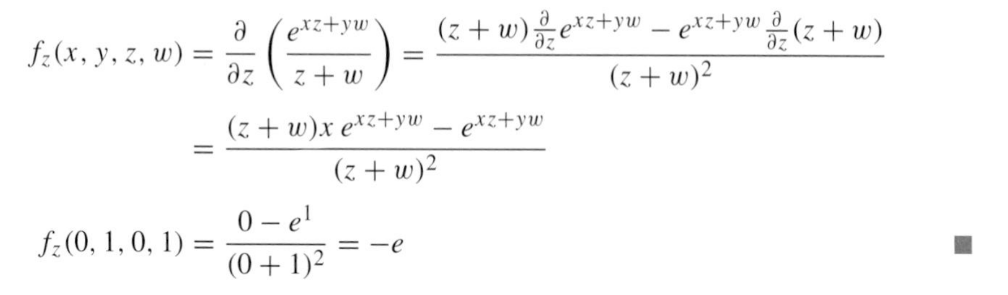

$$f_z(x, y, z, w) = \frac{\partial}{\partial z}\left(\frac{e^{xz+yw}}{z+w}\right) = \frac{(z+w)\frac{\partial}{\partial z}e^{xz+yw} - e^{xz+yw}\frac{\partial}{\partial z}(z+w)}{(z+w)^2}$$

$$= \frac{(z+w)x\, e^{xz+yw} - e^{xz+yw}}{(z+w)^2}$$

$$f_z(0, 1, 0, 1) = \frac{0 - e^1}{(0+1)^2} = -e$$ ■

Since $f_x(a, b)$ is the derivative of the function $f(x, b)$, we can use the partial derivative to estimate the change in $f(x, b)$ when x changes from a to $a + \Delta x$ (but $y = b$ is fixed) as in the single-variable case. Similarly, we can estimate the change in $f(a, y)$:

$$f(a + \Delta x, b) - f(a, b) \approx f_x(a, b)\Delta x$$

$$f(a, b + \Delta y) - f(a, b) \approx f_y(a, b)\Delta y$$

This approximation applies in any number of variables.

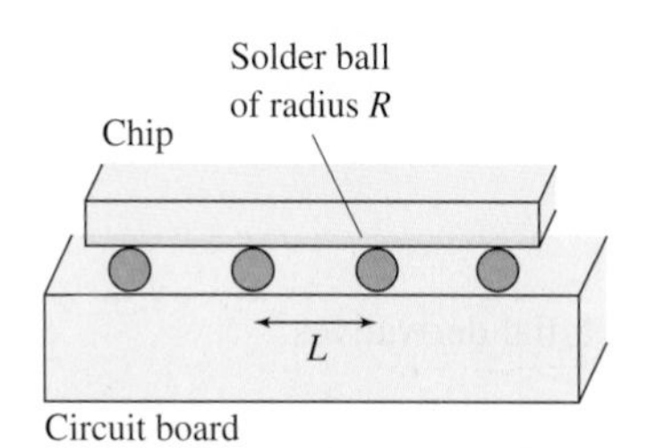

FIGURE 3 A BGA package. Temperature variations strain the BGA and may cause it to fail because the chip and board expand at different rates.

■ **EXAMPLE 5** **Testing Microchips for Reliability** A **ball grid array** (BGA) consists of a microchip joined to a circuit board by small solder balls of radius R mm separated by a distance L mm (Figure 3). Manufacturers test the reliability of BGA's by subjecting them to repeated cycles in which the temperature is varied from 0° to 100° over a 40-min period. According to one model, the average number N of cycles before the chip fails is

$$N = \left(\frac{2{,}200R}{Ld}\right)^{1.9}$$

where d is the difference between the coefficients of expansion of the chip and the board. Estimate the change in N when $d = 10$, $R = 0.12$, and L is increased from 0.4 to 0.42.

Solution Since we are interested in the effect on N of a change in L, we calculate the partial derivative with respect to L at the point $(L, R, d) = (0.4, 0.12, 10)$:

$$\frac{\partial N}{\partial L} = \frac{\partial}{\partial L}\left(\frac{2{,}200R}{Ld}\right)^{1.9} = \left(\frac{2{,}200R}{d}\right)^{1.9}\frac{\partial}{\partial L}L^{-1.9} = -1.9\left(\frac{2{,}200R}{d}\right)^{1.9}L^{-2.9}$$

$$\left.\frac{\partial N}{\partial L}\right|_{(L,R,d)=(0.4,0.12,10)} = -1.9\left(\frac{2{,}200(0.12)}{10}\right)^{1.9}(0.4)^{-2.9} \approx -13{,}609$$

The change ΔN in N is approximately

$$\Delta N \approx \frac{\partial N}{\partial L}\Delta L = -13{,}609(0.02) \approx -272 \text{ cycles}$$ ■

In the next example, we estimate a partial derivative numerically. Since f_x and f_y are limits of difference quotients, we have the following approximations when h and k

are "small" (just how small depends on the function and the accuracy required):

$$f_x(a, b) \approx \frac{\Delta f}{\Delta x} = \frac{f(a+h, b) - f(a, b)}{h}$$

$$f_y(a, b) \approx \frac{\Delta f}{\Delta y} = \frac{f(a, b+k) - f(a, b)}{k}$$

A similar approximation is valid in any number of variables.

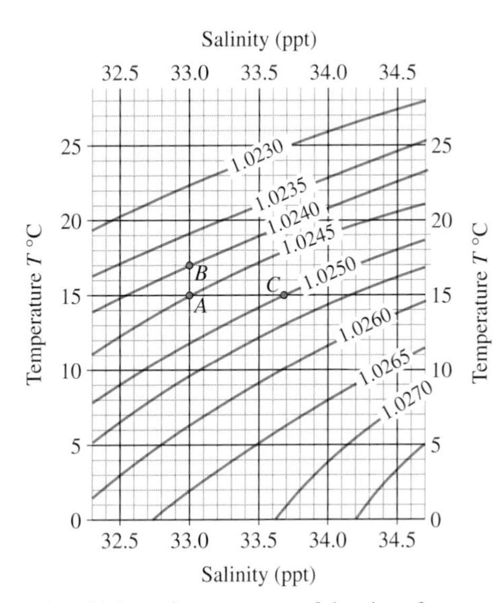

FIGURE 4 Contour map of density of seawater as a function of temperature and salinity.

■ **EXAMPLE 6 Estimating Partial Derivatives Using Contour Maps** The density of seawater ρ (kg/m^3) depends on salinity S (parts per thousand or ppt) and the temperature T (°C). Use the contour map in Figure 4 to estimate $\frac{\partial \rho}{\partial T}$ and $\frac{\partial \rho}{\partial S}$ at A.

Solution Point A has coordinates $(S, T) = (33, 15)$ and lies on the level curve $\rho = 1.0245$. We estimate $\frac{\partial \rho}{\partial T}$ at A in two steps.

Step 1. **Move vertically from A to the next level curve.**
Since T varies in the vertical direction, we move vertically from point A to the point B on the next higher level curve, where $\rho = 1.0240$. The point B has coordinates $(S, T) = (33, 17)$. Note that in moving from A to B, we have kept S constant since both points have salinity $S = 33$.

Step 2. **Compute the changes and the difference quotient.**

$$\Delta\rho = \text{change in } \rho \text{ from } A \text{ to } B = 1.0240 - 1.0245 = -0.0005$$

$$\Delta T = 17 - 15 = 2$$

$$\left.\frac{\partial \rho}{\partial T}\right|_A \approx \frac{\Delta\rho}{\Delta T} \approx \frac{-0.0005}{2} = -0.00025 \text{ kg-m}^{-3}/°\text{C}$$

To estimate $\frac{\partial \rho}{\partial S}$, we move horizontally to point C on the next level curve where $\rho = 1.0250$, and compute the changes as before. Since C has coordinates $(S, T) \approx (33.7, 15)$,

$$\Delta\rho = \text{change in } \rho \text{ from } A \text{ to } C = 1.0250 - 1.0245 = 0.0005$$

$$\Delta S \approx 33.7 - 33 = 0.7$$

$$\left.\frac{\partial \rho}{\partial S}\right|_A \approx \frac{\Delta\rho}{\Delta S} \approx \frac{0.0005}{0.7} \approx 0.0007 \text{ kg-m}^{-3}/\text{ppt}$$ ■

Higher-Order Partial Derivatives

The higher-order partial derivatives of a function $f(x, y)$ are the derivatives of the partial derivatives f_x and f_y. For instance, the *second-order* partial derivatives are the derivatives of f_x and f_y with respect to x or y. We write f_{xx} for the x-derivative of f_x and f_{yy} for the y-derivative of f_y:

$$f_{xx} = \frac{\partial}{\partial x}\left(\frac{\partial f}{\partial x}\right), \qquad f_{yy} = \frac{\partial}{\partial y}\left(\frac{\partial f}{\partial y}\right)$$

We also have the "mixed partials," which are the derivatives

$$f_{xy} = \frac{\partial}{\partial y}\left(\frac{\partial f}{\partial x}\right), \qquad f_{yx} = \frac{\partial}{\partial x}\left(\frac{\partial f}{\partial y}\right)$$

The process can be continued. For example, f_{xyx} is the x-derivative of f_{xy} and f_{xyy} is the y-derivative of f_{xy} (perform the differentiation in the order of the subscripts from left to right). Leibniz notation for the higher-order partial derivatives is

$$f_{xx} = \frac{\partial^2 f}{\partial x^2}, \qquad f_{xy} = \frac{\partial^2 f}{\partial y \partial x}, \qquad f_{yx} = \frac{\partial^2 f}{\partial x \partial y}, \qquad f_{yy} = \frac{\partial^2 f}{\partial y^2}$$

Higher partial derivatives are defined for functions of three or more variables in a similar manner.

EXAMPLE 7 Calculate the second-order partials and f_{xyy} of $f(x, y) = x^3 + y^2 e^x$.

Solution

Remember how the subscripts are used in partial derivatives. The notation f_{xyy} means "first differentiate with respect to x and then differentiate twice with respect to y."

$$f_x(x, y) = \frac{\partial}{\partial x}(x^3 + y^2 e^x) = 3x^2 + y^2 e^x \qquad f_y(x, y) = \frac{\partial}{\partial y}(x^3 + y^2 e^x) = 2ye^x$$

$$f_{xx}(x, y) = \frac{\partial}{\partial x} f_x = \frac{\partial}{\partial x}(3x^2 + y^2 e^x) = 6x + y^2 e^x \qquad f_{yy}(x, y) = \frac{\partial}{\partial y} f_y = \frac{\partial}{\partial y} 2ye^x = 2e^x$$

$$f_{xy}(x, y) = \frac{\partial f_x}{\partial y} = \frac{\partial}{\partial y}(3x^2 + y^2 e^x) = 2ye^x \qquad f_{yx}(x, y) = \frac{\partial f_y}{\partial x} = \frac{\partial}{\partial x} 2ye^x = 2ye^x \tag{2}$$

Now we have $f_{xyy} = \dfrac{\partial}{\partial y} f_{xy} = \dfrac{\partial}{\partial y} 2ye^x = 2e^x$. ■

Observe in Eq. (2) that $f_{xy} = f_{yx}$ for $f(x, y) = x^3 + y^2 e^x$. It is a pleasant circumstance that the equality $f_{xy} = f_{yx}$ holds for all $f(x, y)$, provided that the mixed partials are continuous. See Appendix D for a proof of the following theorem.

The hypothesis of Clairaut's Theorem, that f_{xy} and f_{yx} are continuous, is almost always satisfied in practice, but see Exercise 81 for an example where the mixed partials are not equal.

THEOREM 1 Clairaut's Theorem: Equality of Mixed Partials If f_{xy} and f_{yx} are both continuous functions on a disk D, then $f_{xy}(a, b) = f_{yx}(a, b)$ for all $(a, b) \in D$. In other words,

$$\frac{\partial^2 f}{\partial x \, \partial y} = \frac{\partial^2 f}{\partial y \, \partial x}$$

Although Clairaut's Theorem is stated for f_{xy} and f_{yx}, it implies more generally that partial differentiation may be carried out any order, provided that the derivatives in question are continuous (see Exercise 73). For example, we may compute f_{xyxy} by differentiating f twice with respect to x and twice with respect to y, in any order. Thus,

$$f_{xyxy} = f_{xxyy} = f_{yyxx} = f_{yxyx} = f_{xyyx} = f_{yxxy}$$

EXAMPLE 8 Check that $\dfrac{\partial^3 W}{\partial V \partial T^2} = \dfrac{\partial^3 W}{\partial T^2 \partial V}$ for $W = PV^{4/3}T^{-1}$.

FIGURE 5 Alexis Clairaut (1713–1765) was a brilliant French mathematician who presented his first paper to the Paris Academy of Sciences at the age of 13. In 1752, Clairaut won a prize for an essay on lunar motion that Euler praised (surely an exaggeration) as "the most important and profound discovery that has ever been made in mathematics."

Solution We compute both derivatives and observe that they are equal:

$$\frac{\partial W}{\partial T} = -PV^{4/3}T^{-2} \qquad \frac{\partial W}{\partial V} = \frac{4}{3}PV^{1/3}T^{-1}$$

$$\frac{\partial^2 W}{\partial T^2} = \frac{\partial}{\partial T}\frac{\partial W}{\partial T} = 2PV^{4/3}T^{-3} \qquad \frac{\partial^2 W}{\partial T \partial V} = \frac{\partial}{\partial T}\frac{\partial W}{\partial V} = -\frac{4}{3}PV^{1/3}T^{-2}$$

$$\frac{\partial^3 W}{\partial V \partial T^2} = \frac{\partial}{\partial V}\frac{\partial^2 W}{\partial T^2} = \frac{8}{3}PV^{1/3}T^{-3} \qquad \frac{\partial^3 W}{\partial T^2 \partial V} = \frac{\partial}{\partial T}\frac{\partial^2 W}{\partial T \partial V} = \frac{8}{3}PV^{1/3}T^{-3}$$

EXAMPLE 9 Choosing the Order Wisely Calculate the derivative g_{zzwx}, where $g(x, y, z, w) = x^3w^2z^2 + \sin\left(\frac{xy}{z^2}\right)$.

Solution Let's take advantage of the fact that the derivatives may be calculated in any order to differentiate with respect to w first. This causes the second term, which does not depend on w, to disappear:

$$g_w = \frac{\partial}{\partial w}\left(x^3w^2z^2 + \sin\left(\frac{xy}{z^2}\right)\right) = 2x^3wz^2$$

Next, differentiate twice with respect to z and once with respect to x:

$$g_{wz} = \frac{\partial}{\partial z}2x^3wz^2 = 4x^3wz$$

$$g_{wzz} = \frac{\partial}{\partial z}4x^3wz = 4x^3w$$

$$g_{wzzx} = \frac{\partial}{\partial x}4x^3w = 12x^2w$$

We conclude that $g_{zzwx} = g_{wzzx} = 12x^2w$. ■

Maxwell's equations of electromagnetism and Schrödinger's equation in quantum mechanics are famous examples of PDE's in physics. Partial differential equations are an active field of research with applications in such diverse fields as weather prediction and financial risk analysis.

A **partial differential equation** (PDE) is a differential equation involving functions of several variables and their partial derivatives. The heat equation in the next example is a PDE that models the change in temperature as heat spreads through an object (Figure 6).

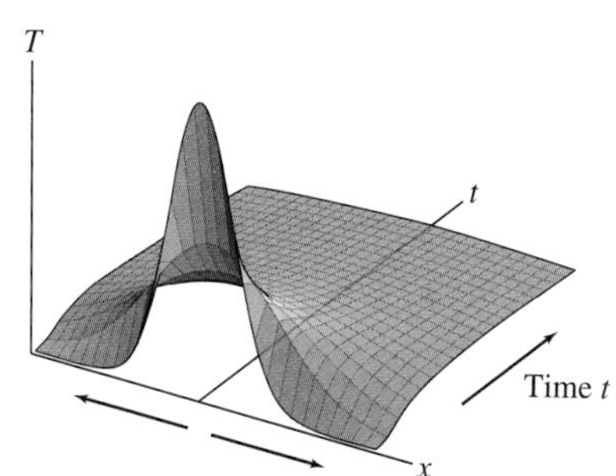

FIGURE 6 The plot of $f(x, t) = \frac{1}{\sqrt{t}}e^{-(x^2/4t)}$ illustrates the diffusion of heat over time.

EXAMPLE 10 The Heat Equation Show that $f(x, t) = \frac{1}{\sqrt{t}}e^{-(x^2/4t)}$ satisfies the heat equation

$$\frac{\partial f}{\partial t} = \frac{\partial^2 f}{\partial x^2}$$

Solution We compute $\frac{\partial^2 f}{\partial x^2}$ and $\frac{\partial f}{\partial t}$ using the Product Rule:

$$\frac{\partial f}{\partial x} = \frac{\partial}{\partial x}t^{-1/2}e^{-(x^2/4t)} = -\frac{1}{2}xt^{-3/2}e^{-(x^2/4t)}$$

$$\frac{\partial^2 f}{\partial x^2} = \frac{\partial}{\partial x}\left(-\frac{1}{2}xt^{-3/2}e^{-(x^2/4t)}\right) = -\frac{1}{2}t^{-3/2}e^{-(x^2/4t)} + \frac{1}{4}x^2t^{-5/2}e^{-(x^2/4t)}$$

As required, $\dfrac{\partial^2 f}{\partial x^2}$ is equal to

$$\frac{\partial f}{\partial t} = \frac{\partial}{\partial t}\left(t^{-1/2}e^{-(x^2/4t)}\right) = -\frac{1}{2}t^{-3/2}e^{-(x^2/4t)} + \frac{1}{4}x^2t^{-5/2}e^{-(x^2/4t)}$$ ■

14.3 SUMMARY

- The partial derivatives of $f(x, y)$ are defined as the limits

$$f_x(a,b) = \left.\frac{\partial f}{\partial x}\right|_{(a,b)} = \lim_{h\to 0}\frac{f(a+h,b) - f(a,b)}{h}$$

$$f_y(a,b) = \left.\frac{\partial f}{\partial y}\right|_{(a,b)} = \lim_{k\to 0}\frac{f(a,b+k) - f(a,b)}{k}$$

- Compute f_x by holding y constant and compute f_y by holding x constant.
- The partial derivative $f_x(a, b)$ is the slope of the tangent line to the trace of the graph of $z = f(x, y)$ in the vertical plane $y = b$ at $x = a$. Similarly, $f_y(a, b)$ is the slope of the tangent line to the trace of the graph in the vertical plane $x = a$ at $y = b$.
- The second-order partial derivatives are

$$\frac{\partial^2}{\partial x^2}f = f_{xx}, \qquad \frac{\partial}{\partial y\,\partial x}f = f_{xy}, \qquad \frac{\partial}{\partial x\,\partial y}f = f_{yx}, \qquad \frac{\partial}{\partial y^2}f = f_{yy}$$

- Clairaut's Theorem states that mixed partials are equal, that is, $f_{xy} = f_{yx}$ (provided that f_{xy} and f_{yx} are continuous). More generally, higher partial derivatives may be computed in any order. For example, $f_{xyyz} = f_{yxzy}$ if f is a function of x, y, z whose fourth-order partial derivatives are continuous.

14.3 EXERCISES

Preliminary Questions

1. Patricia derived the following *incorrect* formula by misapplying the Product Rule:

$$\frac{\partial}{\partial x}(x^2y^2) = x^2(2y) + y^2(2x)$$

What was her mistake and what is the correct calculation?

2. Explain why it is not necessary to use the Quotient Rule to compute $\dfrac{\partial}{\partial x}\left(\dfrac{x+y}{y+1}\right)$. Should the Quotient Rule be used to compute $\dfrac{\partial}{\partial y}\left(\dfrac{x+y}{y+1}\right)$?

3. Which of the following partial derivatives should be evaluated without using the Quotient Rule?

(a) $\dfrac{\partial}{\partial x}\dfrac{xy}{y^2+1}$ **(b)** $\dfrac{\partial}{\partial y}\dfrac{xy}{y^2+1}$ **(c)** $\dfrac{\partial}{\partial x}\dfrac{y^2}{y^2+1}$

4. What is f_x, where $f(x, y, z) = (\sin yz)e^{z^3 - z^{-1}\sqrt{y}}$?

5. Which of the following partial derivatives are equal to f_{xxy}?

(a) f_{xyx} **(b)** f_{yyx} **(c)** f_{xyy} **(d)** f_{yxx}

Exercises

1. Use the limit definition of the partial derivative to verify the formulas

$$\frac{\partial}{\partial x}xy^2 = y^2, \qquad \frac{\partial}{\partial y}xy^2 = 2xy$$

2. Use the Product Rule to compute $\dfrac{\partial}{\partial y}(x^2 + y)(x + y^4)$.

3. Use the Quotient Rule to compute $\dfrac{\partial}{\partial y}\dfrac{y}{x+y}$.

4. Use the Chain Rule to compute $\frac{\partial}{\partial u}\ln(u^2 + uv)$.

5. Calculate $f_z(2, 3, 1)$, where $f(x, y, z) = xyz$.

6. Explain the relation between the two formulas (c is a constant)

$$\frac{d}{dx}\sin(cx) = c\cos(cx), \qquad \frac{\partial}{\partial x}\sin(xy) = y\cos(xy)$$

7. The plane $y = 1$ intersects the surface $z = x^4 + 6xy - y^4$ in a certain curve (Figure 7). Find the slope of the tangent line to this curve at the point $P = (1, 1, 6)$.

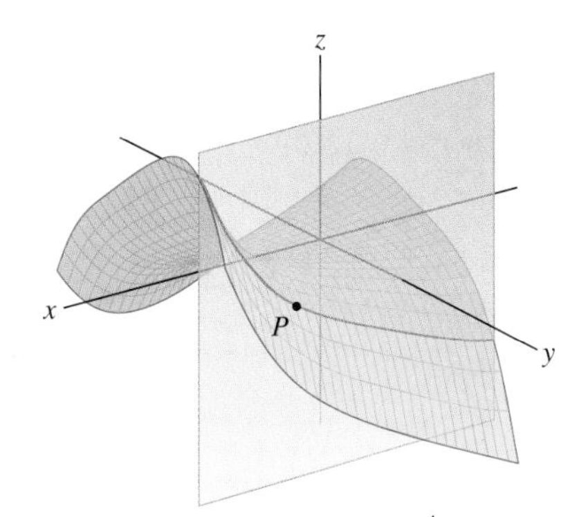

FIGURE 7 Graph of $f(x, y) = x^4 + 6xy - y^4$.

8. Determine if the partial derivatives $\frac{\partial f}{\partial x}$ and $\frac{\partial f}{\partial y}$ are positive or negative at the point P in Figure 7.

In Exercises 9–11, refer to Figure 8.

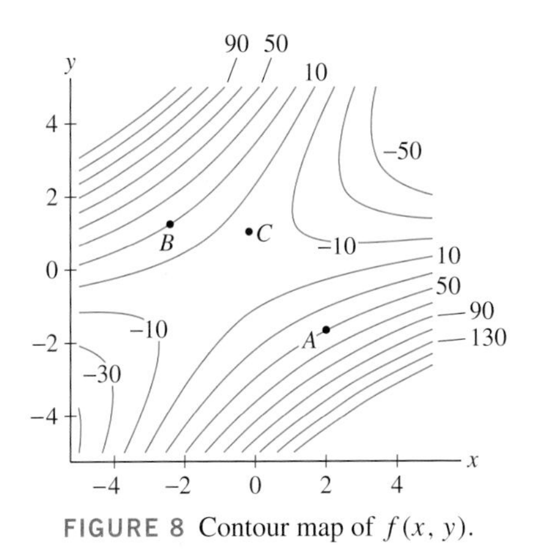

FIGURE 8 Contour map of $f(x, y)$.

9. Estimate f_x and f_y at point A.

10. Starting at point B, in which direction does f increase most rapidly?

11. At which of A, B, or C is f_y smallest?

In Exercises 12–39, compute the partial derivatives.

12. $z = x^2 + y^2$

13. $z = x^4 y^3$

14. $z = x^4 y + xy^{-2}$

15. $V = \pi r^2 h$

16. $z = \frac{x}{y}$

17. $z = \frac{x}{x - y}$

18. $z = \sqrt{9 - x^2 - y^2}$

19. $z = \frac{x}{\sqrt{x^2 + y^2}}$

20. $z = (\sin x)(\sin y)$

21. $z = \sin(u^2 v)$

22. $z = \tan\frac{x}{y}$

23. $S = \tan^{-1}(wz)$

24. $z = \ln(x + y)$

25. $z = \ln(x^2 + y^2)$

26. $W = e^{r+s}$

27. $z = e^{xy}$

28. $R = e^{-v^2/k}$

29. $z = e^{-x^2-y^2}$

30. $P = e^{\sqrt{y^2+z^2}}$

31. $z = y^x$

32. $z = \cosh(3x + 2y)$

33. $z = \sinh(x^2 y)$

34. $w = xy^2 z^3$

35. $w = \frac{x}{y + z}$

36. $Q = re^{\theta}$

37. $w = \frac{x}{(x^2 + y^2 + z^2)^{3/2}}$

38. $A = \sin(4\theta - 9t)$

39. $U = \frac{e^{-rt}}{r}$

In Exercises 40–44, compute the given partial derivatives.

40. $f(x, y) = 3x^2 y + 4x^3 y^2 - 7xy^5$, $\quad f_x(1, 4)$

41. $f(x, y) = \sin(x^2 - y)$, $\quad f_y(0, \pi)$

42. $g(u, v) = u\ln(u + v)$, $\quad g_u(1, 2)$

43. $h(x, z) = e^{xz - x^2 z^3}$, $\quad h_z(2, 1)$

44. $h(x, z) = e^{xz - x^2 z^3}$, $\quad h_z(1, 0)$

45. Calculate $\frac{\partial W}{\partial E}$ and $\frac{\partial W}{\partial T}$, where $W = e^{-E/kT}$.

46. According to the ideal gas law, $PV = nRT$, where P, V, and T are the pressure, volume, and temperature (R and n are constants). View P as a function of V and T, and compute $\frac{\partial P}{\partial T}$ and $\frac{\partial P}{\partial V}$.

47. The volume of a right-circular cone of radius r and height h is $V = \frac{\pi}{3}r^2 h$. Calculate $\frac{\partial V}{\partial r}$ and $\frac{\partial V}{\partial h}$.

48. A right-circular cone has $r = h = 12$ cm. What leads to a greater increase in V, a 1-cm increase in r or 1-cm increase in h? Argue using partial derivatives.

49. Use the contour map of $f(x, y)$ in Figure 9 to explain why the following are true:

(a) f_x and f_y are both larger at P than at Q.

(b) $f_x(x, y)$ is decreasing as a function of y, that is, for any x, $f_x(x, b_1) > f_x(x, b_2)$ if $b_1 < b_2$.

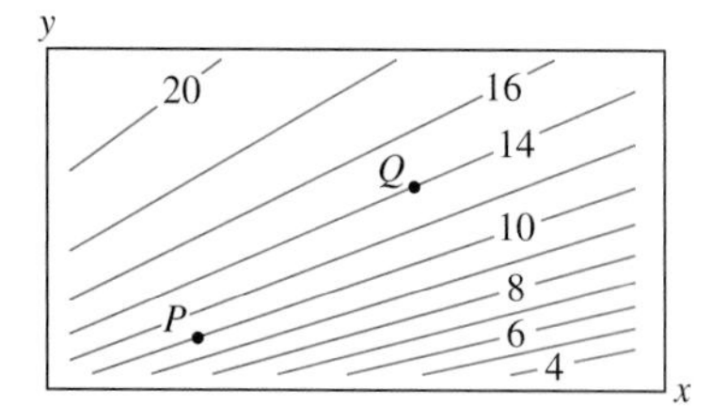

FIGURE 9 Contour interval 2.

50. Over most of the earth, a magnetic compass does not point to true (geographic) north; instead, it points at some angle east or west of true north. The angle D between magnetic north and true north is called the **magnetic declination**. Use Figure 10 to determine which of the following statements is true. Note that because of the way longitude is measured, the x-axis increases from right to left.

(a) $\left.\frac{\partial D}{\partial y}\right|_A > \left.\frac{\partial D}{\partial y}\right|_B$ **(b)** $\left.\frac{\partial D}{\partial x}\right|_C > 0$ **(c)** $\left.\frac{\partial D}{\partial y}\right|_C > 0$

(d) D is 28° larger at A than at C.

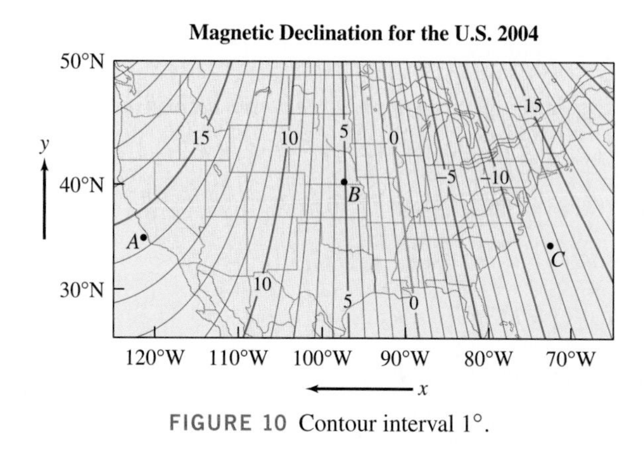

FIGURE 10 Contour interval 1°.

51. Seawater Density Refer to Example 6 and Figure 4.

(a) Estimate $\frac{\partial \rho}{\partial T}$ and $\frac{\partial \rho}{\partial S}$ at points B and C.

(b) Which has a greater effect on density ρ at B, a 1° increase in temperature or a 1-ppt increase in salinity?

(c) True or false: The density of warm seawater is more sensitive to a change in temperature than the density of cold seawater? Explain.

52. Estimate the partial derivatives at P of the function whose contour map is shown in Figure 11.

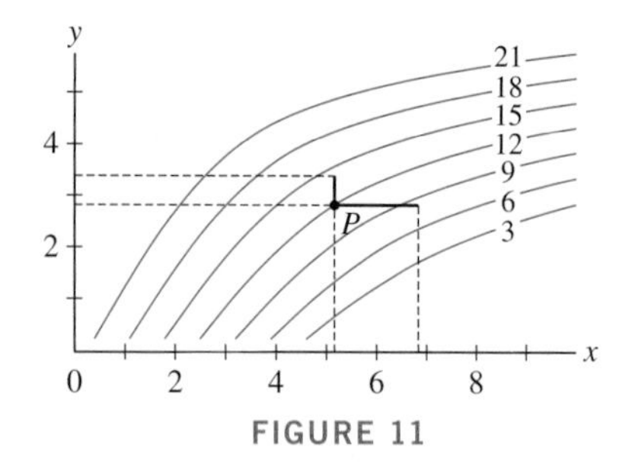

FIGURE 11

In Exercises 53–58, compute the derivative indicated.

53. $f(x, y) = 3x^2y - 6xy^4$, $\frac{\partial^2 f}{\partial x^2}$ and $\frac{\partial^2 f}{\partial y^2}$

54. $f(x, y) = x\ln(y^2)$, $f_{yy}(2, 3)$

55. $g(x, y) = \frac{xy}{x - y}$, $\frac{\partial^2 g}{\partial x\,\partial y}$

56. $g(x, y) = xye^{-y}$, $g_{yy}(1, 0)$

57. $h(u, v) = \frac{u}{u + v}$, $h_{uv}(u, v)$

58. $h(x, y) = \ln(x^3 + y^3)$, $h_{xy}(x, y)$

In Exercises 59–60, use Table 1 to answer the following questions.

TABLE 1 ***Seawater Density ρ as a Function of Temperature T and Salinity S***

T \ S	30	31	32	33	34	35	36
12	22.75	23.51	24.27	25.07	25.82	26.6	27.36
10	23.07	23.85	24.62	25.42	26.17	26.99	27.73
8	23.36	24.15	24.93	25.73	26.5	27.28	29.09
6	23.62	24.44	25.22	26	26.77	27.55	28.35
4	23.85	24.62	25.42	26.23	27	27.8	28.61
2	24	24.78	25.61	26.38	27.18	28.01	28.78
0	24.11	24.92	25.72	26.5	27.34	28.12	28.91

59. Estimate $\frac{\partial \rho}{\partial T}$ and $\frac{\partial \rho}{\partial S}$ at $(S, T) = (34, 2)$ and $(35, 10)$.

60. For fixed salinity $S = 33$, is ρ concave up or concave down as a function of T? *Hint:* Determine if the quotients $\Delta\rho/\Delta T$ are increasing or decreasing. What can you conclude about the sign of $\frac{\partial^2 \rho}{\partial T^2}$?

61. Compute f_{xyz} for

$$f(x, y, z) = \sin(yx) + \tan\left(\frac{z + z^{-1}}{x - x^{-1}}\right)$$

Hint: Use a well-chosen order of differentiation on each term.

62. Let

$$f(x, y, u, v) = \frac{x^2 + e^y v}{3y^2 + \ln(2 + u^2)}$$

What is the fastest way to show that $f_{uvxyvu}(x, y, u, v) = 0$ for all (x, y, u, v)?

In Exercises 63–70, compute the derivative indicated.

63. $f(u, v) = \cos(u + v^2), \quad f_{uuv}$

64. $g(x, y, z) = x^4 y^5 z^6, \quad g_{xxyz}$

65. $F(r, s, t) = r(s^2 + t^2), \quad F_{rst}$

66. $g(x, y, z) = x^4 y^5 z^3, \quad g_{xxyz}$

67. $u(x, t) = t^{-1/2} e^{-(x^2/4t)}, \quad u_{xx}$

68. $R(u, v, w) = \dfrac{u}{v + w}, \quad R_{uvw}$

69. $g(x, y, z) = \sqrt{x^2 + y^2 + z^2}, \quad g_{xyz}$

70. $u(x, t) = \operatorname{sech}^2(x - t), \quad u_{xxx}$

71. Find (by guessing) a function such that $\dfrac{\partial f}{\partial x} = 2xy$ and $\dfrac{\partial f}{\partial y} = x^2$.

72. Prove that there does not exist any function $f(x, y)$ such that $\dfrac{\partial f}{\partial x} = xy$ and $\dfrac{\partial f}{\partial y} = x^2$. *Hint:* Show that f cannot satisfy Clairaut's Theorem.

73. Assume that f_{xy} and f_{yx} are continuous and that f_{yxx} exists. Show that f_{xyx} also exists and that $f_{yxx} = f_{xyx}$.

74. Show that $u(x, t) = \sin(nx)\, e^{-n^2 t}$ satisfies the heat equation (Example 10) for any constant n:

$$\frac{\partial f}{\partial t} = \frac{\partial^2 f}{\partial x^2} \qquad \boxed{3}$$

75. Find all values of A and B such that $f(x, t) = e^{Ax + Bt}$ satisfies Eq. (3).

*In Exercises 76–79, the **Laplace operator** Δ is defined by $\Delta f = f_{xx} + f_{yy}$. A function $u(x, y)$ satisfying the Laplace equation $\Delta u = 0$ is called **harmonic**.*

76. Show that the following functions are harmonic:

(a) $u(x, y) = x$

(b) $u(x, y) = e^x \cos y$

(c) $u(x, y) = \tan^{-1} \dfrac{y}{x}$

(d) $u(x, y) = \ln(x^2 + y^2)$

77. Find all harmonic polynomials $u(x, y)$ of degree three, that is, $u(x, y) = ax^3 + bx^2 y + cxy^2 + dy^3$.

78. Show that if $u(x, y)$ is harmonic, then the partial derivatives $\dfrac{\partial u}{\partial x}$ and $\dfrac{\partial u}{\partial y}$ are harmonic.

79. Find all constants a, b such that $u(x, y) = \cos(ax)e^{by}$ is harmonic.

80. Show that $u(x, t) = \operatorname{sech}^2(x - t)$ satisfies the **Korteweg–deVries equation** (which arises in the study of water waves)

$$4u_t + u_{xxx} + 12uu_x = 0$$

Further Insights and Challenges

81. CAS **Assumptions Matter** This exercise shows that the hypotheses of Clairaut's Theorem are needed. Let $f(x, y) = xy\dfrac{x^2 - y^2}{x^2 + y^2}$ for $(x, y) \neq (0, 0)$ and $f(0, 0) = 0$.

(a) Use a computer algebra system to verify the following formulas for $(x, y) \neq (0, 0)$:

$$f_x(x, y) = \frac{y(x^4 + 4x^2y^2 - y^4)}{(x^2 + y^2)^2}$$

$$f_y(x, y) = \frac{x(x^4 - 4x^2y^2 - y^4)}{(x^2 + y^2)^2}$$

(b) Use the limit definition of the partial derivative to show that $f_x(0, 0) = f_y(0, 0) = 0$ and that $f_{yx}(0, 0)$ and $f_{xy}(0, 0)$ both exist but are not equal.

(c) Use a computer algebra system to show that for $(x, y) \neq (0, 0)$:

$$f_{xy}(x, y) = f_{yx}(x, y) = \frac{x^6 + 9x^4y^2 - 9x^2y^4 - y^6}{(x^2 + y^2)^3}$$

Show that f_{xy} is not continuous at $(0, 0)$. *Hint:* Show that $\lim_{h \to 0} f_{xy}(h, 0) \neq \lim_{h \to 0} f_{xy}(0, h)$.

(d) Explain why the result of (b) does not contradict Clairaut's Theorem.

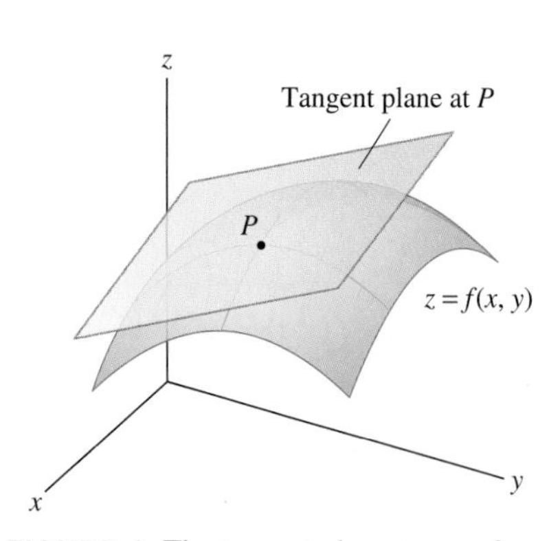

FIGURE 1 The tangent plane to a surface.

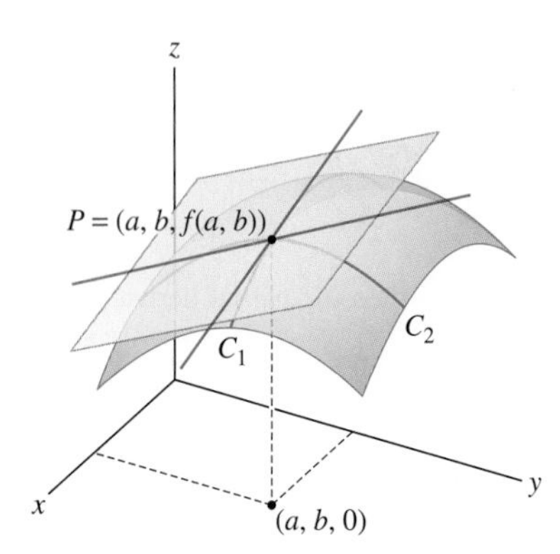

FIGURE 2 The tangent lines to the trace curves C_1 and C_2 lie in the tangent plane (and span it).

14.4 Differentiability, Linear Approximation, and Tangent Planes

In this section, we generalize two key concepts from single-variable calculus: the linear approximation and the tangent line (which becomes the *tangent plane* for functions of two variables; see Figure 1). Recall that in one variable:

- The linearization of $f(x)$ at $x = a$ is the linear function

$$L(x) = f(a) + f'(a)(x - a)$$

- The tangent line to $y = f(x)$ at $x = a$ is the line $y = L(x)$.

The **linearization** of a function $f(x, y)$ of two variables at (a, b), assuming that $f_x(a, b)$ and $f_y(a, b)$ exist, is the linear function

$$\boxed{L(x, y) = f(a, b) + f_x(a, b)(x - a) + f_y(a, b)(y - b)}$$

To justify this definition, we observe first that the point $P = (a, b, f(a, b))$ lies on the plane $z = L(x, y)$ since $L(a, b) = f(a, b)$. Furthermore, $z = L(x, y)$ is the unique plane containing the tangent lines to the two vertical trace curves through P (Figure 2). For instance, the trace in $y = b$ is the curve $z = f(x, b)$ whose tangent line at P is

$$\underbrace{z = L(x, b) = f(a, b) + f_x(a, b)(x - a)}_{\text{Equation of tangent line to vertical trace } z = f(x, b)}$$

■ **EXAMPLE 1** Find the linearization of $f(x, y) = 5x + 4y^2$ at $(a, b) = (2, 1)$.

Solution First evaluate $f(x, y)$ and its partial derivatives at $(a, b) = (2, 1)$:

$$f(x, y) = 5x + 4y^2, \qquad f_x(x, y) = 5, \qquad f_y(x, y) = 8y$$

$$f(2, 1) = 14, \qquad f_x(2, 1) = 5, \qquad f_y(2, 1) = 8$$

$$L(x, y) = \underbrace{14 + 5(x - 2) + 8(y - 1)}_{f(a,b)+f_x(a,b)(x-a)+f_y(a,b)(y-b)} = -4 + 5x + 8y \qquad ■$$

Observe that the graph $z = L(x, y)$ is the tangent plane at $P = (2, 1, 14)$ (see Figure 3). Our next goal is to define differentiability. It might seem appropriate to say that $f(x, y)$ is differentiable at (a, b) if the partial derivatives $f_x(a, b)$ and $f_y(a, b)$ exist. It turns out, however, that this is not sufficient to guarantee that $f(x, y)$ has the smoothness we expect of a differentiable function. What is needed is the additional requirement of local linearity.

Recall that in one variable, local linearity at a point $x = a$ means that the linear approximation $L(x)$ approximates the function $f(x)$ at a to *first-order*. In other words, the error $E(x) = f(x) - L(x)$ tends to zero faster than the distance $|x - a|$ as $x \to a$. We express this in a precise fashion by saying that there exists a function $\epsilon(x)$ such that

$$f(x) = L(x) + \epsilon(x)|x - a|, \qquad \text{where } \lim_{x \to a} \epsilon(x) = 0$$

The multivariable case is similar. A function $f(x, y)$ is locally linear at (a, b) if $|f(x, y) - L(x, y)|$ tends to zero faster than the distance $\sqrt{(x - a)^2 + (y - b)^2}$ from (x, y) to (a, b).

DEFINITION Local Linearity Assume that $f(x, y)$ is defined in a disk D containing (a, b) and that $f_x(a, b)$ and $f_y(a, b)$ exist. We say that $f(x, y)$ is **locally linear** at (a, b) if $L(x, y)$ approximates $f(x, y)$ at (a, b) to **first order**, that is, if

$$f(x, y) = L(x, y) + \underbrace{\epsilon(x, y)\sqrt{(x-a)^2 + (y-b)^2}}_{\text{The error } E(x,y)} \quad \text{for } (x, y) \in D \qquad \boxed{1}$$

where $\epsilon(x, y)$ is a function such that $\lim_{(x,y)\to(a,b)} \epsilon(x, y) = 0$.

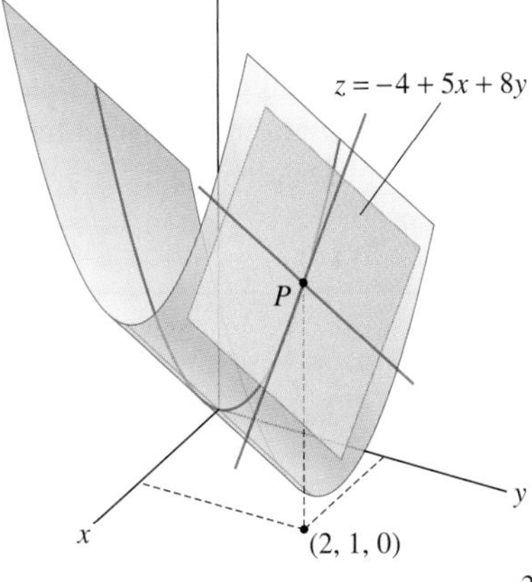

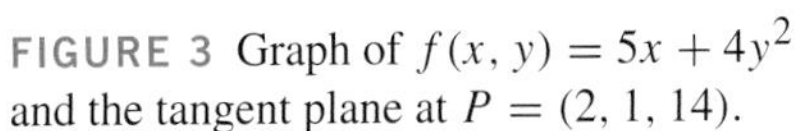
FIGURE 3 Graph of $f(x, y) = 5x + 4y^2$ and the tangent plane at $P = (2, 1, 14)$.

EXAMPLE 2 Verifying a First-Order Approximation Show that $f(x, y) = 5x + 4y^2$ is locally linear at $P = (2, 1)$.

Solution In Example 1, we showed that the linearization of $f(x, y)$ at $P = (2, 1)$ is $L(x, y) = -4 + 5x + 8y$. To verify local linearity, we compute the error

$$E(x, y) = f(x, y) - L(x, y) = (5x + 4y^2) - (-4 + 5x + 8y)$$
$$= 4y^2 - 8y + 4 = 4(y-1)^2$$

For $(x, y) \neq (2, 1)$, this may be written

$$E(x, y) = 4(y-1)^2 = \frac{4(y-1)^2}{\sqrt{(x-2)^2 + (y-1)^2}}\sqrt{(x-2)^2 + (y-1)^2}$$

So if we set

$$\epsilon(x, y) = \frac{4(y-1)^2}{\sqrt{(x-2)^2 + (y-1)^2}}$$

then

$$f(x, y) = L(x, y) + \epsilon(x, y)\sqrt{(x-2)^2 + (y-1)^2}$$

To check that $L(x, y)$ yields a first-order approximation, we must verify that

$$\lim_{(x,y)\to(2,1)} \epsilon(x, y) = 0$$

Observe that

The inequality (2) is valid because

$$\sqrt{(x-2)^2 + (y-1)^2} \geq \sqrt{(y-1)^2}$$

$$\epsilon(x, y) = \frac{4(y-1)^2}{\sqrt{(x-2)^2 + (y-1)^2}} \leq \frac{4(y-1)^2}{\sqrt{(y-1)^2}} = 4|y-1| \qquad \boxed{2}$$

Thus, $0 \leq \epsilon(x, y) \leq 4|y-1|$, and since $\lim_{(x,y)\to(2,1)} 4|y-1| = 0$, the Squeeze Theorem implies that $\lim_{(x,y)\to(2,1)} \epsilon(x, y) = 0$ as desired. ■

Does the linearization $L(x, y)$ always approximate $f(x, y)$ to first order? The answer is *"yes" for most functions that arise in practice*. However, the mere existence of $f_x(a, b)$ and $f_y(a, b)$ does not guarantee that $f(x, y)$ is locally linear at (a, b). For this reason, we define differentiability by requiring that $f(x, y)$ be locally linear in addition to having partial derivatives at (a, b).

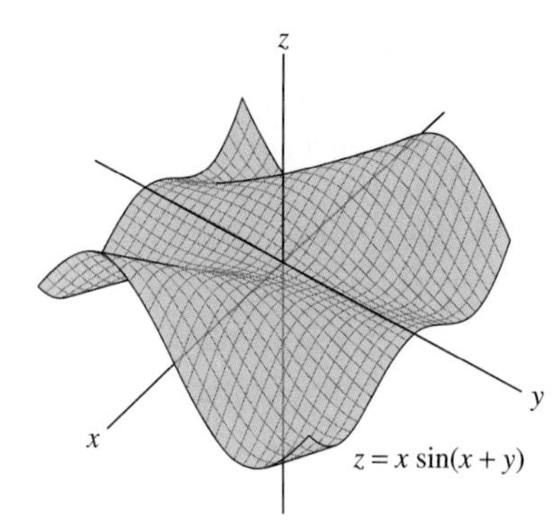

FIGURE 4 The function $f(x, y) = x\sin(x + y)$ is differentiable.

DEFINITION Differentiability and the Tangent Plane Assume that $f(x, y)$ is defined in a disk D containing (a, b). We say that $f(x, y)$ is **differentiable** at (a, b) if:

- $f_x(a, b)$ and $f_y(a, b)$ exist, and
- $f(x, y)$ is locally linear at (a, b).

In this case, the **tangent plane** to the surface $z = f(x, y)$ at $(a, b, f(a, b))$ is the plane with equation $z = L(x, y)$. Explicitly,

$$z = f(a, b) + f_x(a, b)(x - a) + f_y(a, b)(y - b) \qquad \boxed{3}$$

If $f(x, y)$ is differentiable at all points in a domain $\mathcal{D}$, we say that $f(x, y)$ is differentiable on $\mathcal{D}$.

It is cumbersome to check differentiability using the above definition, but fortunately, this is usually not necessary. The following theorem provides a criterion for differentiability that is easy to apply. It assures us that most functions built out of elementary functions are differentiable on their domains. See Appendix D for a proof.

The definition of differentiability extends to functions of n-variables and Theorem 1 holds in this setting: If all of the partial derivatives of $f(x_1, \ldots, x_n)$ exist and are continuous on an open domain $\mathcal{D}$, then $f(x_1, \ldots, x_n)$ is differentiable on $\mathcal{D}$.

THEOREM 1 Criterion for Differentiability If $f_x(x, y)$ and $f_y(x, y)$ exist and are continuous on an open disk D, then $f(x, y)$ is differentiable on D.

■ **EXAMPLE 3 Checking Differentiability** Show that $f(x, y) = x\sin(x + y)$ is differentiable.

Solution The function $f(x, y)$ (Figure 4) is differentiable by Theorem 1 because its partial derivatives are continuous:

$$f_x(x, y) = x\cos(x + y) + \sin(x + y), \qquad f_y(x, y) = x\cos(x + y)$$ ■

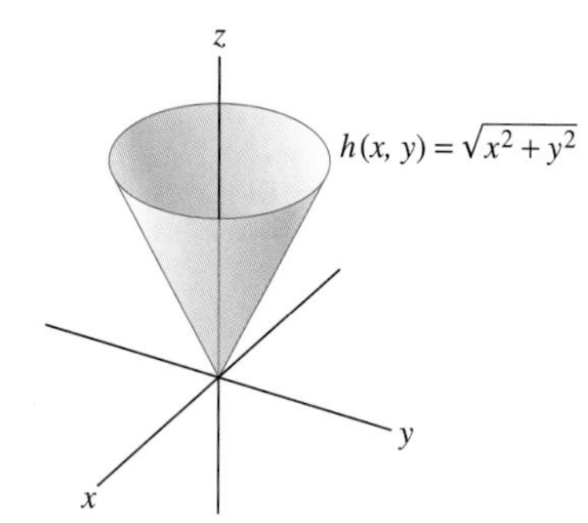

FIGURE 5 The function $h(x, y) = \sqrt{x^2 + y^2}$ is differentiable except at the origin.

■ **EXAMPLE 4 Checking Differentiability** Where is $h(x, y) = \sqrt{x^2 + y^2}$ differentiable?

Solution The function $h(x, y) = \sqrt{x^2 + y^2}$ is differentiable for $(x, y) \neq (0, 0)$ because the partial derivatives exist and are continuous except at $(0, 0)$ (Figure 5):

$$h_x(x, y) = \frac{x}{\sqrt{x^2 + y^2}}, \qquad h_y(x, y) = \frac{y}{\sqrt{x^2 + y^2}}$$

However, $h(x, y)$ is not differentiable at $(0, 0)$ because $h_x(0, 0)$ and $h_y(0, 0)$ do not exist. ■

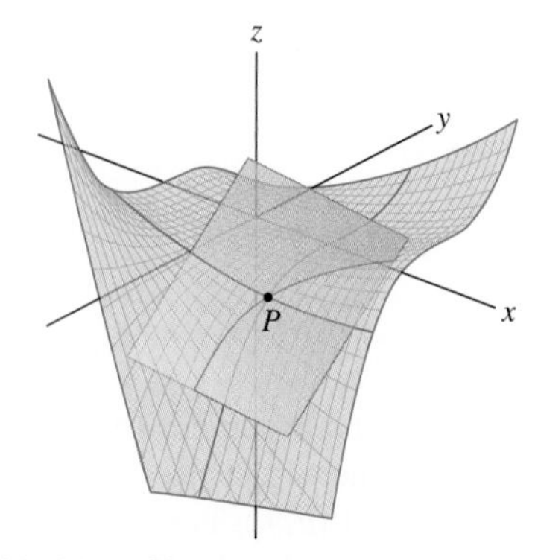

FIGURE 6 Tangent plane to the surface $f(x, y) = xy^3 + x^2$ at $(2, -2)$.

■ **EXAMPLE 5** Find the tangent plane to the graph of $f(x, y) = xy^3 + x^2$ at $(2, -2)$.

Solution We have

$$f_x(x, y) = y^3 + 2x, \qquad f_x(2, -2) = -4$$

$$f_y(x, y) = 3xy^2, \qquad f_y(2, -2) = 24$$

Since $f(2, -2) = -12$, the tangent plane has equation

$$z = -12 - 4(x - 2) + 24(y + 2)$$

This can be rewritten as $z = 44 - 4x + 24y$ (Figure 6). ■

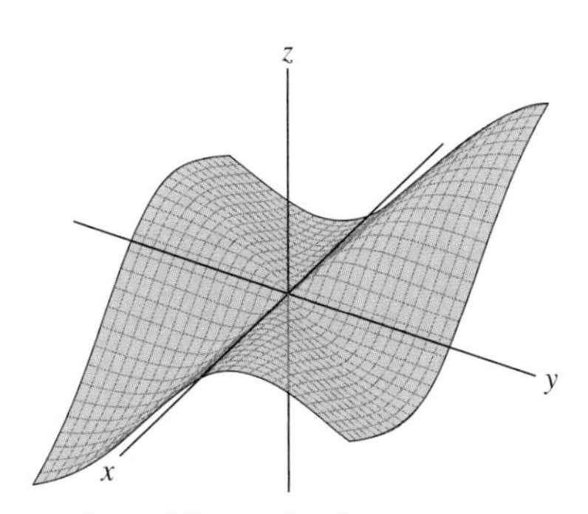

FIGURE 7 The graph of $g(x, y) = \dfrac{2xy(x+y)}{x^2+y^2}$. The partial derivatives at (0, 0) exist, but the graph is not locally linear at (0, 0).

Assumptions Matter Differentiability is a stricter requirement in several variables than in one variable. In one variable, if $f'(a)$ exists, then $f(x)$ is automatically locally linear at $x = a$. As noted above, however, in two variables, the existence of the partial derivatives $f_x(a, b)$ and $f_y(a, b)$ does not guarantee that $f(x, y)$ is locally linear at (a, b). Figure 7 shows the graph of a continuous function $g(x, y)$ that is not locally linear at $P = (0, 0)$ (and hence not differentiable), even though both $g_x(0, 0)$ and $g_y(0, 0)$ exist (see Exercise 42).

There are two useful ways of rewriting the linearization as a **linear approximation**. First, set $x = a + h$ and $y = b + k$. Then $h = x - a$, $k = y - b$, and we have

$$f(a+h, b+k) \approx f(a, b) + f_x(a, b)h + f_y(a, b)k \tag{4}$$

The linear approximation can also be written in terms of the *change in* f:

$$\Delta f = f(x, y) - f(a, b), \qquad \Delta x = x - a, \qquad \Delta y = y - b$$

$$\Delta f \approx f_x(a, b)\Delta x + f_y(a, b)\Delta y \tag{5}$$

EXAMPLE 6 Estimating Change: Body Mass Index A person's Body Mass Index is $I = \dfrac{W}{H^2}$, where W is the body weight (in kilograms) and H is the body height (in meters). The Body Mass Index (BMI) for a child of mass $W = 17$ kg and height $H = 1.05$ m is $I = 17/1.05^2 \approx 15.4$. Estimate ΔI if (W, H) changes to (18.5, 1.07).

BMI is one factor used by health care professionals to assess the risk of certain diseases such as diabetes or high blood pressure. The range $18.5 \le I \le 24.9$ is considered normal for adults over 20 years of age.

Solution We compute the partial derivatives at $(W, H) = (17, 1.05)$:

$$\frac{\partial I}{\partial W} = \frac{\partial}{\partial W}\left(\frac{W}{H^2}\right) = \frac{1}{H^2}, \qquad \frac{\partial I}{\partial H} = \frac{\partial}{\partial H}\left(\frac{W}{H^2}\right) = -\frac{2W}{H^3}$$

$$\left.\frac{\partial I}{\partial W}\right|_{(17,1.05)} = \frac{1}{1.05^2} \approx 0.91, \qquad \left.\frac{\partial I}{\partial H}\right|_{(17,1.05)} = -\frac{2(17)}{1.05^3} \approx -29.37$$

The linear approximation in delta form is

$$\Delta I \approx 0.91\Delta W - 29.37\Delta H$$

If (W, H) changes from (17, 1.05) to (18.5, 1.07), then $\Delta W = 1.5$ and $\Delta H = 0.02$. Thus,

$$\Delta I \approx 0.91(1.5) - 29.37(0.02) \approx 0.8$$

The linear approximation applies to functions of more than two variables. In three variables, the linear approximation at (a, b, c) is

$$f(a+h, b+k, c+\ell) \approx f(a, b, c) + f_x(a, b, c)h + f_y(a, b, c)k + f_z(a, b, c)\ell$$

Furthermore, the linear approximation is often expressed in terms of differentials:

$$df = f_x(x, y, z)\,dx + f_y(x, y, z)\,dy + f_z(x, y, z)\,dz = \frac{\partial f}{\partial x}dx + \frac{\partial f}{\partial y}dy + \frac{\partial f}{\partial x}dz$$

In this text, we use Δ-notation rather than differentials:

$$\Delta f \approx f_x(x, y, z)\Delta x + f_y(x, y, z)\Delta y + f_z(x, y, z)\Delta z$$

← REMINDER the percentage error is equal to

$$\left|\frac{\text{error} \times 100}{\text{actual value}}\right| \%$$

EXAMPLE 7 Estimate $(3.99)^3(1.01)^4(1.98)^{-1}$ using the linear approximation and use a calculator to find the percentage error.

Solution Think of $(3.99)^3(1.01)^4(1.98)^{-1}$ as a value of $f(x, y, z) = x^3y^4z^{-1}$. Then

$$f(3.99, 1.01, 1.98) = (3.99)^3(1.01)^4(1.98)^{-1}$$

and it makes sense to use the linear approximation at $(4, 1, 2)$. We have

$$f(x, y, z) = x^3y^4z^{-1}, \qquad f(4, 1, 2) = (4^3)(1^4)(2^{-1}) = 32$$

$$f_x(x, y, z) = 3x^2y^4z^{-1}, \qquad f_x(4, 1, 2) = 24$$

$$f_y(x, y, z) = 4x^3y^3z^{-1}, \qquad f_y(4, 1, 2) = 128$$

$$f_z(x, y, z) = -x^3y^4z^{-2}, \qquad f_z(4, 1, 2) = -16$$

The linear approximation yields

$$f(4 + h, 1 + k, 2 + \ell) \approx 32 + 24h + 128k - 16\ell \qquad \boxed{6}$$

We apply (6) with $h = -0.01$, $k = 0.01$, and $\ell = -0.02$:

$$(3.99)^3(1.01)^4(1.98)^{-1} \approx 32 + 24(-0.01) + 128(0.01) - 16(-0.02) = 33.36$$

Using a calculator, we find that $(3.99)^3(1.01)^4(1.98)^{-1} \approx 33.384$, so the error is less than 0.025 and the percentage error is at most

$$\text{Percentage error} \approx \frac{(0.025)}{33.384} \times 100 \approx 0.075\%$$ ■

14.4 SUMMARY

- The *linearization* of $f(x, y)$ at (a, b) is the linear function

$$L(x, y) = f(a, b) + f_x(a, b)(x - a) + f_y(a, b)(y - b)$$

- The linearization in three variables is

$$f(x, y, z) \approx f(a, b, c) + f_x(a, b, c)(x - a) + f_y(a, b, c)(y - b) + f_z(a, b, c)(z - c)$$

- A function $f(x, y)$ is *locally linear* at (a, b) if $f_x(a, b)$ and $f_y(a, b)$ exist and $L(x, y)$ approximates $f(x, y)$ to first order at (a, b). This means that the error $E(x, y) = f(x, y) - L(x, y)$ has the form

$$E(x, y) = \epsilon(x, y)\sqrt{(x - a)^2 + (y - b)^2}, \quad \text{where} \quad \lim_{(x,y)\to(a,b)} \epsilon(x, y) = 0$$

- There are three equivalent ways of writing the linear approximation:

$$f(x, y) \approx f(a, b) + f_x(a, b)(x - a) + f_y(a, b)(y - b)$$

$$f(a + h, b + k) \approx f(a, b) + f_x(a, b)h + f_y(a, b)k$$

Or setting $\Delta f = f(x, y) - f(a, b)$, $\Delta x = x - a$, $\Delta y = y - b$,

$$\boxed{\Delta f \approx f_x(a, b)\, \Delta x + f_y(a, b)\, \Delta y}$$

- By definition, $f(x, y)$ is differentiable at (a, b) if $f_x(a, b)$ and $f_y(a, b)$ exist and $f(x, y)$ is locally linear at (a, b).
- In practice, the following result is used to check differentiability: *If $f_x(x, y)$ and $f_y(x, y)$ exist and are continuous in a disk D containing (a, b), then $f(x, y)$ is differentiable at (a, b).*
- An equation of the tangent plane to $z = f(x, y)$ at (a, b) is

$$z = f(a, b) + f_x(a, b)(x - a) + f_y(a, b)(y - b)$$

14.4 EXERCISES

Preliminary Questions

1. How is the linearization of $f(x, y)$ at (a, b) defined?

2. Define local linearity for functions of two variables.

In Questions 3–5, assume that

$$f(2, 3) = 8, \qquad f_x(2, 3) = 5, \qquad f_y(2, 3) = 7$$

3. Which of (a)–(b) is the linearization of f at $(2, 3)$?

(a) $L(x, y) = 8 + 5x + 7y$

(b) $L(x, y) = 8 + 5(x - 2) + 7(y - 3)$

4. Estimate $f(2, 3.1)$.

5. Estimate Δf at $(2, 3)$ if $\Delta x = -0.3$ and $\Delta y = 0.2$.

6. Which theorem allows us to conclude that $f(x, y) = x^3 y^8$ is differentiable?

Exercises

1. Let $f(x, y) = x^2 y^3$.

(a) Find the linearization of f at $(a, b) = (2, 1)$.

(b) Use the linear approximation to estimate $f(2.01, 1.02)$ and $f(1.97, 1.01)$, and compare your estimates with the values obtained using a calculator.

2. Write the linear approximation to $f(x, y) = x(1 + y)^{-1}$ at $(a, b) = (1, 5)$ in the form

$$f(a + h, b + k) \approx f(a, b) + f_x(a, b)h + f_y(a, b)k$$

Use it to estimate $0.95/5.02$ and compare the estimate with the value obtained using a calculator.

3. Let $f(x, y) = x^3 y^{-4}$. Use Eq. (5) to estimate the change $\Delta f = f(2.03, 0.9) - f(2, 1)$.

4. Assume that $f(3, 2) = 2$, $f_x(3, 2) = -1$, and $f_y(3, 2) = 3$. Use the linear approximation to estimate $f(3.1, 1.8)$.

5. Use the linear approximation of $f(x, y) = e^{x^2 + y}$ at $(0, 0)$ to estimate $f(0.01, -0.02)$. Compare with the value obtained using a calculator.

6. Let $f(x, y) = \dfrac{x^2}{y^2 + 1}$. Use the linear approximation at an appropriate point (a, b) to estimate $f(4.01, 0.98)$.

7. Use Eq. (3) to find an equation of the tangent plane to the graph of $f(x, y) = 2x^2 - 4xy^2$ at the point $(-1, 2)$.

8. Let $f(x, y) = e^{x/y}$ and $P = (2, 1, e^2)$.

(a) Find an equation of the tangent plane to the graph of $f(x, y)$ at P.

(b) CAS Plot $f(x, y)$ and the tangent plane at P on the same screen to illustrate local linearity.

9. Find the linear approximation to $f(x, y, z) = \dfrac{xy}{z}$ at the point $(2, 1, 2)$.

10. Assume that $f(1, 0, 0) = -3$, $f_x(1, 0, 0) = -2$, $f_y(1, 0, 0) = 4$, and $f_z(1, 0, 0) = 2$. Use the linear approximation to estimate $f(1.02, 0.01, -0.03)$.

In Exercises 11–18, find an equation of the tangent plane at the given point.

11. $f(x, y) = x^2 y + xy^3$, $(2, 1)$

12. $f(x, y) = \dfrac{x}{\sqrt{y}}$, $(4, 4)$

13. $f(x, y) = x^2 + y^{-2}$, $(4, 1)$

14. $G(u, w) = \sin(uw)$, $(\frac{\pi}{6}, 1)$

15. $F(r, s) = r^2 s^{-1/2} + s^{-3}$, $(2, 1)$

16. $H(s, t) = se^{st}$, $(0, 0)$

17. $f(x, y) = e^x \ln y$, $(0, 1)$

18. $f(x, y) = \ln(x^2 + y^2)$, $(1, 1)$

19. Find the points on the graph of $z = 3x^2 - 4y^2$ at which the vector $\mathbf{n} = \langle 3, 2, 2 \rangle$ is normal to the tangent plane.

20. Find the points on the graph of $z = x^2e^y$ at which the tangent plane is parallel to $5x - 2y + \frac{1}{2}z = 0$.

21. CAS Use a computer algebra system to plot the graph of $f(x, y) = x^2 + 3xy + y$ together with the tangent plane at $(x, y) = (1, 1)$ on the same screen.

22. Suppose that the plane tangent to the surface $z = f(x, y)$ at $(-2, 3, 4)$ has equation $z + 4x + 2y = 2$. Estimate $f(-2.1, 3.1)$.

23. The following values are given:

$$f(1, 2) = 10, \qquad f(1.1, 2.01) = 10.3, \qquad f(1.04, 2.1) = 9.7$$

Find an approximation to the equation of the tangent plane to the graph of f at $(1, 2, 10)$.

24. Use the linear approximation to $f(x, y) = \sqrt{x/y}$ at $(9, 4)$ to estimate $\sqrt{9.1/3.9}$.

In Exercises 25–30, use the linear approximation to estimate the value. Compare with the value given by a calculator.

25. $(2.01)^3(1.02)^2$

26. $4.1/7.9$

27. $\sqrt{3.01^2 + 3.99^2}$

28. $\dfrac{0.98^2}{2.01^3 + 1}$

29. $\sqrt{(1.9)(2.02)(4.05)}$

30. $\dfrac{8.01}{\sqrt{(1.99)(2.01)}}$

31. Estimate $f(2.1, 3.8)$ given that $f(2, 4) = 5$, $f_x(2, 4) = 0.3$, and $f_y(2, 4) = -0.2$.

In Exercises 32–34, let $I = W/H^2$ denote the BMI described in Example 6.

32. A boy has weight $W = 34$ kg and height $H = 1.3$. Use the linear approximation to estimate the change in I if (W, H) changes to $(36, 1.32)$.

33. Suppose that $(W, H) = (34, 1.3)$ and W increases to 35. Use the linear approximation to estimate the increase in H required to keep I constant.

34. (a) Show that $\Delta I \approx 0$ if $\Delta H/\Delta W \approx H/2W$.

(b) Suppose that $(W, H) = (25, 1.1)$. What increase in H will leave I (approximately) constant if W is increased by 1 kg?

35. The volume of a cylinder of radius r and height h is $V = \pi r^2 h$.

(a) Use the linear approximation to show that

$$\frac{\Delta V}{V} \approx \frac{2\Delta r}{r} + \frac{\Delta h}{h}$$

(b) Calculate the percentage increase in V if r and h are each increased by 2%.

(c) The volume of a certain cylinder V is determined by measuring r and h. Which will lead to a greater error in V: a 1% error in r or a 1% error in h?

36. Use the linear approximation to show that if $I = x^a y^b$, then

$$\frac{\Delta I}{I} \approx a\frac{\Delta x}{x} + b\frac{\Delta y}{y}$$

37. The monthly payment for a home loan is given by a function $f(P, r, N)$, where P is the principal (the initial size of the loan), r the interest rate, and N the length of the loan in months. Interest rates are expressed as a decimal: A 6% interest rate is denoted by $r = 0.06$. If $P = \$100{,}000$, $r = 0.06$, and $N = 240$ (a 20-year loan), then the monthly payment is $f(100{,}000, 0.06, 240) = 716.43$. Furthermore, with these values, we have

$$\frac{\partial f}{\partial P} = 0.0071, \qquad \frac{\partial f}{\partial r} = 5{,}769, \qquad \frac{\partial f}{\partial N} = -1.5467$$

Estimate:

(a) The change in monthly payment per \$1,000 increase in loan principal.

(b) The change in monthly payment if the interest rate increases to $r = 6.5\%$ and $r = 7\%$.

(c) The change in monthly payment if the length of the loan increases to 24 years.

38. Automobile traffic passes a point P on a road of width w ft at an average rate of R vehicles per second. The arrival of automobiles is irregular, however, and traffic engineers have found that the formula $T = te^{Rt}$ gives a good model for the average waiting time T until there is a gap in traffic of at least t seconds. A pedestrian walking at a speed of 3.5 ft/s (5.1 mph) requires $t = w/3.5$ s to cross the road. Therefore, the average time the pedestrian will have to wait before crossing is $f(w, R) = (w/3.5)e^{wR/3.5}$ s.

(a) What is the pedestrian's average waiting time if $w = 25$ ft and $R = 0.2$ vehicles per second?

(b) Use the linear approximation to estimate the increase in waiting time if w is increased to 27 ft.

(c) Estimate the waiting time if the width is increased to 27 ft and R decreases to 0.18.

(d) What is the rate of increase in waiting time per 1-ft increase in width when $w = 30$ and $R = 0.3$?

39. The volume V of a cylinder is computed using the values 3.5 m for diameter and 6.2 m for height. Use the linear approximation to estimate the maximum error in V if each of these values has a possible error of at most 5%.

Further Insights and Challenges

40. Show that if $f(x, y)$ is differentiable at (a, b), then the function of one variable $f(x, b)$ is differentiable at $x = a$. Use this to prove that $f(x, y) = \sqrt{x^2 + y^2}$ is *not* differentiable at $(0, 0)$.

41. Assume that $f(0) = 0$. By the discussion in this section, $f(x)$ is differentiable at $x = 0$ if there is a constant M such that

$$f(h) = Mh + h\epsilon(h) \qquad \boxed{7}$$

where $\lim_{h\to 0} \epsilon(h) = 0$ [in this case, $M = f'(0)$].

(a) Use this definition to verify that $f(x) = 1/(x+1)$ is differentiable with $M = -1$ and $E(h) = h/(h+1)$.

(b) Use this definition to verify that $f(x) = x^{3/2}$ is differentiable with $M = 0$.

(c) Show that $f(x) = x^{1/2}$ is not differentiable at $x = 0$ by showing that if M is any constant, and if we write $\sqrt{h} = Mh + h\epsilon(h)$, then the function $\epsilon(h)$ does not approach zero as $h \to 0$.

42. Assumptions Matter Recall that if $f(x, y)$ is differentiable at (a, b), then the partial derivatives $f_x(a, b)$ and $f_y(a, b)$ exist. This exercise shows that the mere existence of $f_x(a, b)$ and $f_y(a, b)$ does not imply $f(x, y)$ is differentiable. Define $g(x, y) = \dfrac{2xy(x+y)}{x^2+y^2}$ for $(x, y) \neq 0$ and $g(0, 0) = 0$.

(a) Prove that $|g(x, y)| \le |x + y|$. *Hint:* Use the inequality

$$(x \pm y)^2 \ge 0$$

to show that $\left|\dfrac{2xy}{x^2+y^2}\right| \le 1$.

(b) Use (a) to show that $g(x, y)$ is continuous at $(0, 0)$.

(c) Use the limit definitions to show that $g_x(0, 0)$ and $g_y(0, 0)$ exist and both are equal to zero.

(d) Show that if a function $f(x, y)$ is locally linear at $(0, 0)$ and $f(0, 0) = 0$, then $f(h, h)/h$ approaches 0 as $h \to 0$. Show that $g(x, y)$ does not have this property and hence is not differentiable at $(0, 0)$.

(e) Explain why $g_x(x, y)$ and $g_y(x, y)$ cannot both be continuous at $(0, 0)$. In fact, since $g(x, y)$ is symmetric in x and y, both partial derivatives are discontinuous at $(0, 0)$.

(f) Show directly that $\lim_{(x,y)\to(0,0)} g_x(x, y)$ does not exist by verifying that $g_x(x, 0) = 0$ for $x \neq 0$ and $g_x(0, y) = 2$ for $y \neq 0$.

14.5 The Gradient and Directional Derivatives

In previous sections, we have emphasized that the rate of change of a function f of several variables depends on the direction of change. Since directions are indicated by vectors, it is natural to use vectors to describe the derivative of f in a specified direction.

To do this, we first introduce the **gradient** ∇f_P, which is the vector whose components are the partial derivatives of f at P.

The gradient of a function of n variables is the vector

$$\nabla f = \left\langle \frac{\partial f}{\partial x_1}, \frac{\partial f}{\partial x_2}, \dots, \frac{\partial f}{\partial x_n} \right\rangle$$

DEFINITION The Gradient The gradient of a function $f(x, y)$ at a point $P = (a, b)$ is the vector

$$\nabla f_P = \langle f_x(a, b), f_y(a, b) \rangle$$

In three variables, if $P = (a, b, c)$,

$$\nabla f_P = \langle f_x(a, b, c), f_y(a, b, c), f_z(a, b, c) \rangle$$

The symbol ∇, called "del," is an upside-down Greek delta. The use of ∇ for the gradient was popularized by the Scottish physicist P. G. Tait (1831–1901) who called the symbol "nabla." Nabla means harp in Hebrew and refers to ∇'s resemblance to an ancient ten-string harp. The great physicist James Clerk Maxwell was reluctant to adopt this term and referred to the gradient simply as the "slope." He wrote jokingly to his friend Tait in 1871: "Still harping on that nabla?"

We also write $\nabla f_{(a,b)}$ or $\nabla f(a, b)$ for the gradient. Sometimes, we omit reference to the point P and write

$$\nabla f = \left\langle \frac{\partial f}{\partial x}, \frac{\partial f}{\partial y} \right\rangle \qquad \text{or} \qquad \nabla f = \left\langle \frac{\partial f}{\partial x}, \frac{\partial f}{\partial y}, \frac{\partial f}{\partial z} \right\rangle$$

The gradient ∇f "assigns" a vector ∇f_P to each point in the domain of f.

EXAMPLE 1 Drawing Gradient Vectors Find the gradient of $f(x, y) = x^2 + y^2$ at $P = (1, 1)$ and draw several gradient vectors.

Solution The partial derivatives are $f_x(x, y) = 2x$ and $f_y(x, y) = 2y$, so

$$\nabla f = \langle 2x, 2y \rangle$$

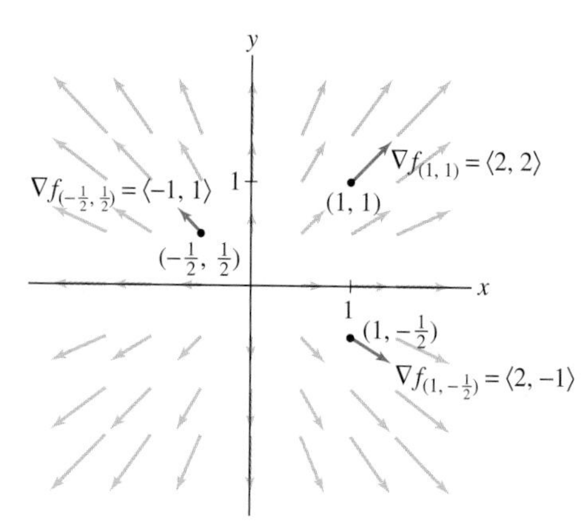

FIGURE 1 Gradient vectors of $f(x, y) = x^2 + y^2$ at several points (vectors not drawn to scale).

At $P = (1, 1)$, we have $\nabla f_P = \nabla f(1, 1) = \langle 2, 2\rangle$. Figure 1 shows gradient vectors of $f(x, y)$ at several points. ■

■ **EXAMPLE 2 Gradient in Three Variables** Calculate the vector $\nabla f_{(3,-2,4)}$, where $f(x, y, z) = ze^{2x+3y}$.

Solution The partial derivatives are

$$\frac{\partial f}{\partial x} = 2ze^{2x+3y}, \qquad \frac{\partial f}{\partial y} = 3ze^{2x+3y}, \qquad \frac{\partial f}{\partial z} = e^{2x+3y}$$

Therefore, $\nabla f = \langle 2ze^{2x+3y}, 3ze^{2x+3y}, e^{2x+3y}\rangle$, and (see Figure 2)

$$\nabla f_{(3,-2,4)} = \langle 2 \cdot 4e^0, 3 \cdot 4e^0, e^0\rangle = \langle 8, 12, 1\rangle$$ ■

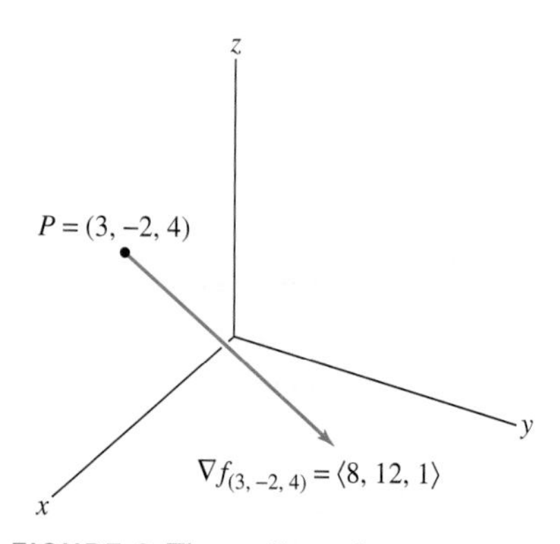

FIGURE 2 The gradient of $f(x, y, z) = ze^{2x+3y}$ at $P = (3, -2, 4)$.

The following theorem lists some useful properties of the gradient. The proofs are left as exercises (see Exercises 61–63).

THEOREM 1 Properties of the Gradient If $f(x, y, z)$ and $g(x, y, z)$ are differentiable and c is a constant, then:

(i) $\nabla(f + g) = \nabla f + \nabla g$

(ii) $\nabla(cf) = c\nabla f$

(iii) Product Rule for Gradients: $\nabla(fg) = f\nabla g + g\nabla f$

(iv) Chain Rule for Gradients: If $F(t)$ is a differentiable function of one variable, then

$$\nabla(F(f(x, y, z)) = F'(f(x, y, z))\nabla f \qquad (1)$$

■ **EXAMPLE 3 Using the Chain Rule for Gradients** Find the gradient of $g(x, y, z) = (x^2 + y^2 + z^2)^8$.

Solution Apply Eq. (1) with $F(t) = t^8$ and $f(x, y, z) = x^2 + y^2 + z^2$. Then $g(x, y, z) = F(f(x, y, z))$ and

$$\nabla g = \nabla\big((x^2 + y^2 + z^2)^8\big) = 8(x^2 + y^2 + z^2)^7 \nabla(x^2 + y^2 + z^2)$$
$$= 8(x^2 + y^2 + z^2)^7 \langle 2x, 2y, 2z\rangle = 16(x^2 + y^2 + z^2)^7 \langle x, y, z\rangle$$ ■

The Chain Rule for Paths

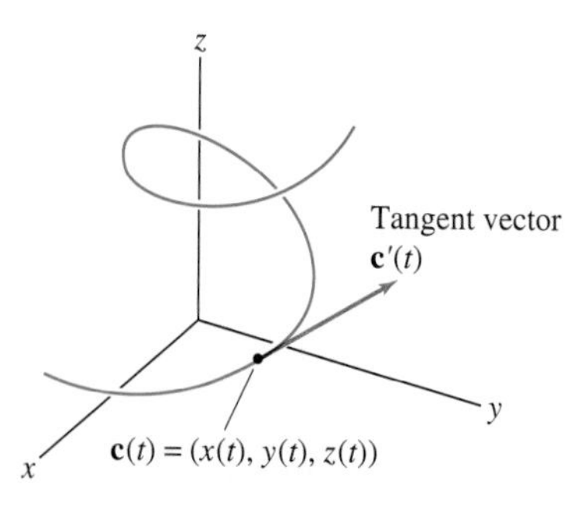

FIGURE 3 Tangent vector $\mathbf{c}'(t)$ to a path $\mathbf{c}(t) = (x(t), y(t), z(t))$.

As a first application of the gradient, we discuss the Chain Rule for Paths. Recall that in Chapter 13, we represented a path (or parametrized curve) in $\mathbf{R}^3$ by a vector-valued function $\mathbf{r}(t) = \langle x(t), y(t), z(t)\rangle$. In this chapter, we change notation and represent such a path instead by a function $\mathbf{c}(t) = (x(t), y(t), z(t))$ (Figure 3). We think of $\mathbf{c}(t)$ as a moving point rather than a moving vector. As in Chapter 13, the derivative $\mathbf{c}'(t)$ is the tangent or "velocity" vector that points in the direction of motion tangent to the path:

$$\mathbf{c}'(t) = \langle x'(t), y'(t), z'(t)\rangle$$

We use similar notation for paths in $\mathbf{R}^2$.

The Chain Rule for Paths is different from the more elementary Chain Rule for Gradients stated in Eq. (1) above. Suppose that $f(x, y)$ is a function of two variables and let

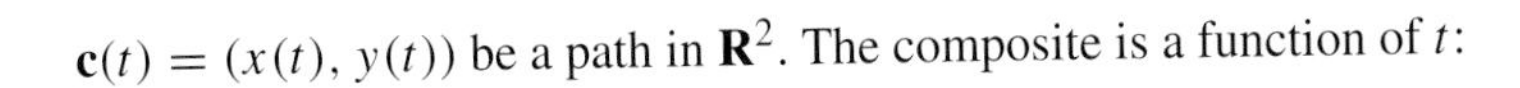

$\mathbf{c}(t) = (x(t), y(t))$ be a path in $\mathbf{R}^2$. The composite is a function of t:

$$f(\mathbf{c}(t)) = f(x(t), y(t))$$

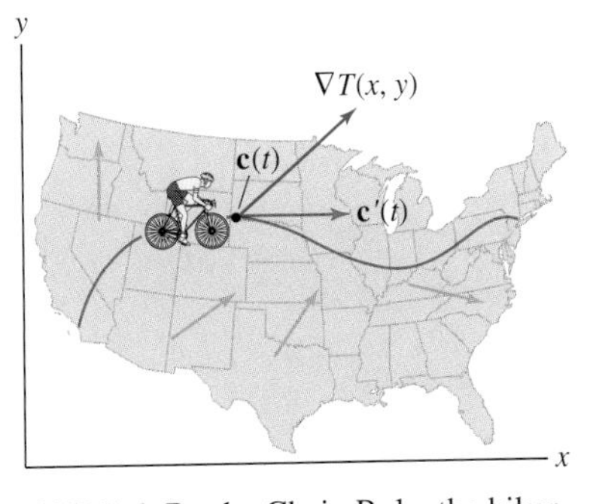

FIGURE 4 By the Chain Rule, the biker feels the temperature T changing at a rate equal to the dot product of the temperature gradient ∇T and the velocity vector $\mathbf{c}'(t)$.

What is the idea behind this type of composite function? Imagine riding a bike along the path $\mathbf{c}(t) = (x(t), y(t))$ and let $T(x, y)$ be the outside temperature at location (x, y) (Figure 4). Then $T(\mathbf{c}(t))$ is the temperature at your location $\mathbf{c}(t)$ at time t. The derivative $\frac{d}{dt}T(\mathbf{c}(t))$ is the rate at which the outside temperature changes as you ride along the path. The Chain Rule states that you will observe the temperature changing at a rate equal to the dot product of the temperature gradient ∇T and the velocity vector $\mathbf{c}'(t)$.

THEOREM 2 Chain Rule for Paths If f is a differentiable function and $\mathbf{c}(t)$ is a differentiable path, then

$$\frac{d}{dt} f(\mathbf{c}(t)) = \nabla f_{\mathbf{c}(t)} \cdot \mathbf{c}'(t)$$

Explicitly, in the case of two variables, if $\mathbf{c}(t) = (x(t), y(t))$ is a differentiable path in $\mathbf{R}^2$, then

$$\frac{d}{dt} f(\mathbf{c}(t)) = \left\langle \frac{\partial f}{\partial x}, \frac{\partial f}{\partial y} \right\rangle \cdot \left\langle x'(t), y'(t) \right\rangle = \frac{\partial f}{\partial x}\frac{dx}{dt} + \frac{\partial f}{\partial y}\frac{dy}{dt}$$

Proof We derive this result using the linear approximation. By definition,

$$\frac{d}{dt} f(\mathbf{c}(t)) = \lim_{h \to 0} \frac{f(x(t+h), y(t+h)) - f(x(t), y(t))}{h}$$

To calculate this derivative, set

$$\Delta x = x(t+h) - x(t), \qquad \Delta y = y(t+h) - y(t)$$

Since $f(x, y)$ is differentiable, we may apply the linear approximation (Section 14.4):

$$\begin{aligned} &f(x(t+h), y(t+h)) - f(x(t), y(t)) \\ &\qquad = f_x(x(t), y(t))\Delta x + f_y(x(t), y(t))\Delta y + \epsilon(\Delta x, \Delta y)\sqrt{\Delta x^2 + \Delta y^2} \end{aligned}$$

where $\epsilon(\Delta x, \Delta y) \to 0$ as Δx and Δy tend to zero. Now set $h = \Delta t$ and for $\Delta t \neq 0$, divide by Δt:

$$\begin{aligned} &\frac{f(x(t+h), y(t+h)) - f(x(t), y(t))}{h} \\ &\qquad = f_x(x(t), y(t))\frac{\Delta x}{\Delta t} + f_y(x(t), y(t))\frac{\Delta y}{\Delta t} + \epsilon(\Delta x, \Delta y)\sqrt{\left(\frac{\Delta x}{\Delta t}\right)^2 + \left(\frac{\Delta y}{\Delta t}\right)^2} \end{aligned}$$

As $h = \Delta t$ tends to zero, the left-hand side approaches $\frac{d}{dt} f(\mathbf{c}(t))$ and the difference quotients $\Delta x/\Delta t$ and $\Delta y/\Delta t$ on the right approach dx/dt and dy/dt, respectively. Since $\epsilon(\Delta x, \Delta y)$ approaches zero, we obtain the desired result:

$$\frac{d}{dt}f(\mathbf{c}(t)) = \lim_{h\to 0}\frac{f(x(t+h), y(t+h)) - f(x(t), y(t))}{h}$$

$$= f_x(x(t), y(t))\lim_{\Delta t\to 0}\frac{\Delta x}{\Delta t} + f_y(x(t), y(t))\lim_{\Delta t\to 0}\frac{\Delta y}{\Delta t}$$

$$= f_x(x(t), y(t))\frac{dx}{dt} + f_y(x(t), y(t))\frac{dy}{dt} = \nabla f_{\mathbf{c}(t)}\cdot \mathbf{c}'(t)$$ ■

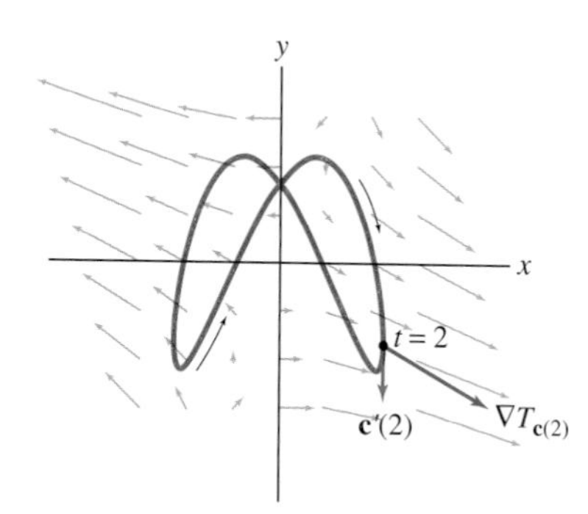

FIGURE 5 The path $\mathbf{c}(t) = (\cos(t-2), \sin 2t)$.

EXAMPLE 4 A bug carries a tiny thermometer along the path

$$\mathbf{c}(t) = (\cos(t-2), \sin 2t)$$

(t in seconds) as in Figure 5. How fast is the temperature changing at $t = 2$ if the temperature at location (x, y) is $T(x, y) = x^2 - xy + 60\,^\circ\text{F}$?

Solution The bug's location at $t = 2$ is $\mathbf{c}(2) = (\cos 0, \sin 4) \approx (1, -0.76)$. We compute the gradient ∇T and evaluate it at $\mathbf{c}(2)$:

$$\nabla T = \langle 2x - y, -x\rangle, \qquad \nabla T_{\mathbf{c}(2)} = \langle 2 - (-0.76), -1\rangle = \langle 2.76, -1\rangle$$

Next, we compute the tangent vector

$$\mathbf{c}'(t) = \langle -\sin(t-2), 2\cos 2t\rangle, \qquad \mathbf{c}'(2) = \langle 0, 2\cos 4\rangle \approx \langle 0, -1.31\rangle$$

According to the Chain Rule, the temperature changes at the rate

$$\left.\frac{d}{dt}T(\mathbf{c}(t))\right|_{t=2} = \nabla T_{\mathbf{c}(2)}\cdot \mathbf{c}'(2) = \langle 2.76, -1\rangle \cdot \langle 0, -1.31\rangle = 1.31^\circ\text{F/s}$$ ■

The Chain Rule holds for functions of n variables. If $f(x_1, \dots, x_n)$ is differentiable and if $\mathbf{c}(t) = (x_1(t), \dots, x_n(t))$ is a differentiable path, then

$$\frac{d}{dt}f(\mathbf{c}(t)) = \nabla f \cdot \mathbf{c}'(t) = \frac{\partial f}{\partial x_1}\frac{dx_1}{dt} + \frac{\partial f}{\partial x_2}\frac{dx_2}{dt} + \cdots + \frac{\partial f}{\partial x_n}\frac{dx_n}{dt}$$

EXAMPLE 5 Calculate $\left.\frac{d}{dt}f(\mathbf{c}(t))\right|_{t=\pi/2}$, where

$$f(x, y, z) = xy + z^2 \quad \text{and} \quad \mathbf{c}(t) = (\cos t, \sin t, t)$$

Solution First, note that $\mathbf{c}(\frac{\pi}{2}) = (\cos\frac{\pi}{2}, \sin\frac{\pi}{2}, \frac{\pi}{2}) = (0, 1, \frac{\pi}{2})$. Next, evaluate ∇f and $\mathbf{c}'(t)$ at $t = \frac{\pi}{2}$:

$$\nabla f = \left\langle \frac{\partial f}{\partial x}, \frac{\partial f}{\partial y}, \frac{\partial f}{\partial z}\right\rangle = \langle y, x, 2z\rangle, \quad \nabla f_{\mathbf{c}(\pi/2)} = \nabla f\left(0, 1, \frac{\pi}{2}\right) = \langle 1, 0, \pi\rangle$$

$$\mathbf{c}'(t) = \langle -\sin t, \cos t, 1\rangle, \qquad \mathbf{c}'\left(\frac{\pi}{2}\right) = \left\langle -\sin\frac{\pi}{2}, \cos\frac{\pi}{2}, 1\right\rangle = \langle -1, 0, 1\rangle$$

By the Chain Rule,

$$\left.\frac{d}{dt}f(\mathbf{c}(t))\right|_{t=\pi/2} = \nabla f_{\mathbf{c}(\pi/2)}\cdot \mathbf{c}'\left(\frac{\pi}{2}\right) = \langle 1, 0, \pi\rangle \cdot \langle -1, 0, 1\rangle = \pi - 1$$ ■

Directional Derivatives

As mentioned at the beginning of this section, we are interested in studying the derivative of a function f in the direction of an arbitrary vector $\mathbf{v}$. To define this derivative, let $P = (a, b)$ and $\mathbf{v} = \langle h, k \rangle$. The path

$$\mathbf{c}(t) = (a + th, b + tk)$$

parametrizes the line through P in the direction of $\mathbf{v}$ with $\mathbf{c}(0) = P$ (Figure 6). The derivative of f with respect to $\mathbf{v}$, denoted $D_{\mathbf{v}}f$, is the derivative of f along this path at $t = 0$:

$$D_{\mathbf{v}}f(P) = D_{\mathbf{v}}f(a, b) = \frac{d}{dt}f(\mathbf{c}(t))\Big|_{t=0} = \lim_{t\to 0}\frac{f(\mathbf{c}(t)) - f(\mathbf{c}(0))}{t}$$

$$= \lim_{t\to 0}\frac{f(a + th, b + tk) - f(a, b)}{t}$$

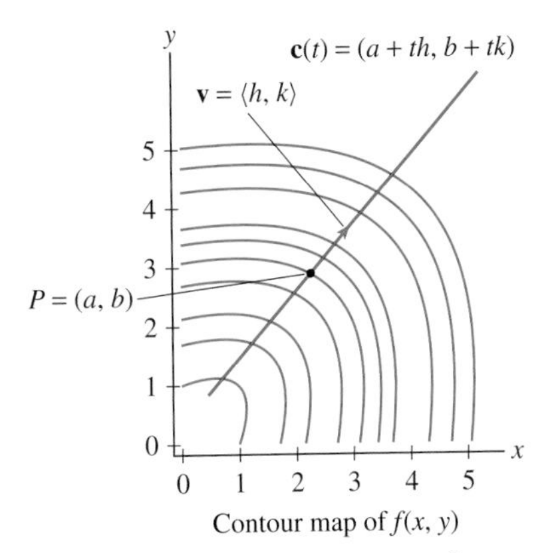

FIGURE 6 $D_{\mathbf{v}}f(a, b)$ is the rate of change of $f(\mathbf{c}(t))$ at $t = 0$.

The derivative $D_{\mathbf{v}}f$ is defined for any vector $\mathbf{v}$, but when $\mathbf{u}$ is a unit vector, $D_{\mathbf{u}}f$ is called the *directional derivative* in the direction of $\mathbf{u}$. In this case, $D_{\mathbf{u}}f$ is the rate of change of f per *unit change* in the direction of $\mathbf{u}$. We can also interpret $D_{\mathbf{u}}f(a, b)$ as the slope of the trace curve through $P' = (a, b, f(a, b))$ in the vertical plane through $P = (a, b)$ in the direction $\mathbf{u}$ (Figure 7).

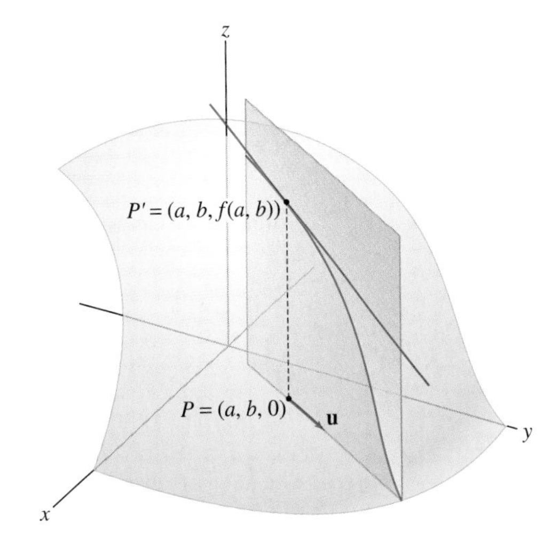

FIGURE 7 For a unit vector $\mathbf{u}$, $D_{\mathbf{u}}f(a, b)$ is the slope of the trace curve through $P' = (a, b, f(a, b))$ in the vertical plane through $P = (a, b)$ in the direction $\mathbf{u}$.

DEFINITION Directional Derivative The derivative of $f(x, y)$ at $P = (a, b)$ with respect to a nonzero vector $\mathbf{v} = \langle h, k \rangle$ is the limit (assuming it exists):

$$D_{\mathbf{v}}f(P) = D_{\mathbf{v}}f(a, b) = \lim_{t\to 0}\frac{f(a + th, b + tk) - f(a, b)}{t}$$

If $\mathbf{u}$ is a unit vector, $D_{\mathbf{u}}f$ is called the **directional derivative** in the direction $\mathbf{u}$.

By definition, $D_{\mathbf{v}}f(a, b)$ is the derivative at $t = 0$ of the composite function $f(\mathbf{c}(t))$ where $\mathbf{c}(t) = (a + th, b + tk)$. Therefore, if $f(x, y)$ is differentiable, we may compute $D_{\mathbf{v}}f(a, b)$ using the Chain Rule for Paths. We have $\mathbf{c}'(t) = \langle h, k \rangle = \mathbf{v}$, so

$$D_{\mathbf{v}} f(a, b) = \frac{d}{dt} f(\mathbf{c}(t))\Big|_{t=0} = \nabla f_{(a,b)} \cdot \mathbf{c}'(0) = \nabla f_{(a,b)} \cdot \mathbf{v}$$

We also see that for any scalar λ, $D_{\lambda\mathbf{v}} f(P) = \nabla f_P \cdot (\lambda\mathbf{v}) = \lambda \nabla f_P \cdot \mathbf{v}$. Therefore,

$$D_{\lambda\mathbf{v}} f(P) = \lambda D_{\mathbf{v}} f(P)$$

In particular, if $\mathbf{v} \neq \mathbf{0}$, then $\mathbf{u} = \dfrac{\mathbf{v}}{\|\mathbf{v}\|}$ is the unit vector in the direction $\mathbf{v}$ and the directional derivative in the direction of $\mathbf{v}$ is

$$D_{\mathbf{u}} f(P) = \frac{1}{\|\mathbf{v}\|} D_{\mathbf{v}} f(P) \qquad \boxed{2}$$

THEOREM 3 Evaluating the Derivative with Respect to v If $f(x, y)$ is differentiable, then $D_{\mathbf{v}} f(a, b)$ exists for any vector $\mathbf{v}$ and

$$D_{\mathbf{v}} f(a, b) = \nabla f_{(a,b)} \cdot \mathbf{v}$$

Similarly, in three variables, $D_{\mathbf{v}} f(a, b, c) = \nabla f_{(a,b,c)} \cdot \mathbf{v}$. These formulas can be written $D_{\mathbf{v}} f(P) = \nabla f_P \cdot \mathbf{v}$.

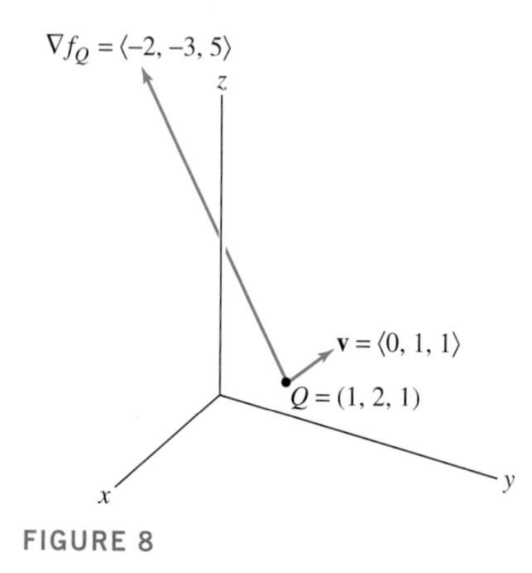

FIGURE 8

■ **EXAMPLE 6** Find the rate of change (in millibars per meter) of pressure at the point $Q = (1, 2, 1)$ in the direction of $\mathbf{v} = \langle 0, 1, 1 \rangle$ if the pressure is given by

$$f(x, y, z) = 1{,}000 + yz^2 + x^2z - xy^2 \qquad (x, y, z \text{ in meters})$$

Solution We compute the gradient at $Q = (1, 2, 1)$ and apply Theorem 3:

$$\nabla f = \left\langle \frac{\partial f}{\partial x}, \frac{\partial f}{\partial y}, \frac{\partial f}{\partial z} \right\rangle = \left\langle 2xz - y^2, z^2 - 2xy, 2yz + x^2 \right\rangle$$

$$\nabla f_Q = \nabla f_{(1,2,1)} = \langle -2, -3, 5 \rangle$$

$$D_{\mathbf{v}} f(Q) = \nabla f_Q \cdot \mathbf{v} = \langle -2, -3, 5 \rangle \cdot \langle 0, 1, 1 \rangle = -3 + 5 = 2$$

The rate of change per unit meter is the directional derivative with respect to the unit vector $\mathbf{u} = \dfrac{\mathbf{v}}{\|\mathbf{v}\|}$ (Figure 8). Since $\|\mathbf{v}\| = \sqrt{2}$, Eq. (2) yields

$$D_{\mathbf{u}} f(Q) = \frac{1}{\|\mathbf{v}\|} D_{\mathbf{v}} f(Q) = \frac{2}{\sqrt{2}} \approx 1.4 \text{ mb/m} \qquad \blacksquare$$

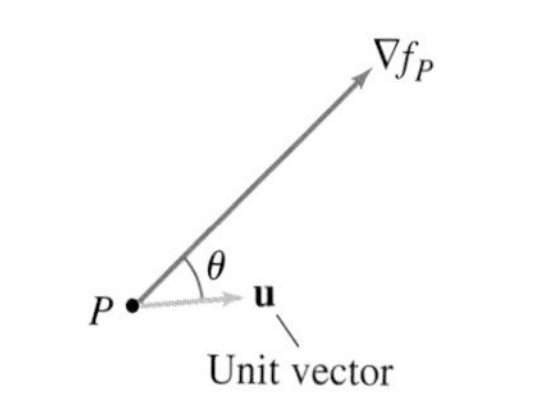

FIGURE 9 $\nabla f_P \cdot \mathbf{u} = \|\nabla f_P\| \cos\theta$.

REMINDER For any vectors $\mathbf{u}$ *and* $\mathbf{v}$, *the dot product is*

$$\mathbf{v} \cdot \mathbf{u} = \|\mathbf{v}\| \|\mathbf{u}\| \cos\theta$$

where θ *is the angle between* $\mathbf{v}$ *and* $\mathbf{u}$.

We now discuss two important geometric properties of the gradient. First, suppose that $\nabla f_P \neq 0$ and let $\mathbf{u}$ be a unit vector (Figure 9). By the properties of the dot product,

$$D_{\mathbf{u}} f(P) = \nabla f_P \cdot \mathbf{u} = \|\nabla f_P\| \cos\theta \qquad \boxed{3}$$

where θ is the angle between ∇f_P and $\mathbf{u}$. Note that $\cos\theta$ takes its maximum value when $\theta = 0$. Therefore, $D_{\mathbf{u}} f(P)$ has the largest possible value when $\theta = 0$, that is, when $\mathbf{u}$ points in the direction of ∇f_P. It follows that the *gradient vector points in the direction of the maximum rate of increase*. Similarly, f decreases most rapidly in the opposite direction, $-\nabla f_P$, since $\cos\theta = -1$ for $\theta = \pi$. In the earlier scenario of a biker measuring outside temperature, the rate of change of temperature depends on the angle between the temperature gradient ∇T and the biker's direction of motion. Temperature increases most rapidly in the direction of the gradient ∇T (Figure 10).

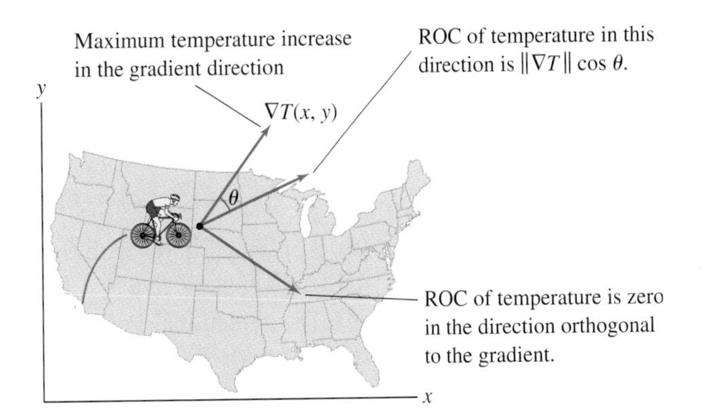

FIGURE 10 The biker should travel in the direction of the gradient to increase temperature most rapidly.

Next, suppose that $f(P) = k$, so that P lies on the level curve $f(x, y) = k$, and let us parametrize this level curve by a path $\mathbf{c}(t)$ such that $\mathbf{c}(0) = P$ and $\mathbf{c}'(0) \neq \mathbf{0}$ (this is possible whenever $\nabla f_P \neq \mathbf{0}$). Then $f(\mathbf{c}(t)) = k$ for all t, and by the Chain Rule,

$$\nabla f_P \cdot \mathbf{c}'(0) = \frac{d}{dt} f(\mathbf{c}(t))\Big|_{t=0} = \frac{d}{dt}k = 0$$

This shows that the vector ∇f_P is perpendicular to $\mathbf{c}'(0)$. However, $\mathbf{c}'(0)$ is tangent to the level curve, so we may conclude that ∇f_P is perpendicular to the level curve through P. For functions of three variables, ∇f_P is normal to the level surface $f(x, y, z) = k$ through P.

THEOREM 4 Interpretation of the Gradient Vector Assume that $\nabla f_P \neq \mathbf{0}$. Let $\mathbf{u}$ be a unit vector making an angle θ with ∇f_P. Then

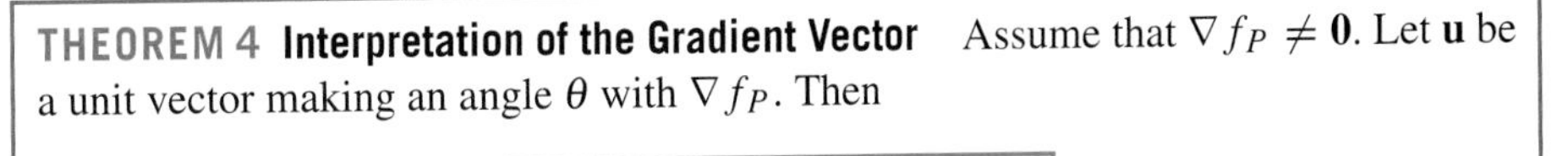

$$D_{\mathbf{u}} f(P) = \|\nabla f_P\| \cos\theta \qquad \boxed{4}$$

- ∇f_P points in the direction of maximum rate of increase of f and $-\nabla f_P$ points in the direction of maximum rate of decrease.
- ∇f_P is normal to the level curve (or surface) of f at P.

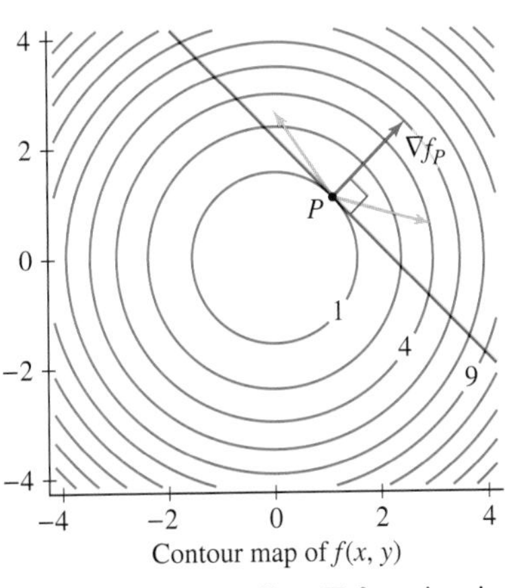

FIGURE 11 The gradient ∇f_P points in the direction of maximal increase and is perpendicular to the level curve of f at P.

GRAPHICAL INSIGHT On a contour map, the gradient direction at P is perpendicular to the level curve through P. This is the direction along which we cross the largest number of contour lines per unit distance. In Figure 11, ∇f_P spans three contour lines, while vectors of the same length pointing in other directions cross fewer lines. Thus, f increases more slowly in these other directions.

■ **EXAMPLE 7** Figure 12 shows the contour map of $f(x, y) = 25 + (x + y^2)e^{-0.3y^2}$ describing the temperature (in degrees Celsius) of a rectangular metal plate (x, y in inches). Find the directional derivative of temperature at $P = (-1, -1)$ in the direction of:

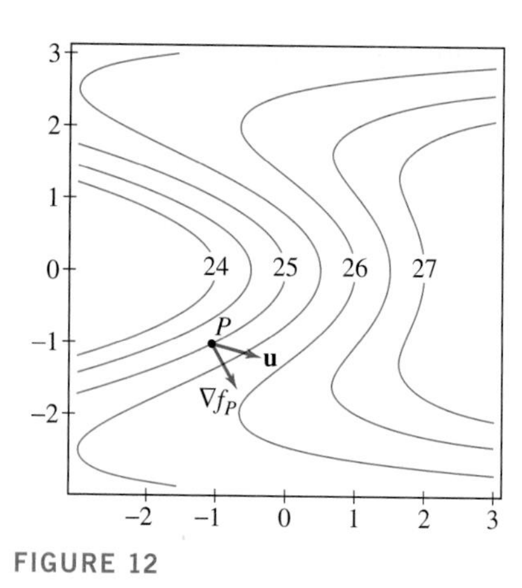

FIGURE 12

By Eq. (4), the directional derivative in the direction of the gradient itself ($\theta = 0$) is equal to $\|\nabla f_P\|$.

(a) The gradient.

(b) The unit vector **u** making an angle of $\theta = 45^\circ$ with the gradient.

Solution To apply Eq. (4), we compute the length of the gradient at $P = (-1, -1)$:

$$f_x(x, y) = e^{-0.3y^2}, \qquad f_y(x, y) = y(2 - 0.6x - 0.6y^2)e^{-0.3y^2}$$

$$f_x(P) = e^{-0.3} \approx 0.74, \qquad f_y(P) = -2e^{-0.3} \approx -1.48$$

Hence, $\nabla f_P \approx \langle 0.74, -1.48 \rangle$ and

$$\|\nabla f_P\| \approx \sqrt{0.74^2 + (-1.48)^2} \approx 1.65^\circ\text{C/in.}$$

(a) Applying Eq. (4) with $\theta = 0$, we see that the directional derivative in the direction of the gradient is $\|\nabla f_P\| \approx 1.65^\circ$C/in.

(b) The directional derivative in the direction of a unit vector **u** making an angle of $\theta = 45^\circ$ with ∇f_P is

$$D_{\mathbf{u}} f(P) = \|\nabla f_P\| \cos 45^\circ \approx 1.65 \left(\frac{\sqrt{2}}{2}\right) \approx 1.17^\circ\text{C/in.} \qquad \blacksquare$$

Another application of the gradient is to finding an equation for the tangent plane to a surface defined implicitly by an equation $F(x, y, z) = k$, where k is a constant. Let $P = (a, b, c)$ and assume that $\nabla F_P \neq \mathbf{0}$. According to Theorem 4, ∇F_P is a normal vector to the level surface $F(x, y, z) = k$. Therefore, the tangent plane has equation

$$\nabla F_P \cdot \langle x - a, y - b, z - c \rangle = 0$$

or

$$F_x(a, b, c)(x - a) + F_y(a, b, c)(y - b) + F_z(a, b, c)(z - c) = 0$$

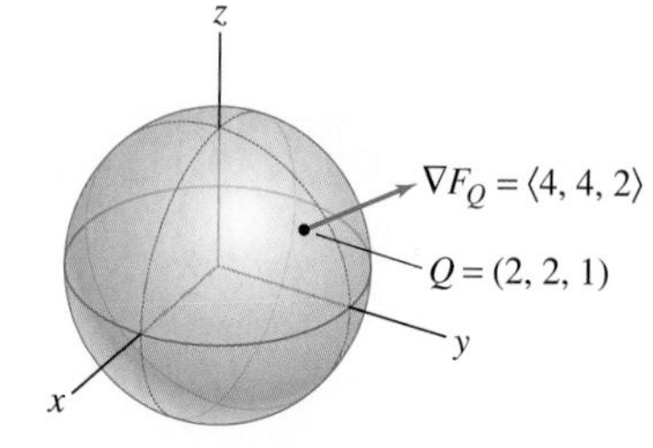

FIGURE 13 The sphere $F(x, y, z) = x^2 + y^2 + z^2 - 9 = 0$.

■ **EXAMPLE 8 Finding a Normal Vector** Let P be a point on the sphere with equation $x^2 + y^2 + z^2 = 9$. Show that the radial vector $\overrightarrow{OP}$ is normal to the sphere. Find a vector normal at $Q = (2, 2, 1)$.

Solution Let $F(x, y, z) = x^2 + y^2 + z^2$ and $P = (x, y, z)$. Then

$$\nabla F_P = \langle 2x, 2y, 2z \rangle = 2\langle x, y, z \rangle$$

Since ∇F_P is normal to the sphere at P and is a multiple of the radial vector $\overrightarrow{OP} = \langle x, y, z \rangle$, we conclude that $\overrightarrow{OP}$ is normal to the sphere. A normal vector at $Q = (2, 2, 1)$ is $\nabla F_Q = \langle 4, 4, 2 \rangle$ (Figure 13). $\blacksquare$

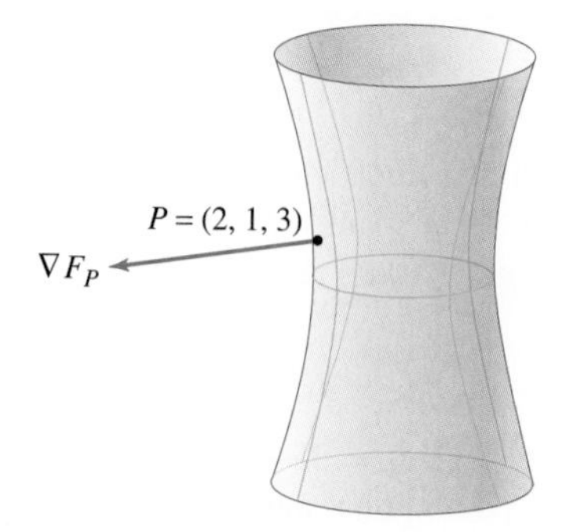

FIGURE 14 The gradient vector ∇F is normal to $F(x, y, z) = 16$.

■ **EXAMPLE 9 Finding the Tangent Plane** Find an equation of the tangent plane to the surface $4x^2 + 9y^2 - z^2 = 16$ at $P = (2, 1, 3)$.

Solution Let $F(x, y, z) = 4x^2 + 9y^2 - z^2$. Then

$$\nabla F = \langle 8x, 18y, -2z \rangle, \qquad \nabla F_P = \nabla F_{(2,1,3)} = \langle 16, 18, -6 \rangle$$

The vector ∇F_P is normal to surface $F(x, y, z) = 16$ (Figure 14), so the tangent plane has equation

$$16(x - 2) + 18(y - 1) - 6(z - 3) = 0 \qquad \text{or} \qquad 16x + 8y - 6z = 32 \qquad \blacksquare$$

The symbol ψ (pronounced "p-sigh" or "p-see") is the lowercase Greek letter psi.

CONCEPTUAL INSIGHT The directional derivative has an interpretation as $\tan\psi$, where ψ is the **angle of inclination**. Think of the graph of $z = f(x, y)$ as a mountain lying over the xy-plane (Figure 15). Let $P = (a, b)$ be a point in the domain of $f(x, y)$ and $P' = (a, b, f(a, b))$ the point on the mountain above P. Given a unit vector $\mathbf{u} = \langle u_1, u_2 \rangle$, you can move up the mountain "in the direction $\mathbf{u}$," that is, such that your horizontal displacement occurs in the direction $\mathbf{u}$. This amounts to moving along the trace curve obtained by intersecting the graph with the vertical plane through P' in the direction $\mathbf{u}$. The directional derivative $D_{\mathbf{u}} f$ is the slope of the tangent line to this trace curve at P' and the vector $\mathbf{w} = \langle u_1, u_2, D_{\mathbf{u}} f \rangle$ is tangent to the trace curve. The angle of inclination ψ, which measures the steepness of the terrain in the direction $\mathbf{u}$, is defined by

$$\tan\psi = \text{slope of tangent in direction } \mathbf{u} = D_{\mathbf{u}} f \qquad \boxed{5}$$

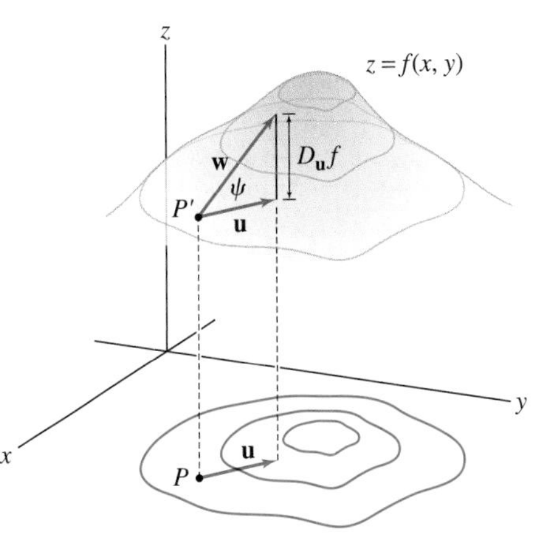

FIGURE 15 The vector $\mathbf{w}$ is tangent to $z = f(x, y)$ at P'.

■ **EXAMPLE 10 Angle of Inclination** You are standing on the side of a mountain [a surface $z = f(x, y)$] at a point $P' = (a, b, f(a, b))$, where $\nabla f_{(a,b)} = \langle 0.4, 0.02 \rangle$. Find the angle of inclination:

(a) In the direction of the gradient.

(b) In a direction making an angle of $\theta = 60^\circ$ with the gradient.

Solution

(a) Let $\mathbf{u}$ be the unit vector in the direction of $\nabla f_{(a,b)}$. Then

$$D_{\mathbf{u}} f(a, b) = \|\nabla f_{P'}\| \cos 0 = \sqrt{(0.4)^2 + (0.02)^2} \approx 0.4$$

The angle of inclination ψ satisfies $\tan\psi = D_{\mathbf{u}} f(a, b) \approx 0.4$. Therefore,

$$\psi \approx \tan^{-1} 0.4 \approx 21.8^\circ$$

(b) Now let $\mathbf{u}$ be a unit vector making an angle of $\theta = 60^\circ$ with $\nabla f_{(a,b)}$. Then

$$D_{\mathbf{u}} f(a, b) = \|\nabla f_{(a,b)}\| \cos 60^\circ \approx (0.4)(0.5) \approx 0.2$$

The angle of inclination ψ in the direction of $\mathbf{u}$ satisfies $\tan\psi = 0.2$, so $\psi = \tan^{-1} 0.2 \approx 11.3^\circ$. ■

14.5 SUMMARY

- The *gradient* of a function f is the vector of partial derivatives:

$$\nabla f = \left\langle \frac{\partial f}{\partial x}, \frac{\partial f}{\partial y} \right\rangle \quad \text{or} \quad \nabla f = \left\langle \frac{\partial f}{\partial x}, \frac{\partial f}{\partial y}, \frac{\partial f}{\partial z} \right\rangle$$

- The *Chain Rule for Paths:* If $\mathbf{c}(t)$ is a path, then the derivative of the composite function $f(\mathbf{c}(t))$ is $\frac{d}{dt} f(\mathbf{c}(t)) = \nabla f_{\mathbf{c}(t)} \cdot \mathbf{c}'(t)$.
- For any vector $\mathbf{v} = \langle v_1, v_2 \rangle$, we define the *derivative of f with respect to* $\mathbf{v}$:

$$D_{\mathbf{v}} f(a, b) = \lim_{t \to 0} \frac{f(a + tv_1, b + tv_2) - f(a, b)}{t}$$

If $\mathbf{u}$ is a *unit vector*, we call $D_{\mathbf{u}} f$ the *directional derivative* of f in the direction $\mathbf{u}$. These definitions extend to three or more variables.
- For any vector $\mathbf{v}$, $D_{\mathbf{v}} f(a, b) = \nabla f_{(a,b)} \cdot \mathbf{v}$.
- The gradient has two basic geometric properties: If $\nabla f_P \neq \mathbf{0}$, then ∇f_P points in the direction of maximum rate of increase and ∇f_P is orthogonal to the level curve (or surface) through P.
- We can write the equation of the tangent plane to the level surface $F(x, y, z) = k$ at a point $P = (a, b, c)$ in the following ways:

$$\nabla F_P \cdot \langle x - a, y - b, z - c \rangle = 0$$

$$F_x(a, b, c)(x - a) + F_y(a, b, c)(y - b) + F_z(a, b, c)(z - c) = 0$$

14.5 EXERCISES

Preliminary Questions

1. Which of the following is a possible value of the gradient ∇f of a function $f(x, y)$ of two variables?

(a) 5 **(b)** $\langle 3, 4 \rangle$ **(c)** $\langle 3, 4, 5 \rangle$

2. True or false: A differentiable function increases at the rate $\|\nabla f_P\|$ in the direction of ∇f_P?

3. Describe the two main geometric properties of the gradient ∇f.

4. Express the partial derivative $\frac{\partial f}{\partial x}$ as a directional derivative $D_{\mathbf{u}} f$ for some unit vector $\mathbf{u}$.

5. You are standing at point where the temperature gradient vector is pointing in the northeast (NE) direction. In which direction(s) should you walk to avoid a change in temperature?

(a) NE **(b)** NW **(c)** SE **(d)** SW

6. What is the rate of change of $f(x, y)$ at $(0, 0)$ in the direction making an angle of $45°$ with the x-axis if $\nabla f(0, 0) = \langle 2, 4 \rangle$?

Exercises

1. Let $f(x, y) = xy^2$ and $\mathbf{c}(t) = (\frac{1}{2}t^2, t^3)$.

(a) Calculate $\nabla f \cdot \mathbf{c}'(t)$.

(b) Use the Chain Rule for Paths to evaluate $\frac{d}{dt} f(\mathbf{c}(t))$ at $t = 1$ and $t = -1$.

2. Let $f(x, y) = e^{xy}$ and $\mathbf{c}(t) = (t^3, 1 + t)$.

(a) Calculate ∇f and $\mathbf{c}'(t)$.

(b) Use the Chain Rule for Paths to calculate $\frac{d}{dt} f(\mathbf{c}(t))$ as $\nabla f \cdot \mathbf{c}'(t)$.

(c) Write out the composite $f(\mathbf{c}(t))$ as a function of t and differentiate. Check that the result agrees with (b).

3. Let $f(x, y) = x^2 + y^2$ and $\mathbf{c}(t) = (\cos t, \sin t)$.

(a) Find $\frac{d}{dt} f(\mathbf{c}(t))$ without making any calculations. Explain.

(b) Verify your answer to (a) using the Chain Rule.

4. Figure 16 shows the level curves of a function $f(x, y)$ and a path $\mathbf{c}(t)$, traversed in the direction indicated. State whether the derivative $\frac{d}{dt} f(\mathbf{c}(t))$ is positive, negative, or zero at points A–D.

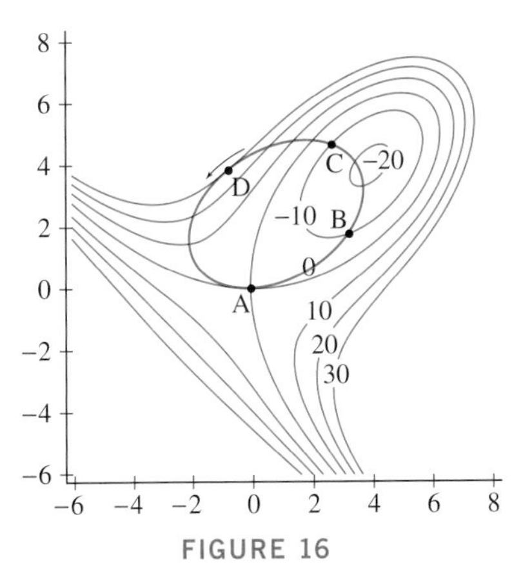

FIGURE 16

In Exercises 5–8, calculate the gradient.

5. $f(x, y) = \cos(x^2 + y)$

6. $g(x, y) = \dfrac{x}{x^2 + y^2}$

7. $h(x, y, z) = xyz^{-3}$

8. $r(x, y, z, w) = xze^{yw}$

In Exercises 9–20, use the Chain Rule to calculate $\frac{d}{dt} f(\mathbf{c}(t))$.

9. $f(x, y) = 3x - 7y$, $\mathbf{c}(t) = (\cos t, \sin t)$, $t = 0$

10. $f(x, y) = 3x - 7y$, $\mathbf{c}(t) = (t^2, t^3)$, $t = 2$

11. $f(x, y) = x^2 - 3xy$, $\mathbf{c}(t) = (\cos t, \sin t)$, $t = 0$

12. $f(x, y) = x^2 - 3xy$, $\mathbf{c}(t) = (\cos t, \sin t)$, $t = \frac{\pi}{2}$

13. $f(x, y) = \sin(xy)$, $\mathbf{c}(t) = (e^{2t}, e^{3t})$, $t = 0$

14. $f(x, y) = \cos(y - x)$, $\mathbf{c}(t) = (e^t, e^{2t})$, $t = \ln 3$

15. $f(x, y) = x - xy$, $\mathbf{c}(t) = (t^2, t^2 - 4t)$, $t = 4$

16. $f(x, y) = xe^y$, $\mathbf{c}(t) = (t^2, t^2 - 4t)$, $t = 0$

17. $f(x, y) = \ln x + \ln y$, $\mathbf{c}(t) = (\cos t, t^2)$, $t = \frac{\pi}{4}$

18. $g(x, y, z) = xye^z$, $\mathbf{c}(t) = (t^2, t^3, t - 1)$, $t = 1$

19. $g(x, y, z) = xyz^{-1}$, $\mathbf{c}(t) = (e^t, t, t^2)$, $t = 1$

20. $g(x, y, z, w) = x + 2y + 3z + 5w$, $\mathbf{c}(t) = (t^2, t^3, t, t-2)$, $t = 1$

In Exercises 21–30, calculate the directional derivative in the direction of $\mathbf{v}$ *at the given point. Remember to normalize the direction vector or use Eq. (2).*

21. $f(x, y) = x^2 + y^3$, $\mathbf{v} = \langle 4, 3 \rangle$, $P = (1, 2)$

22. $f(x, y) = x^2y^3$, $\mathbf{v} = \mathbf{i} + \mathbf{j}$, $P = (-2, 1)$

23. $f(x, y) = x^2y^3$, $\mathbf{v} = \mathbf{i} + \mathbf{j}$, $P = (\frac{1}{6}, 3)$

24. $f(x, y) = \sin(x - y)$, $\mathbf{v} = \langle 1, 1 \rangle$, $P = (\frac{\pi}{2}, \frac{\pi}{6})$

25. $f(x, y) = \tan^{-1}(xy)$, $\mathbf{v} = \langle 1, 1 \rangle$, $P = (3, 4)$

26. $f(x, y) = e^{xy - y^2}$, $\mathbf{v} = \langle 12, -5 \rangle$, $P = (2, 2)$

27. $f(x, y) = \ln(x^2 + y^2)$, $\mathbf{v} = 3\mathbf{i} - 2\mathbf{j}$, $P = (1, 0)$

28. $g(x, y, z) = z^2 - xy^2$, $\mathbf{v} = \langle -1, 2, 2 \rangle$, $P = (2, 1, 3)$

29. $g(x, y, z) = xe^{-yz}$, $\mathbf{v} = \langle 1, 1, 1 \rangle$, $P = (1, 2, 0)$

30. $g(x, y, z) = x \ln(y + z)$, $\mathbf{v} = 2\mathbf{i} - \mathbf{j} + \mathbf{k}$, $P = (2, e, e)$

31. Find the directional derivative of $f(x, y) = x^2 + 4y^2$ at $P = (3, 2)$ in the direction pointing to the origin.

32. Find the directional derivative of $f(x, y, z) = xy + z^3$ at $P = (3, -2, -1)$ in the direction pointing to the origin.

33. A bug located at $(3, 9, 4)$ begins walking in a straight line toward $(5, 7, 3)$. At what rate is the bug's temperature changing if the temperature is $T(x, y, z) = xe^{y-z}$? Units are in meters and degrees Celsius.

34. Suppose that $\nabla f_P = \langle 2, -4, 4 \rangle$. Is f increasing or decreasing at P in the direction $\mathbf{v} = \langle 2, 1, 3 \rangle$?

35. Let $f(x, y) = xe^{x^2 - y}$ and $P = (1, 1)$.

(a) Calculate $\|\nabla f_P\|$.

(b) Find the rate of change of f in the direction ∇f_P.

(c) Find the rate of change of f in the direction of a vector making an angle of 45° with ∇f_P.

36. Let $f(x, y, z) = \sin(xy + z)$ and $P = (0, -1, \pi)$. Calculate $D_{\mathbf{u}} f(P)$ where $\mathbf{u}$ is a unit vector making an angle $\theta = 30°$ with ∇f_P.

37. Let $T(x, y)$ be the temperature at location (x, y). Assume that $\nabla T = \langle y - 4, x + 2y \rangle$. Let $\mathbf{c}(t) = (t^2, t)$ be a path in the plane. Find the values of t such that

$$\frac{d}{dt} T(\mathbf{c}(t)) = 0$$

38. Find a vector normal to the surface $x^2 + y^2 - z^2 = 6$ at $P = (3, 1, 2)$.

39. Find a vector normal to the surface $3z^3 + x^2y - y^2x = 1$ at $P = (1, -1, 1)$.

40. Find the two points on the ellipsoid $\dfrac{x^2}{4} + \dfrac{y^2}{9} + z^2 = 1$ where the tangent plane is normal to $\mathbf{v} = \langle 1, 1, -2 \rangle$.

In Exercises 41–44, find an equation of the tangent plane to the surface at the given point.

41. $x^2 + 3y^2 + 4z^2 = 20$, $P = (2, 2, 1)$

42. $xz + 2x^2y + y^2z^3 = 11$, $P = (2, 1, 1)$

43. $x^2 + z^2 e^{y-x} = 13, \quad P = \left(2, 3, \frac{3}{\sqrt{e}}\right)$

44. $\ln[1 + 4x^2 + 9y^4] - z^2 = 0, \quad P = (3, 1, 1)$

45. Verify what is clear from Figure 17: Every tangent plane to the cone $x^2 + y^2 - z^2 = 0$ passes through the origin.

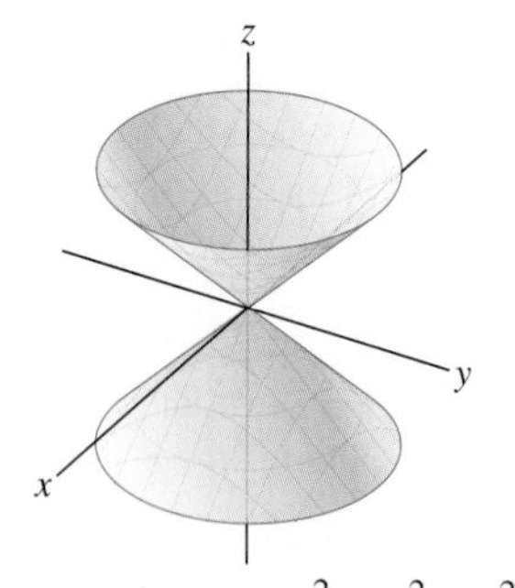

FIGURE 17 Graph of $x^2 + y^2 - z^2 = 0$.

46. CAS Use a computer algebra system to produce a contour plot of $f(x, y) = x^2 - 3xy + y - y^2$ together with its gradient vector field on the domain $[-4, 4] \times [-4, 4]$.

47. Find a function $f(x, y, z)$ such that ∇f is the constant vector $\langle 1, 3, 1 \rangle$.

48. Find a function $f(x, y, z)$ such that $\nabla f = \langle 2x, 1, 2 \rangle$.

49. Find a function $f(x, y, z)$ such that $\nabla f = \langle x, y^2, z^3 \rangle$.

50. Find a function $f(x, y, z)$ such that $\nabla f = \langle z, 2y, x \rangle$.

51. Find a function $f(x, y)$ such that $\nabla f = \langle y, x \rangle$.

52. Show that there does not exist a function $f(x, y)$ such that $\nabla f = \langle y^2, x \rangle$. *Hint:* Use Clairaut's Theorem $f_{xy} = f_{yx}$.

53. Let $\Delta f = f(a + h, b + k) - f(a, b)$ be the change in f at $P = (a, b)$. Set $\Delta \mathbf{v} = \langle h, k \rangle$. Show that the linear approximation can be written

$$\Delta f \approx \nabla f_P \cdot \Delta \mathbf{v} \qquad \boxed{6}$$

54. Use Eq. (6) to estimate

$$\Delta f = f(3.53, 8.98) - f(3.5, 9)$$

assuming that $\nabla f_{(3.5,9)} = \langle 2, -1 \rangle$.

55. Find the unit vector $\mathbf{n}$ normal to the surface $z^2 - 2x^4 - y^4 = 16$ at $P = (2, 2, 8)$ that points in the direction of the xy-plane.

56. Suppose, in the previous exercise, that a particle located at the point $P = (2, 2, 8)$ travels toward the xy-plane in the direction normal to the surface.

(a) Through which point Q on the xy-plane will the particle pass?

(b) Suppose the axes are calibrated in centimeters. Determine the path $\mathbf{c}(t)$ of the particle if it travels at a constant speed of 8 cm/s. How long will it take the particle to reach Q?

57. Let $f(x, y) = \tan^{-1} \frac{x}{y}$ and $\mathbf{u} = \left\langle \frac{\sqrt{2}}{2}, \frac{\sqrt{2}}{2} \right\rangle$.

(a) Calculate the gradient of f.

(b) Calculate $D_{\mathbf{u}} f(1, 1)$ and $D_{\mathbf{u}} f(\sqrt{3}, 1)$.

(c) Show that the lines $y = mx$ for $m \neq 0$ are level curves for f.

(d) Verify that ∇f_P is orthogonal to the level curve through P for $P = (x, y) \neq (0, 0)$.

58. Suppose that the intersection of two surfaces $F(x, y, z) = 0$ and $G(x, y, z) = 0$ is a curve $\mathcal{C}$ and let P be a point on $\mathcal{C}$. Explain why the vector $\mathbf{v} = \nabla F_P \times \nabla G_P$ is a direction vector for the tangent line to $\mathcal{C}$ at P.

59. Let $\mathcal{C}$ be the curve of intersection of the two surfaces $x^2 + y^2 + z^2 = 3$ and $(x - 2)^2 + (y - 2)^2 + z^2 = 3$. Use the result of Exercise 58 to find parametric equations of the tangent line to $\mathcal{C}$ at $P = (1, 1, 1)$.

60. Let $\mathcal{C}$ be the curve of intersection of the two surfaces $x^3 + 2xy + yz = 7$ and $3x^2 - yz = 1$. Find the parametric equations of the tangent line to $\mathcal{C}$ at $P = (1, 2, 1)$.

61. Verify the linearity relations for gradients:

(a) $\nabla(f + g) = \nabla f + \nabla g$

(b) $\nabla(cf) = c\nabla f$

62. Prove the Chain Rule for gradients (Theorem 1).

63. Prove the Product Rule for gradients (Theorem 1).

Further Insights and Challenges

64. Let $\mathbf{u}$ be a unit vector. Show that the directional derivative $D_{\mathbf{u}} f$ is equal to the component of ∇f along $\mathbf{u}$.

65. Let $f(x, y) = (xy)^{1/3}$.

(a) Use the limit definition to show that $f_x(0, 0) = f_y(0, 0) = 0$.

(b) Use the limit definition to show that the directional derivative $D_{\mathbf{u}} f(0, 0)$ does not exist for any unit vector $\mathbf{u}$ other than $\mathbf{i}$ and $\mathbf{j}$.

(c) Is f differentiable at $(0, 0)$?

66. Use the definition of differentiability to show that if $f(x, y)$ is differentiable at $(0, 0)$ and

$$f(0, 0) = f_x(0, 0) = f_y(0, 0) = 0$$

then

$$\lim_{(x,y)\to(0,0)} \frac{f(x, y)}{\sqrt{x^2 + y^2}} = 0 \qquad \boxed{7}$$

67. This exercise shows that there exists a function which is not differentiable at (0, 0) even though all directional derivatives at (0, 0) exist. Define $f(x, y) = \dfrac{x^2 y}{x^2 + y^2}$ for $(x, y) \neq 0$ and $f(0, 0) = 0$.

(a) Use the limit definition to show that $D_{\mathbf{v}} f(0, 0)$ exists for all vectors **v**. Show that $f_x(0, 0) = f_y(0, 0) = 0$.

(b) Prove that f is *not* differentiable at (0, 0) by showing that Eq. (7) does not hold.

68. Prove that if $f(x, y)$ is differentiable and $\nabla f_{(x,y)} = \mathbf{0}$ for all (x, y), then f is constant.

69. Prove the following Quotient Rule where f, g are differentiable:

$$\nabla\left(\frac{f}{g}\right) = \frac{g\nabla f - f\nabla g}{g^2}$$

In Exercises 70–72, a path $\mathbf{c}(t) = (x(t), y(t))$ *follows the gradient of a function* $f(x, y)$ *if the tangent vector* $\mathbf{c}'(t)$ *points in the direction of* ∇f *for all* t*. In other words,* $\mathbf{c}'(t) = k(t)\nabla f_{\mathbf{c}(t)}$ *for some positive function* $k(t)$*. Note that in this case,* $\mathbf{c}(t)$ *crosses each level curve of* $f(x, y)$ *at a right angle.*

70. Show that if the path $\mathbf{c}(t) = (x(t), y(t))$ follows the gradient of $f(x, y)$, then $\dfrac{y'(t)}{x'(t)} = \dfrac{f_y}{f_x}$.

71. Find a path of the form $\mathbf{c}(t) = (t, g(t))$ passing through (1, 2) that follows the gradient of $f(x, y) = 2x^2 + 8y^2$ (Figure 18). *Hint:* Use Separation of Variables.

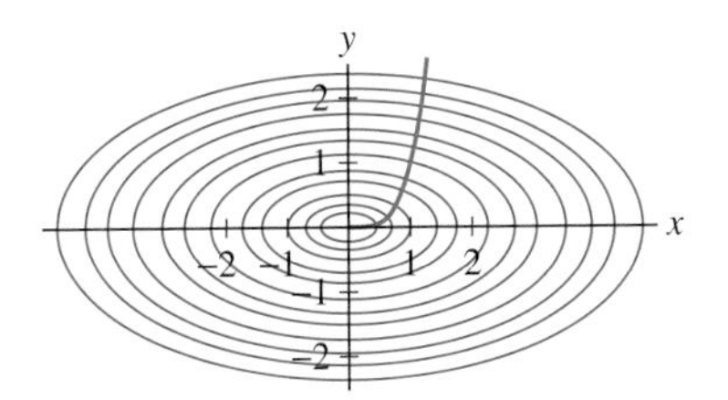

FIGURE 18 The path $\mathbf{c}(t)$ is orthogonal to the level curves of $f(x, y) = 2x^2 + 8y^2$.

72. CAS Find the curve $y = g(x)$ passing through (0, 1) that crosses each level curve of $f(x, y) = y \sin x$ at a right angle. If you have a computer algebra system, graph $y = g(x)$ together with the level curves of f.

14.6 The Chain Rule

In the previous section, we proved the Chain Rule for Paths and used it to derive the main properties of the gradient. There is a more general version of the Chain Rule that applies to general composite functions. Suppose, for example, that x, y, z are differentiable functions of s and t, say, $x = x(s, t)$, $y = y(s, t)$, and $z = z(s, t)$. Then the composite

$$f(x(s, t), y(s, t), z(s, t)) \qquad \boxed{1}$$

is a function of s and t.

■ **EXAMPLE 1** Find the composite function where $f(x, y, z) = xy + z$ and $x = s^2$, $y = st$, $z = t^2$.

Solution The composite function is

$$f(x(s, t), y(s, t), z(s, t)) = xy + z = (s^2)(st) + t^2 = s^3 t + t^2$$ ■

For the composite function in Eq. (1), the general Chain Rule has the form (1)

$$\frac{\partial f}{\partial s} = \frac{\partial f}{\partial x}\frac{\partial x}{\partial s} + \frac{\partial f}{\partial y}\frac{\partial y}{\partial s} + \frac{\partial f}{\partial z}\frac{\partial z}{\partial s} \qquad \boxed{2}$$

$$\frac{\partial f}{\partial t} = \frac{\partial f}{\partial x}\frac{\partial x}{\partial t} + \frac{\partial f}{\partial y}\frac{\partial y}{\partial t} + \frac{\partial f}{\partial z}\frac{\partial z}{\partial t} \qquad \boxed{3}$$

We prove these formulas by showing that they follow from the Chain Rule for Paths. Fix a point (s_0, t_0). The partial derivative $\partial f/\partial s$ at (s_0, t_0) is the derivative of the function $f(x(s, t_0), y(s, t_0), z(s, t_0))$ with respect to s. To evaluate it, we define a path

$\mathbf{c}(s) = (x(s, t_0), y(s, t_0), z(s, t_0))$. The tangent vector is

$$\mathbf{c}'(s) = \left\langle \frac{\partial x}{\partial s}(s, t_0), \frac{\partial y}{\partial s}(s, t_0), \frac{\partial z}{\partial s}(s, t_0) \right\rangle$$

By the Chain Rule for Paths,

$$\left.\frac{\partial f}{\partial s}\right|_{(s_0,t_0)} = \left.\frac{d}{ds} f(\mathbf{c}(s))\right|_{s=s_0} = \nabla f \cdot \mathbf{c}'(s) = \frac{\partial f}{\partial x}\frac{\partial x}{\partial s} + \frac{\partial f}{\partial y}\frac{\partial y}{\partial s} + \frac{\partial f}{\partial z}\frac{\partial z}{\partial s}$$

where the partial derivatives on the right are evaluated at (s_0, t_0). Equation (3) is proved similarly. Furthermore, this argument generalizes to functions $f(x_1, \ldots, x_n)$ of n variables that depend on other variables $t_1, \ldots, t_m$. We call $t_1, \ldots, t_m$ the **independent variables**.

THEOREM 1 General Version of Chain Rule Let $f(x_1, \ldots, x_n)$ be a differentiable function of n variables. Suppose that each of the variables $x_1, \ldots, x_n$ is a differentiable function of m independent variables $t_1, \ldots, t_m$. Then for $k = 1, \ldots, m$,

$$\frac{\partial f}{\partial t_k} = \frac{\partial f}{\partial x_1}\frac{\partial x_1}{\partial t_k} + \frac{\partial f}{\partial x_2}\frac{\partial x_2}{\partial t_k} + \cdots + \frac{\partial f}{\partial x_n}\frac{\partial x_n}{\partial t_k} \qquad \boxed{4}$$

The term "primary derivative" is not standard. We use it in this section to clarify the structure of the Chain Rule.

As an aid to remembering the general Chain Rule, we will refer to the derivatives $\frac{\partial f}{\partial x_1}, \ldots, \frac{\partial f}{\partial x_n}$ as the **primary derivatives**. If $t = t_k$ is any one of the independent variables, then $\frac{\partial f}{\partial t_k}$ is equal to a sum of n terms, one for each primary derivative:

$$j\text{th term:} \quad \frac{\partial f}{\partial x_j}\frac{\partial x_j}{\partial t_k} \quad \text{for } j = 1, 2, \ldots, n$$

Note that the primary derivatives are the components of the gradient, so we can express Eq. (4) as a dot product:

$$\frac{\partial f}{\partial t_k} = \underbrace{\left\langle \frac{\partial f}{\partial x_1}, \frac{\partial f}{\partial x_2}, \ldots, \frac{\partial f}{\partial x_n} \right\rangle}_{\nabla f} \cdot \left\langle \frac{\partial x_1}{\partial t_k}, \frac{\partial x_2}{\partial t_k}, \ldots, \frac{\partial x_n}{\partial t_k} \right\rangle$$

■ **EXAMPLE 2 Using the Chain Rule** Let $f(x, y, z) = xy + z$. Calculate $\frac{\partial f}{\partial s}$, where $x = s^2$, $y = st$, $z = t^2$.

Solution

***Step 1.* Compute the primary derivatives.**

$$\frac{\partial f}{\partial x} = y, \qquad \frac{\partial f}{\partial y} = x \qquad \frac{\partial f}{\partial z} = 1$$

***Step 2.* Apply the Chain Rule:**

$$\frac{\partial f}{\partial s} = \frac{\partial f}{\partial x}\frac{\partial x}{\partial s} + \frac{\partial f}{\partial y}\frac{\partial y}{\partial s} + \frac{\partial f}{\partial z}\frac{\partial z}{\partial s} = y\frac{\partial}{\partial s}s^2 + x\frac{\partial}{\partial s}st + \frac{\partial}{\partial s}t^2$$

$$= (y)(2s) + (x)(t) + 0 = 2sy + xt$$

Step 3. **Express the answer in terms of the independent variables.**

If desired, we may use $x = s^2$ and $y = st$ to express the derivative in terms of s and t:

$$\frac{\partial f}{\partial s} = 2ys + xt = 2(st)s + (s^2)t = 3s^2t$$

As a check of this result, recall that in Example 1, we computed the composite function

$$f(x(s,t), y(s,t), z(s,t)) = f(s^2, st, t^2) = s^3t + t^2$$

From this we see directly that $\frac{\partial f}{\partial s} = 3s^2t$, confirming our result obtained using the Chain Rule. ■

■ **EXAMPLE 3 Evaluating the Derivative at a Point** Let $f(x, y) = e^{xy}$. Evaluate $\frac{\partial f}{\partial t}$ at $(s, t, u) = (-1, 3, 5)$, where

$$x = st, \qquad y = u - t$$

Solution First, we calculate the primary derivatives:

$$\frac{\partial f}{\partial x} = ye^{xy}, \qquad \frac{\partial f}{\partial y} = xe^{xy}$$

Since $\frac{\partial x}{\partial t} = s$ and $\frac{\partial y}{\partial t} = -1$, using the Chain Rule, we have

$$\frac{\partial f}{\partial t} = \frac{\partial f}{\partial x}\frac{\partial x}{\partial t} + \frac{\partial f}{\partial y}\frac{\partial y}{\partial t} = ye^{xy}(s) + xe^{xy}(-1) = (ys - x)e^{xy}$$

It is not necessary to express $\frac{\partial f}{\partial t}$ in terms of s, t, u. For $(s, t, u) = (-1, 3, 5)$, we have $x = st = -3$ and $y = u - t = 2$. Therefore,

$$\left.\frac{\partial f}{\partial t}\right|_{(-1,3,5)} = \left.(ys - x)e^{xy}\right|_{(-1,3,5)} = (2(-1) - (-3))e^{(-3)(2)} = e^{-6}$$ ■

■ **EXAMPLE 4 Polar Coordinates** Let $f(x, y)$ be a function of two variables and let (r, θ) be polar coordinates.

(a) Express $\frac{\partial f}{\partial \theta}$ in terms of $\frac{\partial f}{\partial x}$ and $\frac{\partial f}{\partial y}$.

(b) Evaluate $\frac{\partial f}{\partial \theta}$ at $(x, y) = (1, 1)$ if $f(x, y) = yx^{-1}$.

Solution

(a) Since $x = r\cos\theta$ and $y = r\sin\theta$,

$$\frac{\partial x}{\partial \theta} = -r\sin\theta, \qquad \frac{\partial y}{\partial \theta} = r\cos\theta$$

By the Chain Rule,

$$\frac{\partial f}{\partial \theta} = \frac{\partial f}{\partial x}\frac{\partial x}{\partial \theta} + \frac{\partial f}{\partial y}\frac{\partial y}{\partial \theta} = -r\sin\theta\frac{\partial f}{\partial x} + r\cos\theta\frac{\partial f}{\partial y} \qquad \boxed{5}$$

(b) Applying Eq. (5) to $f(x, y) = yx^{-1}$, we obtain

$$\frac{\partial f}{\partial \theta} = -r\sin\theta\frac{\partial f}{\partial x} + r\cos\theta\frac{\partial f}{\partial y} = (-r\sin\theta)(-yx^{-2}) + (r\cos\theta)(x^{-1})$$

The point $(x, y) = (1, 1)$ has polar coordinates $r = \sqrt{2}$ and $\theta = \tan^{-1} 1 = \frac{\pi}{4}$, so

$$\left.\frac{\partial f}{\partial \theta}\right|_{(1,1)} = (-r \sin \theta)(-yx^{-2}) + (r \cos \theta)(x^{-1})$$

$$= \left(-\sqrt{2} \sin \frac{\pi}{4}\right)(-1) + \left(\sqrt{2} \cos \frac{\pi}{4}\right)(1)$$

$$= \sqrt{2}\left(\frac{\sqrt{2}}{2}\right) + \sqrt{2}\left(\frac{\sqrt{2}}{2}\right) = 2$$

Implicit Differentiation

In calculus of one variable, implicit differentiation is used to compute $\frac{dy}{dx}$ in cases where y is defined implicitly as a function of x through an equation $f(x, y) = 0$. This extends to functions of several variables. Consider a surface defined by an equation

$$F(x, y, z) = 0$$

where $F(x, y, z)$ is differentiable. We may differentiate both sides with respect to x using the Chain Rule:

$$\frac{\partial F}{\partial x}\frac{\partial x}{\partial x} + \frac{\partial F}{\partial y}\frac{\partial y}{\partial x} + \frac{\partial F}{\partial z}\frac{\partial z}{\partial x} = 0$$

Since $\partial x/\partial x = 1$ and $\partial y/\partial x = 0$, this equation reduces to

$$\frac{\partial F}{\partial x} + \frac{\partial F}{\partial z}\frac{\partial z}{\partial x} = F_x + F_z\frac{\partial z}{\partial x} = 0$$

If $F_z \neq 0$, we may solve for $\frac{\partial z}{\partial x}$ $\left(\frac{\partial z}{\partial y} \text{ is computed similarly}\right)$:

$$\frac{\partial z}{\partial x} = -\frac{F_x}{F_z}, \qquad \frac{\partial z}{\partial y} = -\frac{F_y}{F_z} \tag{6}$$

EXAMPLE 5 Calculate $\partial z/\partial x$ and $\partial z/\partial y$ at $P = (1, 1, 1)$, where (Figure 1)

$$F(x, y, z) = x^2 + y^2 - 2z^2 + 12x - 8z - 4 = 0$$

What is the graphical interpretation of these partial derivatives?

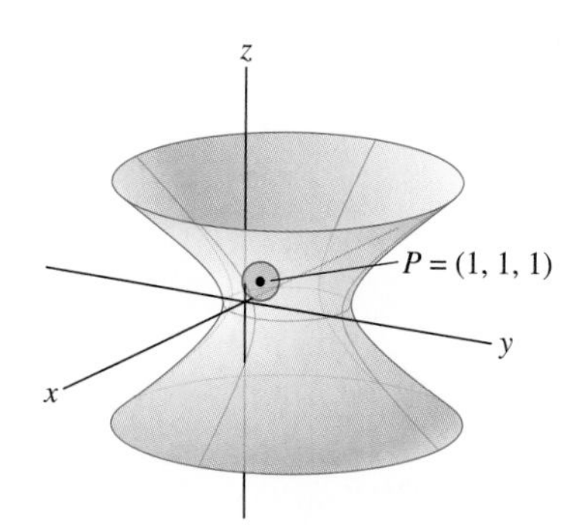

FIGURE 1 The surface $x^2 + y^2 - 2z^2 + 12x - 8z - 4 = 0$. In a small patch around P, the surface can be represented as the graph of a function.

Solution We have

$$F_x = 2x + 12, \qquad F_y = 2y, \qquad F_z = -4z - 8$$

and hence,

$$\frac{\partial z}{\partial x} = -\frac{F_x}{F_z} = \frac{2x + 12}{4z + 8}, \qquad \frac{\partial z}{\partial y} = -\frac{F_y}{F_z} = \frac{2y}{4z + 8}$$

$$\left.\frac{\partial z}{\partial x}\right|_{(1,1,1)} = \frac{14}{12} = \frac{7}{6}, \qquad \left.\frac{\partial z}{\partial y}\right|_{(1,1,1)} = \frac{2}{12} = \frac{1}{6}$$

Figure 1 shows the surface $F(x, y, z) = 0$. The surface as a whole is not the graph of a function (it fails the Vertical Line Test). However, a small patch near P may be represented as a graph of some function $z = f(x, y)$, and the partial derivatives $\frac{\partial z}{\partial x}$ and $\frac{\partial z}{\partial y}$ are equal to the partial derivatives f_x and f_y. Implicit differentiation has allowed us to compute these partial derivatives without explicitly finding $f(x, y)$. ■

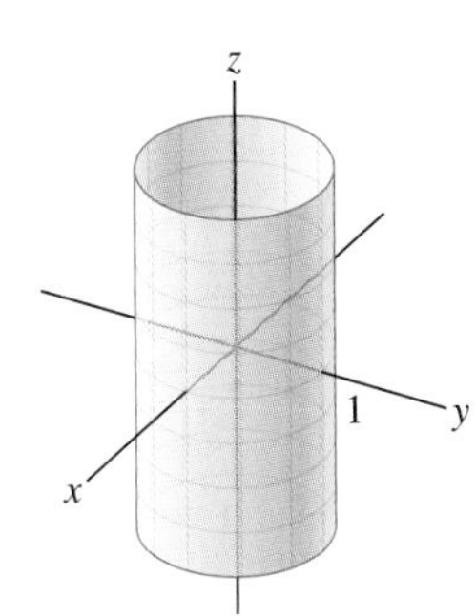

FIGURE 2 Graph of the cylinder $x^2 + y^2 - 1 = 0$.

Assumptions Matter Implicit differentiation is based on the assumption that the equation $F(x, y, z) = 0$ defines z as a function $z = f(x, y)$, at least near a given point $P = (a, b, c)$. According to the Implicit Function Theorem of advanced calculus, such a function $f(x, y)$ exists if F has continuous partial derivatives and if $F_z(P) \neq 0$ [this is also the condition under which the formulas (6) are valid]. Note that $F_z(P) \neq 0$ implies that the tangent plane at P is nonvertical. To see what can go wrong when $F_z(P) = 0$, consider the cylinder (Figure 2)

$$F(x, y, z) = x^2 + y^2 - 1 = 0$$

In this case, $F_z = 0$. The z-coordinate on the cylinder does not depend on x or y, so it is impossible to represent the cylinder as a graph $z = f(x, y)$.

14.6 SUMMARY

- If $f(x, y, z)$ is a function of variables x, y, z, and x, y, z depend on two other variables, say, s and t, then

$$f(x, y, z) = f(x(s, t), y(s, t), z(s, t))$$

is a composite function of s and t. We refer to s and t as the *independent variables*.
- The *Chain Rule* expresses the partial derivatives with respect to the independent variables s and t in terms of the *primary derivatives:*

$$\frac{\partial f}{\partial x}, \quad \frac{\partial f}{\partial y}, \quad \frac{\partial f}{\partial z}$$

Namely,

$$\frac{\partial f}{\partial s} = \frac{\partial f}{\partial x}\frac{\partial x}{\partial s} + \frac{\partial f}{\partial y}\frac{\partial y}{\partial s} + \frac{\partial f}{\partial z}\frac{\partial z}{\partial s}, \qquad \frac{\partial f}{\partial t} = \frac{\partial f}{\partial x}\frac{\partial x}{\partial t} + \frac{\partial f}{\partial y}\frac{\partial y}{\partial t} + \frac{\partial f}{\partial z}\frac{\partial z}{\partial t}$$

- The general case is similar. Let $f(x_1, \dots, x_n)$ be a function of n variables. If $x_1, \dots, x_n$ depend on the independent variables $t_1, \dots, t_m$, then

$$\frac{\partial f}{\partial t_k} = \frac{\partial f}{\partial x_1}\frac{\partial x_1}{\partial t_k} + \frac{\partial f}{\partial x_2}\frac{\partial x_2}{\partial t_k} + \cdots + \frac{\partial f}{\partial x_n}\frac{\partial x_n}{\partial t_k}$$

This relation can be expressed as a dot product:

$$\frac{\partial f}{\partial t_k} = \underbrace{\left\langle \frac{\partial f}{\partial x_1}, \frac{\partial f}{\partial x_2}, \dots, \frac{\partial f}{\partial x_n} \right\rangle}_{\nabla f} \cdot \left\langle \frac{\partial x_1}{\partial t_k}, \frac{\partial x_2}{\partial t_k}, \dots, \frac{\partial x_n}{\partial t_k} \right\rangle$$

- Implicit differentiation is used to find the partial derivatives $\frac{\partial z}{\partial x}$ and $\frac{\partial z}{\partial y}$ when z is defined implicitly by an equation $F(x, y, z) = 0$:

$$\frac{\partial z}{\partial x} = -\frac{F_x}{F_z}, \qquad \frac{\partial z}{\partial y} = -\frac{F_y}{F_z}$$

14.6 EXERCISES

Preliminary Questions

1. Consider a function $f(x, y)$ where $x = uv$ and $y = u/v$.

(a) What are the primary derivatives of f?

(b) What are the independent variables?

In Questions 2–3, suppose that $f(u, v) = ue^v$, where $u = rs$ and $v = r + s$.

2. The composite function $f(u, v)$ is equal to:

(a) rse^{r+s} **(b)** re^s **(c)** rse^{rs}

3. What is the value of $f(u, v)$ at $(r, s) = (1, 1)$?

4. According to the Chain Rule, $\frac{\partial f}{\partial r}$ is equal to (choose correct answer):

(a) $\frac{\partial f}{\partial x}\frac{\partial x}{\partial r} + \frac{\partial f}{\partial x}\frac{\partial x}{\partial s}$

(b) $\frac{\partial f}{\partial x}\frac{\partial x}{\partial r} + \frac{\partial f}{\partial y}\frac{\partial y}{\partial r}$

(c) $\frac{\partial f}{\partial r}\frac{\partial r}{\partial x} + \frac{\partial f}{\partial s}\frac{\partial s}{\partial x}$

5. Suppose that x, y, z are functions of the independent variables u, v, w. Given a function $f(x, y, z)$, which of the following terms appear in the Chain Rule expression for $\frac{\partial f}{\partial w}$?

(a) $\frac{\partial f}{\partial v}\frac{\partial x}{\partial v}$ **(b)** $\frac{\partial f}{\partial w}\frac{\partial w}{\partial x}$

(c) $\frac{\partial f}{\partial z}\frac{\partial z}{\partial w}$ **(d)** $\frac{\partial f}{\partial v}\frac{\partial v}{\partial w}$

6. With notation as in the previous question, does $\frac{\partial x}{\partial v}$ appear in the Chain Rule expression for $\frac{\partial f}{\partial u}$?

Exercises

1. Let $f(x, y, z) = x^2y^3 + z^4$ and $x = s^2$, $y = st^2$, and $z = s^2t$.

(a) Calculate the primary derivatives $\frac{\partial f}{\partial x}, \frac{\partial f}{\partial y}, \frac{\partial f}{\partial z}$.

(b) Calculate $\frac{\partial x}{\partial s}, \frac{\partial y}{\partial s}, \frac{\partial z}{\partial s}$.

(c) Compute $\frac{\partial f}{\partial s}$ using the Chain Rule:

$$\frac{\partial f}{\partial s} = \frac{\partial f}{\partial x}\frac{\partial x}{\partial s} + \frac{\partial f}{\partial y}\frac{\partial y}{\partial s} + \frac{\partial f}{\partial z}\frac{\partial z}{\partial s}$$

Express the answer in terms of the independent variables s, t.

2. Let $f(x, y) = x\cos(y)$ and $x = u^2 + v^2$ and $y = u - v$.

(a) Calculate the primary derivatives $\frac{\partial f}{\partial x}, \frac{\partial f}{\partial y}$.

(b) Use the Chain Rule to calculate $\frac{\partial f}{\partial v}$. Leave the answer in terms of both the dependent and independent variables.

(c) Determine (x, y) for $(u, v) = (2, 1)$ and evaluate $\frac{\partial f}{\partial v}$ at $(u, v) = (2, 1)$.

In Exercises 3–10, use the Chain Rule to calculate the partial derivatives. Express the answer in terms of the independent variables.

3. $\frac{\partial f}{\partial s}, \frac{\partial f}{\partial r}$: $f(x, y, z) = xy + z^2$, $x = s^2$, $y = 2rs$, $z = r^2$

4. $\frac{\partial f}{\partial r}, \frac{\partial f}{\partial t}$: $f(x, y, z) = xy + z^2$, $x = r + s - 2t$, $y = 3rt$, $z = s^2$

5. $\frac{\partial g}{\partial u}, \frac{\partial g}{\partial v}$: $g(x, y) = \cos(x^2 - y^2)$, $x = 2u - 3v$, $y = -5u + 8v$

6. $\frac{\partial h}{\partial t_2}$: $h(x, y) = \frac{x}{y}$, $x = t_1t_2$, $y = t_1^2t_2$

7. $\frac{\partial F}{\partial x}, \frac{\partial F}{\partial y}$: $F(u, v) = e^{u+v}$, $u = x^2$, $v = xy$

8. $\frac{\partial f}{\partial u}$: $f(x, y) = x^2 + y^2$, $x = e^{u+v}$, $y = u + v$

9. $\frac{\partial f}{\partial x}, \frac{\partial f}{\partial y}$: $f(r, \theta) = r\sin^2\theta$, $x = r\cos\theta$, $y = r\sin\theta$

10. $\frac{\partial f}{\partial \theta}, \frac{\partial f}{\partial r}$: $f(x, y, z) = xy - z^2$, $x = r\cos\theta$, $y = \cos^2\theta$, $z = r$

In Exercises 11–16, use the Chain Rule to evaluate the partial derivative at the point specified.

11. $\dfrac{\partial f}{\partial u}$ and $\dfrac{\partial f}{\partial v}$ at $(u, v) = (-1, -1)$, where $f(x, y, z) = x^3 + yz^2$, $x = u^2 + v$, $y = u + v^2$, $z = uv$.

12. $\dfrac{\partial f}{\partial s}$ at $(r, s) = (1, 0)$, where $f(x, y) = \ln(xy)$, $x = 3r + 2s$, $y = 5r + 3s$.

13. $\dfrac{\partial g}{\partial \theta}$ at $(r, \theta) = (2\sqrt{2}, \frac{\pi}{4})$, where $g(x, y) = \dfrac{1}{x + y^2}$, $x = r \sin \theta$, $y = r \cos \theta$.

14. $\dfrac{\partial g}{\partial s}$ at $s = 4$, where $g(x, y) = x^2 - y^2$, $x = s^2 + 1$, $y = 1 - 2s$.

15. $\dfrac{\partial g}{\partial u}$ at $(u, v) = (0, 1)$, where $g(x, y) = x^2 - y^2$, $x = e^u \cos v$, $y = e^u \sin v$.

16. $\dfrac{\partial h}{\partial q}$ at $(q, r) = (3, 2)$, where $h(u, v) = ue^v$, $u = q^3$, $v = qr^2$.

17. The temperature at a point (x, y) is $T(x, y) = 20 + 0.1(x^2 - xy)$ (degrees Celsius). A particle moves clockwise along the unit circle at unit speed (1 cm/s). How fast is the particle's temperature changing at time $t = \pi$?

18. Let $u = u(x, y)$ and let (r, θ) be polar coordinates. Express u_x and u_y in terms of u_r and u_θ. Then show that

$$\|\nabla u\|^2 = u_r^2 + \frac{1}{r^2}u_\theta^2 \qquad \boxed{7}$$

19. Let $u(r, \theta) = r^2 \cos^2 \theta$. Use Eq. (7) to compute $\|\nabla u\|^2$. Then compute $\|\nabla u\|^2$ directly by observing that $u(x, y) = x^2$ and compare.

20. Let $x = s + t$ and $y = s - t$. Show that for any differentiable function $f(x, y)$,

$$\left(\frac{\partial f}{\partial x}\right)^2 - \left(\frac{\partial f}{\partial y}\right)^2 = \frac{\partial f}{\partial s}\frac{\partial f}{\partial t}$$

21. Express the derivatives $\dfrac{\partial f}{\partial \rho}, \dfrac{\partial f}{\partial \theta}, \dfrac{\partial f}{\partial \phi}$ in terms of $\dfrac{\partial f}{\partial x}, \dfrac{\partial f}{\partial y}, \dfrac{\partial f}{\partial z}$, where (ρ, θ, ϕ) are spherical coordinates.

22. Suppose that z is defined implicitly as a function of x and y by the equation $F(x, y, z) = xz^2 + y^2z + xy - 1 = 0$.

(a) Calculate F_x, F_y, F_z.

(b) Use Eq. (6) to calculate $\dfrac{\partial z}{\partial x}$ and $\dfrac{\partial z}{\partial y}$.

23. Calculate $\dfrac{\partial z}{\partial x}$ and $\dfrac{\partial z}{\partial y}$ at the points $(3, 2, 1)$ and $(3, 2, -1)$, where z is defined implicitly by the equation $z^4 + z^2x^2 - y - 8 = 0$.

In Exercises 24–29, calculate the derivative using implicit differentiation.

24. $\dfrac{\partial z}{\partial x}$, $\quad x^2y + y^2z + xz^2 = 10$

25. $\dfrac{\partial w}{\partial z}$, $\quad x^2w + w^3 + wz^2 + 3yz = 0$

26. $\dfrac{\partial z}{\partial y}$, $\quad e^{xy} + \sin(xz) + y = 0$

27. $\dfrac{\partial r}{\partial t}$ and $\dfrac{\partial t}{\partial r}$, $\quad r^2 = te^{s/r}$

28. $\dfrac{\partial w}{\partial y}$, $\quad \dfrac{1}{w^2 + x^2} + \dfrac{1}{w^2 + y^2} = 1$ at $(x, y, w) = (1, 1, 1)$

29. $\dfrac{\partial U}{\partial T}$ and $\dfrac{\partial T}{\partial U}$, $\quad (TU - V)^2 \ln(W - UV) = 1$ at $(T, U, V, W) = (1, 1, 2, 4)$

30. The pressure P, volume V, and temperature T of a van der Waals gas with n molecules (n constant) are related by the equation

$$\left(P + \frac{an^2}{V^2}\right)(V - nb) = nRT$$

where a, b, and R are constant. Calculate $\dfrac{\partial P}{\partial T}$ and $\dfrac{\partial V}{\partial P}$.

31. Let $f(x, y, z) = F(r)$, where $r = \sqrt{x^2 + y^2 + z^2}$. Show that

$$\nabla f = F'(r)\mathbf{e_r} \qquad \boxed{8}$$

where $\mathbf{e_r} = \dfrac{\mathbf{r}}{\|\mathbf{r}\|}$ and $\mathbf{r} = \langle x, y, z\rangle$.

32. Let $f(x, y, z) = e^{-x^2-y^2-z^2} = e^{-r^2}$, with r as in Exercise 31. Compute ∇f directly and using Eq. (8).

33. Use Eq. (8) to compute $\nabla\left(\dfrac{1}{r}\right)$ and $\nabla(\ln r)$.

34. Show that if $f(x)$ is differentiable and $c \neq 0$ is a constant, then $u(x, t) = f(x - ct)$ satisfies the so-called **advection equation**

$$\frac{\partial u}{\partial t} + c\frac{\partial u}{\partial x} = 0$$

35. Jessica and Matthew are running toward the point P along the straight paths that make a fixed angle of θ (Figure 3). Suppose that Matthew runs with velocity v_a m/s and Jessica with velocity v_b m/s.

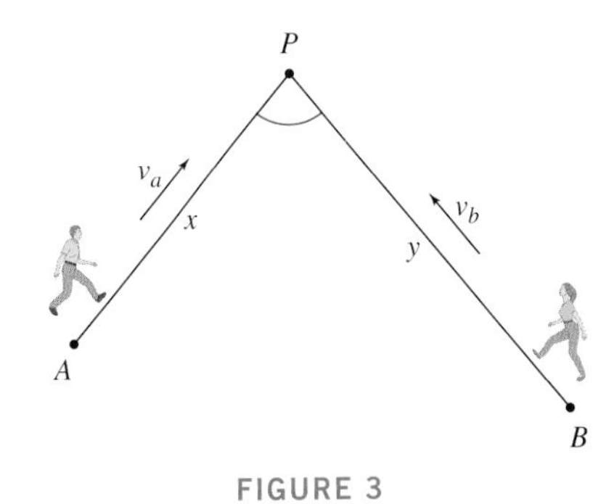

FIGURE 3

Let $f(x, y)$ be the distance from Matthew to Jessica when Matthew is x meters from P and Jessica is y meters from P.

(a) Show that $f(x, y) = \sqrt{x^2 + y^2 - 2xy\cos\theta}$.

(b) Use the Chain Rule to determine the rate at which the distance between Matthew and Jessica is changing when $x = 30$, $y = 20$, $v_a = 4$ m/s, and $v_b = 3$ m/s.

Further Insights and Challenges

36. The Law of Cosines states that $c^2 = a^2 + b^2 - 2ab\cos\theta$, where a, b, c are the sides of a triangle and θ is the angle opposite the side of length c.

(a) Use implicit differentiation to compute the derivatives $\dfrac{\partial\theta}{\partial a}$, $\dfrac{\partial\theta}{\partial b}$, and $\dfrac{\partial\theta}{\partial c}$.

(b) Suppose that $a = 10$, $b = 16$, $c = 22$. Estimate the change in θ if a and b are increased by 1 and c is increased by 2.

In Exercises 37–40, a function $f(x, y, z)$ is called ***homogeneous of degree n*** *if $f(\lambda x, \lambda y, \lambda z) = \lambda^n f(x, y, z)$ for all $\lambda \in \mathbf{R}$.*

37. Show that the following functions are homogeneous and determine their degree.

(a) $f(x, y, z) = x^2y + xyz$

(b) $f(x, y, z) = 3x + 2y - 8z$

(c) $f(x, y, z) = \ln\left(\dfrac{xy}{z^2}\right)$

(d) $f(x, y, z) = z^4$

38. Prove that if $f(x, y, z)$ is homogeneous of degree n, then $f_x(x, y, z)$ is homogeneous of degree $n - 1$. *Hint:* Either use the limit definition or apply the Chain Rule to $f(\lambda x, \lambda y, \lambda z)$.

39. Prove that if $f(x, y, z)$ is homogeneous of degree n, then

$$x\frac{\partial f}{\partial x} + y\frac{\partial f}{\partial y} + z\frac{\partial f}{\partial z} = nf \tag{9}$$

Hint: Let $F(t) = f(tx, ty, tz)$ and calculate $F'(1)$ using the Chain Rule.

40. Verify Eq. (9) for the functions in Exercise 37.

41. Suppose that $x = g(t, s)$, $y = h(t, s)$. Show that f_{tt} is equal to

$$f_{xx}\left(\frac{\partial x}{\partial t}\right)^2 + 2f_{xy}\left(\frac{\partial x}{\partial t}\right)\left(\frac{\partial y}{\partial t}\right) + f_{yy}\left(\frac{\partial y}{\partial t}\right)^2 + f_x\frac{\partial^2 x}{\partial t^2} + f_y\frac{\partial^2 y}{\partial t^2} \tag{10}$$

42. Let $r = \sqrt{x_1^2 + \cdots + x_n^2}$ and let $g(r)$ be a function of r. Prove the formulas

$$\frac{\partial g}{\partial x_i} = \frac{x_i}{r}g_r, \qquad \frac{\partial^2 g}{\partial x_i^2} = \frac{x_i^2}{r^2}g_{rr} + \frac{r^2 - x_i^2}{r^3}g_r$$

43. Prove that if $g(r)$ is a function of r as in Exercise 42, then

$$\frac{\partial^2 g}{\partial x_1^2} + \cdots + \frac{\partial^2 g}{\partial x_n^2} = g_{rr} + \frac{n-1}{r}g_r$$

In Exercises 44–48, the ***Laplace operator*** *is defined by $\Delta f = f_{xx} + f_{yy}$. A function $f(x, y)$ satisfying the Laplace equation $\Delta f = 0$ is called* ***harmonic****. A function $f(x, y)$ is called* ***radial*** *if $f(x, y) = g(r)$, where $r = \sqrt{x^2 + y^2}$.*

44. Use Eq. (10) to prove that in polar coordinates (r, θ),

$$\Delta f = f_{rr} + \frac{1}{r^2}f_{\theta\theta} + \frac{1}{r}f_r \tag{11}$$

45. Use Eq. (11) to show that $f(x, y) = \ln r$ is harmonic.

46. Verify that $f(x, y) = x$ and $f(x, y) = y$ are harmonic using both the rectangular and polar expressions for Δf.

47. Verify that $f(x, y) = \tan^{-1}\dfrac{y}{x}$ is harmonic using both the rectangular and polar expressions for Δf.

48. Use the Product Rule to show that

$$f_{rr} + \frac{1}{r}f_r = r^{-1}\frac{\partial}{\partial r}\left(r\frac{\partial f}{\partial r}\right)$$

Use this formula to show that if f is a radial harmonic function, then $rf_r = C$ for some constant C. Conclude that $f(x, y) = C\ln r + b$ for some constant b.

49. Figure 4 shows the graph of the equation

$$F(x, y, z) = x^2 + y^2 - z^2 - 12x - 8z - 4 = 0$$

(a) Use the quadratic formula to solve for z as a function of x and y. This gives two formulas, depending on the choice of a sign.

(b) Which formula defines the portion of the surface satisfying $z \geq -4$? Which formula defines the portion satisfying $z \leq -4$?

(c) Calculate $\dfrac{\partial z}{\partial x}$ using the formula $z = f(x, y)$ (for both choices of sign) and again via implicit differentiation. Verify that the two answers agree.

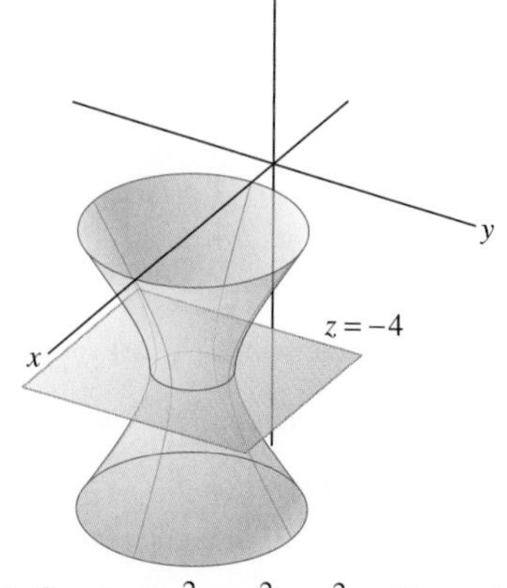

FIGURE 4 Graph of $x^2 + y^2 - z^2 - 12x - 8z - 4 = 0$.

50. When x, y, and z are related by an equation $F(x, y, z) = 0$, we sometimes write $\left(\frac{\partial z}{\partial x}\right)_y$ in place of $\frac{\partial z}{\partial x}$ to indicate that in the differentiation, z is treated as a function of x with y held constant (and similarly for the other variables).

(a) Use Eq. (6) to prove the **cyclic relation**

$$\left(\frac{\partial z}{\partial x}\right)_y \left(\frac{\partial x}{\partial y}\right)_z \left(\frac{\partial y}{\partial z}\right)_x = -1 \qquad \boxed{12}$$

(b) Verify Eq. (12) for $F(x, y, z) = x + y + z = 0$.

(c) Verify the cyclic relation for the variables P, V, T in the ideal gas law $PV - nRT = 0$ (n and R are constants).

14.7 Optimization in Several Variables

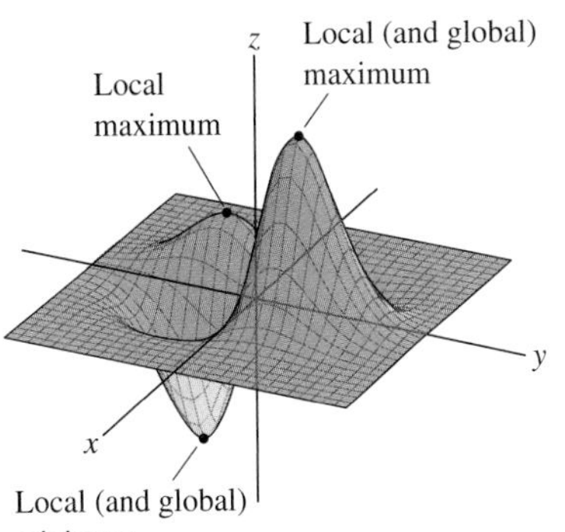

FIGURE 1 Graph of $f(x, y) = (x^2 + 40xy + 20y + 5y^2)e^{-0.4(2x^2+y^2)}$.

Recall that optimization is the process of finding the extreme values of a function. This amounts to finding the highest and lowest points on the graph over a given domain. As we saw in the one-variable case, it is important to distinguish between *local* and *global* extreme values. A local extreme value is a value $f(a, b)$ that is a maximum or minimum in some small open disk around (a, b) (Figure 1).

DEFINITION Local Extreme Values A function $f(x, y)$ has a **local extremum** at $P = (a, b)$ if there exists an open disk $D(P, r)$ such that:

- **Local maximum:** $f(x, y) \le f(a, b)$ for all $(x, y) \in D(P, r)$.
- **Local minimum:** $f(x, y) \ge f(a, b)$ for all $(x, y) \in D(P, r)$.

REMINDER The term "extremum" (the plural is "extrema") means a minimum or maximum value.

In one variable, Fermat's Theorem states that if $f(a)$ is a local extreme value, then a is a critical point [either $f'(a) = 0$ or $f'(a)$ does not exist]. Thus, the tangent line at a local extremum is horizontal (if the tangent line exists). We can expect a similar result for functions of two variables, but in this case, it is the *tangent plane* that must be horizontal at a local extremum (Figure 2). Recall that the tangent plane to the surface $z = f(x, y)$ at $P = (a, b)$ has equation

$$z = f(a, b) + f_x(a, b)(x - a) + f_y(a, b)(y - b)$$

We see that the tangent plane is horizontal if $f_x(a, b) = f_y(a, b) = 0$, that is, if the equation reduces to $z = f(a, b)$.

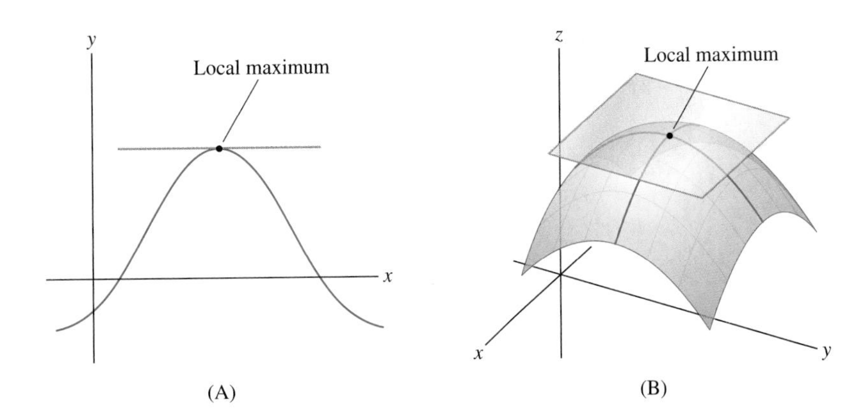

FIGURE 2 The tangent line or plane is horizontal at a local extremum.

DEFINITION Critical Point A point $P = (a, b)$ in the domain of $f(x, y)$ is called a **critical point** if:

- $f_x(a, b) = 0$ and $f_y(a, b) = 0$, or
- One of the partial derivatives $f_x(a, b)$, $f_y(a, b)$ does not exist.

According to the next theorem, local extrema occur at critical points, as in the single-variable case.

> **THEOREM 1 Fermat's Theorem in Several Variables** If $f(x, y)$ has a local minimum or maximum at $P = (a, b)$, then $P = (a, b)$ is a critical point of $f(x, y)$.

Proof Assume that $f(x, y)$ has a local minimum at $P = (a, b)$ (a local maximum is similar). Then $f(x, y) \geq f(a, b)$ for (x, y) in a small disk $D(r, P)$. In particular, $f(x, b) \geq f(a, b)$ if $|x - a| < r$. This shows that the function $g(x) = f(x, b)$ has a local minimum at $x = a$, and therefore, either $g'(a) = 0$ or $g'(a)$ does not exist. Since $g'(a) = f_x(a, b)$, we conclude that either $f_x(a, b) = 0$ or $f_x(a, b)$ does not exist. Similarly, $f_y(a, b) = 0$ or $f_y(a, b)$ does not exist. Therefore, (a, b) is a critical point. ■

More generally, we say that $(a_1, \ldots, a_n)$ is a critical point of $f(x_1, \ldots, x_n)$ if all the partial derivatives are zero:

$$f_{x_j}(a_1, \ldots, a_n) = 0$$

or if one of the partial derivatives $f_{x_j}(a_1, \ldots, a_n)$ does not exist. Theorem 1 holds in any number of variables: Local extrema occur at critical points.

■ **EXAMPLE 1 Finding Critical Points** Show that $f(x, y) = 11x^2 - 2xy + 2y^2 + 3y$ has one critical point. Use Figure 3 to determine whether it corresponds to local minima or maxima.

Solution We set the partial derivatives equal to zero and solve the resulting pair of equations:

$$f_x(x, y) = 22x - 2y = 0$$

$$f_y(x, y) = -2x + 4y + 3 = 0$$

By the first equation, $y = 11x$. Substituting $y = 11x$ in the second equation gives

$$-2x + 4y + 3 = -2x + 4(11x) + 3 = 42x + 3 = 0$$

Thus $x = -\frac{1}{14}$ and $y = -\frac{11}{14}$. There is just one critical point, $P = (-\frac{1}{14}, -\frac{11}{14})$. Figure 3 shows that $f(x, y)$ has a local minimum at P. ■

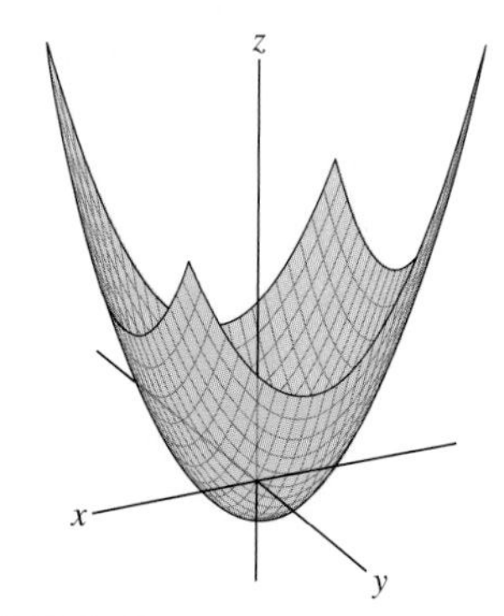

FIGURE 3 Graph of $f(x, y) = 11x^2 - 2xy + 2y^2 + 3y$.

As we saw in the previous example, finding the critical points involves solving the simultaneous equations $f_x(x, y) = 0$ and $f_x(x, y) = 0$. It is not always possible to find the solutions exactly, but we can use a computer to find numerical approximations.

■ **EXAMPLE 2** CAS **Numerical Example** Use a computer algebra system to find numerical approximations to the critical points of

$$f(x, y) = \frac{x - y}{2x^2 + 8y^2 + 3}$$

Refer to Figure 4 to determine whether they correspond to local minima or maxima.

Solution We use a CAS to compute the partial derivatives and solve

$$f_x(x, y) = \frac{-2x^2 + 8y^2 + 4xy + 3}{(2x^2 + 8y^2 + 3)^2} = 0$$

$$f_x(x, y) = \frac{-2x^2 + 8y^2 - 16xy - 3}{(2x^2 + 8y^2 + 3)^2} = 0$$

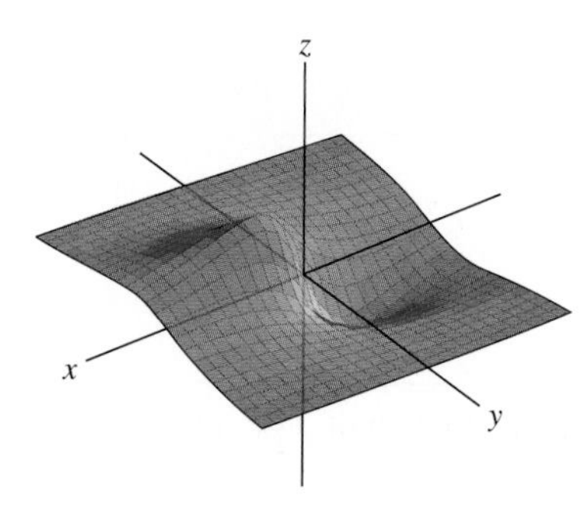

FIGURE 4 Plot of $f(x, y) = \dfrac{x - y}{2x^2 + 8y^2 + 3}$.

Figure 4 suggests that $f(x, y)$ has two critical points near $(1, 0)$ and $(-1, 0)$. We use a CAS to search for solutions near these points. It suffices to set numerators of f_x and f_y equal to zero, and we do so to simplify the problem. The Mathematica command

```
[FindRoot[{-2x^2+8y^2+4xy+3 == 0, -2x^2+8y^2-16xy-3 == 0},
 {{x,1},{y,0}}]
```

returns the answer

```
{x -> 1.095, y -> -0.274}
```

Thus, $(1.095, -0.274)$ is an approximate critical point at which f has a local maximum, as we see in Figure 4. A search for the second critical point near $(-1, 0)$ yields $(-1.095, 0.274)$, which approximates the critical point where $f(x, y)$ takes on a local minimum. ■

A basic problem in optimization is determining whether a critical point yields a local minimum or maximum or neither. In one variable, $f(x)$ may have a point of inflection rather than a local extremum at a critical point. A similar phenomenon occurs in several variables. Figure 5(C) shows a saddle point. The origin is a critical point, but it yields neither a local minimum nor a local maximum. If you stand at the saddle point and begin walking, some directions take you uphill and other directions take you downhill.

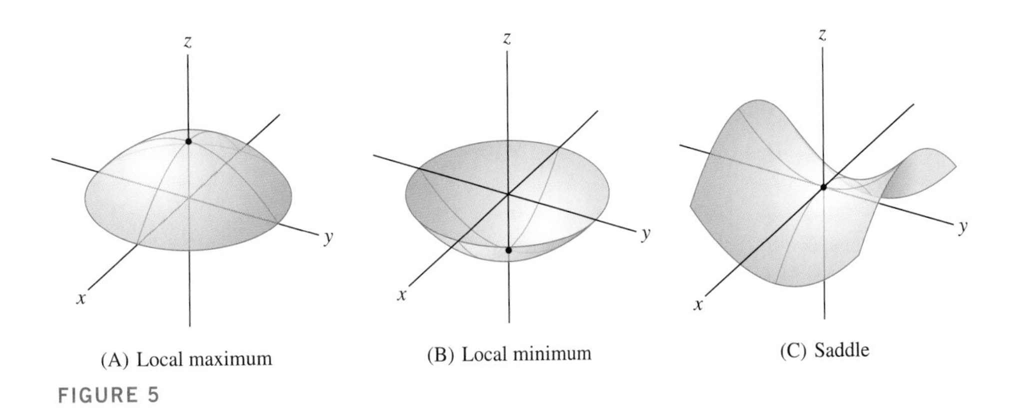

(A) Local maximum (B) Local minimum (C) Saddle

FIGURE 5

■ **EXAMPLE 3** Find the critical points of $f(x, y) = x^3 + y^3 - 12xy$ and use Figure 6 to determine whether they correspond to extrema.

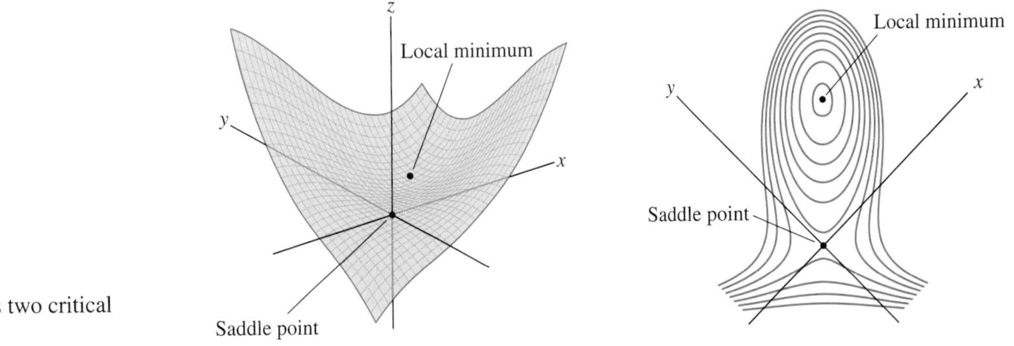

FIGURE 6 The function $f(x, y) = x^3 + y^3 - 12xy$ has two critical points.

Solution Again, we set the partial derivatives equal to zero and solve:

$$f_x(x, y) = 3x^2 - 12y = 0 \quad \Rightarrow \quad y = \frac{1}{4}x^2$$

$$f_y(x, y) = 3y^2 - 12x = 0$$

Substituting $y = \frac{1}{4}x^2$ in the second equation yields

$$3y^2 - 12x = 3\left(\frac{1}{4}x^2\right)^2 - 12x = \frac{3}{16}x(x^3 - 64) = 0 \quad \Rightarrow \quad x = 0, 4$$

Since $y = \frac{1}{4}x^2$, the critical points are $(0, 0)$ and $(4, 4)$. Figure 6 shows that $f(x, y)$ has a local minimum at $(4, 4)$ and saddle point at $(0, 0)$. ■

GRAPHICAL INSIGHT We can read off the type of a critical point from the contour map. Typically, if $f(P)$ is a local minimum or maximum, then the nearby level curves are closed curves encircling P as in Figure 7(A). The contour plot shows that $f(x, y)$ increases in all directions emanating from P. In Figure 7(B), $g(x, y)$ has a saddle point at Q. The level curves of $g(x, y)$ through Q consist of two intersecting lines that divide the neighborhood near Q into four regions. The contour map shows that $g(x, y)$ is decreasing in the x-direction and increasing in the y-direction. There also exist more general types of saddle points. The graph of $h(x, y)$ in Figure 7(C) is called a "monkey saddle" (because a monkey can sit on this saddle with room for his tail in the back).

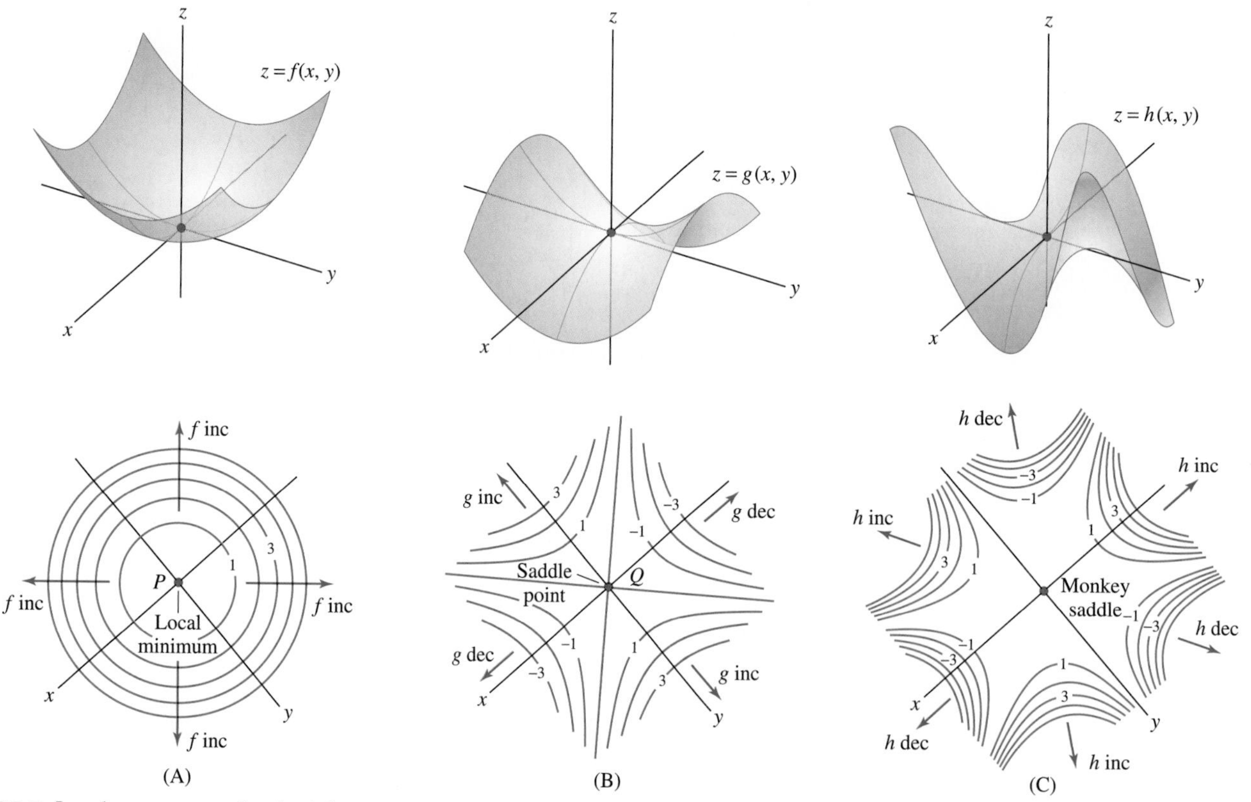

FIGURE 7 Level curves near a local minimum or saddle point.

Recall that in one variable, if $f''(c) \neq 0$, then the sign of $f''(c)$ at a critical point $x = c$ determines whether $f(c)$ is a local minimum or maximum. There is also a Second Derivative Test for determining the type of a critical point (a, b) of a function $f(x, y)$ in two variables. It relies on the sign of the **discriminant** $D = D(a, b)$, defined as follows:

$$D = D(a, b) = f_{xx}(a, b)f_{yy}(a, b) - f_{xy}(a, b)^2$$

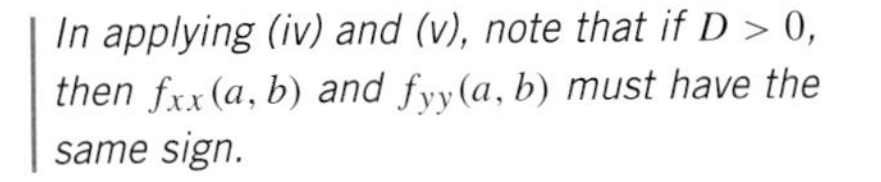
In applying (iv) and (v), note that if $D > 0$, then $f_{xx}(a, b)$ and $f_{yy}(a, b)$ must have the same sign.

THEOREM 2 Second Derivative Test Let $P = (a, b)$ be a critical point of $f(x, y)$. Assume that the second derivatives f_{xx}, f_{yy}, f_{xy} are continuous near P. Then:

(i) If $D > 0$, then $f(a, b)$ is a local minimum or local maximum.
(ii) If $D < 0$, then (a, b) is a saddle point.
(iii) If $D = 0$, the test is inconclusive.

Furthermore,

(iv) If $D > 0$ and $f_{xx}(a, b) > 0$, then $f(a, b)$ is a local minimum.
(v) If $D > 0$ and $f_{xx}(a, b) < 0$, then $f(a, b)$ is a local maximum.

A proof of this theorem is discussed at the end of this section.

EXAMPLE 4 Applying the Second Derivative Test Find the critical points of the function $f(x, y) = (x^2 + y^2)e^{-x}$ and analyze them using the Second Derivative Test.

Solution

***Step 1.* Find the critical points.**
Set the partial derivatives equal to zero and solve:

$$f_x(x, y) = -(x^2 + y^2)e^{-x} + 2xe^{-x} = (2x - x^2 - y^2)e^{-x} = 0$$

$$f_y(x, y) = 2ye^{-x} = 0 \quad \Rightarrow \quad y = 0$$

Substituting $y = 0$ in the first equation then gives

$$(2x - x^2 - y^2)e^{-x} = (2x - x^2)e^{-x} = 0 \quad \Rightarrow \quad x = 0, 2$$

The critical points are $(0, 0)$ and $(2, 0)$ (Figure 8).

***Step 2.* Compute the second-order partials and the discriminant.**

$$f_{xx}(x, y) = \frac{\partial}{\partial x}(2x - x^2 - y^2)e^{-x} = (2 - 4x + x^2 + y^2)e^{-x}$$

$$f_{yy}(x, y) = \frac{\partial}{\partial y}2ye^{-x} = 2e^{-x}$$

$$f_{xy}(x, y) = f_{yx}(x, y) = \frac{\partial}{\partial x}2ye^{-x} = -2ye^{-x}$$

$$D(x, y) = f_{xx}f_{yy} - f_{xy}^2 = 2(2 - 4x + x^2 + y^2)e^{-2x} - 4y^2e^{-2x}$$

$$= (4 - 8x + 2x^2 - 2y^2)e^{-2x}$$

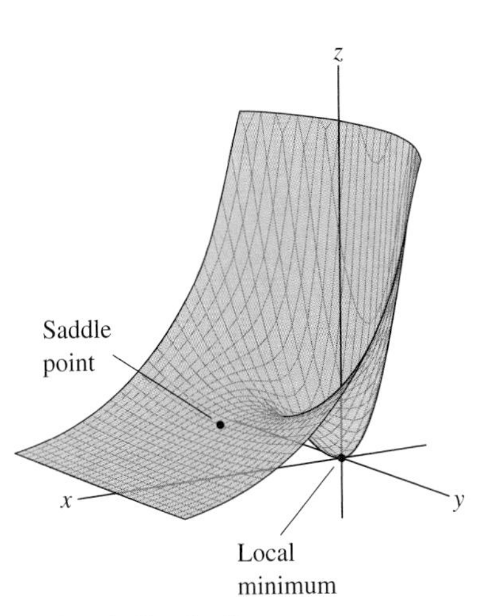

FIGURE 8 Graph of $f(x, y) = (x^2 + y^2)e^{-x}$.

Step 3. **Apply the Second Derivative Test.**

Critical Point	f_{xx}	f_{yy}	f_{xy}	Discriminant D	Type
$(0, 0)$	2	2	0	$2(2) - 0^2 = 4$	Local minimum since $D > 0$ and $f_{xx} > 0$
$(2, 0)$	$-2e^{-2}$	$2e^{-2}$	0	$-2e^{-2}(2e^{-2}) - 0^2 = -4e^{-2}$	Saddle since $D < 0$

■

Global Extrema

In many optimization problems, we are interested in finding the minimum or maximum value of a function f on a given domain $\mathcal{D}$. These values are called the **global** or **absolute extreme values** of f on $\mathcal{D}$. The existence of global extrema depends on the domain $\mathcal{D}$. For example, $f(x, y) = x + y$ has a maximum value $f(1, 1) = 2$ on the square unit $\mathcal{D}_1$ in Figure 9, but it has no maximum value on the entire plane $\mathbf{R}^2$.

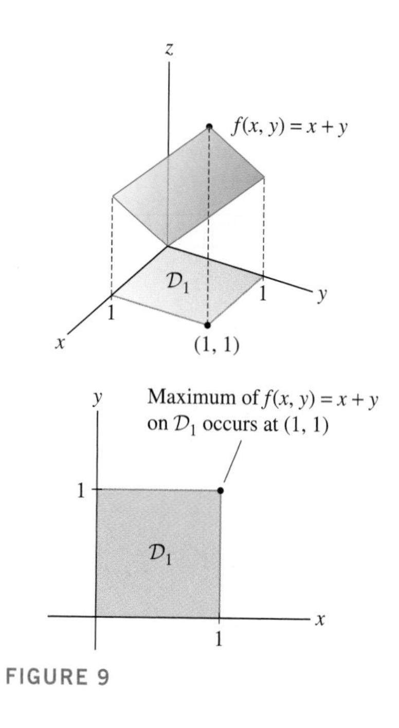

FIGURE 9

To state conditions that guarantee the existence of global extrema, a few definitions are needed. First, we say that a domain $\mathcal{D}$ is **bounded** if it is contained in some disk of radius M centered at the origin. In other words, no point of $\mathcal{D}$ is more than a distance M from the origin [Figures 10(A) and (C)]. Next, a point P is called an:

- **Interior point** of $\mathcal{D}$ if $\mathcal{D}$ contains some open disk $D(P, r)$ centered at P.
- **Boundary point** of $\mathcal{D}$ if every disk centered at P contains points in $\mathcal{D}$ and points not in $\mathcal{D}$.

The **interior** of $\mathcal{D}$ is the set of all interior points and the **boundary** of $\mathcal{D}$ is the set of all boundary points. In Figure 10(C), the boundary is the curve surrounding the domain and the interior consists of all points in the domain not lying on the boundary curve.

A domain $\mathcal{D}$ is called **closed** if $\mathcal{D}$ contains all its boundary points; $\mathcal{D}$ is called **open** if every point of $\mathcal{D}$ is an interior point. The domain in Figure 10(A) is closed because the domain includes its boundary curve. In Figure 10(C), some of the boundary points are excluded and the domain is neither open nor closed.

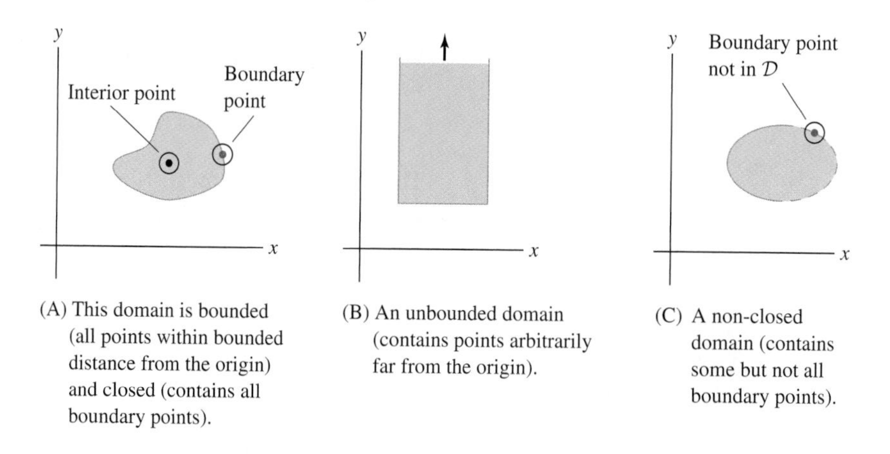

FIGURE 10 Domains in $\mathbf{R}^2$.

In Section 4.2, we stated two basic results on optimization. First, every continuous function $f(x)$ on a *closed, bounded interval* $[a, b]$ takes on both a minimum and a maximum value on $[a, b]$. Second, these extreme values occur either at critical points in the interior (a, b) or at the endpoints. Analogous results hold in several variables.

> **THEOREM 3 Existence of Global Extrema** Assume that $f(x, y)$ is defined and continuous on a closed, bounded domain $\mathcal{D}$ in $\mathbf{R}^2$. Then:
>
> **(i)** $f(x, y)$ takes on both a minimum and a maximum value on $\mathcal{D}$.
>
> **(ii)** The extreme values occur either at critical points in the interior of $\mathcal{D}$ or at points on the boundary of $\mathcal{D}$.

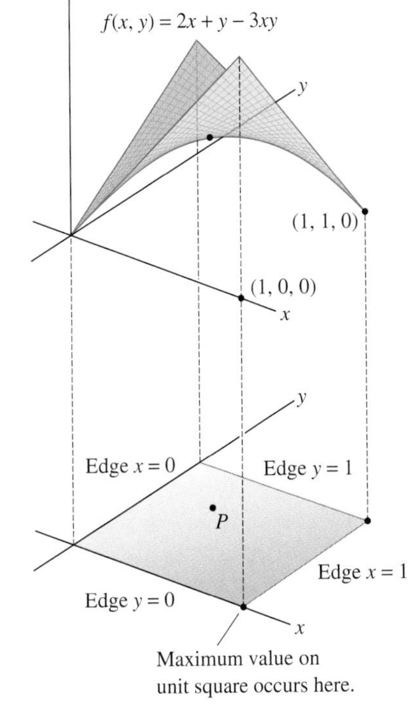

FIGURE 11

■ **EXAMPLE 5** Find the maximum of $f(x, y) = 2x + y - 3xy$ on the unit square $\mathcal{D} = \{(x, y) : 0 \le x, y \le 1\}$ (Figure 11).

Solution By Theorem 3, the maximum occurs either at a critical point or on the boundary of the square.

***Step 1.* Examine the critical points.**

Set the partial derivatives equal to zero and solve:

$$f_x(x, y) = 2 - 3y = 0 \Rightarrow y = \frac{2}{3}, \qquad f_y(x, y) = 1 - 3x = 0 \Rightarrow x = \frac{1}{3}$$

There is a unique critical point $P = (\frac{1}{3}, \frac{2}{3})$ and

$$f(P) = f\left(\frac{1}{3}, \frac{2}{3}\right) = 2\left(\frac{1}{3}\right) + \left(\frac{2}{3}\right) - 3\left(\frac{1}{3}\right)\left(\frac{2}{3}\right) = \frac{2}{3}$$

***Step 2.* Find the maximum on the boundary.**

We must find the maximum value on the boundary of the square. We do this by checking each of the four edges separately. The bottom edge is described by $y = 0$, $0 \le x \le 1$, and $f(x, 0) = 2x$. Hence, the maximum value of f on this edge occurs at $x = 1$, where $f(1, 0) = 2$. Proceeding in a similar fashion with the other edges, we obtain the following:

Edge	Restriction of $f(x, y)$ to Edge	Maximum of $f(x, y)$ on Edge
Lower: $y = 0, 0 \le x \le 1$	$f(x, 0) = 2x$	$f(1, 0) = 2$
Upper: $y = 1, 0 \le x \le 1$	$f(1, x) = 1 - x$	$f(0, 0) = 1$
Left: $x = 0, 0 \le y \le 1$	$f(0, y) = y$	$f(0, 1) = 1$
Right: $x = 1, 0 \le y \le 1$	$f(1, y) = 2 - 2y$	$f(1, 0) = 2$

The maximum of f on the boundary is $f(1, 0) = 2$. This is larger than the value $f(P) = \frac{2}{3}$ at the critical point, so the maximum of f on the unit square is 2. ■

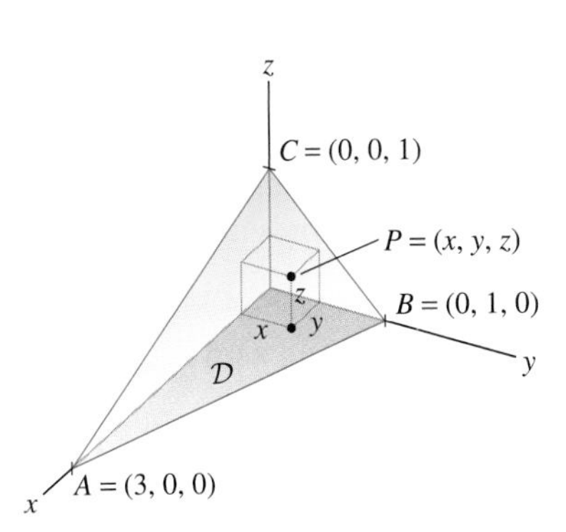

FIGURE 12 The shaded triangle is the domain of $V(x, y)$.

■ **EXAMPLE 6 Box of Maximum Volume** Find the volume of the largest box that can be inscribed in the tetrahedron bounded by the coordinate planes and the plane

$$\frac{1}{3}x + y + z = 1$$

Solution

***Step 1.* Find a function to be maximized.**

Let $P = (x, y, z)$ be the corner of the box lying on the front face of the tetrahedron (Figure 12). Then the box has sides of lengths x, y, z and volume $V = xyz$. We use the equation $\frac{1}{3}x + y + z = 1$, or $z = 1 - \frac{1}{3}x - y$, to express V in terms of x and y:

$$V(x, y) = xyz = xy\left(1 - \frac{1}{3}x - y\right) = xy - \frac{1}{3}x^2y - xy^2$$

Step 2. **Determine the domain.**
Let $\mathcal{D}$ be the shaded triangle $\triangle OAB$ in the xy-plane in Figure 12. The corner point $P = (x, y, z)$ projects to the point (x, y) in $\mathcal{D}$, so our problem is to find the maximum of V on $\mathcal{D}$. Since $\mathcal{D}$ is a closed, bounded region, this maximum occurs at a critical point inside $\mathcal{D}$ or on the boundary of $\mathcal{D}$.

Step 3. **Examine the critical points.**
First, set the partial derivatives equal to zero and solve:

$$\frac{\partial V}{\partial x} = y - \frac{2}{3}xy - y^2 = y\left(1 - \frac{2}{3}x - y\right) = 0$$

$$\frac{\partial V}{\partial y} = x - \frac{1}{3}x^2 - 2xy = x\left(1 - \frac{1}{3}x - 2y\right) = 0$$

If $x = 0$ or $y = 0$, then (x, y) lies on the boundary of $\mathcal{D}$, so assume that x and y are both nonzero. Then we may divide by y and x in the above equations to obtain

$$1 - \frac{2}{3}x - y = 0 \quad \Rightarrow \quad y = 1 - \frac{2}{3}x$$

$$1 - \frac{1}{3}x - 2y = 0$$

Substituting $y = 1 - \frac{2}{3}x$ in the second equation yields

$$1 - \frac{1}{3}x - 2\left(1 - \frac{2}{3}x\right) = 0 \quad \Rightarrow \quad x - 1 = 0 \quad \Rightarrow \quad x = 1$$

Therefore, $(1, \frac{1}{3})$ is a critical point, and

$$V\left(1, \frac{1}{3}\right) = (1)\frac{1}{3} - \frac{1}{3}(1)^2\frac{1}{3} - (1)\left(\frac{1}{3}\right)^2 = \frac{1}{9}$$

Step 4. **Check the boundary.**
We have $V(x, y) = 0$ for all points (x, y) on the boundary of $\mathcal{D}$ (because the three edges of the boundary are defined by $x = 0$, $y = 0$, or $1 - \frac{1}{3}x - y = 0$). Clearly, then, the maximum occurs at the critical point and the maximum volume is $\frac{1}{9}$. ■

Proof of the Second Derivative Test We discuss the main ideas in the proof of the Second Derivative Test, leaving the full details to the Internet supplement. The proof is based on an algebraic fact about quadratic forms. A **quadratic form** is a function in two variables of the type

$$Q(h, k) = ah^2 + 2bhk + ck^2 \quad (a, b, c \text{ constants not all zero})$$

A quadratic form with a nonzero discriminant D is often called "nondegenerate."

The discriminant of Q is the quantity $D = ac - b^2$. For example,

$$\text{If } Q(h, k) = h^2 - k^2, \qquad \text{then } D = (1)(-1) - 0 = -1$$

$$\text{If } Q(h, k) = h^2 + 2hk + 5k^2, \qquad \text{then } D = (1)(5) - 1 = 4$$

Some quadratic forms, such as $Q(h,k) = h^2 - k^2$, take on both positive and negative values. Other quadratic forms, such as $Q(h,k) = h^2 + k^2$, satisfy $Q(h,k) > 0$ for $(h,k) \neq 0$. According to the next theorem, the sign of the discriminant determines which of these two possibilities occurs (Figure 13).

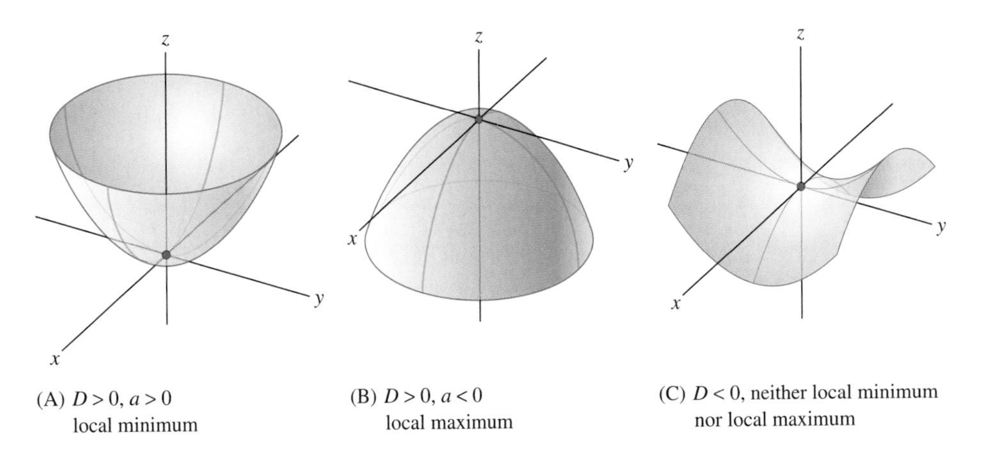

FIGURE 13 Graphs of $Q(x, y)$ with $D > 0$ (local minimum) and $D < 0$ (local maximum).

THEOREM 4 Let $Q(h,k) = ah^2 + 2bhk + ck^2$ be a quadratic form with discriminant $D = ac - b^2$. Then:

(i) If $D > 0$ and $a > 0$, then $Q(h,k) > 0$ for $(h,k) \neq (0,0)$.
(ii) If $D > 0$ and $a < 0$, then $Q(h,k) < 0$ for $(h,k) \neq (0,0)$.
(iii) If $D < 0$, then $Q(h,k)$ takes on both positive and negative values.

Proof Assume first that $a \neq 0$. Then we may rewrite $Q(h,k)$ by "completing the square":

$$Q(h,k) = ah^2 + 2bhk + ck^2 = a\left(h + \frac{b}{a}k\right)^2 + \left(c - \frac{b^2}{a}\right)k^2$$

$$= a\left(h + \frac{b}{a}k\right)^2 + \frac{D}{a}k^2 \qquad \boxed{1}$$

If $D > 0$ and $a > 0$, then $D/a > 0$ and both terms in (1) are positive. Hence, $Q(h,k) \geq 0$. Furthermore, if $Q(h,k) = 0$, then $k = 0$ and $h + \frac{b}{a}k = 0$, in which case $h = 0$. Thus, $Q(h,k) > 0$ if $(h,k) \neq 0$. This proves (i) and (ii) follows similarly. To prove (iii), note that if $a \neq 0$ and $D < 0$, then coefficients of the squared terms in Eq. (1) have opposite signs and $Q(h,k)$ takes on both positive and negative values. If $a = 0$ and $D < 0$, then $Q(h,k) = 2bhk + ck^2$ with $b \neq 0$, and $Q(h,k)$ again takes on both positive and negative values. ■

The Second Derivative Test is based on the second-order approximation of a function $f(x,y)$ of two variables (see the Internet supplement):

$$f(a+h, b+k) \approx f(a,b) + f_x(a,b)h + f_y(a,b)k$$
$$+ \frac{1}{2}\left(f_{xx}(a,b)h^2 + 2f_{xy}(a,b)hk + f_{yy}(a,b)k^2\right) + E_2(h,k) \qquad \boxed{2}$$

where $E_2(h,k)$ is the error. If (a,b) is a critical point, then the linear term is zero since $f_x(a,b) = f_y(a,b) = 0$. The behavior of f near (a,b) is then determined by the quadratic term $Q(h,k)$:

$$f(a+h, b+k)) \approx f(a,b) + \frac{1}{2}\overbrace{\left(f_{xx}(a,b)h^2 + 2f_{xy}(a,b)hk + f_{yy}(a,b)k^2\right)}^{Q(h,k)}$$

The discriminant of $Q(h,k)$ is equal to the discriminant of f at (a,b):

$$D = f_{xx}(a,b)f_{yy}(a,b) - f_{xy}(a,b)^2$$

Now we can apply Theorem 4. If $D > 0$ and $f_{xx}(a,b) > 0$, then $Q(h,k) \geq 0$ and Eq. (2) shows that $f(a+h, b+k) - f(a,b) > 0$ when $|h|$ and $|k|$ are small. Thus, $f(a,b)$ is a local minimum. Similarly, $f(a,b)$ is a local maximum if $D > 0$ and $f_{xx}(a,b) < 0$. If $D < 0$, then $Q(h,k)$ takes on both positive and negative values, in which case (a,b) is a saddle point. To complete this argument, we would have to prove the second-order approximation in the multivariable setting and verify that the error is sufficiently small to ignore.

14.7 SUMMARY

- We say that $P = (a,b)$ is a *critical point* of $f(x,y)$ if either

$$f_x(a,b) = f_y(a,b) = 0$$

or one of the two partial derivatives $f_x(a,b)$, $f_y(a,b)$ does not exist. In n-variables, $P = (a_1, \dots, a_n)$ is a critical point of $f(x_1, \dots, x_n)$ if $f_{x_j}(a_1, \dots, a_n) = 0$ for $j = 1, \dots, n$, or if $f_{x_j}(a_1, \dots, a_n)$ does not exist for some j.
- If $f(P)$ is a local minimum or maximum value of f, then P is a critical point.
- The *discriminant* of $f(x,y)$ at $P = (a,b)$ is the quantity

$$D(a,b) = f_{xx}(a,b)f_{yy}(a,b) - f_{xy}(a,b)^2$$

- *Second Derivative Test:* If (a,b) is a critical point of $f(x,y)$ and the second derivatives f_{xx}, f_{yy}, f_{xy} are continuous near P, then

$$D(a,b) > 0 \quad \Rightarrow \quad f(a,b) \text{ is a local minimum or maximum}$$
$$D(a,b) < 0 \quad \Rightarrow \quad (a,b) \text{ is a saddle point}$$
$$D(a,b) = 0 \quad \Rightarrow \quad \text{test inconclusive}$$

Furthermore, if $D(a,b) > 0$, then $f(a,b)$ is a local minimum if $f_{xx}(a,b) > 0$, and it is a local maximum if $f_{xx}(a,b) < 0$.
- Let $\mathcal{D}$ be a domain. A point P in $\mathcal{D}$ is an *interior* point if $\mathcal{D}$ contains some open disk $D(P,r)$ centered at P. A point P is a *boundary point* of $\mathcal{D}$ if every open disk $D(P,r)$ contains points in $\mathcal{D}$ and points not in $\mathcal{D}$. The *interior* of $\mathcal{D}$ is the set of all interior points and the *boundary* is the set of all boundary points. A domain is *closed* if it contains all its boundary points and *open* if it is equal to its interior.
- *Existence of Global Extrema:* A continuous function f takes on both a minimum and a maximum value on any closed, bounded domain $\mathcal{D}$.

- *Location of Extreme Values:* The extreme values of f on a closed, bounded domain $\mathcal{D}$ occur either at critical points in the interior of $\mathcal{D}$ or at points on the boundary of $\mathcal{D}$. To determine the extreme values, first find the critical points in the interior of $\mathcal{D}$. Then compare the values of f at the critical points with the minimum and maximum values of f on the boundary.

14.7 EXERCISES

Preliminary Questions

1. The functions $f(x, y) = x^2 + y^2$ and $g(x, y) = x^2 - y^2$ both have a critical point at $(0, 0)$. How is the behavior of the two functions at the critical point different?

2. Identify the points indicated in the contour maps as local minima, maxima, saddle points, or neither (Figure 14).

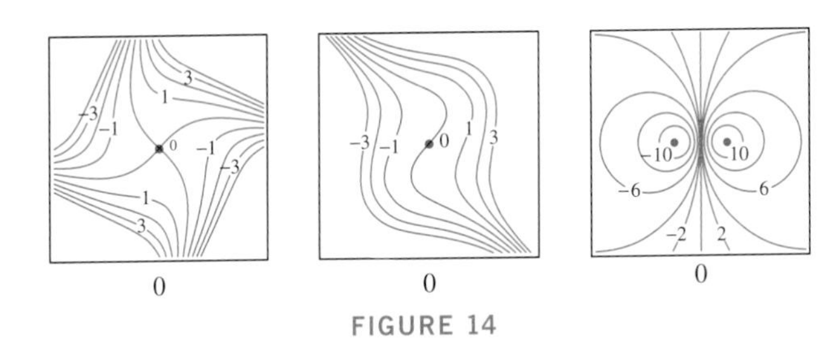

FIGURE 14

3. Let $f(x, y)$ be a continuous function on a domain $\mathcal{D}$ in $\mathbf{R}^2$. Determine which of the following statements are true:

(a) If $\mathcal{D}$ is closed and bounded, then f takes on a maximum value on $\mathcal{D}$.

(b) If $\mathcal{D}$ is neither closed nor bounded, then f does not take on a maximum value of $\mathcal{D}$.

(c) $f(x, y)$ need not have a maximum value on $\mathcal{D} = \{(x, y) : 0 \le x, y \le 1\}$.

(d) A continuous function takes on neither a minimum nor a maximum value on the open quadrant $\{(x, y) : x > 0, y > 0\}$.

Exercises

1. Let $P = (a, b)$ be a critical point of $f(x, y) = x^2 + y^4 - 4xy$.

(a) First use $f_x(,x, y) = 0$ to show that $a = 2b$. Then use $f_y(x, y) = 0$ to show that $P = (0, 0)$, $(2\sqrt{2}, \sqrt{2})$, or $(-2\sqrt{2}, -\sqrt{2})$.

(b) Referring to Figure 15, determine the local minima and saddle points of $f(x, y)$ and find the absolute minimum value of $f(x, y)$.

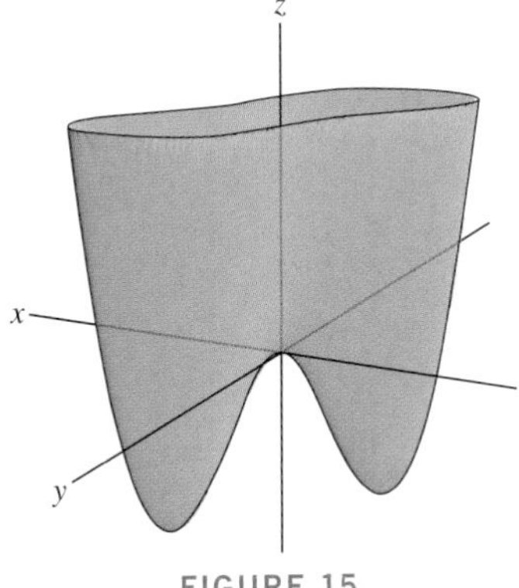

FIGURE 15

2. Find the critical points of the functions

$$f(x, y) = x^2 + 2y^2 - 4y + 6x, \quad g(x, y) = x^2 - 12xy + y$$

Use the Second Derivative Test to determine the local minimum, maximum, and saddle points. Match $f(x, y)$ and $g(x, y)$ with their graphs in Figure 16.

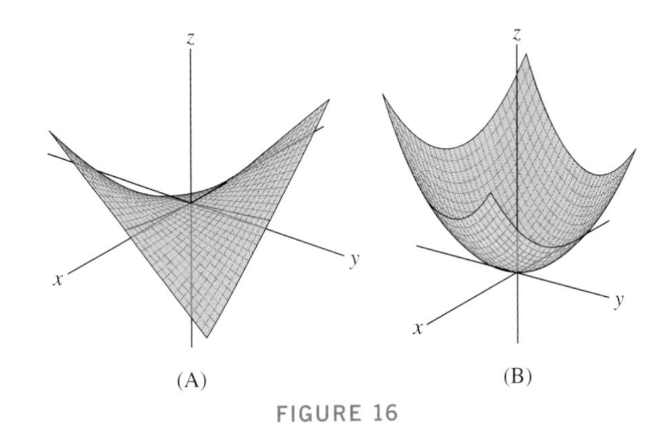

FIGURE 16

3. Find the critical points of

$$f(x, y) = 8y^4 + x^2 + xy - 3y^2 - y^3$$

Use the contour map in Figure 17 to determine their nature (minimum, maximum, saddle point).

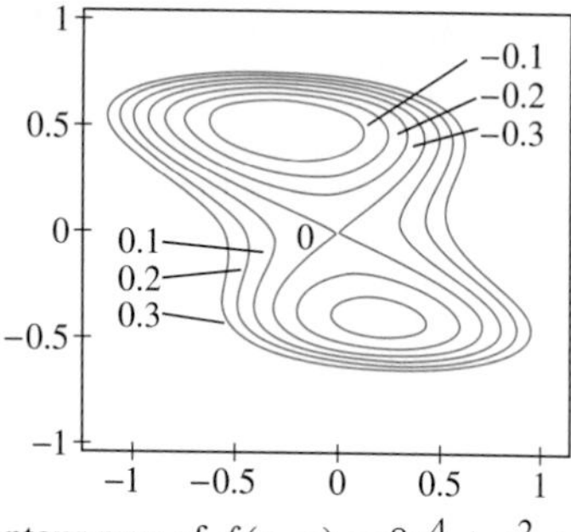

FIGURE 17 Contour map of $f(x, y) = 8y^4 + x^2 + xy - 3y^2 - y^3$.

4. Let $f(x, y) = y^2x - yx^2 + xy$.

(a) Show that the critical points (x, y) satisfy the equations

$$y(y - 2x + 1) = 0, \qquad x(2y - x + 1) = 0$$

(b) Show that there are three critical points with either $x = 0$ or $y = 0$ (or both) and one critical point with $x, y \neq 0$.

(c) Use the second derivative to determine the nature of the critical points.

In Exercises 5–20, find the critical points of the function. Then use the Second Derivative Test to determine whether they are local minima or maxima (or state that the test fails).

5. $f(x, y) = x^2 + y^2 - xy + x$

6. $f(x, y) = x^3 - xy + y^3$

7. $f(x, y) = x^3y + 12x^2 - 8y$

8. $f(x, y) = x^3 + 2xy - 2y^2 - 10x$

9. $f(x, y) = 4x - 3x^3 - 2xy^2$

10. $f(x, y) = x^3 + y^4 - 6x - 2y^2$

11. $f(x, y) = x^4 + y^4 - 4xy$

12. $f(x, y) = e^{x^2 - y^2 + 4y}$

13. $f(x, y) = xye^{-x^2 - y^2}$

14. $f(x, y) = x \ln(x + y)$

15. $f(x, y) = e^x - xe^y$

16. $f(x, y) = (x + 3y)e^{y - x^2}$

17. $f(x, y) = \ln x + 2\ln y - x - 4y$

18. $f(x, y) = (x^2 + y^2)e^{-x^2 - y^2}$

19. $f(x, y) = x - y^2 - \ln(x + y)$

20. $f(x, y) = (x - y)e^{x^2 - y^2}$

21. Show that $f(x, y) = \sqrt{x^2 + y^2}$ has one critical point P and that f is nondifferentiable at P. Show that $f(P)$ is an absolute minimum value.

22. Let $f(x, y) = (x^2 + y^2 - 1)^2$.

(a) Find the critical points of $f(x, y)$.

(b) CAS Use a computer algebra system to graph $f(x, y)$ and identify the critical points on the graph.

(c) Find all points where f takes on a minimum value. Does the Second Derivative Test apply?

23. CAS Use a computer algebra system to find numerical approximations to the critical points of

$$f(x, y) = (1 - x + x^2)e^{y^2} + (1 - y + y^2)e^{x^2}$$

Use Figure 18 to determine whether they correspond to local minima or maxima.

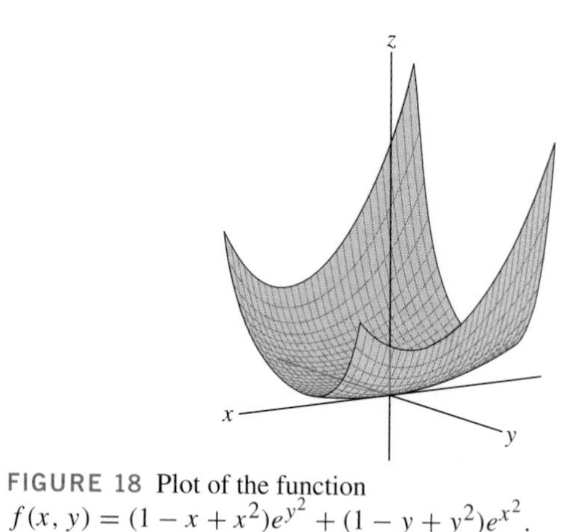

FIGURE 18 Plot of the function
$f(x, y) = (1 - x + x^2)e^{y^2} + (1 - y + y^2)e^{x^2}$.

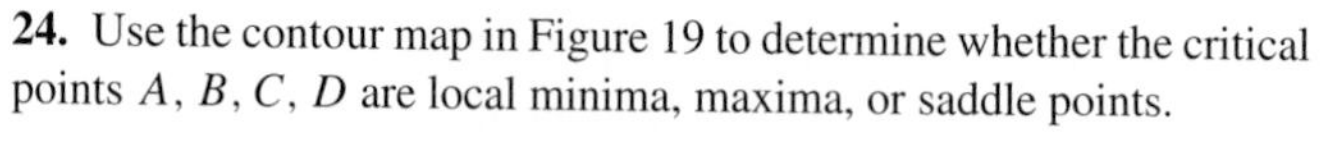

24. Use the contour map in Figure 19 to determine whether the critical points A, B, C, D are local minima, maxima, or saddle points.

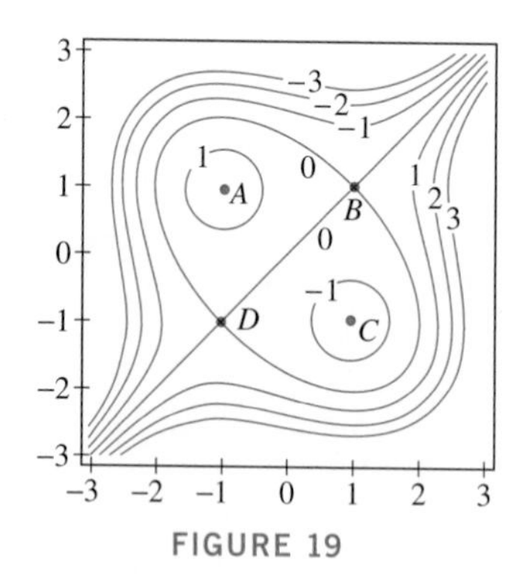

FIGURE 19

25. Which of the following domains are closed and which are bounded?

(a) $\{(x, y) \in \mathbf{R}^2 : x^2 + y^2 \leq 1\}$

(b) $\{(x, y) \in \mathbf{R}^2 : x^2 + y^2 < 1\}$

(c) $\{(x, y) \in \mathbf{R}^2 : x \geq 0\}$

(d) $\{(x, y) \in \mathbf{R}^2 : x > 0, y > 0\}$

(e) $\{(x, y) \in \mathbf{R}^2 : 1 \le x \le 4, 5 \le y \le 10\}$

(f) $\{(x, y) \in \mathbf{R}^2 : x > 0, x^2 + y^2 \le 10\}$

In Exercises 26–29, determine the global extreme values of the function on the given set without using calculus.

26. $f(x, y) = x + y, \quad 0 \le x, y \le 1$

27. $f(x, y) = 2x - y, \quad 0 \le x \le 1, 0 \le y \le 3$

28. $f(x, y) = (x^2 + y^2 + 1)^{-1}, \quad 0 \le x \le 3, 0 \le y \le 5$

29. $f(x, y) = e^{-x^2-y^2}, \quad x^2 + y^2 \le 1$

30. Assumptions Matter Show that $f(x, y) = x + y$ has no global minimum or maximum on the domain $0 < x, y < 1$. Does this contradict Theorem 3?

31. Let $\mathcal{D} = \{(x, y) : x > 0, y > 0\}$. Show that $\mathcal{D}$ is not closed. Find a continuous function that does not have a global minimum value on $\mathcal{D}$.

32. The goal is to find the extreme values of $f(x, y) = x^2 - 2xy + 2y$ on the square $0 \le x, y \le 2$ (Figure 20).

(a) Evaluate f at the critical points that lie in the square.

(b) On the bottom edge of the square, $y = 0$ and $f(x, 0) = x^2$. Find the extreme values of f on the bottom edge.

(c) Find the extreme values of f on the remaining edges of the square.

(d) Find the largest and smallest values of f among the values computed in (a), (b), and (c).

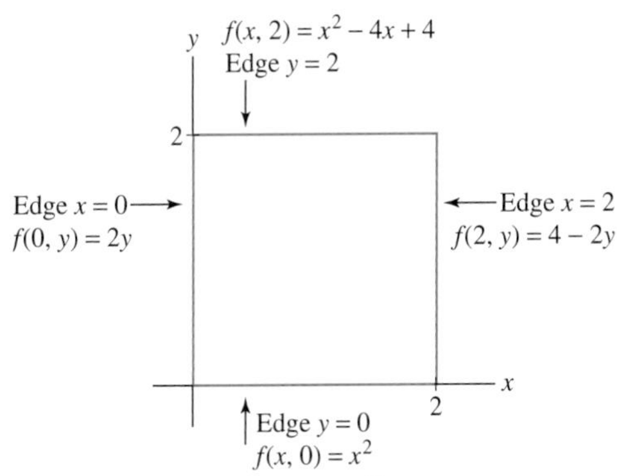

FIGURE 20 The function $f(x, y) = x^2 - 2xy + 2y$ on the four boundary segments of the square $0 \le x \le 2$.

33. Let $f(x, y) = (x + y)\ln(x^2 + y^2)$, defined for $(x, y) \ne (0, 0)$.

(a) Show that if (x, y) is a critical point of $f(x, y)$, then either $x = y$ or $x = -y$. Determine the critical points (four in all).

(b) CAS Plot a graph of $f(x, y)$ and use it to determine whether these critical points are maxima, minima, or neither.

34. Let $f(x, y) = x^4 - 4xy + 2y^2$.

(a) Find the critical points of $f(x, y)$.

(b) CAS Use a plot to determine whether the critical points are maxima, minima, or saddle points. Then confirm this result using the Second Derivative Test.

In Exercises 35–41, determine the global extreme values of the function on the given domain.

35. $f(x, y) = x^3 - 2y, \quad 0 \le x, y \le 1$

36. $f(x, y) = 4x + 3y, \quad 0 \le x, y \le 1$

37. $f(x, y) = x^2 + 2y^2, \quad 0 \le x, y \le 1$

38. $f(x, y) = (4y^2 - x^2)e^{-x^2-y^2}, \quad x^2 + y^2 \le 2$

39. $f(x, y) = x^3 + x^2y + 2y^2, \quad x, y \ge 0, x + y \le 1$

40. $f(x, y) = x^3 + y^3 - 3xy, \quad 0 \le x, y \le 1$

41. CAS $f(x, y) = x^3y^5$, the set bounded by $x = 0$, $y = 0$, and $y = 1 - \sqrt{x}$. *Hint:* Use a computer algebra system to find the minimum along the boundary curve $y = 1 - \sqrt{x}$, which is parametrized by $(t, 1 - \sqrt{t})$ for $0 \le t \le 1$.

42. Show that the rectangular box (including a top and bottom) with fixed volume V with the smallest possible surface area is a cube (Figure 21).

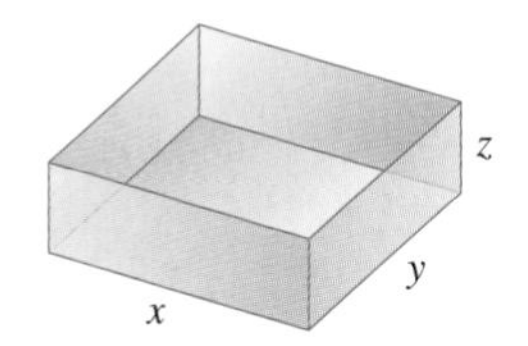

FIGURE 21 Rectangular box with sides x, y, z.

43. Consider a rectangular box B with a bottom and sides but no top such that B has minimal surface area among all boxes with fixed volume V.

(a) Do you think B is a cube as in the solution to Exercise 42? If not, how would you make its shape differ from a cube?

(b) Find the dimensions of B and compare with your response to (a).

Further Insights and Challenges

44. Given n data points $(x_1, y_1), \dots, (x_n, y_n)$, we may seek a linear function $f(x) = mx + b$ that best fits the data. The **linear least-squares fit** is the linear function $f(x) = mx + b$ that minimizes the sum of the squares (Figure 22)

$$E(m, b) = \sum_{j=1}^{n} (y_j - f(x_j))^2$$

Show that E is minimized for m and b satisfying

$$m\sum_{j=1}^{n} x_j + bn = \sum_{j=1}^{n} y_j$$

$$m\sum_{j=1}^{n} x_j^2 + b\sum_{j=1}^{n} x_j = \sum_{j=1}^{n} x_j y_j$$

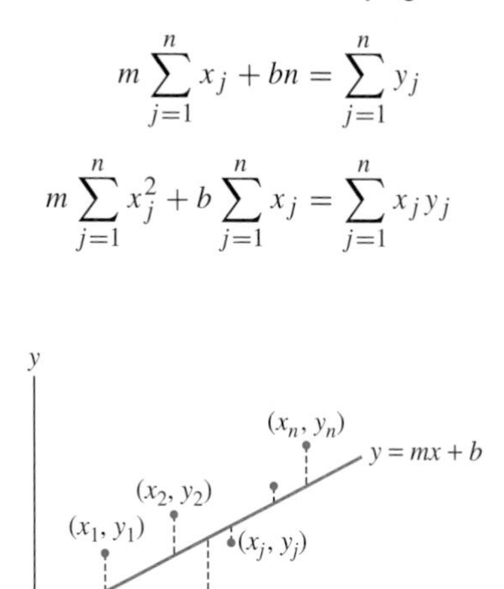

FIGURE 22 The linear least-squares fit minimizes the sum of the squares of the vertical distances from the data points to the line.

45. The power (in microwatts) of a laser is measured as a function of current (in milliamps). Find the linear least-squares fit (Exercise 44) for the data points.

Current (mA)	1.0	1.1	1.2	1.3	1.4	1.5
Laser power (μW)	0.52	0.56	0.82	0.78	1.23	1.50

46. Prove that if $\alpha, \beta \geq 1$ are numbers such that $\alpha^{-1} + \beta^{-1} = 1$, then the following inequality holds for all $x, y \geq 0$:

$$\frac{1}{\alpha}x^{\alpha} + \frac{1}{\beta}x^{\beta} \geq xy$$

For example,

$$\frac{1}{3}x^3 + \frac{2}{3}y^{3/2} \geq xy, \qquad \frac{2}{5}x^{5/2} + \frac{3}{5}y^{5/3} \geq xy, \qquad \text{etc.}$$

Hint: Find the critical points of $f(x, y) = \alpha^{-1}x^{\alpha} + \beta^{-1}x^{\beta} - xy$.

47. The following problem was proposed by Fermat as a challenge to the Italian scientist Evangelista Torricelli (1608–1647), a student of Galileo and inventor of the barometer. Given three points $A = (a_1, a_2)$, $B = (b_1, b_2)$, and $C = (c_1, c_2)$ in the plane, find the point $P = (x, y)$ that minimizes the sum of the distances

$$f(x, y) = AP + BP + CP$$

(a) Write out $f(x, y)$ as a function of x and y, and show that $f(x, y)$ is differentiable except at the points A, B, C.

(b) Define the unit vectors

$$\mathbf{e} = \frac{\overrightarrow{AP}}{\|\overrightarrow{AP}\|}, \qquad \mathbf{f} = \frac{\overrightarrow{BP}}{\|\overrightarrow{BP}\|}, \qquad \mathbf{g} = \frac{\overrightarrow{CP}}{\|\overrightarrow{CP}\|}$$

Show that the condition $\nabla f = 0$ is equivalent to

$$\mathbf{e} + \mathbf{f} + \mathbf{g} = 0 \qquad \boxed{3}$$

Prove that Eq. (3) holds if and only if the mutual angles between the unit vectors are all 120°.

(c) Define the Fermat point to be the point P such that angles between the segments $\overline{AP}$, $\overline{BP}$, $\overline{CP}$ are all 120°. Conclude that the **Fermat point** solves the minimization problem (Figure 23).

(d) Show that the Fermat point does not exist if one of the angles in $\triangle ABC$ is $\geq 120°$. Where does the minimum occur in this case? *Hint:* The minimum must occur at a point where $f(x, y)$ is not differentiable.

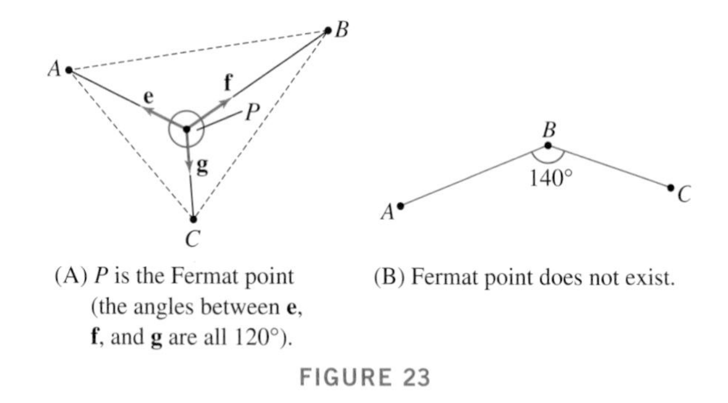

(A) P is the Fermat point (the angles between **e**, **f**, and **g** are all 120°).

(B) Fermat point does not exist.

FIGURE 23

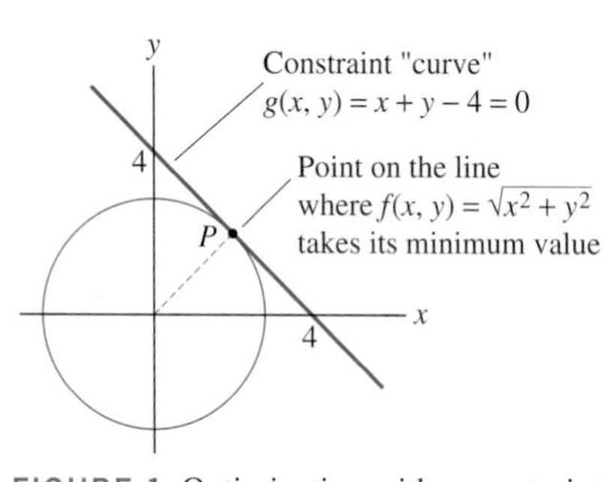

FIGURE 1 Optimization with a constraint: find the minimum of $f(x, y) = \sqrt{x^2 + y^2}$ on the line $x + y = 4$.

14.8 Lagrange Multipliers: Optimizing with a Constraint

Some optimization problems involve finding the minimum or maximum of a function $f(x, y)$ subject to a constraint $g(x, y) = 0$ (Figure 1). An example is the problem of minimizing

$$f(x, y) = \sqrt{x^2 + y^2} \quad \text{subject to} \quad g(x, y) = x + y - 4 = 0$$

Since $f(x, y)$ is the distance from (x, y) to the origin, this problem asks us to find the point $P = (x, y)$ on the line $x + y = 4$ closest to the origin. The method of **Lagrange**

multipliers is a general procedure for finding extreme values of $f(x, y)$ on a constraint curve $g(x, y) = 0$. The main idea is as follows.

GRAPHICAL INSIGHT Suppose we are standing at point Q in Figure 2(A), and we wish to move along the constraint curve in a direction that increases the value of f. The gradient vector ∇f_Q points in the direction of maximal increase of f (the northeast direction). Although moving in the gradient direction would take us off the constraint curve, we can still increase f by moving to the right along the constraint curve.

We keep moving to the right until we arrive at the point P, where ∇f_P is orthogonal to the constraint curve [Figure 2(B)]. Once at P, we cannot increase f further by moving to the right or left along the constraint curve, and thus, $f(P)$ is a local maximum (subject to the constraint). At the point P, the level curve of f is tangent to the constraint curve. Now observe that ∇f_P and ∇g_P are both normal to the constraint curve, and hence they must point in the same or opposite directions. In other words, $\nabla f_P = \lambda \nabla g_P$ for some scalar λ (called a **Lagrange multiplier**). Note also that level curves of f and g through P are tangent to each other at P.

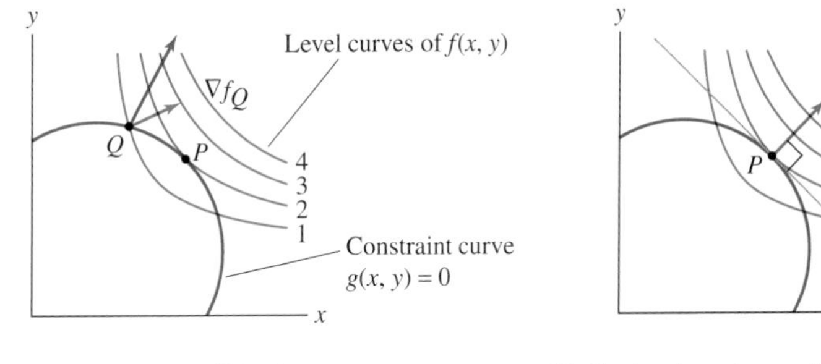

(A) The gradient vector ∇f_Q shows that f increases as we move to the right along the constraint curve.

(B) The local maximum of f on the constraint curve $g(x, y) = 0$ occurs at a point P where ∇f_P and ∇g_P point in the same direction.

FIGURE 2

THEOREM 1 Lagrange Multipliers Assume that $f(x, y)$ and $g(x, y)$ are differentiable functions. If $f(x, y)$ has a local minimum or maximum on the constraint curve $g(x, y) = 0$ at $P = (a, b)$, and if $\nabla g_P \neq \mathbf{0}$, then there is a scalar λ such that

$$\nabla f_P = \lambda \nabla g_P \qquad \boxed{1}$$

In Theorem 1, the assumption $\nabla g_P \neq \mathbf{0}$ guarantees (by the Implicit Function Theorem of advanced calculus) that we can parametrize the curve $g(x, y) = 0$ near P.

Proof Let $\mathbf{c}(t)$ be a parametrization of the constraint curve $g(x, y) = 0$ near P such that $\mathbf{c}(0) = P$ and (a, b) and $\mathbf{c}'(0) \neq \mathbf{0}$. Then $f(\mathbf{c}(0)) = f(P)$, and $f(\mathbf{c}(t))$ has a local minimum or maximum at $t = 0$. Thus, $t = 0$ is a critical point of $f(\mathbf{c}(t))$ and

$$\underbrace{\frac{d}{dt} f(\mathbf{c}(t))\bigg|_{t=0} = \nabla f_P \cdot \mathbf{c}'(0)}_{\text{Chain Rule}} = 0$$

This shows that ∇f_P is orthogonal to the tangent vector $\mathbf{c}'(0)$ to the curve $g(x, y) = 0$. The gradient ∇g_P is also orthogonal to $\mathbf{c}'(0)$ since ∇g_P is orthogonal to the level curve $g(x, y) = 0$ at P. We conclude that ∇f_P and ∇g_P are proportional as claimed. ■

We refer to Eq. (1) as the **Lagrange condition**. When we write this condition in terms of components, we obtain the **Lagrange equations**:

$$f_x(a,b) = \lambda g_x(a,b)$$
$$f_y(a,b) = \lambda g_y(a,b)$$

The point $P = (a,b)$ is called a critical point for the optimization problem with constraint and $f(a,b)$ is called a critical value.

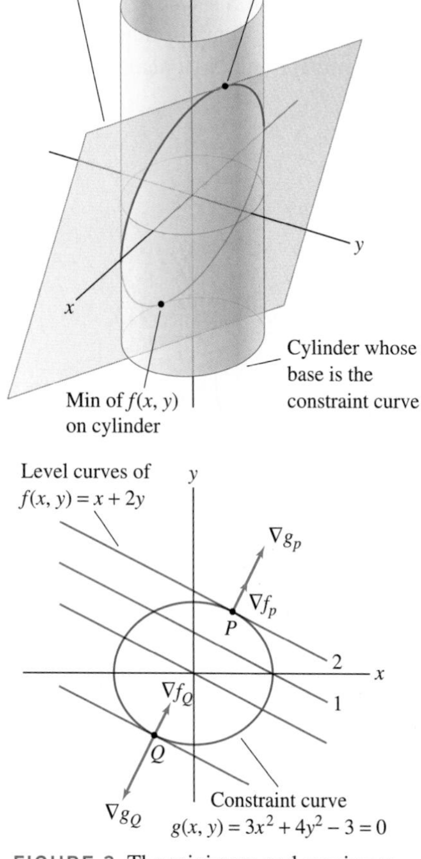

FIGURE 3 The minimum and maximum occur where the level curve of $f(x,y) = x + 2y$ is tangent to the constraint curve $3x^2 + 4y^2 = 3$.

■ **EXAMPLE 1** Find the extreme values of the function $f(x,y) = x + 2y$ on the ellipse $3x^2 + 4y^2 = 3$.

Solution

***Step 1.* Write out the Lagrange equations.**
The constraint curve is $g(x,y) = 0$, where $g(x,y) = 3x^2 + 4y^2 - 3$. The gradients are $\nabla f = \langle 1, 2\rangle$ and $\nabla g = \langle 6x, 8y\rangle$, so the Lagrange condition gives us

$$\langle 1, 2\rangle = \lambda\,\langle 6x, 8y\rangle \quad\Rightarrow\quad 1 = \lambda(6x), \qquad 2 = \lambda(8y) \tag{2}$$

***Step 2.* Solve for λ in terms of x and y.**
We use (2) to obtain two equations for λ:

$$\lambda = \frac{1}{6x}, \qquad \lambda = \frac{2}{8y} \tag{3}$$

To justify dividing by x and y, we observe that x and y must be nonzero since $x = 0$ or $y = 0$ would violate (2).

***Step 3.* Solve for x and y using the constraint.**
The two expressions for λ must be equal, so we obtain $\frac{1}{6x} = \frac{2}{8y}$ or $y = \frac{3}{2}x$. Now substitute $y = \frac{3}{2}x$ in the constraint equation:

$$g(x,y) = 3x^2 + 4y^2 - 3 = 3x^2 + 4\left(\frac{3}{2}x\right)^2 - 3 = 12x^2 - 3 = 0$$

Thus, $x = \frac{1}{2}$ or $-\frac{1}{2}$ and $y = \frac{3}{2}x = \frac{3}{4}$ or $-\frac{3}{4}$.

***Step 4.* Calculate the critical values.**
Now evaluate $f(x,y)$ at the critical points $P = (\frac{1}{2}, \frac{3}{4})$ and $Q = (-\frac{1}{2}, -\frac{3}{4})$:

$$f(P) = \frac{1}{2} + 2\left(\frac{3}{4}\right) = 2, \qquad f(Q) = -\frac{1}{2} + 2\left(-\frac{3}{4}\right) = -2$$

We conclude that the maximum of $f(x,y)$ on the ellipse is 2 and the minimum is -2 (Figure 3). ■

GRAPHICAL INSIGHT In an ordinary optimization problem without a constraint, the maximum value of $f(x,y)$ corresponds to the highest point on the surface $z = f(x,y)$. In the case of optimization subject to a constraint $g(x,y) = 0$, consider the cylinder $\mathcal{S}$ whose base is the constraint curve, that is, the cylinder consisting of all points (x,y,z) such that $g(x,y) = 0$ (Figure 3). The maximum subject to the constraint corresponds to the highest point on the intersection of the surface $z = f(x,y)$ and the cylinder $\mathcal{S}$.

FIGURE 4 Economist Paul Douglas and mathematician Charles Cobb arrived at the production functions $P(x, y) = Cx^a y^b$ in 1927 by fitting data gathered on the relationship between labor, capital, and the output in an industrial economy. The photo shows Douglas, who was a professor at the University of Chicago and also served as U.S. Senator from Illinois from 1949–1967.

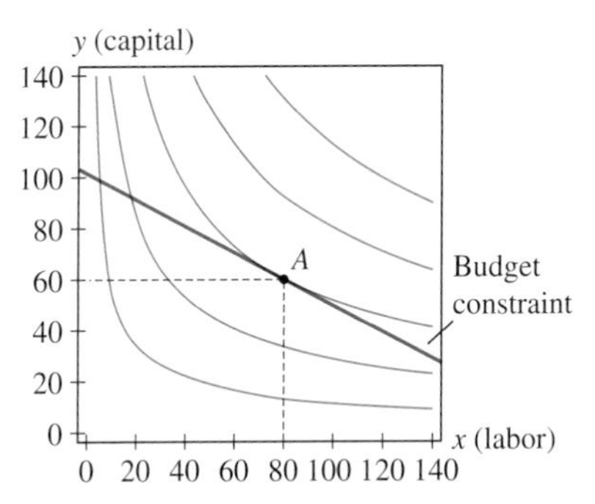

FIGURE 5 Contour plot of the Cobb–Douglas production function $P(x, y) = 50x^{0.4}y^{0.6}$.

■ **EXAMPLE 2 Cobb–Douglas Production Function** By investing x units of labor and y units of capital, a low-end watch manufacturer can produce $P(x, y) = 50x^{0.4}y^{0.6}$ watches. Find the maximum number of watches that can be produced on a budget of \$20,000 if labor costs \$100 per unit and capital costs \$200 per unit.

Solution The total cost of x units of labor and y units of capital is $100x + 200y$, so the budget constraint is $100x + 200y = 20{,}000$. Thus, our task is to maximize the function $P(x, y) = 50x^{0.4}y^{0.6}$ subject to the constraint (Figure 5)

$$g(x, y) = 100x + 200y - 20{,}000 = 0 \qquad \boxed{4}$$

***Step 1.* Write out the Lagrange equations.**
The Lagrange equations are

$$P_x(x, y) = \lambda g_x(x, y): \quad 20x^{-0.6}y^{0.6} = 100\lambda$$

$$P_y(x, y) = \lambda g_y(x, y): \quad 30x^{0.4}y^{-0.4} = 200\lambda$$

***Step 2.* Solve for λ in terms of x and y.**
These equations yield two expressions for λ that must be equal:

$$\lambda = \frac{1}{5}x^{-0.6}y^{0.6} = \frac{3}{20}x^{0.4}y^{-0.4} \qquad \boxed{5}$$

***Step 3.* Solve for x and y using the constraint.**
Multiply Eq. (5) by $5x^{0.6}y^{0.4}$ to obtain $y = \frac{3}{4}x$, and substitute in Eq. (4):

$$100x + 200y = 100x + 200\left(\frac{3}{4}x\right) = 20{,}000 \quad \Rightarrow \quad 250x = 20{,}000$$

We obtain $x = 20{,}000/250 = 80$ and $y = \frac{3}{4}x = 60$. Therefore, the critical point is $A = (80, 60)$.

***Step 4.* Calculate the Critical Values.**
Since $P(x, y)$ is increasing as a function of x and y, ∇P points to the northeast and it is clear that $P(x, y)$ takes on a maximum value at A (Figure 5). The maximum is $P(80, 60) = 50(80)^{0.6}(60)^{0.4} = 3{,}365.87$, or roughly 3,365 watches, with a cost per watch of $20{,}000/3{,}365$ or \$5.94. ■

The method of Lagrange multipliers is valid in any number of variables. The next example illustrates the method for three variables.

■ **EXAMPLE 3 Lagrange Multipliers in Three Variables** Find the point on the plane $\frac{x}{2} + \frac{y}{4} + \frac{z}{4} = 1$ closest to the origin in $\mathbf{R}^3$.

Solution Our task is to minimize the distance $d = \sqrt{x^2 + y^2 + z^2}$ subject to the constraint $\frac{x}{2} + \frac{y}{4} + \frac{z}{4} = 1$. When the distance d is at a minimum, the *square* d^2 is also at a minimum (since the function d^2 is increasing). Thus, it suffices to find the minimum of the square of the distance $f(x, y, z) = x^2 + y^2 + z^2$.

***Step 1.* Write out the Lagrange equations.**
The Lagrange condition is

$$\underbrace{\langle 2x, 2y, 2z\rangle}_{\nabla f} = \lambda \underbrace{\left\langle \frac{1}{2}, \frac{1}{4}, \frac{1}{4}\right\rangle}_{\nabla g}$$

***Step 2.* Solve for λ in terms of x, y, and z.**
We obtain

$$\lambda = 4x = 8y = 8z \quad \Rightarrow \quad z = y = \frac{x}{2}$$

***Step 3.* Solve for x, y, and z using the constraint.**
Substitute in the constraint equation:

$$\frac{x}{2} + \frac{y}{4} + \frac{z}{4} = \frac{2z}{2} + \frac{z}{4} + \frac{z}{4} = \frac{3z}{2} = 1 \quad \Rightarrow \quad z = \frac{2}{3}$$

We obtain $x = 2z = \frac{4}{3}$ and $y = z = \frac{2}{3}$. It is clear geometrically that the critical point corresponds to a minimum of f. Hence, the point on the plane closest to the origin is $P = (\frac{4}{3}, \frac{2}{3}, \frac{2}{3})$ (Figure 6). ■

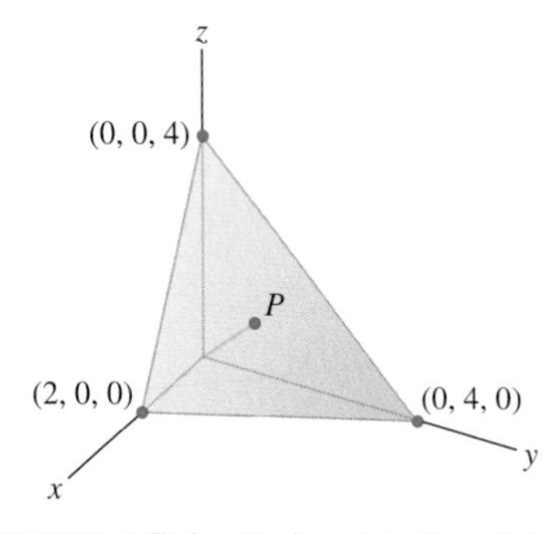

FIGURE 6 Point P closest to the origin on the plane.

The method of Lagrange multipliers applies when there is more than one constraint equation, but we must add another multiplier for each constraint. For example, if the problem is to minimize $f(x, y, z)$ subject to constraints $g(x, y, z) = 0$ and $h(x, y, z) = 0$, the Lagrange condition is

$$\nabla f = \lambda \nabla g + \mu \nabla h$$

■ **EXAMPLE 4 Lagrange Multipliers with Multiple Constraints** Find the minimum of $f(x, y, z) = x^2y^2 + z^2$ subject to $x + y = 2$ and $y + z = 4$.

Solution The constraint equations are

$$g(x, y) = x + y - 2 = 0, \qquad h(x, y) = y + z - 4 = 0$$

***Step 1.* Write out the Lagrange equations.**
We have $\nabla f = \langle 2xy^2, 2x^2y, 2z\rangle$, so the Lagrange condition is

$$\underbrace{\left\langle 2xy^2, 2x^2y, 2z\right\rangle}_{\nabla f} = \lambda \underbrace{\langle 1, 1, 0\rangle}_{\nabla g} + \mu \underbrace{\langle 0, 1, 1\rangle}_{\nabla h} = \langle \lambda, \lambda + \mu, \mu\rangle$$

***Step 2.* Solve for λ and μ.**
We obtain $\lambda = 2xy^2$, $\lambda + \mu = 2x^2y$, and $\mu = 2z$. Using the first and third equation in the middle equation, we obtain

$$2xy^2 + 2z = 2x^2y \quad \Rightarrow \quad z = x^2y - xy^2 \tag{6}$$

***Step 3.* Solve for x, y, and z using the constraints.**
First, use the constraint equations to express y and z in terms of x:

$$y = 2 - x \qquad \text{and} \qquad z = 4 - y = 4 - (2 - x) = x + 2$$

Then substitute in Eq. (6) to obtain an equation for x:

$$x + 2 = x^2(2 - x) - x(2 - x)^2 \quad \Rightarrow \quad 2x^3 - 6x^2 + 5x + 2 = 0$$

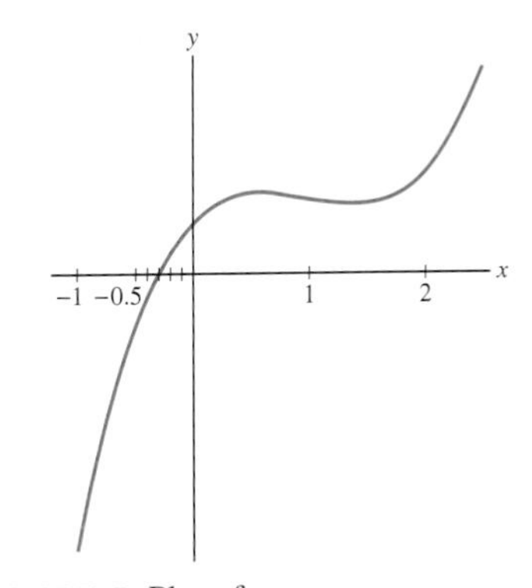

FIGURE 7 Plot of $F(x) = 2x^3 - 6x^2 + 5x + 2$.

The plot of the function $F(x) = 2x^3 - 6x^2 + 5x + 2$ in Figure 7 shows that $F(x) = 0$ has a unique real root $x \approx -0.3$. We then have

$$y = 2 - x \approx 2 - (-0.3) = 2.3, \qquad z = x + 2 \approx -0.3 + 2 = 1.7$$

The unique critical point is $P \approx (-0.3, 2.3, 1.7)$.

***Step 4.* Calculate the critical values.**

The value of f at the unique critical point is approximately $f(-0.3, 2.3, 1.7) \approx 3.4$. To show that this is a minimum value, observe that $y = 2 - x$ and $z = x + 2$ and $f(x, y, z) = x^2(2 - x)^2 + (x + 2)^2$. Thus $f(x, y, z)$ tends to ∞ as x grows large and the critical point must yield a minimum. In fact, we could have found this minimum by optimizing $g(x) = x^2(2 - x)^2 + (x + 2)^2$, but our purpose was to illustrate the method of Lagrange multipliers. ■

Assumptions Matter Constrained optimization problems do not always have solutions. For example, $f(x, y) = x$ has no maximum or minimum subject to the constraint $g(x, y) = x - y = 0$ because $P = (a, a)$ satisfies the constraint and $f(a, a) = a$ can be arbitrarily large or small. However, a minimum and maximum are guaranteed to exist if the constraint curve $g(x, y) = 0$ is *bounded*. For example, extreme values exist if the constraint curve is an ellipse. This follows from Theorem 3 in Section 14.7, which states that a continuous function on a closed, bounded domain takes on extreme values.

14.8 SUMMARY

- According to the method of *Lagrange multipliers*, the extreme values of $f(x, y)$ subject to a constraint $g(x, y) = 0$ occur at points P (called critical points) satisfying the Lagrange condition $\nabla f_P = \lambda \nabla g_P$. This condition is equivalent to the *Lagrange equations*

$$f_x(x, y) = \lambda g_x(x, y), \qquad f_y(x, y) = \lambda g_y(x, y)$$

- Theorem 1 does not guarantee that a minimum or maximum value exists, but if it does exist, it must occur at a critical point.
- If the constraint curve $g(x, y) = 0$ is bounded [e.g., if $g(x, y) = 0$ is a circle or ellipse], then minimum and maximum values exist and occur at a critical point.
- The Lagrange condition for the extreme values of a function of three variables $f(x, y, z)$ subject to two constraints $g(x, y, z) = 0$ and $h(x, y, z) = 0$ is

$$\nabla f = \lambda \nabla g + \mu \nabla h$$

14.8 EXERCISES

Preliminary Questions

1. Suppose that the maximum of $f(x, y)$ subject to the constraint $g(x, y) = 0$ occurs at a point $P = (a, b)$ such that $\nabla f_P \neq 0$. Which of the following are true?

(a) ∇f_P is tangent to $g(x, y) = 0$ at P.

(b) ∇f_P is orthogonal to $g(x, y) = 0$ at P.

2. Figure 8 shows a constraint $g(x, y) = 0$ and the level curves of a function f. In each case, determine whether f has a local minimum, local maximum, or neither at the labeled point.

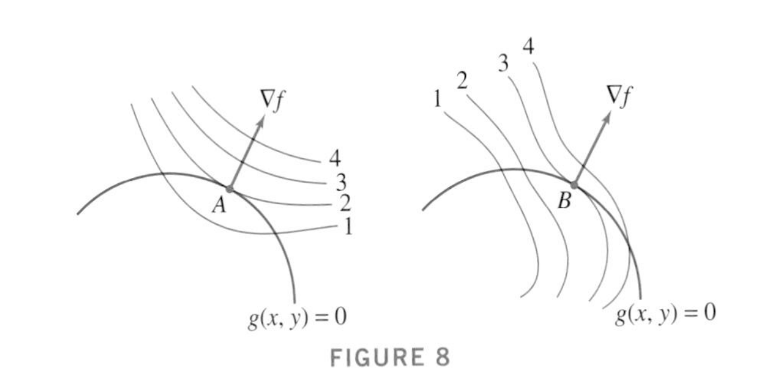

FIGURE 8

3. On the contour map in Figure 9:

(a) Identify the points where $\nabla f = \lambda \nabla g$ for some scalar λ.

(b) Identify the minimum and maximum values of $f(x, y)$ subject to $g(x, y) = 0$.

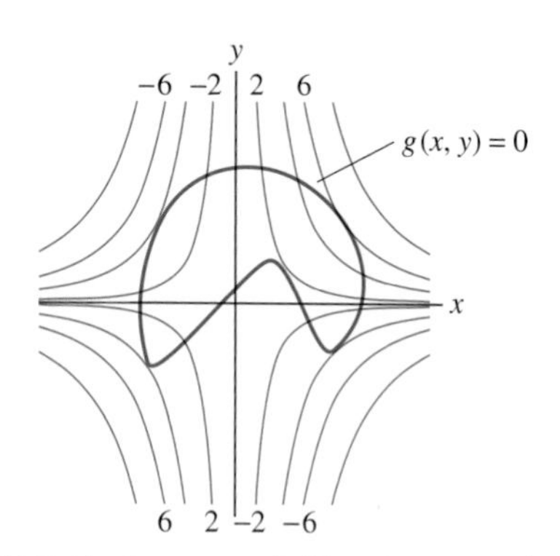

FIGURE 9 Contour map of $f(x, y)$; contour interval 2.

Exercises

1. Use Lagrange multipliers to find the extreme values of the function $f(x, y) = 2x + 4y$ subject to the constraint $g(x, y) = x^2 + y^2 - 5 = 0$.

(a) Show that the Lagrange equation $\nabla f = \lambda \nabla g$ gives $\lambda x = 1$ and $\lambda y = 2$.

(b) Show that these equations imply $\lambda \neq 0$ and $y = 2x$.

(c) Use the constraint equation to determine the possible critical points (x, y).

(d) Evaluate $f(x, y)$ at the critical points and determine the minimum and maximum values.

2. Find the extreme values of $f(x, y) = x^2 + 2y^2$ subject to the constraint $g(x, y) = 4x - 6y = 25$.

(a) Show that the Lagrange equations yield $2x = 4\lambda$, $4y = -6\lambda$.

(b) Show that if $x = 0$ or $y = 0$, then $\lambda = 0$ and the Lagrange equations give $x = y = 0$. Since $(0, 0)$ does not satisfy the constraint, you may assume that x and y are nonzero.

(c) Use the Lagrange equations to show that $y = -\frac{3}{4}x$.

(d) Substitute in the constraint equation to show that there is a unique critical point P.

(e) Does P correspond to a minimum or maximum value of f? Refer to Figure 10 to justify your answer. *Hint:* Do the values of $f(x, y)$ increase or decrease as (x, y) moves away from P along the line $g(x, y) = 0$?

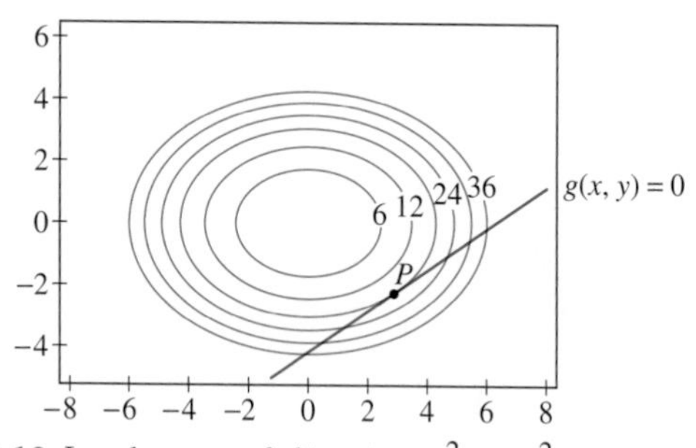

FIGURE 10 Level curves of $f(x, y) = x^2 + 2y^2$ and constraint $g(x, y) = 4x - 6y - 25 = 0$.

3. Apply the method of Lagrange multipliers to the function $f(x, y) = (x^2 + 1)y$ subject to the constraint $x^2 + y^2 = 5$. *Hint:* First show that $y \neq 0$; then treat the cases $x = 0$ and $x \neq 0$ separately.

In Exercises 4–13, find the minimum and maximum values of the function subject to the given constraint.

4. $f(x, y) = 2x + 3y, \quad x^2 + y^2 = 4$

5. $f(x, y) = x^2 + y^2, \quad 2x + 3y = 6$

6. $f(x, y) = 4x^2 + 9y^2, \quad xy = 4$

7. $f(x, y) = xy, \quad 4x^2 + 9y^2 = 32$

8. $f(x, y) = x^2y + x + y, \quad xy = 4$

9. $f(x, y) = x^2 + y^2, \quad x^4 + y^4 = 1$

10. $f(x, y) = x^2y^4, \quad x^2 + 2y^2 = 6$

11. $f(x, y, z) = 3x + 2y + 4z, \quad x^2 + 2y^2 + 6z^2 = 1$

12. $f(x, y, z) = x^2 - y - z, \quad x^2 - y^2 + z = 0$

13. $f(x, y, z) = xy + 3xz + 2yz, \quad 5x + 9y + z = 10$

14. Let $f(x, y) = x^3 + xy + y^3$, $g(x, y) = x^3 - xy + y^3$.

(a) Show that there is a unique point $P = (a, b)$ on $g(x, y) = 1$ where $\nabla f_P = \lambda \nabla g_P$ for some scalar λ.

(b) Refer to Figure 11 to determine whether $f(P)$ is a local minimum or maximum of f subject to the constraint.

(c) Does Figure 11 suggest that $f(P)$ is a global extremum subject to the constraint?

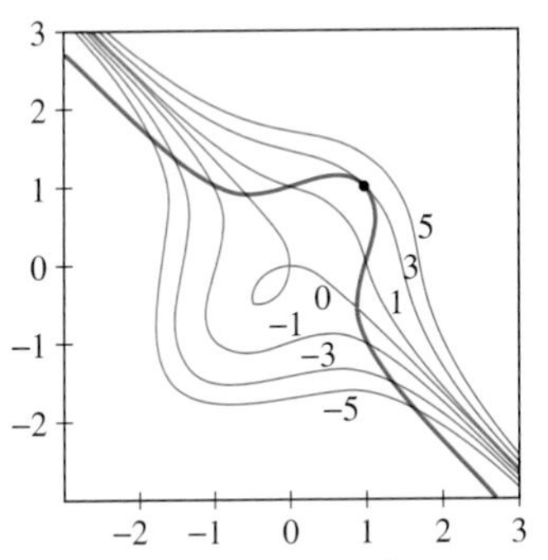

FIGURE 11 Contour map of $f(x, y) = x^3 + xy + y^3$ and graph of $g(x, y) = x^3 - xy + y^3 = 1$.

15. Use Lagrange multipliers to find the point (a, b) on the graph of $y = e^x$, where the value ab is as small as possible.

16. Find the rectangular box of maximum volume if the sum of the lengths of the edges is 300 cm.

17. The surface area of a right-circular cone of radius r and height h is $S = \pi r\sqrt{r^2 + h^2}$, and its volume is $V = \frac{1}{3}\pi r^2 h$.

(a) Determine the ratio h/r for the cone with given surface area S and maximal volume V.

(b) What is the ratio h/r for a cone with given volume V and minimal surface area S?

(c) Does a cone with given volume V and maximal surface area exist?

18. In Example 1, we found the maximum of $f(x, y) = x + 2y$ on the ellipse $3x^2 + 4y^2 = 3$. Solve this problem again without using Lagrange multipliers. First, show that the ellipse is parametrized by $x = \cos t$, $y = \frac{\sqrt{3}}{2}\sin t$. Then find the maximum value of $f\left(\cos t, \frac{\sqrt{3}}{2}\sin t\right)$ using single-variable calculus. Which method do you find easier?

19. Use Lagrange multipliers to find the maximum area of a rectangle inscribed in the ellipse (Figure 12):

$$\frac{x^2}{a^2} + \frac{y^2}{b^2} = 1$$

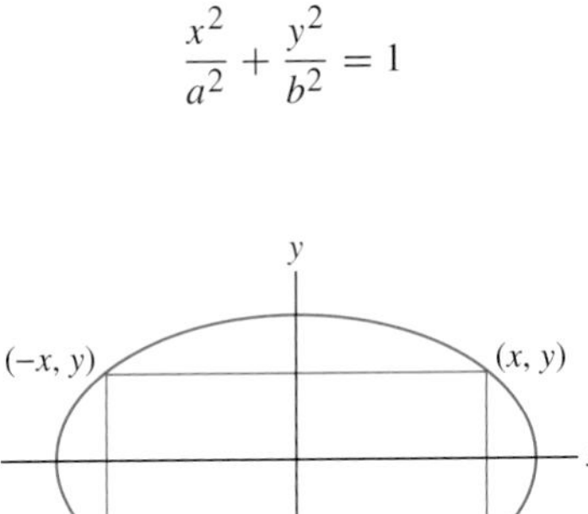

FIGURE 12 Rectangle inscribed in the ellipse $\frac{x^2}{a^2} + \frac{y^2}{b^2} = 1$.

20. Show that the point (x_0, y_0) closest to the origin on the line $ax + by = c$ has coordinates

$$x_0 = \frac{ac}{a^2 + b^2}, \qquad y_0 = \frac{bc}{a^2 + b^2}$$

21. Find the maximum value of $f(x, y) = x^a y^b$ for $x, y \geq 0$ on the unit circle, where $a, b > 0$ are constants.

22. Find the maximum value of $f(x, y) = x^a y^b$ for $x, y \geq 0$ on the line $x + y = 1$, where $a, b > 0$ are constants.

23. Find the maximum value of $f(x, y, z) = x^a y^b z^c$ for $x, y, z \geq 0$ on the unit sphere, where $a, b, c > 0$ are constants.

24. Show that the Lagrange equations for $f(x, y, z) = x^2 y + zy^2$ subject to the constraint $g(x, y) = x + yz = 4$ have no solution. What can you conclude about the minimum and maximum values of f subject to $g = 0$?

25. Let $f(x, y, z) = y + z - x^2$.

(a) Find the solutions to the Lagrange equations for f subject to the constraint $g(x, y, z) = x^2 - y^2 + z^3 = 0$. *Hint:* Show that at a critical point, λ, y, z must be nonzero and $x = 0$.

(b) Show that f has no minimum or maximum subject to the constraint. *Hint:* Consider the values of f at the points $(0, y^3, y^2)$, which satisfy the constraint.

(c) Does (b) contradict Theorem 1?

26. Use Lagrange multipliers to find the point $P = (x_0, y_0, z_0)$ on the plane $ax + by + cz = d$ closest to the origin. Then calculate the distance from P to O.

27. Let Q be the point on an ellipse closest to a given point P outside the ellipse. It was known to the Greek mathematician Apollonius (third century BCE) that $\overline{PQ}$ is perpendicular to the tangent at Q (Figure 13). Explain in words why this conclusion is a consequence of the method of Lagrange multipliers. *Hint:* The circles centered at P are level curves of the function to be minimized.

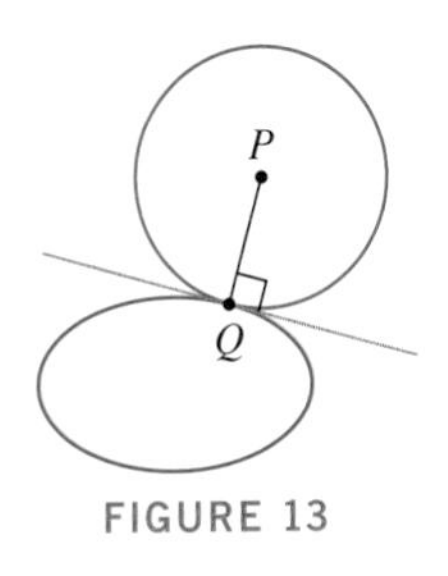

FIGURE 13

28. Antonio has \$5.00 to spend on a lunch consisting of hamburgers (\$1.50 each) and French fries (\$1.00 per order). Antonio's satisfaction from eating x_1 hamburgers and x_2 orders of French fries is measured by a function $U(x_1, x_2) = \sqrt{x_1 x_2}$. How much of each type of food should he purchase to maximize his satisfaction (assume that fractional amounts of each food can be purchased)?

29. Find the maximum value of $f(x, y, z) = xy + xz + yz - 4xyz$ subject to the constraints $x + y + z = 1$ and $x, y, z \geq 0$.

30. A plane with equation $\frac{x}{a} + \frac{y}{b} + \frac{z}{c} = 1$ $(a, b, c > 0)$ together with the positive coordinate planes forms a tetrahedron of volume $V = \frac{1}{6}abc$ (Figure 14). Find the plane that minimizes V if the plane is constrained to pass through the point $P = (1, 1, 1)$.

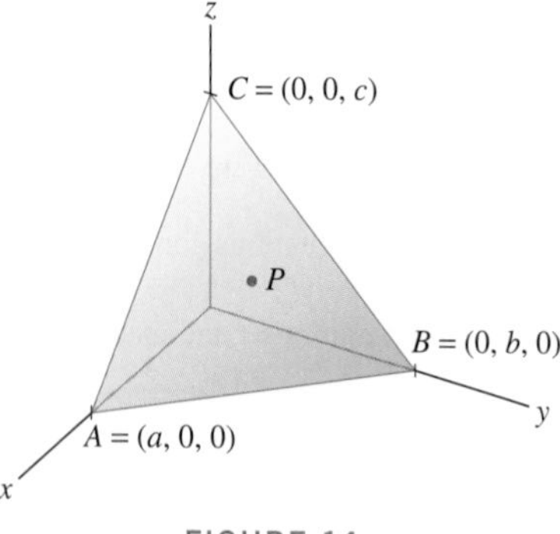

FIGURE 14

31. With the same set-up as in the previous problem, find the plane that minimizes V if the plane is constrained to pass through a point $P = (\alpha, \beta, \gamma)$ with $\alpha, \beta, \gamma > 0$.

32. In a contest, a runner starting at A must touch a point P along a river and then run to B in the shortest time possible (Figure 15). The runner should choose the point P minimizing the total length of the path.

(a) Define a function $f(x, y) = AP + PB$, where $P = (x, y)$. Rephrase the runner's problem as a constrained optimization problem, assuming that the river is given by an equation $g(x, y) = 0$.

(b) Explain why the level curves of $f(x, y)$ are ellipses.

(c) Use Lagrange multipliers to justify the following statement: The ellipse through the point P minimizing the length of the path is tangent to the river.

(d) Identify the point on the river in Figure 15 for which the length is minimal.

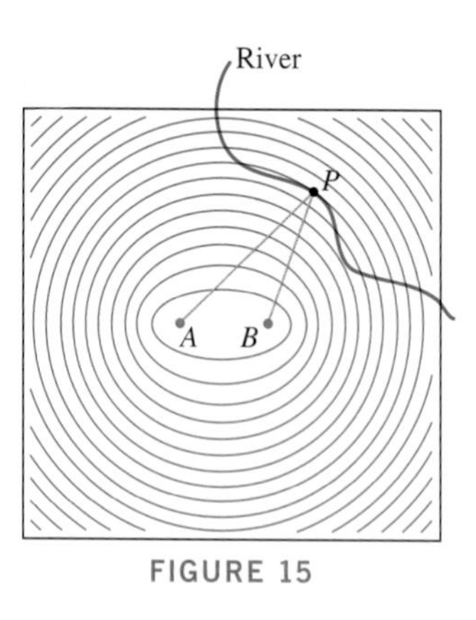

FIGURE 15

33. Let L be the minimum length of a ladder that can reach over a fence of height h to a wall located a distance b behind the wall.

(a) Use Lagrange multipliers to show that $L = (h^{2/3} + b^{2/3})^{3/2}$ (Figure 16). *Hint:* Show that the problem amounts to minimizing $f(x, y) = (x + b)^2 + (y + h)^2$ subject to $y/b = h/x$ or $xy = bh$.

(b) Show that the value of L is also equal to the radius of the circle with center $(-b, -h)$ that is tangent to the graph of $xy = bh$.

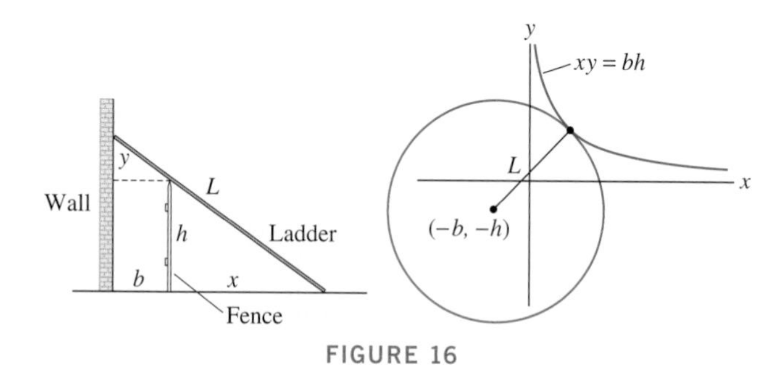

FIGURE 16

34. Find the minimum and maximum of $f(x, y, z) = y + 2z$ subject to two constraints, $2x + y = 4$ and $x^2 + y^2 = 1$.

35. Find the minimum and maximum of $f(x, y, z) = x^2 + y^2 + z^2$ subject to two constraints, $x + 2y + z = 3$ and $x - y = 4$.

Further Insights and Challenges

36. Suppose that both $f(x, y)$ and the constraint function $g(x, y)$ are linear. Use contour maps to explain why $f(x, y)$ does not have a maximum subject to $g(x, y) = 0$ unless $g = af + b$ for some constants a, b.

37. Assumptions Matter CAS Consider the problem of minimizing $f(x, y) = x$ subject to $g(x, y) = (x - 1)^3 - y^2 = 0$.

(a) Show, without using calculus, that the minimum occurs at $P = (1, 0)$.

(b) Show that the Lagrange condition $\nabla f_P = \lambda \nabla g_P$ is not satisfied for any value of λ.

(c) Does this contradict Theorem 1?

38. Marginal Utility Goods 1 and 2 are available at prices (in dollars) of p_1 per unit of good 1 and p_2 per unit of good 2. A utility function $U(x_1, x_2)$ is a function representing the **utility** or benefit of consuming x_j units of good j. The **marginal utility** of the jth good is $\partial U/\partial x_j$, the rate of increase in utility per unit increase in the jth good. Prove the following law of economics: Given a budget of L dollars, utility is maximized at the consumption level (a, b) where the ratio of marginal utility is equal to the ratio of prices:

$$\frac{\text{Marginal utility of good 1}}{\text{Marginal utility of good 2}} = \frac{U_{x_1}(a, b)}{U_{x_2}(a, b)} = \frac{p_1}{p_2}$$

39. Consider the utility function $U(x_1, x_2) = x_1x_2$ with budget constraint $p_1x_1 + p_2x_2 = c$.

(a) Show that the maximum of $U(x_1, x_2)$ subject to the budget constraint is equal to $c^2/(4p_1p_2)$.

(b) Calculate the value of the Lagrange multiplier λ occurring in (a).

(c) Prove the following interpretation: λ is the rate of increase in utility per unit increase of total budget c.

40. This exercise shows that the multiplier λ may be interpreted as a rate of change in general. Assume that the maximum value of $f(x, y)$ subject to $g(x, y) = c$ occurs at a point P. Then P depends on the value of c, so we may write $P = (x(c), y(c))$ and we have $g(x(c), y(c)) = c$.

(a) Show that

$$\nabla g(x(c), y(c)) \cdot \langle x'(c), y'(c)\rangle = 1$$

Hint: Use the Chain Rule to differentiate the equation $g(x(c), y(c)) = c$ with respect to c.

(b) Show that

$$\frac{d}{dc} f(x(c), y(c)) = \lambda$$

Hint: Use the Chain Rule and the Lagrange condition $\nabla f_P = \lambda \nabla g_P$.

(c) Conclude that λ is the rate of increase in f per unit increase in the "budget level" c.

41. Let $B > 0$. Show that the maximum of $f(x_1, \dots, x_n) = x_1x_2\cdots x_n$ subject to the constraints $x_1 + \cdots + x_n = B$ and $x_j \geq 0$ for $j = 1, \dots, n$ occurs for $x_1 = \cdots = x_n = B/n$. Use this to conclude that

$$(a_1a_2\cdots a_n)^{1/n} \leq \frac{a_1 + \cdots + a_n}{n}$$

for all positive numbers $a_1, \dots, a_n$.

42. Let $B > 0$. Show that the maximum of $f(x_1, \dots, x_n) = x_1 + \cdots + x_n$ subject to $x_1^2 + \cdots + x_n^2 = B^2$ is $\sqrt{n}B$. Conclude that

$$|a_1| + \cdots + |a_n| \leq \sqrt{n}(a_1^2 + \cdots + a_n^2)^{1/2}$$

for all numbers $a_1, \dots, a_n$.

43. Given constants E, E_1, E_2, E_3, consider the maximum of

$$S(x_1, x_2, x_3) = x_1 \ln x_1 + x_2 \ln x_2 + x_3 \ln x_3$$

subject to two constraints:

$$x_1 + x_2 + x_3 = N, \quad E_1x_1 + E_2x_2 + E_3x_3 = E$$

Show that there is a constant μ such that $x_i = A^{-1}e^{\mu E_i}$ for $i = 1, 2, 3$, where $A = N^{-1}(e^{\mu E_1} + e^{\mu E_2} + e^{\mu E_3})$.

44. Boltzmann Distribution Generalize Exercise 43 to n variables: Show that there is a constant μ such that the maximum of

$$S = x_1 \ln x_1 + \cdots + x_n \ln x_n$$

subject to the constraints

$$x_1 + \cdots + x_n = N, \qquad E_1x_1 + \cdots + E_nx_n = E$$

occurs for $x_i = A^{-1}e^{\mu E_i}$, where

$$A = N^{-1}(e^{\mu E_1} + \cdots + e^{\mu E_n})$$

This result is used in physics to determine the distribution of velocities of gas molecules at temperature T; x_i is the number of molecules with kinetic energy E_i; $\mu = -(kT)^{-1}$, where k is Boltzmann's constant. The quantity S is called the **entropy**.

CHAPTER REVIEW EXERCISES

1. Given $f(x, y) = \dfrac{\sqrt{x^2 - y^2}}{x + 3}$,

(a) Sketch the domain of f.

(b) Calculate $f(3, 1)$ and $f(-5, -3)$.

(c) Find a point satisfying $f(x, y) = 1$.

2. Find the domain and range of:

(a) $f(x, y, z) = \sqrt{x - y} + \sqrt{y - z}$

(b) $f(x, y) = \ln(4x^2 - y)$

3. Sketch the graph $f(x, y) = x^2 - y + 1$ and describe its vertical and horizontal traces.

4. CAS Use a graphing utility to draw the graph of the function $\cos(x^2 + y^2)e^{1-xy}$ in the domains $[-1, 1] \times [-1, 1]$, $[-2, 2] \times [-2, 2]$, and $[-3, 3] \times [-3, 3]$, and explain its behavior.

5. Match the functions (a)–(d) with their graphs in Figure 1.

(a) $f(x, y) = x^2 + y$

(b) $f(x, y) = x^2 + 4y^2$

(c) $f(x, y) = \sin(4xy)e^{-x^2-y^2}$

(d) $f(x, y) = \sin(4x)e^{-x^2-y^2}$

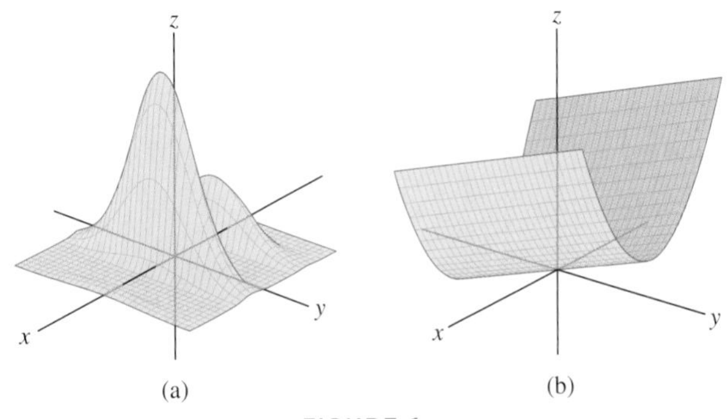

FIGURE 1

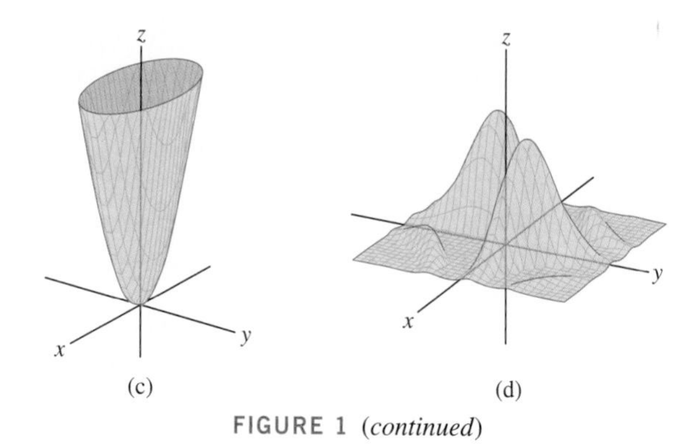

FIGURE 1 (*continued*)

6. Referring to the contour map in Figure 2, estimate:

(a) The average rate of change of elevation from A to C and from A to D.

(b) The directional derivative at A in the direction $\mathbf{v}$.

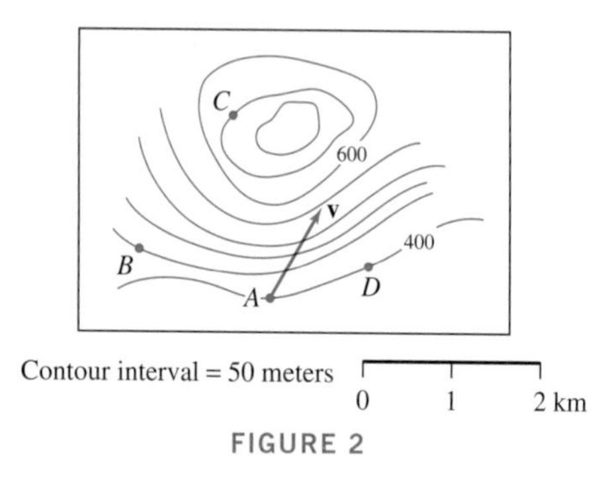

FIGURE 2

7. Describe the level curves of:

(a) $f(x, y) = e^{4x-y}$

(b) $f(x, y) = \ln(4x - y)$

(c) $f(x, y) = 3x^2 - 4y^2$

(d) $f(x, y) = x + y^2$

8. Match each function (a)–(c) with its contour graph (i)–(iii) in Figure 3:

(a) $f(x, y) = xy$

(b) $f(x, y) = e^{xy}$

(c) $f(x, y) = \sin(xy)$

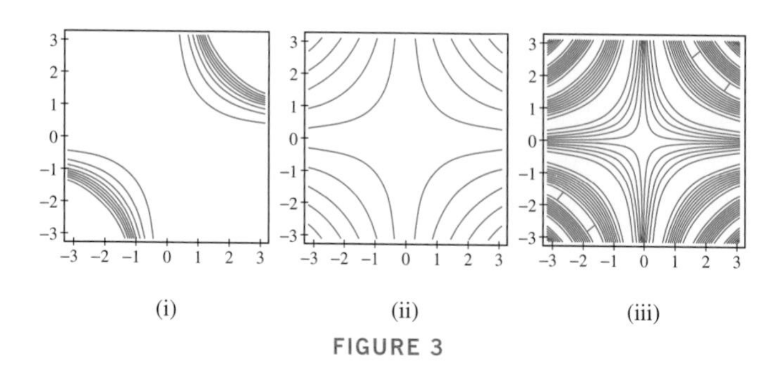

FIGURE 3

In Exercises 9–14, evaluate the limit or state that it does not exist.

9. $\displaystyle\lim_{(x,y)\to(1,-3)} (xy + y^2)$

10. $\displaystyle\lim_{(x,y)\to(1,-3)} \ln(3x + y)$

11. $\displaystyle\lim_{(x,y)\to(0,0)} \frac{xy + xy^2}{x^2 + y^2}$

12. $\displaystyle\lim_{(x,y)\to(0,0)} \frac{x^3y^2 + x^2y^3}{x^4 + y^4}$

13. $\displaystyle\lim_{(x,y)\to(1,-3)} (2x + y)e^{-x+y}$

14. $\displaystyle\lim_{(x,y)\to(0,2)} \frac{(e^x - 1)(e^y - 1)}{x}$

15. Let

$$f(x, y) = \begin{cases} \dfrac{(xy)^p}{x^4 + y^4} & (x, y) \neq (0, 0) \\ 0 & (x, y) = (0, 0) \end{cases}$$

Use polar coordinates to show that $f(x, y)$ is continuous at all (x, y) if $p > 2$, but discontinuous at $(0, 0)$ if $p \leq 2$.

16. Calculate $f_x(1, 3)$ and $f_y(1, 3)$ for $f(x, y) = \sqrt{7x + y^2}$.

In Exercises 17–20, compute f_x and f_y.

17. $f(x, y) = 2x + y^2$

18. $f(x, y) = 4xy^3$

19. $f(x, y) = \sin(xy)e^{-x-y}$

20. $f(x, y) = \ln(x^2 + xy^2)$

21. Calculate f_{xxyz} for $f(x, y, z) = y\sin(x + z)$.

22. Fix $c > 0$. Show that for any constants α, β, the function $u(t, x) = \sin(\alpha ct + \beta)\sin(\alpha x)$ satisfies the wave equation

$$\frac{\partial^2 u}{\partial t^2} = c^2 \frac{\partial^2 u}{\partial x^2}$$

23. Find an equation of the tangent to the graph of $f(x, y) = xy^2 - xy + 3x^3y$ at $P = (1, 3)$.

24. Suppose that $f(4, 4) = 3$ and $f_x(4, 4) = f_y(4, 4) = -1$. Use the linear approximation to estimate $f(4.1, 4)$ and $f(3.88, 4.03)$.

25. Estimate $\sqrt{7.1^2 + 4.9^2 + 70.1}$ using the linear approximation. Compare with a calculator value.

26. Suppose that the plane $z = 2x - y - 3$ is tangent to the graph of $z = f(x, y)$ at $P = (2, 4)$.

(a) Determine $f(2, 4)$, $f_x(2, 4)$, and $f_y(2, 4)$.

(b) Approximate $f(2.2, 3.9)$.

27. Jason earns $S(h, c) = 20h\left(1 + \dfrac{c}{100}\right)^{1.5}$ dollars per month at a used car lot, where h is the number of hours worked and c is the number of cars sold. He has already worked 160 hours and sold 69 cars. Right now Jason wants to go home but wonders how much more he might earn if he stays another 10 minutes with a customer who is considering buying a car. Use the linear approximation to estimate how much extra money Jason will earn if he sells his 70th car during these 10 minutes.

In Exercises 28–31, compute $\frac{d}{dt} f(\mathbf{c}(t))$ *at the given value of* t.

28. $f(x, y) = x + e^y, \quad \mathbf{c}(t) = (3t - 1, t^2)$ at $t = 2$

29. $f(x, y, z) = xz - y^2, \quad \mathbf{c}(t) = (t, t^3, 1 - t)$

30. $f(x, y) = xe^{3y} - ye^{3x}, \quad \mathbf{c}(t) = (e^t, \ln t)$ at $t = 1$

31. $f(x, y) = \tan^{-1}\frac{y}{x}, \quad \mathbf{c}(t) = (\cos t, \sin t), t = \frac{\pi}{3}$

In Exercises 32–35, compute the directional derivative at P in the direction $\mathbf{v}$.

32. $f(x, y) = x^3y^4, \quad P = (3, -1), \quad \mathbf{v} = 2\mathbf{i} + \mathbf{j}$

33. $f(x, y, z) = zx - xy^2, \quad P = (1, 1, 1), \quad \mathbf{v} = \langle 2, -1, 2\rangle$

34. $f(x, y) = e^{x^2+y^2}, \quad P = \left(\frac{\sqrt{2}}{2}, \frac{\sqrt{2}}{2}\right), \quad \mathbf{v} = \langle 3, -4\rangle$

35. $f(x, y, z) = \sin(xy + z), \quad P = (0, 0, 0), \quad \mathbf{v} = \mathbf{j} + \mathbf{k}$

36. Find the unit vector $\mathbf{e}$ at $P = (0, 0, 1)$ pointing in the direction along which $f(x, y, z) = xz + e^{-x^2+y}$ increases most rapidly.

37. Find an equation of the tangent plane at $P = (0, 3, -1)$ to the surface with equation

$$ze^x + e^{z+1} = xy + y - 3$$

38. Let $n \neq 0$ be an integer and r an arbitrary constant. Show that the tangent plane to the surface $x^n + y^n + z^n = r$ at $P = (a, b, c)$ has equation

$$a^{n-1}x + b^{n-1}y + c^{n-1}z = r$$

39. Let $f(x, y) = (x - y)e^x$. Use the Chain Rule to calculate $\frac{\partial f}{\partial u}$ and $\frac{\partial f}{\partial v}$, where $x = u - v$ and $y = u + v$.

40. Let $f(x, y) = x^2y + y^2z$. Use the Chain Rule to calculate $\frac{\partial f}{\partial s}$ and $\frac{\partial f}{\partial t}$, where

$$x = s + t, \quad y = st, \quad z = 2s - t$$

41. Express the partial derivatives $\frac{\partial f}{\partial r}$ and $\frac{\partial f}{\partial \theta}$ of a function $f(x, y, z)$ in terms of $\frac{\partial f}{\partial x}$, $\frac{\partial f}{\partial y}$, and $\frac{\partial f}{\partial z}$, where (r, θ, z) are cylindrical coordinates.

42. Let $f(x, y, z) = (x^2 + y^2)e^{-xz}$ and $P = (1, 0, 2)$. Use the result of Exercise 41 to calculate $\frac{\partial f}{\partial r}$ and $\frac{\partial f}{\partial \theta}$ at the point P.

43. Let $g(u, v) = f(u^3 - v^3, v^3 - u^3)$. Prove that

$$v^2\frac{\partial g}{\partial u} - u^2\frac{\partial g}{\partial v} = 0$$

44. Let $f(x, y) = g(u)$, where $u = x^2 + y^2$ and $g(u)$ is a differentiable function in one variable. Prove that

$$\left(\frac{\partial f}{\partial x}\right)^2 + \left(\frac{\partial f}{\partial y}\right)^2 = 4u\left(\frac{\partial g}{\partial u}\right)^2$$

45. Calculate $\frac{\partial z}{\partial x}$, where $xe^z + ze^y = x + y$.

46. Let $f(x, y) = x^4 - 2x^2 + y^2 - 6y$.

(a) Find the critical points of f and use the Second Derivative Test to determine whether they are local minima or maxima.

(b) Find the minimum value of f without calculus by completing the square.

In Exercises 47–50, find the critical points of the function and analyze them using the Second Derivative Test.

47. $f(x, y) = x^2 + 2y^2 - 4xy + 6x$

48. $f(x, y) = x^3 + 2y^3 - xy$

49. $f(x, y) = e^{x+y} - xe^{2y}$

50. $f(x, y) = \sin(x + y) - \frac{1}{2}(x + y^2)$

51. Prove that $f(x, y) = (x + 2y)e^{xy}$ has no critical points.

52. Find the global extrema of $f(x, y) = x^3 - xy - y^2 + y$ on the square $[0, 1] \times [0, 1]$.

53. Find the global extrema of $f(x, y) = 2xy - x - y$ on the domain $\{y \le 4, y \ge x^2\}$.

54. Find the maximum of $f(x, y, z) = xyz$ subject to the constraint $g(x, y, z) = 2x + y + 4z = 1$.

55. Use Lagrange multipliers to find the minimum and maximum value of $f(x, y) = 3x - 2y$ on the circle $x^2 + y^2 = 4$.

56. Find the minimum value of $f(x, y) = xy$ subject to the constraint $5x - y = 4$ in two ways: using Lagrange multipliers and setting $y = 5x - 4$ in $f(x, y)$.

57. Find the minimum and maximum values of $f(x, y) = x^2y$ on the ellipse $4x^2 + 9y^2 = 36$.

58. Find the point in the first quadrant on the curve $y = x + x^{-1}$ closest to the origin.

59. Find the extreme values of $f(x, y, z) = x + 2y + 3z$ subject to the two constraints $x + y + z = 1$ and $x^2 + y^2 + z^2 = 1$.

60. Use Lagrange multipliers to find the dimensions of a cylindrical can of fixed volume V with minimal surface area (including the top and bottom of the can).

61. Find the dimensions of the box of maximum volume with its sides parallel to the coordinate planes that can be inscribed in the ellipsoid (Figure 4)

$$\left(\frac{x}{a}\right)^2 + \left(\frac{y}{b}\right)^2 + \left(\frac{z}{c}\right)^2 = 1$$

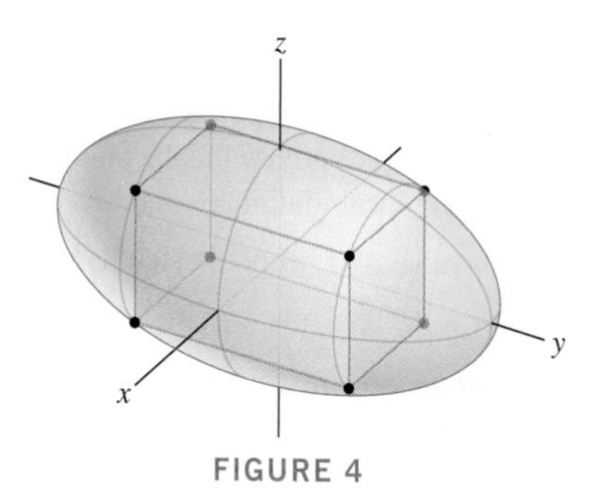

FIGURE 4

62. Given n nonzero numbers $\sigma_1, \dots, \sigma_n$, show that the minimum value of

$$f(x_1, \dots, x_n) = x_1^2\sigma_1^2 + \cdots + x_n^2\sigma_n^2$$

subject to $x_1 + \cdots + x_n = 1$ is c, where $c = \left(\sum_{j=1}^{n} \sigma_j^{-2}\right)^{-1}$.

63. A bead hangs on a string of length ℓ whose ends are fixed by thumbtacks located at points $(0, 0)$ and (a, b) on a bulletin board (Figure 5). The bead rests in the position that minimizes its height y. Use Lagrange multipliers to show that the two sides of the string make equal angles with the horizontal, that is, show that $\theta_1 = \theta_2$.

As an aside, note that the locus of the bead when pulled taut is an ellipse with foci at O and P, and the tangent line at the lowest point is horizontal. Therefore, this exercise provides another proof of the reflective property of the ellipse (a light ray emanating from one focus and bouncing off the ellipse is reflected to the other focus).

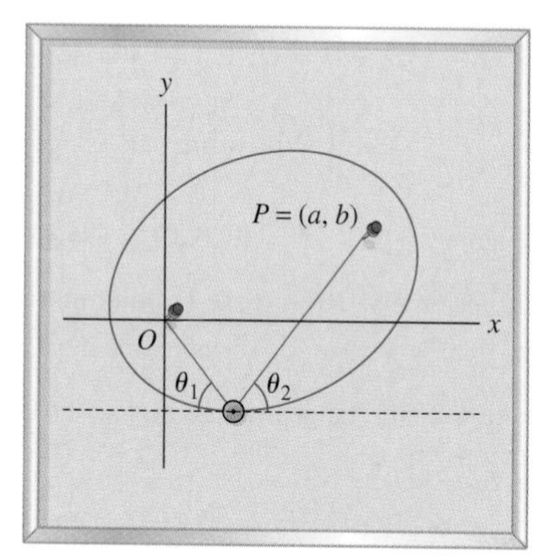

FIGURE 5

15 MULTIPLE INTEGRATION

Ranger Uranium Mine in Kakadu National Park, Australia. The volume of ore removed is one type of quantity that is expressed by a multiple integral.

The integral of a function of several variables is called a **multiple integral**. In this chapter, we define multiple integrals and develop techniques for computing them. We also discuss some of their applications. Like the one-variable definite integral, multiple integrals are used to compute many important and diverse quantities, such as volumes, surface areas, centers of mass, probabilities, and average values.

15.1 Integration in Several Variables

In the first three sections of this chapter, we study integration of functions $f(x, y)$ of two variables. An integral of $f(x, y)$ is called a **double integral**.

There are many similarities between double integrals and the single integrals with which you are familiar:

- Double and single integrals are both defined as limits of Riemann sums.
- A double integral represents signed volume, just as a single integral represents signed area.
- We evaluate double integrals using the Fundamental Theorem of Calculus (FTC), but we have to apply it twice (see the discussion of iterated integrals below).
- Double integrals can be approximated using numerical methods similar to the Endpoint and Midpoint Rules for single integrals.

In the context of multiple integration, we often refer to the definite integral of a function $f(x)$ of one variable as a "single integral."

An important difference in the multivariable case is that the domain of integration plays a more prominent role. The domain of integration of a single integral $\int_a^b f(x)\,dx$ is an interval $[a, b]$, but in two variables, the domain is a region $\mathcal{D}$ in the plane with a more general boundary curve (Figure 1).

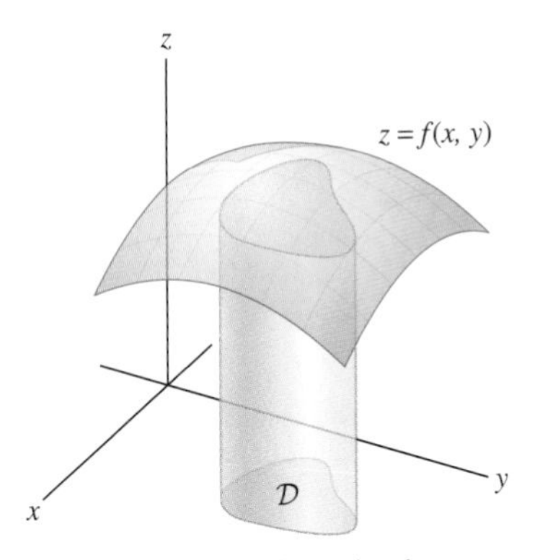

FIGURE 1 Volume of a region between the graph of $z = f(x, y)$ and the domain $\mathcal{D}$.

In this section, we restrict our attention to the simplest case in which $\mathcal{D}$ is a rectangle, leaving more general domains for discussion in Section 15.2. The notation

$$\mathcal{R} = [a, b] \times [c, d]$$

denotes the rectangle (Figure 2) consisting of all points (x, y) such that

$$\mathcal{R}: \quad a \le x \le b, \qquad c \le y \le d$$

As mentioned above, the double integral is a limit of Riemann sums. To form a Riemann sum, we choose positive integers N and M and partitions of the intervals $[a, b]$ and $[c, d]$:

$$a = x_0 < x_1 < \cdots < x_N = b, \qquad c = y_0 < y_1 < \cdots < y_M = d$$

Let $\Delta x_i = x_i - x_{i-1}$ and $\Delta y_j = y_j - y_{j-1}$ be the widths of the subintervals defined by these partitions. We then partition $\mathcal{R}$ into NM subrectangles $\mathcal{R}_{ij}$, whose area we denote

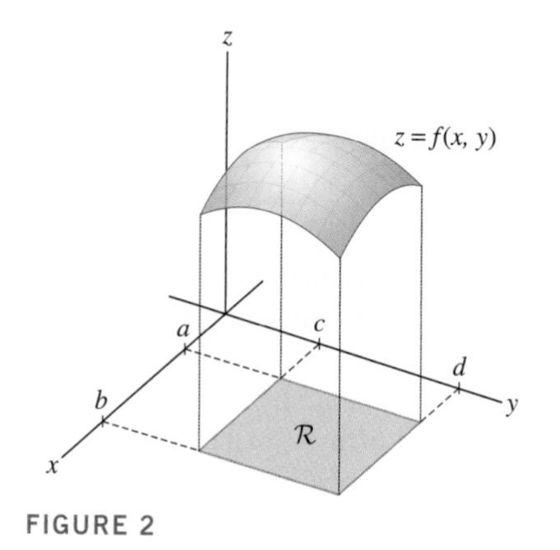

FIGURE 2

by ΔA_{ij} [Figure 3(B)]:

$$\mathcal{R}_{ij} = [x_{i-1}, x_i] \times [y_{j-1}, y_j], \qquad \text{Area}(\mathcal{R}_{ij}) = \Delta A_{ij} = \Delta x_i \, \Delta y_j$$

The partition is called **regular** if the intervals $[a, b]$ and $[c, d]$ are both divided into subintervals of equal length. In other words, the partition is regular if $\Delta x_i = \Delta x$ and $\Delta y_j = \Delta y$ where

$$\Delta x = \frac{b-a}{N}, \qquad \Delta y = \frac{d-c}{M}$$

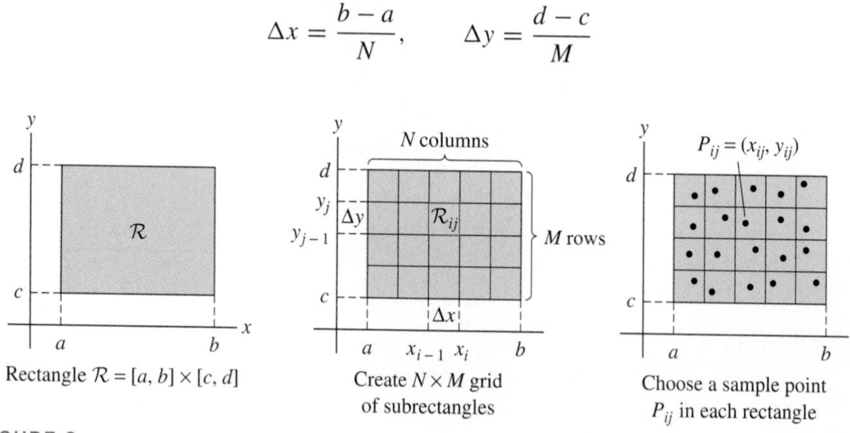

FIGURE 3

Next, we choose a **sample point** $P_{ij} = (x_{ij}, y_{ij})$ in each subrectangle $\mathcal{R}_{ij}$ [Figure 3(C)] and construct a box of height $f(P_{ij})$ above $\mathcal{R}_{ij}$ as in Figure 4(B). This box has signed volume

$$f(P_{ij}) \, \Delta A_{ij} = f(P_{ij}) \, \Delta x_i \, \Delta y_j = \underbrace{\text{height} \times \text{area}}_{\text{Signed volume of box}}$$

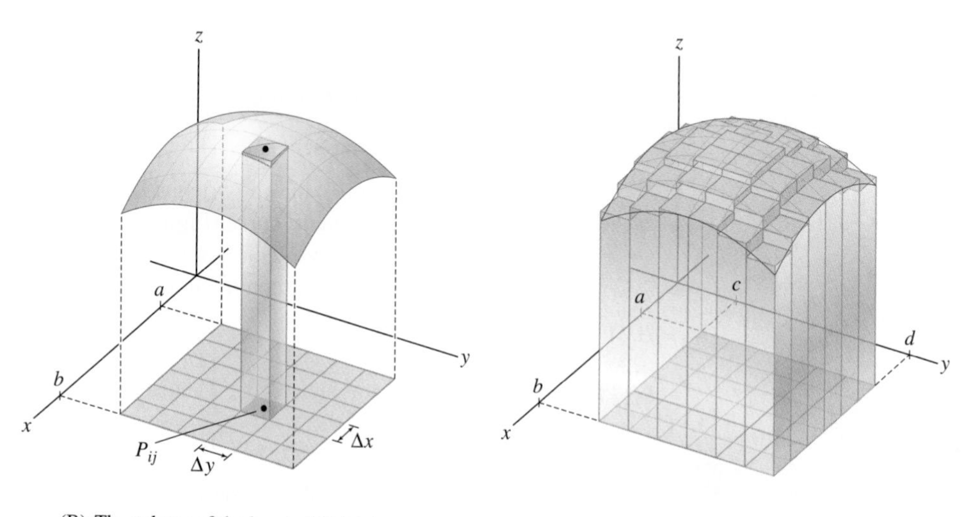

(A) In one variable, a Riemann sum approximates the area under the curve by a sum of areas of rectangles.

(B) The volume of the box is $f(P_{ij})\Delta A$ where $\Delta A = \Delta x \Delta y$.

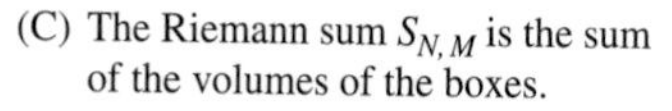

(C) The Riemann sum $S_{N,M}$ is the sum of the volumes of the boxes.

FIGURE 4

Here, by definition, the signed volume of a region is positive if it lies above the xy-plane and negative if it lies below the xy-plane. If $f(P_{ij}) < 0$, the box extends below the xy-

Keep in mind that a Riemann sum depends on the choice of partition and sample points. It would be more proper to write

$$S_{N,M}(\{P_{ij}\}, \{x_i\}, \{y_j\})$$

but we write $S_{N,M}$ to simplify the notation.

plane and the signed volume is negative. The Riemann sum $S_{N,M}$ is the sum of the signed volumes of the boxes [Figure 4(C)]:

$$S_{N,M} = \sum_{i=1}^{N}\sum_{j=1}^{M} f(P_{ij})\,\Delta A_{ij} = \sum_{i=1}^{N}\sum_{j=1}^{M} f(P_{ij})\,\Delta x_i\,\Delta x_j$$

This double summation runs over all i and j in the ranges $1 \le i \le N$ and $1 \le j \le M$, a total of NM terms.

The Riemann sums $S_{N,M}$ yield approximations to the (signed) volume of the region between the graph of $z = f(x, y)$ and the xy-plane. They are a natural generalization of the rectangular approximations to area provided by Riemann sums in the single variable case [Figure 4(A)].

Let $\mathcal{P} = \{\{x_i\}, \{y_j\}\}$ be the set of endpoints of the partition and let $\|\mathcal{P}\|$ be the maximum of the widths Δx_i, Δy_j. As $\|\mathcal{P}\|$ tends to zero, the boxes become thinner and they approximate the signed volume more closely (Figure 5). Therefore, we may expect $S_{N,M}$ to converge to the signed volume under the graph as $\|\mathcal{P}\|$ approaches zero. Note that N and M both approach ∞ as $\|\mathcal{P}\| \to 0$. The precise definition of the limit is as follows:

The Riemann sum $S_{N,M}$ approaches a limit L as $\|\mathcal{P}\| \to 0$ if, for all $\epsilon > 0$, there exists $\delta > 0$ such that $|L - S_{N,M}| < \epsilon$ for all partitions satisfying $\|\mathcal{P}\| < \delta$ and all choices of sample points.

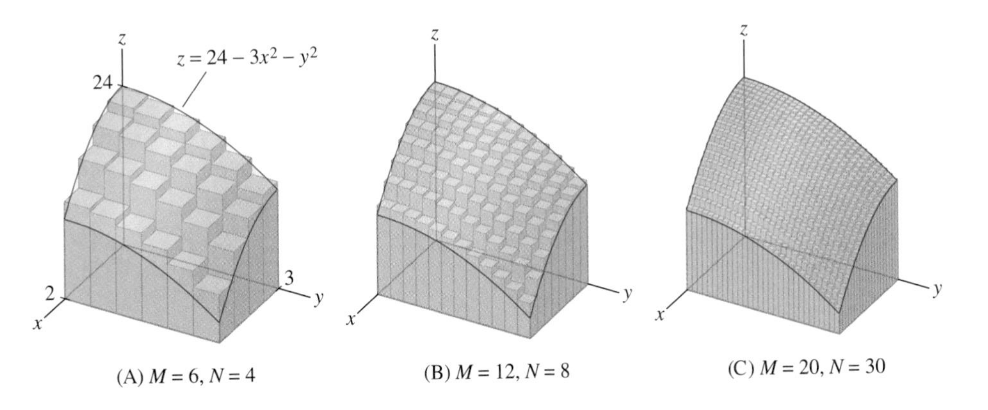

FIGURE 5 Midpoint approximations to the volume under $z = 12 + 3x^2 + 5y^2$.

In this case, we write

$$\lim_{\|\mathcal{P}\|\to 0} S_{N,M} = \lim_{\|\mathcal{P}\|\to 0} \sum_{i=1}^{N}\sum_{j=1}^{M} f(P_{ij})\,\Delta A_{ij} = L$$

This limit L, if it exists, is called the double integral of $f(x, y)$ over $\mathcal{R}$ (Figure 6), and is denoted $\iint_{\mathcal{R}} f(x, y)\,dA$.

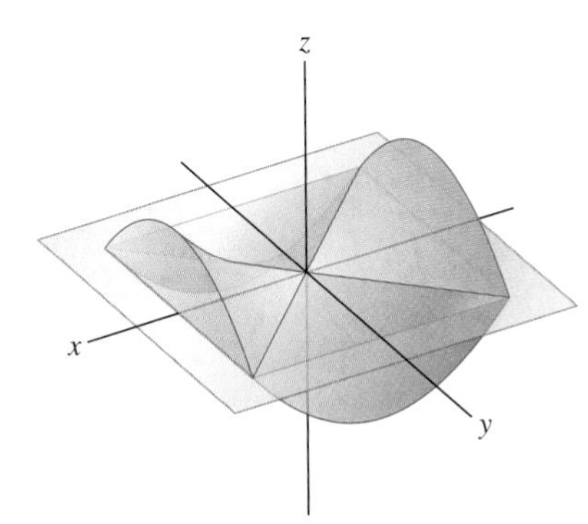

FIGURE 6 The double integral $\iint_{\mathcal{R}} f(x, y)\,dA$ is the signed volume between the surface $z = f(x, y)$ and the xy-plane.

DEFINITION Double Integral over a Rectangle The double integral of $f(x, y)$ over a rectangle $\mathcal{R}$ is defined as the limit

$$\iint_{\mathcal{R}} f(x, y)\,dA = \lim_{\|\mathcal{P}\|\to 0} \sum_{i=1}^{N}\sum_{j=1}^{M} f(P_{ij})\Delta A_{ij}$$

If this limit exists, we say that $f(x, y)$ is **integrable** over $\mathcal{R}$.

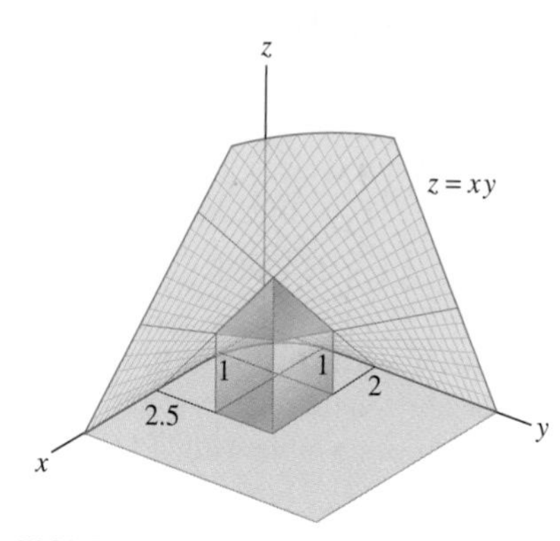

FIGURE 7 Graph of $z = xy$.

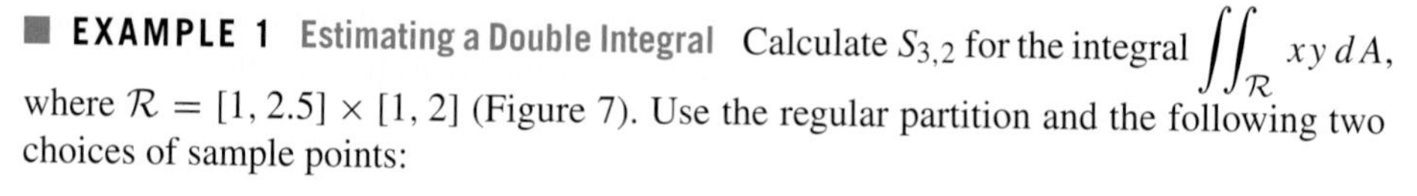

■ **EXAMPLE 1 Estimating a Double Integral** Calculate $S_{3,2}$ for the integral $\iint_{\mathcal{R}} xy\, dA$, where $\mathcal{R} = [1, 2.5] \times [1, 2]$ (Figure 7). Use the regular partition and the following two choices of sample points:

(a) Lower-left vertex **(b)** Midpoint of rectangle

Solution Figures 8 and 9 show a 3×2 grid of subrectangles on the rectangle $\mathcal{R}$. Since we use the regular partition, each subrectangle has sides of length

$$\Delta x = \frac{2.5 - 1}{3} = \frac{1}{2}, \qquad \Delta y = \frac{2 - 1}{2} = \frac{1}{2}$$

and area $\Delta A = \Delta x\, \Delta y = \frac{1}{4}$. The corresponding Riemann sum is

$$S_{3,2} = \sum_{i=1}^{3} \sum_{j=1}^{2} f(P_{ij})\, \Delta A = \frac{1}{4} \sum_{i=1}^{3} \sum_{j=1}^{2} f(P_{ij})$$

(a) If we use the lower-left vertices shown in Figure 8, the Riemann sum is

$$S_{3,2} = \frac{1}{4}\big(f(1, 1) + f(1, 1.5) + f(1.5, 1) + f(1.5, 1.5) + f(2, 1) + f(2, 1.5)\big)$$
$$= \frac{1}{4}\big(1 + 1.5 + 1.5 + 2.25 + 2 + 3\big) = \frac{1}{4}(11.25) = 2.8125$$

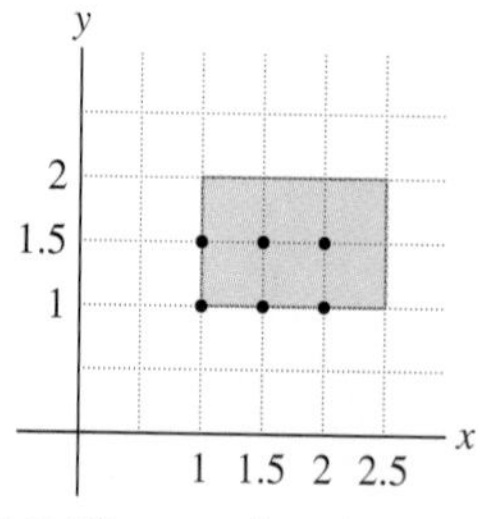

FIGURE 8 The sample points are the lower-left vertices of each square.

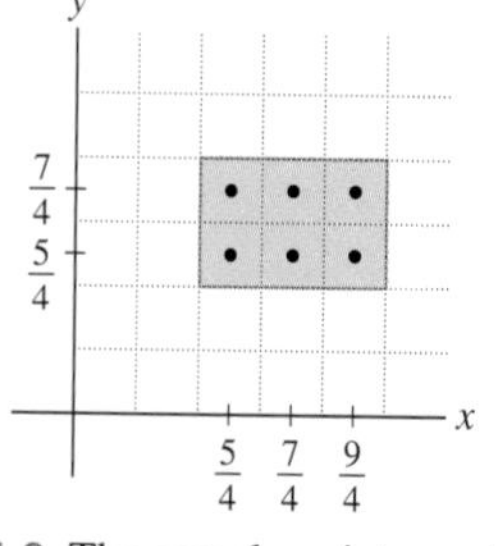

FIGURE 9 The sample points are the centers of each square.

(b) Using the midpoints of the rectangles shown in Figure 9, we obtain

$$S_{3,2} = \frac{1}{4}\left(f\left(\frac{5}{4}, \frac{5}{4}\right) + f\left(\frac{5}{4}, \frac{7}{4}\right) + f\left(\frac{7}{4}, \frac{5}{4}\right) + f\left(\frac{7}{4}, \frac{7}{4}\right) + f\left(\frac{9}{4}, \frac{5}{4}\right) + f\left(\frac{9}{4}, \frac{7}{4}\right)\right)$$
$$= \frac{1}{4}\left(\frac{25}{16} + \frac{35}{16} + \frac{35}{16} + \frac{49}{16} + \frac{45}{16} + \frac{63}{16}\right) = \frac{1}{4}\left(\frac{252}{16}\right) = 3.9375$$ ■

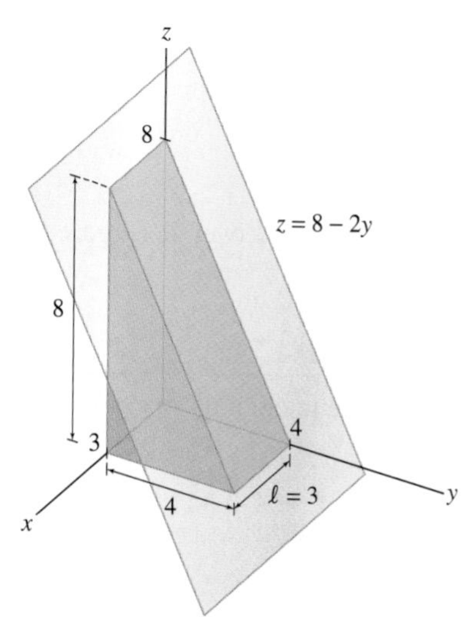

FIGURE 10 Graph of $z = 8 - 2y$.

■ **EXAMPLE 2** Evaluate $\iint_{\mathcal{R}} (8 - 2y)\, dA$, where $\mathcal{R} = [0, 3] \times [0, 4]$.

Solution The graph of $z = 8 - 2y$ is the plane shown in Figure 10. The double integral represents the volume V of a solid of length $\ell = 3$ whose vertical cross section is a triangle of height 8 and base 4, and area $A = \frac{1}{2}(8)4 = 16$. The volume of the solid is $V = \ell A = 3(16) = 48$. Therefore,

$$\iint_{\mathcal{R}} (8 - 2y)\, dA = 48$$ ■

The following theorem states that continuous functions are integrable. The proof is similar to the single-variable case and is omitted.

THEOREM 1 Continuous Functions Are Integrable If $f(x, y)$ is continuous on a rectangle $\mathcal{R}$, then $f(x, y)$ is integrable over $\mathcal{R}$.

As in the single-variable case, we often make use of the linearity properties of the double integral. These follow from the definition of the double integral as a limit of Riemann sums.

THEOREM 2 Linearity of the Double Integral Assume that $f(x, y)$ and $g(x, y)$ are integrable over a rectangle $\mathcal{R}$. Then:

(i) $\displaystyle\iint_{\mathcal{R}} \big(f(x, y) + g(x, y)\big)\, dA = \iint_{\mathcal{R}} f(x, y)\, dA + \iint_{\mathcal{R}} g(x, y)\, dA$

(ii) For any constant C, $\displaystyle\iint_{\mathcal{R}} Cf(x, y)\, dA = C \iint_{\mathcal{R}} f(x, y)\, dA$

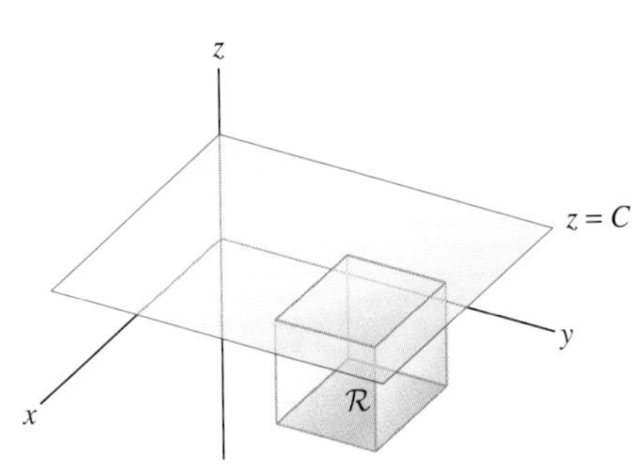

FIGURE 11 The double integral of $f(x, y) = C$ over a rectangle $\mathcal{R}$ is $C \cdot \text{Area}(\mathcal{R})$.

If $f(x, y) = C$ is a constant function, then

$$\iint_{\mathcal{R}} C\, dA = C \cdot \text{Area}(\mathcal{R})$$

In this case, the double integral is the signed volume of the box of base $\mathcal{R}$ and height $|C|$ (Figure 11).

■ **EXAMPLE 3 Arguing by Symmetry** Use symmetry to justify $\displaystyle\iint_{\mathcal{R}} xy^2\, dA = 0$, where $\mathcal{R} = [-1, 1] \times [-1, 1]$.

Solution The double integral is the signed volume of the region between the graph of $f(x, y) = xy^2$ and the xy-plane. Our function satisfies

$$f(-x, y) = -xy^2 = -f(x, y)$$

so the region below the xy-plane (where $-1 \le x \le 0$) cancels with the region above the xy-plane (where $0 \le x \le 1$) (Figure 12). ■

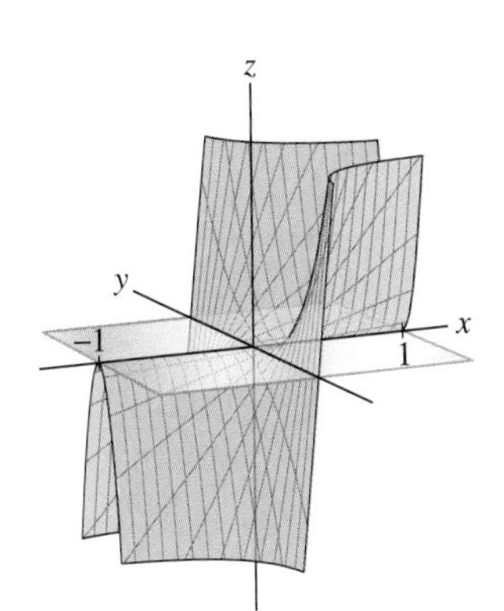

FIGURE 12

Iterated Integrals

Our main technique for evaluating double integrals is based on the FTC, as in the single-variable case. To use the FTC, we must first express the double integral as an iterated integral. An **iterated integral** is an expression of the form

$$\int_a^b \left(\int_c^d f(x, y)\, dy \right) dx$$

Notice that the inner integral $\int_c^d$ specifies the limits of integration for the y-variable, and the outer integral $\int_a^b$ specifies the limits for the x-variable.

It is evaluated in two steps. First, hold x constant and evaluate the inner integral with respect to y:

$$S(x) = \int_c^d f(x, y)\, dy$$

Then integrate the resulting function $S(x)$ with respect to x.

■ **EXAMPLE 4** Evaluate $\int_2^4 \left(\int_1^9 ye^x\,dy\right) dx$.

Solution First evaluate the inner integral, treating x as a constant:

$$S(x) = \int_1^9 ye^x\,dy = e^x \int_1^9 y\,dy = e^x \left(\frac{1}{2}y^2\right)\bigg|_{y=1}^9 = e^x\left(\frac{81-1}{2}\right) = 40e^x$$

Then integrate $S(x)$ with respect to x:

$$\int_2^4 \left(\int_1^9 ye^x\,dy\right) dx = \int_2^4 40e^x\,dx = 40e^x\bigg|_2^4 = 40(e^4 - e^2)$$ ■

If an iterated integral is written with dx preceding dy, then we integrate first with respect to x:

$$\int_c^d \int_a^b f(x,y)\,dx\,dy = \int_{y=c}^d \left(\int_{x=a}^b f(x,y)\,dx\right) dy$$

Here, for clarity, we have included the variables in the limits of integration of the iterated integral on the right.

We often omit the parentheses in the notation for an iterated integral:

$$\int_a^b \int_c^d f(x,y)\,dy\,dx$$

The order of the variables in $dy\,dx$ tells us to integrate first with respect to y between the limits $y = c$ and $y = d$.

■ **EXAMPLE 5** Evaluate $\int_0^{\pi/2} \int_0^{\pi/2} \sin(2x+y)\,dx\,dy$.

Solution Compute the inner integral, treating y as a constant:

$$\int_0^{\pi/2} \sin(2x+y)\,dx = -\frac{1}{2}\cos(2x+y)\bigg|_{x=0}^{\pi/2} = -\frac{1}{2}\cos(\pi+y) + \frac{1}{2}\cos y$$

$$\int_0^{\pi/2}\int_0^{\pi/2} \sin(2x+y)\,dx\,dy = -\frac{1}{2}\int_0^{\pi/2}\big(\cos(\pi+y) - \cos y\big)\,dy$$

$$= -\frac{1}{2}\big(\sin(\pi+y) - \sin(y)\big)\bigg|_{y=0}^{\pi/2}$$

$$= -\frac{1}{2}\left(\sin\frac{3\pi}{2} - \sin\frac{\pi}{2}\right) + \frac{1}{2}\big(\sin\pi - \sin 0\big) = 1$$ ■

■ **EXAMPLE 6 Reversing the Order of Integration** Verify that

$$\int_0^1 \int_2^4 x^2y^3\,dy\,dx = \int_2^4 \int_0^1 x^2y^3\,dx\,dy$$

Solution Both integrals have the value 20:

$$\int_0^1 \int_2^4 x^2y^3\,dy\,dx = \int_0^1 \left(\frac{1}{4}x^2y^4\bigg|_{y=2}^4\right) dx = \int_0^1 \left(64x^2 - 4x^2\right) dx$$

$$= \int_0^1 60x^2\,dx = 20x^3\bigg|_0^1 = 20$$

$$\int_2^4 \int_0^1 x^2 y^3 \, dx \, dy = \int_2^4 \left(\frac{1}{3} x^3 y^3 \bigg|_{x=0}^{1} \right) dy = \frac{1}{3} \int_2^4 y^3 \, dy$$

$$= \frac{1}{12} y^4 \bigg|_2^4 = \frac{1}{12}(256 - 16) = 20$$

The previous example illustrates a general fact: The value of an iterated integral does not depend on the order in which the integration is performed. This is part of Fubini's Theorem. Even more important, Fubini's Theorem states that a double integral over a rectangle can be evaluated as an iterated integral.

CAUTION When you reverse the order of integration in an iterated integral, remember to interchange the limits of integration (the inner limits become the outer limits).

THEOREM 3 Fubini's Theorem The double integral of a continuous function $f(x, y)$ over a rectangle $\mathcal{R} = [a, b] \times [c, d]$ is equal to the iterated integral (in either order):

$$\iint_{\mathcal{R}} f(x, y) \, dA = \int_a^b \int_c^d f(x, y) \, dy \, dx = \int_c^d \int_a^b f(x, y) \, dx \, dy$$

Proof We sketch the proof. Consider a Riemann sum $S_{N,M}$ for a regular partition. Then $S_{N,M}$ is the sum of values $f(P_{ij}) \, \Delta x \, \Delta y$, where

$$\Delta x = \frac{b - a}{N} \qquad \text{and} \qquad \Delta y = \frac{d - c}{M}$$

3	$f(P_{13})$	$f(P_{23})$	$f(P_{33})$
2	$f(P_{12})$	$f(P_{22})$	$f(P_{32})$
1	$f(P_{11})$	$f(P_{21})$	$f(P_{31})$
j / i	1	2	3

Fubini's Theorem ultimately stems from the elementary fact that we may sum these values in any order. Think of the values $f(P_{ij})$ as listed in an $N \times M$ array. For example, the values for $S_{3,3}$ are listed in the array in the margin. If we first sum the columns and then add up the column sums, we obtain

$$S_{N,M} = \underbrace{\sum_{i=1}^{N} \sum_{j=1}^{M} f(P_{ij}) \, \Delta x \, \Delta y}_{\text{Sum in any order}} = \underbrace{\sum_{i=1}^{N} \left(\sum_{j=1}^{M} f(P_{ij}) \, \Delta y \right) \Delta x}_{\text{First sum the columns, then add up the column sums}}$$

Let us choose sample points of the form $P_{ij} = (x_i, y_j)$, where $\{x_i\}$ are sample points for the regular partition on $[a, b]$ and $\{y_j\}$ are sample points for the regular partition of $[c, d]$. We observe that Δx and Δy approach zero as M and N tend to infinity. Therefore, the limit of the sums $S_{N,M}$ as $\|\mathcal{P}\| \to 0$ may be expressed as a limit as $N, M \to \infty$:

$$\iint_{\mathcal{R}} f(x, y) \, dA = \lim_{N,M \to \infty} \sum_{i=1}^{N} \left(\sum_{j=1}^{M} f(x_i, y_j) \Delta y \right) \Delta x$$

The inner sum on the right is a Riemann sum that approaches the single integral $\int_c^d f(x_i, y) \, dy$ as M approaches ∞:

$$\lim_{M \to \infty} S_{N,M} = \lim_{M \to \infty} \sum_{j=1}^{M} f(x_i, y_j) = \int_c^d f(x_i, y) \, dy$$

To complete the proof, we take two facts for granted. First is the fact that the function $S(x) = \int_c^d f(x, y) \, dy$ is continuous for $a \le x \le b$. Second, the limit $\lim_{N,M \to \infty} S_{N,M}$ may be computed by first taking the limit with respect to M and then with respect to N.

Then

$$\iint_{\mathcal{R}} f(x, y)\, dA = \lim_{N,M\to\infty} S_{N,M} = \lim_{N\to\infty}\left(\lim_{M\to\infty} S_{N,M}\right)$$

$$= \lim_{N\to\infty} \sum_{i=1}^{N} \left(\int_c^d f(x_i, y)\, dy\right) \Delta x \qquad \boxed{1}$$

$$= \int_a^b \left(\int_c^d f(x, y)\, dy\right) dx \qquad \boxed{2}$$

Notice that (2) follows from (1) because the sum in (1) is a Riemann sum for $S(x)$ over the interval $[a, b]$. This proves Fubini's Theorem for the order $dy\,dx$. A similar argument applies to the order $dx\,dy$. ■

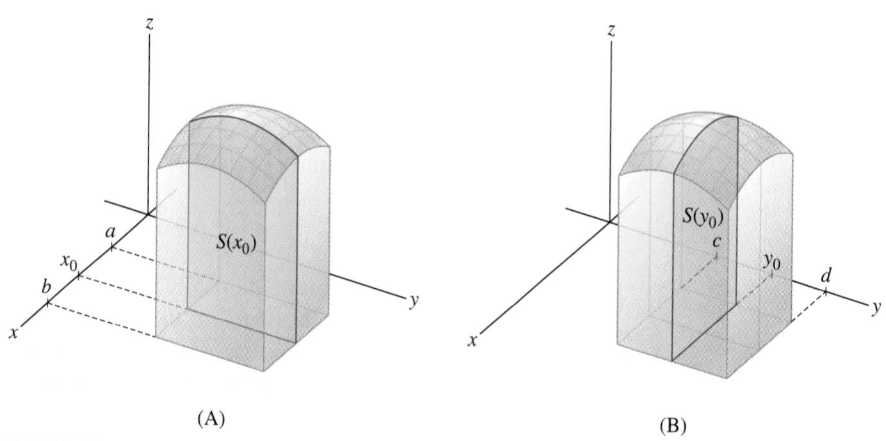

FIGURE 13

GRAPHICAL INSIGHT The double integral $\iint_{\mathcal{R}} f(x, y)\, dA$ is the signed volume V of the solid region S between the xy-plane and the graph of $f(x, y)$. When we write a double integral as an iterated integral in the order $dy\,dx$, then for each fixed value $x = x_0$, the inner integral is the area of the cross section of S in the vertical plane $x = x_0$ perpendicular to the x-axis (Figure 13):

$$S(x_0) = \int_c^d f(x_0, y)\, dy = \begin{array}{l}\text{(area of cross section in vertical plane}\\ x = x_0 \text{ perpendicular to the } x\text{-axis)}\end{array}$$

Fubini's Theorem thus asserts that the volume V of S may be calculated as the integral of cross-sectional area $S(x)$:

$$V = \int_a^b \int_c^d f(x, y)\, dy\, dx = \int_a^b S(x)\, dx = \text{integral of cross-sectional area}$$

Similarly, when we write V as an iterated integral in the order $dx\,dy$, we are calculating V as the integral of cross sections perpendicular to the y-axis.

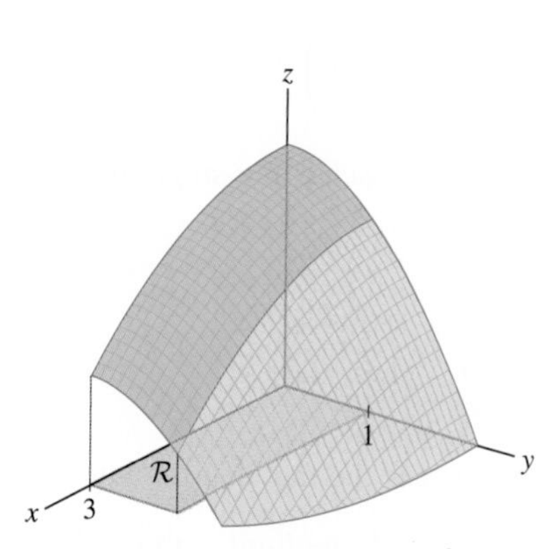

FIGURE 14 Graph of $z = 16 - x^2 - 3y^2$ over $\mathcal{R} = [0, 3] \times [0, 1]$.

■ **EXAMPLE 7** Find the volume V between the surface $z = 16 - x^2 - 3y^2$ and the rectangle $\mathcal{R} = [0, 3] \times [0, 1]$ (Figure 14).

Exercises

1. Compute the Riemann sum $S_{4,3}$ to estimate the double integral of $f(x, y) = xy$ over $\mathcal{R} = [1, 3] \times [1, 2.5]$. Use the regular partition and upper-right vertices of the subrectangles as sample points.

2. Compute the Riemann sum with $N = M = 2$ to estimate the integral of $\sqrt{x + y}$ over $\mathcal{R} = [0, 1] \times [0, 1]$. Use the regular partition and midpoints of the subrectangles as sample points.

In Exercises 3–6, compute the Riemann sums for the double integral $\iint_{\mathcal{R}} f(x, y)\, dA$, *where* $\mathcal{R} = [1, 4] \times [1, 3]$, *for the grid and two choices of sample points shown in Figure 17.*

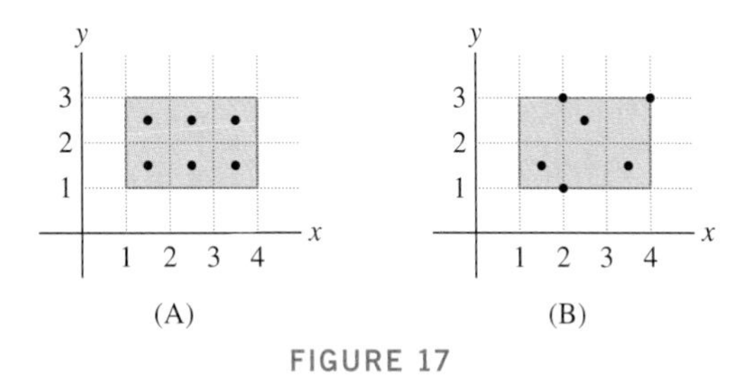

FIGURE 17

3. $f(x, y) = 2x + y$

4. $f(x, y) = 7$

5. $f(x, y) = 4x$

6. $f(x, y) = x - 2y$

7. Let $\mathcal{R} = [0, 1] \times [0, 1]$. Estimate $\iint_{\mathcal{R}} (x + y)\, dA$ by computing two different Riemann sums, each with at least six rectangles.

8. Evaluate $\iint_{\mathcal{R}} 4\, dA$, where $\mathcal{R} = [2, 5] \times [4, 7]$.

9. Evaluate $\iint_{\mathcal{R}} (-5)\, dA$, where $\mathcal{R} = [2, 5] \times [4, 7]$.

10. Use the following table to compute a Riemann sum $S_{3,3}$ for $f(x, y)$ on the square $\mathcal{R} = [0, 1.5] \times [0.5, 2]$. Use the regular partition and sample points of your choosing.

Values of $f(x, y)$

y \ x	0	0.5	1	1.5	2
2	2.6	2.17	1.86	1.62	1.44
1.5	2.2	1.83	1.57	1.37	1.22
1	1.8	1.5	1.29	1.12	1
0.5	1.4	1.17	1	0.87	0.78
0	1	0.83	0.71	0.62	0.56

11. The following table gives the approximate height at 1-ft intervals of a 5 × 4-ft mound of gravel. Estimate the volume of the mound by computing the average of the four Riemann sums $S_{5,4}$ with lower left, lower right, upper left, and upper right vertices of the subrectangles as sample points.

y \ x	0	1	2	3	4	5
4	0.4	0.4	0.6	0.8	0.6	0.6
3	0.8	1.6	1.8	2.5	2.1	0.8
2	0.5	1.5	3.2	3.5	2.1	0.6
1	0.4	0.8	1.3	1.5	1.4	0.6
0	0.3	0.3	0.5	0.8	0.5	0.4

12. Figure 18 shows a grid with values of $f(x, y)$ at sample points in each square. Estimate the double integral of $f(x, y)$ over the rectangles:

(a) $[0.25, 1] \times [0.5, 1]$

(b) $[-0.5, 0.5] \times [0, 1]$

(c) $[-1, -0.5] \times [-1, -0.5]$

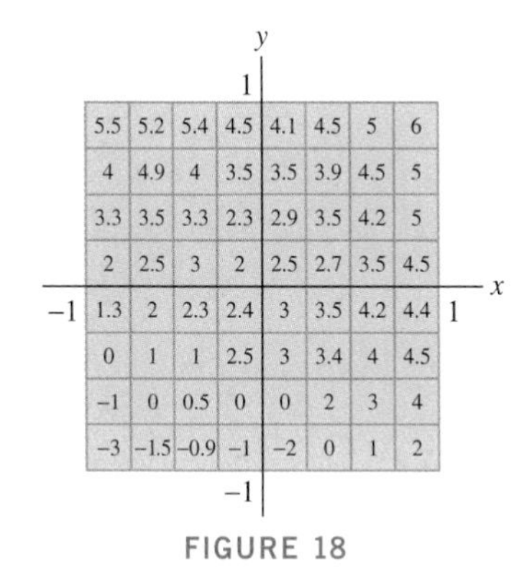

FIGURE 18

13. CAS Let $S_{N,N}$ be the Riemann sum for $\int_0^1 \int_0^1 e^{x^3 - y^3}\, dy\, dx$ using the regular partition and the lower left-hand vertex of each subrectangle as sample points. Use a computer algebra system to calculate $S_{N,N}$ for $N = 25, 50, 100$.

14. CAS Let $S_{N,M}$ be the Riemann sum for

$$\int_0^4 \int_0^2 \ln(1 + x^2 + y^2)\, dy\, dx$$

using the regular partition and the upper right-hand vertex of each subrectangle as sample points. Use a computer algebra system to calculate $S_{2N,N}$ for $N = 25, 50, 100$.

In Exercises 15–16, use symmetry to evaluate the double integral.

15. $\iint_{\mathcal{R}} \sin x\, dA, \quad \mathcal{R} = [0, 2\pi] \times [0, 2\pi]$

16. $\iint_{\mathcal{R}} (2 + x^2 y)\, dA, \quad \mathcal{R} = [0, 1] \times [-1, 1]$

In Exercises 17–32, evaluate the iterated integral.

17. $\int_1^3 \int_0^2 x^3 y\, dy\, dx$

18. $\int_0^2 \int_1^3 x^3 y\, dx\, dy$

19. $\displaystyle\int_0^2 \int_1^3 x^3 y \, dy \, dx$

20. $\displaystyle\int_{-1}^1 \int_0^\pi x^2 \sin y \, dy \, dx$

21. $\displaystyle\int_2^6 \int_1^4 x^2 \, dx \, dy$

22. $\displaystyle\int_2^6 \int_1^4 y^2 \, dx \, dy$

23. $\displaystyle\int_0^1 \int_0^2 (x + 4y^3) \, dx \, dy$

24. $\displaystyle\int_0^2 \int_0^2 (x^2 - y^2) \, dy \, dx$

25. $\displaystyle\int_0^1 \int_2^3 \sqrt{x + 4y} \, dx \, dy$

26. $\displaystyle\int_0^{\pi/4} \int_0^{\pi/2} \cos(2x + y) \, dy \, dx$

27. $\displaystyle\int_1^2 \int_0^4 \frac{dy \, dx}{x + y}$

28. $\displaystyle\int_0^1 \int_0^2 e^{2x+3y} \, dy \, dx$

29. $\displaystyle\int_0^1 \int_2^3 \frac{1}{(x + 4y)^3} \, dx \, dy$

30. $\displaystyle\int_0^1 \int_0^1 \frac{y}{x^2 + 1} \, dy \, dx$

31. $\displaystyle\int_1^2 \int_1^2 \ln(xy) \, dy \, dx$

32. $\displaystyle\int_1^4 \int_0^2 y^{-2} e^{2x} \, dx \, dy$

33. Let $f(x, y) = mxy^2$, where m is a constant. Find a value of m such that $\displaystyle\iint_{\mathcal{R}} f(x, y) \, dx \, dy = 1$, where $\mathcal{R} = [0, 1] \times [0, 2]$.

In Exercises 34–37, evaluate the double integral of the function over the rectangle.

34. $\displaystyle\iint_{\mathcal{R}} (2x + 6y) \, dA, \quad \mathcal{R} = [0, 3] \times [0, 2]$

35. $\displaystyle\iint_{\mathcal{R}} x^2 y \, dA, \quad \mathcal{R} = [-1, 1] \times [0, 2]$

36. $\displaystyle\iint_{\mathcal{R}} \cos x \sin 2y \, dA, \quad \mathcal{R} = [0, \tfrac{\pi}{2}] \times [0, \tfrac{\pi}{2}]$

37. $\displaystyle\iint_{\mathcal{R}} \frac{y}{x + 1} \, dA, \quad \mathcal{R} = [0, 2] \times [0, 4]$

In Exercises 38–41, use Eq. (2) to evaluate the integral.

38. $\displaystyle\iint_{\mathcal{R}} e^x \sin y \, dA, \quad \mathcal{R} = [0, 2] \times [0, \tfrac{\pi}{4}]$

39. $\displaystyle\iint_{\mathcal{R}} e^{2x+3y} \, dA, \quad \mathcal{R} = [0, 1] \times [0, 2]$

40. $\displaystyle\int_0^1 \int_0^2 x^4 y^3 \, dy \, dx$

41. $\displaystyle\int_1^2 \int_0^2 \frac{x}{y} \, dx \, dy$

42. Evaluate $I = \displaystyle\int_1^3 \int_0^1 y e^{xy} \, dy \, dx$. You will need Integration by Parts and the formula

$$\int e^x (x^{-1} - x^{-2}) \, dx = x^{-1} e^x + C$$

Then evaluate I again using Fubini's Theorem to change the order of integration (i.e., integrate first with respect to y). Which method is easier?

43. CAS Use the inequality $0 \le \sin x \le x$ for $x \ge 0$ to show that

$$\int_0^1 \int_0^1 \sin(xy) \, dx \, dy \le \frac{1}{4}$$

Then use a computer algebra system to evaluate the double integral to three decimal places.

44. Evaluate $\displaystyle\int_0^1 \int_0^1 \frac{y}{1 + xy} \, dy \, dx$. *Hint:* Change the order of integration.

45. Calculate a Riemann sum $S_{3,3}$ on the square $\mathcal{R} = [0, 3] \times [0, 3]$ for the function $f(x, y)$ whose contour plot is shown in Figure 19. Choose sample points and use the plot to find the values of $f(x, y)$ at these points.

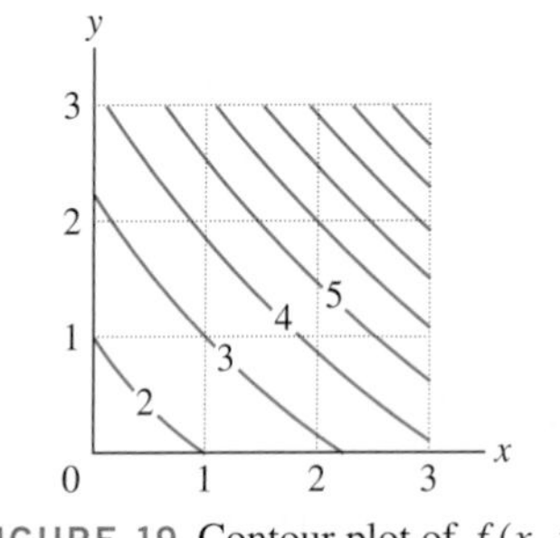

FIGURE 19 Contour plot of $f(x, y)$.

46. Using Fubini's Theorem, argue that the solid in Figure 20 has volume AL.

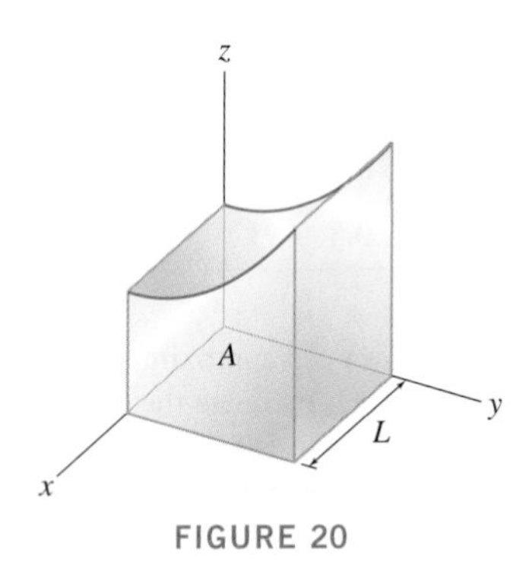

FIGURE 20

Further Insights and Challenges

47. Prove the following extension of the FTC to two variables: If $\dfrac{\partial^2 F}{\partial x\, \partial y} = f(x, y)$, then

$$\iint_{\mathcal{R}} f(x, y)\, dA = F(b, d) - F(a, d) - F(b, c) + F(a, c)$$

where $\mathcal{R} = [a, b] \times [c, d]$.

48. Let $F(x, y) = x^{-1}e^{xy}$. Show that $\dfrac{\partial^2 F}{\partial x\, \partial y} = ye^{xy}$ and use the result of Exercise 47 to evaluate $\iint_{\mathcal{R}} ye^{xy}\, dA$ for the rectangle $\mathcal{R} = [1, 3] \times [0, 1]$.

49. Find a function $F(x, y)$ satisfying $\dfrac{\partial^2 F}{\partial x\, \partial y} = 6x^2 y$ and use the result of Exercise 47 to evaluate $\iint_{\mathcal{R}} 6x^2 y\, dA$ for the rectangle $\mathcal{R} = [0, 1] \times [0, 4]$.

50. In this exercise, we use double integration to evaluate the following improper integral for $a > 0$ a positive constant:

$$I(a) = \int_0^\infty \frac{e^{-x} - e^{-ax}}{x}\, dx$$

(a) Use L'Hôpital's Rule to show that $f(x) = \dfrac{e^{-x} - e^{-ax}}{x}$, though not defined at $x = 0$, can be made continuous by assigning the value $f(0) = a - 1$. Then use the Comparison Theorem to show that $I(a)$ converges. *Hint:* $f(x) \le e^{-x} + e^{-ax}$ for $x > 1$.

(b) Show that $I(a) = \displaystyle\int_0^\infty \int_1^a e^{-xy}\, dy\, dx$.

(c) Prove, by interchanging the order of integration, that

$$I(a) = \ln a - \lim_{T \to \infty} \int_1^a \frac{e^{-Ty}}{y}\, dy \qquad \boxed{4}$$

(d) Use the Comparison Theorem to show that the limit in Eq. (4) is zero. *Hint:* If $a \ge 1$, use $e^{-Ty} \le e^{-T}$ for $y \ge a$, and if $a < 1$, use $e^{-Ty} \le e^{-aT}$ for $a \le y \le 1$. Conclude that $I(a) = \ln a$ (Figure 21).

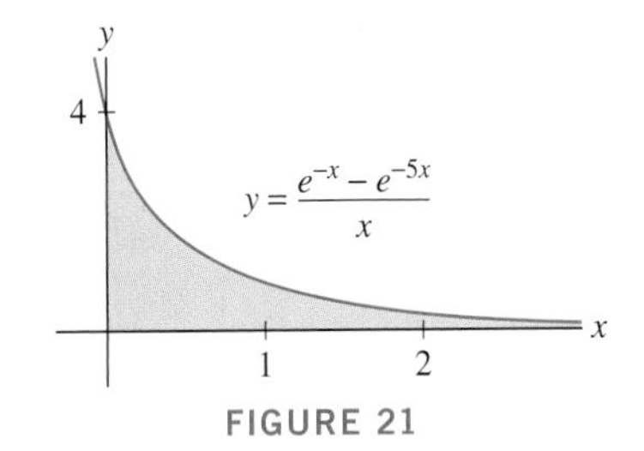

FIGURE 21

15.2 Double Integrals over More General Regions

In the previous section, we restricted our attention to rectangular domains. We now treat double integrals over more general domains $\mathcal{D}$ whose boundaries are simple closed curves (a curve is *simple* if it does not intersect itself). We assume that $\mathcal{D}$ is closed, that is, $\mathcal{D}$ contains its boundary, and that the boundary is either smooth as in Figure 1(A) or has at most finitely many points of nondifferentiability as in Figure 1(B). A boundary curve of this type is called **piecewise smooth**.

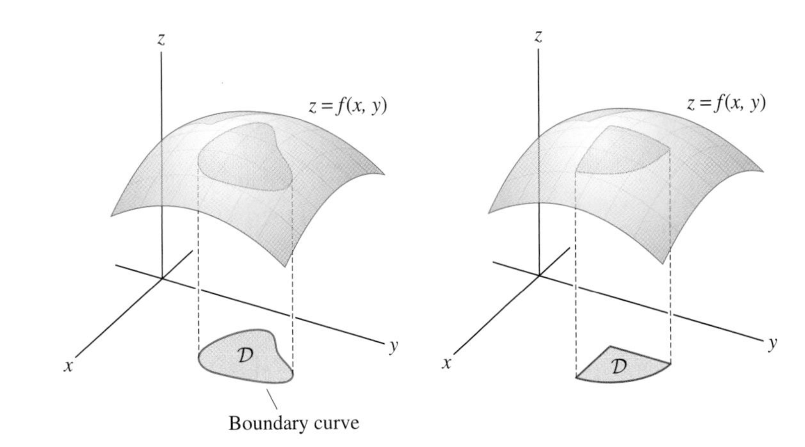

FIGURE 1

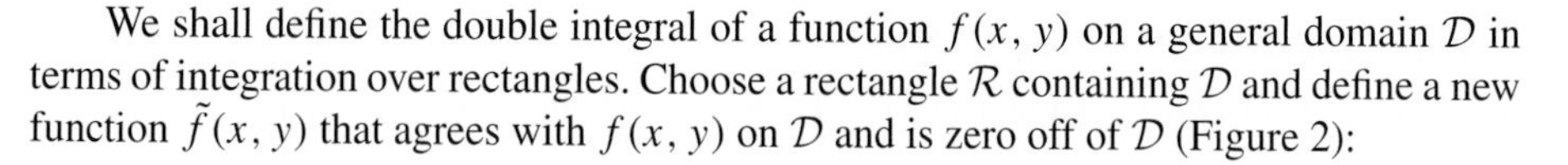

We shall define the double integral of a function $f(x, y)$ on a general domain $\mathcal{D}$ in terms of integration over rectangles. Choose a rectangle $\mathcal{R}$ containing $\mathcal{D}$ and define a new function $\tilde{f}(x, y)$ that agrees with $f(x, y)$ on $\mathcal{D}$ and is zero off of $\mathcal{D}$ (Figure 2):

$$\tilde{f}(x, y) = \begin{cases} f(x, y) & \text{if } (x, y) \in \mathcal{D} \\ 0 & \text{if } (x, y) \notin \mathcal{D} \end{cases}$$

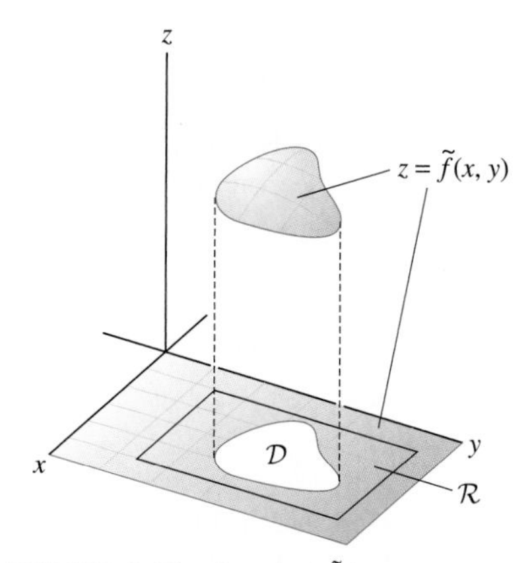

FIGURE 2 The function $\tilde{f}$ is zero outside of $\mathcal{D}$.

We then define the double integral over $\mathcal{D}$ as the integral of $\tilde{f}$ over $\mathcal{R}$:

$$\iint_{\mathcal{D}} f(x, y)\, dA = \iint_{\mathcal{R}} \tilde{f}(x, y)\, dA \qquad \boxed{1}$$

We say that f is integrable over $\mathcal{D}$ if the integral of $\tilde{f}$ over $\mathcal{R}$ exists. The value of the integral does not depend on the choice of $\mathcal{R}$ because $\tilde{f}$ is zero outside of $\mathcal{D}$.

This definition seems reasonable because the integral of $\tilde{f}$ only "picks up" the values of f on $\mathcal{D}$. There is a possible difficulty, however. The function $\tilde{f}$ is usually discontinuous along the boundary of $\mathcal{D}$ since its values suddenly jump to zero when we cross the boundary. The following theorem, which we state without proof, tells us that these possible discontinuities do not affect the integrability of $\tilde{f}$ if f is continuous.

THEOREM 1 Let $\mathcal{D}$ be a domain whose boundary is a simple, closed, piecewise smooth curve. If $f(x, y)$ is continuous on $\mathcal{D}$, then $\iint_{\mathcal{D}} f(x, y)\, dA$ exists.

We may approximate the double integral of $f(x, y)$ over a nonrectangular domain $\mathcal{D}$ by Riemann sums. Choose a rectangle $\mathcal{R}$ containing $\mathcal{D}$ and subdivide $\mathcal{R}$ into MN subrectangles $\mathcal{R}_{ij}$ of size $\Delta A = \Delta x \Delta y$ using a regular partition, as in Section 15.1. Then choose a sample point P_{ij} in each $\mathcal{R}_{ij}$. Since $\tilde{f}(P_{ij}) = 0$ unless P_{ij} lies in $\mathcal{D}$, the Riemann sum reduces to a sum over those sample points that lie in $\mathcal{D}$:

$$S_{N,M} = \sum_{i=1}^{N} \sum_{j=1}^{M} \tilde{f}(P_{ij})\, \Delta x\, \Delta y = \underbrace{\sum f(P_{ij})\, \Delta x\, \Delta y}_{\substack{\text{Sum only over points} \\ P_{ij} \text{ that lie in } \mathcal{D}.}} \approx \iint_{\mathcal{D}} f(x, y)\, dA \qquad \boxed{2}$$

■ **EXAMPLE 1** Let $\mathcal{D}$ be the shaded region in Figure 3 and let $f(x + y) = x + y$. Approximate $\iint_{\mathcal{D}} f(x, y)\, dA$ by computing $S_{4,4}$ for $\iint_{\mathcal{R}} \tilde{f}(x, y)\, dA$ using the regular partition of $\mathcal{R} = [0, 2] \times [0, 2]$ and the upper right-hand corners of the squares as sample points.

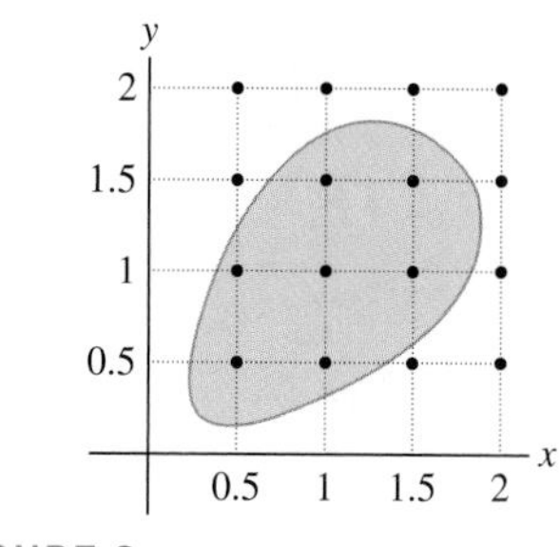

FIGURE 3

Solution The subrectangles in Figure 3 have sides of length $\Delta x = \Delta y = \frac{1}{2}$ and area $\Delta A = \frac{1}{4}$. Of the 16 sample points in Figure 3, only seven lie in $\mathcal{D}$, so

$$S_{4,4} = \sum_{i=1}^{4} \sum_{j=1}^{4} \tilde{f}(P_{ij})\, \Delta x\, \Delta y = \frac{1}{4}\big(f(0.5, 0.5) + f(1, 0.5) + f(0.5, 1) + f(1, 1)$$

$$+ f(1.5, 1) + f(1, 1.5) + f(1.5, 1.5)\big)$$

$$= \frac{1}{4}\big(1 + 1.5 + 1.5 + 2 + 2.5 + 2.5 + 3\big) = \frac{7}{2}$$

■

The linearity properties of the double integral carry over to general domains: If $f(x, y)$ and $g(x, y)$ are integrable and C is a constant, then

$$\iint_{\mathcal{D}} (f(x, y) + g(x, y))\, dA = \iint_{\mathcal{D}} f(x, y)\, dA + \iint_{\mathcal{D}} g(x, y)\, dA$$

$$\iint_{\mathcal{D}} Cf(x, y)\, dA = C \iint_{\mathcal{D}} f(x, y)\, dA$$

We define the area of a domain $\mathcal{D}$ to be the double integral of the constant function $f(x, y) = 1$:

$$\boxed{\text{Area}(\mathcal{D}) = \iint_{\mathcal{D}} 1\, dA} \qquad \boxed{3}$$

More generally, for any constant C,

$$\iint_{\mathcal{D}} C\, dA = C\, \text{Area}(\mathcal{D}) \qquad \boxed{4}$$

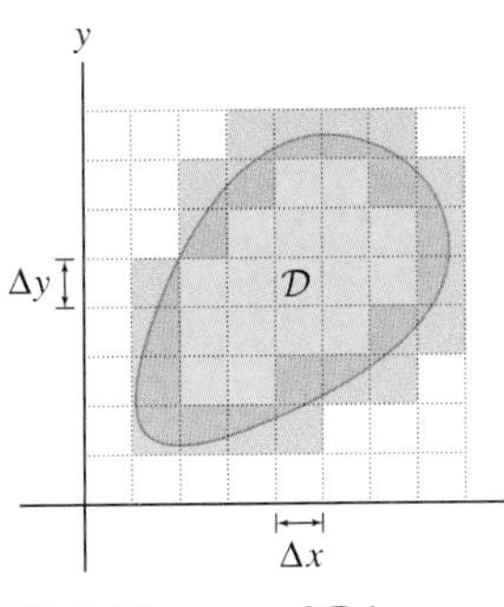

FIGURE 4 The area of $\mathcal{D}$ is approximated by the sum of the areas of rectangles that intersect $\mathcal{D}$.

CONCEPTUAL INSIGHT According to Eq. (3), the area of a domain $\mathcal{D}$ is equal to the limit of Riemann sums for the function $f(x, y) = 1$. By Eq. (2), such a Riemann sum is obtained by covering $\mathcal{D}$ with a grid of rectangles (Figure 4) and summing up the areas of those rectangles that intersect $\mathcal{D}$ (the rectangles on the boundary are included only if the sample point lies in $\mathcal{D}$). In Exercise 35, we show that Eq. (3) yields the same value for the area as a single integral in the case of a region between two graphs.

Regions Between Curves

When the domain $\mathcal{D}$ is the region between two curves, we can evaluate the double integral as an iterated integral. A region of this type is called **simple**. More precisely, we say that $\mathcal{D}$ is **vertically simple** if $\mathcal{D}$ is the region between the graphs of continuous functions $y = \alpha(x)$ and $y = \beta(x)$ (Figure 5):

$$\mathcal{D} = \{(x, y) : a \le x \le b \quad \alpha(x) \le y \le \beta(x)\}$$

Similarly, $\mathcal{D}$ is **horizontally simple** if

$$\mathcal{D} = \{(x, y) : c \le y \le d \quad \alpha(y) \le x \le \beta(y)\}$$

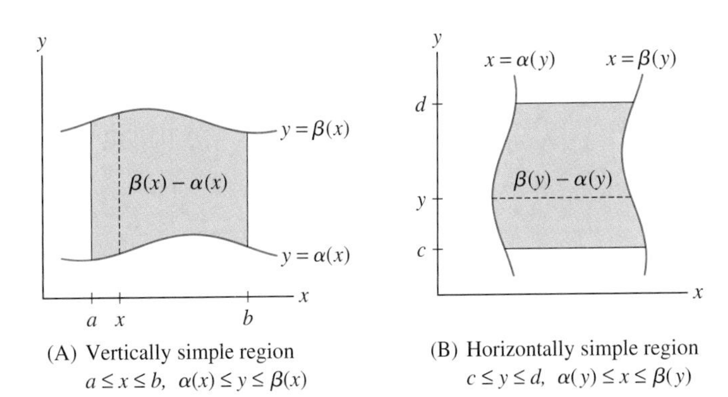

(A) Vertically simple region $a \le x \le b,\ \alpha(x) \le y \le \beta(x)$

(B) Horizontally simple region $c \le y \le d,\ \alpha(y) \le x \le \beta(y)$

FIGURE 5

THEOREM 2 If $\mathcal{D}$ is a vertically simple region with description

$$a \le x \le b, \qquad \alpha(x) \le y \le \beta(x)$$

where $\alpha(x)$ and $\beta(x)$ are continuous, then

$$\iint_{\mathcal{D}} f(x,y)\,dA = \int_a^b \int_{\alpha(x)}^{\beta(x)} f(x,y)\,dy\,dx$$

If $\mathcal{D}$ is a horizontally simple region with description

$$c \le y \le d, \qquad \alpha(y) \le x \le \beta(y)$$

then

$$\iint_{\mathcal{D}} f(x,y)\,dA = \int_c^d \int_{\alpha(y)}^{\beta(y)} f(x,y)\,dx\,dy$$

Proof We sketch the proof. Assume that $\mathcal{D}$ is vertically simple (the horizontally simple case is similar). Let $\mathcal{R} = [a,b] \times [c,d]$ be a rectangle containing $\mathcal{D}$. Then

Although $\tilde{f}$ need not be continuous, the use of Fubini's Theorem in Eq. (5) can be justified. In particular, the integral $\int_c^d \tilde{f}(x,y)\,dy$ exists and is a continuous function of x.

$$\iint_{\mathcal{D}} f(x,y)\,dA = \iint_{\mathcal{R}} \tilde{f}(x,y)\,dA = \int_a^b \int_c^d \tilde{f}(x,y)\,dy\,dx \qquad \boxed{5}$$

By definition, $\tilde{f}(x,y)$ is zero outside $\mathcal{D}$, so for fixed x, $\tilde{f}(x,y)$ is zero unless y satisfies $\alpha(x) \le y \le \beta(x)$. Therefore,

$$\int_c^d \tilde{f}(x,y)\,dy = \int_{\alpha(x)}^{\beta(x)} f(x,y)\,dy$$

Substituting in Eq. (5), we obtain the desired equality:

$$\iint_{\mathcal{D}} f(x,y)\,dA = \int_a^b \int_{\alpha(x)}^{\beta(x)} f(x,y)\,dy\,dx \qquad \blacksquare$$

Integration over a simple region is similar to integration over a rectangle with one difference: The limits of the inner integral may be functions instead of constants.

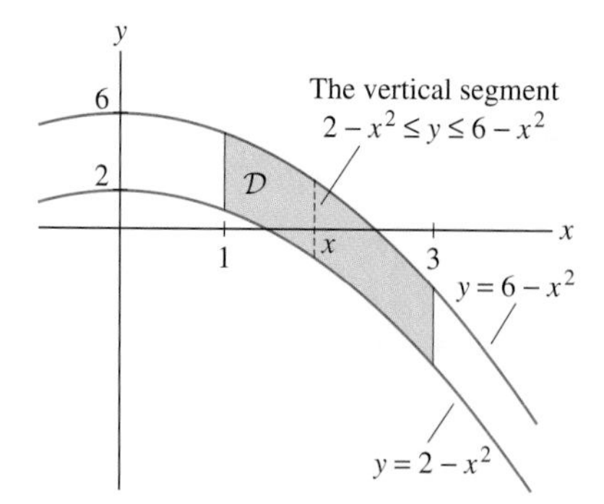

FIGURE 6 The region $1 \le x \le 3$, $2 - x^2 \le y \le 6 - x^2$. The inner integral in Eq. (6) is the integral of $f(x,y) = x^3$ over the vertical segment between the curves at x.

■ **EXAMPLE 2** Evaluate $\iint_{\mathcal{D}} x^3\,dA$, where $\mathcal{D}$ is the region in Figure 6.

Solution We may describe $\mathcal{D}$ as a vertically simple region:

$$\underbrace{1 \le x \le 3}_{\text{Limits of outer integral}}, \qquad \underbrace{2 - x^2 \le y \le 6 - x^2}_{\text{Limits of inner integral}}$$

Step 1. Set up the iterated integral.

$$\iint_{\mathcal{D}} x^3\,dA = \int_1^3 \int_{y=2-x^2}^{6-x^2} x^3\,dy\,dx \qquad \boxed{6}$$

so we may rewrite our integral and evaluate:

$$\int_1^9 \int_{\sqrt{y}}^3 xe^y\,dx\,dy = \int_1^3 \int_1^{x^2} xe^y\,dy\,dx = \int_1^3 \left(\int_{y=1}^{x^2} xe^y\,dy\right) dx$$

$$= \int_1^3 \left(xe^y\Big|_{y=1}^{x^2}\right) dx = \int_1^3 (xe^{x^2} - ex)\,dx = \frac{1}{2}(e^{x^2} - ex^2)\Big|_1^3$$

$$= \frac{1}{2}(e^9 - 9e) - 0 = \frac{1}{2}(e^9 - 9e)$$ ■

The **average value** of a function $f(x, y)$ on a domain $\mathcal{D}$ is defined by

$$\overline{f} = \frac{1}{\text{Area}(\mathcal{D})} \iint_{\mathcal{D}} f(x, y)\,dA = \frac{\iint_{\mathcal{D}} f(x, y)\,dA}{\iint_{\mathcal{D}} 1\,dA} \qquad \boxed{7}$$

← REMINDER *Equation (7) is similar to the definition of an average value in one variable:*

$$\overline{f} = \frac{1}{b-a}\int_a^b f(x)\,dx = \frac{\int_a^b f(x)\,dx}{\int_a^b 1\,dx}$$

Note that the integral of $f(x, y)$ over $\mathcal{D}$ has the same value as the integral of the constant function $\overline{f}(x, y) = \overline{f}$ (Figure 11).

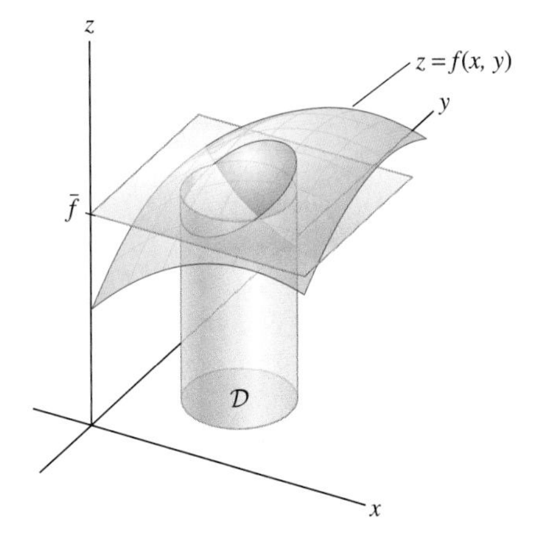

FIGURE 11

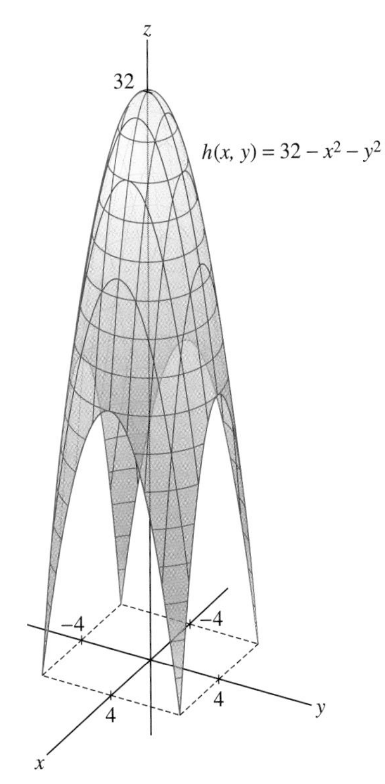

FIGURE 12 Pagoda with ceiling $H(x, y) = 32 - x^2 - y^2$.

■ **EXAMPLE 7 Computing Average Value** An architect needs to know the average height $\overline{H}$ of the ceiling of a pagoda whose base $\mathcal{D}$ is the square $[-4, 4] \times [-4, 4]$ and roof is the graph of $H(x, y) = 32 - x^2 - y^2$ where distances are in feet (Figure 12). Find this average height.

Solution First, we compute the integral of $H(x, y)$ over $\mathcal{D}$:

$$\iint_{\mathcal{D}} (32 - x^2 - y^2)\,dA = \int_{-4}^4 \int_{-4}^4 (32 - x^2 - y^2)\,dy\,dx$$

$$= \int_{-4}^4 \left(\left(32y - x^2y - \frac{1}{3}y^3\right)\Big|_{-4}^4\right) dx = \int_{-4}^4 \left(\frac{640}{3} - 8x^2\right) dx$$

$$= \left(\frac{640}{3}x - \frac{8}{3}x^3\right)\Big|_{-4}^4 = \frac{4{,}096}{3}$$

The area of $\mathcal{D}$ is $8 \times 8 = 64$, so the average height of the pagoda's ceiling is

$$\overline{H} = \frac{1}{\text{Area}(\mathcal{D})} \iint_{\mathcal{D}} H(x, y)\, dA = \frac{1}{64}\left(\frac{4{,}096}{3}\right) = \frac{64}{3} \approx 21.3 \text{ ft}$$

In the previous example, the average $\overline{H}$ lies between the minimum height $H = 0$ and maximum height $H = 32$. According to the first part of the next theorem, it is true in general that the average value lies between the minimum and maximum values of the function on the domain. The second part tells us that integration "preserves order" for functions f and g such that $f(x, y) \le g(x, y)$ on $\mathcal{D}$.

THEOREM 3 Let $f(x, y)$ and $g(x, y)$ be integrable functions on $\mathcal{D}$. Assume that $m \le f(x, y) \le M$ for all $(x, y) \in \mathcal{D}$, then

$$m \,\text{Area}(\mathcal{D}) \le \iint_{\mathcal{D}} f(x, y)\, dA \le M \,\text{Area}(\mathcal{D}) \tag{8}$$

If $f(x, y) \le g(x, y)$ for all $(x, y) \in \mathcal{D}$, then

$$\iint_{\mathcal{D}} f(x, y)\, dA \le \iint_{\mathcal{D}} g(x, y)\, dA \tag{9}$$

Proof We prove (9) first. If $f(x, y) \le g(x, y)$, then every Riemann sum for $f(x, y)$ is less than or equal to the corresponding Riemann sum for g:

$$\sum f(P_{ij})\, \Delta x_i\, \Delta y_j \le \sum g(P_{ij})\, \Delta x_i\, \Delta y_j$$

We obtain (9) by taking the limit. Now suppose that $f(x, y) \le M$ and apply (9) with $g(x, y) = M$:

$$\iint_{\mathcal{D}} f(x, y)\, dA \le \iint_{\mathcal{D}} M\, dA = M \,\text{Area}(\mathcal{D})$$

This proves half of Eq. (8). The other half follows similarly. ■

If $\mathcal{D}$ is a nonconnected domain consisting of two or more subdomains bounded by simple closed curves [as in Figure 13(B)], then the integral $\iint_{\mathcal{D}} f(x, y)\, dA$ is defined as the sum of the integrals over the subdomains.

According to the Mean Value Theorem for Integrals (Section 6.2), if $f(x)$ is a continuous function with average value $\overline{f}$ on $[a, b]$, then there exists $c \in [a, b]$ such that $f(c) = \overline{f}$. A similar statement holds for functions of two or more variables on a domain $\mathcal{D}$, provided that $\mathcal{D}$ is closed, bounded, and connected. By definition, $\mathcal{D}$ is **connected** if any two points can be joined by a curve lying entirely in $\mathcal{D}$ [Figure 13(A)]. The domain in Figure 13(B) is not connected. See Exercise 72 for a proof of the MVT for Double Integrals.

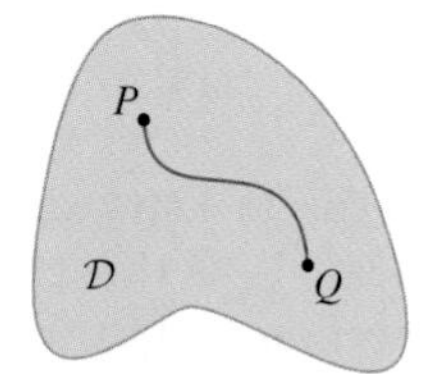

(A) Connected domain: Any two points can be joined by a curve lying entirely in $\mathcal{D}$.

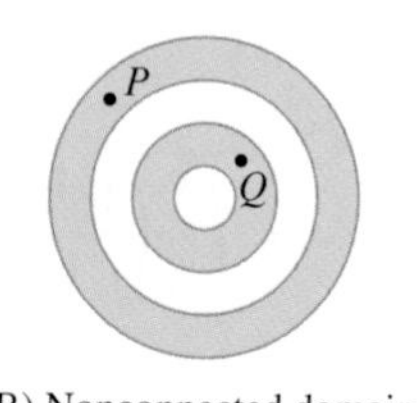

(B) Nonconnected domain.

FIGURE 13

THEOREM 4 Mean Value Theorem for Double Integrals If $f(x, y)$ is continuous and $\mathcal{D}$ is closed, bounded, and connected, then there exists a point $P \in \mathcal{D}$ such that

$$\iint_{\mathcal{D}} f(x, y)\, dA = f(P)\, \text{Area}(\mathcal{D}) \tag{10}$$

Equivalently, $f(P) = \overline{f}$, where $\overline{f}$ is the average value of f on $\mathcal{D}$.

Decomposing the Domain into Smaller Domains

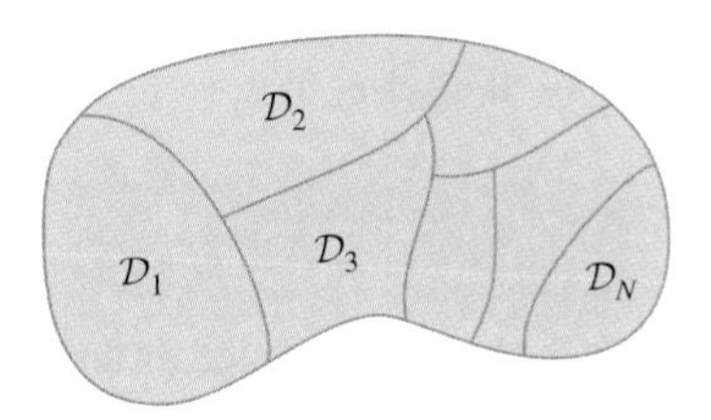

FIGURE 14 The region $\mathcal{D}$ is a union of smaller domains.

Double integrals are additive with respect to the domain: If $\mathcal{D}$ is the union of domains $\mathcal{D}_1, \mathcal{D}_2, \ldots, \mathcal{D}_N$ that do not overlap except possibly on boundary curves (Figure 14), then

$$\boxed{\iint_{\mathcal{D}} f(x, y)\, dA = \iint_{\mathcal{D}_1} f(x, y)\, dA + \cdots + \iint_{\mathcal{D}_N} f(x, y)\, dA}$$

Additivity may be used to evaluate double integrals over domains $\mathcal{D}$ that are not simple but can be decomposed into finitely many simple domains.

We close this section with a simple remark that will be useful later on. If $f(x, y)$ is a continuous function on a *small* domain $\mathcal{D}$, then

In general, the approximation (11) is useful only if $\mathcal{D}$ is small in both width and length, that is, if $\mathcal{D}$ is contained in a circle of small radius. If $\mathcal{D}$ has small area but is very long and thin, then f may be far from constant on $\mathcal{D}$.

$$\iint_{\mathcal{D}} f(x, y)\, dx\, dy \approx \underbrace{f(P)\,\text{Area}(\mathcal{D})}_{\text{Function value} \times \text{area}} \qquad \boxed{11}$$

where P is any sample point in $\mathcal{D}$. In fact, by the MVT for Integrals, it is possible to choose P so that (11) holds with equality. If $\mathcal{D}$ is small, then f is nearly constant on $\mathcal{D}$ and (11) holds as a good approximation for all $P \in \mathcal{D}$.

If the domain $\mathcal{D}$ is not small, we may partition it into N smaller subdomains $\mathcal{D}_1, \ldots, \mathcal{D}_N$ and choose sample points P_j in $\mathcal{D}_j$. By additivity,

$$\iint_{\mathcal{D}} f(x, y)\, dx\, dy = \sum_{j=1}^{N} \iint_{\mathcal{D}_j} f(x, y)\, dx\, dy = \sum_{j=1}^{N} f(P_j)\,\text{Area}(\mathcal{D}_j)$$

and thus we have the approximation

$$\boxed{\iint_{\mathcal{D}} f(x, y)\, dx\, dy \approx \sum_{j=1}^{N} f(P_j)\,\text{Area}(\mathcal{D}_j)} \qquad \boxed{12}$$

Observe that Eq. (12) is a generalization of the Riemann sum approximation. In the case of a Riemann sum, the $\mathcal{D}_j$ are rectangles of area $\Delta A_{ij} = \Delta x_i\, \Delta y_j$.

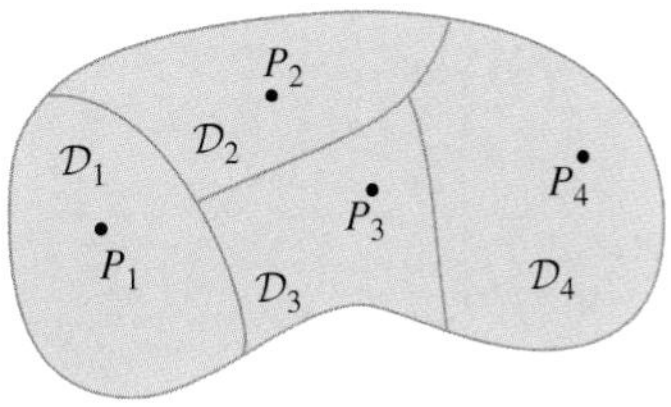

■ **EXAMPLE 8** Estimate $\iint_{\mathcal{D}} f(x, y)\, dA$ for the domain $\mathcal{D}$ in Figure 15, using the areas and function values given there and the accompanying table.

FIGURE 15

j	1	2	3	4
Area($\mathcal{D}_j$)	1	1	0.9	1.2
$f(P_j)$	1.8	2.2	2.1	2.4

Solution

$$\iint_{\mathcal{D}} f(x, y)\, dA \approx \sum_{j=1}^{4} f(P_j)\,\text{Area}(\mathcal{D}_j)$$

$$= (1.8)(1) + (2.2)(1) + (2.1)(0.9) + (2.4)(1.2) \approx 8.8 \qquad ■$$

15.2 SUMMARY

- Let $\mathcal{D}$ be a domain whose boundary is a simple closed curve that is either smooth or has a finite number of corners. To define the integral of $f(x, y)$ over $\mathcal{D}$, choose a rectangle $\mathcal{R}$ containing $\mathcal{D}$ and set

$$\iint_{\mathcal{D}} f(x, y)\, dA = \iint_{\mathcal{R}} \tilde{f}(x, y)\, dA$$

Here, $\tilde{f}(x, y) = f(x, y)$ if $(x, y) \in \mathcal{D}$ and $\tilde{f}(x, y) = 0$ otherwise. The value of the integral does not depend on the choice of $\mathcal{R}$.

- For any constant C, $\iint_{\mathcal{D}} C\, dA = C \cdot \text{Area}(\mathcal{D})$.
- If $\mathcal{D}$ is vertically or horizontally simple, $\iint_{\mathcal{D}} f(x, y)\, dA$ may be evaluated as an iterated integral:

Vertically simple domain $a \le x \le b, \quad \alpha(x) \le y \le \beta(x)$	$\int_a^b \int_{\alpha(x)}^{\beta(x)} f(x, y)\, dy\, dx$
Horizontally simple domain $c \le y \le d, \quad \alpha(y) \le x \le \beta(y)$	$\int_c^d \int_{\alpha(y)}^{\beta(y)} f(x, y)\, dx\, dy$

- The *average value* of f on $\mathcal{D}$ is $\overline{f} = \dfrac{1}{\text{Area}(\mathcal{D})} \iint_{\mathcal{D}} f(x, y)\, dA$.
- If $f(x, y) \le g(x, y)$ on $\mathcal{D}$, then $\iint_{\mathcal{D}} f(x, y)\, dA \le \iint_{\mathcal{D}} g(x, y)\, dA$.
- *Additivity with respect to the domain:* If $\mathcal{D}$ is a union of nonoverlapping (except possibly on their boundaries) domains $\mathcal{D}_1, \dots, \mathcal{D}_N$, then

$$\iint_{\mathcal{D}} f(x, y)\, dA = \sum_{j=1}^{N} \iint_{\mathcal{D}_j} f(x, y)\, dA$$

- If the domains $\mathcal{D}_1, \dots, \mathcal{D}_N$ are small and P_j is a sample point in $\mathcal{D}_j$, then

$$\iint_{\mathcal{D}} f(x, y)\, dA \approx \sum_{j=1}^{N} f(P_j)\text{Area}(\mathcal{D}_j)$$

15.2 EXERCISES

Preliminary Questions

1. Which of the following expressions do not make sense?

(a) $\int_0^1 \int_1^y f(x, y)\, dy\, dx$ **(b)** $\int_0^1 \int_1^x f(x, y)\, dy\, dx$

(c) $\int_1^y \int_0^1 f(x, y)\, dy\, dx$ **(d)** $\int_0^1 \int_x^1 f(x, y)\, dy\, dx$

2. Draw a domain in the plane that is neither vertically nor horizontally simple.

3. Which of the four regions in Figure 16 is the domain of integration for $\int_{-\sqrt{2}/2}^{0} \int_{x}^{\sqrt{1-x^2}} f(x, y)\, dy\, dx$?

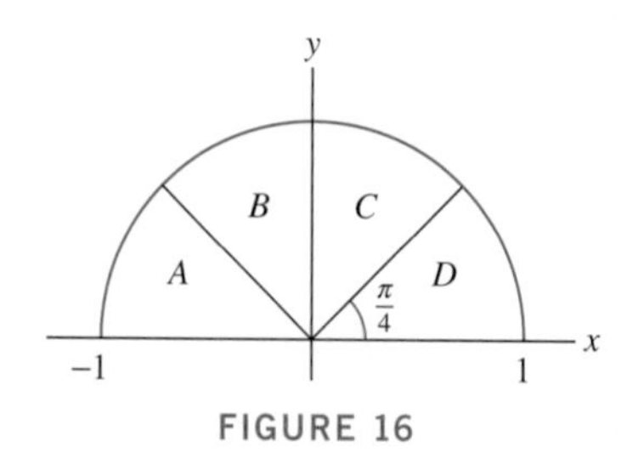

FIGURE 16

4. Let $\mathcal{D}$ be the unit circle. If the maximum value of $f(x, y)$ on $\mathcal{D}$ is 4, then the largest possible value of $\iint_{\mathcal{D}} f(x, y)\, dx\, dy$ is (choose the correct answer):

(a) 4 **(b)** 4π **(c)** $\dfrac{4}{\pi}$

Exercises

1. Calculate the Riemann sum for $f(x, y) = x - y$ and domain $\mathcal{D}$ in Figure 17 with two choices of sample points, • and ∘. Which do you think is a better approximation to the integral of f over $\mathcal{D}$? Why?

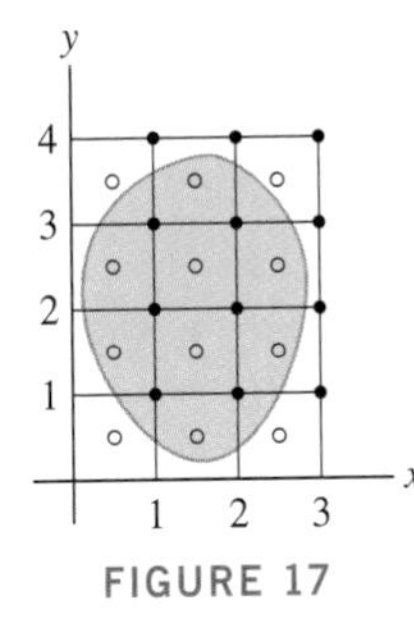

FIGURE 17

2. Approximate values of $f(x, y)$ at sample points on a grid are given in Figure 18. Estimate $\iint_{\mathcal{D}} f(x, y)\,dx\,dy$ by computing the Riemann sum with the given sample points.

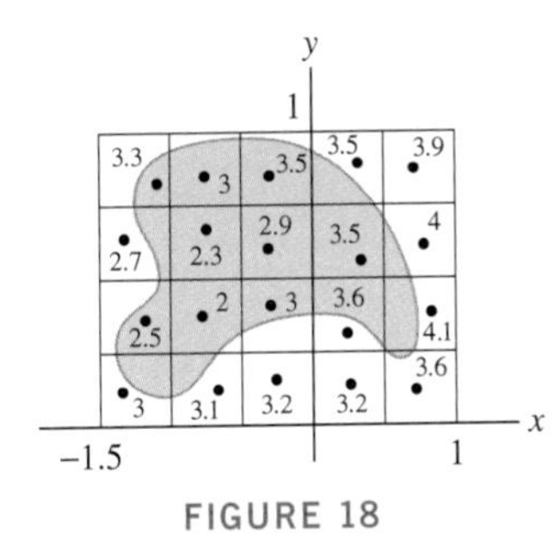

FIGURE 18

3. Let $\mathcal{D}$ be the domain defined by $0 \le x \le 1$, $x^2 \le y \le 4 - x^2$. Sketch $\mathcal{D}$ and express $\iint_{\mathcal{D}} y\,dA$ as an iterated integral and evaluate the result.

4. Express the domain $\mathcal{D}$ in Figure 19 as both a vertically and horizontally simple region and evaluate the integral of $f(x, y) = xy$ over $\mathcal{D}$ as an iterated integral in two ways.

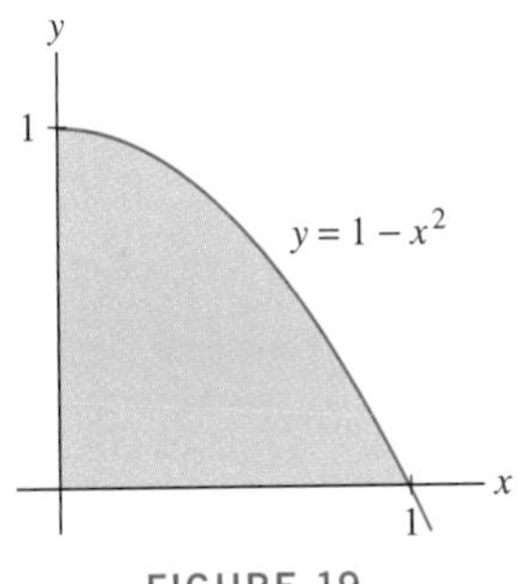

FIGURE 19

In Exercises 5–7, compute the double integral of $f(x, y) = x^2y$ over the given shaded domain in Figure 20.

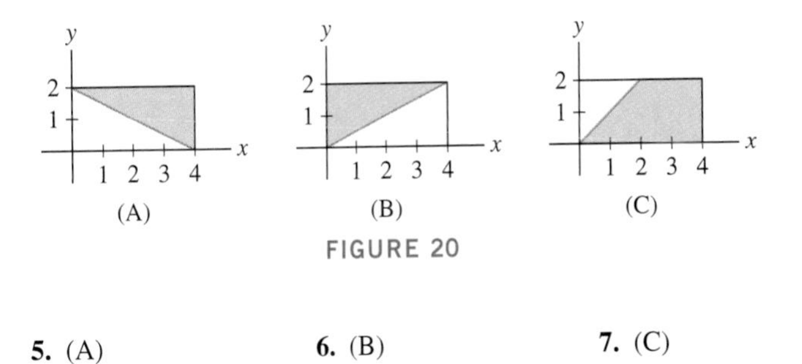

FIGURE 20

5. (A) **6.** (B) **7.** (C)

8. Sketch the domain $\mathcal{D}$ defined by $x + y \le 12$, $x \ge 4$, $y \ge 4$ and compute $\iint_{\mathcal{D}} e^{x+y}\,dA$.

9. Integrate $f(x, y) = x$ over the region bounded by $y = x^2$ and $y = x + 2$.

10. Calculate the double integral of $f(x, y) = y^2$ over the rhombus $\mathcal{R}$ in Figure 21.

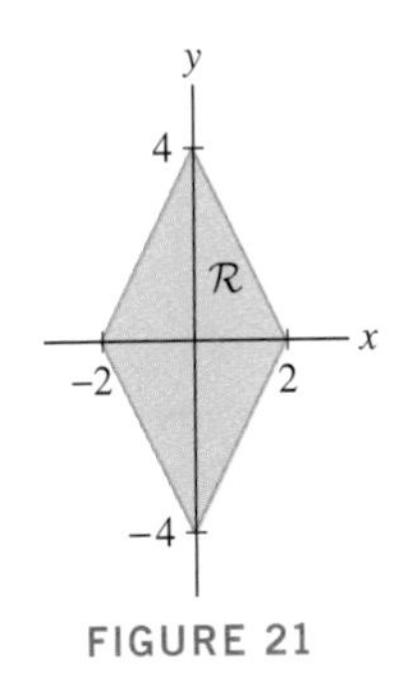

FIGURE 21

11. Sketch the region $\mathcal{D}$ between $y = x^2$ and $y = x(1 - x)$. Express $\mathcal{D}$ as an elementary region and calculate the integral of $f(x, y) = 2y$ over $\mathcal{D}$.

12. Find the double integral of $f(x, y) = x^3y$ over the region between the curves $y = x^2$ and $y = x(1 - x)$.

13. Calculate the integral of $f(x, y) = x$ over the region $\mathcal{D}$ bounded above by $y = x(2 - x)$ and below by $x = y(2 - y)$. *Hint:* Apply the quadratic formula to the lower boundary curve to solve for y as a function of x.

14. Evaluate $\iint_{\mathcal{D}} (x + y)\,dx\,dy$, where $\mathcal{D} = \{(x, y) : x^2 + y^2 \le 4, y \ge 0\}$.

15. Integrate $f(x, y) = (x + y + 1)^{-2}$ over the triangle with vertices $(0, 0)$, $(4, 0)$, and $(0, 8)$.

16. Integrate $f(x, y) = x + y$ over:

(a) The region between $y = x$ and $y = 4x - x^2$.

(b) The region with boundary $y = x$, $y = 4x - x^2$, and $y = 0$. *Hint:* Write the answer as a sum of two integrals.

In Exercises 17–30, compute the integral of the function over the given region. Sketch the region.

17. $f(x, y) = x^3, \quad 0 \le x \le 2, \quad x^2 \le y \le 4$

18. $f(x, y) = xy, \quad 0 \le x \le 2, \quad 0 \le y \le x^2$

19. $f(x, y) = x^2 y, \quad 1 \le x \le 3, \quad x \le y \le 2x + 1$

20. $f(x, y) = x + y, \quad 0 \le x \le 3, \quad 0 \le y \le \sqrt{9 - x^2}$

21. $f(x, y) = 1, \quad 0 \le x \le 1, \quad 1 \le y \le e^x$

22. $f(x, y) = \sin x, \quad 0 \le x \le \frac{\pi}{2}, \quad 0 \le y \le \cos x$

23. $f(x, y) = x, \quad 0 \le x \le 1, \quad 1 \le y \le e^{x^2}$

24. $f(x, y) = \sin(x + y), \quad 0 \le x \le \frac{\pi}{2}, \quad 0 \le y \le 2x$

25. $f(x, y) = \cos(2x + y), \quad -\frac{\pi}{2} \le x \le \frac{\pi}{2}, \quad 1 \le y \le 2x$

26. $f(x, y) = \frac{y}{x}, \quad 0 \le y \le 1, \quad 1 \le x \le e^y$

27. $f(x, y) = 2xy, \quad 0 \le y \le 1, \quad y^2 \le x \le y$

28. $f(x, y) = e^{x+y}, \quad 2 \le y \le 4, \quad y + 1 \le x \le 12 - y$

29. $f(x, y) = (x + y)^{-1}, \quad 1 \le y \le e, \quad 0 \le x \le y$

30. $f(x, y) = \dfrac{\ln y}{x^2}, \quad 1 \le y \le 2, \quad y \le x \le 2y$

In Exercises 31–34, sketch the domain of integration and express the iterated integral in the opposite order.

31. $\displaystyle\int_0^4 \int_x^4 f(x, y)\, dy\, dx$

32. $\displaystyle\int_4^9 \int_{\sqrt{y}}^3 f(x, y)\, dx\, dy$

33. $\displaystyle\int_4^9 \int_2^{\sqrt{y}} f(x, y)\, dx\, dy$

34. $\displaystyle\int_0^1 \int_{e^x}^e f(x, y)\, dy\, dx$

35. According to Eq. (3), the area of a domain $\mathcal{D}$ is equal to $\iint_{\mathcal{D}} 1\, dA$. Prove that if $\mathcal{D}$ is the region between two curves $y = \alpha(x)$ and $y = \beta(x)$ for $a \le x \le b$, then

$$\iint_{\mathcal{D}} 1\, dA = \int_a^b (\beta(x) - \alpha(x))\, dx$$

The integral on the right is the area of $\mathcal{D}$ as defined in Chapter 6.

36. Sketch the domain $\mathcal{D}$ corresponding to

$$\int_0^4 \int_{\sqrt{y}}^2 \sqrt{x^2 + y}\, dx\, dy$$

Then change the order of integration and evaluate.

In Exercises 37–44, for each double integral, sketch the region. Then change the order of integration and evaluate. Explain the simplification achieved by interchanging the order.

37. $\displaystyle\int_0^1 \int_y^1 \frac{\sin x}{x}\, dx\, dy$

38. $\displaystyle\int_0^4 \int_{\sqrt{y}}^2 \sqrt{x^3 + 1}\, dx\, dy$

39. $\displaystyle\int_0^1 \int_{y=x}^1 x e^{y^3}\, dy\, dx$

40. $\displaystyle\int_0^1 \int_{y=x^{2/3}}^1 x e^{y^4}\, dy\, dx$

41. $\displaystyle\int_0^1 \int_0^{\pi/2} x \cos(xy)\, dx\, dy$

42. $\displaystyle\int_0^2 \int_0^3 y^3 \cos(xy^2)\, dy\, dx$

43. $\displaystyle\int_0^9 \int_0^{\sqrt{y}} \frac{x^3\, dx\, dy}{(3x^2 + y)^{1/2}}$

44. $\displaystyle\int_{-1}^1 \int_{-\sqrt{1-x^2}}^{\sqrt{1-x^2}} \sqrt{1 - y^2}\, dy\, dx$

45. Sketch the region $\mathcal{D}$ bounded by the curves $y = e^x$, $y = e^{\sqrt{x}}$, $y = 1$, and $y = 2$, and compute $\iint_{\mathcal{D}} (\ln y)^{-1}\, dA$ by writing $\mathcal{D}$ as a horizontally simple region.

46. Sketch $\mathcal{D}$ where $0 \le x, y \le 2$ and x *or* y is greater than 1, and compute $\iint_{\mathcal{D}} e^{x+y}\, dA$.

In Exercises 47–50, calculate the double integral of $f(x, y)$ over the triangle indicated in Figure 22.

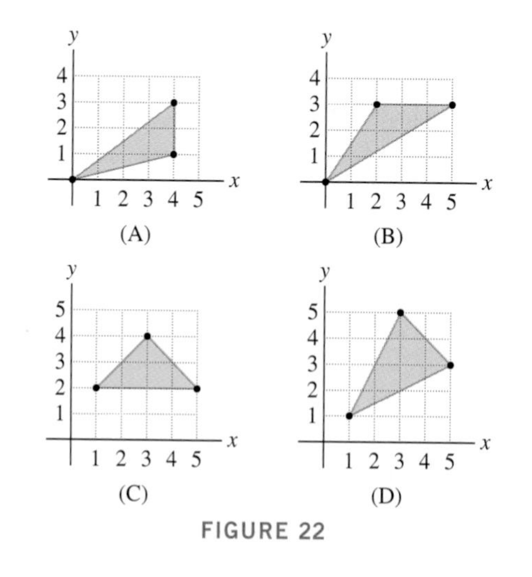

FIGURE 22

47. $f(x, y) = ye^x$, (A)

48. $f(x, y) = 1 - 2x$, (B)

49. $f(x, y) = \dfrac{x}{y^2}$, (C)

50. $f(x, y) = \dfrac{x}{y^2}$, (D)

51. Calculate the double integral $f(x, y) = \dfrac{\sin y}{y}$ over the region $\mathcal{D}$ in Figure 23.

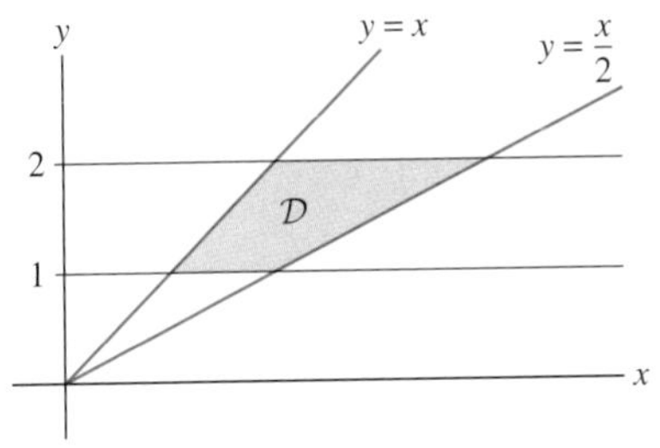

FIGURE 23

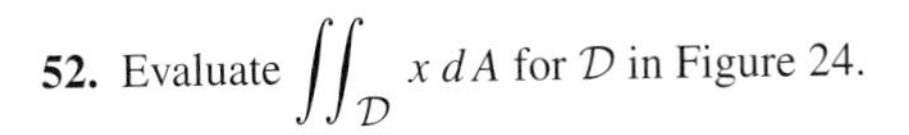
52. Evaluate $\iint_{\mathcal{D}} x\,dA$ for $\mathcal{D}$ in Figure 24.

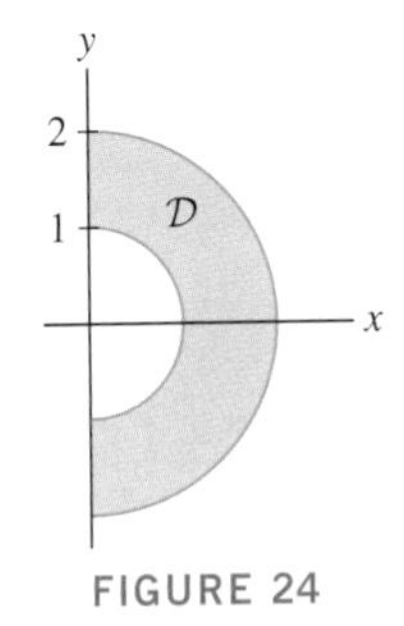

FIGURE 24

53. Find the volume of the region in Figure 25 bounded by the paraboloids $z = x^2 + y^2$ and $z = 8 - x^2 - y^2$.

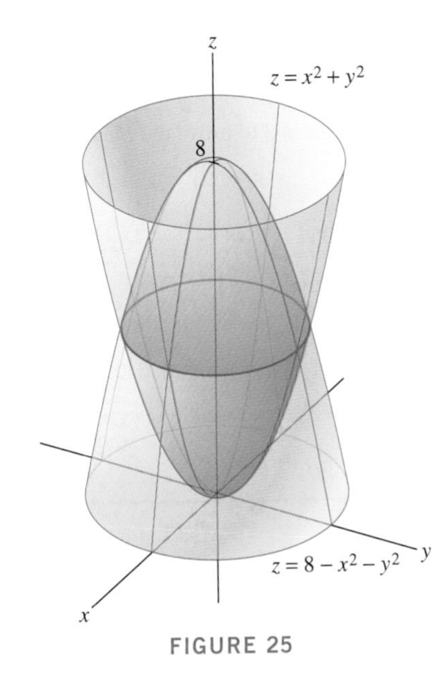

FIGURE 25

54. Find the volume of the region bounded by $z = 8 - y^2$, $y = 8 - x^2$, $x = 0$, $y = 0$, and $z = 0$.

55. Find the volume of the region enclosed by $z = 1 - y^2$ and $z = y^2 - 1$ for $0 \le x \le 2$.

56. Find the total mass of the square disk $0 \le x, y \le 1$ if the mass density is $\rho(x, y) = x^2 + y^2$.

57. A plate in the shape of the region bounded by $y = x^{-1}$ and $y = 0$ for $1 \le x \le 4$ has mass density $\rho(x, y) = y/x$. Calculate the total mass of the plate.

58. Calculate the average value of the function $f(x, y) = e^{x+y}$ on the square $[0, 1] \times [0, 1]$.

59. Calculate the average height above the x-axis of a point in the region $0 \le x \le 1$, $0 \le y \le x^2$.

60. Find the average height of the "ceiling" in Figure 26 defined by $z = y^2 \sin x$ for $0 \le x \le \pi$, $0 \le y \le 1$.

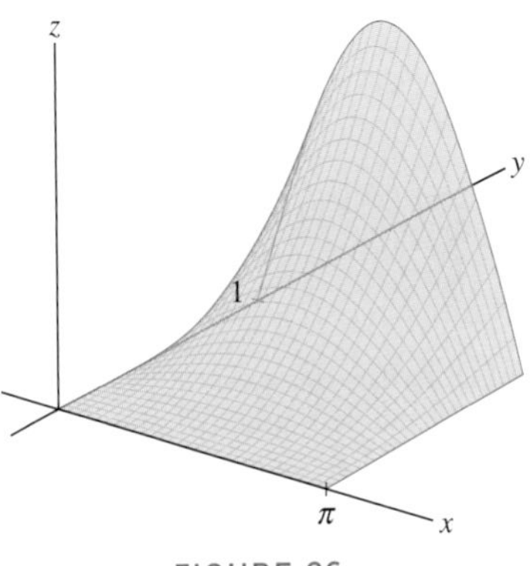

FIGURE 26

61. Calculate the average value of the x-coordinate of a point on the semicircle $x^2 + y^2 \le R^2$, $x \ge 0$. What is the average value of the y-coordinate?

62. What is the average value of a constant function $f(x, y) = C$ over a domain $\mathcal{D}$?

63. What is the average value of the linear function $f(x, y) = mx + ny + p$ on the ellipse $\left(\frac{x}{a}\right)^2 + \left(\frac{y}{b}\right)^2 \le 1$? Argue by symmetry rather than calculation.

64. Find the average square distance from the origin to a point in the domain $\mathcal{D}$ in Figure 27.

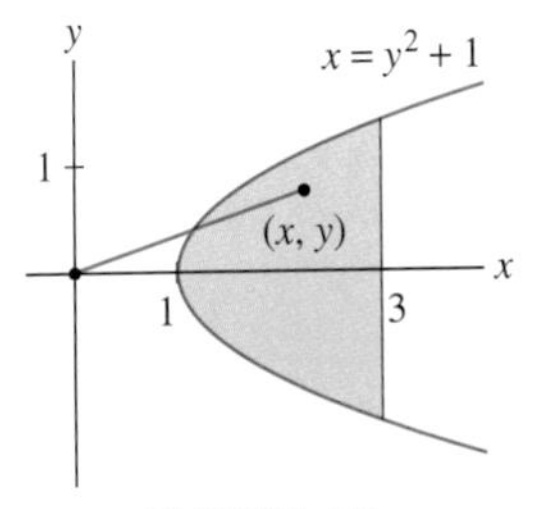

FIGURE 27

65. Show that the average value of $f(x, y) = y$ (the y-coordinate of the centroid) of the sector in Figure 28 is equal to $\overline{y} = \left(\frac{2R}{3}\right)\left(\frac{\sin\theta}{\theta}\right)$.

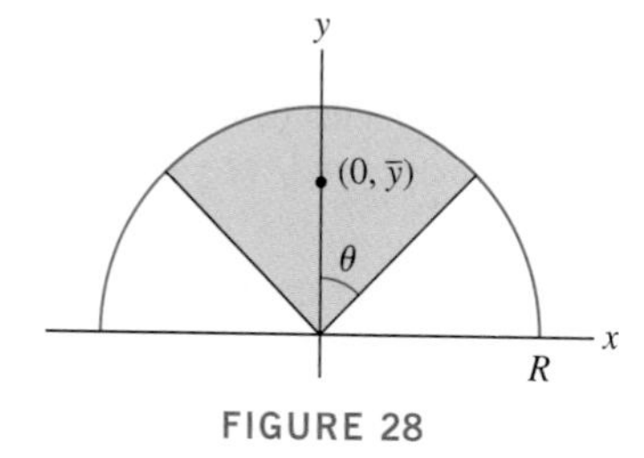

FIGURE 28

66. Prove the inequality $\iint_{\mathcal{D}} \frac{dA}{4 + x^2 + y^2} \le \pi$, where $\mathcal{D}$ is the disk $x^2 + y^2 \le 4$.

67. Find a point P in $\mathcal{D} = [0, 1] \times [0, 4]$ such that $f(P) = \overline{f}$, where $\overline{f}$ is the average of $f(x, y) = xy^2$ on $\mathcal{D}$ (the existence of such a point is guaranteed by the MVT for Double Integrals).

68. Verify the MVT for Double Integrals for $f(x, y) = e^{x-y}$ on the triangle bounded by $y = 0$, $x = 1$, and $y = x$.

In Exercises 69–70, use Eq. (12) to estimate the double integral.

69. The following table lists the areas of the subdomains $\mathcal{D}_j$ of the domain $\mathcal{D}$ in Figure 29 and the values of a function $f(x, y)$ at sample points $P_j \in \mathcal{D}_j$. Estimate $\iint_{\mathcal{D}} f(x, y)\, dA$.

j	1	2	3	4	5	6
Area($\mathcal{D}_j$)	1.2	1.1	1.4	0.6	1.2	0.8
$f(P_j)$	9	9.1	9.3	9.1	8.9	8.8

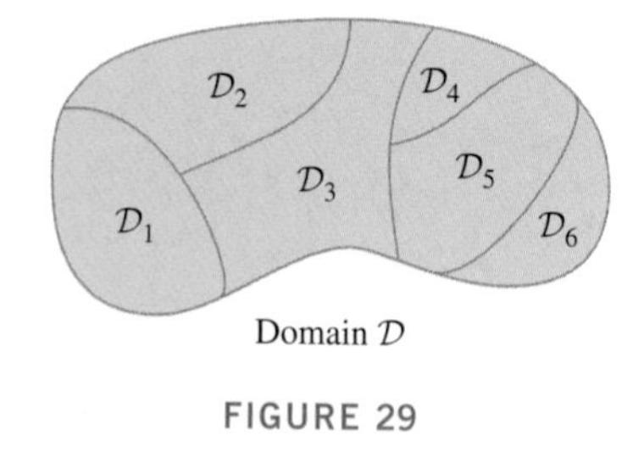

FIGURE 29

70. The domain $\mathcal{D}$ between the circles of radii 5 and 5.2 in the first quadrant in Figure 30 is divided into six subdomains of angular width $\Delta\theta = \frac{\pi}{12}$, and the values of a function $f(x, y)$ at sample points are given. Compute the area of the subdomains and estimate $\iint_{\mathcal{D}} f(x, y)\, dA$.

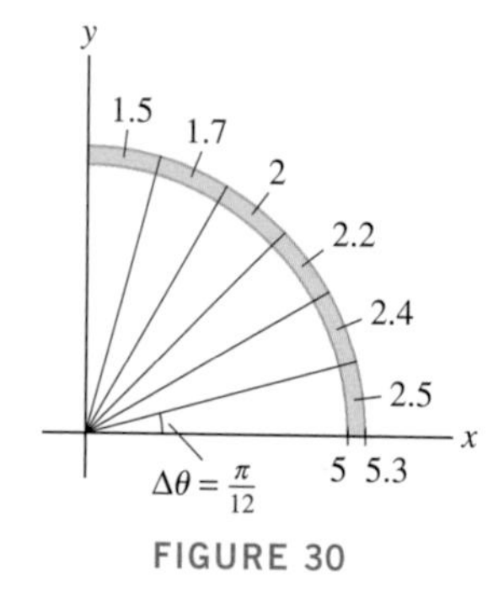

FIGURE 30

Further Insights and Challenges

71. Suppose that $f(x, y)$ is a continuous function on a closed connected domain $\mathcal{D}$. The Intermediate Value Theorem (IVT) states that if $f(P) = a$ and $f(Q) = b$, where $P, Q \in \mathcal{D}$, then $f(x, y)$ takes on every value between a and b at some point in $\mathcal{D}$.

(a) Show, by constructing a counterexample, that the IVT is false if $\mathcal{D}$ is not connected.

(b) Prove the IVT in two variables as follows: Let $\mathbf{c}(t)$ be a path such that $\mathbf{c}(0) = P$ and $\mathbf{c}(1) = Q$ (such a path exists because $\mathcal{D}$ is connected). Apply the IVT in one variable to the composite function $f(\mathbf{c}(t))$.

72. Use the IVT and Theorem 3 to prove the Mean Value Theorem for Double Integrals.

73. Let $f(y)$ be a function of y alone and set $G(t) = \int_0^t \int_0^x f(y)\, dy\, dx$.

(a) Use the FTC to prove that $G''(t) = f(t)$.

(b) Show, by changing the order in the double integral, that $G(t) = \int_0^t (t - y) f(y)\, dy$. This shows that the "second antiderivative" of $f(y)$ can be expressed as a single integral.

15.3 Triple Integrals

Triple integrals of functions $f(x, y, z)$ of three variables are a fairly straightforward generalization of double integrals. Instead of a rectangle in the plane, we consider a box

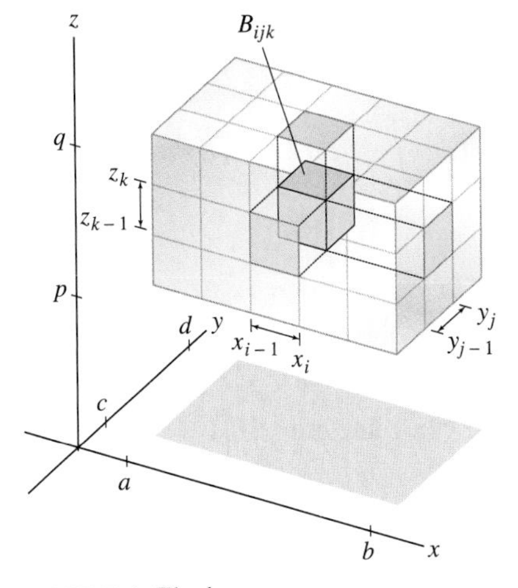

FIGURE 1 The box $\mathcal{B} = [a, b] \times [c, d] \times [p, q]$ decomposed into subboxes.

(Figure 1)

$$\mathcal{B} = [a, b] \times [c, d] \times [p, q]$$

consisting of all points (x, y, z) in $\mathbf{R}^3$ such that

$$a \le x \le b, \qquad c \le y \le d, \qquad p \le z \le q$$

To form a Riemann sum, we choose positive integers N, M, L and partitions of the three intervals

$$a = x_0 < x_1 < \cdots < x_N = b$$

$$c = y_0 < y_1 < \cdots < y_M = d$$

$$p = z_0 < z_1 < \cdots < z_L = q$$

Let $\Delta x_i = x_i - x_{i-1}$, $\Delta y_j = y_j - y_{j-1}$, $\Delta z_k = z_k - z_{k-1}$ be the widths of the subintervals defined by these partitions. We then partition $\mathcal{B}$ into NML smaller boxes $\mathcal{B}_{ijk}$, whose volume we denote by ΔV_{ijk} (Figure 1):

$$\mathcal{B}_{ijk} = [x_{i-1}, x_i] \times [y_{j-1}, y_j] \times [z_{k-1}, z_k],$$

$$\text{Volume}(\mathcal{B}_{ijk}) = \Delta V_{ijk} = \Delta x_i \, \Delta y_j \, \Delta z_k$$

This partition is called **regular** if $[a, b]$, $[c, d]$, and $[p, q]$ are each divided into subintervals of equal lengths Δx, Δy, and Δz, where

$$\Delta x = \frac{b-a}{N}, \qquad \Delta y = \frac{d-c}{M}, \qquad \Delta z = \frac{q-p}{L}$$

For any choice of sample points P_{ijk} in $\mathcal{B}_{ijk}$, we define the Riemann sum:

$$S_{N,M,L} = \sum_{i=1}^{N} \sum_{j=1}^{M} \sum_{k=1}^{L} f(P_{ijk}) \, \Delta V_{ijk}$$

As in the previous section, let $\mathcal{P} = \{\{x_i\}, \{y_j\}, \{z_k\}\}$ be the set of endpoints and let $\|\mathcal{P}\|$ be the maximum of the widths $\Delta x_i, \Delta y_j, \Delta z_k$. If the sums $S_{N,M,L}$ approach a limit as $\|\mathcal{P}\| \to 0$ for arbitrary choices of sample points, we say that f is **integrable** over $\mathcal{B}$. The limit value is denoted

$$\iiint_{\mathcal{B}} f(x, y, z) \, dV = \lim_{\|\mathcal{P}\| \to 0} S_{N,M,L}$$

Triple integrals obey the same linearity properties as double and single integrals. The following theorem assures us that continuous functions are integrable over a box and that the triple integral can be evaluated as an iterated integral.

The notation dA, used in the previous section, suggests area and occurs in double integrals over domains in the plane. Similarly, dV suggests volume and occurs in triple integrals over regions in $\mathbf{R}^3$.

THEOREM 1 Fubini's Theorem for Triple Integrals If $f(x, y, z)$ is continuous on $\mathcal{B} = [a, b] \times [c, d] \times [p, q]$, then the triple integral exists and is equal to the iterated integral:

$$\iiint_{\mathcal{B}} f(x, y, z) \, dV = \int_{x=a}^{b} \int_{y=c}^{d} \int_{z=p}^{q} f(x, y, z) \, dz \, dy \, dx$$

Furthermore, the iterated integral may be evaluated in any order.

As noted in the theorem, we are free to evaluate the iterated integral in any order. For instance,

$$\int_{x=a}^{b}\int_{y=c}^{d}\int_{z=p}^{q} f(x, y, z)\, dz\, dy\, dx = \int_{z=p}^{q}\int_{y=c}^{d}\int_{x=a}^{b} f(x, y, z)\, dx\, dy\, dz$$

There are in total six different orders given by

$$dz\,dy\,dx \quad dy\,dz\,dx \quad dz\,dx\,dy \quad dx\,dz\,dy \quad dy\,dx\,dz \quad dx\,dy\,dz$$

When the integrand is a product $f(x, y, z) = a(x)b(y)c(z)$ and the domain of integration is a box, then the triple integral factors as a product of three integrals. For example, the function in Example 1

$$f(x, y, z) = x^2e^{y+3z} = x^2e^ye^{3z}$$

is a product, and

$$\iiint_{\mathcal{B}} x^2e^ye^{3z}\, dV$$

can be evaluated more simply as the product

$$\left(\int_1^4 x^2\, dx\right)\left(\int_0^3 e^y\, dy\right)\left(\int_2^6 e^{3z}\, dz\right)$$
$$= (21)(e^3 - 1)\left(\frac{e^{18} - e^6}{3}\right)$$

■ **EXAMPLE 1 Integration over a Box** Calculate the integral $\iiint_{\mathcal{B}} x^2e^{y+3z}\, dV$, where $\mathcal{B} = [1, 4] \times [0, 3] \times [2, 6]$.

Solution We write this triple integral as an iterated integral:

$$\iiint_{\mathcal{B}} x^2e^{y+3z}\, dV = \int_1^4\int_0^3\int_2^6 x^2e^{y+3z}\, dz\, dy\, dx$$

***Step 1.* Evaluate the inner integral with respect to z, holding x and y constant.**

$$\int_{z=2}^{6} x^2e^{y+3z}\, dz = \frac{1}{3}x^2e^{y+3z}\Big|_2^6 = \frac{1}{3}x^2e^{y+18} - \frac{1}{3}x^2e^{y+6} = \frac{1}{3}(e^{18} - e^6)x^2e^y$$

***Step 2.* Evaluate the middle integral with respect to y, holding x constant.**

$$\int_{y=0}^{3} \frac{1}{3}(e^{18} - e^6)x^2e^y\, dy = \frac{1}{3}(e^{18} - e^6)x^2\int_{y=0}^{3} e^y\, dy = \frac{1}{3}(e^{18} - e^6)(e^3 - 1)x^2$$

***Step 3.* Evaluate the outer integral respect to x.**

$$\iiint_{\mathcal{B}} (x^2e^{y+3z})\, dV = \frac{1}{3}(e^{18} - e^6)(e^3 - 1)\int_{x=1}^{4} x^2\, dx = 7(e^{18} - e^6)(e^3 - 1) \quad ■$$

Next, we consider triple integrals over solid regions $\mathcal{W}$ that are *simple* in the following sense: $\mathcal{W}$ consists of all points $P = (x, y, z)$ that lie between the graphs of continuous functions $\psi(x, y)$ and $\phi(x, y)$ over a domain $\mathcal{D}$ in the xy-plane (Figure 2). Assume that $\psi(x, y) \le \phi(x, y)$. Then

$$\mathcal{W} = \{(x, y, z) : (x, y) \in \mathcal{D} \quad \text{and} \quad \psi(x, y) \le z \le \phi(x, y)\} \qquad \boxed{1}$$

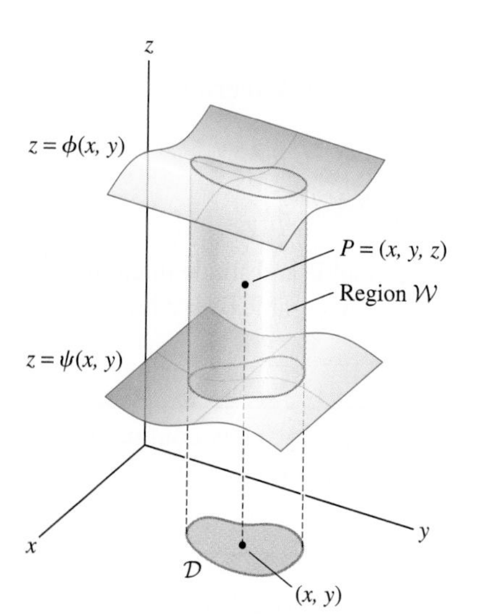

FIGURE 2 The point $P = (x, y, z)$ in the simple region $\mathcal{W}$ if $(x, y) \in \mathcal{D}$ and $\psi(x, y) \le z \le \phi(x, y)$.

The domain $\mathcal{D}$ is the **projection** of $\mathcal{W}$ onto the xy-plane. As in the case of double integrals, we define the triple integral of $f(x, y, z)$ over $\mathcal{W}$ by

$$\iiint_{\mathcal{W}} f(x, y, z)\, dV = \iiint_{\mathcal{B}} \tilde{f}(x, y, z)\, dV$$

where $\mathcal{B}$ is a box containing $\mathcal{W}$ and $\tilde{f}$ is the function that is equal to f on $\mathcal{W}$ and equal to zero outside of $\mathcal{W}$. In practice, instead of using this formal definition, we evaluate triple integrals as iterated integrals. This is justified by the following theorem, whose proof is similar to that of Theorem 2 in Section 15.2.

THEOREM 2 Let $\mathcal{D}$ be a region in the xy-plane. Assume that $\psi(x, y)$ and $\phi(x, y)$ are continuous with $\psi(x, y) \le \phi(x, y)$ for $(x, y) \in \mathcal{D}$. Then the triple integral of a continuous function $f(x, y, z)$ over the domain

$$\mathcal{W} = \{(x, y, z) : (x, y) \in \mathcal{D} \quad \text{and} \quad \psi(x, y) \le z \le \phi(x, y)\} \qquad \boxed{2}$$

exists and is equal to the iterated integral:

$$\iiint_{\mathcal{W}} f(x, y, z)\, dV = \iint_{\mathcal{D}} \left(\int_{z=\psi(x,y)}^{\phi(x,y)} f(x, y, z)\, dz \right) dA$$

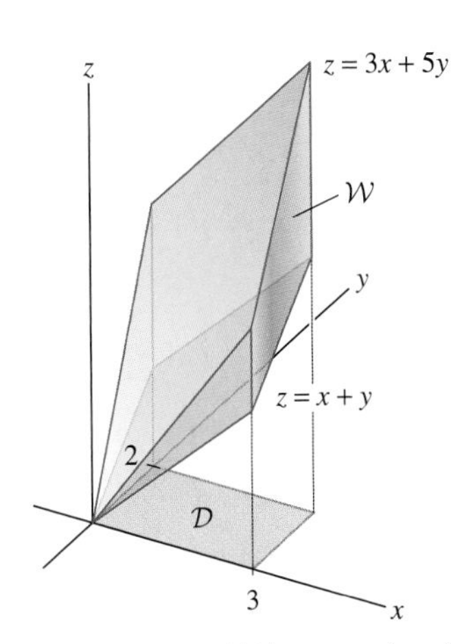

FIGURE 3 Region $\mathcal{W}$ between the planes $z = x + y$ and $z = 3x + 5y$ lying over $\mathcal{D} = [0, 3] \times [0, 2]$.

EXAMPLE 2 Integrating over a Solid Region with a Rectangular Base Evaluate $\iiint_{\mathcal{W}} z\, dV$, where $\mathcal{W}$ is the region between the planes $z = x + y$ and $z = 3x + 5y$ lying over the rectangle $\mathcal{D} = [0, 3] \times [0, 2]$ (Figure 3).

Solution We apply Theorem 2 with $\psi(x, y) = x + y$ and $\phi(x, y) = 3x + 5y$ to write the triple integral as an iterated integral and evaluate:

$$\iiint_{\mathcal{W}} z\, dV = \iint_{\mathcal{D}} \left(\int_{z=x+y}^{3x+5y} z\, dz \right) dA = \int_{x=0}^{3} \int_{y=0}^{2} \int_{z=x+y}^{3x+5y} z\, dz\, dy\, dx$$

Note that the integral over $\mathcal{D}$ has been converted to an iterated integral.

***Step 1.* Evaluate the inner integral with respect to z, holding x and y constant.**

$$\int_{z=x+y}^{3x+5y} z\, dz = \frac{1}{2} z^2 \bigg|_{z=x+y}^{3x+5y} = \frac{1}{2}(3x + 5y)^2 - \frac{1}{2}(x + y)^2 = 4x^2 + 14xy + 12y^2 \qquad \boxed{3}$$

***Step 2.* Evaluate the middle integral with respect to y, holding x constant.**

$$\int_{y=0}^{2} (4x^2 + 14xy + 12y^2)\, dy = (4x^2 y + 7xy^2 + 4y^3)\bigg|_{y=0}^{2} = 8x^2 + 28x + 32$$

***Step 3.* Evaluate the outer integral with respect to x.**

$$\iiint_{\mathcal{W}} z\, dV = \int_{x=0}^{3} (8x^2 + 28x + 32)\, dx = \left(\frac{8}{3}x^3 + 14x^2 + 32x \right) \bigg|_0^3$$

$$= 72 + 126 + 96 = 294$$

■

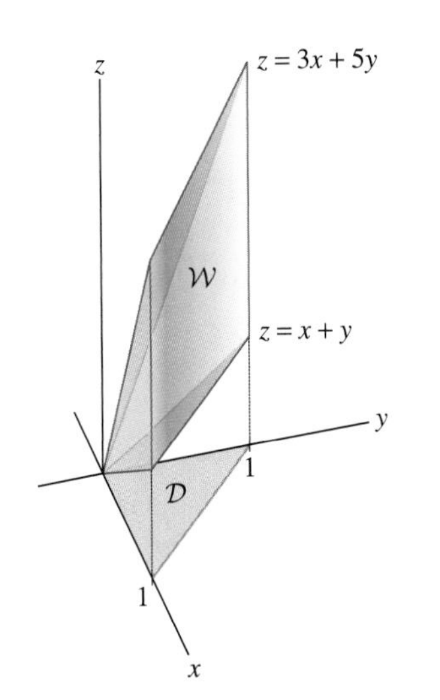

FIGURE 4 Region $\mathcal{W}$ between the planes $z = x + y$ and $z = 3x + 5y$ lying over the triangle $\mathcal{D}$.

EXAMPLE 3 Integrating over a Solid Region with a Triangular Base Evaluate $\iiint_{\mathcal{W}} z\, dV$, where $\mathcal{W}$ is the region in Figure 4:

$$\mathcal{W} = \{(x, y, z) : x + y \le z \le 3x + 5y, x \ge 0, y \ge 0, x + y \le 1\}$$

Solution $\mathcal{W}$ is the region between the two planes $z = x + y$ and $z = 3x + 5y$ lying over the triangle $\mathcal{D}$ in the xy-plane defined by

$$0 \le x \le 1, \qquad 0 \le y \le 1 - x$$

Thus, the triple integral is equal to the iterated integral:

$$\iiint_{\mathcal{W}} z\,dV = \iint_{\mathcal{D}} \left(\int_{z=x+y}^{3x+5y} z\,dz \right) dA = \underbrace{\int_{x=0}^{1} \int_{y=0}^{1-x}}_{\text{Integral over triangle}} \int_{z=x+y}^{3x+5y} z\,dz\,dy\,dx$$

The inner integral is the same as in the previous example [see Eq. (3)]:

$$\int_{z=x+y}^{3x+5y} z\,dz = \frac{1}{2}z^2 \Big|_{x+y}^{3x+5y} = 4x^2 + 14xy + 12y^2$$

Next, we integrate with respect to y:

$$\begin{aligned} \int_{y=0}^{1-x} (4x^2 + 14xy + 12y^2)\,dy &= (4x^2y + 7xy^2 + 4y^3)\Big|_{y=0}^{1-x} \\ &= 4x^2(1-x) + 7x(1-x)^2 + 4(1-x)^3 \\ &= 4 - 5x + 2x^2 - x^3 \end{aligned}$$

and finally,

$$\iiint_{\mathcal{W}} z\,dV = \int_{x=0}^{1} (4 - 5x + 2x^2 - x^3)\,dx = 4 - \frac{5}{2} + \frac{2}{3} - \frac{1}{4} = \frac{23}{12}$$ ■

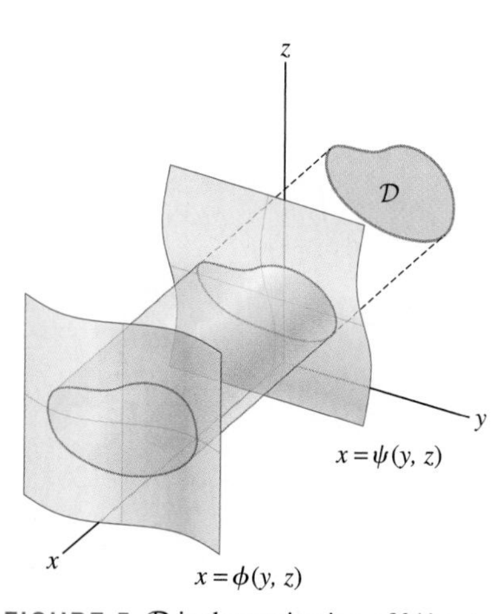

FIGURE 5 $\mathcal{D}$ is the projection of $\mathcal{W}$ onto the yz-plane.

In the examples so far, we have evaluated triple integrals by projecting the region $\mathcal{W}$ onto a domain $\mathcal{D}$ in the xy-plane. We can integrate equally well by projecting onto domains in the xz- or yz-planes. For example, if $\mathcal{W}$ is the region between the graphs of $x = \psi(y, z)$ and $x = \phi(y, z)$, where $\psi(y, z) \le \phi(y, z)$, lying over a domain $\mathcal{D}$ in the yz-plane (Figure 5), then

$$\iiint_{\mathcal{W}} f(x, y, z)\,dV = \iint_{\mathcal{D}} \left(\int_{x=\psi(y,z)}^{x=\phi(y,z)} f(x, y, z)\,dx \right) dA$$

■ **EXAMPLE 4 Writing a Triple Integral in Three Ways** Let $\mathcal{W}$ be the region bounded by (Figure 6)

$$z = 4 - y^2, \qquad y = 2x, \qquad z = 0, \qquad x = 0$$

Express $\iiint_{\mathcal{W}} xyz\,dV$ as an iterated integral in three ways, by projecting onto each of the three coordinate planes (but do not evaluate).

Solution We consider each coordinate plane separately.

***Step 1.* The *xy*-plane.**

The upper face $z = 4 - y^2$ intersects the first quadrant of the xy-plane ($z = 0$) in the line $y = 2$ [Figure 6(A)]. Therefore, the projection of $\mathcal{W}$ onto the xy-plane is a triangle $\mathcal{D}$ defined by $0 \le x \le 1$, $2x \le y \le 2$, and

$$\mathcal{W}: \quad 0 \le x \le 1, \qquad 2x \le y \le 2, \qquad 0 \le z \le 4 - y^2$$

$$\iiint_{\mathcal{W}} xyz\,dV = \int_{x=0}^{1} \int_{y=2x}^{2} \int_{z=0}^{4-y^2} xyz\,dz\,dy\,dx \tag{4}$$

You can check that all three ways of writing the triple integral in Example 4 as an iterated integral yield the same answer:

$$\iiint xyz\,dV = \frac{2}{3}$$

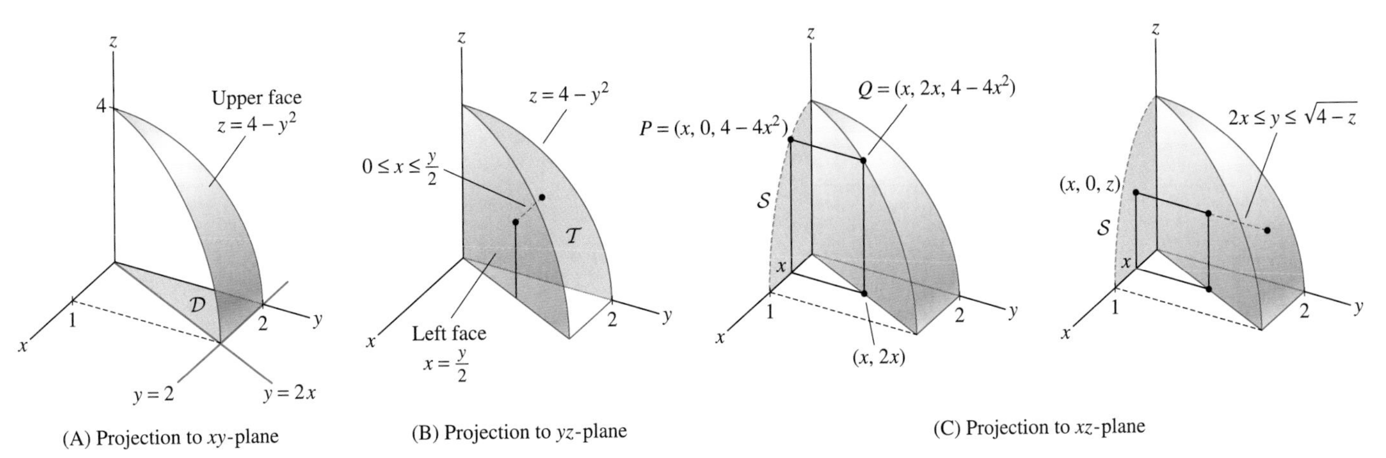

(A) Projection to xy-plane (B) Projection to yz-plane (C) Projection to xz-plane

FIGURE 6 Projections of $\mathcal{W}$ onto the coordinate planes.

***Step 2.* The yz-plane.**

The projection of $\mathcal{W}$ onto the yz-plane is the domain $\mathcal{T}$ [Figure 6(B)]:

$$\mathcal{T}: \quad 0 \le y \le 2, \qquad 0 \le z \le 4 - y^2$$

The region $\mathcal{W}$ consists of all points lying between $\mathcal{T}$ and the "left face" $y = 2x$. In other words, the x-coordinate must satisfy $0 \le x \le \frac{1}{2}y$. Thus,

$$\mathcal{W}: \quad 0 \le y \le 2, \qquad 0 \le z \le 4 - y^2, \qquad 0 \le x \le \frac{1}{2}y$$

$$\iiint_{\mathcal{W}} xyz\, dV = \int_{y=0}^{2} \int_{z=0}^{4-y^2} \int_{x=0}^{y/2} xyz\, dx\, dz\, dy$$

***Step 3.* The xz-plane.**

This is the most difficult case. First, we must describe the projection $\mathcal{S}$ of $\mathcal{W}$ onto the xz-plane. To do this, we need the equation of the dashed curve in the xz-plane in Figure 6(C). We claim that a point P on this dashed curve has coordinates $P = (x, 0, 4 - 4x^2)$. To check this, observe that P is the projection of a point $Q = (x, y, z)$ on the boundary of the left face. Since Q lies on both the plane $y = 2x$ and the surface $z = 4 - y^2$, $Q = (x, 2x, 4 - 4x^2)$ and $P = (x, 0, 4 - 4x^2)$, as claimed. We see that the projection of $\mathcal{W}$ onto the xz-plane is the domain

$$\mathcal{S}: \quad 0 \le x \le 1, \qquad 0 \le z \le 4 - 4x^2$$

We have limits for x and z variables, so the triple integral can be written

$$\iiint_{\mathcal{W}} xyz\, dV = \int_{x=0}^{1} \int_{z=0}^{4-4x^2} \int_{y=??}^{??} xyz\, dy\, dz\, dx$$

What are the limits for y? The equation of the upper face $z = 4 - y^2$ may be written $y = \sqrt{4 - z}$. Thus, $\mathcal{W}$ is bounded by the left face $y = 2x$ and the upper face $y = \sqrt{4 - z}$, and the y-coordinate of a point in $\mathcal{W}$ satisfies

$$2x \le y \le \sqrt{4 - z}$$

Now we can write the triple integral as the following iterated integral:

$$\iiint_{\mathcal{W}} xyz\, dV = \int_{x=0}^{1} \int_{z=0}^{4-4x^2} \int_{y=2x}^{\sqrt{4-z}} xyz\, dy\, dz\, dx$$ ■

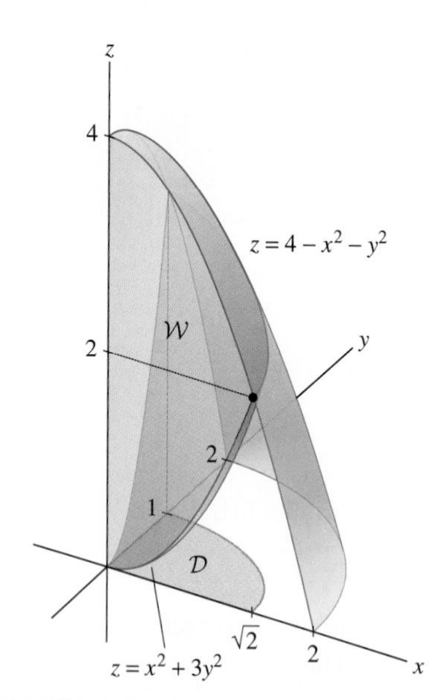

FIGURE 7 Region $x^2 + 3y^2 \le z \le 4 - x^2 - y^2$ lying over the ellipse $x^2 + 2y^2 = 2$ in the octant $x, y, z \ge 0$.

■ **EXAMPLE 5 Integration over the Region Between Intersecting Surfaces** Integrate $f(x, y, z) = x$ over the region $\mathcal{W}$ bounded above by $z = 4 - x^2 - y^2$ and below by $z = x^2 + 3y^2$ in the octant $x, y, z \ge 0$.

Solution First, we must determine the projection of $\mathcal{W}$ onto the xy-plane. The upper and lower surfaces intersect in the curve

$$x^2 + 3y^2 = 4 - x^2 - y^2 \qquad \text{or} \qquad x^2 + 2y^2 = 2$$

Therefore, as we see in Figure 7, $\mathcal{W}$ projects onto the domain $\mathcal{D}$ in the first quadrant bounded by the ellipse $x^2 + 2y^2 = 2$, and we have

$$\mathcal{W} = \left\{(x, y) \in \mathcal{D},\ x^2 + 3y^2 \le z \le 4 - x^2 - y^2, x \ge 0, y \ge 0\right\}$$

$$\iiint_{\mathcal{W}} x\, dV = \iint_{\mathcal{D}} \int_{z=x^2+3y^2}^{4-x^2-y^2} x\, dz\, dA$$

Next, we express the quarter ellipse $\mathcal{D}$ in the first quadrant as a simple domain:

$$0 \le y \le 1, \qquad 0 \le x \le \sqrt{2 - 2y^2}$$

and write the triple integral as an iterated integral:

$$\iiint_{\mathcal{W}} x\, dV = \int_0^1 \int_{x=0}^{\sqrt{2-2y^2}} \int_{z=x^2+3y^2}^{4-x^2-y^2} x\, dz\, dx\, dy$$

Here are the results of evaluating the integrals in order:

Inner integral: $\displaystyle\int_{z=x^2+y^2}^{4-x^2-y^2} x\, dz = xz\Big|_{z=x^2+3y^2}^{4-x^2-y^2} = 4x - 2x^3 - 4y^2x$

Middle integral:

$$\int_{x=0}^{\sqrt{2-2y^2}} (4x - 2x^3 - 4y^2x)\, dx = \left(2x^2 - \frac{1}{2}x^4 - 2x^2y^2\right)\Bigg|_{x=0}^{\sqrt{2-2y^2}}$$
$$= 2 - 4y^2 + 2y^4$$

Triple integral: $\displaystyle\iiint_{\mathcal{W}} x\, dV = \int_0^1 (2 - 4y^2 + 2y^4)\, dy = 2 - \frac{4}{3} + \frac{2}{5} = \frac{16}{15}$ ■

We conclude this section by mentioning some applications of triple integrals. First, we note that the triple integral of the function $f(x, y, z) = 1$ is the volume

$$\text{Volume}(\mathcal{W}) = \iiint_{\mathcal{W}} 1\, dV$$

More generally, if we imagine $\mathcal{W}$ as a solid object with variable but continuous mass density $\rho(x, y, z)$ (in units of mass per unit volume), then the total mass of $\mathcal{W}$ is

$$\text{Total mass of } \mathcal{W} = \iiint_{\mathcal{W}} \rho(x, y, z)\, dV$$

We may justify this formula when $\mathcal{W}$ is a box by noting that the triple integral is a limit of Riemann sums:

$$S_{N,M,L} = \sum_{i=1}^{N}\sum_{j=1}^{M}\sum_{k=1}^{L} \underbrace{\rho(P_{ijk})\Delta V}_{\text{Approximate mass of } \mathcal{B}_{ijk}}$$

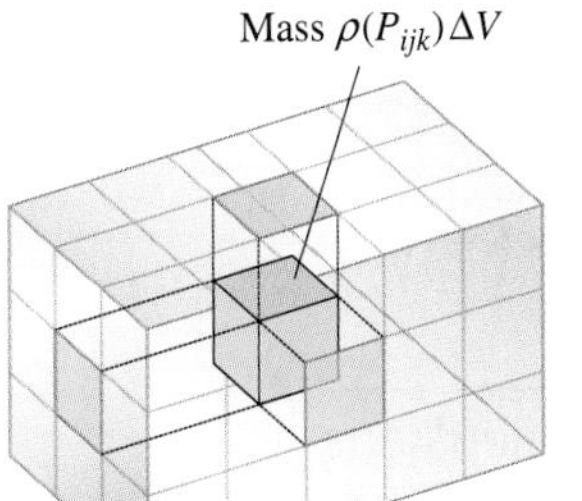

FIGURE 8 The mass of a small cube is approximately $\rho(P)\,\Delta V$.

Here, $\mathcal{W}$ is divided into small boxes $\mathcal{B}_{ijk}$ with sides of lengths Δx, Δy, Δz and volume $\Delta V = \Delta x\,\Delta y\,\Delta z$, and P_{ijk} is a sample point in $\mathcal{B}_{ijk}$. If $\rho(x, y, z)$ is continuous, then the density is nearly constant on each small box, so the mass of $\mathcal{B}_{ijk}$ is approximately $\rho(P_{ijk})\,\Delta V$ (Figure 8). In the limit, we obtain the exact mass of $\mathcal{W}$. This argument applies in a straightforward way to more general regions.

The **average value** of $f(x, y, z)$ over a region $\mathcal{W}$ with volume V is defined by

$$\overline{f} = \frac{1}{V}\iiint_{\mathcal{W}} f(x, y, z)\, dV = \frac{\iiint_{\mathcal{W}} f(x, y, z)\, dV}{\iiint_{\mathcal{W}} 1\, dV}$$

The **centroid** of a region $\mathcal{W}$ is the point $P = (\overline{x}, \overline{y}, \overline{z})$ whose coordinates are the averages of x, y, and z over $\mathcal{W}$. More generally, if $\mathcal{W}$ represents a solid with mass density $\rho(x, y, z)$, then the **center of mass** of $\mathcal{W}$ is the point $P_{\mathrm{CM}} = (x_{\mathrm{CM}}, y_{\mathrm{CM}}, z_{\mathrm{CM}})$, where x_{CM} is defined as follows (and y_{CM}, z_{CM} are defined similarly):

$$x_{\mathrm{CM}} = \frac{\iiint_{\mathcal{W}} x\rho(x, y, z)\, dV}{\iiint_{\mathcal{W}} \rho(x, y, z)\, dV}$$

■ **EXAMPLE 6 Center of Mass** Find the z-coordinate of the center of mass of the first octant $\mathcal{W}$ of the unit sphere, assuming a mass density of $\rho(x, y, z) = y$ (Figure 9).

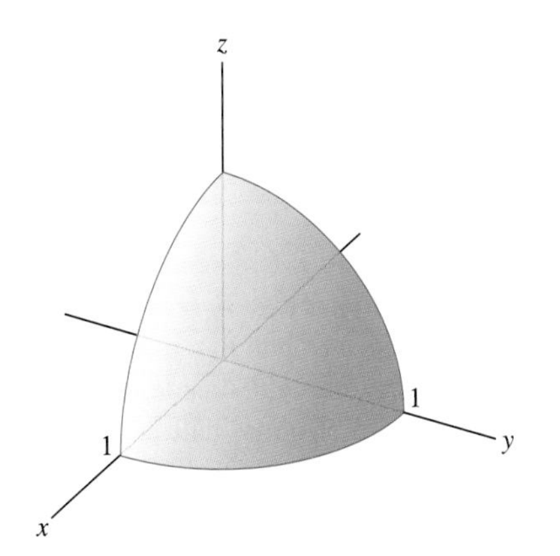

FIGURE 9

Solution The numerator of z_{CM} is

$$\iiint_{\mathcal{W}} z\,\rho(x, y, z)\, dV = \int_{x=0}^{1}\int_{y=0}^{\sqrt{1-x^2}}\int_{z=0}^{\sqrt{1-x^2-y^2}} zy\, dz\, dy\, dx$$

$$= \int_{x=0}^{1}\int_{y=0}^{\sqrt{1-x^2}} \left(\frac{1}{2}yz^2\bigg|_{z=0}^{\sqrt{1-x^2-y^2}}\right) dy\, dx$$

$$= \int_{x=0}^{1}\int_{y=0}^{\sqrt{1-x^2}} \frac{1}{2}y\left(1 - x^2 - y^2\right) dy\, dx$$

$$= \int_{x=0}^{1} \left(\frac{1}{4}y^2 - \frac{1}{4}x^2y^2 - \frac{1}{8}y^4\right)\bigg|_{y=0}^{\sqrt{1-x^2}} dx$$

$$= \int_{x=0}^{1} \left(\frac{1}{8}x^4 - \frac{1}{4}x^2 + \frac{1}{8}\right) dx = \frac{1}{15}$$

The total mass of the $\mathcal{W}$ is equal to the integral of the mass density $\rho(x, y, z)$ over $\mathcal{W}$. In the last step of the following computation, we use the formula stated in the margin:

The following formula may be derived using trigonometric substitution $x = \sin\theta$:

$$\int (1-x^2)^{3/2}\,dx = \frac{1}{8}\left(x(5-x^2)\sqrt{1-x^2} + 3\sin^{-1}x\right) + C$$

$$m = \iiint_{\mathcal{W}} \rho(x,y,z)\,dV = \int_{x=0}^{1}\int_{y=0}^{\sqrt{1-x^2}}\int_{z=0}^{\sqrt{1-x^2-y^2}} y\,dz\,dy\,dx$$

$$= \int_{x=0}^{1}\int_{y=0}^{\sqrt{1-x^2}} y\sqrt{1-x^2-y^2}\,dy\,dx$$

$$= \int_{x=0}^{1}\left(-\frac{1}{3}(1-x^2-y^2)^{3/2}\right)\Bigg|_{y=0}^{\sqrt{1-x^2}} dx$$

$$= \frac{1}{3}\int_{x=0}^{1}(1-x^2)^{3/2}\,dx = \frac{\pi}{16}$$

We conclude that

$$z_{\text{CM}} = \frac{1}{m}\iiint_{\mathcal{W}} z\,\rho(x,y,z)\,dV = \frac{16}{\pi}\left(\frac{1}{15}\right) = \frac{16}{15\pi}$$ ■

Excursion: Volume of the Sphere in Higher Dimensions

Archimedes (287–212 BCE) discovered the beautiful formula $V = \frac{4}{3}\pi r^3$ for the volume of a sphere nearly 2,000 years before calculus was invented. Archimedes gave a brilliant geometric argument showing that the volume of a sphere is equal to two-thirds the volume of the circumscribed cylinder. He valued this achievement so highly that he requested that a sphere with circumscribed cylinder be engraved on his tomb. We can use integration to generalize Archimedes's formula to n dimensions. The ball of radius r in $\mathbf{R}^n$, denoted $B_n(r)$, is the set of points $(x_1, \dots, x_n)$ in $\mathbf{R}^n$ such that

$$x_1^2 + x_2^2 + \cdots + x_n^2 \le r^2$$

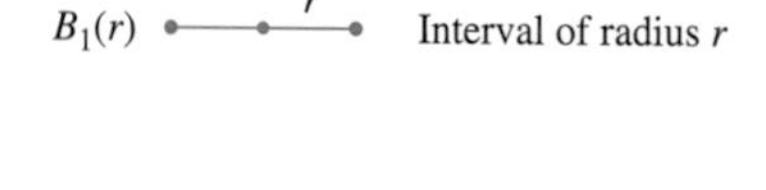

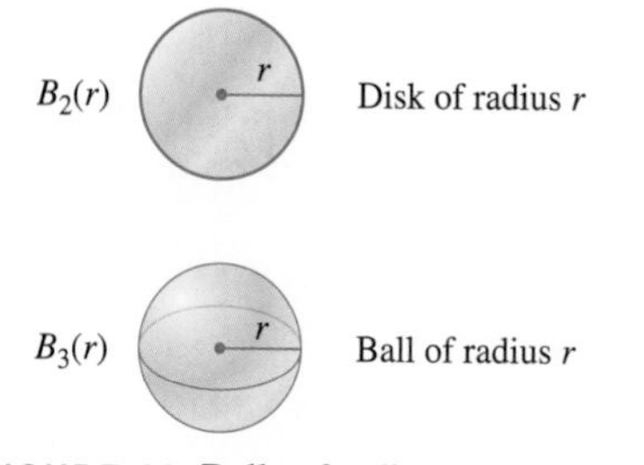

FIGURE 10 Balls of radius r in dimensions $n = 1, 2, 3$.

The balls $B_n(r)$ in dimensions 1, 2, and 3 are the interval, disk, and ball shown in Figure 10. In dimensions $n \ge 4$, the ball $B_n(r)$ is hard, if not impossible, to visualize.

Let $V_n(r)$ be the volume (also called the **hypervolume**) of $B_n(r)$. We know that

$$V_1(r) = 2r, \qquad V_2(r) = \pi r^2, \qquad V_3(r) = \frac{4}{3}\pi r^3$$

It is reasonable to assume that $V_n(r)$ is proportional to r^n for all n (this can be proved using the Change of Variables Formula for multiple integrals). Thus $V_n(r) = A_n r^n$, where A_n is a constant. In fact, $A_n = V_n(1)$ is the volume of the unit ball in n dimensions.

To compute A_n, we shall integrate cross-sectional hypervolume. Consider the case $n = 3$, where the horizontal slice at height $z = c$ is a two-dimensional ball (a disk) of radius $\sqrt{r^2 - c^2}$ (Figure 11). The volume $V_3(r)$ is equal to the integral of these horizontal slices:

$$V_3(r) = \int_{z=-r}^{r} V_2\left(\sqrt{r^2 - z^2}\right) dz = \int_{z=-r}^{r} \pi(r^2 - z^2)\,dz = \frac{4}{3}\pi r^3$$

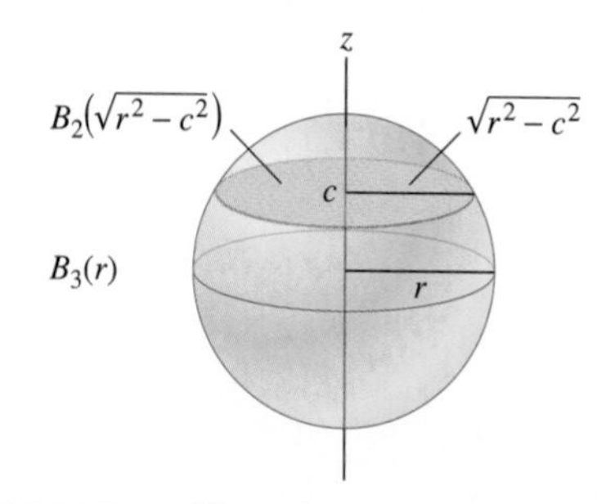

FIGURE 11 The volume $V_3(r)$ is the integral of cross-sectional area $V_2(\sqrt{r^2 - c^2})$.

The same approach works in higher dimensions. The slice of $B_n(r)$ at height $x_n = c$ has equation

$$x_1^2 + x_2^2 + \cdots + x_{n-1}^2 + c^2 = r^2$$

This slice is the ball $B_{n-1}\left(\sqrt{r^2 - c^2}\right)$ of radius $\sqrt{r^2 - c^2}$, and $V_n(r)$ is obtained by integrating the hypervolume of these slices:

$$V_n(r) = \int_{x_n=-r}^{r} V_{n-1}\left(\sqrt{r^2 - x_n^2}\right) dx_n = A_{n-1} \int_{x_n=-r}^{r} \left(\sqrt{r^2 - x_n^2}\right)^{n-1} dx_n$$

Using the substitution $x_n = r\sin\theta$ and $dx_n = r\cos\theta\, d\theta$, we have

$$V_n(r) = A_{n-1} r^n \int_{-\pi/2}^{\pi/2} \cos^n\theta\, d\theta = A_{n-1} C_n r^n$$

where $C_n = \int_{\theta=-\pi/2}^{\pi/2} \cos^n\theta\, d\theta$. Setting $r = 1$, we obtain [since $V_n(1) = A_n$]

$$\boxed{A_n = A_{n-1} C_n} \qquad (5)$$

In Exercise 41, you are asked to use Integration by Parts to verify the relation

$$C_n = \left(\frac{n-1}{n}\right) C_{n-2} \qquad (6)$$

It is easy to check directly that $C_0 = \pi$ and $C_1 = 2$. By Eq. (6), $C_2 = \frac{1}{2}C_0 = \frac{\pi}{2}$, $C_3 = \frac{2}{3}(2) = \frac{4}{3}$, and so on. Here are the first few values of C_n:

n	0	1	2	3	4	5	6	7
C_n	π	2	$\frac{\pi}{2}$	$\frac{4}{3}$	$\frac{3\pi}{8}$	$\frac{16}{15}$	$\frac{5\pi}{16}$	$\frac{32}{35}$

TABLE 1

n	A_n
1	2
2	$\pi \approx 3.14$
3	$\frac{4}{3}\pi \approx 4.19$
4	$\frac{\pi^2}{2} \approx 4.93$
5	$\mathbf{\frac{8\pi^2}{15} \approx 5.26}$
6	$\frac{\pi^3}{6} \approx 5.17$
7	$\frac{16\pi^3}{105} \approx 4.72$

We also have $A_1 = 2$ and $A_2 = \pi$, so we can use the values of C_n together with Eq. (5) to obtain the values of A_n in Table 1. Using induction and formulas (5) and (6), one can verify the following general formulas:

$$\boxed{A_{2m} = \frac{\pi^m}{m!}, \qquad A_{2m+1} = \frac{2^{m+1}\pi^m}{1 \cdot 3 \cdot 5 \cdot \cdots \cdot (2m+1)}}$$

This sequence of numbers A_n has a curious property. Recall that A_n is the volume of the unit ball in n dimensions. From Table 1, it appears that the hypervolumes increase up to dimension 5 and then begin to decrease. In Exercise 42, you are asked to verify that the five-dimensional ball has the largest volume. Furthermore, the volumes A_n tend to 0 as $n \to \infty$.

15.3 SUMMARY

- The triple integral over a box $\mathcal{B} = [a, b] \times [c, d] \times [p, q]$ is equal to the iterated integral

$$\iiint_{\mathcal{B}} f(x, y, z)\, dV = \int_{x=a}^{b} \int_{y=c}^{d} \int_{z=p}^{q} f(x, y, z)\, dz\, dy\, dx$$

The iterated integral may be written in any one of six possible orders, for example,

$$\int_{z=p}^{q} \int_{y=c}^{d} \int_{x=a}^{b} f(x, y, z)\, dx\, dy\, dz$$

- A *simple region* $\mathcal{W}$ is a region consisting of the points (x, y, z) between two surfaces $z = \psi(x, y)$ and $z = \phi(x, y)$, where $\psi(x, y) \le \phi(x, y)$, lying over a domain $\mathcal{D}$ in the xy-plane. In other words, $\mathcal{W}$ is defined by

$$(x, y) \in \mathcal{D}, \qquad \psi(x, y) \le z \le \phi(x, y)$$

The triple integral over $\mathcal{W}$ is equal to an iterated integral:

$$\iiint_{\mathcal{W}} f(x, y, z)\, dV = \iint_{\mathcal{D}} \left(\int_{z=\psi(x,y)}^{\phi(x,y)} f(x, y, z)\, dz \right) dA$$

- The *average value* of $f(x, y, z)$ on a region $\mathcal{W}$ of volume V is the quantity

$$\overline{f} = \frac{1}{V} \iiint_{\mathcal{W}} f(x, y, z)\, dV, \qquad V = \iiint_{\mathcal{W}} 1\, dV$$

- If a solid $\mathcal{W}$ has continuous mass density $\rho(x, y, z)$, then the *center of mass* of $\mathcal{W}$ is $P_{\rm CM} = (x_{\rm CM}, y_{\rm CM}, z_{\rm CM})$, where $x_{\rm CM}$ is ($y_{\rm CM}, z_{\rm CM}$ are defined similarly):

$$x_{\rm CM} = \frac{\iiint_{\mathcal{W}} x\rho(x, y, z)\, dV}{\iiint_{\mathcal{W}} \rho(x, y, z)\, dV}$$

The *centroid* of $\mathcal{W}$ is the center of mass with $\rho = 1$.

15.3 EXERCISES

Preliminary Questions

1. Which of (a)–(c) is not equal to $\int_0^1 \int_3^4 \int_6^7 f(x, y, z)\, dz\, dy\, dx$?

(a) $\int_6^7 \int_0^1 \int_3^4 f(x, y, z)\, dy\, dx\, dz$

(b) $\int_3^4 \int_0^1 \int_6^7 f(x, y, z)\, dz\, dx\, dy$

(c) $\int_3^4 \int_3^4 \int_6^7 f(x, y, z)\, dx\, dz\, dy$

2. Which of the following does not represent a meaningful triple integral?

(a) $\int_0^1 \int_0^x \int_{x+y}^{2x+y} e^{x+y+z}\, dz\, dy\, dx$

(b) $\int_0^1 \int_0^z \int_{x+y}^{2x+y} e^{x+y+z}\, dz\, dy\, dx$

3. Describe the projection of the region of integration $\mathcal{W}$ onto the xy-plane:

(a) $\int_0^1 \int_0^x \int_0^{x^2+y^2} f(x, y, z)\, dz\, dy\, dx$

(b) $\int_0^1 \int_0^{\sqrt{1-x^2}} \int_2^4 f(x, y, z)\, dz\, dy\, dx$

Exercises

In Exercises 1–10, evaluate $\iiint_{\mathcal{B}} f(x, y, z)\, dV$ *for the specified function f and box* $\mathcal{B}$.

1. $f(x, y, z) = z^4$; $2 \le x \le 8, 0 \le y \le 5, 0 \le z \le 1$

2. $f(x, y, z) = xyz$; $0 \le x, y, z \le 1$

3. $f(x, y, z) = xz^2$; $[0, 2] \times [1, 6] \times [3, 4]$

4. $f(x, y, z) = (x - y)(y - z)$; $[0, 1] \times [0, 3] \times [0, 3]$

5. $f(x, y, z) = xe^{y-2z}$; $0 \le x \le 2, 0 \le y, z \le 1$

6. $f(x, y, z) = \dfrac{x}{(y+z)^2}$; $[0, 2] \times [2, 4] \times [-1, 1]$

7. $f(x, y, z) = \dfrac{z}{x}$; $1 \le x \le 3, 0 \le y \le 2, 0 \le z \le 4$

8. $f(x, y, z) = xy^2e^{xyz}$; $0 \le x, z \le 1, 0 \le y \le 2$

9. $f(x, y, z) = (x + z)^3$; $[0, a] \times [0, b] \times [0, c]$

10. $f(x, y, z) = (2x + 3y - z)^4$; $[0, a] \times [0, b] \times [0, c]$

In Exercises 11–14, evaluate $\iiint_{\mathcal{W}} f(x, y, z)\, dV$ *for the function f and region* $\mathcal{W}$ *specified.*

11. $f(x, y, z) = x + y$; $\mathcal{W}: y \le z \le x, 0 \le y \le x, 0 \le x \le 1$

12. $f(x, y, z) = e^{x+y+z}$; $\mathcal{W}: 0 \le z \le 1, 0 \le y \le x, 0 \le x \le 1$

13. $f(x, y, z) = xyz$; $\mathcal{W}: 0 \le z \le 1, 0 \le y \le \sqrt{1 - x^2}, 0 \le x \le 1$

14. $f(x, y, z) = e^{2x-z}$; $\mathcal{W}: x + y + z \le 1, x, y, z \ge 0$

15. Calculate the integral of $f(x, y, z) = z$ over the region $\mathcal{W}$ below the upper hemisphere of the sphere $x^2 + y^2 + z^2 = 9$ lying over the unit square $0 \le x, y \le 1$ in Figure 12.

16. Integrate $f(x, y, z) = z$ over the region $\mathcal{W}$ below the upper hemisphere of radius 3 as in Figure 12, but lying over the triangle in the xy-plane bounded by the lines $x = 1$, $y = 0$, and $x = y$.

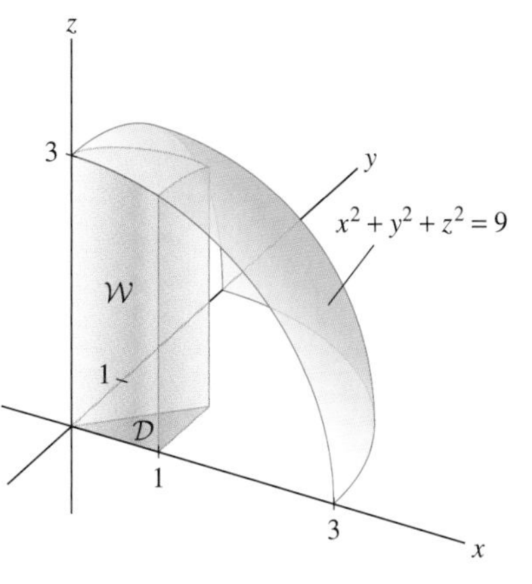

FIGURE 12

17. Calculate the integral of $f(x, y, z) = e^{x+y+z}$ over the tetrahedron $\mathcal{W}$ in Figure 13.

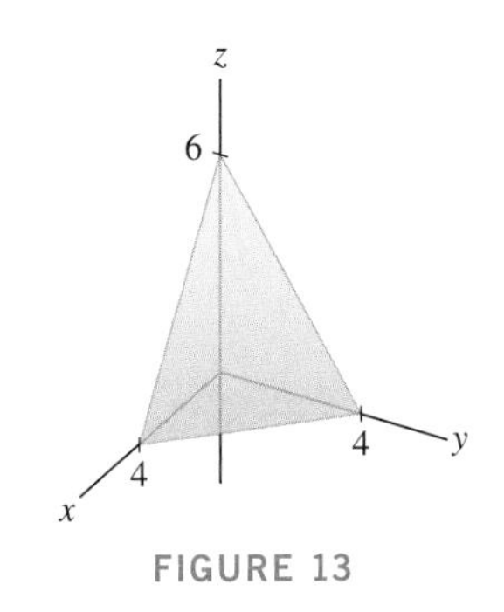

FIGURE 13

18. Describe the domain of integration and evaluate:

$$\int_0^9 \int_0^{\sqrt{9-x^2}} \int_0^{\sqrt{9-x^2-y^2}} xy \, dz \, dy \, dx$$

19. Integrate $f(x, y, z) = xz$ over the region in the first octant $(x, y, z \ge 0)$ above the parabolic cylinder $z = y^2$ and below the paraboloid $z = 8 - 2x^2 - y^2$.

20. Compute the integral of $f(x, y, z) = (y - x)z$ over the region within the cylinder $x^2 + y^2 = 16$ between the planes $z = 0$ and $z = y$.

21. Find the triple integral of the function z over the ramp in Figure 14. Here, z is the height above the ground.

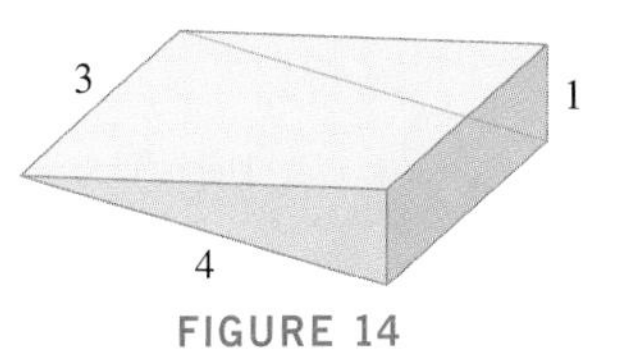

FIGURE 14

22. Let W be the region in the first octant ($x \ge 0, y \ge 0, z \ge 0$) of $\mathbf{R}^3$ bounded by $x + y + z = 1$ and $x^2 + y^2 = z - 2$. Express the triple integral

$$\iiint_{\mathcal{W}} f(x, y, z) \, dV$$

as an iterated integral.

23. Find the volume of the solid in $\mathbf{R}^3$ bounded by $y = x^2$, $x = y^2$, $z = x + y + 5$, and $z = 0$.

24. Find the volume of the solid in the octant $x, y, z \ge 0$ bounded by $x + y + z = 1$ and $x + y + 2z = 1$.

25. Calculate $\iiint_{\mathcal{W}} y \, dV$, where $\mathcal{W}$ is the region above $z = x^2 + y^2$ and below $z = 5$, and bounded by $y = 0$ and $y = 1$.

26. Evaluate $\iiint_{\mathcal{W}} y \, dV$, where W is the domain bounded by the elliptic cylinder $\dfrac{x^2}{4} + \dfrac{y^2}{9} = 1$, the sphere $x^2 + y^2 + z^2 = 16$ in the first octant $x \ge 0$, $y \ge 0$, $z \ge 0$ (Figure 15).

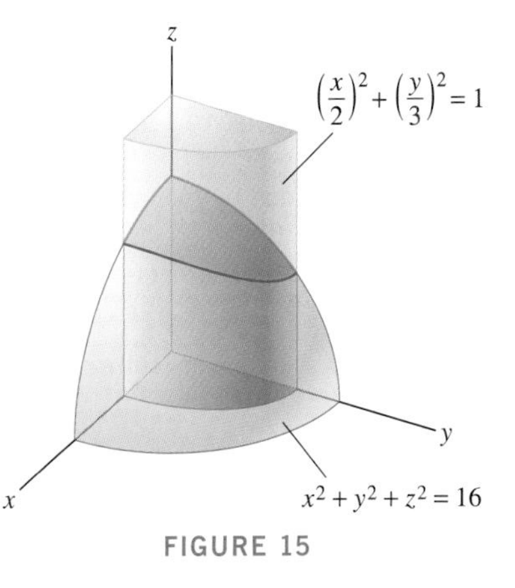

FIGURE 15

27. The solution to Example 4 expresses a triple integral in the three orders $dz\,dy\,dx$, $dx\,dz\,dy$, and $dy\,dz\,dx$. Write the integral in the three other orders: $dz\,dx\,dy$, $dx\,dy\,dz$, and $dy\,dx\,dz$.

28. Let $\mathcal{W}$ be the region bounded by $y + z = 2$, $2x = y$, $x = 0$, and $z = 0$ (Figure 16). Express and evaluate the triple integral of xe^z by projecting $\mathcal{W}$ onto the:

(a) xy-plane **(b)** yz-plane **(c)** xz-plane

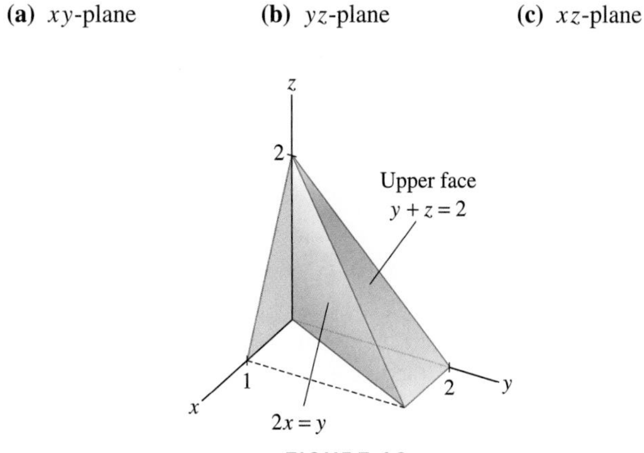

FIGURE 16

29. Express the triple integral $\iiint_{\mathcal{W}} f(x, y, z)\, dV$ as an iterated integral in the two orders $dz\, dy\, dx$ and $dx\, dy\, dz$, where (Figure 17)

$$\mathcal{W} = \left\{(x, y, z) : \sqrt{x^2 + y^2} \le z \le 1\right\}$$

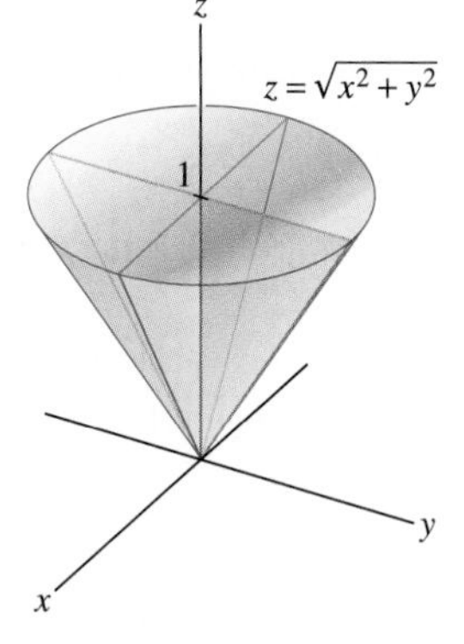

FIGURE 17

30. Let $\mathcal{W}$ be the region bounded by the cylinders $z = 1 - y^2$ and $y = x^2$, and the planes $y = 0$ and $y = 1$. Calculate the volume of $\mathcal{W}$ as a triple integral in the three orders $dz\, dy\, dx$, $dx\, dz\, dy$, and $dy\, dz\, dx$.

31. Find the average of $f(x, y, z) = xy \sin(\pi z)$ over the cube $0 \le x, y, z \le 1$.

32. Let $\mathcal{W}$ be the region bounded by the planes $y = z$, $2y + z = 3$, and $z = 0$ for $0 \le x \le 4$. Express the triple integral $\iiint_{\mathcal{W}} f(x, y, z)\, dV$ as an iterated integral in the order $dx\, dz\, dy$ (project $\mathcal{W}$ onto the yz-plane). Then find the centroid of $\mathcal{W}$.

33. Find the average of $f(x, y, z) = x^2 + y^2 + z^2$ over the region bounded by the planes $2y + z = 1$, $x = 0$, $x = 1$, $z = 0$, and $y = 0$.

34. Find the total mass of the part of the solid cylinder $x^2 + y^2 \le 4$ such that $x^2 \le z \le 9 - x^2$, assuming that the mass density is $\rho(x, y, z) = |y|$.

35. Calculate the centroid of the tetrahedron in Figure 18.

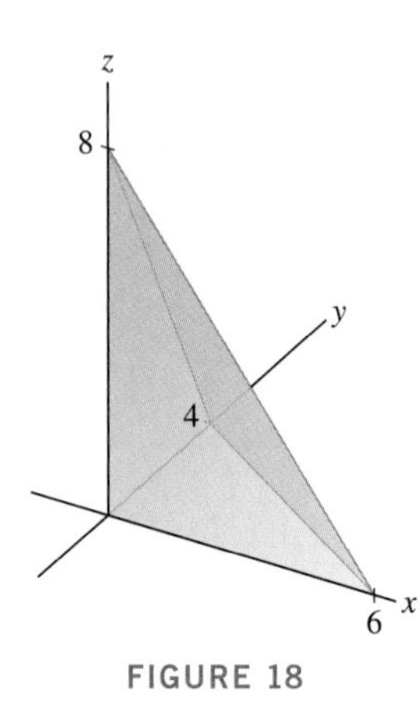

FIGURE 18

36. Find the centroid of the region described in Exercise 30.

37. Find the center mass of the solid bounded by planes $x + y + z = 1$, $x = 0$, $y = 0$, and $z = 0$, assuming a mass density of $\rho(x, y, z) = \sqrt{z}$.

38. Find the center mass of the cylinder $x^2 + y^2 = 1$ for $0 \le z \le 1$, assuming a mass density of $\rho(x, y, z) = x^2 + y^2$.

In Exercises 39–40, let $I = \int_0^1 \int_0^1 \int_0^1 f(x, y, z)\, dV$ *and let* $S_{N,N,N}$ *be the Riemann sum approximation*

$$S_{N,N,N} = \frac{1}{N^3} \sum_{i=1}^{N} \sum_{j=1}^{N} \sum_{k=1}^{N} f\left(\frac{i}{N}, \frac{j}{N}, \frac{k}{N}\right)$$

39. CAS Calculate $S_{N,N,N}$ for $f(x, y, z) = e^{x^2 - y - z}$ for $N = 10$, 20, 30. Then evaluate I and find an N such that $S_{N,N,N}$ approximates I to two decimal places.

40. CAS Calculate $S_{N,N,N}$ for $f(x, y, z) = \sin(xyz)$ for $N = 10$, 20, 30. Then use a computer algebra system to calculate I numerically and estimate the error $|I - S_{N,N,N}|$.

Further Insights and Challenges

41. Use Integration by Parts to verify Eq. (6).

42. Compute the volume A_n of the unit ball in $\mathbf{R}^n$ for $n = 8, 9, 10$. Show that $C_n \le 1$ for $n \ge 6$ and use this to prove that of all unit balls, the five-dimensional ball has the largest volume. Can you explain why A_n tends to 0 as $n \to \infty$?

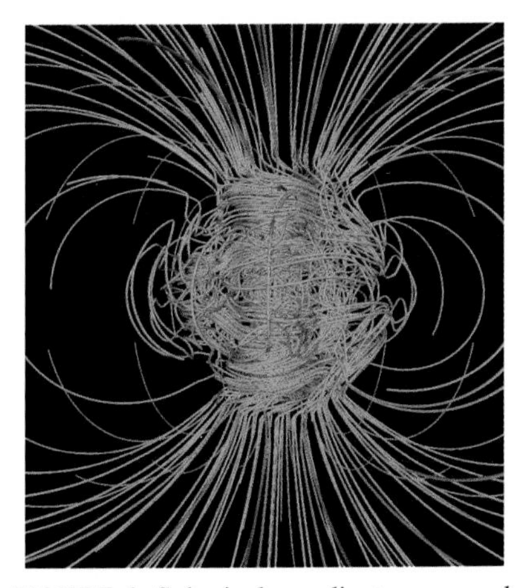

FIGURE 1 Spherical coordinates are used in mathematical models of the earth's magnetic field. This computer simulation, based on the Glatzmaier–Roberts model, shows the magnetic lines of force, representing inward and outward directed field lines in blue and yellow, respectively.

15.4 Integration in Polar, Cylindrical, and Spherical Coordinates

We have seen many examples in single-variable calculus where a well-chosen change of variables transforms a complicated integral into a simpler one. Change of variables is also useful in multivariable calculus, but with a slight change of emphasis. In the single-variable case, the goal is to find a change of variables that simplifies the integrand, whereas in the multivariable case, we are usually interested in simplifying the domain of integration. In this section, we treat the change of variables for polar, cylindrical, and spherical coordinates. These are the three most important cases. They are especially useful when the domain has radial or spherical symmetry (for example, a disk, ball, or cylinder). The general Change of Variables Formula is discussed in the next section.

Double Integrals in Polar Coordinates

Polar coordinates are convenient when the domain of integration is an angular sector or a **polar rectangle** (Figure 2):

$$\text{Polar rectangle } \mathcal{R}: \quad \theta_1 \le \theta \le \theta_2, \qquad r_1 \le r \le r_2 \tag{1}$$

We assume throughout that $r_1 \ge 0$ and that all radial coordinates are nonnegative. Recall that rectangular and polar coordinates are related by

$$x = r\cos\theta, \qquad y = r\sin\theta$$

Thus, we write a function $f(x, y)$ in polar coordinates as $f(r\cos\theta, r\sin\theta)$. We will prove the following Change of Variables Formula for the polar rectangle $\mathcal{R}$:

Equation (2) expresses the integral of $f(x, y)$ over a sector as the integral of a new function $f(r\cos\theta, \sin\theta)$ over the rectangle $[\theta_1, \theta_2] \times [r_1, r_2]$. In this sense, the change of variables "simplifies" the domain of integration.

$$\iint_{\mathcal{R}} f(x, y)\, dA = \int_{\theta_1}^{\theta_2} \int_{r_1}^{r_2} f(r\cos\theta, r\sin\theta)\, r\, dr\, d\theta \tag{2}$$

Notice the extra factor r in the integrand on the right.

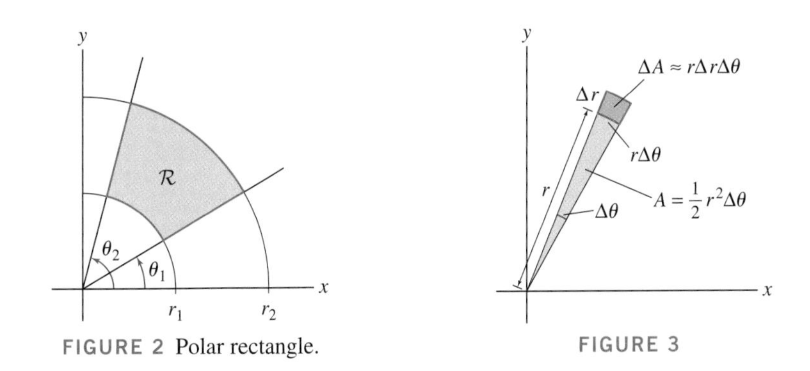

FIGURE 2 Polar rectangle.

FIGURE 3

To verify Eq. (2), we first compute the area ΔA of the small polar rectangle as in Figure 3 (recall that the area of a sector is $\frac{1}{2}r^2\Delta\theta$):

$$\Delta A = \frac{1}{2}(r + \Delta r)^2\,\Delta\theta - \frac{1}{2}r^2\,\Delta\theta = r(\Delta r\,\Delta\theta) + \frac{1}{2}(\Delta r)^2\Delta\theta \approx r(\Delta r\,\Delta\theta)$$

Next, we decompose $\mathcal{R}$ into a grid of $N \times M$ small polar subrectangles $\mathcal{R}_{ij}$, each of angular width $\Delta\theta = (\theta_2 - \theta_1)/N$ and radial width $\Delta r = (r_2 - r_1)/M$. If $\mathcal{R}_{ij}$ is small and $f(x, y)$ is continuous, then

$$\iint_{\mathcal{R}_{ij}} f(x, y)\,dx\,dy \approx f(P_{ij})\,\text{Area}(\mathcal{R}_{ij}) \approx f(P_{ij})\,r_{ij}\,\Delta r\,\Delta\theta \qquad \boxed{3}$$

REMINDER According to (11) in Section 15.2, if f is continuous and $\mathcal{D}$ is a small domain,

$$\iint_{\mathcal{D}} f(x, y)\,dA \approx f(P)\,\text{Area}(\mathcal{D})$$

where P is any sample point in $\mathcal{D}$.

where $P_{ij} = (r_{ij}, \theta_{ij})$ is a sample point in $\mathcal{R}_{ij}$ (Figure 4). [In fact, we may choose P_{ij} so that (3) is an equality by the Mean Value Theorem for Double Integrals.] The integral over $\mathcal{R}$ is the sum of integrals (3); thus,

$$\iint_{\mathcal{R}} f(x, y)\,dx\,dy = \sum_{i=1}^{N}\sum_{j=1}^{M} \iint_{\mathcal{R}_{ij}} f(x, y)\,dx\,dy$$

$$\approx \sum_{i=1}^{N}\sum_{j=1}^{M} f(P_{ij})\,\text{Area}(\mathcal{R}_{ij})$$

$$\approx \sum_{i=1}^{N}\sum_{j=1}^{M} f(r_{ij}\cos\theta_{ij}, r_{ij}\sin\theta_{ij})\,r_{ij}\,\Delta r\,\Delta\theta \qquad \boxed{4}$$

Observe that (4) is a Riemann sum for the integral of $rf(r\sin\theta, r\cos\theta)$ over the region $r_1 \le r \le r_2$, $\theta_1 \le \theta \le \theta_2$. As $N, M \to \infty$, this Riemann sum approaches the integral and we obtain Eq. (2).

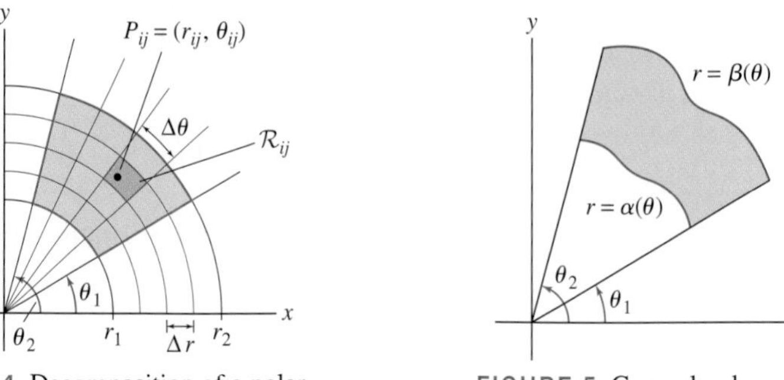

FIGURE 4 Decomposition of a polar rectangle into subrectangles.

FIGURE 5 General polar region.

A similar derivation is valid for more general polar regions that may be described as the region between two polar curves $r = \alpha(\theta)$ and $r = \beta(\theta)$ (Figure 5).

Equation (5) is summarized in the symbolic expression for the "area element" dA in polar coordinates:

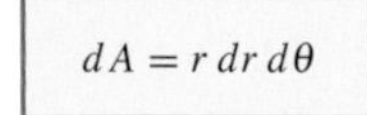

THEOREM 1 Double Integral in Polar Coordinates Let $\mathcal{D}$ be a domain of the form $\theta_1 \le \theta \le \theta_2$, $\alpha(\theta) \le r \le \beta(\theta)$, and assume that $f(x, y)$ is continuous on $\mathcal{D}$. Then

$$\iint_{\mathcal{D}} f(x, y)\,dA = \int_{\theta_1}^{\theta_2}\int_{r=\alpha(\theta)}^{\beta(\theta)} f(r\cos\theta, r\sin\theta)\,r\,dr\,d\theta \qquad \boxed{5}$$

■ **EXAMPLE 1** Use polar coordinates to compute $\iint_{\mathcal{D}} (x + y)\,dA$, where $\mathcal{D}$ is the quarter annulus $4 \le x^2 + y^2 \le 16$, $x, y \ge 0$.

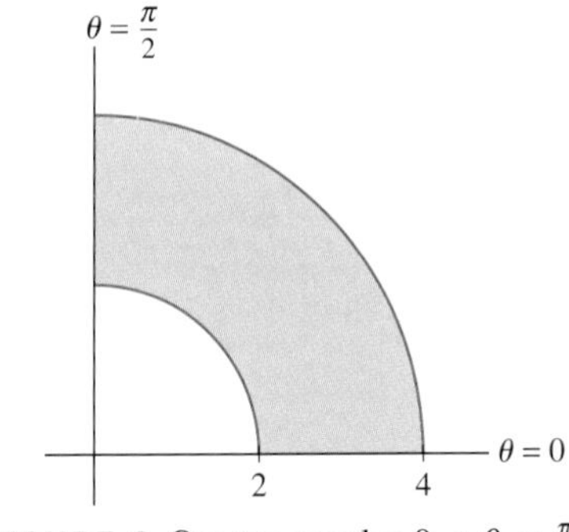

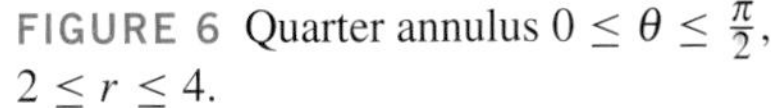
FIGURE 6 Quarter annulus $0 \le \theta \le \frac{\pi}{2}$, $2 \le r \le 4$.

Solution

***Step 1.* Describe $\mathcal{D}$ and f in polar coordinates.**
The quarter annulus $\mathcal{D}$ is defined by the inequalities (Figure 6)

$$0 \le \theta \le \frac{\pi}{2}, \qquad 2 \le r \le 4$$

$$f(x, y) = x + y = r\cos\theta + r\sin\theta = r(\cos\theta + \sin\theta)$$

***Step 2.* Change variables in the integral and evaluate.**
In converting to polar coordinates, we replace dA by $r\,dr\,d\theta$:

$$\iint_{\mathcal{D}} (x + y)\,dA = \int_0^{\pi/2} \int_2^4 r(\cos\theta + \sin\theta)\,r\,dr\,d\theta$$

The inner integral is

$$\int_2^4 (\cos\theta + \sin\theta)\,r^2\,dr = (\cos\theta + \sin\theta)\left(\frac{4^3}{3} - \frac{2^3}{3}\right) = \frac{56}{3}(\cos\theta + \sin\theta)$$

and

$$\iint_{\mathcal{D}} (x + y)\,dA = \frac{56}{3}\int_0^{\pi/2} (\cos\theta + \sin\theta)\,d\theta = \frac{56}{3}(\sin\theta - \cos\theta)\bigg|_0^{\pi/2} = \frac{112}{3} \quad ■$$

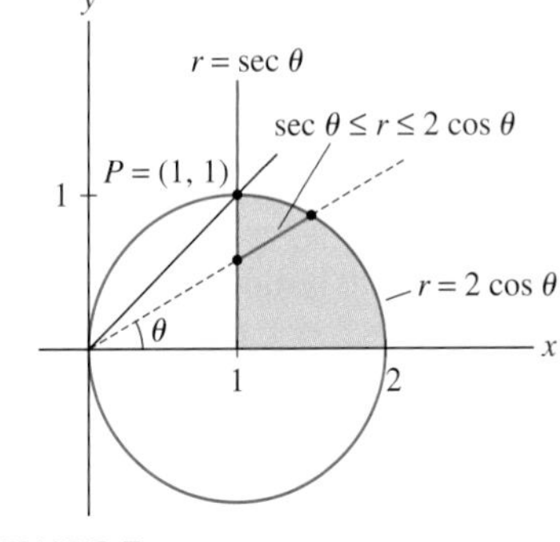

FIGURE 7

■ **EXAMPLE 2** Calculate $\iint_{\mathcal{D}} (x^2 + y^2)^{-2}\,dA$ for the shaded domain $\mathcal{D}$ in Figure 7.

Solution

***Step 1.* Describe $\mathcal{D}$ and f in polar coordinates.**
Recall from Section 11.3 (Example 7) that:

- The vertical line $x = 1$ has polar equation $r\cos\theta = 1$ or $r = \sec\theta$.
- The circle of radius 1 and center $(1, 0)$ has polar equation $r = 2\cos\theta$.

The quarter circle lies in the angular sector $0 \le \theta \le \frac{\pi}{4}$ [note that the point $P = (1, 1)$ has polar angle $\frac{\pi}{4}$]. A ray of angle θ intersects $\mathcal{D}$ for r between $\sec\theta$ and $2\cos\theta$. Therefore,

$$\mathcal{D}: \quad 0 \le \theta \le \frac{\pi}{4}, \qquad \sec\theta \le r \le 2\cos\theta$$

On the other hand, $f(x, y) = (x^2 + y^2)^{-2} = (r^2)^{-2} = r^{-4}$.

***Step 2.* Change variables in the integral and evaluate.**
By the Change of Variables Formula,

$$\iint_{\mathcal{D}} (x^2 + y^2)^{-2}\,dA = \int_0^{\pi/4} \int_{\sec\theta}^{2\cos\theta} r^{-4}\,r\,dr\,d\theta = \int_0^{\pi/4} \int_{\sec\theta}^{2\cos\theta} r^{-3}\,dr\,d\theta$$

The inner integral is

$$\int_{\sec\theta}^{2\cos\theta} r^{-3}\,dr = -\frac{1}{2}r^{-2}\bigg|_{\sec\theta}^{2\cos\theta} = -\frac{1}{8}\sec^2\theta + \frac{1}{2}\cos^2\theta$$

← REMINDER

$$\cos^2\theta = \frac{1}{2}(1 + \cos 2\theta)$$

$$\int \cos^2\theta\,d\theta = \frac{1}{2}\left(\theta + \frac{1}{2}\sin 2\theta\right) + C$$

$$\int \sec^2\theta\,d\theta = \tan\theta + C$$

Therefore,

$$\iint_{\mathcal{D}} (x^2 + y^2)^{-2}\, dA = \int_0^{\pi/4} \left(\frac{1}{2}\cos^2\theta - \frac{1}{8}\sec^2\theta \right) d\theta$$

$$= \left(\frac{1}{4}\left(\theta + \frac{1}{2}\sin 2\theta\right) - \frac{1}{8}\tan\theta \right)\Bigg|_0^{\pi/4}$$

$$= \frac{1}{4}\left(\frac{\pi}{4} + \frac{1}{2}\sin\frac{\pi}{2} \right) - \frac{1}{8}\tan\frac{\pi}{4} = \frac{\pi}{16}$$

■

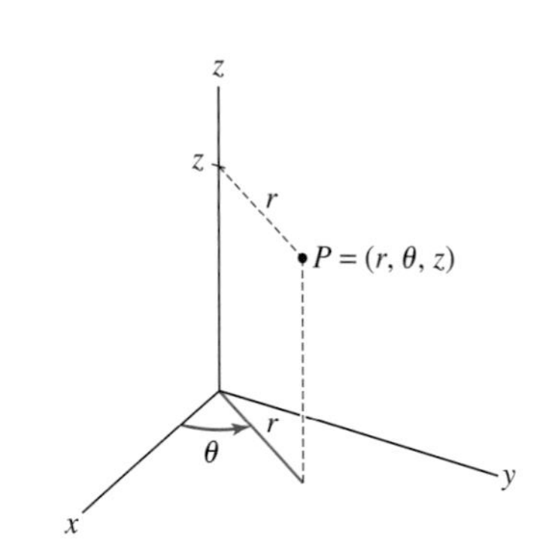

FIGURE 8 Cylindrical coordinates.

Triple Integrals in Cylindrical Coordinates

Cylindrical coordinates, introduced in Section 12.7, are useful when the domain has **axial symmetry**, that is, symmetry with respect to an axis. In cylindrical coordinates (r, θ, z), the axis of symmetry is the z-axis. Recall the relations (Figure 8)

$$x = r\cos\theta, \quad y = r\sin\theta, \quad z = z$$

To set up a triple integral in cylindrical coordinates, we assume that the domain of integration $\mathcal{W}$ can be described as the region between two surfaces (Figure 9)

$$z_1(r, \theta) \le z \le z_2(r, \theta)$$

lying over a domain $\mathcal{D}$ in the xy-plane with polar description

$$\mathcal{D}: \quad \theta_1 \le \theta \le \theta_2, \qquad \alpha(\theta) \le r \le \beta(\theta)$$

By Fubini's Theorem,

$$\iiint_{\mathcal{W}} f(x, y, z)\, dV = \iint_{\mathcal{D}} \left(\int_{z=z_1(r,\theta)}^{z=z_2(r,\theta)} f(x, y, z)\, dz \right) dA$$

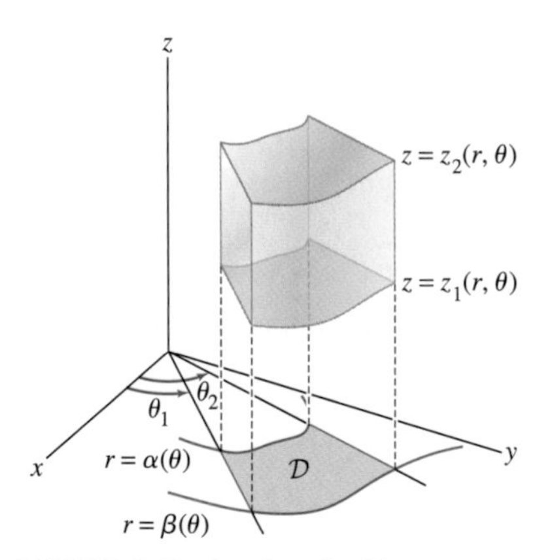

FIGURE 9 Region described in cylindrical coordinates.

Expressing the integral over $\mathcal{D}$ in polar coordinates, we obtain the following Change of Variables Formula.

We summarize Eq. (6) in a formula for the volume element (Figure 10):

$$dV = r\, dz\, dr\, d\theta$$

THEOREM 2 Triple Integrals in Cylindrical Coordinates For a region $\mathcal{W}$ of the form

$$\theta_1 \le \theta \le \theta_2, \qquad \alpha(\theta) \le r \le \beta(\theta), \qquad z_1(r, \theta) \le z \le z_2(r, \theta)$$

the triple integral $\iiint_{\mathcal{W}} f(x, y, z)\, dV$ is equal to:

$$\int_{\theta_1}^{\theta_2} \int_{r=\alpha(\theta)}^{\beta(\theta)} \int_{z=z_1(r,\theta)}^{z=z_2(r,\theta)} f(r\cos\theta, r\sin\theta, z)\, r\, dz\, dr\, d\theta \qquad \boxed{6}$$

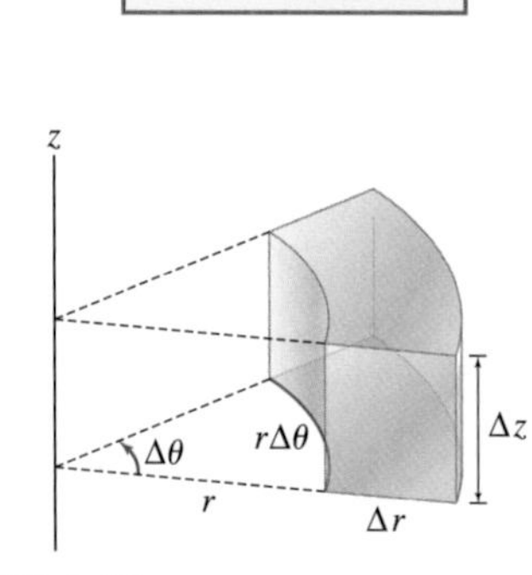

FIGURE 10

■ **EXAMPLE 3** Integrate $f(x, y, z) = z\sqrt{x^2 + y^2}$ over the cylinder $x^2 + y^2 \le 4$ for $1 \le z \le 5$ (Figure 11).

Solution The domain of integration $\mathcal{W}$ lies above the circle of radius 2, so in cylindrical coordinates,

$$\mathcal{W}: \quad 0 \le \theta \le 2\pi, \qquad 0 \le r \le 2, \qquad 1 \le z \le 5$$

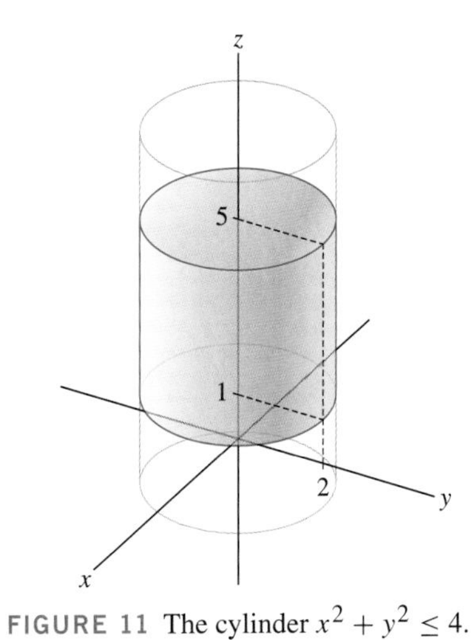

FIGURE 11 The cylinder $x^2 + y^2 \le 4$.

We write the function in cylindrical coordinates:

$$f(x, y, z) = z\sqrt{x^2 + y^2} = zr$$

and apply Eq. (6), noting that the resulting triple integral is equal to a product of single integrals:

$$\begin{aligned}\iiint_{\mathcal{W}} z\sqrt{x^2 + y^2}\, dV &= \int_0^{2\pi} \int_{r=0}^{2} \int_{z=1}^{5} (zr)r\, dz\, dr\, d\theta \\ &= \left(\int_0^{2\pi} d\theta\right)\left(\int_{r=0}^{2} r^2\, dr\right)\left(\int_{z=1}^{5} z\, dz\right) \\ &= (2\pi)\left(\frac{2^3}{3}\right)\left(\frac{5^2 - 1^2}{2}\right) = 64\pi\end{aligned}$$ ■

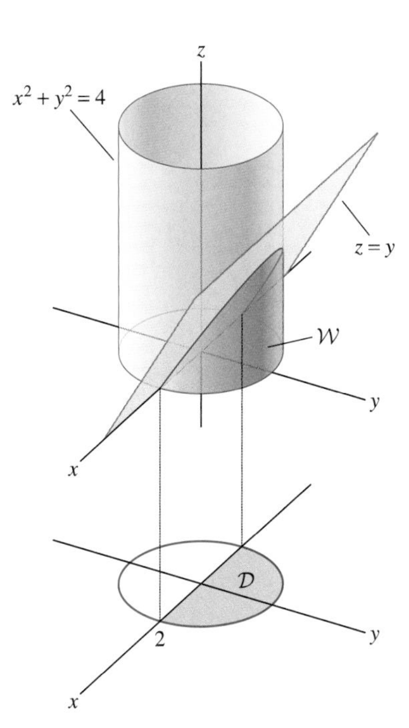

FIGURE 12

■ **EXAMPLE 4** Compute the total mass of the region $\mathcal{W}$ within the cylinder $x^2 + y^2 \le 4$ where $0 \le z \le y$, assuming a mass density of $\rho(x, y, z) = z$.

Solution

***Step 1.* Express $\mathcal{W}$ in cylindrical coordinates.**

The condition $0 \le z \le y$ tells us that $y \ge 0$, so $\mathcal{W}$ projects onto the semicircle $\mathcal{D}$ in the xy-plane of radius 2 where $y \ge 0$, as in Figure 12. In polar coordinates,

$$\mathcal{D}: \quad 0 \le \theta \le \pi, \qquad 0 \le r \le 2$$

The upper boundary of $\mathcal{W}$ is the surface $z = y = r\sin\theta$ and the lower boundary is $z = 0$, so

$$\mathcal{W}: \quad 0 \le \theta \le \pi, \qquad 0 \le r \le 2, \qquad 0 \le z \le r\sin\theta$$

***Step 2.* Set up the integral in cylindrical coordinates and evaluate.**

The total mass is the integral of mass density:

$$\begin{aligned}\iiint_{\mathcal{W}} \rho(x, y, z)\, dV &= \int_0^{\pi} \int_{r=0}^{2} \int_{z=0}^{r\sin\theta} zr\, dz\, dr\, d\theta \\ &= \int_0^{\pi} \int_0^{2} \frac{1}{2}(r\sin\theta)^2 r\, dr\, d\theta = \frac{1}{2}\int_0^{\pi} \int_0^{2} r^3 \sin^2\theta\, dr\, d\theta \\ &= \frac{1}{2}\left(\frac{2^4}{4}\right)\int_0^{\pi} \sin^2\theta\, d\theta = \left(\theta - \frac{1}{2}\sin 2\theta\right)\Big|_0^{\pi} = \pi\end{aligned}$$ ■

← REMINDER

$$\int \sin^2\theta\, d\theta = \frac{1}{2}\left(\theta - \frac{1}{2}\cos 2\theta\right) + C$$

Triple Integrals in Spherical Coordinates

We noted above that the Change of Variables Formula in cylindrical coordinates is summarized by the symbolic equation $dV = r\, dr\, d\theta\, dz$. The analog in spherical coordinates (Figure 13) is the formula

$$\boxed{dV = \rho^2 \sin\phi\, d\rho\, d\phi\, d\theta}$$

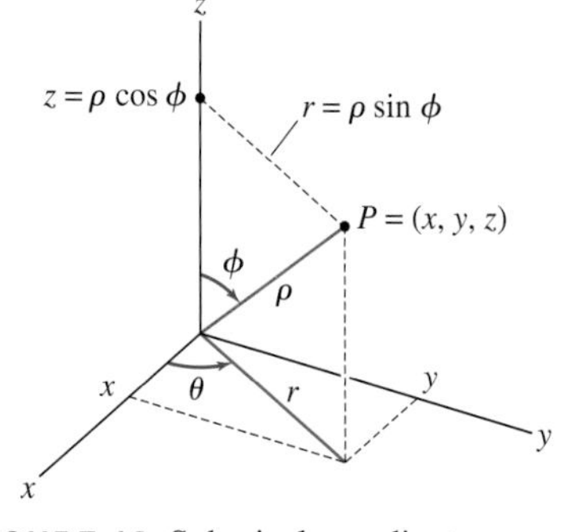

FIGURE 13 Spherical coordinates.

where we recall (Section 12.7)

$$x = \rho\cos\theta\sin\phi, \qquad y = \rho\sin\theta\sin\phi, \qquad z = \rho\cos\phi$$

The key step in deriving this formula is to compute the volume of a small **spherical wedge** $\mathcal{W}$, defined by the inequalities

$$\mathcal{W}: \quad \theta_1 \le \theta \le \theta_2, \quad \phi_1 \le \phi \le \phi_2, \quad \rho_1 \le \rho \le \rho_2 \tag{7}$$

Referring to Figure 14, we see that when the increments

$$\Delta\theta = \theta_2 - \theta_1, \quad \Delta\phi = \phi_2 - \phi_1, \quad \Delta\rho = \rho_2 - \rho_1$$

are small, the spherical wedge is nearly a box with sides $\Delta\rho$, $\rho_1\Delta\phi$, and $\rho_1\sin\phi_1\Delta\theta$ and volume

$$\text{Volume}(\mathcal{W}) \approx \rho_1^2\sin\phi_1\,\Delta\rho\,\Delta\phi\,\Delta\theta \tag{8}$$

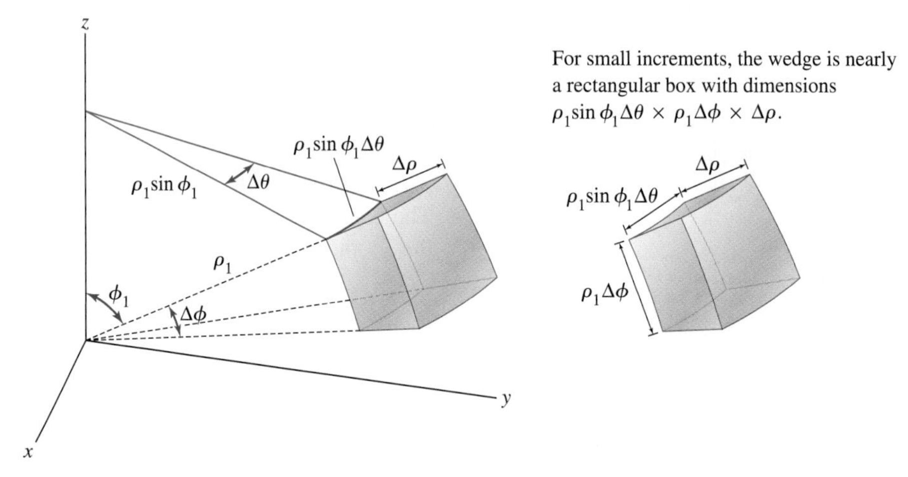

FIGURE 14 Spherical wedge.

Now we may proceed as in our discussion of polar coordinates. Decompose $\mathcal{W}$ into N^3 spherical subwedges $\mathcal{W}_i$ (Figure 15) with increments

$$\Delta\theta = \frac{\theta_2 - \theta_1}{N}, \qquad \Delta\phi = \frac{\phi_2 - \phi_1}{N}, \qquad \Delta\rho = \frac{\rho_2 - \rho_1}{N}$$

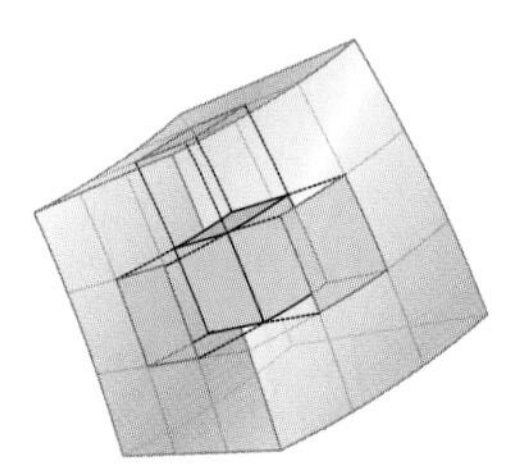
FIGURE 15 Decomposition of a spherical wedge into subwedges.

If N is large, then $\mathcal{W}_i$ is small and if $f(x, y, z)$ is continuous, we have the approximation

$$\iiint_{\mathcal{W}_i} f(x, y, z)\,dx\,dy\,dz \approx f(P_i)\text{Volume}(\mathcal{W}_i) \approx f(P_i)\rho_i^2\sin\phi_i\,\Delta\rho_i\,\Delta\theta_i\,\Delta\phi_i$$

where $P_i = (\rho_i, \theta_i, \phi_i)$ is any sample point in $\mathcal{W}_i$. Thus,

$$\iiint_{\mathcal{W}} f(x, y, z)\,dx\,dy\,dz \approx \sum_i f(P_i)\rho_i^2\sin\phi_i\,\Delta\rho_i\,\Delta\theta_i\,\Delta\phi_i \tag{9}$$

The sum on the right is a Riemann sum for the function

$$f(\rho\cos\theta\sin\phi, \rho\sin\theta\sin\phi, \rho\cos\phi)\,\rho^2\sin\phi$$

on the domain $\mathcal{W}$. The Change of Variables Formula follows by passing to the limit as $N \to \infty$. This argument applies more generally to regions in which the radial variable ρ is defined by an inequality $\rho_1(\theta, \phi) \le \rho \le \rho_2(\theta, \phi)$.

THEOREM 3 Triple Integrals in Spherical Coordinates For a region $\mathcal{W}$ defined by

$$\theta_1 \le \theta \le \theta_2, \qquad \phi_1 \le \phi \le \phi_2, \qquad \rho_1(\theta, \phi) \le \rho \le \rho_2(\theta, \phi)$$

the triple integral $\iiint_{\mathcal{W}} f(x, y, z)\, dV$ is equal to

$$\int_{\theta_1}^{\theta_2} \int_{\phi_1}^{\phi_2} \int_{\rho_1(\theta,\phi)}^{\rho_2(\theta,\phi)} f(\rho\cos\theta\sin\phi, \rho\sin\theta\sin\phi, \rho\cos\phi)\, \rho^2 \sin\phi\, d\rho\, d\phi\, d\theta \tag{10}$$

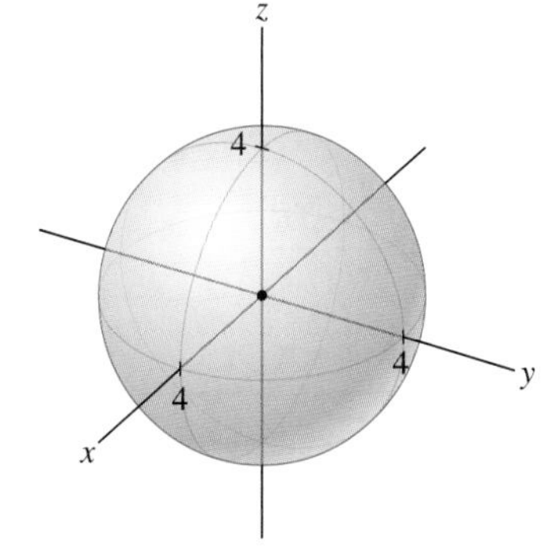

FIGURE 16 Sphere of radius 4.

■ **EXAMPLE 5** Find the mass of a sphere S of radius 4 and center at the origin with mass density $f(x, y, z) = x^2 + y^2$ (Figure 16).

Solution We compute total mass by integrating mass density in spherical coordinates. First, we write $f(x, y, z)$ in spherical coordinates:

$$\begin{aligned} f(x, y, z) = x^2 + y^2 &= (\rho\cos\theta\sin\phi)^2 + (\rho\sin\theta\sin\phi)^2 \\ &= \rho^2\sin^2\phi(\cos^2\theta + \sin^2\theta) = \rho^2\sin^2\phi \end{aligned}$$

Since we are integrating over the entire sphere S of radius 4, r varies from 0 to 4, θ from 0 to 2π, and ϕ from 0 to π. In the following computation, we carry out the integration first with respect to θ:

$$\begin{aligned} \iiint_S (x^2 + y^2)\, dV &= \int_0^{2\pi} \int_0^{\pi} \int_0^4 (\rho^2\sin^2\phi)\, \rho^2 \sin\phi\, d\rho\, d\phi\, d\theta \\ &= 2\pi \int_0^{\pi} \int_0^4 \rho^4 \sin^3\phi\, d\rho\, d\phi = 2\pi \int_0^{\pi} \left(\frac{\rho^5}{5}\bigg|_0^4 \right) \sin^3\phi\, d\phi \\ &= \frac{2{,}048\pi}{5} \int_0^{\pi} \sin^3\phi\, d\phi \\ &= \frac{2{,}048\pi}{5} \left(\frac{1}{3}\cos^3\phi - \cos\phi \right)\bigg|_0^{\pi} = \frac{8{,}192\pi}{15} \end{aligned}$$

■

REMINDER

$$\int \sin^3\phi\, d\phi = \frac{1}{3}\cos^3\phi - \cos\phi + C$$

[write $\sin^3\phi = \sin\phi(1 - \cos^2\phi)$].

■ **EXAMPLE 6** Find the centroid of the ice cream cone $\mathcal{W}$ consisting of the region below the hemisphere $x^2 + y^2 + z^2 = R^2$, $z \ge 0$, and above the cone $x^2 + y^2 = z^2$.

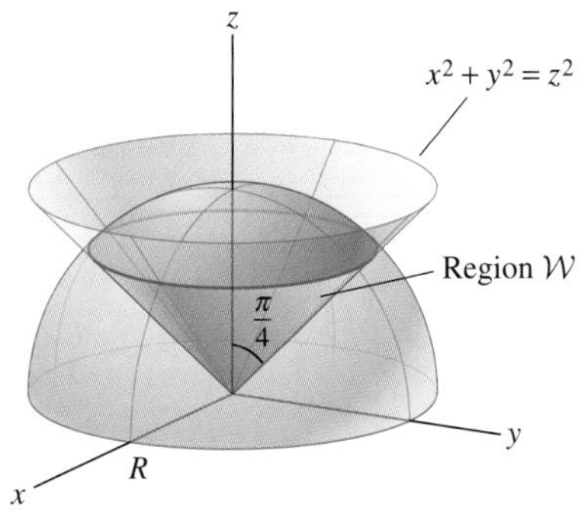

Solution First, we write equation $x^2 + y^2 = z^2$ in spherical coordinates:

$$\begin{aligned} (\rho\cos\theta\sin\phi)^2 + (\rho\sin\theta\sin\phi)^2 &= (\rho\cos\phi)^2 \\ \rho^2\sin^2\phi(\cos^2\theta + \sin^2\theta) &= \rho^2\cos^2\phi \\ \sin^2\phi &= \cos^2\phi \end{aligned}$$

FIGURE 17 Ice cream cone defined by $0 \le \rho \le R$, $0 \le \phi \le \pi/4$.

Thus, $\sin\phi = \pm\cos\phi$ or $\phi = \frac{\pi}{4}, \frac{3\pi}{4}$, and the equation of the upper branch of the cone is $\phi = \frac{\pi}{4}$ (Figure 17). The ice cream cone has the description

$$\mathcal{W}: \quad 0 \le \theta \le 2\pi, \qquad 0 \le \phi \le \frac{\pi}{4}, \qquad 0 \le \rho \le R$$

To find the centroid, we first compute the volume V by integrating $f(x, y, z) = 1$ over $\mathcal{W}$. In the following computation, we carry out the integration in the order θ, ρ, ϕ:

$$V = \iiint_{\mathcal{W}} 1\, dV = \int_0^{2\pi} \int_{\phi=0}^{\pi/4} \int_0^R \rho^2 \sin\phi\, d\rho\, d\phi\, d\theta$$

$$= 2\pi \int_{\phi=0}^{\pi/4} \int_0^R \rho^2 \sin\phi\, d\rho\, d\phi = \frac{2\pi R^3}{3} \int_{\phi=0}^{\pi/4} \sin\phi\, d\phi = \frac{\pi R^3(2-\sqrt{2})}{3}$$

The region $\mathcal{W}$ is symmetric with respect to the z-axis, so its centroid lies on the z-axis and we need only find the z-coordinate z_{CM}. We have

$$\iiint_{\mathcal{W}} z\, dV = \int_0^{2\pi} \int_{\phi=0}^{\pi/4} \int_0^R (\rho\cos\phi)\rho^2 \sin\phi\, d\rho\, d\phi\, d\theta$$

$$= 2\pi \int_{\phi=0}^{\pi/4} \int_0^R \rho^3 \cos\phi \sin\phi\, d\rho\, d\phi = \frac{\pi R^4}{2} \int_0^{\pi/4} \sin\phi\cos\phi\, d\phi = \frac{\pi R^4}{8}$$

$$z_{\mathrm{CM}} = \frac{\iiint_{\mathcal{W}} z\, dV}{\iiint_{\mathcal{W}} 1\, dV} = \left(\frac{3}{\pi R^3(2-\sqrt{2})}\right)\left(\frac{\pi R^4}{8}\right) = \frac{3R}{8(2-\sqrt{2})}$$ ■

15.4 SUMMARY

- Double integral in *polar coordinates:*

$$\iint_{\mathcal{D}} f(x, y)\, dA = \int_{\theta_1}^{\theta_2} \int_{r=\alpha(\theta)}^{\beta(\theta)} f(r\cos\theta, r\sin\theta)\, r\, dr\, d\theta$$

- Triple integral $\iiint_{\mathcal{R}} f(x, y, z)\, dV$

  – *Cylindrical coordinates:*

$$\int_{\theta_1}^{\theta_2} \int_{r=\alpha(\theta)}^{\beta(\theta)} \int_{z=z_1(r,\theta)}^{z=z_2(r,\theta)} f(r\cos\theta, r\sin\theta, z)\, r\, dz\, dr\, d\theta$$

  – *Spherical coordinates:*

$$\int_{\theta_1}^{\theta_2} \int_{\phi_1}^{\phi_2} \int_{\rho_1(\theta,\phi)}^{\rho_2(\theta,\phi)} f(\rho\cos\theta\sin\phi, \rho\sin\theta\sin\phi, \rho\cos\phi)\, \rho^2 \sin\phi\, d\rho\, d\phi\, d\theta$$

15.4 EXERCISES

Preliminary Questions

1. Which of the following represent the integral of $f(x, y) = x^2 + y^2$ over the unit circle?

(a) $\int_0^1 \int_0^{2\pi} r^2\, dr\, d\theta$ **(b)** $\int_0^{2\pi} \int_0^1 r^2\, dr\, d\theta$

(c) $\int_0^1 \int_0^{2\pi} r^3\, dr\, d\theta$ **(d)** $\int_0^{2\pi} \int_0^1 r^3\, dr\, d\theta$

2. What are the limits of integration in $\iiint f(r, \theta, z) r\, dr\, d\theta\, dz$ if the integration extends over the following region?

(a) $x^2 + y^2 \le 4,\ -1 \le z \le 2$

(b) Lower hemisphere of the sphere of radius 2, center at origin

3. What are the limits of integration in

$$\iiint f(\rho, \phi, \theta)\rho^2 \sin\phi \, d\rho \, d\phi \, d\theta$$

if the integration extends over the following spherical regions centered at the origin?

(a) Sphere of radius 4

(b) Region between the spheres of radii 4 and 5

(c) Lower hemisphere of the sphere of radius 2

4. An ordinary rectangle of sides Δx and Δy has area $\Delta x \, \Delta y$, no matter where it is located in the plane. However, the area of a polar rectangle of sides Δr and $\Delta\theta$ depends on its distance from the origin. How is this difference reflected in the Change of Variables Formula for polar coordinates?

Exercises

In Exercises 1–6, sketch the $\mathcal{D}$ indicated and integrate $f(x, y)$ over $\mathcal{D}$ using polar coordinates.

1. $f(x, y) = \sqrt{x^2 + y^2}, \quad x^2 + y^2 \le 2$

2. $f(x, y) = x^2 + y^2, \quad 1 \le x^2 + y^2 \le 4$

3. $f(x, y) = xy; \quad x \ge 0, \quad y \ge 0, \quad x^2 + y^2 \le 4$

4. $f(x, y) = y(x^2 + y^2)^3; \quad y \ge 0, \quad x^2 + y^2 \le 1$

5. $f(x, y) = y(x^2 + y^2)^{-1}; \quad y \ge \frac{1}{2}, \quad x^2 + y^2 \le 1$

6. $f(x, y) = e^{x^2+y^2}, \quad x^2 + y^2 \le R$

7. Find the volume of the wedge-shaped region (Figure 18) contained in the cylinder $x^2 + y^2 = 9$ and bounded above by the plane $z = x$ and below by the xy-plane.

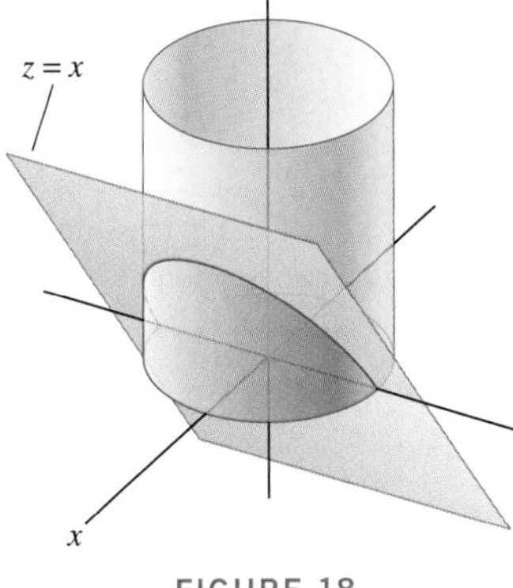

FIGURE 18

8. Let $\mathcal{W}$ be the region above the unit sphere $x^2 + y^2 + z^2 = 6$ and below the paraboloid $z = 4 - x^2 - y^2$.

(a) Show that the projection of $\mathcal{W}$ on the xy-plane is the disk $x^2 + y^2 \le 2$ (Figure 19).

(b) Compute the volume of $\mathcal{W}$ using polar coordinates.

9. Use polar coordinates to compute the volume of the region defined by $4 - x^2 - y^2 \le z \le 10 - 4x^2 - 4y^2$.

10. Compute the total mass of the plate in Figure 20, assuming a mass density of $f(x, y) = x^{-1}y$ g/cm^2.

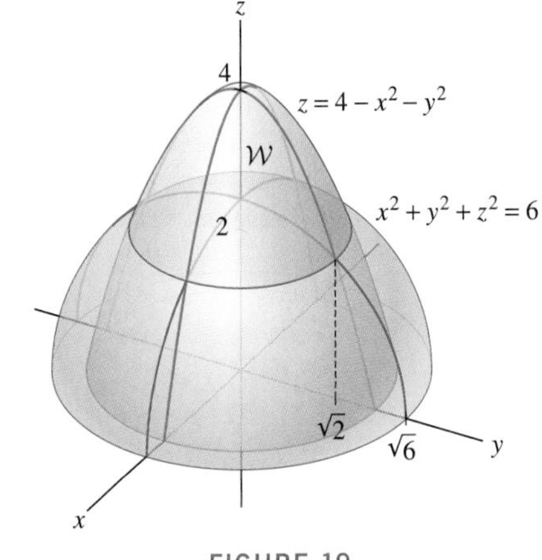

FIGURE 19

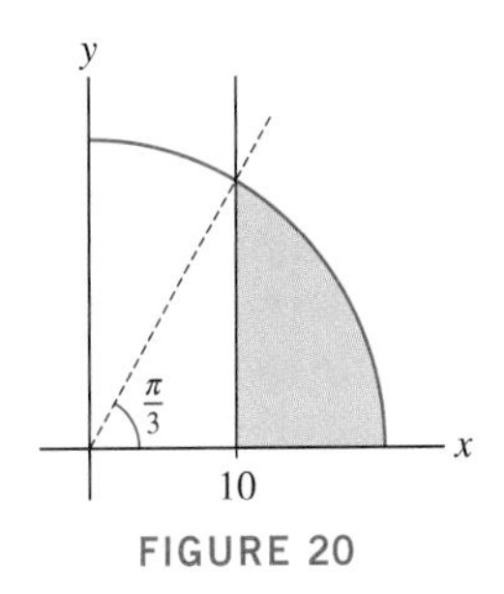

FIGURE 20

11. Use polar coordinates to find the centroid of a quarter circle of radius R.

12. Find the center of mass of the semicircle $x^2 + y^2 \le R^2$, $y \ge 0$, if the mass density is $\rho(x, y) = y$.

In Exercises 13–20, sketch the region of integration and evaluate by changing to polar coordinates.

13. $\displaystyle\int_{-2}^{2}\int_{0}^{\sqrt{4-x^2}} (x^2 + y^2)\, dy\, dx$

14. $\displaystyle\int_{0}^{3}\int_{0}^{\sqrt{9-y^2}} \sqrt{x^2 + y^2}\, dx\, dy$

15. $\displaystyle\int_{0}^{1/2}\int_{\sqrt{3}x}^{\sqrt{1-x^2}} x\, dy\, dx$

16. $\displaystyle\int_{-1}^{2}\int_{0}^{\sqrt{4-x^2}} (x^2+y^2)\,dy\,dx$

17. $\displaystyle\int_{0}^{4}\int_{0}^{\sqrt{16-x^2}} \tan^{-1}\frac{y}{x}\,dy\,dx$

18. $\displaystyle\int_{0}^{2}\int_{x}^{\sqrt{3}x} x\,dy\,dx$

19. $\displaystyle\int_{1}^{2}\int_{0}^{\sqrt{2x-x^2}} \frac{1}{\sqrt{x^2+y^2}}\,dy\,dx$

20. $\displaystyle\int_{0}^{2}\int_{x}^{\sqrt{3}x} y\,dy\,dx$

In Exercises 21–26, calculate the integral over the given region by changing to polar coordinates.

21. $f(x,y)=(x^2+y^2)^{-2}; \quad x^2+y^2\le 2, \quad x\ge 1$

22. $f(x,y)=x, \quad 2\le x^2+y^2\le 4$

23. $f(x,y)=|xy|, \quad x^2+y^2\le 1$

24. $f(x,y)=(x^2+y^2)^{-3/2}; \quad x^2+y^2\le 1, \quad x+y\ge 1$

25. $f(x,y)=x-y; \quad x^2+y^2\le 1, \quad x+y\ge 1$

26. $f(x,y)=y; \quad x^2+y^2\le 1, \quad (x-1)^2+y^2\le 1$

27. Evaluate $\displaystyle\iint_{\mathcal{D}} \sqrt{x^2+y^2}\,dA$, where $\mathcal{D}$ is the domain in Figure 21.
Hint: Find the equation of the inner circle in polar coordinates and treat the right and left parts of the region separately.

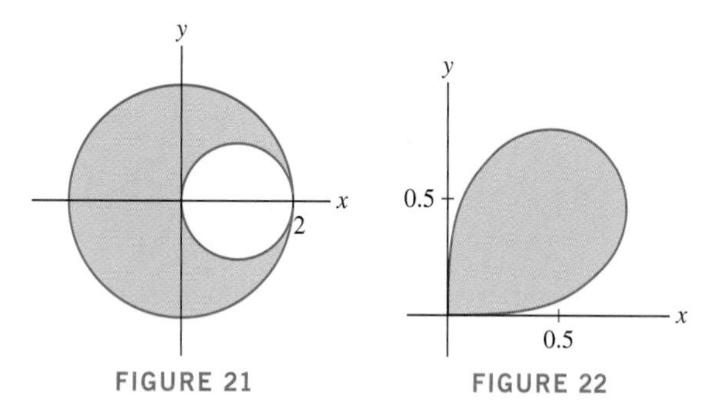

FIGURE 21 FIGURE 22

28. Evaluate $\displaystyle\iint_{\mathcal{D}} x^2\,dA$, where $\mathcal{D}$ is the region in the first quadrant enclosed by the lemniscate $r^2=\sin 2\theta$ (Figure 22).

29. Let $\mathcal{W}$ be the region between the paraboloids $z=x^2+y^2$ and $z=8-x^2-y^2$.

(a) Describe $\mathcal{W}$ in cylindrical coordinates.

(b) Use cylindrical coordinates to compute the volume of $\mathcal{W}$.

30. Use cylindrical coordinates to calculate the integral of $f(x,y,z)=z$ over the region above the disk $x^2+y^2=1$ in the xy-plane and below the surface $z=4+x^2+y^2$.

In Exercises 31–36, use cylindrical coordinates to calculate $\displaystyle\iiint_{\mathcal{W}} f(x,y,z)\,dV$ *for the given function and region.*

31. $f(x,y,z)=x^2+y^2; \quad x^2+y^2\le 9, \quad 0\le z\le 5$

32. $f(x,y,z)=xz; \quad x^2+y^2\le 1, \quad x\ge 0, \quad 0\le z\le 2$

33. $f(x,y,z)=y; \quad x^2+y^2\le 1, \quad x\ge 0, \quad y\ge 0, \quad 0\le z\le 2$

34. $f(x,y,z)=z\sqrt{x^2+y^2}, \quad x^2+y^2\le z\le 8-(x^2+y^2)$

35. $f(x,y,z)=z, \quad x^2+y^2\le z\le 9$

36. $f(x,y,z)=z, \quad 0\le z\le x^2+y^2\le 9$

37. Find the height of the centroid (average value of the y-coordinate) for the region $\mathcal{W}$ in Figure 19, lying above the unit sphere $x^2+y^2+z^2=6$ and below the paraboloid $z=4-x^2-y^2$.

In Exercises 38–41, express the triple integral in cylindrical coordinates.

38. $\displaystyle\int_{-1}^{1}\int_{y=-\sqrt{1-x^2}}^{y=\sqrt{1-x^2}}\int_{z=0}^{4} f(x,y,z)\,dz\,dy\,dx$

39. $\displaystyle\int_{0}^{1}\int_{y=-\sqrt{1-x^2}}^{y=\sqrt{1-x^2}}\int_{z=0}^{4} f(x,y,z)\,dz\,dy\,dx$

40. $\displaystyle\int_{-1}^{1}\int_{y=0}^{y=\sqrt{1-x^2}}\int_{z=0}^{x^2+y^2} f(x,y,z)\,dz\,dy\,dx$

41. $\displaystyle\int_{0}^{2}\int_{y=0}^{y=\sqrt{2x-x^2}}\int_{z=0}^{\sqrt{x^2+y^2}} f(x,y,z)\,dz\,dy\,dx$

42. Use cylindrical coordinates to find the mass of a cylinder of radius 4 and height 10 if the mass density at a point is equal to the square of the distance from the cylinder's central axis.

43. Find the equation of the right-circular cone in Figure 23 in cylindrical coordinates and compute its volume.

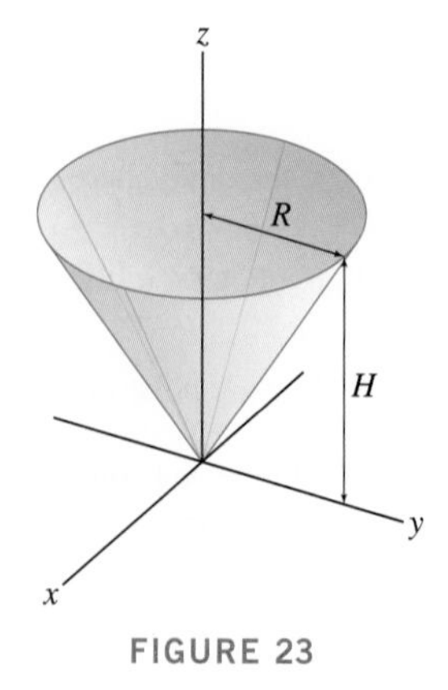

FIGURE 23

44. Let $\mathcal{W}$ be the portion of the half-cylinder $x^2 + y^2 \le 4, x \ge 0$ such that $0 \le z \le 3y$. Use cylindrical coordinates to compute the total mass of $\mathcal{W}$ if the mass density is $\rho(x, y, z) = z^2$.

45. Use cylindrical coordinates to integrate $f(x, y, z) = z$ over the intersection of the sphere $x^2 + y^2 + z^2 = 4$ and the cylinder $x^2 + y^2 = 1$.

46. Use spherical coordinates to evaluate the triple integral of $f(x, y, z) = z$ over the region

$$0 \le \theta \le \frac{\pi}{3}, \qquad 0 \le \phi \le \frac{\pi}{2}, \qquad 1 \le \rho \le 2 \qquad \boxed{11}$$

47. Evaluate the triple integral of $f(x, y, z) = x^2$ for region (11).

48. Calculate the volume of the sphere $x^2 + y^2 + z^2 = a^2$, using both spherical and cylindrical coordinates.

49. Find the volume of the region lying above the cone $\phi = \phi_0$ and below the sphere $\rho = R$.

50. Let $\mathcal{W}$ be the region within the cylinder $x^2 + y^2 = 2$ between $z = 0$ and the cone $z = \sqrt{x^2 + y^2}$. Calculate the integral of $f(x, y, z) = x^2 + y^2$ over $\mathcal{W}$, using both spherical and cylindrical coordinates.

In Exercises 51–56, use spherical coordinates to calculate the triple integral of $f(x, y, z)$ over the given region.

51. $f(x, y, z) = y; \quad x^2 + y^2 + z^2 \le 1, \quad x, y, z \le 0$

52. $f(x, y, z) = \sqrt{x^2 + y^2 + z^2}, \quad x^2 + y^2 + z^2 \le 2z$

53. $f(x, y, z) = \rho^{-3}, \quad 2 \le x^2 + y^2 + z^2 \le 4$

54. $f(x, y, z) = x^2 + y^2, \quad \rho \le 1$

55. $f(x, y, z) = 1; \quad x^2 + y^2 + z^2 \le 4z, \quad z \ge \sqrt{x^2 + y^2}$

56. $f(x, y, z) = \rho; \quad x^2 + y^2 + z^2 \le 4, \quad z \le 1 \; x \ge 0$

57. Find the centroid of the region $\mathcal{W}$ bounded by the cone $\phi = \phi_0$ and the sphere $\rho = R$ as a function of ϕ_0 and R.

58. Find the total mass of the portion of the sphere $x^2 + y^2 + z^2 \le 12$ with $x, y, z \ge 0$ if the mass density is $f(x, y, z) = x(x^2 + y^2 + z^2)^{-1/2}$.

59. Evaluate the triple integral of $f(x, y, z) = z(x^2 + y^2 + z^2)^{-3/2}$ over the part of the ball $x^2 + y^2 + z^2 \le 16$ defined by $z \ge 2$.

60. Calculate the volume of the cone in Figure 23 using spherical coordinates.

61. Find the center of mass of a cylinder of radius 2 and height 4 and mass density e^{-z}, where z is the height above the base.

62. Find the center of mass of a cylinder defined by $0 \le r \le 2$ and $2 \le z \le 4$ with mass density ρ, where ρ is the distance to the origin.

In Exercises 63–65, compute the centroid of the shapes in Figure 24.

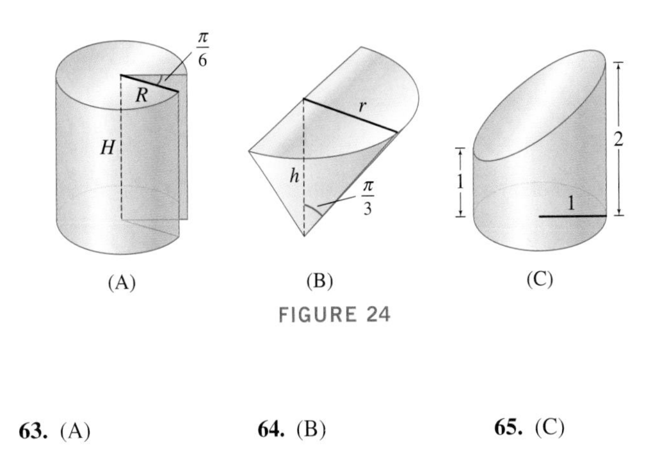

FIGURE 24

63. (A) **64.** (B) **65.** (C)

66. According to Coulomb's Law, the force between two electric charges of magnitude q_1 and q_2 separated by a distance r is Cq_1q_2/r^2 acting along the line through the two charges (C is a constant). Suppose that a circular disk of radius R has a uniformly distributed charge of density k Coulombs per square centimeter (Figure 25). Let F be the net force on a charged particle P of magnitude Q coulombs located at a distance d cm vertically above the center of the disk (by symmetry, F acts in the vertical direction).

(a) Let $\mathcal{R}$ be a small polar rectangle of size $\Delta r \times \Delta\theta$ located at distance r. Show that $\mathcal{R}$ exerts a force on P whose vertical component is

$$\left(\frac{kCQd}{(r^2 + d^2)^{3/2}}\right) r\,\Delta r\,\Delta\theta$$

(b) Explain why F is equal to the following double integral and evaluate:

$$F = kCQd \int_0^{2\pi} \int_0^R \frac{r\,dr\,d\theta}{(r^2 + d^2)^{3/2}}$$

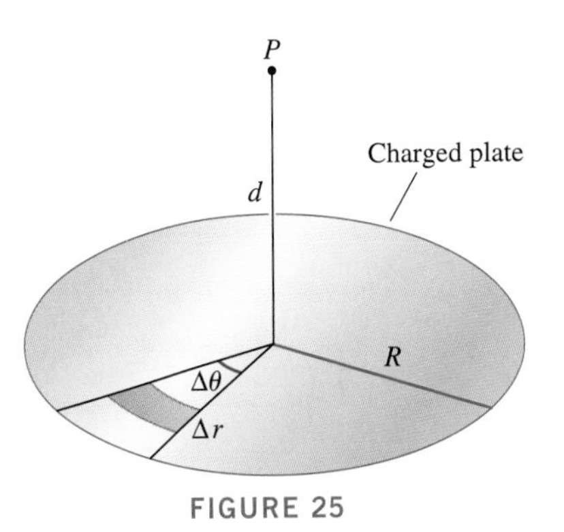

FIGURE 25

Further Insights and Challenges

67. Area Under the Bell-Shaped Curve Let $I = \int_{-\infty}^{\infty} e^{-x^2}\,dx$ be the area under the bell-shaped curve (Figure 26). By Fubini's Theorem, $I^2 = J$, where J is the improper double integral:

$$J = \int_{-\infty}^{\infty}\int_{-\infty}^{\infty} e^{-x^2-y^2}\,dx\,dy$$

Write J in polar coordinates and evaluate, showing that $J = \pi$ and $I = \sqrt{\pi}$.

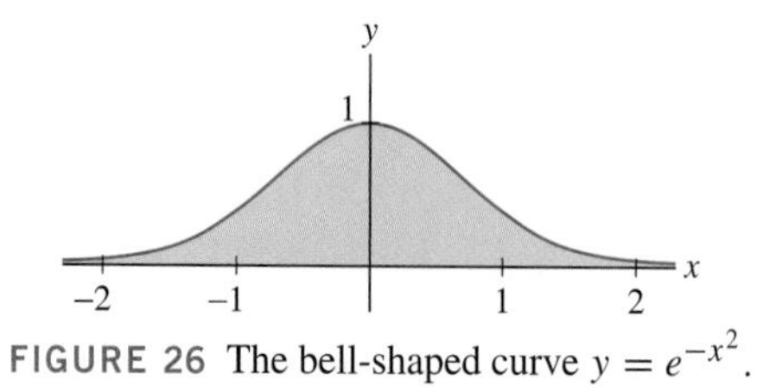

FIGURE 26 The bell-shaped curve $y = e^{-x^2}$.

68. Let D be a plate in the shape of an ellipse with the polar equation

$$r^2 = \left(\frac{1}{6}\sin^2\theta + \frac{1}{9}\cos^2\theta\right)^{-1}$$

with the disk $x^2 + y^2 \le 1$ removed (Figure 27). Assume that D is charged with a charge density $\rho(r,\theta) = 3r^{-4}$ Coulombs per square centimeter. Calculate the total charge on D.

69. Calculate the integral of $\cos z$ over the ball $x^2 + y^2 + z^2 \le 1$. *Hint:* Use cylindrical coordinates and integrate in the order $d\theta\,dz\,dr$.

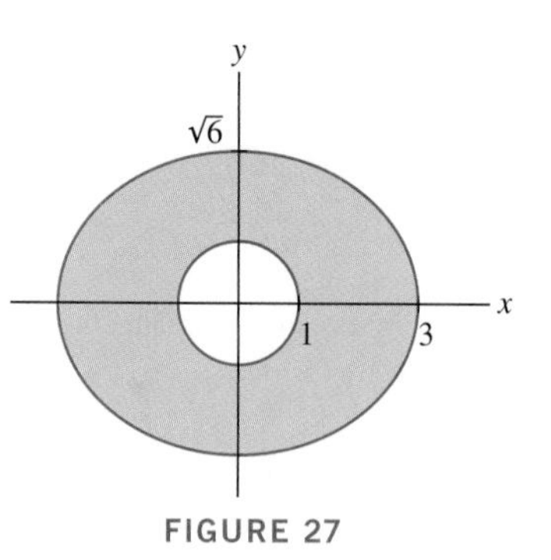

FIGURE 27

70. An Improper Multiple Integral Show that a triple integral of $(x^2 + y^2 + z^2 + 1)^{-2}$ over all of $\mathbf{R}^3$ is equal to π^2. This is an improper integral, so integrate first over $\rho \le R$ and let $R \to \infty$. *Hint:* Use trigonometric substitution to evaluate the integral with respect to ρ.

71. Prove the formula

$$\iint_{\mathcal{D}} \ln r\,dA = -\frac{\pi}{2}$$

where $r = \sqrt{x^2 + y^2}$ and $\mathcal{D}$ is the unit disk $x^2 + y^2 \le 1$. This is an improper integral since $\ln r$ is not defined at $(0, 0)$, so integrate first over the annulus $a \le r \le 1$ and let $a \to 0$.

72. Recall that the improper integral $\int_0^1 x^{-a}\,dx$ converges if $a < 1$. For which values of a does $\iint_{\mathcal{D}} r^{-a}\,dA$ converge, where $r = \sqrt{x^2 + y^2}$ and $\mathcal{D}$ is the unit disk $x^2 + y^2 \le 1$?

15.5 Change of Variables

In the previous section, we discussed the Change of Variables Formulas for the three most common non-Cartesian coordinate systems: polar, cylindrical, and spherical. These formulas are instances of the general Change of Variables Formula for multiple integrals. In this section, we shall use the term **map**, which is another name for a function. It is commonly used when the domain of the function is a domain in $\mathbf{R}^n$ or some set other than a subset of real numbers. A map is also called a **mapping**.

Maps from $\mathbf{R}^2$ to $\mathbf{R}^2$

A map or mapping is function $\Phi : X \to Y$ from a set X (the domain) to a set Y. For $P \in X$, the element $\Phi(P)$ is called the **image** of P. If Z is a subset of X, the image of Z is the set $\Phi(Z)$ consisting of all images $\Phi(z)$ for $z \in Z$. The set $\mathcal{R} = \Phi(X)$ is called the image or **range** of Φ.

We will study maps $\Phi : \mathcal{D} \to \mathbf{R}^2$ defined on a domain $\mathcal{D}$ in $\mathbf{R}^2$. To prevent confusion, we use different variables for the domain and range. For instance, we may view Φ as a map from the uv-plane to the xy-plane (Figure 1). Explicitly, the map Φ has the form $\Phi(u, v) = (\phi(u, v), \psi(u, v))$, where ϕ and ψ are the **component functions**. We also

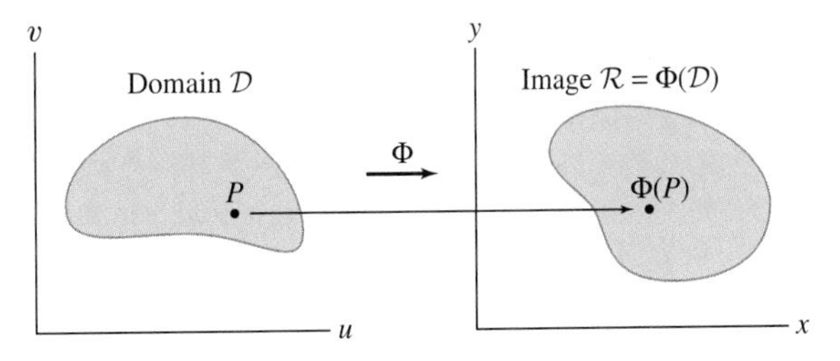

FIGURE 1 Φ maps $\mathcal{D}$ to $\mathcal{R}$.

write

$$x = \phi(u, v), \qquad y = \psi(u, v)$$

One map with which we are already familiar is the map defining polar coordinates. For this map, we use variables r, θ instead of u, v.

■ **EXAMPLE 1 Polar Coordinates Map** Describe the image of a rectangle $[r_1, r_2] \times [\theta_1, \theta_2]$ under the polar coordinates map $\Phi : \mathbf{R}^2 \to \mathbf{R}^2$ defined by

$$\Phi(r, \theta) = (r \sin\theta, r\cos\theta)$$

Solution First, we determine the images of horizontal and vertical lines. Referring to Figure 2, we see that:

- The vertical line $r = r_0$ is mapped to the set of points with radial coordinate r_0, that is, the circle of radius r_0.
- The horizontal line $\theta = \theta_0$ is mapped to the line through the origin of angle θ_0.

It follows that the image of the rectangle $[r_1, r_2] \times [\theta_1, \theta_2]$ is the polar rectangle in the xy-plane defined by $r_1 \le r \le r_2$, $\theta_1 \le \theta \le \theta_2$. ■

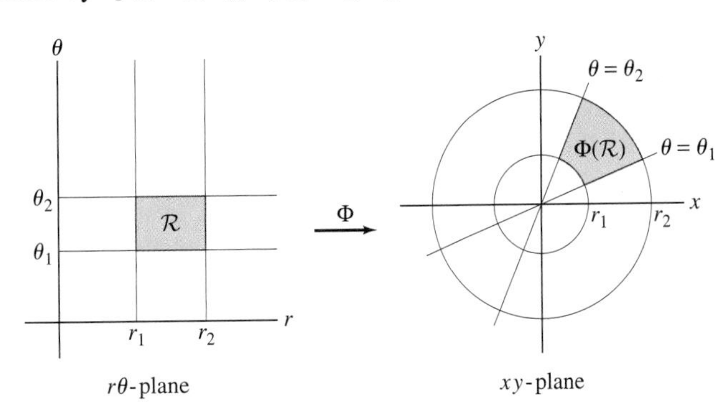

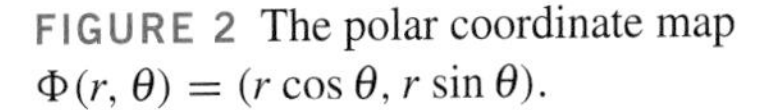

FIGURE 2 The polar coordinate map $\Phi(r, \theta) = (r\cos\theta, r\sin\theta)$.

General mappings may be quite complicated, so it is useful to study the simplest type of maps, the linear maps. A map $\Phi(u, v)$ is called **linear** if it has the form

$$\Phi(u, v) = (Au + Cv, Bu + Dv) \quad (A, B, C, D \text{ constants})$$

A linear map satisfies the following linearity properties (see Exercise 46). For all points (u_1, v_1), (u_2, v_2), (u, v) and constants c,

$$\Phi(u_1 + u_2, v_1 + v_2) = \Phi(u_1, v_1) + \Phi(u_2, v_2) \tag{1}$$

$$\Phi(cu, cv) = c\Phi(u, v) \tag{2}$$

A key property of linear maps is that they map lines to lines. More precisely, *the segment through any two points P and Q is mapped to the segment through $\Phi(P)$ and $\Phi(Q)$* (Figure 3).

To check this, it is convenient to think of Φ as a map from vectors in the uv-plane to vectors in the xy-plane. In vector notation, the segment $\overline{PQ}$ is parametrized as

$$(1-t)\overrightarrow{OP} + t\overrightarrow{OQ} \quad \text{for} \quad 0 \le t \le 1$$

Using linearity, we see that the image of $\overline{PQ}$ is the segment from $\Phi(P)$ to $\Phi(Q)$:

$$\Phi\left((1-t)\overrightarrow{OP} + t\overrightarrow{OQ}\right) = \Phi\left((1-t)\overrightarrow{OP}\right) + \Phi\left(t\overrightarrow{OQ}\right) = (1-t)\Phi(\overrightarrow{OP}) + t\Phi(\overrightarrow{OQ})$$

This gives us a clear picture of the linear mapping $\Phi(u, v) = (Au + Cv, Bu + Dv)$. The images of the basis vectors $\mathbf{i} = \langle 1, 0\rangle$ and $\mathbf{j} = \langle 0, 1\rangle$ are

$$\mathbf{r} = \Phi(\langle 1, 0\rangle) = \langle A, B\rangle, \qquad \mathbf{s} = \Phi(\langle 0, 1\rangle) = \langle C, D\rangle$$

Thus, Φ maps the grid of horizontal and vertical lines in the uv-plane to the grid generated by the vectors $\mathbf{r}$ and $\mathbf{s}$ in the xy-plane (Figure 3).

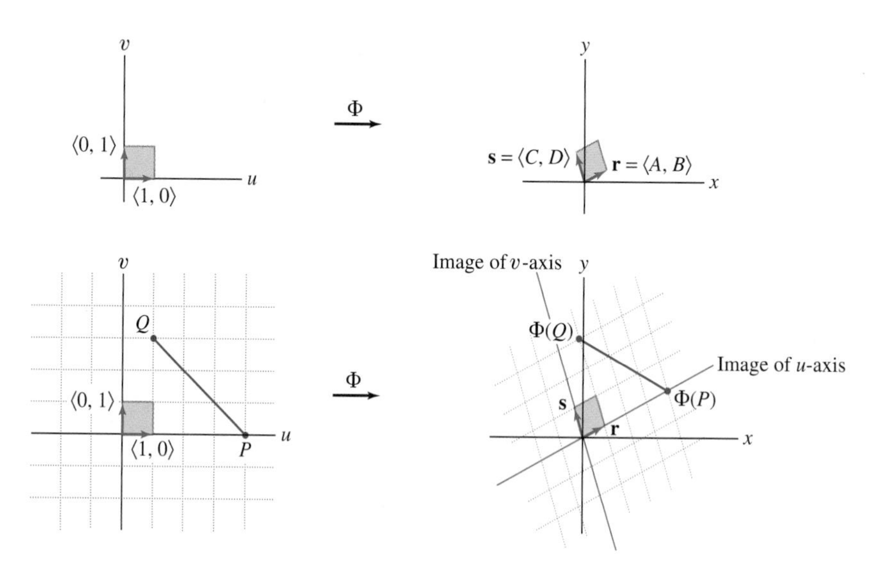

FIGURE 3 A linear mapping Φ maps the grid of horizontal and vertical lines in the uv-plane to the grid generated by **r** and **s** in the xy-plane.

EXAMPLE 2 Image of a Triangle Under a Linear Map Find the image of the triangle $\mathcal{T}$ with vertices $(1, 2)$, $(2, 1)$, $(3, 4)$ under the mapping $\Phi(u, v) = (2u - v, u + v)$.

Solution The image of any polygon $\mathcal{P}$ under a linear map Φ is the polygon whose vertices are the images of the vertices of $\mathcal{P}$. Thus, to find the image of $\mathcal{T}$, we compute the images of the vertices under $\Phi(u, v) = (2u - v, u + v)$:

$$\Phi(1, 2) = (0, 3), \qquad \Phi(2, 1) = (3, 3), \qquad \Phi(3, 4) = (2, 7)$$

These are the vertices of $\Phi(\mathcal{T})$ as shown in Figure 4. ■

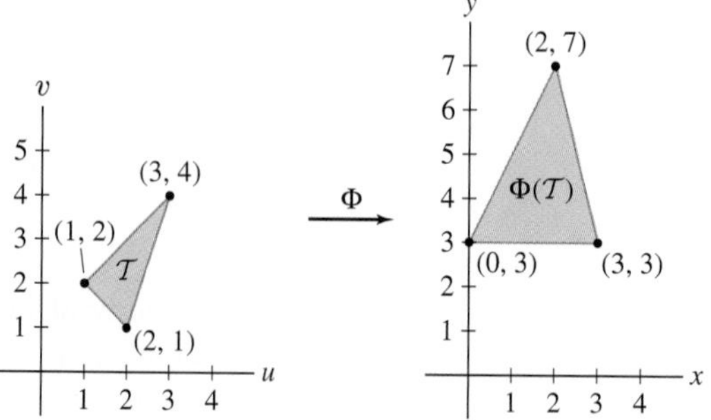

FIGURE 4 The map $\Phi(u, v) = (2u - v, u + v)$.

To understand a nonlinear map, it is usually helpful to determine the images of horizontal and vertical lines as we did for the polar coordinate mapping.

EXAMPLE 3 Let $\Phi(u, v) = (uv^{-1}, uv)$ for $u, v > 0$. Determine the images of:

(a) The lines $u = c$ and $v = c$ **(b)** $[1, 2] \times [1, 2]$

Find the inverse map Φ^{-1}.

Solution

(a) In this map, we have $x = uv^{-1}$ and $y = uv$. Thus $xy = u^2$ and $y/x = v^2$. It follows that Φ maps the vertical line $u = c$ to the hyperbola $xy = c^2$. Similarly, Φ maps the horizontal line $v = c$ to the line $y/x = c^2$, or $y = c^2 x$.

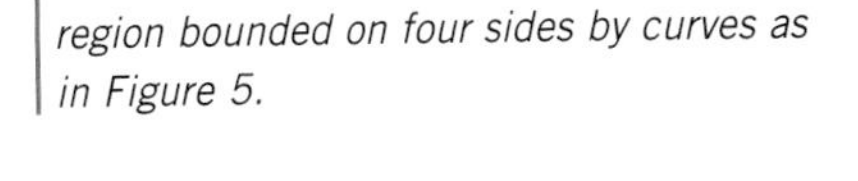
The term "curvilinear rectangle" refers to a region bounded on four sides by curves as in Figure 5.

(b) The image of $[1, 2] \times [1, 2]$ is the *curvilinear* rectangle bounded by the four curves which are the images of the lines $u = 1$, $u = 2$, and $v = 1$, $v = 2$. By (a), the lines $u = 1$ and $u = 2$ are mapped by Φ to the hyperbolas $xy = 1$ and $xy = 4$. Similarly, the lines $v = 1$ and $v = 2$ are mapped to the lines through the origin $y = x$ and $y = 4x$. Therefore, Φ maps $[1, 2] \times [1, 2]$ to the shaded curvilinear rectangle on the right in Figure 5. This image is defined by the inequalities

$$1 \le xy \le 4, \qquad 1 \le \frac{y}{x} \le 4$$

To find Φ^{-1}, we solve for u and v in terms of x and y. We noted that $xy = u^2$ and $y/x = v^2$. Therefore, $u = \sqrt{xy}$ and $v = \sqrt{y/x}$ and the inverse map is $\Phi^{-1}(x, y) = \left(\sqrt{xy}, \sqrt{y/x}\right)$.

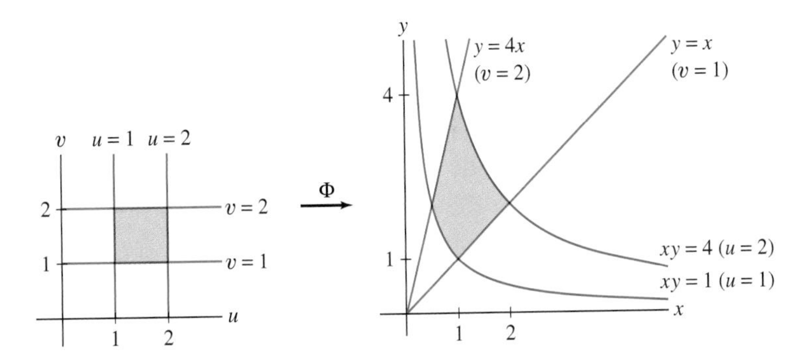

FIGURE 5 The mapping $\Phi(u, v) = (uv^{-1}, uv)$.

How Area Changes Under a Mapping: The Jacobian Determinant

The **Jacobian determinant** (or simply "Jacobian") of a map

$$\Phi(u, v) = (\phi(u, v), \psi(u, v))$$

is the determinant

$$\operatorname{Jac}(\Phi) = \begin{vmatrix} \dfrac{\partial \phi}{\partial u} & \dfrac{\partial \phi}{\partial v} \\[2mm] \dfrac{\partial \psi}{\partial u} & \dfrac{\partial \psi}{\partial v} \end{vmatrix}$$

REMINDER The definition of a 2×2 determinant is

$$\begin{vmatrix} a & b \\ c & d \end{vmatrix} = ad - bc \qquad \boxed{3}$$

Since x and y are functions of u and v:

$$x = \phi(u, v), \qquad y = \psi(u, v)$$

we may also write the Jacobian determinant in the notation

$$\frac{\partial(x, y)}{\partial(u, v)} = \begin{vmatrix} \dfrac{\partial x}{\partial u} & \dfrac{\partial x}{\partial v} \\ \dfrac{\partial y}{\partial u} & \dfrac{\partial y}{\partial v} \end{vmatrix} = \frac{\partial x}{\partial u}\frac{\partial y}{\partial v} - \frac{\partial x}{\partial v}\frac{\partial y}{\partial u}$$

Keep in mind that $\text{Jac}(\Phi)$ is a function of u and v.

EXAMPLE 4 Evaluate the Jacobian of $\Phi(u, v) = (u^3 + v, uv)$ at $(u, v) = (2, 1)$.

Solution We have $x = u^3 + v$ and $y = uv$, so

$$\text{Jac}(\Phi) = \frac{\partial(x, y)}{\partial(u, v)} = \begin{vmatrix} \dfrac{\partial x}{\partial u} & \dfrac{\partial x}{\partial v} \\ \dfrac{\partial y}{\partial u} & \dfrac{\partial y}{\partial v} \end{vmatrix} = \begin{vmatrix} 3u^2 & 1 \\ v & u \end{vmatrix} = 3u^3 - v$$

The value of the Jacobian at $(2, 1)$ is

$$\text{Jac}(\Phi)(2, 1) = 3(2)^3 - 1 = 23$$

The Jacobian of a linear map $\Phi(u, v) = (Au + Cv, Bu + Dv)$ has two special properties:

- The Jacobian is *constant* with value $\text{Jac}(\Phi) = \begin{vmatrix} A & C \\ B & D \end{vmatrix} = AD - BC$.
- Under Φ, the area of a region is multiplied by the factor $|\text{Jac}(\Phi)|$.

The first property follows by direct calculation: Since $x = Au + Cv$ and $y = Bu + Dv$, the partials in Jacobians are the constants A, B, C, D as indicated.

The claim in the second property is that if $\mathcal{D}$ is a domain in the uv-plane, then the area of its image $\Phi(\mathcal{D})$ is (Figure 6):

$$\text{Area}(\Phi(\mathcal{D})) = |\text{Jac}(\Phi)|\text{Area}(\mathcal{D}) \tag{4}$$

This is certainly true for the unit rectangle $\mathcal{D} = [1, 0] \times [0, 1]$ because $\Phi(\mathcal{D})$ is the parallelogram spanned by $\langle A, B\rangle$ and $\langle C, D\rangle$ (Figure 6). This parallelogram has area $|\text{Jac}(\Phi)| = |AD - BC|$ by Theorem 3 in Section 12.4. Similarly, we can check that Eq. (4) holds for arbitrary parallelograms (see Exercise 47). However, Eq. (4) also holds for general domains because a general domain may be approximated as closely as desired by a union of rectangles in a fine grid of lines parallel to the u- and v-axes.

We cannot expect Eq. (4) to hold for a nonlinear map $\Phi(u, v) = (\phi(u, v), \psi(u, v))$. In fact, it would not make sense as stated because $\text{Jac}(\Phi)(P)$ may vary from point to

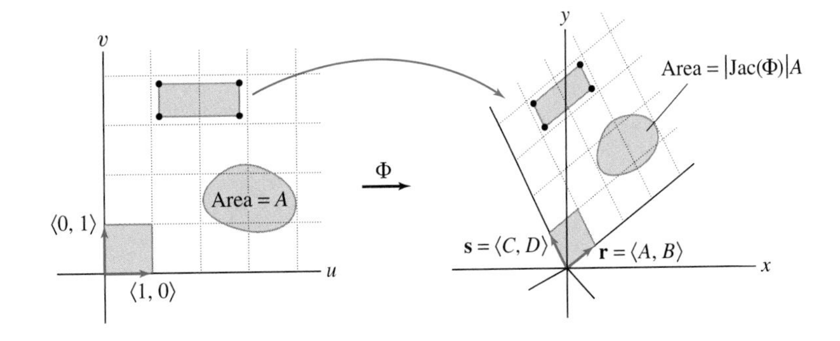

FIGURE 6 A linear map Φ expands (or shrinks) area by the factor $|\text{Jac}(\Phi)|$.

point. However, it is *approximately true* if the domain $\mathcal{D}$ is small and P is a sample point in $\mathcal{D}$:

$$\text{Area}(\Phi(\mathcal{D})) \approx |\text{Jac}(\Phi)(P)|\text{Area}(\mathcal{D}) \quad \boxed{5}$$

This result may be stated more precisely as the limit relation:

$$|\text{Jac}(\Phi)(P)| = \lim_{|\mathcal{D}|\to 0} \frac{\text{Area}(\Phi(\mathcal{D}))}{\text{Area}(\mathcal{D})} \quad \boxed{6}$$

The notation $|\mathcal{D}| \to 0$ means that the diameter of $\mathcal{D}$ (the maximum distance between two points in $\mathcal{D}$) tends to zero.

CONCEPTUAL INSIGHT Although a rigorous proof of Eq. (6) is somewhat technical, we can understand the underlying idea as an application of linear approximation. Consider a rectangle $\mathcal{R}$ with vertex at $P = (u, v)$ and sides of lengths Δu and Δv, assumed to be small as in Figure 7. The image $\Phi(\mathcal{R})$ is not a parallelogram, but it is approximated well by the parallelogram spanned by the vectors $\mathbf{A}$ and $\mathbf{B}$ in the figure:

$$\mathbf{A} = \Phi(u + \Delta u, v) - \Phi(u, v)$$

$$\mathbf{B} = \Phi(u, v + \Delta v) - \Phi(u, v)$$

The linear approximation applied to the components of Φ yields

$$\begin{aligned} \mathbf{A} &= \langle \phi(u + \Delta u, v), \psi(u + \Delta u, v)\rangle - \langle \phi(u, v), \psi(u, v)\rangle \\ &= \langle \phi(u + \Delta u, v) - \phi(u, v), \psi(u + \Delta u, v) - \psi(u, v)\rangle \\ &\approx \langle \phi_u(u, v)\Delta u, \psi_u(u, v)\Delta u\rangle \\ &= \langle \phi_u(u, v), \psi_u(u, v)\rangle \Delta u \end{aligned} \quad \boxed{7}$$

← *REMINDER Equation (7) uses the linear approximation for $\phi(u, v)$ and $\psi(u, v)$:*

$$\phi(u + \Delta u, v) \approx \phi_u(u, v)\Delta u$$

$$\psi(u + \Delta u, v) \approx \psi_u(u, v)\Delta u$$

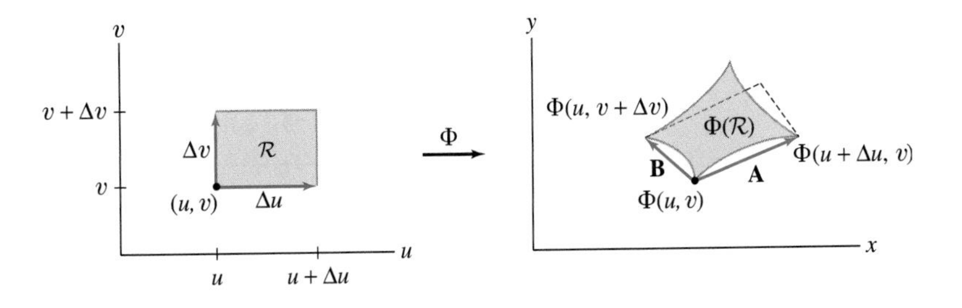

FIGURE 7 The image of a small rectangle under a nonlinear map can be approximated by a parallelogram whose sides are determined by the linear approximation.

Similarly,

$$\mathbf{B} = \Phi(u, v + \Delta v) - \Phi(u, v) \approx \langle \phi_v(u, v), \psi_v(u, v) \rangle \Delta v$$

Since the area of $\mathcal{R}$ is $\Delta u \, \Delta v$, we obtain the desired approximation:

$$\text{Area}(\Phi(\mathcal{R})) \approx \left| \det \begin{pmatrix} \mathbf{A} \\ \mathbf{B} \end{pmatrix} \right| = \left| \det \begin{pmatrix} \phi_u(u, v)\Delta u & \psi_u(u, v)\Delta u \\ \phi_v(u, v)\Delta v & \psi_v(u, v)\Delta v \end{pmatrix} \right|$$

$$= |\phi_u(u, v)\psi_v(u, v) - \psi_u(u, v)\phi_v(u, v)| \, \Delta u \, \Delta v$$

$$= |\text{Jac}(\Phi)(P)| \text{Area}(\mathcal{R})$$

This justifies (5) in the case of a rectangle. For more general regions $\mathcal{D}$, we justify (5) by approximating $\mathcal{D}$ by rectangles.

The Change of Variables Formula

Recall the formula for integration in polar coordinates:

$$\iint_{\mathcal{R}} f(x, y) \, dx \, dy = \int_{\theta_1}^{\theta_2} \int_{r_1}^{r_2} f(r \cos\theta, r \sin\theta) \, r \, dr \, d\theta \qquad \boxed{8}$$

where $\mathcal{R} = [\theta_1, \theta_2] \times [r_1, r_2]$ is a rectangle in the $r\theta$-plane (Figure 8).

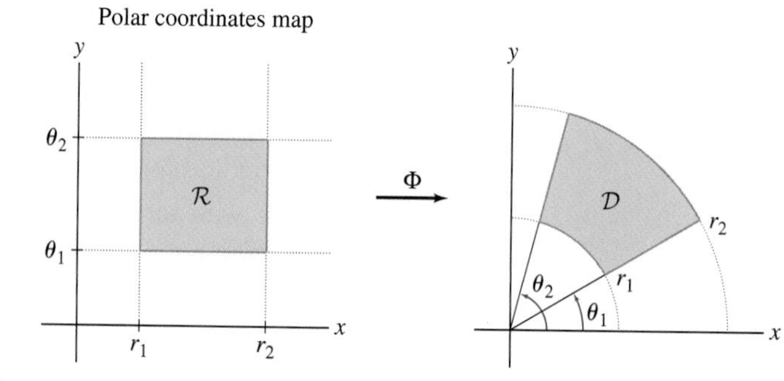

FIGURE 8

The general Change of Variables Formula has a similar form. We consider a one-to-one map $\Phi : \mathcal{D}_0 \to \mathcal{D}$ from a domain in the uv-plane to a domain in the xy-plane (Figure 9). The Change of Variables Formula expresses an integral over $\mathcal{D}$ as an integral over $\mathcal{D}_0$. The Jacobian appears instead of the factor r on the right-hand side of Eq. (8):

REMINDER Φ *is called "one-to-one" if* $\Phi(P) = \Phi(Q)$ *only for* $P = Q$.

$$\iint_{\mathcal{D}} f(x, y) \, dx \, dy = \iint_{\mathcal{D}_0} f(x(u, v), y(u, v)) \left| \frac{\partial(x, y)}{\partial(u, v)} \right| du \, dv$$

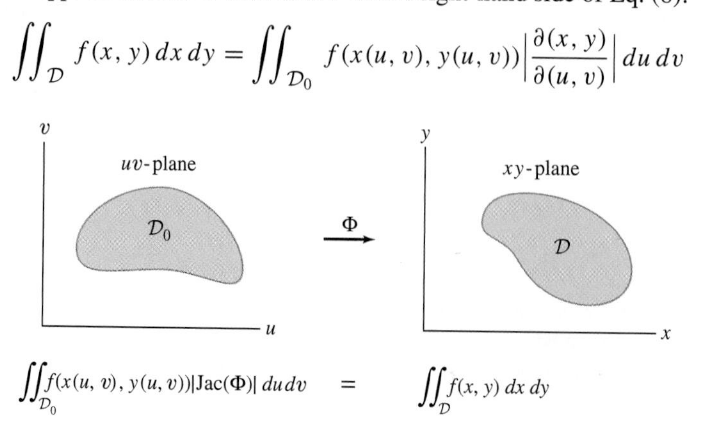

FIGURE 9 The Change of Variables Formula expresses a double integral over $\mathcal{D}$ as a double integral over $\mathcal{D}_0$.

To justify this formula (without attempting to give a fully rigorous proof), suppose first that the domain $\mathcal{D}$ is small and let $P \in \mathcal{D}$ be a sample point. There is a unique point $P_0 \in \mathcal{D}_0$ such that $\Phi(P_0) = P$ since Φ is one-to-one. Assume $f(x, y)$ is continuous and consider the following sequence of approximations:

REMINDER If $\mathcal{D}$ is a domain of small diameter, $P \in \mathcal{D}$ is a sample point, and $f(x, y)$ is continuous, then (see Section 15.2)

$$\iint_{\mathcal{D}} f(x, y)\, dA \approx f(P)\text{Area}(\mathcal{D})$$

We use this in (9) and (11).

$$\iint_{\mathcal{D}} f(x, y)\, dx\, dy \approx f(P)\text{Area}(\mathcal{D}) \qquad \boxed{9}$$

$$\approx f(\Phi(P_0))\, |\text{Jac}(\Phi)(P_0)|\, \text{Area}(\mathcal{D}_0) \qquad \boxed{10}$$

$$\approx \iint_{\mathcal{D}_0} f(\Phi(u, v))\, |\text{Jac}(\Phi)(u, v)|\, du\, dv \qquad \boxed{11}$$

If $\mathcal{D}$ is not small, divide it into small subdomains $D_j = \Phi(\mathcal{D}_{0j})$ and choose sample points $P_j = \Phi(P_{0j})$ where $P_{0j} \in \mathcal{D}_{0j}$ (Figure 10). Then apply the approximation to each subdomain and take the sum:

$$\iint_{\mathcal{D}} f(x, y)\, dx\, dy = \sum_j \iint_{\mathcal{D}_j} f(x, y)\, dx\, dy$$

$$\approx \sum_j \iint_{\mathcal{D}_{0j}} f(\Phi(u, v)))\, |\text{Jac}(\Phi)(u, v)|\, du\, dv$$

$$= \iint_{\mathcal{D}_0} f(\Phi(u, v))\, |\text{Jac}(\Phi)(u, v)|\, du\, dv$$

By carefully estimating the error in each step of this argument, we can show that the error tends to zero as maximum diameters of the subdomains $\mathcal{D}_j$ tend to zero. This yields the Change of Variables Formula. To state it precisely, we assume that Φ is a **C^1 map**, by which we mean that the component functions ϕ and ψ have continuous partial derivatives. We also assume that the domains are simple, that is, either vertically or horizontally simple (see Section 15.2). Although we assumed earlier that Φ is one-to-one, in fact it suffices to assume that Φ is one-to-one on the interior of $\mathcal{D}_0$.

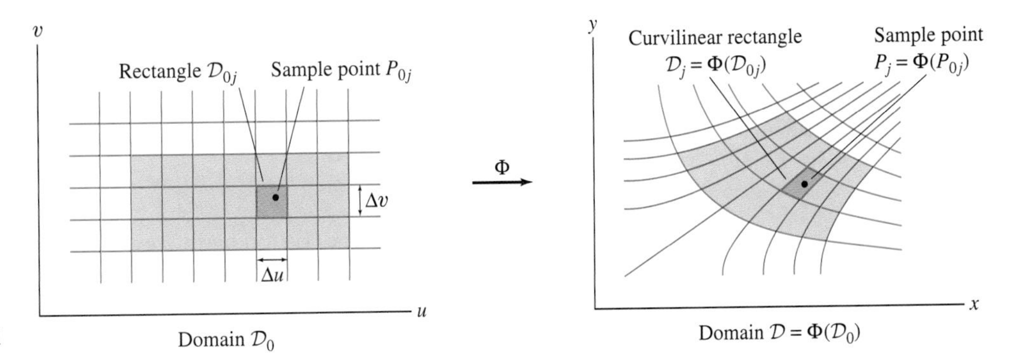

FIGURE 10 Φ maps a rectangular grid on $\mathcal{D}_0$ to a curvilinear grid on $\mathcal{D}$.

THEOREM 1 General Change of Variables Let $\Phi : \mathcal{D}_0 \to \mathcal{D}$ be a C^1 mapping of vertically or horizontally simple regions and assume that Φ is one-to-one on the interior of $\mathcal{D}_0$. Let $f(x, y)$ be a continuous function on $\mathcal{D}$. Then

$$\iint_{\mathcal{D}} f(x, y)\, dx\, dy = \iint_{\mathcal{D}_0} f(x(u, v), y(u, v)) \left| \frac{\partial(x, y)}{\partial(u, v)} \right| du\, dv \qquad \boxed{12}$$

Equation (12) is summarized in the symbolic equality

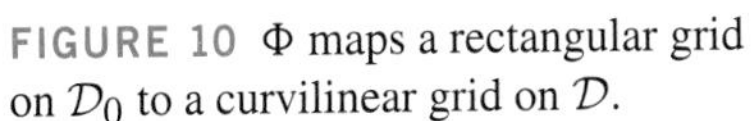

$$dx\, dy = \left| \frac{\partial(x, y)}{\partial(u, v)} \right| du\, dv$$

EXAMPLE 5 Polar Coordinates Revisited Use the general Change of Variables Formula to derive the formula for integration in polar coordinates.

Solution The Jacobian of the polar coordinate map $\Phi(r, \theta) = (r\cos\theta, r\sin\theta)$ is

$$\text{Jac}(\Phi) = \begin{vmatrix} \dfrac{\partial x}{\partial r} & \dfrac{\partial x}{\partial \theta} \\ \dfrac{\partial y}{\partial r} & \dfrac{\partial y}{\partial \theta} \end{vmatrix} = \begin{vmatrix} \cos\theta & -r\sin\theta \\ \sin\theta & r\cos\theta \end{vmatrix} = r(\cos^2\theta + \sin^2\theta) = r$$

If $\mathcal{D}$ is the polar rectangle defined by $r_0 \le r \le r_1$, $\theta_0 \le \theta \le \theta_1$, then Eq. (12) yields

$$\iint_{\mathcal{D}} f(x, y)\, dA = \int_{\theta_0}^{\theta_1} \int_{r_0}^{r_1} f(r\cos\theta, r\sin\theta) r\, d\theta \qquad \boxed{13}$$

■

Assumptions Matter In Theorem 1, we assume that Φ is one-to-one on the interior but not necessarily on the boundary of the domain. Thus, it is valid to apply Eq. (12) to the polar coordinates map Φ on the rectangle $\mathcal{D}_0 = [0, 1] \times [0, 2\pi]$. In this case, Φ is one-to-one on the interior but not on the boundary of $\mathcal{D}_0$ since $\Phi(0, \theta) = (0, 0)$ for all θ and $\Phi(r, 0) = \Phi(r, 2\pi)$ for all r. On the other hand, Eq. (13) cannot be applied to Φ on the rectangle $[0, 1] \times [0, 4\pi]$ because it is not one-to-one on the interior.

EXAMPLE 6 Use the Change of Variables Formula to calculate $J = \iint_{\mathcal{P}} e^{4x-y}\, dA$, where $\mathcal{P}$ is the parallelogram spanned by the vectors $\langle 4, 1\rangle$, $\langle 3, 3\rangle$ in Figure 11.

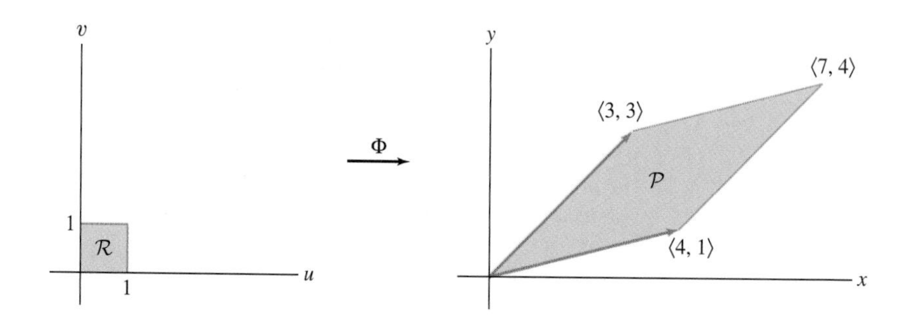

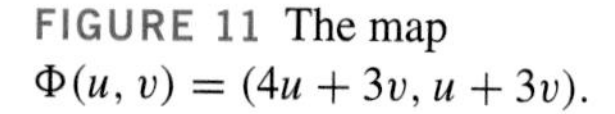
FIGURE 11 The map $\Phi(u, v) = (4u + 3v, u + 3v)$.

In general (Figure 3), the map

$$\Phi(u, v) = (Au + Cv, Bu + Dv)$$

satisfies

$$\Phi(1, 0) = (A, B), \quad \Phi(1, 0) = (C, D)$$

Solution It would not be too hard to evaluate this double integral directly, but our purpose here is to illustrate the Change of Variables Formula.

***Step 1.* Define a map.**
We look for a linear map Φ that sends the unit square $\mathcal{R} = [0, 1] \times [0, 1]$ to $\mathcal{P}$. Observe that (Figure 11)

$$\Phi(u, v) = (4u + 3v, u + 3v)$$

satisfies $\Phi(1, 0) = (4, 1)$ and $\Phi(0, 1) = (3, 3)$. Thus, Φ maps the vectors $\langle 1, 0\rangle$ and $\langle 0, 1\rangle$ to $\langle 4, 1\rangle$ and $\langle 3, 3\rangle$. By linearity, Φ maps the unit square $\mathcal{R}$ to $\mathcal{P}$.

***Step 2.* Express $f(x, y)$ in terms of the new variables.**
Since $x = 4u + 3v$ and $y = u + 3v$, we have

$$e^{4x-y} = e^{4(4u+3v)-(u+3v)} = e^{15u+9v}$$

Step 3. **Compute the Jacobian.**

$$\text{Jac}(\Phi) = \begin{vmatrix} \dfrac{\partial x}{\partial u} & \dfrac{\partial x}{\partial v} \\ \dfrac{\partial y}{\partial u} & \dfrac{\partial y}{\partial v} \end{vmatrix} = \begin{vmatrix} 4 & 3 \\ 1 & 3 \end{vmatrix} = 9$$

Step 4. **Apply the Change of Variables Formula.**

The Change of Variables Formula tells us that $dA = 9\,du\,dv$:

$$\iint_{\mathcal{P}} e^{4x-y}\,dA = \iint_{\mathcal{R}} e^{15u+9v}\,|\text{Jac}(\Phi)|\,du\,dv = 9\int_0^1\int_0^1 e^{15u+9v}\,du\,dv$$

$$= 9\left(\int_0^1 e^{15u}\,du\right)\left(\int_0^1 e^{9v}\,dv\right) = \frac{1}{15}(e^{15}-1)(e^9-1)$$

■ **EXAMPLE 7** Use the Change of Variables $x = uv^{-1}$, $y = uv$ to compute

$$\iint_{\mathcal{D}} (x^2 + y^2)\,dA$$

where $\mathcal{D}$ is the domain $1 \le xy \le 4$, $1 \le y/x \le 4$.

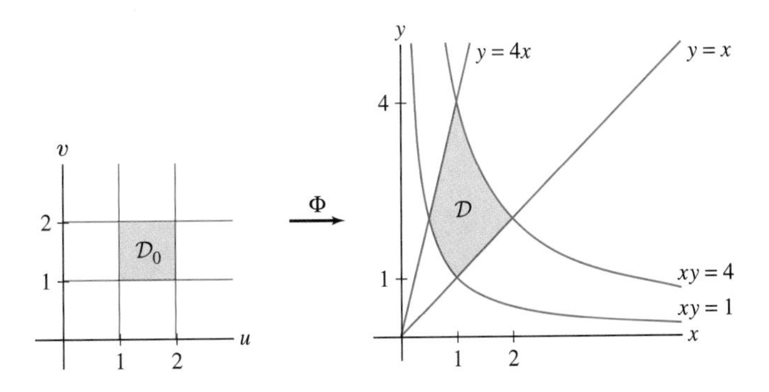

FIGURE 12 The map $\Phi(u, v) = (uv^{-1}, uv)$.

Solution We studied the map $\Phi(u, v) = (uv^{-1}, uv)$ in Example 3 (Figure 12), where we showed that $u = \sqrt{xy}$ and $v = \sqrt{y/x}$. The inequalities $1 \le xy \le 4$ and $1 \le y/x \le 4$ are thus equivalent to $1 \le u \le 2$ and $1 \le v \le 2$. Therefore, Φ maps the rectangle $\mathcal{D}_0 = [1, 2] \times [1, 2]$ to the domain $\mathcal{D}$. The Jacobian of Φ is

$$\frac{\partial(x, y)}{\partial(u, v)} = \begin{vmatrix} \dfrac{\partial x}{\partial u} & \dfrac{\partial x}{\partial v} \\ \dfrac{\partial y}{\partial u} & \dfrac{\partial y}{\partial v} \end{vmatrix} = \begin{vmatrix} v^{-1} & -uv^{-2} \\ v & u \end{vmatrix} = \frac{2u}{v}$$

To apply the Change of Variables, we write $f(x, y)$ in terms of u and v:

$$f(x, y) = x^2 + y^2 = \left(\frac{u}{v}\right)^2 + (uv)^2 = u^2(v^{-2} + v^2)$$

By the Change of Variables Formula,

$$\iint_{\mathcal{D}} (x^2 + y^2)\,dA = \iint_{\mathcal{D}_0} u^2(v^{-2} + v^2)\left|\frac{2u}{v}\right|\,du\,dv$$

$$= 2\int_{v=1}^{2}\int_{u=1}^{2} u^3(v^{-3} + v)\,du\,dv$$

$$= 2\left(\int_{u=1}^{2} u^3\,du\right)\left(\int_{v=1}^{2} (v^{-3} + v)\,dv\right)$$

$$= 2\left(\frac{2^4 - 1^4}{4}\right)\left(\left(\frac{1}{2}v^2 - \frac{1}{2}v^{-2}\right)\Big|_1^2\right) = \frac{225}{16}$$ ■

The Jacobian of the inverse map Φ^{-1} is the reciprocal of Jac(Φ). This can be verified using the Chain Rule (see Exercises 48–50). Thus, we have $\text{Jac}(\Phi^{-1}) = \text{Jac}(\Phi)^{-1}$, which may be written

$$\frac{\partial(x, y)}{\partial(u, v)} = \left(\frac{\partial(u, v)}{\partial(x, y)}\right)^{-1} \qquad \boxed{14}$$

■ **EXAMPLE 8 Using the Inverse Map** Use Change of Variables to integrate the function $f(x, y) = x^2 + y^2$ over the domain

$$\mathcal{D}: \quad -3 \le x^2 - y^2 \le 3, \qquad 1 \le xy \le 4$$

Solution This is different from Example 7 because it is not clear which map to use to change variables. What we can do is define a map Φ in the wrong direction (Figure 13), from $\mathcal{D}$ to the rectangle $\mathcal{R} = [-3, 3] \times [1, 4]$ in the uv-plane:

$$\Phi : \mathcal{D} \to \mathcal{R}$$

$$(x, y) \to (x^2 - y^2, xy)$$

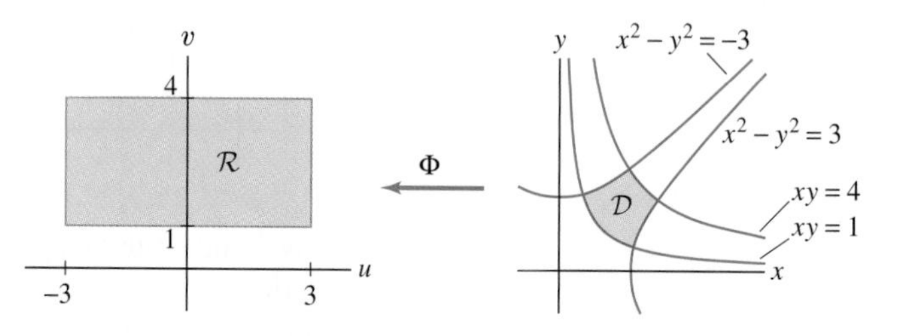

FIGURE 13 The map Φ. Note that Φ goes in the "wrong" direction.

Since we want to convert an integral over $\mathcal{D}$ into an integral over the rectangle $\mathcal{R}$, we must apply the Change of Variables Formula to the inverse mapping:

$$\Phi^{-1} : \mathcal{R} \to \mathcal{D}$$

To compute the Jacobian of Φ^{-1}, we first compute the Jacobian of Φ. Since $u = x^2 - y^2$ and $v = xy$, we have

$$\text{Jac}(\Phi) = \frac{\partial(u, v)}{\partial(x, y)} = \begin{vmatrix} \frac{\partial u}{\partial x} & \frac{\partial u}{\partial y} \\ \frac{\partial v}{\partial x} & \frac{\partial v}{\partial y} \end{vmatrix} = \begin{vmatrix} 2x & -2y \\ y & x \end{vmatrix} = 2(x^2 + y^2)$$

Then, by Eq. (14),

$$\text{Jac}(\Phi^{-1}) = \frac{\partial(x, y)}{\partial(u, v)} = \left(\frac{\partial(u, v)}{\partial(x, y)}\right)^{-1} = \frac{1}{2(x^2 + y^2)}$$

Normally, the next step would be to express $f(x, y)$ in terms of u and v. To do so, we would have to express x and y in terms of u and v (this amounts to finding Φ^{-1}). Fortunately, this is not necessary in our case because the Jacobian cancels with $f(x, y)$:

$$\iint_{\mathcal{D}} (x^2 + y^2)\, dx\, dy = \iint_{\mathcal{R}} f(x(u, v), y(u, v)) \left|\text{Jac}(\Phi^{-1})\right| du\, dv$$

$$= \iint_{\mathcal{R}} (x^2 + y^2) \frac{1}{2(x^2 + y^2)}\, du\, dv$$

$$= \frac{1}{2} \iint_{\mathcal{R}} du\, dv = \frac{1}{2} \int_{-3}^{3} \int_{1}^{4} dv\, du = \frac{1}{2}(6)(3) = 9 \quad \blacksquare$$

Change of Variables in Three Variables

The Change of Variables Formula has the same form in three (or more) variables as in two variables. Let

$$\Phi : \mathcal{W}_0 \to \mathcal{W}$$

be a mapping from a three-dimensional region $\mathcal{W}_0$ in (u, v, w)-space to a region $\mathcal{W}$ in (x, y, z)-space, say,

$$x = \phi_1(u, v, w), \qquad y = \phi_2(u, v, w), \qquad z = \phi_3(u, v, w)$$

← REMINDER 3×3-determinants are defined in Eq. (2) of Section 12.4.

The Jacobian $\text{Jac}(\Phi)$ is the 3×3-determinant:

$$\text{Jac}(\Phi) = \left|\frac{\partial(x, y, z)}{\partial(u, v, w)}\right| = \begin{vmatrix} \frac{\partial x}{\partial u} & \frac{\partial x}{\partial v} & \frac{\partial x}{\partial w} \\ \frac{\partial y}{\partial u} & \frac{\partial y}{\partial v} & \frac{\partial y}{\partial w} \\ \frac{\partial z}{\partial u} & \frac{\partial z}{\partial v} & \frac{\partial z}{\partial w} \end{vmatrix} \qquad \boxed{15}$$

The Change of Variables Formula amounts to the relation

$$\boxed{dx\, dy\, dz = \left|\frac{\partial(x, y, z)}{\partial(u, v, w)}\right| du\, dv\, dw}$$

Assume that Φ is C^1 and one-to-one on the interior of $\mathcal{W}_0$, and that f is continuous. Then

$$\iiint_{\mathcal{W}} f(x, y, z)\, dx\, dy\, dz$$

$$= \iiint_{\mathcal{W}_0} f(x(u, v, w), y(u, v, w), z(u, v, w)) \left|\frac{\partial(x, y, z)}{\partial(u, v, w)}\right| du\, dv\, dw \qquad \boxed{16}$$

In Exercises 42 and 43, you are asked to use the general Change of Variables Formula to derive the formulas for integration in cylindrical and spherical coordinates developed in Section 15.4.

15.5 SUMMARY

- Let $\Phi(u, v) = (\phi(u, v), \psi(u, v))$ be a mapping. We also write $x = \phi(u, v)$ and $y = \psi(u, v)$. The Jacobian of Φ is the determinant

$$\text{Jac}(\Phi) = \left|\frac{\partial(x, y)}{\partial(u, v)}\right| = \begin{vmatrix} \dfrac{\partial x}{\partial u} & \dfrac{\partial x}{\partial v} \\ \dfrac{\partial y}{\partial u} & \dfrac{\partial y}{\partial u} \end{vmatrix}$$

- If $\Phi : \mathcal{D}_0 \to \mathcal{D}$ is C^1 and one-to-one on the interior of $\mathcal{D}_0$, and if f is continuous, then

$$\iint_{\mathcal{D}} f(x, y)\, dA = \iint_{\mathcal{D}_0} f(x(u, v), y(u, v)) \left|\frac{\partial(x, y)}{\partial(u, v)}\right| du\, dv$$

- $\dfrac{\partial(u, v)}{\partial(x, y)} = \left(\dfrac{\partial(x, y)}{\partial(u, v)}\right)^{-1}$

15.5 EXERCISES

Preliminary Questions

1. Which of the following maps is linear?

(a) (uv, v) **(b)** $(u + v, u)$ **(c)** $(3, e^u)$

2. Suppose that Φ is a linear map such that $\Phi(2, 0) = (4, 0)$ and $\Phi(0, 3) = (-3, 9)$. Find the images of:

(a) $\Phi(1, 0)$ **(b)** $\Phi(1, 1)$ **(c)** $\Phi(2, 1)$

3. What is the area of $\Phi(\mathcal{R})$ if $\mathcal{R}$ is a rectangle of area 9 and Φ is a mapping whose Jacobian has constant value 4?

4. Estimate the area of $\Phi(\mathcal{R})$, where $\mathcal{R} = [1, 1.2] \times [3, 3.1]$ and Φ is a mapping such that $\text{Jac}(\Phi)(1, 3) = 3$.

Exercises

1. Determine the image under $\Phi(u, v) = (2u, u + v)$ of the following sets:

(a) The u- and v-axes

(b) The rectangle $\mathcal{R} = [0, 5] \times [0, 7]$

(c) The line segment joining $(1, 2)$ and $(5, 3)$

(d) The triangle with vertices $(0, 1)$, $(1, 0)$, and $(1, 1)$

2. Describe [in the form $y = f(x)$] the images of the lines $u = c$ and $v = c$ under the mapping $\Phi(u, v) = (u/v, u^2 - v^2)$.

3. Let $\Phi(u, v) = (u^2, v)$. Is Φ one-to-one? If not, determine a domain on which Φ is one-to-one. Find the image under Φ of:

(a) The u- and v-axes

(b) The rectangle $\mathcal{R} = [-1, 1] \times [-1, 1]$

(c) The line segment joining $(0, 0)$ and $(1, 1)$

(d) The triangle with vertices $(0, 0)$, $(0, 1)$, and $(1, 1)$

4. Let $\Phi(u, v) = (e^u, e^{u+v})$.

(a) Is Φ one-to-one? What is the image of Φ?

(b) Describe the images of the vertical lines $u = c$ and horizontal lines $v = c$.

In Exercises 5–12, let $\Phi(u, v) = (2u + v, 5u + 3v)$ be a map from the uv-plane to the xy-plane.

5. Show that the image of the horizontal line $v = c$ is the line $y = \frac{5}{2}x + \frac{1}{2}c$. What is the image (in slope-intercept form) of the vertical line $u = c$?

6. Describe the image of the line through the points $(u, v) = (1, 1)$ and $(u, v) = (1, -1)$ under Φ in slope-intercept form.

7. Describe the image of the line $v = 4u$ under Φ in slope-intercept form.

8. Show that Φ maps the line $v = mu$ to the line of slope $(5 + 3m)/(2 + m)$ through the origin in the xy-plane.

9. Show that the inverse of Φ is

$$\Phi^{-1}(x, y) = (3x - y, -5x + 2y)$$

Hint: Show that $\Phi(\Phi^{-1}(x, y)) = (x, y)$ and $\Phi^{-1}(\Phi(u, v)) = (u, v)$.

10. Use the inverse in Exercise 9 to find:

(a) A point in the uv-plane mapping to $(2, 1)$.

(b) A segment in the uv-plane mapping to the segment joining $(-2, 1)$ and $(3, 4)$.

11. Calculate $\text{Jac}(\Phi) = \dfrac{\partial(x, y)}{\partial(u, v)}$.

12. Calculate $\text{Jac}(\Phi^{-1}) = \dfrac{\partial(u, v)}{\partial(x, y)}$.

In Exercises 13–18, compute the Jacobian (at the point, if indicated).

13. $\Phi(u, v) = (3u + 4v, u - 2v)$

14. $\Phi(r, s) = (rs, r + s)$

15. $\Phi(u, v) = (ue^v, ve^{3u})$, $(u, v) = (1, 2)$

16. $\Phi(r, t) = (r \sin t, r - \cos t)$, $(r, t) = (1, \pi)$

17. $\Phi(r, \theta) = (r\cos\theta, r\sin\theta)$

18. $\Phi(u, v) = (ue^v, e^u)$

19. Find a linear mapping Φ that maps $[0, 1] \times [0, 1]$ to the parallelogram in the xy-plane spanned by the vectors $\langle 2, 3\rangle$ and $\langle 4, 1\rangle$.

20. Find a linear mapping Φ that maps $[0, 1] \times [0, 1]$ to the parallelogram in the xy-plane spanned by the vectors $\langle -2, 5\rangle$ and $\langle 1, 7\rangle$.

21. Let $\mathcal{D}$ be the parallelogram in Figure 14. Apply the Change of Variables Formula to the $\Phi(u, v) = (5u + 3v, u + 4v)$ to evaluate $\iint_{\mathcal{D}} xy\,dA$ as an integral over $\mathcal{D}_0 = [0, 1] \times [0, 1]$.

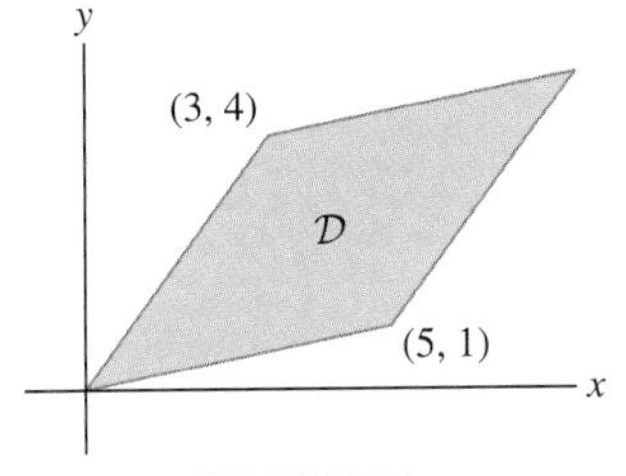

FIGURE 14

22. Let $\Phi(u, v) = (u - uv, uv)$ of the following sets:

(a) Determine the images of the horizontal and vertical lines in the uv-plane.

(b) Compute the Jacobian of Φ.

(c) Observe that by the formula for the area of a triangle, the region $\mathcal{D}$ in Figure 15 has area $\frac{1}{2}(b^2 - a^2)$. Compute this area again using Change of Variables Formula applied to Φ.

(d) Calculate $\iint_{\mathcal{D}} xy\,dA$.

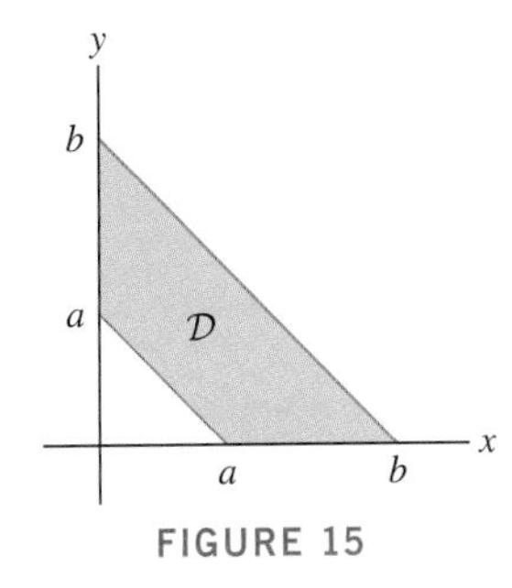

FIGURE 15

23. Let $\Phi(u, v) = (3u + v, u - 2v)$. Use the Jacobian to determine the area of $\Phi(\mathcal{R})$ for:

(a) $\mathcal{R} = [0, 3] \times [0, 5]$ **(b)** $\mathcal{R} = [2, 5] \times [1, 7]$

24. Find a linear map T that maps $[0, 1] \times [0, 1]$ to the parallelogram $\mathcal{P}$ in the xy-plane with vertices $(0, 0)$, $(2, 2)$, $(1, 4)$, $(3, 6)$. Then calculate the double integral of e^{2x-y} over $\mathcal{P}$ via change of variables.

25. With Φ as in Example 3, use the Change of Variables Formula to compute the area of the image of $[1, 4] \times [1, 4]$.

In Exercises 26–28, let $\mathcal{R}_0 = [0, 1] \times [0, 1]$ be the unit square. The translate of a map $\Phi_0(u, v) = (\phi(u, v), \psi(u, v))$ is a map

$$\Phi(u, v) = (a + \phi(u, v), b + \psi(u, v))$$

where a, b are constants. Observe that the map Φ_0 in Figure 16 maps $\mathcal{R}_0$ to the parallelogram $\mathcal{P}_0$ and the translate

$$\Phi_1(u, v) = (2 + 4u + 2v, 1 + u + 3v)$$

maps $\mathcal{R}_0$ to $\mathcal{P}_1$.

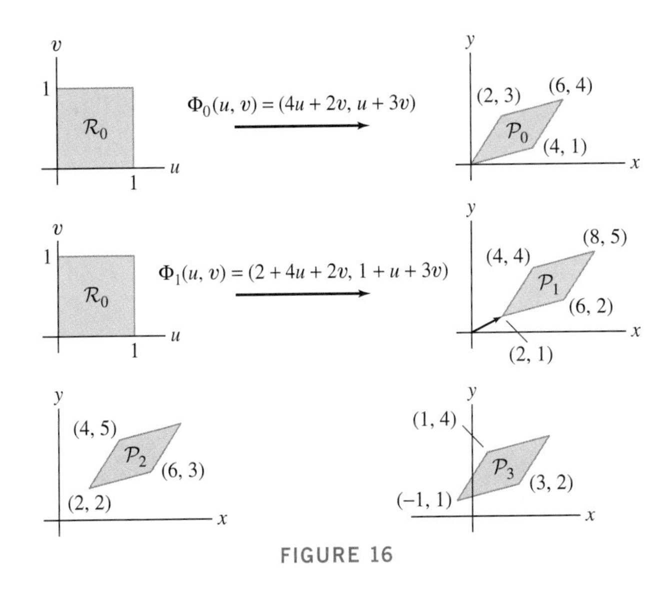

FIGURE 16

26. Find translates Φ_2 and Φ_3 of the mapping Φ_0 in Figure 16 that maps the unit square $\mathcal{R}_0$ to the parallelograms $\mathcal{P}_2$ and $\mathcal{P}_3$.

27. Sketch the parallelogram $\mathcal{P}$ with vertices $(1, 1)$, $(2, 4)$, $(3, 6)$, $(4, 9)$ and find the translate of a linear mapping that maps $\mathcal{R}_0$ to $\mathcal{P}$.

28. Find the translate of a linear mapping that maps $\mathcal{R}_0$ to the parallelogram spanned by the vectors $\langle 3, 9\rangle$ and $\langle -4, 6\rangle$ based at $(4, 2)$.

29. Let $\mathcal{D} = \Phi(\mathcal{R})$, where $\Phi(u, v) = (u^2, u + v)$ and $\mathcal{R} = [1, 2] \times [0, 6]$. Calculate $\iint_{\mathcal{D}} y\,dA$. *Note:* It is not necessary to describe $\mathcal{D}$.

30. Let $\mathcal{D}$ be the image of $\mathcal{R} = [1, 4] \times [1, 4]$ under the map $\Phi(u, v) = (u^2/v, v^2/u)$.

(a) Compute $\text{Jac}(\Phi)$.

(b) Sketch $\mathcal{D}$.

(c) Use the Change of Variables Formula to compute Area($\mathcal{D}$) and $\iint_{\mathcal{D}} f(x, y)\,dA$, where $f(x, y) = x^2 + y^2$.

31. Compute $\iint_{\mathcal{D}} (x + 3y)\, dx\, dy$, where $\mathcal{D}$ is the shaded region in Figure 17. *Hint:* Use the map $\Phi(u, v) = (u - 2v, v)$.

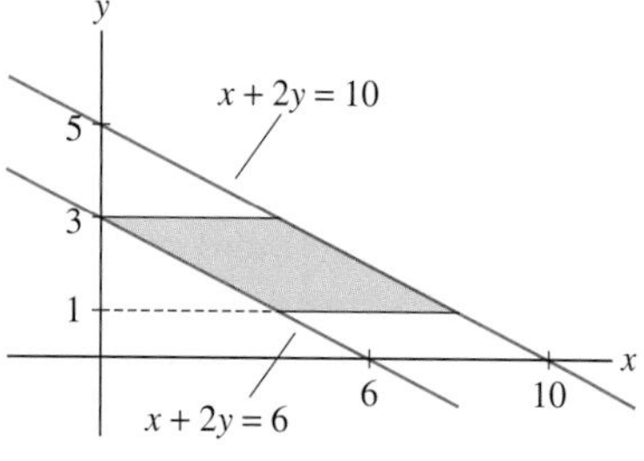

FIGURE 17

32. Use the map $\Phi(u, v) = \left(\dfrac{u}{v+1}, \dfrac{uv}{v+1}\right)$ to compute

$$\iint_{\mathcal{D}} (x + y)\, dx\, dy$$

where $\mathcal{D}$ is the shaded region in Figure 18.

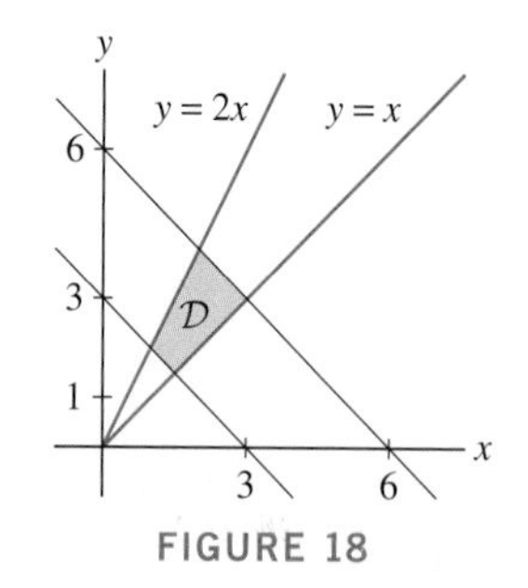

FIGURE 18

33. Show that $T(u, v) = (u^2 - v^2, 2uv)$ maps the triangle $\mathcal{D}_0 = \{(u, v) : 0 \le v \le u \le 1\}$ to the domain $\mathcal{D}$ bounded by $x = 0$, $y = 0$, and $y^2 = 4 - 4x$. Use T to evaluate

$$\iint_{\mathcal{D}} \sqrt{x^2 + y^2}\, dx\, dy$$

34. Find a mapping Φ that maps the disk $u^2 + v^2 \le 1$ onto the interior of the ellipse $\left(\frac{x}{a}\right)^2 + \left(\frac{y}{b}\right)^2 \le 1$. Then use the Change of Variables Formula to prove that the area of the ellipse is πab.

35. Calculate $\iint_{\mathcal{D}} e^{9x^2+4y^2}\, dA$, where $\mathcal{D}$ is the interior of the ellipse $\left(\frac{x}{2}\right)^2 + \left(\frac{y}{3}\right)^2 \le 1$.

36. Compute the area of the region enclosed by the ellipse $x^2 + 2xy + 2y^2 - 4y = 8$ as an integral in the variables $u = x + y$, $v = y$.

37. Sketch the domain $\mathcal{D}$ bounded by $y = x^2$, $y = \frac{1}{2}x^2$, and $y = x$. Use a change of variables with the map $x = uv$, $y = u^2$ to calculate

$$\iint_{\mathcal{D}} y^{-1}\, dx\, dy$$

This is an improper integral since $f(x, y) = y^{-1}$ is undefined at $(0, 0)$, but it becomes proper after changing variables.

38. Find an appropriate change of variables to evaluate

$$\iint_{\mathcal{R}} (x + y)^2 e^{x^2 - y^2}\, dx\, dy$$

where $\mathcal{R}$ is the square with vertices $(1, 0)$, $(0, 1)$, $(-1, 0)$, $(0, -1)$.

39. Let Φ be the inverse of the map $T(x, y) = (xy, x^2 y)$ from the xy-plane to the uv-plane. Let $\mathcal{D}$ be the domain in Figure 19. Show, by applying the Change of Variables Formula to the inverse $\Phi = T^{-1}$, that

$$\iint_{\mathcal{D}} e^{xy}\, dA = \int_{10}^{20} \int_{20}^{40} e^u v^{-1}\, dv\, du$$

and evaluate this result. *Hint:* Use Eq. (14) to compute $\text{Jac}(\Phi)$. It is not necessary to determine Φ explicitly.

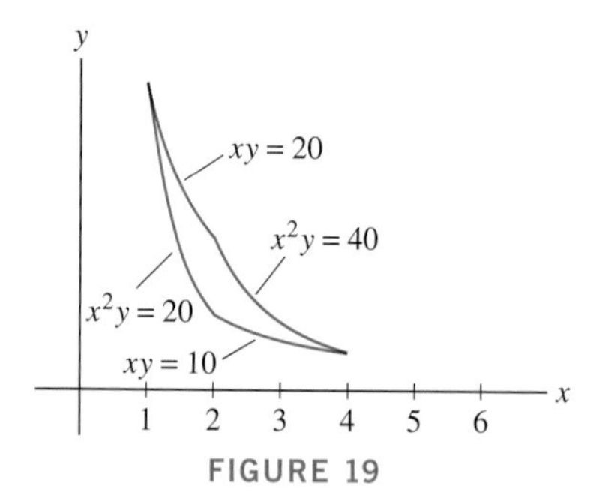

FIGURE 19

40. Sketch the domain

$$\mathcal{D} = \{(x, y) : 1 \le x + y \le 4,\ -4 \le y - 2x \le 1\}$$

(a) Let T be the map $u = x + y$, $v = y - 2x$ from the xy-plane to the uv-plane and let Φ be its inverse. Use Eq. (14) to compute $\text{Jac}(\Phi)$.

(b) Compute $\iint_{\mathcal{D}} e^{x+y}\, dA$ using Change of Variables with the map Φ. *Hint:* It is not necessary to solve for Φ explicitly.

41. Let $I = \iint_{\mathcal{D}} (x^2 - y^2)\, dx\, dy$, where

$$\mathcal{D} = \{(x, y) : 2 \le xy \le 4, 0 \le x - y \le 3, x \ge 0, y \ge 0\}$$

(a) Show that the mapping $u = xy$, $v = x - y$ maps $\mathcal{D}$ to the rectangle $\mathcal{R} = [2, 4] \times [0, 3]$.

(b) Compute $\partial(x, y)/\partial(u, v)$ by first computing $\partial(u, v)/\partial(x, y)$.

(c) Use the Change of Variables Formula to show that I is equal to the integral of $f(u, v) = v$ over $\mathcal{R}$ and evaluate.

42. Derive formula (6) in Section 15.4 for integration in cylindrical coordinates from the general Change of Variables Formula.

43. Derive formula (10) in Section 15.4 for integration in spherical coordinates from the general Change of Variables Formula.

44. Use the Change of Variables Formula in three variables to prove that the volume of the ellipsoid $\left(\frac{x}{a}\right)^2 + \left(\frac{y}{b}\right)^2 + \left(\frac{z}{c}\right)^2 = 1$ is equal to $abc \times$ the volume of the unit sphere.

Further Insights and Challenges

45. Use the map

$$x = \frac{\sin u}{\cos v}, \qquad y = \frac{\sin v}{\cos u}$$

to evaluate the integral

$$\int_0^1 \int_0^1 \frac{dx\,dy}{1 - x^2 y^2}$$

This is an improper integral since the integrand is infinite if $x = \pm 1$, $y = \pm 1$, but the Change of Variables shows that the result is finite.

46. Verify properties (1) and (2) for linear functions and show that any map satisfying these two properties is linear.

47. Let Φ be a linear map. Prove Eq. (4) in the following steps.

(a) For any set $\mathcal{D}$ in the uv-plane and any vector $\mathbf{u}$, let $\mathcal{D} + \mathbf{u}$ be the set obtained by translating all points in $\mathcal{D}$ by $\mathbf{u}$. By linearity, Φ maps $\mathcal{D} + \mathbf{u}$ to the translate $\Phi(\mathcal{D}) + \Phi(\mathbf{u})$ [Figure 20(C)]. Therefore, if Eq. (4) holds for $\mathcal{D}$, it also holds for $\mathcal{D} + \mathbf{u}$.

(b) In the text, we verified Eq. (4) for the unit rectangle. Use linearity to show that Eq. (4) also holds for all rectangles with vertex at the origin and sides parallel to the axes. Then argue that it also holds for each triangular half of such a rectangle, as in Figure 20(A).

(c) Figure 20(B) shows that the area of a parallelogram is a difference of the areas of rectangles and triangles covered by steps (a) and (b). Use this to prove Eq. (4) for arbitrary parallelograms.

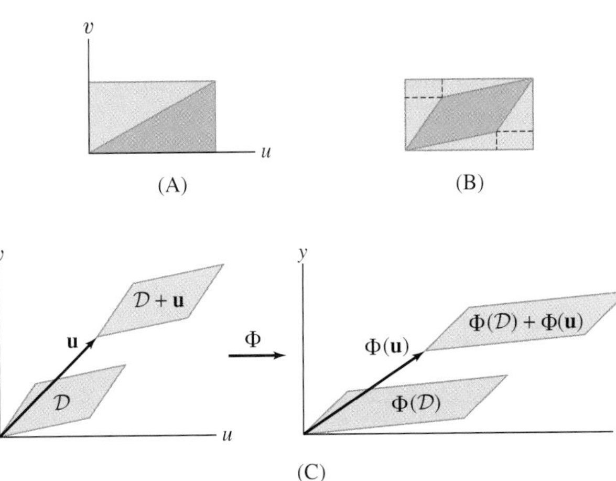

FIGURE 20

48. The product of 2×2 matrices A and B is the matrix AB defined by

$$\underbrace{\begin{pmatrix} a & b \\ c & d \end{pmatrix}}_{A} \underbrace{\begin{pmatrix} a' & b' \\ c' & d' \end{pmatrix}}_{B} = \underbrace{\begin{pmatrix} aa' + bc' & ab' + bd' \\ ca' + dc' & cb' + dd' \end{pmatrix}}_{AB}$$

The (i, j)-entry of A is the **dot product** of the ith row of A and jth column of B. Prove that $\det(AB) = \det(A)\det(B)$.

49. Let $\Phi_1 : \mathcal{D}_1 \to \mathcal{D}_2$ and $\Phi_2 : \mathcal{D}_2 \to \mathcal{D}_3$ be C^1 maps, and let $\Phi_2 \circ \Phi_1 : \mathcal{D}_1 \to \mathcal{D}_3$ be the composite map. Use the Multivariable Chain Rule and Exercise 48 to show that

$$\text{Jac}(\Phi_2 \circ \Phi_1) = \text{Jac}(\Phi_2)\text{Jac}(\Phi_1)$$

50. Use Exercise 49 to prove that $\text{Jac}(\Phi^{-1}) = \text{Jac}(\Phi)^{-1}$. *Hint:* Verify that $\text{Jac}(I) = 1$, where I is the identity map $I(u, v) = (u, v)$.

51. Let $(\overline{x}, \overline{y})$ be the centroid of a domain $\mathcal{D}$. For $\lambda > 0$, let $\lambda\mathcal{D}$ be the **dilate** of $\mathcal{D}$, defined by

$$\lambda\mathcal{D} = \{(\lambda x, \lambda y) : (x, y) \in \mathcal{D}\}$$

Use the Change of Variables Formula to prove that the centroid of $\lambda\mathcal{D}$ is $(\lambda\overline{x}, \lambda\overline{y})$.

CHAPTER REVIEW EXERCISES

1. Calculate the Riemann sum $S_{3,4}$ for $\displaystyle\int_1^2 \int_2^3 x^2 y\, dx\, dy$ using two choices of sample points:

(a) Lower-left vertex

(b) Midpoint of rectangle

Then calculate the exact value of the double integral.

2. Let $S_{N,N}$ be the Riemann sum for $\displaystyle\int_0^1 \int_0^1 \cos(xy)\, dx\, dy$ using midpoints as sample points.

(a) Calculate $S_{4,4}$.

(b) CAS Use a computer algebra system to calculate $S_{N,N}$ for $N = 10, 50, 100$.

3. Let $\mathcal{D}$ be the shaded domain in Figure 1.

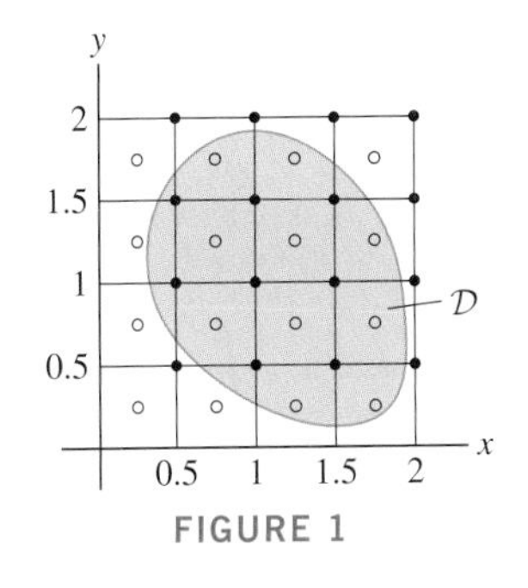

FIGURE 1

Estimate $\iint_{\mathcal{D}} xy\,dA$ by the Riemann sum whose sample points are the midpoints of the squares in the grid.

4. Explain the following:

(a) $\int_{-1}^{1}\int_{-1}^{1} \sin(xy)\,dx\,dy = 0$

(b) $\int_{-1}^{1}\int_{-1}^{1} \cos(xy)\,dx\,dy > 0$

In Exercises 5–8, evaluate the iterated integral.

5. $\int_{0}^{2}\int_{3}^{5} y(x-y)\,dx\,dy$

6. $\int_{1/2}^{0}\int_{0}^{\pi/6} e^{2y}\sin 3x\,dx\,dy$

7. $\int_{0}^{\pi/3}\int_{0}^{\pi/6} \sin(x+y)\,dx\,dy$

8. $\int_{1}^{2}\int_{1}^{2} \frac{y\,dx\,dy}{x+y^2}$

In Exercises 9–14, sketch the domain $\mathcal{D}$ and calculate $\iint_{\mathcal{D}} f(x,y)\,dA$.

9. $\mathcal{D} = \{0 \le x \le 4,\ 0 \le y \le x\}$, $\quad f(x,y) = \cos y$

10. $\mathcal{D} = \{0 \le x \le 2,\ 0 \le y \le 2x - x^2\}$, $\quad f(x,y) = \sqrt{x}y$

11. $\mathcal{D} = \{0 \le x \le 1,\ 1 - x \le y \le 2 - x\}$ $\quad f(x,y) = e^{x+2y}$

12. $\mathcal{D} = \{1 \le x \le 2,\ 0 \le y \le \frac{1}{x}\}$, $\quad f(x,y) = \cos(xy)$

13. $\mathcal{D} = \{0 \le y \le 1,\ 0.5y^2 \le x \le y^2\}$ $\quad f(x,y) = ye^{1+x}$

14. $\mathcal{D} = \{1 \le y \le e,\ y \le x \le 2y\}$, $\quad f(x,y) = \ln(x+y)$

15. Express $\int_{-3}^{3}\int_{0}^{9-x^2} f(x,y)\,dy\,dx$ as an iterated integral in the order $dx\,dy$.

16. Let $\mathcal{D}$ be the domain between $y = x$ and $y = \sqrt{x}$. Calculate $\iint_{\mathcal{D}} xy\,dA$ as an iterated integral in the order $dx\,dy$ and $dy\,dx$.

17. Verify directly that

$$\int_{2}^{3}\int_{0}^{2} \frac{dy\,dx}{1+x-y} = \int_{0}^{2}\int_{2}^{3} \frac{dx\,dy}{1+x-y}$$

18. Prove the formula

$$\int_{0}^{1}\int_{0}^{y} f(x)\,dx\,dy = \int_{0}^{1} (1-x)f(x)\,dx$$

Then use it to calculate $\int_{0}^{1}\int_{0}^{y} \frac{\sin x}{1-x}\,dx\,dy$.

19. Rewrite $\int_{0}^{1}\int_{-\sqrt{1-y^2}}^{\sqrt{1-y^2}} \frac{y\,dx\,dy}{(1+x^2+y^2)^2}$ by interchanging the order of integration and evaluate.

20. Evaluate $\iint_{\mathcal{D}} x\,dA$, where $\mathcal{D}$ is the shaded domain in Figure 2.

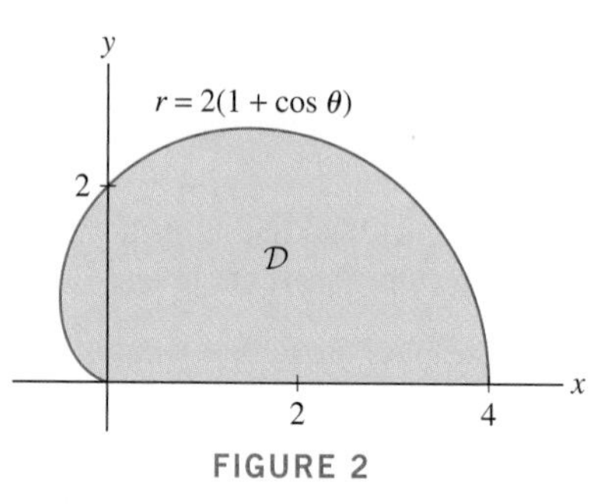

FIGURE 2

21. Find the center of mass of the sector of central angle 2θ (symmetric with respect to the y-axis) in Figure 3, assuming that the mass density is $\rho(x,y) = x^2$.

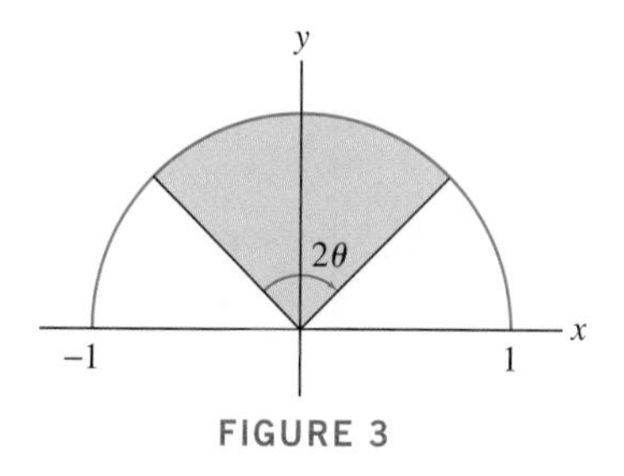

FIGURE 3

22. Find the volume of the region between the graph of the function $f(x,y) = 1 - (x^2 + y^2)$ and the xy-plane.

23. CAS Express the average value of $f(x,y) = e^{xy}$ over the ellipse $\frac{x^2}{2} + y^2 = 1$ as an iterated integral and evaluate numerically using a computer algebra system.

24. Evaluate $\int_{0}^{3}\int_{1}^{4}\int_{2}^{5} (x^3 + y^2 + z)\,dx\,dy\,dz$.

25. Calculate $\iiint_{\mathcal{B}} (xy + z)\,dV$, where

$$\mathcal{B} = \{0 \le x \le 2,\ 0 \le y \le 1,\ 1 \le z \le 3\}$$

as an iterated integral in two different ways.

26. Calculate $\iiint_{\mathcal{W}} xyz\,dV$, where

$$\mathcal{W} = \{0 \le x \le 1,\ x \le y \le 1,\ x \le z \le x + y\}$$

27. Evaluate $I = \int_{-1}^{1} \int_{0}^{\sqrt{1-x^2}} \int_{0}^{1} (x + y + z)\,dz\,dy\,dx$. Then rewrite I as an iterated integral in the order $dy\,dz\,dx$ and evaluate a second time.

28. Find the average value of $f(x, y, z) = xy^2z^3$ on the box $[0, 1] \times [0, 2] \times [0, 3]$.

29. Find the center of mass of the first octant of the ball $x^2 + y^2 + z^2 = 1$, assuming a mass density of $\rho(x, y, z) = x$.

30. Find the centroid of the tetrahedron with vertices $(0, 0, 0)$, $(2, 0, 0)$, $(0, 4, 0)$, and $(0, 0, 6)$.

31. Use polar coordinates to calculate $\iint_{\mathcal{D}} \sqrt{x^2 + y^2}\,dA$, where $\mathcal{D}$ is the region in the first quadrant bounded by the spiral $r = \theta$, the circle $r = 1$, and the x-axis.

32. Calculate $\iint_{\mathcal{D}} \sin(x^2 + y^2)\,dA$, where

$$\mathcal{D} = \left\{\frac{\pi}{2} \le x^2 + y^2 \le \pi\right\}$$

33. Use cylindrical coordinates to find the total mass of the solid bounded by $z = 8 - x^2 - y^2$ and $z = x^2 + y^2$, assuming a mass density of $f(x, y, z) = (x^2 + y^2)^{1/2}$.

34. Describe a region whose volume is equal to:

(a) $\int_{0}^{2\pi} \int_{0}^{\pi/2} \int_{4}^{9} \rho^2 \sin\phi\,d\rho\,d\phi\,d\theta$

(b) $\int_{-2}^{1} \int_{\pi/3}^{\pi/4} \int_{0}^{2} r\,dr\,d\theta\,dz$

(c) $\int_{0}^{2\pi} \int_{0}^{3} \int_{-\sqrt{9-r^2}}^{0} r\,dz\,dr\,d\theta$

35. Find the volume of the solid contained in the cylinder $x^2 + y^2 = 1$ below the curve $z = (x + y)^2$ and above the curve $z = -(x - y)^2$.

36. Use polar coordinates to evaluate $\iint_{\mathcal{D}} x\,dA$, where $\mathcal{D}$ is the shaded region between the two circles of radius 1 in Figure 4.

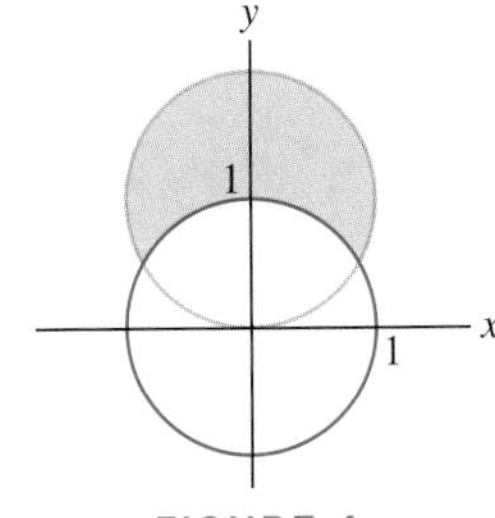

FIGURE 4

37. Express in cylindrical coordinates and evaluate:

$$\int_{0}^{1} \int_{0}^{\sqrt{1-x^2}} \int_{0}^{\sqrt{x^2+y^2}} z\,dz\,dy\,dx$$

38. Use spherical coordinates to calculate the triple integral of $f(x, y, z) = x^2 + y^2 + z^2$ over the region

$$1 \le x^2 + y^2 + z^2 \le 4$$

39. Convert to spherical coordinates and evaluate:

$$\int_{-2}^{2} \int_{-\sqrt{4-x^2}}^{\sqrt{4-x^2}} \int_{0}^{\sqrt{4-x^2-y^2}} e^{-(x^2+y^2+z^2)^{3/2}}\,dz\,dy\,dx$$

40. Using cylindrical coordinates, prove that the centroid of a right-circular cone of height h and radius R is located at height $\frac{h}{4}$ on the central axis.

41. Compute the Jacobian of the map

$$\Phi(r, s) = \left(e^r \cosh(s), e^r \sinh(s)\right)$$

42. Find a linear mapping $\Phi(u, v)$ that maps the unit square to the parallelogram in the xy-plane spanned by the vectors $\langle 3, -1\rangle$ and $\langle 1, 4\rangle$. Then, use the Jacobian to find the area of $\Phi(\mathcal{R})$, where $\mathcal{R} = [0, 4] \times [0, 3]$.

43. Use the map

$$\Phi(u, v) = \left(\frac{u + v}{2}, \frac{u - v}{2}\right)$$

to compute $\iint_{\mathcal{R}} \left((x - y)\sin(x + y)\right)^2 dx\,dy$, where $\mathcal{R}$ is the square with vertices $(\pi, 0)$, $(2\pi, \pi)$, $(\pi, 2\pi)$, and $(0, \pi)$.

44. Let $\mathcal{D}$ be the shaded region in Figure 5 and let Φ be the map

$$u = -y + x^2, \qquad v = y - x^3$$

(a) Show that Φ maps $\mathcal{D}$ to a rectangle $\mathcal{R}$ in the uv-plane.

(b) Apply Eq. (5) in Section 15.5 with $P = (1, 7)$ to estimate Area($\mathcal{D}$).

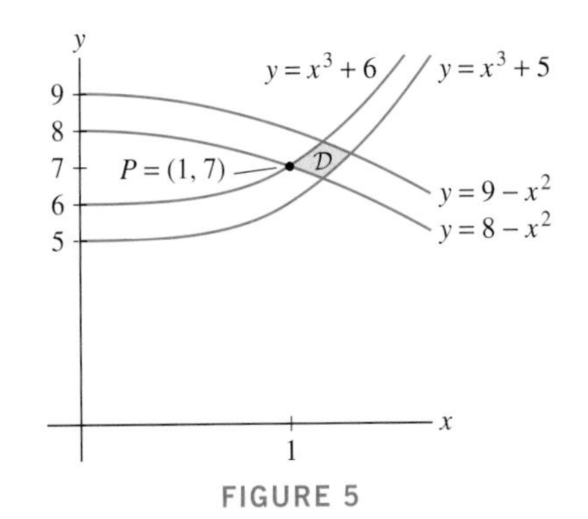

FIGURE 5

45. Calculate the integral of $f(x, y) = e^{3x-2y}$ over the parallelogram in Figure 6.

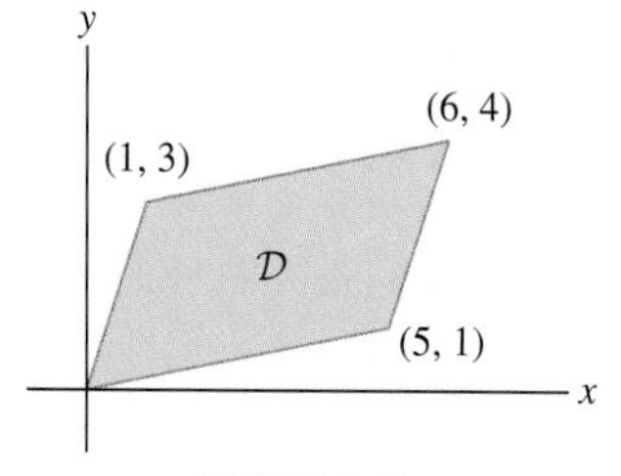

FIGURE 6

46. Sketch the region $\mathcal{D}$ bounded by the curves $y = \frac{2}{x}$, $y = \frac{1}{2x}$, $y = 2x$, $y = \frac{x}{2}$ in the first quadrant. Let Φ be the map $u = xy$, $v = y/x$ from the xy-plane to the uv-plane.

(a) Find the image of $\mathcal{D}$ under Φ.

(b) Show that $\left|\text{Jac}(\Phi^{-1})\right| = \frac{1}{2|v|}$.

(c) Apply the Change of Variables Formula to prove the formula

$$\iint_{\mathcal{D}} f\left(\frac{y}{x}\right) dA = \frac{3}{4}\int_{1/2}^{2} \frac{f(v)\,dv}{v}$$

(d) Apply (c) to evaluate $\displaystyle\iint_{\mathcal{D}} \frac{ye^{y/x}}{x}\,dA$.

16 LINE AND SURFACE INTEGRALS

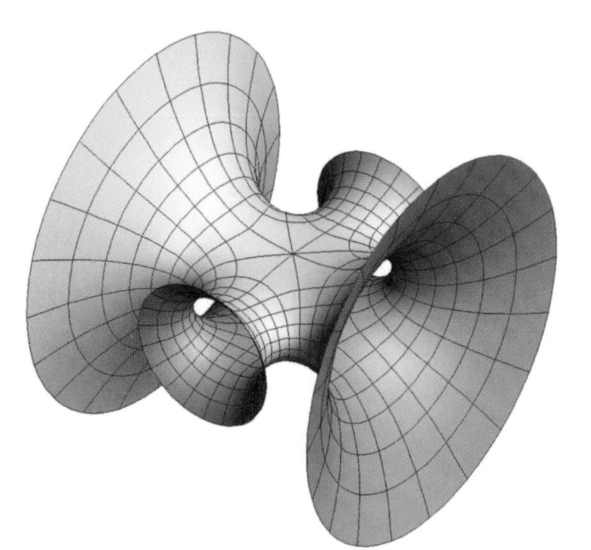

This image of the Symmetric Four-Noid surface was produced by R. Palais using the mathematical visualization program 3D-XplorMath.

In the previous chapter, we generalized integration from one variable to several variables. Our goal in this chapter is to generalize integration still further to include integration over curves or paths and surfaces. We will define integration not just of functions but also of vector fields. Integrals of vector fields are particularly important in applications involving the "field theories" of physics, such as the theory of electromagnetism, heat transfer, fluid dynamics, and aerodynamics. To lay the groundwork, this chapter begins with a discussion of vector fields.

16.1 Vector Fields

There are situations in which it is natural to consider a vector attached to each point in a given region. Such a collection of vectors is called a **vector field**. For example, Figure 2 shows the velocity vector field of wind off the coast of California at Los Angeles. The vector at each point is the velocity vector of the wind at that point.

We represent a vector field in three-space by a vector whose components are functions:

$$\mathbf{F}(x, y, z) = \langle F_1(x, y, z), F_2(x, y, z), F_3(x, y, z)\rangle$$

We also write

$$\mathbf{F} = F_1\mathbf{i} + F_2\mathbf{j} + F_3\mathbf{k}$$

and we denote the value of $\mathbf{F}$ at a point P by $\mathbf{F}(P)$. A vector field in the plane has the form $\mathbf{F}(x, y) = \langle F_1(x, y), F_2(x, y)\rangle$ or $F_1\mathbf{i} + F_2\mathbf{j}$. *In this chapter, we assume that the components F_j are smooth functions (i.e., have partial derivatives of all orders) on their domains.*

"Fluid dynamics" is the study of the equations that must be satisfied by the components of velocity fields of fluids. Many interesting problems remain unsolved, especially the problem of describing turbulent flow (Figure 1). Researchers use high-speed computers to simulate the complex dynamics of turbulence.

FIGURE 1 Turbulent flow of water around the hull of a submarine.

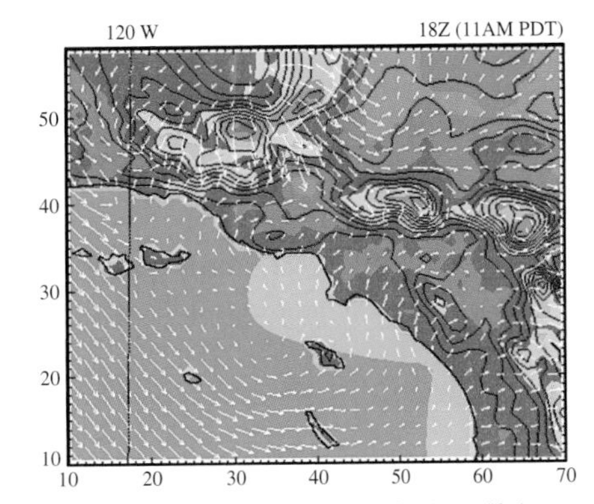

FIGURE 2 Vector field representing horizontal wind velocity off the coast of California at Los Angeles.

EXAMPLE 1 Sketch the vector fields $\mathbf{F} = \langle -y, x \rangle$ and $\mathbf{G} = \mathbf{i} + x\mathbf{j}$.

Solution These vector fields are sketched in Figure 3. Observe that $\mathbf{F} = \langle -y, x \rangle$ is perpendicular to the position vector $\langle x, y \rangle$ and has the same length. Thus, we may describe $\mathbf{F}$ as follows: The vectors along a circle of radius r centered at the origin are tangent to the circle and have length r.

The vector field $\mathbf{G} = \mathbf{i} + x\mathbf{j}$ depends only on x, so it assigns the same vector to points with the same x-coordinate. In other words, the vectors do not change along vertical lines. As $|x|$ increases, the vector $\langle 1, x \rangle$ grows longer and its slope is positive or negative, depending on the sign of x. ■

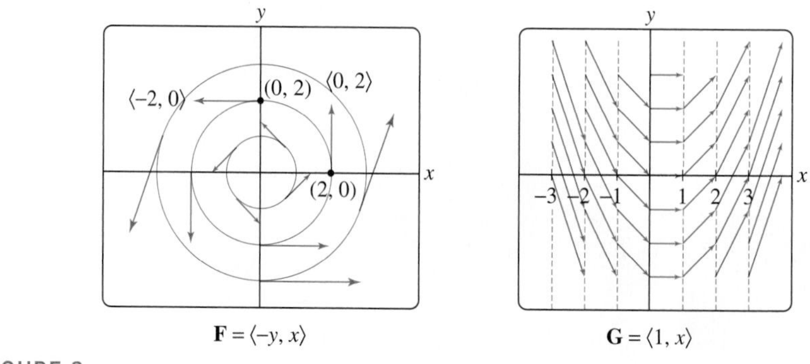

$\mathbf{F} = \langle -y, x \rangle$ $\quad$ $\mathbf{G} = \langle 1, x \rangle$

FIGURE 3

Although it is generally not practical to sketch vector fields in three-space by hand, a computer algebra system can be used to produce useful visual representations (Figure 4). Note that the vector field in Figure 4(B) is a **constant vector field** whose value at every point is the vector $\langle 1, -1, 3 \rangle$.

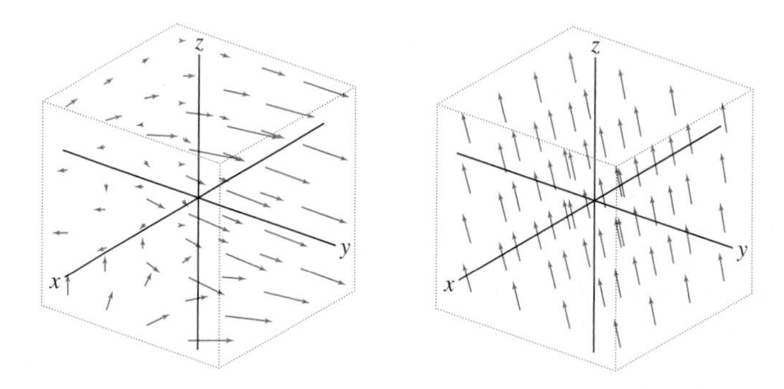

(A) $\mathbf{F} = \langle x \sin z, y^2, x/(z^2+1) \rangle$ $\quad$ (B) Constant vector field $\mathbf{F} = \langle 1, -1, 3 \rangle$

FIGURE 4

A **unit vector field** is a vector field $\mathbf{F}$ such that $\|\mathbf{F}(P)\| = 1$ for all points P. Important examples are the unit radial vector fields in three and two dimensions [Figures 5(A) and 5(B)]:

$$\mathbf{e}_r = \left\langle \frac{x}{r}, \frac{y}{r}, \frac{z}{r} \right\rangle \quad \text{or} \quad \mathbf{e}_r = \left\langle \frac{x}{r}, \frac{y}{r} \right\rangle \tag{1}$$

Here $r = (x^2 + y^2 + z^2)^{1/2}$ if $n = 3$ and $r = (x^2 + y^2)^{1/2}$ if $n = 2$. At each point P, $\mathbf{e}_r(P)$ is a unit vector pointing in the direction of the radial vector $\overrightarrow{OP}$.

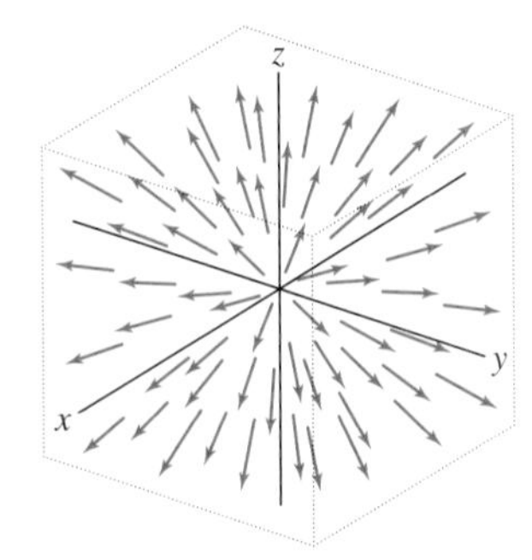

(A) Unit radial vector field in 3-space
$\mathbf{e}_r = \langle x/r, y/r, z/r \rangle$

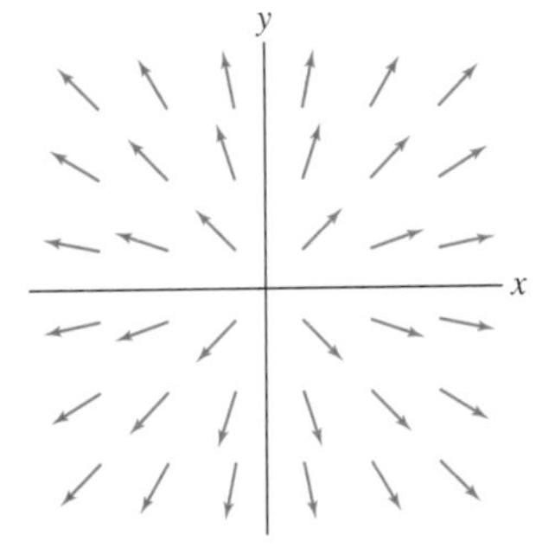

(B) Unit radial vector field in the plane
$\mathbf{e}_r = \langle x/r, y/r \rangle$

FIGURE 5

Gradient Vector Fields

One way of producing a vector field is to take the gradient of a differentiable function φ:

$$\mathbf{F} = \nabla\varphi = \left\langle \frac{\partial\varphi}{\partial x}, \frac{\partial\varphi}{\partial y}, \frac{\partial\varphi}{\partial z} \right\rangle$$

A vector field of this type is called a **gradient vector field** and the function φ is called a **potential function** for $\mathbf{F}$. The same definitions apply in two variables or, more generally, in n variables. We will see that gradient vector fields have important properties when we study line integrals in Section 16.3.

EXAMPLE 2 Show that $\varphi(x, y, z) = xy + yz^2$ is a potential function for the vector field $\mathbf{F} = \langle y, x + z^2, 2yz \rangle$.

Solution We compute the gradient of φ:

$$\frac{\partial\varphi}{\partial x} = y, \quad \frac{\partial\varphi}{\partial y} = x + z^2, \quad \frac{\partial\varphi}{\partial z} = 2yz \qquad \Rightarrow \qquad \nabla\varphi = \langle y, x + z^2, 2yz \rangle$$

Thus, $\nabla\varphi = \mathbf{F}$ as claimed. ■

The English physicist Paul Dirac (1902–1984), who received the Nobel Prize in 1933, introduced a generalization of vectors called "spinors" to unify the special theory of relativity with quantum mechanics. This led to the discovery of the positron, an elementary particle used in the PET-scan imaging to detect cancer and study metabolic processes in the body.

Is every vector field equal to the gradient of some function? The answer is no, as we will show in Example 3; the reason is that the components of a gradient vector field have the special property that the "cross partials are equal." Suppose that $\mathbf{F} = \nabla\varphi$, where φ has continuous second partial derivatives:

$$\mathbf{F} = \langle F_1, F_2, F_3 \rangle = \left\langle \frac{\partial\varphi}{\partial x}, \frac{\partial\varphi}{\partial y}, \frac{\partial\varphi}{\partial z} \right\rangle$$

Now differentiate the x-component with respect to y and the y-component with respect to x:

$$\frac{\partial F_1}{\partial y} = \frac{\partial}{\partial y}\left(\frac{\partial\varphi}{\partial x}\right) = \frac{\partial^2\varphi}{\partial y\partial x}, \qquad \frac{\partial F_2}{\partial x} = \frac{\partial}{\partial x}\left(\frac{\partial\varphi}{\partial y}\right) = \frac{\partial^2\varphi}{\partial x\partial y}$$

Recall that by Clairaut's Theorem in Section 14.3, the mixed partial derivatives are equal: $\dfrac{\partial^2\varphi}{\partial y\,\partial x} = \dfrac{\partial^2\varphi}{\partial x\,\partial y}$. It follows that for a gradient vector field,

$$\frac{\partial F_1}{\partial y} = \frac{\partial F_2}{\partial x}$$

Similarly, the other cross partials of **F** are equal:

$$\frac{\partial F_1}{\partial z} = \frac{\partial F_3}{\partial x}, \qquad \frac{\partial F_2}{\partial z} = \frac{\partial F_3}{\partial y}$$

A vector **F** that does not satisfy these equations cannot be a gradient vector field. This proves the following theorem.

THEOREM 1 Cross Partials of a Gradient Vector Field Are Equal Let

$$\mathbf{F} = \langle F_1, F_2, F_3 \rangle$$

be a gradient vector field whose components have continuous partial derivatives. Then the cross partials are equal:

$$\frac{\partial F_1}{\partial y} = \frac{\partial F_2}{\partial x}, \qquad \frac{\partial F_2}{\partial z} = \frac{\partial F_3}{\partial y}, \qquad \frac{\partial F_3}{\partial x} = \frac{\partial F_1}{\partial z}$$

Similarly, if the vector field in the plane $\mathbf{F} = \langle F_1, F_2 \rangle$ is a gradient vector field, then $\frac{\partial F_1}{\partial y} = \frac{\partial F_2}{\partial x}$.

EXAMPLE 3 Show that $\mathbf{F} = \langle e^x, y^2, xz \rangle$ is not a gradient vector field.

Solution The cross partials for x and z are not equal:

$$\frac{\partial F_1}{\partial z} = \frac{\partial}{\partial z} e^x = 0 \quad \text{is not equal to} \quad \frac{\partial F_3}{\partial x} = \frac{\partial}{\partial x} xz = z$$

Thus, **F** is not a gradient vector field, even though the other cross partials agree:

$$\frac{\partial F_1}{\partial y} = \frac{\partial F_2}{\partial x} = 0 \qquad \text{and} \qquad \frac{\partial F_2}{\partial z} = \frac{\partial F_3}{\partial y} = 0$$

■

Although not every vector field has a potential function, the next theorem shows that if a potential function does exist, then it is unique up to a constant.

THEOREM 2 Uniqueness of Potential Functions If **F** is a gradient vector field on a ball $\mathcal{D}$ in $\mathbf{R}^3$ (or a disk $\mathcal{D}$ in $\mathbf{R}^2$), then any two potential functions differ by a constant.

Proof If both φ_1 and φ_2 are potential functions of **F**, then the gradient of the difference $\varphi = \varphi_1 - \varphi_2$ is zero:

$$\nabla\varphi = \nabla\varphi_1 - \nabla\varphi_2 = \mathbf{F} - \mathbf{F} = \mathbf{0}$$

However, a function whose gradient is zero on a ball is a constant function (this generalizes the fact from single-variable calculus that a function on an interval with zero derivative is a constant function—see Exercise 33). Therefore, $\varphi(x, y, z) = C$ and $\varphi_1 = \varphi_2 + C$ for some constant C.

■

REMINDER In $\mathbf{R}^3$,

$$\mathbf{e}_r = \left\langle \frac{x}{r}, \frac{y}{r}, \frac{z}{r} \right\rangle$$

where $r = (x^2 + y^2 + z^2)^{1/2}$. In $\mathbf{R}^2$,

$$\mathbf{e}_r = \left\langle \frac{x}{r}, \frac{y}{r} \right\rangle$$

where $r = (x^2 + y^2)^{1/2}$.

In the next example, we show that the inverse-square radial field is a gradient vector field. This field is of great importance because it describes the gravitational force due to a point mass or the electrostatic force due to a point charge. If the mass or charge is located at the origin, then there is a constant k such that the force field is

$$\mathbf{F} = \frac{k}{r^2}\mathbf{e}_r = k\left\langle \frac{x}{r^3}, \frac{y}{r^3}, \frac{z}{r^3} \right\rangle$$

This vector field is not defined at the origin $r = 0$.

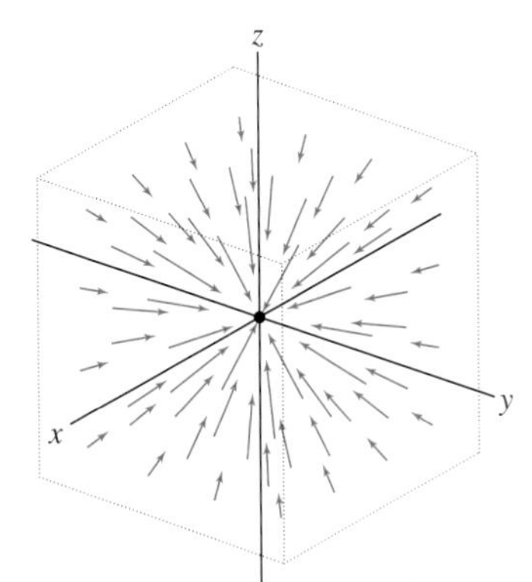

FIGURE 6 The vector field $\frac{-\mathbf{e}_r}{r^2}$, which represents the gravitational force field due to a point mass (up to a scalar multiple).

EXAMPLE 4 Potentials for the Unit and Inverse Square Radial Vector Fields Show that $\mathbf{e}_r = \nabla r$ and $\frac{\mathbf{e}_r}{r^2} = \nabla\left(\frac{-1}{r}\right)$.

Solution We treat the case of radial vector fields in $\mathbf{R}^3$ (Figure 6). The computation is similar for $\mathbf{R}^2$. Since $r = (x^2 + y^2 + z^2)^{1/2}$, we have

$$\frac{\partial r}{\partial x} = \frac{\partial}{\partial x}(x^2 + y^2 + z^2)^{1/2} = \frac{1}{2}(x^2 + y^2 + z^2)^{-1/2}(2x) = \frac{x}{r}$$

Similarly, $\frac{\partial r}{\partial y} = \frac{y}{r}$ and $\frac{\partial r}{\partial z} = \frac{z}{r}$. Therefore,

$$\nabla r = \left\langle \frac{x}{r}, \frac{y}{r}, \frac{z}{r} \right\rangle = \mathbf{e}_r$$

Using the Chain Rule for gradients (Theorem 1 in Section 14.5), we obtain

$$\nabla(-r^{-1}) = r^{-2}\nabla r = r^{-2}\mathbf{e}_r$$

16.1 SUMMARY

- A *vector field* assigns a vector to each point in a domain in $\mathbf{R}^3$ or $\mathbf{R}^2$. A vector field in $\mathbf{R}^3$ is represented by a triple of functions: $\mathbf{F} = \langle F_1, F_2, F_3 \rangle$ and in $\mathbf{R}^2$ by a pair of functions $\mathbf{F} = \langle F_1, F_2 \rangle$. We assume that the components F_j are smooth functions on their domains.
- A *gradient vector field* is a vector field of the form $\mathbf{F} = \nabla\varphi$, where $\varphi(x, y, z)$ is a differentiable function.
- If $\mathbf{F} = \nabla\varphi$, then φ is called a *potential function* for $\mathbf{F}$. Any two potential functions for a gradient vector field differ by a constant.
- If $\mathbf{F} = \langle F_1, F_2, F_3 \rangle = \nabla\varphi$ is a gradient vector field, then the cross partials are equal:

$$\frac{\partial F_1}{\partial y} = \frac{\partial F_2}{\partial x}, \qquad \frac{\partial F_1}{\partial z} = \frac{\partial F_3}{\partial x}, \qquad \frac{\partial F_2}{\partial z} = \frac{\partial F_3}{\partial y}$$

If the components of $\mathbf{F}$ do not satisfy these equations, then $\mathbf{F}$ is not a gradient vector field.
- The unit and inverse square radial vector fields are gradient vector fields

$$\mathbf{e}_r = \left\langle \frac{x}{r}, \frac{y}{r}, \frac{z}{r} \right\rangle = \nabla r, \qquad \frac{\mathbf{e}_r}{r^2} = \left\langle \frac{x}{r^3}, \frac{y}{r^3}, \frac{z}{r^3} \right\rangle = \nabla(-r^{-1})$$

where $r = (x^2 + y^2 + z^2)^{1/2}$. In $\mathbf{R}^2$, $\mathbf{e}_r = \left\langle \frac{x}{r}, \frac{y}{r} \right\rangle$, where $r = (x^2 + y^2)^{1/2}$.

16.1 EXERCISES

Preliminary Questions

1. Which of the following is a unit vector field in the plane?

(a) $\mathbf{F} = \langle y, x \rangle$

(b) $\mathbf{F} = \left\langle \dfrac{y}{\sqrt{x^2+y^2}}, \dfrac{x}{\sqrt{x^2+y^2}} \right\rangle$

(c) $\mathbf{F} = \left\langle \dfrac{y}{x^2+y^2}, \dfrac{x}{x^2+y^2} \right\rangle$

2. Sketch an example of a nonconstant vector field in the plane in which each vector is parallel to $\langle 1, 1 \rangle$.

3. Show that the vector field $\mathbf{F} = \langle -z, 0, x \rangle$ is orthogonal to the position vector $\overrightarrow{OP}$ at each point P. Give an example of another vector field with this property.

4. Give an example of a potential function for $\langle yz, xz, xy \rangle$ other than $\varphi(x, y, z) = xyz$.

Exercises

1. Compute and sketch the vector assigned to the points $P = (1, 2)$ and $Q = (-1, -1)$ by the vector field $\mathbf{F} = \langle x^2, x \rangle$.

2. Compute and sketch the vector assigned to the points $P = (1, 2)$ and $Q = (-1, -1)$ by the vector field $\mathbf{F} = \langle -y, x \rangle$.

3. Compute and sketch the vector assigned to the points $P = (0, 1, 1)$ and $Q = (2, 1, 0)$ by the vector field $\mathbf{F} = \langle xy, z^2, x \rangle$.

4. Compute the vector assigned to the points $P = (1, 1, 0)$ and $Q = (2, 1, 2)$ by the vector fields $\mathbf{e}_r$, $\dfrac{\mathbf{e}_r}{r}$, and $\dfrac{\mathbf{e}_r}{r^2}$.

In Exercises 5–13, sketch the following planar vector fields by drawing the vectors attached to points with integer coordinates in the rectangle $-3 \le x, y \le 3$. Instead of drawing the vectors with their true lengths, scale them if necessary to avoid overlap.

5. $\mathbf{F} = \langle 1, 0 \rangle$

6. $\mathbf{F} = x\mathbf{i}$

7. $\mathbf{F} = \langle 0, x \rangle$

8. $\mathbf{F} = y\mathbf{j}$

9. $\mathbf{F} = \langle 1, 1 \rangle$

10. $\mathbf{F} = x^2\mathbf{i} + y\mathbf{j}$

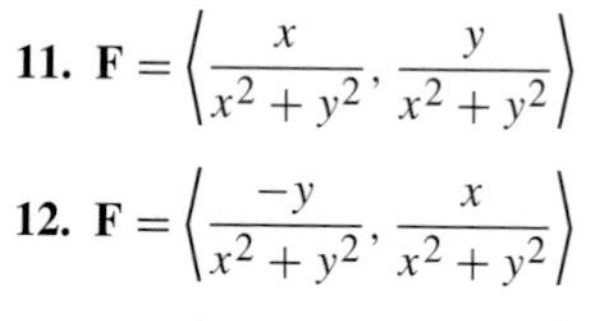

11. $\mathbf{F} = \left\langle \dfrac{x}{x^2+y^2}, \dfrac{y}{x^2+y^2} \right\rangle$

12. $\mathbf{F} = \left\langle \dfrac{-y}{x^2+y^2}, \dfrac{x}{x^2+y^2} \right\rangle$

13. $\mathbf{F} = \left\langle \dfrac{y}{x^2+y^2}, \dfrac{-x}{x^2+y^2} \right\rangle$

In Exercises 14–17, match the planar vector field with the corresponding plot in Figure 7.

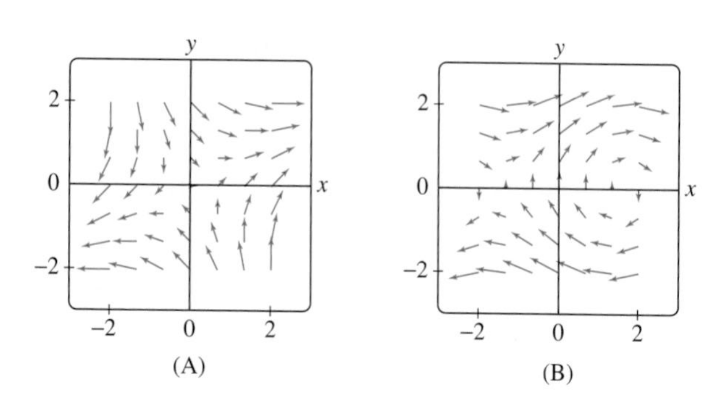

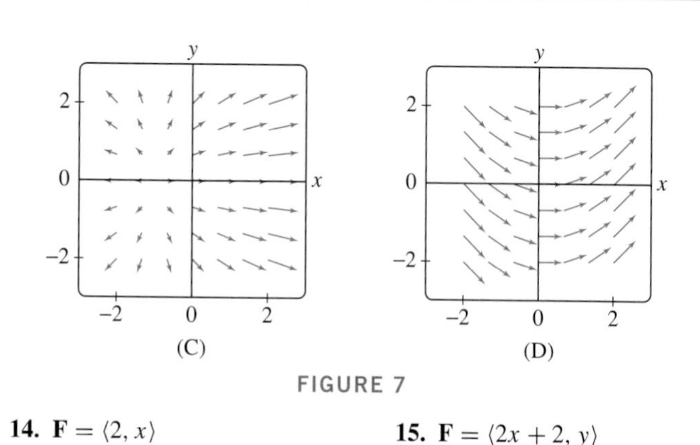

FIGURE 7

14. $\mathbf{F} = \langle 2, x \rangle$

15. $\mathbf{F} = \langle 2x + 2, y \rangle$

16. $\mathbf{F} = \langle y, \cos x \rangle$

17. $\mathbf{F} = \langle x + y, x - y \rangle$

In Exercises 18–21, match the three-dimensional vector field with the corresponding plot in Figure 8.

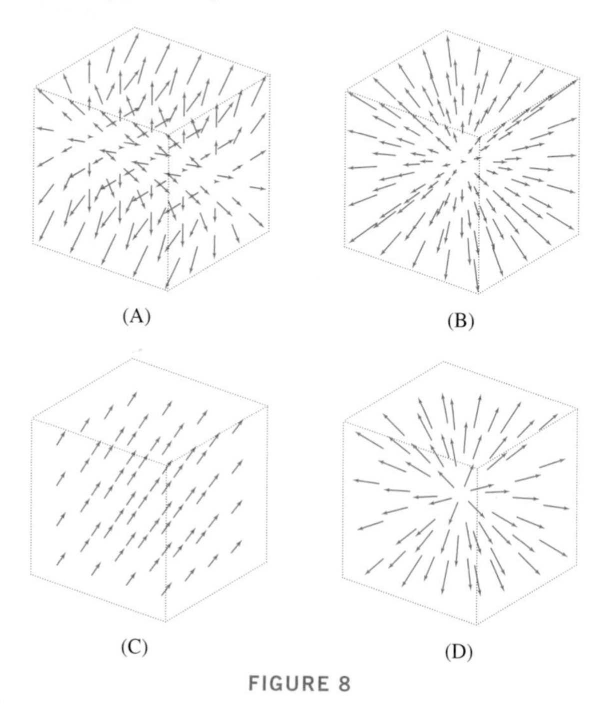

FIGURE 8

18. $\mathbf{F} = \langle x, y, z \rangle$

19. $\mathbf{F} = \langle x, 0, z \rangle$

20. $\mathbf{F} = \left\langle \dfrac{x}{\sqrt{x^2+y^2+z^2}}, \dfrac{y}{\sqrt{x^2+y^2+z^2}}, \dfrac{z}{\sqrt{x^2+y^2+z^2}} \right\rangle$

21. $\mathbf{F} = \langle 1, 1, 1 \rangle$

In Exercises 22–25, find a potential function for the vector field $\mathbf{F}$ *by inspection.*

22. $\mathbf{F} = \langle x, y \rangle$

23. $\mathbf{F} = \langle ye^{xy}, xe^{xy} \rangle$

24. $\mathbf{F} = \langle yz^2, xz^2, 2xyz \rangle$

25. $\mathbf{F} = \langle 2xze^{x^2}, 0, e^{x^2} \rangle$

26. Find potential functions for the vector fields $\mathbf{F} = \dfrac{\mathbf{e}_r}{r^3}$ and $\mathbf{G} = \dfrac{\mathbf{e}_r}{r^4}$ in $\mathbf{R}^3$.

27. Which of (A) or (B) in Figure 9 is the contour plot of the vector field $\mathbf{F}$? Recall that the gradient vectors are perpendicular to the level curves.

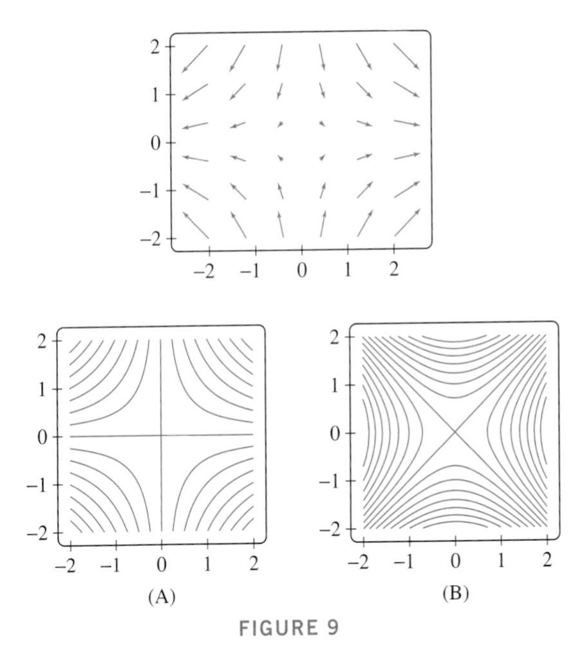

FIGURE 9

28. Which of (A) or (B) in Figure 10 is the contour plot of a potential function of the vector field $\mathbf{F}$?

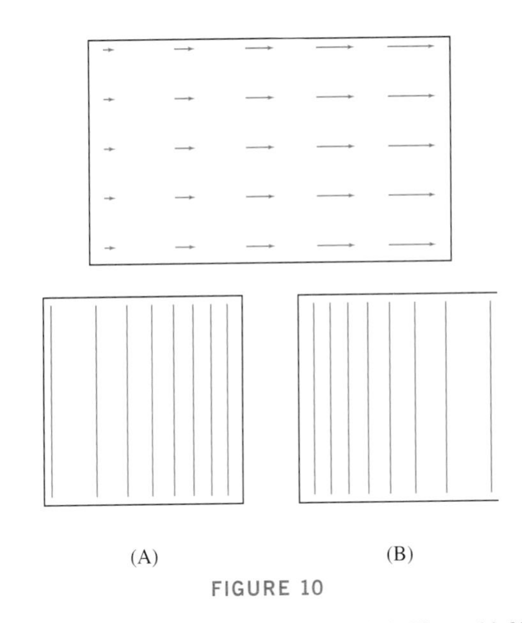

FIGURE 10

29. Let $\varphi = \ln r$, where $r = \sqrt{x^2+y^2}$. Express $\nabla\varphi$ in terms of the unit radial vector $\mathbf{e}_r$ in $\mathbf{R}^2$.

30. Match the force fields (a)–(c) with (A)–(C) in Figure 11. Note that the vectors (A)–(C) indicate direction but are not drawn to indicate magnitude.

(a) One positive and one negative charge

(b) Two positive charges

(c) Two positive charges and one negative charge

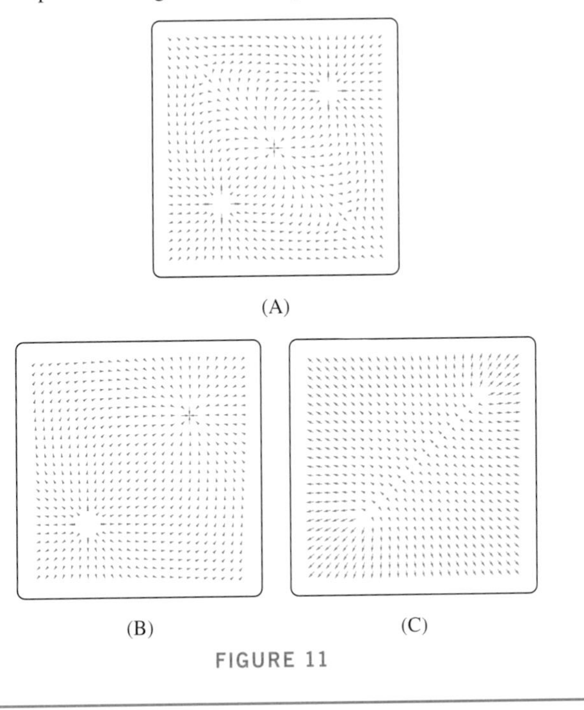

FIGURE 11

Further Insights and Challenges

31. Show that any vector field of the form $\mathbf{F} = \langle f(x), g(y), h(z) \rangle$ has a potential function. Assume that f, g, and h are continuous.

32. In this exercise, we show that the vector field $\mathbf{F}$ in Figure 12 is not a gradient vector field. Suppose that a potential function φ did exist.

(a) Argue that the level curves of φ would have to be vertical lines.

(b) Argue that the level curves would have to grow farther apart as y increases.

(c) Explain why the conclusions of (a) and (b) are incompatible.

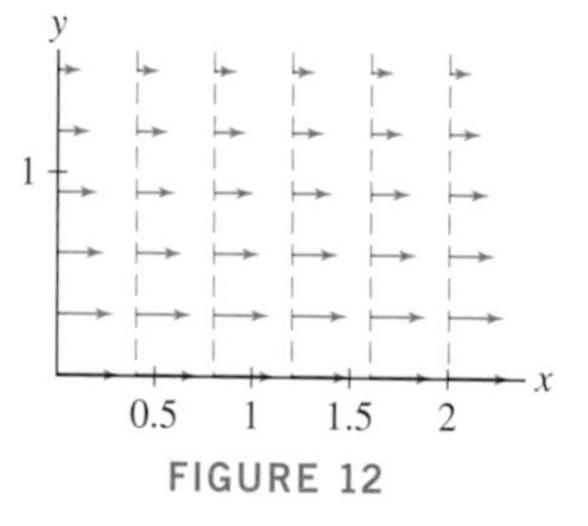

FIGURE 12

33. Show that if $\nabla\varphi(x, y) = \mathbf{0}$ for all (x, y) in a disk $\mathcal{D}$ in $\mathbf{R}^2$, then φ is constant on $\mathcal{D}$. *Hint:* Given points $P = (a, b)$ and $Q = (c, d)$ in $\mathcal{D}$, let $R = (c, b)$ (Figure 13). Use single-variable calculus to show that φ is constant along the segments $\overline{PR}$ and $\overline{RQ}$ and conclude that $\varphi(P) = \varphi(R)$.

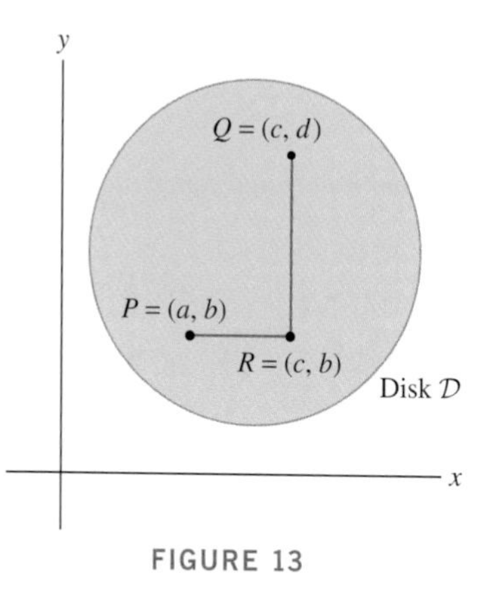

FIGURE 13

16.2 Line Integrals

In this section, we introduce integration over curves in $\mathbf{R}^3$. Integrals over curves are traditionally called **line integrals**, although it would be more appropriate to call them "curve integrals." As mentioned in the introduction to this chapter, we define line integrals both of functions and vector fields. The line integral of a function $f(x, y, z)$ over a curve $\mathcal{C}$ is called a **scalar line integral** and is denoted $\int_{\mathcal{C}} f(x, y, z)\, ds$.

All integrals, line integrals included, are defined as limits of suitable Riemann sums. To define the scalar line integral, we form a Riemann sum by dividing $\mathcal{C}$ into N consecutive arcs $\mathcal{C}_1, \ldots, \mathcal{C}_N$ (Figure 1). Denote the length of the arc $\mathcal{C}_i$ by Δs_i. Within each arc $\mathcal{C}_i$, choose a sample point $P_i \in \mathcal{C}_i$ and consider the Riemann sum:

$$\sum_{i=1}^{N} f(P_i)\,\text{length}(\mathcal{C}_i) = \sum_{i=1}^{N} f(P_i)\,\Delta s_i$$

The line integral of f over $\mathcal{C}$ is the limit (if it exists) of these Riemann sums as the maximum of the lengths Δs_i approaches zero:

$$\int_{\mathcal{C}} f(x, y, z)\, ds = \lim_{\{\Delta s_i\}\to 0} \sum_{i=1}^{N} f(P_i)\,\Delta s_i \qquad \boxed{1}$$

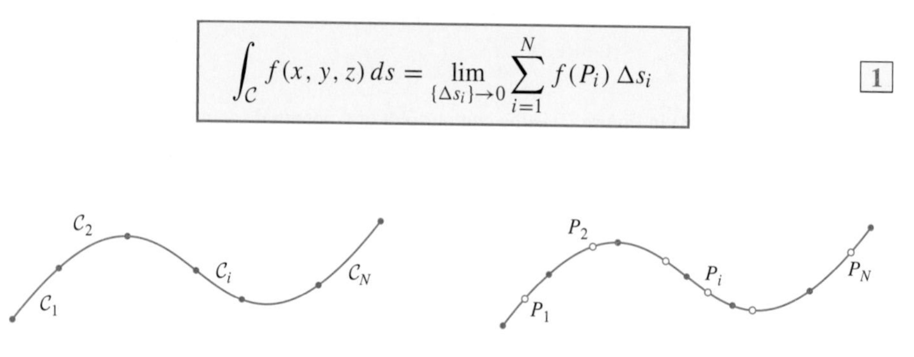

FIGURE 1 The curve $\mathcal{C}$ divided into N small arcs.

Here we write $\{\Delta s_i\} \to 0$ to indicate that the limit is taken over all Riemann sums as the maximum of the lengths Δs_i tends to zero.

The scalar line integral is a generalization of the arc length integral discussed in Section 13.3. If we take the constant function $f(x, y, z) = 1$, then all Riemann sums have the same value:

$$\sum_{i=1}^{N} 1\,\Delta s_i = \sum_{i=1}^{N} \text{length}(\mathcal{C}_i) = \text{length}(\mathcal{C})$$

and thus $\int_{\mathcal{C}} 1\,ds = \text{length}(\mathcal{C})$.

The definition of the scalar line integral as a limit in Eq. (1) is rarely if ever used in computations. As we now show, line integrals may be computed using parametrizations of the curve.

In this chapter, we use the notation $\mathbf{c}(t) = (x(t), y(t), z(t))$ to denote a path in $\mathbf{R}^3$. This notation was used in Section 14.5. We think of $\mathbf{c}(t)$ as a point moving in space as a function of time t. The derivative $\mathbf{c}'(t)$ is the tangent vector (also called the velocity vector)

$$\mathbf{c}'(t) = \langle x'(t), y'(t), z'(t) \rangle$$

Recall that if $\mathbf{c}'(t) \neq \mathbf{0}$, then $\mathbf{c}'(t)$ is tangent to the path, points in the direction of motion, and its length $\|\mathbf{c}'(t)\|$ is the speed at time t.

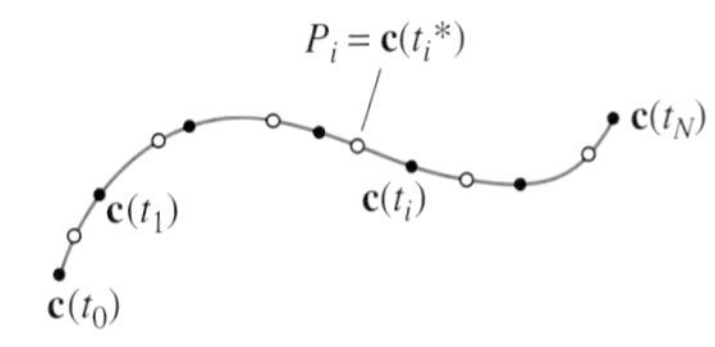

FIGURE 2 Partition of parametrized curve $\mathbf{c}(t)$.

The arc length formula from Section 13.3 states that the length of $\mathbf{c}(t)$ for $a \le t \le b$ is given by the integral

$$\int_a^b \|\mathbf{c}'(t)\|\,dt$$

Now assume that $\mathcal{C}$ has a parametrization $\mathbf{c}(t)$ for $a \le t \le b$ with continuous derivative $\mathbf{c}'(t)$. Choose a partition of the interval $[a, b]$:

$$a = t_0 < t_1 < \cdots < t_{N-1} < t_N = b$$

and let $\mathcal{C}_i$ be the portion of $\mathcal{C}$ parametrized by $\mathbf{c}(t)$ for $t_{i-1} \le t \le t_i$ (Figure 2). According to the arc length formula,

$$\text{Length}(\mathcal{C}_i) = \Delta s_i = \int_{t_{i-1}}^{t_i} \|\mathbf{c}'(t)\|\,dt$$

By the Mean Value Theorem for Integrals, there exists an intermediate point t_i^* in $[t_{i-1}, t_i]$ such that

$$\Delta s_i = \int_{t_{i-1}}^{t_i} \|\mathbf{c}'(t)\|\,dt = \|\mathbf{c}'(t_i^*)\|\,\Delta t_i \qquad \Delta t_i = t_i - t_{i-1}$$

Taking $P_i = \mathbf{c}(t_i^*)$ as our intermediate point in $\mathcal{C}_i$, we see from Eq. (1) that

$$\int_{\mathcal{C}} f(x, y, z)\,ds = \lim_{\{\Delta t_i\}\to 0} \sum_{i=1}^{N} f(\mathbf{c}(t_i^*))\|\mathbf{c}'(t_i^*)\|\,\Delta t_i \qquad \boxed{2}$$

The limit is taken over all partitions as the maximum of the lengths Δt_i tend to zero. Since the sums on the right are Riemann sums for the integral

$$\int_a^b f(\mathbf{c}(t))\|\mathbf{c}'(t)\|\,dt$$

we obtain Eq. (3) in the following theorem. Furthermore, our proof does not depend on the choice of parametrization. It follows that you can use any parametrization to compute a scalar line integral.

THEOREM 1 Computing a Scalar Line Integral Let $\mathbf{c}(t)$ be a parametrization of a curve $\mathcal{C}$ for $a \le t \le b$. Assume that $f(x, y, z)$ and $\mathbf{c}'(t)$ are continuous. Then

$$\int_{\mathcal{C}} f(x, y, z)\, ds = \int_a^b f(\mathbf{c}(t))\|\mathbf{c}'(t)\|\, dt \qquad \boxed{3}$$

The value of the integral on the right does not depend on the choice of parametrization. For $f(x, y, z) = 1$, we obtain the length of $\mathcal{C}$:

$$\text{Length of } \mathcal{C} = \int_{\mathcal{C}} \|\mathbf{c}'(t)\|\, dt \qquad \boxed{4}$$

We also note that if $\mathbf{c}(t) = (x(t), y(t), z(t))$, then

$$\|\mathbf{c}'(t)\| = \sqrt{x'(t)^2 + y'(t)^2 + z'(t)^2}$$

and the scalar line integral may be written explicitly as

$$\int_{\mathcal{C}} f(x, y, z)\, ds = \int_a^b f(\mathbf{c}(t))\sqrt{x'(t)^2 + y'(t)^2 + z'(t)^2}\, dt$$

The symbol ds is intended to suggest arc length s and is often referred to as the **line element** or **arc length differential**. When we evaluate the line integral in terms of a parametrization, we replace ds by $\|\mathbf{c}'(t)\|\, dt$.

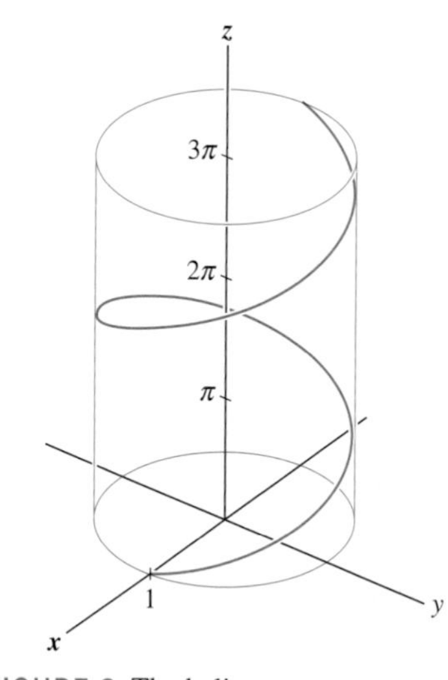

FIGURE 3 The helix $\mathbf{c}(t) = (\cos t, \sin t, t)$.

■ **EXAMPLE 1 Integrating along a Helix** Calculate $\int_{\mathcal{C}} (x + y + z)\, ds$, where $\mathcal{C}$ is the helix $\mathbf{c}(t) = (\cos t, \sin t, t)$ for $0 \le t \le \pi$ (Figure 3).

Solution

***Step 1.* Compute $ds = \|\mathbf{c}'(t)\|\, dt$.**

We have $\mathbf{c}'(t) = \langle -\sin t, \cos t, 1\rangle$, so

$$\|\mathbf{c}'(t)\| = \sqrt{(-\sin t)^2 + \cos^2 t + 1} = \sqrt{2}$$

$$ds = \|\mathbf{c}'(t)\| dt = \sqrt{2}\, dt$$

***Step 2.* Write out $f(\mathbf{c}(t))$ and evaluate the line integral.**

Let $f(x, y, z) = x + y + z$. Then

$$f(\mathbf{c}(t)) = f(\cos t, \sin t, t) = \cos t + \sin t + t$$

By Eq. (3),

$$\int_{\mathcal{C}} f(x, y, z)\, ds = \int_0^\pi f(\mathbf{c}(t))\, \|\mathbf{c}'(t)\|\, dt = \int_0^\pi (\cos t + \sin t + t)\sqrt{2}\, dt$$

$$= \sqrt{2}\left(\sin t - \cos t + \frac{1}{2}t^2\right)\Big|_0^\pi$$

$$= \sqrt{2}\left(0 + 1 + \frac{1}{2}\pi^2\right) - \sqrt{2}\,(0 - 1 + 0) = 2\sqrt{2} + \frac{\sqrt{2}}{2}\pi^2 \qquad ■$$

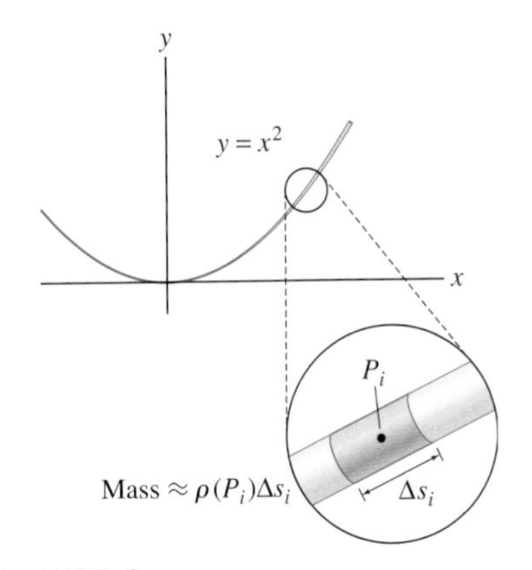

FIGURE 4

The scalar line integral can be used to express the **total mass** of a curve (think of the curve as a wire) in terms of its **mass density** $\rho(x, y, z)$, given in units of mass per unit length:

$$\text{Total mass of } \mathcal{C} = \int_{\mathcal{C}} \rho(x, y, z)\, ds \qquad \boxed{5}$$

To justify this interpretation, divide $\mathcal{C}$ into N arcs $\mathcal{C}_i$ of length Δs_i as above. If Δs_i is small, then the mass density along $\mathcal{C}_i$ is nearly constant and the mass of $\mathcal{C}_i$ is approximately equal to $\rho(P_i)\,\Delta s_i$, where P_i is any sample point on $\mathcal{C}_i$ (Figure 4). Thus, we have the following approximation for the total mass:

$$\text{Total mass of } \mathcal{C} = \sum_{i=1}^{N} \text{mass of } \mathcal{C}_i \approx \sum_{i=1}^{N} \rho(P_i)\,\Delta s_i$$

In the limit as the maximum of the lengths Δs_i tends to zero, the sums on the right approach the line integral of $\rho(x, y, z)$ and we obtain Eq. (5).

EXAMPLE 2 Scalar Line Integrals as Total Mass A wire in the shape of the parabola $y = x^2$ for $1 \le x \le 4$ has mass density $\rho(x, y) = y/x$ in units of grams per centimeter. Find the total mass of the wire.

We defined line integrals for functions of three variables. The line integral of a function $f(x, y)$ of two variables over a curve in $\mathbf{R}^2$ is defined and evaluated in a similar fashion (but the z-component does not appear in the formulas).

Solution The arc of the parabola is parametrized by $\mathbf{c}(t) = (t, t^2)$ for $1 \le t \le 4$.

***Step 1.* Compute $ds = \|\mathbf{c}'(t)\|\, dt$.**

$$\mathbf{c}'(t) = \langle 1, 2t \rangle$$

$$ds = \|\mathbf{c}'(t)\|\, dt = \sqrt{1 + 4t^2}\, dt$$

***Step 2.* Write out $\rho(\mathbf{c}(t))$ and evaluate the line integral.**

$$\rho(\mathbf{c}(t)) = \rho(t, t^2) = \frac{t^2}{t} = t$$

$$\int_{\mathcal{C}} \rho(x, y)\, ds = \int_1^4 \rho(\mathbf{c}(t))\|\mathbf{c}'(t)\|\, dt = \int_1^4 t\sqrt{1 + 4t^2}\, dt$$

Now use the substitution $u = 1 + 4t^2$, $du = 8t\, dt$. Then the new limits of integration are $u(1) = 5$ and $u(4) = 65$, and

$$\int_1^4 t\sqrt{1 + 4t^2} = \frac{1}{8}\int_5^{65} \sqrt{u}\, du = \frac{1}{12}u^{3/2}\bigg|_5^{65} = \frac{1}{12}(65^{3/2} - 5^{3/2}) \approx 42.74$$

The total mass of the wire is approximately 42.7 g. ■

The Vector Line Integral

We now discuss the line integral of a vector field $\mathbf{F}$ along a curve $\mathcal{C}$, denoted $\int_{\mathcal{C}} \mathbf{F} \cdot d\mathbf{s}$. We make the standing assumption that $\mathbf{F} = \langle F_1, F_2, F_3 \rangle$ is smooth, that is, the components F_1, F_2, F_3 are smooth and, unless otherwise stated, that the curve $\mathcal{C}$ is smooth.

Intuitively speaking, the vector line integral is the integral of the **tangential component** of the vector field along the curve. When the context is clear, we refer to a vector line integral more simply as a line integral. However, there is an important difference between vector and scalar line integrals: To define a vector line integral, we must specify a *direction* along the path or curve. A curve $\mathcal{C}$ can be traversed in one of two directions, and we say that $\mathcal{C}$ is **oriented** if one of these two directions is specified. We refer to this direction as the **forward direction** along the curve (Figure 5).

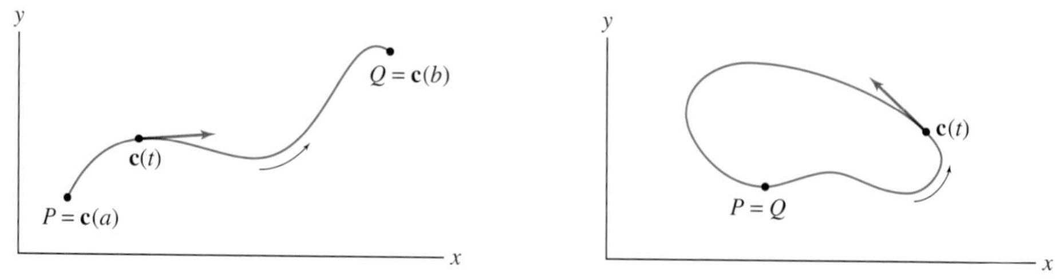

FIGURE 5 An oriented curve is a curve with a specified direction.

Given an oriented curve $\mathcal{C}$ in the plane or in 3-space, let $\mathbf{T} = \mathbf{T}(P)$ denote the unit tangent vector pointing in the forward direction along the curve. If $\mathbf{F}$ is a vector field and P a point on $\mathcal{C}$, the dot product $\mathbf{F}(P) \cdot \mathbf{T}(P)$ represents the tangential component of $\mathbf{F}$ along $\mathcal{C}$ at P, that is, the length of the projection of $\mathbf{F}(P)$ along $\mathbf{T}(P)$ (Figure 6). Recall that, since $\|\mathbf{T}(P)\| = 1$, we have

The unit tangent vector $\mathbf{T}$ varies from point to point along the curve. When it is necessary to stress this dependence, we write $\mathbf{T}(P)$ or $\mathbf{T}(\mathbf{c}(t))$.

$$\mathbf{F}(P) \cdot \mathbf{T}(P) = \|\mathbf{F}(P)\| \cos\theta$$

where θ is the angle between $\mathbf{F}(P)$ and $\mathbf{T}(P)$. As P varies along the curve, $\mathbf{F} \cdot \mathbf{T}$ defines a scalar function on $\mathcal{C}$ and the scalar line integral of this function is the vector line integral of $\mathbf{F}$.

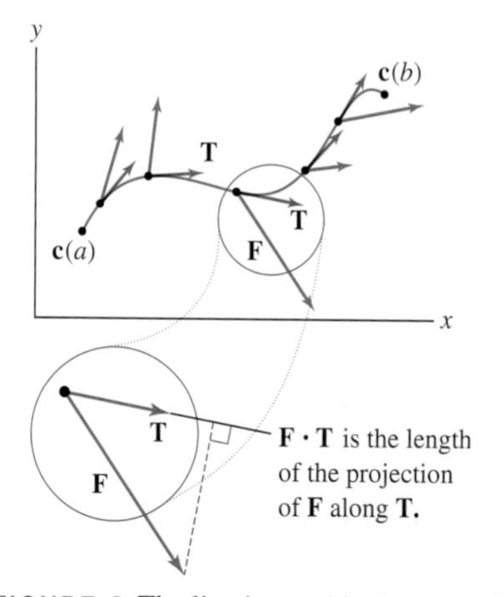

FIGURE 6 The line integral is the integral of the tangential component of $\mathbf{F}$ along $\mathcal{C}$.

DEFINITION Vector Line Integral Let $\mathcal{C}$ be an oriented curve and let $\mathbf{T}$ denote the unit tangent vector pointing in the forward direction along $\mathcal{C}$. The line integral of a vector field $\mathbf{F}$ along $\mathcal{C}$ is the integral of the tangential component of $\mathbf{F}$:

$$\int_{\mathcal{C}} \mathbf{F} \cdot d\mathbf{s} = \int_{\mathcal{C}} (\mathbf{F} \cdot \mathbf{T})\, ds \qquad \boxed{6}$$

As in the scalar case, we evaluate vector line integrals by using parametrizations. However, in the vector case, we must choose a parametrization $\mathbf{c}(t)$ that traces $\mathcal{C}$ in the forward direction. We also assume that $\mathbf{c}(t)$ is regular, that is, $\mathbf{c}'(t) \neq 0$ for $a \le t \le b$. In this case, $\mathbf{c}'(t)$ is a nonzero tangent vector pointing in the forward direction and the unit tangent vector is

$$\mathbf{T} = \text{unit tangent vector} = \frac{\mathbf{c}'(t)}{\|\mathbf{c}'(t)\|}$$

The tangential component of $\mathbf{F}$ at $\mathbf{c}(t)$ is $\mathbf{F}(\mathbf{c}(t)) \cdot \mathbf{T}(\mathbf{c}(t))$, and thus,

$$\begin{aligned}\int_{\mathcal{C}} \mathbf{F} \cdot d\mathbf{s} = \int_{\mathcal{C}} (\mathbf{F} \cdot \mathbf{T})\, ds &= \int_a^b \mathbf{F}(\mathbf{c}(t)) \cdot \mathbf{T}(\mathbf{c}(t))\, \|\mathbf{c}'(t)\|\, dt \\ &= \int_a^b \mathbf{F}(\mathbf{c}(t)) \cdot \left(\frac{\mathbf{c}'(t)}{\|\mathbf{c}'(t)\|}\right) \|\mathbf{c}'(t)\|\, dt \\ &= \int_a^b \mathbf{F}(\mathbf{c}(t)) \cdot \mathbf{c}'(t)\, dt\end{aligned}$$

THEOREM 2 Computing a Vector Line Integral Let $\mathbf{c}(t)$ be a regular parametrization of an oriented curve $\mathcal{C}$ for $a \le t \le b$. The line integral of a vector field $\mathbf{F}$ over a curve $\mathcal{C}$ is equal to

$$\int_{\mathcal{C}} \mathbf{F} \cdot d\mathbf{s} = \int_a^b \mathbf{F}(\mathbf{c}(t)) \cdot \mathbf{c}'(t)\, dt \tag{7}$$

It is useful to think of $d\mathbf{s}$ as a "vector line element" or "vector differential" whose expression is

$$d\mathbf{s} = \mathbf{c}'(t)\, dt$$

when we choose a parametrization. The vector line integral is the integral of the dot product of $\mathbf{F}$ and the vector differential $d\mathbf{s}$.

EXAMPLE 3 Let $\mathbf{F} = \langle z, y^2, x\rangle$. Evaluate $\int_{\mathcal{C}} \mathbf{F} \cdot d\mathbf{s}$, where $\mathcal{C}$ is parametrized by $\mathbf{c}(t) = (t+1, e^t, t^2)$ for $0 \le t \le 2$.

Solution There are two steps in evaluating a line integral.

Step 1. **Calculate the integrand $\mathbf{F}(\mathbf{c}(t)) \cdot \mathbf{c}'(t)$.**

$$\mathbf{c}(t) = (t+1, e^t, t^2)$$

$$\mathbf{F}(\mathbf{c}(t)) = \langle z, y^2, x\rangle = \langle t^2, e^{2t}, t+1\rangle$$

$$\mathbf{c}'(t) = \langle 1, e^t, 2t\rangle$$

The integrand is the dot product:

$$\mathbf{F}(\mathbf{c}(t)) \cdot \mathbf{c}'(t) = \langle t^2, e^{2t}, t+1\rangle \cdot \langle 1, e^t, 2t\rangle = e^{3t} + 3t^2 + 2t$$

Step 2. **Evaluate the line integral.**

$$\int_{\mathcal{C}} \mathbf{F} \cdot d\mathbf{s} = \int_0^2 \mathbf{F}(\mathbf{c}(t))\mathbf{c}'(t)\, dt$$

$$= \int_0^2 (e^{3t} + 3t^2 + 2t)\, dt = \left(\frac{1}{3}e^{3t} + t^3 + t^2\right)\Big|_0^2$$

$$= \left(\frac{1}{3}e^6 + 8 + 4\right) - \frac{1}{3} = \frac{1}{3}\left(e^6 + 35\right)$$

■

GRAPHICAL INSIGHT To reinforce the concept of the line integral as the integral of the tangential component, consider the line integral of the vector field $\mathbf{F} = \langle 2y, -3\rangle$ around the ellipse in Figure 7. Observe in part (A) that the dot product of $\mathbf{F} \cdot \mathbf{T}$ at each point along the top part of the ellipse is negative because $\mathbf{F} \cdot \mathbf{T} = \|\mathbf{F}\| \cos\theta$ and the angle θ between $\mathbf{F}$ and $\mathbf{T}$ is obtuse. Therefore, the line integral in part (A) is negative. Similarly, in Figure 7(B), $\mathbf{F} \cdot \mathbf{T}$ is positive and the line integral is positive. The line integral around the entire ellipse in Figure 7(C) appears to be negative because the negative tangential components from the upper part of the curve appear to dominate the positive contribution of the tangential components from the lower part. We verify this in the next example.

Before proceeding, we introduce another standard notation for the line integral $\int_{\mathcal{C}} \mathbf{F} \cdot d\mathbf{s}$ of a vector field $\mathbf{F} = \langle F_1, F_2, F_3\rangle$:

$$\int_{\mathcal{C}} F_1\, dx + F_2\, dy + F_3\, dz$$

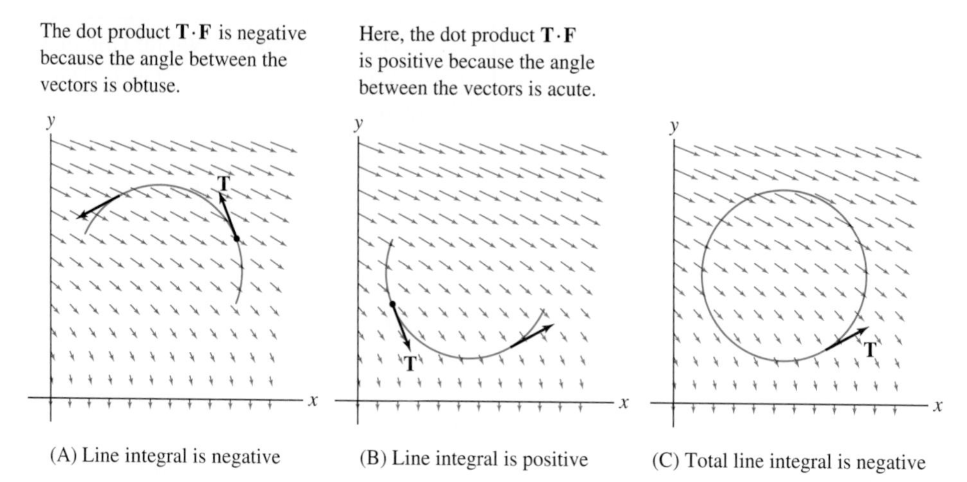

FIGURE 7 The vector field $\mathbf{F} = \langle 2y, -3\rangle$.

To evaluate the vector line integral in this notation, we choose a parametrization $\mathbf{c}(t) = (x(t), y(t), z(t))$ and replace dx by $\frac{dx}{dt}\,dt$, dy by $\frac{dy}{dt}\,dt$, and dz by $\frac{dz}{dt}\,dt$:

$$\int_a^b \underbrace{\left(F_1(\mathbf{c}(t))\frac{dx}{dt} + F_2(\mathbf{c}(t))\frac{dy}{dt} + F_3(\mathbf{c}(t))\frac{dz}{dt}\right)}_{\mathbf{F}(\mathbf{c}(t))\cdot\mathbf{c}'(t)} dt$$

EXAMPLE 4 Calculate $\int_C 2y\,dx - 3\,dy$, where C is the ellipse in Figure 7 parametrized counterclockwise by $\mathbf{c}(t) = (4 + 3\cos\theta, 3 + 2\sin\theta)$ for $0 \le \theta < 2\pi$.

In Example 4, keep in mind that

$$\int_C 2y\,dx - 3\,dy$$

is another notation for the line integral of the vector field $\mathbf{F} = \langle 2y, -3\rangle$ over C.

Solution We have $x(\theta) = 4 + 3\cos\theta$ and $y(\theta) = 3 + 2\sin\theta$, and

$$\frac{dx}{d\theta} = -3\sin\theta\,d\theta, \qquad \frac{dy}{d\theta} = 2\cos\theta\,d\theta$$

The integrand of the line integral is

$$\begin{aligned} 2y\,dx - 3\,dy &= \left(2y\frac{dx}{d\theta} - 3\frac{dy}{d\theta}\right)d\theta \\ &= \big(2(3 + 2\sin\theta)(-3\sin\theta) - 3(2\cos\theta)\big)\,d\theta \\ &= -\big(6\cos\theta + 18\sin\theta + 12\sin^2\theta\big)\,d\theta \end{aligned}$$

We evaluate the line integral, noting that the integrals of $\cos\theta$ and $\sin\theta$ over $[0, \pi]$ are zero:

$$\begin{aligned} \int_C 2y\,dx - 3\,dy &= -\int_0^{2\pi}\big(6\cos\theta + 18\sin\theta + 12\sin^2\theta\big)\,d\theta \\ &= -12\int_0^{2\pi}\sin^2\theta\,d\theta = -12\int_0^{2\pi}\left(\frac{1-\cos 2\theta}{2}\right)d\theta \\ &= (-12)(\pi) = -12\pi \end{aligned}$$

■

We now state some basic properties of vector line integrals. First, we note that on every curve there are two possible orientations corresponding to the two directions along the curve (Figure 8). We write $-\mathcal{C}$ to denote the curve $\mathcal{C}$ with the opposite orientation. If we reverse the orientation of the curve, the unit tangent vector changes sign from $\mathbf{T}$ to $-\mathbf{T}$, and it follows that the vector line integral changes sign:

$$\int_{-\mathcal{C}} \mathbf{F} \cdot d\mathbf{s} = -\int_{\mathcal{C}} \mathbf{F} \cdot d\mathbf{s}$$

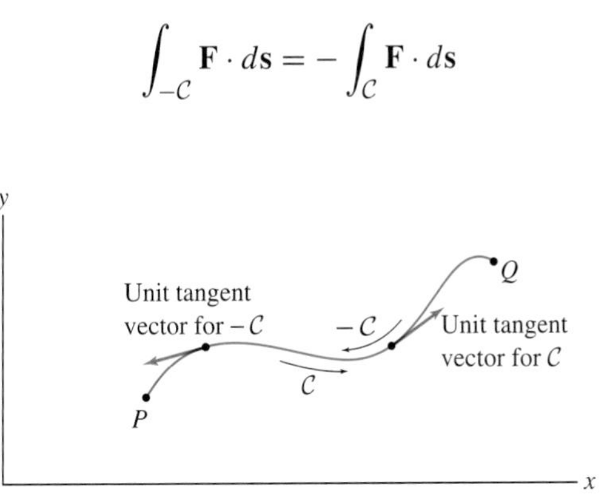

FIGURE 8 The path from P to Q has two possible orientations.

Next, suppose that $\mathcal{C}$ is the disjoint union of n curves $\mathcal{C}_1, \dots, \mathcal{C}_n$. We indicate this using the notation

$$\mathcal{C} = \mathcal{C}_1 + \cdots + \mathcal{C}_n$$

In this case, the line integral over $\mathcal{C}$ is equal to the sum of line integrals:

$$\int_{\mathcal{C}} \mathbf{F} \cdot d\mathbf{s} = \int_{\mathcal{C}_1} \mathbf{F} \cdot d\mathbf{s} + \cdots + \int_{\mathcal{C}_n} \mathbf{F} \cdot d\mathbf{s}$$

We use this formula to define the line integral when $\mathcal{C}$ is not smooth, but can be written as a sum of smooth curves $\mathcal{C}_1, \dots, \mathcal{C}_n$. In this case, $\mathcal{C}$ is called **piecewise smooth**. The next theorem summarizes the main properties of line integrals.

THEOREM 3 Properties of Line Integrals Let $\mathcal{C}$ be a smooth oriented curve and let $\mathbf{F}$ and $\mathbf{G}$ be vector fields.

(i) Linearity: $\int_{\mathcal{C}} (\mathbf{F} + \mathbf{G}) \cdot d\mathbf{s} = \int_{\mathcal{C}} \mathbf{F} \cdot d\mathbf{s} + \int_{\mathcal{C}} \mathbf{G} \cdot d\mathbf{s}$

$$\int_{\mathcal{C}} k\mathbf{F} \cdot d\mathbf{s} = k \int_{\mathcal{C}} \mathbf{F} \cdot d\mathbf{s} \quad (k \text{ a constant})$$

(ii) Reversing orientation: $\int_{-\mathcal{C}} \mathbf{F} \cdot d\mathbf{s} = -\int_{\mathcal{C}} \mathbf{F} \cdot d\mathbf{s}$

(iii) Additivity: If $\mathcal{C}$ is a sum of n smooth curves $\mathcal{C}_1 + \cdots + \mathcal{C}_n$, then

$$\int_{\mathcal{C}} \mathbf{F} \cdot d\mathbf{s} = \int_{\mathcal{C}_1} \mathbf{F} \cdot d\mathbf{s} + \cdots + \int_{\mathcal{C}_n} \mathbf{F} \cdot d\mathbf{s}$$

EXAMPLE 5 Compute $\int_{\mathcal{C}} \mathbf{F} \cdot d\mathbf{s}$, where $\mathbf{F} = \langle e^z, e^y, x + y \rangle$ and $\mathcal{C}$ is the triangle joining $(1, 0, 0)$, $(0, 1, 0)$, and $(0, 0, 1)$ oriented in the counterclockwise direction when viewed from above.

Solution With vertices labeled as in Figure 9, we evaluate the line integral as the sum of line integrals over the edges of the triangle:

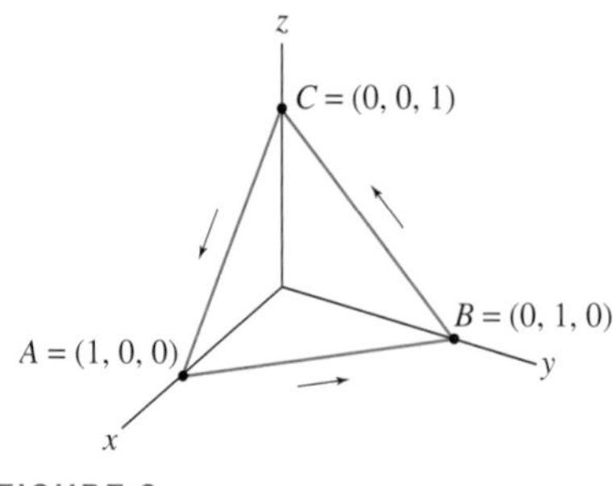

FIGURE 9

$$\int_C \mathbf{F} \cdot d\mathbf{s} = \int_{\overline{AB}} \mathbf{F} \cdot d\mathbf{s} + \int_{\overline{BC}} \mathbf{F} \cdot d\mathbf{s} + \int_{\overline{CA}} \mathbf{F} \cdot d\mathbf{s}$$

The segment $\overline{AB}$ is parametrized by the path $\mathbf{c}(t) = (1 - t, t, 0)$ for $0 \le t \le 1$. We have $\mathbf{c}'(t) = \langle -1, 1, 0 \rangle$ and

$$\mathbf{F}(\mathbf{c}(t)) \cdot \mathbf{c}'(t) = \mathbf{F}(1 - t, t, 0) \cdot \langle -1, 1, 0 \rangle$$
$$= \langle e^0, e^t, 1 \rangle \cdot \langle -1, 1, 0 \rangle = -1 + e^t$$

Thus,

$$\int_{\overline{AB}} \mathbf{F} \cdot d\mathbf{s} = \int_0^1 (e^t - 1)\, dt = (e^t - t)\Big|_0^1 = (e - 1) - 1 = \boxed{e - 2}$$

Similarly, $\overline{BC}$ is parametrized by $\mathbf{c}(t) = (0, 1 - t, t)$ for $0 \le t \le 1$ and

$$\mathbf{F}(\mathbf{c}(t)) \cdot \mathbf{c}'(t) = \langle e^t, e^{1-t}, 1 - t \rangle \cdot \langle 0, -1, 1 \rangle = -e^{1-t} + 1 - t$$

$$\int_{\overline{BC}} \mathbf{F} \cdot d\mathbf{s} = \int_0^1 (-e^{1-t} + 1 - t)\, dt = \left(e^{1-t} + t - \frac{1}{2}t^2\right)\Big|_0^1 = \boxed{\frac{3}{2} - e}$$

Finally, $\overline{CA}$ is parametrized by $\mathbf{c}(t) = (t, 0, 1 - t)$ for $0 \le t \le 1$ and

$$\mathbf{F}(\mathbf{c}(t)) \cdot \mathbf{c}'(t) = \langle e^{1-t}, 1, t \rangle \cdot \langle 1, 0, -1 \rangle = e^{1-t} - t$$

$$\int_{\overline{CA}} \mathbf{F} \cdot d\mathbf{s} = \int_0^1 (e^{1-t} - t)\, dt = \left(-e^{1-t} - \frac{1}{2}t^2\right)\Big|_0^1 = \boxed{-\frac{3}{2} + e}$$

The total line integral is equal to

$$\int_C \mathbf{F} \cdot d\mathbf{s} = (e - 2) + \left(\frac{3}{2} - e\right) + \left(-\frac{3}{2} + e\right) = e - 2$$

■

Physical Interpretation of the Line Integral

Recall that in physics, the term "work" refers to the energy expended when a force is applied to an object to move it along a path. If we move the object along the straight line from P to Q by applying a constant force $\mathbf{F}$ [Figure 10(A)], then the work W is equal to the product:

$$W = (\text{tangential component of force}) \text{ times } (\text{distance from } P \text{ to } Q)$$

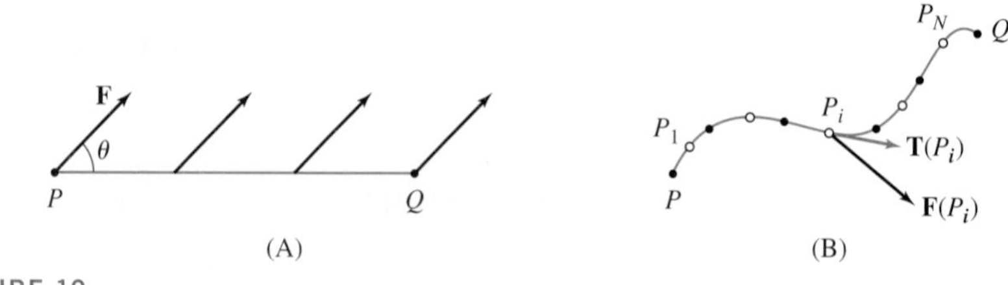

FIGURE 10

The distance from P to Q is the length $\|\mathbf{v}\|$ of the vector $\mathbf{v} = \overrightarrow{PQ}$, and the tangential component of the force is $\mathbf{F} \cdot (\mathbf{v}/\|\mathbf{v}\|)$. We see that the work is simply the dot product

$$W = \left(\mathbf{F} \cdot \frac{\mathbf{v}}{\|\mathbf{v}\|}\right) \|\mathbf{v}\| = \mathbf{F} \cdot \mathbf{v}$$

When we move the object along a curved path $\mathcal{C}$ and when the applied force $\mathbf{F}$ is no longer constant, it makes sense to define the work performed by $\mathbf{F}$ as the line integral of the tangential component of $\mathbf{F}$ along the path [Figure 10(B)]:

$$\text{Work } W = \int_{\mathcal{C}} \mathbf{F} \cdot d\mathbf{s}$$

In other words, work is equal to the integral of the tangential component of force along the direction of motion. To justify this definition, we divide $\mathcal{C}$ into N consecutive arcs $\mathcal{C}_1, \ldots, \mathcal{C}_N$, where $\mathcal{C}_i$ has length Δs_i. The work W_i required to move the object along $\mathcal{C}_i$ is approximately equal to $(\mathbf{F}(P_i) \cdot \mathbf{T}(P_i))\,\Delta s_i$, where P_i is a sample point in $\mathcal{C}_i$. This gives us the approximation to the total work:

$$W = \sum_{i=1}^{N} W_i \approx \sum_{i=1}^{N} (\mathbf{F}(P_i) \cdot \mathbf{T}(P_i))\Delta s_i$$

The right-hand side approaches the line integral $\int_{\mathcal{C}} \mathbf{F} \cdot d\mathbf{s}$ as the maximum of the lengths Δs_i tends to zero.

REMINDER *Work has units of energy. The SI unit of force is the newton and the unit of energy is the joule, defined as 1 newton-meter. The British unit is the foot-pound.*

We often wish to calculate the work required to move an object along a path in the presence of a force field $\mathbf{F}$ (such as an electrical or gravitational field). In this case, $\mathbf{F}$ acts on the object and we must work *against* the force field to move the object. The work required to move the object is the negative of the line integral:

$$\text{Work performed against } \mathbf{F} = -\int_{\mathcal{C}} \mathbf{F} \cdot d\mathbf{s}$$

EXAMPLE 6 Calculating Work The force field $\mathbf{F} = \langle x^2, -z, -yz^{-1}\rangle$ acts on a particle. Calculate the work performed moving the particle from $P = (0,0,0)$ to $Q = (1,1,1)$ along the path $\mathbf{c}(t) = (t^2, t^3, t)$ for $0 \le t \le 1$. Assume that $\mathbf{F}$ is given in newtons and the unit of distance is the meter.

Solution In this case, $\mathbf{F}(\mathbf{c}(t)) = \langle t^4, -t, -t^2\rangle$ and $\mathbf{c}'(t) = \langle 2t, 3t^2, 1\rangle$, so

$$\mathbf{F}(\mathbf{c}(t)) \cdot \mathbf{c}'(t) = \langle t^4, -t, -t^2\rangle \cdot \langle 2t, 3t^2, 1\rangle = 2t^5 - 3t^3 - t^2$$

Since we must work against the force field, the work performed in joules is

$$W = \int_{\mathcal{C}} \mathbf{F} \cdot d\mathbf{s} = -\int_0^1 (2t^5 - 3t^3 - t^2)\,dt = -\left(\frac{1}{3} - \frac{3}{4} - \frac{1}{3}\right) = \frac{3}{4} \text{ J}$$

16.2 SUMMARY

- We write $\mathbf{c}(t) = (x(t), y(t), z(t))$ for the parametrization of a curve in $\mathbf{R}^3$. The derivative $\mathbf{c}'(t) = \langle x'(t), y'(t), z'(t)\rangle$ is the tangent or velocity vector, which points in the direction of motion.
- The *line integral of a scalar function* $f(x, y, z)$ over a curve $\mathcal{C}$ is evaluated using the formula

$$\int_C f(x, y, z)\, ds = \int_a^b f(\mathbf{c}(t))\, \|\mathbf{c}'(t)\|\, dt$$

where $\mathbf{c}(t)$ is a parametrization of $\mathcal{C}$ for $a \le t \le b$. We have the symbolic equality $ds = \|\mathbf{c}'(t)\|\, dt$.

- An *oriented curve* $\mathcal{C}$ is a curve in which one of the two possible directions along $\mathcal{C}$ (called the *forward direction*) is chosen.
- The *line integral of a vector field* $\mathbf{F}$ along an oriented curve $\mathcal{C}$ is defined as the integral of the tangential component $\mathbf{F} \cdot \mathbf{T}$, where $\mathbf{T}$ is the unit tangent vector pointing in the forward direction of $\mathcal{C}$:

$$\int_C \mathbf{F} \cdot d\mathbf{s} = \int_C (\mathbf{F} \cdot \mathbf{T})\, ds = \int_a^b \mathbf{F}(\mathbf{c}(t)) \cdot \mathbf{c}'(t)\, dt$$

Here $\mathbf{c}(t)$ is a regular parametrization ($\mathbf{c}'(t) \neq \mathbf{0}$) that traces the oriented curve $\mathcal{C}$ in the forward direction. We have the symbolic equality of vector differentials, $d\mathbf{s} = \mathbf{c}'(t)\, dt$.

- Let $-\mathcal{C}$ be the curve $\mathcal{C}$ with the opposite orientation. Then

$$\int_C \mathbf{F} \cdot d\mathbf{s} = -\int_{-C} \mathbf{F} \cdot d\mathbf{s}$$

- When a variable force $\mathbf{F}$ is exerted on an object to move it along a curve $\mathcal{C}$, the work performed by $\mathbf{F}$ is equal to $\int_C \mathbf{F} \cdot d\mathbf{s}$.

16.2 EXERCISES

Preliminary Questions

1. What is the line integral of the constant function $f(x, y, z) = 10$ over a curve $\mathcal{C}$ of length 5?

2. Which of the following have a zero line integral over the vertical segment from $(0, 0)$ to $(0, 1)$?

(a) $f(x, y) = x$ **(b)** $f(x, y) = y$

(c) $\mathbf{F} = \langle x, 0\rangle$ **(d)** $\mathbf{F} = \langle y, 0\rangle$

(e) $\mathbf{F} = \langle 0, x\rangle$ **(f)** $\mathbf{F} = \langle 0, y\rangle$

3. State whether true or false. If false, give the correct statement.

(a) The scalar line integral does not depend on how you parametrize the curve.

(b) If you reverse the orientation of the curve, neither the vector nor the scalar line integral changes sign.

4. Let $\mathcal{C}$ be a curve of length 5. What is the value of $\int_C \mathbf{F} \cdot d\mathbf{s}$ if

(a) $\mathbf{F}(P)$ is normal to $\mathcal{C}$ at all points P on $\mathcal{C}$?

(b) $\mathbf{F}(P) = \mathbf{T}(P)$ at all points P on $\mathcal{C}$, where $\mathbf{T}(P)$ is the unit tangent vector pointing in the forward direction along the curve?

Exercises

1. Let $f(x, y, z) = x + yz$ and let $\mathcal{C}$ be the line segment from $P = (0, 0, 0)$ to $(6, 2, 2)$.

(a) Calculate $f(\mathbf{c}(t))$ and $ds = \|\mathbf{c}'(t)\|\, dt$ for the parametrization $\mathbf{c}(t) = (6t, 2t, 2t)$ for $0 \le t \le 1$.

(b) Evaluate $\int_C f(x, y, z)\, ds$.

2. Repeat Exercise 1 with the parametrization $\mathbf{c}(t) = (3t, t, t)$ for $0 \le t \le 2$.

3. Let $\mathbf{F} = \langle y^2, x^2\rangle$ and let $\mathcal{C}$ be the $y = x^{-1}$ for $1 \le x \le 2$, oriented from left to right.

(a) Calculate $\mathbf{F}(\mathbf{c}(t))$ and $d\mathbf{s} = \mathbf{c}'(t)\, dt$ for the parametrization $\mathbf{c}(t) = (t, t^{-1})$.

(b) Calculate the dot product $\mathbf{F}(\mathbf{c}(t)) \cdot \mathbf{c}'(t)\, dt$ and evaluate $\int_C \mathbf{F} \cdot d\mathbf{s}$.

4. Let $\mathbf{F} = \langle z^2, x, y\rangle$ and let $\mathcal{C}$ be the path $\mathbf{c}(t) = \langle 3 + 5t^2, 3 - t^2, t\rangle$ for $0 \le t \le 2$.

(a) Calculate $\mathbf{F}(\mathbf{c}(t))$ and $d\mathbf{s} = \mathbf{c}'(t)\, dt$.

(b) Calculate the dot product $\mathbf{F}(\mathbf{c}(t)) \cdot \mathbf{c}'(t)\, dt$ and evaluate $\int_C \mathbf{F} \cdot d\mathbf{s}$.

In Exercises 5–8, calculate the integral of the given scalar function or vector field over the curve $\mathbf{c}(t) = (\cos t, \sin t, t)$ *for* $0 \le t \le \pi$.

5. $f(x, y, z) = x^2 + y^2 + z^2$

6. $f(x, y, z) = xy + z$

7. $\mathbf{F} = \langle x, y, z\rangle$

8. $\mathbf{F} = \langle xy, 2, z^3\rangle$

9. Calculate the total mass of a circular piece of wire of radius 4 cm centered at the origin whose mass density is $\rho(x, y) = x^2$ g/cm.

10. Calculate the total mass of a metal tube in the helical shape $\mathbf{c}(t) = (\cos t, \sin t, t^2)$ (distance in centimeters) for $0 \le t \le 2\pi$ if the mass density is $\rho(x, y, z) = \sqrt{z}$ g/cm.

11. The values of a function $f(x, y, z)$ and vector field $\mathbf{F}(x, y, z)$ are given at six sample points along the path ABC in Figure 11. Estimate the line integrals of f and $\mathbf{F}$ along ABC.

Point	$f(x, y, z)$	$\mathbf{F}(x, y, z)$
$(1, \frac{1}{6}, 0)$	3	$\langle 1, 0, 2\rangle$
$(1, \frac{1}{2}, 0)$	3.3	$\langle 1, 1, 3\rangle$
$(1, \frac{5}{6}, 0)$	3.6	$\langle 2, 1, 5\rangle$
$(1, 1, \frac{1}{6})$	4.2	$\langle 3, 2, 4\rangle$
$(1, 1, \frac{1}{2})$	4.5	$\langle 3, 3, 3\rangle$
$(1, 1, \frac{5}{6})$	4.2	$\langle 5, 3, 3\rangle$

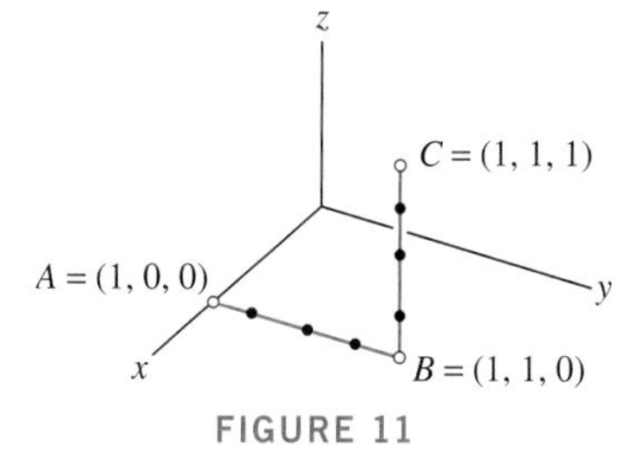

FIGURE 11

12. Estimate the line integrals of $f(x, y)$ and $\mathbf{F}(x, y)$ along the quarter circle (oriented counterclockwise) in Figure 12 using the values at the three sample points along each path.

Point	$f(x, y)$	$\mathbf{F}(x, y)$
A	1	$\langle 1, 2\rangle$
B	-2	$\langle 1, 3\rangle$
C	4	$\langle -2, 4\rangle$

FIGURE 12

13. Figure 13 shows three vector fields. In each case, determine whether the line integral around the circle (oriented counterclockwise) is positive, negative, or zero.

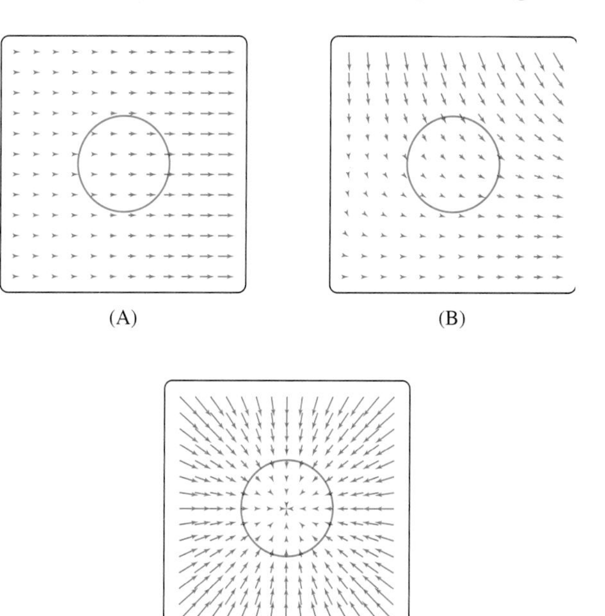

FIGURE 13

14. Determine whether the line integrals around the curves (oriented counterclockwise) in Figures 14(A) and (B) are positive or negative. What is the value of the line integral in Figure 14(C)? Explain.

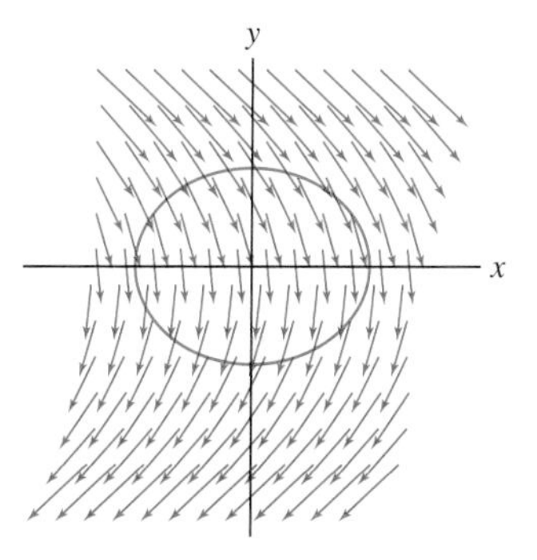

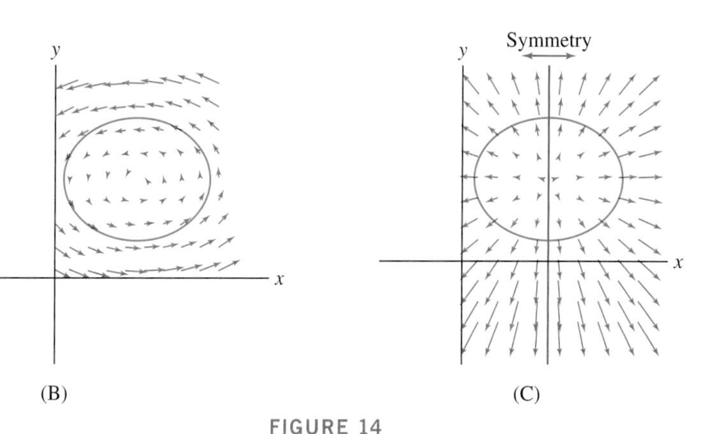

FIGURE 14

In Exercises 15–22, compute the line integral of the scalar function over the curve.

15. $f(x, y, z) = z^2$, $\mathbf{c}(t) = (2t, 3t, 4t)$ for $0 \le t \le 2$

16. $f(x, y, z) = 3x - 2y + z$, $\mathbf{c}(t) = (2 + t, 2 - t, 2t)$ for $-2 \le t \le 1$

17. $f(x, y) = \sqrt{1 + 9xy}$, $y = x^3$ for $0 \le x \le 1$

18. $f(x, y) = \dfrac{y^3}{x^7}$, $y = \frac{1}{4}x^4$ for $1 \le x \le 2$

19. $f(x, y, z) = xe^{z^2}$, piecewise linear path from $(0, 0, 1)$ to $(0, 2, 0)$ to $(1, 1, 1)$.

20. $f(x, y, z) = x^2 z$, $\mathbf{c}(t) = (e^t, \sqrt{2}t, e^{-t})$ for $0 \le t \le 1$

21. $f(x, y, z) = 2x^2 + 8z$, $\mathbf{c}(t) = (e^t, t^2, t)$, $0 \le t \le 1$

22. $f(x, y, z) = 3zy - 1 + 12xz$, $\mathbf{c}(t) = (t, \frac{t^2}{\sqrt{2}}, \frac{t^3}{3})$, $0 \le t \le 2$

In Exercises 23–35, compute the line integral of the vector field over the oriented curve.

23. $\mathbf{F} = \langle x^2, xy \rangle$, line segment from $(0, 0)$ to $(2, 2)$

24. $\mathbf{F} = \langle 4, y \rangle$, quarter circle $x^2 + y^2 = 1$ with $x, y \le 0$ oriented counterclockwise

25. $\mathbf{F} = \langle x^2, xy \rangle$, circle $x^2 + y^2 = 9$ oriented clockwise

26. $\mathbf{F} = \langle e^{y-x}, e^{2x} \rangle$, piecewise linear path from $(1, 1)$ to $(2, 2)$ to $(0, 2)$

27. $\mathbf{F} = \langle xy, x + y \rangle$, $\mathbf{c}(t) = (1 + t^{-1}, t^2)$ for $1 \le t \le 2$

28. $\mathbf{F} = \langle y + z, x, y \rangle$, $\mathbf{c}(t) = (t + t^2, \frac{1}{3}t^3, 2 + t)$ for $0 \le t \le 2$

29. $\mathbf{F} = \langle 3zy^{-1}, 4x, -y \rangle$, $\mathbf{c}(t) = (e^t, e^t, t)$ for $-1 \le t \le 1$

30. $\mathbf{F} = \langle x - y, y - z, z \rangle$, line segment from $(0, 0, 0)$ to $(1, 4, 4)$

31. $\mathbf{F} = \left\langle \dfrac{-y}{x^2 + y^2}, \dfrac{x}{x^2 + y^2} \right\rangle$, circle of radius R with center at the origin oriented counterclockwise

32. $\mathbf{F} = \left\langle \dfrac{-y}{x^2 + y^2}, \dfrac{x}{x^2 + y^2} \right\rangle$, the square with vertices $(1, 1)$, $(1, -1)$, $(-1, -1)$, and $(-1, 1)$ in the counterclockwise direction

33. $\mathbf{F} = \langle z^2, x, y \rangle$, $\mathbf{c}(t) = (\cos t, \tan t, t)$ for $0 \le t \le \frac{\pi}{4}$

34. $\mathbf{F} = \left\langle \dfrac{1}{y^3 + 1}, \dfrac{1}{z + 1}, 1 \right\rangle$, $\mathbf{c}(t) = (t^3, 2t, t^2)$ for $0 \le t \le 1$

35. $\mathbf{F} = \langle z^3, yz, x \rangle$, circle of radius 2 in the yz-plane with center at the origin oriented clockwise when viewed from the positive x-axis

36. CAS Let $f(x, y, z) = x^{-1}yz$ and let C be the curve parametrized by $\mathbf{c}(t) = (\ln t, t, t^2)$ for $2 \le t \le 4$. Use a computer algebra system to calculate $\int_C f(x, y, z)\, ds$ to four decimal places.

37. CAS Use a CAS to calculate $\int_C \langle e^{x-y}, e^{x+y} \rangle \cdot d\mathbf{s}$ to four decimal places, where C is the curve $y = \sin x$ for $0 \le x \le \pi$, oriented from left to right.

In Exercises 38–39, calculate the line integral of $\mathbf{F} = \langle e^z, e^{x-y}, e^y \rangle$ *over the given path.*

38. The path from P to Q in Figure 15

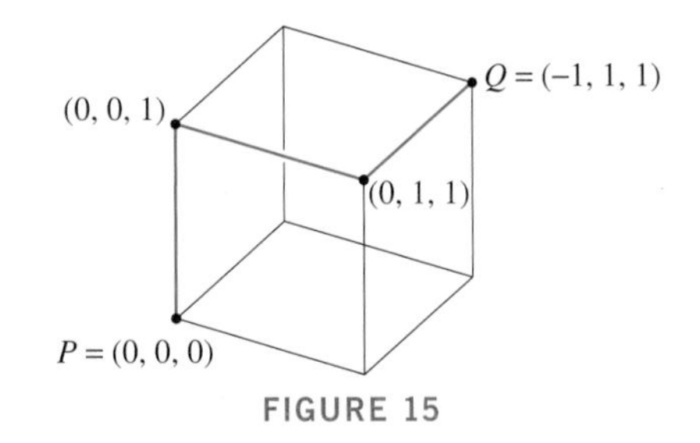

FIGURE 15

39. The path ABC in Figure 16

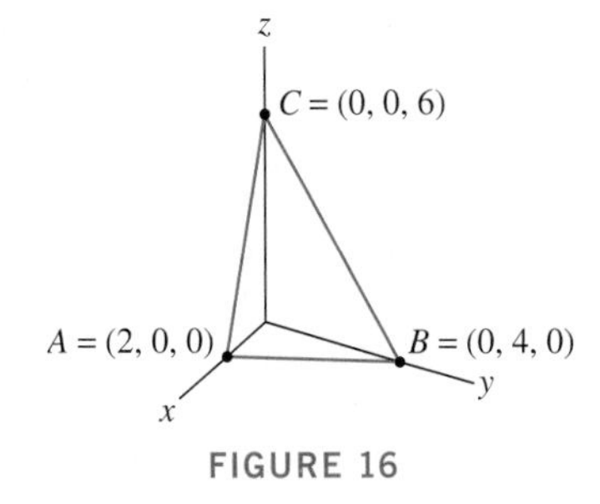

FIGURE 16

40. Let C be the path from P to Q in Figure 17 that traces C_1, C_2, and C_3 in the orientation indicated. Suppose that

$$\int_C \mathbf{F} \cdot d\mathbf{s} = 5, \qquad \int_{C_1} \mathbf{F} \cdot d\mathbf{s} = 8, \qquad \int_{C_3} \mathbf{F} \cdot d\mathbf{s} = 8$$

Determine:

(a) $\displaystyle\int_{-C_3} \mathbf{F} \cdot d\mathbf{s}$ **(b)** $\displaystyle\int_{C_2} \mathbf{F} \cdot d\mathbf{s}$ **(c)** $\displaystyle\int_{-C_1 - C_3} \mathbf{F} \cdot d\mathbf{s}$

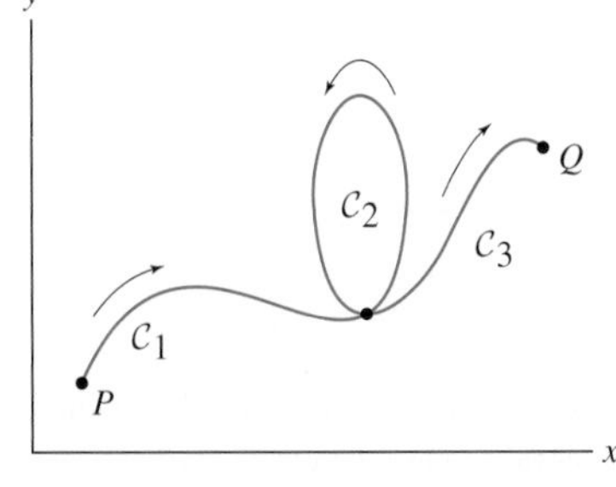

FIGURE 17

(d) What is the value of the line integral of $\mathbf{F}$ over the path that traverses the loop $\mathcal{C}_2$ four times in the clockwise direction?

In Exercises 41–44, let $\mathbf{F}$ *be the* **vortex vector field** *(so-called because it swirls around the origin as shown in Figure 18)*

$$\mathbf{F} = \left\langle \frac{-y}{x^2+y^2}, \frac{x}{x^2+y^2} \right\rangle$$

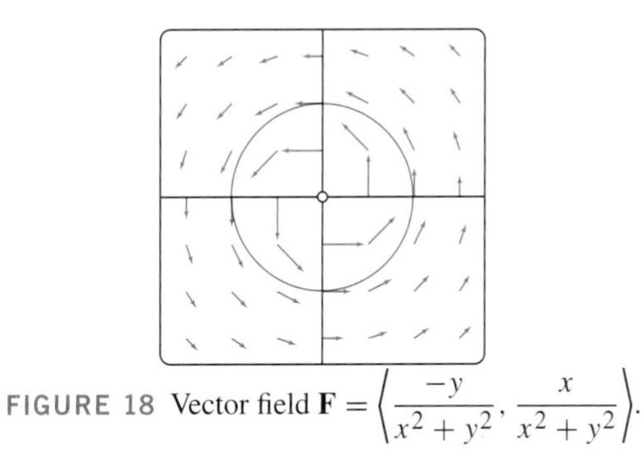

FIGURE 18 Vector field $\mathbf{F} = \left\langle \frac{-y}{x^2+y^2}, \frac{x}{x^2+y^2} \right\rangle$.

41. Let $I = \int_{\mathcal{C}} \mathbf{F} \cdot d\mathbf{s}$, where $\mathcal{C}$ is the circle of radius 2 centered at the origin oriented counterclockwise (Figure 18).

(a) Do you expect I to be positive, negative, or zero?

(b) Evaluate I.

(c) Verify that I changes sign when $\mathcal{C}$ is oriented in the clockwise direction.

42. Compute $\int_{\mathcal{C}_R} \mathbf{F} \cdot d\mathbf{s}$, where $\mathcal{C}_R$ is the circle of radius R centered at the origin oriented counterclockwise. Show that the result is independent of R.

43. Calculate $\int_{\mathcal{A}} \mathbf{F} \cdot d\mathbf{s}$, where $\mathcal{A}$ is the arc of angle θ_0 on the circle of radius R centered at the origin oriented counterclockwise. *Note:* $\mathcal{A}$ begins at $(R, 0)$ and ends at $(R\cos\theta_0, R\sin\theta_0)$.

44. Let $a > 0$, $b < c$. Show that the integral of $\mathbf{F}$ along the segment from $P = (a, b)$ to $Q = (a, c)$ is equal to the angle $\angle POQ$ (O is the origin).

45. Calculate the line integral of the constant vector field $\mathbf{F} = \langle 2, -1, 4 \rangle$ along the segment $\overline{PQ}$, where:

(a) $P = (0, 0, 0)$, $Q = (1, 0, 0)$

(b) $P = (0, 0, 0)$, $Q = (4, 3, 5)$

(c) $P = (3, 2, 3)$, $Q = (4, 8, 12)$

46. Show that if $\mathbf{F}$ is a constant vector field and $\mathcal{C}$ is any oriented path from P to Q, then

$$\int_{\mathcal{C}} \mathbf{F} \cdot d\mathbf{s} = \mathbf{F} \cdot \overrightarrow{PQ}$$

47. Figure 19 shows the vector field $\mathbf{F}(x, y) = \langle x, x \rangle$.

(a) Are $\int_{\overline{AB}} \mathbf{F} \cdot d\mathbf{s}$ and $\int_{\overline{DC}} \mathbf{F} \cdot d\mathbf{s}$ equal? If not, which is larger?

(b) Which is smaller: the line integral of $\mathbf{F}$ over the path ADC or ABC?

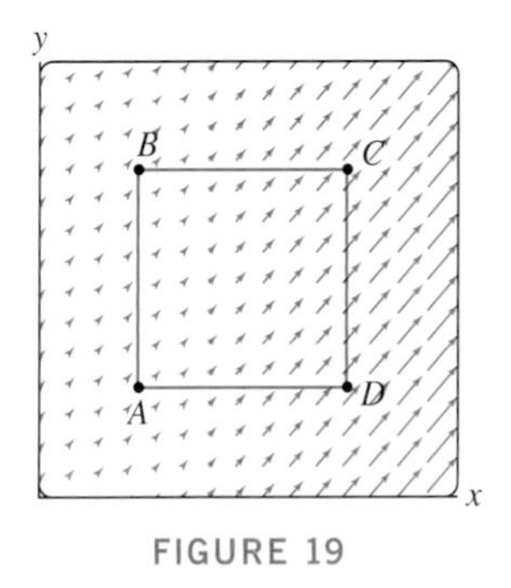

FIGURE 19

48. Calculate the work done by a field $\mathbf{F} = \langle x + y, x - y \rangle$ when an object moves from $(0, 0)$ to $(1, 1)$ along each of the paths $y = x^2$ and $x = y^2$.

49. Calculate the work done by the force field $\mathbf{F} = \langle x, y, z \rangle$ along the path $(\cos t, \sin t, t)$ for $0 \le t \le 3\pi$.

50. Let $I = \int_{\mathcal{C}} f(x, y, z)\, ds$. Assume that $f(x, y, z) \ge m$ for some number m and all points (x, y, z) on $\mathcal{C}$. Which of the following conclusions is correct? Explain.

(a) $I \ge m$

(b) $I \ge mL$, where L is the length of $\mathcal{C}$

Further Insights and Challenges

51. As observed in the text, the value of a scalar line integral does not depend on the choice of parametrization. Prove this directly. Namely, suppose that $\mathbf{c}_1(t)$ and $\mathbf{c}(t)$ are two parametrizations of $\mathcal{C}$ and that $\mathbf{c}_1(t) = \mathbf{c}(\varphi(t))$, where $\varphi(t)$ is an increasing function. Use the Change of Variables Formula to verify that

$$\int_c^d f(\mathbf{c}_1(t)) \|\mathbf{c}_1'(t)\|\, dt = \int_a^b f(\mathbf{c}(t)) \|\mathbf{c}'(t)\|\, dt$$

where $a = \varphi(c)$ and $b = \varphi(d)$.

52. Suppose we wish to compute the *average value* of a continuous function $f(x, y, z)$ along a curve $\mathcal{C}$ of length L. Given a large number N, divide $\mathcal{C}$ into N consecutive arcs $\mathcal{C}_1, \dots, \mathcal{C}_N$, each of length L/N, and let P_i be a sample point in $\mathcal{C}_i$ (Figure 20). Then the points P_i are somewhat evenly spaced along the curve and the sum $\frac{1}{N} \sum_{i=1} f(P_i)$ may be considered an approximation to the average value of f along $\mathcal{C}$. We define the average to be the limit (if it exists) as $N \to \infty$:

$$\text{Av}(f) = \lim_{N \to \infty} \frac{1}{N} \sum_{i=1} f(P_i)$$

Prove that

$$\text{Av}(f) = \frac{1}{L}\int_{C} f(x, y, z)\, ds \qquad \boxed{8}$$

Hint: Show that $\frac{L}{N}\sum_{i=1} f(P_i)$ is a Riemann sum approximation to the line integral of f along $\mathcal{C}$.

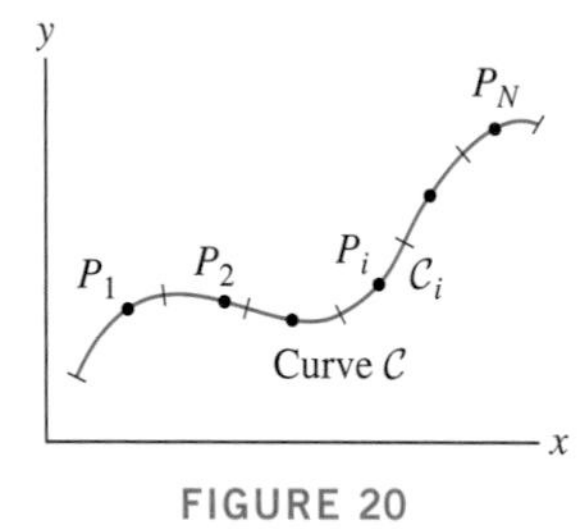

FIGURE 20

53. Use Eq. (8) to calculate the average value of $f(x, y) = x - y$ along the segment from $P = (2, 1)$ to $Q = (5, 5)$.

54. Use Eq. (8) to calculate the average value of xy along the curve $y = x^2$ for $0 \le x \le 1$.

55. The temperature (in degrees centigrade) at a point P on a circular wire of radius 2 cm centered at the origin is equal to the square of the distance from P to $P_0 = (2, 0)$. Compute the average temperature along the wire.

56. Let $\mathbf{F} = \langle x, 0\rangle$. Prove that if $\mathcal{C}$ is any path from (a, b) to (c, d), then

$$\int_{\mathcal{C}} \mathbf{F}\cdot d\mathbf{s} = \frac{1}{2}(c^2 - a^2)$$

57. Let $\mathbf{F} = \langle y, x\rangle$. Prove that if $\mathcal{C}$ is any path from (a, b) to (c, d), then

$$\int_{\mathcal{C}} \mathbf{F}\cdot d\mathbf{s} = cd - ab$$

16.3 Conservative Vector Fields

Some vector fields possess a special property called **path independence**. By this we mean that when the vector field is integrated along a path from P to Q, the result depends only on the endpoints P and Q, not on the path followed. A vector field with this property is called **conservative**.

If $\mathbf{c}(t)$ is a path from P to Q, we write $\int_{\mathbf{c}} \mathbf{F}\cdot d\mathbf{s}$ for the line integral $\int_{\mathcal{C}} \mathbf{F}\cdot d\mathbf{s}$ along the curve $\mathcal{C}$ parametrized by $\mathbf{c}(t)$. We use these two notations for the line integral interchangeably. Then a vector field $\mathbf{F}$ defined on a domain $\mathcal{D}$ is conservative if, for all P and Q in $\mathcal{D}$ and for any two paths $\mathbf{c}_1(t)$ and $\mathbf{c}_2(t)$ in $\mathcal{D}$ from P to Q (Figure 1), we have

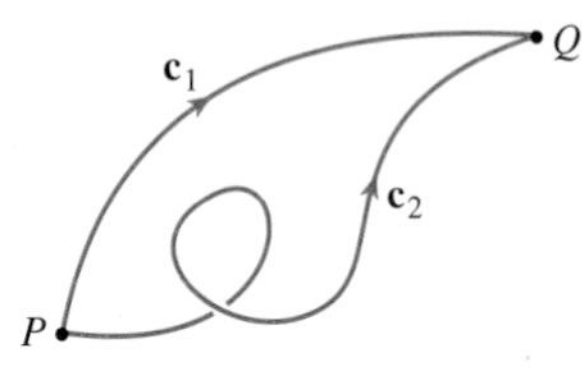

FIGURE 1 If $\mathbf{F}$ is conservative, then the line integrals over $\mathbf{c}_1$ and $\mathbf{c}_2$ are equal.

$$\int_{\mathbf{c}_1} \mathbf{F}\cdot d\mathbf{s} = \int_{\mathbf{c}_2} \mathbf{F}\cdot d\mathbf{s}$$

When $\mathcal{C}$ is a closed curve, we sometimes use the symbol $\oint$ and write the line integral as $\oint_{\mathcal{C}} \mathbf{F}\cdot d\mathbf{s}$.

Our first result states that every gradient vector field is conservative. Recall that $\mathbf{F}$ is a gradient vector field if $\mathbf{F} = \nabla\varphi$ for some function φ, called a *potential function*. We continue to assume without further mention that the components of all vector fields and all potential functions are smooth.

THEOREM 1 Fundamental Theorem for Gradient Vector Fields If $\mathbf{F} = \nabla\varphi$ on a domain $\mathcal{D}$, then for every oriented curve $\mathcal{C}$ in $\mathcal{D}$ with initial point P and terminal point Q,

$$\int_{\mathcal{C}} \mathbf{F}\cdot d\mathbf{s} = \varphi(Q) - \varphi(P) \qquad \boxed{1}$$

If $\mathcal{C}$ is closed (i.e., if $P = Q$), then $\oint_{\mathcal{C}} \mathbf{F}\cdot d\mathbf{s} = 0$.

Proof Let $\mathbf{c}(t)$ be a parametrization of $\mathcal{C}$ for $a \le t \le b$. Then $\mathbf{c}(a) = P$ and $\mathbf{c}(b) = Q$. According to the Chain Rule for Paths (Theorem 2 in Section 14.5),

$$\frac{d}{dt}\varphi(\mathbf{c}(t)) = \nabla\varphi(\mathbf{c}(t)) \cdot \mathbf{c}'(t) = \mathbf{F}(\mathbf{c}(t)) \cdot \mathbf{c}'(t)$$

Therefore,

$$\int_{\mathcal{C}} \mathbf{F} \cdot d\mathbf{s} = \int_a^b \mathbf{F}(\mathbf{c}(t)) \cdot \mathbf{c}'(t)\, dt = \int_a^b \frac{d}{dt}\varphi(\mathbf{c}(t))\, dt$$

Equation (1) follows from the Fundamental Theorem of Calculus:

$$\int_a^b \frac{d}{dt}\varphi(\mathbf{c}(t))\, dt = \varphi(\mathbf{c}(t))\Big|_a^b = \varphi(\mathbf{c}(b)) - \varphi(\mathbf{c}(a)) = \varphi(Q) - \varphi(P)$$

If $\mathcal{C}$ is a closed curve, then the initial and terminal points coincide, that is, $P = Q$. In this case, the line integral is equal to $\varphi(Q) - \varphi(P) = 0$. ■

■ **EXAMPLE 1** Let $\mathbf{F} = \langle 2xy + z, x^2, x\rangle$. Verify that $\varphi(x, y, z) = x^2y + xz$ is a potential function for $\mathbf{F}$ and evaluate the line integrals of $\mathbf{F}$ over the following paths:

(a) $\mathbf{c}(t) = (e^t, e^{2t}, t^2)$ for $0 \le t \le 2$

(b) Any path from $(1, -1, 2)$ to $(2, 2, 2)$

Solution The partial derivatives of φ are the components of $\mathbf{F}$:

$$\frac{\partial\varphi}{\partial x} = 2xy + z, \qquad \frac{\partial\varphi}{\partial y} = x^2, \qquad \frac{\partial\varphi}{\partial z} = x$$

Therefore, $\nabla\varphi = \langle 2xy + z, x^2, x\rangle = \mathbf{F}$.

(a) The initial and terminal points of the path $\mathbf{c}(t) = (e^t, e^{2t}, t^2)$ for $0 \le t \le 2$ are

$$P = \mathbf{c}(0) = (1, 1, 0), \qquad Q = \mathbf{c}(2) = (e^2, e^4, 4)$$

The values of φ at the endpoints are

$$\varphi(Q) = \varphi(e^2, e^4, 4) = e^8 + 4e^2, \qquad \varphi(P) = \varphi(1, 1, 0) = 1$$

By the Fundamental Theorem for Gradient Vector Fields,

$$\int_{\mathbf{c}} \mathbf{F} \cdot d\mathbf{s} = \varphi(Q) - \varphi(P) = (e^8 + 4e^2) - 1 = e^8 + 4e^2 - 1$$

(b) The line integral of $\mathbf{F}$ along any path $\mathbf{c}$ from $(1, -1, 2)$ to $(2, 2, 2)$ is

$$\int_{\mathbf{c}} \mathbf{F} \cdot d\mathbf{s} = \varphi(Q) - \varphi(P) = \varphi(2, 2, 2) - \varphi(1, -1, 2) = 12 - 1 = 11$$ ■

■ **EXAMPLE 2** Let $\mathbf{F} = \langle 2x + y, x\rangle$. Evaluate $\int_{\mathbf{c}} \mathbf{F} \cdot d\mathbf{s}$, where $\mathbf{c}$ is a path from $(1, 1)$ to $(3, -1)$, by finding a potential function for $\mathbf{F}$.

Solution A general method for finding potential functions is discussed later in this section. For $\mathbf{F} = \langle 2x + y, x\rangle$, we see by inspection that $\varphi(x, y) = x^2 + xy$ is a potential function. Indeed,

$$\frac{\partial\varphi}{\partial x} = \frac{\partial}{\partial x}(x^2 + xy) = 2x + y, \qquad \frac{\partial\varphi}{\partial y} = \frac{\partial}{\partial y}(x^2 + xy) = x$$

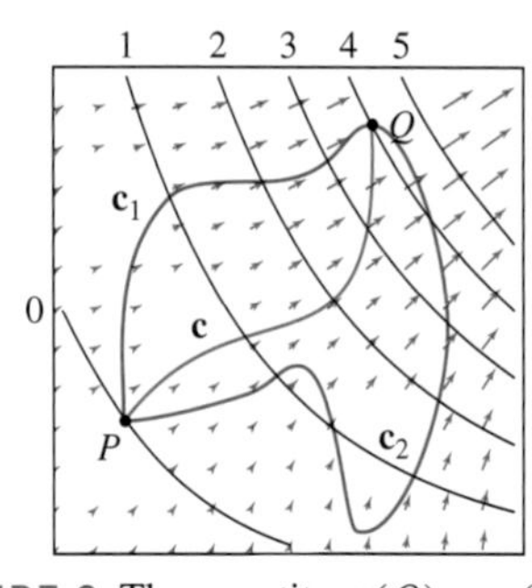

FIGURE 2 The quantity $\varphi(Q) - \varphi(P)$ is the "net" number of level curves you have to cross to get from P to Q.

Therefore, for any path $\mathbf{c}$ from $(1, 1)$ to $(3, -1)$,

$$\int_{\mathbf{c}} \mathbf{F} \cdot d\mathbf{s} = \int_{\mathbf{c}} \nabla\varphi \cdot d\mathbf{s} = \varphi(3, -1) - \varphi(1, 1) = (3^2 - 3) - (1^2 + 1) = 4 \qquad \blacksquare$$

CONCEPTUAL INSIGHT We have seen that if φ is a potential function for a gradient vector field $\mathbf{F} = \nabla\varphi$ and $\mathbf{c}(t)$ is a path from P to Q, then

$$\int_{\mathbf{c}} \nabla\varphi \cdot d\mathbf{s} = \varphi(Q) - \varphi(P) \qquad \boxed{2}$$

We shall interpret this formula in terms of the contour map of φ. If we assume that the contour interval is 1, as in Figure 2, then the potential φ changes by 1 when we move from one level curve to the next. Therefore, the net change in φ along the path $\mathbf{c}(t)$ is equal to the number of level curves crossed by $\mathbf{c}(t)$. More precisely,

$$\varphi(Q) - \varphi(P) = \text{net number of level curves crossed by } \mathbf{c} \qquad \boxed{3}$$

For example, in Figure 2, $\varphi(P) = 4$ and $\varphi(Q) = 4$, so every path from P to Q begins at the level curve $c = 0$ and ends at the level curve $c = 4$. Along the way, the path makes four crossings from one level curve to the next higher level curve. This is true even if the curve backtracks, provided that we count the "net" number of crossings. In the figure, $\mathbf{c}$ and $\mathbf{c}_1$ travel consistently in the direction of increasing potential whereas $\mathbf{c}_2$ backtracks.

Now we can explain Eq. (2). For any path $\mathbf{c}(t)$ from P to Q for $a \le t \le b$, the composite function $\varphi(\mathbf{c}(t))$ tells us which level curve of φ we are on at time t, and its derivative is the rate at which we cross level curves:

$$\frac{d}{dt}\varphi(\mathbf{c}(t)) = \nabla\varphi(\mathbf{c}(t)) \cdot \mathbf{c}'(t) = \begin{array}{l}\text{number of level curves}\\ \text{crossed per unit time}\end{array}$$

The line integral is nothing more than the integral of this rate, which is equal to the net number of crossings:

$$\int_{\mathbf{c}} \nabla\varphi \cdot d\mathbf{s} = \int_a^b \nabla\varphi(\mathbf{c}(t)) \cdot \mathbf{c}'(t)\, dt = \int_a^b \frac{d}{dt}\varphi(\mathbf{c}(t))\, dt = \begin{array}{l}\text{net number of}\\ \text{level curves crossed}\end{array} \qquad \boxed{4}$$

The net number of crossings depends only on the endpoints of the path, not on the path itself, and hence the line integral is path-independent. However, this interpretation of the line integral is valid only for a gradient vector field—otherwise, we cannot speak of the level curves of a potential function.

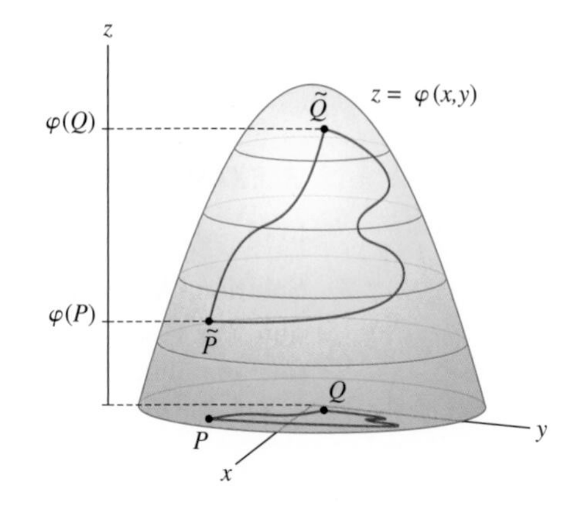

FIGURE 3 The line integral is independent of path.

In terms of the graph of a potential function $z = \varphi(x, y)$, the line integral is simply the change in height from the point $\tilde{P}$ above P to the point $\tilde{Q}$ above Q (Figure 3).

In the Conceptual Insight, we discussed gradient vector fields in the plane. A standard example of a conservative field in $\mathbf{R}^3$ is the force field $\mathbf{F} = -mg\mathbf{k}$ due to gravity acting on an object of mass m near the surface of the earth. This force field is constant, and $\mathbf{F} = -\nabla\varphi$ where $\varphi(x, y, z) = mgz$. When you hike up a mountain along a path $\mathbf{c}(t)$ beginning at altitude z_1 and ending at altitude z_2, your work against gravity is independent of the particular path and is equal to

$$-\int_{\mathbf{c}} \mathbf{F} \cdot d\mathbf{s} = \varphi(z_2) - \varphi(z_1) = mg(z_2 - z_1)$$

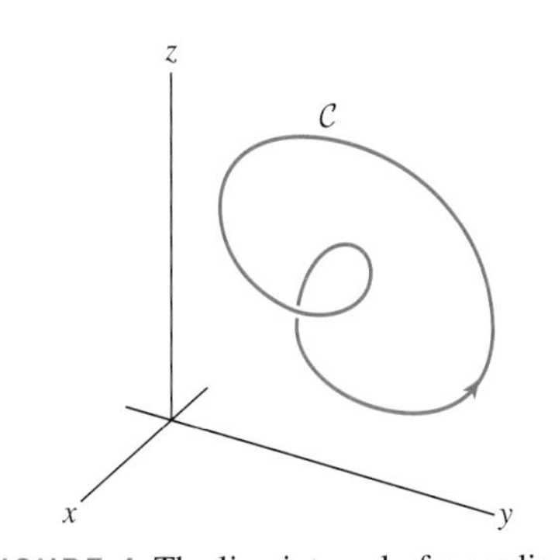

FIGURE 4 The line integral of a gradient vector field around a closed curve is zero.

EXAMPLE 3 **Integral around a Closed Path** Let $\varphi(x, y, z) = xy\sin(yz)$. Evaluate $\int_C \nabla\varphi \cdot d\mathbf{s}$, where C is the closed curve in Figure 4.

Solution By Theorem 1, the integral of a gradient vector around any closed path is zero. In other words, $\int_C \nabla\varphi \cdot d\mathbf{s} = 0$. ■

In the next theorem, we show that the condition of path independence can be stated in terms of closed paths.

THEOREM 2 Equivalent Conditions for Path Independence Let $\mathbf{F}$ be a vector field defined on a domain $\mathcal{D}$. The following two conditions are equivalent:

(a) For any two paths $\mathcal{C}_1$ and $\mathcal{C}_2$ in $\mathcal{D}$ with the same initial and terminal points,

$$\int_{\mathcal{C}_1} \mathbf{F} \cdot d\mathbf{s} = \int_{\mathcal{C}_2} \mathbf{F} \cdot d\mathbf{s}$$

(b) For all closed paths $\mathcal{C}$ in $\mathcal{D}$, $\oint_{\mathcal{C}} \mathbf{F} \cdot d\mathbf{s} = 0$.

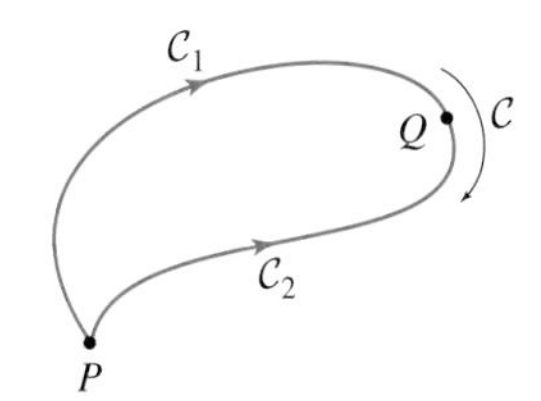

FIGURE 5 The closed loop $\mathcal{C}$ is represented as $\mathcal{C}_1 - \mathcal{C}_2$.

Proof Given any closed path $\mathcal{C}$, we may choose distinct points P and Q along $\mathcal{C}$ and let $\mathcal{C}_1$ and $\mathcal{C}_2$ be the two paths from P to Q as in Figure 5. Then

$$\mathcal{C} = \mathcal{C}_1 - \mathcal{C}_2$$

since $-\mathcal{C}_2$ is the path from Q back to P along $\mathcal{C}_2$ in the opposite direction. Thus,

$$\oint_{\mathcal{C}} \mathbf{F} \cdot d\mathbf{s} = \int_{\mathcal{C}_1} \mathbf{F} \cdot d\mathbf{s} - \int_{\mathcal{C}_2} \mathbf{F} \cdot d\mathbf{s} \qquad \boxed{5}$$

If we assume (a), then the right-hand term in Eq. (5) is zero. Thus the left-hand term is zero and (b) holds. On the other hand, Eq. (5) is valid for any two paths $\mathcal{C}_1$ and $\mathcal{C}_2$ from P to Q. If we assume (b), then the left-hand side of Eq. (5) is zero. Thus, the right-hand term is zero and (a) holds as well. ■

The Fundamental Theorem (Theorem 1) tells us that gradient vector fields are conservative. We now prove the converse statement: *Every* conservative vector field is a gradient

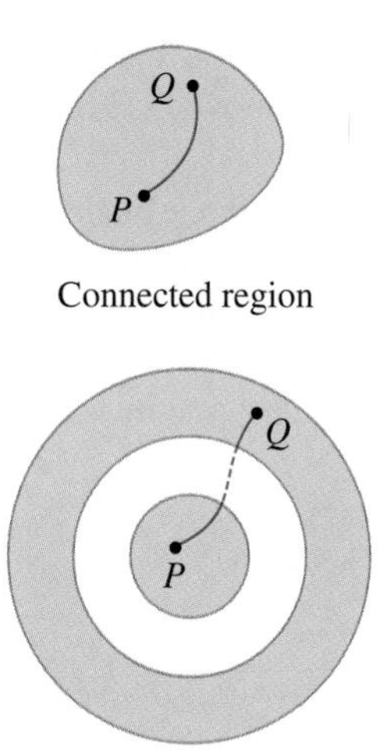

FIGURE 6 $\mathcal{D}$ is connected if any two points in $\mathcal{D}$ can be joined by a path entirely contained in $\mathcal{D}$.

vector field. Recall that a domain $\mathcal{D}$ is said to be connected if any two points in $\mathcal{D}$ can be joined by a path in $\mathcal{D}$ (Figure 6) and that $\mathcal{D}$ is open if for all $P \in \mathcal{D}$, there exists an open disk centered at P contained in $\mathcal{D}$.

> **THEOREM 3** Every conservative vector field $\mathbf{F}$ on an open connected domain is a gradient vector field, that is, $\mathbf{F} = \nabla\varphi$ for some potential function φ.

Proof We treat the case of a planar vector field $\mathbf{F} = \langle F_1, F_2 \rangle$. The proof for vector fields in $\mathbf{R}^3$ is similar. Choose a point P_0 in $\mathcal{D}$ and for any point $P = (x, y) \in \mathcal{D}$, define

$$\varphi(P) = \varphi(x, y) = \int_{\mathcal{C}} \mathbf{F} \cdot d\mathbf{s}$$

where $\mathcal{C}$ is any curve beginning at P_0 and ending at P (Figure 7). For a general vector field, this definition would not be meaningful because the line integral on the right would depend on the choice of $\mathcal{C}$. However, since $\mathbf{F}$ is assumed conservative, the integral is path-independent and φ is defined unambiguously. We must show that $\dfrac{\partial\varphi}{\partial x} = F_1$ and $\dfrac{\partial\varphi}{\partial y} = F_2$. We prove the first equation since the second may be verified in a similar manner.

To compute $\dfrac{\partial\varphi}{\partial x}$, let $\mathcal{C}_1$ be the horizontal segment from (x, y) to $(x+h, y)$. For $|h|$ small enough, $\mathcal{C}_1$ lies inside $\mathcal{D}$ and the sum $\mathcal{C} + \mathcal{C}_1$ is a curve beginning at P_0 and ending at $(x+h, y)$ (Figure 7). Therefore,

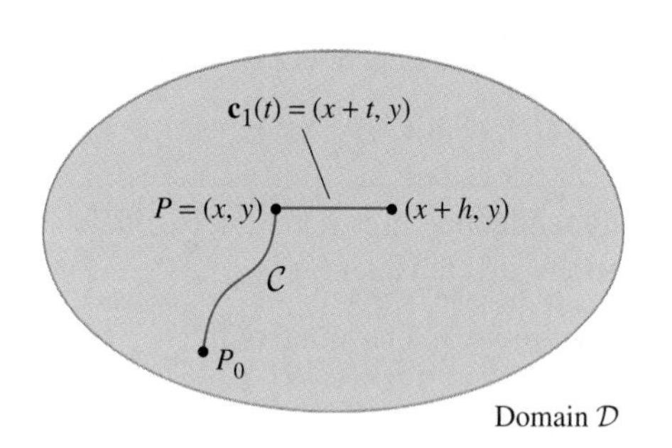

FIGURE 7

$$\varphi(x+h, y) - \varphi(x, y) = \int_{\mathcal{C}+\mathcal{C}_1} \mathbf{F} \cdot d\mathbf{s} - \int_{\mathcal{C}} \mathbf{F} \cdot d\mathbf{s}$$

$$\left(\int_{\mathcal{C}} \mathbf{F} \cdot d\mathbf{s} + \int_{\mathcal{C}_1} \mathbf{F} \cdot d\mathbf{s}\right) - \int_{\mathcal{C}} \mathbf{F} \cdot d\mathbf{s} = \int_{\mathcal{C}_1} \mathbf{F} \cdot d\mathbf{s}$$

We may parametrize $\mathcal{C}_1$ by $\mathbf{c}_1(t) = (x+t, y)$ for $0 \le t \le h$. Then $\mathbf{c}'(t) = \langle 1, 0 \rangle$ and

$$\mathbf{F}(\mathbf{c}(t)) \cdot \mathbf{c}'(t) = \langle F_1(x+t, y), F_2(x+t, y) \rangle \cdot \langle 1, 0 \rangle = F_1(x+t, y)$$

$$\varphi(x+h, y) - \varphi(x, y) = \int_{\mathbf{c}_1} \mathbf{F} \cdot d\mathbf{s} = \int_0^h F_1(x+t, y)\, dt$$

Therefore, making the substitution $u = x + t$, we obtain

$$\frac{\varphi(x+h, y) - \varphi(x, y)}{h} = \frac{1}{h}\int_0^h F_1(x+t, y)\, dt = \frac{1}{h}\int_x^{x+h} F_1(u, y)\, du$$

Now take the limit as $h \to 0$:

$$\frac{\partial\varphi}{\partial x} = \lim_{h\to 0} \frac{\varphi(x+h, y) - \varphi(x, y)}{h} = \underbrace{\lim_{h\to 0} \frac{1}{h}\int_x^{x+h} F_1(u, y)\, du}_{\text{Equal to } F_1(x, y) \text{ by the FTC}}$$

The limit on the right is equal to the derivative of the integral of $F_1(x, y)$ (for fixed y) and is therefore equal to $F_1(x, y)$ itself by the Fundamental Theorem of Calculus. This completes the proof. ∎

Finding a Potential Function

Having shown that a vector field $\mathbf{F}$ is conservative if and only if $\mathbf{F} = \nabla\varphi$ for some potential function φ, we now address the following two problems:

- How to determine whether a vector field is conservative
- Given a conservative vector field, how to find a potential function

Suppose that $\mathbf{F} = \langle F_1, F_2, F_3 \rangle$. We already know, by Theorem 1 of Section 16.1, that if $\mathbf{F}$ has a potential function, then the cross partials of the components are equal (Figure 8):

$$\frac{\partial F_1}{\partial y} = \frac{\partial F_2}{\partial x}, \qquad \frac{\partial F_2}{\partial z} = \frac{\partial F_3}{\partial y}, \qquad \frac{\partial F_3}{\partial x} = \frac{\partial F_1}{\partial z} \tag{6}$$

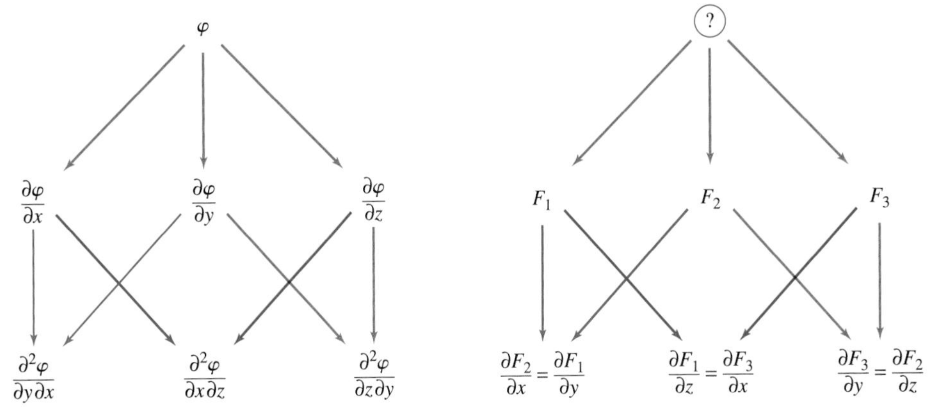

FIGURE 8

Now suppose that $\mathbf{F}$ satisfies the condition of equal cross partials. Does it follow that $\mathbf{F}$ has a potential function? The answer is yes, provided that the domain $\mathcal{D}$ has a property called simple connectedness. Roughly speaking, a domain $\mathcal{D}$ in the plane is **simply connected** if it does not have any "holes" (Figure 9). More precisely, in a simply connected domain $\mathcal{D}$, every loop in $\mathcal{D}$ can be shrunk or "contracted" down to a point while staying within $\mathcal{D}$. On the other hand, the disk with a point removed in Figure 10 is not simply connected. The loop shown in the figure cannot be shrunk down to a point without passing through the point that was removed. Standard examples of simply-

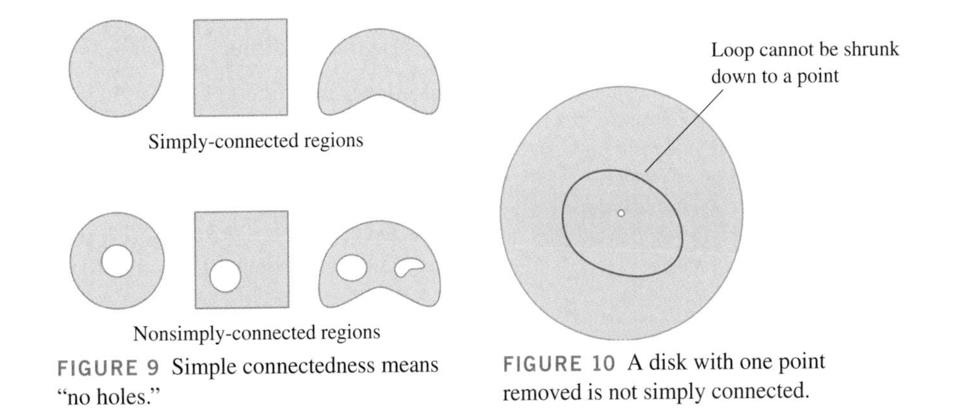

FIGURE 9 Simple connectedness means "no holes."

FIGURE 10 A disk with one point removed is not simply connected.

connected domains in $\mathbf{R}^2$ are disks, rectangles, and the entire xy-plane. In three-space, the interiors of balls and boxes are simply connected, as is the entire space $\mathbf{R}^3$.

Rather than prove Theorem 4, we indicate a procedure for finding a potential function in the next two examples. This procedure always works "locally"; that is, it allows us to find a potential function near a given point in the domain $\mathcal{D}$. The more subtle issue of proving that it always works on the entire domain relies on the fact that $\mathcal{D}$ is simply connected. A rigorous proof uses the theorems of Green and Stokes from Chapter 17.

THEOREM 4 Existence of a Potential Function Let $\mathbf{F} = \langle F_1, F_2, F_3 \rangle$ be a vector field on a simply-connected domain $\mathcal{D}$. If the cross partials of $\mathbf{F}$ are equal,

$$\frac{\partial F_1}{\partial y} = \frac{\partial F_2}{\partial x}, \qquad \frac{\partial F_2}{\partial z} = \frac{\partial F_3}{\partial y}, \qquad \frac{\partial F_3}{\partial x} = \frac{\partial F_1}{\partial z} \tag{7}$$

then $\mathbf{F} = \nabla\varphi$ for some function φ.

EXAMPLE 4 Finding a Potential Function Verify that the cross partials are equal for the vector field

$$\mathbf{F} = \left\langle 2xy + y^3 + 2, x^2 + 3xy^2 + 2y \right\rangle$$

and find a potential function in the domain $\mathcal{D} = \mathbf{R}^2$.

Solution Let $\mathbf{F} = \langle F_1, F_2 \rangle$. We first check that the cross partials are equal:

$$\frac{\partial F_1}{\partial y} = \frac{\partial}{\partial y}(2xy + y^3 + 2) \quad = 2x + 3y^2$$

$$\frac{\partial F_2}{\partial x} = \frac{\partial}{\partial x}(x^2 + 3xy^2 + 2y) = 2x + 3y^2$$

By Theorem 4, this guarantees that a potential function exists. We find it in three steps.

***Step 1.* Use the condition $\dfrac{\partial\varphi}{\partial x} = F_1$.**

The condition $\dfrac{\partial\varphi}{\partial x} = F_1$ tells us that φ is an antiderivative of $F_1(x, y)$ when y is fixed:

$$\varphi(x, y) = \int F_1(x, y)\,dx = \underbrace{\int (2xy + y^3 + 2)\,dx}_{\text{Integrate with respect to } x} = x^2y + xy^3 + 2x + g(y)$$

Note that the antiderivative is only determined up to a constant of integration. This "constant" may depend on y (but not on x), so we write it as $g(y)$. Thus,

$$\varphi(x, y) = x^2y + xy^3 + 2x + g(y)$$

***Step 2.* Use the condition $\dfrac{\partial\varphi}{\partial y} = F_2$.**

This allows us to find $g(y)$:

In summary, to find a potential function for $\mathbf{F} = \langle F_1, F_2 \rangle$, integrate $F_1(x, y)$ with respect to x and "adjust" the result by adding on a function $g(y)$ of y alone. Alternatively, integrate $F_2(x, y)$ with respect to y and adjust the result by adding on a function $f(x)$ of x alone.

$$F_2(x, y) = \frac{\partial\varphi}{\partial y} = \frac{\partial}{\partial y}\left(x^2y + xy^3 + 2x + g(y)\right) = x^2 + 3xy^2 + g'(y)$$

Since $F_2(x, y) = x^2 + 3xy^2 + 2y$, we obtain $g'(y) = 2y$ or $g(y) = y^2 + C$. Therefore, the general potential function for $\mathbf{F}$ is

$$\varphi(x, y) = x^2y + xy^3 + 2x + y^2 + C$$ ■

The same method works for vector fields in three-space.

■ **EXAMPLE 5** **Finding a Potential Function** Verify that the following vector field has a potential function and find one:

$$\mathbf{F} = \langle 2xyz^{-1}, z + x^2z^{-1}, y - x^2yz^{-2} \rangle$$

Solution First, we check that the cross partials are equal by verifying that their differences are zero:

$$\frac{\partial F_1}{\partial y} - \frac{\partial F_2}{\partial x} = \frac{\partial}{\partial y}2xyz^{-1} - \frac{\partial}{\partial x}(z + x^2z^{-1}) = 2xz^{-1} - 2xz^{-1} = 0$$

$$\frac{\partial F_2}{\partial z} - \frac{\partial F_3}{\partial y} = \frac{\partial}{\partial z}(z + x^2z^{-1}) - \frac{\partial}{\partial y}(y - x^2yz^{-2}) = (1 - x^2z^{-2}) - (1 - x^2z^{-2}) = 0$$

$$\frac{\partial F_3}{\partial x} - \frac{\partial F_1}{\partial z} = \frac{\partial}{\partial x}(y - x^2yz^{-2}) - \frac{\partial}{\partial z}2xyz^{-1} = -2xyz^{-2} - (-2xyz^{-2}) = 0$$

Step 1. **Use the condition** $\frac{\partial \varphi}{\partial x} = F_1$.

The function φ is an antiderivative of F_1 when y and z are fixed:

$$\varphi(x, y, z) = \underbrace{\int F_1(x, y, z)\,dx}_{\text{Integrate with respect to } x} = \int 2xyz^{-1}\,dx = x^2yz^{-1} + \underbrace{g(y, z)}_{\text{Constant of integration}} \qquad \boxed{8}$$

Therefore, $\varphi(x, y, z) = x^2yz^{-1} + g(y, z)$, where $g(y, z)$ does not depend on x.

Step 2. **Use the condition** $\frac{\partial \varphi}{\partial y} = F_2$.

This allows us to determine $g(y, z)$ up to a function depending on z alone:

$$F_2(x, y, z) = \frac{\partial \varphi}{\partial y} = \frac{\partial}{\partial y}(x^2yz^{-1} + g(y, z)) = x^2z^{-1} + \frac{\partial g}{\partial y} \qquad \boxed{9}$$

Since $F_2(x, y, z) = z + x^2z^{-1}$, we conclude that $\frac{\partial g}{\partial y} = z$ and hence

$$g(y, z) = \int z\,dy = yz + h(z)$$

Step 3. **Use the condition** $\frac{\partial \varphi}{\partial z} = F_3$.

At this point, by Eqs. (8) and (9), we know that

$$\varphi(x, y, z) = x^2yz^{-1} + g(y, z) = x^2yz^{-1} + yz + h(z)$$

We use this last condition to identify $h(z)$:

$$\frac{\partial \varphi}{\partial z} = \frac{\partial}{\partial z}\left(x^2yz^{-1} + zy + h(z)\right) = -x^2yz^{-2} + y + \frac{dh}{dz}$$

Since $F_3(x, y, z) = -x^2yz^{-2} + y$, it follows that $\frac{dh}{dz} = 0$. Therefore, $h(z)$ is a constant C and the general potential function is

$$\varphi(x, y, z) = x^2yz^{-1} + zy + C$$

■

Assumptions Matter We cannot expect the above method for finding a potential function to work if $\mathbf{F}$ does not satisfy the cross-partials condition. After all, in this case, no potential function exists. For example, $\mathbf{F} = \langle y, 0 \rangle$ does not satisfy the cross-partials condition. If we attempt to find a potential function, we first use the condition $\partial\varphi/\partial x = y$ to obtain $\varphi(x, y) = xy + g(y)$. But the second condition yields an equation with no solution:

$$\frac{\partial \varphi}{\partial y} = x + \frac{dg}{dy} = 0 \quad \Rightarrow \quad \frac{dg}{dy} = -x \quad \text{(no solution)}$$

Indeed, $g(y)$ depends only on y, so its derivative cannot depend on x.

CONCEPTUAL INSIGHT In physics, two important conservative force fields $\mathbf{F}$ are the gravitation force and electrostatic force. It is customary to change sign and write $\mathbf{F} = -\nabla\varphi$ and to define the **potential energy** (PE) of an object at location (x, y, z) to be $\varphi(x, y, z)$. The work required to move an object from point P to point Q along any path $\mathbf{c}$ is equal to the change in potential energy:

$$\text{Work against } \mathbf{F} = -\int_{\mathbf{c}} \mathbf{F} \cdot d\mathbf{s} = \underbrace{\varphi(Q) - \varphi(P)}_{\text{Change in potential energy}}$$

The name "conservative field" stems from the Law of Conservation of Energy. Suppose that an object subject to a conservative force $\mathbf{F}$ travels along a path $\mathbf{c}(t)$ with velocity $\mathbf{v} = \mathbf{c}'(t)$. By definition, the object's **kinetic energy** (KE) is $\frac{1}{2}\mathbf{v} \cdot \mathbf{v}$ and its total energy at time t is the sum

$$E = KE + PE = \frac{1}{2}m\mathbf{v} \cdot \mathbf{v} + \varphi(\mathbf{c}(t))$$

The Law of Conservation of Energy states that E is constant in time. To prove it, we use Newton's Law $\mathbf{F} = m\mathbf{a}$ (where $\mathbf{a} = \mathbf{v}'(t)$):

$$\begin{aligned}
\frac{dE}{dt} &= \frac{d}{dt}\left(\frac{1}{2}m\mathbf{v} \cdot \mathbf{v} + \varphi(\mathbf{c}(t))\right) \\
&= m\mathbf{v} \cdot \mathbf{a} + \nabla\varphi(\mathbf{c}(t)) \cdot \mathbf{c}'(t) && \text{(by the Chain Rule)} \\
&= m\mathbf{v} \cdot \mathbf{a} + \nabla\varphi(\mathbf{c}(t)) \cdot \mathbf{v} \\
&= m\mathbf{v} \cdot \mathbf{a} - \mathbf{F} \cdot \mathbf{v} && \text{(since } \mathbf{F} = -\nabla\varphi) \\
&= \mathbf{v} \cdot (m\mathbf{a} - \mathbf{F}) = 0 && \text{(since } \mathbf{F} = m\mathbf{a})
\end{aligned}$$

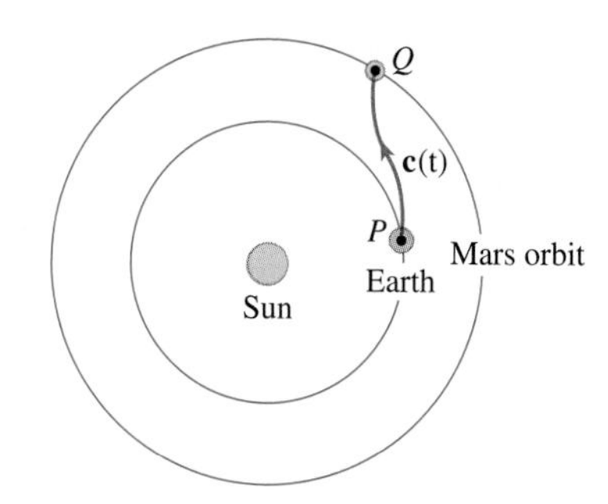

FIGURE 11

The gravitational force exerted by the sun (located at the origin) on a mass the size of the earth is $\mathbf{F} = -k\frac{\mathbf{e}_r}{r^2}$, *where*

$$\mathbf{e}_r = \frac{1}{r}\langle x, y, z\rangle \qquad \left(r = \sqrt{x^2 + y^2 + z^2}\right)$$

and $k \approx 7.9 \times 10^{44}$ N-m^2.

■ **EXAMPLE 6 Work against Gravity** Compute the work W (in joules) required to move the earth, along any path, from its current orbit to the orbit of Mars. The average distance from the sun to the earth is 1.5×10^{11} m and to Mars is 2.28×10^{11} m.

Solution The gravitational force of the sun on the earth is described by the inverse square field $\mathbf{F} = -k\frac{\mathbf{e}_r}{r^2}$ (see the marginal note). The computation in Example 4 of Section 16.1 showed that $\mathbf{F} = -\nabla\varphi$ where $\varphi = -kr^{-1}$. If $\mathbf{c}(t)$ is a path from a point P on the orbit of the earth to a point Q on the orbit of Mars (Figure 11), the work W performed *against* gravity in moving the earth along $\mathbf{c}$ is the line integral

$$W = -\int_{\mathbf{c}} \mathbf{F} \cdot d\mathbf{s} = \int_{\mathbf{c}} \nabla\varphi \cdot d\mathbf{s} = \varphi(Q) - \varphi(P) = -kr^{-1}\Big|_P^Q$$

$$= -kr^{-1}\Big|_{1.5\times 10^{11}}^{2.28\times 10^{11}} = -(7.9\times 10^{44})\left(\frac{1}{2.28\times 10^{11}} - \frac{1}{1.5\times 10^{11}}\right) \approx 1.8\times 10^{33}\ \mathrm{J}$$ ■

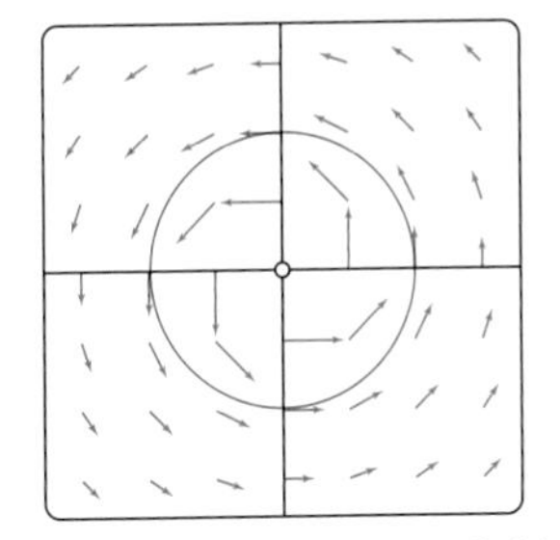
FIGURE 12 The vortex vector field.

Assumptions Matter **A Close Examination of the Vortex Vector Field** Theorem 4 states that a vector field **F** is conservative if the cross partials of its components are equal. However, Theorem 4 assumes that the domain $\mathcal{D}$ is simply connected. To see why this assumption is necessary, consider the vortex vector field **F** defined by (Figure 12)

$$\mathbf{F} = \left\langle \frac{-y}{x^2+y^2}, \frac{x}{x^2+y^2} \right\rangle$$

This vector field has three properties that seem contradictory.

(i) The cross partials are equal:

$$\frac{\partial}{\partial x}\left(\frac{x}{x^2+y^2}\right) = \frac{(x^2+y^2) - x(\partial/\partial x)(x^2+y^2)}{(x^2+y^2)^2} = \frac{y^2-x^2}{(x^2+y^2)^2}$$

$$\frac{\partial}{\partial y}\left(\frac{-y}{(x^2+y^2)}\right) = \frac{-(x^2+y^2) + y(\partial/\partial y)(x^2+y^2)}{(x^2+y^2)^2} = \frac{y^2-x^2}{(x^2+y^2)^2}$$

(ii) But **F** is not conservative because the line integral around the unit circle $\mathcal{C}$ is not zero. Indeed, using the parametrization $\mathbf{c}(\theta) = (\cos\theta, \sin\theta)$, we have

$$F(\mathbf{c}(\theta)) \cdot \mathbf{c}'(\theta) = \langle -\sin\theta, \cos\theta\rangle \cdot \langle -\sin\theta, \cos\theta\rangle = \sin^2\theta + \cos^2\theta = 1$$

$$\int_{\mathcal{C}} \mathbf{F} \cdot d\mathbf{s} = \int_0^{2\pi} \mathbf{F}(\mathbf{c}(\theta)) \cdot \mathbf{c}'(\theta)\, d\theta = \int_0^{2\pi} d\theta = 2\pi \neq 0 \qquad \boxed{10}$$

(iii) Yet **F** seems to have a potential function, namely $\varphi(x, y) = \tan^{-1}\frac{y}{x}$ (defined for $x \neq 0$). Indeed, the formula

$$\frac{d}{dt}\tan^{-1} t = \frac{1}{1+t^2}$$

and the Chain Rule give us:

$$\frac{\partial\varphi}{\partial x} = \frac{\partial}{\partial x}\tan^{-1}\frac{y}{x} = \frac{1}{1+(y/x)^2}\left(-\frac{y}{x^2}\right) = \frac{-y}{x^2+y^2}$$

$$\frac{\partial\varphi}{\partial y} = \frac{\partial}{\partial y}\tan^{-1}\frac{y}{x} = \frac{1}{1+(y/x)^2}\left(\frac{1}{x}\right) = \frac{x}{x^2+y^2}$$

In other words, $\nabla\varphi = \mathbf{F}$.

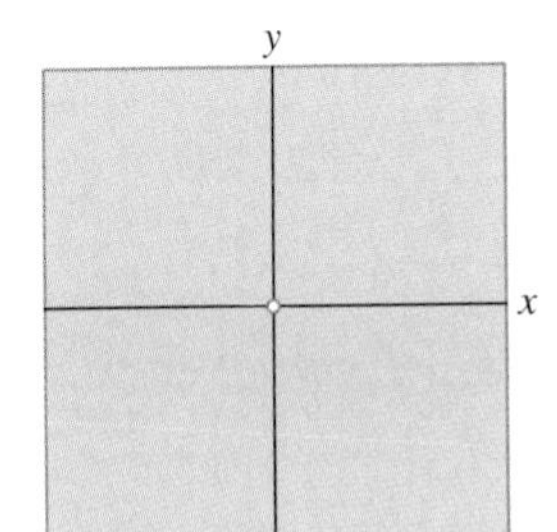

FIGURE 13 The domain $\mathcal{D}$ of the vortex vector **F** is the plane with the origin removed. This domain is not simply connected.

These contradictions are resolved when we take the domain $\mathcal{D}$ of **F** into account (Figure 13). The vector field **F** is not defined at the origin, so $\mathcal{D} = \{(x, y) \neq (0, 0)\}$. Since this domain is not simply connected, Theorem 4 does not apply and the cross-partial condition does not guarantee that **F** is conservative.

Now observe that the potential function may be interpreted as the angular coordinate of polar coordinates:

$$\varphi(x, y) = \theta = \tan^{-1}\frac{y}{x}$$

Therefore, the line integral of $\mathbf{F}$ along a curve $\mathcal{C}$ from P to Q as in Figure 14(A) is equal to the change in angle along the path:

$$\int_{\mathcal{C}} \mathbf{F} \cdot d\mathbf{s} = \int_{\mathcal{C}} \nabla\theta \cdot d\mathbf{s} = \theta_2 - \theta_1 = \text{the change in angle along } \mathcal{C}$$

But recall that the angle θ is only defined up to integer multiples of 2π. If we move along a path that goes all the way around the origin, the angle increases by 2π. This explains why the integral of $\mathbf{F}$ around the unit circle is 2π rather than zero. The key point is that $\varphi = \theta$ may be defined locally, but it cannot be defined as a continuous potential function on the entire domain $\mathcal{D}$ of $\mathbf{F}$. In general, if a closed curve $\mathcal{C}$ winds around the origin n times (where n is negative if the curve winds in the clockwise direction), then [see Figures 14(C) and (D)]:

$$\oint \mathbf{F} \cdot d\mathbf{s} = 2\pi n$$

The number n is called the *winding number* of the curve.

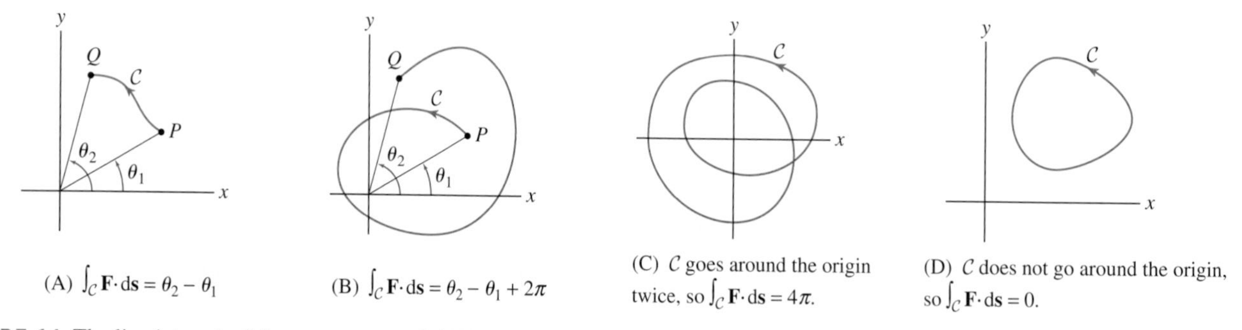

(A) $\int_{\mathcal{C}} \mathbf{F} \cdot d\mathbf{s} = \theta_2 - \theta_1$

(B) $\int_{\mathcal{C}} \mathbf{F} \cdot d\mathbf{s} = \theta_2 - \theta_1 + 2\pi$

(C) $\mathcal{C}$ goes around the origin twice, so $\int_{\mathcal{C}} \mathbf{F} \cdot d\mathbf{s} = 4\pi$.

(D) $\mathcal{C}$ does not go around the origin, so $\int_{\mathcal{C}} \mathbf{F} \cdot d\mathbf{s} = 0$.

FIGURE 14 The line integral of the vortex vector field $\mathbf{F} = \nabla\theta$ is equal to the change in θ along the curve.

16.3 SUMMARY

- A vector field $\mathbf{F}$ on a domain $\mathcal{D}$ is called *conservative* or *path-independent* if

$$\int_{\mathbf{c}_1} \mathbf{F} \cdot d\mathbf{s} = \int_{\mathbf{c}_2} \mathbf{F} \cdot d\mathbf{s}$$

for any two paths $\mathbf{c}_1$ and $\mathbf{c}_2$ in $\mathcal{D}$ with the same initial and terminal points.
- A vector field $\mathbf{F}$ on a domain $\mathcal{D}$ is called a *gradient vector field* if there exists a function φ such that $\nabla\varphi = \mathbf{F}$ on $\mathcal{D}$. The function φ is called a *potential function* of $\mathbf{F}$.
- The *Fundamental Theorem for Gradient Vector Fields*: If $\mathbf{F} = \nabla\varphi$, then

$$\int_{\mathbf{c}} \mathbf{F} \cdot d\mathbf{s} = \varphi(Q) - \varphi(P)$$

where $\mathbf{c}$ is a path from P to Q in the domain of $\mathbf{F}$. If $\mathbf{c}$ is a *closed path* ($P = Q$), then

$$\oint_{\mathbf{c}} \mathbf{F} \cdot d\mathbf{s} = 0$$

In particular, gradient vector fields are conservative.
- Also, the converse is true: For an open connected domain, a conservative vector field is a gradient vector field.

- All conservative vector fields satisfy the cross partials condition

$$\frac{\partial F_1}{\partial y} = \frac{\partial F_2}{\partial x}, \qquad \frac{\partial F_2}{\partial z} = \frac{\partial F_3}{\partial y}, \qquad \frac{\partial F_3}{\partial x} = \frac{\partial F_1}{\partial z} \qquad \boxed{11}$$

- Equality of the cross partials guarantees that $\mathbf{F}$ is conservative if the domain $\mathcal{D}$ is simply connected; that is, any loop in $\mathcal{D}$ can be shrunk to a point.

16.3 EXERCISES

Preliminary Questions

1. The following statement is false. *If* $\mathbf{F}$ *is a gradient vector field, then the line integral of* $\mathbf{F}$ *along every curve is zero.* Which single word must be added to make it true?

2. Which of the following statements are true for all vector fields, which are true only for conservative vector fields?

(a) The line integral along a path from P to Q does not depend on which path is chosen.
(b) The line integral over an oriented curve C does not depend on how the C is parametrized.
(c) The line integral around a closed curve is zero.
(d) The line integral changes sign if the orientation is reversed.
(e) The line integral is equal to the difference of a potential function at the two endpoints.
(f) The line integral is equal to the integral of the tangential component along the curve.
(g) The cross partials of the components are equal.

3. Let $\mathbf{F}$ be a vector field on an open, connected domain $\mathcal{D}$. Which of the following statements are always true, and which are true under additional hypotheses on $\mathcal{D}$?

(a) If $\mathbf{F}$ has a potential function, then $\mathbf{F}$ is conservative.
(b) If $\mathbf{F}$ is conservative, then the cross partials of $\mathbf{F}$ are equal.
(c) If the cross partials of $\mathbf{F}$ are equal, then $\mathbf{F}$ is conservative.

4. Let C, $\mathcal{D}$, and $\mathcal{E}$ be the oriented curves in Figure 15 and let $\mathbf{F} = \nabla\varphi$ be a gradient vector field such that $\int_C \mathbf{F}\cdot d\mathbf{s} = 4$. What are the values of the following integrals?

(a) $\int_{\mathcal{D}} \mathbf{F}\cdot d\mathbf{s}$ **(b)** $\int_{\mathcal{E}} \mathbf{F}\cdot d\mathbf{s}$

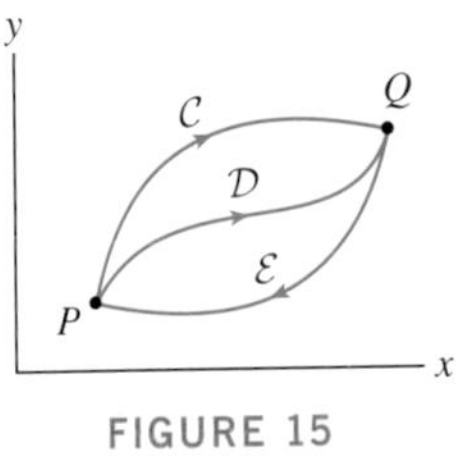

FIGURE 15

Exercises

1. Let $\varphi(x, y, z) = xy\sin(yz)$. Evaluate $\int_{\mathbf{c}} \nabla\varphi\cdot d\mathbf{s}$, where $\mathbf{c}$ is any path from $(0, 0, 0)$ to $(1, 1, \pi)$.

2. Let $\mathbf{F}$ be the gradient of $\varphi(x, y, z) = xy + z^2$. Compute the line integral of $\mathbf{F}$ along:

(a) The line segment from $(1, 1, 1)$ to $(1, 2, 2)$

(b) A circle in the xz-plane

(c) The upper half of a circle of radius 4 with its center $(0, 1, 1)$ in the yz-plane, oriented clockwise

In Exercises 3–8, verify that $\mathbf{F} = \nabla\varphi$ *and evaluate the line integral of* $\mathbf{F}$ *over the given path.*

3. $\mathbf{F} = \langle 3, 6y\rangle$, $\varphi(x, y, z) = 3x + 3y^2$; $\mathbf{c}(t) = (t, 2t^{-1})$ for $1 \le t \le 4$

4. $\mathbf{F} = \langle \sin y^2, 2xy\cos y^2\rangle$, $\varphi(x, y) = x\sin y^2$; line segment from $(2, 0)$ to $(3, \sqrt{\pi})$

5. $\mathbf{F} = \langle xy^2, x^2y\rangle$, $\varphi(x, y) = \frac{1}{2}x^2y^2$; upper half of the unit circle centered at the origin oriented counterclockwise

6. $\mathbf{F} = \langle y + 2xz, x, x^2\rangle$, $\varphi(x, y, z) = xy + x^2z$; any path from $(1, 2, 1)$ to $(3, 1, 0)$

7. $\mathbf{F} = \langle ye^x, xe^z, xye^z\rangle$, $\varphi(x, y, z) = xye^z$; $\mathbf{c}(t) = (t^2, t^3, t - 1)$ for $1 \le t \le 2$

8. $\mathbf{F} = \left\langle \frac{x}{x-y}, \frac{z}{y-z}, \ln(x-y)\right\rangle$, $\varphi(x, y, z) = z\ln(x-y)$; ellipse $2x^2 + 3(y-4)^2 = 12$ in the clockwise direction

9. Find a potential function for $\mathbf{F} = \langle 2xy + 5, x^2 - 4z, -4y\rangle$ and evaluate

$$\int_{\mathbf{c}} \mathbf{F}\cdot d\mathbf{s}$$

where $\mathbf{c}(t) = (t^2, \sin(\pi t), e^{t^2-2t})$ for $0 \le t \le 2$.

10. Verify the cross-partials condition and find a potential function for

$$\mathbf{F} = \langle 2xyz^{-1} + yz, x^2z^{-1} + xz, -x^2yz^{-2} + xy\rangle$$

on the domain $\{z \ne 0\}$.

11. Find the line integral of $\mathbf{F} = \langle 2xyz, x^2z, x^2y \rangle$ over any path from $(0, 0, 0)$ to $(3, 2, 1)$.

In Exercises 12–17, determine whether the vector field is conservative and, if so, find a potential function.

12. $\mathbf{F} = \langle z, 1, x \rangle$

13. $\mathbf{F} = \langle 0, x, y \rangle$

14. $\mathbf{F} = \langle y^2, 2xy + e^z, ye^z \rangle$

15. $\mathbf{F} = \langle y, x, z^3 \rangle$

16. $\mathbf{F} = \langle \cos(xz), \sin(yz), xy \sin z \rangle$

17. $\mathbf{F} = \langle \cos z, 2y, -x \sin z \rangle$

18. Calculate the work expended when a particle is moved from O to Q along segments $\overline{OP}$ and $\overline{PQ}$ in Figure 16 in the presence of the force field $\mathbf{F} = \langle x^2, y^2 \rangle$. How much work is expended moving in a complete circuit around the square?

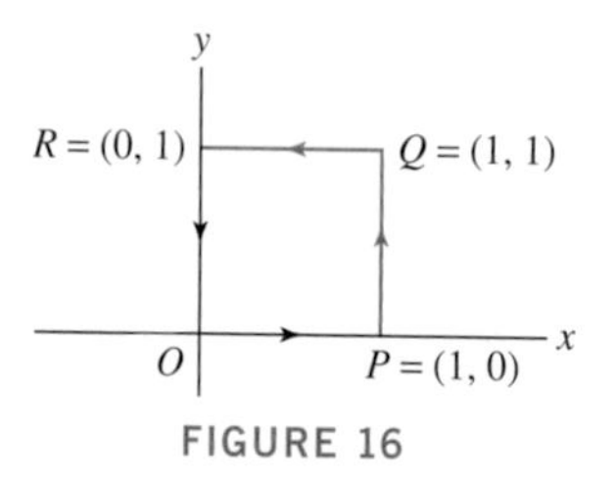

FIGURE 16

19. Let $\mathbf{F} = \left\langle \dfrac{1}{x}, \dfrac{-1}{y} \right\rangle$. Calculate the work against F required to move an object from $(1, 1)$ to $(3, 4)$ along any path in the first quadrant.

20. Let $\mathbf{F}$ be the vector field in Figure 17. Let $\overline{PQ}$, $\overline{QR}$, and $\overline{PR}$ be the segments in the figure with the orientations indicated.

(a) Is $\displaystyle\int_{\overline{PR}} \mathbf{F} \cdot d\mathbf{s}$ positive, negative, or zero?

(b) What is the value of $\displaystyle\int_{\overline{QR}} \mathbf{F} \cdot d\mathbf{s}$?

21. The vector field $\mathbf{F}$ in Figure 17 is horizontal and appears to depend on only the x-coordinate. Suppose that $\mathbf{F} = \langle g(x), 0 \rangle$. Prove that

$$\int_{\overline{PR}} \mathbf{F} \cdot d\mathbf{s} = \int_{\overline{PQ}} \mathbf{F} \cdot d\mathbf{s}$$

by showing that both integrals are equal to $\displaystyle\int_a^b g(x)\, dx$.

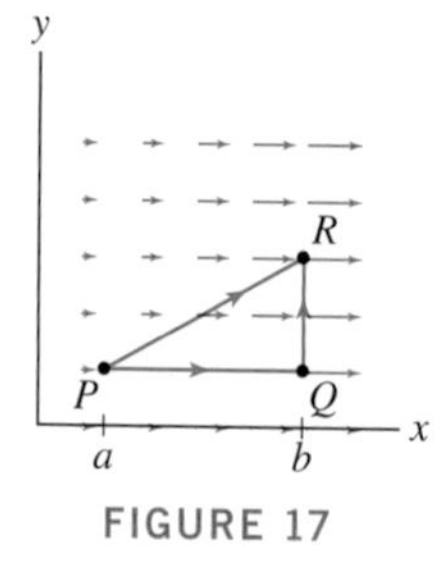

FIGURE 17

22. Let $\mathbf{F} = \langle x^{-1}z^2, y^{-1}z^2, 2z \log(xy) \rangle$.

(a) Verify that $\mathbf{F} = \nabla\varphi$, where $\varphi(x, y, z) = z^2 \log(xy)$.

(b) Evaluate $\displaystyle\int_{\mathbf{c}} \mathbf{F} \cdot d\mathbf{s}$, where $\mathbf{c}(t) = \langle e^t, e^{2t}, t^2 \rangle$ for $1 \le t \le 3$.

(c) Evaluate $\displaystyle\int_{\mathbf{c}} \mathbf{F} \cdot d\mathbf{s}$ for any path $\mathbf{c}$ from $P = (\frac{1}{2}, 4, 2)$ to $Q = (2, 3, 3)$ contained in the region $x > 0$, $y > 0$, $z > 0$.

(d) Why was it necessary to specify that the path lie in the region where x, y, and z are positive?

23. How much energy (in joules) does it take to carry a 2-kg object from sea level along any path to the top of a hill that is 1,000 m high? Assume that the force of gravity $\mathbf{F}$ is constant $-mg$ in the vertical direction, where $g = 9.8$ m/s^2. *Hint:* Find a potential function for $\mathbf{F}$.

24. Let $\mathbf{F} = \left\langle \dfrac{-y}{x^2 + y^2}, \dfrac{x}{x^2 + y^2} \right\rangle$ be the vortex vector field. Determine $\displaystyle\int_{\mathbf{c}} \mathbf{F} \cdot d\mathbf{s}$ for each of the paths in Figure 18.

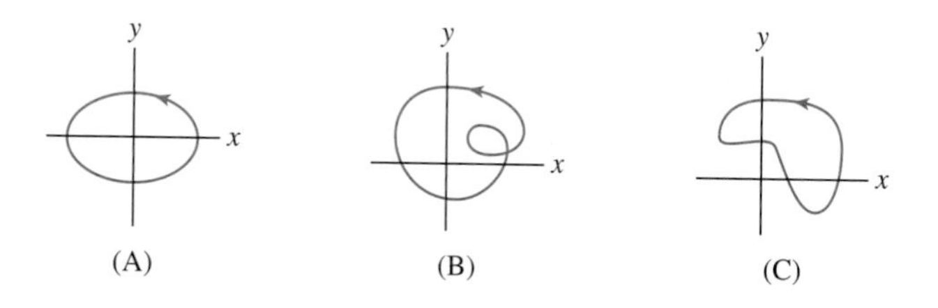

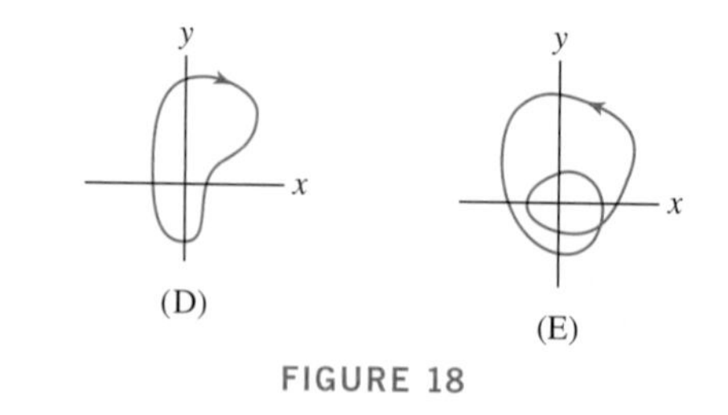

FIGURE 18

Further Insights and Challenges

25. The vector field $\mathbf{F} = \left\langle \dfrac{x}{x^2 + y^2}, \dfrac{y}{x^2 + y^2} \right\rangle$ is defined on the domain $\mathcal{D} = \{(x, y) \neq (0, 0)\}$.

(a) Show that $\mathbf{F}$ satisfies the cross-partials condition on $\mathcal{D}$.

(b) Show that $\varphi(x, y) = \frac{1}{2} \ln(x^2 + y^2)$ is a potential function for $\mathbf{F}$.

(c) Is $\mathcal{D}$ simply connected?

(d) Do these results contradict Theorem 4?

16.4 Parametrized Surfaces and Surface Integrals

We now develop the concept of a surface integral, focusing in this section on the scalar case. The first step is to introduce parametrized surfaces which play a role analogous to parametrized curves in line integrals.

A **parametrized surface** $\mathcal{S}$ is a surface whose points are described in the form

$$\Phi(u, v) = (x(u, v), y(u, v), z(u, v))$$

where the variables u, v (called parameters) vary in a region $\mathcal{D}$ called the **parameter domain**. Note that two parameters u and v are needed to parametrize a surface because the surface is two-dimensional, whereas just one parameter is needed for a curve. We shall always assume that Φ is **continuously differentiable**, by which we mean that the functions $x(u, v)$, $y(u, v)$, and $z(u, v)$ have continuous partial derivatives.

Figure 1 shows a plot of the surface $\mathcal{S}$ with the parametrization

$$\Phi = (u + v, u^3 - v, v^3 - u)$$

This surface consists of all points (x, y, z) in $\mathbf{R}^3$ such that

$$x = u + v, \qquad y = u^3 - v, \qquad z = v^3 - u$$

for (u, v) in $\mathcal{D} = \mathbf{R}^2$.

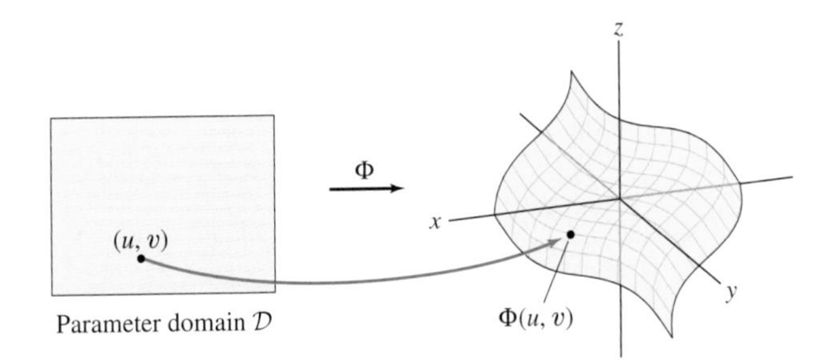

FIGURE 1 The parametric surface $\Phi(u, v) = (u + v, u^3 - v, v^3 - u)$.

■ **EXAMPLE 1 Parametrization of a Cone** Find a parametrization of the portion $\mathcal{S}$ of the cone with equation $x^2 + y^2 = z^2$ lying above or below the disk $x^2 + y^2 \le 4$. Specify the domain $\mathcal{D}$ of the parametrization.

Solution This surface $x^2 + y^2 = z^2$ is a cone whose horizonal cross section at height $z = u$ is the circle $x^2 + y^2 = u^2$ of radius u (Figure 2). So the coordinates of a point on the cone at height u are of the form $(u \cos v, u \sin v, u)$ for some angle v. Thus, the cone has the parametrization

$$\Phi(u, v) = (u \cos v, u \sin v, u)$$

Since we are interested in the portion of the cone where $x^2 + y^2 = u^2 \le 4$, the height variable u satisfies $-2 \le u \le 2$. The angular variable v varies in the interval $[0, 2\pi)$, and therefore, the parameter domain is $\mathcal{D} = [-2, 2] \times [0, 2\pi)$. ■

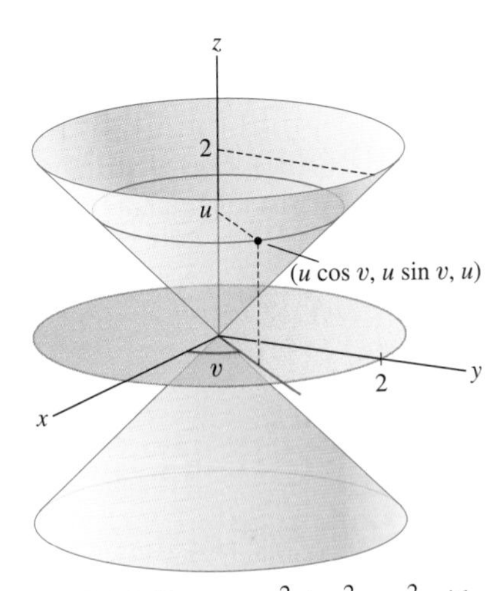

FIGURE 2 The cone $x^2 + y^2 = z^2$ with coordinates (u, v).

Three standard parametrizations arise often in computations. First, the **cylinder of radius** R with equation $x^2 + y^2 = R^2$ is conveniently parametrized in cylindrical coordinates (Figure 3). The cylinder consists of all points with cylindrical coordinates (R, θ, z), so it makes sense to parametrize the cylinder using θ and z as parameters (with fixed R):

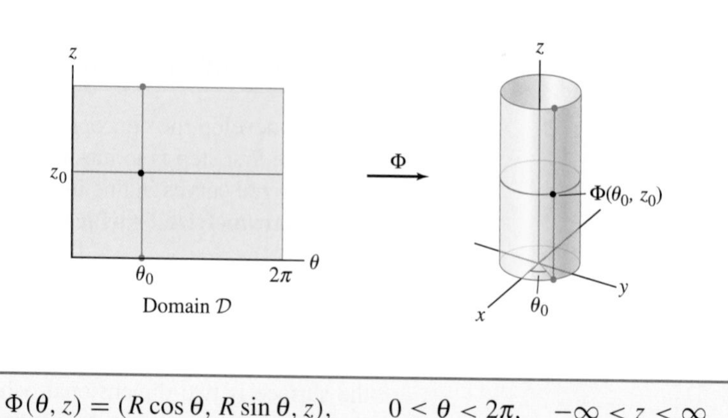

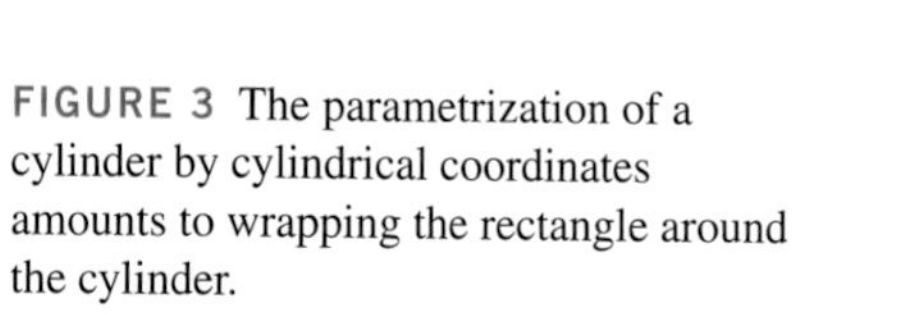

FIGURE 3 The parametrization of a cylinder by cylindrical coordinates amounts to wrapping the rectangle around the cylinder.

$$\Phi(\theta, z) = (R\cos\theta, R\sin\theta, z), \qquad 0 \le \theta < 2\pi, \qquad -\infty < z < \infty$$

If necessary, review cylindrical and spherical coordinates in Section 12.7. They are used often in surface calculations.

The sphere of radius R with center at the origin may be parametrized using spherical coordinates (ρ, θ, ϕ) with $\rho = R$ (Figure 4):

$$\Phi(\theta, \phi) = (R\cos\theta\sin\phi, R\sin\theta\sin\phi, R\cos\phi), \qquad 0 \le \theta < 2\pi, \qquad 0 \le \phi \le \pi$$

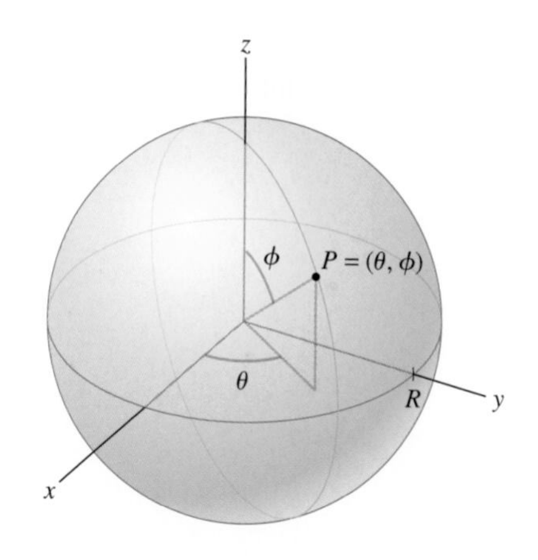

FIGURE 4 Spherical coordinates on a sphere of radius R.

The North and South Poles correspond to $\phi = 0$ and $\phi = \pi$ with any value of θ (the map Φ fails to be one-to-one at the poles):

$$\text{North Pole: } \Phi(\theta, 0) = (0, 0, R), \qquad \text{South Pole: } \Phi(\theta, \pi) = (0, 0, -R)$$

As shown in Figure 5, Φ maps each horizontal segment $\phi = c$ $(0 < c < \pi)$ to a latitude (a circle parallel to the equator) and each vertical segment $\theta = c$ to a longitudinal circle passing through the North and South Poles.

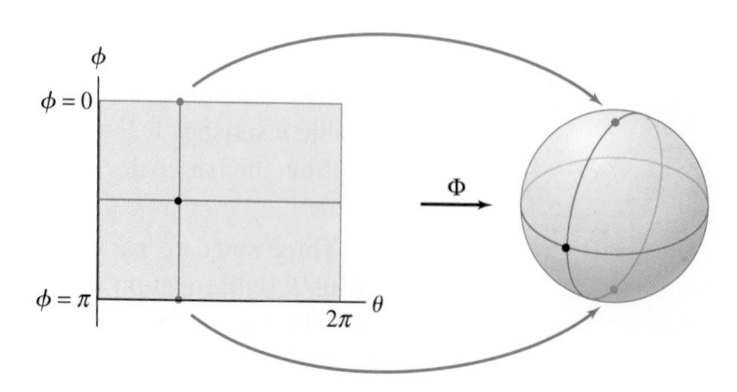

FIGURE 5 The parametrization of the sphere by spherical coordinates amounts to wrapping the rectangle around the sphere. The top and bottom edges of the rectangle are collapsed to the North and South Poles.

Finally, there is a simple way to parametrize the **graph of a function** $z = f(x, y)$ (Figure 6). We set

$$\boxed{\Phi(x, y) = (x, y, f(x, y))}$$

In this case, the parameters are x and y.

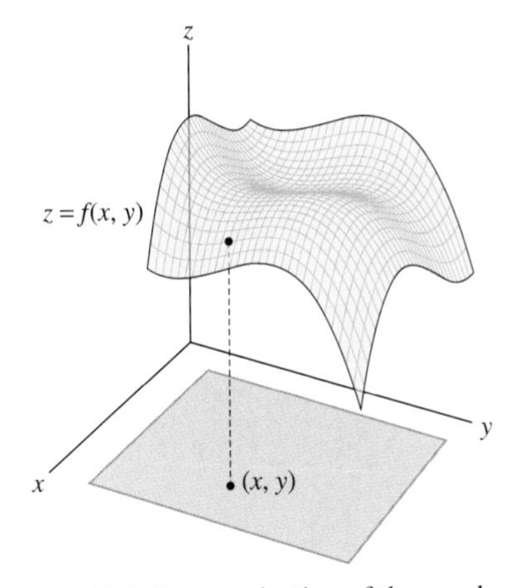

FIGURE 6 Parametrization of the graph $z = f(x, y)$.

Grid Curves, Normal Vectors, and the Tangent Plane

Suppose that a surface $\mathcal{S}$ has a parametrization

$$\Phi(u, v) = (x(u, v), y(u, v), z(u, v))$$

that is one-to-one on a domain $\mathcal{D}$. Then each point P on $\mathcal{S}$ corresponds to a unique pair (u_0, v_0) such that $P = \Phi(u_0, v_0)$ and it is useful to think of (u_0, v_0) as the "coordinates" of P determined by the parametrization. Coordinates of this type are sometimes called **curvilinear coordinates**.

In the uv-plane, we have a grid formed by the families of lines parallel to the coordinates axes. The parametrization maps these grid lines to a system of **grid curves** on the surface (Figure 7). The horizontal line $v = v_0$ maps to the curve $\Phi(u, v_0)$ (the grid curve in the u-direction) and the vertical line $u = u_0$ maps to the curve $\Phi(u_0, v)$ (the grid curve in the v-direction).

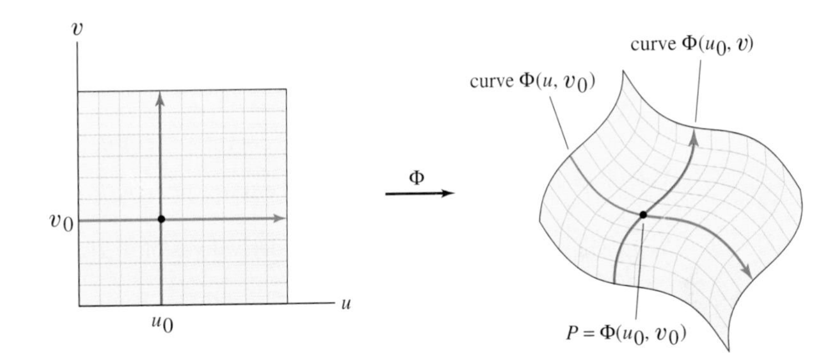

FIGURE 7 Grid curves.

Through each point $P = \Phi(u_0, v_0)$ on $\mathcal{S}$ there pass two grid curves, namely, the curve $\Phi(u, v_0)$ in the u-direction and the curve $\Phi(u_0, v)$ in the v-direction. The tangent vectors to these grid curves are

$$\mathbf{T}_u(P) = \frac{\partial \Phi}{\partial u}(u_0, v_0) = \left\langle \frac{\partial x}{\partial u}(u_0, v_0), \frac{\partial y}{\partial u}(u_0, v_0), \frac{\partial z}{\partial u}(u_0, v_0) \right\rangle$$

$$\mathbf{T}_v(P) = \frac{\partial \Phi}{\partial v}(u_0, v_0) = \left\langle \frac{\partial x}{\partial v}(u_0, v_0), \frac{\partial y}{\partial v}(u_0, v_0), \frac{\partial z}{\partial v}(u_0, v_0) \right\rangle$$

Observe that these vectors are tangent to the surface $\mathcal{S}$ at P because they are tangent to grid curves that lie on $\mathcal{S}$ (Figure 8). Set

$$\mathbf{n}(P) = \mathbf{n}(u_0, v_0) = \mathbf{T}_u(P) \times \mathbf{T}_v(P)$$

If this cross product is nonzero, then it is normal (perpendicular) to $\mathcal{S}$. In this case, $\mathbf{T}_u$ and $\mathbf{T}_v$ span the tangent plane at P and $\mathbf{n}(u, v)$ is a **normal vector** to the tangent plane. We often write $\mathbf{n}$ instead of $\mathbf{n}(u, v)$ or $\mathbf{n}(P)$, but it is understood that the vector $\mathbf{n}$ varies from point to point on the surface. Similarly, we often denote the tangent vectors by $\mathbf{T}_u$ and $\mathbf{T}_v$. The parametrization Φ is called **regular** at P if $\mathbf{n}(u_0, v_0) \neq 0$. Note that $\mathbf{T}_u$, $\mathbf{T}_v$, and $\mathbf{n}$ need not be unit vectors (thus the notation here differs from that in Sections 13.4, 13.5, and 16.2, where $\mathbf{T}$ and $\mathbf{n}$ denote unit vectors).

At each point on a surface, the normal vector points in one of two opposite directions. If we change the parametrization, the length of $\mathbf{n}$ *may change and its direction may be reversed.*

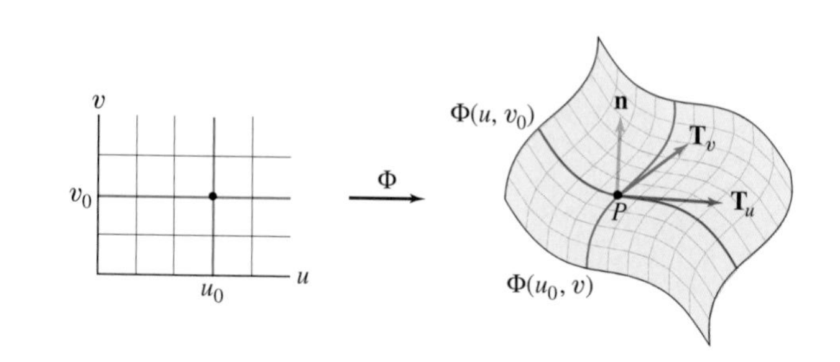

FIGURE 8 The vectors $\mathbf{T}_u$ and $\mathbf{T}_v$ are tangent to the grid curves through $P = \Phi(u_0, v_0)$.

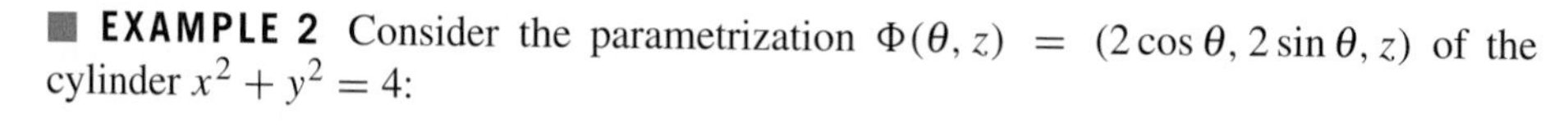

EXAMPLE 2 Consider the parametrization $\Phi(\theta, z) = (2\cos\theta, 2\sin\theta, z)$ of the cylinder $x^2 + y^2 = 4$:

(a) Describe the grid curves.

(b) Compute $\mathbf{T}_\theta$, $\mathbf{T}_z$, and $\mathbf{n}(\theta, z)$.

(c) Find an equation of the tangent plane at $P = \Phi(\frac{\pi}{4}, 5)$.

Solution

(a) The grid curves on the cylinder through $P = (\theta_0, z_0)$ are (Figure 9)

$$\theta\text{-grid curve:} \quad (2\cos\theta, 2\sin\theta, z_0) = \text{circle of radius 2 at height } z_0$$

$$z\text{-grid curve:} \quad (2\cos\theta_0, 2\sin\theta_0, z) = \text{vertical line through } P$$

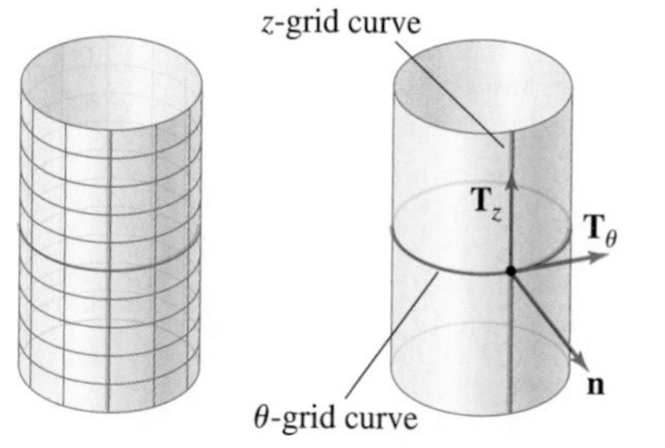

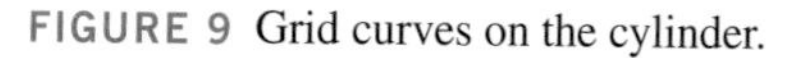

FIGURE 9 Grid curves on the cylinder.

(b) The partial derivatives of Φ give us the tangent vectors at P:

$$\mathbf{T}_\theta = \frac{\partial\Phi}{\partial\theta} = \frac{\partial}{\partial\theta}(2\cos\theta, 2\sin\theta, z) = \langle -2\sin\theta, 2\cos\theta, 0\rangle$$

$$\mathbf{T}_z = \frac{\partial\Phi}{\partial z} = \frac{\partial}{\partial z}(2\cos\theta, 2\sin\theta, z) = \langle 0, 0, 1\rangle$$

As shown in Figure 9, $\mathbf{T}_\theta$ is tangent to the θ-grid curve (a circle through P) and $\mathbf{T}_z$ points in the direction of the z-grid curve (the vertical line through P). The normal vector is

$$\mathbf{n}(\theta, z) = \mathbf{T}_\theta \times \mathbf{T}_z = \begin{vmatrix} \mathbf{i} & \mathbf{j} & \mathbf{k} \\ -2\sin\theta & 2\cos\theta & 0 \\ 0 & 0 & 1 \end{vmatrix} = 2\cos\theta\mathbf{i} + 2\sin\theta\mathbf{j}$$

Observe that $\mathbf{n}$ is horizontal and points directly out of the cylinder.

(c) For $u = \frac{\pi}{4}$, $v = 5$,

$$P = \Phi\left(\frac{\pi}{4}, 5\right) = \langle\sqrt{2}, \sqrt{2}, 5\rangle, \qquad \mathbf{n} = \mathbf{n}\left(\frac{\pi}{4}, 5\right) = \langle\sqrt{2}, \sqrt{2}, 0\rangle$$

REMINDER An equation of the plane through $P = (x_0, y_0, z_0)$ *with normal vector* $\mathbf{n}$ *is*

$$\langle x - x_0, y - y_0, z - z_0\rangle \cdot \mathbf{n} = 0$$

The tangent plane through P has normal vector $\mathbf{n}$ and thus has equation

$$\langle x - \sqrt{2}, y - \sqrt{2}, z - 5\rangle \cdot \langle\sqrt{2}, \sqrt{2}, 0\rangle = 0$$

This can be written

$$\sqrt{2}(x - \sqrt{2}) + \sqrt{2}(y - \sqrt{2}) = 0 \qquad \text{or} \qquad x + y = 2\sqrt{2}$$

As we might have expected, the tangent plane is vertical (since z does not appear in the equation). ■

EXAMPLE 3 CAS **Helicoid Surface** Describe the surface $\mathcal{S}$ with parametrization

$$\Phi(u, v) = (u\cos v, u\sin v, v), \qquad 0 \le u \le 1, \quad 0 \le v < 2\pi$$

(a) Use a computer algebra system to plot $\mathcal{S}$.

(b) Compute $\mathbf{n}(u, v)$ at $u = \frac{1}{2}$, $v = \frac{\pi}{2}$.

Solution For each fixed value $u = a$, the curve $\Phi(a, v) = (a\cos v, a\sin v, v)$ is a helix of radius a. Therefore, as u varies from 0 to 1, $\Phi(u, v)$ describes a family of helices of radius u. The resulting surface is a "helical ramp."

(a) A typical command for plotting a parametric surface with a computer algebra system is

```
ParametricPlot3D[{u*Cos[v],u*Sin[v],v},{u,0,1},{v,0,2Pi}]
```

We obtain the plot in Figure 10.

(b) The tangent and normal vectors are

$$\mathbf{T}_u = \frac{\partial \Phi}{\partial u} = \langle \cos v, \sin v, 0 \rangle$$

$$\mathbf{T}_v = \frac{\partial \Phi}{\partial v} = \langle -u\sin v, u\cos v, 1 \rangle$$

$$\mathbf{n}(u, v) = \mathbf{T}_u \times \mathbf{T}_v = \begin{vmatrix} \mathbf{i} & \mathbf{j} & \mathbf{k} \\ \cos v & \sin v & 0 \\ -u\sin v & u\cos v & 1 \end{vmatrix}$$

$$= (\sin v)\mathbf{i} - (\cos v)\mathbf{j} + (u\cos^2 v + u\sin^2 v)\mathbf{k}$$

$$= (\sin v)\mathbf{i} - (\cos v)\mathbf{j} + u\mathbf{k}$$

At $u = \frac{1}{2}$, $v = \frac{\pi}{2}$, we have $\mathbf{n} = \mathbf{i} + \frac{1}{2}\mathbf{k}$. ■

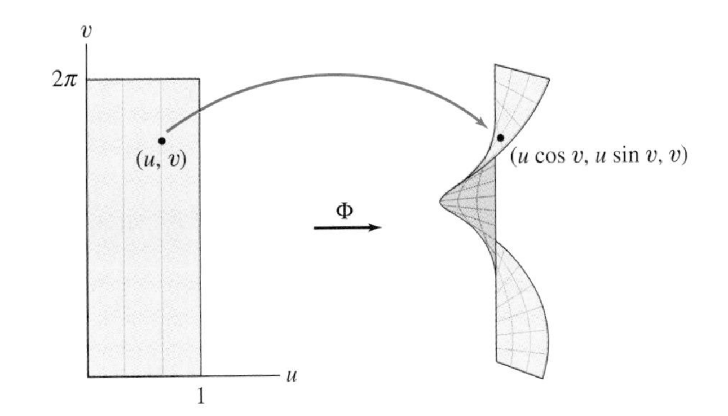

FIGURE 10 Helicoid.

Surface Area

Note that we require only that Φ be one-to-one on the interior of $\mathcal{D}$. Many common parametrizations (such as the parametrizations by cylindrical and spherical coordinates) fail to be one-to-one on the boundary of their domains.

Let $\Phi(u, v)$ be a parametrization of a surface $\mathcal{S}$ with domain $\mathcal{D}$. Since we want Φ to cover the surface $\mathcal{S}$ just once, we assume that Φ is one-to-one, except possibly on the boundary of $\mathcal{D}$. We also assume that Φ is regular, except possibly on the boundary of $\mathcal{D}$. Recall that Φ is regular at (u, v) if the normal vector $\mathbf{n} = \mathbf{T}_u \times \mathbf{T}_v$ is nonzero.

The length $\|\mathbf{n}\|$ of the normal vector has an important interpretation in terms of area. To discuss this interpretation, we assume, for simplicity, that the domain $\mathcal{D}$ is a rectangle but our argument applies to more general domains. Let us divide $\mathcal{D}$ into a grid of small rectangles $\mathcal{R}_{ij}$ of size $\Delta u \times \Delta v$, as in Figure 11. Each rectangle $\mathcal{R}_{ij}$ is mapped by Φ to a "curved" parallelogram $\mathcal{S}_{ij}$ on the surface $\mathcal{S}$. Referring to the figure, suppose that $\mathcal{R}_{ij}$ is the rectangle with vertices $P'Q'R'S'$, where $P' = (u_{ij}, v_{ij})$. Let $\mathcal{S}_{ij}$ be the image $\Phi(\mathcal{R}_{ij})$ of $\mathcal{R}_{ij}$. Let P, Q, R, and S be the images of P', Q', R', and S'. Thus, $P = \Phi(P') = \Phi(u_{ij}, v_{ij})$.

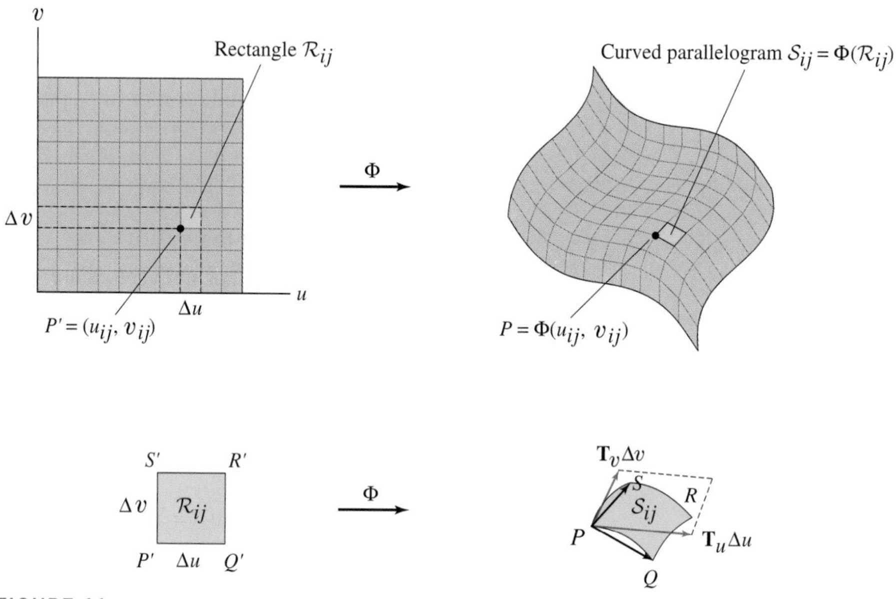

FIGURE 11

Let's examine how the area of $\mathcal{S}_{ij}$ is related to the area of $\mathcal{R}_{ij}$. First, we observe that if Δu and Δv are small, then the "curved" parallelogram $\mathcal{S}_{ij}$ has approximately the same area as the "genuine" parallelogram with sides $\overrightarrow{PQ}$ and $\overrightarrow{PS}$. Since the area of the parallelogram spanned by two vectors is the length of their cross product, we obtain the approximation

$$\text{Area}(\mathcal{S}_{ij}) \approx \text{area of parallelogram spanned by } \overrightarrow{PQ} \text{ and } \overrightarrow{PS}$$
$$= \|\overrightarrow{PQ} \times \overrightarrow{PS}\|$$

*← **REMINDER** The area of the parallelogram spanned by vectors $\mathbf{v}$ and $\mathbf{w}$ in $\mathbf{R}^3$ is equal to $\|\mathbf{v} \times \mathbf{w}\|$ (Theorem 3 in Section 12.4).*

Next, we use the linear approximation to estimate the vectors $\overrightarrow{PQ}$ and $\overrightarrow{PS}$:

$$\overrightarrow{PQ} = \Phi(u_{ij} + \Delta u, v_{ij}) - \Phi(u_{ij}, v_{ij}) \approx \frac{\partial \Phi}{\partial u}(u_{ij}, v_{ij})\Delta u = \mathbf{T}_u \Delta u$$

$$\overrightarrow{PS} = \Phi(u_{ij}, v_{ij} + \Delta v) - \Phi(u_{ij}, v_{ij}) \approx \frac{\partial \Phi}{\partial v}(u_{ij}, v_{ij})\Delta v = \mathbf{T}_v \Delta v$$

We obtain

$$\text{Area}(\mathcal{S}_{ij}) \approx \|\overrightarrow{PQ} \times \overrightarrow{PS}\| \approx \|\mathbf{T}_u \Delta u \times \mathbf{T}_v \Delta v\| = \|\mathbf{T}_u \times \mathbf{T}_v\| \, \Delta u \, \Delta v$$

Observing that $\mathbf{n}(u_{ij}, v_{ij}) = \mathbf{T}_u \times \mathbf{T}_v$ and $\text{Area}(\mathcal{R}_{ij}) = \Delta u \Delta v$, we may write our result as follows:

$$\boxed{\text{Area}(\mathcal{S}_{ij}) \approx \|\mathbf{n}(u_{ij}, v_{ij})\| \text{Area}(\mathcal{R}_{ij})} \qquad \boxed{1}$$

The approximation (1) is valid for any small region $\mathcal{R}$ in the uv-plane:

$$\text{Area}(\mathcal{S}) \approx \|\mathbf{n}(u_0, v_0)\| \text{Area}(\mathcal{R})$$

where $\mathcal{S} = \Phi(\mathcal{R})$ and (u_0, v_0) is any sample point in $\mathcal{R}$. Here, "small" means contained in a small disk. We do not allow $\mathcal{R}$ to be very thin and wide.

Thus, $\|\mathbf{n}\|$ *is a distortion factor that tells us how the area of a small rectangle $\mathcal{R}_{ij}$ is altered under the map Φ.*

The entire surface $\mathcal{S}$ is the union of the small patches $\mathcal{S}_{ij}$, so we may approximate the total area of $\mathcal{S}$ by the sum:

$$\text{Area}(\mathcal{S}) = \sum_{ij} \text{Area}(\mathcal{S}_{ij}) \approx \sum_{ij} \|\mathbf{n}(u_{ij}, v_{ij})\| \Delta u \Delta v \qquad \boxed{2}$$

The sum on the right is a Riemann sum for the double integral of $\|\mathbf{n}(u, v)\|$ over the parameter domain $\mathcal{D}$. As Δu and Δv tend to zero, these Riemann sums converge to the following double integral, which we take as the definition of surface area:

$$\text{Area}(\mathcal{S}) = \iint_{\mathcal{D}} \|\mathbf{n}(u, v)\| \, du \, dv$$

A modification of this formula leads to the notion of the **surface integral** of a function $f(x, y, z)$ over $\mathcal{S}$. We consider sums of the form

$$\sum_{i,j} f(P_{ij}) \text{Area}(\mathcal{S}_{ij}) \qquad \boxed{3}$$

where $P_{ij} = \Phi(u_{ij}, v_{ij})$ is a sample point in $\mathcal{S}_{ij}$. If we think of $f(P)$ as a continuous mass density on the surface, then

$$\text{Mass of } \mathcal{S}_{ij} \approx \text{density} \times \text{area} \approx f(P_{ij}) \text{Area}(\mathcal{S}_{ij})$$

and (3) is an approximation to the total mass of the surface. In any case, the limit of the sums as Δu and Δv tend to zero (if it exists) is called the surface integral of f over $\mathcal{S}$; it is denoted

$$\iint_{\mathcal{S}} f(x, y, z) \, dS = \lim_{\Delta u, \Delta v \to 0} \sum_{i,j} f(P_{ij}) \text{Area}(\mathcal{S}_{ij})$$

Applications of surface integrals: If $\mathcal{S}$ is a surface with spherical mass density $\rho(x, y, z)$, then

$$\text{Mass of } \mathcal{S} = \iint_{\mathcal{S}} \rho(x, y, z) \, dS$$

Similarly, if an electric charge is distributed over $\mathcal{S}$ with charge density $\rho(x, y, z)$, then this integral yields the total charge on $\mathcal{S}$.

To evaluate the surface integral, let $P_{ij} = \Phi(u_{ij}, v_{ij})$ and apply the approximation (1):

$$\sum_{ij} f(P_{ij}) \text{Area}(\mathcal{S}_{ij}) \approx \sum_{i,j} f(\Phi(u_{ij}, v_{ij})) \|\mathbf{n}(u_{ij}, v_{ij})\| \, \Delta u \, \Delta v$$

The sum on the right is a Riemann sum for the double integral of

$$f(\Phi(u, v)) \|\mathbf{n}(u, v)\|$$

over the parameter domain $\mathcal{D}$. Under our assumption that Φ is continuously differentiable, the Riemann sums on the right approach the same limit as the sums on the left (the proof is omitted), yielding the formula in the next theorem.

It is interesting to note that Eq. (4) includes the Change of Variables Formula for double integrals (Theorem 1 in Section 15.5) as a special case. If the surface $\mathcal{S}$ is a domain in the xy-plane [in other words, $z(u, v) = 0$], then the integral over $\mathcal{S}$ reduces to the double integral of the function $f(x, y, 0)$. We may view $\Phi(u, v)$ as a mapping from the uv-plane to the xy-plane, and we find that $\|\mathbf{n}(u, v)\|$ is the Jacobian of this mapping.

THEOREM 1 Surface Integrals and Surface Area Let $\Phi(u, v)$ be a parametrization of a surface $\mathcal{S}$ with parameter domain $\mathcal{D}$. Assume that Φ is continuously differentiable, one-to-one, and regular (except possibly at the boundary of $\mathcal{D}$). Then

$$\iint_{\mathcal{S}} f(x, y, z)\, dS = \iint_{\mathcal{D}} f(\Phi(u, v))\|\mathbf{n}(u, v)\|\, du\, dv \qquad \boxed{4}$$

For $f(x, y, z) = 1$, we obtain the surface area of $\mathcal{S}$:

$$\text{Area}(\mathcal{S}) = \iint_{\mathcal{D}} \|\mathbf{n}(u, v)\|\, du\, dv$$

Equation (4) is summarized by the symbolic expression for the "surface element":

$$dS = \|\mathbf{n}(u, v)\|\, du\, dv$$

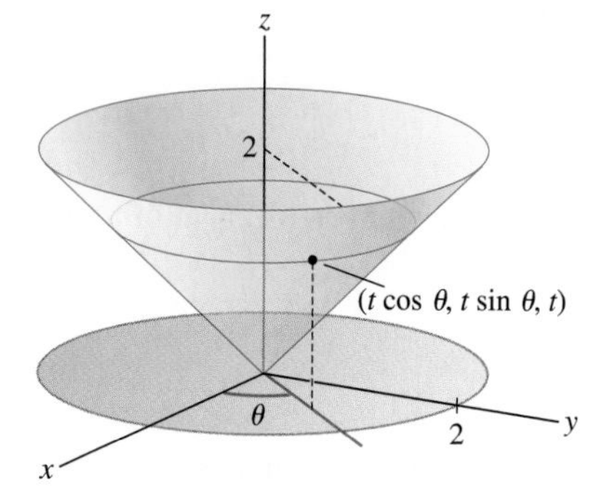

FIGURE 12 Portion $\mathcal{S}$ of the cone $x^2 + y^2 = z^2$ lying over the disk $x^2 + y^2 \le 4$.

EXAMPLE 4 Calculate Area($\mathcal{S}$) and $\iint_{\mathcal{S}} x^2 z\, dS$, where $\mathcal{S}$ is the portion of the cone $x^2 + y^2 = z^2$ lying above the disk $x^2 + y^2 \le 4$ (Figure 12).

Solution A parametrization of the cone was found in Example 1. Using the variables θ and t, this parametrization is

$$\Phi(\theta, t) = (t\cos\theta, t\sin\theta, t), \qquad 0 \le t \le 2, \quad 0 \le \theta < 2\pi$$

***Step 1.* Compute the tangent and normal vectors.**

$$\mathbf{T}_\theta = \frac{\partial\Phi}{\partial\theta} = \langle -t\sin\theta, t\cos\theta, 0\rangle, \qquad \mathbf{T}_t = \frac{\partial\Phi}{\partial t} = \langle \cos\theta, \sin\theta, 1\rangle$$

$$\mathbf{n} = \mathbf{T}_\theta \times \mathbf{T}_t = \begin{vmatrix} \mathbf{i} & \mathbf{j} & \mathbf{k} \\ -t\sin\theta & t\cos\theta & 0 \\ \cos\theta & \sin\theta & 1 \end{vmatrix}$$
$$= t\cos\theta\mathbf{i} + t\sin\theta\mathbf{j} - t\mathbf{k}$$

The normal vector has length

$$\|\mathbf{n}\| = \sqrt{t^2\cos^2\theta + t^2\sin^2\theta + t^2} = \sqrt{2t^2} = \sqrt{2}\,|t|$$

Thus, $dS = \sqrt{2}|t|\, d\theta\, dt$. In our case, we are integrating over a region where $t \ge 0$, so we may drop the absolute value.

***Step 2.* Calculate the surface area.**

$$\text{Area}(\mathcal{S}) = \iint_{\mathcal{D}} \|\mathbf{n}\|\, du\, dv = \int_0^2 \int_0^{2\pi} \sqrt{2}\,t\, d\theta\, dt = \sqrt{2}\pi t^2 \Big|_0^2 = 4\sqrt{2}\pi$$

***Step 3.* Calculate the surface integral.**
We express $f(x, y, z) = x^2 z$ in terms of the parameters t and θ:

$$f(\Phi(\theta, t)) = f(t\cos\theta, t\sin\theta, t) = (t\cos\theta)^2 t = t^3\cos^2\theta$$

← REMINDER

$$\int_0^{2\pi} \cos^2\theta\, d\theta = \int_0^{2\pi} \frac{1+\cos 2\theta}{2}\, d\theta = \pi$$

Then

$$\iint_S f(x, y, z)\, dS = \int_{t=0}^{2} \int_{\theta=0}^{2\pi} f(\Phi(\theta, t))\, \|\mathbf{n}(\theta, t)\|\, d\theta\, dt$$
$$= \int_{t=0}^{2} \int_{\theta=0}^{2\pi} (t^3 \cos^2\theta)(\sqrt{2}t)\, d\theta\, dt$$
$$= \sqrt{2}\left(\int_0^2 t^4\, dt\right)\left(\int_0^{2\pi} \cos^2\theta\, d\theta\right)$$
$$= \sqrt{2}\left(\frac{32}{5}\right)(\pi) = \frac{32\sqrt{2}\pi}{5}$$ ■

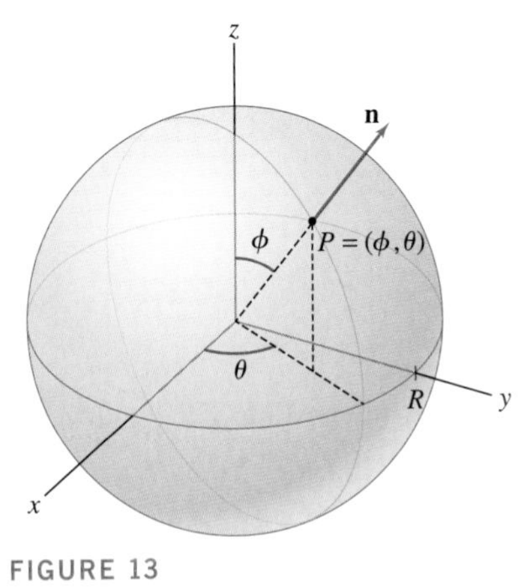

FIGURE 13

For use in the next example and for future reference, we compute the normal vector in the parametrization of the sphere of radius R centered at the origin by spherical coordinates, with outward-pointing normal (Figure 13):

$$\boxed{\Phi(\phi, \theta) = (R\cos\theta\sin\phi,\ R\sin\theta\sin\phi,\ R\cos\phi)}$$

We have

$$\mathbf{T}_\phi = \langle R\cos\theta\cos\phi,\ R\sin\theta\cos\phi,\ -R\sin\phi\rangle$$
$$\mathbf{T}_\theta = \langle -R\sin\theta\sin\phi,\ R\cos\theta\sin\phi,\ 0\rangle$$
$$\mathbf{n} = \mathbf{T}_\phi \times \mathbf{T}_\theta = \begin{vmatrix} \mathbf{i} & \mathbf{j} & \mathbf{k} \\ R\cos\theta\cos\phi & R\sin\theta\cos\phi & -R\sin\phi \\ -R\sin\theta\sin\phi & R\cos\theta\sin\phi & 0 \end{vmatrix}$$
$$= R^2\cos\theta\sin^2\phi\,\mathbf{i} + R^2\sin\theta\sin^2\phi\,\mathbf{j} + R^2\cos\phi\sin\phi\,\mathbf{k}$$
$$= R^2\sin\phi\,\langle\cos\theta\sin\phi,\ \sin\theta\sin\phi,\ \cos\phi\rangle \qquad \boxed{5}$$

Note that the vector $\langle\cos\theta\sin\phi, \sin\theta\sin\phi, \cos\phi\rangle$ lies on the unit sphere and is, in fact, the radial unit vector $\mathbf{e}_r$ that points in the direction from the origin to $\Phi(\phi, \theta)$. Therefore, for the parametrization by spherical coordinates with outward-pointing normal, we have

$$\boxed{\mathbf{n} = \mathbf{T}_\phi \times \mathbf{T}_\theta = (R^2\sin\phi)\,\mathbf{e}_r, \qquad \text{and} \qquad \|\mathbf{n}\| = R^2\sin\phi} \qquad \boxed{6}$$

The normal vector $\mathbf{T}_\theta \times \mathbf{T}_\phi$ is an inward-pointing normal.

■ **EXAMPLE 5** **Total Charge on a Surface** Find the total charge (in coulombs) on a sphere S of radius 5 cm whose charge density in spherical coordinates is $\rho(\phi, \theta) = 0.003\cos^2\phi\ \mathrm{C/cm^2}$.

Solution Integrals over the sphere are usually easier to evaluate in spherical coordinates, so we parametrize S by

$$\Phi(\phi, \theta) = (5\cos\theta\sin\phi,\ 5\sin\theta\sin\phi,\ 5\cos\phi)$$

By Eq. (6), $\|\mathbf{n}\| = 5^2\sin\phi$ and

$$\text{Total charge} = \iint_S \rho(\phi, \theta)\, dS = \int_{\theta=0}^{2\pi} \int_{\phi=0}^{\pi} \rho(\phi, \theta) \|\mathbf{n}\|\, d\phi\, d\theta$$

$$= \int_{\theta=0}^{2\pi} \int_{\phi=0}^{\pi} (0.003 \cos^2 \phi)(25 \sin \phi)\, d\phi\, d\theta$$

$$= (0.075)(2\pi) \int_{\phi=0}^{\pi} \cos^2 \phi \sin \phi\, d\phi$$

$$= 0.15\pi \left(-\frac{\cos^3 \phi}{3} \right) \Bigg|_0^{\pi} = 0.15\pi \left(\frac{2}{3} \right) \approx 0.3 \text{ C}$$

■

When a graph $z = g(x, y)$ is parametrized by $\Phi(x, y) = (x, y, g(x, y))$, the tangent vectors are

$$\mathbf{T}_x = (1, 0, g_x), \qquad \mathbf{T}_y = (0, 1, g_y)$$

and the normal vector is

$$\mathbf{n} = \mathbf{T}_x \times \mathbf{T}_y = \begin{vmatrix} \mathbf{i} & \mathbf{j} & \mathbf{k} \\ 1 & 0 & g_x \\ 0 & 1 & g_y \end{vmatrix} = -g_x \mathbf{i} - g_y \mathbf{j} + \mathbf{k} \qquad \boxed{7}$$

Thus, $\|\mathbf{n}\| = \sqrt{1 + g_x^2 + g_y^2}$, and the surface integral over the portion of a graph lying over a domain $\mathcal{D}$ in the xy-plane is

$$\text{Surface integral over a graph} = \iint_{\mathcal{D}} f(x, y, g(x, y)) \sqrt{1 + g_x^2 + g_y^2}\, dx\, dy \qquad \boxed{8}$$

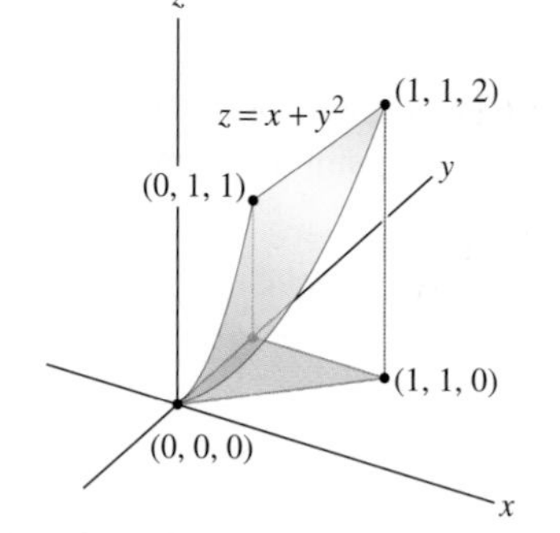

FIGURE 14 The surface $z = x + y^2$ over $0 \le x \le y \le 1$.

■ **EXAMPLE 6** Calculate the surface integral $\iint_{\mathcal{S}} (z - x)\, dS$, where $\mathcal{S}$ is the portion of the graph of $z = x + y^2$ where $0 \le x \le y \le 1$ (Figure 14).

Solution Let $z = g(x, y) = x + y^2$. Then $g_x = 1$ and $g_y = 2y$, and

$$dS = \sqrt{1 + g_x^2 + g_y^2}\, dx\, dy = \sqrt{1 + 1 + 4y^2}\, dx\, dy = \sqrt{2 + 4y^2}\, dx\, dy$$

We express $f(x, y, z) = z - x$ in terms of the parameters x and y:

$$f(x, y, z) = f(x, y, x + y^2) = z - x = (x + y^2) - x = y^2$$

and use Eq. (8) with the domain $0 \le x \le y$, $0 \le y \le 1$:

$$\iint_{\mathcal{S}} f(x, y, z)\, dS = \int_{y=0}^{1} \int_{x=0}^{y} y^2 \sqrt{2 + 4y^2}\, dx\, dy$$

$$= \int_{y=0}^{1} \left(y^2 \sqrt{2 + 4y^2} \right) x \Bigg|_{x=0}^{y} dy = \int_0^1 y^3 \sqrt{2 + 4y^2}\, dy$$

Let $u = 2 + 4y^2$, $du = 8y\, dy$. Then $y^2 = \frac{1}{4}(u - 2)$, and

$$\int_0^1 y^3\sqrt{2+4y^2}\,dy = \frac{1}{8}\int_2^6 \frac{1}{4}(u-2)\sqrt{u}\,du = \frac{1}{32}\int_2^6 (u^{3/2}-2u^{1/2})\,du$$

$$= \frac{1}{32}\left(\frac{2}{5}u^{5/2}-\frac{4}{3}u^{3/2}\right)\Bigg|_2^6 = \frac{1}{30}(6\sqrt{6}+\sqrt{2}) \approx 0.54 \qquad \blacksquare$$

Excursion

The French mathematician Pierre Simon, Marquis de Laplace (1749–1827) showed that the gravitational potential satisfies the Laplace equation $\Delta\varphi = 0$, *where* Δ *is the* Laplace operator

$$\Delta\varphi = \frac{\partial^2\varphi}{\partial x^2} + \frac{\partial^2\varphi}{\partial y^2} + \frac{\partial^2\varphi}{\partial z^2}$$

This equation plays an important role in more advanced branches of math and physics.

The concept of a potential function was first introduced by the French mathematician Joseph-Louis Lagrange (1736–1813) who observed that the gravitational force field $\mathbf{F}$ can be written as a gradient $\mathbf{F} = -\nabla\varphi$ (recall that the minus sign is a convention of physicists, chosen so that positive work is required to achieve a gain in potential energy). We call φ the **gravitational potential**. If a mass m is located at a point Q, then the force on a unit mass at a point P is $\mathbf{F} = -\dfrac{Gm}{r^2}\mathbf{e}_r$, where $\mathbf{e}_r$ is the unit vector pointing from Q to P and $r = |P-Q|$ is the distance from P to Q. We saw in Example 4 of Section 16.1 that $\mathbf{F} = -\nabla\varphi$, where $\varphi(P) = -\dfrac{Gm}{r} = -\dfrac{Gm}{|P-Q|}$. If, instead of a single mass, we have N point masses $m_1, \dots, m_N$ located at $Q_1, \dots, Q_N$, then the gravitational potential is the sum

$$\varphi(P) = -G\sum_{i=1}^{N} \frac{m_i}{|P-Q_i|} \qquad \boxed{9}$$

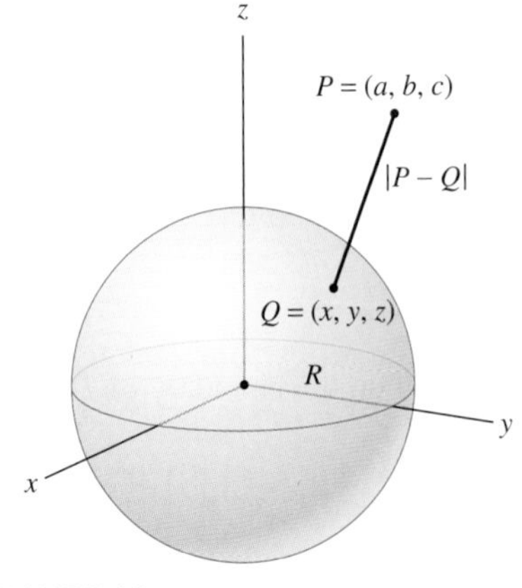

FIGURE 15

It is often more realistic to think of mass as distributed continuously in a region according to some density function $\rho(x, y, z)$. If the mass is distributed over a thin surface $\mathcal{S}$, we obtain the potential function by replacing the sum in Eq. (9) with a surface integral:

$$\varphi(P) = -G\iint_{\mathcal{S}} \frac{\rho(x,y,z)\,dS}{|P-Q|} = -G\iint_{\mathcal{S}} \frac{\rho(x,y,z)\,dS}{\sqrt{(x-a)^2+(y-b)^2+(z-c)^2}} \qquad \boxed{10}$$

where $P = (a, b, c)$. This surface integral is hard or impossible to evaluate explicitly unless the surface and mass distribution are sufficiently symmetrical. A basic case is that of a sphere of uniform mass density (Figure 15).

> **THEOREM 2 Gravitational Potential of a Uniform Sphere** The gravitation potential φ due to a sphere of radius R with uniform mass distribution of total mass m at a point P located at a distance r from the center is
>
> $$\varphi(P) = \begin{cases} \dfrac{-Gm}{r} & \text{if } r > R \quad (P \text{ outside the sphere}) \\[2ex] \dfrac{-Gm}{R} & \text{if } r < R \quad (P \text{ inside the sphere}) \end{cases} \qquad \boxed{11}$$

We leave this calculation as an exercise (Exercise 50). We will prove it again with much less calculation using Gauss's Law in Section 17.3.

In his magnum opus, *Principia Mathematica*, Isaac Newton proved that a sphere of uniform mass density attracts a particle outside the sphere as if the entire mass were concentrated at the center. In other words, a uniform sphere behaves like a point mass as far as gravity is concerned. Furthermore, if the sphere is hollow (i.e., the mass is spread uniformly on the surface of the sphere), then the shell exerts no gravitation force on a particle inside it. This result follows immediately from Eq. (11). Indeed, outside the sphere, φ has the same formula as the potential due to a point mass. Inside the sphere, the potential is *constant* with a value of $-Gm/R$. But constant potential means zero force because the force is the gradient of the potential.

This discussion applies equally well to the electrostatic force between charged objects because this force also obeys an inverse square law (Coulomb's Law). Thus, for example, a uniformly charged sphere behaves like a point charge (when seen from outside the sphere).

16.4 SUMMARY

- A *parametrized surface* is a surface $\mathcal{S}$ whose points are described in the form $\Phi(u, v) = (x(u, v), y(u, v), z(u, v))$, where the *parameters* u and v vary in a domain $\mathcal{D}$ in the uv-plane. We assume that Φ is continuously differentiable.
- At each point on $\mathcal{S}$, we define the tangent and normal vectors

$$\mathbf{T}_u = \frac{\partial \Phi}{\partial u} = \left\langle \frac{\partial x}{\partial u}, \frac{\partial y}{\partial u}, \frac{\partial z}{\partial u} \right\rangle, \qquad \mathbf{T}_v = \frac{\partial \Phi}{\partial v} = \left\langle \frac{\partial x}{\partial v}, \frac{\partial y}{\partial v}, \frac{\partial z}{\partial v} \right\rangle$$

$$\mathbf{n} = \mathbf{n}(u, v) = \mathbf{T}_u \times \mathbf{T}_v$$

The parametrization is called *regular* at (u, v) if $\mathbf{n}(u, v) \neq 0$.

- The quantity $\|\mathbf{n}\|$ may be thought of as a "distortion factor." If $\mathcal{D}$ is a small region in the uv-plane and $\mathcal{S} = \Phi(\mathcal{D})$, then

$$\text{Area}(\mathcal{S}) \approx \|\mathbf{n}(u_0, v_0)\| \text{Area}(\mathcal{D})$$

where (u_0, v_0) is any sample point in $\mathcal{D}$.

- Assume that Φ is one-to-one and regular, except possibly on the boundary of $\mathcal{D}$. The surface area and surface integral are given by the formulas

$$\text{Area}(\mathcal{S}) = \iint_{\mathcal{D}} \|\mathbf{n}(u, v)\| \, du \, dv$$

$$\iint_{\mathcal{S}} f(x, y, z) \, dS = \iint_{\mathcal{D}} f(\Phi(u, v)) \, \|\mathbf{n}(u, v)\| \, du \, dv$$

- Some standard parametrizations:
  - Cylinder of radius R with outward-pointing normal:

$$\Phi(\theta, z) = (R\cos\theta, R\sin\theta, z)$$

$$\mathbf{n} = \mathbf{T}_\theta \times \mathbf{T}_z = R\langle \cos\theta, \sin\theta, 0 \rangle$$

$$dS = \|\mathbf{n}\| \, d\theta \, dz = R \, d\theta \, dz$$

  - Sphere of radius R with outward-pointing normal:

$$\Phi(\phi, \theta) = (R\cos\theta\sin\phi, R\sin\theta\sin\phi, R\cos\phi)$$

$$\mathbf{n} = \mathbf{T}_\phi \times \mathbf{T}_\theta = R^2 \sin\phi \langle \cos\theta\sin\phi, \sin\theta\sin\phi, \cos\phi \rangle = (R^2\sin\phi)\,\mathbf{e}_r$$

$$dS = \|\mathbf{n}\| \, d\phi \, d\theta = R^2 \sin\phi \, d\phi \, d\theta$$

– Graph of $z = g(x, y)$:

$$\Phi(x, y) = (x, y, g(x, y))$$

$$\mathbf{n} = \mathbf{T}_x \times \mathbf{T}_y = \langle -g_x, -g_y, 1 \rangle$$

$$dS = \|\mathbf{n}\|\, dx\, dy = \sqrt{1 + g_x^2 + g_y^2}\, dx\, dy$$

16.4 EXERCISES

Preliminary Questions

1. What is the surface integral of the function $f(x, y, z) = 10$ over a surface of total area 5?

2. What interpretation can we give to the length $\|\mathbf{n}\|$ of the normal vector for a parametrization $\Phi(u, v)$?

3. A parametrization maps a rectangle of size 0.01×0.02 in the uv-plane onto a small patch $\mathcal{S}$ of a surface. Estimate Area($\mathcal{S}$) if $\mathbf{T}_u \times \mathbf{T}_v = \langle 1, 2, 2 \rangle$ at a sample point in the rectangle.

4. A small surface $\mathcal{S}$ is divided into three small pieces, each of area 0.2. Estimate $\iint_{\mathcal{S}} f(x, y, z)\, dS$ if $f(x, y, z)$ takes the values 0.9, 1, and 1.1 at sample points in these three pieces.

5. A surface $\mathcal{S}$ has a parametrization whose domain is the square $0 \le u, v \le 2$ such that $\|\mathbf{n}(u, v)\| = 5$ for all (u, v). What is Area($\mathcal{S}$)?

6. What is the outward-pointing unit normal to the sphere of radius 3 centered at the origin at $P = (2, 2, 1)$?

Exercises

1. Match the parametrization with the surface in Figure 16.

(a) $(u, \cos v, \sin v)$ **(b)** $(u, u + v, v)$

(c) (u, v^3, v)

(d) $(\cos u \sin v, 3 \cos u \sin v, \cos v)$

(e) $(u, u(2 + \cos v), u(2 + \sin v))$

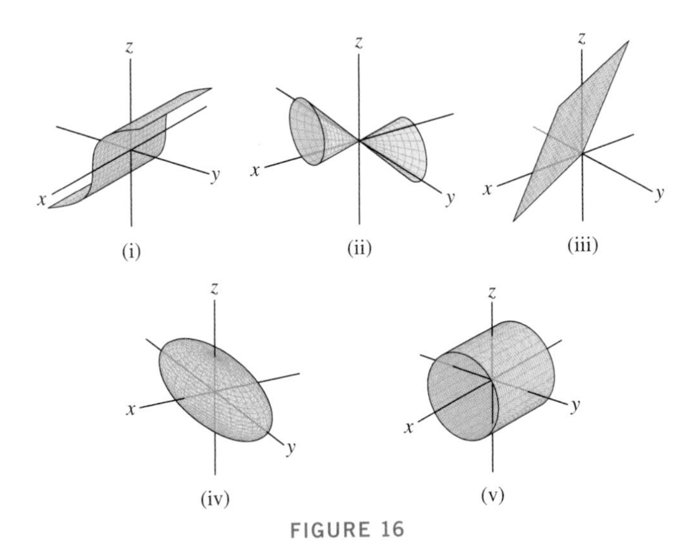

FIGURE 16

2. Show that $\Phi(r, \theta) = (r \cos\theta, r \sin\theta, 1 - r^2)$ parametrizes the paraboloid $z = 1 - x^2 - y^2$. Describe the grid curves of this parametrization.

3. Show that $\Phi(u, v) = (2u + 1, u - v, 3u + v)$ parametrizes the plane $2x - y - z = 2$. Then:

(a) Calculate $\mathbf{T}_u$, $\mathbf{T}_v$, and $\mathbf{n}(u, v)$.

(b) Find the area of $\mathcal{S} = \Phi(\mathcal{D})$, where $\mathcal{D} = \{(u, v) : 0 \le u \le 2, 0 \le v \le 1\}$.

(c) Express $f(x, y, z) = yz$ in terms of u and v and evaluate $\iint_{\mathcal{S}} f(x, y, z)\, dS$.

4. Let $\mathcal{S} = \Phi(\mathcal{D})$, where $\mathcal{D} = \{(u, v) : u^2 + v^2 \le 1, u \ge 0, v \ge 0\}$ and Φ is as defined in Exercise 3.

(a) Calculate the surface area of $\mathcal{S}$.

(b) Evaluate $\iint_{\mathcal{S}} (x - y)\, dS$. *Hint:* Use polar coordinates.

5. Let $\Phi(x, z) = (x, y, xy)$.

(a) Calculate $\mathbf{T}_x$, $\mathbf{T}_y$, and $\mathbf{n}(x, y)$.

(b) Let S be the part of the surface with parameter domain $\mathcal{D} = \{(x, y) : x^2 + y^2 \le 1, x \ge 0, y \ge 0\}$. Verify the following formula and evaluate using polar coordinates:

$$\iint_S 1\, dS = \iint_{\mathcal{D}} \sqrt{1 + x^2 + y^2}\, dx\, dy$$

(c) Verify the following formula and evaluate:

$$\iint_S z\, dS = \int_0^{\pi/2} \int_0^1 (\sin\theta \cos\theta) r^3 \sqrt{1 + r^2}\, dr\, d\theta$$

6. A surface $\mathcal{S}$ has a parametrization $\Phi(u, v)$ whose domain $\mathcal{D}$ is the square in Figure 17. Suppose that Φ has the following normal vectors:

$$\mathbf{n}(A) = \langle 2, 10 \rangle, \quad \mathbf{n}(B) = \langle 1, 3, 0 \rangle$$

$$\mathbf{n}(C) = \langle 3, 0, 1 \rangle, \quad \mathbf{n}(D) = \langle 2, 0, 1 \rangle$$

Estimate $\int_{\mathcal{S}} f(x, y, z)\, dS$, where f is a function such that $f(\Phi(u, v)) = u + v$.

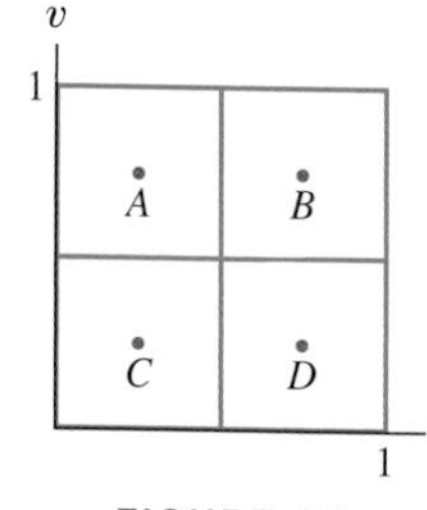

FIGURE 17

In Exercises 7–10, calculate $\mathbf{T}_u$, $\mathbf{T}_v$, *and* $\mathbf{n}(u, v)$ *for the parametrized surface at the given point. Then find the equation of the tangent plane to the surface at that point.*

7. $\Phi(u, v) = (2u + v, u - 4v, 3u)$; $\quad u = 1, \quad v = 4$

8. $\Phi(u, v) = (u^2 - v^2, u + v, u - v)$; $\quad u = 2, \quad v = 3$

9. $\Phi(\theta, \phi) = (\cos\theta \sin\phi, \sin\theta \sin\phi, \cos\phi)$; $\quad \theta = \frac{\pi}{2}, \quad \phi = \frac{\pi}{4}$

10. $\Phi(r, \theta) = (r\cos\theta, r\sin\theta, 1 - r^2)$; $\quad r = \frac{1}{2}, \quad \theta = \frac{\pi}{4}$

11. Use the normal vector computed in Exercise 8 to estimate the area of the small patch of the surface $\Phi(u, v) = (u^2 - v^2, u + v, u - v)$ defined by

$$2 \le u \le 2.1, \qquad 3 \le v \le 3.2$$

12. Sketch the small patch of the sphere whose spherical coordinates satisfy

$$\frac{\pi}{2} - 0.15 \le \theta \le \frac{\pi}{2} + 0.15, \qquad \frac{\pi}{4} - 0.1 \le \phi \le \frac{\pi}{4} + 0.1$$

Use the normal vector computed in Exercise 9 to estimate its area.

13. A surface $\mathcal{S}$ has a parametrization $\Phi(u, v)$ with domain $0 \le u \le 2$, $0 \le v \le 4$ such that the following partial derivatives are constant:

$$\frac{\partial\Phi}{\partial u} = \langle 2, 0, 1\rangle, \qquad \frac{\partial\Phi}{\partial v} = \langle 4, 0, 3\rangle$$

What is the surface area of $\mathcal{S}$?

14. Show that the ellipsoid

$$\left(\frac{x}{a}\right)^2 + \left(\frac{y}{b}\right)^2 + \left(\frac{z}{c}\right)^2 = 1$$

is parametrized by

$$\Phi(\phi, \theta) = (a\cos\theta\sin\phi, b\sin\theta\sin\phi, c\cos\phi)$$

Express the surface area of the ellipsoid as an integral but do not attempt to evaluate it.

15. CAS Let $\mathcal{S}$ be the surface with parametrization

$$\Phi(u, v) = \big((3 + \sin v)\cos u, (3 + \sin v)\sin u, v\big)$$

for $0 \le u, v \le 2\pi$. Using a computer algebra system:

(a) Plot $\mathcal{S}$ from several different viewpoints. Is $\mathcal{S}$ best described as a "vase that holds water" or a "bottomless vase"?

(b) Calculate the normal vector $\mathbf{n}(u, v)$.

(c) Calculate the surface area of $\mathcal{S}$ to four decimal places.

16. CAS Let $\mathcal{S}$ be the surface $z = \ln(5 - x^2 - y^2)$ for $0 \le x, y \le 1$. Using a computer algebra system:

(a) Calculate the surface area of $\mathcal{S}$ to four decimal places.

(b) Calculate $\iint_{\mathcal{S}} x^2 y^3\, dS$ to four decimal places.

17. Use spherical coordinates to compute the surface area of a sphere of radius R.

18. Compute the integral of z over the upper hemisphere of a sphere of radius R centered at the origin.

19. Compute the integral of x^2 over the octant of the unit sphere centered at the origin, where $x, y, z \ge 0$.

20. Show that the hemisphere $x^2 + y^2 + z^2 = R^2$, $z \ge 0$ is parametrized by

$$\Phi(r, \theta) = (r\cos\theta, r\sin\theta, \sqrt{R^2 - r^2})$$

for $0 \le \theta \le 2\pi$, $0 \le r \le R$. Compute the surface area of the hemisphere using this parametrization.

In Exercises 21–32, calculate $\iint_{\mathcal{S}} f(x, y, z)\, dS$ *for the given surface and function.*

21. $\Phi(u, v) = (u\cos v, u\sin v, u)$, $\quad 0 \le u, v \le 1$; $\quad f(x, y, z) = z(x^2 + y^2)$

22. $\Phi(r, \theta) = (r\cos\theta, r\sin\theta, \theta)$, $\quad 0 \le r \le 1$, $\quad 0 \le \theta \le 2\pi$; $f(x, y, z) = \sqrt{x^2 + y^2}$

23. $x^2 + y^2 = 4$, $\quad 0 \le z \le 4$; $\quad f(x, y, z) = e^{-z}$

24. $z = 4 - x^2 - y^2$, $\quad z \ge 0$; $\quad f(x, y, z) = z$

25. $z = 4 - x^2 - y^2$, $\quad z \ge 0$; $\quad f(x, y, z) = z(x^2 + y^2)$

26. $y = 9 - z^2$, $\quad 0 \le x \le z \le 3$; $\quad f(x, y, z) = 1$

27. $y = 9 - z^2$, $\quad 0 \le x, z \le 3$; $\quad f(x, y, z) = z$

28. $\Phi(u, v) = (u, v^3, u + v)$, $\quad 0 \le u, v \le 1$; $\quad f(x, y, z) = y$

29. Part of the plane $x + y + z = 1$, where $x, y, z \ge 0$; $\quad f(x, y, z) = z$

30. Region in the plane $x + y + z = 0$ contained in the cylinder $x^2 + y^2 = 1$; $\quad f(x, y, z) = xz$

31. Part of the surface $x = z^3$, where $0 \le x, y \le 1$; $\quad f(x, y, z) = x$

32. Part of the unit sphere centered at the origin, where $x \ge 0$ and $|y| \le x$; $\quad f(x, y, z) = x$

33. Let S be the sphere of radius R centered at the origin. Explain the following equalities using symmetry:

(a) $\iint_S x\,dS = \iint_S y\,dS = \iint_S z\,dS = 0$

(b) $\iint_S x^2\,dS = \iint_S y^2\,dS = \iint_S z^2\,dS$

Then show, by adding the three integrals in part (b), that $\iint_S x^2\,dS = \frac{4}{3}\pi R^4$.

34. Calculate $\iint_{\mathcal{S}} (xy + e^z)\,dS$, where $\mathcal{S}$ is the triangle in Figure 18 with vertices $(0, 0, 3)$, $(1, 0, 2)$, and $(0, 4, 1)$. *Hint:* Find the equation of the plane containing the triangle.

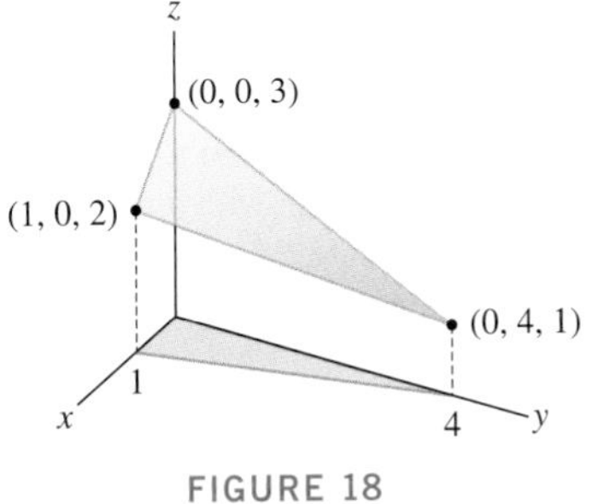

FIGURE 18

35. Find the area of the portion of the plane $2x + 3y + 4z = 28$ lying above the rectangle $1 \le x \le 3$, $2 \le y \le 5$ in the xy-plane.

36. What is the area of the portion of the plane $2x + 3y + 4z = 28$ lying above the domain $\mathcal{D}$ in the xy-plane in Figure 19 if Area$(\mathcal{D}) = 5$?

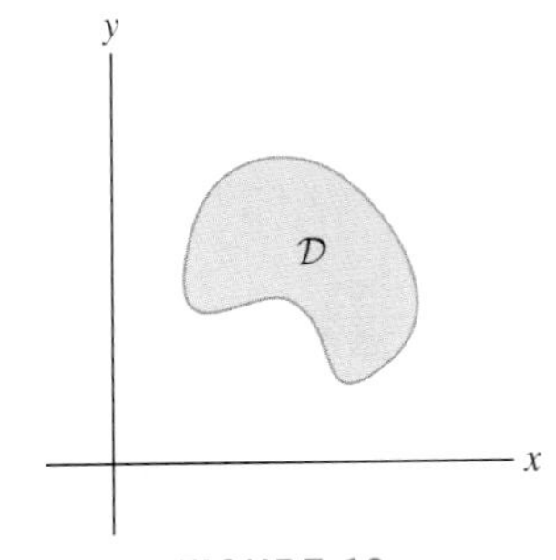

FIGURE 19

37. Compute the integral of $f(x, y, z) = z^2(x^2 + y^2 + z^2)^{-1}$ over the cap of the sphere $x^2 + y^2 + z^2 = 4$ defined by $z \ge 1$.

38. Calculate $\iint_{\mathcal{S}} f(x, y, z) = z(x^2 + y^2)\,dS$, where $\mathcal{S}$ is the hemisphere $x^2 + y^2 + z^2 = R^2$, $z \ge 0$.

39. Let S be the portion of the sphere $x^2 + y^2 + z^2 = 9$, where $1 \le x^2 + y^2 \le 4$ and $z \ge 0$ (Figure 20). Find a parametrization of S in polar coordinates and use it to compute:

(a) The area of S

(b) $\iint_S z^{-1}\,dS$

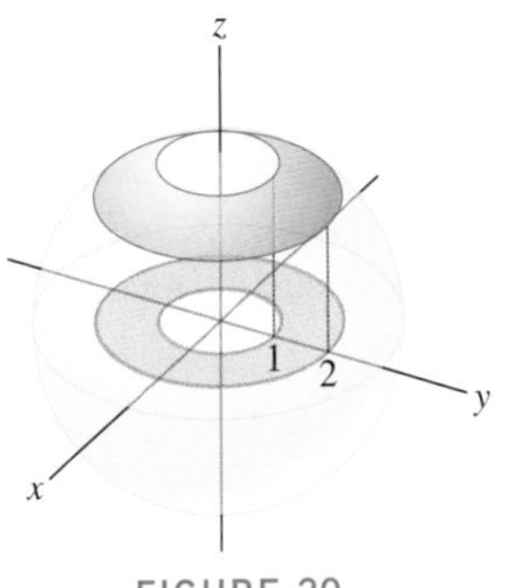

FIGURE 20

40. Find the surface area of the part of the cone $x^2 + y^2 = z^2$ between the planes $z = 2$ and $z = 5$.

41. Find the surface area of the portion S of the cone $z^2 = x^2 + y^2$, where $z \ge 0$, contained within the cylinder $y^2 + z^2 \le 1$.

42. Find the integral of e^{2x+yz} over the following faces of the cube of side 2 in Figure 21.

(a) The top face

(b) The face $PQRS$

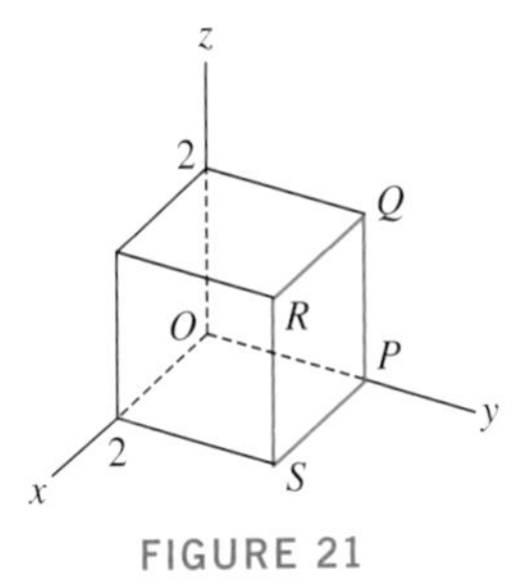

FIGURE 21

43. Prove a famous result of Archimedes: The surface area of the portion of the sphere of radius r between two horizontal planes $z = a$ and $z = b$ is equal to the surface area of the corresponding portion of the circumscribed cylinder (Figure 22).

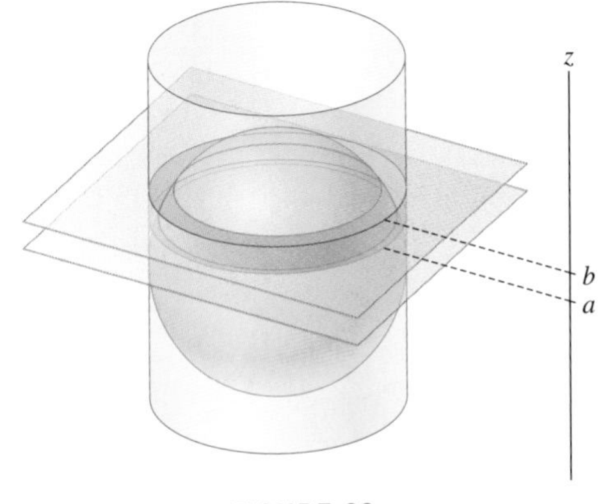

FIGURE 22

Further Insights and Challenges

44. Surfaces of Revolution Let $\mathcal{S}$ be the surface formed by revolving the region underneath the graph $z = g(y)$ in the yz-plane for $c \le y \le d$ about the z-axis (Figure 23). Assume that $c \ge 0$.

(a) Show that the circle generated by rotating a point $(0, a, b)$ about the z-axis is parametrized by

$$(a\cos\theta, a\sin\theta, b), \quad 0 \le \theta \le 2\pi$$

(b) Show that $\mathcal{S}$ is parametrized by

$$\Phi(y, \theta) = (y\cos\theta, y\sin\theta, g(y)) \qquad \boxed{12}$$

for $c \le y \le d$, $0 \le \theta \le 2\pi$.

(c) Show that

$$\text{Area}(\mathcal{S}) = 2\pi \int_c^d y\sqrt{1 + g'(y)^2}\, dy \qquad \boxed{13}$$

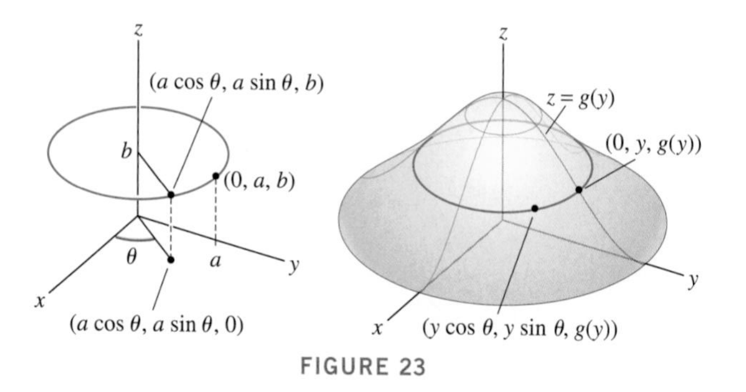

FIGURE 23

45. Use Eq. (13) to compute the surface area of $z = 4 - y^2$ for $0 \le y \le 2$ rotated about the z-axis.

46. Describe the upper half of the cone $x^2 + y^2 = z^2$ for $0 \le z \le d$ as a surface of revolution (Figure 2) and use Eq. (13) to compute its surface area.

47. Area of a Torus Let T be the torus obtained by rotating the circle in the yz-plane of radius a centered at $(0, b, 0)$ about the z-axis (Figure 24). We assume that $b > a > 0$.

(a) Use Eq. (13) to show that

$$\text{Area(T)} = 4\pi \int_{b-a}^{b+a} \frac{ay}{\sqrt{a^2 - (b-y)^2}}\, dy$$

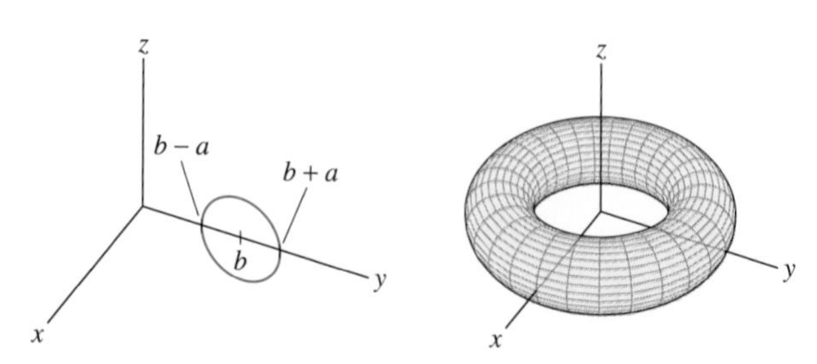

FIGURE 24 The torus obtained by rotating a circle of radius a.

(b) Show that Area(T) $= 4\pi^2 ab$. *Hint:* Rewrite the integral using substitution.

48. Pappus's Theorem Also called **Guldin's Rule**, Pappus's Theorem states that the area of a surface of revolution $\mathcal{S}$ is equal to the length L of the generating curve times the distance traversed by the center of mass. Use Eq. (13) to prove Pappus's Theorem. If $\mathcal{C}$ is the graph $z = g(y)$ for $c \le y \le d$, then the center of mass is defined as the point $(\overline{y}, \overline{z})$ with

$$\overline{y} = \frac{1}{L}\int_{\mathcal{C}} y\, ds, \qquad \overline{z} = \frac{1}{L}\int_{\mathcal{C}} z\, ds$$

49. Compute the surface area of the torus in Exercise 47 using Pappus's Theorem.

50. **Potential due to a Uniform Sphere** Let $\mathcal{S}$ be a sphere of radius R with center at the origin with a uniform mass distribution of total mass m [since $\mathcal{S}$ has surface area $4\pi R^2$, the mass density is $\rho = m/(4\pi R^2)$]. The gravitational potential $\varphi(P)$ due to $\mathcal{S}$ at a point $P = (a, b, c)$ is equal to

$$-G \iint_{\mathcal{S}} \frac{\rho\, dS}{\sqrt{(x-a)^2 + (y-b)^2 + (z-c)^2}}$$

(a) Use symmetry to conclude that the potential depends only on the distance r from P to the center of the sphere. Therefore, it suffices to compute $\varphi(P)$ for a point $P = (0, 0, r)$ on the z-axis (with $r \ne R$).

(b) Use spherical coordinates to show that $\varphi(0, 0, r)$ is equal to

$$\frac{-Gm}{4\pi} \int_0^{\pi} \int_0^{2\pi} \frac{\sin\phi\, d\theta\, d\phi}{\sqrt{R^2 + r^2 - 2Rr\cos\phi}}$$

(c) Use the substitution $u = R^2 + r^2 - 2Rr\cos\phi$ to show that

$$\varphi(0, 0, r) = \frac{-mG}{2Rr}\big(|R + r| - |R - r|\big)$$

(d) Verify formula (11) for φ.

51. Calculate the gravitational potential φ for a hemisphere of radius R with uniform mass distribution.

52. The surface of a cylinder of radius R and length L has a uniform mass distribution ρ (the top and bottom of the cylinder are excluded). Use Eq. (10) to find the gravitational potential at a point P located along the axis of the cylinder.

53. Let S be the part of the graph $z = g(x, y)$ lying over a domain $\mathcal{D}$ in the xy-plane. Let $\phi = \phi(x, y)$ be the angle between the normal to S and the vertical. Prove the formula

$$\text{Area}(S) = \iint_{\mathcal{D}} \frac{dA}{|\cos\phi|}$$

16.5 Surface Integrals of Vector Fields

We now study the integrals of vector fields over surfaces. These integrals represent the quantity flux, which arises often in physics and engineering as the rate at which a quantity flows across a surface.

As a preliminary step, it is necessary to introduce the concept of an **oriented surface.** An *orientation* of a surface $\mathcal{S}$ is a continuously varying choice of unit normal vector $\mathbf{e_n}(P)$ at each point P on the surface (Figure 1). Since there are two normal directions at each point on $\mathcal{S}$, an orientation serves to specify one of the two "sides" of the surface. For example, a sphere has two orientations, depending on whether the unit normal vector points to the outside or to the inside of the sphere. If $\mathbf{e_n}(P)$ is the unit normal vector determined by an orientation, then $-\mathbf{e_n}(P)$ is the unit normal vector determined by the *opposite orientation.*

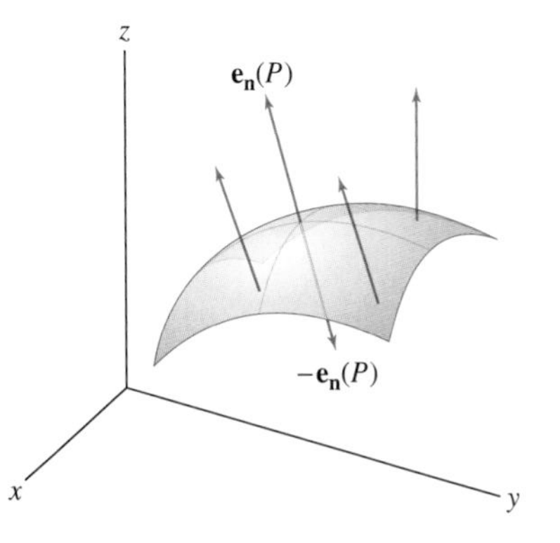

FIGURE 1 Two possible unit normal vectors. An orientation is a continuously varying choice of normal direction at each point.

The **normal component** of a vector field $\mathbf{F}$ at a point P on an oriented surface is the dot product $\mathbf{F}(P) \cdot \mathbf{e_n}(P) = \|\mathbf{F}(P)\| \cos\theta$, where θ is the angle between $\mathbf{F}(P)$ and $\mathbf{e_n}(P)$ (Figure 2). We usually write $\mathbf{e_n}$ instead of $\mathbf{e_n}(P)$, but it is understood that $\mathbf{e_n}$ varies from point to point on the surface. The surface integral of $\mathbf{F}$ is defined as the integral of this normal component:

$$\text{Vector surface integral:} \quad \iint_{\mathcal{S}} \mathbf{F} \cdot d\mathbf{S} = \iint_{\mathcal{S}} (\mathbf{F} \cdot \mathbf{e_n})\, dS$$

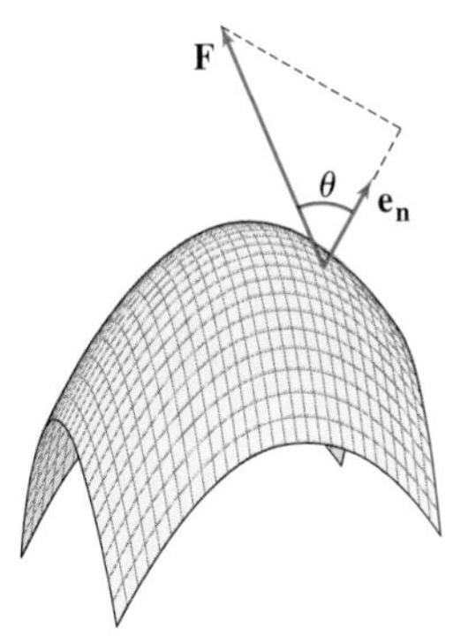

FIGURE 2 The normal component of a vector to a surface.

The surface integral of $\mathbf{F}$ is also called the **flux** of $\mathbf{F}$ across $\mathcal{S}$.

As in the scalar case, we use a parametrization $\Phi(u, v)$ of $\mathcal{S}$ to evaluate the surface integral. Assume that Φ is regular, so that $\mathbf{n} \neq 0$, and define a unit normal vector:

$$\mathbf{e_n} = \mathbf{e_n}(u, v) = \frac{\mathbf{n}}{\|\mathbf{n}\|}$$

We assume that Φ parametrizes the oriented surface $\mathcal{S}$, that is, we assume that $\mathbf{e_n}$ points in the normal direction specified by the orientation of $\mathcal{S}$. In this case,

$$\begin{aligned} \iint_{\mathcal{S}} \mathbf{F} \cdot d\mathbf{S} &= \iint_{\mathcal{D}} (\mathbf{F} \cdot \mathbf{e_n}) \|\mathbf{n}(u, v)\|\, du\, dv \\ &= \iint_{\mathcal{D}} \mathbf{F}(\Phi(u, v)) \cdot \left(\frac{\mathbf{n}(u, v)}{\|\mathbf{n}(u, v)\|} \right) \|\mathbf{n}(u, v)\|\, du\, dv \\ &= \iint_{\mathcal{D}} \mathbf{F}(\Phi(u, v)) \cdot \mathbf{n}(u, v)\, du\, dv \end{aligned} \qquad \boxed{1}$$

Compare Eq. (1) with the formula for a scalar surface integral:

$$\iint_{\mathcal{S}} f(x, y, z)\, dS = \iint f(\Phi(u, v)) \|\mathbf{n}(u, v)\|\, du\, dv$$

Orientation plays no role in a scalar surface integral because the integral involves the length $\|\mathbf{n}\|$ *but not the direction of* $\mathbf{n}$.

This formula remains valid even if $\mathbf{n}(u, v)$ is zero at points on the boundary of the parameter domain. If we reverse the orientation of $\mathcal{S}$ in a vector surface integral, $\mathbf{n}(u, v)$ is replaced by $-\mathbf{n}(u, v)$ and the integral changes sign.

We may interpret the symbol $d\mathbf{S}$ as a "vector surface element" that is expressed in terms of a parametrization by

$$d\mathbf{S} = \mathbf{n}(u, v)\, du\, dv$$

THEOREM 1 Vector Surface Integral Let $\Phi(u, v)$ be a parametrization of an oriented surface $\mathcal{S}$ with parameter domain $\mathcal{D}$. Assume that Φ is one-to-one and regular, except possibly at points on the boundary of $\mathcal{D}$. Let $\mathbf{n}(u, v)$ be the normal vector defined by Φ. Then

$$\iint_{\mathcal{S}} \mathbf{F} \cdot d\mathbf{S} = \iint_{\mathcal{D}} \mathbf{F}(\Phi(u, v)) \cdot \mathbf{n}(u, v)\, du\, dv \qquad \boxed{2}$$

If the orientation of $\mathcal{S}$ is reversed, the surface integral changes sign.

■ **EXAMPLE 1** Calculate $\iint_{\mathcal{S}} \mathbf{F} \cdot d\mathbf{S}$, where $\mathbf{F} = \langle 0, 0, x \rangle$ and $\mathcal{S}$ is the surface with parametrization

$$\Phi(u, v) = (u^2, v, u^3 - v^2), \qquad 0 \le u, v \le 1$$

oriented with an upward-pointing normal.

Solution

***Step 1.* Compute the tangent and normal vectors.**

$$\mathbf{T}_u = \langle 2u, 0, 3u^2 \rangle, \qquad T_v = \langle 0, 1, -2v \rangle$$

$$\mathbf{n}(u, v) = \mathbf{T}_u \times \mathbf{T}_v = \begin{vmatrix} \mathbf{i} & \mathbf{j} & \mathbf{k} \\ 2u & 0 & 3u^2 \\ 0 & 1 & -2v \end{vmatrix}$$

$$= -3u^2\mathbf{i} + 4uv\mathbf{j} + 2u\mathbf{k} = \langle -3u^2, 4uv, 2u \rangle$$

The z-component of $\mathbf{n}$ is positive on the domain $0 \le u \le 1$, so $\mathbf{n}$ is the upward-pointing normal (Figure 3).

***Step 2.* Evaluate the dot product F · n.**

Write $\mathbf{F}$ in terms of the parameters $x = u^2$, $y = v$, and $z = u^3 - v^2$:

$$\mathbf{F}(\Phi(u, v)) = \langle 0, 0, x \rangle = \langle 0, 0, u^2 \rangle$$

and compute the dot product:

$$\mathbf{F}(\Phi(u, v)) \cdot \mathbf{n}(u, v) = \langle 0, 0, u^2 \rangle \cdot \langle -3u^2, 4uv, 2u \rangle = 2u^3$$

***Step 3.* Evaluate the surface integral.**

The surface integral is equal to a double integral over the rectangle $0 \le u, v \le 1$:

$$\iint_{\mathcal{S}} \mathbf{F} \cdot d\mathbf{S} = \int_{u=0}^{1} \int_{v=0}^{1} \mathbf{F}(\Phi(u, v)) \cdot \mathbf{n}(u, v)\, dv\, du$$

$$= \int_{u=0}^{1} \int_{v=0}^{1} 2u^3\, dv\, du = \int_{u=0}^{1} 2u^3\, du = \frac{1}{2}$$

■

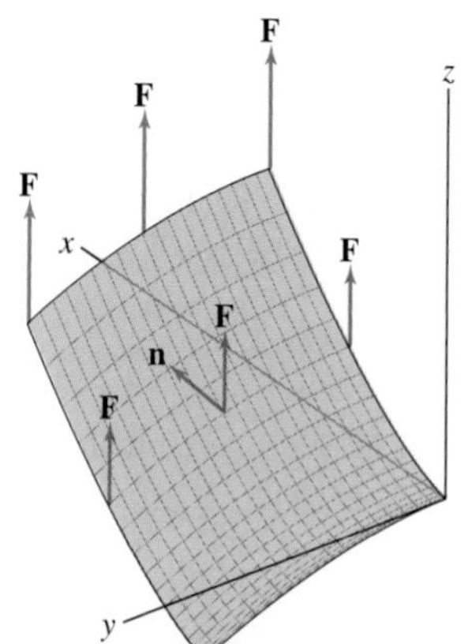

FIGURE 3 The surface $\Phi(u, v) = (u^2, v, u^3 - v^2)$ with an upward-pointing normal. The vector field $\mathbf{F} = \langle 0, 0, x \rangle$ points in the vertical direction.

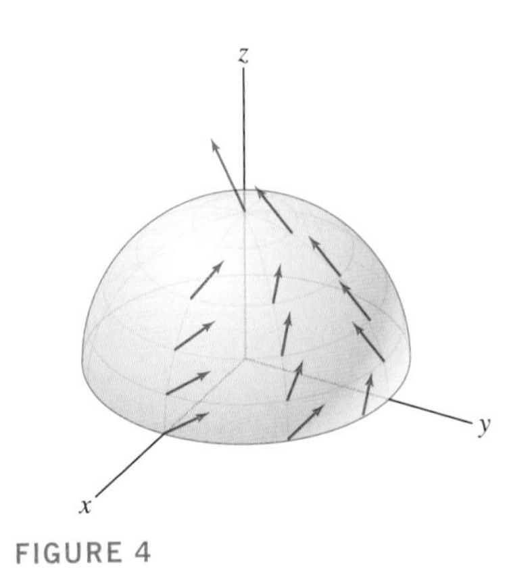

FIGURE 4

■ **EXAMPLE 2** Integral over a Hemisphere Calculate the flux of $\mathbf{F} = \langle z, x, 1 \rangle$ across the upper hemisphere $\mathcal{S}$ of the sphere $x^2 + y^2 + z^2 = 1$, oriented with normal vectors pointing to the outside of the sphere (Figure 4).

Solution The angle of declination ϕ varies between 0 and $\frac{\pi}{2}$ in the upper hemisphere, so we may parametrize S by

$$\Phi(\phi, \theta) = (\cos\theta \sin\phi, \sin\theta \sin\phi, \cos\phi), \qquad 0 \le \phi \le \frac{\pi}{2}, \quad 0 \le \theta < 2\pi$$

***Step 1.* Compute the normal vector.**
According to Eq. (5) in Section 16.4, the outward-pointing normal vector is

$$\mathbf{n} = \mathbf{T}_\phi \times \mathbf{T}_\theta = \sin\phi \langle \cos\theta \sin\phi, \sin\theta \sin\phi, \cos\phi \rangle$$

***Step 2.* Evaluate the dot product $\mathbf{F} \cdot \mathbf{n}$.**
Express the vector field in terms of the parameters and evaluate the dot product:

$$\mathbf{F}(\Phi(\phi, \theta)) = \langle z, x, 1 \rangle = \langle \cos\phi, \cos\theta \sin\phi, 1 \rangle$$

$$\mathbf{F}(\Phi(\phi, \theta)) \cdot \mathbf{n}(\phi, \theta) = \langle \cos\phi, \cos\theta \sin\phi, 1 \rangle \cdot \langle \cos\theta \sin^2\phi, \sin\theta \sin^2\phi, \cos\phi \sin\phi \rangle$$
$$= \cos\theta \sin^2\phi \cos\phi + \cos\theta \sin\theta \sin^3\phi + \cos\phi \sin\phi$$

***Step 3.* Evaluate the surface integral.**

$$\iint_{\mathcal{S}} \mathbf{F} \cdot d\mathbf{S} = \int_{\phi=0}^{\pi/2} \int_{\theta=0}^{2\pi} \mathbf{F}(\Phi(\phi, \theta)) \cdot \mathbf{n}(\phi, \theta)\, d\theta\, d\phi$$
$$= \int_{\phi=0}^{\pi/2} \int_{\theta=0}^{2\pi} (\underbrace{\cos\theta \sin^2\phi \cos\phi + \cos\theta \sin\theta \sin^3\phi}_{\text{Integral over } \theta \text{ is zero}} + \cos\phi \sin\phi)\, d\theta\, d\phi$$

The integrals of $\cos\theta$ and $\cos\theta \sin\theta$ over $0 \le \theta < 2\pi$ are both zero, so we are left with

$$\int_{\phi=0}^{\pi/2} \int_{\theta=0}^{2\pi} \cos\phi \sin\phi\, d\theta\, d\phi = 2\pi \int_{\phi=0}^{\pi/2} \cos\phi \sin\phi\, d\phi = 2\pi \frac{\sin^2\phi}{2}\bigg|_0^{\pi/2} = \pi \quad ■$$

■ **EXAMPLE 3** Surface Integral over a Graph Calculate the flux of $\mathbf{F} = 2\mathbf{i} - y\mathbf{j} + x\mathbf{k}$ through the surface $y = 1 + x^2 + z^2$ for $1 \le y \le 5$, oriented with normal pointing in the negative y-direction.

Solution This surface is the graph of the function $y = 1 + x^2 + z^2$, where x and z are the independent variables.

***Step 1.* Find a parametrization.**
Since y is given explicitly in terms of x and z, we may use x and z as parameters. The surface is parametrized by

$$\Phi(x, z) = (x, 1 + x^2 + z^2, z)$$

What is the parameter domain? Since $y = 1 + x^2 + z^2$, the condition $1 \le y \le 5$ is equivalent to $1 \le 1 + x^2 + z^2 \le 5$ or $0 \le x^2 + z^2 \le 4$. In other words, the parameter domain is $\mathcal{D} = \{(x, z) : x^2 + z^2 \le 4\}$.

Step 2. **Compute the tangent and normal vectors.**

$$\mathbf{T}_x = \langle 1, 2x, 0\rangle, \qquad \mathbf{T}_z = \langle 0, 2z, 1\rangle$$

$$\mathbf{n} = \mathbf{T}_x \times \mathbf{T}_z = \begin{vmatrix} \mathbf{i} & \mathbf{j} & \mathbf{k} \\ 1 & 2x & 0 \\ 0 & 2z & 1 \end{vmatrix} = 2x\mathbf{i} - \mathbf{j} + 2z\mathbf{k}$$

Observe that $\mathbf{n}$ points in the negative y-direction.

Step 3. **Evaluate the dot product $\mathbf{F} \cdot \mathbf{n}$.**

Express the vector field in terms of the parametrization and compute the dot product:

$$\mathbf{F}(\Phi(x,z)) = \langle 2, -y, x\rangle = \langle 2, -(1+x^2+z^2), x\rangle$$

$$\mathbf{F}(\Phi(x,z)) \cdot \mathbf{n} = \langle 2, -(1+x^2+z^2), x\rangle \cdot \langle 2x, -1, 2z\rangle$$

$$= 4x + (1+x^2+z^2) + 2xz$$

Step 4. **Evaluate the surface integral.**

$$\iint_{\mathcal{S}} \mathbf{F} \cdot d\mathbf{S} = \iint_{\mathcal{D}} (x^2 + z^2 + 2xz + 4x + 1)\, dx\, dz$$

Since $\mathcal{D}$ is the circle of radius 2 in the xz-plane, it is convenient to use polar coordinates

$$x = r\cos\theta, \qquad z = r\sin\theta$$

Then $x^2 + z^2 = r^2$ and the integrand becomes

$$x^2 + z^2 + 2xz + 4x + 1 = r^2 + 2r^2 \sin\theta\cos\theta + 4r\cos\theta + 1$$

By the Change of Variables Formula for polar coordinates, $dx\, dz$ becomes $r\, dr\, d\theta$ and

$$\iint_{\mathcal{S}} \mathbf{F} \cdot d\mathbf{S} = \int_{\theta=0}^{2\pi} \int_{r=0}^{2} (r^2 + 2r^2 \sin\theta\cos\theta + 4r\cos\theta + 1)\, r\, dr\, d\theta$$

$$= 2\pi \int_0^2 (r^2+1)\, r\, dr = 2\pi\left(\frac{2^4}{4} + \frac{2^2}{2}\right) = 12\pi$$

■

← REMINDER The Change of Variables Formula for polar coordinates reads

$$\iint f(x,y)\, dx\, dy$$

$$= \iint f(r\sin\theta, r\cos\theta)\, r\, dr\, d\theta$$

We apply it in this example with variables x and z.

Fluid Flux

The word "flux" is derived from the Latin word fluere meaning "to flow."

The term *flux* comes from physics, where it is often used to denote the rate of transfer of fluid, particles, or energy across a given surface. Imagine dipping a net into a stream of flowing water (Figure 5). The flux is the rate at which water flows through the net per unit time.

FIGURE 5 Velocity field of a fluid flow.

To compute this flux or flow rate, let $\mathbf{v}$ be the velocity vector field (Figure 5). At each point P, $\mathbf{v}(P)$ gives the speed and direction of the water particle located at the point P. If we think of the net as a surface $\mathcal{S}$, then the rate of flow through $\mathcal{S}$ (in volume per time) is equal to the surface integral of $\mathbf{v}$ over $\mathcal{S}$.

To explain why, suppose first that $\mathbf{v}$ is constant with value $\mathbf{v}_0$ and that our "net" $\mathcal{S}$ is a rectangle of area A (Figures 6 and 7). Then each water particle flows in the same direction with the same speed $\|\mathbf{v}_0\|$ (say, in meters per second). If $\mathbf{v}_0$ is perpendicular to $\mathcal{S}$, then a water particle passes through $\mathcal{S}$ within a 1-s time interval if its distance to $\mathcal{S}$ is at most $\|\mathbf{v}_0\|$ meters. In other words, the water flowing through $\mathcal{S}$ in a 1-s time interval is

FIGURE 6 Water flowing at constant velocity $\mathbf{v}_0$ perpendicular to a rectangular surface $\mathcal{S}$.

a box of volume $\|\mathbf{v}_0\|A$, and thus,

$$\text{Flow rate} = \text{velocity} \times \text{area} = \|\mathbf{v}_0\|A$$

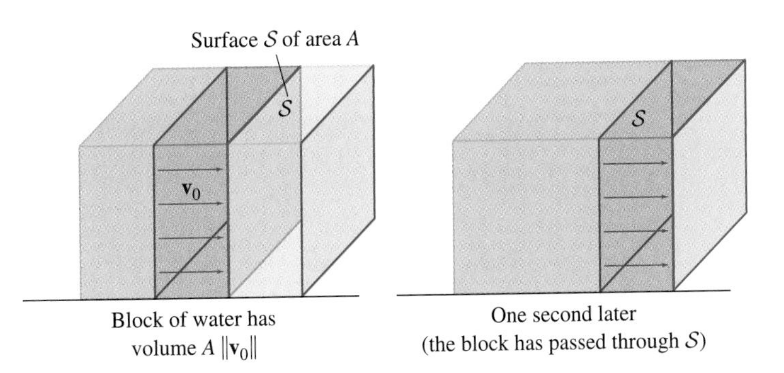

FIGURE 7

If the fluid has constant velocity $\mathbf{v}_0$ but flows at an angle θ relative to $\mathcal{S}$, then the block of water flowing past $\mathcal{S}$ in a 1-s interval is no longer a rectangular box but a parallelepiped of volume $A\|\mathbf{v}_0\|\cos\theta$ (Figure 8). If $\mathbf{n}$ is a vector normal to $\mathcal{S}$ of length equal to the area A, then we may write the flow rate as a dot product:

$$\text{Flow rate} = A\|\mathbf{v}_0\|\cos\theta = \mathbf{v}_0 \cdot \mathbf{n}$$

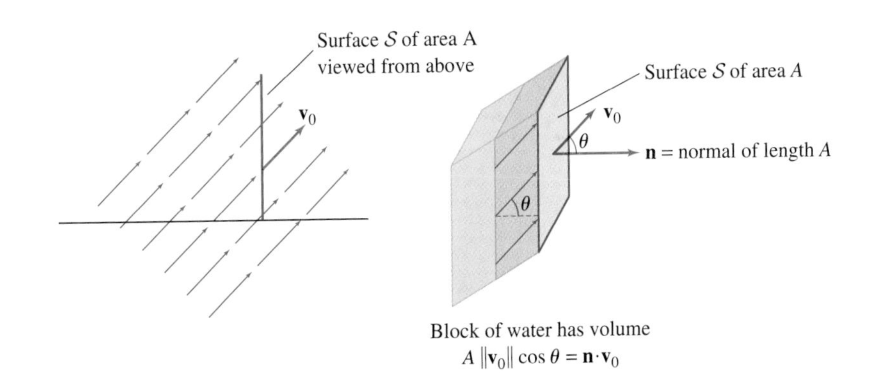

FIGURE 8 Water flowing at constant velocity $\mathbf{v}_0$, making an angle θ with a rectangular surface.

In the general case, the velocity field $\mathbf{v} = \mathbf{v}(u, v)$ need not be constant and the surface $\mathcal{S}$ may be curved. However, suppose that $\mathcal{S}$ has a parametrization $\Phi(u, v)$ that maps a small rectangle of size $\Delta u \times \Delta v$ to a small patch $\mathcal{S}_0$ of $\mathcal{S}$. For any sample point $\Phi(u_0, v_0)$ in $\mathcal{S}_0$, the vector $\mathbf{n}(u_0, v_0)\,\Delta u\,\Delta v$ is a normal vector whose length is approximately equal to the area of $\mathcal{S}_0$ [Eq. (1) in Section 16.4]. This patch is nearly rectangular, and thus, the flow rate through $\mathcal{S}_0$ is approximately $\mathbf{v}(u_0, v_0) \cdot \mathbf{n}(u_0, v_0)\,\Delta u\,\Delta v$. We find that the total flux is equal to the integral of $\mathbf{v}(u, v) \cdot \mathbf{n}(u, v)$, which is the surface integral of $\mathbf{v}$ over $\mathcal{S}$.

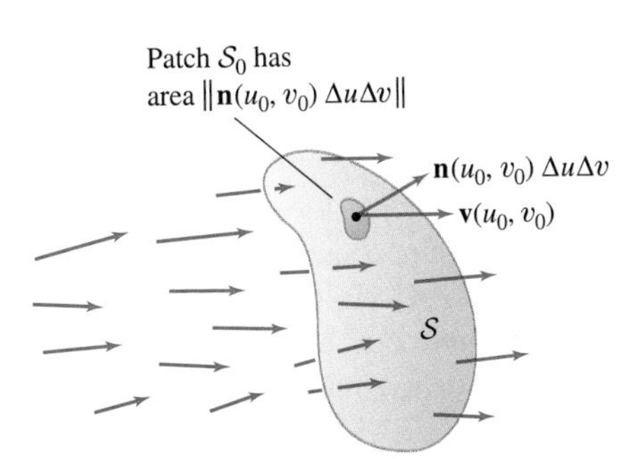

FIGURE 9 The fluid flow across a small patch $\mathcal{S}_0$ is approximately equal to $\mathbf{v}(u_0, v_0) \cdot \mathbf{n}(u_0, v_0)\,\Delta u\,\Delta v$.

Flow Rate of a Fluid Through a Surface The flow rate (in volume per unit time) of a fluid across a surface $\mathcal{S}$ is equal to the surface integral of the velocity vector field $\mathbf{v}$:

$$\text{Flow rate across } \mathcal{S} \text{ (volume per unit time)} = \iint_{\mathcal{S}} \mathbf{v} \cdot d\mathbf{S} \qquad \boxed{3}$$

■ **EXAMPLE 4** Let $\mathbf{v} = \langle x^2 + y^2, 0, z^2 \rangle$ be the velocity field (in centimeters per second) of a fluid in $\mathbf{R}^3$. Compute the flow rate through the upper hemisphere $\mathcal{S}$ of the unit sphere.

Solution We use spherical coordinates:

$$x = \cos\theta \sin\phi, \qquad y = \sin\theta\sin\phi, \qquad z = \cos\phi$$

The upper hemisphere corresponds to the ranges $0 \le \phi \le \frac{\pi}{2}$ and $0 \le \theta \le 2\pi$. By Eq. (5) in Section 16.4, the outward-pointing normal is

$$\mathbf{n} = \mathbf{T}_\phi \times \mathbf{T}_\theta = \sin\phi\langle \cos\theta\sin\phi, \sin\theta\sin\phi, \cos\phi\rangle$$

We have

$$x^2 + y^2 = (\cos\theta\sin\phi)^2 + (\sin\theta\sin\phi)^2 = \sin^2\phi$$

Therefore,

$$\mathbf{v} = \langle x^2 + y^2, 0, z^2\rangle = \langle \sin^2\phi, 0, \cos^2\phi\rangle$$

$$\mathbf{v}\cdot\mathbf{n} = \sin\phi\langle \sin^2\phi, 0, \cos^2\phi\rangle \cdot \langle\cos\theta\sin\phi, \sin\theta\sin\phi, \cos\phi\rangle$$

$$= \sin^4\phi\cos\theta + \sin\phi\cos^3\phi$$

The flux is equal to

$$\iint_{\mathcal{S}} \mathbf{v}\cdot d\mathbf{S} = \int_{\phi=0}^{\pi/2}\int_{\theta=0}^{2\pi} (\sin^4\phi\cos\theta + \sin\phi\cos^3\phi)\, d\theta\, d\phi$$

The integral of the first term, $\sin^4\phi\cos\theta$, with respect to θ is zero, so we are left with

$$\int_{\phi=0}^{\pi/2}\int_{\theta=0}^{2\pi} \sin\phi\cos^3\phi\, d\theta\, d\phi = 2\pi\int_{\phi=0}^{\pi/2}\cos^3\phi\sin\phi\, d\phi$$

$$= 2\pi\left(-\frac{\cos^4\phi}{4}\right)\Bigg|_{\phi=0}^{\pi/2} = \frac{\pi}{2}\ \text{cm}^3/\text{s} \qquad ■$$

Electric and Magnetic Fields

The laws of electricity and magnetism are expressed in terms of two vector fields, the electric field **E** and the magnetic field **B**, whose properties are summarized by four equations known as Maxwell's Equations. One of these equations expresses **Faraday's Law of Induction**, which can be formulated either as a partial differential equation or in the following "integral form":

$$\int_{\mathcal{C}} \mathbf{E}\cdot d\mathbf{s} = -\frac{d}{dt}\iint_{\mathcal{S}} \mathbf{B}\cdot d\mathbf{S} \qquad \boxed{4}$$

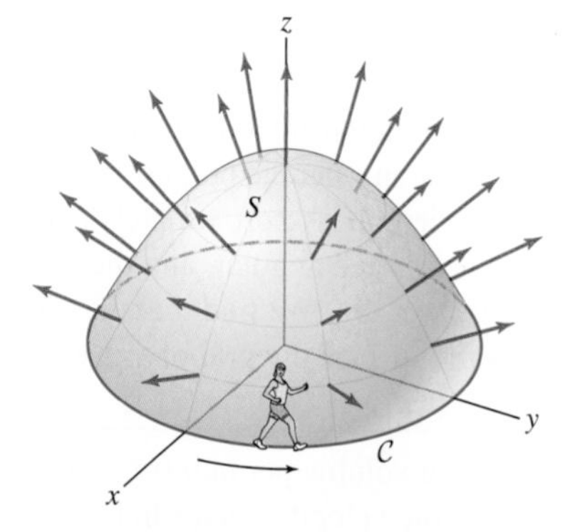

FIGURE 10 If she walks along $\mathcal{C}$ in the positive direction, with the surface to her left, then her head points in the outward direction.

In this equation, $\mathcal{S}$ is a surface with boundary curve $\mathcal{C}$. We orient $\mathcal{S}$ using the unit normal vectors whose direction is determined by the orientation of $\mathcal{C}$ via the right-hand rule (Figure 10). The line integral of **E** around the boundary curve $\mathcal{C}$ is equal to the voltage drop around the curve (the work performed by **E** moving a positive unit charge around $\mathcal{C}$).

■ **EXAMPLE 5** A varying current of magnitude $i = 28\cos\left(\frac{30t}{\pi}\right)$ amperes (t in seconds) flows through a straight wire, producing a magnetic field **B** whose magnitude at

We noted in Section 16.3 that the electric field **E** *is conservative when the charges are stationary (or more generally, when the magnetic field is constant). When the magnetic field* **B** *varies in time, the integral on the right in Eq. (4) is nonzero for some surface, and hence the circulation of* **E** *around* C *is nonzero. This shows that* **E** *is not conservative when the magnetic field varies.*

a distance r m from the wire is $\|\mathbf{B}\| = \dfrac{\mu_0 i}{2\pi r}$ teslas (where $\mu_0 = 4\pi \cdot 10^{-7}$ T-m/A). Furthermore, at any point P, $\mathbf{B}$ points in the direction tangent to the circle through P perpendicular to the wire as in Figure 11.

(a) Calculate the flux $\Phi(t)$ at time t of $\mathbf{B}$ through the rectangle $\mathcal{R}$ shown in Figure 11, oriented with the normal vector pointing out of the page. Assume that $L = 1.2$ m, $H = 0.7$ m, and $d = 0.1$ m.

(b) Use Faraday's Law to determine the voltage drop (in volts) around the rectangular loop $\mathcal{C}$. Assume that $L = 1.2$ m.

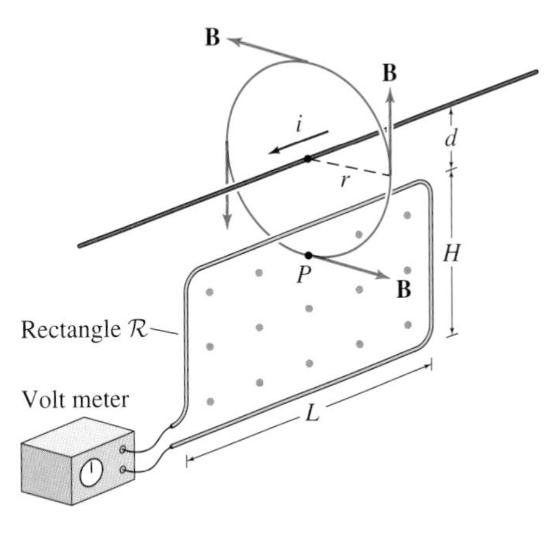

FIGURE 11

Solution Referring to Figure 11, we choose coordinates (x, y) so that the rectangle $\mathcal{R}$ is the region

$$0 \le x \le L, \qquad d \le y \le H + d$$

The distance from a point $P = (x, y)$ in $\mathcal{R}$ to the wire is y. The vector field $\mathbf{B}$ points directly out of the page, so if we orient $\mathcal{R}$ with the unit normal $\mathbf{n}$ pointing out of the page, then $\mathbf{B}$ and $\mathbf{n}$ point in the same direction and

$$\mathbf{B} \cdot \mathbf{n} = \frac{\mu_0 i}{2\pi y}$$

(a) The flux $\Phi(t)$ of $\mathbf{B}$ through $\mathcal{R}$ at time t is then

$$\begin{aligned}
\Phi(t) &= \iint_{\mathcal{R}} \mathbf{B} \cdot \mathbf{n}\, dS = \int_{x=0}^{L} \int_{y=d}^{H+d} \frac{\mu_0 i}{2\pi y}\, dy\, dx = \frac{\mu_0 L i}{2\pi} \int_{y=d}^{H+d} \frac{dy}{y} \\
&= \frac{\mu_0 L}{2\pi} \left(\ln \frac{H+d}{d} \right) i \\
&= \frac{4\pi \cdot 10^{-7}(1.2)}{2\pi} \left(\ln \frac{0.8}{0.1} \right) 28 \cos\left(\frac{30t}{\pi} \right) \\
&\approx 1.4 \times 10^{-5} \cos\left(\frac{30t}{\pi} \right) \text{ T-m}^2
\end{aligned}$$

(b) By Faraday's Law [Eq. (4)], the voltage drop around the rectangular loop $\mathcal{C}$, oriented in the counterclockwise direction, is

$$\int_{\mathcal{C}} \mathbf{E} \cdot d\mathbf{s} = -\frac{d\Phi}{dt} \approx -(1.4 \times 10^{-5}) \left(\frac{30}{\pi} \right) \sin\left(\frac{30t}{\pi} \right) \approx -0.00013 \sin\left(\frac{30t}{\pi} \right) \text{ V} \quad \blacksquare$$

CONCEPTUAL INSIGHT We have seen that a vector surface integral depends on the choice of orientation of the surface. However, some surfaces, such as the Möbius strip (discovered in 1858), cannot be oriented because they are one-sided. You can construct a Möbius strip M with a rectangular strip of paper: Join the two ends of the strip together with a 180° twist. Unlike an ordinary two-sided strip, the Möbius strip M has only one side and it is impossible to specify an outward direction in a consistent manner (Figure 12). If you choose a unit normal vector at a point P and carry that unit vector around M, when you return to P, the vector will point in the opposite direction. Therefore, we cannot integrate a vector field over a Möbius strip. On the other hand, it is possible to integrate a scalar function. For example, the integral of mass density would equal the total mass of the Möbius strip.

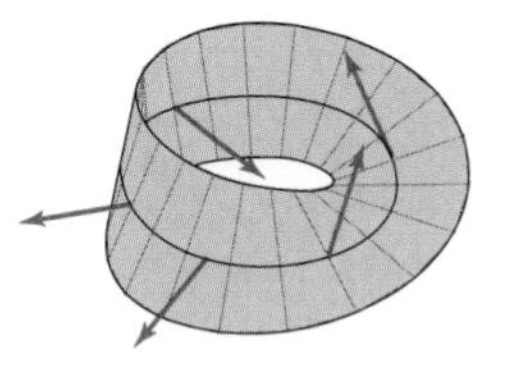
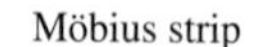

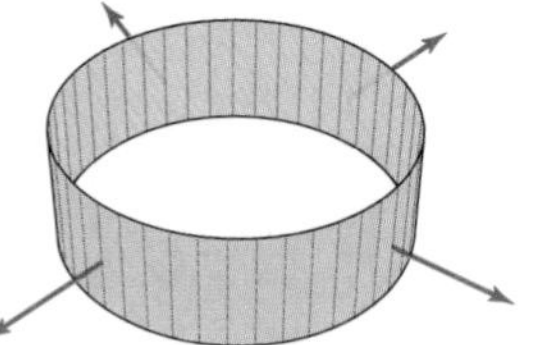

FIGURE 12 It is not possible to choose a continuously varying unit normal vector on a Möbius strip.

16.5 SUMMARY

- A surface $\mathcal{S}$ is *oriented* if a continuously varying unit normal vector $\mathbf{e_n}(P)$ is specified at each point on $\mathcal{S}$. This distinguishes an "outward" direction on the surface.
- The integral of a vector field $\mathbf{F}$ over an oriented surface $\mathcal{S}$ is defined as the integral of the normal component $\mathbf{F} \cdot \mathbf{e_n}$ over $\mathcal{S}$.
- Vector surface integrals are computed using the formula

$$\iint_{\mathcal{S}} \mathbf{F} \cdot d\mathbf{S} = \iint_{\mathcal{D}} \mathbf{F}(\Phi(u, v)) \cdot \mathbf{n}(u, v)\, du\, dv$$

Here, $\Phi(u, v)$ is a parametrization of $\mathcal{S}$ such that $\mathbf{n}(u, v) = \mathbf{T}_u \times \mathbf{T}_v$ points in the direction of the unit normal vector specified by the orientation.
- The surface integral of a vector field $\mathbf{F}$ over $\mathcal{S}$ is also called the *flux* of $\mathbf{F}$ through Φ. If $\mathbf{F}$ is the velocity field of a fluid, then the flux $\iint_{\mathcal{S}} \mathbf{F} \cdot d\mathbf{S}$ is the rate at which fluid flows through $\mathcal{S}$ per unit time.

16.5 EXERCISES

Preliminary Questions

1. Let $\mathbf{F}$ be a vector field and $\Phi(u, v)$ a parametrization of a surface $\mathcal{S}$, and set $\mathbf{n} = \mathbf{T}_u \times \mathbf{T}_v$. Which of the following is the normal component of $\mathbf{F}$?

(a) $\mathbf{F} \cdot \mathbf{n}$ **(b)** $\mathbf{F} \cdot \mathbf{e_n}$

2. The vector surface integral $\iint_{\mathcal{S}} \mathbf{F} \cdot d\mathbf{S}$ is equal to the scalar surface integral of the function (choose the correct answer):

(a) $\|\mathbf{F}\|$

(b) $\mathbf{F} \cdot \mathbf{n}$, where $\mathbf{n}$ is a normal vector

(c) $\mathbf{F} \cdot \mathbf{e_n}$, where $\mathbf{e_n}$ is the unit normal vector

3. $\iint_{\mathcal{S}} \mathbf{F} \cdot d\mathbf{S}$ is zero if (choose the correct answer):

(a) $\mathbf{F}$ is tangent to $\mathcal{S}$ at every point.

(b) $\mathbf{F}$ is perpendicular to $\mathcal{S}$ at every point.

4. If $\mathbf{F}(P) = \mathbf{e_n}(P)$ at each point on $\mathcal{S}$, then $\iint_{\mathcal{S}} \mathbf{F} \cdot d\mathbf{S}$ is equal to (choose the correct answer):

(a) Zero **(b)** Area($\mathcal{S}$) **(c)** Neither

5. Let $\mathcal{S}$ be the disk $x^2 + y^2 \le 1$ in the xy-plane oriented with normal in the positive z-direction. Determine $\iint_{\mathcal{S}} \mathbf{F} \cdot d\mathbf{S}$ for each of the following vector constant fields:

(a) $\mathbf{F} = \langle 1, 0, 0\rangle$

(b) $\mathbf{F} = \langle 0, 0, 1\rangle$

(c) $\mathbf{F} = \langle 1, 1, 1\rangle$

6. Estimate $\iint_{\mathcal{S}} \mathbf{F} \cdot d\mathbf{S}$, where $\mathcal{S}$ is a tiny oriented surface of area 0.05 and the value of $\mathbf{F}$ at a sample point in $\mathcal{S}$ is a vector of length 2 making an angle $\frac{\pi}{4}$ with the normal to the surface.

7. A small surface $\mathcal{S}$ is divided into three pieces of area 0.2. Estimate $\iint_{\mathcal{S}} \mathbf{F} \cdot d\mathbf{S}$ if $\mathbf{F}$ is a unit vector field making angles of 85, 90, and 95° with the normal at sample points in these three pieces.

Exercises

1. Let $\mathbf{F} = \langle y, z, x\rangle$ and let $\mathcal{S}$ be the oriented surface parametrized by $\Phi(u, v) = (u^2 - v, u + v, v^2)$ for $0 \le u \le 2$, $-1 \le v \le 1$. Calculate:

(a) $\mathbf{n}$ and $\mathbf{F} \cdot \mathbf{n}$ as functions of u and v

(b) The normal component of $\mathbf{F}$ to the surface at $P = (3, 3, 1) = \Phi(2, 1)$

(c) $\iint_{\mathcal{S}} \mathbf{F} \cdot d\mathbf{S}$

2. Compute the surface integral of the vector field $\mathbf{F} = \langle x, y, x + y\rangle$ over the portion $\mathcal{S}$ of the paraboloid $z = x^2 + y^2$ lying over the disk $x^2 + y^2 \le 1$.

3. Let $\mathcal{S}$ be the square in the xy-plane shown in Figure 13, oriented with the normal pointing in the positive z-direction. Estimate

$$\iint_{\mathcal{S}} \mathbf{F} \cdot d\mathbf{S}$$

where $\mathbf{F}$ is a vector field whose values at the labeled points are

$$\mathbf{F}(A) = \langle 2, 6, 4\rangle, \qquad \mathbf{F}(B) = \langle 1, 1, 7\rangle$$

$$\mathbf{F}(C) = \langle 3, 3, -3\rangle, \qquad \mathbf{F}(D) = \langle 0, 1, 8\rangle$$

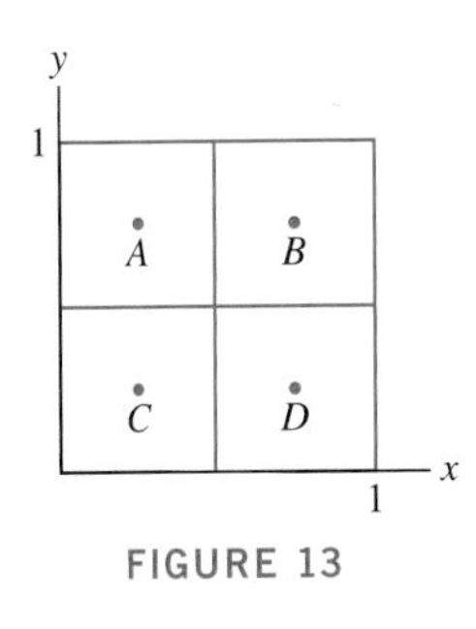

FIGURE 13

4. Suppose that $\mathcal{S}$ is a surface in $\mathbf{R}^3$ with a parametrization Φ whose domain $\mathcal{D}$ is the square in Figure 13. The values of a function f, a vector field $\mathbf{F}$, and the normal vector $\mathbf{n} = \mathbf{T}_u \times \mathbf{T}_v$ at $\Phi(P)$ are given for the four sample points in $\mathcal{D}$ in the following table. Estimate the surface integrals of f and $\mathbf{F}$ over $\mathcal{S}$.

Point P in $\mathcal{D}$	f	$\mathbf{F}$	$\mathbf{n}$
A	3	$\langle 2, 6, 4\rangle$	$\langle 1, 1, 1\rangle$
B	1	$\langle 1, 1, 7\rangle$	$\langle 1, 1, 0\rangle$
C	2	$\langle 3, 3, -3\rangle$	$\langle 1, 0, -1\rangle$
D	5	$\langle 0, 1, 8\rangle$	$\langle 2, 1, 0\rangle$

In Exercises 5–17, compute the surface integral over the given oriented surface.

5. $\mathbf{F} = \langle y, z, x\rangle$, plane $3x - 4y + z = 1$, $0 \le x, y \le 1$, upward-pointing normal

6. $\mathbf{F} = \langle e^z, z, x\rangle$, $\Phi(r, s) = (rs, r + s, r)$, $0 \le r \le 1, 0 \le s \le 1$, oriented by $\mathbf{T}_r \times \mathbf{T}_s$

7. $\mathbf{F} = \langle 0, 3, x^2\rangle$, hemisphere $x^2 + y^2 + z^2 = 9$, $z \ge 0$, outward-pointing normal

8. $\mathbf{F} = \langle x, y, z\rangle$, part of sphere $x^2 + y^2 + z^2 = 1$, where $\frac{1}{2} \le z \le \frac{\sqrt{3}}{2}$, inward-pointing normal

9. $\mathbf{F} = \langle e^z, z, x\rangle$, $z = 9 - x^2 - y^2$, $z \ge 0$, upward-pointing normal

10. $\mathbf{F} = \langle \sin y, \sin z, yz\rangle$, rectangle $0 \le y \le 2$, $0 \le z \le 3$ in the (y, z)-plane, normal pointing in negative x-direction

11. $\mathbf{F} = y^2\mathbf{i} + 2\mathbf{j} - x\mathbf{k}$, portion of the plane $x + y + z = 1$ in the octant $x, y, z \ge 0$, upward-pointing normal

12. $\mathbf{F} = \langle x, y, e^z\rangle$, cylinder $x^2 + y^2 = 4$, $1 \le z \le 5$, outward-pointing normal

13. $\mathbf{F} = \langle xz, yz, z^{-1}\rangle$, disk of radius 3 at height 4 parallel to the xy-plane, upward-pointing normal

14. $\mathbf{F} = \langle xy, y, 0\rangle$, cone $z^2 = x^2 + y^2$, $x^2 + y^2 \le 4$, $z \ge 0$, downward-pointing normal

15. $\mathbf{F} = \langle 0, 0, e^{y+z} \rangle$, boundary of unit cube $0 \le x, y, z \le 1$, outward-pointing normal

16. $\mathbf{F} = \langle 0, 0, z^2 \rangle$, $\Phi(u, v) = (u \cos v, u \sin v, v)$, $0 \le u \le 1$, $0 \le v \le 2\pi$, upward-pointing normal

17. $\mathbf{F} = \langle y, z, 0 \rangle$, $\Phi(u, v) = (u^3 - v, u + v, v^2)$, $0 \le u \le 2$, $0 \le v \le 3$, downward-pointing normal

18. Let $\mathcal{S}$ be the oriented half-cylinder in Figure 14. In (a)–(f), determine whether $\iint_{\mathcal{S}} \mathbf{F} \cdot d\mathbf{S}$ is positive, negative, or zero. Explain your reasoning.

(a) $\mathbf{F} = \mathbf{i}$ **(b)** $\mathbf{F} = \mathbf{j}$

(c) $\mathbf{F} = \mathbf{k}$ **(d)** $\mathbf{F} = y\mathbf{i}$

(e) $\mathbf{F} = -y\mathbf{j}$ **(f)** $\mathbf{F} = x\mathbf{j}$

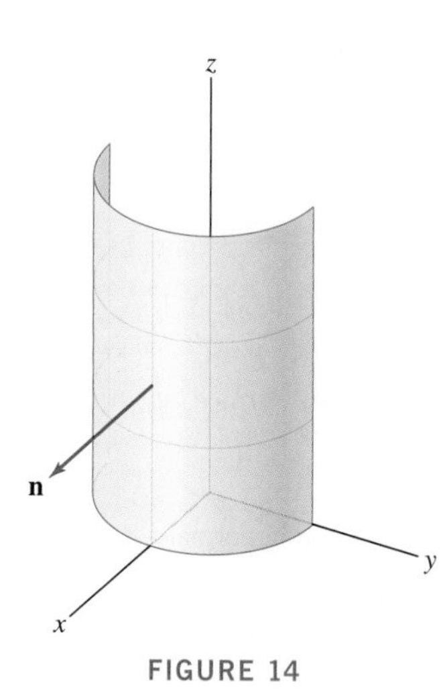

FIGURE 14

19. Let $\mathbf{e_r}$ be the unit radial vector and $r = \sqrt{x^2 + y^2 + z^2}$. Calculate the integral of $\mathbf{F} = e^{-r}\mathbf{e_r}$ over:

(a) The upper-hemisphere of $x^2 + y^2 + z^2 = 9$, outward-pointing normal

(b) The octant $x, y, z \ge 0$ of the unit sphere centered at the origin

20. Show that the flux of $\mathbf{F} = \dfrac{\mathbf{e}_r}{r^2}$ through a sphere centered at the origin does not depend on the radius of the sphere.

21. The electric field due to a point charge located at the origin is $\mathbf{E} = k\dfrac{\mathbf{e}_r}{r^2}$, where k is a constant. Calculate the flux of $\mathbf{E}$ through the disk D of radius 2 parallel to the xy-plane with center $(0, 0, 3)$.

22. Let $\mathcal{S}$ be the ellipsoid $\left(\frac{x}{4}\right)^2 + \left(\frac{y}{3}\right)^2 + \left(\frac{z}{2}\right)^2 = 1$. Calculate the flux of $\mathbf{F} = \langle z, 1, 0 \rangle$ over the portion of $\mathcal{S}$ where $x, y, z \le 0$ with upward-pointing normal. *Hint:* Parametrize $\mathcal{S}$ using a modified form of spherical coordinates (θ, ϕ).

23. Let $\mathbf{v} = \langle x, 0, z \rangle$ be the velocity field (in ft/s) of a fluid in $\mathbf{R}^3$. Calculate the flow rate (in ft^3/s) through the upper hemisphere of the sphere $x^2 + y^2 + z^2 = 1$ $(z \ge 0)$.

24. Calculate the flow rate through the upper hemisphere of the sphere $x^2 + y^2 + z^2 = R^2$ $(z \ge 0)$ for $\mathbf{v}$ as in Exercise 23.

25. Calculate the flow rate of a fluid with velocity field $\mathbf{v} = \langle x, y, x^2 y \rangle$ (in ft/s) through the portion of the ellipse $\left(\frac{x}{2}\right)^2 + \left(\frac{y}{3}\right)^2 = 1$ in the xy-plane, where $x, y \ge 0$, oriented with the normal in the positive z-direction.

In Exercises 26–27, let $\mathcal{T}$ be the triangular region with vertices $(1, 0, 0)$, $(0, 1, 0)$, and $(0, 0, 1)$ oriented with upward-pointing normal vector (Figure 15). Assume distances are in meters.

26. A fluid flows with constant velocity field $\mathbf{v} = 2\mathbf{k}$ (m/s). Calculate:

(a) The flow rate through $\mathcal{T}$

(b) The flow rate through the projection of $\mathcal{T}$ onto the xy-plane [the triangle with vertices $(0, 0, 0)$, $(1, 0, 0)$, and $(0, 1, 0)$]

27. Calculate the flow rate through $\mathcal{T}$ if $\mathbf{v} = -\mathbf{j}$ m/s.

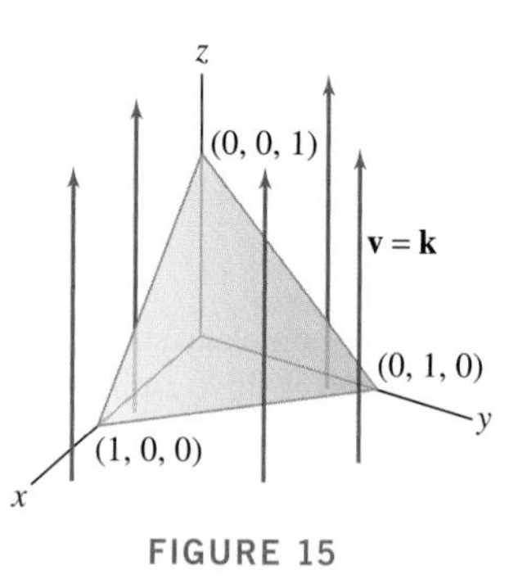

FIGURE 15

28. Prove that if $\mathcal{S}$ is the part of a graph $z = g(x, y)$ lying over a domain $\mathcal{D}$ in the xy-plane, then

$$\iint_{\mathcal{S}} \mathbf{F} \cdot d\mathbf{S} = \iint_{\mathcal{D}} \left(-F_1 \frac{\partial g}{\partial x} - F_2 \frac{\partial g}{\partial y} + F_3 \right) dx\, dy$$

In Exercises 29–30, a varying current $i(t)$ flows through a long straight wire in the xy-plane as in Example 5. The current produces a magnetic field $\mathbf{B}$ whose magnitude at a distance r from the wire is $\|\mathbf{B}\| = \dfrac{\mu_0 i}{2\pi r}$ T, where $\mu_0 = 4\pi \cdot 10^{-7}$ T-m/A. Furthermore, $\mathbf{B}$ points into the page at points P in the xy-plane.

29. Assume that $i(t) = t(12 - t)$ A (t in seconds). Calculate the flux $\Phi(t)$, at time t, of $\mathbf{B}$ through a rectangle of dimensions $L \times H = 3 \times 2$ m, whose top and bottom edges are parallel to the wire and whose bottom edge is located $d = 0.5$ m above the wire (similar to Figure 11). Then use Faraday's Law to determine the voltage drop around the rectangular loop (the boundary of the rectangle) at time t.

30. Assume that $i = 10e^{-0.1t}$ A (t in seconds). Calculate the flux $\Phi(t)$, at time t, of $\mathbf{B}$ through the isosceles triangle of base 12 cm and height 6 cm, whose bottom edge is 3 cm from the wire, as in Figure 16. Assume the triangle is oriented with normal vector pointing out of the page. Use Faraday's Law to determine the voltage drop around the triangular loop (the boundary of the triangle) at time t.

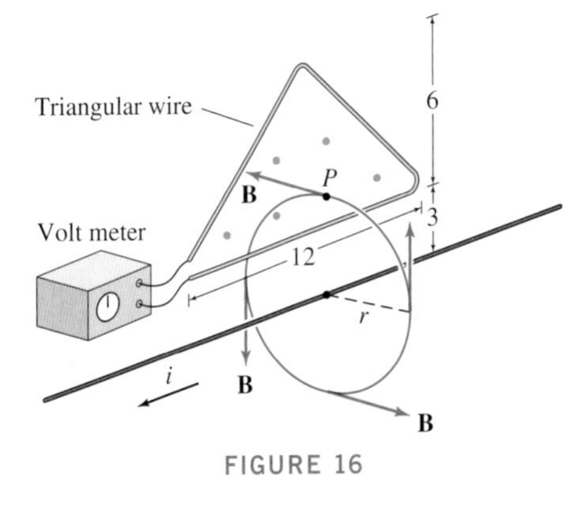

FIGURE 16

Further Insights and Challenges

31. A point mass m is located at the origin. Let Q be the flux of the gravitational field $\mathbf{F} = -Gm\frac{\mathbf{e}_r}{r^2}$ through the cylinder $x^2 + y^2 = R^2$ for $a \le z \le b$, including the top and bottom (Figure 17). Show that $Q = -4\pi Gm$ if $a < 0 < b$ (m lies inside the cylinder) and $Q = 0$ if $0 < a < b$ (m lies outside the cylinder).

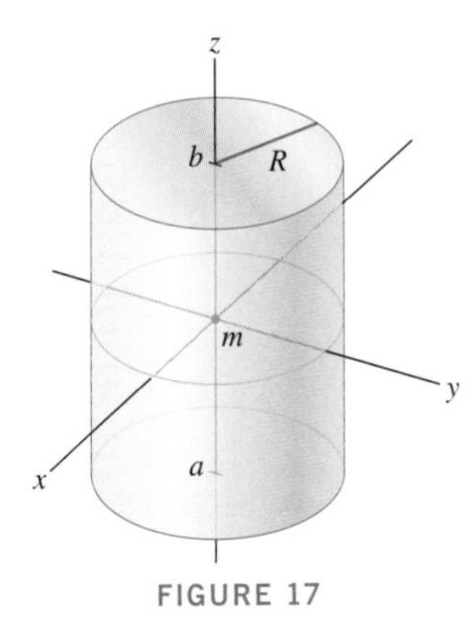

FIGURE 17

In Exercises 32–33, let $\mathcal{S}$ be the surface with parametrization

$$\Phi(u, v) = \left(\left(1 + v\cos\frac{u}{2}\right)\cos u,\ \left(1 + v\cos\frac{u}{2}\right)\sin u,\ v\sin\frac{u}{2}\right)$$

for $0 \le u \le 2\pi$, $-\frac{1}{2} \le v \le \frac{1}{2}$.

32. CAS Use a computer algebra system.

(a) Plot $\mathcal{S}$ and confirm visually that $\mathcal{S}$ is a Möbius strip.

(b) The intersection of $\mathcal{S}$ with the xy-plane is the unit circle $\Phi(u, 0) = (\cos u, \sin u, 0)$. Verify that the normal vector along this circle is

$$\mathbf{n}(u, 0) = \left\langle \cos u \sin\frac{u}{2},\ \sin u \sin\frac{u}{2},\ -\cos\frac{u}{2}\right\rangle$$

(c) As u varies from 0 to 2π, the point $\Phi(u, 0)$ moves once around the unit circle, beginning and ending at $\Phi(0, 0) = \Phi(2\pi, 0) = (1, 0, 0)$. Verify that $\mathbf{n}(u, 0)$ is a unit vector which varies continuously but that $\mathbf{n}(2\pi, 0) = -\mathbf{n}(u, 0)$. This shows that $\mathcal{S}$ is not orientable, that is, it is not possible to choose a nonzero normal vector at each point on $\mathcal{S}$ in a continuously varying manner (if it were possible, the unit normal vector returns to itself rather than its negative when carried around the circle).

33. CAS It is not possible to integrate a vector field over $\mathcal{S}$ because $\mathcal{S}$ is not orientable. However, it is possible to integrate functions over $\mathcal{S}$. Using a computer algebra system:

(a) Verify that

$$\|\mathbf{n}(u, v)\| = 1 + \frac{3}{4}v^2 + 2v\cos\frac{u}{v} + \frac{1}{2}v^2\cos u$$

(b) Compute the surface area of $\mathcal{S}$ to four decimal places.

(c) Compute $\iint_{\mathcal{S}} (x^2 + y^2 + z^2)\, dS$ to four decimal places.

CHAPTER REVIEW EXERCISES

1. Compute the vector assigned to the point $P = (-3, 5)$ by the vector field:

(a) $\mathbf{F} = \langle xy, y - x\rangle$

(b) $\mathbf{F} = \langle 4, 8\rangle$

(c) $\mathbf{F} = \langle 3^{x+y}, \log_2(x + y)\rangle$

2. Find a vector field $\mathbf{F}$ in the plane such that $\|\mathbf{F}(x, y)\| = 1$ and $\mathbf{F}(x, y)$ is orthogonal to $\mathbf{G}(x, y) = \langle x, y\rangle$ for all x, y.

In Exercises 3–6, sketch the vector field.

3. $\mathbf{F}(x, y) = \langle y, 1\rangle$

4. $\mathbf{F}(x, y) = \langle 4, 1\rangle$

5. $\nabla\varphi$, where $\varphi(x, y) = x^2 - y$

6. $\mathbf{F}(x, y) = \left\langle \dfrac{4y}{\sqrt{x^2 + 16y^2}}, \dfrac{-x}{\sqrt{x^2 + 16y^2}} \right\rangle$

Hint: Show that $\mathbf{F}$ is a unit vector field tangent to the family of ellipses $x^2 + 4y^2 = c^2$.

In Exercises 7–14, determine whether or not the vector field is conservative and, if so, find a potential function.

7. $\mathbf{F}(x, y, z) = \langle \sin x, e^y, z \rangle$

8. $\mathbf{F}(x, y, z) = \langle 2, 4, e^z \rangle$

9. $\mathbf{F}(x, y, z) = \langle xyz, \frac{1}{2}x^2 z, 2z^2 y \rangle$

10. $\mathbf{F}(x, y, z) = \langle xy^2 z, x^2 yz, \frac{1}{2}x^2 y^2 \rangle$

11. $\mathbf{F}(x, y, z) = \left\langle \dfrac{y}{1 + x^2}, \tan^{-1} x, 2z \right\rangle$

12. $\mathbf{F}(x, y) = \langle x^2 y, y^2 x \rangle$

13. $\mathbf{F}(x, y, z) = \langle xe^{2x}, ye^{2z}, ze^{2y} \rangle$

14. $\mathbf{F}(x, y) = \langle y^4 x^3, x^4 y^3 \rangle$

15. Calculate $\int_{\mathbf{c}} \nabla\varphi \cdot d\mathbf{s}$, where $\varphi(x, y, z) = x^4 y^3 z^2$ and $\mathbf{c}(t) = (t^2, 1 + t, t^{-1})$ for $1 \le t \le 3$.

16. Find a gradient vector field of the form $\mathbf{F} = \langle g(y), h(x) \rangle$ such that $\mathbf{F}(0, 0) = \langle 1, 1 \rangle$, where $g(y)$ and $h(x)$ are differentiable functions. Determine all such vector fields.

In Exercises 17–20, compute the line integral $\int_C f(x, y)\, ds$ for the given function and path or curve.

17. $f(x, y) = xy$, the path $\mathbf{c}(t) = (t, 2t - 1)$ for $0 \le t \le 1$

18. $f(x, y) = x - y$, the unit semicircle $x^2 + y^2 = 1$, $y \ge 0$

19. $f(x, y, z) = e^x - \dfrac{y}{2\sqrt{2z}}$, the path $\mathbf{c}(t) = \left(\ln t, \sqrt{2}t, \frac{1}{2}t^2\right)$ for $1 \le t \le 2$

20. $f(x, y, z) = x + 2y + z$, the helix $\mathbf{c}(t) = (\cos t, \sin t, t)$ for $-1 \le t \le 3$

21. Find the total mass of an L-shaped rod consisting of the segments $(2t, 2)$ and $(2, 2 - 2t)$ for $0 \le t \le 1$ (length in centimeters) with mass density $\rho(x, y) = x^2 y$ g/cm.

22. Calculate $\mathbf{F} = \nabla\varphi$, where $\varphi(x, y, z) = xye^z$, and compute $\int_C \mathbf{F} \cdot d\mathbf{s}$, where

(a) C is any curve from $(1, 1, 0)$ to $(3, e, -1)$.

(b) C is the boundary of the square $0 \le x, y \le 1$ oriented counterclockwise.

23. Calculate $\int_{C_1} y^3\, dx + x^2 y\, dy$, where C_1 is the oriented curve in Figure 1(A).

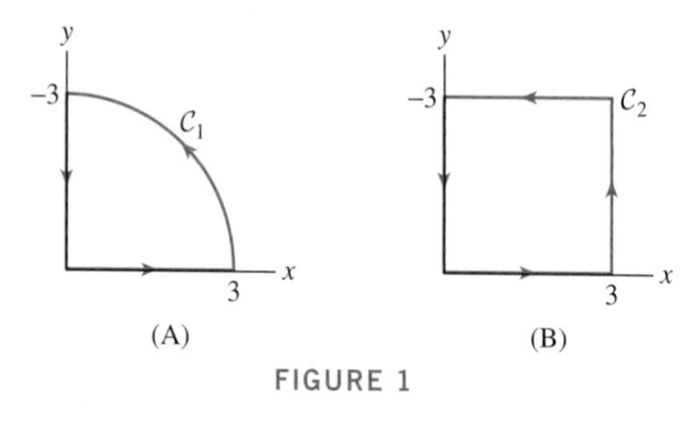

FIGURE 1

24. Let $\mathbf{F}(x, y) = \langle 9y - y^3, e^{\sqrt{y}}(x^2 - 3x) \rangle$ and let C_2 be the oriented curve in Figure 1(B).

(a) Show that $\mathbf{F}$ is not conservative.

(b) Show that $\int_{C_2} \mathbf{F} \cdot d\mathbf{s} = 0$ without explicitly computing the integral. *Hint:* Consider the direction of $\mathbf{F}$ along the edges of the square.

In Exercises 25–28, compute the line integral $\int_{\mathbf{c}} \mathbf{F} \cdot d\mathbf{s}$ for the given vector field and path.

25. $\mathbf{F}(x, y) = \left\langle \dfrac{2y}{x^2 + 4y^2}, \dfrac{x}{x^2 + 4y^2} \right\rangle$,

the path $\mathbf{c}(t) = \left(\cos t, \frac{1}{2}\sin t\right)$ for $0 \le t \le 2\pi$

26. $\mathbf{F}(x, y) = \langle 2xy, x^2 + y^2 \rangle$, the part of the unit circle in the first quadrant oriented counterclockwise.

27. $\mathbf{F}(x, y) = \langle x^2 y, y^2 z, z^2 x \rangle$, the path $\mathbf{c}(t) = \left(e^{-t}, e^{-2t}, e^{-3t}\right)$ for $0 \le t < \infty$

28. $\mathbf{F} = \nabla\varphi$, where $\varphi(x, y, z) = 4x^2 \ln(1 + y^4 + z^2)$, the path $\mathbf{c}(t) = \left(t^3, \ln(1 + t^2), e^t\right)$ for $0 \le t \le 1$

29. Consider the line integrals $\int_{\mathbf{c}} \mathbf{F} \cdot d\mathbf{s}$ for the vector fields $\mathbf{F}$ and paths $\mathbf{c}$ in Figure 2. Which two of the line integrals appear to have a value of zero? Which of the other two is negative?

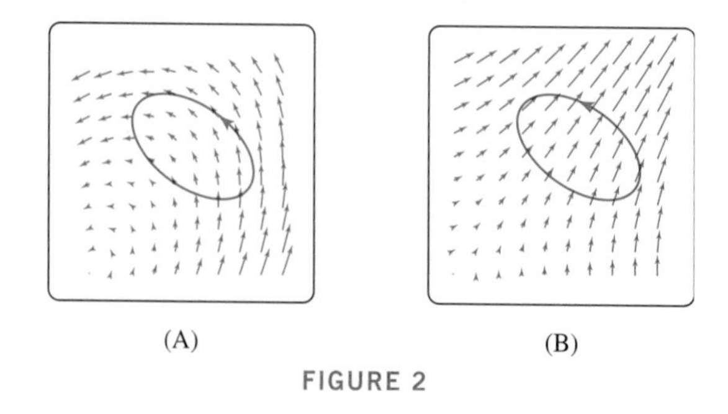

FIGURE 2

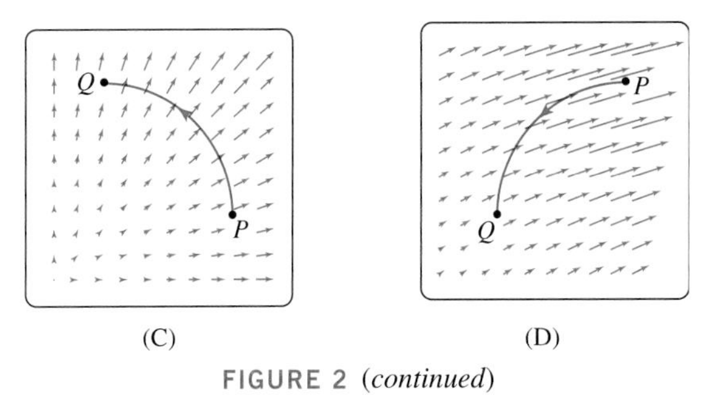

FIGURE 2 (*continued*)

30. Calculate the work required to move an object from $P = (1, 1, 1)$ to $Q = (3, -4, -2)$ against the force field $\mathbf{F}(x, y, z) = -12r^{-4}\langle x, y, z\rangle$ (distance in meters, force in Newtons), where $r = \sqrt{x^2 + y^2 + z^2}$. *Hint:* Find a potential function for $\mathbf{F}$.

31. Find constants a, b, c such that

$$\Phi(u, v) = (u + av, bu + v, 2u - c)$$

parametrizes the plane $3x - 4y + z = 5$. Calculate $\mathbf{T}_u$, $\mathbf{T}_v$, and $\mathbf{n}(u, v)$.

32. Calculate the integral of $f(x, y, z) = e^{z-y}$ over the portion of the plane $6x + 4y + z = 24$, where $x, y, z \geq 0$.

33. Let $\mathcal{S}$ be the surface parametrized by

$$\Phi(u, v) = \left(2u \sin\frac{v}{2}, 2u \cos\frac{v}{2}, 3v\right)$$

for $0 \leq u \leq 1$ and $0 \leq v \leq 2\pi$.

(a) Calculate the tangent vectors $\mathbf{T}_u$ and $\mathbf{T}_v$, and normal vector $\mathbf{n}(u, v)$ at $P = \Phi(1, \frac{\pi}{3})$.

(b) Find the equation of the tangent plane at P.

(c) Compute the surface area of $\mathcal{S}$.

34. CAS Plot the surface with parametrization

$$\Phi(u, v) = (u + 4v, 2u - v, 5uv)$$

for $-1 \leq v \leq 1$, $-1 \leq u \leq 1$. Express the surface area as a double integral and use a computer algebra system to compute the area numerically.

35. CAS Express the surface area of the surface $z = 10 - x^2 - y^2$, $-1 \leq x \leq 1$, $-3 \leq y \leq 3$ as a double integral. Evaluate the integral numerically using a CAS.

36. Evaluate $\iint_{\mathcal{S}} x^2 y\, dS$, where $\mathcal{S}$ is the surface $z = \sqrt{3}x + y^2$, $-1 \leq x \leq 1$, $0 \leq y \leq 1$.

37. Calculate $\iint_{\mathcal{S}} \left(x^2 + y^2\right) e^{-z}\, dS$, where $\mathcal{S}$ is the cylinder with equation $x^2 + y^2 = 9$ for $0 \leq z \leq 10$.

38. Let $\mathcal{S}$ be the upper hemisphere $x^2 + y^2 + z^2 = 1$, $z \geq 0$. For each of the functions (a)–(d), determine whether $\iint_{\mathcal{S}} f\, dS$ is positive, zero, or negative (without evaluating the integral). Explain your reasoning.

(a) $f(x, y, z) = y^3$ **(b)** $f(x, y, z) = z^3$

(c) $f(x, y, z) = xyz$ **(d)** $f(x, y, z) = z^2 - 2$

39. Let $\mathcal{S}$ be a small patch of surface with a parametrization $\Phi(u, v)$ for $0 \leq u, v \leq 0.1$ such that the normal vector $\mathbf{n}(u, v)$ for $(u, v) = (0, 0)$ is $\mathbf{n} = \langle 2, -2, 4\rangle$. Use Eq. (1) in Section 16.4 to estimate the surface area of $\mathcal{S}$.

40. Let $\mathcal{S}$ be the upper hemisphere of the sphere $x^2 + y^2 + z^2 = 9$ and let $\Phi(r, \theta) = (r\cos\theta, r\sin\theta, \sqrt{9 - r^2})$ be its parametrization by cylindrical coordinates (Figure 3).

(a) Calculate the normal vector $\mathbf{n} = \mathbf{T}_r \times \mathbf{T}_\theta$ at the point $\Phi(2, \frac{\pi}{3})$.

(b) Use Eq. (1) in Section 16.4 to estimate the surface area of $\Phi(\mathcal{R})$, where $\mathcal{R}$ is the small domain defined by

$$2 \leq r \leq 2.1, \qquad \frac{\pi}{3} \leq \theta \leq \frac{\pi}{3} + 0.05$$

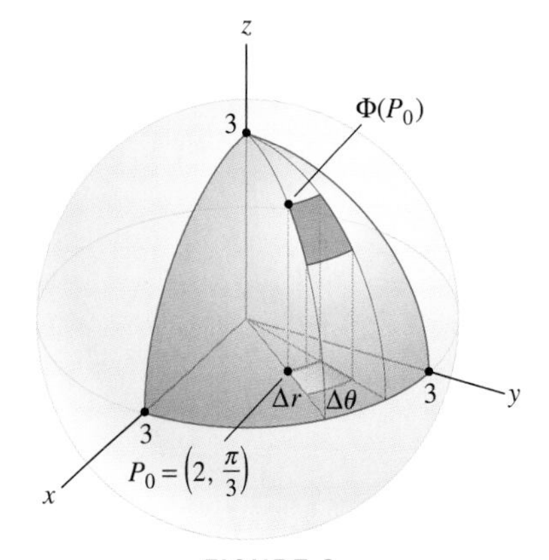

FIGURE 3

In Exercises 41–46, compute $\iint_{\mathcal{S}} \mathbf{F} \cdot d\mathbf{S}$ *for the given oriented surface or parametrized surface.*

41. $\mathbf{F}(x, y, z) = \langle y, x^2 y, e^{xz}\rangle$, $x^2 + y^2 = 9$, $-3 \leq z \leq 3$, outward-pointing normal

42. $\mathbf{F}(x, y, z) = \langle -y, z, -x\rangle$, $\Phi(u, v) = (u + 3v, v - 2u, 2v + 5)$, $0 \leq u, v \leq 1$, upward-pointing normal

43. $\mathbf{F}(x, y, z) = \langle x^2, y^2, x^2 + y^2\rangle$, $x^2 + y^2 + z^2 = 4$, $z \geq 0$, outward-pointing normal

44. $\mathbf{F}(x, y, z) = \langle z, 0, z^2\rangle$, $\Phi(u, v) = (v\cosh u, v\sinh u, v)$, $0 \leq u, v \leq 1$, upward-pointing normal

45. $\mathbf{F}(x, y, z) = \langle 0, 0, xze^{xy}\rangle$, $z = xy$, $0 \leq x, y \leq 1$, upward-pointing normal

46. $\mathbf{F}(x, y, z) = \langle 0, 0, z\rangle$, $3x^2 + 2y^2 + z^2 = 1$, $z \geq 0$, upward-pointing normal

47. Calculate the total charge on the cylinder

$$x^2 + y^2 = R^2, \qquad 0 \le z \le H$$

if the charge density in cylindrical coordinates is $\rho(\theta, z) = Kz^2 \cos^2 \theta$, where K is a constant.

48. Find the flow rate of a fluid with velocity field $\mathbf{v} = \langle 2x, y, xy \rangle$ m/s across the part of the cylinder $x^2 + y^2 = 9$ where $x, y \ge 0$, and $0 \le z \le 4$ (distance in meters).

49. With $\mathbf{v}$ as in Exercise 48, calculate the flow rate across the part of the elliptic cylinder $\frac{x^2}{4} + y^2 = 1$ where $x, y \ge 0$, and $0 \le z \le 4$.

50. Calculate the flux of the vector field $\mathbf{E}(x, y, z) = \langle x, 0, z \rangle$ through the part of the ellipsoid

$$4x^2 + 9y^2 + z^2 = 36$$

where $z \ge 3$. *Hint:* Use the parametrization

$$\Phi(r, \theta) = \left(3r \cos\theta, 2r \sin\theta, 6\sqrt{1 - r^2}\right)$$

17 FUNDAMENTAL THEOREMS OF VECTOR ANALYSIS

A kayaker on the Kananaskis River in Alberta, Canada, contends with the velocity vector field of a turbulent vortex.

In this final chapter, we study the three main theorems of vector analysis: Green's Theorem, Stokes' Theorem, and the Divergence Theorem. This is a fitting conclusion to the text because each of these theorems is a vector generalization of the Fundamental Theorem of Calculus. This chapter is thus the culmination of our efforts to extend the concepts and methods of single-variable calculus to the multivariable setting. However, far from being a terminal point, vector analysis is the gateway to the field theories of mathematics, physics, and engineering. This includes, first and foremost, the theory of electricity and magnetism as expressed by the famous Maxwell equations. It also includes fluid dynamics, aerodynamics, analysis of continuous matter, and at a more advanced level, fundamental physical theories such as general relativity and the theory of elementary particles.

17.1 Green's Theorem

Green's Theorem provides a new perspective on line integrals. In Section 16.3, we proved that the circulation of a gradient vector field $\nabla\varphi$ around a closed curve is equal to zero. Green's Theorem expresses the circulation of a general vector field $\mathbf{F}$ as a double integral, which need not be zero if $\mathbf{F}$ is not a gradient vector field.

Recall that "circulation" is another term for the line integral of a vector field over a closed curve.

Before stating Green's Theorem, we recall the notation for line integrals introduced in Section 16.2. The line integral of $\mathbf{F} = \langle P, Q\rangle$ along an oriented curve $\mathcal{C}$ is denoted by

$$\underbrace{\int_{\mathcal{C}} P\,dx + Q\,dy}_{\text{Another notation for } \int_{\mathcal{C}} \mathbf{F}\cdot d\mathbf{s}}$$

The symbolic differential $P\,dx + Q\,dy$ is interpreted in terms of a parametrization as follows. If $\mathbf{c}(t) = (x(t), y(t))$ parametrizes $\mathcal{C}$ (in the direction of its orientation) for $a \le t \le b$, we use the symbolic relations

$$dx = \frac{dx}{dt}\,dt \qquad \text{and} \qquad dy = \frac{dy}{dt}\,dt$$

$$P\,dx + Q\,dy = P(x(t), y(t))\frac{dx}{dt}\,dt + Q(x(t), y(t))\frac{dy}{dt}\,dt$$

to express the line integral as an ordinary integral with respect to t:

$$\int_{\mathcal{C}} P\,dx + Q\,dy = \int_a^b \left(P(x(t), y(t))\frac{dx}{dt} + Q(x(t), y(t))\frac{dy}{dt}\right) dt$$

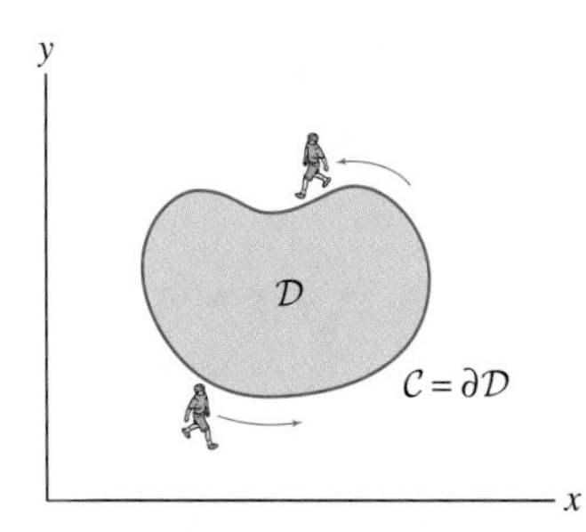

FIGURE 1 The boundary of the domain $\mathcal{D}$ is a simple closed curve $\mathcal{C}$, oriented in the counterclockwise direction.

The integrand on the right is $\mathbf{F} \cdot \mathbf{c}'(t)$, so this coincides with our previous definition of the line integral $\int_{\mathcal{C}} \mathbf{F} \cdot d\mathbf{s}$.

To state Green's Theorem, recall that a **simple closed curve** is a closed curve $\mathcal{C}$ that does not intersect itself. In this case, $\mathcal{C}$ encloses a domain $\mathcal{D}$ (Figure 1). We assume that $\mathcal{C}$ has a parametrization $\mathbf{c}(t)$ such that $\mathbf{c}'(t)$ exists and is continuous with the exception of at most finitely many values of t where the components of $\mathbf{c}'(t)$ may have jump discontinuities (and $\mathcal{C}$ may have a corner). There are two ways to orient $\mathcal{C}$, namely clockwise or counterclockwise. We refer to the counterclockwise orientation as the **boundary orientation**. When you traverse $\mathcal{C}$ in the counterclockwise direction, the domain $\mathcal{D}$ enclosed by $\mathcal{C}$ lies to your left. We write $\partial\mathcal{D}$ to denote the boundary $\mathcal{C}$ with its counterclockwise orientation.

Recall that if $\mathbf{F} = \nabla\varphi$, then

$$\frac{\partial Q}{\partial x} - \frac{\partial P}{\partial y} = 0$$

In this case, Green's Theorem merely confirms what we already know, namely that the circulation of a gradient vector field around any closed curve is zero.

THEOREM 1 Green's Theorem Let $\mathcal{D}$ be a domain whose boundary $\partial\mathcal{D}$ is a simple closed curve, oriented counterclockwise. If $P(x, y)$ and $Q(x, y)$ are differentiable and have continuous first partial derivatives, then

$$\oint_{\partial\mathcal{D}} P\,dx + Q\,dy = \iint_{\mathcal{D}} \left(\frac{\partial Q}{\partial x} - \frac{\partial P}{\partial y}\right) dA \qquad \boxed{1}$$

Proof A complete proof of Green's Theorem for a general domain is beyond the scope of this text. To illustrate the main idea, we prove it for a domain $\mathcal{D}$ as in Figure 2, whose boundary can be described both as the union of two graphs $y = f(x)$ and $y = g(x)$ with $f(x) \geq g(x)$ and also as the union of two graphs $x = f_1(y)$ and $x = g_1(y)$, with $f_1(y) \geq g_1(y)$.

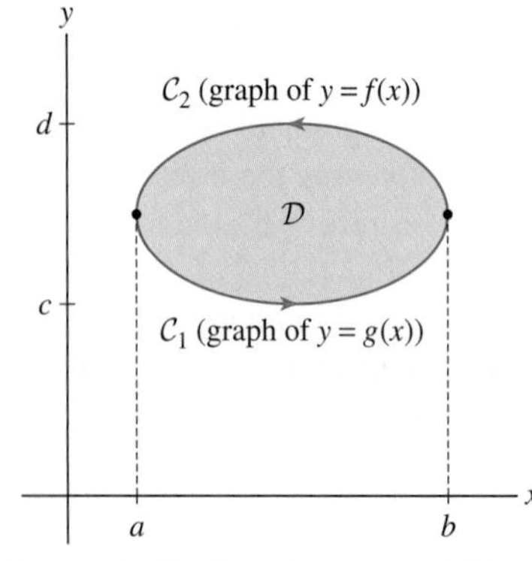

FIGURE 2 The boundary curve $\partial\mathcal{D}$ is the union of the graph of $y = g(x)$ and the graph of $y = f(x)$ oriented counterclockwise.

Equation (1) in Green's Theorem breaks up into two separate equalities, one involving P and one involving Q. Indeed, by additivity,

$$\oint_{\partial\mathcal{D}} P\,dx + Q\,dy = \oint_{\partial\mathcal{D}} \langle P(x, y), 0\rangle \cdot d\mathbf{s} + \oint_{\partial\mathcal{D}} \langle 0, Q(x, y)\rangle \cdot d\mathbf{s}$$

Therefore, Green's Theorem follows if we prove

$$\oint_{\partial\mathcal{D}} \langle P(x, y), 0\rangle \cdot d\mathbf{s} = -\iint_{\mathcal{D}} \frac{\partial P}{\partial y}\,dA \qquad \boxed{2}$$

$$\oint_{\partial\mathcal{D}} \langle 0, Q(x, y)\rangle \cdot d\mathbf{s} = \iint_{\mathcal{D}} \frac{\partial Q}{\partial x}\,dA \qquad \boxed{3}$$

To prove Eq. (2), we write the line integral as a sum:

$$\oint_{\partial\mathcal{D}} \langle P(x, y), 0\rangle \cdot d\mathbf{s} = \oint_{\mathcal{C}_1} \langle P(x, y), 0\rangle \cdot d\mathbf{s} + \oint_{\mathcal{C}_2} \langle P(x, y), 0\rangle \cdot d\mathbf{s}$$

where $\mathcal{C}_1$ is the graph of $y = g(x)$ and $\mathcal{C}_2$ is the graph of $y = f(x)$, oriented as in Figure 2. We parameterize the graphs from left to right using x as parameter:

Graph of $y = g(x)$: $\quad \mathbf{c}_1(x) = (x, g(x)), \quad a \leq x \leq b$

Graph of $y = f(x)$: $\quad \mathbf{c}_2(x) = (x, f(x)), \quad a \leq x \leq b$

Since $\mathcal{C}_2$ is oriented from right to left, the line integral over $\partial\mathcal{D}$ is the difference of the integral over $\mathbf{c}_1$ and $\mathbf{c}_2$:

$$\oint_{\partial\mathcal{D}} \langle P(x,y),0\rangle \cdot d\mathbf{s} = \int_{\mathbf{c}_1} \langle P(x,y),0\rangle \cdot d\mathbf{s} - \int_{\mathbf{c}_2} \langle P(x,y),0\rangle \cdot d\mathbf{s}$$

$$= \int_{x=a}^{b} \langle P(x,g(x)),0\rangle \cdot \mathbf{c}_1{}'(x)\,dx$$

$$- \int_{x=a}^{b} \langle P(x,f(x)),0\rangle \cdot \mathbf{c}_2{}'(x)\,dx$$

Now we have

$$\langle P(x,g(x)),0\rangle \cdot \mathbf{c}_1{}'(x) = \langle P(x,g(x)),0\rangle \cdot \langle 1, g'(x)\rangle = P(x,g(x))$$
$$\langle P(x,f(x)),0\rangle \cdot \mathbf{c}_2{}'(x) = \langle P(x,f(x)),0\rangle \cdot \langle 1, f'(x)\rangle = P(x,f(x))$$

and thus

$$\oint_{\partial\mathcal{D}} \langle P(x,y),0\rangle \cdot d\mathbf{s} = \int_{x=a}^{b} P(x,g(x))\,dx - \int_{x=a}^{b} P(x,f(x))\,dx \qquad \boxed{4}$$

The key step is to apply the Fundamental Theorem of Calculus to $\frac{\partial P}{\partial y}(x,y)$ as a function of y with x held constant:

$$P(x,f(x)) - P(x,g(x)) = \int_{y=g(x)}^{f(x)} \frac{\partial P}{\partial y}(x,y)\,dy$$

We obtain Eq. (2) by substituting the integral on the right in Eq. (4):

$$\oint_{\partial\mathcal{D}} \langle P(x,y),0\rangle \cdot d\mathbf{s} = -\int_{x=a}^{b}\int_{y=g(x)}^{f(x)} \frac{\partial P}{\partial y}(x,y)\,dy\,dx = -\iint_{\mathcal{D}} \frac{\partial P}{\partial y}\,dA$$

Equation (3) is proved in a similar fashion, by expressing $\partial\mathcal{D}$ as the union of the graphs of $x = g_1(y)$ and $x = g_2(y)$. ■

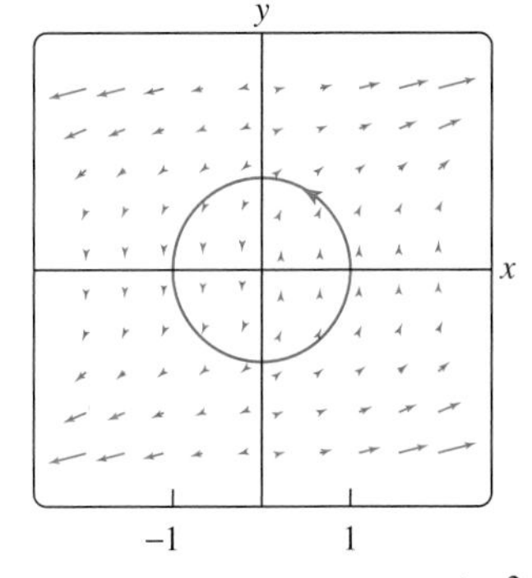

FIGURE 3 The vector field $\mathbf{F} = \langle xy^2, x\rangle$.

■ **EXAMPLE 1 Verifying Green's Theorem** Verify Green's Theorem for the line integral

$$\oint_{\mathcal{C}} xy^2\,dx + x\,dy$$

where $\mathcal{C}$ is the unit circle, oriented counterclockwise (Figure 3).

Solution

***Step 1.* Evaluate the line integral.**

We use the standard parametrization of the unit circle:

$$x = \cos\theta, \qquad y = \sin\theta$$
$$dx = -\sin\theta\,d\theta, \qquad dy = \cos\theta\,d\theta$$

The integrand in the line integral becomes

$$xy^2\,dx + x\,dy = \cos\theta\sin^2\theta(-\sin\theta\,d\theta) + \cos\theta(\cos\theta\,d\theta)$$
$$= \left(-\cos\theta\sin^3\theta + \cos^2\theta\right)d\theta$$

REMINDER To integrate $\cos^2\theta$, use the identity $\cos^2\theta = \frac{1}{2}(1+\cos 2\theta)$.

and

$$\oint_C xy^2\,dx + x\,dy = \int_0^{2\pi} (-\cos\theta\sin^3\theta + \cos^2\theta)\,d\theta$$
$$= -\frac{\sin^4\theta}{4}\Big|_0^{2\pi} + \frac{1}{2}\left(\theta + \frac{1}{4}\sin 2\theta\right)\Big|_0^{2\pi}$$
$$= 0 + \frac{1}{2}(2\pi + 0) = \boxed{\pi}$$

***Step 2.* Evaluate the double integral.**

In this example, $P = xy^2$ and $Q = x$, so

$$\frac{\partial Q}{\partial x} - \frac{\partial P}{\partial y} = \frac{\partial}{\partial x}x - \frac{\partial}{\partial y}xy^2 = 1 - 2xy$$

The double integral in Green's Theorem is taken over the domain bounded by $\mathcal{C}$, which is the disk $\mathcal{D}$ defined by $x^2 + y^2 \le 1$:

$$\iint_{\mathcal{D}} \left(\frac{\partial Q}{\partial x} - \frac{\partial P}{\partial y}\right) dA = \iint_{\mathcal{D}} (1 - 2xy)\,dA$$

By symmetry, the integral of $2xy$ over $\mathcal{D}$ is zero because the contributions for positive and negative x cancel each other (you can also check this directly). Therefore,

$$\iint_{\mathcal{D}} \left(\frac{\partial Q}{\partial x} - \frac{\partial P}{\partial y}\right) dA = \iint_{\mathcal{D}} 1\,dA = \text{Area}(\mathcal{D}) = \boxed{\pi}$$

***Step 3.* Compare.**

Both the line integral and double integral have the value π, so Green's Theorem is verified in this case. ■

It is convenient to refer to the quantity appearing in the double integral in Green's Theorem as the **curl** or **scalar curl**. For $\mathbf{F} = \langle P, Q\rangle$, we write

$$\boxed{\text{curl}_z(\mathbf{F}) = \frac{\partial Q}{\partial x} - \frac{\partial P}{\partial y}}$$

Notice the attached subscript z in curl_z. In the next section, we will define a vector quantity $\text{curl}(\mathbf{F})$ for vector fields in $\mathbf{R}^3$. Our scalar quantity $\text{curl}_z(\mathbf{F})$ is the z-component of the vector curl. With this notation, we may write Green's Theorem as

$$\boxed{\oint_{\partial\mathcal{D}} \mathbf{F}\cdot d\mathbf{s} = \iint_{\mathcal{D}} \text{curl}_z(\mathbf{F})\,dA}$$

■ **EXAMPLE 2 Computing a Line Integral Using Green's Theorem** Use Green's Theorem to compute $\oint_{\mathcal{C}} \sin x\,dx + x^2y^3\,dy$, where $\mathcal{C}$ is the triangular path in Figure 4.

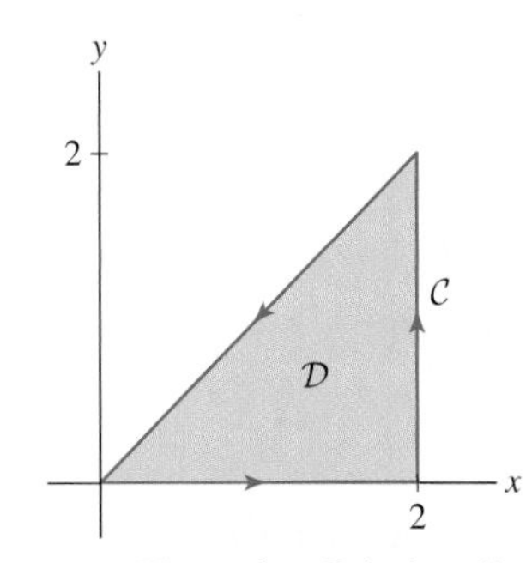

FIGURE 4 The region $\mathcal{D}$ is described by $0 \le x \le 2$, $0 \le y \le x$.

Solution We apply Green's Theorem to $\mathbf{F} = \langle P, Q\rangle = \langle \sin x, x^2y^3\rangle$:

$$\text{curl}_z(\mathbf{F}) = \frac{\partial Q}{\partial x} - \frac{\partial P}{\partial y} = \frac{\partial}{\partial x}x^2y^3 - \frac{\partial}{\partial y}\sin x = 2xy^3$$

The domain $\mathcal{D}$ enclosed by the triangle is described by $0 \le x \le 2$, $0 \le y \le x$, and thus,

$$\oint_{\mathcal{C}} \sin x\,dx + x^2 y\,dy = \iint_{\mathcal{D}} \operatorname{curl}_z(\mathbf{F})\,dA$$

$$= \iint_{\mathcal{D}} 2xy^3\,dA = \int_0^2 \int_{y=0}^{x} 2xy^3\,dy\,dx$$

$$= \int_0^2 \left(\frac{1}{2}xy^4 \Big|_0^x \right) dx = \frac{1}{2}\int_0^2 x^5\,dx = \frac{1}{12}x^6 \Big|_0^2 = \frac{16}{3} \qquad \blacksquare$$

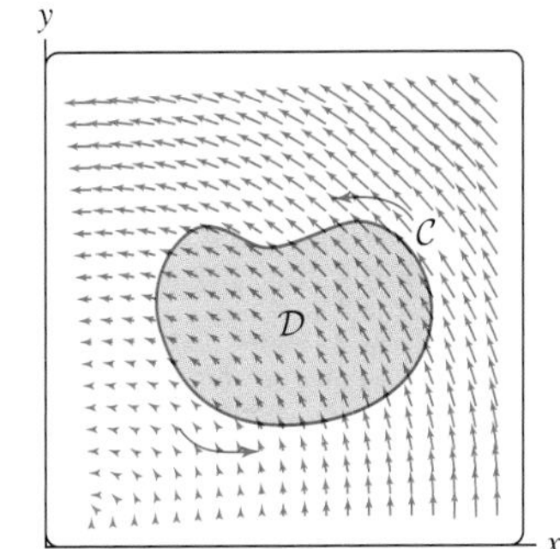

FIGURE 5 The line integral $\oint_{\mathcal{C}} x\,dy - y\,dx$ is equal to twice the area enclosed by $\mathcal{C}$.

Green's Theorem applied to the vector field $\mathbf{F} = \langle -y, x \rangle$ leads to an interesting formula for the area of the region $\mathcal{D}$ enclosed by a simple closed curve $\mathcal{C}$ (Figure 5). In this case, $P = -y$ and $Q = x$, and

$$\operatorname{curl}_z(\mathbf{F}) = \frac{\partial Q}{\partial x} - \frac{\partial P}{\partial y} = \frac{\partial}{\partial x}x - \frac{\partial}{\partial y}(-y) = 2$$

$$\oint_{\mathcal{C}} -y\,dx + x\,dy = \iint_{\mathcal{D}} 2\,dx\,dy = 2\,\text{Area}(\mathcal{D})$$

We obtain the formula

$$\text{Area enclosed by } \mathcal{C} = \frac{1}{2}\oint_{\mathcal{C}} x\,dy - y\,dx \qquad \boxed{5}$$

A practical application of this relation is the **planimeter**, a device that computes the area of an irregular shape when you trace the boundary with a pointer at the end of a movable arm (Figure 6). The mathematical principle behind the planimeter is Eq. (5).

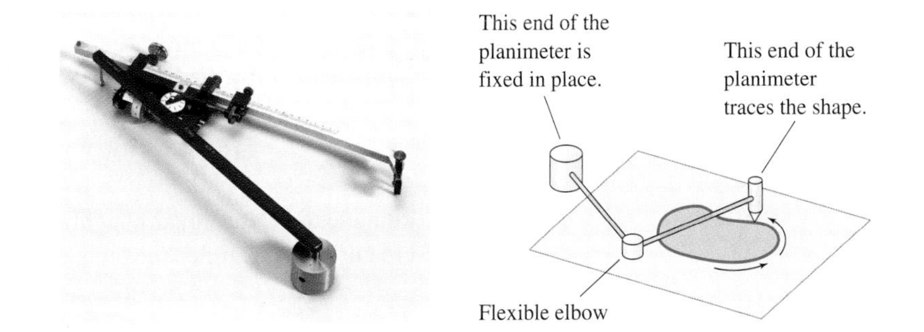

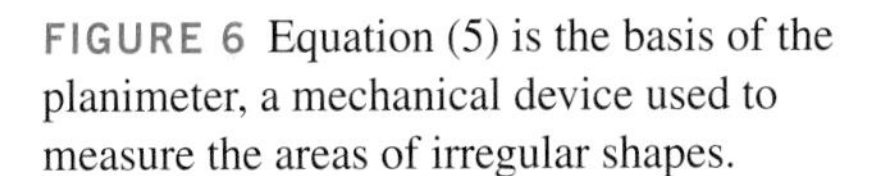

FIGURE 6 Equation (5) is the basis of the planimeter, a mechanical device used to measure the areas of irregular shapes.

■ **EXAMPLE 3 Computing Area via Green's Theorem** Compute the area of the ellipse $\left(\frac{x}{a}\right)^2 + \left(\frac{y}{b}\right)^2 = 1$ as a line integral.

Solution We parametrize the boundary of the ellipse by

$$x = a\cos\theta, \qquad y = b\sin\theta, \qquad 0 \le \theta < 2\pi$$

and compute the area using Eq. (5):

$$x\,dy - y\,dx = (a\cos\theta)(b\cos\theta\,d\theta) - (b\sin\theta)(-a\sin\theta\,d\theta)$$

$$= ab(\cos^2\theta + \sin^2\theta)\,d\theta = ab\,d\theta$$

$$\text{Enclosed area} = \frac{1}{2}\oint_{\mathcal{C}} x\,dy - y\,dx = \frac{1}{2}\int_0^{2\pi} ab\,d\theta = \pi ab$$

This is the well-known formula for the area of an ellipse. ■

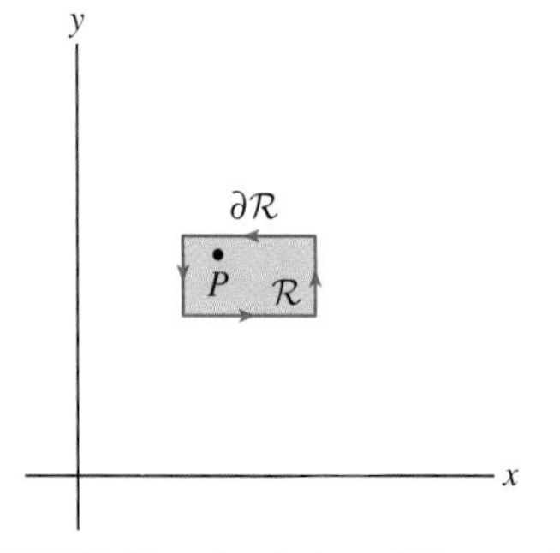

FIGURE 7 The circulation of **F** around $\partial\mathcal{R}$ is approximately $\text{curl}_z(\mathbf{F})(P)\cdot\text{Area}(\mathcal{R})$.

CONCEPTUAL INSIGHT What does the quantity $\text{curl}_z(\mathbf{F})$ represent? To answer this question, let us apply Green's Theorem to a small region $\mathcal{R}$ whose boundary $\mathcal{C} = \partial\mathcal{R}$ is a simple closed curve. Since $\mathcal{R}$ is small and **F** is continuous, $\text{curl}_z(\mathbf{F})$ is nearly constant on $\mathcal{R}$, and to a first approximation, we may replace the function $\text{curl}_z(\mathbf{F})$ by the constant value $\text{curl}_z(\mathbf{F})(P)$, where P is any sample point in $\mathcal{R}$ (Figure 7). Green's Theorem yields the approximation

$$\oint_{\mathcal{C}} \mathbf{F}\cdot d\mathbf{s} = \iint_{\mathcal{R}} \text{curl}_z(\mathbf{F})\,dA \approx \text{curl}_z(\mathbf{F})(P)\cdot\text{Area}(\mathcal{R})$$

In other words, *the circulation around a small region $\mathcal{R}$ is, to a first-order approximation, equal to the curl times the area.* We may think of $\text{curl}_z(\mathbf{F})(P)$ as the **circulation per unit area**.

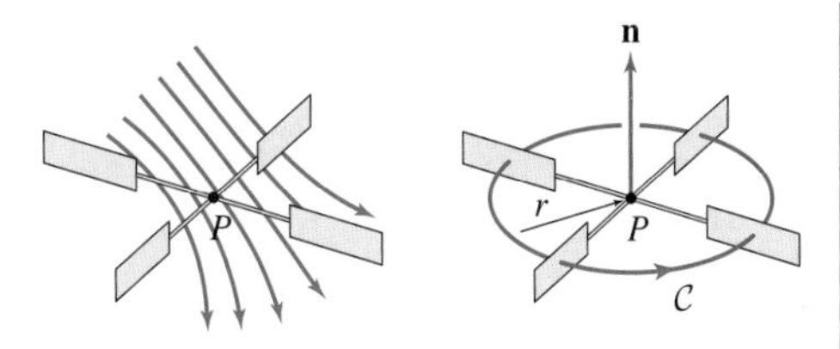

FIGURE 8 The curl at P is approximately equal to one-half the angular velocity (in radians per unit time) of a small paddle wheel placed at P.

GRAPHICAL INSIGHT More geometrically, we may think of the curl as a measure of a vector field's "tendency to rotate." To make this idea precise, let's think of **F** as the velocity field of a fluid. The particles of fluid flow along curves called **integral curves** (or **stream lines**) that are tangent to the vector field at each point. We can measure the curl at a point P by placing a small paddle wheel in the stream at P and observing how fast it rotates (Figure 8).

To justify this statement, suppose that the paddle wheel has radius r and let $\mathcal{C}_r$ be a circle of radius r centered at P. Each paddle on the wheel is constrained to move along the circle $\mathcal{C}_r$ and is pushed to move with a velocity equal to the tangential component of **F**. We may assume that the paddle wheel itself rotates with a velocity v_a equal to the average of the tangential components along the circle. The average is equal to the circulation divided by the length $2\pi r$ of the circle:

$$v_a = \text{velocity of paddle wheel} = \frac{1}{2\pi r}\oint_{\mathcal{C}_r} \mathbf{F}\cdot d\mathbf{s}$$

According to the Conceptual Insight, if r is small, then the circulation is approximately equal to $\text{curl}_z(\mathbf{F})(P)$ times the area πr^2 of the circle. Thus we obtain

$$v_a \approx \frac{1}{2\pi r}(\pi r^2)\text{curl}_z(\mathbf{F})(P) = \left(\frac{1}{2}r\right)\text{curl}_z(\mathbf{F})(P)$$

Angular Velocity *On a circle of radius r meters, an arc of ℓ meters has radian measure ℓ/r. Therefore, an object moving along the circle with a speed of v meters per second travels v/r radians per second. In other words, the object has angular velocity v/r.*

Now if an object moves in a circle with speed v_a, then its angular velocity (in radians per unit time) is $v_a/r \approx \frac{1}{2}\text{curl}_z(\mathbf{F})(P)$. Thus, *the angular velocity of the paddle wheel is approximately one-half the curl.*

Figures 9(A)–(C) show vector fields with nonzero constant curl. The vector field in (A) has positive curl and describes a fluid rotating counterclockwise around the origin. The vector field in (B) has negative curl and describes a fluid that spirals into the origin in the clockwise direction. However, a nonzero curl does not mean that the fluid is necessarily rotating—only that a small paddle wheel would rotate if placed in the fluid. The fluid in Figure 9(C) does not rotate even though its curl is nonzero (this is an example of **shear flow**, also known as a Couette flow). Compare with the vector fields in Figures 9(D) and (E), which have zero curl.

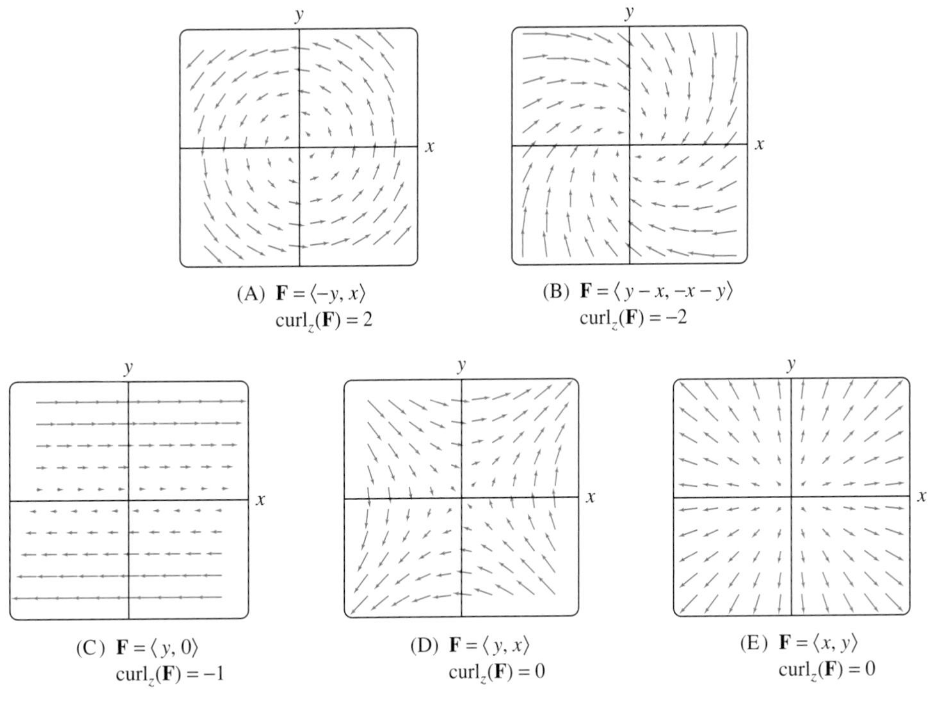

(A) $\mathbf{F} = \langle -y, x \rangle$
$\text{curl}_z(\mathbf{F}) = 2$

(B) $\mathbf{F} = \langle y - x, -x - y \rangle$
$\text{curl}_z(\mathbf{F}) = -2$

(C) $\mathbf{F} = \langle y, 0 \rangle$
$\text{curl}_z(\mathbf{F}) = -1$

(D) $\mathbf{F} = \langle y, x \rangle$
$\text{curl}_z(\mathbf{F}) = 0$

(E) $\mathbf{F} = \langle x, y \rangle$
$\text{curl}_z(\mathbf{F}) = 0$

FIGURE 9

Circulation around a closed curve has an important additivity property: If we decompose a domain $\mathcal{D}$ into two (or more) domains $\mathcal{D}_1$ and $\mathcal{D}_2$ that intersect only on part of their boundaries as in Figure 10, then

$$\oint_{\partial\mathcal{D}} \mathbf{F} \cdot d\mathbf{s} = \oint_{\partial\mathcal{D}_1} \mathbf{F} \cdot d\mathbf{s} + \oint_{\partial\mathcal{D}_2} \mathbf{F} \cdot d\mathbf{s} \qquad \boxed{6}$$

To verify this equation, note first that

$$\oint_{\partial\mathcal{D}} \mathbf{F} \cdot d\mathbf{s} = \int_{\mathcal{C}_{\text{top}}} \mathbf{F} \cdot d\mathbf{s} + \int_{\mathcal{C}_{\text{bot}}} \mathbf{F} \cdot d\mathbf{s}$$

with $\mathcal{C}_{\text{top}}$ and $\mathcal{C}_{\text{bot}}$ as in Figure 10. Then observe that the dashed segment $\mathcal{C}_{\text{middle}}$ occurs in both $\partial\mathcal{D}_1$ and $\partial\mathcal{D}_2$ but with opposite orientations. Therefore,

$$\oint_{\partial\mathcal{D}_1} \mathbf{F} \cdot d\mathbf{s} = \int_{\mathcal{C}_{\text{top}}} \mathbf{F} \cdot d\mathbf{s} - \int_{\mathcal{C}_{\text{middle}}} \mathbf{F} \cdot d\mathbf{s}$$

$$\oint_{\partial\mathcal{D}_2} \mathbf{F} \cdot d\mathbf{s} = \int_{\mathcal{C}_{\text{bot}}} \mathbf{F} \cdot d\mathbf{s} + \int_{\mathcal{C}_{\text{middle}}} \mathbf{F} \cdot d\mathbf{s}$$

FIGURE 10

We obtain Eq. (6) by adding these two equations:

$$\oint_{\partial\mathcal{D}_1} \mathbf{F}\cdot d\mathbf{s} + \oint_{\partial\mathcal{D}_2} \mathbf{F}\cdot d\mathbf{s} = \int_{\mathcal{C}_{\text{top}}} \mathbf{F}\cdot d\mathbf{s} + \int_{\mathcal{C}_{\text{bot}}} \mathbf{F}\cdot d\mathbf{s} = \oint_{\partial\mathcal{D}} \mathbf{F}\cdot \mathbf{s}$$

More General Form of Green's Theorem

Green's Theorem remains valid when $\mathcal{D}$ is a region whose boundary consists of more than one simple closed curve as in Figure 11, provided that we orient the boundary in the correct way. *Each boundary curve must be oriented so that the region lies to your left as you traverse the curve in the direction specified by the orientation.* We write $\partial\mathcal{D}$ for the boundary of $\mathcal{D}$ with this "boundary orientation." For the domains in Figure 11,

$$\partial\mathcal{D}_1 = \mathcal{C}_1 + \mathcal{C}_2, \qquad \partial\mathcal{D}_2 = \mathcal{C}_1 + \mathcal{C}_2 - \mathcal{C}_3$$

The curve $\mathcal{C}_3$ occurs with a minus sign because it is oriented counterclockwise, but the boundary orientation requires a clockwise orientation.

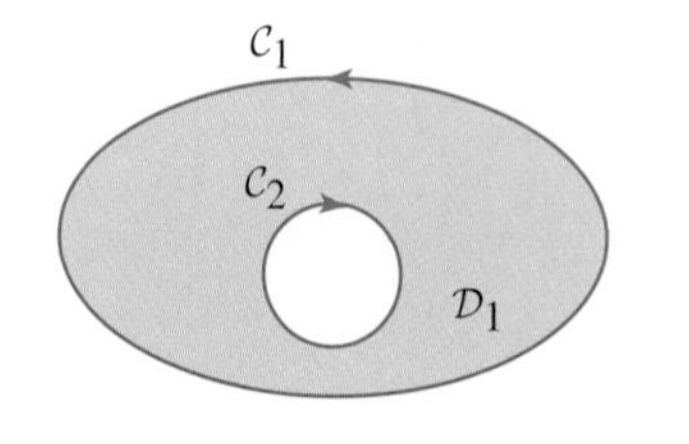

(A) Oriented boundary of $\mathcal{D}_1$ is $\mathcal{C}_1 + \mathcal{C}_2$

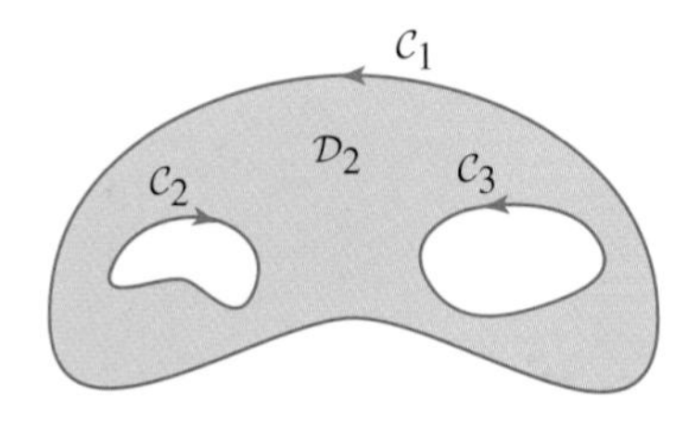

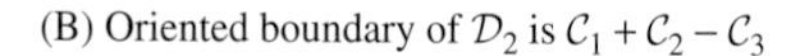

(B) Oriented boundary of $\mathcal{D}_2$ is $\mathcal{C}_1 + \mathcal{C}_2 - \mathcal{C}_3$

FIGURE 11

With this definition of $\partial\mathcal{D}$, Green's Theorem takes the form

$$\oint_{\partial\mathcal{D}} \mathbf{F}\cdot d\mathbf{s} = \iint_{\mathcal{D}} \text{curl}_z(\mathbf{F})\, dA \qquad \boxed{7}$$

This general form of Green's Theorem may be proved using additivity. Consider the region $\mathcal{D}$ in Figure 12. We may decompose $\mathcal{D}$ into domains $\mathcal{D}_1$ and $\mathcal{D}_2$, each of which is bounded by a simple closed curve. Then,

$$\partial\mathcal{D} = \partial\mathcal{D}_1 + \partial\mathcal{D}_2$$

because the edges common to $\partial\mathcal{D}_1$ and $\partial\mathcal{D}_2$ occur with opposite orientation and therefore cancel. Our previous version of Green's Theorem applies to both $\mathcal{D}_1$ and $\mathcal{D}_2$ and thus

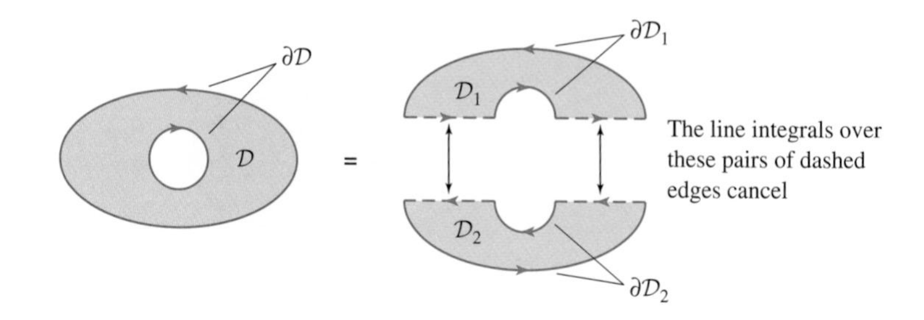

FIGURE 12 The boundary of $\partial\mathcal{D}$ is the sum $\partial\mathcal{D}_1 + \partial\mathcal{D}_2$ because the straight edges cancel.

$$\oint_{\partial \mathcal{D}} \mathbf{F} \cdot d\mathbf{s} = \int_{\partial \mathcal{D}_1} \mathbf{F} \cdot d\mathbf{s} + \int_{\partial \mathcal{D}_2} \mathbf{F} \cdot d\mathbf{s}$$

$$= \iint_{\mathcal{D}_1} \operatorname{curl}_z(\mathbf{F})\, dA + \iint_{\mathcal{D}_2} \operatorname{curl}_z(\mathbf{F})\, dA = \iint_{\mathcal{D}} \operatorname{curl}_z(\mathbf{F})\, dA$$

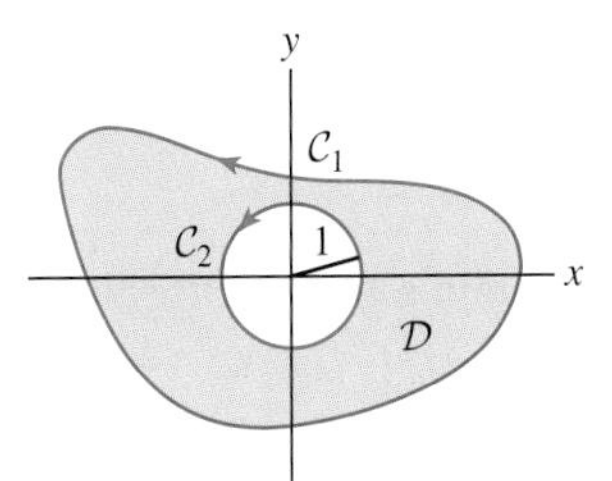

FIGURE 13 $\mathcal{D}$ has area 8 and C_1 is a circle of radius 1.

EXAMPLE 4 Using the General Form of Green's Theorem The domain $\mathcal{D}$ in Figure 13 has area 8. Calculate $\oint_{C_1} \mathbf{F} \cdot d\mathbf{s}$, where $\mathbf{F} = \langle x - y, x + y^3 \rangle$ and C_1 is the outer boundary curve of $\mathcal{D}$ oriented counterclockwise.

Solution We cannot compute the line integral over C_1 directly because the curve C_1 is not specified. However $\partial \mathcal{D} = C_1 - C_2$, so Green's Theorem yields

$$\oint_{C_1} \mathbf{F} \cdot d\mathbf{s} - \oint_{C_2} \mathbf{F} \cdot d\mathbf{s} = \iint_{\mathcal{D}} \operatorname{curl}_z(\mathbf{F})\, dA$$

or

$$\boxed{\oint_{C_1} \mathbf{F} \cdot d\mathbf{s} = \oint_{C_2} \mathbf{F} \cdot d\mathbf{s} + \iint_{\mathcal{D}} \operatorname{curl}_z(\mathbf{F})\, dA} \qquad \boxed{8}$$

We compute each of the integrals on the right. We have

$$\operatorname{curl}_z(\mathbf{F}) = \frac{\partial}{\partial x}(x + y^3) - \frac{\partial}{\partial y}(x - y) = 1 - (-1) = 2$$

and therefore,

$$\iint_{\mathcal{D}} \operatorname{curl}_z(\mathbf{F})\, dA = 2 \iint_{\mathcal{D}} dA = 2\,\text{Area}(\mathcal{D}) = 2(8) = 16$$

To compute the line integral over C_2, we parametrize the circle by $\mathbf{c}(t) = (\cos\theta, \sin\theta)$. Then

$$\mathbf{F} \cdot \mathbf{c}'(t) = \langle \cos\theta - \sin\theta, \cos\theta + \sin^3\theta \rangle \cdot \langle -\sin\theta, \cos\theta \rangle$$

$$= -\sin\theta\cos\theta + \sin^2\theta + \cos^2\theta + \sin^3\theta\cos\theta$$

$$= 1 - \sin\theta\cos\theta + \sin^3\theta\cos\theta$$

The integrals of $\sin\theta\cos\theta$ and $\sin^3\theta\cos\theta$ over $[0, 2\pi]$ are both zero, so

$$\oint_{C_2} \mathbf{F} \cdot d\mathbf{s} = \int_0^{2\pi} (1 - \sin\theta\cos\theta + \sin^3\theta\cos\theta)\, d\theta = \int_0^{2\pi} d\theta = 2\pi$$

Thus, Eq. (8) yields $\oint_{C_1} \mathbf{F} \cdot d\mathbf{s} = 16 + 2\pi$. ■

17.1 SUMMARY

- We have two notations for the line integral of a vector field $\mathbf{F} = \langle P, Q \rangle$ over a curve C:

$$\int_C \mathbf{F} \cdot d\mathbf{s} \qquad \text{and} \qquad \int_C P\, dx + Q\, dy$$

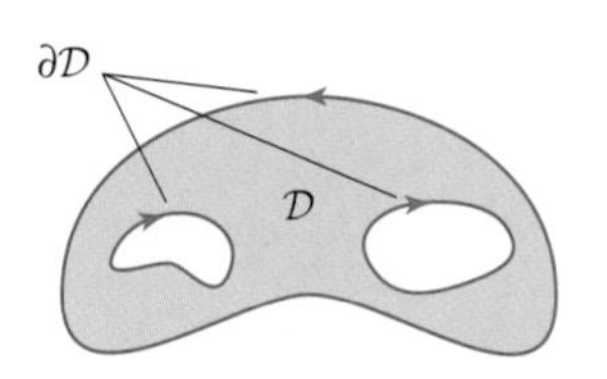

FIGURE 14 The boundary orientation is chosen so that the region lies to your left as you walk along the curve.

- We write $\partial\mathcal{D}$ for the boundary of a domain $\mathcal{D}$ with its boundary orientation (Figure 14).
- Green's Theorem is the equality

$$\oint_{\partial\mathcal{D}} P\,dx + Q\,dy = \iint_{\mathcal{D}} \left(\frac{\partial Q}{\partial x} - \frac{\partial P}{\partial y}\right) dA$$

- The quantity appearing in the double integral of Green's Theorem is called the *curl* or *scalar curl:*

$$\text{curl}_z(\mathbf{F}) = \frac{\partial Q}{\partial x} - \frac{\partial P}{\partial y}$$

- Taking $P = -y$ and $Q = x$ in Green's Theorem, we obtain a formula for the area of the region $\mathcal{D}$ enclosed by $\mathcal{C}$:

$$\text{Area}(\mathcal{D}) = \frac{1}{2}\oint_{\mathcal{C}} x\,dy - y\,dx$$

17.1 EXERCISES

Preliminary Questions

1. Which vector field **F** is being integrated in the line integral $\oint x^2\,dy - e^y\,dx$?

2. Draw a domain in the shape of an ellipse and indicate with an arrow the boundary orientation of the boundary curve. Do the same for the annulus (the region between two concentric circles).

3. The circulation of a gradient vector field around a closed curve is zero. Is this fact consistent with Green's Theorem? Explain.

4. Which of the following vector fields possess the following property: For every simple closed curve $\mathcal{C}$, $\int_{\mathcal{C}} \mathbf{F}\cdot d\mathbf{s}$ is equal to the area enclosed by $\mathcal{C}$?

(a) $\mathbf{F} = \langle -y, 0\rangle$

(b) $\mathbf{F} = \langle x, y\rangle$

(c) $\mathbf{F} = \langle \sin(x^2), x + e^{y^2}\rangle$

Exercises

1. Verify Green's Theorem for the line integral $\oint_{\mathcal{C}} xy\,dx + y\,dy$, where $\mathcal{C}$ is the unit circle, oriented counterclockwise.

2. Let $I = \oint_{\mathcal{C}} \mathbf{F}\cdot d\mathbf{s}$, where $\mathbf{F} = \langle y + \sin x^2, x^2 + e^{y^2}\rangle$ and $\mathcal{C}$ is the circle of radius 4 centered at the origin.

(a) Which is easier: evaluating I directly or using Green's Theorem?

(b) Evaluate I using the easier method.

In Exercises 3–11, use Green's Theorem to evaluate the line integral. Orient the curve counterclockwise unless otherwise indicated.

3. $\oint_{\mathcal{C}} y^2\,dx + x^2\,dy$, where $\mathcal{C}$ is the boundary of the unit square $0 \le x \le 1, 0 \le y \le 1$

4. $\oint_{\mathcal{C}} e^{2x+y}\,dx + e^{-y}\,dy$, where $\mathcal{C}$ is the triangle with vertices (0, 0), (1, 0), and (1, 1)

5. $\oint_{\mathcal{C}} x^2 y\,dx$, where $\mathcal{C}$ is the unit circle centered at the origin

6. $\oint_{\mathcal{C}} \mathbf{F}\cdot d\mathbf{s}$, where $\mathbf{F} = \langle x + y, x^2 - y\rangle$ and $\mathcal{C}$ is the boundary of the region enclosed by $y = x^2$ and $y = \sqrt{x}$ for $0 \le x \le 1$.

7. $\oint_{\mathcal{C}} \mathbf{F}\cdot d\mathbf{s}$, where $\mathbf{F} = \langle x^2, x^2\rangle$ and $\mathcal{C}$ consists of the arcs $y = x^2$ and $y = x$ for $0 \le x \le 1$

8. $\oint_{\mathcal{C}} (\ln x + y)\,dx - x^2\,dy$, where $\mathcal{C}$ is the rectangle with vertices (1, 1), (3, 1), (1, 4), and (3, 4)

9. The line integral of $\mathbf{F} = \langle x^3, 4x\rangle$ around the boundary of the parallelogram in Figure 15 (note the orientation)

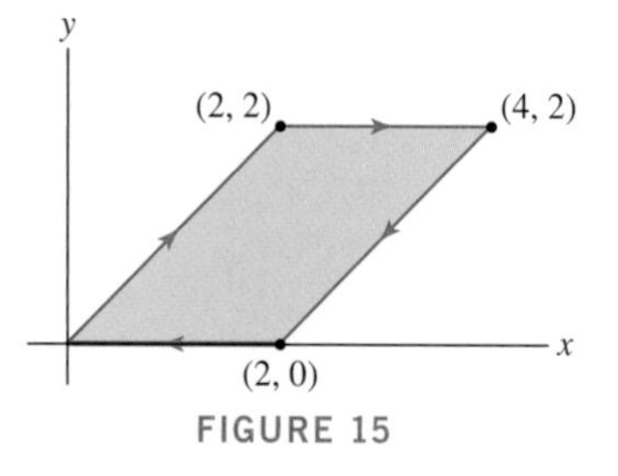

FIGURE 15

10. The line integral of $\mathbf{F} = \langle e^{x+y}, e^{x-y}\rangle$ around the boundary of the parallelogram in Figure 15

11. $\int_C xy\,dx + (x^2 + x)\,dy$, where C is the path in Figure 16

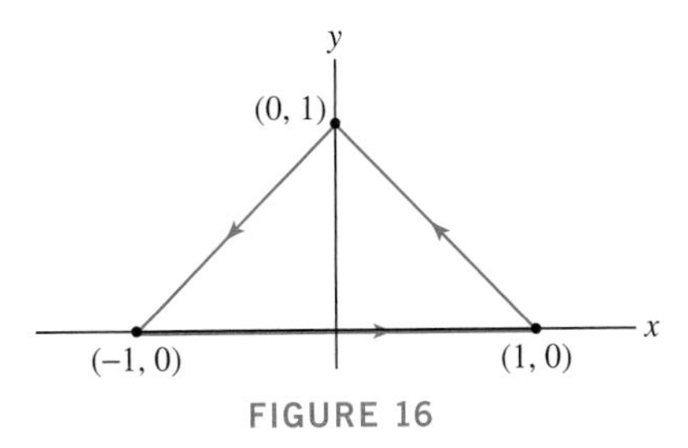

FIGURE 16

12. Let $\mathcal{C}_R$ be the circle of radius R centered at the origin. Use Green's Theorem to determine $\oint_{\mathcal{C}_2} \mathbf{F} \cdot d\mathbf{s}$, where $\mathbf{F}$ is a vector field such that $\oint_{\mathcal{C}_1} \mathbf{F} \cdot d\mathbf{s} = 9$ and $\text{curl}_z(\mathbf{F}) = x^2 + y^2$ for $1 \le x^2 + y^2 \le 4$.

In Exercises 13–16, use Eq. (5) to calculate the area of the given region.

13. The circle of radius 3 centered at the origin

14. The triangle with vertices $(0, 0)$, $(1, 0)$, and $(1, 1)$

15. The region between the x-axis and the cycloid parametrized by $\mathbf{c}(t) = (t - \sin t, 1 - \cos t)$ for $0 \le t \le 2\pi$ (Figure 17)

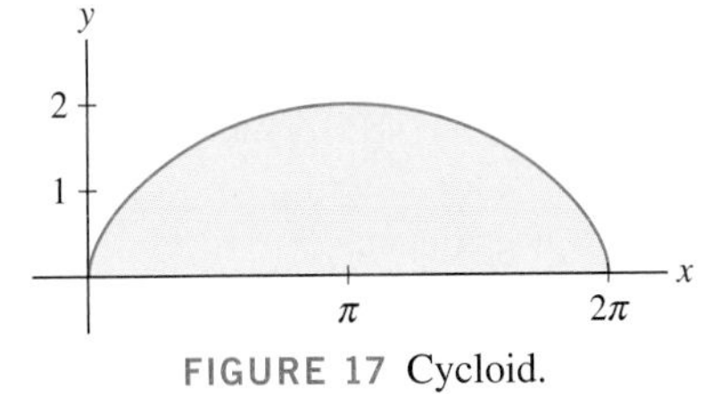

FIGURE 17 Cycloid.

16. The region between the graph of $y = x^2$ and the x-axis for $0 \le x \le 2$

17. Let $x^3 + y^3 = 3xy$ be the **folium of Descartes** (Figure 18).

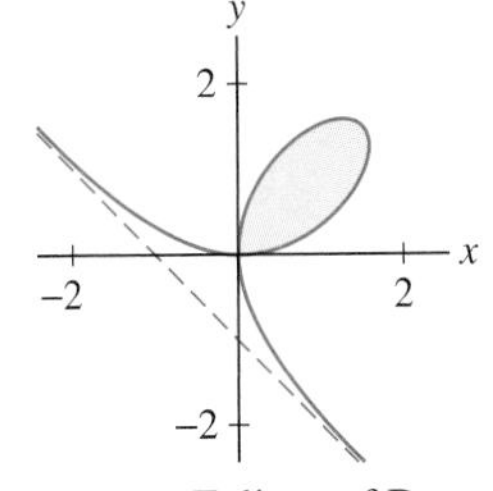

FIGURE 18 Folium of Descartes.

(a) Show that the folium has a parametrization in terms of $t = y/x$ given by

$$x = \frac{3t}{1+t^3}, \qquad y = \frac{3t^2}{1+t^3} \qquad (-\infty < t < \infty) \quad (t \neq -1)$$

(b) Show that

$$x\,dy - y\,dx = \frac{9t^2}{(1+t^3)^2}\,dt$$

Hint: By the Quotient Rule,

$$x^2\,d\left(\frac{y}{x}\right) = x\,dy - y\,dx$$

(c) Find the area of the loop of the folium.

18. Follow the procedure of Exercise 17 to find the area of the loop of the lemniscate curve with equation $(x^2 + y^2)^2 = xy$ (Figure 19).

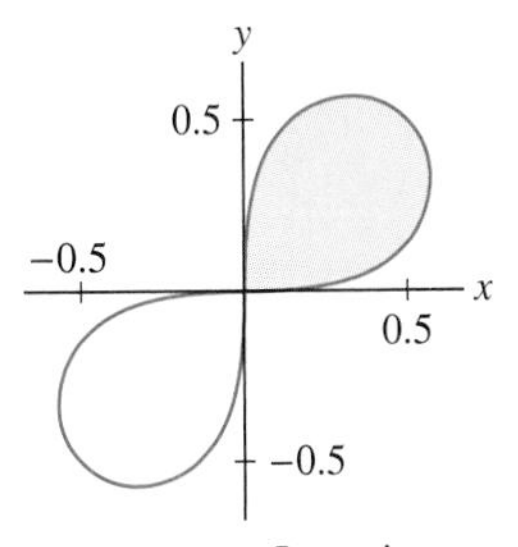

FIGURE 19 Lemniscate.

19. Show that if $\mathcal{C}$ is a simple closed curve, then

$$\oint_{\mathcal{C}} -y\,dx = \oint_{\mathcal{C}} x\,dy$$

and both integrals are equal to the area enclosed by $\mathcal{C}$.

20. For the vector fields (A)–(D) in Figure 20, state whether the curl at the origin appears to be positive, negative, or zero.

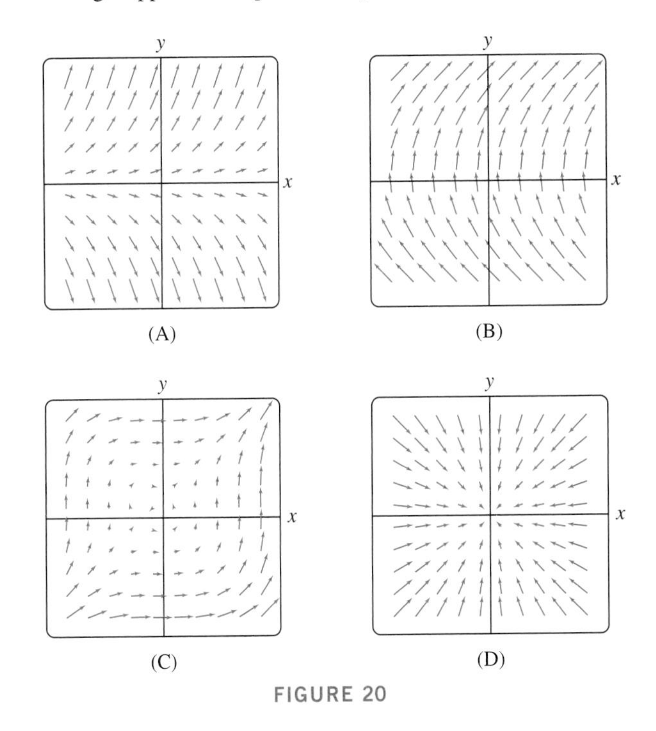

FIGURE 20

21. Let $\mathbf{F} = \langle 2xe^y, x + x^2e^y \rangle$ and let $\mathcal{C}$ be the quarter-circle path from A to B in Figure 21. Evaluate $I = \oint_{\mathcal{C}} \mathbf{F} \cdot d\mathbf{s}$ as follows:

(a) Find a function $\varphi(x, y)$ such that $\mathbf{F} = \mathbf{G} + \nabla\varphi$, where $\mathbf{G} = \langle 0, x \rangle$.

(b) Show that the line integrals of $\mathbf{G}$ along the segments $\overline{OA}$ and $\overline{OB}$ are zero.

(c) Use Green's Theorem to show that

$$I = \varphi(B) - \varphi(A) + 4\pi$$

and evaluate I.

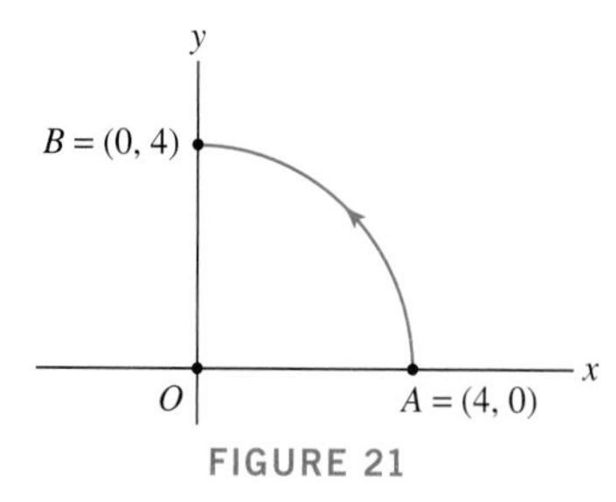

FIGURE 21

22. Compute the line integral of $\mathbf{F} = \langle x^3, 4x \rangle$ along the path from A to B in Figure 22. *Hint:* To save work, use Green's Theorem to relate this line integral to the line integral along the vertical path from B to A.

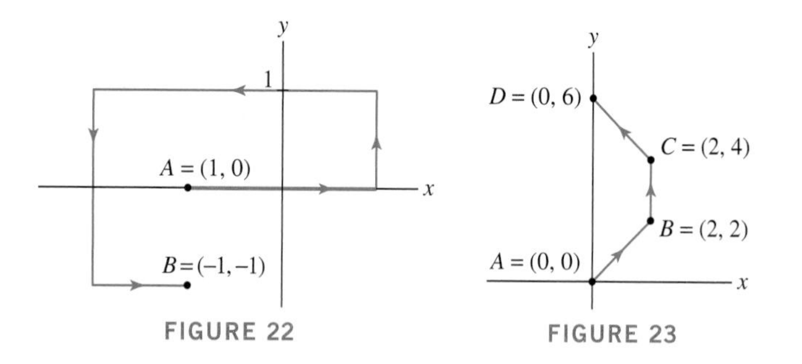

23. Evaluate $I = \int_{\mathcal{C}} (\sin x + y)\, dx + (3x + y)\, dy$ for the nonclosed path $ABCD$ in Figure 23. *Hint:* Use the method of Exercise 22.

24. Estimate the circulation of a vector field $\mathbf{F}$ around a circle of radius $R = 0.1$, assuming that $\text{curl}_z(\mathbf{F})$ takes the value 4 at the center of the circle.

25. Let $\mathbf{F}$ be the velocity field. Estimate the circulation of $\mathbf{F}$ around a circle of radius $R = 0.05$ with center P, assuming that $\text{curl}_z(\mathbf{F})(P) = -3$. In which direction would a small paddle placed at P rotate? How fast would it rotate (in radians per second) if $\mathbf{F}$ is expressed in meters per second?

26. Referring to Figure 24, suppose that $\oint_{\mathcal{C}_2} \mathbf{F} \cdot d\mathbf{s} = 12$. Use Green's Theorem to determine $\int_{\mathcal{C}_1} \mathbf{F} \cdot d\mathbf{s}$, assuming that $\text{curl}_z(\mathbf{F}) = -3$ in $\mathcal{D}$.

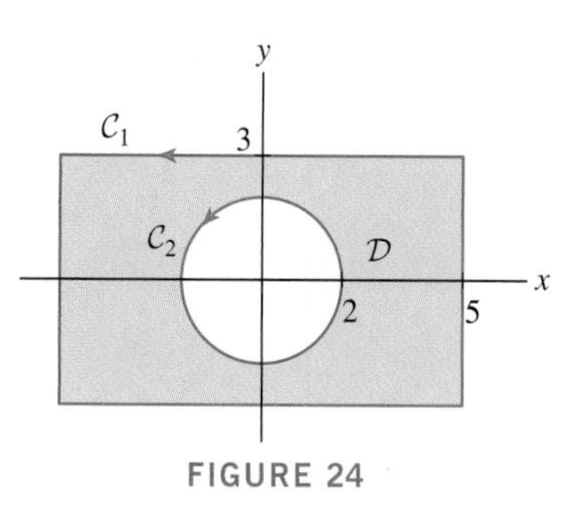

FIGURE 24

27. Referring to Figure 25, suppose that

$$\oint_{\mathcal{C}_2} \mathbf{F} \cdot d\mathbf{s} = 3\pi, \qquad \oint_{\mathcal{C}_3} \mathbf{F} \cdot d\mathbf{s} = 4\pi$$

Use Green's Theorem to determine the circulation of $\mathbf{F}$ around $\mathcal{C}_1$, assuming that $\text{curl}_z(\mathbf{F}) = 9$ on the shaded region.

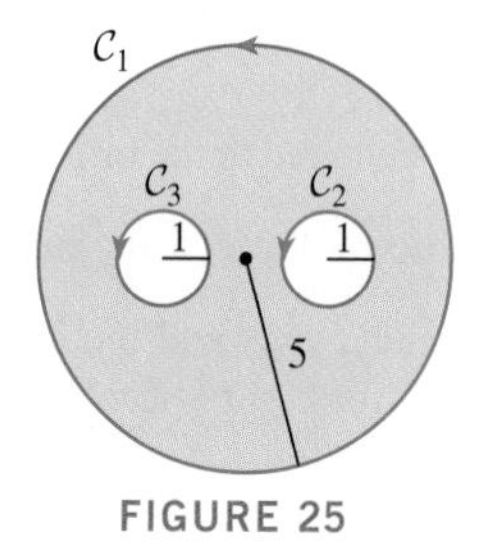

FIGURE 25

28. Area of a Polygon Green's Theorem leads to a convenient formula for the area of a polygon.

(a) Let $\mathcal{C}$ be the line segment joining (x_1, y_1) to (x_2, y_2). Show that

$$\frac{1}{2}\oint_{\mathcal{C}} -y\, dx + x\, dy = \frac{1}{2}(x_1y_2 - x_2y_1)$$

(b) Prove that the area of the polygon with vertices $(x_1, y_1), (x_2, y_2), \ldots, (x_n, y_n)$ is equal [where we set $(x_{n+1}, y_{n+1}) = (x_1, y_1)$] to

$$\frac{1}{2}\sum (x_iy_{i+1} - x_{i+1}y_i)$$

29. Use the result of Exercise 28 to compute the areas of the polygons in Figure 26. Check your result for the area of the triangle in (A) using geometry.

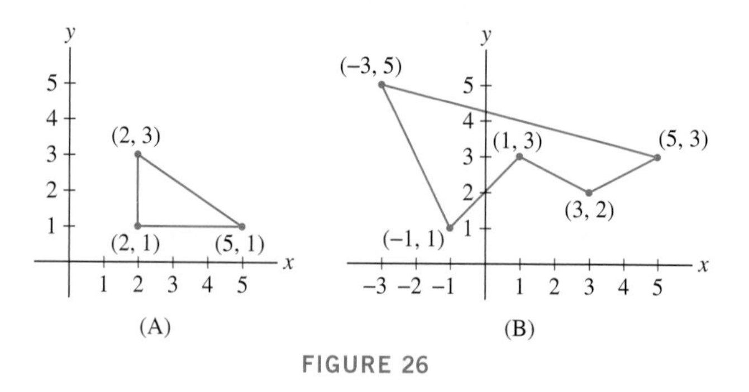

FIGURE 26

Further Insights and Challenges

In Exercises 30–31, let **F** *be the vortex vector field [defined for* $(x, y) \neq (0, 0)$*]:*

$$\mathbf{F} = \left\langle \frac{-y}{x^2 + y^2}, \frac{x}{x^2 + y^2} \right\rangle$$

30. **(a)** Show that $\text{curl}_z(\mathbf{F})(P) = 0$ for all $P \neq (0, 0)$.

(b) Show that the circulation of **F** around the circle $\mathcal{C}_R$ of radius R centered at the origin is equal to 2π for all R.

31. Prove that if $\mathcal{C}$ is a simple closed curve whose interior contains the origin, then $\oint_{\mathcal{C}} \mathbf{F} \cdot d\mathbf{s} = 2\pi$ (Figure 27). *Hint:* Apply Green's Theorem to the domain between $\mathcal{C}$ and $\mathcal{C}_R$ where R is so small that $\mathcal{C}_R$ is contained in $\mathcal{C}$.

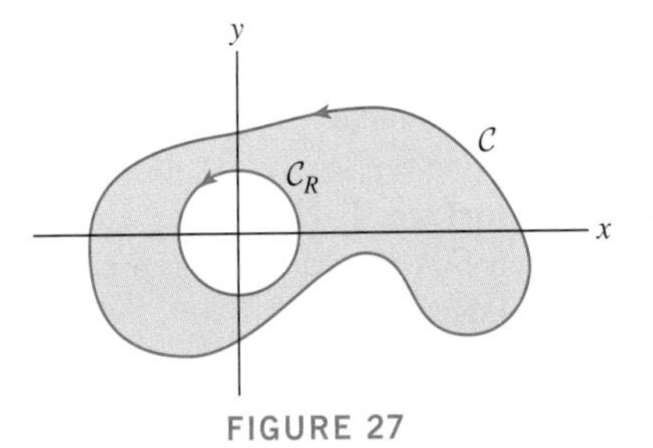

FIGURE 27

In Exercises 32–34, the conjugate *of a vector field* $\mathbf{F} = \langle P, Q \rangle$ *is the vector field* $\mathbf{F}^* = \langle -Q, P \rangle$.

32. Explain the following statement: $\mathbf{F}^*$ is the vector field obtained by rotating the vectors of **F** counterclockwise through an angle of $\frac{\pi}{2}$.

33. The normal component of **F** at a point P on a simple closed $\mathcal{C}$ is the quantity $\mathbf{F}(P) \cdot \mathbf{n}(P)$, where $\mathbf{n}(P)$ is the outward-pointing unit normal vector. The **flux** of **F** across curve $\mathcal{C}$ is defined as the line integral of the normal component around $\mathcal{C}$ (Figure 28). Show that the flux across $\mathcal{C}$ is equal to $\oint_{\mathcal{C}} \mathbf{F}^* \cdot d\mathbf{s}$.

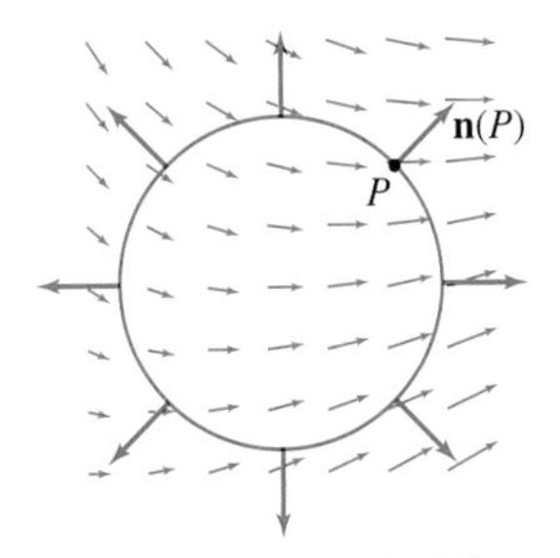

FIGURE 28 The flux of **F** is the integral of the normal component $\mathbf{F} \cdot \mathbf{n}$ around the curve.

34. Define $\text{div}(\mathbf{F}) = \dfrac{\partial P}{\partial x} + \dfrac{\partial Q}{\partial y}$. Use Green's Theorem to prove that for any simple closed curve $\mathcal{C}$,

$$\text{Flux across } \mathcal{C} = \iint_{\mathcal{D}} \text{div}(\mathbf{F})\, dA$$

where $\mathcal{D}$ is the region enclosed by $\mathcal{C}$. This is a two-dimensional version of the **Divergence Theorem** discussed in Section 17.3.

In Exercises 35–38, the **Laplace operator** Δ *is defined by*

$$\Delta\varphi = \frac{\partial^2 \varphi}{\partial x^2} + \frac{\partial^2 \varphi}{\partial y^2} \qquad \boxed{9}$$

35. Let $\mathbf{F} = \nabla\varphi$. Show that $\text{curl}_z(\mathbf{F}^*) = \Delta\varphi$, where $\mathbf{F}^*$ is the conjugate vector field (defined in Exercises 32–34).

36. Let **n** denote the outward-pointing unit normal vector to a simple closed $\mathcal{C}$. The **normal derivative** of a function φ, denoted $\dfrac{\partial \varphi}{\partial \mathbf{n}}$, is the directional derivative $D_{\mathbf{n}}(\varphi) = \nabla\varphi \cdot \mathbf{n}$. Prove the formula

$$\oint_{\mathcal{C}} \frac{\partial \varphi}{\partial \mathbf{n}}\, ds = \iint_{\mathcal{D}} \Delta\varphi\, dA$$

where $\mathcal{D}$ is the domain enclosed by $\mathcal{C}$. *Hint:* Let $\mathbf{F} = \nabla\varphi$. Show that $\dfrac{\partial \varphi}{\partial \mathbf{n}} = \mathbf{F}^* \cdot \mathbf{T}$, where **T** is the unit tangent vector pointing in the counterclockwise direction along $\mathcal{C}$, and apply Green's Theorem.

37. Let $P = (a, b)$ and let $\mathcal{C}(r)$ be the circle of radius r centered at P. The average value of a continuous function φ on $\mathcal{C}(r)$ is defined as the integral

$$I_\varphi(r) = \frac{1}{2\pi} \int_0^{2\pi} \varphi(a + r\cos\theta, b + r\sin\theta)\, d\theta$$

(a) Show that

$$\frac{\partial \varphi}{\partial \mathbf{n}}(a + r\cos\theta, b + r\sin\theta) = \frac{\partial \varphi}{\partial r}(a + r\cos\theta, b + r\sin\theta)$$

(b) Use differentiation under the integral sign to prove that

$$\frac{d}{dr} I_\varphi(r) = \frac{1}{2\pi r} \int_{\mathcal{C}(r)} \frac{\partial \varphi}{\partial \mathbf{n}}\, ds$$

(c) Use Exercise 36 to conclude that

$$\frac{d}{dr} I_\varphi(r) = \frac{1}{2\pi r} \iint_{\mathcal{D}(r)} \Delta\varphi\, dA$$

where $\mathcal{D}(r)$ is the interior of $\mathcal{C}(r)$.

38. Prove that $m(r) \leq I_\varphi(r) \leq M(r)$, where $m(r)$ and $M(r)$ are the minimum and maximum values of φ on $\mathcal{C}(r)$. Then use the continuity of φ to prove that $\lim_{r \to 0} I_\varphi(r) = \varphi(P)$.

In Exercises 39–40, let $\mathcal{D}$ *be the region bounded by a simple closed curve* $\mathcal{C}$. *A function* $\varphi(x, y)$ *on* $\mathcal{D}$ *(whose second-order partial derivatives exist and are continuous) is called* **harmonic** *if* $\Delta\varphi = 0$, *where* $\Delta\varphi$ *is the Laplace operator defined in Eq. (9).*

39. Use the results of Exercises 37 and 38 to prove the **mean-value property** of harmonic functions: If φ is harmonic, then $I_\varphi(r) = \varphi(P)$ for all r.

40. Show that $f(x, y) = x^2 - y^2$ is harmonic. Verify the mean-value property for $f(x, y)$ directly [expand $f(a + r\cos\theta, b + r\sin\theta)$ as a function of θ and compute $I_\varphi(r)$]. Show that $x^2 + y^2$ is not harmonic and does not satisfy the mean-value property.

17.2 Stokes' Theorem

Stokes' Theorem is a generalization of Green's Theorem to three dimensions. Whereas Green's Theorem relates a line integral to a double integral in the plane, Stokes' Theorem relates a line integral in $\mathbf{R}^3$ to a surface integral.

A few preliminaries are needed before we can state Stokes' Theorem. Recall that when applying Green's Theorem to a domain $\mathcal{D}$, we must orient the boundary curve $\partial\mathcal{D}$ so that $\mathcal{D}$ lies to the left when you traverse the boundary in the positive direction (i.e., counterclockwise if $\mathcal{D}$ is bounded by a simple closed curve). In Stokes' Theorem, we must take into account both the orientation of the surface $\mathcal{S}$ and the orientation of its boundary $\partial\mathcal{S}$. First, let us clarify what is meant by the boundary of a surface.

Figure 1 illustrates several possibilities: The boundary may be a simple closed curve as in (A), or it may consist of several closed curves as in (B). If the surface has *no boundary* as in (C), we say that $\mathcal{S}$ is a **closed surface**. To determine whether $\mathcal{S}$ is closed or not, ask yourself if it can hold air. If so, then $\mathcal{S}$ is closed. If it has a boundary, then air can escape through the hole cut off by the boundary curve.

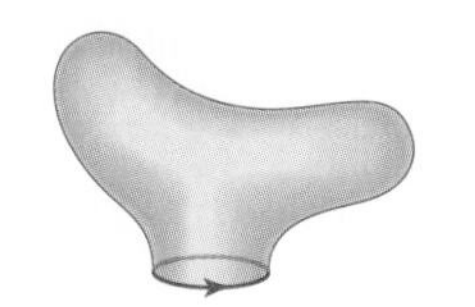

(A) Boundary consists of a single closed curve

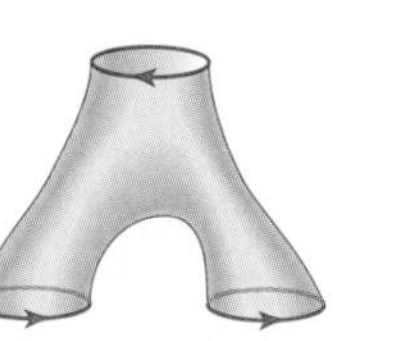

(B) Boundary consists of three closed curves

(C) Closed surface (the boundary is empty)

FIGURE 1 Surfaces and their boundaries.

Recall from Section 16.5 that a surface $\mathcal{S}$ is *oriented* if a unit normal vector at each point has been specified in a continuously varying manner. An oriented surface has two sides and a choice of normal vectors specifies one of these two sides. Once we fix an orientation on $\mathcal{S}$, we may induce a **boundary orientation** of $\partial\mathcal{S}$ as follows: If you imagine yourself as the normal vector walking along the boundary curve, then the surface lies to your left. Figure 2 shows two orientations of a surface whose boundary consists of two curves, $\mathcal{C}_1$ and $\mathcal{C}_2$. Note how the orientation of the boundary depends on the orientation of the surface.

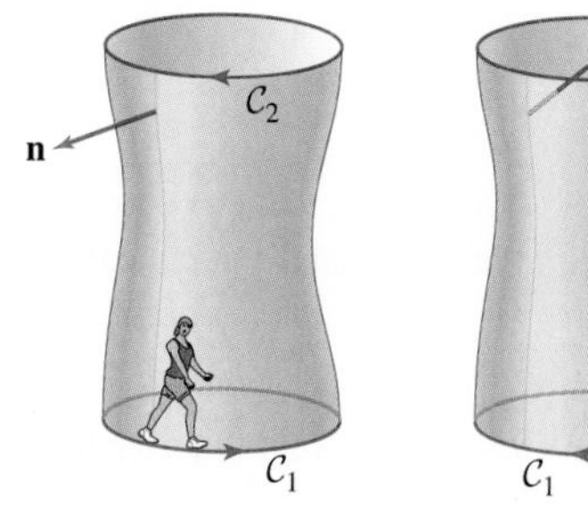

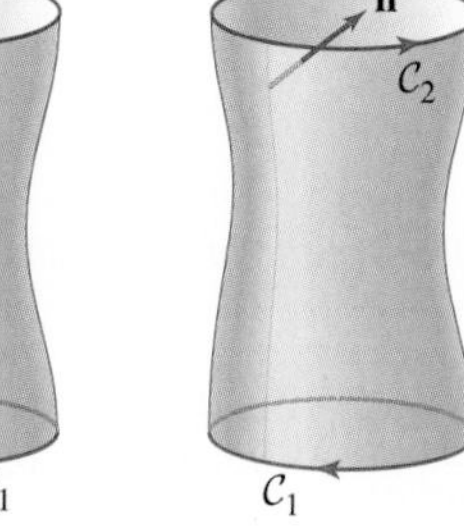

FIGURE 2 The orientation of the boundary ∂S for each of the two possible orientations of the surface $\mathcal{S}$.

The last ingredient in Stokes' Theorem is the **curl** of a vector field $\mathbf{F} = \langle F_1, F_2, F_3 \rangle$, defined by the symbolic determinant

$$\operatorname{curl}(\mathbf{F}) = \begin{vmatrix} \mathbf{i} & \mathbf{j} & \mathbf{k} \\ \frac{\partial}{\partial x} & \frac{\partial}{\partial y} & \frac{\partial}{\partial z} \\ F_1 & F_2 & F_3 \end{vmatrix}$$

$$= \left(\frac{\partial F_3}{\partial y} - \frac{\partial F_2}{\partial z}\right)\mathbf{i} - \left(\frac{\partial F_3}{\partial x} - \frac{\partial F_1}{\partial z}\right)\mathbf{j} + \left(\frac{\partial F_2}{\partial x} - \frac{\partial F_1}{\partial y}\right)\mathbf{k}$$

In terms of components,

$$\operatorname{curl}(\mathbf{F}) = \left\langle \frac{\partial F_3}{\partial y} - \frac{\partial F_2}{\partial z}, \frac{\partial F_1}{\partial z} - \frac{\partial F_3}{\partial x}, \frac{\partial F_2}{\partial x} - \frac{\partial F_1}{\partial y} \right\rangle$$

Keep in mind that curl($\mathbf{F}$) is again a vector field. The scalar curl defined in the previous section is equal to the z-component of the vector curl. It is convenient to define the del "operator" (also called "nabla"):

$$\nabla = \left\langle \frac{\partial}{\partial x}, \frac{\partial}{\partial y}, \frac{\partial}{\partial z} \right\rangle$$

We may think of curl($\mathbf{F}$) as the symbolic cross product of ∇ and $\mathbf{F}$:

$$\operatorname{curl}(\mathbf{F}) = \nabla \times \mathbf{F}$$

It is straightforward to check that taking the curl is a **linear operation** in the following sense:

$$\operatorname{curl}(\mathbf{F} + \mathbf{G}) = \operatorname{curl}(\mathbf{F}) + \operatorname{curl}(\mathbf{G})$$

$$\operatorname{curl}(a\mathbf{F}) = a \operatorname{curl}(\mathbf{F}) \quad (a \text{ any constant})$$

EXAMPLE 1 Calculating the Curl Calculate the curl of $\mathbf{F} = \langle xy, e^x, y + z \rangle$.

Solution We compute the curl as a symbolic determinant:

$$\operatorname{curl}(\mathbf{F}) = \begin{vmatrix} \mathbf{i} & \mathbf{j} & \mathbf{k} \\ \frac{\partial}{\partial x} & \frac{\partial}{\partial y} & \frac{\partial}{\partial z} \\ xy & e^x & y + z \end{vmatrix}$$

$$= \left(\frac{\partial}{\partial y}(y + z) - \frac{\partial}{\partial z}e^x\right)\mathbf{i} - \left(\frac{\partial}{\partial x}(y + z) - \frac{\partial}{\partial z}xy\right)\mathbf{j} + \left(\frac{\partial}{\partial x}e^x - \frac{\partial}{\partial y}xy\right)\mathbf{k}$$

$$= \mathbf{i} + (e^x - x)\mathbf{k}$$

We now state Stokes' Theorem for an oriented surface $\mathcal{S}$ in $\mathbf{R}^3$ bounded by a closed curve or union of closed curves. We must require that $\mathcal{S}$ be smooth or piecewise smooth, in a suitable sense, but it is rather technical to make this requirement precise. It suffices to assume that $\mathcal{S}$ has a one-to-one, regular, continuously differentiable parametrization

$\Phi : \mathcal{D} \to \mathcal{S}$, where $\mathcal{D}$ is a domain in the plane bounded by a smooth, simple closed curve. More generally, Stokes' Theorem is valid for smooth surfaces such as a sphere or torus or piecewise smooth surfaces such as the surface of a cube or tetrahedron.

THEOREM 1 Stokes' Theorem Assume that $\mathcal{S}$ is an oriented, piecewise smooth surface bounded by a closed curve or union of closed curves as described above. Let $\partial\mathcal{S}$ be the boundary of $\mathcal{S}$ with its boundary orientation. Let $\mathbf{F}$ be a vector field whose components have continuous partial derivatives. Then

$$\oint_{\partial\mathcal{S}} \mathbf{F} \cdot d\mathbf{s} = \iint_{\mathcal{S}} \operatorname{curl}(\mathbf{F}) \cdot d\mathbf{S} \qquad \boxed{1}$$

If $\mathcal{S}$ is closed (i.e., $\partial\mathcal{S}$ is empty), then the surface integral on the right is zero.

Proof A complete proof of Stokes' Theorem is beyond the scope of this text. We prove it here in the special case whereby $\mathcal{S}$ is the graph of a function $z = f(x, y)$ lying over a domain $\mathcal{D}$ in the xy-plane. Orient $\mathcal{S}$ with upward-pointing normal.

Let $\mathcal{C} = \partial\mathcal{S}$ be the boundary curve of $\mathcal{S}$. We may write each side of Eq. (1) as a sum over components:

$$\oint_{\mathcal{C}} (F_1\mathbf{i} + F_2\mathbf{j} + F_3\mathbf{k}) \cdot d\mathbf{s} = \oint_{\mathcal{C}} F_1\mathbf{i} \cdot d\mathbf{s} + \oint_{\mathcal{C}} F_2\mathbf{j} \cdot d\mathbf{s} + \oint_{\mathcal{C}} F_3\mathbf{k} \cdot d\mathbf{s}$$

$$\iint_{\mathcal{S}} \operatorname{curl}(F_1\mathbf{i} + F_2\mathbf{j} + F_3\mathbf{k}) \cdot d\mathbf{S} = \iint_{\mathcal{S}} \operatorname{curl}(F_1\mathbf{i}) \cdot d\mathbf{S} + \iint_{\mathcal{S}} \operatorname{curl}(F_2\mathbf{j}) \cdot d\mathbf{S} + \iint_{\mathcal{S}} \operatorname{curl}(F_3\mathbf{k}) \cdot d\mathbf{S}$$

It suffices to show that the terms corresponding to the **i**-, **j**-, and **k**-components are separately equal. We carry this out for the **i**-component. The calculations for the **j**-components are similar and we leave the equality for the **k**-components as an exercise (Exercise 31). Thus, we shall prove that

$$\oint_{\mathcal{C}} F_1(x, y, z)\mathbf{i} \cdot d\mathbf{s} = \iint_{\mathcal{S}} \operatorname{curl}(F_1(x, y, z)\mathbf{i}) \cdot d\mathbf{S} \qquad \boxed{2}$$

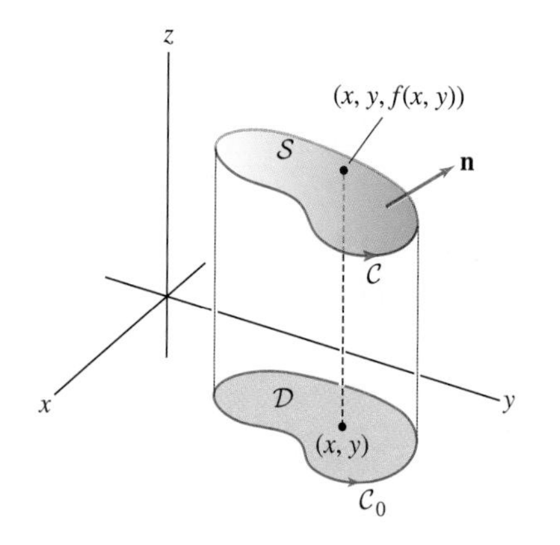

FIGURE 3

Let $\mathcal{C}_0$ be the boundary of the domain $\mathcal{D}$ in the xy-plane, oriented counterclockwise (Figure 3). Observe that the boundary $\mathcal{C}$ of $\mathcal{S}$ projects onto $\mathcal{C}_0$. Let $\mathbf{c}_0(t) = (x(t), y(t))$ (for $a \le t \le b$) be a counterclockwise parametrization of $\mathcal{C}_0$. Then $\mathcal{C}$ with its boundary orientation is parametrized by the path $\mathbf{c}(t) = \big(x(t), y(t), f(x(t), y(t))\big)$. Thus, we have

$$\mathbf{c}(t) = \big(x(t), y(t), f(x(t), y(t))\big)$$

$$\mathbf{c}'(t) = \left\langle x'(t), y'(t), \frac{d}{dt} f(x(t), y(t)) \right\rangle$$

and so

$$\begin{aligned} F_1\mathbf{i} \cdot \mathbf{c}'(t) &= \big\langle F_1\big(x(t), y(t), f(x(t), y(t))\big), 0, 0\big\rangle \cdot \left\langle x'(t), y'(t), \frac{d}{dt} f(x(t), y(t)) \right\rangle \\ &= F_1\big(x(t), y(t), f(x(t), y(t))\big)x'(t) \end{aligned}$$

Therefore, the left-hand side of Eq. (2) is equal to

$$\oint_{\mathcal{C}} F_1\mathbf{i}\cdot d\mathbf{s} = \int_a^b F_1\big(x(t), y(t), f(x(t), y(t))\big)x'(t)\,dt$$

Now we observe that the integral on the right is equal to the line integral of the planar vector field $F_1(x, y, f(x, y))\mathbf{i}$ over $\mathcal{C}_0$, so we may apply Green's Theorem. By the Chain Rule,

$$\frac{\partial}{\partial y}F_1(x, y, f(x, y)) = F_{1y}(x, y, f(x, y)) + F_{1z}(x, y, f(x, y))f_y(x, y)$$

Green's Theorem yields

$$\oint_{\mathcal{C}} F_1\mathbf{i}\cdot d\mathbf{s} = -\iint_{\mathcal{D}} \Big(F_{1y}(x, y, f(x, y)) + F_{1z}(x, y, f(x, y))f_y(x, y)\Big)\,dA \qquad \boxed{3}$$

To finish the proof, we compute the surface integral of $\operatorname{curl}(F_1\mathbf{i})$ using the parametrization $\Phi(x, y) = (x, y, f(x, y))$ of $\mathcal{S}$. We obtain

$$\mathbf{n} = \big\langle -f_x(x, y), -f_y(x, y), 1\big\rangle$$

$$\operatorname{curl}(F_1\mathbf{i})\cdot\mathbf{n} = \big\langle 0, F_{1z}, -F_{1y}\big\rangle\cdot\big\langle -f_x(x, y), -f_y(x, y), 1\big\rangle$$

$$= -F_{1z}(x, y, f(x, y))f_y(x, y) - F_{1y}(x, y, f(x, y))$$

$$\iint_{\mathcal{S}} \operatorname{curl}(F_1\mathbf{i})\cdot d\mathbf{S} = -\iint_{\mathcal{D}} \Big(F_{1z}(x, y, z)f_y(x, y) + F_{1y}(x, y, f(x, y))\Big)\,dA \qquad \boxed{4}$$

Equations (3) and (4) yield Eq. (2) as desired.

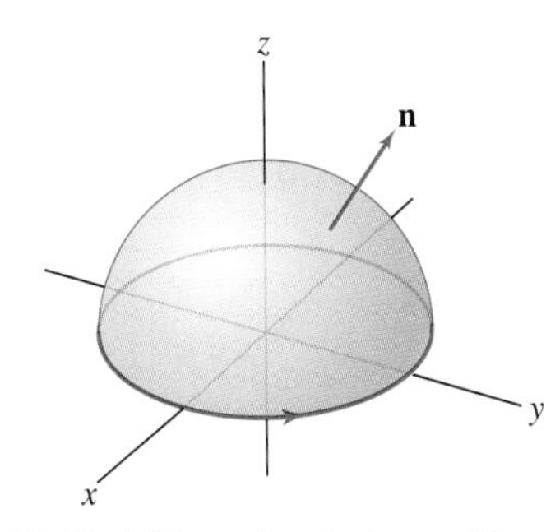

FIGURE 4 Upper hemisphere with oriented boundary.

EXAMPLE 2 Verifying Stokes' Theorem Verify Stokes' Theorem for

$$\mathbf{F} = \langle -y, 2x, x + z\rangle$$

and the upper hemisphere:

$$\mathcal{S} = \{(x, y, z) : x^2 + y^2 + z^2 = 1, z \ge 0\}$$

oriented by outward-pointing normal vectors (Figure 4).

Solution We compute both sides of Eq. (1) in Stokes' Theorem.

***Step 1.* Compute the line integral around the boundary curve.**
The boundary of $\mathcal{S}$ is the unit circle oriented in the counterclockwise direction with parametrization $\mathbf{c}(t) = (\cos t, \sin t, 0)$. Thus,

$$\mathbf{c}'(t) = \langle -\sin t, \cos t, 0\rangle$$

$$\mathbf{F}(\mathbf{c}(t)) = \langle -\sin t, 2\cos t, \cos t\rangle$$

$$\mathbf{F}(\mathbf{c}(t))\cdot\mathbf{c}'(t) = \langle -\sin t, 2\cos t, \cos t\rangle\cdot\langle -\sin t, \cos t, 0\rangle$$

$$= \sin^2 t + 2\cos^2 t = 1 + \cos^2 t$$

$$\oint_{\partial\mathcal{S}} \mathbf{F}\cdot d\mathbf{s} = \int_0^{2\pi} (1 + \cos^2 t)\,dt = 2\pi + \pi = \boxed{3\pi} \qquad \boxed{5}$$

← REMINDER *In Eq. (5), we use*

$$\int_0^{2\pi} \cos^2 t\,dt = \int_0^{2\pi} \frac{1 + \cos 2t}{2}\,dt = \pi$$

***Step 2.* Compute the curl.**

$$\begin{aligned}
\operatorname{curl}(\mathbf{F}) &= \begin{vmatrix} \mathbf{i} & \mathbf{j} & \mathbf{k} \\ \dfrac{\partial}{\partial x} & \dfrac{\partial}{\partial y} & \dfrac{\partial}{\partial z} \\ -y & 2x & x+z \end{vmatrix} \\
&= \left(\frac{\partial}{\partial y}(x+z) - \frac{\partial}{\partial z}2x\right)\mathbf{i} - \left(\frac{\partial}{\partial x}(x+z) - \frac{\partial}{\partial z}(-y)\right)\mathbf{j} \\
&\quad + \left(\frac{\partial}{\partial x}2x - \frac{\partial}{\partial y}(-y)\right)\mathbf{k} \\
&= \langle 0, -1, 3\rangle
\end{aligned}$$

***Step 3.* Compute the flux of the curl through the surface.**

We parametrize the hemisphere using spherical coordinates:

$$\Phi(\phi, \theta) = (\cos\theta\sin\phi, \sin\theta\sin\phi, \cos\phi)$$

As we have seen in previous examples [Eq. (5) in Section 16.4], the outward-pointing normal vector at the point $\Phi(\phi, \theta)$ is

$$\mathbf{n} = \sin\phi\,\langle\cos\theta\sin\phi, \sin\theta\sin\phi, \cos\phi\rangle$$

We have

$$\begin{aligned}
\operatorname{curl}(\mathbf{F})\cdot\mathbf{n} &= \sin\phi\,\langle 0, -1, 3\rangle\cdot\langle\cos\theta\sin\phi, \sin\theta\sin\phi, \cos\phi\rangle \\
&= -\sin\theta\sin^2\phi + 3\cos\phi\sin\phi
\end{aligned}$$

The upper hemisphere $\mathcal{S}$ corresponds to $0 \le \phi \le \frac{\pi}{2}$, so the flux of curl(**F**) through $\mathcal{S}$ is equal to

$$\begin{aligned}
\iint_{\mathcal{S}} \operatorname{curl}(\mathbf{F})\cdot d\mathbf{S} &= \int_{\phi=0}^{\pi/2}\int_{\theta=0}^{2\pi} (-\sin\theta\sin^2\phi + 3\cos\phi\sin\phi)\,d\theta\,d\phi \\
&= 0 + 2\pi\int_{\phi=0}^{\pi/2} 3\cos\phi\sin\phi\,d\phi = 2\pi\left(\frac{3}{2}\sin^2\phi\right)\Big|_{\phi=0}^{\pi/2} \\
&= \boxed{3\pi}
\end{aligned}$$

***Step 4.* Compare the answers.**

Both the line integral and flux of the curl are equal to 3π. Thus Stokes' Theorem is verified in this case. ■

In computing the curl of a vector field $\mathbf{F} = \langle F_1, F_2, F_3\rangle$, it is helpful to observe that the partial derivatives $\dfrac{\partial F_1}{\partial y}$ and $\dfrac{\partial F_1}{\partial z}$ appear in curl(**F**) but $\dfrac{\partial F_1}{\partial x}$ does not. Thus, if $F_1 = F_1(x)$ is a function of x alone, then $\dfrac{\partial F_1}{\partial y} = \dfrac{\partial F_1}{\partial z} = 0$ and F_1 does not contribute to the curl. The same holds for the other components. In other words, if F_1, F_2, and F_3 depend only on their corresponding variables x, y, and z, then

$$\operatorname{curl}(\langle F_1(x), F_2(y), F_3(z)\rangle) = 0$$

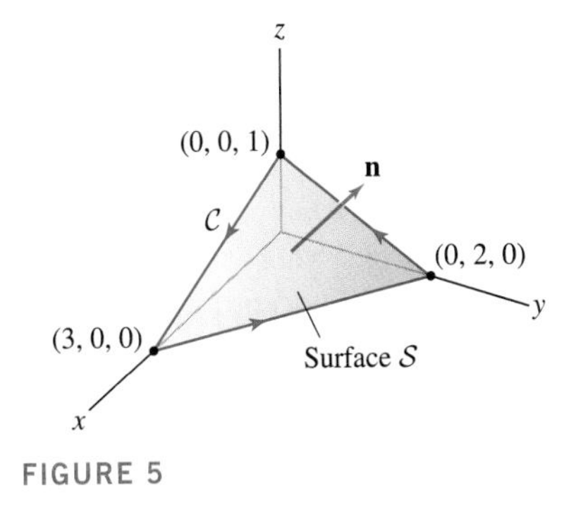

FIGURE 5

EXAMPLE 3 Using Stokes' Theorem to Compute a Line Integral Use Stokes' Theorem to compute $I = \oint_C \mathbf{F} \cdot d\mathbf{s}$, where

$$\mathbf{F} = \langle \sin x^2, e^{y^2} + x^2, z^4 + 2x^2 \rangle$$

and C is the boundary of the triangle in Figure 5 with the indicated orientation.

Solution We apply Stokes' Theorem to the triangular surface $\mathcal{S}$ in Figure 5. The plane containing $\mathcal{S}$ has equation

$$\frac{x}{3} + \frac{y}{2} + z = 1$$

This plane is the graph of $z = f(x, y) = 1 - \frac{x}{3} - \frac{y}{2}$ and $\mathcal{S}$ is the part of the graph lying over a triangle $\mathcal{D}$ in the xy-plane. It is not hard to describe $\mathcal{D}$ explicitly, but we do not need this to solve our problem. The normal vector to $\mathcal{S}$ in the parametrization $\Phi(x, y) = (x, y, f(x, y))$ is

$$\mathbf{n} = \langle -f_x, -f_y, 1 \rangle = \left\langle \frac{1}{3}, \frac{1}{2}, 1 \right\rangle$$

To simplify the computation of curl($\mathbf{F}$), we use the remark in the preceding paragraph that told us the terms $\sin x^2$, e^{y^2}, and z^4 may be dropped:

$$\operatorname{curl}\left(\langle \sin x^2, e^{y^2} + x^2, z^4 + 2x^2 \rangle\right)$$

$$= \overbrace{\operatorname{curl}\left(\langle \sin x^2, e^{y^2}, z^4 \rangle\right)}^{\text{Automatically zero}} + \operatorname{curl}\left(\langle 0, x^2, 2x^2 \rangle\right) = \operatorname{curl}\left(\langle 0, x^2, 2x^2 \rangle\right)$$

It is then straightforward to compute

$$\operatorname{curl}\left(\langle 0, x^2, 2x^2 \rangle\right) = \left\langle 0, -\frac{\partial}{\partial x} 2x^2, \frac{\partial}{\partial x} x^2 \right\rangle = \langle 0, -4x, 2x \rangle$$

Now we observe that curl($\mathbf{F}$) is orthogonal to $\mathbf{n}$:

$$\operatorname{curl}(\mathbf{F}) \cdot \mathbf{n} = \langle 0, -4x, 2x \rangle \cdot \left\langle \frac{1}{3}, \frac{1}{2}, 1 \right\rangle = 0$$

Therefore, by Stokes' Theorem,

$$I = \oint_C \mathbf{F} \cdot d\mathbf{s} = \iint_{\mathcal{S}} \operatorname{curl}(\mathbf{F}) \cdot d\mathbf{S} = \iint_{\mathcal{D}} \overbrace{\operatorname{curl}(\mathbf{F}) \cdot \mathbf{n}}^{\text{Equal to 0}} \, dA = 0$$ ■

CONCEPTUAL INSIGHT There is an interesting and important analogy between Stokes' Theorem and the Fundamental Theorem for Gradient Vector Fields stated in Section 16.3.

- If $\mathbf{F} = \nabla \varphi$, then the Fundamental Theorem tells us that

$$\int_{C_1} \mathbf{F} \cdot d\mathbf{s} = \int_{C_2} \mathbf{F} \cdot d\mathbf{s} = \varphi(Q) - \varphi(P)$$

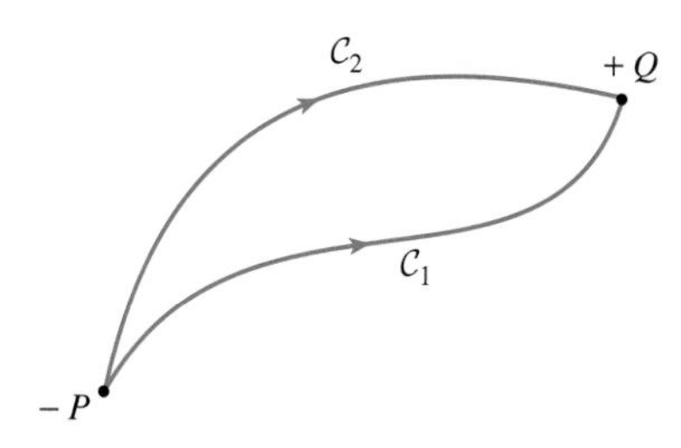

FIGURE 6 Two paths with the same boundary $Q - P$.

for any two paths C_1 and C_2 from P to Q (Figure 6). As we have noted, this shows that the line integral of a gradient vector field is independent of the path and depends only on the *oriented boundary* of the path (think of the oriented boundary as the difference of the endpoints $Q - P$).

- For surface integrals, instead of asking whether $\mathbf{F} = \nabla\varphi$ for some potential function φ, we ask if $\mathbf{F} = \text{curl}(\mathbf{A})$ for some vector field $\mathbf{A}$. The vector field, when it exists, is often called a **vector potential** for $\mathbf{F}$. If $\mathbf{F}$ has a vector potential $\mathbf{A}$, then according to Stokes' Theorem applied to $\mathbf{A}$,

$$\iint_{\mathcal{S}} \mathbf{F} \cdot d\mathbf{S} = \oint_{\partial\mathcal{S}} \mathbf{A} \cdot d\mathbf{s}$$

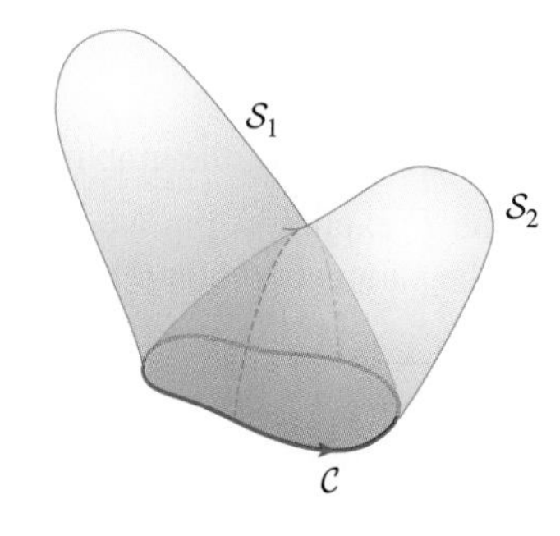

FIGURE 7 Surfaces $\mathcal{S}_1$ and $\mathcal{S}_2$ have the same oriented boundary.

The right-hand side depends only on the oriented boundary $\partial\mathcal{S}$, not on $\mathcal{S}$ itself. In other words, if $\mathcal{S}_1$ and $\mathcal{S}_2$ are any two surfaces with the same oriented boundary curve $\mathcal{C}$ (Figure 7), then

$$\iint_{\mathcal{S}_1} \mathbf{F} \cdot d\mathbf{S} = \iint_{\mathcal{S}_2} \mathbf{F} \cdot d\mathbf{S} \qquad \text{if} \quad \mathbf{F} = \text{curl}(\mathbf{A})$$

since both surface integrals are equal to $\oint_{\mathcal{C}} \mathbf{A} \cdot d\mathbf{s}$. This is the analog of path independence for gradient vector fields. In particular, if $\mathcal{S}$ is closed, then the boundary curve $\partial\mathcal{S}$ is empty. The circulation of $\mathbf{A}$ around the empty curve is zero, so

$$\iint_{\mathcal{S}} \mathbf{F} \cdot d\mathbf{S} = 0 \qquad \text{if} \quad \mathbf{F} = \text{curl}(\mathbf{A}) \text{ and } \mathcal{S} \text{ is closed}$$

A vector potential of a vector field $\mathbf{F}$ *is a kind of "antiderivative," and, like all antiderivatives, it is not unique. If* $\mathbf{F} = \text{curl}(\mathbf{A})$*, then* $\mathbf{F} = \text{curl}(\mathbf{A} + \mathbf{B})$ *for any vector field* $\mathbf{B}$ *such that* $\text{curl}(\mathbf{B}) = \mathbf{0}$.

THEOREM 2 Surface Independence for Curl Vector Fields If $\mathbf{F} = \text{curl}(\mathbf{A})$, then the flux of $\mathbf{F}$ through a surface $\mathcal{S}$ depends only on the oriented boundary $\partial\mathcal{S}$ and not the surface itself:

$$\iint_{\mathcal{S}} \mathbf{F} \cdot d\mathbf{S} = \oint_{\partial\mathcal{S}} \mathbf{A} \cdot d\mathbf{s} \quad \text{if } \mathbf{F} = \text{curl}(\mathbf{A})$$

In particular, if $\mathcal{S}$ is *closed* (i.e., $\partial\mathcal{S}$ is empty), then

$$\iint_{\mathcal{S}} \mathbf{F} \cdot d\mathbf{S} = 0$$

EXAMPLE 4 Let $\mathbf{F} = \text{curl}(\mathbf{A})$, where $\mathbf{A} = \langle y + z, \sin(xy), e^{xyz} \rangle$. Determine the flux of $\mathbf{F}$ through the surfaces S_1 and S_2 whose common boundary $\mathcal{C}$ is the unit circle in the xz-plane as in Figure 8.

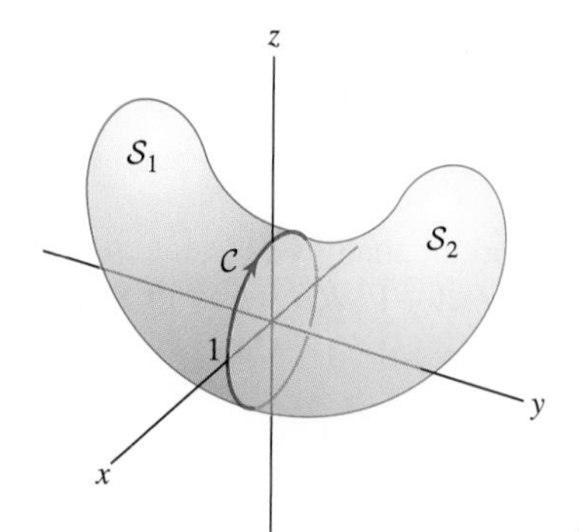

FIGURE 8

Solution When you traverse the circle $\mathcal{C}$ in the direction of the arrow, the surface S_1 lies to the left. Therefore, $\mathcal{C}$ is oriented as the boundary of S_1, and by Stokes' Theorem,

$$\oint_{\mathcal{C}} \mathbf{A} \cdot d\mathbf{s} = \iint_{S_1} \mathbf{F} \cdot d\mathbf{S}$$

The parametrization $\mathbf{c}(t) = (\cos t, 0, \sin t)$ traces $\mathcal{C}$ in the direction of the arrow because it begins at $\mathbf{c}(0) = (1, 0, 0)$ and moves in the direction of $\mathbf{c}(\frac{\pi}{2}) = (0, 0, 1)$. We have

REMINDER We use

$$\int_0^{2\pi} \sin^2 t\, dt = \int_0^{2\pi} \frac{1-\cos 2t}{2}\, dt = \pi$$

$$\int_0^{2\pi} \cos t\, dt = 0$$

$$\mathbf{A}(\mathbf{c}(t)) = \langle 0 + \sin t, \sin(0), e^0 \rangle = \langle \sin t, 0, 1 \rangle$$

$$\mathbf{A}(\mathbf{c}(t)) \cdot \mathbf{c}'(t) = \langle \sin t, 0, 1 \rangle \cdot \langle -\sin t, 0, \cos t \rangle = -\sin^2 t + \cos t$$

$$\oint_{\mathcal{C}} \mathbf{A} \cdot d\mathbf{s} = \int_0^{2\pi} (-\sin^2 t + \cos t)\, dt = -\pi$$

We conclude that $\iint_{\mathcal{S}_1} \mathbf{F} \cdot d\mathbf{S} = -\pi$. On the other hand, $\mathcal{S}_2$ lies on the right as you traverse $\mathcal{C}$ and therefore $-\mathcal{C}$ is oriented as the boundary of $\mathcal{S}_2$. We obtain

$$\iint_{\mathcal{S}_2} \mathbf{F} \cdot d\mathbf{S} = \oint_{-\mathcal{C}} \mathbf{A} \cdot d\mathbf{s} = -\oint_{\mathcal{C}} \mathbf{A} \cdot d\mathbf{s} = \pi \quad \blacksquare$$

CONCEPTUAL INSIGHT In Section 17.1, we interpreted the scalar curl as a measure of the "tendency to rotate." We saw that if $\mathbf{F}$ is the velocity field of a fluid, then the curl of $\mathbf{F}$ at a point P is measured by placing a small paddle wheel at P and determining how fast it rotates. Can we extend this interpretation to three dimensions? Notice an important difference between two and three dimensions. In the plane, there is only one way to place the paddle wheel in the fluid. In three dimensions, we can orient the paddle wheel at P in any direction and the speed of rotation depends on how the wheel is oriented [Figure 9(A)]. Let us show that the vector curl($\mathbf{F}$) simultaneously encodes the tendency to rotate in all possible directions.

Let $\mathcal{C}$ be the boundary circle of the paddle wheel and $\mathcal{R}$ the enclosed region [Figure 9(B)]. If $\mathcal{C}$ is sufficiently small and curl($\mathbf{F}$) is continuous, then curl($\mathbf{F}$) is nearly constant on $\mathcal{R}$ with value $\mathbf{v}_P = \text{curl}(\mathbf{F})(P)$. If $\mathbf{e_n}$ is a unit normal vector to $\mathcal{R}$ at P, then Stokes' Theorem yields the approximation

$$\oint_{\mathcal{C}} \mathbf{F} \cdot d\mathbf{s} = \iint_{\mathcal{R}} \text{curl}(\mathbf{F}) \cdot d\mathbf{S} \approx \iint_{\mathcal{R}} \mathbf{v}_P \cdot d\mathbf{S} = (\mathbf{v}_P \cdot \mathbf{e_n})\, \text{Area}(\mathcal{R})$$

This approximation holds for any small region $\mathcal{R}$ with boundary $\mathcal{C}$. Our result may be expressed in terms of the angle θ between $\mathbf{v}_P$ and $\mathbf{e_n}$:

$$\int_{\mathcal{C}} \mathbf{F} \cdot d\mathbf{s} \approx (\mathbf{v}_P \cdot \mathbf{e_n})\, \text{Area}(\mathcal{R}) = \|\mathbf{v}_P\|\, \text{Area}(\mathcal{R}) \cos\theta \qquad \boxed{6}$$

This is a remarkable result. If we fix the size of the circle $\mathcal{C}$ but change its orientation in space, then the circulation around $\mathcal{C}$ varies, but (to a first-order approximation) it varies in a simple way, namely as the cosine of the angle θ. In particular, the maximum circulation occurs when $\mathbf{e_n}$ and $\mathbf{v}_P$ point in the same direction, while the circulation is zero (to a first-order approximation) when $\mathbf{e_n}$ is perpendicular to $\mathbf{v}_P$.

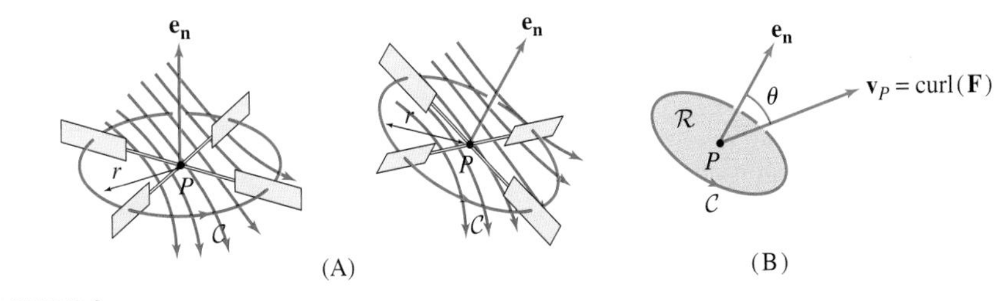

FIGURE 9

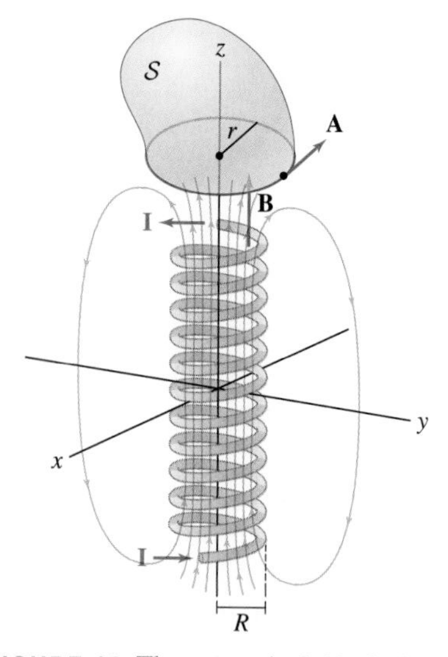

FIGURE 10 The magnetic field of a long solenoid is nearly uniform inside and weak outside. In practice, we treat the solenoid as "infinitely long" if it is very long in comparison with its radius.

EXAMPLE 5 Vector Potential for a Solenoid A magnetic field $\mathbf{B}$ is created when a current flows through a solenoid (a tightly wound spiral of wire; see Figure 10). If we assume that the solenoid is infinitely long, with radius R and a central z-axis, then $\mathbf{B}$ is zero outside the solenoid and has value $B\mathbf{k}$ inside the solenoid, where B is a constant that depends on the current strength and spacing of the turns of wire.

(a) Show that a vector potential for $\mathbf{B}$ is

$$\mathbf{A} = \begin{cases} \frac{1}{2}R^2B\left\langle -\frac{y}{r^2}, \frac{x}{r^2}, 0\right\rangle & \text{if } r > R \\ \frac{1}{2}B\,\langle -y, x, 0\rangle & \text{if } r < R \end{cases}$$

where $r = (x^2 + y^2)^{1/2}$ is the distance to the z-axis.

(b) Calculate the flux of $\mathbf{B}$ through the surface $\mathcal{S}$ in Figure 10 whose boundary is a circle of radius $r > R$.

Solution

(a) We must verify that $\mathbf{B} = \text{curl}(\mathbf{A})$. To simplify the calculation, note that for a vector field whose z-component is zero, the curl reduces to

$$\text{curl}(\langle f, g, 0\rangle) = \langle -g_z, f_z, g_x - f_y\rangle$$

In our case, for $r > R$, we have

$$\text{curl}(\mathbf{A}) = \frac{1}{2}R^2B\left\langle 0, 0, \frac{\partial}{\partial x}xr^{-2} - \frac{\partial}{\partial y}(-yr^{-2})\right\rangle$$

However, the z-component is also zero:

$$\begin{aligned} \frac{\partial}{\partial x}xr^{-2} + \frac{\partial}{\partial y}yr^{-2} &= \frac{\partial}{\partial x}\left(\frac{x}{x^2+y^2}\right) + \frac{\partial}{\partial y}\left(\frac{y}{x^2+y^2}\right) \\ &= \frac{(x^2+y^2) - x(2x)}{(x^2+y^2)^2} + \frac{(x^2+y^2) - y(2y)}{(x^2+y^2)^2} \\ &= 0 \end{aligned}$$

The vector potential $\mathbf{A}$ is not differentiable on the cylinder $r = R$, that is, on the solenoid itself (Figure 11). The magnetic field $\mathbf{B} = \text{curl}(\mathbf{A})$ has a jump discontinuity where $r = R$. We take for granted the fact that Stokes' Theorem remains valid in this setting.

Thus, $\text{curl}(\mathbf{A}) = \mathbf{0}$ when $r > R$ as required. For $r < R$, we have

$$\text{curl}(\mathbf{A}) = \frac{1}{2}B\left\langle 0, 0, \frac{\partial}{\partial x}x - \frac{\partial}{\partial y}(-y)\right\rangle = \langle 0, 0, B\rangle = B\mathbf{k}$$

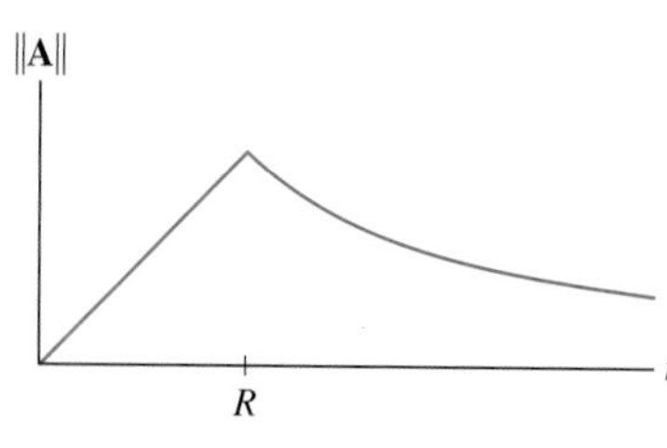

FIGURE 11 The length $\|\mathbf{A}\|$ of the vector potential as a function of distance r to the z-axis.

(b) By Stokes' Theorem, the flux of $\mathbf{B}$ through $\mathcal{S}$ is equal to the circulation of $\mathbf{A}$ through the boundary $\partial\mathcal{S}$, which is a horizontal circle of radius r oriented counterclockwise. We parametrize the circle by $\mathbf{c}(t) = (r\cos t, r\sin t, 0)$. Then

$$\mathbf{c}'(t) = \langle -r\sin t, r\cos t, 0\rangle$$

$$\mathbf{A}(\mathbf{c}(t)) = \frac{1}{2}R^2Br^{-1}\,\langle -\sin t, \cos t, 0\rangle$$

$$\mathbf{A}(\mathbf{c}(t)) \cdot \mathbf{c}'(t) = \frac{1}{2}R^2B\left((-\sin t)^2 + \cos^2 t\right) = \frac{1}{2}R^2B$$

Therefore, the flux of **B** through $\mathcal{S}$ is equal to

$$\iint_{\mathcal{S}} \mathbf{B} \cdot d\mathbf{S} = \oint_{\partial\mathcal{S}} \mathbf{A} \cdot d\mathbf{s} = \int_0^{2\pi} \mathbf{A}(\mathbf{c}(t)) \cdot \mathbf{c}'(t)\, dt$$

$$= \frac{1}{2} R^2 B \int_0^{2\pi} dt = \pi R^2 B \qquad \blacksquare$$

CONCEPTUAL INSIGHT There is an interesting difference between scalar and vector potentials. If $\mathbf{F} = \nabla\varphi$, then the scalar potential φ is constant in regions where the field **F** is zero (since a function with zero gradient is constant). This is not true for vector potentials. As we saw in Example 4, the magnetic field **B** produced by a solenoid is zero everywhere outside the solenoid, but the vector potential **A** is not constant outside the solenoid. In fact, **A** is proportional to $\left\langle -\frac{y}{r^2}, \frac{x}{r^2}, 0 \right\rangle$. This is related to an intriguing phenomenon in physics called the *Aharonov-Bohm (AB) effect*, first proposed on theoretical grounds in the 1940s.

According to electromagnetic theory, a magnetic field **B** exerts a force on a moving electron, causing a deflection in the electron's path. We do not expect any deflection when an electron moves past a solenoid because **B** is zero outside the solenoid (in practice, the field is not actually zero but it is very small—we ignore this difficulty). However, according to quantum mechanics, electrons have both particle and wave properties. In a double slit experiment, a stream of electrons passing through two small slits creates a wavelike interference pattern on a detection screen (Figure 12). The AB effect predicts that if we place a small solenoid between the slits as in the figure (the solenoid is so small that the electrons never pass through it), then the interference pattern will shift slightly. It is as if the elections are "aware" of the magnetic field inside the solenoid even though they never encounter the field directly.

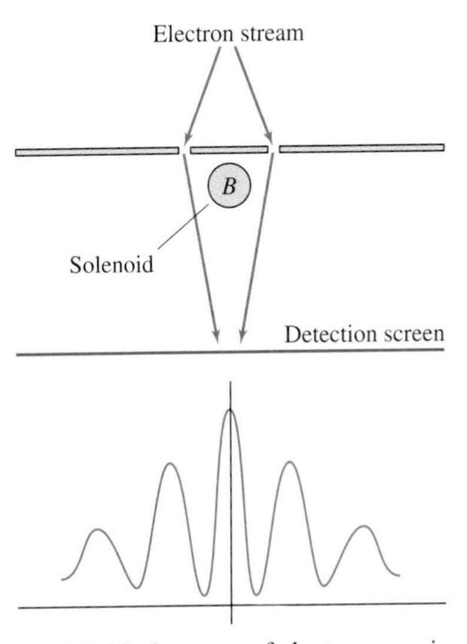

FIGURE 12 A stream of electrons passing through a double slit produces an interference pattern on the detecting screen. A small solenoid is placed between the slits.

The existence of the AB effect was hotly debated until it was confirmed definitively in 1985, in experiments carried out by a team of Japanese physicists led by Akira Tonomura. It appeared to contradict "classical" electromagnetic theory, where the trajectory of an electron is determined by **B** alone. There is no such contradiction in quantum mechanics, because the behavior of the electrons is governed not by **B** but by a "wave function" that is derived from the nonconstant vector potential **A**.

17.2 SUMMARY

- The *boundary* of a surface $\mathcal{S}$ is denoted $\partial\mathcal{S}$. We say that $\mathcal{S}$ is *closed* if $\partial\mathcal{S}$ is empty.
- Suppose that $\mathcal{S}$ is oriented (a continuously varying unit normal is specified at each point of $\mathcal{S}$). We define the *boundary orientation* of $\partial\mathcal{S}$ by the following condition: If you walk along the boundary in the positive direction with your head pointing in the normal direction, then the surface is on your left.
- Stokes' Theorem relates the circulation around the boundary to the surface integral of the curl:

$$\boxed{\oint_{\partial\mathcal{S}} \mathbf{F} \cdot d\mathbf{s} = \iint_{\mathcal{S}} \operatorname{curl}(\mathbf{F}) \cdot d\mathbf{S}}$$

- If $\mathcal{R}$ is a small region with boundary $\mathcal{C}$ and P is a sample point in $\mathcal{R}$ with normal vector $\mathbf{e_n}$ at P, then

$$\int_{\mathcal{C}} \mathbf{F} \cdot d\mathbf{s} \approx (\text{curl}(\mathbf{F})(P) \cdot \mathbf{e_n})\,\text{Area}(\mathcal{R})$$

17.2 EXERCISES

Preliminary Questions

1. Indicate with an arrow the boundary orientation of the boundary curves of the surfaces in Figure 13, oriented by the outward-pointing normal vectors.

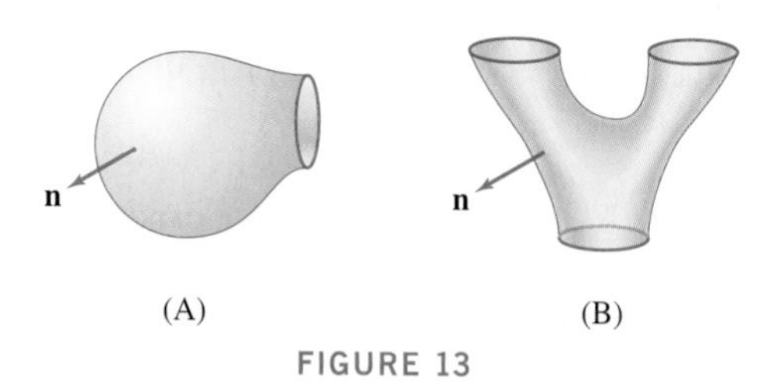

FIGURE 13

2. Let $\mathbf{F} = \text{curl}(\mathbf{A})$. Which of the following are related by Stokes' Theorem?

(a) The circulation of $\mathbf{A}$ and flux of $\mathbf{F}$.

(b) The circulation of $\mathbf{F}$ and flux of $\mathbf{A}$.

3. What is the definition of a vector potential?

4. Which of the following statements is correct?

(a) The flux of $\text{curl}(\mathbf{F})$ through every oriented surface is zero.

(b) The flux of $\text{curl}(\mathbf{F})$ through every closed, oriented surface is zero.

5. Which condition on $\mathbf{F}$ guarantees that the flux through $\mathcal{S}_1$ is equal to the flux through $\mathcal{S}_2$ for any two oriented surfaces $\mathcal{S}_1$ and $\mathcal{S}_2$ with the same oriented boundary?

Exercises

In Exercises 1–4, calculate $\text{curl}(\mathbf{F})$.

1. $\mathbf{F} = \langle z - y^2, x + z^3, y + x^2 \rangle$

2. $\mathbf{F} = \left\langle \frac{y}{x}, \frac{y}{z}, \frac{z}{x} \right\rangle$

3. $\mathbf{F} = \langle e^y, \sin x, \cos x \rangle$

4. $\mathbf{F} = \left\langle \frac{x}{x^2 + y^2}, \frac{y}{x^2 + y^2}, 0 \right\rangle$

In Exercises 5–8, verify Stokes' Theorem for the given vector field and surface, oriented with an upward-pointing normal.

5. $\mathbf{F} = \langle 2xy, x, y + z \rangle$, the surface $z = 1 - x^2 - y^2$ for $x^2 + y^2 \le 1$

6. $\mathbf{F} = \langle yz, 0, x \rangle$, the portion of the plane $\frac{x}{2} + \frac{y}{3} + z = 1$ where $x, y, z \ge 0$

7. $\mathbf{F} = \langle e^{y-z}, 0, 0 \rangle$, the square with vertices $(1, 0, 1)$, $(1, 1, 1)$, $(0, 1, 1)$, and $(0, 0, 1)$

8. $\mathbf{F} = \langle -y, 2x, x + z \rangle$, the upper hemisphere $x^2 + y^2 + z^2 = 1$, $z \ge 0$

In Exercises 9–10, use Stokes' Theorem to compute the flux of $\text{curl}(\mathbf{F})$ *through the given surface.*

9. $\mathbf{F} = \langle z, y, x \rangle$, the hemisphere $x^2 + y^2 + z^2 = 1$, $x \ge 0$

10. $\mathbf{F} = \langle x^2 + y^2, x + z^2, 0 \rangle$, the part of the cone $z^2 = x^2 + y^2$ such that $2 \le z \le 4$

11. Let $\mathcal{S}$ be the surface of the cylinder (not including the top and bottom) of radius 2 for $1 \le z \le 6$, oriented with outward-pointing normal (Figure 14).

(a) Indicate with an arrow the orientation of $\partial\mathcal{S}$ (the top and bottom circles).

(b) Verify Stokes' Theorem for $\mathcal{S}$ and $\mathbf{F} = \langle yz^2, 0, 0 \rangle$.

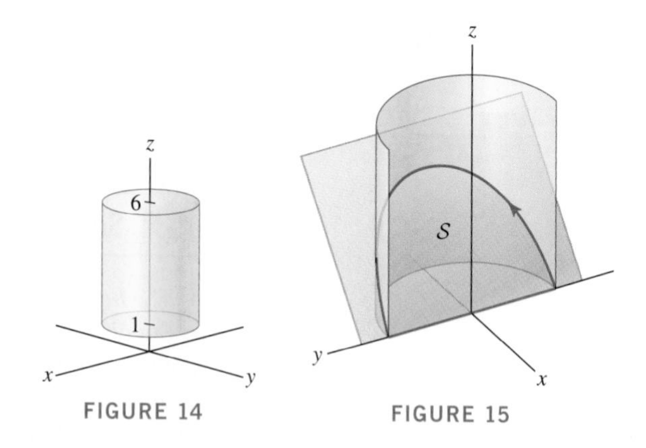

FIGURE 14 FIGURE 15

12. Let $\mathcal{S}$ be the portion of the plane $z = x$ contained in the half-cylinder of radius R depicted in Figure 15. Use Stokes' Theorem to calculate the circulation of $\mathbf{F} = \langle z, x, y + 2z \rangle$ around the boundary of $\mathcal{S}$ (a half-ellipse) in the counterclockwise direction when viewed from above. *Hint:* Show that $\text{curl}(\mathbf{F})$ is orthogonal to the normal vector to the plane.

13. Let I be the flux of $\mathbf{F} = \langle e^y, 2xe^{x^2}, z^2 \rangle$ through the upper hemisphere $\mathcal{S}$ of the unit sphere.

(a) Let $\mathbf{G} = \langle e^y, 2xe^{x^2}, 0 \rangle$. Find a vector field $\mathbf{A}$ such that $\text{curl}(\mathbf{A}) = \mathbf{G}$.

(b) Use Stokes' Theorem to show that the flux of $\mathbf{G}$ through $\mathcal{S}$ is zero. *Hint:* Calculate the circulation of $\mathbf{A}$ around $\partial\mathcal{S}$.

(c) Calculate I. *Hint:* Use (b) to show that I is equal to the flux of $\langle 0, 0, z^2 \rangle$ through $\mathcal{S}$.

14. Let $\mathbf{F} = \langle 0, -z, 1 \rangle$ and let $\mathcal{S}$ be the spherical cap $x^2 + y^2 + z^2 \leq 1$, where $z \geq \frac{1}{2}$. Evaluate $\iint_{\mathcal{S}} \mathbf{F} \cdot d\mathbf{S}$ directly as a surface integral. Then verify that $\mathbf{F} = \text{curl}(\mathbf{A})$, where $\mathbf{A} = \langle 0, x, xz \rangle$ and evaluate the surface integral again using Stokes' Theorem.

15. Let $\mathbf{A}$ be the vector potential and $\mathbf{B}$ the magnetic field of the infinite solenoid of radius R in Example 5. Use Stokes' Theorem to compute:

(a) The flux of $\mathbf{B}$ through a circle in the xy-plane of radius $r < R$

(b) The circulation of $\mathbf{A}$ around the boundary $\mathcal{C}$ of a surface lying outside the solenoid

16. The magnetic field $\mathbf{B}$ due to a small current loop (which we place at the origin) is called a magnetic dipole (Figure 16). Let $\rho = (x^2 + y^2 + z^2)^{1/2}$. For ρ large, $\mathbf{B} = \text{curl}(\mathbf{A})$, where

$$\mathbf{A} = \left\langle -\frac{y}{\rho^3}, \frac{x}{\rho^3}, 0 \right\rangle$$

(a) Let $\mathcal{C}$ be a horizontal circle of radius R located far from the origin with center on z-axis. Show that $\mathbf{A}$ is tangent to $\mathcal{C}$.

(b) Use Stokes' Theorem to calculate the flux of $\mathbf{B}$ through $\mathcal{C}$.

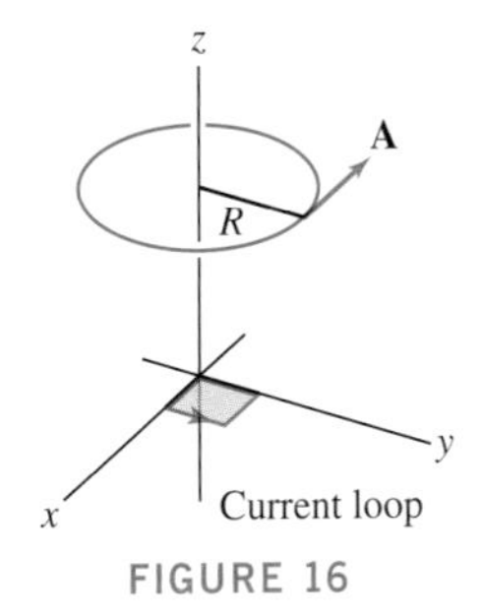

FIGURE 16

17. A uniform magnetic field $\mathbf{B}$ has constant strength b teslas in the z-direction [i.e., $\mathbf{B} = \langle 0, 0, b \rangle$].

(a) Verify that $\mathbf{A} = \frac{1}{2}\mathbf{B} \times \mathbf{r}$ is a vector potential for $\mathbf{B}$, where $\mathbf{r} = \langle x, y, 0 \rangle$.

(b) Calculate the flux of $\mathbf{B}$ through the rectangle with vertices A, B, C, and D in Figure 17.

18. Let $\mathbf{F} = \langle -x^2y, x, 0 \rangle$. Referring to Figure 17, let $\mathcal{C}$ be the closed path $ABCD$. Use Stokes' Theorem to evaluate $\int_{\mathcal{C}} \mathbf{F} \cdot d\mathbf{s}$ in two ways. First, regard $\mathcal{C}$ as the boundary of the rectangle with vertices A, B, C, and D. Then treat $\mathcal{C}$ as the boundary of the wedge-shaped box with open top.

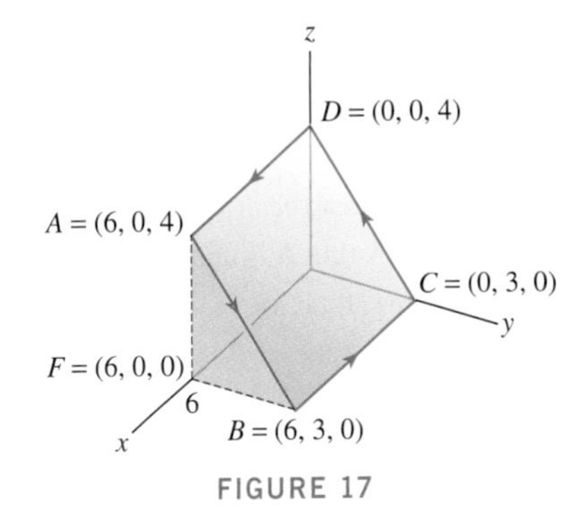

FIGURE 17

19. Let $\mathbf{F} = \langle -z^2, 2zx, 4y - x^2 \rangle$ and let $\mathcal{C}$ be a simple closed curve in the plane $x + y + z = 4$ that encloses a region of area 16 (Figure 18). Calculate $\oint_{\mathcal{C}} \mathbf{F} \cdot d\mathbf{s}$, where $\mathcal{C}$ is oriented in the counterclockwise direction (when viewed from above the plane).

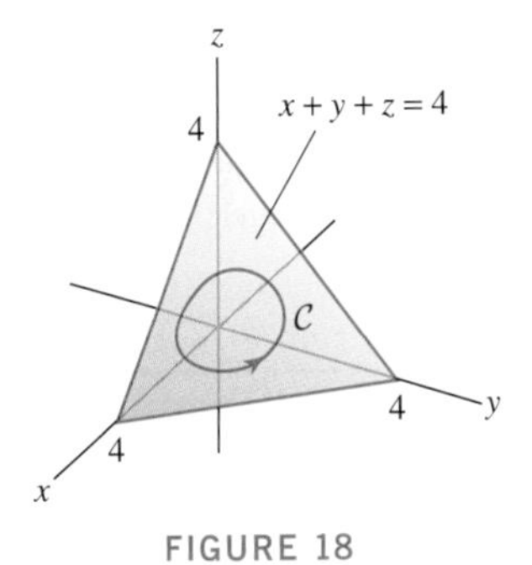

FIGURE 18

20. Let $\mathbf{F} = \langle y^2, x^2, z^2 \rangle$. Show that

$$\int_{\mathcal{C}_1} \mathbf{F} \cdot d\mathbf{s} = \int_{\mathcal{C}_2} \mathbf{F} \cdot d\mathbf{s}$$

for any two closed curves lying on a cylinder whose central axis is the z-axis (Figure 19).

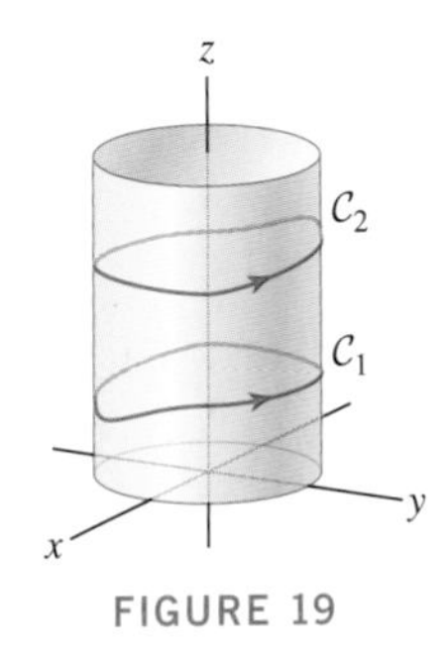

FIGURE 19

21. Let $\mathcal{C}$ be the triangular boundary of the portion of the plane $\frac{x}{a} + \frac{y}{b} + \frac{z}{c} = 1$ lying in the octant $x, y, z \geq 0$. Use Stokes'

Theorem to find positive constants a, b, c such that the line integral of $\mathbf{F} = \langle y^2, 2z + x, 2y^2 \rangle$ around $\mathcal{C}$ is zero. *Hint:* Choose constants so that $\text{curl}(\mathbf{F})$ is orthogonal to the normal vector.

22. The curl of a vector field $\mathbf{F}$ at the origin is $\mathbf{v}_0 = \langle 3, 1, 4 \rangle$. Estimate the circulation around the small parallelogram spanned by the vectors $\mathbf{A} = \langle 0, \frac{1}{2}, \frac{1}{2} \rangle$ and $\mathbf{B} = \langle 0, 0, \frac{1}{3} \rangle$.

23. You know two things about a vector field $\mathbf{F}$:

(a) $\mathbf{F}$ has a vector potential $\mathbf{A}$ (but $\mathbf{A}$ is unknown).

(b) $\mathbf{F}(x, y, 0) = \langle 0, 0, 1 \rangle$ for all (x, y).

Determine the flux of $\mathbf{F}$ through the surface $\mathcal{S}$ in Figure 20.

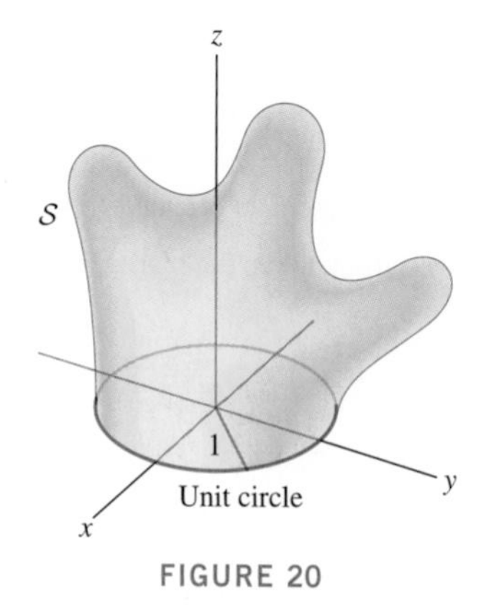

FIGURE 20

24. Find the flux of $\mathbf{F}$ through the surface $\mathcal{S}$ in Figure 20, assuming that $\mathbf{F}$ has a vector potential and $\mathbf{F}(x, y, 0) = \langle \cos x, 0, 0 \rangle$.

25. Use Eq. (8) to prove that if $\mathbf{a}$ is a constant vector, then $\text{curl}(\varphi\mathbf{a}) = \nabla\varphi \times \mathbf{a}$.

26. A vector field $\mathbf{F}$ is called **radial** if it is of the form $\mathbf{F} = \varphi(\rho)\langle x, y, z \rangle$ for some function $\varphi(\rho)$, where $\rho = \sqrt{x^2 + y^2 + z^2}$. Show that the curl of a radial vector field is zero. *Hint:* It is enough to show that one component of the curl is zero, since the calculation for the other two components is similar by symmetry.

27. Verify the identity

$$\text{curl}(\nabla(\varphi)) = \mathbf{0} \qquad \boxed{7}$$

28. Prove the Product Rule

$$\text{curl}(\varphi\mathbf{F}) = \varphi\,\text{curl}(\mathbf{F}) + \nabla\varphi \times \mathbf{F} \qquad \boxed{8}$$

29. Assume that the second partial derivatives of φ and ψ exist and are continuous. Use (7) and (8) to prove that

$$\oint_{\partial\mathcal{S}} \varphi\nabla(\psi) \cdot d\mathbf{s} = \int_{\mathcal{S}} \nabla(\varphi) \times \nabla(\psi) \cdot d\mathbf{s}$$

where $\mathcal{S}$ is a smooth surface with boundary $\partial\mathcal{S}$.

30. Explain carefully why Green's Theorem is a special case of Stokes' Theorem.

Further Insights and Challenges

31. Complete the proof of Theorem 1 by proving the equality

$$\oint_{\mathcal{C}} F_3(x, y, z)\mathbf{k} \cdot d\mathbf{s} = \int_{\mathcal{S}} \text{curl}(F_3(x, y, z)\mathbf{k}) \cdot d\mathbf{S}$$

where $\mathcal{S}$ is the graph of a function $z = f(x, y)$ over a domain $\mathcal{D}$ in the xy-plane whose boundary is a simple closed curve.

32. Let $\mathbf{F}$ be a continuously differentiable vector field in $\mathbf{R}^3$, Q a point, and $\mathcal{S}$ a plane containing Q with unit normal vector $\mathbf{e}$. Let $\mathcal{C}_r$ be a circle of radius r centered at Q in $\mathcal{S}$ and let $\mathcal{S}_r$ be the disk enclosed by $\mathcal{C}_r$. Assume $\mathcal{S}_r$ is oriented with unit normal vector $\mathbf{e}$.

(a) Let $m(r)$ and $M(r)$ be the minimum and maximum values of $\text{curl}(\mathbf{F}(P)) \cdot \mathbf{e}$ for $P \in \mathcal{S}_r$. Prove that

$$m(r) \le \frac{1}{\pi r^2} \iint_{\mathcal{S}_r} \text{curl}(\mathbf{F}) \cdot d\mathbf{S} \le M(r)$$

(b) Prove that

$$\text{curl}(\mathbf{F}(Q)) \cdot \mathbf{e} = \lim_{r \to 0} \frac{1}{\pi r^2} \int_{\mathcal{C}_r} \mathbf{F} \cdot d\mathbf{s}$$

This proves that $\text{curl}(\mathbf{F}(Q)) \cdot \mathbf{e}$ is the circulation per unit area in the plane $\mathcal{S}$.

17.3 Divergence Theorem

Before stating the third and last of the fundamental theorems of vector analysis, we take a moment to put matters in perspective. First of all, we observe that each main theorem is a relation of the type:

Integral of a derivative on a domain	=	Integral over the *oriented boundary* of the domain

In single-variable calculus, the Fundamental Theorem of Calculus (FTC) relates the integral of $f'(x)$ over an interval to the "integral" of $f(x)$ over the boundary:

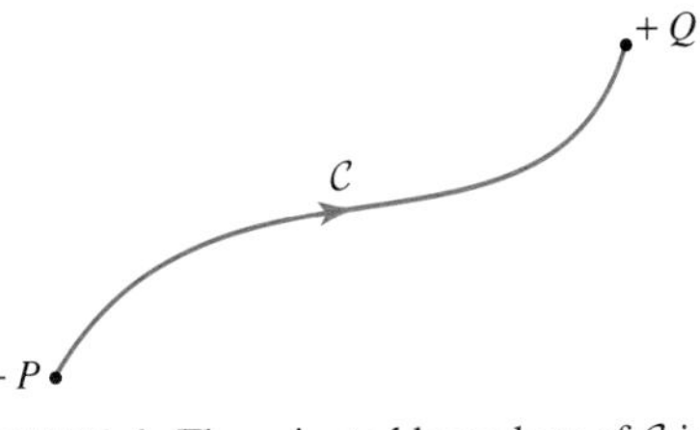

FIGURE 1 The oriented boundary of $\mathcal{C}$ is $Q - P$.

$$\underbrace{\int_a^b f'(x)\,dx}_{\text{Integral of derivative over } [a,b]} = \underbrace{f(b) - f(a)}_{\text{"Integral" over the boundary of } [a,b]}$$

We orient the boundary of the interval $[a, b]$ by assigning a plus sign to b and a minus sign to a.

The Fundamental Theorem for Line Integrals is a direct generalization, where we integrate over a path from P to Q (Figure 1). Instead of the ordinary derivative, we have the gradient:

$$\underbrace{\int_{\mathcal{C}} \nabla\varphi \cdot d\mathbf{s}}_{\text{Integral of derivative over a curve}} = \underbrace{\varphi(Q) - \varphi(P)}_{\text{Integral of } \varphi \text{ over the boundary } Q-P}$$

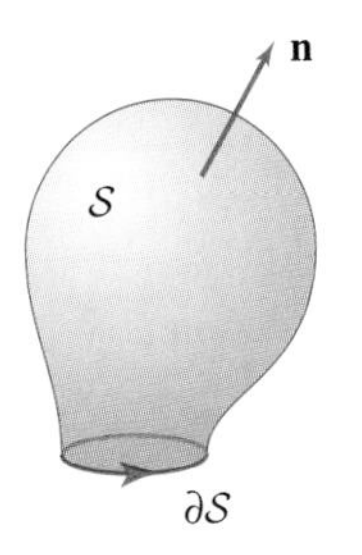

FIGURE 2 The oriented boundary of $\mathcal{S}$ is $\partial\mathcal{S}$.

Stokes' Theorem is a two-dimensional version of the FTC that relates the integral over a surface to an integral over its boundary (Figure 2). In this case, the appropriate derivative is the curl:

$$\underbrace{\iint_{\mathcal{S}} \operatorname{curl}(\mathbf{F}) \cdot d\mathbf{S}}_{\text{Integral of derivative over surface}} = \underbrace{\int_{\partial\mathcal{S}} \mathbf{F} \cdot d\mathbf{s}}_{\text{Integral over boundary}}$$

The Divergence Theorem follows the same pattern, where the domain is a region $\mathcal{W}$ in $\mathbf{R}^3$ whose boundary is a surface $\partial\mathcal{W}$. Figure 3 shows two examples: a ball whose boundary is a sphere and a cube whose boundary is the surface of the cube.

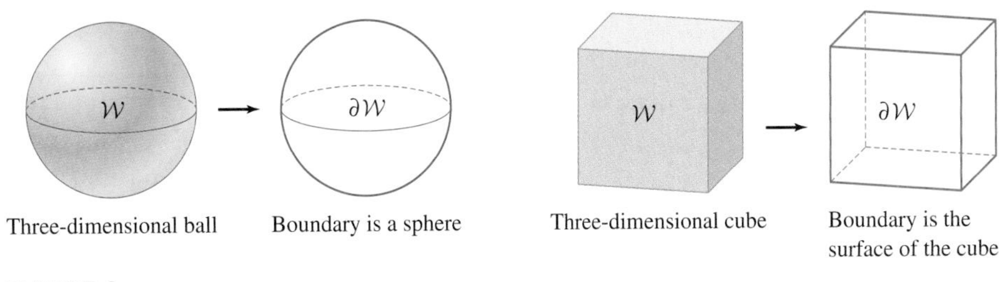

FIGURE 3

The derivative appearing in the Divergence Theorem is the **divergence** of a vector field $\mathbf{F} = \langle F_1, F_2, F_3 \rangle$, defined by

$$\operatorname{div}(\mathbf{F}) = \frac{\partial F_1}{\partial x} + \frac{\partial F_2}{\partial y} + \frac{\partial F_3}{\partial z} \qquad \boxed{1}$$

More advanced treatments of vector calculus use the theory of "differential forms" to formulate a general version of Stokes' Theorem that is valid in all dimensions and includes each of our main theorems (Green, Stokes, Divergence) as a special case.

The divergence is often denoted $\nabla \cdot \mathbf{F}$, which we view as a symbolic dot product:

$$\nabla \cdot \mathbf{F} = \left\langle \frac{\partial}{\partial x}, \frac{\partial}{\partial y}, \frac{\partial}{\partial z} \right\rangle \cdot \langle F_1, F_2, F_3 \rangle = \frac{\partial F_1}{\partial x} + \frac{\partial F_2}{\partial y} + \frac{\partial F_3}{\partial z}$$

Keep in mind that, unlike the gradient and curl, the divergence of a vector field is a scalar function. Furthermore, taking the divergence is a linear operation:

$$\operatorname{div}(\mathbf{F} + \mathbf{G}) = \operatorname{div}(\mathbf{F}) + \operatorname{div}(\mathbf{G}), \qquad \operatorname{div}(c\mathbf{F}) = c\operatorname{div}(\mathbf{F}) \quad (\text{for any constant } c)$$

■ **EXAMPLE 1** Evaluate the divergence of $\mathbf{F} = \langle e^{xy}, xy, z^4 \rangle$ at $P = (1, 0, 2)$.

Solution The divergence is the function

$$\operatorname{div}(\mathbf{F}) = \frac{\partial}{\partial x} e^{xy} + \frac{\partial}{\partial y} xy + \frac{\partial}{\partial z} z^4 = ye^{xy} + x + 4z^3$$

and $\operatorname{div}(\mathbf{F})(P) = 0 \cdot e^0 + 1 + 4 \cdot 2^3 = 33$. ■

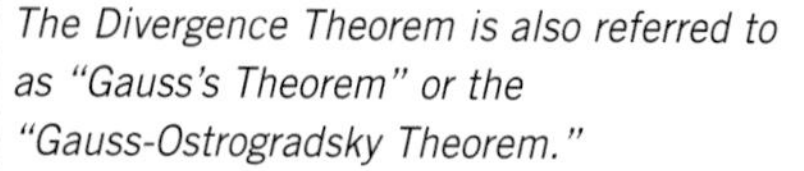

The Divergence Theorem is also referred to as "Gauss's Theorem" or the "Gauss-Ostrogradsky Theorem."

THEOREM 1 Divergence Theorem Let $\mathcal{W}$ be a region in $\mathbf{R}^3$ whose boundary $\partial\mathcal{W}$ is a piecewise smooth surface, oriented so that the normal vectors to $\partial\mathcal{W}$ point outside of $\mathcal{W}$. Let $\mathbf{F}$ be a vector field whose domain contains $\mathcal{W}$ and whose components have continuous partial derivatives. Then

$$\iint_{\partial\mathcal{W}} \mathbf{F} \cdot d\mathbf{S} = \iiint_{\mathcal{W}} \operatorname{div}(\mathbf{F})\, dV \qquad \boxed{2}$$

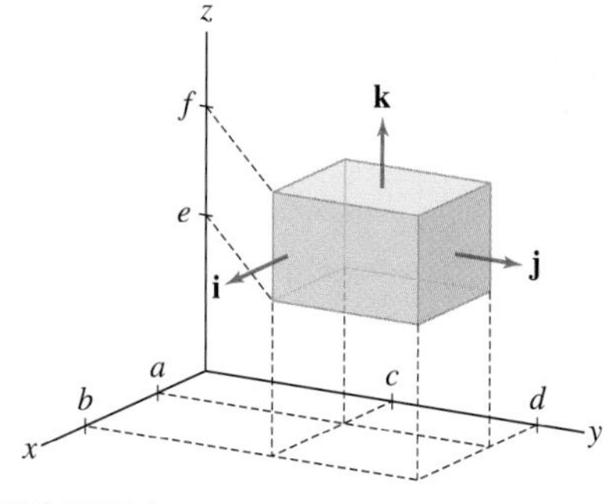

FIGURE 4

Proof We prove the Divergence Theorem for the special case in which $\mathcal{W}$ is a box $[a, b] \times [c, d] \times [e, f]$ as in Figure 4. The general case is technically more difficult but can eventually be reduced to the case of a box.

Both sides of the equality in the Divergence Theorem are additive in the following sense. If $\mathbf{F} = \langle F_1, F_2, F_3 \rangle$, then

$$\iint_{\partial\mathcal{W}} (F_1\mathbf{i} + F_2\mathbf{j} + F_3\mathbf{k}) \cdot d\mathbf{S} = \iint_{\partial\mathcal{W}} F_1\mathbf{i} \cdot d\mathbf{S} + \iint_{\partial\mathcal{W}} F_2\mathbf{j} \cdot d\mathbf{S} + \iint_{\partial\mathcal{W}} F_3\mathbf{k} \cdot d\mathbf{S}$$

$$\iiint_{\mathcal{W}} \operatorname{div}(F_1\mathbf{i} + F_2\mathbf{j} + F_3\mathbf{k}) \cdot dV = \iiint_{\mathcal{W}} \operatorname{div}(F_1\mathbf{i}) \cdot dV + \iiint_{\mathcal{W}} \operatorname{div}(F_2\mathbf{j}) \cdot dV + \iiint_{\mathcal{W}} \operatorname{div}(F_3\mathbf{k}) \cdot dV$$

As in the proof of Green's and Stokes' Theorems, we may prove the Divergence Theorem by showing that the terms corresponding to the **i**-, **j**-, and **k**-components are separately equal. We carry out the argument for the **i**-component (the other two components are similar).

Assume that $\mathbf{F} = F_1\mathbf{i}$ and let $\mathcal{S} = \partial\mathcal{W}$ be the boundary of the box. Then $\mathcal{S}$ consists of six faces. The integral of $\mathbf{F}$ over $\mathcal{S}$ is the sum of the integrals over the six faces. However, $\mathbf{F} = F_1\mathbf{i}$ is orthogonal to the normal vectors to the top and bottom as well as the two side faces since

$$\mathbf{F} \cdot \mathbf{j} = \mathbf{F} \cdot \mathbf{k} = 0$$

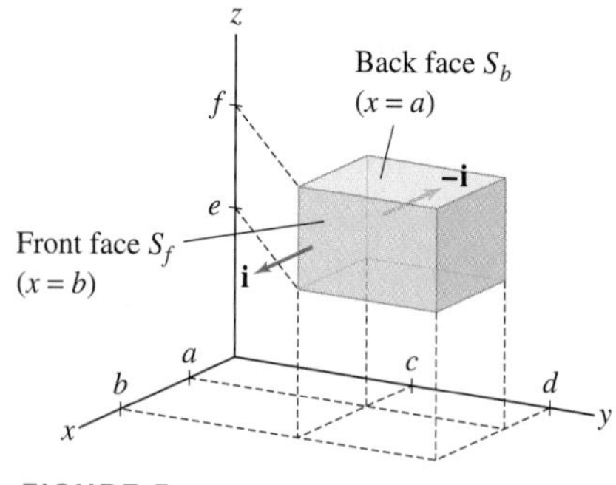

FIGURE 5

Therefore, the integrals over the top, bottom, and two side faces are zero. Nonzero contributions come only from the front and back faces, which we denote $\mathcal{S}_f$ and $\mathcal{S}_b$ (Figure 5):

$$\iint_{\mathcal{S}} \mathbf{F} \cdot d\mathbf{S} = \iint_{\mathcal{S}_f} \mathbf{F} \cdot d\mathbf{S} + \iint_{\mathcal{S}_b} \mathbf{F} \cdot d\mathbf{S}$$

To evaluate these integrals, we parametrize $\mathcal{S}_f$ and $\mathcal{S}_b$ by

$$\Phi_f(y, z) = (b, y, z), \qquad c \le y \le d, \ e \le z \le f$$

$$\Phi_b(y, z) = (a, y, z), \qquad c \le y \le d, \ e \le z \le f$$

The normal vectors for these parametrizations are

$$\frac{\partial \Phi_f}{\partial y} \times \frac{\partial \Phi_f}{\partial z} = \mathbf{j} \times \mathbf{k} = \mathbf{i}$$

$$\frac{\partial \Phi_b}{\partial y} \times \frac{\partial \Phi_b}{\partial z} = \mathbf{j} \times \mathbf{k} = \mathbf{i}$$

However, the outward-pointing normal for the face $\mathcal{S}_b$ is the vector $-\mathbf{i}$ and hence a minus sign occurs when we compute the surface integral of $\mathbf{F}$ over $\mathcal{S}_b$ using the parametrization Φ_b:

$$\iint_{\mathcal{S}_f} \mathbf{F} \cdot d\mathbf{S} + \iint_{\mathcal{S}_b} \mathbf{F} \cdot d\mathbf{S} = \int_e^f \int_c^d F_1(b, y, z)\, dy\, dz - \int_e^f \int_c^d F_1(a, y, z)\, dy\, dz$$
$$= \int_e^f \int_c^d \Big(F_1(b, y, z) - F_1(a, y, z) \Big)\, dy\, dz$$

By the FTC in one variable,

$$F_1(b, y, z) - F_1(a, y, z) = \int_a^b \frac{\partial F_1}{\partial x}(x, y, z)\, dx$$

Since $\operatorname{div}(\mathbf{F}) = \operatorname{div}(F_1\mathbf{i}) = \dfrac{\partial F_1}{\partial x}$, we obtain the desired result:

$$\iint_{\mathcal{S}} \mathbf{F} \cdot d\mathbf{S} = \int_e^f \int_c^d \int_a^b \frac{\partial F_1}{\partial x}(x, y, z)\, dx\, dy\, dz = \iiint_{\mathcal{W}} \operatorname{div}(\mathbf{F})\, dV \qquad \blacksquare$$

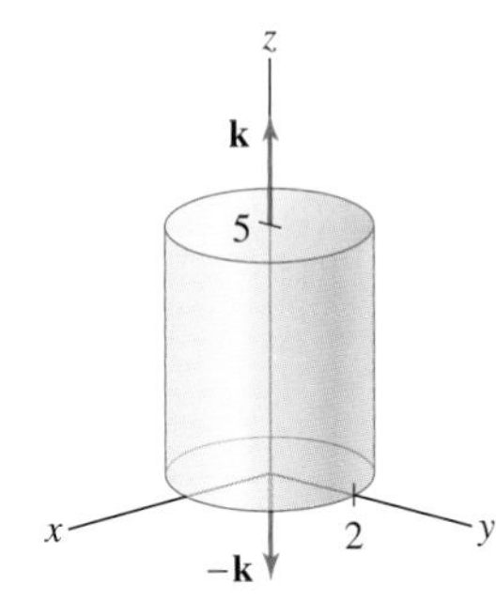

FIGURE 6 Cylinder of radius 2 and height 5.

EXAMPLE 2 Verifying the Divergence Theorem Verify the Divergence Theorem for the vector field $\mathbf{F} = \langle y, yz, z^2 \rangle$ and the portion $\mathcal{S}$ of the cylinder $x^2 + y^2 = 4$ between $z = 0$ and $z = 5$ (Figure 6).

Solution We calculate both sides of Eq. (2).

***Step 1.* Integrate over the side of the cylinder.**

We parametrize the side of the cylinder by $\Phi(\theta, z) = (2\cos\theta, 2\sin\theta, z)$ for $0 \le \theta < 2\pi$ and $0 \le z \le 5$. Then

$$\mathbf{n} = \mathbf{T}_\theta \times \mathbf{T}_z = \langle -2\sin\theta, 2\cos\theta, 0 \rangle \times \langle 0, 0, 1 \rangle = \langle 2\cos\theta, 2\sin\theta, 0 \rangle$$

$$\mathbf{F}(\Phi(\theta, z)) \cdot \mathbf{n} = \langle 2\sin\theta, 2z\sin\theta, z^2 \rangle \cdot \langle 2\cos\theta, 2\sin\theta, 0 \rangle$$
$$= 4\cos\theta\sin\theta + 4z\sin^2\theta$$

$$\iint_{\text{side}} \mathbf{F} \cdot d\mathbf{S} = \int_0^5 \int_0^{2\pi} (4\cos\theta\sin\theta + 4z\sin^2\theta)\, d\theta\, dz$$
$$= 0 + 4\pi \int_0^5 z\, dz = 4\pi\left(\frac{25}{2}\right) = 50\pi \qquad \boxed{3}$$

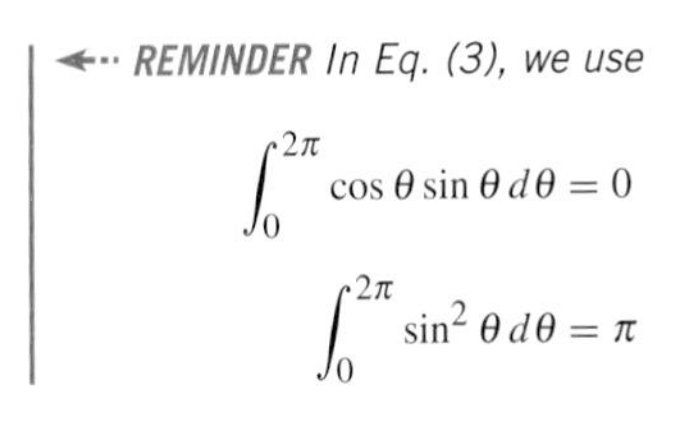
REMINDER In Eq. (3), we use

$$\int_0^{2\pi} \cos\theta\sin\theta\, d\theta = 0$$

$$\int_0^{2\pi} \sin^2\theta\, d\theta = \pi$$

***Step 2.* Integrate over the top and bottom of the cylinder.**

The top of the cylinder is parametrized by $\Phi(x, y) = (x, y, 5)$ with parameter domain $\mathcal{D} = \{(x, y) : x^2 + y^2 \le 4\}$. We obtain

$$\mathbf{F}(\Phi(x, y)) = \langle y, 5y, 5^2 \rangle$$
$$\mathbf{n} = \mathbf{T}_x \times \mathbf{T}_y = \langle 1, 0, 0 \rangle \times \langle 0, 1, 0 \rangle = \langle 0, 0, 1 \rangle$$
$$\mathbf{F}(\Phi(x, y)) \cdot \mathbf{n} = \langle y, 5y, 5^2 \rangle \cdot \langle 0, 0, 1 \rangle = 25$$

Since $\mathcal{D}$ has area 4π,

$$\iint_{\text{top}} \mathbf{F} \cdot d\mathbf{S} = \iint_{\mathcal{D}} 25\, dA = 25\, \text{Area}(\mathcal{D}) = 25(4\pi) = 100\pi$$

The integral over the bottom of the cylinder is zero. Indeed, along the bottom where $z = 0$, the vector field $\mathbf{F}(x, y, 0) = \langle y, 5y, 0 \rangle$ is orthogonal to the normal vector $-\mathbf{k}$.

***Step 3.* Find the total flux.**

$$\iint_{\mathcal{S}} \mathbf{F} \cdot d\mathbf{S} = \text{sides} + \text{top} + \text{bottom} = 50\pi + 100\pi + 0 = \boxed{150\pi}$$

***Step 4.* Compare with integral of divergence.**

$$\text{div}(\mathbf{F}) = \text{div}(\langle y, yz, z^2 \rangle) = \frac{\partial}{\partial x} y + \frac{\partial}{\partial y}(yz) + \frac{\partial}{\partial z} z^2 = 0 + z + 2z = 3z$$

The region $\mathcal{W}$ consists of points (x, y, z), where $(x, y) \in \mathcal{D}$ and $0 \le z \le 5$:

$$\iiint_{\mathcal{W}} \text{div}(\mathbf{F})\, dV = \iint_{\mathcal{D}} \int_{z=0}^{5} 3z\, dV = \iint_{\mathcal{D}} \frac{75}{2}\, dV$$
$$= \left(\frac{75}{2}\right)(\text{Area}(\mathcal{D})) = \left(\frac{75}{2}\right)(4\pi) = \boxed{150\pi}$$

The flux is equal to the integral of divergence, thus verifying the Divergence Theorem. ■

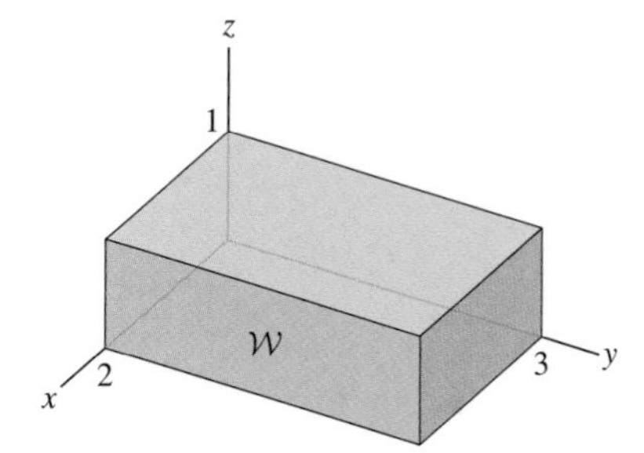

FIGURE 7

■ **EXAMPLE 3 Using the Divergence Theorem** Use the Divergence Theorem to evaluate $\iint_{\mathcal{S}} \langle x^2, z^4, e^z \rangle \cdot d\mathbf{S}$, where $\mathcal{S}$ is the boundary of the box $\mathcal{W} = [0, 2] \times [0, 3] \times [0, 1]$ (Figure 7).

Solution

$$\text{div}(\langle x^2, z^4, e^z \rangle) = \frac{\partial}{\partial x} x^2 + \frac{\partial}{\partial y} z^4 + \frac{\partial}{\partial z} e^z = 2x + e^z$$

By the Divergence Theorem,

$$\iint_{\mathcal{S}} \langle x^2, z^4, e^z \rangle \cdot d\mathbf{S} = \iiint_{\mathcal{W}} (2x + e^z)\, dV = \int_0^2 \int_0^3 \int_0^1 (2x + e^z)\, dz\, dy\, dx$$
$$= 3\int_0^2 2x\, dx + 6\int_0^1 e^z\, dz = 12 + 6(e - 1) = 6e + 6$$ ■

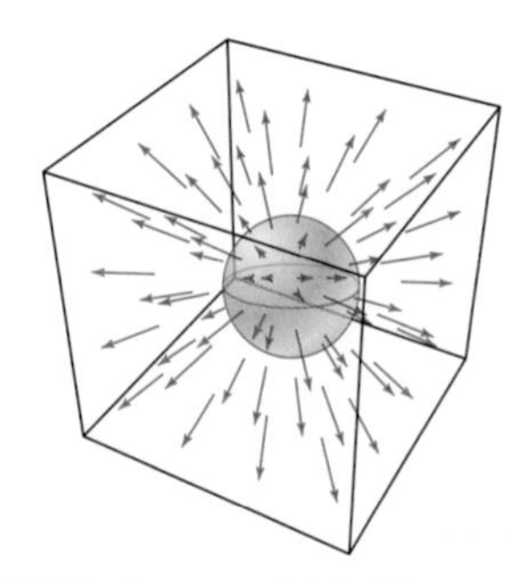
FIGURE 8 A vector field with nonzero divergence.

GRAPHICAL INSIGHT The Interpretation of Divergence As in our discussion of curl, let's assume that $\mathbf{F}$ is the velocity field of a fluid. Then the flux of $\mathbf{F}$ through a surface $\mathcal{S}$ is the volume of fluid passing through $\mathcal{S}$ per unit time (Figure 8). Suppose that $\mathcal{S}$ encloses a small region $\mathcal{W}$ containing a point P (e.g., a ball of small radius). Then div($\mathbf{F}$)

is nearly constant on $\mathcal{W}$ with value $\operatorname{div}(\mathbf{F})(P)$, and the Divergence Theorem yields the approximation

$$\text{Flux across } \mathcal{S} = \iiint_{\mathcal{W}} \operatorname{div}(\mathbf{F})\, dV \approx \operatorname{div}(\mathbf{F})(P)\operatorname{Vol}(\mathcal{W})$$

In other words, the *flux is approximately equal to the divergence times the enclosed volume*. We may think of $\operatorname{div}(\mathbf{F})(P)$ as the "flux per unit volume" at P. In particular,

- If $\operatorname{div}(\mathbf{F})(P) > 0$, there is a net outflow across any small closed surface enclosing P. Thus, fluid is "produced" at P.
- If $\operatorname{div}(\mathbf{F})(P) < 0$, there is a net inflow across any small closed surface enclosing P. Thus, fluid is "consumed" at P.
- If $\operatorname{div}(\mathbf{F})(P) = 0$, then to a first-order approximation, there is no net flow, in or out, across any small closed surface enclosing P.

These cases are more easily visualized in two dimensions, where we define $\operatorname{div}(\langle P, Q\rangle) = \frac{\partial P}{\partial x} + \frac{\partial Q}{\partial y}$. The fluid in Figure 9(A) has positive divergence and there is a positive net flow of fluid across every circle per unit time. Similarly, Figure 9(B) shows a fluid with negative divergence. By contrast, the fluid in Figure 9(C) has zero divergence. The fluid flowing into every circle is balanced by the fluid flowing out. In general, a fluid of constant density whose velocity field satisfies $\operatorname{div}(\mathbf{F}) = 0$ is called **incompressible**. In this case, the net flow across every closed surface is zero, and thus fluid is neither created nor destroyed.

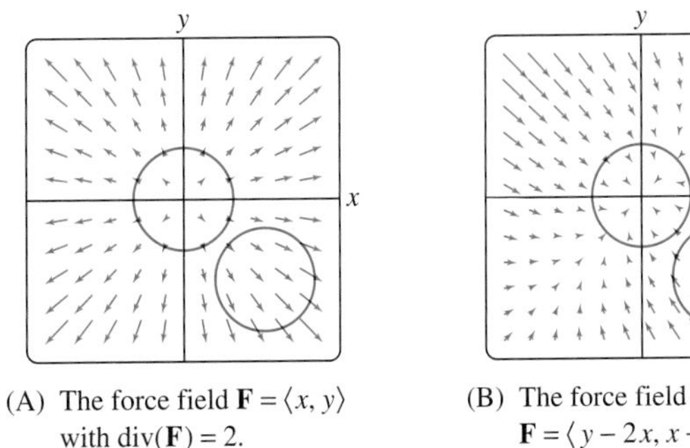

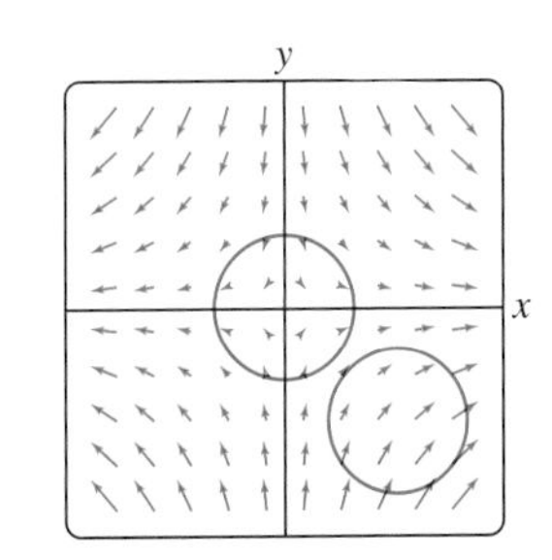

(A) The force field $\mathbf{F} = \langle x, y\rangle$ with $\operatorname{div}(\mathbf{F}) = 2$. There is a net outflow through every circle.

(B) The force field $\mathbf{F} = \langle y - 2x, x - 2y\rangle$ with $\operatorname{div}(\mathbf{F}) = -4$. There is a net inflow into every circle.

(C) The force field $\mathbf{F} = \langle x, -y\rangle$ with $\operatorname{div}(\mathbf{F}) = 0$. The flux through every circle is zero.

FIGURE 9

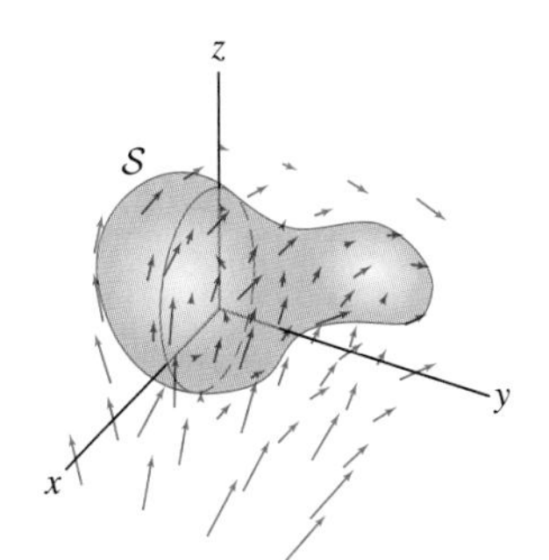

FIGURE 10

■ **EXAMPLE 4 A Vector Field with Zero Divergence** Compute the flux of

$$\mathbf{F} = \left\langle z^2 + xy^2, \cos(x+z), e^{-y} - zy^2\right\rangle$$

through the boundary of the surface $\mathcal{S}$ in Figure 10.

Solution The divergence of $\mathbf{F}$ is zero:

$$\operatorname{div}(\mathbf{F}) = \frac{\partial}{\partial x}(z^2 + xy^2) + \frac{\partial}{\partial y}\cos(x+z) + \frac{\partial}{\partial z}(e^{-y} - zy^2) = y^2 - y^2 = 0$$

By the Divergence Theorem, the flux of **F** through *every* closed surface $\mathcal{S}$ enclosing a region $\mathcal{W}$ is zero:

$$\iint_{\mathcal{S}} \mathbf{F} \cdot d\mathbf{S} = \iiint_{\mathcal{W}} \operatorname{div}(\mathbf{F})\, dV = \iiint_{\mathcal{W}} 0\, dV = 0$$

In particular, the flux through $\mathcal{S}$ in Figure 10 is zero. ■

CONCEPTUAL INSIGHT The Divergence Theorem and Stokes' Theorem give us two ways of showing that a vector field **F** has zero flux through closed surfaces:

- If $\operatorname{div}(\mathbf{F}) = 0$, then $\iint_{\mathcal{S}} \mathbf{F} \cdot d\mathbf{S} = 0$ for every closed surface $\mathcal{S}$ that is the boundary of a region $\mathcal{W}$ contained in the domain of **F**.
- If $\mathbf{F} = \operatorname{curl}(\mathbf{A})$ for some vector potential **A**, then $\iint_{\mathcal{S}} \mathbf{F} \cdot d\mathbf{S} = 0$ for every closed surface contained in the domain of **F** (by Stokes' Theorem; see Theorem 2 in Section 17.2).

This raises the following question: Are the two conditions $\operatorname{div}(\mathbf{F}) = 0$ and $\mathbf{F} = \operatorname{curl}(\mathbf{A})$ related? Half of this question is easy to answer. In Exercise 22, you are asked to verify the identity

$$\boxed{\operatorname{div}(\operatorname{curl}(\mathbf{A})) = 0}$$

So if $\mathbf{F} = \operatorname{curl}(\mathbf{A})$, then $\operatorname{div}(\mathbf{F})$ is automatically zero. The other half is not as straightforward. If $\operatorname{div}(\mathbf{F}) = 0$, we cannot be sure that **F** has a vector potential unless the domain of **F** is a region with "no holes," such as a sphere or box or all of $\mathbf{R}^3$. A precise definition of "no holes" would lead us into the area of mathematics called topology, but roughly speaking, if a domain has no holes, then every vector field with zero divergence has a vector potential (see Exercise 34). By contrast, if the domain has one or more holes (such as $\mathbf{R}^3$ with a sphere or even just a point removed), then there exist vector fields with zero divergence but nonzero flux through some closed surface. This is illustrated in the next example. Such a vector field cannot have a vector potential because $\mathbf{F} = \operatorname{curl}(\mathbf{A})$ implies that the flux through *every* closed surface is zero.

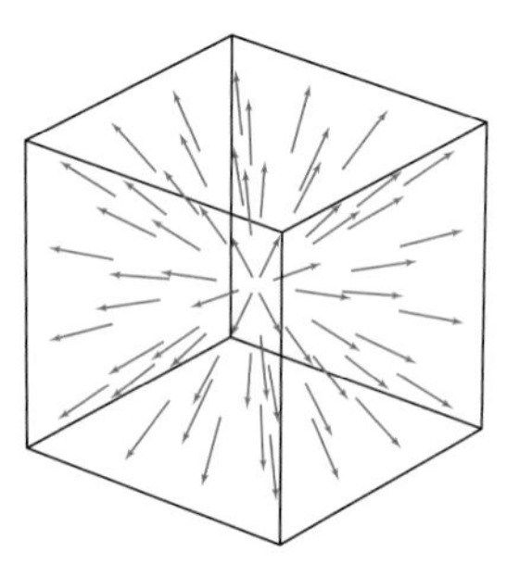

FIGURE 11 Unit radial vector field $\mathbf{e}_r$.

Let $\rho = \sqrt{x^2 + y^2 + z^2}$ and let $\mathbf{e}_r$ denote the unit radial vector field (Figure 11):

$$\boxed{\mathbf{e}_r = \frac{\langle x, y, z\rangle}{\rho} = \frac{\langle x, y, z\rangle}{\sqrt{x^2 + y^2 + z^2}}} \qquad \boxed{4}$$

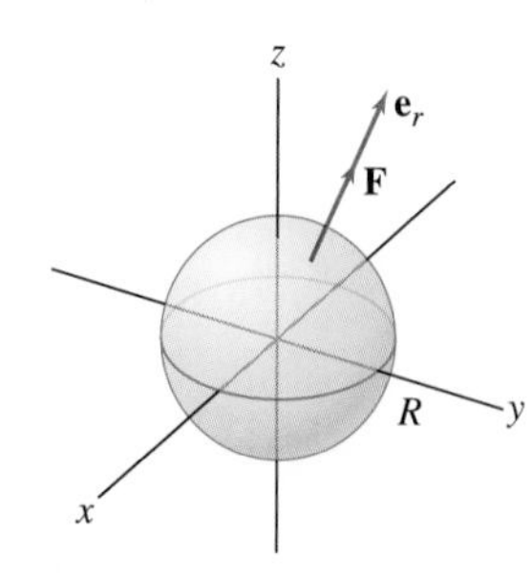

FIGURE 12

■ **EXAMPLE 5 The Inverse-Square Radial Vector Field** Let $\mathbf{F} = \dfrac{\mathbf{e}_r}{\rho^2}$ (defined for $\rho \neq 0$). Verify that:

(a) $\operatorname{div}(\mathbf{F}) = 0$

(b) $\iint_{\mathcal{S}_R} \mathbf{F} \cdot d\mathbf{S} = 4\pi$, where $\mathcal{S}_R$ is the sphere of radius R centered at the origin (Figure 12).

(c) $\iint_{\mathcal{S}} \mathbf{F} \cdot d\mathbf{S} = 0$ for any closed surface $\mathcal{S}$ that does not contain the origin.

Solution Write the field as $\mathbf{F} = \langle F_1, F_2, F_3 \rangle = \langle x\rho^{-3}, y\rho^{-3}, z\rho^{-3} \rangle$.

(a) To compute the divergence, we first note that

$$\frac{\partial \rho}{\partial x} = \frac{\partial}{\partial x}(x^2 + y^2 + z^2)^{1/2} = \frac{1}{2}(x^2 + y^2 + z^2)^{-1/2}(2x) = \frac{x}{\rho}$$

Therefore,

$$\frac{\partial F_1}{\partial x} = \frac{\partial}{\partial x} x\rho^{-3} = \rho^{-3} - 3x\rho^{-4}\frac{\partial \rho}{\partial x} = \rho^{-3} - (3x\rho^{-4})\frac{x}{\rho} = \frac{\rho^2 - 3x^2}{\rho^5}$$

The derivatives $\frac{\partial F_2}{\partial y}$ and $\frac{\partial F_3}{\partial z}$ are similar, so we have

$$\text{div}(\mathbf{F}) = \frac{\rho^2 - 3x^2}{\rho^5} + \frac{\rho^2 - 3y^2}{\rho^5} + \frac{\rho^2 - 3z^2}{\rho^5} = \frac{3\rho^2 - 3(x^2 + y^2 + z^2)}{\rho^5} = 0$$

(b) We compute the flux through $\mathcal{S}_R$ using the parametrization by spherical coordinates $\Phi(\phi, \theta) = (R\cos\theta\sin\phi, R\sin\theta\sin\phi, R\cos\phi)$. As we saw in Section 16.4, Eq. (5), the outward-pointing normal is

$$\mathbf{n} = \mathbf{T}_\phi \times \mathbf{T}_\theta = (R^2 \sin\phi)\mathbf{e}_r$$

Furthermore, on the sphere of radius R, $\mathbf{F}(\Phi(\phi, \theta)) = R^{-2}\mathbf{e}_r$ and therefore,

$$\mathbf{F} \cdot \mathbf{n} = (R^{-2}\mathbf{e}_r) \cdot (R^2 \sin\phi\, \mathbf{e}_r) = \sin\phi(\mathbf{e}_r \cdot \mathbf{e}_r) = \sin\phi$$

Since the flux of $\mathbf{F} = \rho^{-2}\mathbf{e}_r$ through a sphere centered at the origin is nonzero, $\mathbf{F}$ cannot have a vector potential even though $\text{div}(\mathbf{F}) = 0$ on its domain.

Now we can verify that the flux is 4π:

$$\iint_{\mathcal{S}_R} \mathbf{F} \cdot d\mathbf{S} = \int_0^{2\pi}\int_0^{\pi} \mathbf{F} \cdot \mathbf{n}\, d\phi\, d\theta = 2\pi \int_0^{\pi} \sin\phi\, d\phi = 4\pi$$

(c) If $\mathcal{S}$ is a closed surface not containing the origin, then $\mathbf{F}$ is defined at every point in the region $\mathcal{W}$ enclosed by $\mathcal{S}$ and $\text{div}(\mathbf{F}) = 0$ on $\mathcal{W}$. The Divergence Theorem yields

$$\iint_{\mathcal{S}} \mathbf{F} \cdot d\mathbf{S} = \iiint_{\mathcal{W}} \text{div}(\mathbf{F})\, dV = 0$$ ■

In the previous example, we showed that the flux of $\mathbf{F} = \frac{\mathbf{e}_r}{\rho^2}$ through $\mathcal{S}_R$ is equal to 4π. We can prove that this result remains true for any surface $\mathcal{S}$ containing the origin as follows. Choose $R > 0$ small enough so that $\mathcal{S}_R$ is contained inside $\mathcal{S}$ and let $\mathcal{W}$ be the region between $\mathcal{S}_R$ and $\mathcal{S}$. Then the oriented boundary of $\mathcal{W}$ is the difference $\mathcal{S} - \mathcal{S}_R$, that is, the boundary consists of $\mathcal{S}$ with an outward-pointing normal and $\mathcal{S}_R$ with an inward-pointing normal (Figure 13). The Divergence Theorem is valid for the region $\mathcal{W}$, and since $\text{div}(\mathbf{F}) = 0$ on $\mathcal{W}$, we have

$$\iint_{\mathcal{S}} \mathbf{F} \cdot d\mathbf{S} - \iint_{\mathcal{S}_R} \mathbf{F} \cdot d\mathbf{S} = \iiint_{\mathcal{W}} \text{div}(\mathbf{F})\, dV = 0$$

It follows that the flux of $\mathbf{F}$ through $\mathcal{S}$ is equal to the flux through $\mathcal{S}_R$.

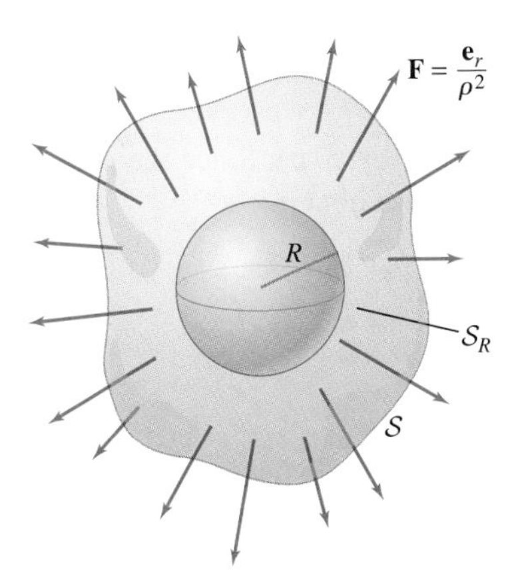

FIGURE 13

This result plays a fundamental role in the theory of electrostatics. The electric field $\mathbf{E}$ due to a point charge of magnitude q coulombs placed at the origin in a vacuum is

$$\mathbf{E} = \left(\frac{q}{4\pi\epsilon_0}\right)\frac{\mathbf{e}_r}{\rho^2}$$

where $\epsilon_0 = 8.85 \times 10^{-12}$ $\mathrm{C}^2/\mathrm{N}\text{-}\mathrm{m}^2$ is the permittivity constant. Since $\mathbf{E}$ and $\mathbf{e}_r/\rho^2$ differ by the constant $q/(4\pi\epsilon_0)$, we have the following general result for any closed surface $\mathcal{S}$:

$$\text{Flux of } \mathbf{E} \text{ through } \mathcal{S} = \begin{cases} 0 & \text{if } q \text{ is outside } \mathcal{S} \\ \dfrac{q}{\epsilon_0} & \text{if } q \text{ is inside } \mathcal{S} \end{cases}$$

Now, instead of placing just one point charge at the origin, we may distribute a finite number N of point charges q_i at different points in space. The resulting electric field $\mathbf{E}$ is the sum of the fields $\mathbf{E}_i$ due to the individual charges and we have

$$\iint_{\mathcal{S}} \mathbf{E}\cdot d\mathbf{S} = \iint_{\mathcal{S}} \mathbf{E}_1\cdot d\mathbf{S} + \cdots + \iint_{\mathcal{S}} \mathbf{E}_N\cdot d\mathbf{S}$$

Each integral on the right is either 0 or $\dfrac{q_i}{\epsilon_0}$, according to whether or not $\mathcal{S}$ contains q_i, so we conclude that

$$\iint_{\mathcal{S}} \mathbf{E}\cdot d\mathbf{S} = \frac{\text{total charge enclosed by } \mathcal{S}}{\epsilon_0} \qquad \boxed{5}$$

This relation is called **Gauss's Law**. A limiting argument may be used to show that Eq. (5) remains valid for the electric field due to a *continuous* distribution of charge.

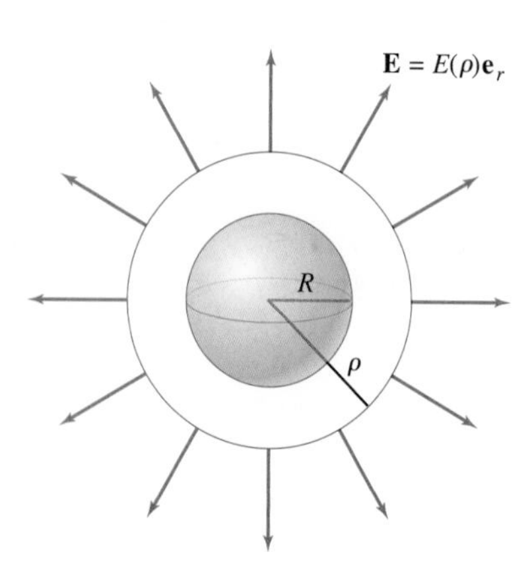

FIGURE 14 The electric field due to a uniformly charged sphere.

Let's examine Gauss's Law for a uniformly charged sphere of radius R centered at the origin (Figure 14). Let Q be the total quantity of charge on the sphere. Let P be a point not lying on the sphere. By symmetry, the electric field $\mathbf{E}$ must be directed radially away from the origin and its magnitude depends only on the distance ρ from P to the origin. Thus, $\mathbf{E} = E(\rho)\mathbf{e}_r$ for some function $E(\rho)$. The flux of $\mathbf{E}$ through the sphere $\mathcal{S}_\rho$ is

$$\iint_{\mathcal{S}_\rho} \mathbf{E}\cdot d\mathbf{S} = E(\rho)\underbrace{\iint_{\mathcal{S}_\rho} \mathbf{e}_r\cdot d\mathbf{S}}_{\text{Surface area of sphere}} = 4\pi\rho^2 E(\rho)$$

By Gauss's Law, this flux is equal to the charge enclosed by $\mathcal{S}(\rho)$. Since the enclosed charge is either Q or zero, according as $\rho > R$ or $\rho < R$, we obtain

$$E(\rho) = \begin{cases} 0 & \text{if } \rho < R \\ \dfrac{Q}{4\pi\epsilon_0\rho^2} & \text{if } \rho > R \end{cases} \qquad \boxed{6}$$

In other words, $\mathbf{E}$ coincides with the electric field due to a point charge Q at the origin outside the sphere, and inside the sphere the field $\mathbf{E}$ is zero. We obtained this same result for the gravitational field by a laborious calculation in Section 16.4. Here, we have derived it from Gauss's Law and a simple appeal to symmetry.

HISTORICAL PERSPECTIVE

James Clerk Maxwell (1831–1879)

The theorems of vector analysis that we have studied in this chapter were developed in the nineteenth century, in large part, to express the laws of electricity and magnetism. Electromagnetism was studied intensively in the period 1750–1890, culminating in the famous Maxwell Equations, which provide a unified understanding in terms of two vector fields: the electric field **E** and the magnetic field **B**. In a region of empty space (where there are no charged particles), Maxwell's Equations are

$$\operatorname{div}(\mathbf{E}) = 0, \qquad \operatorname{div}(\mathbf{B}) = 0$$

$$\operatorname{curl}(\mathbf{E}) = -\frac{\partial \mathbf{B}}{\partial t}, \qquad \operatorname{curl}(\mathbf{B}) = \mu_0\epsilon_0\frac{\partial \mathbf{E}}{\partial t}$$

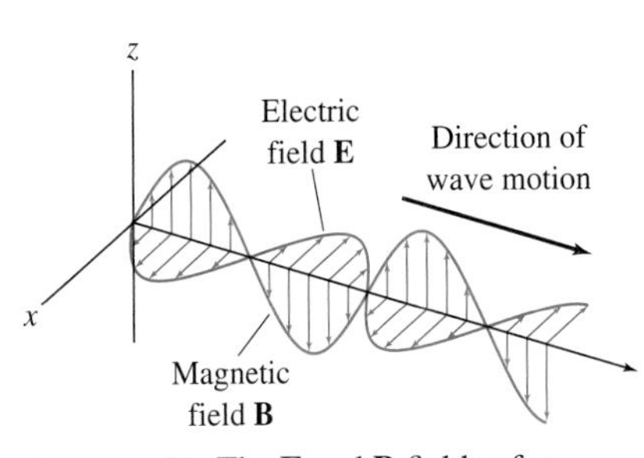

FIGURE 15 The **E** and **B** fields of an electromagnetic wave along an axis of motion.

where $\mu_0\epsilon_0$ is an experimentally determined constant. In MKS units,

$$\mu_0 = 4\pi \times 10^{-7} \text{ henries/m}$$

and

$$\epsilon_0 \approx 8.85 \times 10^{-12} \text{ farads/m}$$

On the basis of these four equations, Maxwell was led to make two predictions of fundamental importance: (1) That electromagnetic waves exist (this was confirmed by H. Hertz in 1887), and (2) that light is an electromagnetic wave.

How do Maxwell's equations suggest that electromagnetic waves exist? And why did Maxwell conclude that light is an electromagnetic wave? It was known to mathematicians in the eighteenth century that waves traveling with velocity c may be described by functions $\varphi(x, y, z, t)$ that satisfy the *wave equation*

$$\Delta\varphi = \frac{1}{c^2}\frac{\partial^2\varphi}{\partial t^2} \qquad \boxed{7}$$

where Δ is the Laplace operator (or "Laplacian") $\Delta\varphi = \frac{\partial^2\varphi}{\partial x^2} + \frac{\partial^2\varphi}{\partial y^2} + \frac{\partial^2\varphi}{\partial z^2}$.

We will show that the components of **E** satisfy this wave equation. Take the curl of both sides of Maxwell's third equation:

$$\operatorname{curl}(\operatorname{curl}(\mathbf{E})) = \operatorname{curl}\left(-\frac{\partial \mathbf{B}}{\partial t}\right) = -\frac{\partial}{\partial t}\operatorname{curl}(\mathbf{B})$$

Then apply Maxwell's fourth equation to obtain

$$\begin{aligned}\operatorname{curl}(\operatorname{curl}(\mathbf{E})) &= -\frac{\partial}{\partial t}\left(\mu_0\epsilon_0\frac{\partial \mathbf{E}}{\partial t}\right)\\ &= -\mu_0\epsilon_0\frac{\partial^2\mathbf{E}}{\partial t^2}\end{aligned} \qquad \boxed{8}$$

Finally, let us define the Laplacian of a vector field

$$\mathbf{F} = \langle F_1, F_2, F_3\rangle$$

by applying Δ to each component, $\Delta\mathbf{F} = \langle \Delta F_1, \Delta F_2, \Delta F_3\rangle$. Then the following identity holds (see Exercise 26):

$$\operatorname{curl}(\operatorname{curl}(\mathbf{F})) = \nabla(\operatorname{div}(\mathbf{F})) - \Delta\mathbf{F}$$

Applying this identity to **E**, we obtain $\operatorname{curl}(\operatorname{curl}(\mathbf{E})) = -\Delta\mathbf{E}$ since $\operatorname{div}(\mathbf{E}) = 0$ by Maxwell's first equation. Thus, Eq. (8) yields

$$\Delta\mathbf{E} = \mu_0\epsilon_0\frac{\partial^2\mathbf{E}}{\partial t^2}$$

In other words, each component of the electric field satisfies the wave equation (7), with $c = (\mu_0\epsilon_0)^{-1/2}$. This tells us that the **E**-field (and similarly the **B**-field) can propagate through space like a wave, giving rise to electromagnetic radiation (Figure 15). Maxwell noticed that the velocity of an electromagnetic wave is

$$c = (\mu_0\epsilon_0)^{-1/2} \approx 3 \times 10^8 \text{ m/s}$$

This value of c turned out be suspiciously close to the velocity of light (first measured by Olaf Römer in 1676). As Maxwell wrote in 1862, "We can scarcely avoid the conclusion that light consists in the transverse undulations of the same medium which is the cause of electric and magnetic phenomena." Needless to say, our modern world relies at every turn on the unseen electromagnetic radiation whose existence was first predicted by Maxwell on mathematical grounds.

17.3 SUMMARY

- The *divergence* of a vector field $\mathbf{F} = \langle F_1, F_2, F_3 \rangle$ is defined by

$$\operatorname{div}(\mathbf{F}) = \frac{\partial F_1}{\partial x} + \frac{\partial F_2}{\partial y} + \frac{\partial F_3}{\partial z}$$

- The Divergence Theorem applies to a region $\mathcal{W}$ in $\mathbf{R}^3$ whose boundary $\partial\mathcal{W}$ is a surface, oriented by normal vectors pointing outside $\mathcal{W}$. Assume that $\mathbf{F}$ is defined and has continuous partial derivatives on $\mathcal{W}$. Then

$$\iint_{\partial\mathcal{W}} \mathbf{F} \cdot d\mathbf{S} = \iiint_{\mathcal{W}} \operatorname{div}(\mathbf{F})\, dV$$

- The Divergence Theorem has the following corollary: If $\operatorname{div}(\mathbf{F}) = 0$ and $\mathcal{W}$ is a region contained in the domain of $\mathbf{F}$, then the flux of $\mathbf{F}$ through the boundary $\partial\mathcal{W}$ is zero.
- The vector field $\mathbf{F} = \dfrac{\mathbf{e}_r}{\rho^2}$ in Example 5, defined for $\rho \neq 0$, has zero divergence. However, the flux of $\mathbf{F}$ through every closed surface $\mathcal{S}$ containing the origin is equal to 4π. This does not contradict the Divergence Theorem because $\mathbf{F}$ is not defined at the origin.

17.3 EXERCISES

Preliminary Questions

1. What is the flux of $\mathbf{F} = \langle 1, 0, 0 \rangle$ through a closed surface?

2. Justify the following statement: The flux of $\mathbf{F} = \langle x^3, y^3, z^3 \rangle$ through every closed surface is positive.

3. Which of the following expressions are meaningful (where $\mathbf{F}$ is a vector field and φ is a function)? Of those that are meaningful, which are automatically zero?

(a) $\operatorname{div}(\nabla\varphi)$ **(b)** $\operatorname{curl}(\nabla\varphi)$ **(c)** $\nabla\operatorname{curl}(\varphi)$

(d) $\operatorname{div}(\operatorname{curl}(\mathbf{F}))$ **(e)** $\operatorname{curl}(\operatorname{div}(\mathbf{F}))$ **(f)** $\nabla(\operatorname{div}(\mathbf{F}))$

4. Which of the following statements is correct (where $\mathbf{F}$ is a continuously differentiable vector field defined everywhere)?

(a) The flux of $\operatorname{curl}(\mathbf{F})$ through all surfaces is zero.

(b) If $\mathbf{F} = \nabla\varphi$, then the flux of $\mathbf{F}$ through all surfaces is zero.

(c) The flux of $\operatorname{curl}(\mathbf{F})$ through all closed surfaces is zero.

5. How does the Divergence Theorem imply that the flux of $\mathbf{F} = \langle x^2, y - e^z, y - 2zx \rangle$ through a closed surface is equal to the enclosed volume?

Exercises

In Exercises 1–4, compute the divergence of the vector field.

1. $\mathbf{F} = \langle xy, yz, y^2 - x^3 \rangle$

2. $x\mathbf{i} + y\mathbf{j} + z\mathbf{k}$

3. $\mathbf{F} = \langle x - 2zx^2, z - xy, z^2x^2 \rangle$

4. $\sin(x + z)\mathbf{i} - ye^{xz}\mathbf{k}$

In Exercises 5–8, verify the Divergence Theorem for the vector field and region.

5. $\mathbf{F} = \langle z, x, y \rangle$ and the box $[0, 4] \times [0, 2] \times [0, 3]$

6. $\mathbf{F} = \langle y, x, z \rangle$ and the region $x^2 + y^2 + z^2 \le 4$

7. $\mathbf{F} = \langle 2x, 3z, 3y \rangle$ and the region $x^2 + y^2 \le 1, 0 \le z \le 2$

8. $\mathbf{F} = \langle x, 0, 0 \rangle$ and the region $x^2 + y^2 \le z \le 4$

In Exercises 9–16, use the Divergence Theorem to evaluate the surface integral $\iint_{\mathcal{S}} \mathbf{F} \cdot d\mathbf{S}$.

9. $\mathbf{F} = \langle x, y, z \rangle$, $\mathcal{S}$ is the sphere $x^2 + y^2 + z^2 = 1$.

10. $\mathbf{F} = \langle y, z, x \rangle$, $\mathcal{S}$ is the sphere $x^2 + y^2 + z^2 = 1$.

11. $\mathbf{F} = \langle x^3, 0, z^3 \rangle$, $\mathcal{S}$ is the sphere $x^2 + y^2 + z^2 = 4$.

12. $\mathbf{F} = \langle x, -y, z \rangle$, $\mathcal{S}$ is the boundary of the unit cube $0 \le x, y, z \le 1$.

13. $\mathbf{F} = \langle x, y^2, z + y \rangle$, $\mathcal{S}$ is the boundary of the region contained in the cylinder $x^2 + y^2 = 4$ between the planes $z = x$ and $z = 8$.

14. $\mathbf{F} = \langle x^2 - z^2, e^{z^2} - \cos x, y^3 \rangle$, $\mathcal{S}$ is the boundary of the region bounded by $x + 2y + 4z = 12$ and the coordinate planes in the first octant.

15. $\mathbf{F} = \langle x + y, z, z - x \rangle$, $\mathcal{S}$ is the boundary of the region between the paraboloid $z = 9 - x^2 - y^2$ and the xy-plane.

16. $\mathbf{F} = \left\langle e^{z^2}, \sin(x^2 z), \sqrt{x^2 + 9y^2} \right\rangle$, $\mathcal{S}$ is the region $x^2 + y^2 \le z \le 8 - x^2 - y^2$.

17. Let $\mathcal{W}$ be the region in Figure 16 bounded by the cylinder $x^2 + y^2 = 9$, the plane $z = x + 1$, and the xy-plane. Use the Divergence Theorem to compute the flux of $\mathbf{F} = \langle z, x, y + 2z \rangle$ through the boundary of $\mathcal{W}$.

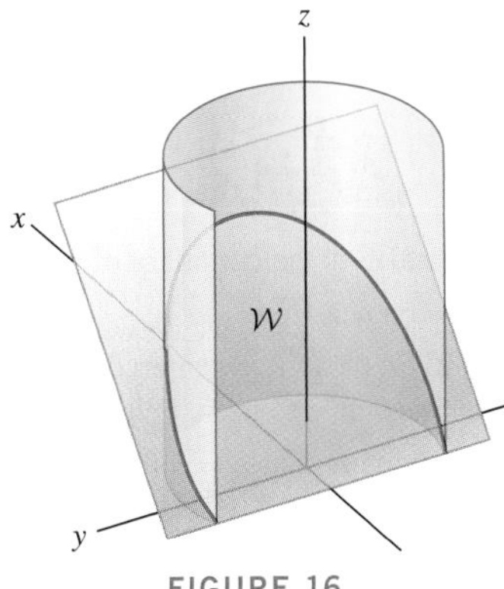

FIGURE 16

18. Find a constant c for which the velocity field

$$\mathbf{F} = (cx - y)\mathbf{i} + (y - z)\mathbf{j} + (3x + 4cz)\mathbf{k}$$

of a fluid is incompressible [meaning that $\mathrm{div}(\mathbf{F}) = 0$].

19. Volume as a Surface Integral Let $\mathbf{F} = \langle x, y, z \rangle$. Prove that if $\mathcal{W}$ is a region $\mathbf{R}^3$ with a smooth boundary $\mathcal{S}$, then

$$\text{Volume}(\mathcal{W}) = \frac{1}{3} \iint_{\mathcal{S}} \mathbf{F} \cdot d\mathbf{S} \qquad \boxed{9}$$

20. Use Eq. (9) to calculate the volume of the unit ball as a surface integral over the unit sphere.

21. Show that $\Phi(a \cos\theta \sin\phi, b \sin\theta \sin\phi, c \cos\phi)$ is a parametrization of the ellipsoid

$$\left(\frac{x}{a}\right)^2 + \left(\frac{y}{b}\right)^2 + \left(\frac{z}{c}\right)^2 = 1$$

Then use Eq. (9) to calculate the volume of the ellipsoid as a surface integral over its boundary.

22. Prove the identity

$$\boxed{\mathrm{div}(\mathrm{curl}(\mathbf{F})) = 0}$$

23. Find and prove a Product Rule expressing $\mathrm{div}(f\mathbf{F})$ in terms of $\mathrm{div}(\mathbf{F})$ and ∇f.

24. Prove the identity

$$\mathrm{div}(\mathbf{F} \times \mathbf{G}) = \mathrm{curl}(\mathbf{F}) \cdot \mathbf{G} - \mathbf{F} \cdot \mathrm{curl}(\mathbf{G})$$

Then prove that the cross product of two irrotational vector fields is source-free [$\mathbf{F}$ is called **irrotational** if $\mathrm{curl}(\mathbf{F}) = 0$ and **source-free** if $\mathrm{div}(\mathbf{F}) = 0$].

25. Prove that $\mathrm{div}(\nabla f \times \nabla g) = 0$.

In Exercises 26–28, let Δ denote the Laplace operator defined by

$$\boxed{\Delta\varphi = \frac{\partial\varphi^2}{\partial x^2} + \frac{\partial\varphi^2}{\partial y^2} + \frac{\partial\varphi^2}{\partial z^2}}$$

26. Prove the identity

$$\mathrm{curl}(\mathrm{curl}(\mathbf{F})) = \nabla(\mathrm{div}(\mathbf{F})) - \Delta\mathbf{F}$$

where $\Delta\mathbf{F}$ denotes $\langle \Delta F_1, \Delta F_2, \Delta F_3 \rangle$.

27. A function φ satisfying $\Delta\varphi = 0$ is called **harmonic**.

(a) Show that $\Delta\varphi = \mathrm{div}(\nabla\varphi)$ for any function φ.

(b) Show that φ is harmonic if and only if $\mathrm{div}(\nabla\varphi) = 0$.

(c) Show that if $\mathbf{F}$ is the gradient of a harmonic function, then $\mathrm{curl}(F) = 0$ and $\mathrm{div}(F) = 0$.

(d) Show $\mathbf{F} = \left\langle xz, -yz, \frac{1}{2}(x^2 - y^2) \right\rangle$ is the gradient of a harmonic function. What is the flux of $\mathbf{F}$ through a closed surface?

28. Let $\mathbf{F} = \rho^n \mathbf{e}_r$, where n is any number, $\rho = (x^2 + y^2 + z^2)^{1/2}$, and $\mathbf{e}_r = \rho^{-1}\langle x, y, z \rangle$ is the unit radial vector.

(a) Calculate $\mathrm{div}(\mathbf{F})$.

(b) Use the Divergence Theorem to calculate the flux of $\mathbf{F}$ through the surface of a sphere of radius R centered at the origin. For which values of n is this flux independent of R?

(c) Prove that $\nabla(\rho^n) = n\rho^{n-1}\mathbf{e}_r$.

(d) Use (c) to show that $\mathbf{F}$ is a gradient vector field for $n \ne -1$. Then show that $\mathbf{F} = \rho^{-1}\mathbf{e}_r$ is also a gradient vector field by computing the gradient of $\ln \rho$.

(e) What is the value of $\int_{\mathcal{C}} \mathbf{F} \cdot d\mathbf{s}$, where $\mathcal{C}$ is a closed curve?

(f) Find the values of n for which the function $\varphi = \rho^n$ is harmonic.

29. The electric field due to a unit electric dipole oriented in the $\mathbf{k}$ direction is $\mathbf{E} = \nabla\left(\frac{z}{\rho^3}\right)$, where $\rho = (x^2 + y^2 + z^2)^{1/2}$ (Figure 17). Let $\mathbf{e}_r = \rho^{-1}\langle x, y, z \rangle$.

(a) Show that $\mathbf{E} = \rho^{-3}\mathbf{k} - 3z\rho^{-4}\mathbf{e}_r$.

(b) Calculate the flux of $\mathbf{E}$ through a sphere centered at the origin.

(c) Calculate $\mathrm{div}(\mathbf{E})$.

(d) Can we use the Divergence Theorem to compute the flux of $\mathbf{E}$ through a sphere centered at the origin?

FIGURE 17 The dipole vector field restricted to the xz-plane.

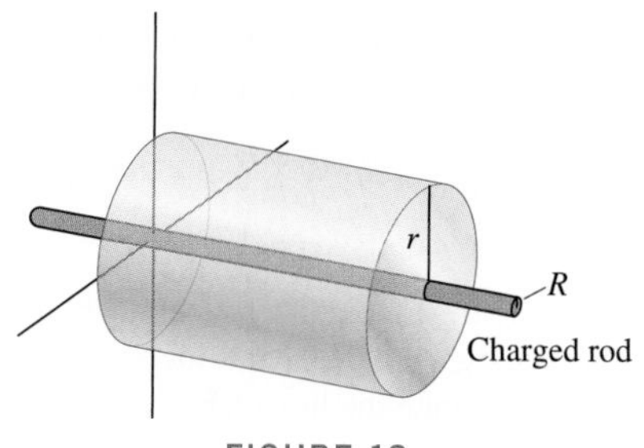

FIGURE 18

30. Let **E** be the electric field due to a long, uniformly charged rod of radius R with charge density δ per unit length (Figure 18). By symmetry, we may assume that **E** is everywhere perpendicular to the rod and its magnitude $E(r)$ depends only on the distance r to the rod (strictly speaking, this would hold only if the rod were infinite, but it is nearly true if the rod is long enough). Show that $E(r) = \dfrac{\delta}{2\pi\epsilon_0 r}$ for $r > R$.
Hint: Apply Gauss's Law to a cylinder of radius R and of unit length with its axis along the rod.

31. Let $I = \iint_{\mathcal{S}} \mathbf{F} \cdot d\mathbf{S}$, where

$$\mathbf{F} = \left\langle \frac{2yz}{\rho^2}, -\frac{xz}{\rho^2}, -\frac{xy}{\rho^2} \right\rangle$$

($\rho = \sqrt{x^2 + y^2 + z^2}$) and $\mathcal{S}$ is the boundary of a region $\mathcal{W}$.

(a) Check that **F** is divergence-free.

(b) Show that $I = 0$ if $\mathcal{S}$ is a sphere centered at the origin. Note that the Divergence Theorem cannot be used. Why not?

(c) Give an argument showing that $I = 0$ for all $\mathcal{S}$.

Further Insights and Challenges

32. Let $\mathcal{S}$ be the boundary surface of a region $\mathcal{W}$ in $\mathbf{R}^3$ and let $D_{\mathbf{e}_n}\varphi$ denote the directional derivative of φ, where $\mathbf{e}_n$ is the outward unit normal vector. Let Δ be the Laplace operator defined earlier.

(a) Use the Divergence Theorem to prove

$$\iint_{\mathcal{S}} D_{\mathbf{e}_n}\varphi\, dS = \iiint_{\mathcal{W}} \Delta\varphi\, dV$$

(b) Show that if φ is a harmonic function (defined in Exercise 27), then

$$\iint_{\mathcal{S}} D_{\mathbf{e}_n}\varphi\, dS = 0$$

33. Assume that φ is harmonic. Show that $\text{div}(\varphi\nabla\varphi) = \|\nabla\varphi\|^2$ and conclude that

$$\iint_{\mathcal{S}} \varphi D_{\mathbf{e}_n}\varphi\, dS = \iiint_{\mathcal{W}} \|\nabla\varphi\|^2\, dV$$

34. Let $\mathbf{F} = \langle P, Q, R\rangle$ be a vector field defined on $\mathbf{R}^3$ such that $\text{div}(\mathbf{F}) = 0$. Use the following steps to show that **F** has a vector potential.

(a) Let $\mathbf{A} = \langle f, 0, g\rangle$. Show that

$$\text{curl}(\mathbf{A}) = \left\langle \frac{\partial g}{\partial y}, \frac{\partial f}{\partial z} - \frac{\partial g}{\partial x}, -\frac{\partial f}{\partial y} \right\rangle$$

(b) Fix any value y_0 and show that if we define

$$f(x, y, z) = -\int_{y_0}^{y} R(x, t, z)\, dt + \alpha(x, z)$$

$$g(x, y, z) = \int_{y_0}^{y} P(x, t, z)\, dt + \beta(x, z)$$

where α and β are any functions of x and z, then $\partial g/\partial y = P$ and $-\partial f/\partial y = R$.

(c) It remains for us to show that α and β can be chosen so $Q = \partial f/\partial z - \partial g/\partial x$. Verify that the following choice works (for any choice of z_0):

$$\alpha(x, z) = \int_{z_0}^{z} Q(x, y_0, t)\, dt, \qquad \beta(x, z) = 0$$

Hint: You will need to use the relation $\text{div}(\mathbf{F}) = 0$.

35. Show that $\mathbf{F} = \langle 2y - 1, 3z^2, 2xy\rangle$ has a vector potential and find one.

36. Show that $\mathbf{F} = \langle 2ye^z - xy, y, yz - z\rangle$ has a vector potential and find one.

37. A vector field with a vector potential has zero flux through every closed surface in its domain. In the text, we observed that although the inverse-square radial vector field $\mathbf{F} = \dfrac{\mathbf{e}_r}{\rho^2}$ satisfies $\text{div}(\mathbf{F}) = 0$, **F** cannot have a vector potential on its domain $\{(x, y, z) \neq (0, 0, 0)\}$ because the flux of **F** through a sphere containing the origin is nonzero.

(a) Show that the method of Exercise 34 produces a vector potential **A** such that $\mathbf{F} = \text{curl}(\mathbf{A})$ on the restricted domain $\mathcal{D}$ consisting of $\mathbf{R}^3$ with the y-axis removed.

(b) Show that **F** also has a vector potential on the domains obtained by removing either the x-axis or the z-axis from $\mathbf{R}^3$.

(c) Does the existence of a vector potential on these restricted domains contradict the fact that the flux of **F** through a sphere containing the origin is nonzero?

CHAPTER REVIEW EXERCISES

1. Let $\mathbf{F}(x, y) = \langle x + y^2, x^2 - y \rangle$ and let $\mathcal{C}$ be the unit circle, oriented counterclockwise. Evaluate $\oint_{\mathcal{C}} \mathbf{F} \cdot d\mathbf{s}$ directly as a line integral and using Green's Theorem.

2. Let $\partial\mathcal{R}$ be the boundary of the rectangle in Figure 1 and let $\partial\mathcal{R}_1$ and $\partial\mathcal{R}_2$ be the boundaries of the two triangles, all oriented counterclockwise.

(a) Determine $\oint_{\partial\mathcal{R}_1} \mathbf{F} \cdot d\mathbf{s}$ if $\oint_{\partial\mathcal{R}} \mathbf{F} \cdot d\mathbf{s} = 4$ and $\oint_{\partial\mathcal{R}_2} \mathbf{F} \cdot d\mathbf{s} = -2$.

(b) What is the value of $\oint_{\partial\mathcal{R}} \mathbf{F}\, d\mathbf{s}$ if $\partial\mathcal{R}$ is oriented clockwise?

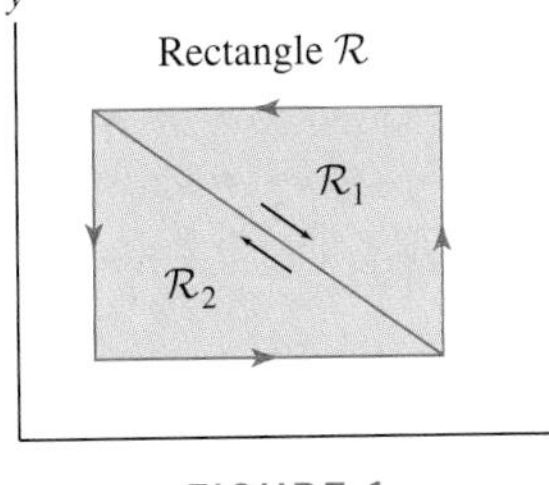

FIGURE 1

In Exercises 3–6, use Green's Theorem to evaluate the line integral around the given closed curve.

3. $\oint_{\mathcal{C}} xy^3\, dx + x^3 y\, dy$, where $\mathcal{C}$ is the rectangle $-1 \le x \le 2$, $-2 \le y \le 3$, oriented counterclockwise.

4. $\oint_{\mathcal{C}} (3x + 5y - \cos y)\, dx + x \sin y\, dy$, where $\mathcal{C}$ is any closed curve enclosing a region with area 4, oriented counterclockwise.

5. $\oint_{\mathcal{C}} y^2\, dx - x^2\, dy$, where $\mathcal{C}$ consists of the arcs $y = x^2$ and $y = \sqrt{x}$, $0 \le x \le 1$, oriented clockwise.

6. $\oint_{\mathcal{C}} ye^x\, dx + xe^y\, dy$, where $\mathcal{C}$ is the triangle with vertices $(-1, 0)$, $(0, 4)$, and $(0, 1)$, oriented counterclockwise.

7. Let $\mathbf{c}(t) = (t^2(1 - t), t(t - 1)^2)$.

(a) GU Plot the path $\mathbf{c}(t)$ for $0 \le t \le 1$.

(b) Calculate the area of the region enclosed by $\mathbf{c}(t)$ for $0 \le t \le 1$.

8. In (a)–(d), state whether the equation is an identity (valid for all $\mathbf{F}$ or φ). If it is not, provide an example in which the equation does not hold.

(a) $\text{curl}(\nabla\varphi) = 0$ **(b)** $\text{div}(\nabla\varphi) = 0$

(c) $\text{div}(\text{curl}(\mathbf{F})) = 0$ **(d)** $\nabla(\text{div}(\mathbf{F})) = 0$

In Exercises 9–12, calculate the curl and divergence of the vector field.

9. $\mathbf{F} = y\mathbf{i} - z\mathbf{k}$

10. $\mathbf{F} = \langle e^{x+y}, e^{y+z}, xyz \rangle$

11. $\mathbf{F} = \nabla(e^{-x^2-y^2-z^2})$

12. $\mathbf{e}_\rho = \rho^{-1}\langle x, y, z \rangle$ $(\rho = \sqrt{x^2 + y^2 + z^2})$

13. Recall that if F_1, F_2, and F_3 are differentiable functions of one variable, then

$$\text{curl}\,(\langle F_1(x), F_2(y), F_3(z) \rangle) = \mathbf{0}$$

Use this to calculate the curl of

$$\mathbf{F} = \langle x^2 + y^2, \ln y + z^2, z^3 \sin(z^2) e^{z^3} \rangle$$

14. Give an example of a nonzero vector field $\mathbf{F}$ such that $\text{curl}(\mathbf{F}) = \mathbf{0}$ and $\text{div}(\mathbf{F}) = 0$.

15. Verify the identities of Exercises 22 and 24 in Section 17.3 for the vector fields $\mathbf{F} = \langle xz, ye^x, yz \rangle$ and $\mathbf{G} = \langle z^2, xy^3, x^2 y \rangle$.

16. Suppose that $\mathcal{S}_1$ and $\mathcal{S}_2$ are surfaces with the same oriented boundary curve $\mathcal{C}$. Which of the following conditions guarantees that the flux of $\mathbf{F}$ through $\mathcal{S}_1$ is equal to the flux of $\mathbf{F}$ through $\mathcal{S}_2$?

(a) $\mathbf{F} = \nabla\varphi$ for some function φ

(b) $\mathbf{F} = \text{curl}(\mathbf{G})$ for some vector field $\mathbf{G}$

17. Prove that if $\mathbf{F}$ is a gradient vector field, then the flux of $\text{curl}(\mathbf{F})$ through a smooth surface $\mathcal{S}$ (whether closed or not) is equal to zero.

18. Verify Stokes' Theorem for $\mathbf{F} = \langle y, z - x, 0 \rangle$ and the surface $z = 4 - x^2 - y^2$, $z \ge 0$, oriented by outward-pointing normals.

In Exercises 19–20, let $\mathbf{F} = \langle z^2, x + z, y^2 \rangle$ and let $\mathcal{S}$ be the upper half of the ellipsoid $\frac{x^2}{4} + y^2 + z^2 = 1$, oriented by outward-pointing normals.

19. Use Stokes' Theorem to compute $\iint_{\mathcal{S}} \text{curl}(\mathbf{F}) \cdot d\mathbf{S}$.

20. Compute $\iint_{\mathcal{S}} \mathbf{F} \cdot d\mathbf{S}$. *Hint:* Find a vector potential $\mathbf{A}$, that is, a vector field $\mathbf{A}$ such that $\mathbf{F} = \text{curl}\,(\mathbf{A})$, and use Stokes' Theorem.

21. Use Stokes' Theorem to evaluate $\oint_{\mathcal{C}} \langle y, z, x \rangle \cdot d\mathbf{s}$, where $\mathcal{C}$ is the curve in Figure 2.

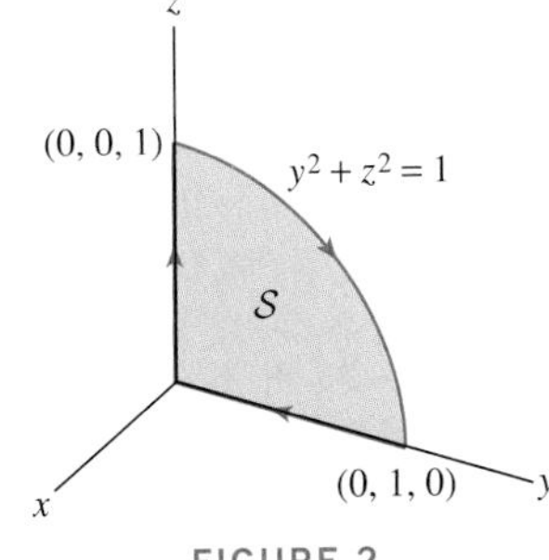

FIGURE 2

22. Verify the Divergence Theorem for $\mathbf{F} = \langle 0, 0, z \rangle$ and the region $x^2 + y^2 + z^2 = 1$.

In Exercises 23–26, use the Divergence Theorem to calculate $\iint_S \mathbf{F} \cdot d\mathbf{S}$ *for the given vector field and surface.*

23. $\mathbf{F} = \langle xy, yz, x^2z + z^2 \rangle$, $\mathcal{S}$ is the boundary of the box $[0, 1] \times [2, 4] \times [1, 5]$.

24. $\mathbf{F} = \langle xy, yz, x^2z + z^2 \rangle$, $\mathcal{S}$ is the boundary of the unit sphere.

25. $\mathbf{F} = \langle xyz + xy, \frac{1}{2}y^2(1 - z) + e^x, e^{x^2+y^2} \rangle$, $\mathcal{S}$ is the boundary of the solid bounded by the cylinder $x^2 + y^2 = 16$ and the planes $z = 0$ and $z = y - 4$.

26. $\mathbf{F} = \langle \sin(yz), \sqrt{x^2 + z^4}, x\cos(x - y) \rangle$, $\mathcal{S}$ is any smooth closed surface that is the boundary of a region in $\mathbf{R}^3$.

27. Find the volume of a region $\mathcal{W}$ if

$$\iint_{\partial \mathcal{W}} \left\langle x + xy + z, x + 3y - \frac{1}{2}y^2, 4z \right\rangle \cdot d\mathbf{S} = 16$$

28. Show that the circulation of $\mathbf{F} = \langle x^2, y^2, z(x^2 + y^2) \rangle$ around any curve C on the surface of the cone $z^2 = x^2 + y^2$ is equal to zero (Figure 3).

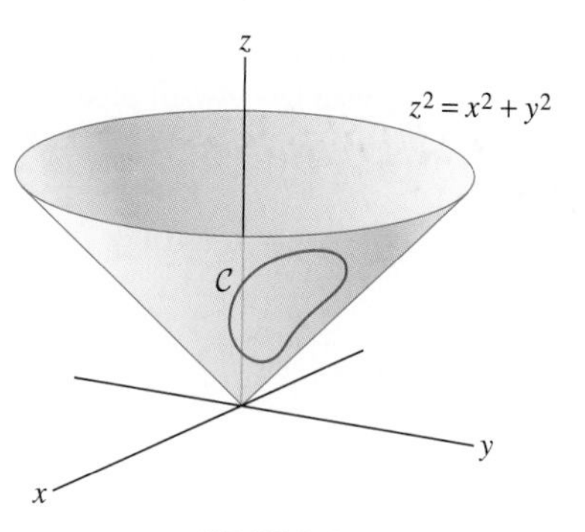

FIGURE 3

In Exercises 29–32, let $\mathbf{F}$ *be a vector field whose curl and divergence at the origin are*

$$\text{curl}(\mathbf{F})(0, 0, 0) = \langle 2, -1, 4 \rangle, \qquad \text{div}(\mathbf{F})(0, 0, 0) = -2$$

29. Estimate $\oint_C \mathbf{F} \cdot d\mathbf{s}$, where C is the circle of radius 0.03 in the xy-plane centered at the origin.

30. Estimate $\oint_C \mathbf{F} \cdot d\mathbf{s}$, where C is the boundary of the square of side 0.03 in the yz-plane centered at the origin. Does the estimate depend on how the square is oriented within the yz-plane? Might the actual circulation depend on how it is oriented?

31. Suppose that $\mathbf{F}$ is the velocity field of a fluid and imagine placing a small paddle wheel at the origin. Find the equation of the plane in which the paddle wheel should be placed to make it rotate as quickly as possible.

32. Estimate the flux of $\mathbf{F}$ through the box of side 0.5 in Figure 4. Does the result depend on how the box is oriented relative to the coordinate axes?

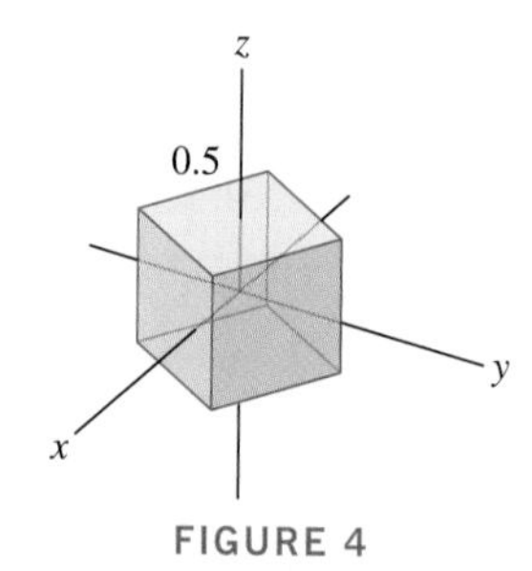

FIGURE 4

33. The velocity field of a fluid (in meters per second) is

$$\mathbf{F} = \langle x^2 + y^2, 0, z^2 \rangle$$

Let $\mathcal{W}$ be the region between the hemisphere

$$\mathcal{S} = \{(x, y, z) : x^2 + y^2 + z^2 = 1, \quad x, y, z \geq 0\}$$

and the disk

$$\mathcal{D} = \{(x, y, 0) : x^2 + y^2 \leq 1\}$$

in the xy-plane.

(a) Show that no fluid flows across $\mathcal{D}$.

(b) Use (a) to show that the rate of fluid flow across $\mathcal{S}$ is equal to $\iiint_{\mathcal{W}} \text{div}(\mathbf{F})\, dV$. Compute this triple integral using spherical coordinates.

34. The velocity field of a fluid (in meters per second) is

$$\mathbf{F} = (3y - 4)\mathbf{i} + e^{-y(z+1)}\mathbf{j} + (x^2 + y^2)\mathbf{k}$$

(a) Estimate the flow rate (in cubic meters per second) through a small surface $\mathcal{S}$ around the origin if $\mathcal{S}$ encloses a region of volume 0.01 m^3.

(b) Estimate the circulation of $\mathbf{F}$ about a circle in the xy-plane of radius $r = 0.1$ m centered at the origin (oriented counterclockwise when viewed from above).

(c) Estimate the circulation of $\mathbf{F}$ about a circle in the yz-plane of radius $r = 0.1$ m centered at the origin (oriented counterclockwise when viewed from the positive x-axis).

35. Let $\varphi(x, y) = x + \dfrac{x}{x^2 + y^2}$. The vector field $\mathbf{F} = \nabla\varphi$ (Figure 5) provides a model in the plane of the velocity field of an incompressible, irrotational fluid flowing past a cylindrical obstacle (in this case, the obstacle is the unit circle $x^2 + y^2 = 1$).

(a) Verify that $\mathbf{F}$ is irrotational [by definition, $\mathbf{F}$ is irrotational if $\text{curl}(\mathbf{F}) = \mathbf{0}$].

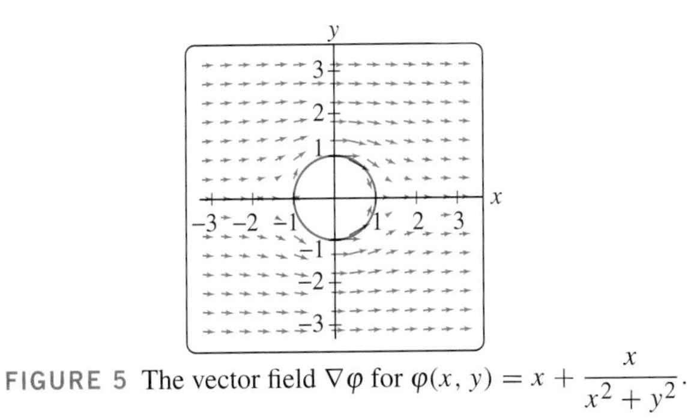

FIGURE 5 The vector field $\nabla\varphi$ for $\varphi(x, y) = x + \dfrac{x}{x^2 + y^2}$.

(b) Verify that $\mathbf{F}$ is tangent to the unit circle at each point along the unit circle except $(1, 0)$ and $(-1, 0)$ (where $\mathbf{F} = \mathbf{0}$).

(c) What is the circulation of $\mathbf{F}$ around the unit circle?

(d) Calculate the line integral of $\mathbf{F}$ along the upper and lower halves of the unit circle separately.

36. Figure 6 shows the vector field $\mathbf{F} = \nabla\varphi$, where $\varphi(x, y) = \ln(x^2 + (y - 1)^2) + \ln(x^2 + (y + 1)^2)$, which is the velocity field for the flow of a fluid with sources of equal strength at $(0, \pm 1)$ (note that φ is undefined at these two points). Show that $\mathbf{F}$ is both irrotational and incompressible, that is, $\text{curl}_z(\mathbf{F}) = 0$ and $\text{div}(\mathbf{F}) = 0$ [in computing $\text{div}(\mathbf{F})$, treat $\mathbf{F}$ as a vector field in $\mathbf{R}^3$ with a zero z-component]. Is it necessary to compute $\text{curl}_z(\mathbf{F})$ to conclude that it is zero?

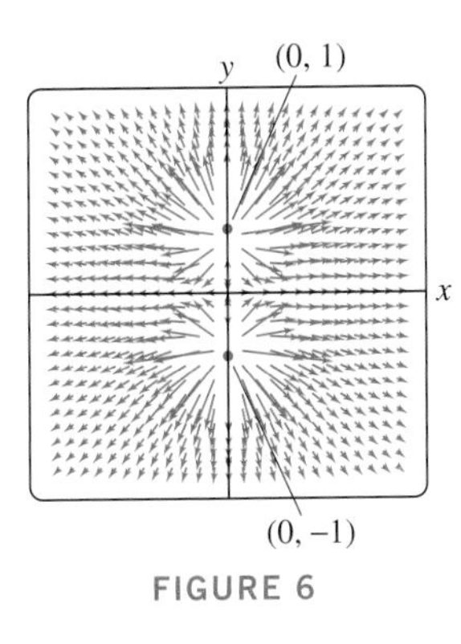

FIGURE 6

37. In Section 17.1, we showed that if $\mathcal{C}$ is a simple closed curve, oriented counterclockwise, then the line integral is

$$\text{Area enclosed by } \mathcal{C} = \frac{1}{2}\oint_{\mathcal{C}} x\,dy - y\,dx \qquad \boxed{1}$$

Suppose that $\mathcal{C}$ is a path from P to Q that is not closed but has the property that every line through the origin intersects $\mathcal{C}$ in at most one point, as in Figure 7. Let $\mathcal{R}$ be the region enclosed by $\mathcal{C}$ and the two radial segments joining P and Q to the origin. Show that the line integral in Eq. (1) is equal to the area of $\mathcal{R}$. *Hint:* Show that the line integral of $\mathbf{F} = \langle -y, x\rangle$ along the two radial segments is zero and apply Green's Theorem.

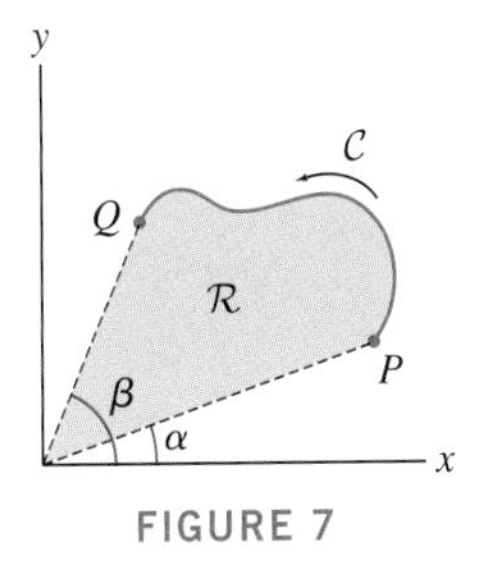

FIGURE 7

38. Suppose that the curve $\mathcal{C}$ in Figure 7 has the polar equation $r = f(\theta)$.

(a) Show that $\mathbf{c}(\theta) = (f(\theta)\cos\theta, f(\theta)\sin\theta)$ is a counterclockwise parametrization of $\mathcal{C}$.

(b) In Section 11.4, we showed that the area of the region $\mathcal{R}$ is given by the formula

$$\text{Area of } \mathcal{R} = \frac{1}{2}\int_{\alpha}^{\beta} f(\theta)^2\,d\theta$$

Use the result of Exercise 37 to give a new proof of this formula. *Hint:* Evaluate the line integral in (1) using $\mathbf{c}(\theta)$.

39. Prove the following generalization of Eq. (1). Let $\mathcal{C}$ be a simple closed curve in the plane (Figure 8)

$$\mathcal{S}: \quad ax + by + cz + d = 0$$

Then the area of the region R enclosed by $\mathcal{C}$ is equal to

$$\frac{1}{2\|\mathbf{n}\|}\oint_{\mathcal{C}} (bz - cy)\,dx + (cx - az)\,dy + (ay - bx)\,dz$$

where $\mathbf{n} = \langle a, b, c\rangle$ is the normal to $\mathcal{S}$ and $\mathcal{C}$ is oriented as the boundary of $\mathcal{R}$ (relative to the normal vector $\mathbf{n}$). *Hint:* Apply Stokes' Theorem to $\mathbf{F} = \langle bz - cy, cx - az, ay - bx\rangle$.

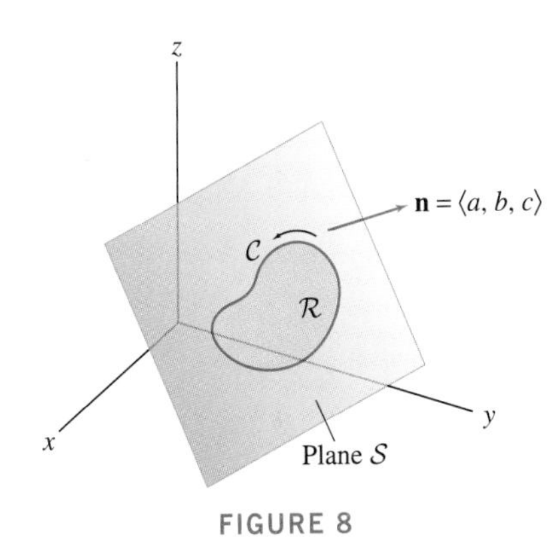

FIGURE 8

40. Use the result of Exercise 39 to calculate the area of the triangle with vertices $(1, 0, 0)$, $(0, 1, 0)$, and $(0, 0, 1)$ as a line integral. Verify your result using geometry.

A THE LANGUAGE OF MATHEMATICS

One of the challenges in learning calculus is growing accustomed to its precise language and terminology, especially in the statements of theorems. In this section, we analyze a few details of logic that are helpful, and indeed essential, in understanding and applying theorems properly.

Many theorems in mathematics involve an **implication**. If A and B are statements, then the implication $A \Longrightarrow B$ is the assertion that A implies B:

$$A \Longrightarrow B: \qquad \textit{If } A \textit{ is true, then } B \textit{ is true.}$$

Statement A is called the **hypothesis** (or premise) and statement B the **conclusion** of the implication. Here is an example: *If m and n are even integers, then $m+n$ is an even integer*. This statement may be divided into a hypothesis and conclusion:

$$\underbrace{m \text{ and } n \text{ are even integers}}_{A} \quad \Longrightarrow \quad \underbrace{m+n \text{ is an even integer}}_{B}$$

In everyday speech, implications are often used in a less precise way. An example is: *If you work hard, then you will succeed*. Furthermore, some statements that do not initially have the form $A \Longrightarrow B$ may be restated as implications. For example, the statement "Cats are mammals" can be rephrased:

$$\text{Let } X \text{ be an animal.} \quad \underbrace{X \text{ is a cat}}_{A} \quad \Longrightarrow \quad \underbrace{X \text{ is a mammal}}_{B}$$

When we say that an implication $A \Longrightarrow B$ is true, we do not claim that A or B is necessarily true. Rather, we are making the conditional statement that *if* A happens to be true, *then* B is also true. In the above, if X does not happen to be a cat, the implication tells us nothing.

The **negation** of a statement A is the assertion that A is false and is denoted $\neg A$.

Statement A	**Negation $\neg A$**
X lives in California.	X does not live in California.
$\triangle ABC$ is a right triangle.	$\triangle ABC$ is not a right triangle.

The negation of the negation is the original statement: $\neg(\neg A) = A$. To say that X does *not not live in California* is the same as saying that X *lives in California*.

EXAMPLE 1 State the negation of:

(a) The door is open and the dog is barking.

(b) The door is open or the dog is barking (or both).

Solution

(a) The first statement is true if two conditions are satisfied (door open and dog barking), and it is false if at least one of these conditions is not satisfied. So the negation is

Either the door is not open *OR* the dog is not barking *(or both)*.

(b) The second statement is true if at least one of the conditions (door open or dog barking) is satisfied, and it is false if neither condition is satisfied. So the negation is

The door is not open *AND* the dog is not barking. ■

Contrapositive and Converse

Two important operations are the formation of the contrapositive and converse of a statement. The **contrapositive** of $A \Longrightarrow B$ is the statement "If B is false, then A is false":

Keep in mind that when we form the contrapositive, we reverse the order of A and B. The contrapositive of $A \Longrightarrow B$ is NOT $\neg A \Longrightarrow \neg B$.

The contrapositive of $A \Longrightarrow B$ is $\neg B \Longrightarrow \neg A$.

Here are some examples:

Statement	Contrapositive
If X is a cat, then X is a mammal.	If X is not a mammal, then X is not a cat.
If you work hard, then you will succeed.	If you did not succeed, then you did not work hard.
If m and n are both even, then $m+n$ is even.	If $m+n$ is not even, then m and n are not both even.

A key observation is this:

The contrapositive and the original implication are equivalent.

The fact that $A \Longrightarrow B$ is equivalent to its contrapositive is a general rule of logic that does not depend on what A and B happen to mean. This rule belongs to the subject of "formal logic," which deals with logical relations between statements without concern for the actual content of these statements.

In other words, if an implication is true, then its contrapositive is automatically true and vice versa. In essence, an implication and its contrapositive are two ways of saying the same thing. For example, the contrapositive "If X is not a mammal, then X is not a cat" is a roundabout way of saying that cats are mammals.

The **converse** of $A \Longrightarrow B$ is the *reverse* implication $B \Longrightarrow A$:

Implication: $A \Longrightarrow B$	Converse $B \Longrightarrow A$
If A is true, then B is true.	If B is true, then A is true.

The converse plays a very different role than the contrapositive because *the converse is NOT equivalent to the original implication*. The converse may be true or false, even if the original implication is true. Here are some examples:

True Statement	Converse	Converse True or False?
If X is a cat, then X is a mammal.	If X is a mammal, then X is a cat.	False
If m is even, then m^2 is even.	If m^2 is even, then m is even.	True

■ **EXAMPLE 2 An Example Where the Converse Is False** Show that the converse of "If m and n are even, then $m+n$ is even" is false.

A counterexample is an example that satisfies the hypothesis but not the conclusion of a statement. If a single counterexample exists, then the statement is false. However, we cannot prove that a statement is true merely by giving an example.

Solution The converse is "If $m + n$ is even, then m and n are even." To show that the converse is false, we display a counterexample. Take $m = 1$ and $n = 3$ (or any other pair of odd numbers). The sum is even (since $1 + 3 = 4$) but neither 1 nor 3 is even. Therefore, the converse is false. ■

■ **EXAMPLE 3** **An Example Where the Converse Is True** State the contrapositive and converse of the Pythagorean Theorem. Are either or both of these true?

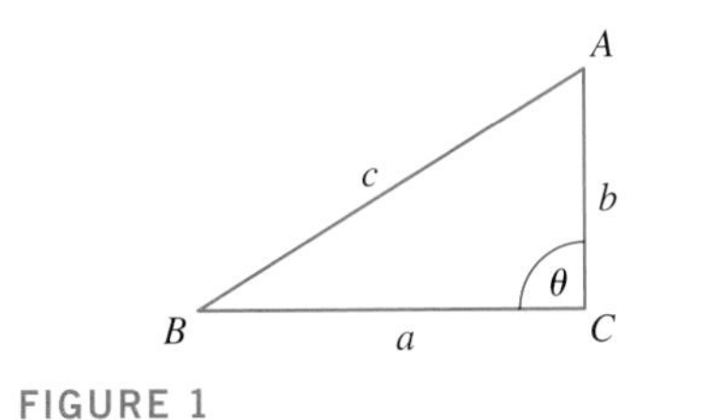

FIGURE 1

Solution Consider a triangle with sides a, b, and c, and let θ be the angle opposite the side of length c as in Figure 1. The Pythagorean Theorem states that if $\theta = 90°$, then $a^2 + b^2 = c^2$. Here are the contrapositive and converse:

Pythagorean Theorem	$\theta = 90° \Longrightarrow a^2 + b^2 = c^2$	True
Contrapositive	$a^2 + b^2 \neq c^2 \Longrightarrow \theta \neq 90°$	Automatically true
Converse	$a^2 + b^2 = c^2 \Longrightarrow \theta = 90°$	True (but not automatic)

The contrapositive is automatically true because it is just another way of stating the original theorem. The converse is not automatically true since there could conceivably exist a nonright triangle that satisfies $a^2 + b^2 = c^2$. However, the converse of the Pythagorean Theorem is, in fact, true. This follows from the Law of Cosines (see Exercise 38). ■

When both a statement $A \Longrightarrow B$ and its converse $B \Longrightarrow A$ are true, we write $A \Longleftrightarrow B$. In this case, A and B are **equivalent**. We often express this with the phrase

$$A \Longleftrightarrow B \qquad A \text{ is true } \textit{if and only if } B \text{ is true.}$$

For example,

$a^2 + b^2 = c^2$	if and only if	$\theta = 90°$
It is morning	if and only if	the sun is rising.

We mention the following variations of terminology involving implications that you may come across:

Statement	Is Another Way of Saying
A is true if B is true.	$B \Longrightarrow A$
A is true only if B is true.	$A \Longrightarrow B$ (A cannot be true unless B is also true.)
For A to be true, it is necessary that B be true.	$A \Longrightarrow B$ (A cannot be true unless B is also true.)
For A to be true, it is sufficient that B be true.	$B \Longrightarrow A$
For A to be true, it is necessary and sufficient that B be true.	$B \Longleftrightarrow A$

Analyzing a Theorem

To see how these rules of logic arise in the study of calculus, consider the following result from Section 4.2.

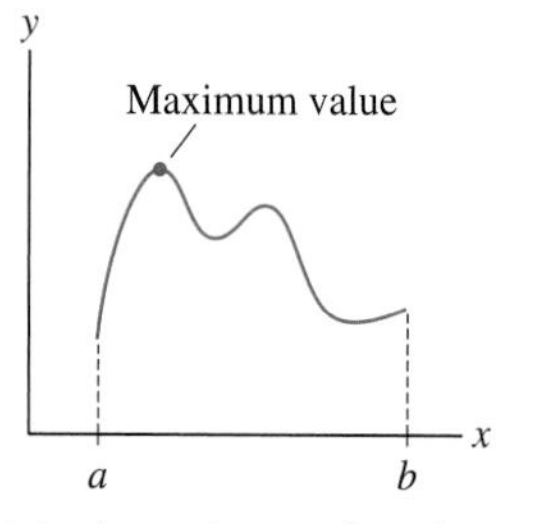

FIGURE 2 A continuous function on a closed interval $I = [a, b]$ has a maximum value.

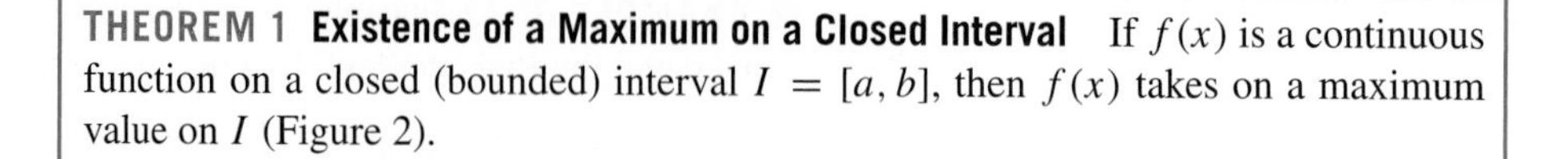

THEOREM 1 Existence of a Maximum on a Closed Interval If $f(x)$ is a continuous function on a closed (bounded) interval $I = [a, b]$, then $f(x)$ takes on a maximum value on I (Figure 2).

To analyze this theorem, let's write out the hypotheses and conclusion separately:

Hypotheses A: $f(x)$ is continuous and I is closed.

Conclusion B: $f(x)$ takes on a maximum value on I.

A first question to ask is: "Are the hypotheses necessary?" Is the conclusion still true if we drop one or both assumptions? To show that both hypotheses are necessary, we provide the counterexamples:

- **The continuity of $f(x)$ is a necessary hypothesis.** Figure 3(A) shows the graph of a function on a closed interval $[a, b]$ that is not continuous. This function has no maximum value on $[a, b]$, which shows that the conclusion may fail if the continuity hypothesis is not satisfied.
- **The hypothesis that I is closed is necessary.** Figure 3(B) shows the graph of a continuous function on an *open interval* (a, b). This function has no maximum value, which shows that the conclusion may fail if the interval is not closed.

We see that both hypotheses in Theorem 1 are necessary. In stating this, we do not claim that the conclusion *always* fails when one or both of the hypotheses are not satisfied. We claim only that the conclusion *may* fail when the hypotheses are not satisfied. Next, let's analyze the contrapositive and converse:

- **Contrapositive $\neg B \Rightarrow \neg A$ (automatically true):** If $f(x)$ does not have a maximum value on I, then either $f(x)$ is not continuous or I is not closed (or both).
- **Converse $B \Rightarrow A$ (in this case, false):** If $f(x)$ has a maximum value on I, then $f(x)$ is continuous and I is closed.

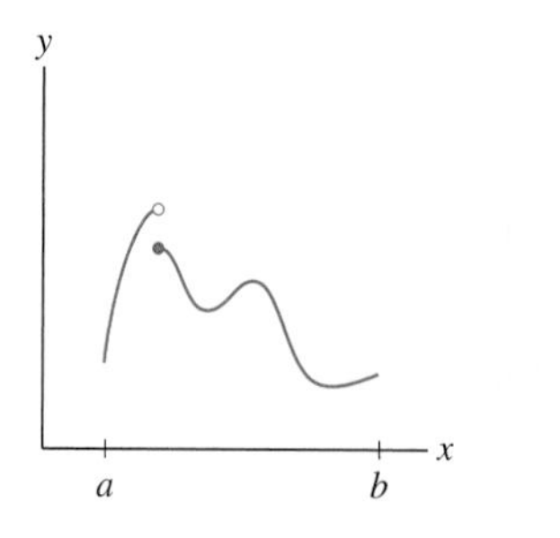

(A) The interval is closed but the function is not continuous. The function has no maximum value.

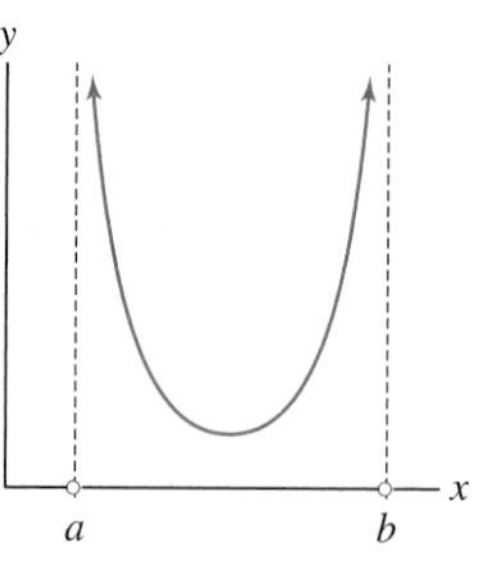

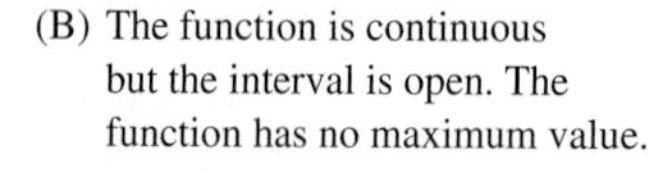

(B) The function is continuous but the interval is open. The function has no maximum value.

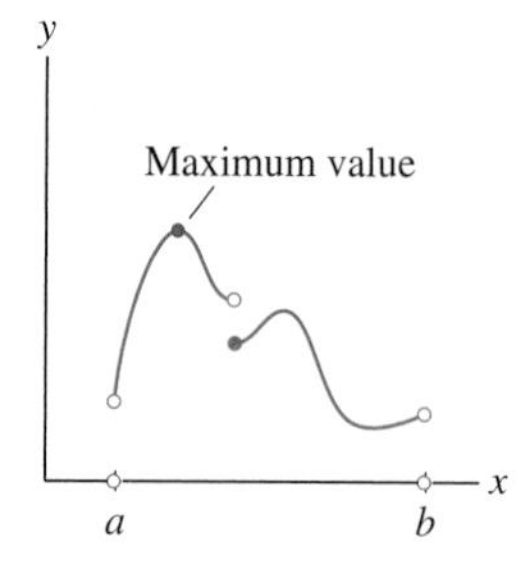

(C) This function is not continuous and the interval is not closed, but the function does have a maximum value.

FIGURE 3

The technique of proof by contradiction *is also known by its Latin name* reductio ad absurdum *or "reduction to the absurd." The ancient Greek mathematicians used proof by contradiction as early as the fifth century* BC*, and Euclid (325–265* BC*) employed it in his classic treatise on geometry entitled* The Elements*. A famous example is the proof that $\sqrt{2}$ is irrational in Example 4. The philosopher Plato (427–347* BC*) wrote: "He is unworthy of the name of man who is ignorant of the fact that the diagonal of a square is incommensurable with its side."*

As we know, the contrapositive is merely a way of restating the theorem, so it is automatically true. The converse is not automatically true, and in fact, in this case it is false. The function in Figure 3(C) provides a counterexample to the converse: $f(x)$ has a maximum value on $I = (a, b)$, but $f(x)$ is not continuous and I is not closed.

Mathematicians have devised various general strategies and methods for proving theorems. The method of proof by induction is discussed in Appendix C. Another important method is **proof by contradiction**, also called **indirect proof**. Suppose our goal is to prove that $A \Longrightarrow B$. In a proof by contradiction, we start by assuming that A is true but B is false, and we then show that this leads to a contradiction. Therefore, B must be true (to avoid the contradiction).

EXAMPLE 4 Proof by Contradiction The number $\sqrt{2}$ is irrational (Figure 4).

Solution Assume that the theorem is false, namely that $\sqrt{2} = p/q$, where p and q are whole numbers. We may assume that p/q is in lowest terms and, therefore, at most one of p and q is even.

The relation $\sqrt{2} = p/q$ implies that $2 = p^2/q^2$ or $p^2 = 2q^2$. This shows that p must be even. But if p is even, say, $p = 2m$, then $4m^2 = 2q^2$ or $q^2 = 2m^2$. This shows that q is also even. This contradicts our assumption that p/q is in lowest terms. We conclude that our original assumption, that $\sqrt{2} = p/q$, must be false. Therefore, $\sqrt{2}$ is irrational. ■

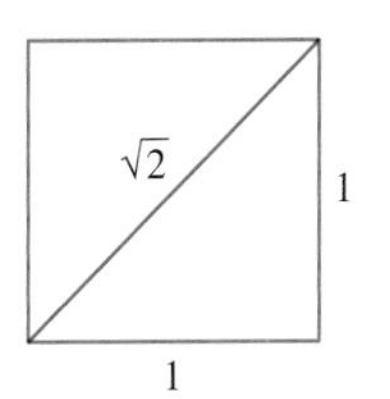

FIGURE 4 The diagonal of the unit square has length $\sqrt{2}$.

CONCEPTUAL INSIGHT The hallmark of mathematics is precision and rigor. A theorem is established, not through observation or experimentation, but by a proof that consists of a chain of reasoning with no gaps.

This approach to mathematics comes down to us from the ancient Greek mathematicians, especially Euclid, and it remains the standard in contemporary research. In recent decades, the computer has become a powerful tool for mathematical experimentation and data analysis. Researchers may use experimental data to discover potential new mathematical facts, but the title "theorem" is not bestowed until someone writes down a proof.

This insistence on theorems and proofs distinguishes mathematics from the other sciences. In the natural sciences, facts are established through experiment and are subject to change or modification as more knowledge is acquired. In mathematics, theories are also developed and expanded, but previous results are not invalidated. The Pythagorean Theorem was discovered in antiquity and is a cornerstone of plane geometry. In the nineteenth century, mathematicians began to study more general types of geometry (of the type that eventually led to Einstein's four-dimensional space-time geometry in the Theory of Relativity). Pythagoras's Theorem does not hold in these more general geometries, but its status in plane geometry is unchanged.

One of the most famous problems in mathematics is known as "Fermat's Last Theorem." It states that the equation

$$x^n + y^n = z^n$$

has no solutions in positive integers if $n \geq 3$. In a marginal note written around 1630, Fermat claimed to have a proof, and over the centuries, that assertion was verified for many values of the exponent n. However, only in 1994 did the British-American mathematician Andrew Wiles, working at Princeton University, find a complete proof.

A. SUMMARY

- The implication $A \Longrightarrow B$ is the assertion "If A is true, then B is true."
- The *contrapositive* of $A \Longrightarrow B$ is the implication $\neg B \Longrightarrow \neg A$, which says "If B is false, then A is false." An implication and its contrapositive are equivalent (one is true if and only if the other is true).
- The converse of $A \Longrightarrow B$ is $B \Longrightarrow A$. An implication and its converse are not necessarily equivalent. One may be true and the other false.
- A and B are *equivalent* if $A \Longrightarrow B$ and $B \Longrightarrow A$ are both true.

- In a proof by contradiction (in which the goal is to prove statement A), we start by assuming that A is false and show that this assumption leads to a contradiction.

A. EXERCISES

Preliminary Questions

1. Which is the contrapositive of $A \Longrightarrow B$?

(a) $B \Longrightarrow A$ **(b)** $\neg B \Longrightarrow A$

(c) $\neg B \Longrightarrow \neg A$ **(d)** $\neg A \Longrightarrow \neg B$

2. Which of the choices in Question 1 is the converse of $A \Longrightarrow B$?

3. Suppose that $A \Longrightarrow B$ is true. Which is then automatically true, the converse or the contrapositive?

4. Restate as an implication: "A triangle is a polygon."

Exercises

1. Which is the negation of the statement "The car and the shirt are both blue"?

(a) Neither the car nor the shirt is blue.

(b) The car is not blue and/or the shirt is not blue.

2. Which is the contrapositive of the implication "If the car has gas, then it will run"?

(a) If the car has no gas, then it will not run.

(b) If the car will not run, then it has no gas.

In Exercises 3–6, state the negation.

3. The time is 4 o'clock.

4. $\triangle ABC$ is an isosceles triangle.

5. m and n are odd integers.

6. Either m is odd or n is odd.

7. x is a real number and y is an integer.

8. $f(x)$ is a linear function.

In Exercises 9–14, state the contrapositive and converse.

9. If m and n are odd integers, then mn is odd.

10. If today is Tuesday, then we are in Belgium.

11. If today is Tuesday, then we are not in Belgium.

12. If $x > 4$, then $x^2 > 16$.

13. If m^2 is divisible by 3, then m is divisible by 3.

14. If $x^2 = 2$, then x is irrational.

In Exercise 15–18, give a counterexample to show that the converse of the statement is false.

15. If m is odd, then $2m + 1$ is also odd.

16. If $\triangle ABC$ is equilateral, then it is an isosceles triangle.

17. If m is divisible by 9 and 4, then m is divisible by 12.

18. If m is odd, then $m^3 - m$ is divisible by 3.

In Exercise 19–22, determine whether the converse of the statement is false.

19. If $x > 4$ and $y > 4$, then $x + y > 8$.

20. If $x > 4$, then $x^2 > 16$.

21. If $|x| > 4$, then $x^2 > 16$.

22. If m and n are even, then mn is even.

In Exercises 23–24, state the contrapositive and converse (it is not necessary to know what these statements mean).

23. If $f(x)$ and $g(x)$ are differentiable, then $f(x)g(x)$ is differentiable.

24. If the force field is radial and decreases as the inverse square of the distance, then all closed orbits are ellipses.

*In Exercises 25–28, the **inverse** of $A \Longrightarrow B$ is the implication $\neg A \Longrightarrow \neg B$.*

25. Which of the following is the inverse of the implication "If she jumped in the lake, then she got wet"?

(a) If she did not get wet, then she did not jump in the lake.

(b) If she did not jump in the lake, then she did not get wet.

Is the inverse true?

26. State the inverses of these implications:

(a) If X is a mouse, then X is a rodent.

(b) If you sleep late, you will miss class.

(c) If a star revolves around the sun, then it's a planet.

27. Explain why the inverse is equivalent to the converse.

28. State the inverse of the Pythagorean Theorem. Is it true?

29. Theorem 1 in Section 2.4 states the following: "If $f(x)$ and $g(x)$ are continuous functions, then $f(x) + g(x)$ is continuous." Does it follow logically that if $f(x)$ and $g(x)$ are not continuous, then $f(x) + g(x)$ is not continuous?

30. Write out a proof by contradiction for this fact: There is no smallest positive rational number. Base your proof on the fact that if $r > 0$, then $0 < r/2 < r$.

31. Use proof by contradiction to prove that if $x + y > 1$, then $x > 1$ or $y > 1$ (or both).

In Exercises 32–35, use proof by contradiction to show that the number is irrational.

32. $\sqrt{\frac{1}{2}}$ **33.** $\sqrt{3}$ **34.** $\sqrt[3]{2}$ **35.** $\sqrt[4]{11}$

36. An isosceles triangle is a triangle with two equal sides. The following theorem holds: If $\triangle$ is a triangle with two equal angles, then $\triangle$ is an isosceles triangle.

(a) What are the hypotheses?

(b) Show by providing a counterexample that the hypothesis is necessary.

(c) What is the contrapositive?

(d) What is the converse? Is it true?

37. Consider the following theorem: Let $f(x)$ be a quadratic polynomial with a positive leading coefficient. Then $f(x)$ has a minimum value.

(a) What are the hypotheses?

(b) What is the contrapositive?

(c) What is the converse? Is it true?

Further Insights and Challenges

38. Let a, b, and c be the sides of a triangle and let θ be the angle opposite c. Use the Law of Cosines (Theorem 1 in Section 1.4) to prove the converse of the Pythagorean Theorem.

39. Carry out the details of the following proof by contradiction, due to R. Palais, that $\sqrt{2}$ is irrational. If $\sqrt{2}$ is rational, then $n\sqrt{2}$ is a whole number for some whole number n. Let n be the smallest such whole number and let $m = n\sqrt{2} - n$.

(a) Prove that $m < n$.

(b) Prove that $m\sqrt{2}$ is a whole number.

Explain why (a) and (b) imply that $\sqrt{2}$ is irrational.

40. Generalize the argument of Exercise 39 to prove that $\sqrt{A}$ is irrational if A is a whole number but not a perfect square. *Hint:* Choose n as before and let $m = n\sqrt{A} - n[\sqrt{A}]$, where $[x]$ is the greatest integer function.

41. Generalize further and show that for any whole r, the rth root $\sqrt[r]{A}$ is irrational unless A is an rth power. *Hint:* Let $x = \sqrt[r]{A}$. Show that if x is rational, we may choose a smallest whole number n such that nx^j is a whole number for $j = 1, \dots, r - 1$. Then consider $m = nx - n[x]$ as before.

42. Given a finite list of prime numbers $p_1, \dots, p_N$, let $M = p_1 \cdot p_2 \cdots p_N + 1$. Show that M is not divisible by any of the primes $p_1, \dots, p_N$. Use this and the fact that every number has a prime factorization to prove that there exist infinitely many prime numbers. This argument was advanced by Euclid in *The Elements*.

B | PROPERTIES OF REAL NUMBERS

"The ingenious method of expressing every possible number using a set of ten symbols (each symbol having a place value and an absolute value) emerged in India. The idea seems so simple nowadays that its significance and profound importance is no longer appreciated. Its simplicity lies in the way it facilitated calculation and placed arithmetic foremost amongst useful inventions. The importance of this invention is more readily appreciated when one considers that it was beyond the two greatest men of Antiquity, Archimedes and Apollonius."

—Pierre-Simon Laplace,
one of the great French mathematicians
of the seventeenth century

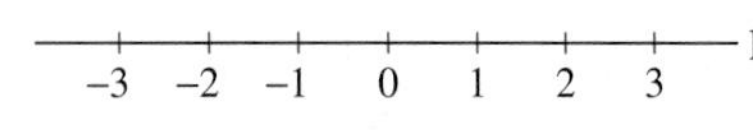

FIGURE 1 The real number line.

In this appendix, we discuss the basic properties of real numbers. First, let us recall that a real number is a number that may be represented by a finite or infinite decimal expansion. The set of all real numbers is denoted **R** and is often visualized as the "number line" (Figure 1). The two algebraic operations of addition and multiplication are defined on **R** and satisfy the Commutative, Associative, and Distributive Laws (Table 1).

TABLE 1 Algebraic Laws

Commutative Laws:	$a + b = b + a, \quad ab = ba$
Associative Laws:	$(a + b) + c = a + (b + c), \quad (ab)c = a(bc)$
Distributive Law:	$a(b + c) = ab + ac$

Every real number x has an additive inverse $-x$ such that $x + (-x) = 0$ and every nonzero real number x has a multiplicative inverse x^{-1} such that $x(x^{-1}) = 1$. We do not regard subtraction and division as separate algebraic operations because they are defined in terms of inverses. By definition, the difference $x - y$ is equal to $x + (-y)$ and the quotient x/y is equal to $x(y^{-1})$ for $y \neq 0$.

In addition to the algebraic operations, there is an **order relation** on **R**: For any two real numbers a and b, precisely one of the following is true:

$$\text{Either} \quad a = b, \quad \text{or} \quad a < b, \quad \text{or} \quad a > b$$

To distinguish between the conditions $a \leq b$ and $a < b$, we often refer to $a < b$ as a **strict inequality**. Similar conventions hold for $>$ and $\geq$. The rules given in Table 2 allow us to manipulate inequalities.

TABLE 2 Order Properties

If $a < b$ and $b < c$,	then $a < c$.
If $a < b$ and $c < d$,	then $a + c < b + d$.
If $a < b$ and $c > 0$,	then $ac < bc$.
If $a < b$ and $c < 0$,	then $ac > bc$.

The last order property says that an inequality reverses direction when multiplied by a negative number c. For example,

$$-2 < 5 \quad \text{but} \quad (-3)(-2) > (-3)5$$

The algebraic and order properties of real numbers are certainly familiar. We now discuss the less familiar **Least Upper Bound (LUB) Property** of the real numbers. This property is one way of expressing the so-called **completeness** of the real numbers. There are other ways of formulating completeness (such as the so-called nested interval property discussed in any book on analysis) that are equivalent to the LUB Property and serve the same purpose. Completeness is used in calculus to construct rigorous proofs of basic theorems about continuous functions such as the Intermediate Value Theorem (IVT) or the existence of extreme values on a closed interval. The underlying idea is that the real number line "has no holes." We elaborate on this idea below. First, we introduce the necessary definitions.

Suppose that S is a nonempty set of real numbers. A number M is called an **upper bound** for S if

$$x \le M \qquad \text{for all } x \in S$$

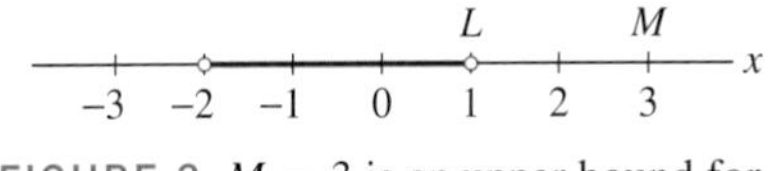

FIGURE 2 $M = 3$ is an upper bound for the set $S = (-2, 1)$ and the LUB is $L = 1$.

If S has an upper bound, we say that S is **bounded above**. A **least upper bound** L is an upper bound for S such that every other upper bound M satisfies $M \ge L$. For example (Figure 2),

- $M = 3$ is an upper bound for the open interval $S = (-2, 1)$.
- $L = 1$ is the LUB for $S = (-2, 1)$.

We now state the LUB Property of the real numbers.

THEOREM 1 Existence of a Least Upper Bound Let S be a nonempty set of real numbers that is bounded above. Then S has an LUB.

In a similar fashion, we say that a number B is a **lower bound** for S if $x \ge B$ for all $x \in S$. We say that S is **bounded below** if S has a lower bound. A **greatest lower bound** (GLB) is a lower bound M such that every other lower bound B satisfies $B \le M$. The set of real numbers also has the GLB Property: If S is a nonempty set of real numbers that is bounded below, then S has a GLB. This may be deduced immediately from Theorem 1. For any nonempty set of real numbers S, let $-S$ be the set of numbers of the form $-x$ for $x \in S$. Then $-S$ has an upper bound if S has a lower bound. Consequently, $-S$ has an LUB L by Theorem 1, and $-L$ is a GLB for S.

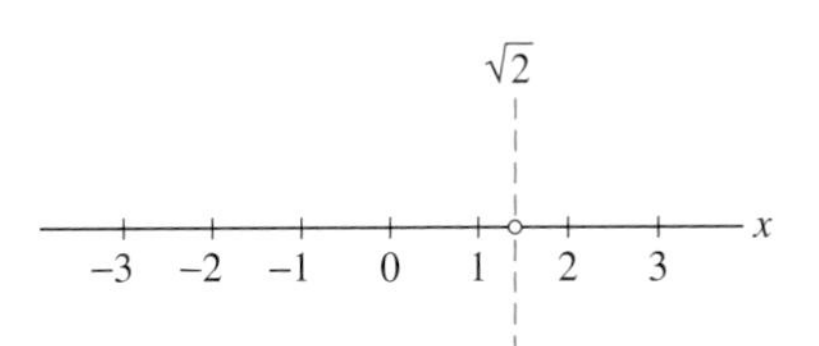

FIGURE 3 The rational numbers have a "hole" at the location $\sqrt{2}$.

CONCEPTUAL INSIGHT Theorem 1 may appear quite reasonable, but perhaps it is not clear why it is useful. We suggested above that the LUB Property expresses the idea that **R** is "complete" or "has no holes." To illustrate this idea, let's compare **R** to the set of rational numbers, denoted **Q**. Intuitively, **Q** is not complete because the irrational numbers are missing. For example, **Q** has a "hole" where the irrational number $\sqrt{2}$ should be located (Figure 3). This hole divides **Q** into two disconnected halves (the half to the left and the half to the right of $\sqrt{2}$). Because of this, **Q** is not a truly "continuous" object. In the following example, we will see that the existence of $\sqrt{2}$ as a real number is directly related to the LUB Property. Another consequence of completeness is the IVT, whose proof below relies on the LUB Property.

EXAMPLE 1 Show that 2 has a square root by applying the LUB Property to the set

$$S = \{x : x^2 < 2\}$$

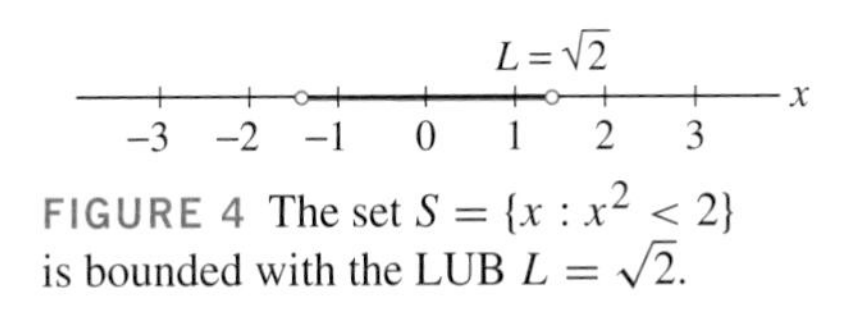

FIGURE 4 The set $S = \{x : x^2 < 2\}$ is bounded with the LUB $L = \sqrt{2}$.

Solution First, we note that S is bounded with the upper bound $M = 2$. Indeed, if $x > 2$, then satisfies $x^2 > 4$, and hence, x does not belong to S. By the LUB Property, S has a least upper bound. Call it L. We claim that $L = \sqrt{2}$ or, equivalently, that $L^2 = 2$ (Figure 4). We prove this by showing that the inequalities $L^2 < 2$ or $L^2 > 2$ are both false.

If $L^2 < 2$, let $b = L + h$, where $h > 0$. Then

$$b^2 = L^2 + 2Lh + h^2 = L^2 + h(2L + h) \qquad \boxed{1}$$

We can make the quantity $h(2L+h)$ as small as desired by choosing $h>0$ small enough. In particular, we may choose a positive h so that $h(2L+h)<2-L^2$. For this choice, $b^2<L^2-(2-L)^2=2$ by Eq. (1). Therefore, $b \in S$. But $b>L$ since $h>0$, and thus L is not an upper bound for S, in contradiction to our hypothesis on L. We conclude that $L^2 \geq 2$.

If $L^2>2$, let $b=L-h$, where $h>0$. Then

$$b^2=L^2-2Lh+h^2=L^2-h(2L+h)$$

Now choose h positive but small enough so that $h(2L+h)<L^2-2$. Then $b<L$ and $b^2>L^2-(L^2-2)=2$. But in this case, b is a smaller lower bound for S. Indeed, if $x \geq b$, then $x^2 \geq b^2>2$, and x does not belong to S. This contradicts our hypothesis that L is the LUB. We conclude that $L^2 \leq 2$, and since we have already shown that $L^2 \geq 2$, we have $L^2=2$ as claimed. ■

We now prove three important theorems, the third of which is used in the proof of the LUB Property below.

THEOREM 2 Bolzano–Weierstrass Theorem Let S be a bounded, infinite set of real numbers. Then there exists a sequence of elements $\{a_n\}$ in S such that the limit $L=\lim_{n\to\infty} a_n$ exists.

Proof For simplicity of notation, we assume that S is contained in the unit interval $[0,1]$ (a similar proof works in general). If $k_1, k_2, \ldots, k_n$ is a sequence of n digits (i.e., each k_j is a whole number and $0 \leq k_j \leq 9$), let

$$S(k_1, k_2, \ldots, k_n)$$

be the set of $x \in S$ whose decimal expansion begins $0.k_1k_2\ldots k_n$. The set S is the union of the subsets $S(0), S(1), \ldots, S(9)$, and since S is infinite, at least one of these subsets must be infinite. Therefore, we may choose k_1 so that $S(k_1)$ is infinite. In a similar fashion, at least one of the set $S(k_1,0), S(k_2,1), \ldots, S(k_1,9)$ must be infinite, so we may choose k_2 so that $S(k_1,k_2)$ is infinite. Continuing in this way, we obtain an infinite sequence $\{k_n\}$ such that $S(k_1,k_2,\ldots,k_n)$ is infinite for all n. Let a_n be an arbitrary element of $S(k_1,k_2,\ldots,k_n)$ and let L be the infinite decimal $0.k_1k_2k_3\ldots$. Then $\lim_{n\to\infty} a_n=L$ since $|L-a_n|<10^{-n}$ for all n. ■

We use the Bolzano–Weierstrass Theorem to prove two important results about sequences $\{a_n\}$. Recall that an upper bound for $\{a_n\}$ is a number M such that $a_j \leq M$ for all j. If an upper bound exists, $\{a_n\}$ is said to be bounded from above. Lower bounds are defined similarly and $\{a_n\}$ is said to be bounded from below if a lower bound exists. A sequence is bounded if it is bounded from above and below. A **subsequence** of $\{a_n\}$ is a sequence of elements $a_{n_1}, a_{n_2}, a_{n_3}, \ldots$, where $n_1<n_2<n_3<\cdots$.

Now consider a set $S=\{a_1,a_2,a_3,\ldots\}$. In this case, the Bolzano–Weierstrass Theorem asserts that there exists a subsequence of elements $a_{n_1}, a_{n_2}, \ldots$ in S such that $\lim_{k\to\infty} a_{n_k}$ exists. Thus, we obtain the following result.

| Section 10.1

THEOREM 3 Every bounded sequence has a convergent subsequence.

THEOREM 4 Bounded Monotonic Sequences Converge

- If $\{a_n\}$ is increasing and $a_n \le M$ for all n, then $\{a_n\}$ converges and $\lim_{n\to\infty} a_n \le M$.
- If $\{a_n\}$ is decreasing and $a_n \ge M$ for all n, then $\{a_n\}$ converges and $\lim_{n\to\infty} a_n \ge M$.

Proof Suppose that $\{a_n\}$ is increasing and bounded above by M. Then $\{a_n\}$ is automatically bounded below by $m = a_1$ since $a_1 \le a_2 \le a_3 \cdots$. Hence, $\{a_n\}$ is bounded, and by Theorem 3, we may choose a convergent subsequence $a_{n_1}, a_{n_2}, \ldots$. Let

$$L = \lim_{k\to\infty} a_{n_k}$$

Observe that $a_n \le L$ for all n. For if not, then $a_n > L$ for some n and then $a_{n_k} \ge a_n > L$ for all k such that $n_k \ge n$. But this contradicts that $a_{n_k} \to L$. Now, by definition, for any $\epsilon > 0$, there exists $N_\epsilon > 0$ such that

$$|a_{n_k} - L| < \epsilon \qquad \text{if } n_k > N_\epsilon$$

Choose k_0 such that $n_{k_0} > N_\epsilon$. If $n \ge n_{k_0} > N_\epsilon$, then $n_{k_0} \le a_n \le L$, and therefore,

$$|a_n - L| \le |a_{n_k} - L| < \epsilon \qquad \text{for all } n \ge n_{k_0}$$

This proves that $\lim_{n\to\infty} a_n = L$ as desired. It remains to prove that $L \le M$. If $L > M$, let $\epsilon = (L - M)/2$ and choose N so that

$$|a_n - L| < \epsilon \qquad \text{if } k > N$$

Then $a_n > L - \epsilon = M + \epsilon$. This contradicts our assumption that M is an upper bound for $\{a_n\}$. Therefore, $L \le M$ as claimed. ■

Proof of Theorem 1 We now use Theorem 4 to prove the LUB Property (Theorem 1). If x is a real number, we write $x(d)$ for the real number obtained by truncating the decimal expansion after the dth digit to the right of the decimal point. We call $x(d)$ the truncation of x of length d. Thus,

$$\text{If } x = 1.41569, \text{ then } x(3) = 1.415.$$

We say that x is a *decimal of length* d if $x = x(d)$. Any two distinct decimals of length d differ by at least 10^{-d}. It follows that for any two real numbers $A < B$, there are at most finitely many decimals of length d between A and B.

Now let S be a nonempty set of real numbers with an upper bound M. We shall prove that S has an LUB. Let $S(d)$ be the set of truncations of length d:

$$S(d) = \{x(d) : x \in S\}$$

We claim that $S(d)$ has a maximum element. To verify this, choose any $a \in S$. If $x \in S$ and $x(d) > a(d)$, then

$$a(d) \le x(d) \le M$$

Thus, by the remark of the previous paragraph, there are at most finitely many values of $x(d)$ in $S(d)$ larger than $a(d)$. The largest of these is the maximum element in $S(d)$.

For $d = 1, 2, \ldots$, choose an element x_d such that $x_d(d)$ is the maximum element in $S(d)$. By construction, $\{x_d(d)\}$ is an increasing sequence (since the largest dth truncation can only become larger as d increases). Furthermore, $x_d(d) \le M$ for all d. We now apply

Theorem 4 to conclude that $\{x_d(d)\}$ converges to a limit L. We claim that L is the LUB of S. Observe first that L is an upper bound for S. Indeed, if $x \in S$, then $x(d) \le L$ for all d and thus $x \le L$. To show that L is the LUB, suppose that M is an upper bound such that $M < L$. Then $x_d \le M$ for all d and hence $x_d(d) \le M$ for all d. But then

$$L = \lim_{d \to \infty} x_d(d) \le M$$

This contradiction shows that L is the LUB. ■

As mentioned above, the LUB Property is used in calculus to establish certain basic theorems about continuous functions. As an example, we prove the IVT. Another example is the theorem on the existence of extrema on a closed interval (see Appendix D).

THEOREM 5 Intermediate Value Theorem If $f(x)$ is continuous on a closed interval $[a, b]$, then for every value M between $f(a)$ and $f(b)$, there exists at least one value $c \in [a, b]$ such that $f(c) = M$.

Proof We may replace $f(x)$ by $f(x) - M$ to reduce to the case $M = 0$. Similarly, replacing $f(x)$ by $-f(x)$ if necessary, we may assume that $f(a) < 0$ and $f(b) > 0$. Now let

$$S = \{x \in [a, b] : f(x) < 0\}$$

Then $a \in S$ since $f(a) < 0$ and thus S is nonempty. Clearly, b is an upper bound for S. Therefore, by the LUB Property, S has an LUB L. We claim that $f(L) = 0$. If not, set $r = f(L)$. We may assume that $r > 0$ (the case $r < 0$ is similar).

Since $f(x)$ is continuous, there exists a number $\delta > 0$ such that

$$|f(x) - f(L)| = |f(x) - r| < \frac{1}{2}r \qquad \text{if} \qquad |x - L| < \delta$$

Equivalently,

$$\frac{1}{2}r < f(x) < \frac{3}{2}r \qquad \text{if} \qquad |x - L| < \delta$$

The number $\frac{1}{2}r$ is positive so we conclude that

$$f(x) > 0 \qquad \text{if} \qquad L - \delta < x < L + \delta$$

But $f(x) \ge 0$ for all $x > L$ since L is an upper bound for S, so we may conclude that $f(x) \ge 0$ for $x \ge L - \delta$. Therefore, $L - \delta$ is an upper bound for S, contradicting our definition of L as the LUB. We conclude that $f(L) = 0$ as desired. ■

C INDUCTION AND THE BINOMIAL THEOREM

The Principle of Induction is a method of proof that is widely used to prove that a given statement $P(n)$ is valid for all natural numbers $n = 1, 2, 3, \ldots$. Here are two statements of this kind:

- $P(n)$: The sum of the first n odd numbers is equal to n^2.
- $P(n)$: $\dfrac{d}{dx}x^n = nx^{n-1}$.

The first statement claims that for all natural numbers n,

$$\underbrace{1 + 3 + \cdots + (2n - 1)}_{\text{Sum of first } n \text{ odd numbers}} = n^2 \qquad \boxed{1}$$

We can check directly that $P(n)$ is true for the first few values of n:

$$P(1) \text{ is the equality:} \qquad 1 = 1^2 \quad \text{(true)}$$

$$P(2) \text{ is the equality:} \qquad 1 + 3 = 2^2 \quad \text{(true)}$$

$$P(3) \text{ is the equality:} \qquad 1 + 3 + 5 = 3^2 \quad \text{(true)}$$

The Principle of Induction may be used to establish $P(n)$ for all n.

The Principle of Induction applies if $P(n)$ is an assertion defined for $n \geq n_0$, where n_0 is a fixed integer. Assume that

*(i) **Initial step:** $P(n_0)$ is true.*

*(ii) **Induction step:** If $P(n)$ is true for $n = k$, then $P(n)$ is also true for $n = k + 1$.*

Then $P(n)$ is true for all $n \geq n_0$.

THEOREM 1 Principle of Induction Let $P(n)$ be an assertion that depends on a natural number n. Assume that:

(i) Initial step: $P(1)$ is true.

(ii) Induction step: If $P(n)$ is true for $n = k$, then $P(n)$ is also true for $n = k + 1$.

Then $P(n)$ is true for all natural numbers $n = 1, 2, 3, \ldots$.

EXAMPLE 1 Prove that $1 + 3 + \cdots + (2n - 1) = n^2$ for all natural numbers n.

Solution As above, we let $P(n)$ denote the equality

$$P(n): \qquad 1 + 3 + \cdots + (2n - 1) = n^2$$

***Step 1.* Initial step: Show that $P(1)$ is true.**

We checked this above. $P(1)$ is the equality $1 = 1^2$.

***Step 2.* Induction step: Show that if $P(n)$ is true for $n = k$, then $P(n)$ is also true for $n = k + 1$.**

Assume that $P(k)$ is true. Then

$$1 + 3 + \cdots + (2k - 1) = k^2$$

Add $2k + 1$ to both sides:

$$\big[1 + 3 + \cdots + (2k - 1)\big] + (2k + 1) = k^2 + 2k + 1 = (k + 1)^2$$

$$1 + 3 + \cdots + (2k + 1) = (k + 1)^2$$

This is precisely the statement $P(k + 1)$. Thus, $P(k + 1)$ is true whenever $P(k)$ is true. By the Principle of Induction, $P(k)$ is true for all k. ■

The intuition behind the Principle of Induction is the following. If $P(n)$ were not true for all n, then there would exist a smallest natural number k such that $P(k)$ is false. Furthermore, $k > 1$ since $P(1)$ is true. Thus $P(k - 1)$ is true [otherwise, $P(k)$ would not be the smallest "counterexample"]. On the other hand, if $P(k - 1)$ is true, then $P(k)$ is also true by the induction step. This is a contradiction. So $P(k)$ must be true for all k.

■ **EXAMPLE 2** Use Induction and the Product Rule to prove that for all whole numbers n,

$$\frac{d}{dx}x^n = nx^{n-1}$$

Solution Let $P(n)$ be the formula $\frac{d}{dx}x^n = nx^{n-1}$.

Step 1. **Initial step: Show that $P(1)$ is true.**

We use the limit definition to verify $P(1)$:

$$\frac{d}{dx}x = \lim_{h\to 0}\frac{(x + h) - x}{h} = \lim_{h\to 0}\frac{h}{h} = \lim_{h\to 0} 1 = 1$$

Step 2. **Induction step: Show that if $P(n)$ is true for $n = k$, then $P(n)$ is also true for $n = k + 1$.**

To carry out the induction step, assume that $\frac{d}{dx}x^k = kx^{k-1}$, where $k \geq 1$. Then, by the Product Rule,

$$\frac{d}{dx}x^{k+1} = \frac{d}{dx}(x \cdot x^k) = x\frac{d}{dx}x^k + x^k\frac{d}{dx}x = x(kx^{k-1}) + x^k$$

$$= kx^k + x^k = (k + 1)x^k$$

This shows that $P(k + 1)$ is true.

By the Principle of Induction, $P(n)$ is true for all $n \geq 1$. ■

In Pascal's Triangle, *the* n*th row displays the coefficients in the expansion of* $(a + b)^n$*:*

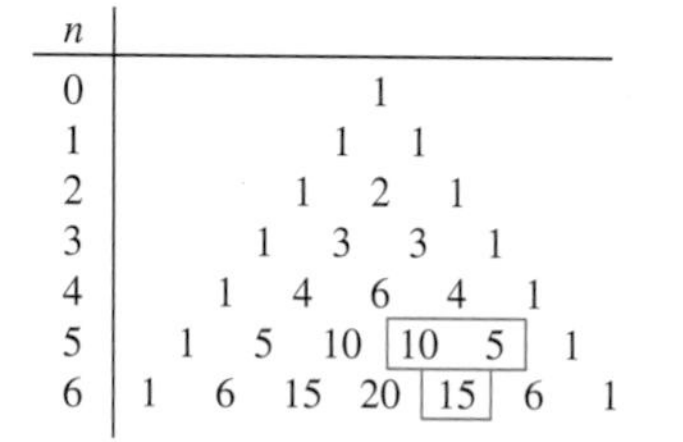

n	
0	1
1	1 1
2	1 2 1
3	1 3 3 1
4	1 4 6 4 1
5	1 5 10 10 5 1
6	1 6 15 20 15 6 1

The triangle is constructed as follows: Each entry is the sum of the two entries above it in the previous line. For example, the entry 15 in line $n = 6$ *is the sum* $10 + 5$ *of the entries above it in line* $n = 5$*. The recursion relation guarantees that the entries in the triangle are the binomial coefficients.*

As another application of induction, we prove the Binomial Theorem, which describes the expansion of the binomial $(a + b)^n$. The first few expansions are familiar:

$$(a + b)^1 = a + b$$

$$(a + b)^2 = a^2 + 2ab + b^2$$

$$(a + b)^3 = a^3 + 3a^2b + 3ab^2 + b^3$$

In general, we have an expansion

$$(a + b)^n = a^n + \binom{n}{1}a^{n-1}b + \binom{n}{2}a^{n-2}b^2 + \binom{n}{3}a^{n-3}b^3 + \cdots + \binom{n}{n-1}ab^{n-1} + b^n \quad \boxed{2}$$

where the coefficient of $x^{n-k}x^k$, denoted $\binom{n}{k}$, is called the **binomial coefficient**. Note that the first term in (2) corresponds to $k = 0$ and the last term to $k = n$; thus, $\binom{n}{0} = \binom{n}{n} = 1$. In summation notation,

$$(a+b)^n = \sum_{k=0}^{n} \binom{n}{k} a^k b^{n-k}$$

Pascal's Triangle (described in the marginal note) can be used to compute binomial coefficients if n and k are not too large. The Binomial Theorem provides the following general formula:

$$\binom{n}{k} = \frac{n!}{k!\,(n-k)!} = \frac{n(n-1)(n-2)\cdots(n-k+1)}{k(k-1)(k-2)\cdots 2\cdot 1} \qquad \boxed{3}$$

Before proving this formula, we prove a recursion relation for binomial coefficients. Note, however, that (3) is certainly correct for $k = 0$ and $k = n$ (recall that by convention, $0! = 1$):

$$\binom{n}{0} = \frac{n!}{(n-0)!\,0!} = \frac{n!}{n!} = 1, \qquad \binom{n}{n} = \frac{n!}{(n-n)!\,n!} = \frac{n!}{n!} = 1$$

THEOREM 2 Recursion Relation for Binomial Coefficients

$$\binom{n}{k} = \binom{n-1}{k} + \binom{n-1}{k-1} \qquad \text{for } 1 \le k \le n-1$$

Proof We write $(a+b)^n$ as $(a+b)(a+b)^{n-1}$ and expand in terms of binomial coefficients:

$$(a+b)^n = (a+b)(a+b)^{n-1}$$

$$\sum_{k=0}^{n} \binom{n}{k} a^{n-k} b^k = (a+b) \sum_{k=0}^{n-1} \binom{n-1}{k} a^{n-1-k} b^k$$

$$= a \sum_{k=0}^{n-1} \binom{n-1}{k} a^{n-1-k} b^k + b \sum_{k=0}^{n-1} \binom{n-1}{k} a^{n-1-k} b^k$$

$$= \sum_{k=0}^{n-1} \binom{n-1}{k} a^{n-k} b^k + \sum_{k=0}^{n-1} \binom{n-1}{k} a^{n-(k+1)} b^{k+1}$$

Replacing k by $k-1$ in the second sum, we obtain

$$\sum_{k=0}^{n} \binom{n}{k} a^{n-k} b^k = \sum_{k=0}^{n-1} \binom{n-1}{k} a^{n-k} b^k + \sum_{k=1}^{n} \binom{n-1}{k-1} a^{n-k} b^k$$

On the right-hand side, the first term in the first sum is a^n and the last term in the second sum is b^n. Thus, we have

$$\sum_{k=0}^{n} \binom{n}{k} a^{n-k} b^k = a^n + \left(\sum_{k=1}^{n-1} \left(\binom{n-1}{k} + \binom{n-1}{k-1} \right) a^{n-k} b^k \right) + b^n$$

The recursion relation follows because the coefficients of $a^{n-k}b^k$ on the two sides of the equation must be equal. ■

We now use induction to prove Eq. (3). Let $P(n)$ be the claim:

$$\binom{n}{k} = \frac{n!}{k!\,(n-k)!} \qquad \text{for } 0 \le k \le n$$

We have $\binom{1}{0} = \binom{1}{1} = 1$ since $(a+b)^1 = a+b$, so $P(1)$ is true. Furthermore, $\binom{n}{n} = 1$ for all n since we can see directly that b^n has coefficient 1 in the expansion of $(a+b)^n$. For the inductive step, assume that $P(n)$ is true. By the recursion relation, for $0 \le k \le n$, we have

$$\binom{n+1}{k} = \binom{n}{k} + \binom{n}{k-1} = \frac{n!}{k!\,(n-k)!} + \frac{n!}{(k-1)!\,(n-k+1)!}$$

$$= n!\left(\frac{n+1-k}{k!\,(n+1-k)!} + \frac{k}{k!\,(n+1-k)!}\right) = n!\left(\frac{n+1}{k!\,(n+1-k)!}\right)$$

$$= \frac{(n+1)!}{k!\,(n+1-k)!}$$

Thus, $P(n+1)$ is also true and the Binomial Theorem follows by induction.

■ **EXAMPLE 3** Use the Binomial Theorem to expand $(x+y)^5$ and $(x+2)^3$.

Solution The fifth row in Pascal's Triangle yields

$$(x+y)^5 = x^5 + 5x^4y + 10x^3y^2 + 10x^2y^3 + 5xy^4 + y^5$$

The third row in Pascal's Triangle yields

$$(x+2)^3 = x^3 + 3x^2(2) + 3x(2)^2 + 2^3 = x^3 + 6x^2 + 12x + 8$$ ■

C. EXERCISES

In Exercises 1–4, use the Principle of Induction to prove the formula for all natural numbers n.

1. $1 + 2 + 3 + \cdots + n = \dfrac{n(n+1)}{2}$

2. $1^3 + 2^3 + 3^3 + \cdots + n^3 = \dfrac{n^2(n+1)^2}{4}$

3. $\dfrac{1}{1\cdot 2} + \dfrac{1}{2\cdot 3} + \cdots + \dfrac{1}{n(n+1)} = \dfrac{n}{n+1}$

4. $1 + x + x^2 + \cdots + x^n = \dfrac{1-x^{n+1}}{1-x}$ for any $x \ne 1$

5. Let $P(n)$ be the statement $2^n > n$.

(a) Show that $P(1)$ is true.

(b) Observe that if $2^n > n$, then $2^n + 2^n > 2n$. Use this to show that if $P(n)$ is true for $n = k$, then $P(n)$ is true for $n = k+1$. Conclude that $P(n)$ is true for all n.

6. Use induction to prove that $n! > 2^n$ for $n \ge 4$.

Let $\{F_n\}$ be the Fibonacci sequence, defined by the recursion formula

$$F_n = F_{n-1} + F_{n-2}, \qquad F_1 = F_2 = 1$$

The first few terms are 1, 1, 2, 3, 5, 8, 13, ... In Exercises 7–10, use induction to prove the identity.

7. $F_1 + F_2 + \cdots + F_n = F_{n+2} - 1$

8. $F_1^2 + F_2^2 + \cdots + F_n^2 = F_{n+1}F_n$

9. $F_n = \dfrac{R_+^n - R_-^n}{\sqrt{5}}$, where $R_\pm = \dfrac{1 \pm \sqrt{5}}{2}$

10. $F_{n+1}F_{n-1} = F_n^2 + (-1)^n$. *Hint:* For the induction step, show that

$$F_{n+2}F_n = F_{n+1}F_n + F_n^2$$

$$F_{n+1}^2 = F_{n+1}F_n + F_{n+1}F_{n-1}$$

11. Use induction to prove that $f(n) = 8^n - 1$ is divisible by 7 for all natural numbers n. *Hint:* For the induction step, show that

$$8^{k+1} - 1 = 7 \cdot 8^k + (8^k - 1)$$

12. Use induction to prove that $n^3 - n$ is divisible by 3 for all natural numbers n.

13. Use induction to prove that $5^{2n} - 4^n$ is divisible by 7 for all natural numbers n.

14. Use Pascal's Triangle to write out the expansions of $(a + b)^6$ and $(a - b)^4$.

15. Expand $(x + x^{-1})^4$.

16. What is the coefficient of x^9 in $(x^3 + x)^5$?

17. Let $S(n) = \sum_{k=0}^{n} \binom{n}{k}$.

(a) Use Pascal's Triangle to compute $S(n)$ for $n = 1, 2, 3, 4$.

(b) Prove that $S(n) = 2^n$ for all $n \geq 1$. *Hint:* Expand $(a + b)^n$ and evaluate at $a = b = 1$.

18. Let $T(n) = \sum_{k=0}^{n} (-1)^k \binom{n}{k}$.

(a) Use Pascal's Triangle to compute $T(n)$ for $n = 1, 2, 3, 4$.

(b) Prove that $T(n) = 0$ for all $n \geq 1$. *Hint:* Expand $(a + b)^n$ and evaluate at $a = 1, b = -1$.

D ADDITIONAL PROOFS

In this appendix, we provide proofs of several theorems that were stated or used in the text.

| Section 10.1

THEOREM 1 If $f(x)$ is continuous and $\{a_n\}$ is a sequence such that the limit $\lim_{n\to\infty} a_n = L$ exists, then

$$\lim_{n\to\infty} f(a_n) = f(L)$$

Proof Choose any $\epsilon > 0$. Since $f(x)$ is continuous, there exists $\delta > 0$ such that

$$|f(x) - f(L)| < \epsilon \qquad \text{if } 0 < |x - L| < \delta$$

Since $\lim_{n\to\infty} a_n = L$, there exists $N > 0$ such that $|a_n - L| < \delta$ for $n > N$. Thus,

$$|f(a_n) - f(L)| < \epsilon \qquad \text{for } n > N$$

It follows that $\lim_{n\to\infty} f(a_n) = f(L)$. ■

| Section 14.3

THEOREM 2 Clairaut's Theorem If f_{xy} and f_{yx} are both continuous functions on a disk D, then $f_{xy}(a, b) = f_{yx}(a, b)$ for all $(a, b) \in D$.

Proof We prove that both $f_{xy}(a, b)$ and $f_{yx}(a, b)$ are equal to the limit

$$L = \lim_{h\to 0} \frac{f(a+h, b+h) - f(a+h, b) - f(a, b+h) + f(a, b)}{h^2}$$

Let $F(x) = f(x, b+h) - f(x, b)$. The numerator in the limit is equal to

$$F(a+h) - F(a)$$

and $F'(x) = f_x(x, b+h) - f_x(x, b)$. By the MVT, there exists a_1 between a and $a+h$ such that

$$F(a+h) - F(a) = hF'(a_1) = h(f_x(a_1, b+h) - f_x(a_1, b))$$

By the MVT applied to f_x, there exists b_1 between b and $b+h$ such that

$$f_x(a_1, b+h) - f_x(a_1, b) = hf_{xy}(a_1, b_1)$$

Thus,

$$F(a+h) - F(a) = h^2 f_{xy}(a_1, b_1)$$

and

$$L = \lim_{h\to 0} \frac{h^2 f_{xy}(a_1, b_1)}{h^2} = \lim_{h\to 0} f_{xy}(a_1, b_1) = f_{xy}(a, b)$$

The last equality follows from the continuity of f_{xy} since (a_1, b_1) approaches (a, b) as $h \to 0$. To prove that $L = f_{yx}(a, b)$, repeat the argument using the function $F(y) = f(a+h, y) - f(a, y)$, with the roles of x and y reversed. ■

| Section 14.4

THEOREM 3 Criterion for Differentiability If $f_x(x, y)$ and $f_y(x, y)$ exist and are continuous on an open disk D, then $f(x, y)$ is differentiable on D.

Proof Let $(a, b) \in D$ and set

$$L(x, y) = f(a, b) + f_x(a, b)(x - a) + f_y(a, b)(y - b)$$

To prove that $f(x, y)$ is differentiable, we must show that

$$f(x, y) = L(x, y) + \epsilon(x, y)\sqrt{(x - a)^2 + (y - b)^2}$$

where $\epsilon(x, y)$ is a function such that $\lim_{(x,y)\to(a,b)} \epsilon(x, y) = 0$. It is convenient to switch to the variables h and k, where $x = a + h$ and $y = b + k$. Set

$$\Delta f = f(a + h, b + k) - f(a, b)$$

Then

$$L(x, y) = f(a, b) + f_x(a, b)h + f_y(a, b)k$$

and we may define the function

$$E(h, k) = f(x, y) - L(x, y) = \Delta f - (f_x(a, b)h + f_y(a, b)k)$$

Thus, $\epsilon(x, y) = E(h, k)/\sqrt{h^2 + k^2}$ and it is necessary to show that

$$\lim_{(h,k)\to(0,0)} \frac{E(h, k)}{\sqrt{h^2 + k^2}} = 0$$

To do this, we write Δf as a sum of two terms:

$$\Delta f = (f(a + h, b + k) - f(a, b + k)) + (f(a, b + k) - f(a, b))$$

and apply the MVT to each term separately. We find that there exist a_1 between a and $a + h$ and b_1 between b and $b + k$ such that

$$f(a + h, b + k) - f(a, b + k) = hf_x(a_1, b + k)$$
$$f(a, b + k) - f(a, b) = kf_y(a, b_1)$$

Therefore,

$$E(h, k) = h(f_x(a_1, b + k) - f_x(a, b)) + k(f_y(a, b_1) - f_y(a, b))$$

and

$$\begin{aligned}
\left|\frac{E(h, k)}{\sqrt{h^2 + k^2}}\right| &= \left|\frac{h(f_x(a_1, b + k) - f_x(a, b)) + k(f_y(a, b_1) - f_y(a, b))}{\sqrt{h^2 + k^2}}\right| \\
&\le \left|\frac{h(f_x(a_1, b + k) - f_x(a, b))}{\sqrt{h^2 + k^2}}\right| + \left|\frac{k(f_y(a, b_1) - f_y(a, b))}{\sqrt{h^2 + k^2}}\right| \\
&\le \left|\frac{h(f_x(a_1, b + k) - f_x(a, b))}{h}\right| + \left|\frac{k(f_y(a, b_1) - f_y(a, b))}{k}\right| \\
&= |f_x(a_1, b + k) - f_x(a, b)| + |f_y(a, b_1) - f_y(a, b)|
\end{aligned}$$

In the second line, we use the Triangle Inequality, and we may pass to the third line because $|h|$ and $|k|$ are both smaller than $\sqrt{h^2 + k^2}$. Both terms in the last line tend to zero as $(h, k) \to (0, 0)$ because f_x and f_y are assumed to be continuous. This completes the proof that $f(x, y)$ is differentiable. ■

E | TAYLOR POLYNOMIALS

English mathematician Brook Taylor (1685–1731) made important contributions to calculus and physics, as well as to the theory of linear perspective used in drawing.

We have seen that approximation is a basic theme in calculus. One example is linearization, where we approximate a differentiable function $f(x)$ at $x = a$ by the linear function

$$L(x) = f(a) + f'(a)(x - a)$$

However, a drawback of the linearization is that it is accurate only in a small interval around $x = a$. Taylor polynomials are higher-degree approximations that generalize the linearization using the higher derivatives $f^{(k)}(a)$. They are useful because, by taking sufficiently high degree, we can approximate transcendental functions such as $\sin x$ and e^x to arbitrary accuracy on any given interval.

Assume that $f(x)$ is defined on an open interval I and that all higher derivatives $f^{(k)}(x)$ exist on I. Fix a number $a \in I$. The nth **Taylor polynomial** for f centered at $x = a$ is the polynomial

$$T_n(x) = f(a) + \frac{f'(a)}{1!}(x - a) + \frac{f''(a)}{2!}(x - a)^2 + \cdots + \frac{f^{(n)}(a)}{n!}(x - a)^n$$

← REMINDER k-factorial *is the number* $k! = k(k-1)(k-2)\cdots(2)(1)$. *Thus,*

$$1! = 1, \quad 2! = (2)1 = 2, \quad 3! = (3)(2)1 = 6$$

By convention, we define $0! = 1$.

It is convenient to regard $f(x)$ itself as the *zeroth* derivative $f^{(0)}(x)$. Then we may write the Taylor polynomial in summation notation,

$$T_n(x) = \sum_{j=0}^{n} \frac{f^{(j)}(a)}{j!}(x - a)^j$$

When $a = 0$, $T_n(x)$ is also called the nth **Maclaurin polynomial**. The first few Taylor polynomials are

$$T_0(x) = f(a)$$

$$T_1(x) = f(a) + f'(a)(x - a)$$

$$T_2(x) = f(a) + f'(a)(x - a) + \frac{1}{2}f''(a)(x - a)^2$$

$$T_3(x) = f(a) + f'(a)(x - a) + \frac{1}{2}f''(a)(x - a)^2 + \frac{1}{6}f'''(a)(x - a)^3$$

Scottish mathematician Colin Maclaurin (1698–1746) was a professor in Edinburgh. Newton was so impressed by his work that he once offered to pay part of Maclaurin's salary.

Note that $T_1(x)$ is the *linear approximation* to $f(x)$ at a. As we will see, the higher-degree Taylor polynomials provide increasingly better approximations to $f(x)$. Before computing some Taylor polynomials, we record two important properties that follow from the definition:

- $T_n(a) = f(a)$ [since all terms in $T_n(x)$ after the first are zero at $x = a$].
- $T_n(x)$ is obtained from $T_{n-1}(x)$ by adding on a term of degree n:

$$T_n(x) = T_{n-1}(x) + \frac{f^{(n)}(a)}{n!}(x - a)^n$$

EXAMPLE 1 Computing Taylor Polynomials Let $f(x) = \sqrt{x + 1}$. Compute $T_n(x)$ at $a = 3$ for $n = 0, 1, 2, 3$, and 4.

Solution First evaluate the derivatives $f^{(j)}(3)$:

$$f(x) = (x+1)^{1/2}, \qquad f(3) = 2$$
$$f'(x) = \frac{1}{2}(x+1)^{-1/2} \qquad f'(3) = \frac{1}{4}$$
$$f''(x) = -\frac{1}{4}(x+1)^{-3/2} \qquad f''(3) = -\frac{1}{32}$$
$$f'''(x) = \frac{3}{8}(x+1)^{-5/2} \qquad f'''(3) = \frac{3}{256}$$
$$f^{(4)}(x) = -\frac{15}{16}(x+1)^{-7/2} \qquad f^{(4)}(3) = -\frac{15}{2{,}048}$$

Then compute the coefficients $\dfrac{f^{(j)}(3)}{j!}$:

The first term $f(a)$ in the Taylor polynomial $T_n(x)$ is called the constant term.

$$\text{Constant term} = f(3) = 2$$
$$\text{Coefficient of } (x-3) = f'(3) = \frac{1}{4}$$
$$\text{Coefficient of } (x-3)^2 = \frac{f''(3)}{2!} = -\frac{1}{32}\cdot\frac{1}{2!} = -\frac{1}{64}$$
$$\text{Coefficient of } (x-3)^3 = \frac{f'''(3)}{3!} = \frac{3}{256}\cdot\frac{1}{3!} = \frac{1}{512}$$
$$\text{Coefficient of } (x-3)^4 = \frac{f^{(4)}(3)}{4!} = -\frac{15}{2{,}048}\cdot\frac{1}{4!} = -\frac{5}{16{,}384}$$

The first four Taylor polynomials centered at $a = 3$ are (see Figure 1):

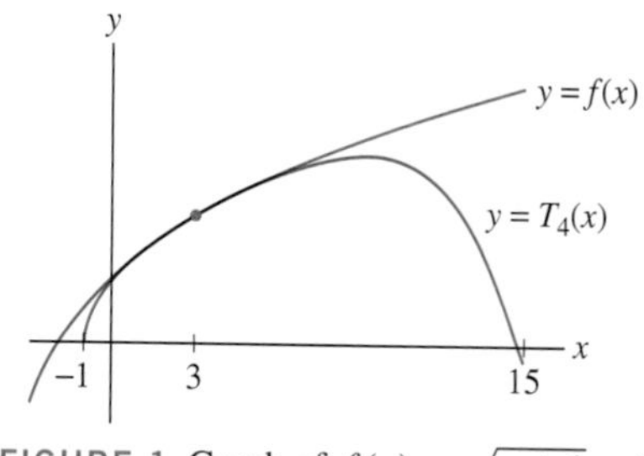

FIGURE 1 Graph of $f(x) = \sqrt{x+1}$ and $T_4(x)$ centered at $x = 3$.

$$T_0(x) = 2$$
$$T_1(x) = 2 + \frac{1}{4}(x-3)$$
$$T_2(x) = 2 + \frac{1}{4}(x-3) - \frac{1}{64}(x-3)^2$$
$$T_3(x) = 2 + \frac{1}{4}(x-3) - \frac{1}{64}(x-3)^2 + \frac{1}{512}(x-3)^3$$
$$T_4(x) = 2 + \frac{1}{4}(x-3) - \frac{1}{64}(x-3)^2 + \frac{1}{512}(x-3)^3 - \frac{5}{16{,}384}(x-3)^4 \quad ■$$

■ **EXAMPLE 2 Finding a General Formula for T_n** Find the Taylor polynomials $T_n(x)$ of $f(x) = \ln x$ at $a = 1$.

After computing several derivatives of $f(x) = \ln x$, we begin to discern the pattern. Often, however, the derivatives follow no simple pattern and there is no simple formula for the general Taylor polynomial.

Solution For $f(x) = \ln x$, the constant term of $T_n(x)$ at $a = 1$ is zero since $f(1) = \ln 1 = 0$. Next, we compute the derivatives:

$$f'(x) = x^{-1}, \qquad f''(x) = -x^{-2}, \qquad f'''(x) = 2x^{-3}, \qquad f^{(4)}(x) = -3\cdot 2x^{-4}$$

Similarly, $f^{(5)}(x) = 4\cdot 3\cdot 2x^{-5}$. The general pattern is that $f^{(k)}(x)$ is a multiple of x^{-k}, with a coefficient $(k-1)!$ that alternates in sign:

$$f^{(k)}(x) = (-1)^{k-1}(k-1)!\,x^{-k} \qquad \boxed{1}$$

Therefore, the coefficient of $(x-1)^k$ in $T_n(x)$ is

$$\frac{f^{(k)}(1)}{k!} = \frac{(-1)^{k-1}(k-1)!}{k!} = \frac{(-1)^{k-1}}{k} \quad (\text{for } k \geq 1)$$

In other words, the coefficients for $k \geq 1$ are $1, -\frac{1}{2}, \frac{1}{3}, -\frac{1}{4}, \ldots$, and

$$T_n(x) = (x-1) - \frac{1}{2}(x-1)^2 + \frac{1}{3}(x-1)^3 - \cdots + (-1)^{n-1}\frac{1}{n}(x-1)^n$$

Taylor polynomials for $\ln x$ *at* $a = 1$:

$T_1(x) = (x-1)$

$T_2(x) = (x-1) - \frac{1}{2}(x-1)^2$

$T_3(x) = (x-1) - \frac{1}{2}(x-1)^2 + \frac{1}{3}(x-1)^3$

■ **EXAMPLE 3 Maclaurin Polynomials for $f(x) = \cos x$** Find the Maclaurin polynomials of $f(x) = \cos x$.

Solution Recall that the Maclaurin polynomials are the Taylor polynomials centered at $a = 0$. The key observation is that the derivatives of $f(x) = \cos x$ form a pattern that repeats with period 4:

$$f'(x) = -\sin x, \qquad f''(x) = -\cos x, \qquad f'''(x) = \sin x, \qquad f^{(4)}(x) = \cos x,$$

and, in general, $f^{(j+4)}(x) = f^{(j)}(x)$. At $x = 0$, the derivatives form the repeating pattern $1, 0, -1$, and 0:

$f(0)$	$f'(0)$	$f''(0)$	$f'''(0)$	$f^{(4)}(0)$	$f^{(5)}(0)$	$f^{(6)}(0)$	$f^{(7)}(0)$	$\ldots$
1	0	-1	0	1	0	-1	0	$\ldots$

In other words, the even derivatives are $f^{(2k)}(0) = (-1)^k$ and the odd derivatives are zero: $f^{(2k+1)}(0) = 0$. Therefore, the coefficient of x^{2k} is $(-1)^k/(2k)!$ and the coefficient of x^{2k+1} is zero. We have

$$T_0(x) = T_1(x) = 1$$

$$T_2(x) = T_3(x) = 1 - \frac{1}{2!}x^2$$

$$T_4(x) = T_5(x) = 1 - \frac{x^2}{2} + \frac{x^4}{4!}$$

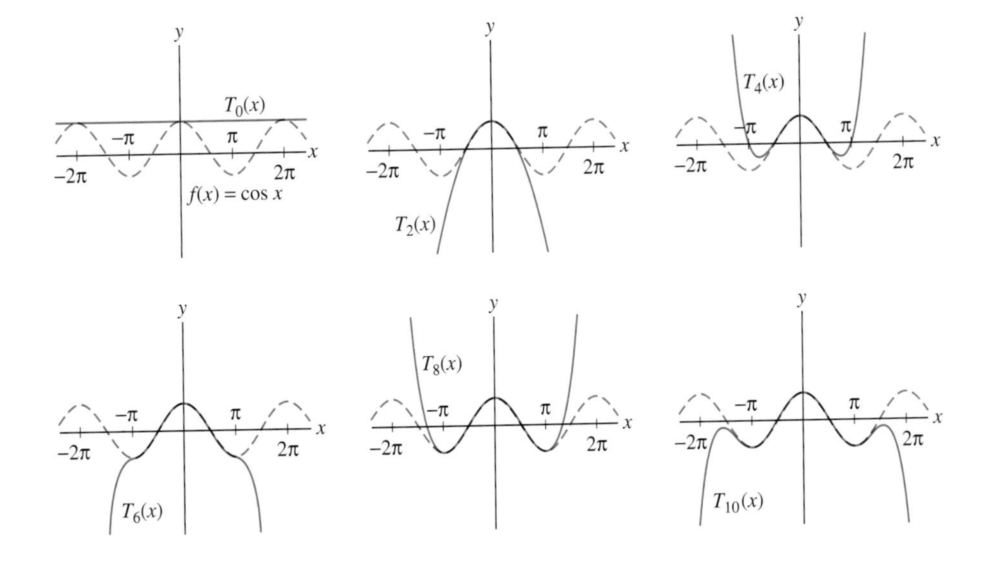

FIGURE 2 Graphs of Taylor polynomials for $f(x) = \cos x$. The graph of $f(x)$ is shown as a dashed curve.

and, in general,

$$T_{2n}(x) = T_{2n+1}(x) = 1 - \frac{1}{2}x^2 + \frac{1}{4!}x^4 - \frac{1}{6!}x^6 + \cdots + (-1)^n \frac{1}{(2n)!}x^{2n}$$ ■

Taylor polynomials $T_n(x)$ are designed to approximate $f(x)$ in an interval around $x = a$. Figure 2 shows the first few Maclaurin polynomials for $f(x) = \cos x$. Observe that as n gets larger, $T_n(x)$ approximates $f(x) = \cos x$ well over larger and larger intervals. Outside this interval, the approximation fails.

CONCEPTUAL INSIGHT To understand how the formula for T_n arises, it is useful to introduce a notion of closeness for two functions. We say that f and g *agree to order n* at $x = a$ if

$$f(a) = g(a), \quad f'(a) = g'(a), \quad f''(a) = g''(a), \ldots, \quad f^{(n)}(a) = g^{(n)}(a)$$

Taylor polynomials are defined so that $T_n(x)$ and $f(x)$ agree to order n at $x = a$ (see Exercise 59) and, in fact, $T_n(x)$ is the only polynomial of degree n with this property. For example, we can check that $T_2(x)$ agrees with $f(x)$ to order $n = 2$:

$$T_2(x) = f(a) + f'(a)(x-a) + \frac{1}{2}f''(a)(x-a)^2, \qquad T_2(a) = f(a)$$

$$T_2'(x) = f'(a) + f''(a)(x-a), \qquad T_2'(a) = f'(a)$$

$$T_2''(x) = f''(a), \qquad T_2''(a) = f''(a)$$

The Remainder Term

Our next goal is to study the error $|T_n(x) - f(x)|$ in the approximation provided by the nth Taylor polynomial centered at $x = a$. Define the ***n*th remainder** for $f(x)$ at $x = a$ by

$$R_n(x) = f(x) - T_n(x)$$

The error is the absolute value of the remainder. Also, $f(x) = T_n(x) + R_n(x)$, so

$$f(x) = f(a) + \frac{f'(a)}{1!}(x-a)^1 + \frac{f''(a)}{2!}(x-a)^2 + \cdots + \frac{f^{(n)}(a)}{n!}(x-a)^n + R_n(x)$$

Taylor's Theorem Assume that $f^{(n+1)}(x)$ exists and is continuous. Then

$$R_n(x) = \frac{1}{n!}\int_a^x (x-u)^n f^{(n+1)}(u)\,du \qquad \boxed{2}$$

Proof Set

$$I_n(x) = \frac{1}{n!}\int_a^x (x-u)^n f^{(n+1)}(u)\,du$$

Our goal is to show that $R_n(x) = I_n(x)$. For $n = 0$, $R_0(x) = f(x) - f(a)$ and the desired result is just a restatement of the Fundamental Theorem of Calculus:

$$I_0(x) = \int_a^x f'(u)\,du = f(x) - f(a) = R_0(x)$$

Exercise 55 reviews this proof for the special case $n = 2$.

To prove the formula for $n > 0$, we apply Integration by Parts to $I_n(x)$ with

$$h(u) = \frac{(x-u)^n}{n!}, \qquad g(u) = f^{(n)}(u)$$

$$\begin{aligned} I_n(x) &= \int_a^x h(u)\,g'(u)\,du = h(u)g(u)\Big|_a^x - \int_a^x h'(u)g(u)\,du \\ &= \frac{1}{n!}(x-u)^n f^{(n)}(u)\Big|_a^x - \frac{1}{n!}\int_a^x (-n)(x-u)^{n-1} f^{(n)}(u)\,du \\ &= -\frac{1}{n!}(x-a)^n f^{(n)}(a) + I_{n-1}(x) \end{aligned}$$

This result can be rewritten as

$$I_{n-1}(x) = \frac{f^{(n)}(a)}{n!}(x-a)^n + I_n(x)$$

Now apply this relation n times:

$$\begin{aligned} f(x) &= f(a) + I_0(x) \\ &= f(a) + \frac{f'(a)}{1!}(x-a) + I_1(x) \\ &= f(a) + \frac{f'(a)}{1!}(x-a) + \frac{f''(a)}{2!}(x-a)^2 + I_2(x) \\ &\ \ \vdots \\ &= f(a) + \frac{f'(a)}{1!}(x-a) + \cdots + \frac{f^{(n)}(a)}{n!}(x-a)^n + I_n(x) \end{aligned}$$

This shows that $f(x) = T_n(x) + I_n(x)$ and hence $I_n(x) = R_n(x)$ as desired. ■

Although Taylor's Theorem gives us an explicit formula for the remainder, we will not use this formula directly. Instead, we will use it to *estimate* the size of the error.

THEOREM 1 Error Bound Assume that $f^{(n+1)}(x)$ exists and is continuous. Let K be a number such that $|f^{(n+1)}(u)| \le K$ for all u between a and x. Then

$$|T_n(x) - f(x)| \le K\frac{|x-a|^{n+1}}{(n+1)!}$$

Proof Assume that $x \ge a$ (the case $x \le a$ is similar). Then, since $|f^{(n+1)}(u)| \le K$ for $a \le u \le x$,

In (3), we use the inequality

$$\left| \int_a^b f(x)\,dx \right| \le \int_a^b |f(x)|\,dx$$

which is valid for all integrable functions.

$$\begin{aligned} |T_n(x) - f(x)| = |R_n(x)| &= \left| \frac{1}{n!} \int_a^x (x-u)^n f^{(n+1)}(u)\,du \right| \\ &\le \frac{1}{n!} \int_a^x \left| (x-u)^n f^{(n+1)}(u) \right| du \qquad \boxed{3} \\ &\le \frac{K}{n!} \int_a^x (x-u)^n\,du \\ &= \frac{K}{n!} \frac{-(x-u)^{n+1}}{n+1} \Bigg|_{u=a}^{x} = K \frac{|x-a|^{n+1}}{(n+1)!} \end{aligned}$$ ■

■ **EXAMPLE 4 Using the Error Bound** Use the Error Bound to find a bound for the error $|T_3(1.2) - \ln 1.2|$, where $T_3(x)$ is the third Taylor polynomial for $f(x) = \ln x$ at $a = 1$. Check your result using a calculator.

Solution To use the Error Bound with $n = 3$, we must find a value of K such that $|f^{(4)}(u)| \le K$ for $1 \le u \le 1.2$. By Eq. (1) in Example 2, $|f^{(4)}(x)| = 6x^{-4}$. Since $|f^{(4)}(x)|$ is decreasing for $x \ge 1$, its maximum value on $[1, 1.2]$ is $|f^{(4)}(1)| = 6$. Therefore, we may apply the Error Bound with $K = 6$ to obtain

$$|T_3(1.2) - \ln 1.2| \le K \frac{|x-a|^{n+1}}{(n+1)!} = 6 \frac{|1.2-1|^4}{4!} \approx 0.0004$$

By Example 2,

$$T_3(x) = (x-1) - \frac{1}{2}(x-1)^2 + \frac{1}{3}(x-1)^3$$

The following values from a calculator confirm that the error is at most 0.0004:

$$|T_3(1.2) - \ln 1.2| \approx |0.182667 - 0.182322| \approx 0.00035 < 0.0004$$ ■

■ **EXAMPLE 5 Approximating with a Given Accuracy** Let $T_n(x)$ be the nth Maclaurin polynomial for $f(x) = \cos x$. Find a value of n such that

$$|T_n(0.2) - \cos(0.2)| \le 10^{-5}$$

Use a calculator to verify that this value of n works.

To use the Error Bound, it is not necessary to find the smallest possible value of K. In this example, we take $K = 1$. This works for all n, but for odd n we could have used the smaller value $K = \sin(0.2) \approx 0.2$.

Solution Since $|f^{(n)}(x)|$ is $|\cos x|$ or $|\sin x|$, depending on whether n is even or odd, we have $|f^{(n)}(u)| \le 1$ for all u. Thus, we may apply the Error Bound with $K = 1$:

$$|T_n(0.2) - \cos(0.2)| \le K \frac{|0.2-0|^{n+1}}{(n+1)!} = \frac{|0.2|^{n+1}}{(n+1)!}$$

To make the error less than 10^{-5}, we must choose n so that

$$\frac{|0.2|^{n+1}}{(n+1)!} < 10^{-5}$$

We find a suitable n by checking several values:

n	2	3	4
$\frac{|0.2|^{n+1}}{(n+1)!}$	$\frac{0.2^3}{3!} \approx 0.0013$	$\frac{0.2^4}{4!} \approx 6.67 \times 10^{-5}$	$\frac{0.2^5}{5!} \approx 2.67 \times 10^{-6} < 10^{-5}$

We see that the error is less than 10^{-5} for $n = 4$. To verify this, recall that $T_4(x) = 1 - \frac{1}{2}x^2 + \frac{1}{4!}x^4$ by Example 3. The following values from a calculator confirm that the error is significantly less than 10^{-5} as required:

$$\text{Actual error} = |T_4(0.2) - \cos(0.2)| \approx |0.98006667 - 0.98006657| = 10^{-7}$$ ■

CONCEPTUAL INSIGHT Recall that functions $g(x)$ and $f(x)$ agree at $x = a$ to order n if $g^{(k)}(a) = f^{(k)}(a)$ for $0 \le k \le n$. The defining property of the Taylor polynomial $T_n(x)$ is that it has degree n and agrees with $f(x)$ at $x = a$ to order n. Now, let us say that a function $g(x)$ *approximates* $f(x)$ *at* $x = a$ *to order* n if the error $E(x) = |f(x) - g(x)|$ satisfies

$$\lim_{x \to a} \frac{E(x)}{|x - a|^n} = 0 \qquad \boxed{4}$$

Thus, in an nth-order approximation, the error tends to zero faster than $|x - a|^n$. The Error Bound tells us that $T_n(x)$ is an nth-order approximation to $f(x)$ at $x = a$, provided we assume that $f^{(n+1)}(x)$ exists and is continuous and that $|f^{(n+1)}(u)| \le K$ for u in some open interval I around a. In fact, the Error Bound gives

$$|f(x) - T_n(x)| \le C|x - a|^{n+1} \qquad \text{for } x \in I$$

where

$$C = \frac{K}{(n+1)!}$$

so

$$\frac{|f(x) - T_n(x)|}{|x - a|^n} \le C|x - a| \to 0 \qquad (\text{as } x \to a)$$

If $f^{(k)}(x)$ exists for all k [as is the case for transcendental functions such as $f(x) = \sin x$ or $f(x) = e^x$], then we can approximate $f(x)$ to arbitrarily high order using Taylor polynomials.

E. SUMMARY

- The nth *Taylor polynomial* centered at $x = a$ for the function $f(x)$ is

$$T_n(x) = f(a) + \frac{f'(a)}{1!}(x - a)^1 + \frac{f''(a)}{2!}(x - a)^2 + \cdots + \frac{f^{(n)}(a)}{n!}(x - a)^n$$

When $a = 0$, $T_n(x)$ is also called the nth *Maclaurin polynomial*.

- If $f^{(n+1)}(x)$ exists and is continuous, then we have the *Error Bound*

$$|T_n(x) - f(x)| \le K \frac{|x - a|^{n+1}}{(n+1)!}$$

where K is a number such that $|f^{(n+1)}(u)| \le K$ for all u between a and x.

- For reference, we include a table of standard Maclaurin and Taylor polynomials.

$f(x)$	a	Maclaurin or Taylor Polynomial
e^x	0	$T_n(x) = 1 + x + \frac{x^2}{2!} + \frac{x^3}{3!} + \cdots + \frac{x^n}{n!}$
$\sin x$	0	$T_{2n+1}(x) = T_{2n+2}(x) = x - \frac{x^3}{3!} + \cdots + (-1)^n \frac{x^{2n+1}}{(2n+1)!}$
$\cos x$	0	$T_{2n}(x) = T_{2n+1}(x) = 1 - \frac{x^2}{2!} + \frac{x^4}{4!} - \cdots + (-1)^n \frac{x^{2n}}{(2n)!}$
$\ln x$	1	$T_n(x) = (x-1) - \frac{1}{2}(x-1)^2 + \cdots + \frac{(-1)^{n-1}}{n}(x-1)^n$
$\frac{1}{1-x}$	0	$T_n(x) = 1 + x + x^2 + \cdots + x^n$

E. EXERCISES

Preliminary Questions

1. What is $T_3(x)$ centered at $a = 3$ for a function $f(x)$ such that $f(3) = 9$, $f'(3) = 8$, $f''(3) = 4$, and $f'''(3) = 12$.

2. The dashed graphs in Figure 3 are Taylor polynomials for a function $f(x)$. Which of the two is a Maclaurin polynomial?

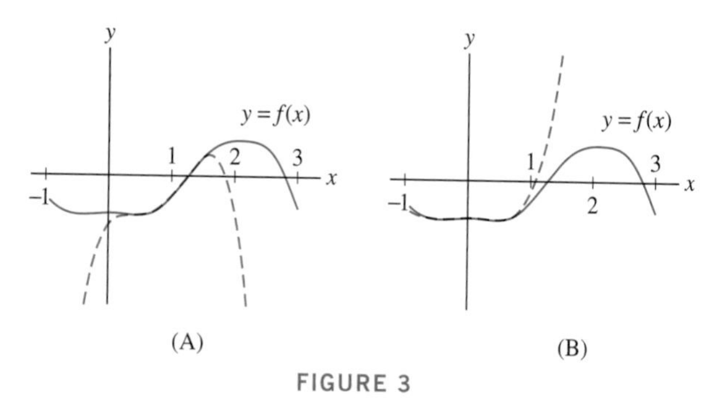

FIGURE 3

3. For which value of x does the Maclaurin polynomial $T_n(x)$ satisfy $T_n(x) = f(x)$, no matter what $f(x)$ is?

4. Let $T_n(x)$ be the Maclaurin polynomial of a function $f(x)$ satisfying $|f^{(4)}(x)| \le 1$ for all x. Which of the following statements follow from the Error Bound?

(a) $|T_4(2) - f(2)| \le 2^4/24$

(b) $|T_3(2) - f(2)| \le 2^3/6$

(c) $|T_3(2) - f(2)| \le 1/3$

Exercises

In Exercises 1–14, calculate the Taylor polynomials $T_2(x)$ and $T_3(x)$ centered at $x = a$ for the given function and value of a.

1. $f(x) = \sin x, \quad a = 0$

2. $f(x) = \sin x, \quad a = \pi/2$

3. $f(x) = \frac{1}{1+x}, \quad a = 0$

4. $f(x) = \frac{1}{1+x}, \quad a = 1$

5. $f(x) = \tan x, \quad a = 0$

6. $f(x) = \tan x, \quad a = \frac{\pi}{4}$

7. $f(x) = \frac{1}{1+x^2}, \quad a = 0$

8. $f(x) = \frac{1}{1+x^2}, \quad a = -1$

9. $f(x) = e^x, \quad a = 0$

10. $f(x) = e^x, \quad a = \ln 2$

11. $f(x) = e^{-x} + e^{-2x}, \quad a = 0$

12. $f(x) = x^2 e^{-x}, \quad a = 1$

13. $f(x) = \ln(x+1), \quad a = 0$

14. $f(x) = \cosh x, \quad a = 0$

In Exercises 15–18, compute $T_2(x)$ at $x = a$ and use a calculator to compute the error $|f(x) - T_2(x)|$ at the given value of x.

15. $y = e^x, \quad x = -0.5, \quad a = 0$

16. $y = \cos x, \quad x = \frac{\pi}{12}, \quad a = 0$

17. $y = x^{-3/2}, \quad x = 0.3, \quad a = 1$

18. $y = e^{\sin x}, \quad x = 1.5, \quad a = \frac{\pi}{2}$

19. Show that the nth Maclaurin polynomial for $f(x) = e^x$ is

$$T_n(x) = 1 + \frac{x}{1!} + \frac{x^2}{2!} + \cdots + \frac{x^n}{n!}$$

20. Show that the nth Taylor polynomial for $\frac{1}{x+1}$ at $a = 1$ is

$$T_n(x) = \frac{1}{2} - \frac{x-1}{4} + \frac{(x-1)^2}{8} + \cdots + (-1)^n \frac{(x-1)^n}{2^{n+1}}$$

In Exercises 21–26, find $T_n(x)$ at $x = a$ for all n.

21. $f(x) = \frac{1}{1-x}, \quad a = 0$

22. $f(x) = \frac{1}{x-1}, \quad a = 4$

23. $f(x) = e^x, \quad a = 1$

24. $f(x) = \sqrt{x}, \quad a = 1$

25. $f(x) = x^{5/2}, \quad a = 2$

26. $f(x) = \cos x, \quad a = \pi/4$

27. CAS Plot $y = e^x$ together with the Maclaurin polynomials $T_n(x)$ for $n = 1, 3, 5$ and then for $n = 2, 4, 6$ on the interval $[-3, 3]$. What difference do you notice between the even and odd Maclaurin polynomials?

28. CAS Plot $f(x) = \frac{1}{1+x}$ together with the Taylor polynomials $T_n(x)$ at $a = 1$ for $1 \le n \le 4$ on the interval $[-2, 8]$ (be sure to limit the upper plot range).

(a) Over which interval does T_4 appear to give a close approximation to $f(x)$?

(b) What happens with $x < -1$?

(c) Use your computer algebra system to produce T_{30} and plot it together with $f(x)$ on $[-2, 8]$. Over which interval does T_{30} appear to give a close approximation?

29. Use the Error Bound to find the maximum possible size of $|\cos 0.3 - T_5(0.3)|$. Verify your result with a calculator.

30. Calculate $T_4(x)$ at $a = 1$ for $f(x) = x^{11/2}$ and use the Error Bound to find the maximum possible size of $|T_4(1.2) - f(1.2)|$. *Hint:* Show that the max $|f^{(5)}(x)|$ on $[1, c]$ with $c > 1$ is $f^{(5)}(c)$.

31. Let T_n be the Maclaurin polynomial of $f(x) = e^x$ and let $c > 0$. Show that we may take $K = e^c$ in the Error Bound for $|T_n(c) - f(c)|$. Then use the Error Bound to determine a number n such that $|T_n(0.1) - e^{0.1}| \le 10^{-5}$.

32. Let $f(x) = \sqrt{1+x}$ and let $T_n(x)$ be the Taylor polynomial centered at $a = 8$.

(a) Find $T_3(x)$ and calculate $T_3(8.02)$.

(b) Use the Error Bound to find a bound for $|T_3(8.02) - \sqrt{9.02}|$.

33. Calculate $T_3(x)$ at $a = 0$ for $f(x) = \tan^{-1} x$. Then compute $T_3(\frac{1}{2})$ and use the Error Bound to find a bound for $|T_3(\frac{1}{2}) - \tan^{-1}(\frac{1}{2})|$. Refer to the graph in Figure 4 to find an acceptable value of K.

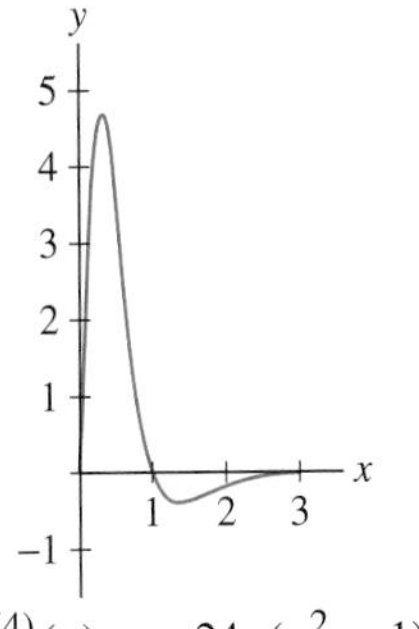

FIGURE 4 Graph of $f^{(4)}(x) = -24x(x^2-1)/(x^2+1)^4$, where $f(x) = \tan^{-1} x$.

34. GU Calculate $T_2(x)$ for $f(x) = \operatorname{sech} x$ at $a = 0$. Then compute $T_2(\frac{1}{2})$ and use the Error Bound to find a bound for

$$\left| T_2\left(\frac{1}{2}\right) - f\left(\frac{1}{2}\right) \right|.$$

Hint: Plot $f'''(x)$ to find an acceptable value of K.

35. Show that the Maclaurin polynomials for $f(x) = \sin x$ are

$$T_{2n-1}(x) = T_{2n} = x - \frac{x^3}{3!} + \frac{x^5}{5!} - \cdots + (-1)^{n-1} \frac{x^{2n-1}}{(2n-1)!}$$

Use the Error Bound with $n = 4$ to show that

$$\left| \sin x - \left(x - \frac{x^3}{6} \right) \right| \le \frac{|x|^5}{120} \quad \text{(for all } x\text{)}$$

36. Find n such that $|T_n(0.1) - \cos(0.1)| \le 10^{-7}$, where T_n is the Maclaurin polynomial for $f(x) = \cos x$ (see Example 5). Calculate $|T_n(0.1) - \cos(0.1)|$ for this value of n and verify the bound on the error.

37. Find n such that $|T_n(1.3) - \ln(1.3)| \le 10^{-4}$, where T_n is the Taylor polynomial for $f(x) = \ln x$ at $a = 1$.

38. Find n such that $|T_n(1) - e| \le 10^{-6}$, where T_n is the Maclaurin polynomial for $f(x) = e^x$.

39. Find n such that $|T_n(1.3) - \sqrt{1.3}| \le 10^{-6}$, where $T_n(x)$ is the Taylor polynomial for $f(x) = \sqrt{x}$ at $a = 1$.

40. Let $T_n(x)$ be the Taylor polynomial for $f(x) = \ln x$ at $a = 1$ and let $c > 1$.

(a) Show that the maximum of $f^{(k+1)}(x)$ on $[1, c]$ is $f^{(k+1)}(1)$ (see Example 4).

(b) Prove $|T_n(c) - \ln c| \le \frac{|c-1|^{n+1}}{n+1}$.

(c) Find n such that $|T_n(1.5) - \ln 1.5| \le 10^{-2}$.

41. Let $n \ge 1$. Show that if $|x|$ is small, then

$$(x+1)^{1/n} \approx 1 + \frac{x}{n} + \frac{1-n}{2n^2} x^2$$

Use this approximation with $n = 6$ to estimate $1.5^{1/6}$.

42. Verify that the third Maclaurin polynomial for $f(x) = e^x \sin x$ is equal to the product of the third Maclaurin polynomials of e^x and $\sin x$ (after discarding terms of degree greater than 3 in the product).

43. Find the fourth Maclaurin polynomial for $f(x) = \sin x \cos x$ by multiplying the fourth Maclaurin polynomials for $f(x) = \sin x$ and $f(x) = \cos x$.

44. Find the Maclaurin polynomials $T_n(x)$ for $f(x) = \cos(x^2)$. You may use the fact that $T_n(x)$ is equal to the sum of the terms up to degree n obtained by substituting x^2 for x in the nth Maclaurin polynomial of $\cos x$.

45. Find the Maclaurin polynomials of $\dfrac{1}{1+x^2}$ by substituting $-x^2$ for x in the Maclaurin polynomials of $\dfrac{1}{1-x}$ (see Exercise 21).

46. The seventeenth-century Dutch scientist Christian Huygens used the approximation $\theta \approx \dfrac{8b-a}{3}$ for the length θ of a circular arc of the unit circle, where a is the length of the chord $\overline{AC}$ of angle θ and b is length of the chord $\overline{AB}$ of angle $\theta/2$ (Figure 5).

(a) Prove that $a = 2\sin(\theta/2)$ and $b = 2\sin(\theta/4)$, and show that the Huygens approximation amounts to the approximation

$$\theta \approx \frac{16}{3}\sin\frac{\theta}{4} - \frac{2}{3}\sin\frac{\theta}{2}$$

(b) Compute the fifth MacLaurin polynomial of the function on the right.

(c) Use the Error Bound to show that the error in the Huygens approximation is less than $0.00022|\theta|^5$.

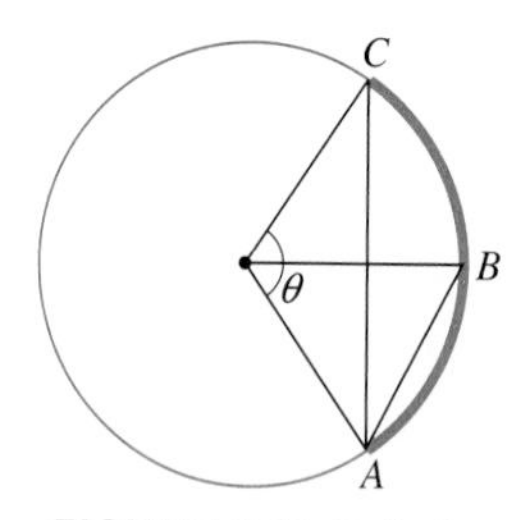

FIGURE 5 Unit circle.

47. Let $f(x) = 3x^3 + 2x^2 - x - 4$. Calculate $T_j(x)$ for $j = 1, 2, 3, 4, 5$ at both $a = 0$ and $a = 1$. Show that $T_3(x) = f(x)$ in both cases.

48. Let $T_n(x)$ be the nth Taylor polynomial at $x = a$ for a polynomial $f(x)$ of degree n. Based on the result of Exercise 47, guess the value of $|f(x) - T_n(x)|$. Prove that your guess is correct using the Error Bound.

49. Taylor polynomials can be used instead of L'Hôpital's Rule to evaluate limits. Consider $L = \lim\limits_{x\to 0} \dfrac{e^x + e^{-x} - 2}{1 - \cos x}$.

(a) Show that the second Maclaurin polynomial for

$$f(x) = e^x + e^{-x} - 2$$

is $T_2(x) = x^2$. Use the Error Bound with $n = 2$ to show that

$$e^x + e^{-x} - 2 = x^2 + g_1(x)$$

where $\lim\limits_{x\to 0} \dfrac{g_1(x)}{x^2} = 0$. Similarly, prove that

$$1 - \cos x = \frac{1}{2}x^2 + g_2(x)$$

where $\lim\limits_{x\to 0} \dfrac{g_2(x)}{x^2} = 0$.

(b) Evaluate L by using (a) to show that $L = \lim\limits_{x\to 0} \dfrac{1 + \frac{g_1(x)}{x^2}}{\frac{1}{2} + \frac{g_2(x)}{x^2}}$.

(c) Evaluate L again using L'Hôpital's Rule.

50. Use the method of Exercise 49 to evaluate

$$\lim_{x\to 0} \frac{1 - \cos x}{x^2} \quad \text{and} \quad \lim_{x\to 0} \frac{\sin x - x}{x^3}.$$

51. A light wave of wavelength λ travels from A to B by passing through an aperture (see Figure 6 for the notation). The aperture (circular region) is located in a plane that is perpendicular to $\overline{AB}$. Let $f(r) = d' + h'$, that is, $f(r)$ is the distance $AC + CB$ as a function of r. The **Fresnel zones**, used to determine the optical disturbance at B, are the concentric bands bounded by the circles of radius R_n such that $f(R_n) = AB + n\lambda/2 = d + h + n\lambda/2$.

(a) Show that $f(r) = \sqrt{d^2 + r^2} + \sqrt{h^2 + r^2}$, and use the Maclaurin polynomial of order 2 to show

$$f(r) \approx d + h + \frac{1}{2}\left(\frac{1}{d} + \frac{1}{h}\right)r^2$$

(b) Deduce that $R_n \approx \sqrt{n\lambda L}$, where $L = (d^{-1} + h^{-1})^{-1}$.

(c) Estimate the radii R_1 and R_{100} for blue light ($\lambda = 475 \times 10^{-7}$ cm) if $d = h = 100$ cm.

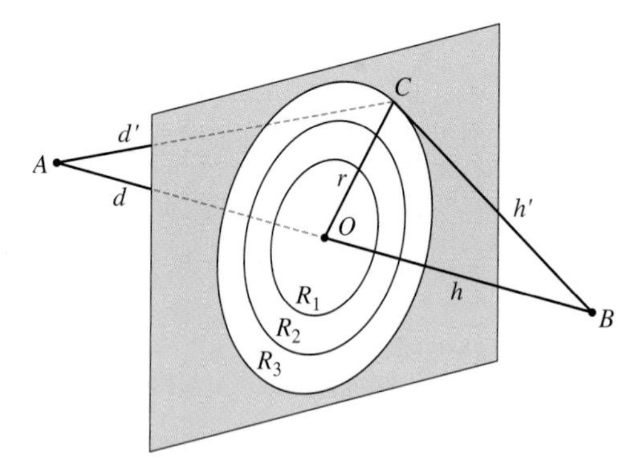

FIGURE 6 The Fresnel zones are the regions between the circles of radius R_n.

Further Insights and Challenges

52. Show that the nth Maclaurin polynomial of $f(x) = \arcsin x$ for n odd is

$$T_n(x) = x + \frac{1}{2}\frac{x^3}{3} + \frac{1 \cdot 3}{2 \cdot 4}\frac{x^5}{5} + \frac{1 \cdot 3 \cdot 7}{2 \cdot 4 \cdot 6}\frac{x^7}{7} + \cdots + \frac{1 \cdot 3 \cdot 7 \cdots (n-1)}{2 \cdot 4 \cdot 6 \cdots (n-2)}\frac{x^n}{n}$$

53. Use Taylor's Theorem to show that if $f^{(n+1)}(t) \ge 0$ for all t, then the nth Maclaurin polynomial $T_n(x)$ satisfies $T_n(x) \le f(x)$ for all $x \ge 0$.

54. Use Exercise 53 to show that for $x \ge 0$ and all n,

$$e^x \ge 1 + x + \frac{x^2}{2!} + \cdots + \frac{x^n}{n!}$$

Sketch the graphs of e^x, $T_1(x)$, and $T_2(x)$ on the same coordinate axes. Does this inequality remain true for $x < 0$?

55. This exercise is intended to reinforce the proof of Taylor's Theorem.

(a) Show that $f(x) = T_0(x) + \int_a^x f'(u)\,du$.

(b) Use Integration by Parts to prove the formula

$$\int_a^x (x-u)f^{(2)}(u)\,du = -f'(a)(x-a) + \int_a^x f'(u)\,du$$

(c) Prove the case $n = 2$ of Taylor's Theorem:

$$f(x) = T_1(x) + \int_a^x (x-u)f^{(2)}(u)\,du.$$

56. Approximating Integrals Using Taylor Polynomials Taylor polynomials can be used to obtain numerical approximations to integrals.

(a) Let $L > 0$. Show that if two functions $f(x)$ and $g(x)$ satisfy $|f(x) - g(x)| < L$ for all $x \in [a, b]$, then

$$\int_a^b \big(f(x) - g(x)\big)\,dx < L(b-a)$$

(b) Show that $|T_3(x) - \sin x| \le (\frac{1}{2})^5/5!$ for all $x \in [0, \frac{1}{2}]$.

(c) Evaluate $\int_0^{1/2} T_3(x)\,dx$ as an approximation to $\int_0^{1/2} \sin x\,dx$. Use (a) and (b) to find a bound for the size of the error.

57. Use the fourth Maclaurin polynomial and the method of Exercise 56 to approximate $\int_0^{1/2} \sin(x^2)\,dx$. Find a bound for the error.

58. The Second Derivative Test for local extrema fails when $f''(a) = 0$. This exercise shows us how to extend the test under the following assumption:

$$f'(a) = f''(a) = f'''(a) = 0 \quad \text{but} \quad f^{(4)}(a) > 0$$

(a) Show that $T_4(x) = f(a) + \frac{1}{24}f^{(4)}(a)(x-a)^4$.

(b) Show that $T_4(x) > f(a)$ for $x \ne a$.

(c) By the Error Bound, there is a constant C such that $|T_4(x) - f(x)| \le C|x-a|^5$ for all x in an interval containing a. Use this and (b) to show that $f(x) > f(a)$ for x sufficiently close to but not equal to a.

(d) Conclude that $f(a)$ is a local minimum. A similar argument shows that $f(a)$ is a local maximum if $f^{(4)}(a) < 0$.

59. Prove by induction that for all k,

$$\frac{d^j}{dx^j}\left(\frac{(x-a)^k}{k!}\right) = \frac{k(k-1)\cdots(k-j+1)(x-a)^{k-j}}{k!}$$

$$\frac{d^j}{dx^j}\left(\frac{(x-a)^k}{k!}\right)\Bigg|_{x=a} = \begin{cases} 1 & \text{for } k = j \\ 0 & \text{for } k \ne j \end{cases}$$

Use this to prove that $T_n(x)$ agrees with $f(x)$ at $x = a$ to order n.

60. The following equation arises in the description of Bose–Einstein condensation (the quantum theory of gases cooled to near absolute zero):

$$A_0 = \frac{4A}{\sqrt{\pi}}\int_0^\infty \frac{x^2 e^{-x^2}\,dx}{1 - Ae^{-x^2}}$$

It is necessary to derive an approximate expression for A_0 in terms of A for $|A|$ small.

(a) Show that the second Maclaurin polynomial for the function $f(A) = \dfrac{x^2 e^{-x^2}}{1 - Ae^{-x^2}}$ (where A is the variable and x is treated as a constant) is

$$T_2(A) = x^2 e^{-x^2} + x^2 e^{-2x^2} A + x^2 e^{-3x^2} A^2$$

(b) Use the approximation $A_0 \approx \dfrac{4A}{\sqrt{\pi}}\displaystyle\int_0^\infty T_2(A)\,dx$ to show

$$A_0 \approx A + \frac{1}{2\sqrt{2}}A^2 + \frac{1}{3\sqrt{3}}A^3$$

You may use the formula (valid for $\lambda > 0$)

$$\int_0^\infty x^2 e^{-\lambda x^2}\,dx = \frac{1}{4}\sqrt{\frac{\pi}{\lambda^3}}$$

ANSWERS TO ODD-NUMBERED EXERCISES

Chapter 10

Section 10.1 Preliminary

1. $a_4 = 12$ **2.** (c) **3.** $\lim_{n\to\infty} a_n = \sqrt{2}$ **4.** (b)
5. **(a)** False. Counterexample: $a_n = \cos \pi n$. **(b)** True
(c) False. Counterexample: $a_n = (-1)^n$.

Section 10.1 Exercises

1. **(a)** (iv) **(b)** (i) **(c)** (iii) **(d)** (ii)
3. $c_1 = 2, c_2 = 2, c_3 = \frac{4}{3}, c_4 = \frac{2}{3}$
5. $a_1 = 3, a_2 = 10, a_3 = 101, a_4 = 10{,}202$
7. $c_1 = 1, c_2 = \frac{3}{2}, c_3 = \frac{11}{6}, c_4 = \frac{25}{12}$
9. $b_1 = 2, b_2 = 5, b_3 = 9, b_4 = 19$
11. **(a)** $a_n = \frac{(-1)^{n+1}}{n^3}$ **(b)** $a_n = \frac{n+1}{5+n}$
13. $\lim_{n\to\infty} a_n = 4$ **15.** $\lim_{n\to\infty} \left(5 - \frac{9}{n^2}\right) = 5$
17. $\lim_{n\to\infty} (-2^{-n}) = 0$ **19.** $\lim_{n\to\infty} \left(\frac{1}{3}\right)^n = 0$
21. The sequence diverges. **23.** $\lim_{n\to\infty} \frac{n}{\sqrt{n^3+1}} = 0$
25. The sequence diverges.
27. **(a)** $M = 999$ **(b)** $M = 99{,}999$
31. $\lim_{n\to\infty} 2^{1/n} = 1$ **33.** $\lim_{n\to\infty} n^{1/n} = 1$ **35.** $\lim_{n\to\infty} \left(1 + \frac{1}{n}\right)^n = e$
37. $\lim_{n\to\infty} a_n = 0$ **39.** $\lim_{n\to\infty} n \sin \frac{1}{n} = 1$ **41.** (b)
43. $\lim_{n\to\infty} \frac{3n^2+n+2}{2n^2-3} = \frac{3}{2}$ **45.** $\lim_{n\to\infty} 3 + \left(-\frac{1}{2}\right)^n = 3$
47. $\lim_{n\to\infty} \tan^{-1}(1 - \frac{2}{n}) = \frac{\pi}{4}$ **49.** $\lim_{n\to\infty} \ln\left(\frac{2n+1}{3n+4}\right) = \ln \frac{2}{3}$
51. $\lim_{n\to\infty} \frac{e^n+3^n}{5^n} = 0$ **53.** $\lim_{n\to\infty} \frac{e^n}{2^{n^2}} = 0$
55. $\lim_{n\to\infty} \frac{n^3+2e^{-n}}{3n^3+4e^{-n}} = \frac{1}{3}$ **57.** The sequence diverges.
59. $\lim_{n\to\infty} \frac{(-4{,}000)^n}{n!} = 0$ **61.** $\lim_{n\to\infty} \cos \frac{\pi}{n} = 1$ **63.** $\lim_{n\to\infty} \sqrt[n]{n} = 1$
65. Any number greater than or equal to 3 is an upper bound.
71. Example: $a_n = -n^2, b_n = n^2 + \frac{1}{n}$
75. $\lim_{n\to\infty} a_n = \frac{1+\sqrt{21}}{2}$ **77.** $\lim_{n\to\infty} c_n = 1$
81. **(a)** $c_1 = \frac{3}{2}; c_2 = \frac{13}{12}; c_3 = \frac{19}{20}; c_4 = \frac{743}{840}$ **(c)** $\lim_{n\to\infty} c_n = \ln 2$

Section 10.2 Preliminary

1. The sum of an infinite series is defined as the limit of the sequence of partial sums.
2. $S = \frac{1}{2}$ **3.** The result is negative hence the formula is invalid.
4. No **5.** No. It may be used to establish divergence. **6.** $N = 13$
7. No **8.** Example: $\sum_{n=1}^{\infty} \frac{1}{n^{9/10}}$.

Section 10.2 Exercises

1. **(a)** $a_n = \frac{1}{3^n}$ **(b)** $a_n = \left(\frac{5}{2}\right)^{n-1}$
(c) $a_n = (-1)^{n+1} \frac{n^n}{n!}$ **(d)** $a_n = \frac{1+\frac{(-1)^{n+1}+1}{2}}{n^2+1}$
3. $S_2 = \frac{5}{4}, S_4 = \frac{205}{144}, S_6 = \frac{5{,}369}{3{,}600}$ **5.** $S_2 = \frac{2}{3}, S_4 = \frac{4}{5}, S_6 = \frac{6}{7}$
7. $S_5 = 0.035672, S_{10} = 0.035352, S_{15} = 0.035413$. Yes.
9. $S_3 = \frac{3}{10}, S_4 = \frac{1}{3}, S_5 = \frac{5}{14}; \sum_{n=1}^{\infty} \left(\frac{1}{n+1} - \frac{1}{n+2}\right) = \frac{1}{2}$
11. $\sum_{n=3}^{\infty} \frac{1}{n(n-1)} = \sum_{n=3}^{\infty} \left(\frac{1}{n-1} - \frac{1}{n}\right) = \frac{1}{2}$
13. $\lim_{n\to\infty} (-1)^n n^2 \neq 0$
17. (a) **19.** $S = \frac{1}{18}$ **21.** $S = \frac{27}{968}$ **23.** $S = \frac{7}{5}$ **25.** $S = \frac{e}{e^2-1}$
27. The series diverges. **29.** $S = \frac{8}{5}$ **31.** $S = \frac{512}{49}$ **33.** (b), (c)
35. **(a)** $\sum_{n=1}^{10} a_n = \frac{249}{50}, \sum_{n=4}^{16} a_n = \frac{494}{2{,}304}$ **(b)** $a_3 = \frac{5}{18}$
(c) $a_n = \frac{2(2n-1)}{n^2(n-1)^2}$ **(d)** $\sum_{n=1}^{\infty} a_n = 5$
39. 50 ft **41.** $S = \frac{1}{2}$ **45.** $\frac{\sqrt{2}}{\sqrt{2}-1}$

Section 10.3 Preliminary

1. (b) **2.** f must be positive, continuous and decreasing
3. Integral Test **4.** Comparison Test
5. The Comparison Test is not efficient for these two series.

Section 10.3 Exercises

1. $\int_1^\infty \frac{dx}{x^4} = \frac{1}{3}$. The series converges.
3. $\int_1^\infty x^{-1/3}\, dx = \infty$. The series diverges.
5. $\int_{25}^\infty \frac{x^2}{(x^3+9)^{5/2}}\, dx = \frac{1}{70{,}353 \cdot 15{,}634^{3/2}}$. The series converges.
7. $\int_1^\infty \frac{dx}{x^2+1} = \frac{\pi}{4}$. The series converges.
9. $\int_1^\infty x e^{-x^2}\, dx = \frac{1}{2e}$. The series converges.
11. $\int_1^\infty \frac{1}{2^{\ln x}}\, dx = \infty$. The series diverges.

13. $\int_1^\infty \frac{\ln x}{x^2}\,dx = \frac{1+\ln 2}{2}$. The series converges.

15. $\frac{1}{n^3+8n} \le \frac{1}{n^3}$. The series converges.

19. $\frac{1}{n2^n} \le \frac{1}{2^n}$. The series converges.

21. $\frac{k^{1/3}}{k^2+k} \le \frac{k^{1/3}}{k^2} = \frac{1}{k^{5/3}}$. The series converges.

23. $\frac{1}{\sqrt{n^3+1}} \le \frac{1}{\sqrt{n^3}} = \frac{1}{n^{3/2}}$. The series converges.

25. $0 \le \frac{\sin^2 k}{k^2} \le \frac{1}{k^2}$. The series converges.

27. $\frac{4}{m!+4^m} \le \frac{4}{4^m} = 4\cdot\left(\frac{1}{4}\right)^m$. The series converges.

29. $\frac{1}{2^{k^2}} \le \frac{1}{2^k} = \left(\frac{1}{2}\right)^k$. The series converges.

31. $0 \le \frac{\ln n}{n^3+3\ln n} \le \frac{\ln n}{n^3}$. The series converges.

33. The general term of the series does not approach zero and the series diverges.

35. The series converges. **37.** The series diverges.

39. The series converges. **41.** The series converges.

47. The series converges for $a > 1$ and diverges for $a \le 1$.

49. The series converges. **51.** The series diverges.

53. The series converges. **55.** The series converges.

57. The series diverges. **59.** The series converges.

61. The series converges. **63.** The series converges.

65. The series converges. **67.** The series converges.

69. The series converges. **71.** The series converges.

73. The series converges. **81.** $\sum_{n=1}^{\infty} \frac{1}{n^5} \approx 1.036895$

87. **(c)** $\sum_{n=1}^{1,000} \frac{1}{n^2} = 1.643934568$; $1 + \sum_{n=1}^{100} \frac{1}{n^2(n+1)} = 1.644884890$; the latter sum is a better approximation to $\pi^2/6$

Section 10.4 Preliminary

1. (c) **2.** (b) **3.** No **4.** Example: $\sum \frac{(-1)^n}{\sqrt[3]{n}}$.

Section 10.4 Exercises

3. Converges conditionally **5.** Converges absolutely

7. Converges conditionally **9.** Converges absolutely

11. Converges absolutely

13. **(a)**

n	S_n
1	1.000000
2	0.875000
3	0.912037
4	0.896412
5	0.904412
6	0.899782
7	0.902698
8	0.900745
9	0.902116
10	0.901116

15. $S_5 = 0.947539$ **17.** $S_{44} = 0.0656746$

19. Converges (by Comparison Test)

21. Converges (by Leibniz Test)

23. Converges (by Linearity of Series)

25. Converges (by Leibniz Test)

27. Converges conditionally

Section 10.5 Preliminary

1. $\rho = \lim_{n\to\infty} \left|\frac{a_{n+1}}{a_n}\right|$

2. The Ratio Test is conclusive for $\sum_{n=1}^{\infty} \frac{1}{2^n}$ and inconclusive for $\sum_{n=1}^{\infty} \frac{1}{n}$.

3. No

Section 10.5 Exercises

1. Converges (by the Ratio Test)

3. Converges absolutely (by the Ratio Test)

5. The Ratio Test is inconclusive. **7.** Diverges (by the Ratio Test)

9. Converges (by the Ratio Test)

11. Converges (by the Ratio Test)

13. Diverges (by the Ratio Test) **15.** The Ratio Test is inconclusive.

17. Converges (by the Ratio Test) **19.** $\lim_{n\to\infty} \left|\frac{a_{n+1}}{a_n}\right| = \frac{1}{3} < 1$

21. $\lim_{n\to\infty} \left|\frac{a_{n+1}}{a_n}\right| = 2|x|$ **23.** $\lim_{n\to\infty} \left|\frac{a_{n+1}}{a_n}\right| = |r|$

27. Converges absolutely (by the Ratio Test)

29. The Ratio Test is inconclusive and the series may diverge/converge, depending on a_n.

31. Converges absolutely (by the Ratio Test)

33. The Ratio Test is inconclusive.

35. Converges (by the Root Test) **37.** Converges (by the Root Test)

39. Converges (by the Root Test)

41. Converges (by the Ratio Test or by Linearity)

43. Converges (by the Ratio Test)

45. Converges (by the Limit Comparison Test)

47. Converges (by the Limit Comparison Test)

49. Converges (by the Limit Comparison Test)

51. Diverges (by the Divergence Test)

Section 10.6 Preliminary

1. Yes. The series converges for both $x = 4$ and $x = -3$.

2. Converges at $x = 8$; cannot determine convergence for $x = 0$, $x = 2$, $x = 12$.

3. $R = 4$ **4.** $\sum_{n=0}^{\infty} (n+1)^2 x^n$; $R = 1$

Section 10.6 Exercises

1. $R = 2$ **3.** $R = 3$ for all series.
7. Converges in $|x| < 1$ and diverges elsewhere.
9. Converges in $-\frac{1}{2} \le x < \frac{1}{2}$ and diverges elsewhere.
11. Converges in $-1 \le x < 1$ and diverges elsewhere.
13. Converges for all x.
15. Converges in $-5 < x < -3$ and diverges elsewhere.
17. Converges in $-2 < x < 2$ and diverges elsewhere.
19. Converges in $3 \le x \le 5$ and diverges elsewhere.
21. Converges for $x = -5$ and diverges for $x \ne -5$.
23. Converges for $2 - \frac{1}{e} < x < 2 + \frac{1}{e}$ and diverges elsewhere.
25. Converges in $-1 \le x < 1$ and diverges elsewhere.
27. Expansion: $\sum_{n=0}^{\infty} 3^n x^n$; the expansion is valid for $|x| < \frac{1}{3}$.
29. Expansion: $\sum_{n=0}^{\infty} \frac{x^n}{3^{n+1}}$; the expansion is valid for $|x| < 3$.
31. Expansion: $\sum_{n=0}^{\infty} (-1)^n x^{9n}$; the expansion is valid for $|x| < 1$.
33. Expansion: $\sum_{n=0}^{\infty} (-1)^n 3^n x^{7n}$; the expansion is valid for $|x| < \frac{1}{3^{1/7}}$.
37. $\frac{1}{4-x} = \sum_{n=0}^{\infty} (-1)^{n+1}(x-5)^n$ on the interval $(4, 6)$.
39. Example: $\sum_{n=1}^{\infty} \frac{(x-4)^n}{n2^n}$ **41.** N is at least 7; $S_7 = 0.405804$.
45. N is at least 6; $S_6 = 0.368056$.
47. $1 - \frac{1}{2}x^2 - \sum_{n=2}^{\infty} \frac{1\cdot 3\cdot 5\cdots(2n-3)}{(2n)!}x^{2n}$; $R = \infty$
53. Converges for $-\frac{1}{3} < x < \frac{1}{3}$

Section 10.7 Preliminary

1. $f(0) = 3$ and $f'''(0) = 30$ **2.** $f(-2) = 0$; $f^{(4)}(-2) = 48$
3. Substituting x^2 for the MacLaurin series for $\sin(x)$.
4. $f(x) = 4 + \sum_{n=1}^{\infty} \frac{(x-3)^{n+1}}{n(n+1)}$ **5.** (a) and (c)

Section 10.7 Exercises

1. $f(x) = 2 + 3x + 2x^2 + 2x^3 + \cdots$
3. $\frac{1}{1-2x} = \sum_{n=0}^{\infty} 2^n x^n$ for $|x| < \frac{1}{2}$
5. $\cos(3x) = \sum_{n=0}^{\infty} \frac{(-1)^n 9^n}{(2n)!} x^{2n}$ for all x
7. $\sin(x^2) = \sum_{n=0}^{\infty} \frac{(-1)^n}{(2n+1)!} x^{4n+2}$ for all x
9. $\ln(1 - x^2) = -\sum_{n=1}^{\infty} \frac{x^{2n}}{n}$ for $|x| < 1$
11. $\tan^{-1}(x^2) = \sum_{n=1}^{\infty} \frac{(-1)^n x^{4n+2}}{2n+1}$ for $|x| \le 1$
13. $e^{x-2} = \sum_{n=0}^{\infty} \frac{1}{n!e^2} x^n$ for all x
15. $\ln(1 - 5x) = -\sum_{n=1}^{\infty} \frac{5^n}{n} x^n$ for $|x| < \frac{1}{5}$ and $x = -\frac{1}{5}$
17. $\sinh(x) = \sum_{n=0}^{\infty} \frac{1}{(2n+1)!} x^{2n+1}$ for all x
19. $\frac{1-\cos(x^2)}{x} = \sum_{n=1}^{\infty} \frac{(-1)^{n+1}}{(2n)!} x^{4n-1}$ for $x \ne 0$
21. $e^x \sin(x) = x + x^2 + \frac{x^3}{3} - \frac{x^5}{30} + \cdots$
23. $e^x \ln(1 - x) = -x - \frac{3x^2}{2} - \frac{4x^3}{3} - x^4 + \cdots$
25. $(1 + x)^{-3/2} = 1 - \frac{3x}{2} + \frac{15x^2}{8} - \frac{35x^3}{16} + \frac{315x^4}{128} - \cdots$
27. $e^{e^x} = e + ex + ex^2 + \frac{5e}{6}x^3 + \cdots$
29. $\sin(x) = \sum_{n=0}^{\infty} \frac{(-1)^n}{(2n)!}\left(x - \frac{\pi}{2}\right)^{2n}$ **31.** $\frac{1}{x} = \sum_{n=0}^{\infty} (-1)^n (x-1)^n$
33. $\frac{1}{1-x} = \sum_{n=0}^{\infty} \frac{(-1)^{n+1}(x-5)^n}{4^{n+1}}$
35. $x^4 + 3x - 1$
$= 21 + 35(x-2) + 24(x-2)^2 + 8(x-2)^3 + (x-2)^4$
37. $e^{3x} = \sum_{n=0}^{\infty} \frac{3^n e^{-3}}{n!}(x+1)^n$
39. $\frac{1}{1-x^2} = \sum_{n=0}^{\infty} \frac{(-1)^{n+1}(2^{n+1}-1)}{2^{2n+3}}(x-3)^n$, for $|x - 3| < 2$
41. $f(x) = 1 + \sum_{n=0}^{\infty} \frac{1\cdot 3\cdot 5\cdots(2n-1)}{2^n n!}(3x)^{2n}$ for $|x| < \frac{1}{3}$
43. $\sin^{-1}\frac{1}{2} \approx 0.52358519539$
47. **(a)** 5 **(b)** $\sum_{n=0}^{4} \frac{(-1)^n}{(2n+1)n!} = 0.747487$
49. $\int_0^1 \cos(x^2)\,dx = \sum_{n=0}^{\infty} \frac{(-1)^n}{(4n+1)(2n)!}$; $\sum_{n=0}^{3} \frac{(-1)^n}{(4n+1)(2n)!} = 0.904522792$
51. $\int_0^2 e^{-x^3}\,dx = \sum_{n=0}^{\infty} \frac{(-1)^n 2^{3n+1}}{(3n+1)n!}$; $\sum_{n=0}^{24} \frac{(-1)^n 2^{3n+1}}{(3n+1)n!} = 0.892953509$
53. $\int_0^x \frac{1-\cos t}{t}\,dt = \sum_{n=1}^{\infty} \frac{(-1)^n}{2n(2n)!} x^{2n}$
55. $\int_0^x \ln(1 + t^2)\,dt = \sum_{n=1}^{\infty} \frac{(-1)^{n-1}}{n(2n+1)} x^{2n+1}$
57. $f(x) = \frac{1}{1+2x}$ **59.** $x^2 - \frac{2}{3}x^6 + \frac{2}{15}x^{10} - \frac{4}{315}x^{14}$
61. $1 + x - \frac{1}{2}x^2 - \frac{1}{8}x^4$ **63.** $f(x) = e^{x^3}$
65. $f(x) = 1 - 5x + \sin(5x)$ **67.** $I(t) = \frac{V}{R}\sum_{n=1}^{\infty} \frac{(-1)^{n+1}}{n!}\left(\frac{Rt}{L}\right)^n$
69. $f(x) = \sum_{n=0}^{\infty} \frac{(-1)^n x^n}{(2n)!}$ and $f^{(5)}(0) = -\frac{1}{30,240}$

71. $f^{(8)}(0) = -1{,}575$

73. The expansion is not valid since the series only converges for $|x| < 1$.

77. $4; \sum_{n=1}^{4} \frac{(-1)^{n-1}(0.2)^n}{n} = 0.182267$

79. $\frac{1}{(1-x)(1-2x)} = \sum_{n=0}^{\infty} (-1)^n \left(1 - \left(\frac{2}{3}\right)^{n+1}\right)(x-2)^n$, for $|x-2| < 1$

83. If a positive whole, $R = \infty$; if negative whole, $R = 1$.

Chapter 10 Review

1. **(a)** $a_1^2 = 4, a_2^2 = \frac{1}{4}, a_3^2 = 0$
(b) $b_1 = \frac{1}{24}, b_2 = \frac{1}{60}, b_3 = \frac{1}{240}$
(c) $a_1b_1 = -\frac{1}{12}, a_2b_2 = -\frac{1}{120}, a_3b_3 = 0$
(d) $2a_2 - 3a_1 = 5, 2a_3 - 3a_2 = 1.5, 2a_4 - 3a_3 = \frac{1}{12}$

3. $\lim_{n\to\infty}(5a_n - 2a_n^2) = 2$ **5.** $\lim_{n\to\infty} e^{a_n} = e^2$

7. The limit does not exist. **9.** $\lim_{n\to\infty} a_n = 0$

11. $\lim_{n\to\infty} a_n = 1$ **13.** The sequence diverges.

15. $\lim_{n\to\infty} b_n = \frac{\pi}{4}$ **17.** $\lim_{n\to\infty} b_n = \frac{1}{2}$ **19.** $\lim_{n\to\infty} a_n = 0$

21. $\lim_{n\to\infty} c_n = e^3$

25. **(a)** $\lim_{n\to\infty} a_n = \infty$ **(b)** $\lim_{n\to\infty} \frac{a_{n+1}}{a_n} = 3$

27. $S_4 = -\frac{11}{60} = -0.183333; S_7 = \frac{41}{630} = 0.0650794$

29. $\sum_{n=2}^{\infty}\left(\frac{2}{3}\right)^n = \frac{4}{3}$ **31.** $\sum_{n=0}^{\infty}\frac{2^{n+1}}{3^n} = 6$

33. Example: $a_n = \left(\frac{1}{2}\right)^n + 1, b_n = -1$

35. The sequence diverges.

37. $\int_1^{\infty}\frac{x^3}{e^{x^4}}dx = \frac{e^{-1}}{4}$. The sequence converges.

39. $\frac{1}{n^2} > \frac{1}{(n+1)^2}$. The sequence converges.

41. $\frac{n^2+1}{n^{3.5}-2} < \frac{4}{n^{1.5}}$. The sequence converges.

43. Since $\frac{\ln n}{1.5^n} < \frac{n}{1.5^n}$ for all $n > 1$. The sequence converges.

45. $\frac{1}{3^n - 2^n} < \frac{1}{2^n}$ for all $n \geq 2$. The sequence converges.

49. $0.3971162690 \leq S \leq 0.3971172688$. The maximum size of the error 10^{-6}.

51. Converges absolutely. **53.** The series diverges.

55. $16; K \approx \sum_{n=0}^{1} 5\frac{(-1)^k}{(2k+1)^2} = 0.915479$

57. **(a)** Converges **(b)** Converges **(c)** Diverges
(d) Converges

59. The Ratio Test is inconclusive. **61.** The series converges.

63. The Ratio Test is inconclusive. **65.** The series converges.

67. The series converges. **69.** The series converges.

71. The series converges. **73.** The series converges.

75. The series converges by Leibniz Test.

77. The series converges by Leibniz Test.

79. The series converges. **81.** The series diverges.

83. The series converges. **85.** The series converges for $x \in \Re$.

87. The series converge for $x \in [2, 4]$.

89. The series converge for $x = 0$.

91. $\frac{2}{4-3x} = \frac{1}{2}\sum_{n=0}^{\infty}\left(\frac{3}{4}\right)^n x^n$. The series converges for $|x| < \frac{4}{3}$.

93. **(c)**

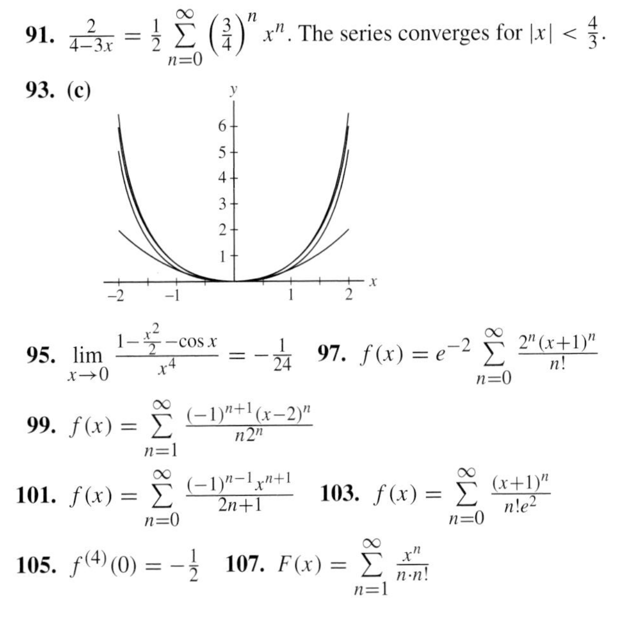

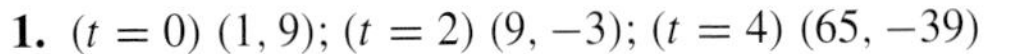

95. $\lim_{x\to 0}\frac{1-\frac{x^2}{2}-\cos x}{x^4} = -\frac{1}{24}$ **97.** $f(x) = e^{-2}\sum_{n=0}^{\infty}\frac{2^n(x+1)^n}{n!}$

99. $f(x) = \sum_{n=1}^{\infty}\frac{(-1)^{n+1}(x-2)^n}{n2^n}$

101. $f(x) = \sum_{n=0}^{\infty}\frac{(-1)^{n-1}x^{n+1}}{2n+1}$ **103.** $f(x) = \sum_{n=0}^{\infty}\frac{(x+1)^n}{n!e^2}$

105. $f^{(4)}(0) = -\frac{1}{2}$ **107.** $F(x) = \sum_{n=1}^{\infty}\frac{x^n}{n \cdot n!}$

Chapter 11

Section 11.1 Preliminary

1. A circle of radius 3 centered at the origin.

2. The center is at (4, 5). **3.** Maximum height: 4

4. Yes; no **5.** (a) ↔ (iii), (b) ↔ (ii), (c) ↔ (i)

Section 11.1 Exercises

1. $(t = 0)$ $(1, 9)$; $(t = 2)$ $(9, -3)$; $(t = 4)$ $(65, -39)$

5. **(a)** **(b)**

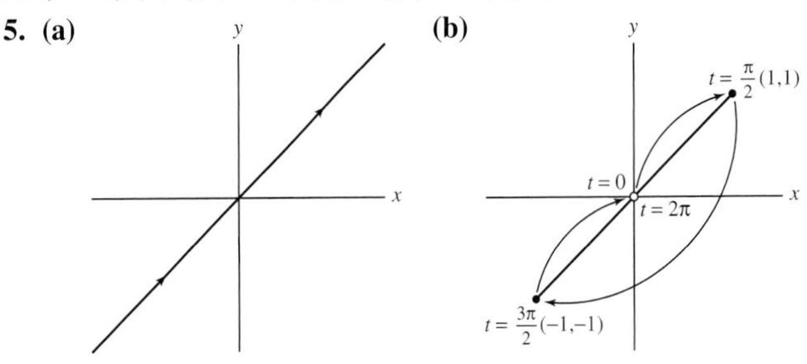

(c) (d)

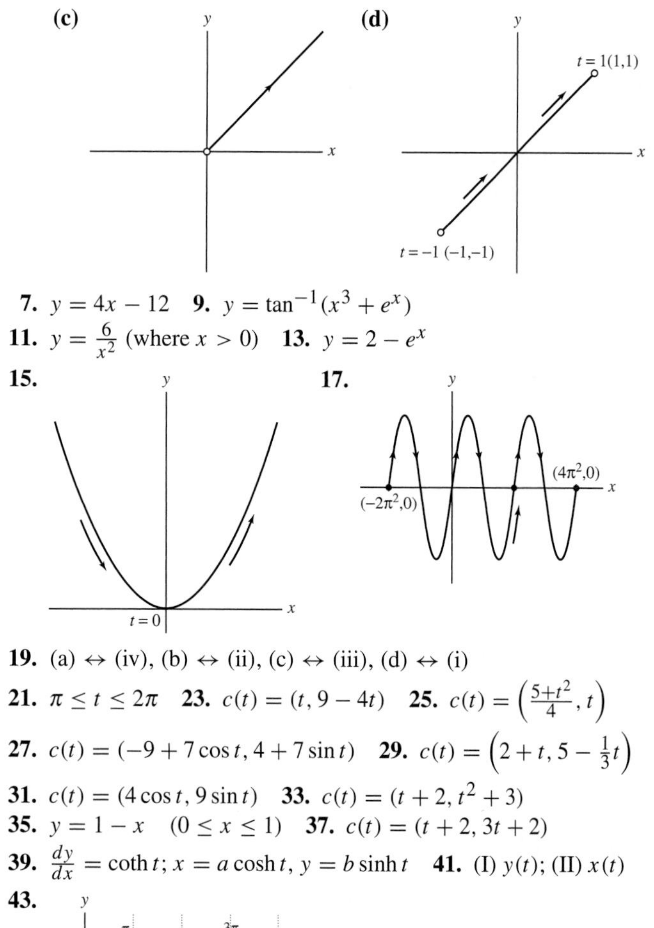

7. $y = 4x - 12$ **9.** $y = \tan^{-1}(x^3 + e^x)$

11. $y = \frac{6}{x^2}$ (where $x > 0$) **13.** $y = 2 - e^x$

15. **17.**

19. (a) $\leftrightarrow$ (iv), (b) $\leftrightarrow$ (ii), (c) $\leftrightarrow$ (iii), (d) $\leftrightarrow$ (i)

21. $\pi \le t \le 2\pi$ **23.** $c(t) = (t, 9 - 4t)$ **25.** $c(t) = \left(\frac{5+t^2}{4}, t\right)$

27. $c(t) = (-9 + 7\cos t, 4 + 7\sin t)$ **29.** $c(t) = \left(2 + t, 5 - \frac{1}{3}t\right)$

31. $c(t) = (4\cos t, 9\sin t)$ **33.** $c(t) = (t + 2, t^2 + 3)$

35. $y = 1 - x \quad (0 \le x \le 1)$ **37.** $c(t) = (t + 2, 3t + 2)$

39. $\frac{dy}{dx} = \coth t$; $x = a\cosh t$, $y = b\sinh t$ **41.** (I) $y(t)$; (II) $x(t)$

43.

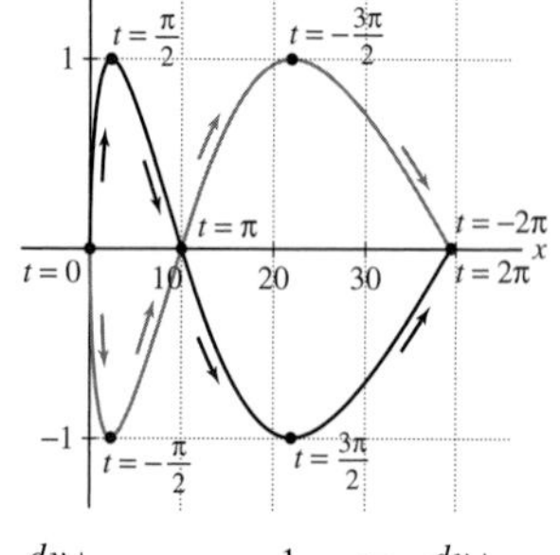

45. $\left.\frac{dy}{dx}\right|_{t=-4} = -\frac{1}{6}$ **47.** $\left.\frac{dy}{dx}\right|_{s=-1} = -\frac{3}{4}$

49. $y = -\frac{9}{2}x + \frac{11}{2}$; $\left.\frac{dy}{dx}\right|_{x=1} = \left.\frac{dy}{dx}\right|_{t=0} = -\frac{9}{2}$

51. $y = x^2 + x^{-1}$; $\left.\frac{dy}{dx}\right|_{x=1} = \left.\frac{dy}{dx}\right|_{s=1} = 1$

53.

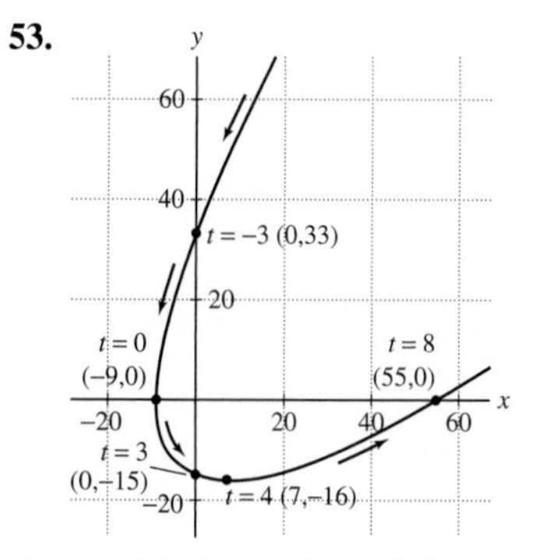

The graph is in: quadrant (i) for $t < -3$ or $t > 8$, quadrant (ii) for $-3 < t < 0$, quadrant (iii) for $0 < t < 3$, quadrant (iv) for $3 < t < 8$.

55. $(55, 0)$ **57.** $\left(\frac{x}{4}\right)^2 + \left(\frac{y}{7}\right)^2 = 1$; $m = -\frac{7}{4}$

59. $c(t) = (3 - 9t + 24t^2 - 16t^3, 2 + 6t^2 - 4t^3)$, $0 \le t \le 1$

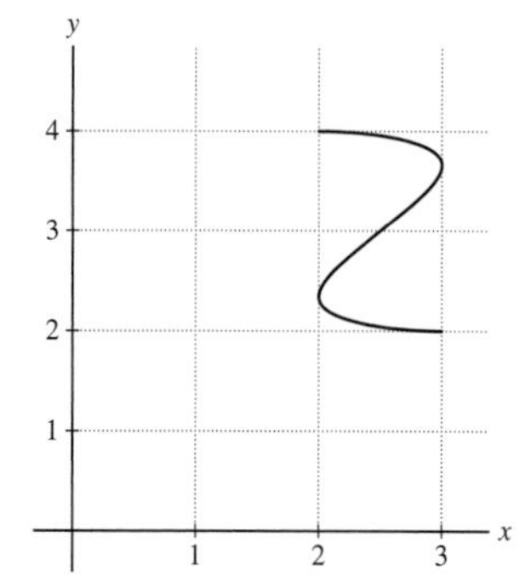

63.

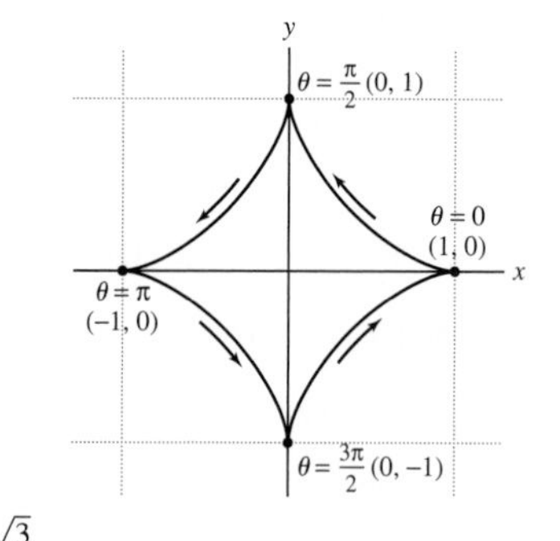

$y = -\sqrt{3}x + \frac{\sqrt{3}}{2}$

65. $((2k - 1)\pi, 2)$, $k = 0, \pm 1, \pm 2, \ldots$

75. The coordinates of P, $(R\cos\theta, r\sin\theta)$ describe an ellipse for $0 \le \theta \le 2\pi$.

79. $\left.\frac{d^2y}{dx^2}\right|_{t=2} = -\frac{21}{512}$ **81.** $\left.\frac{d^2y}{dx^2}\right|_{t=-3} = 0$ **83.** Concave up: $t > 0$

87. Formula for the area under the graph of a positive function.

Section 11.2 Preliminary

1. $S = \int_a^b \sqrt{x'(t)^2 + y'(t)^2}\,dt$ **2.** The speed at time t

3. Displacement: 5; no **4.** $L = 180$ cm

Section 11.2 Exercises

1. $S = 3\pi$ **3.** $S = 10$ **5.** $S = 16\sqrt{13}$

7. $S = \frac{1}{2}(65^{3/2} - 5^{3/2}) \approx 256.43$ **9.** $S = 3\pi$ **11.** $S = 2$

15. $S = \pi\sqrt{1+4\pi^2} + \frac{1}{2}\ln\left(2\pi + \sqrt{1+4\pi^2}\right) \approx 21.256$

17. $\left.\frac{ds}{dt}\right|_{t=\frac{\pi}{4}} = 5\sqrt{9 + 55\sin^2\left(5 \cdot \frac{\pi}{4}\right)} \approx 30.21$ m/s

19. $\left.\frac{ds}{dt}\right|_{t=1} = \sqrt{10} \approx 3.16$ m/s **21.** $\left(\frac{ds}{dt}\right)_{\min} \approx \sqrt{4.89} \approx 2.21$

23. $\frac{ds}{dt} = 8$

25.

$t = \frac{\pi}{2}$ (0, e); $t = \pi$, (−1, 1); $t = 0, t = 2\pi$, (1, 1); $t = \frac{3\pi}{2}$ $(0, \frac{1}{e})$

$M_{10} = 6.903734$, $M_{20} = 6.915035$, $M_{30} = 6.914949$, $M_{50} = 6.914951$

27. $M_{10} = 25.528309$,

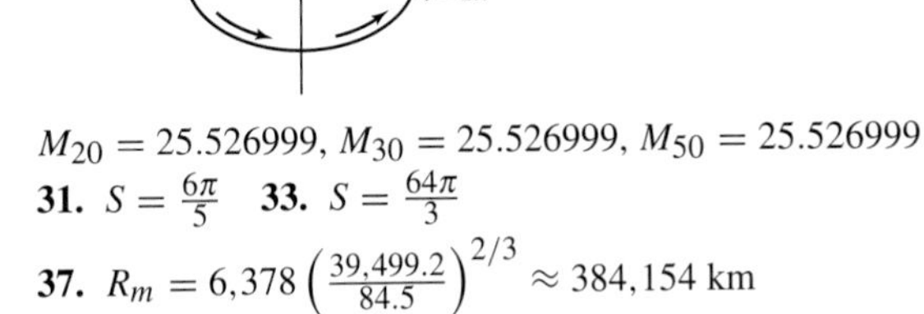

$M_{20} = 25.526999$, $M_{30} = 25.526999$, $M_{50} = 25.526999$

31. $S = \frac{6\pi}{5}$ **33.** $S = \frac{64\pi}{3}$

37. $R_m = 6{,}378\left(\frac{39{,}499.2}{84.5}\right)^{2/3} \approx 384{,}154$ km

Section 11.3 Preliminary

1. (b) **2.** Positive: $(r, \theta) = \left(1, \frac{\pi}{2}\right)$; Negative: $(r, \theta) = \left(-1, \frac{3\pi}{2}\right)$

3. Exactly one

4. **(a)** Equation of the circle of radius 2 centered at the origin.
(b) Equation of the circle of radius $\sqrt{2}$ centered at the origin.
(c) Equation of the vertical line through the point (2, 0).

5. (a)

Section 11.3 Exercises

1. **(a)** $\left(3\sqrt{2}, \frac{3\pi}{4}\right)$ **(b)** $(3, \pi)$ **(c)** $(\sqrt{5}, \pi + 0.46) \approx (\sqrt{5}, 3.6)$
(d) $\left(\sqrt{2}, \frac{5\pi}{4}\right)$ **(e)** $\left(\sqrt{2}, \frac{\pi}{4}\right)$ **(f)** $\left(4, \frac{\pi}{6}\right)$ **(g)** $\left(4, \frac{11\pi}{6}\right)$

3. **(a)** $(1, 0)$ **(b)** $\left(\sqrt{12}, \frac{\pi}{6}\right)$ **(c)** $\left(\sqrt{8}, \frac{3\pi}{4}\right)$ **(d)** $\left(2, \frac{2\pi}{3}\right)$

5. **(a)** $\left(\frac{3\sqrt{3}}{2}, \frac{3}{2}\right)$ **(b)** $\left(-\frac{6}{\sqrt{2}}, \frac{6}{\sqrt{2}}\right)$ **(c)** $(0, -5)$

7. (b), (c), (f) **9.** $\theta \approx 0.46$ **11.** $r = 2\csc\theta$

13. $x^2 + \left(y - \frac{1}{2}\right)^2 = \left(\frac{1}{2}\right)^2$ **15.** $y = 2$ **17.** $\frac{\left(x - \frac{1}{3}\right)^2}{\frac{4}{9}} + \frac{y^2}{\frac{1}{3}} = 1$

19. $r = 5\sec\theta$ **21.** $r^2 = 2\csc 2\theta$

23. A, $\theta = 0$; B, $\theta = \frac{\pi}{4}$; C, $\theta = \frac{\pi}{2}, \frac{3\pi}{2}$; D, $\theta = \frac{7\pi}{4}$

25. (a) ↔ (iv), (b) ↔ (iii), (c) ↔ (i), (d) ↔ (ii)

27. A, $\left(\frac{\pi}{2}, 1\right)$; B, $\left(\frac{3\pi}{4}, 0\right)$; C, $(0, 1)$; D, $\left(\frac{\pi}{4}, \sqrt{2}\right)$

29.

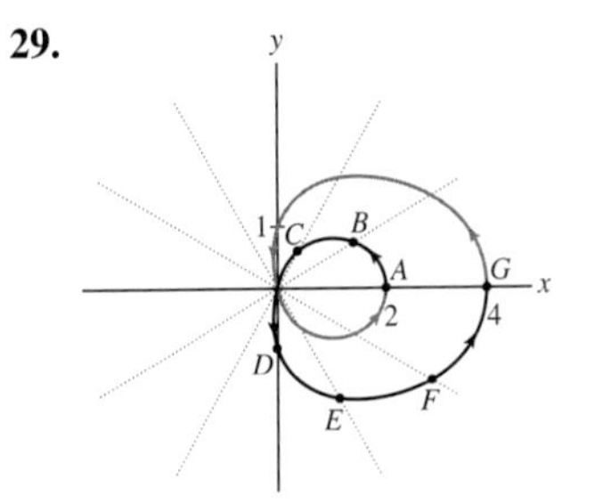

31. **(a)** A, $\theta = 0, r = 0$; B, $\theta = \frac{\pi}{4}, r = \sin\frac{2\pi}{4} = 1$; C, $\theta = \frac{\pi}{2}$, $r = 0$; D, $\theta = \frac{3\pi}{4}, r = \sin\frac{2\cdot 3\pi}{4} = -1$; E, $\theta = \pi, r = 0$; F, $\theta = \frac{5\pi}{4}$, $r = 1$; G, $\theta = \frac{3\pi}{2}, r = 0$; H, $\theta = \frac{7\pi}{4}, r = -1$; I, $\theta = 2\pi, r = 0$
(b) $0 \le \theta \le \frac{\pi}{2}$ is in the first quadrant. $\frac{\pi}{2} \le \theta \le \pi$ is in the fourth quadrant. $\pi \le \theta \le \frac{3\pi}{2}$ is in the third quadrant. $\frac{3\pi}{2} \le \theta \le 2\pi$ is in the second quadrant.

33.

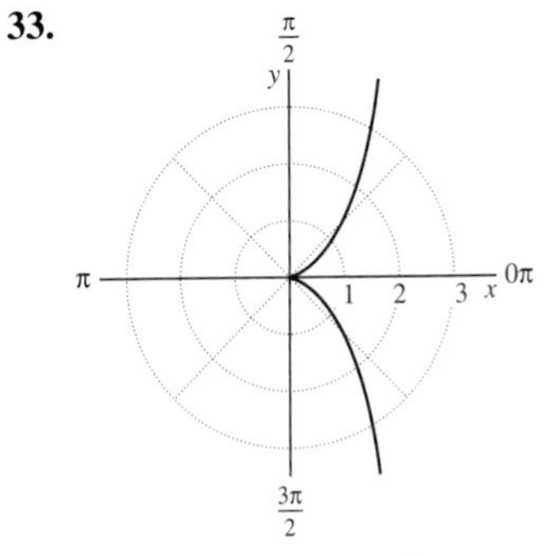

35. $\left(x - \frac{a}{2}\right)^2 + \left(y - \frac{b}{2}\right)^2 = \frac{a^2+b^2}{4}$, $r = \frac{\sqrt{a^2+b^2}}{2}$, centered at the point $\left(\frac{a}{2}, \frac{b}{2}\right)$.

37. $r^2 = \sec 2\theta$ **39.** $\left(x^2 + y^2\right)^2 = x^3 - 3y^2x$

41. $r = 2\sec\left(\theta - \frac{\pi}{9}\right)$ **43.** $r = 2\sqrt{10}\sec(\theta - 4.39)$

45. $r = \frac{9}{4\cos\theta - \sin\theta}$ **53.** $y = \frac{x}{\sqrt{3}}$

Section 11.4 Preliminary

1. False **2.** (b) **3.** False **4.** (c)

Section 11.4 Exercises

1. $A = \frac{1}{2}\int_{\pi/2}^{\pi} r^2\,d\theta = \frac{25\pi}{4}$

$\theta = \frac{\pi}{2}$, y, $\theta = \pi$, x

3. $A = \frac{1}{2}\int_0^{\pi} r^2\,d\theta = 4\pi$ **5.** $A = \frac{3\pi}{2}$

7. $A = \frac{\pi}{8} \approx 0.39$ **9.** $A = 1$ **11.** $A = \frac{3\pi a^2}{2}$

13. $A = \frac{\sqrt{15}}{2} + 7\cos^{-1}\left(\frac{1}{4}\right) \approx 11.163$

15. $A = \pi - \frac{3\sqrt{3}}{2} \approx 0.54$ **17.** $A = \frac{\pi}{8} - \frac{1}{4} \approx 0.14$

19. $A = 4\pi$ **21.** $A = \frac{9\pi}{2} - 4\sqrt{2}$ **23.** $A = \frac{\pi^3}{48} - \frac{\pi}{8} \approx 0.25$

25. $L = \frac{A}{2}\sqrt{A^2+1} + \frac{1}{2}\ln\left|A + \sqrt{A^2+1}\right|$

27. $L = \tan A$

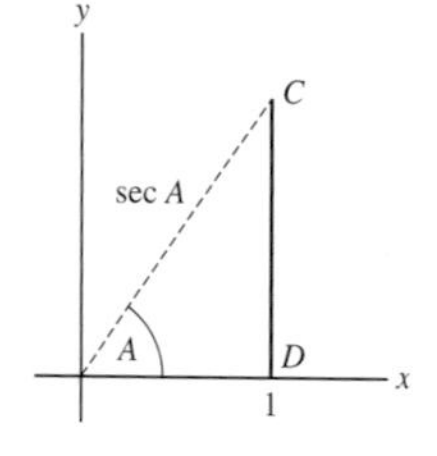

29. $L = 8$ **31.** $L = \sqrt{2}(e^{2\pi} - 1) \approx 755.9$

33. $L = \int_0^{\pi} \sqrt{1 + a^2}e^{a\theta}\, d\theta$

35. $L = \int_0^{2\pi} \sqrt{5 - 4\cos\theta}(2 - \cos\theta)^{-2}\, d\theta$

Section 11.5 Preliminary

1. (a) Hyperbola **(b)** Not a conic section **(c)** Ellipse **(d)** Not a conic section

2. Hyperbola **3.** The points $(0, c)$ and $(0, -c)$

5. $\pm\frac{b}{a}$ are the slopes of the two asymptotes of the hyperbola.

Section 11.5 Exercises

1. $F_1 = (-\sqrt{65}, 0)$, $F_2 = (\sqrt{65}, 0)$; The vertices are $(9, 0)$, $(-9, 0)$, $(0, 4)$, $(0, -4)$.

3. $F_1 = (-\sqrt{5}, 0)$, $F_2 = (\sqrt{5}, 0)$; The vertices are $(3, 0)$, $(-3, 0)$, $(0, 2)$, $(0, -2)$.

5. $F_1 = (\sqrt{97}, 0)$, $F_2 = (-\sqrt{97}, 0)$; The vertices are $(4, 0)$ and $(-4, 0)$.

7. $F_1 = (\sqrt{65} + 3, -1)$, $F_2 = (-\sqrt{65} + 3, -1)$; The vertices are $(10, -1)$ and $(-4, -1)$.

9. $\frac{x^2}{5^2} + \frac{y^2}{7^2} = 1$ **11.** $\frac{(x-12)^2}{5^2} + \frac{(y-6)^2}{7^2} = 1$

13. $\left(\frac{x}{9}\right)^2 + \left(\frac{y}{16}\right)^2 = 1$ **15.** $\left(\frac{x}{4}\right)^2 + \left(\frac{y}{2\sqrt{13}}\right)^2 = 1$

17. $\frac{x^2}{9} + \frac{y^2}{16} = 1$ **19.** $\left(\frac{x}{4}\right)^2 - \left(\frac{y}{12}\right)^2 = 1$ **23.** $x = \frac{y^2}{8}$

25. $y = \frac{1}{20}x^2$ **27.** $y = \frac{x^2}{16}$ **29.** $x = \frac{y^2}{8}$

31. Vertices $(\pm 4, 0)$, $(0, \pm 2)$. Foci $(-\sqrt{12}, 0)$, $(\sqrt{12}, 0)$. Focal axis is the x-axis. Conjugate axis is the y-axis. Centered at the origin.

33. Vertices $(\pm 2, 0)$, $(0, \pm 4)$. Foci $(0, \pm\sqrt{12})$. Focal axis is the y-axis. Conjugate axis is the x-axis. Centered at the origin.

35. Vertices $(5, 5)$, $(-7, 5)$. Foci $(\sqrt{84} - 1, 5)$, $(-\sqrt{84} - 1, 5)$. Focal axis: $y = 5$. Conjugate axis: $x = -1$. Asymptotes $y = -1.15x + 3.85$, $y = 1.15x + 6.15$. Centered at $(-1, 5)$.

37. Vertex $(4, 0)$. Focus $\left(4, \frac{1}{16}\right)$. The axis is the vertical line $x = 4$.

39. Foci $\left(-\frac{\sqrt{21}}{2} + 1, \frac{1}{5}\right)$, $\left(\frac{\sqrt{21}}{2} + 1, \frac{1}{5}\right)$. Focal axis: $y = 0.2$. Conjugate axis: $x = 1$. Centered at $(1, 0.2)$.

41. $D = -87$; ellipse **43.** $D = 88$; hyperbola

47. Focus $(0, c)$. Directrix $y = -c$.

49. $\overline{A'F_0} = a + c = 3.6 + 0.9 = 4.5$ billion mi

51. $r = \frac{3}{2+\cos\theta}$ **53.** $r = \frac{4}{1+\cos\theta}$

55. Hyperbola, $e = 4$; directrix $x = 2$

57. Ellipse, $e = \frac{3}{4}$; directrix $x = \frac{8}{3}$

61. $\left(\frac{x+\frac{4}{3}}{\frac{8}{3}}\right)^2 + \left(\frac{y}{\frac{4}{\sqrt{3}}}\right)^2 = 1$

Chapter 11 Review

1. a, c

3. $c(t) = (1 + 2\cos t, 1 + 2\sin t)$. The intersection points with the y-axis are $(0, 1 \pm \sqrt{3})$. The intersection points with the x-axis are $(1 \pm \sqrt{3}, 0)$.

5. $c(\theta) = (\cos(\theta + \pi), \sin(\theta + \pi))$ **7.** $c(t) = (1 + 2t, 3 + 4t)$

9. $y = -\frac{x}{4} + \frac{37}{4}$ **11.** $y = \frac{8}{(3-x)^3} + \frac{3-x}{2}$

13. $(\pi^2 k^2, 0)$, where $k \in \mathbb{Z}$

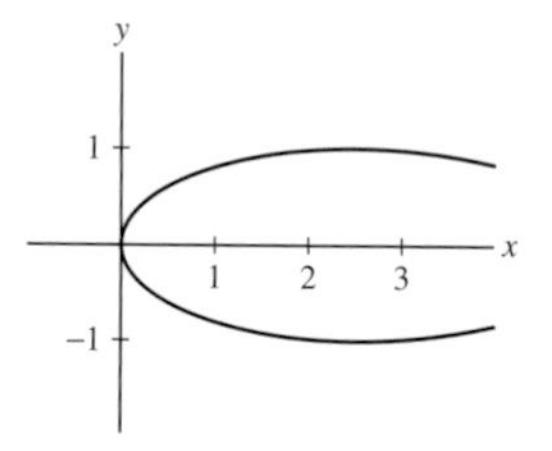

15. $\left.\frac{dy}{dx}\right|_{\theta=\frac{\pi}{4}} - \frac{1}{4\sqrt{2}}$ **17.** $\left.\frac{dy}{dx}\right|_P = 5$

19. Vertical: $t = \pi + 2\pi k$ where $k \in \mathbb{Z}$; horizontal: $t = \pm\frac{\pi}{3} + 2\pi k$ where $k \in \mathbb{Z}$

21. $\left.\frac{ds}{dt}\right|_{t=\frac{\pi}{4}} = \sqrt{20.5} \approx 4.53$ **23.** $L = 5(e - 1)$

25. $t_0 = \frac{3}{4}\pi$; $\left.\frac{ds}{dt}\right|_{t=\frac{3}{4}\pi} = e^{-3\pi/4}\sqrt{2}$

27. $(\sqrt{10}, 5.034)$; $(\sqrt{10}, -0.321)$ **29.** $r^2 = \frac{12}{2+\sin 2\theta}$

31. $7x - y = 4$ **33.** $A = \frac{9}{4}\left(\frac{\pi}{3} + \frac{\sqrt{3}}{2}\right)$ **35.** $A = \frac{\pi}{2}$

37. $A = \frac{5\pi}{8} - 1$ **39.** $L_{\text{inner}} = 7.5087$; $L_{\text{outer}} = 36.121$

41. Foci $\left(\pm\sqrt{6}, 0\right)$. Vertices $(\pm 2, 0)$. Hyperbola centered at $(0, 0)$.

43. Hyperbola shifted 3 units on the y-axis. Foci $\left(\pm\sqrt{\frac{3}{2}}, 3\right)$. Vertices $\left(\pm\frac{1}{\sqrt{2}}, 3\right)$.

45. $\left(\frac{x}{8}\right)^2 - \left(\frac{y}{6}\right)^2 = 1$ **47.** $\left(\frac{x}{64}\right)^2 + \left(\frac{y}{8\sqrt{63}}\right)^2 = 1$

51. $e = \frac{76}{816} = 0.0931372549$

Chapter 12

Section 12.1 Preliminary

1. **(a)** True **(b)** False **(c)** True **(d)** True

2. $\|-3\mathbf{a}\| = 15$ **3.** The components are not changed. **4.** $\langle 0, 0\rangle$

5. **(a)** True **(b)** False

Section 12.1 Exercises

1. $\mathbf{v}_1 = \langle 2, 0\rangle$, $\|\mathbf{v}_1\| = 2$ $\mathbf{v}_2 = \langle 2, 0\rangle$, $\|\mathbf{v}_2\| = 2$

$\mathbf{v}_3 = \langle 3, 1\rangle$, $\|\mathbf{v}_3\| = \sqrt{10}$ $\mathbf{v}_4 = \langle 2, 2\rangle$, $\|\mathbf{v}_4\| = 2\sqrt{2}$

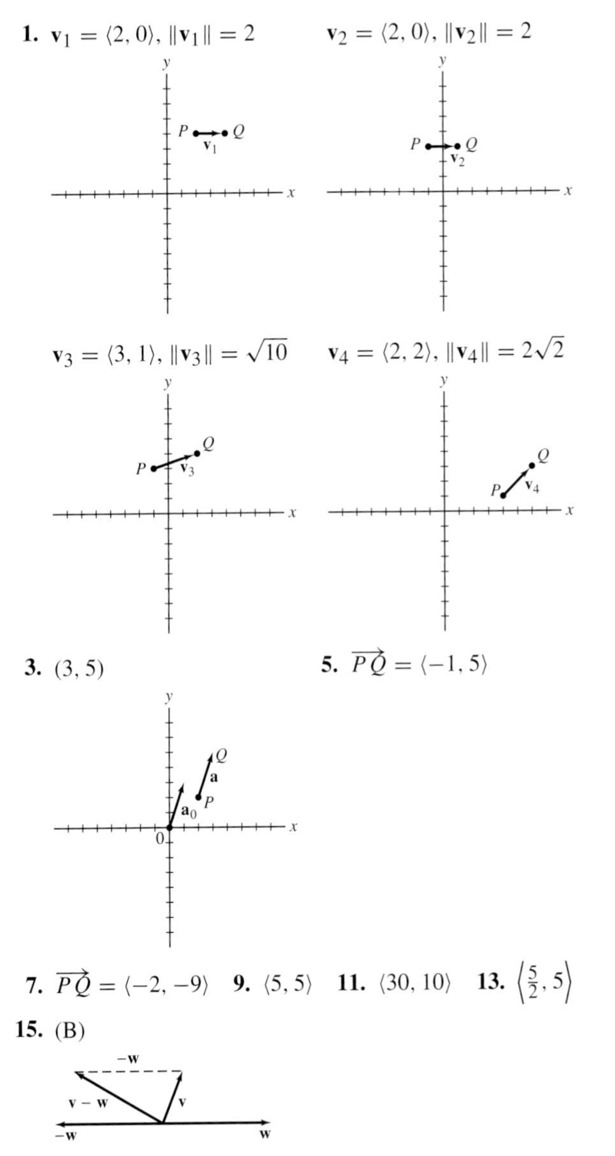

3. $(3, 5)$ **5.** $\overrightarrow{PQ} = \langle -1, 5\rangle$

7. $\overrightarrow{PQ} = \langle -2, -9\rangle$ **9.** $\langle 5, 5\rangle$ **11.** $\langle 30, 10\rangle$ **13.** $\left\langle \frac{5}{2}, 5\right\rangle$

15. (B)

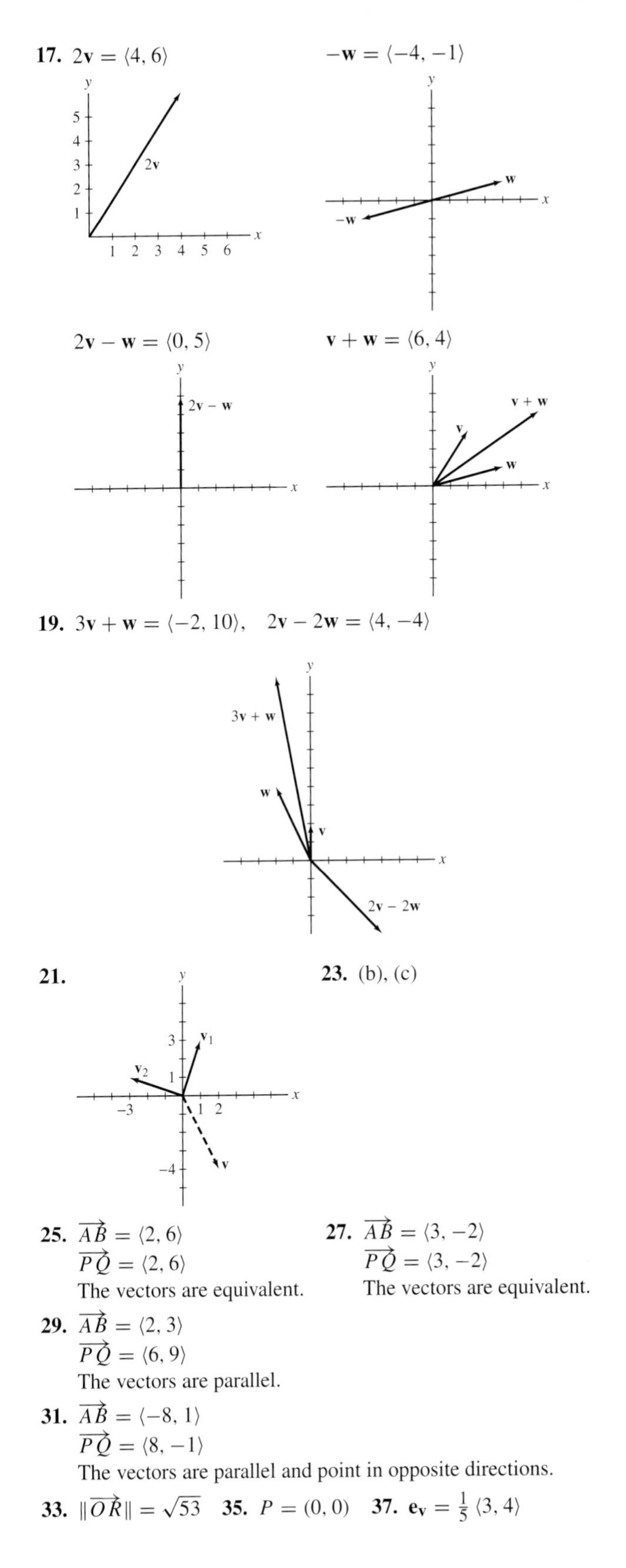

17. $2\mathbf{v} = \langle 4, 6\rangle$ $-\mathbf{w} = \langle -4, -1\rangle$

$2\mathbf{v} - \mathbf{w} = \langle 0, 5\rangle$ $\mathbf{v} + \mathbf{w} = \langle 6, 4\rangle$

19. $3\mathbf{v} + \mathbf{w} = \langle -2, 10\rangle$, $2\mathbf{v} - 2\mathbf{w} = \langle 4, -4\rangle$

21.

23. (b), (c)

25. $\overrightarrow{AB} = \langle 2, 6\rangle$
$\overrightarrow{PQ} = \langle 2, 6\rangle$
The vectors are equivalent.

27. $\overrightarrow{AB} = \langle 3, -2\rangle$
$\overrightarrow{PQ} = \langle 3, -2\rangle$
The vectors are equivalent.

29. $\overrightarrow{AB} = \langle 2, 3\rangle$
$\overrightarrow{PQ} = \langle 6, 9\rangle$
The vectors are parallel.

31. $\overrightarrow{AB} = \langle -8, 1\rangle$
$\overrightarrow{PQ} = \langle 8, -1\rangle$
The vectors are parallel and point in opposite directions.

33. $\|\overrightarrow{OR}\| = \sqrt{53}$ **35.** $P = (0, 0)$ **37.** $\mathbf{e}_{\mathbf{v}} = \frac{1}{5}\langle 3, 4\rangle$

39. $\mathbf{e_u} = \frac{1}{\sqrt{2}}\langle -1, -1\rangle$ **41.** $-\mathbf{e_v} = \left\langle \frac{1}{\sqrt{5}}, -\frac{2}{\sqrt{5}}\right\rangle$

43. $\left\langle \sqrt{3}, 1\right\rangle$ **45.** $\lambda = \pm\frac{1}{\sqrt{13}}$ **47.** $a = 0; b = 4$

49. **(a)** $\langle 4, 3\rangle$; $\|4\mathbf{i}+3\mathbf{j}\| = 5$ **(b)** $\langle 2, -3\rangle$; $\|2\mathbf{i}-3\mathbf{j}\| = \sqrt{13}$ **(c)** $\langle 1, 1\rangle$; $\|\mathbf{i}+\mathbf{j}\| = \sqrt{2}$ **(d)** $\langle 1, -3\rangle$; $\|\mathbf{i}-3\mathbf{j}\| = \sqrt{10}$

51. $8\mathbf{i}+46\mathbf{j}$ **53.** $24\mathbf{i}-4\mathbf{j}$

55. $s = -1, r = 2$ **57.** $\mathbf{u} = \frac{11}{18}\langle 1, 4\rangle + \frac{5}{18}\langle 5, 2\rangle$

59. $f_1 = 27.32$ lb, $f_2 = 24.5$ lb

61. $\mathbf{r} = \langle L_1\sin\theta_1 + L_2\sin\theta_2, L_1\cos\theta_1 - L_2\cos\theta_2\rangle$

63. $\left(\frac{x}{8}\right)^2 + \left(\frac{y}{2}\right)^2 = 1$

Section 12.2 Preliminary

1. $Q = (4, 3, 2)$

2. $\mathbf{v} = \langle 3, 2, 1\rangle$ regardless of the base point.

3. (a) **4.** (c) **5.** Infinitely many direction vectors. **6.** True

Section 12.2 Exercises

1. **3.**

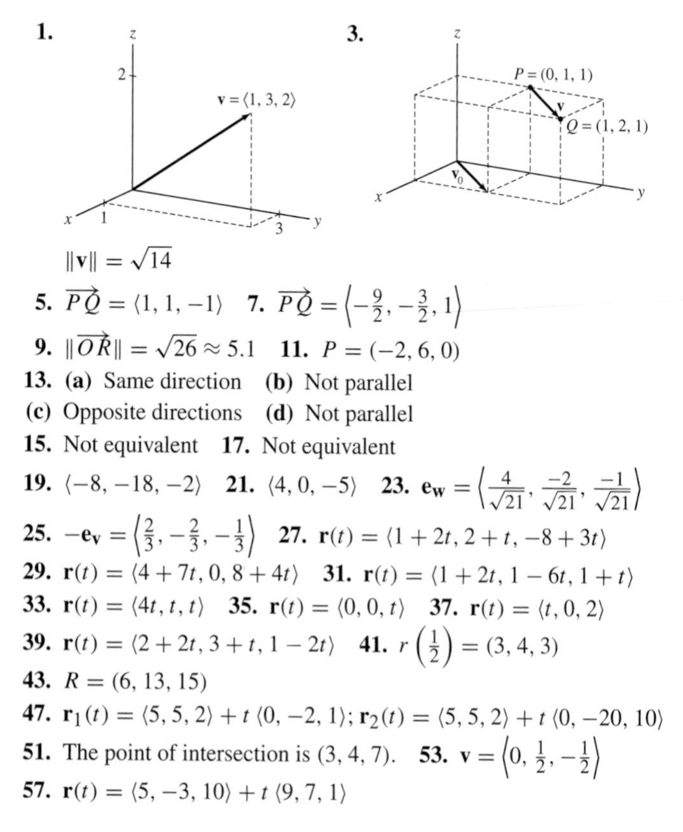

$\|\mathbf{v}\| = \sqrt{14}$

5. $\overrightarrow{PQ} = \langle 1, 1, -1\rangle$ **7.** $\overrightarrow{PQ} = \left\langle -\frac{9}{2}, -\frac{3}{2}, 1\right\rangle$

9. $\|\overrightarrow{OR}\| = \sqrt{26} \approx 5.1$ **11.** $P = (-2, 6, 0)$

13. **(a)** Same direction **(b)** Not parallel **(c)** Opposite directions **(d)** Not parallel

15. Not equivalent **17.** Not equivalent

19. $\langle -8, -18, -2\rangle$ **21.** $\langle 4, 0, -5\rangle$ **23.** $\mathbf{e_w} = \left\langle \frac{4}{\sqrt{21}}, \frac{-2}{\sqrt{21}}, \frac{-1}{\sqrt{21}}\right\rangle$

25. $-\mathbf{e_v} = \left\langle \frac{2}{3}, -\frac{2}{3}, -\frac{1}{3}\right\rangle$ **27.** $\mathbf{r}(t) = \langle 1+2t, 2+t, -8+3t\rangle$

29. $\mathbf{r}(t) = \langle 4+7t, 0, 8+4t\rangle$ **31.** $\mathbf{r}(t) = \langle 1+2t, 1-6t, 1+t\rangle$

33. $\mathbf{r}(t) = \langle 4t, t, t\rangle$ **35.** $\mathbf{r}(t) = \langle 0, 0, t\rangle$ **37.** $\mathbf{r}(t) = \langle t, 0, 2\rangle$

39. $\mathbf{r}(t) = \langle 2+2t, 3+t, 1-2t\rangle$ **41.** $r\left(\frac{1}{2}\right) = (3, 4, 3)$

43. $R = (6, 13, 15)$

47. $\mathbf{r}_1(t) = \langle 5, 5, 2\rangle + t\langle 0, -2, 1\rangle$; $\mathbf{r}_2(t) = \langle 5, 5, 2\rangle + t\langle 0, -20, 10\rangle$

51. The point of intersection is $(3, 4, 7)$. **53.** $\mathbf{v} = \left\langle 0, \frac{1}{2}, -\frac{1}{2}\right\rangle$

57. $\mathbf{r}(t) = \langle 5, -3, 10\rangle + t\langle 9, 7, 1\rangle$

Section 12.3 Preliminary

1. Scalar **2.** The angle between **a** and **b** is obtuse.

3. Distributive Law **4.** $\text{proj}_{\mathbf{v}}(\mathbf{v}) = \mathbf{v}$ **5.** No difference

6. **(a)** $\text{proj}_{2\mathbf{v}}(\mathbf{u}) = \mathbf{v}$ **(b)** $\text{proj}_{\mathbf{v}}(2\mathbf{u}) = 2\mathbf{v}$

7. **(b)** $\mathbf{u}_{\|}$

8. **(c)** $\mathbf{e_u}\cdot\mathbf{e_v}$

Section 12.3 Exercises

1. 15 **3.** 41 **5.** 5 **7.** 0 **9.** 1 **11.** 0 **13.** Obtuse

15. Acute **17.** Obtuse **19.** $\theta = 90°$ **21.** $\cos\theta = \frac{1}{\sqrt{10}}$

23. $\theta = \cos^{-1} 0.997 \approx 4.4°$ **25.** $\theta \approx 70.53°$

27. $\theta = \cos^{-1} 0.724 \approx 43.57°$

29. **(a)** $b = -\frac{1}{2}$ **(b)** $b = 0$ or $b = \frac{1}{2}$

31. $\mathbf{v}_1 = \langle 0, 1, 0\rangle$, $\mathbf{v}_2 = \langle 3, 2, 2\rangle$ **33.** 8 **35.** 2 **37.** $\|\mathbf{v}\|^2$

39. $\|\mathbf{v}\|^2 - \|\mathbf{w}\|^2$ **41.** $\text{proj}_{\mathbf{v}}(\mathbf{u}) = \left\langle -\frac{4}{5}, 0, -\frac{2}{5}\right\rangle$

43. $\text{proj}_{\mathbf{v}}(\mathbf{u}) = \langle 1, 1, 0\rangle$ **45.** $\text{proj}_{\mathbf{v}}(\mathbf{u}) = -4\mathbf{k}$

47. $\text{proj}_{\mathbf{v}}(\mathbf{u}) = a\mathbf{i}$ **49.** $\mathbf{a} = \langle -2, 0, 0\rangle + \langle 0, 1, 1\rangle$

51. $\mathbf{a} = \left\langle \frac{1}{2}, \frac{1}{2}\right\rangle + \left\langle \frac{1}{2}, -\frac{1}{2}\right\rangle$ **53.** $\mathbf{a}_{\|} = \mathbf{0}, \mathbf{a}_{\perp} = \mathbf{a} = \langle 4, 2\rangle$

57. $\|\mathbf{v}+\mathbf{w}\| = \sqrt{2+\sqrt{3}} \approx 1.93$

59. **(a)** $\mathbf{v}\cdot\mathbf{w} = -3$ **(b)** $\|2\mathbf{v}+\mathbf{w}\| = \sqrt{13} \approx 3.61$ **(c)** $\|2\mathbf{v}-3\mathbf{w}\| = \sqrt{133} \approx 11.53$

61. $\theta_1 \approx 47.47°$, $\theta_2 \approx 71.58°$, $\theta_3 \approx 60.95°$ **63.** $\beta = 60°$

65. $\text{proj}_{\overrightarrow{AB}}(\overrightarrow{AD}) = \left\langle \frac{1}{2}, 0, -\frac{1}{2}\right\rangle$

67. Counterexample: $\mathbf{a} = \langle 1, 0, 1\rangle$, $\mathbf{v} = \langle 3, 1, 1\rangle$, $\mathbf{w} = \langle 4, 1, 0\rangle$

69. The force required is 68.07 N.

71. $\mathbf{F}_1 = -62.1\mathbf{i} + 35.9\mathbf{j}$, $\mathbf{F}_2 = 62.1\mathbf{i} + 62.1\mathbf{j}$, in Newtons

85. $\cos\alpha = \frac{3}{7}$, $\cos\beta = \frac{6}{7}$, $\cos\gamma = -\frac{2}{7}$

Section 12.4 Preliminary

1. $\begin{vmatrix} -5 & -1 \\ 4 & 0 \end{vmatrix}$ **2.** $\|\mathbf{e}\times\mathbf{f}\| = \frac{1}{\sqrt{2}}$ **3.** $\mathbf{u}\times\mathbf{w} = \langle -2, -2, -1\rangle$

4. **(a)** 0 **(b)** 0

5. $\mathbf{i}\times\mathbf{j} = \mathbf{k}$ and $\mathbf{i}\times\mathbf{k} = -\mathbf{j}$

6. **v** or **w** (or both) is the zero vector, or if **v** and **w** are parallel vectors.

Section 12.4 Exercises

1. −5 **3.** −15 **5.** −8 **7.** 15 **9.** 1 **11.** $\mathbf{v}\times\mathbf{w} = \mathbf{i}+2\mathbf{j}-5\mathbf{k}$

13. $\mathbf{v}\times\mathbf{w} = 6\mathbf{i}-8\mathbf{k}$ **15.** $\mathbf{v}\times\mathbf{w} = \frac{7}{3}\mathbf{i}-\mathbf{j}+\frac{2}{3}\mathbf{k}$ **17.** $-\mathbf{j}+\mathbf{i}$

19. $-2\mathbf{i}+\mathbf{j}+\mathbf{k}$ **21.** $\langle -1, -1, 0\rangle$ **23.** $\langle 0, 9, 3\rangle$

25. $\langle -2, -2, -2\rangle$

27. $\mathbf{v}\times\mathbf{i} = \langle 0, c, -b\rangle$; $\mathbf{v}\times\mathbf{j} = \langle -c, 0, a\rangle$; $\mathbf{v}\times\mathbf{k} = \langle b, -a, 0\rangle$

29. $\mathbf{v}\times\mathbf{w}$ equals $-\mathbf{u}$ **31.** $\mathbf{u} = \mathbf{v}\times\mathbf{w} = \langle 0, 3, 3\rangle$ **35.** $\mathbf{e}'$

37. $\mathbf{F}_1$ **41.** The area is $8\sqrt{53}$.

43.

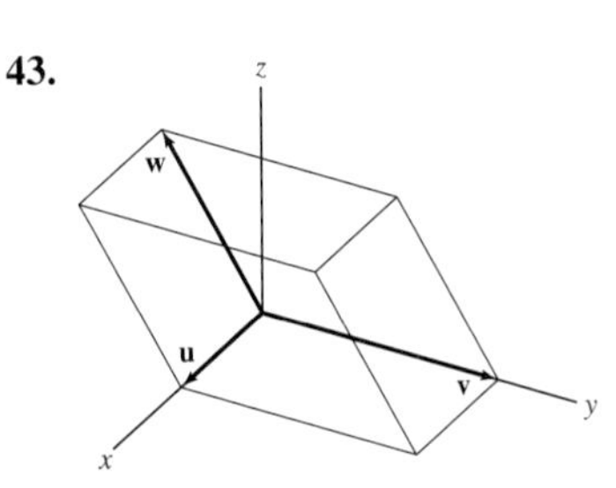

The volume is 4.

45. $A = \sqrt{35} \approx 5.92$. **47.** $S = \frac{9\sqrt{3}}{2} \approx 7.8$.

61. $\mathbf{X} = \langle 1, 1, 2 \rangle$ **65.** $\tau = (204.79 - 2{,}940 \cos\theta)\mathbf{k}$

67. $\theta = 41.39°$ **69.** 107

Section 12.5 Preliminary

1. $3x + 4y - z = 0$ **2.** (c) $z = 1$ **3.** Plane (c)

4. xz-plane **5.** (c) $x + y = 0$ **6.** (a) is correct

Section 12.5 Exercises

1. $\langle 1, 3, 2 \rangle \cdot \langle x, y, z \rangle = 3$
$x + 3y + 2z = 3$
$(x - 4) + 3(y + 1) + 2(z - 1) = 0$

3. $\langle -1, 2, 1 \rangle \cdot \langle x, y, z \rangle = 3$
$-x + 2y + z = 3$
$-(x - 4) + 2(y - 1) + (z - 5) = 0$

5. $\langle 1, 0, 0 \rangle \cdot \langle x, y, z \rangle = 3$
$x = 3$
$(x - 3) + 0 \cdot (y - 1) + 0 \cdot (z + 9) = 0$

7. $\langle 0, 0, 1 \rangle \cdot \langle x, y, z \rangle = 2$
$z = 2$
$0(x - 6) + 0(y - 7) + (z - 2) = 0$

9. $2x + y + z = 1$ **11.** $6x + 9y + 4z = 19$ **13.** $x + 2y - z = 1$

15. $\mathbf{n} = \langle 9, -4, -11 \rangle$ **17.** $\mathbf{n} = \langle 3, 25, 9 \rangle$ **19.** $\mathbf{n} = \langle 1, 0, -1 \rangle$

21. $4x - 9y + z = 0$ **23.** $x = 4$ **25.** $13x + y - 5z = 0$

27. The planes are parallel. **29.** $10x + 15y + 6z = 30$

31. $P = (2, -1, -1)$ **33.** $P = (9, 18, 3)$ **35.** $3x - 4y = 8$

37. $3x + 4z = 5$ **39.** $3x + 4z = -2$ **41.** $y = 4$

43. $4x + 3y + z = 8,\ 4x + 3y - 5z = 8$

45. $ax - 2ay + cz = 0,\ a \neq 0$ **49.** $\theta = \frac{\pi}{2}$ **51.** $\theta = \frac{\pi}{2}$

53. $\theta = 94.80°$ **55.** $x + y + z = 1$

57. $x - y - z = f$ **59.** $x = \frac{9}{5} + 2t,\ y = -\frac{6}{5} - 3t,\ z = 2 + 5t$

61. $x = 3 + 3t,\ y = -1 + 5t,\ z = 1 - 7t$

63. $P = \left(\frac{2}{3}, -\frac{1}{3}, \frac{2}{3}\right)$ **67.** $\ell = \frac{4}{5}$

Section 12.6 Preliminary

1. True, mostly, except at $x = \pm a$, $y = \pm b$, or $z = \pm c$.

2. False **3.** Hyperbolic paraboloid **4.** No **5.** Ellipsoid

6. $-\left(\frac{x}{2}\right)^2 - \left(\frac{y}{3}\right)^2 + \left(\frac{z}{4}\right)^2 = 1$

7. All vertical lines passing through a parabola C in the xy-plane.

Section 12.6 Exercises

1. Ellipsoid **3.** Ellipsoid **5.** Hyperboloid of one sheet

7. Elliptic paraboloid **9.** Hyperbolic paraboloid

11. Hyperbolic paraboloid **13.** Ellipsoid, a circle on the xz-plane.

15. Ellipsoid, the trace is an ellipse.

17. Hyperboloid of one sheet, the trace is a hyperbola.

19. Parabolic cylinder, the trace is the parabola $y = 3x^2$.

21. a $\leftrightarrow$ Figure b; b $\leftrightarrow$ Figure c; c $\leftrightarrow$ Figure a

23. $y = \left(\frac{x}{2}\right)^2 + \left(\frac{z}{4}\right)^2$ **25.** $|h| < 2$

27. **31.**

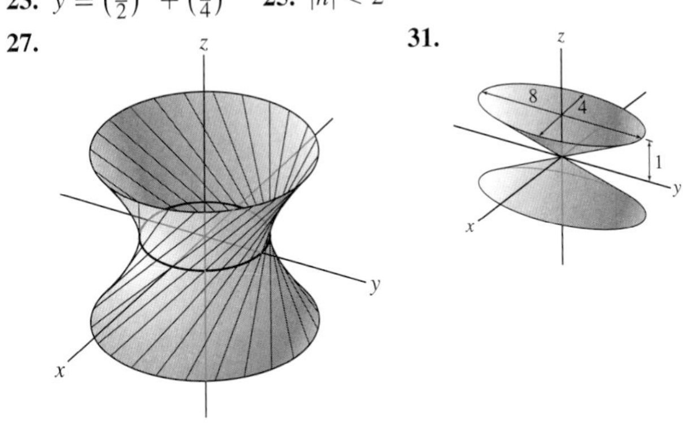

33. $\left(\frac{x}{2}\right)^2 + \left(\frac{y}{4}\right)^2 + \left(\frac{z}{6}\right)^2 = 1$ **35.** $\left(\frac{x}{4}\right)^2 + \left(\frac{y}{6}\right)^2 - \left(\frac{z}{3\sqrt{3}}\right)^2 = 1$

39. The upper part of an elliptic cone

Section 12.7 Preliminary

1. Cylinder of radius R whose axis is the z-axis.
Sphere of radius R centered at the origin.

2. (b) **3.** (a) **4.** $\phi = 0, \pi$ **5.** $\phi = \frac{\pi}{2}$, the xy-plane

Section 12.7 Exercises

1. $(-4, 0, 4)$ **3.** $(x, y, z) = \left(0, 0, \frac{1}{2}\right)$

5. $(r, \theta, z) = \left(\sqrt{2}, \frac{7\pi}{4}, 1\right)$ **7.** $(r, \theta, z) = \left(2, \frac{\pi}{3}, 7\right)$

9. $(r, \theta, z) = \left(5, \frac{\pi}{4}, 2\right)$ **11.** $r^2 \leq 1$

13. $z = \sqrt{4 - r^2},\ 0 \leq \theta \leq \frac{\pi}{2}$

15. $r^2 \sin^2\theta + z^2 \leq 9,\ 0 \leq \theta \leq \frac{\pi}{4}$ and $\frac{5\pi}{4} \leq \theta \leq 2\pi$

17. **19.**

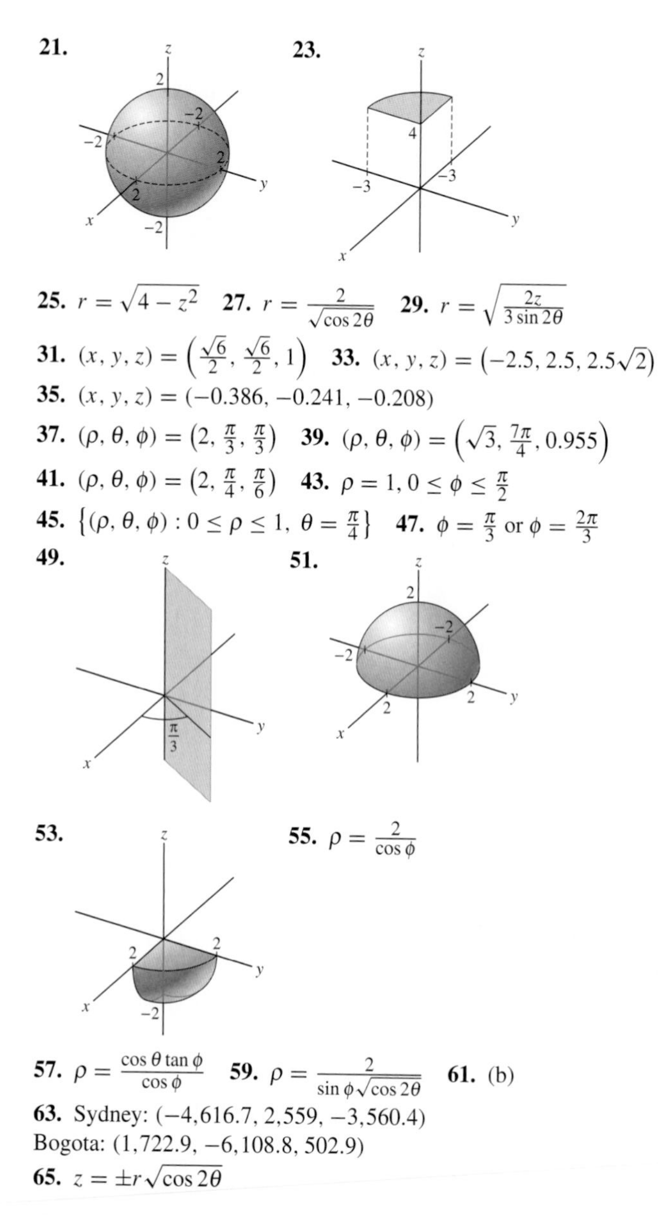

21. **23.**

25. $r = \sqrt{4 - z^2}$ **27.** $r = \dfrac{2}{\sqrt{\cos 2\theta}}$ **29.** $r = \sqrt{\dfrac{2z}{3 \sin 2\theta}}$

31. $(x, y, z) = \left(\frac{\sqrt{6}}{2}, \frac{\sqrt{6}}{2}, 1\right)$ **33.** $(x, y, z) = \left(-2.5, 2.5, 2.5\sqrt{2}\right)$

35. $(x, y, z) = (-0.386, -0.241, -0.208)$

37. $(\rho, \theta, \phi) = \left(2, \frac{\pi}{3}, \frac{\pi}{3}\right)$ **39.** $(\rho, \theta, \phi) = \left(\sqrt{3}, \frac{7\pi}{4}, 0.955\right)$

41. $(\rho, \theta, \phi) = \left(2, \frac{\pi}{4}, \frac{\pi}{6}\right)$ **43.** $\rho = 1, 0 \le \phi \le \frac{\pi}{2}$

45. $\left\{(\rho, \theta, \phi) : 0 \le \rho \le 1,\ \theta = \frac{\pi}{4}\right\}$ **47.** $\phi = \frac{\pi}{3}$ or $\phi = \frac{2\pi}{3}$

49. **51.**

53. **55.** $\rho = \dfrac{2}{\cos\phi}$

57. $\rho = \dfrac{\cos\theta \tan\phi}{\cos\phi}$ **59.** $\rho = \dfrac{2}{\sin\phi\sqrt{\cos 2\theta}}$ **61.** (b)

63. Sydney: $(-4{,}616.7, 2{,}559, -3{,}560.4)$
Bogota: $(1{,}722.9, -6{,}108.8, 502.9)$

65. $z = \pm r\sqrt{\cos 2\theta}$

Chapter 12 Review

1. $\langle 21, -25\rangle$ and $\langle -19, 31\rangle$ **3.** $\left\langle \frac{-2}{\sqrt{29}}, \frac{5}{\sqrt{29}} \right\rangle$ **5.** $\mathbf{i} = \frac{2}{11}\mathbf{v} + \frac{5}{11}\mathbf{w}$

7. $\overrightarrow{PQ} = \langle -4, 1\rangle$; $\|\overrightarrow{PQ}\| = \sqrt{17}$ **9.** $\left\langle \frac{3}{\sqrt{2}}, -\frac{3}{\sqrt{2}} \right\rangle$

11. $\beta = \frac{3}{2}$ **13.** $\mathbf{u} = \left\langle \frac{1}{3}, \frac{-11}{6}, \frac{7}{6} \right\rangle$

15. $\mathbf{r}_1(t) = \langle 1 + 3t, 4 + t, 5 + 6t\rangle$; $\mathbf{r}_2(t) = \langle 1 + 3t, t, 6t\rangle$

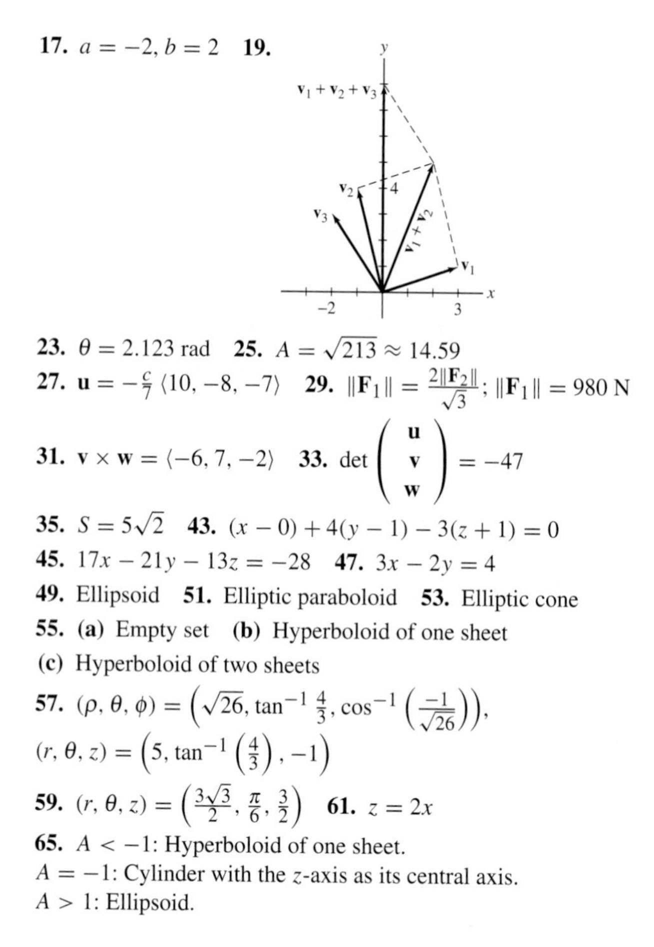

17. $a = -2, b = 2$ **19.**

23. $\theta = 2.123$ rad **25.** $A = \sqrt{213} \approx 14.59$

27. $\mathbf{u} = -\frac{c}{7}\langle 10, -8, -7\rangle$ **29.** $\|\mathbf{F}_1\| = \frac{2\|\mathbf{F}_2\|}{\sqrt{3}}$; $\|\mathbf{F}_1\| = 980$ N

31. $\mathbf{v} \times \mathbf{w} = \langle -6, 7, -2\rangle$ **33.** $\det\begin{pmatrix} \mathbf{u} \\ \mathbf{v} \\ \mathbf{w} \end{pmatrix} = -47$

35. $S = 5\sqrt{2}$ **43.** $(x - 0) + 4(y - 1) - 3(z + 1) = 0$

45. $17x - 21y - 13z = -28$ **47.** $3x - 2y = 4$

49. Ellipsoid **51.** Elliptic paraboloid **53.** Elliptic cone

55. **(a)** Empty set **(b)** Hyperboloid of one sheet
(c) Hyperboloid of two sheets

57. $(\rho, \theta, \phi) = \left(\sqrt{26}, \tan^{-1}\frac{4}{3}, \cos^{-1}\left(\frac{-1}{\sqrt{26}}\right)\right)$,
$(r, \theta, z) = \left(5, \tan^{-1}\left(\frac{4}{3}\right), -1\right)$

59. $(r, \theta, z) = \left(\frac{3\sqrt{3}}{2}, \frac{\pi}{6}, \frac{3}{2}\right)$ **61.** $z = 2x$

65. $A < -1$: Hyperboloid of one sheet.
$A = -1$: Cylinder with the z-axis as its central axis.
$A > 1$: Ellipsoid.

Chapter 13

Section 13.1 Preliminary

1. (c) **2.** The curve $z = e^x$

3. The projection onto the xz-plane **4.** The point $(-2, 2, 3)$

5. As t increases from 0 to 2π, a point on $\sin t\mathbf{i} + \cos t\mathbf{j}$ moves clockwise and a point on $\cos t\mathbf{i} + \sin t\mathbf{j}$ moves counterclockwise.

6. (a), (c), and (d)

Section 13.1 Exercises

1. $D = \{t \in \mathbf{R},\ t \ne 0, t \ne -1\}$

3. $\mathbf{r}(t) = (3 + 3t)\mathbf{i} - 5\mathbf{j} + (7 + t)\mathbf{k}$

5. A $\leftrightarrow$ ii, B $\leftrightarrow$ i, C $\leftrightarrow$ iii

7. $(a) = $ (v), $(b) = $ (i), $(c) = $ (ii), $(d) = $ (vi), $(e) = $ (iv), $(f) = $ (iii)

9. C $\leftrightarrow$ i, A $\leftrightarrow$ ii, B $\leftrightarrow$ iii

11. Plane $x = 7$, circle centered at $(7, 0, 0)$ with radius $\sqrt{144} = 12$.

13. Plane $y = 9$, circle centered at $(6, 9, 4)$ with $r = 3$.

15. Q lies on the curve.

17. $(0,1,0), (0,-1,0), \left(\frac{1}{\sqrt{2}}, \frac{1}{\sqrt{2}}, 0\right), \left(\frac{1}{\sqrt{2}}, -\frac{1}{\sqrt{2}}, 0\right),$
$\left(-\frac{1}{\sqrt{2}}, -\frac{1}{\sqrt{2}}, 0\right), \left(-\frac{1}{\sqrt{2}}, \frac{1}{\sqrt{2}}, 0\right)$

19. $\mathbf{r}(t) = \langle 2 + 9\cos 2t, 3\cos t, 3\sin t\rangle$

23. $\mathbf{r}(t) = \left\langle \cos t, \sin t, 4\cos^2 t\right\rangle$

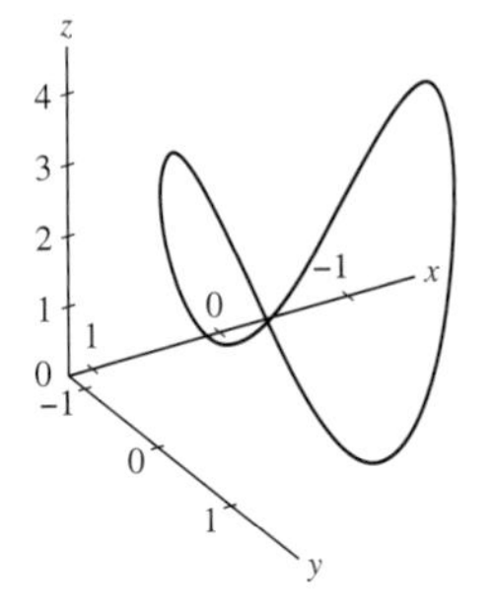

25. $\mathbf{r}(t) = \langle 3, 2, t\rangle, -\infty < t < \infty$ **27.** $\mathbf{r}(t) = \langle t, 3t, 15t\rangle$

29. $\mathbf{r}(t) = \langle 1, 2 + 2\cos t, 5 + 2\sin t\rangle$

31. $\mathbf{r}(t) = \left\langle \frac{\sqrt{3}}{2}\cos t, \frac{1}{2}, \frac{\sqrt{3}}{2}\sin t\right\rangle$

33. $\mathbf{r}(t) = \langle 3 + 2\cos t, 1, 5 + 3\sin t\rangle$

35. **(a)** False **(b)** True **(c)** True

37. The paths intersect. **39.** $z_{\max} = 4$

Section 13.2 Preliminary

1. $\frac{d}{dt} f(t)\mathbf{r}(t) = f(t)\mathbf{r}'(t) + f'(t)\mathbf{r}(t)$

$\frac{d}{dt}\mathbf{r}_1(t) \cdot \mathbf{r}_2(t) = \mathbf{r}_1(t) \cdot \mathbf{r}_2'(t) + \mathbf{r}_1'(t) \cdot \mathbf{r}_2(t)$

$\frac{d}{dt}\mathbf{r}_1(t) \times \mathbf{r}_2(t) = \mathbf{r}_1(t) \times \mathbf{r}_2'(t) + \mathbf{r}_1'(t) \times \mathbf{r}_2(t)$

2. True **3.** False **4.** True **5.** False **6.** False

7. **(a)** Vector **(b)** Scalar **(c)** Vector

Section 13.2 Exercises

1. $\lim_{t\to 3}\left\langle t^2, 4t, \frac{1}{t}\right\rangle = \left\langle 9, 12, \frac{1}{3}\right\rangle$

3. $\lim_{t\to 0}\left(e^{2t}\mathbf{i} + \ln(t+1)\mathbf{j} + 4\mathbf{k}\right) = \mathbf{i} + 4\mathbf{k}$

5. $\lim_{h\to 0}\frac{\mathbf{r}(t+h)-\mathbf{r}(t)}{h} = \left\langle -\frac{1}{t^2}, \cos t, 0\right\rangle$ **7.** $\frac{d\mathbf{r}}{dt} = \left\langle 1, 2t, 3t^2\right\rangle$

9. $\mathbf{w}'(s) = \left\langle e^s, -2e^{-2s}\right\rangle$ **11.** $\mathbf{r}'(t) = \left\langle 1 + t^{-2}, 8t, 0\right\rangle$

13. $\mathbf{a}'(\theta) = (-2\sin 2\theta)\mathbf{i} + (2\cos 2\theta)\mathbf{j} + (4\cos 4\theta)\mathbf{k}$

15. $\mathbf{r}'(t) = \left\langle 1, 2t, 3t^2\right\rangle$
$\mathbf{r}''(t) = \langle 0, 2, 6t\rangle$

17.

19. $\frac{d}{dt}\left(\mathbf{r}_1(t) \cdot \mathbf{r}_2(t)\right) = 4e^t + 18t^2$

21. $\frac{d}{dt}\left(\mathbf{r}_1(t) \cdot \mathbf{r}_2(t)\right) = \left\langle t^2 e^t(t+3), 48, 8e^t(t+1)\right\rangle$

23. $F'(t) = 4t^3 + 3t^2 + 9$ **27.** $\frac{d}{dt}\mathbf{r}(g(t)) = \left\langle 4e^{4t+9}, 8e^{8t+18}, 0\right\rangle$

29. $\frac{d}{dt}\mathbf{r}(g(t)) = \left\langle 3^{\sin t}\cos t \ln 3, \frac{\cos t}{1+\sin^2 t}\right\rangle$

31. $\frac{d}{dt}\mathbf{r}(t) \cdot \mathbf{a}(t)\Big|_{t=2} = 13$

33. $\ell(t) = \langle -3 - 4t, 10 + 5t, 16 + 24t\rangle$

35. $\ell(t) = (2 - s)\mathbf{i} + \left(\frac{3}{2}s - 1\right)\mathbf{k}$

37. $\ell(s) = s\mathbf{i} + (1 - s)\mathbf{j} + (9 + 9s)\mathbf{k}$

41. $(\ln 4)\mathbf{i} + \frac{56}{3}\mathbf{j} - \frac{496}{5}\mathbf{k}$ **43.** $-8\mathbf{i} - \frac{64}{3}\mathbf{j} - 64\mathbf{k}$

45. $\left(\frac{3}{2}t^2\right)\mathbf{i} + (2t^3)\mathbf{j} + (9t)\mathbf{k}$

47. $\mathbf{r}(t) = t\mathbf{i} - t\mathbf{j} + c$, with initial conditions
$\mathbf{r}(t) = (t+2)\mathbf{i} - t\mathbf{j} + 3\mathbf{k}$.

49. $\mathbf{r}(t) = \left\langle -\frac{1}{3}\cos 3t, -\frac{1}{3}\cos 3t, \frac{1}{2}t^2\right\rangle + \mathbf{c}$, with initial conditions
$\mathbf{r}(t) = \left\langle \frac{1}{3}(1 - \cos 3t), \frac{1}{3}(4 - \cos 3t), 8 + \frac{1}{2}t^2\right\rangle$.

51. $\mathbf{r}(t) = (8t^2)\mathbf{k} + \mathbf{c}_1 t + \mathbf{c}_2$, with initial conditions
$\mathbf{r}(t) = \mathbf{i} + t\mathbf{j} + (8t^2)\mathbf{k}$.

53. $\mathbf{r}(t) = \left\langle e^t, -\sin t, -\cos t\right\rangle + \mathbf{c}_1 t + \mathbf{c}_2$, with initial conditions
$\mathbf{r}(t) = \left\langle e^t - t, -\sin t + 3t, -\cos t + 2t + 2\right\rangle$.

55. $\mathbf{r}(4) = \left\langle 33, 2, \frac{256}{3}\right\rangle$ **57.** $\mathbf{r}(t) = \mathbf{c} + t\mathbf{v}$ **59.** $\mathbf{r}(t) = e^{2t}\mathbf{c}$

Section 13.3 Preliminary

1. $2\mathbf{r}' = \langle 50, -70, 20\rangle, -\mathbf{r}' = \langle -25, 35, -10\rangle$

2. **(b)** True

3. **(a)** $L'(2) = 4$
(b) $L(t)$ is the distance along the path traveled, which is usually different from the distance from the origin.

4. 6

Section 13.3 Exercises

1. $L = 3\sqrt{61}$ **3.** $L = 15 + \ln 4$ **5.** $L \approx 6.26$

7. $s(t) = \frac{1}{27}\left((20 + 9t^2)^{3/2} - 20^{3/2}\right)$ **9.** $v(3) \approx 1.05$

11. $v(0) = \sqrt{2}$

13. At the origin. The integral $\int_0^T \|\mathbf{r}'(u)\|du$ is the length of the path traveled in the time interval $0 \le t \le T$.

15. **(a)** $s(t) = \sqrt{29}t$ **(b)** $t = \varphi(s) = \frac{s}{\sqrt{29}}$

17. $\mathbf{r}_1(s) = \langle 2 + 4\cos(2s), 10, -3 + 4\sin(2s)\rangle$

19. $\mathbf{r}(\varphi(s)) =$
$$\left(1 + \frac{s}{\sqrt{3}}\right)\left\langle \sin\left(\ln\left(1 + \frac{s}{\sqrt{3}}\right)\right), \cos\left(\ln\left(1 + \frac{s}{\sqrt{3}}\right)\right), 1\right\rangle$$

21. $L = \int_0^8 \sqrt{1 + 9t^4}\,dt$, $L = \int_0^2 \sqrt{1 + 9t^{12}}\,3t^2\,dt$

23. **(b)** $L_1 \approx 94.3, L_2 \approx 125.7$

25. **(a)** $\mathbf{r}_1(s) = \frac{s}{\sqrt{17}}\left\langle \cos\left(4\ln\frac{s}{\sqrt{17}}\right), \sin\left(4\ln\frac{s}{\sqrt{17}}\right)\right\rangle$

27. $L = 2\pi$

29.

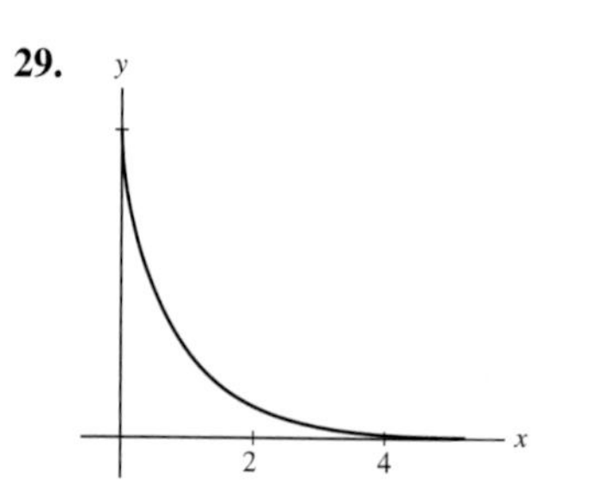

Section 13.4 Preliminary

1. $\mathbf{T}(t) = \frac{\mathbf{v}}{\|\mathbf{v}\|} = \left\langle \frac{2}{3}, \frac{1}{3}, -\frac{2}{3} \right\rangle$ **2.** $\frac{1}{4}$ **3.** A circle of radius 2

4. Zero curvature **5.** $\kappa = \sqrt{14}$ **6.** 4 **7.** $R = \frac{1}{9}$

Section 13.4 Exercises

1. $\|\mathbf{r}'(t)\| = 36\sqrt{t^4 + t^2 + t^6}$, $\mathbf{T}(t) = \frac{t}{|t|\sqrt{1+t^2+t^4}} \left\langle t, 1, t^2 \right\rangle$,
$\mathbf{T}(1) = \left\langle \frac{1}{\sqrt{3}}, \frac{1}{\sqrt{3}}, \frac{1}{\sqrt{3}} \right\rangle$

3. $\|\mathbf{r}'(t)\| = \sqrt{122}$, $\mathbf{T}(t) = \left\langle \frac{4}{\sqrt{122}}, -\frac{5}{\sqrt{122}}, \frac{9}{\sqrt{122}} \right\rangle$, $\mathbf{T}(1) = \mathbf{T}(t)$

5. $\|\mathbf{r}'(t)\| = \sqrt{64t^2 + 81}$, $\mathbf{T}(t) = \frac{1}{\sqrt{64t^2+81}} \langle 8t, 9 \rangle$,
$\mathbf{T}(1) = \left\langle \frac{8}{\sqrt{145}}, \frac{9}{\sqrt{145}} \right\rangle$

7. $\kappa = \frac{e^t}{(1+e^{2t})^{3/2}}$ **9.** $\kappa = \frac{4}{17}$ **11.** $\kappa(t) = \frac{2|t|^3}{(1+|t|^4)^{3/2}}$

13. $\kappa(3) \approx 0.0025$ **15.** $\kappa(2) \approx 0.0015$

17. $\kappa\left(\frac{\pi}{3}\right) \approx 4.54$, $\kappa\left(\frac{\pi}{2}\right) = 0.2$ **21.** $\alpha = \pm\sqrt{2}$

27. $\kappa(2) \approx 0.012$ **29.** $\kappa(\pi) \approx 1.11$

33. $\kappa(t) = t^2$

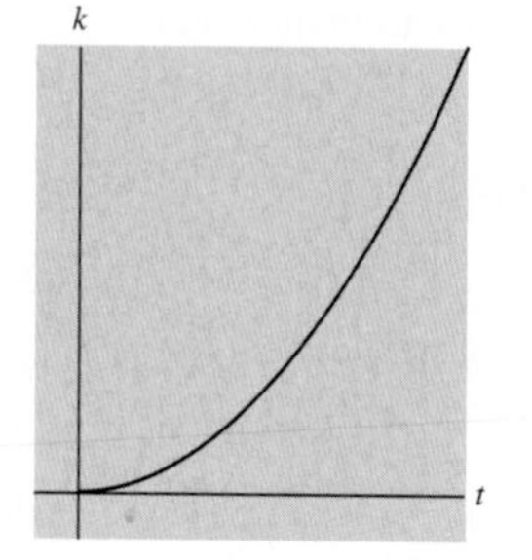

35. $\mathbf{N}(t) = \langle 0, -\sin 2t, -\cos 2t \rangle$

37. $\mathbf{T}'\left(\frac{\pi}{4}\right) = \left\langle -\frac{\sqrt{2}}{3\sqrt{3}}, -\frac{2}{3\sqrt{3}} \right\rangle$, $\mathbf{T}'\left(\frac{3\pi}{4}\right) = \left\langle \frac{\sqrt{2}}{3\sqrt{3}}, \frac{2}{3\sqrt{3}} \right\rangle$

39. $\mathbf{N}\left(\pi^{1/3}\right) = \left\langle \frac{1}{2}, -\frac{\sqrt{3}}{2} \right\rangle$ **41.** $\mathbf{N}(1) = \frac{1}{7\sqrt{17}} \langle -5, -22, 18 \rangle$

43. $\mathbf{N}(\pi) = \frac{\langle 0, -1 \rangle}{1} = \langle 0, -1 \rangle$ **45.** $\mathbf{N}(-1) = \frac{1}{\sqrt{66}} \langle -7, 1, 4 \rangle$

49. $\mathbf{c}(t) = \left\langle \frac{\pi}{2}, 0 \right\rangle + \langle \cos t, \sin t \rangle$ **51.** $\langle \cos t, \sin t \rangle$

53. $\mathbf{c}(t) = \langle \pi, -2 \rangle + 4 \langle \cos t, \sin t \rangle$

55. $\mathbf{c}(t) = \langle 3, 0, 0 \rangle + 2 \langle 1, 0, 0 \rangle \cos t + \frac{2}{\sqrt{2}} \langle 0, 1, 1 \rangle \sin t$

65. $\kappa(\theta) = \frac{\theta^2+2}{(\theta^2+1)^{3/2}}$ **67.** $\kappa(\theta) = \frac{1}{a\sqrt{1+b^2}} e^{-b\theta}$

81. **(a)** $\mathbf{N}(0) = \frac{1}{\sqrt{5}} \langle -1, 0, 2 \rangle$ **(b)** $\mathbf{N}(1) = \frac{1}{\sqrt{66}} \langle 4, 7, -1 \rangle$

Section 13.5 Preliminary

1. No, since the particle may change its direction. **2.** $\mathbf{a}(t)$

3. **(a)** Their velocity vectors point in the same direction.

5. **(b)** Parallel

6. $\|\mathbf{a}(t)\| = 8$ cm/s^2

7. **(b)** $a_{\mathbf{N}}$

8. Increased by a factor of 4

Section 13.5 Exercises

1. $h = -0.2 : \frac{\mathbf{r}(1-0.2) - \mathbf{r}(1)}{-0.2} = \langle -0.085, 1.91, 2.635 \rangle$

$h = -0.1 : \frac{\mathbf{r}(1-0.1) - \mathbf{r}(1)}{-0.1} = \langle -0.19, 2.07, 2.97 \rangle$

$h = 0.1 : \frac{\mathbf{r}(1+0.1) - \mathbf{r}(1)}{0.1} = \langle -0.41, 2.37, 4.08 \rangle$

$h = 0.2 : \frac{\mathbf{r}(1+0.2) - \mathbf{r}(1)}{0.2} = \langle -0.525, 2.505, 5.075 \rangle$

$\mathbf{v}(1) \approx \langle -0.3, 2.2, 3.5 \rangle$, $v(1) \cong 4.1$

3. $\mathbf{v}(t) = \langle 3, -1, 8 \rangle$, $\mathbf{a}(t) = \langle 6, 0, 8 \rangle$, $v(1) = \sqrt{74}$

5. $\mathbf{v}(\theta) = \left\langle \frac{1}{2}, -\frac{\sqrt{3}}{2}, 0 \right\rangle$, $\mathbf{a}(\theta) = \left\langle -\frac{\sqrt{3}}{2}, -\frac{1}{2}, 9 \right\rangle$, $v\left(\frac{\pi}{3}\right) = 1$

7. $\mathbf{a}(t) = -2\left\langle \cos\frac{t}{2}, \sin\frac{t}{2} \right\rangle$

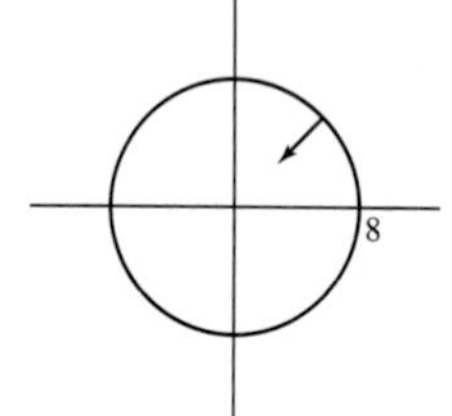

$\mathbf{r}(t) = 8\left\langle \cos\frac{t}{2}, \sin\frac{t}{2} \right\rangle$

9.

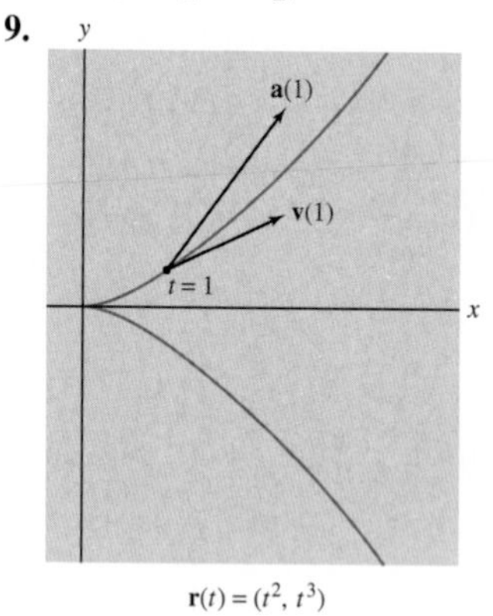

$\mathbf{r}(t) = \langle t^2, t^3 \rangle$

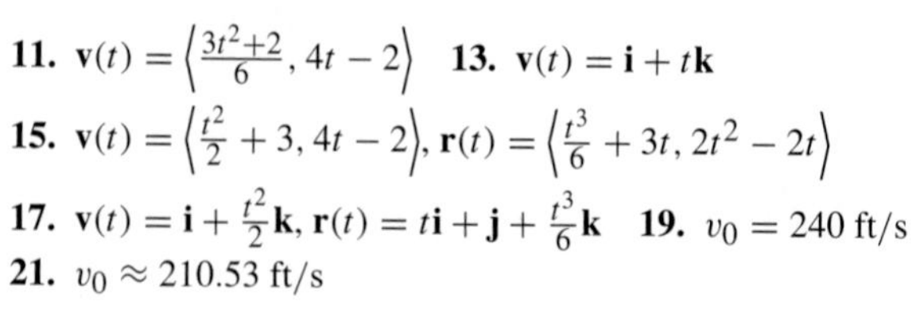

11. $\mathbf{v}(t) = \left\langle \frac{3t^2+2}{6}, 4t - 2 \right\rangle$ **13.** $\mathbf{v}(t) = \mathbf{i} + t\mathbf{k}$

15. $\mathbf{v}(t) = \left\langle \frac{t^2}{2} + 3, 4t - 2 \right\rangle$, $\mathbf{r}(t) = \left\langle \frac{t^3}{6} + 3t, 2t^2 - 2t \right\rangle$

17. $\mathbf{v}(t) = \mathbf{i} + \frac{t^2}{2}\mathbf{k}$, $\mathbf{r}(t) = t\mathbf{i} + \mathbf{j} + \frac{t^3}{6}\mathbf{k}$ **19.** $v_0 = 240$ ft/s

21. $v_0 \approx 210.53$ ft/s

23. $\theta \approx 22.66^\circ$ or $\theta \approx 67.34^\circ$
$\theta = 22.66^\circ$ will get the ball faster to the other player.

25. $\mathbf{T} = \frac{1}{3}\langle 2, 2, -1\rangle$, $\mathbf{N} = \frac{1}{15\sqrt{2}}\langle -5, 13, 16\rangle$, $\mathbf{a} = \frac{5}{3}\mathbf{T} + \frac{10\sqrt{2}}{3}\mathbf{N}$

27. **(a)** At its original position **(b)** No

29. $\|\mathbf{a}\| = 27.6\ \text{ft/s}^2$ **31.** $a_{\mathbf{T}} = 0, a_{\mathbf{N}} = 1$

33. $a_{\mathbf{T}}(t = 0) = 0, a_{\mathbf{N}}(t = 0) = \sqrt{2}$

35. $\mathbf{a}(0) = \sqrt{3}\mathbf{T} + \sqrt{2}\mathbf{N}$, with $\mathbf{T} = \frac{1}{\sqrt{3}}\langle 1, 1, 1\rangle$ and
$\mathbf{N} = \frac{1}{\sqrt{2}}\langle -1, 0, 1\rangle$

37. $\mathbf{a_T}(4) = 4$, $\mathbf{a_N}(4) = 1$, so $\mathbf{a} = 4\mathbf{T} + \mathbf{N}$, with $\mathbf{T} = \left\langle \frac{1}{9}, \frac{4}{9}, \frac{8}{9}\right\rangle$ and
$\mathbf{N} = \left\langle -\frac{4}{9}, -\frac{7}{9}, \frac{4}{9}\right\rangle$

39. $\mathbf{a_T}(t) = 0$, $\mathbf{a_N}(0) = 1$, so $\mathbf{a} = \mathbf{N}$, with $\mathbf{N} = \langle -1, 0, 0\rangle$

41. $\mathbf{a}\left(\frac{\pi}{2}\right) = -\frac{\pi}{2\sqrt{3}}\mathbf{T} + \frac{\pi}{\sqrt{6}}\mathbf{N}$, with $\mathbf{T} = \frac{1}{\sqrt{3}}\langle 1, -1, 1\rangle$ and
$\mathbf{N} = \frac{1}{\sqrt{6}}\langle 1, -1, -2\rangle$

43. $a_{\mathbf{T}} = 0, a_{\mathbf{N}} = 0.25\ \text{cm/s}^2$

45. $\mathbf{a}(3) = 3\mathbf{T} + 0.02025\mathbf{N}$, where $\mathbf{T} \approx \langle -0.0045, 0.9999\rangle$,
$\mathbf{N} \approx \langle -0.9999, -0.0045\rangle$

51. The maximum speed (in case of constant speed) is about 98 ft/s.

53. It is faster to hug the inside curve of radius r, $(r < R)$.

Section 13.6 Preliminary

1. $\frac{dA}{dt} = \frac{1}{2}\|\mathbf{J}\|$ **3.** The period is increased eightfold.

Section 13.6 Exercises

3. $|\mathbf{J}| = 4.78 \times 10^{16}\ \text{ft}^2/\text{s}$ **5.** $M = 1.897 \times 10^{27}$ kg

11. $v = 11{,}060$ km/h

13. $M = 2.6225 \times 10^{41}$ kg (132 billions of sun)

17. $r_{\text{per}} = 4.377 \cdot 10^7$ km, $r_{\text{ap}} = 7.203 \cdot 10^7$ km

Chapter 13 Review

1. **(a)** $-1 < t < 0$ or $0 < t \le 1$ **(b)** $0 < t \le 2$

3. $\mathbf{r}(t) = \left\langle t^2, t, \sqrt[3]{3 - t^4}\right\rangle$ **5.** $\mathbf{r}'(t) = \left\langle -1, -2t^{-3}, \frac{1}{t}\right\rangle$

7. $\mathbf{r}'(0) = \langle 2, 0, 6\rangle$. **9.** $\frac{d}{dt}e^t\left\langle 1, t, t^2\right\rangle = e^t\left\langle 1, 1+t, 2t+t^2\right\rangle$

11. $\frac{d}{dt}(6\mathbf{r}_1(t) - 4\mathbf{r}_2(t))\Big|_{t=3} = \langle 0, -8, -10\rangle$

13. $\frac{d}{dt}(\mathbf{r}_1(t) \cdot \mathbf{r}_2(t))\Big|_{t=3} = 2$

15. $\int_0^3 \left\langle 4t + 3, t^2, -4t^3\right\rangle dt = \langle 27, 9, -81\rangle$

17. $\mathbf{T}(\pi) = \left\langle \frac{-1}{\sqrt{2}}, \frac{1}{\sqrt{2}}, 0\right\rangle$ **19.** $\left(3, 3, 5\frac{1}{3}\right)$ **21.** $L = 2\sqrt{13}$

23. $\mathbf{r}(t) = \left\langle \frac{5}{2}\cos\frac{2\pi t}{5}, \frac{5}{2}\sin\frac{2\pi t}{5}, t\right\rangle$, $0 \le t \le 20$, $L \approx 127.2$

25. $\mathbf{T}(1) = \left\langle -\frac{1}{\sqrt{3}}, \frac{1}{\sqrt{3}}, \frac{1}{\sqrt{3}}\right\rangle$, $\mathbf{N}(1) = \frac{3}{\sqrt{18}}\langle 1, 0, 1\rangle$, $\kappa(t) = \frac{t^2}{t^4+t^2+1}$

27. $\kappa(t) = \frac{|4t^2-2|e^{-t^2}}{\left(1+4t^2e^{-2t^2}\right)^{3/2}}$

29. $\mathbf{T}(0) = \langle 1, 0\rangle$, $\mathbf{T}(1) = \left\langle \frac{e}{\sqrt{e^2+4}}, \frac{-2}{\sqrt{e^2+4}}\right\rangle$, $\mathbf{N}(0) = \langle 0, -1\rangle$,
$\mathbf{N}(1) = \left\langle \frac{2}{\sqrt{e^2+4}}, \frac{e}{\sqrt{e^2+4}}\right\rangle$

31. $\kappa(t) = \frac{\sqrt{2}}{(1+\cos^2 t)^{3/2}}$, $\mathbf{N}(t) = \frac{-\langle \sin t, \sin t, 2\cos t\rangle}{\sqrt{2}\cdot\sqrt{1+\cos^2 t}}$

33. $\mathbf{a} = \frac{1}{\sqrt{2}}\mathbf{T} + 4\mathbf{N}$, where $\mathbf{T} = \langle -1, 0\rangle$ and $\mathbf{N} = \langle 0, -1\rangle$.

35. $\mathbf{c}(t) = \left\langle 0, \frac{1}{2}\right\rangle + \frac{1}{2}\langle \cos t, \sin t\rangle$, $0 \le t \le 2\pi$

Chapter 14

Section 14.1 Preliminary

1. The two curves have the same shape but are located in parallel planes.

2. A parabola $z = x^2$ in the xz-plane.

3. Two different level curves of a function do not intersect.

4. The vertical lines $x = c$ with distance of 1 unit between adjacent lines.

5. In the contour map of $g(x, y) = 2x$, the distance between two adjacent vertical lines is $\frac{1}{2}$.

Section 14.1 Exercises

1. $f(2, 2) = 18$, $f(-1, 4) = -5$, $f\left(6, \frac{1}{2}\right) = 114$

3. $h(3, 8, 2) = 6$; $h(3, -2, -6) = -\frac{1}{6}$

5. The domain is the entire xy-plane.

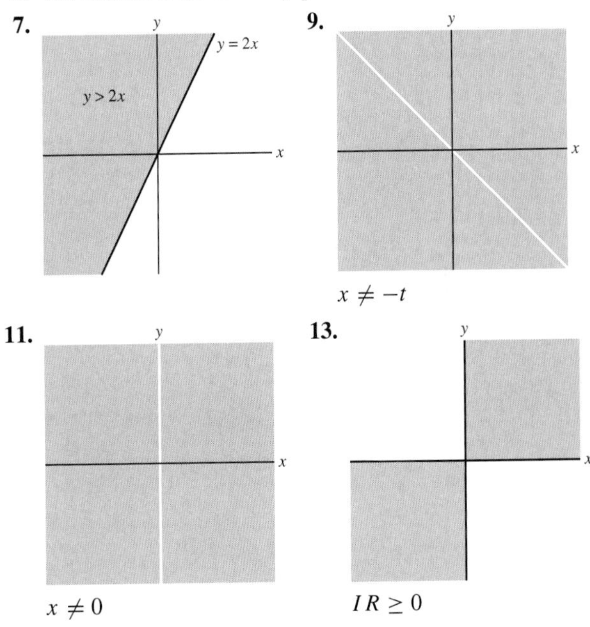

15.

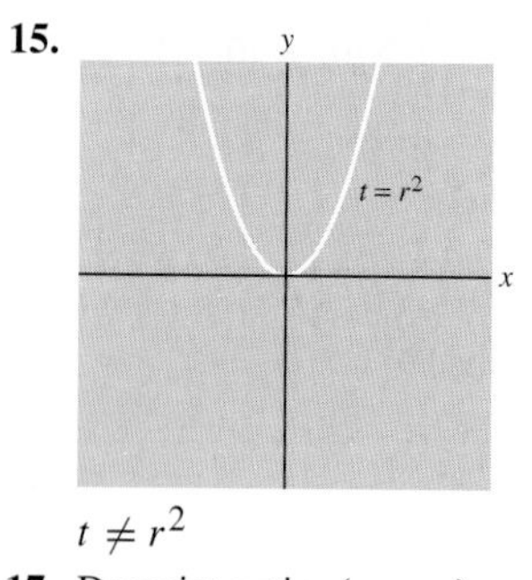

$t \neq r^2$

17. Domain: entire (x, y, z)-space. Range: entire real line.

19. $\left\{(x, y, z)\, x^2 + y^2 + z^2 \leq 9\right\}$, $\{\omega \in R : 0 \leq \omega \leq 3\}$.

21.

Horizontal trace: $3x + 4y = 12 - c$ in the plane $z = c$.
Vertical trace: $z = (12 - 3a) - 4y$ and $z = -3x + (12 - 4a)$ in the planes $x = a$ and $y = a$, respectively.

23.

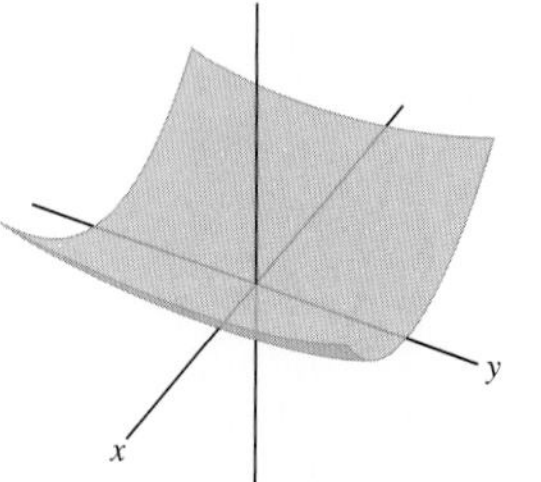

The horizontal traces are ellipses for $c > 0$.
The vertical trace in the plane $x = a$ is the parabola $z = a^2 + 4y^2$.
The vertical trace in the plane $y = a$ is the parabola $z = x^2 + 4a^2$.

25.

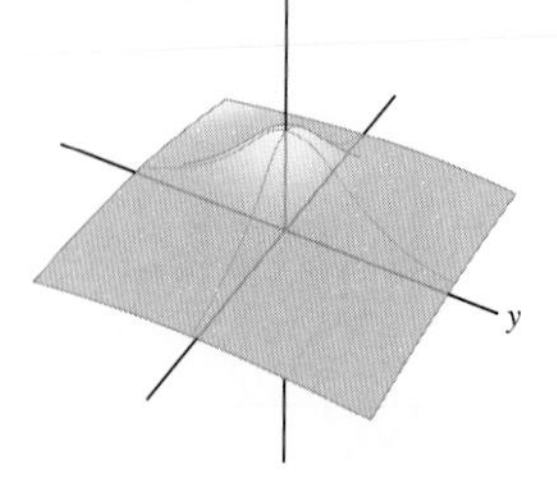

Horizontal trace in the plane $z = c$: $x^2 + y^2 = \frac{1}{c} - 1$.
Vertical trace in the plane $x = a$: $z = \frac{1}{(1+a^2)+y^2}$.
Vertical trace in the plane $y = a$: $z = \frac{1}{x^2+a^2+1}$

27.

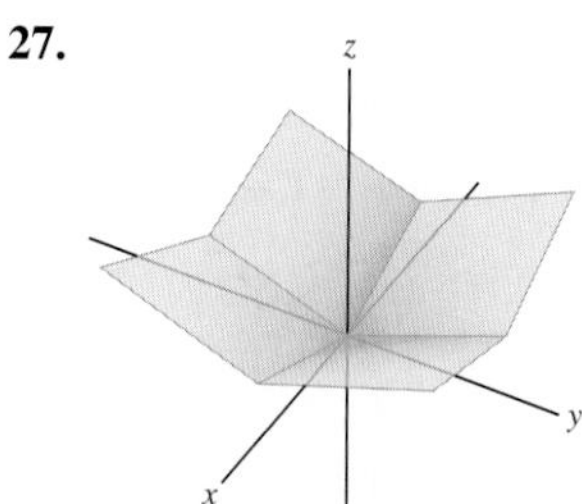

Horizontal trace: $|x| + |y| = c$.
Vertical traces: $z = |a| + |y|$ or $z = |x| + |a|$, in the planes $x = a$ and $y = a$, respectively.

29.

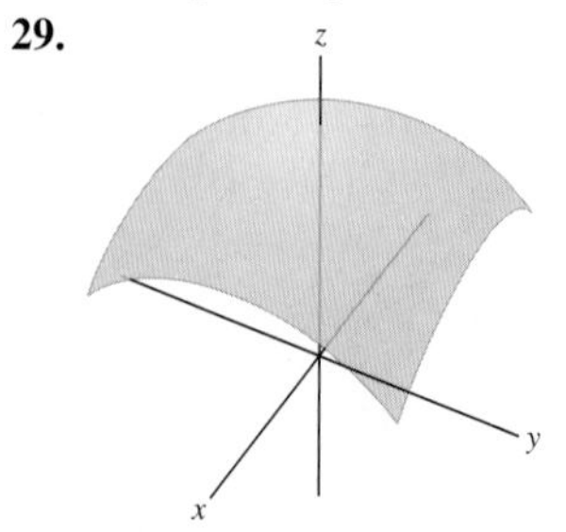

Horizontal trace: $x^2 + y^2 = 1 - c^2$.
Vertical traces: $z = \sqrt{1 - a^2 - y^2}$ in the plane $x = a$ and $z = \sqrt{1 - a^2 - x^2}$ in the plane $y = a$.

31. $m = 1$: $m = 2$:

33. (a)

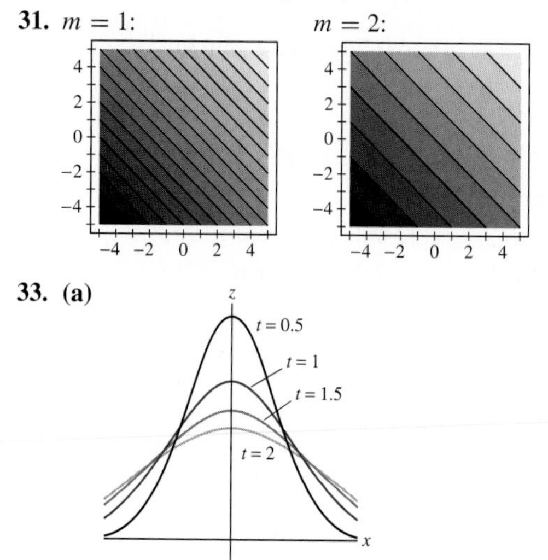

For each t, the temperature decreases as we move away from the center point. Also, as t increases, the temperature at each point in the bar (except at the middle) increases and then decreases.

(b)

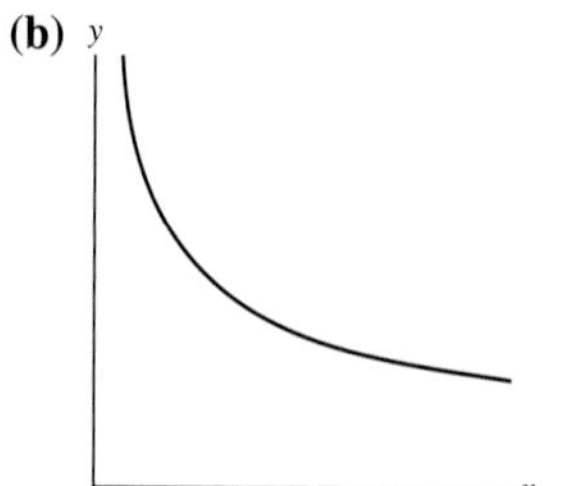

The temperature is decreasing with time at the center point, according to $\frac{1}{\sqrt{t}}$.

(c)

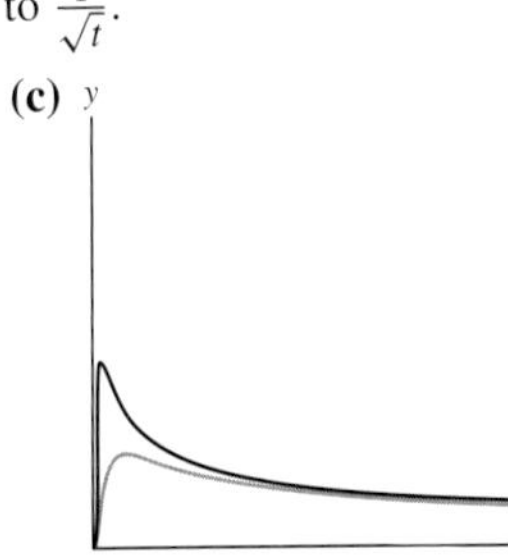

For small values of t, the temperature increases fast and then slowly decreases as t increases.

35. **37.** **39.** **41.** **43.**

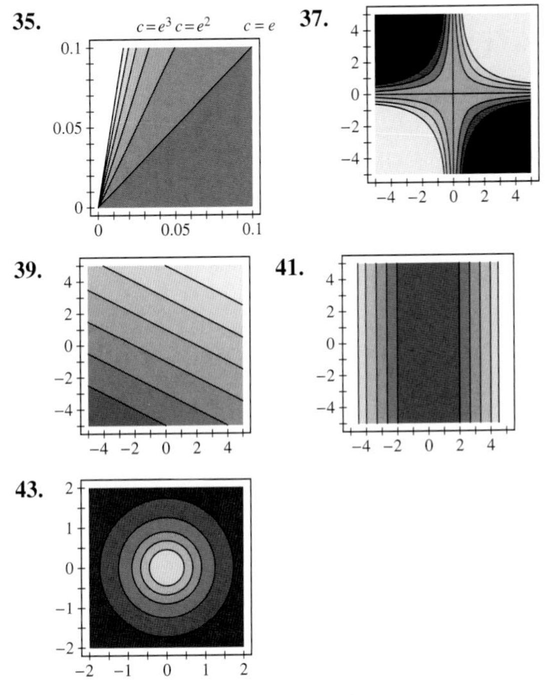

45. $m = 6 : f(x, y) = 2x + 6y + 6$
$m = 3 : f(x, y) = x + 3y + 3$

47. Average ROC from C to $D = 0.000625\text{kg/m}^3 \cdot \text{ppt}$

49. At point A

51. Average ROC from A to $B \approx 0.047$
Average ROC from A to $C \approx 0.023$

53.

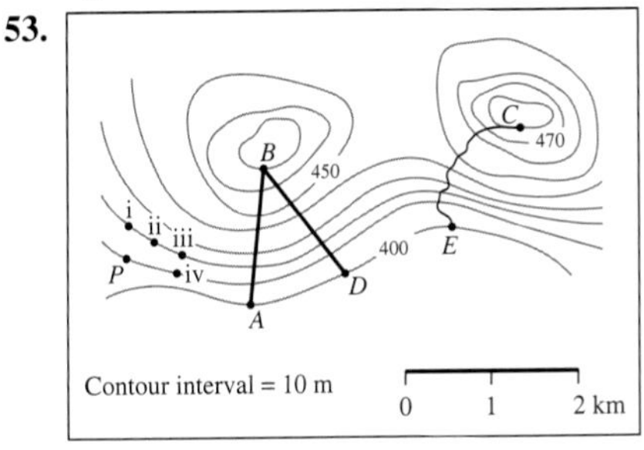

55. $f(r, \theta) = \cos\theta$

Section 14.2 Preliminary

1. $D^*(p, r)$ consists of all points in $D(p, r)$ other than p itself.

2. $f(2, 3) = 27$ **3.** All four statements are true.

4. $\lim_{(x,y)\to(0,0)} f(x, y)$ does not exist.

Section 14.2 Exercises

1. $\lim_{(x,y)\to(1,2)} (x^2 + y) = 3$

3. $\lim_{(x,y)\to(3,-1)} (3x^2y - 2xy^3) = -21$

5. $\lim_{(x,y)\to(1,2)} \frac{x^2+y}{x+y^2} = \frac{3}{5}$ **7.** $\lim_{(x,y)\to(4,4)} \arctan\frac{y}{x} = \frac{\pi}{4}$

9. $\lim_{(x,y)\to(1,1)} \frac{e^{x^2}-e^{-y^2}}{x+y} = \frac{1}{2}(e - e^{-1})$

11. $\lim_{(x,y)\to(0,1)} \frac{x}{y} = 0$ **13.** $\lim_{(x,y)\to(1,1)} e^{xy}\ln(1 + xy) = e\ln 2$

15. $\lim_{(x,y)\to(-1,-2)} x^2|y|^3 = 8$ **17.** $\lim_{(x,y)\to(4,2)} \frac{y-2}{\sqrt{x^2-4}} = 0$

19. $\lim_{(x,y)\to(2,5)} \big(f(x, y) + 4g(x, y)\big) = 31$

21. $\lim_{(x,y)\to(2,5)} e^{f(x,y)} = e^3$ **23.** $\lim_{(x,y)\to(0,0)} \frac{(\sin x)(\sin y)}{xy} = 1$

25. $\lim_{(z,w)\to(-1,2)} (z^2w - 9z) = 11$

27. $\lim_{(h,k)\to(2,0)} h^4\frac{(2+k)^2-4}{k} = 64$

29. The limit does not exist. **31.** The limit does not exist.

39. In (B), $\lim_{(x,y)\to P} g(x, y)$ appears to exist. If so, it must be 6, which is the level curve of P.

41. **(a)** Yes

(b)

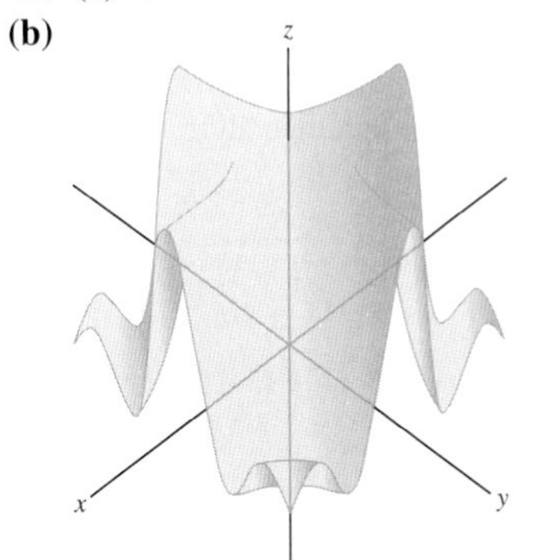

Near the axes, the values of $f(x, y)$ are approaching 1.

43. $f(10^{-1}, 10^{-2}) = \frac{1}{2}$, $f(10^{-5}, 10^{-10}) = \frac{1}{2}$, $f(10^{-20}, 10^{-40}) = \frac{1}{2}$

Section 14.3 Preliminary

1. $\frac{\partial}{\partial x}(x^2y^2) = 2xy^2$

2. In this case, the Constant Multiple Rule can be used. In the second part, since y appears in both the numerator and the denominator, then the Quotient Rule is preferred.

3. (a), (c) **4.** $f_x = 0$ **5.** (a), (d)

Section 14.3 Exercises

3. $\frac{\partial}{\partial y}\frac{y}{x+y} = \frac{x}{(x+y)^2}$ **5.** $f_z(2,3,1) = 6$ **7.** $m = 10$

9. $f_x(A) = 16$, $f_y(A) = -33.3$ **11.** f_y is smallest at C.

13. $\frac{\partial}{\partial x}(x^4y^3) = 4x^3y^3$, $\frac{\partial}{\partial y}(x^4y^3) = 3x^4y^2$

15. $\frac{\partial V}{\partial r} = 2\pi rh$, $\frac{\partial V}{\partial h} = \pi r^2$

17. $\frac{\partial}{\partial x}\left(\frac{x}{x-y}\right) = -\frac{y}{(x-y)^2}$, $\frac{\partial}{\partial y}\left(\frac{x}{x-y}\right) = \frac{x}{(x-y)^2}$

19. $\frac{\partial z}{\partial x} = \frac{y^2}{(x^2+y^2)^{3/2}}$, $\frac{\partial z}{\partial y} = \frac{-xy}{(x^2+y^2)^{3/2}}$

21. $\frac{\partial}{\partial u}\sin(u^2v) = 2uv\cos(u^2v)$, $\frac{\partial}{\partial v}\sin(u^2v) = u^2\cos(u^2v)$

23. $\frac{dS}{dw} = \frac{z}{1+w^2z^2}$, $\frac{dS}{dz} = \frac{w}{1+w^2z^2}$

25. $\frac{\partial z}{\partial x} = \frac{2x}{x^2+y^2}$, $\frac{\partial z}{\partial y} = \frac{2y}{x^2+y^2}$

27. $\frac{\partial}{\partial x}e^{xy} = ye^{xy}$, $\frac{\partial}{\partial y}e^{xy} = xe^{xy}$

29. $\frac{\partial z}{\partial x} = -2xe^{-x^2-y^2}$, $\frac{\partial z}{\partial y} = -2ye^{-x^2-y^2}$

31. $\frac{\partial z}{\partial y} = xy^{x-1}$, $\frac{\partial z}{\partial x} = y^x \ln y$

33. $\frac{\partial}{\partial x}\sinh(x^2y) = 2xy\cosh(x^2y)$, $\frac{\partial}{\partial y}\sinh(x^2y) = x^2\cosh(x^2y)$

35. $\frac{\partial w}{\partial x} = \frac{1}{y+z}$, $\frac{\partial w}{\partial y} = \frac{-x}{(y+z)^2}$, $\frac{\partial w}{\partial z} = \frac{-x}{(y+z)^2}$

37. $\frac{\partial w}{\partial x} = \frac{y^2+z^2-2x^2}{(x^2+y^2+z^2)^{5/2}}$, $\frac{\partial w}{\partial y} = -\frac{3xy}{(x^2+y^2+z^2)^{5/2}}$,
$\frac{\partial w}{\partial z} = -\frac{3xz}{(x^2+y^2+z^2)^{5/2}}$

39. $\frac{\partial U}{\partial t} = -e^{-rt}$, $\frac{\partial U}{\partial r} = \frac{-e^{-rt}(rt+1)}{r^2}$

41. $f_y(0,\pi) = 1$ **43.** $h_z(2,1) = -10e^{-2}$

45. $\frac{\partial W}{\partial E} = -\frac{1}{kT}e^{-\frac{E}{kT}}$, $\frac{\partial W}{\partial T} = \frac{E}{kT^2}e^{-\frac{E}{kT}}$

47. $\frac{\partial V}{\partial r} = \frac{2\pi hr}{3}$, $\frac{\partial V}{\partial h} = \frac{\pi}{3}r^2$

51. **(a)** $\left.\frac{\partial \rho}{\partial T}\right|_B = -2 \cdot 10^{-4}$, $\left.\frac{\partial \rho}{\partial S}\right|_C = 0.001$
(b) At the point C, the effect caused by the change in salinity is greater than the effect caused by the change in temperature.
(c) True

53. $\frac{\partial^2 f}{\partial x^2} = 6y$; $\frac{\partial^2 f}{\partial y^2} = -72xy^2$ **55.** $\frac{\partial^2 g}{\partial x \partial y} = -\frac{2xy}{(x-y)^3}$

57. $h_{uv}(u,v) = \frac{u-v}{(u+v)^3}$

59. $\frac{\partial \rho}{\partial T}(34,2) = -0.09$, $\frac{\partial \rho}{\partial T}(35,10) = -0.195$, $\frac{\partial \rho}{\partial S}(34,2) = 0.83$,
$\frac{\partial \rho}{\partial S}(35,10) = 0.74$

61. $f_{xyz}(x,y,z) = 0$ **63.** $f_{uuv} = 2v\sin(u+v^2)$ **65.** $F_{rst} = 0$

67. $u_{xx} = -\frac{1}{2}t^{-3/2}e^{-x^2/4t}\left(1 - \frac{x^2}{2t}\right)$

69. $g_{xyz} = \frac{3xyz}{(x^2+y^2+z^2)^{5/2}}$ **71.** $f(x,y) = x^2y$ **75.** $B = A^2$

77. $u(x,y) = ax^3 - 3dx^2y - 3axy^2 + dy^3$

Section 14.4 Preliminary

1. $L(x,y) = f(a,b) + f_x(a,b)(x-a) + f_y(a,b)(y-b)$

2. $f(x,y) - L(x,y) = \epsilon(x,y)\sqrt{(x-a)^2 + (y-b)^2}$

3. **(b)** $L(x,y) = 8 + 5(x-2) + 7(y-3)$

4. $f(2,3.1) \approx 8.7$ **5.** $\Delta f \approx -0.1$

6. Criterion for Differentiability

Section 14.4 Exercises

1. **(a)** $f(x,y) = -16 + 4x + 12y$
(b) $f(2.01, 1.02) \approx 4.28$, $f(1.97, 1.01) \approx 4$

3. $\Delta f \approx 3.56$ **5.** $f(0.01, -0.02) \approx 0.98$

7. $z = -34 - 20x + 16y$ **9.** $\frac{xy}{z} \approx \frac{1}{2}x + y - \frac{1}{2}z$

11. $z = 5x + 10y - 14$ **13.** $z = 8x - 2y - 13$

15. $z = 4r - 5s + 2$ **17.** $z = y - 1$ **19.** $\left(-\frac{1}{4}, \frac{1}{8}, \frac{1}{8}\right)$

23. $z = 3x - 3y + 13$ **25.** 8.44 **27.** 4.998

29. 3.945 **31.** 5.07 **33.** 0.0191

35. **(b)** 6% **(c)** 1% error in r

37. **(a)** \$7.1 **(b)** \$28.845, \$57.69 **(c)** −\$74.2416

Section 14.5 Preliminary

1. (b) **2.** False **3.** True **4.** $\frac{\partial f}{\partial x}(a,b) = D_{\mathbf{i}}f(a,b)$

5. **(b)** NW **(c)** SE

6. $3\sqrt{2}$

Section 14.5 Exercises

1. **(a)** $\nabla f = \langle y^2, 2xy\rangle$, $\mathbf{c}'(t) = \langle t, 3t^2\rangle$, $\nabla f \cdot \mathbf{c}'(t) = y^2t + 6xyt^2$
(b) $\nabla f \cdot \mathbf{c}'(t) = 4t^7$
(c) $\left.\frac{d}{dt}(f(\mathbf{c}(t)))\right|_{t=1} = 4$; $\left.\frac{d}{dt}(f(\mathbf{c}(t)))\right|_{t=-1} = -4$

3. **(a)** $\frac{d}{dt}f(\mathbf{c}(t)) = 0$

5. $\nabla f = -\sin(x^2 + y)\langle 2x, 1\rangle$

7. $\nabla h = \langle yz^{-3}, xz^{-3}, -3xyz^{-4}\rangle$

9. $\left.\frac{d}{dt}f(\mathbf{c}(t))\right|_{t=0} = -7$ **11.** $\left.\frac{d}{dt}f(\mathbf{c}(t))\right|_{t=0} = -3$

13. $\left.\frac{d}{dt}f(\mathbf{c}(t))\right|_{t=0} = 43$ **15.** $\left.\frac{d}{dt}f(\mathbf{c}(t))\right|_{t=4} = -56$

17. $\left.\frac{d}{dt}f(\mathbf{c}(t))\right|_{t=\frac{\pi}{4}} \approx 1.546$ **19.** $\left.\frac{d}{dt}g(\mathbf{c}(t))\right|_{t=1} = 0$

21. $D_{\mathbf{u}}f(1,2) = 8.8$, $D_{\mathbf{u}}f(1,2) = \frac{18}{\sqrt{13}}$

23. $D_{\mathbf{u}}f\left(\frac{1}{6}, 3\right) = \frac{39}{4\sqrt{2}}$ **25.** $D_{\mathbf{u}}f(3,4) = \frac{7\sqrt{2}}{290}$

27. $D_{\mathbf{u}}f(1,0) = \frac{6}{\sqrt{13}}$ **29.** $D_{\mathbf{u}}f(1,2,0) = -\frac{1}{\sqrt{3}}$

31. $D_{\mathbf{u}}f(3,2) = \frac{-50}{\sqrt{13}}$ **33.** $D_{\mathbf{u}}f(P) = -\frac{e^5}{3} \approx -49.47$

35. **(a)** $\|\nabla f_P\| = \sqrt{10}$ **(b)** $\|\nabla f_P\| = \sqrt{10}$
(c) $D_{\mathbf{e}_\mathbf{v}}f(P) = \sqrt{5} \approx 2.236$

37. $t_1 = 0$, $t_2 = 2$ **39.** $\nabla f_P = \langle -3, 3, 9\rangle$ **41.** $x + 3y + 2z = 10$

43. $-5x + 9y + 6\sqrt{e}z = 35$ **47.** $f(x,y,z) = x + 3y + z$

49. $f(x,y,z) = \frac{x^2}{2} + \frac{y^3}{3} + \frac{z^4}{4}$ **55.** $\frac{1}{\sqrt{21}}\langle 4, 2, -1\rangle$

59. $x = 1 + 8t$, $y = 1 - 8t$, $z = 1$

65. **(c)** No

71. $\mathbf{c}(t) = (t, 2t^4)$

Section 14.6 Preliminary

1. **(a)** $\frac{\partial f}{\partial x}$ and $\frac{\partial f}{\partial y}$ **(b)** u and v

2. (a) **3.** $f(u, v)|_{(r,s)} = e^2$ **4.** (b) **5.** (c) **6.** No

Section 14.6 Exercises

1. **(a)** $\frac{\partial f}{\partial x} = 2xy^3$, $\frac{\partial f}{\partial y} = 3x^2y^2$, $\frac{\partial f}{\partial z} = 4z^3$
(b) $\frac{\partial x}{\partial s} = 2s$, $\frac{\partial y}{\partial s} = t^2$, $\frac{\partial z}{\partial s} = 2st$ **(c)** $\frac{\partial f}{\partial s} = 4s^5t^6 + 3s^6t^6 + 8s^7t^4$

3. $\frac{\partial f}{\partial s} = 6rs^2$, $\frac{\partial f}{\partial r} = 2s^3 + 4r^3$

5. $\frac{\partial g}{\partial u} = (42u - 68v)\sin(68uv - 21u^2 - 55v^2)$
$\frac{\partial g}{\partial v} = (110v - 68u)\sin(68uv - 21u^2 - 55v^2)$

7. $\frac{\partial F}{\partial x} = (2x + y)e^{x^2+xy}$, $\frac{\partial F}{\partial y} = xe^{x^2+xy}$

9. $\frac{\partial f}{\partial x} = \frac{-xy^2}{(x^2 + y^2)^{3/2}}$
$\frac{\partial f}{\partial y} = \frac{y(y^2 + 2x^2)}{(x^2 + y^2)^{3/2}}$

11. $\left.\frac{\partial f}{\partial u}\right|_{(u,v)=(-1,-1)} = 1$
$\left.\frac{\partial f}{\partial v}\right|_{(u,v)=(-1,-1)} = -2$

13. $\left.\frac{\partial g}{\partial \theta}\right|_{(r,\theta)=\left(2\sqrt{2}, \frac{\pi}{4}\right)} = \frac{1}{6}$ **15.** $\left.\frac{\partial g}{\partial u}\right|_{(u,v)=(0,1)} = 2\cos 2$

17. -0.1 degrees per second **19.** $\|\nabla u\|^2 = 4r^2\cos^2\theta$

21. $\frac{\partial f}{\partial \rho} = (\sin\phi\cos\theta)\frac{\partial f}{\partial x} + (\sin\phi\sin\theta)\frac{\partial f}{\partial y} + (\cos\phi)\frac{\partial f}{\partial z}$
$\frac{\partial f}{\partial \phi} = (\rho\cos\phi\cos\theta)\frac{\partial f}{\partial x} + (\rho\cos\phi\sin\theta)\frac{\partial f}{\partial y} - (\rho\sin\phi)\frac{\partial f}{\partial z}$
$\frac{\partial f}{\partial \theta} = (-\rho\sin\phi\sin\theta)\frac{\partial f}{\partial x} + (\rho\sin\phi\cos\theta)\frac{\partial f}{\partial y}$

23. $\left.\frac{\partial z}{\partial x}\right|_{(3,2,-1)} = \frac{3}{11}$, $\left.\frac{\partial z}{\partial y}\right|_{(3,2,-1)} = -\frac{1}{22}$

25. $\frac{\partial w}{\partial z} = -\frac{2wz+3y}{x^2+3w^2+z^2}$ **27.** $\frac{\partial r}{\partial t} = \frac{r^2e^{s/r}}{2r^3+ste^{s/r}}$, $\frac{\partial t}{\partial r} = 2re^{-s/r} + \frac{st}{r^2}$

29. $\left.\frac{\partial U}{\partial T}\right|_{(1,1,2,4)} = -\frac{2\ln 2}{1+2\ln 2}$, $\left.\frac{\partial T}{\partial U}\right|_{(1,1,2,4)} = -\frac{1+2\ln 2}{2\ln 2}$

33. $\nabla\left(\frac{1}{r}\right) = -\frac{1}{r^3}\mathbf{r}$, $\nabla(\ln r) = \frac{\mathbf{r}}{r^2}$

Section 14.7 Preliminary

1. f has a local (and global) min at $(0, 0)$; g has a saddle point at $(0, 0)$.

2.

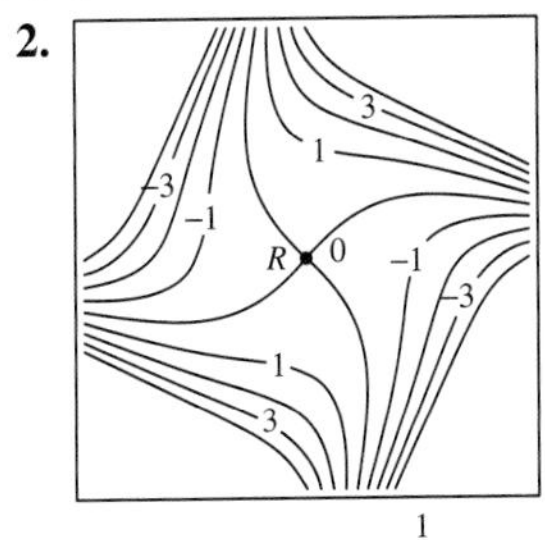

R is a saddle point.

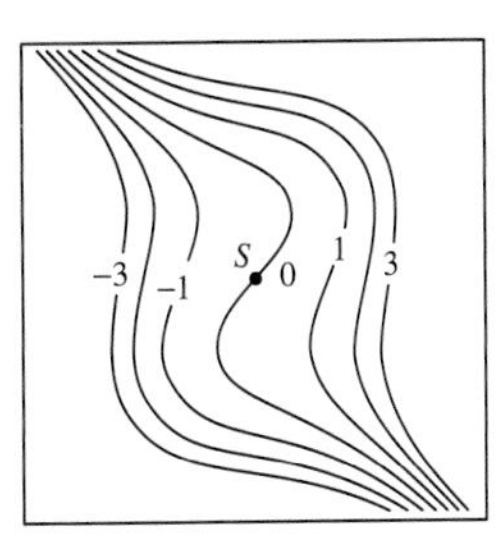

Point S is neither a local extremum nor a saddle point.

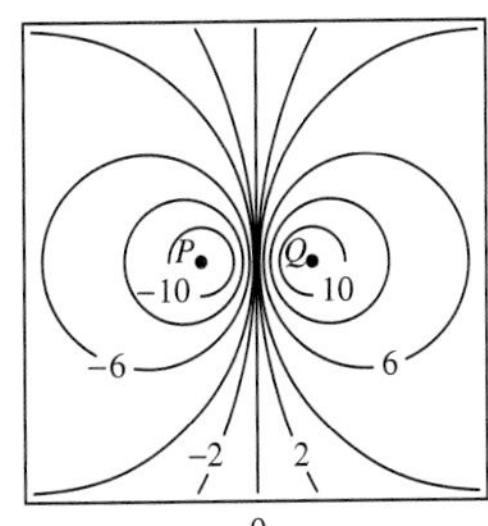

Local minimum at P, local maximum at Q.

3. (a)

Section 14.7 Exercises

1. **(b)** $P_1 = (0, 0)$ saddle point, $P_2 = (2\sqrt{2}, \sqrt{2})$, $P_3 = (-2\sqrt{2}, -\sqrt{2})$ local minima. Absolute minimum value of f is -4.

3. $(0, 0)$, $\left(\frac{13}{64}, -\frac{13}{32}\right)$, $\left(-\frac{1}{4}, \frac{1}{2}\right)$ **5.** $\left(-\frac{2}{3}, -\frac{1}{3}\right)$ local minimum

7. $(2, -4)$ saddle point

9. $(0, \pm\sqrt{2})$ saddle points, $\left(\frac{2}{3}, 0\right)$ local maximum, $\left(-\frac{2}{3}, 0\right)$ local minimum.

11. $(0, 0)$ saddle point, $(1, 1)$ and $(-1, -1)$ local minima.

13. $(0, 0)$ saddle point, $\left(\frac{1}{\sqrt{2}}, \frac{1}{\sqrt{2}}\right)$ local maximum, $\left(\frac{1}{\sqrt{2}}, -\frac{1}{\sqrt{2}}\right)$ local minimum, $\left(-\frac{1}{\sqrt{2}}, \frac{1}{\sqrt{2}}\right)$ local minimum, $\left(-\frac{1}{\sqrt{2}}, -\frac{1}{\sqrt{2}}\right)$ local maximum.

15. $(0, 0)$ saddle point **17.** $\left(1, \frac{1}{2}\right)$ local maximum

19. $\left(\frac{3}{2}, -\frac{1}{2}\right)$ saddle point **23.** $x = y = 0.27788$ local minimum

25. **(a)** Bounded and closed **(b)** Bounded
(c) Closed **(d)** Neither bounded nor closed
(e) Bounded and closed **(f)** Bounded

27. Global maximum 2, global minimum -3

29. Maximum value $f(0, 0) = 1$, minimum value e^{-1}

31. $f(x, y) = -\frac{1}{x+y}$

35. Global minimum $f(0, 1) = -2$, global maximum $f(1, 0) = 1$

37. Global maximum 3, global minimum 0

39. Global minimum $f(0, 0) = 0$, global maximum $f(0, 1) = 2$

41. Minimum value 0, maximum value $f\left(\frac{36}{121}, \frac{5}{11}\right) = 0.0005$

43. **(a)** In the box B with minimal surface area, z is smaller than $\sqrt[3]{V}$, which is the side of a cube with volume V.
(b) Width: $x = (2V)^{1/3}$; length: $y = (2V)^{1/3}$, height: $z = \left(\frac{V}{4}\right)^{1/3}$

45. $f(x) = 1.9629x - 1.5519$

47. (a) $f_x(x, y) =$
$\frac{x-a_1}{\sqrt{(x-a_1)^2+(y-a_2)^2}} + \frac{x-b_1}{\sqrt{(x-b_1)^2+(y-b_2)^2}} + \frac{x-c_1}{\sqrt{(x-c_1)^2+(y-c_2)^2}}$
$f_y(x, y) =$
$\frac{y-a_2}{\sqrt{(x-a_1)^2+(y-a_2)^2}} + \frac{y-b_2}{\sqrt{(x-b_1)^2+(y-b_2)^2}} + \frac{y-c_2}{\sqrt{(x-c_1)^2+(y-c_2)^2}}$
(b) $\mathbf{e} = \frac{\langle x-a_1, y-a_2\rangle}{\sqrt{(x-a_1)^2+(y-a_2)^2}}$, $\mathbf{f} = \frac{\langle x-b_1, y-b_2\rangle}{\sqrt{(x-b_1)^2+(y-b_2)^2}}$,
$\mathbf{g} = \frac{\langle x-c_1, y-c_2\rangle}{\sqrt{(x-c_1)^2+(y-c_2)^2}}$

Section 14.8 Preliminary

1. (b)
2. f has a local maximum 2, under the constraint, at A. $f(B)$ is neither local minimum nor local maximum of f.
3. (a)

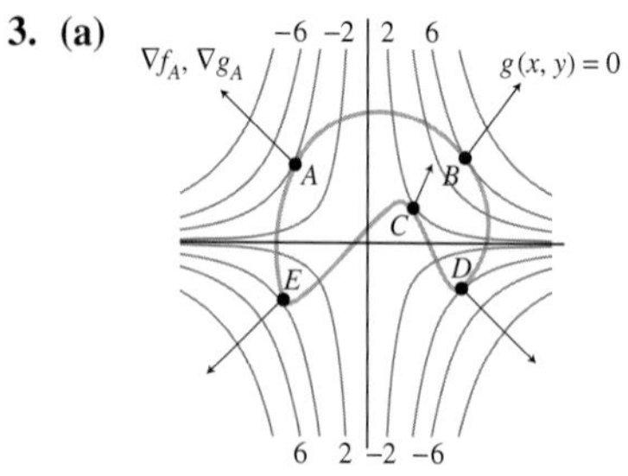

(b) Global minimum −4, global maximum 6

Section 14.8 Exercises

1. (c) $(-1, -2)$ and $(1, 2)$ **(d)** Maximum: 10, minimum: −10
3. Maximum: $4\sqrt{2}$, minimum: $-4\sqrt{2}$
5. Minimum: $\frac{468}{169}$, no maximum value
7. Maximum: $\frac{8}{3}$, minimum: $-\frac{8}{3}$ **9.** Maximum: $\sqrt{2}$, minimum: 1
11. Minimum: −3.7, maximum: 3.7
13. No maximum and minimum values **15.** $(-1, e^{-1})$
17. (a) $h = \sqrt{\frac{2}{\sqrt{3}\pi}} \approx 0.6$, $r = \sqrt{\frac{1}{\sqrt{3}\pi}} \approx 0.43$ **(b)** $\frac{h}{r} = \sqrt{2}$
(c) There is no cone of volume 1 and maximal surface area.
19. $2ab$ **21.** $\sqrt{\frac{a^a b^b}{(a+b)^{a+b}}}$ **23.** $\sqrt{\frac{a^a b^b c^c}{(a+b+c)^{a+b+c}}}$
25. (a) $(0, -\frac{8}{27}, \frac{4}{9})$ **(c)** No
29. $\frac{3}{16}$ **31.** $\frac{x}{\alpha} + \frac{y}{\beta} + \frac{z}{\gamma} = 3$ **35.** Minimum: $\frac{138}{11}$
39. (b) $\lambda = \frac{c}{2p_1p_2}$

Chapter 14 Review

1. (a)

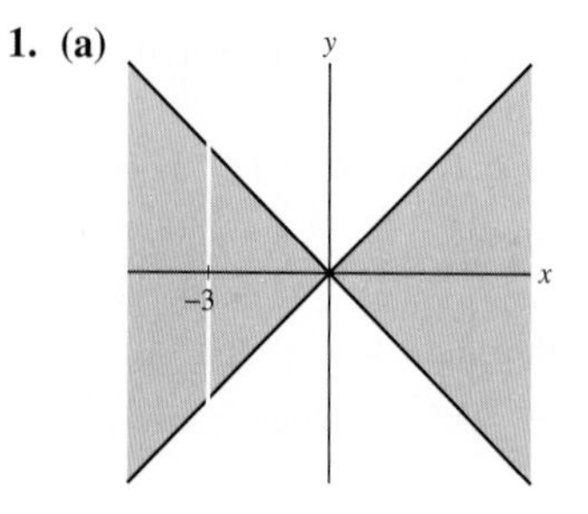

(b) $f(3, 1) = \frac{\sqrt{2}}{3}$, $f(-5, -3) = -2$ **(c)** $\left(-\frac{5}{3}, 1\right)$
3.

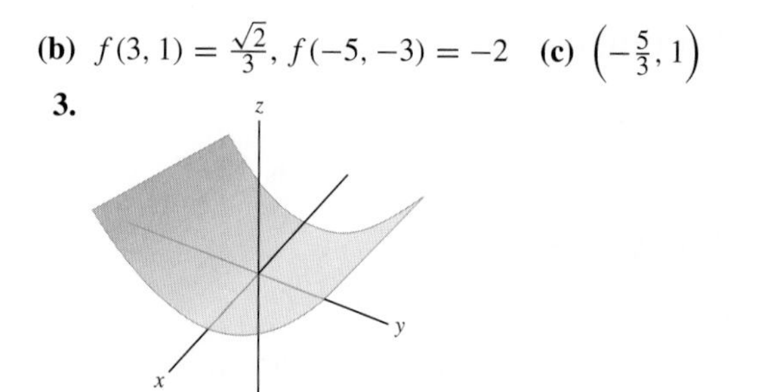

Vertical and horizontal traces: the line $z = (c^2 + 1) - y$ in the plane $x = c$, the parabola $z = x^2 - c + 1$ in the plane $y = c$.
5. (a) (b) **(b)** (c) **(c)** (d) **(d)** (a)
7. (a) Parallel lines $4x - y = \ln c$, $c > 0$, in the xy-plane.
(b) Parallel lines $4x - y = e^c$, in the xy-plane.
(c) Hyperbolas $3x^2 - 4y^2 = c$ in the xy-plane.
(d) Parabolas $x = c - y^2$ in the xy-plane.
9. $\lim_{(x,y)\to(1,-3)} (xy + y^2) = 6$ **11.** The limit does not exist.
13. $\lim_{(x,y)\to(1,-3)} (2x + y)e^{-x+y} = -e^{-4}$ **17.** $f_x = 2$, $f_y = 2y$
19. $f_x = e^{-x-y}\big(y\cos(xy) - \sin(xy)\big)$
$f_y = e^{-x-y}\big(x\cos(yx) - \sin(yx)\big)$
21. $f_{xxyz} = -\cos(x + z)$ **23.** $z = 33x + 8y - 42$
25. $\sqrt{7.1^2 + 4.9^2 + 70.1} \approx 12.020833$
27. Approximately \$69.72 **29.** $\frac{dt}{df}(\mathbf{c}(t)) = -6t^5 - 2t + 1$
31. $\frac{d}{dt} f(\mathbf{c}(t))\Big|_{t=\frac{\pi}{3}} = 1$ **33.** $D_{\mathbf{u}} f(1, 1, 1) = \frac{4}{3}$
35. $D_{\mathbf{v}} f(P) = \frac{1}{\sqrt{2}}$ **37.** $-4x - y + 2z = -5$
39. $\frac{\partial f}{\partial u} = -2ve^{u-v}$, $\frac{\partial f}{\partial v} = 2e^{u-v}(v - 1)$
41. $\frac{\partial f}{\partial r} = \frac{\partial f}{\partial x}\cos\theta + \frac{\partial f}{\partial y}\sin\theta$, $\frac{\partial f}{\partial \theta} = -\frac{\partial f}{\partial x} r\sin\theta + \frac{\partial f}{\partial y} r\cos\theta$
45. $\frac{\partial z}{\partial x} = -\frac{e^z - 1}{xe^z + e^y}$ **47.** $(3, 3)$ saddle point
49. $\left(\frac{1}{2}, \frac{1}{2}\right)$ saddle point
53. $f(2, 4) = 10$ global maximum, $f(-2, 4) = -18$ global minimum
55. Maximum: $\frac{26}{\sqrt{13}}$, minimum: $-\frac{26}{\sqrt{13}}$
57. Maximum: $\frac{12}{\sqrt{3}}$, minimum: $-\frac{12}{\sqrt{3}}$
61. $x = \frac{a}{\sqrt{3}}$, $y = \frac{b}{\sqrt{3}}$, $z = \frac{c}{\sqrt{3}}$

Chapter 15

Section 15.1 Preliminary

1. $\Delta A = 1$, the number of subrectangles is 32.
2. $\iint_{\mathcal{R}} f\,dA \approx S_{1,1} = 0.16$ **3.** $\iint_{\mathcal{R}} 5\,dA = 50$
4. The signed volume between the graph $z = f(x, y)$ and the xy-plane. The region below the xy-plane is treated as negative volume.
5. (b) **6.** (c), (d)

Section 15.1 Exercises

1. $S_{4,3} = 13.5$ **3.** (A) $S_{32} = 42$ (B) $S_{32} = 43.5$

5. (A) $S_{32} = 60$ (B) $S_{32} = 62$

7. (A) $S_{32} \approx 1.069$ (B) $S_{23} \approx 1.097$

9. $\iint_{\mathcal{R}} (-5)\, dA = -45$ **11.** $S_{54} = 24.2$

13. 1.0731, 1.0783, 1.0809 **15.** $\iint_{\mathcal{R}} \sin x\, dA = 0$

17. $\int_1^3 \int_0^2 x^3 y\, dy\, dx = 40$ **19.** $\int_0^2 \int_1^3 x^3 y\, dy\, dx = 16$

21. $\int_2^6 \int_1^4 x^2\, dx\, dy = 84$ **23.** $\int_0^1 \int_0^2 (x + 4y^3)\, dx\, dy = 4$

25. $\int_0^1 \int_2^3 \sqrt{x + 4y}\, dx\, dy = \frac{1}{15}(7^{5/2} - 6^{5/2} - 3^{5/2} + 2^{5/2}) \approx 2.102$

27. $\int_1^2 \int_0^4 \frac{dy\, dx}{x+y} = 6 \ln 6 - 2 \ln 2 - 5 \ln 5 \approx 1.31$

29. $\int_0^1 \int_2^3 \frac{1}{(x+4y)^3}\, dx\, dy = \frac{1}{56}$

31. $\int_1^2 \int_1^2 \ln(xy)\, dy\, dx = 4 \ln 2 - 2 \approx 0.773$ **33.** $m = \frac{3}{4}$

35. $\iint_{\mathcal{R}} x^2 y\, dA = \frac{4}{3}$ **37.** $\iint_{\mathcal{R}} \frac{y}{x+1}\, dA = 8 \ln 3 \approx 8.79$

39. $\iint_{\mathcal{R}} e^{2x+3y}\, dA = \frac{1}{6}(e^2 - 1)(e^6 - 1) \approx 428.52$

41. $\int_1^2 \int_0^2 \frac{x}{y}\, dx\, dy = 2 \ln 2 \approx 1.39$

43. $\int_0^1 \int_0^1 \sin(xy)\, dx\, dy \approx 0.240$ **45.** $S_{33} = 44$

49. $F(x, y) = y^2 x^3$, $\iint_{\mathcal{R}} 6x^2 y\, dA = 16$

Section 15.2 Preliminary

1. (a)

2. **3.**

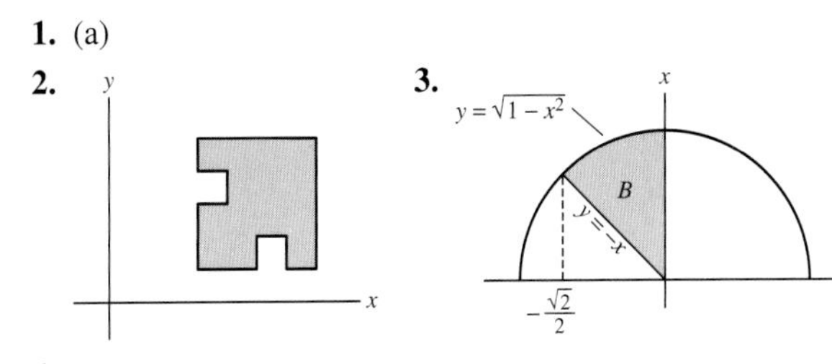

4. (b)

Section 15.2 Exercises

1. (a) Sample points •: $S_{34} = -3$

(b) Sample points ∘: $S_{34} = -4$

3.

$\iint_{\mathcal{D}} y\, dA = \frac{20}{3} \approx 6.67$

5. $\iint_{\mathcal{D}} x^2 y\, dA = 38.4$ **7.** $\iint_{\mathcal{D}} x^2 y\, dA = \frac{608}{15} \approx 40.53$

9. $\iint_{\mathcal{D}} x\, dA = 2\frac{1}{4}$

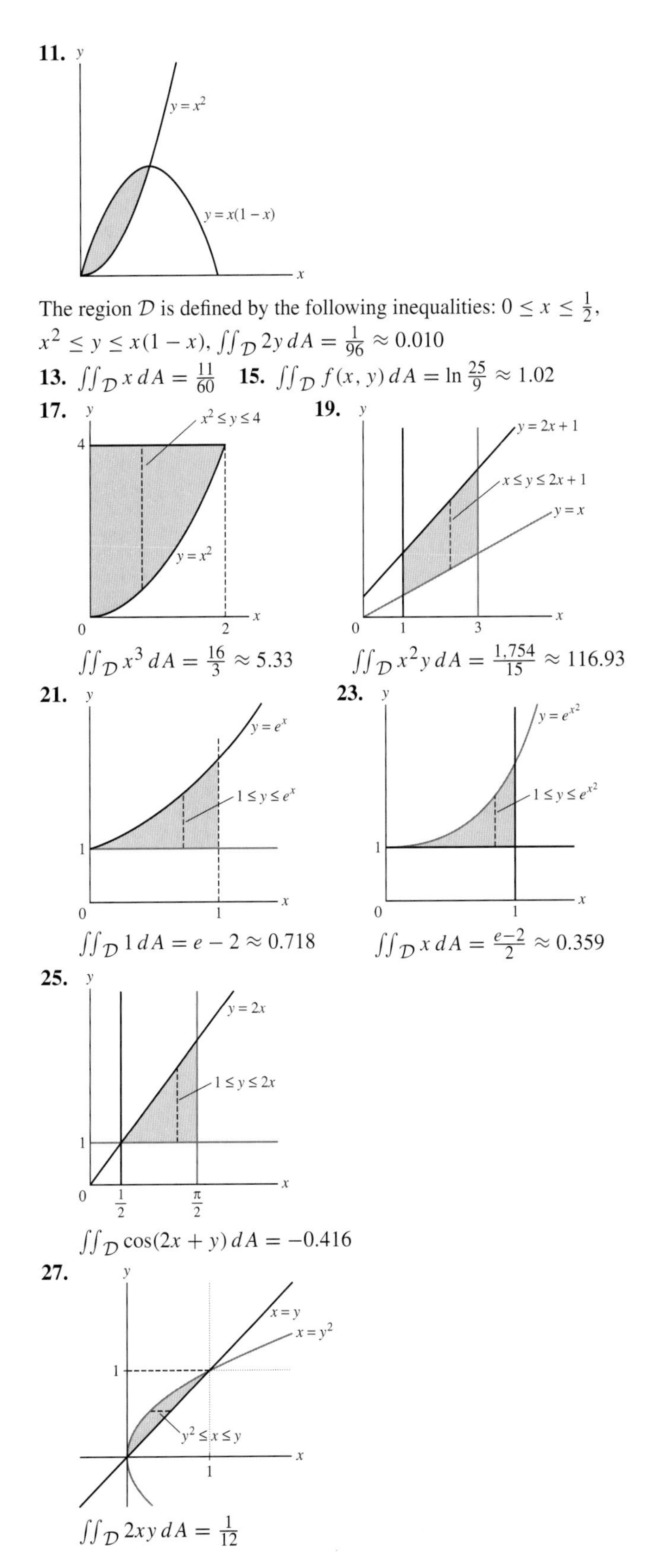

11.

The region $\mathcal{D}$ is defined by the following inequalities: $0 \le x \le \frac{1}{2}$, $x^2 \le y \le x(1 - x)$, $\iint_{\mathcal{D}} 2y\, dA = \frac{1}{96} \approx 0.010$

13. $\iint_{\mathcal{D}} x\, dA = \frac{11}{60}$ **15.** $\iint_{\mathcal{D}} f(x, y)\, dA = \ln \frac{25}{9} \approx 1.02$

17.

$\iint_{\mathcal{D}} x^3\, dA = \frac{16}{3} \approx 5.33$

19.

$\iint_{\mathcal{D}} x^2 y\, dA = \frac{1{,}754}{15} \approx 116.93$

21.

$\iint_{\mathcal{D}} 1\, dA = e - 2 \approx 0.718$

23.

$\iint_{\mathcal{D}} x\, dA = \frac{e-2}{2} \approx 0.359$

25.

$\iint_{\mathcal{D}} \cos(2x + y)\, dA = -0.416$

27.

$\iint_{\mathcal{D}} 2xy\, dA = \frac{1}{12}$

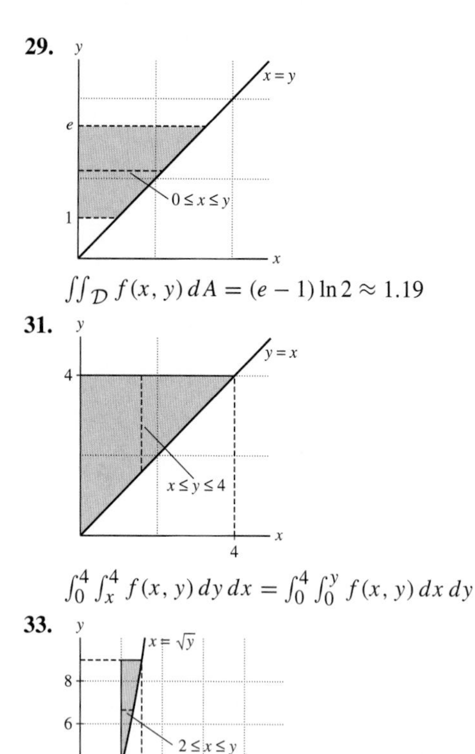

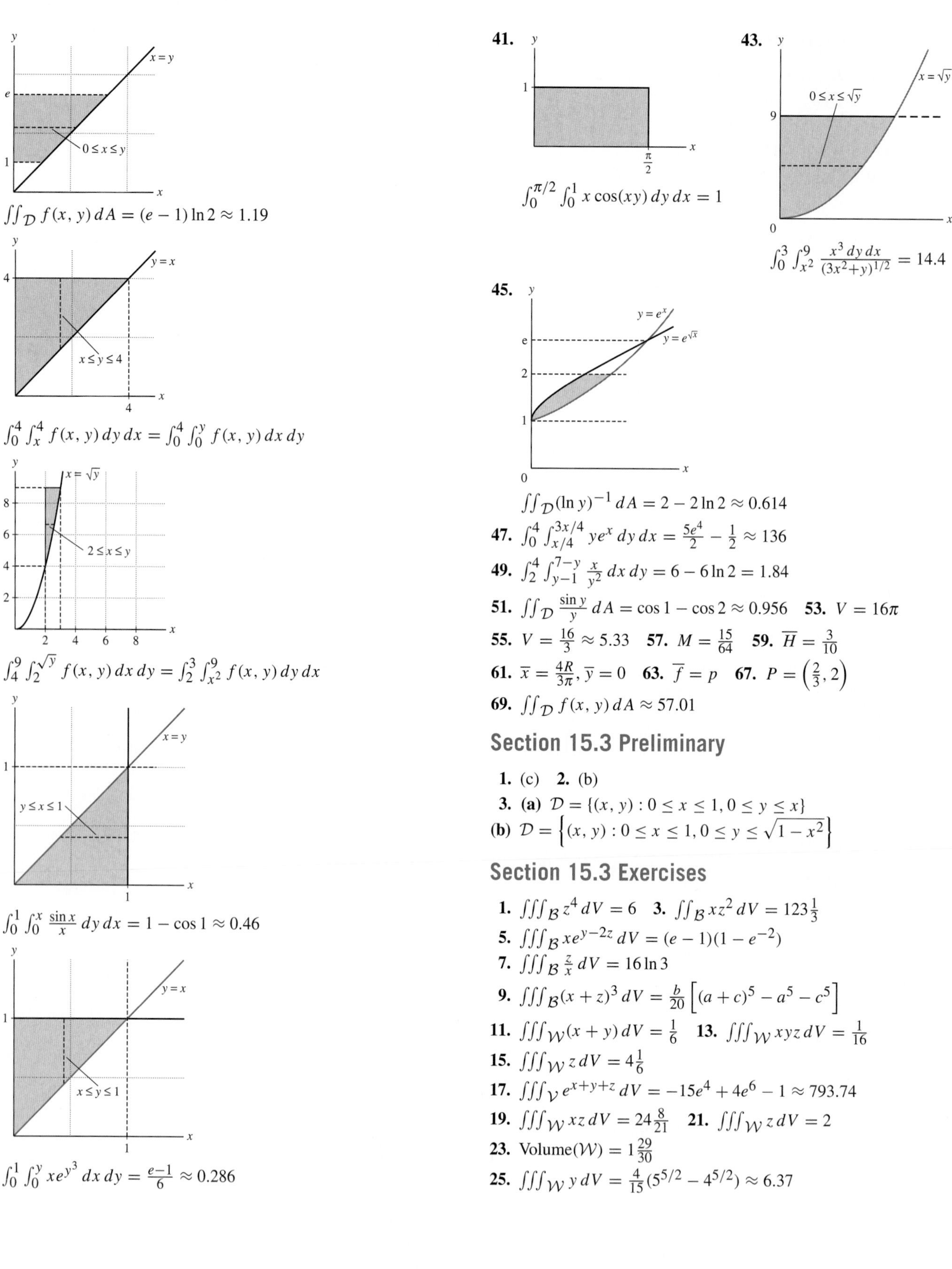

29. $\iint_{\mathcal{D}} f(x, y)\, dA = (e - 1)\ln 2 \approx 1.19$

31. $\int_0^4 \int_x^4 f(x, y)\, dy\, dx = \int_0^4 \int_0^y f(x, y)\, dx\, dy$

33. $\int_4^9 \int_2^{\sqrt{y}} f(x, y)\, dx\, dy = \int_2^3 \int_{x^2}^9 f(x, y)\, dy\, dx$

37. $\int_0^1 \int_0^x \frac{\sin x}{x}\, dy\, dx = 1 - \cos 1 \approx 0.46$

39. $\int_0^1 \int_0^y xe^{y^3}\, dx\, dy = \frac{e-1}{6} \approx 0.286$

41. $\int_0^{\pi/2} \int_0^1 x\cos(xy)\, dy\, dx = 1$

43. $\int_0^3 \int_{x^2}^9 \frac{x^3\, dy\, dx}{(3x^2+y)^{1/2}} = 14.4$

45. $\iint_{\mathcal{D}} (\ln y)^{-1}\, dA = 2 - 2\ln 2 \approx 0.614$

47. $\int_0^4 \int_{x/4}^{3x/4} ye^x\, dy\, dx = \frac{5e^4}{2} - \frac{1}{2} \approx 136$

49. $\int_2^4 \int_{y-1}^{7-y} \frac{x}{y^2}\, dx\, dy = 6 - 6\ln 2 = 1.84$

51. $\iint_{\mathcal{D}} \frac{\sin y}{y}\, dA = \cos 1 - \cos 2 \approx 0.956$ **53.** $V = 16\pi$

55. $V = \frac{16}{3} \approx 5.33$ **57.** $M = \frac{15}{64}$ **59.** $\overline{H} = \frac{3}{10}$

61. $\overline{x} = \frac{4R}{3\pi}, \overline{y} = 0$ **63.** $\overline{f} = p$ **67.** $P = \left(\frac{2}{3}, 2\right)$

69. $\iint_{\mathcal{D}} f(x, y)\, dA \approx 57.01$

Section 15.3 Preliminary

1. (c) **2.** (b)

3. **(a)** $\mathcal{D} = \{(x, y) : 0 \le x \le 1, 0 \le y \le x\}$
(b) $\mathcal{D} = \left\{(x, y) : 0 \le x \le 1, 0 \le y \le \sqrt{1 - x^2}\right\}$

Section 15.3 Exercises

1. $\iiint_{\mathcal{B}} z^4\, dV = 6$ **3.** $\iint_{\mathcal{B}} xz^2\, dV = 123\frac{1}{3}$

5. $\iiint_{\mathcal{B}} xe^{y-2z}\, dV = (e - 1)(1 - e^{-2})$

7. $\iiint_{\mathcal{B}} \frac{z}{x}\, dV = 16\ln 3$

9. $\iiint_{\mathcal{B}} (x + z)^3\, dV = \frac{b}{20}\left[(a + c)^5 - a^5 - c^5\right]$

11. $\iiint_{\mathcal{W}} (x + y)\, dV = \frac{1}{6}$ **13.** $\iiint_{\mathcal{W}} xyz\, dV = \frac{1}{16}$

15. $\iiint_{\mathcal{W}} z\, dV = 4\frac{1}{6}$

17. $\iiint_{\mathcal{V}} e^{x+y+z}\, dV = -15e^4 + 4e^6 - 1 \approx 793.74$

19. $\iiint_{\mathcal{W}} xz\, dV = 24\frac{8}{21}$ **21.** $\iiint_{\mathcal{W}} z\, dV = 2$

23. Volume$(\mathcal{W}) = 1\frac{29}{30}$

25. $\iiint_{\mathcal{W}} y\, dV = \frac{4}{15}(5^{5/2} - 4^{5/2}) \approx 6.37$

27. $\iiint_{\mathcal{W}} xyz^2\,dV = \int_0^2 \int_0^{y/2} \int_0^{4-y^2} xyz^2\,dz\,dx\,dy$

$\iiint_{\mathcal{W}} xyz^2\,dV = \int_0^4 \int_0^{\sqrt{4-z}} \int_0^{y/2} xyz^2\,dx\,dy\,dz$

$\iiint_{\mathcal{W}} xyz^2\,dV = \int_0^4 \int_0^{\sqrt{1-\frac{z}{4}}} \int_{2x}^{\sqrt{4-z}} xyz^2\,dy\,dx\,dz$

29. $\iiint_{\mathcal{W}} f(x,y,z)\,dV$
$= \int_{-1}^1 \int_{-\sqrt{1-x^2}}^{\sqrt{1-x^2}} \int_{\sqrt{x^2+y^2}}^1 f(x,y,z)\,dz\,dy\,dx$

$\iiint_{\mathcal{W}} f(x,y,z)\,dV$
$= \int_0^1 \int_{-z}^{z} \int_{-\sqrt{z^2-y^2}}^{\sqrt{z^2-y^2}} f(x,y,z)\,dx\,dy\,dz$

31. $\overline{f} = \frac{1}{2\pi}$ **33.** $\overline{f} = \frac{13}{24}$ **35.** $P = (1.5, 1, 2)$

37. $(x_{CM}, y_{CM}, z_{CM}) = \left(\frac{2}{9}, \frac{2}{9}, \frac{1}{3}\right)$

39. $S_{N,N,N} \approx 0.561, 0.572$, and 0.576 for $N = 10, 20, 30$, respectively. $I \approx 0.584$, for $N = 100$ $S_{N,N,N} \approx 0.582$.

Section 15.4 Preliminary

1. (d)

2. **(a)** $\int_{-4}^4 \int_0^{2\pi} \int_0^{\sqrt{16-z^2}} f(P)r\,dr\,d\theta\,dz$

(b) $\int_{-5}^{-4} \int_0^{2\pi} \int_0^{\sqrt{25-z^2}} f(P)r\,dr\,d\theta\,dz +$
$\int_{-4}^4 \int_0^{2\pi} \int_{\sqrt{16-z^2}}^{\sqrt{25-z^2}} f(P)r\,dr\,d\theta\,dz +$
$\int_4^5 \int_0^{2\pi} \int_0^{\sqrt{25-z^2}} f(P)r\,dr\,d\theta\,dz$

(c) $\int_{-2}^0 \int_0^{2\pi} \int_0^{\sqrt{4-z^2}} f(P)r\,dr\,d\theta\,dz$

3. **(a)** $\int_0^{2\pi} \int_0^{\pi} \int_0^4 f(P)\rho^2 \sin\phi\,d\rho\,d\phi\,d\theta$

(b) $\int_0^{2\pi} \int_0^{\pi} \int_4^5 f(P)\rho^2 \sin\phi\,d\rho\,d\phi\,d\theta$

(c) $\int_0^{2\pi} \int_{\pi/2}^{\pi} \int_0^2 f(P)\rho^2 \sin\phi\,d\rho\,d\phi\,d\theta$

4. $\Delta A \approx r(\Delta r \Delta\theta)$

Section 15.4 Exercises

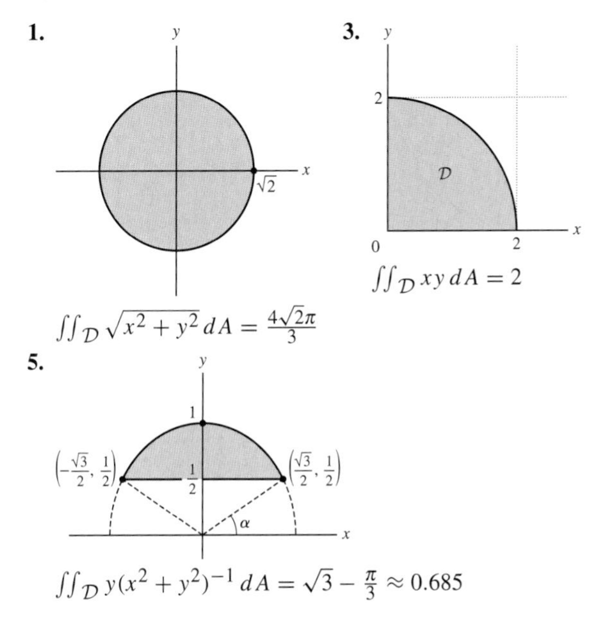

1. $\iint_{\mathcal{D}} \sqrt{x^2+y^2}\,dA = \frac{4\sqrt{2}\pi}{3}$

3. $\iint_{\mathcal{D}} xy\,dA = 2$

5. $\iint_{\mathcal{D}} y(x^2+y^2)^{-1}\,dA = \sqrt{3} - \frac{\pi}{3} \approx 0.685$

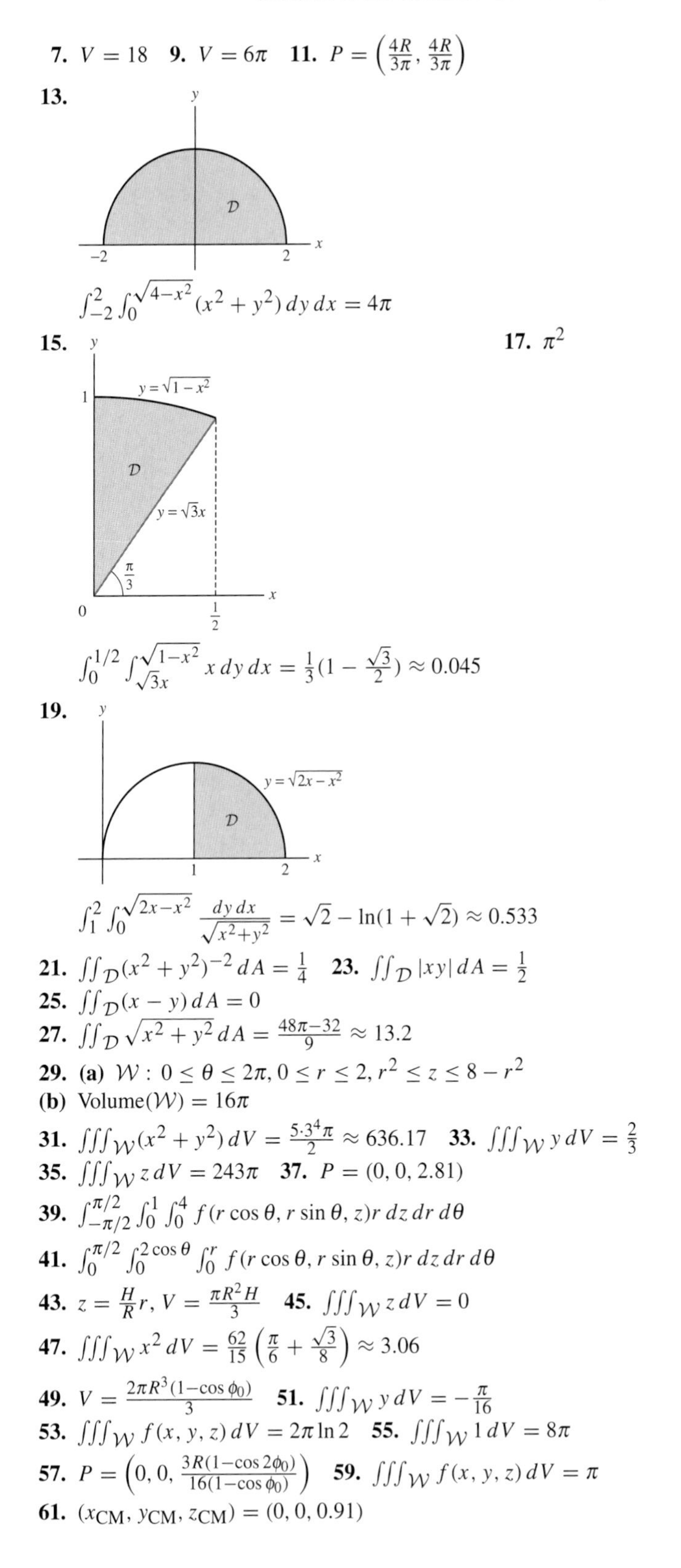

7. $V = 18$ **9.** $V = 6\pi$ **11.** $P = \left(\frac{4R}{3\pi}, \frac{4R}{3\pi}\right)$

13. $\int_{-2}^2 \int_0^{\sqrt{4-x^2}} (x^2+y^2)\,dy\,dx = 4\pi$

15. $\int_0^{1/2} \int_{\sqrt{3}x}^{\sqrt{1-x^2}} x\,dy\,dx = \frac{1}{3}\left(1 - \frac{\sqrt{3}}{2}\right) \approx 0.045$

17. π^2

19. $\int_1^2 \int_0^{\sqrt{2x-x^2}} \frac{dy\,dx}{\sqrt{x^2+y^2}} = \sqrt{2} - \ln(1+\sqrt{2}) \approx 0.533$

21. $\iint_{\mathcal{D}} (x^2+y^2)^{-2}\,dA = \frac{1}{4}$ **23.** $\iint_{\mathcal{D}} |xy|\,dA = \frac{1}{2}$

25. $\iint_{\mathcal{D}} (x-y)\,dA = 0$

27. $\iint_{\mathcal{D}} \sqrt{x^2+y^2}\,dA = \frac{48\pi-32}{9} \approx 13.2$

29. **(a)** $\mathcal{W}: 0 \le \theta \le 2\pi, 0 \le r \le 2, r^2 \le z \le 8 - r^2$

(b) $\text{Volume}(\mathcal{W}) = 16\pi$

31. $\iiint_{\mathcal{W}} (x^2+y^2)\,dV = \frac{5 \cdot 3^4 \pi}{2} \approx 636.17$ **33.** $\iiint_{\mathcal{W}} y\,dV = \frac{2}{3}$

35. $\iiint_{\mathcal{W}} z\,dV = 243\pi$ **37.** $P = (0, 0, 2.81)$

39. $\int_{-\pi/2}^{\pi/2} \int_0^1 \int_0^4 f(r\cos\theta, r\sin\theta, z)r\,dz\,dr\,d\theta$

41. $\int_0^{\pi/2} \int_0^{2\cos\theta} \int_0^r f(r\cos\theta, r\sin\theta, z)r\,dz\,dr\,d\theta$

43. $z = \frac{H}{R}r$, $V = \frac{\pi R^2 H}{3}$ **45.** $\iiint_{\mathcal{W}} z\,dV = 0$

47. $\iiint_{\mathcal{W}} x^2\,dV = \frac{62}{15}\left(\frac{\pi}{6} + \frac{\sqrt{3}}{8}\right) \approx 3.06$

49. $V = \frac{2\pi R^3(1-\cos\phi_0)}{3}$ **51.** $\iiint_{\mathcal{W}} y\,dV = -\frac{\pi}{16}$

53. $\iiint_{\mathcal{W}} f(x,y,z)\,dV = 2\pi\ln 2$ **55.** $\iiint_{\mathcal{W}} 1\,dV = 8\pi$

57. $P = \left(0, 0, \frac{3R(1-\cos 2\phi_0)}{16(1-\cos\phi_0)}\right)$ **59.** $\iiint_{\mathcal{W}} f(x,y,z)\,dV = \pi$

61. $(x_{CM}, y_{CM}, z_{CM}) = (0, 0, 0.91)$

63. $(x_{CM}, y_{CM}, z_{CM}) = \left(\frac{2R}{\pi}, \frac{4R}{\pi}\left(1 - \frac{\sqrt{3}}{2}\right), \frac{H}{2}\right)$

65. $(x_{CM}, y_{CM}, z_{CM}) = \left(0, \frac{1}{12}, \frac{37}{48}\right)$

69. $4\pi(\sin 1 - \cos 1)$ **73.** $a < 2$

Section 15.5 Preliminary

1. (b)

2. **(a)** $\Phi(1, 0) = (2, 0)$ **(b)** $\Phi(1, 1) = (1, 3)$
(c) $\Phi(2, 1) = (3, 3)$

3. $\text{Area}(\Phi(\mathcal{R})) = 36$ **4.** $\text{Area}(\Phi(\mathcal{R})) = 0.06$

Section 15.5 Exercises

1. **(a)** The image of the u-axis is the line $y = \frac{1}{2}x$; the image of the v-axis is the line $x = 0$.
(b) The parallelogram with vertices (0, 0), (10, 5), (10, 12), (0, 7).
(c) A segment in the xy-plane joining the points (2, 3) and (10, 8).
(d) A triangle in the xy-plane whose vertices are at the points (0, 1), (2, 1), and (2, 2).

3. ϕ is one-to-one on the domain $\{(u, v)\, u \geq 0\}$ and on the domain $\{(u, v)\, u \leq 0\}$.
(a) The positive x-axis and the y-axis, respectively.
(b) The rectangle $[0, 1] \times [-1, 1]$
(c) The curve $y = \sqrt{x}$ for $0 \leq x \leq 1$
(d)

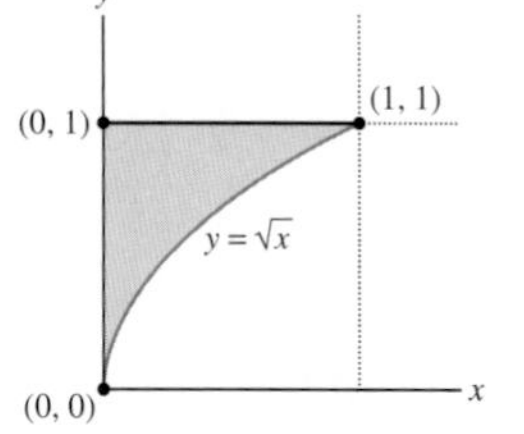

5. $y = \frac{5x}{2} + \frac{c}{2}$ **7.** $y = \frac{17}{6}x$ **11.** $\text{Jac}(\Phi) = 1$

13. $\text{Jac}(\Phi) = -10$ **15.** $\text{Jac}(\Phi) = (1 - 3uv)e^{v+3u}$

17. $\text{Jac}(\Phi) = r$ **19.** $\Phi(u, v) = (4u + 2v, u + 3v)$

21. $\iint_{\mathcal{D}} xy\, dA = \frac{2{,}329}{12} \approx 194.08$

23. **(a)** $\text{Area}(\Phi(\mathcal{R})) = 105$ **(b)** $\text{Area}(\Phi(\mathcal{R})) = 126$

25. $\text{Jac}(T) = \frac{2u}{v}$, $\iint_{\Phi(\mathcal{T})} 1\, dA = 15 \ln 4$

27.

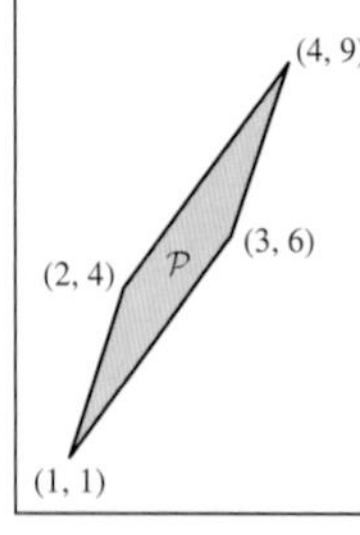

$\Phi(u, v) = (1 + 2u + v, 1 + 5u + 3v)$

29. $\iint_{\mathcal{D}} y\, dA = 82$ **31.** $\iint_{\mathcal{D}} (x + 3y)\, dx\, dy = 80$

33. $\iint_{\mathcal{D}} \sqrt{x^2 + y^2}\, dx\, dy = \frac{56}{45}$

35. $\iint_{\mathcal{D}} e^{9x^2+4y^2}\, dA = \frac{\pi(e^{36} - 1)}{6}$

37.

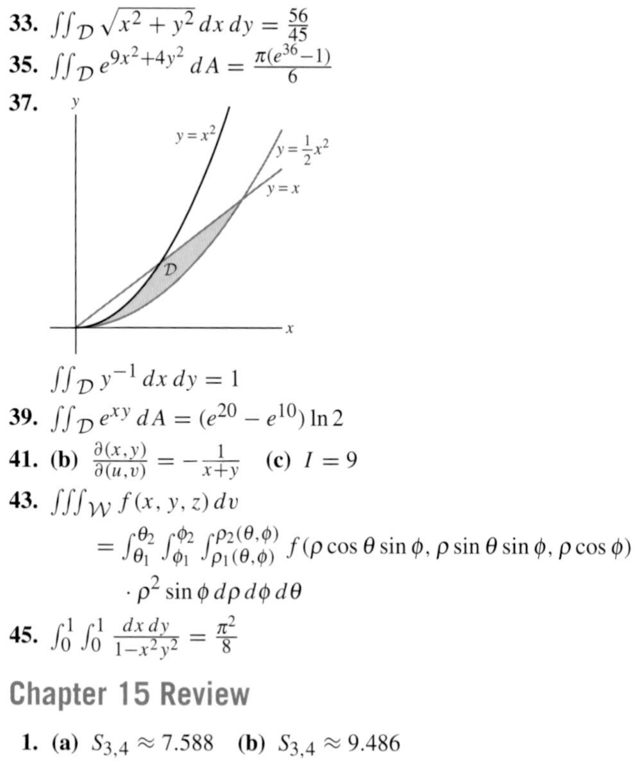

$\iint_{\mathcal{D}} y^{-1}\, dx\, dy = 1$

39. $\iint_{\mathcal{D}} e^{xy}\, dA = (e^{20} - e^{10}) \ln 2$

41. **(b)** $\frac{\partial(x, y)}{\partial(u, v)} = -\frac{1}{x+y}$ **(c)** $I = 9$

43. $\iiint_{\mathcal{W}} f(x, y, z)\, dv$

$$= \int_{\theta_1}^{\theta_2} \int_{\phi_1}^{\phi_2} \int_{\rho_1(\theta,\phi)}^{\rho_2(\theta,\phi)} f(\rho \cos\theta \sin\phi, \rho \sin\theta \sin\phi, \rho \cos\phi) \cdot \rho^2 \sin\phi\, d\rho\, d\phi\, d\theta$$

45. $\int_0^1 \int_0^1 \frac{dx\, dy}{1 - x^2 y^2} = \frac{\pi^2}{8}$

Chapter 15 Review

1. **(a)** $S_{3,4} \approx 7.588$ **(b)** $S_{3,4} \approx 9.486$
(c) $\int_1^2 \int_2^3 x^2 y\, dx\, dy = 9.5$

3. $S_{44} = 2.9375$ **5.** $\int_0^2 \int_3^5 y(x - y)\, dx\, dy = \frac{32}{3}$

7. $\int_0^{\pi/3} \int_0^{\pi/6} \sin(x + y)\, dx\, dy = \frac{\sqrt{3} - 1}{2}$

9.

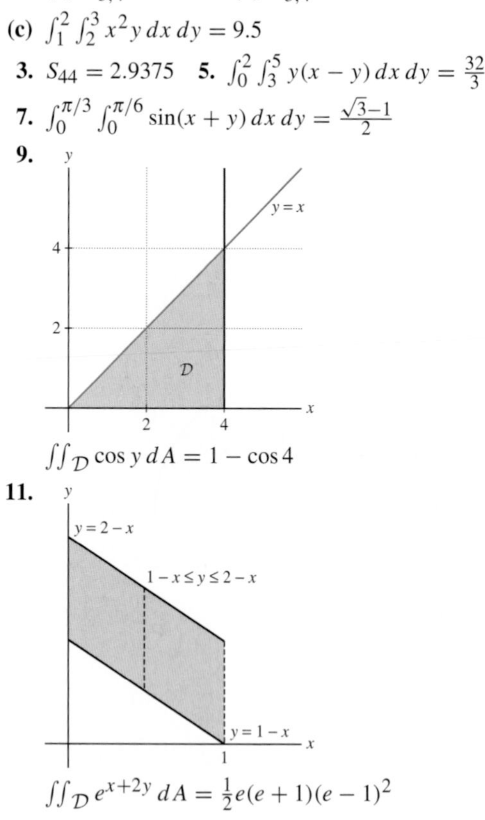

$\iint_{\mathcal{D}} \cos y\, dA = 1 - \cos 4$

11.

$\iint_{\mathcal{D}} e^{x+2y}\, dA = \frac{1}{2}e(e + 1)(e - 1)^2$

13.

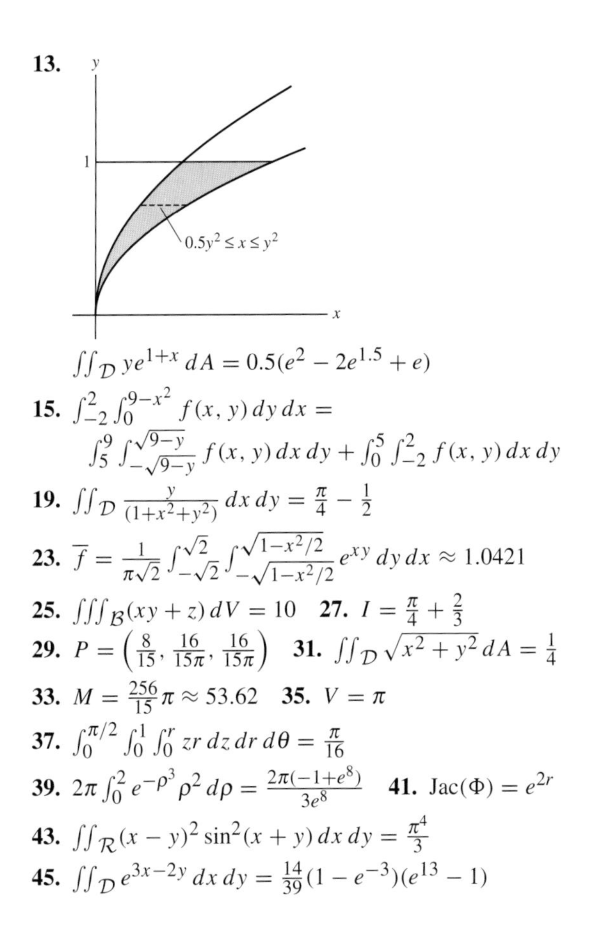

$\iint_{\mathcal{D}} ye^{1+x}\,dA = 0.5(e^2 - 2e^{1.5} + e)$

15. $\int_{-2}^{2}\int_{0}^{9-x^2} f(x,y)\,dy\,dx =$
$\int_5^9 \int_{-\sqrt{9-y}}^{\sqrt{9-y}} f(x,y)\,dx\,dy + \int_0^5 \int_{-2}^{2} f(x,y)\,dx\,dy$

19. $\iint_{\mathcal{D}} \frac{y}{(1+x^2+y^2)}\,dx\,dy = \frac{\pi}{4} - \frac{1}{2}$

23. $\overline{f} = \frac{1}{\pi\sqrt{2}} \int_{-\sqrt{2}}^{\sqrt{2}} \int_{-\sqrt{1-x^2/2}}^{\sqrt{1-x^2/2}} e^{xy}\,dy\,dx \approx 1.0421$

25. $\iiint_{\mathcal{B}} (xy+z)\,dV = 10$ **27.** $I = \frac{\pi}{4} + \frac{2}{3}$

29. $P = \left(\frac{8}{15}, \frac{16}{15\pi}, \frac{16}{15\pi}\right)$ **31.** $\iint_{\mathcal{D}} \sqrt{x^2+y^2}\,dA = \frac{1}{4}$

33. $M = \frac{256}{15}\pi \approx 53.62$ **35.** $V = \pi$

37. $\int_0^{\pi/2}\int_0^1\int_0^r zr\,dz\,dr\,d\theta = \frac{\pi}{16}$

39. $2\pi \int_0^2 e^{-\rho^3}\rho^2\,d\rho = \frac{2\pi(-1+e^8)}{3e^8}$ **41.** $\text{Jac}(\Phi) = e^{2r}$

43. $\iint_{\mathcal{R}} (x-y)^2 \sin^2(x+y)\,dx\,dy = \frac{\pi^4}{3}$

45. $\iint_{\mathcal{D}} e^{3x-2y}\,dx\,dy = \frac{14}{39}(1-e^{-3})(e^{13}-1)$

Chapter 16

Section 16.1 Preliminary

1. (b) **2.** The vector $\langle x, x\rangle$ **3.** $\mathbf{F} = \langle 0, -z, y\rangle$

4. $\phi_1(x,y,z) = xyz + 1$

Section 16.1 Exercises

1. $\mathbf{F}(1,2) = \langle 1, 1\rangle$, $\mathbf{F}(-1,-1) = \langle 1, -1\rangle$

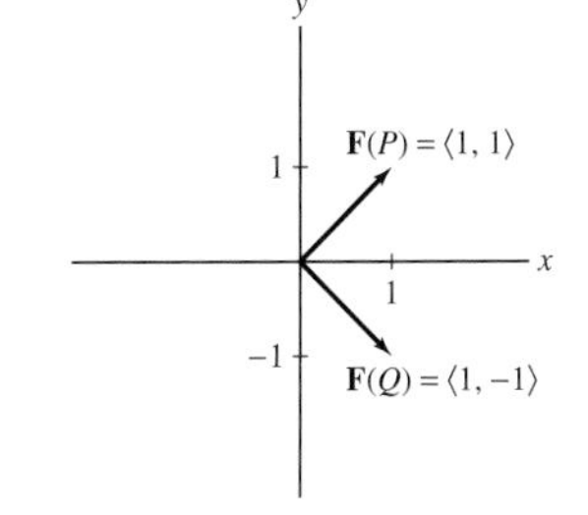

3. $\mathbf{F}(P) = \langle 0, 1, 0\rangle$, $\mathbf{F}(Q) = \langle 2, 0, 2\rangle$

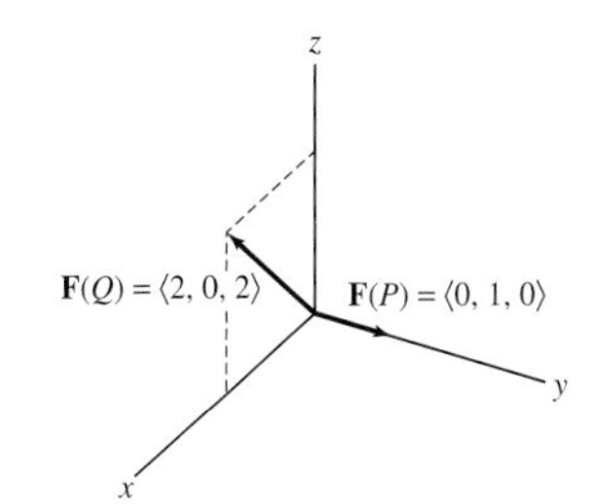

5. $\mathbf{F} = \langle 1, 0\rangle$

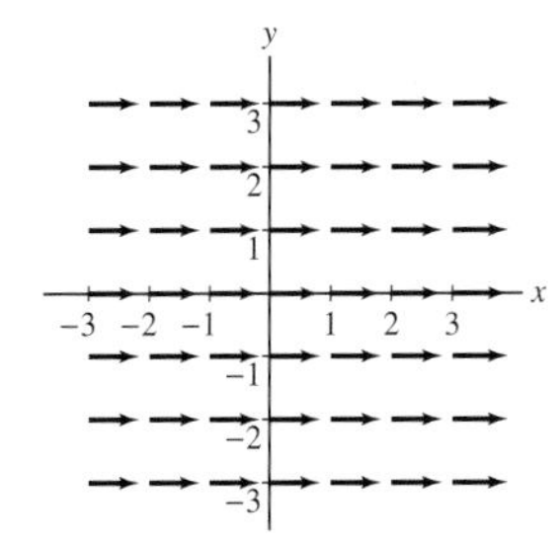

7. $\mathbf{F}(x,y) = \langle 0, x\rangle$

9. $\mathbf{F} = \langle 1, 1\rangle$

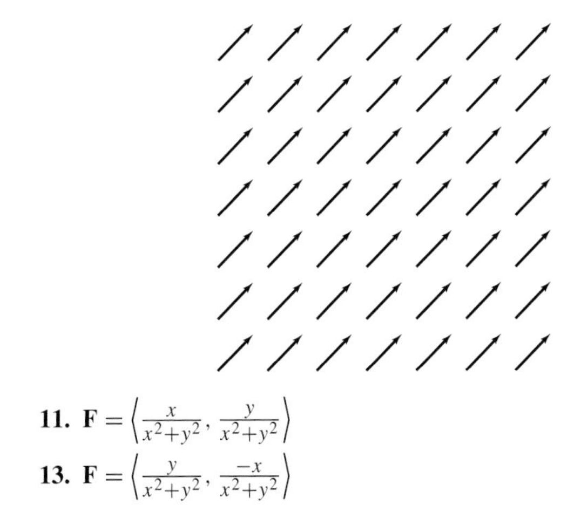

11. $\mathbf{F} = \left\langle \frac{x}{x^2+y^2}, \frac{y}{x^2+y^2}\right\rangle$

13. $\mathbf{F} = \left\langle \frac{y}{x^2+y^2}, \frac{-x}{x^2+y^2}\right\rangle$

15. Plot (C) **17.** Plot (A) **19.** Plot (A) **21.** Plot (C)
23. $\varphi(x, y) = e^{xy}$ **25.** $\varphi(x, y, z) = ze^{x^2}$ **27.** Plot (B)
29. $\nabla\varphi = \frac{\mathbf{e}_r}{r}$

Section 16.2 Preliminary

1. $\int_C 10\,ds = 10\int_C 1\,ds = 50$ **2.** (a), (c), (d), (e)
3. **(a)** True
(b) False. Reversing the orientation of the curve changes the sign of the vector line integral.
4. **(a)** $\int_C \mathbf{F}\cdot d\mathbf{s} = 0$ **(b)** $\int_C \mathbf{F}\cdot d\mathbf{s} = 5$

Section 16.2 Exercises

1. **(a)** $f(\mathbf{c}(t)) = 6t + 4t^2$, $ds = 2\sqrt{11}\,dt$
(b) $\int_C f(x, y, z)\,ds = \frac{26\sqrt{11}}{3}$
3. **(a)** $\mathbf{F}(\mathbf{c}(t)) = \langle t^{-2}, t^2\rangle$, $d\mathbf{s} = \langle 1, -t^{-2}\rangle\,dt$
(b) $\int_C \mathbf{F}\cdot d\mathbf{s} = -\frac{1}{2}$
5. $\int_C (x^2 + y^2 + z^2)\,ds = \sqrt{2}\left(\pi + \frac{\pi^3}{3}\right)$
7. $\int_C \mathbf{F}\,ds = \frac{\pi^2}{2}$ **9.** $M = 64\pi$ g
11. $\int_{ABC} f(x, y, z)\,ds \approx 7.6$, $\int_{ABC} \mathbf{F}\cdot d\mathbf{s} \approx 4\frac{2}{3}$
13. (A) Zero (B) Negative (C) Zero
15. $\int_C f(x, y, z)\,ds = \frac{128\sqrt{29}}{3} \approx 229.8$
17. $\int_C f(x, y)\,ds = 2.8$
19. $\int_C f(x, y, z)\,ds = \int_0^1 te^{2t-t^2}\sqrt{3}\,dt \approx 2.028$
21. $\int_C f(x, y, z)\,ds = \frac{2}{3}\left((e^2 + 5)^{3/2} - 2^{3/2}\right)$
23. $\int_C \mathbf{F}\cdot d\mathbf{s} = \frac{16}{3}$ **25.** $\int_C \mathbf{F}\cdot d\mathbf{s} = 0$ **27.** $\int_C \mathbf{F}\cdot d\mathbf{s} = 11.5 - \ln 2$
29. $\int_C \mathbf{F}\cdot d\mathbf{s} = 2(e^2 - e^{-2}) - (e - e^{-1}) \approx 12.157$
31. $\int_C \mathbf{F}\cdot d\mathbf{s} = 2\pi$
33. $\int_C \mathbf{F}\cdot d\mathbf{s} = \frac{\pi(\pi-8)}{16\sqrt{2}} + \ln(2 + \sqrt{2}) + 2 - \sqrt{2} \approx 1.139$
35. $\int_C \mathbf{F}\cdot d\mathbf{s} = 0$ **37.** $\int_C \langle e^{x-y}, e^{x+y}\rangle\cdot d\mathbf{s} \approx -4.50879$
39. $\int_C \mathbf{F}\cdot d\mathbf{s} = \frac{1}{3}e^6 + \frac{3}{2}e^4 + \frac{2}{3}e^2 - \frac{65}{6} + \frac{1}{3}e^{-4}$
41. **(a)** Positive **(b)** $\int_C \mathbf{F}\cdot d\mathbf{s} = 2\pi$
43. $\int_A \mathbf{F}\cdot d\mathbf{s} = \theta_0$
45. **(a)** $\int_C \mathbf{F}\cdot d\mathbf{s} = 2$ **(b)** $\int_C \mathbf{F}\cdot d\mathbf{s} = 25$ **(c)** $\int_C \mathbf{F}\cdot d\mathbf{s} = 32$
47. **(a)** $\int_{AB} \mathbf{F}\cdot d\mathbf{s} < \int_{DC} \mathbf{F}\cdot d\mathbf{s}$ **(b)** $\int_{ABC} \mathbf{F}\cdot d\mathbf{s} < \int_{ADC} \mathbf{F}\cdot d\mathbf{s}$
49. $W = \frac{9\pi^2}{2}$ **53.** $\text{Av}(f) \approx 3.744$ **55.** $\text{Av}(T) = 8$

Section 16.3 Preliminary

1. Closed
2. **(a)** Conservative vector fields **(b)** All vector fields
(c) Conservative vector fields **(d)** All vector fields
(e) Conservative vector fields **(f)** All vector fields
(g) Conservative vector fields
3. **(a)** Always true
(b) True under additional hypotheses on $\mathcal{D}$
(c) True under additional hypotheses on $\mathcal{D}$
4. **(a)** $\int_D \mathbf{F}\cdot d\mathbf{s} = 4$ **(b)** $\int_E \mathbf{F}\cdot d\mathbf{s} = -4$

Section 16.3 Exercises

1. $\int_{\mathbf{c}} \nabla\varphi\cdot d\mathbf{s} = 0$ **3.** $\int_{\mathbf{c}} \mathbf{F}\cdot d\mathbf{s} = -\frac{9}{4}$ **5.** $\int_{\mathbf{c}} \mathbf{F}\cdot d\mathbf{s} = 0$
7. $\int_{\mathbf{c}} \mathbf{F}\cdot d\mathbf{s} = 32e - 1$
9. $\varphi(x, y, z) = x^2y + 5x - 4zy$, $\int_{\mathbf{c}} \mathbf{F}\cdot d\mathbf{s} = 20$
11. $\int_{\mathbf{c}} \mathbf{F}\cdot d\mathbf{s} = 18$ **13.** $\mathbf{F}$ is not conservative.
15. $\varphi(x, y, z) = yx + \frac{z^4}{4}$ **17.** $\varphi(x, y, z) = x\cos z + y^2$
19. $W = \ln 4 - \ln 3$ **23.** $W = 19{,}600$ joules
25. **(c)** $\mathcal{D}$ is not simply connected. **(d)** No

Section 16.4 Preliminary

1. $\iint_S f(x, y, z)\,dS = 50$
2. A distortion factor that indicates how much the area of R_{ij} is altered under the map ϕ.
3. $\text{Area}(S) \approx 0.0006$ **4.** $\iint_S f(x, y, z)\,dS \approx 0.6$
5. $\text{Area}(S) = 20$ **6.** $\left\langle \frac{2}{3}, \frac{2}{3}, \frac{1}{3}\right\rangle$

Section 16.4 Exercises

1. **(a)** v **(b)** iii **(c)** i **(d)** iv **(e)** ii
3. **(a)** $\mathbf{T}_u = \langle 2, 1, 3\rangle$, $\mathbf{T}_v = \langle 0, -1, 1\rangle$, $\mathbf{n}(u, v) = \langle 4, -2, -2\rangle$
(b) $\text{Area}(S) = 4\sqrt{6}$ **(c)** $\iint_S f(x, y, z)\,dS = \frac{32\sqrt{6}}{3}$
5. **(a)** $\mathbf{T}_x = \langle 1, 0, y\rangle$, $\mathbf{T}_y = \langle 0, 1, x\rangle$, $\mathbf{n}(x, y) = \langle -y, -x, 1\rangle$
(b) $\iint_S 1\,dS = \frac{(2\sqrt{2}-1)\pi}{6}$ **(c)** $\iint_S z\,dS = \frac{\sqrt{2}+1}{15}$
7. $\mathbf{T}_u = \langle 2, 1, 3\rangle$, $\mathbf{T}_v = \langle 1, -4, 0\rangle$, $\mathbf{n}(u, v) = 3\langle 4, 1, -3\rangle$, $4x + y - 3z = 0$
9. $\mathbf{T}_\theta = \langle -\sin\theta\sin\phi, \cos\theta\sin\phi, 0\rangle$
$\mathbf{T}_\phi = \langle \cos\theta\cos\phi, \sin\theta\cos\phi, -\sin\phi\rangle$
$\mathbf{n}(\theta, \phi) = -(\cos\theta\sin^2\phi)\mathbf{i} - (\sin\theta\sin^2\phi)\mathbf{j} - (\sin\phi\cos\phi)\mathbf{k}$,
$y + z = \sqrt{2}$
11. $\text{Area}(S) \approx 0.2078$ **13.** $\text{Area}(S) = 16$
15. **(a)**

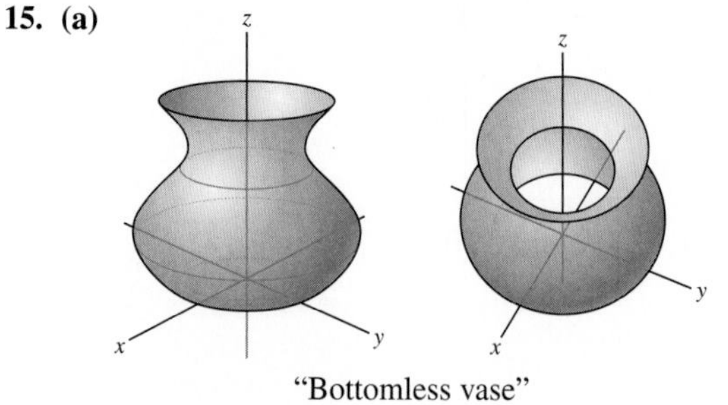

"Bottomless vase"

(b) $\mathbf{n}(u, v) =$
$\big((3+\sin v)\cos u\big)\mathbf{i} + \big((3+\sin v)\sin u\big)\mathbf{j} - \big((3+\sin v)\cos v\big)\mathbf{k}$
(c) $\text{Area}(S) \approx 144.0181$
17. $\text{Area}(S) = 4\pi R^2$ **19.** $\iint_S x^2\, dS = \frac{\pi}{6}$
21. $\iint_S f(x, y, z)\, dS = \frac{\sqrt{2}}{5}$
23. $\iint_S f(x, y, z)\, dS = 4\pi(1 - e^{-4})$
25. $\iint_S f(x, y, z)\, dS = \frac{(3{,}179\sqrt{17}+29)\pi}{420} \approx 98.26$
27. $\iint_S f(x, y, z)\, dS = \frac{37\sqrt{37}-1}{4} \approx 56.02$
29. $\iint_S f(x, y, z)\, dS = \frac{\sqrt{3}}{6}$
31. $\iint_S f(x, y, z)\, dS = \frac{1}{54}(10\sqrt{10} - 1) \approx 0.567$
35. $\text{Area}(S) = \frac{3\sqrt{29}}{2}$ **37.** $\iint_S f(x, y, z)\, dS = \frac{7\pi}{3}$
39. (a) $\text{Area}(S) = 6\pi(\sqrt{8} - \sqrt{5}) \approx 11.166$
(b) $\iint_S z^{-1}\, dS = 3\pi \ln 1.6 \approx 4.43$
41. $\text{Area}(S) = \frac{\pi}{2}$ **45.** $\text{Area}(S) = \frac{\pi}{6}\big(17\sqrt{17} - 1\big) \approx 36.18$
49. $4\pi^2 ab$ **51.** $\varphi(r) = -\frac{Gm}{2Rr}\left(\sqrt{R^2 + r^2} - |R - r|\right)$

Section 16.5 Preliminary

1. (b) **2.** (c) **3.** (a) **4.** (b)
5. (a) $\iint_S \mathbf{F}\cdot d\mathbf{S} = 0$ **(b)** $\iint_S \mathbf{F}\cdot d\mathbf{S} = \pi$ **(c)** $\iint_S \mathbf{F}\cdot d\mathbf{S} = \pi$
6. $\iint_S \mathbf{F}\cdot d\mathbf{S} \approx 0.05\sqrt{2} \approx 0.0707$ **7.** $\iint_S \mathbf{F}\cdot d\mathbf{S} = 0$

Section 16.5 Exercises

1. (a) $\mathbf{n} = \langle 2v, -4uv, 2u + 1\rangle$
$\mathbf{F}\cdot\mathbf{n} = 2u^3 - 4uv^3 + 2v^2 + u^2 - v$
(b) $\mathbf{F}(P)\cdot\mathbf{e_n}(P) = \frac{13}{\sqrt{93}}$ **(c)** $\iint_S \mathbf{F}\cdot d\mathbf{S} = 24$
3. $\iint_S \mathbf{F}\cdot d\mathbf{S} = 4$ **5.** $\iint_S \mathbf{F}\cdot d\mathbf{S} = -4$ **7.** $\iint_S \mathbf{F}\cdot d\mathbf{S} = \frac{9\pi}{4}$
9. $\iint_S \mathbf{F}\cdot d\mathbf{S} = 0$ **11.** $\iint_S \mathbf{F}\cdot d\mathbf{S} = -\frac{11}{12}$
13. $\iint_S \mathbf{F}\cdot d\mathbf{S} = \frac{9\pi}{4}$ **15.** $\iint_S \mathbf{F}\cdot d\mathbf{S} = (e - 1)^2$
17. $\iint_S \mathbf{F}\cdot d\mathbf{S} = 270$
19. (a) $\iint_S \mathbf{F}\cdot d\mathbf{S} = 18\pi e^{-3}$ **(b)** $\iint_S \mathbf{F}\cdot d\mathbf{S} \approx 1.46456$
21. $\iint_{\mathcal{D}} \mathbf{E}\cdot d\mathbf{S} = (2 - \sqrt{2})\pi k$ **23.** $\iint_S \mathbf{v}\cdot d\mathbf{S} = \frac{\pi}{2}\ \text{ft}^3/\text{s}$
25. $\iint_S \mathbf{v}\cdot d\mathbf{S} = 4.8\ \text{ft}^3/\text{s}$ **27.** $\iint_S \mathbf{v}\cdot d\mathbf{S} = -\frac{\sqrt{3}}{2}\ \text{m}^3/\text{s}$
29. $\Phi(t) = (-9.65 \times 10^{-7})\cdot t(12 - t)\ \text{T/m}^2$
$\int_C \mathbf{E}\cdot d\mathbf{S} = 1.93 \times 10^{-6}\cdot(6 - t)\ \text{V}$

Chapter 16 Review

1. (a) $\mathbf{F} = \langle -15, 8\rangle$ **(b)** $\mathbf{F} = \langle 4, 8\rangle$ **(c)** $\mathbf{F} = \langle 9, 1\rangle$
3. $\mathbf{F} = \langle y, 1\rangle$ **5.** $\mathbf{F}(x, y) = \langle 2x, -1\rangle$

$\mathbf{F} = \langle y, 1\rangle$

$\nabla\varphi = \langle 2x, -1\rangle$

7. $\varphi(x, y, z) = -\cos x + e^y + \frac{z^2}{2}$ **9.** $\mathbf{F}$ is not conservative.
11. $\varphi(x, y, z) = y\tan^{-1}(x) + z^2$ **13.** $\mathbf{F}$ is not conservative.
15. $\int_c \nabla\varphi\cdot d\mathbf{s} = 46{,}648$ **17.** $\int_C f(x, y)\, ds = \frac{\sqrt{5}}{6}$
19. $\int_C f(x, y)\, ds = \frac{11}{6}$ **21.** $M = 13\frac{1}{3}$
23. $\int_{C_1} \mathbf{F}\cdot d\mathbf{s} = -\frac{243\pi}{16} + 20.25 \approx -27.463$
25. $\int_C \mathbf{F}\cdot d\mathbf{s} = -\frac{\pi}{2}$ **27.** $\int_C \mathbf{F}\cdot d\mathbf{s} = -\frac{13}{18}$
29. (B) and (C) Zero, (D) Negative
31. $a = \frac{4}{3}, b = \frac{5}{4}, c = -5, \mathbf{T}_u = \left\langle 1, \frac{5}{4}, 2\right\rangle; \mathbf{T}_v = \left\langle \frac{4}{3}, 1, 0\right\rangle,$
$\mathbf{n} = \left\langle -2, \frac{8}{3}, -\frac{2}{3}\right\rangle$
33. (a) $\mathbf{T}_u\left(1, \frac{\pi}{3}\right) = \left\langle 1, \sqrt{3}, 0\right\rangle, \mathbf{T}_v\left(1, \frac{\pi}{3}\right) = \left\langle \frac{\sqrt{3}}{2}, -\frac{1}{2}, 3\right\rangle,$
$\mathbf{n}\left(1, \frac{\pi}{3}\right) = \left\langle 3\sqrt{3}, -3, -2\right\rangle$
(b) $3\sqrt{3}x - 3y - 2z + 2\pi = 0$ **(c)** $\text{Area}(S) \approx 38.4$
35. $\text{Area}(S) = \int_{-3}^{3}\int_{-1}^{1} \sqrt{1 + 4x^2 + 4y^2}\, dx\, dy \approx 41.8525$
37. $\iint_S (x^2 + y^2)e^{-z} dS = 54\pi(-e^{-10} + 1) \approx 54\pi$
39. $\text{Area}(S) = 0.02\sqrt{6} \approx 0.049$ **41.** $\iint_S \mathbf{F}\cdot d\mathbf{S} = \frac{243\pi}{2} \approx 381.7$
43. $\iint_S \mathbf{F}\cdot d\mathbf{S} = 8\pi$ **45.** $\iint_S \mathbf{F}\cdot d\mathbf{S} = 3 - e$
47. $\iint_S \rho\cdot dS = \frac{\pi}{3}KH^3R$ **49.** $\iint_S \mathbf{v}\cdot d\mathbf{S} = 6\pi$

Chapter 17

Section 17.1 Preliminary

1. $\mathbf{F} = \left\langle -e^y, x^2\right\rangle$
2.

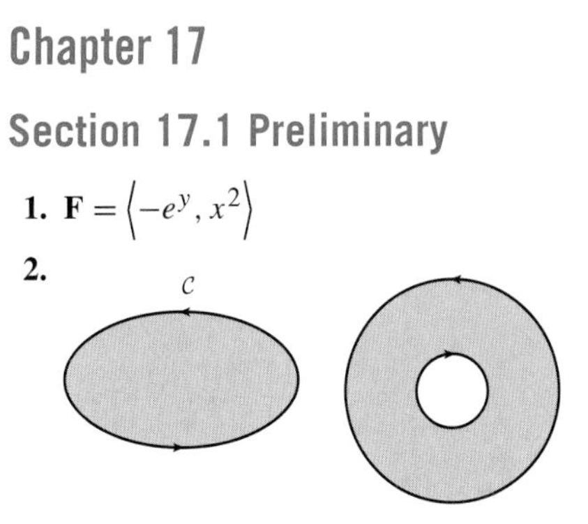

3. Yes **4.** (a), (c)

Section 17.1 Exercises

3. $\int_C y^2\, dx + x^2\, dy = 0$ **5.** $\int_C x^2 y\, dx = -\frac{\pi}{4}$ **7.** $\int_C \mathbf{F}\cdot d\mathbf{s} = \frac{1}{6}$
9. $\int_C x^3\, dx + 4x\, dy = 16$ **11.** $\int_C xy\, dx + (x^2 + x)\, dy = 1$
13. $A = 9\pi$ **15.** $A = 3\pi$
17. (c) $A = \frac{3}{2}$
21. (a) $\varphi(x, y) = x^2 e^y + C$ **(c)** $I = 4\pi - 16$
23. $I = 34$
25. $\int_C \mathbf{F}\cdot d\mathbf{s} \approx -0.024$, CW direction, $\omega = 1.5$ radians per second
27. $\int_{C_1} \mathbf{F}\cdot d\mathbf{s} = 214\pi$ **29.** $A = 12$

Section 17.2 Preliminary

1.

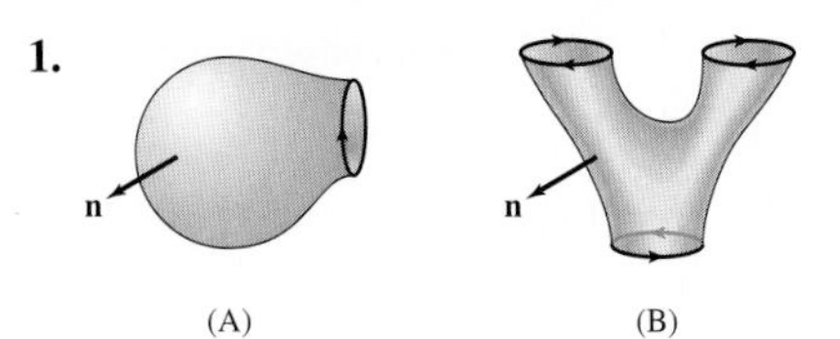

2. (b)

3. A vector field **A** such that $\mathbf{F} = \text{curl}(\mathbf{A})$ is a vector potential for **F**.

4. (b) **5.** $\iint_{S_1} \mathbf{F} \cdot d\mathbf{S} = \iint_{S_2} \mathbf{F} \cdot d\mathbf{S}$ for $\mathbf{F} = \text{curl}(\mathbf{A})$

Section 17.2 Exercises

1. $\text{curl}(\mathbf{F}) = \left\langle 1 - 3z^2, 1 - 2x, 1 + 2y \right\rangle$

3. $\text{curl}(\mathbf{F}) = \langle 0, \sin x, \cos x - e^y \rangle$

5. $\int_C \mathbf{F} \cdot d\mathbf{s} = \iint_S \text{curl}(\mathbf{F}) \cdot d\mathbf{S} = \pi$

7. $e^{-1} - 1$ **9.** $\iint_S \text{curl}(\mathbf{F}) \cdot d\mathbf{S} = 0$

11. (a) **(b)** 140π

13. (a) $\mathbf{A} = \left\langle 0, 0, e^y - e^{x^2} \right\rangle$ **(c)** $\iint_S \mathbf{F} \cdot d\mathbf{S} = \frac{\pi}{2}$

15. (a) $\iint_S \mathbf{B} \cdot dS = r^2 B\pi$ **(b)** $\int_{\partial S} \mathbf{A} \cdot dS = 0$

17. (b) $\iint_S \mathbf{B} \cdot d\mathbf{S} = b\pi$

19. $\int_C \mathbf{F} \cdot d\mathbf{s} = 32$ **21.** $a = 2c$, b any positive number.

23. $\iint_S \mathbf{F} \cdot d\mathbf{S} = 2\pi$

Section 17.3 Preliminary

1. $\iint_S \mathbf{F} \cdot d\mathbf{S} = 0$

2. $\iint_S \mathbf{F} \cdot d\mathbf{S} = \iiint_{\mathcal{W}} (3x^2 + 3y^2 + 3z^2)\, dV$ positive for all $(x, y, z) \neq (0, 0, 0)$.

3. a, b, d, f meaningful; b, d automatically zero vector and zero scalar, respectively.

4. (c)

5. $\text{div}(\mathbf{F}) = 2x + 1 - 2x = 1$ and flux $= \int \text{div}(\mathbf{F})\, dV =$ volume

Section 17.3 Exercises

1. $\text{div}(\mathbf{F}) = y + z$ **3.** $\text{div}(\mathbf{F}) = 1 - 4zx - x + 2zx^2$ **5.** 0

7. 4π **9.** $\iint_S \mathbf{F} \cdot d\mathbf{S} = 4\pi$ **11.** $\iint_S \mathbf{F} \cdot d\mathbf{S} = \frac{256\pi}{5}$

13. $\iint_S \mathbf{F} \cdot d\mathbf{S} = 64\pi$ **15.** $\iint_S \mathbf{F} \cdot d\mathbf{S} = 81\pi$

17. $\iint_S \mathbf{F} \cdot d\mathbf{S} = 70.22$ **21.** $\text{Volume}(\mathcal{W}) = \frac{4\pi abc}{3}$

23. $\text{div}(f\mathbf{F}) = f\,\text{div}(\mathbf{F}) + \mathbf{F} \cdot \nabla f$

29. (b) $\iint_S \mathbf{E} \cdot d\mathbf{S} = 0$ **(c)** $\text{div}(\mathbf{E}) = 0$ **(d)** No

35. $\mathbf{A} = \left\langle z^3 - xy^2, 0, y^2 - y \right\rangle$

Chapter 17 Review

1. $\int_C \mathbf{F} \cdot d\mathbf{s} = 0$ **3.** $\int_C xy^3\, dx + x^3 y\, dy = -30$

5. $\int_C y^2\, dx - x^2\, dy = \frac{3}{5}$

7. (a) **(b)** $A = \frac{1}{60}$

9. $\text{curl}(\mathbf{F}) = -\mathbf{k}$, $\text{div}(\mathbf{F}) = -1$

11. $\text{curl}(\mathbf{F}) = 0$, $\text{div}(\mathbf{F}) = 2e^{-x^2-y^2-z^2}\left(2(x^2 + y^2 + z^2) - 3\right)$

13. $\text{curl}(\mathbf{F}) = \langle -2z, 0, -2y \rangle$ **19.** $\iint_S \text{curl}(\mathbf{F}) \cdot d\mathbf{S} = 2\pi$

21. $\int_C \langle y, z, x \rangle\, d\mathbf{s} = \frac{\pi}{4}$ **23.** $\iint_S \left\langle xy, yz, x^2 z + z^2 \right\rangle \cdot d\mathbf{S} = 142$

25. $\iint_S \mathbf{F} \cdot d\mathbf{S} = -128\pi$ **27.** $\text{Volume}(\mathcal{W}) = 2$

29. $\int_C \mathbf{F} \cdot d\mathbf{s} \approx 0.0113$ **31.** $2x - y + 4z = 0$

33. (b) $\iint_S \mathbf{F} \cdot d\mathbf{S} = \iiint_{\mathcal{W}} \text{div}(\mathbf{F})\, dV = \frac{\pi}{2}$

35. (c) $\int_C \mathbf{F} \cdot d\mathbf{s} = 0$ **(d)** $\int_{C_1} \mathbf{F} \cdot d\mathbf{s} = -4$, $\int_{C_2} \mathbf{F} \cdot d\mathbf{s} = 4$

REFERENCES

The online source MacTutor History of Mathematics Archive (www-history.mcs.st-and.ac.uk) has been a valuable source of historical information.

Section 10.1

40. (EX 77) Adapted from G. Klambauer, *Aspects of Calculus*, Springer-Verlag, New York, 1986, p. 393.

Section 10.2

41. (EX 35) Adapted from *Calculus Problems for a New Century*, Robert Fraga, ed., Mathematical Association of America, Washington, DC, 1993, p. 137.

42. (EX 36) Adapted from *Calculus Problems for a New Century*, Robert Fraga, ed., Mathematical Association of America, Washington, DC, 1993, p. 138.

43. (EX 49) Adapted from George Andrews, "The Geometric Series in Calculus," *American Mathematical Monthly* 105, 1:36–40 (1998).

44. (EX 51) Adapted from Larry E. Knop, "Cantor's Disappearing Table," *The College Mathematics Journal* 16, 5:398–399 (1985).

Section 10.3

45. (EX 83) Adapted from *Calculus Problems for a New Century*, Robert Fraga, ed., Mathematical Association of America, Washington, DC, 1993, p. 141.

Section 10.4

46. (EX 27) Adapted from *Calculus Problems for a New Century*, Robert Fraga, ed., Mathematical Association of America, Washington, DC, 1993, p. 145.

Section 11.2

47. (EX 36) Adapted from Richard Courant and Fritz John, *Differential and Integral Calculus*, Wiley-Interscience, New York, 1965.

Section 11.3

13. (EX 56) Adapted from *Calculus Problems for a New Century*, Robert Fraga, ed., Mathematical Association of America, Washington, DC, 1993.

Section 12.4

48. (EX 66) Adapted from Ethan Berkove and Rich Marchand, "The Long Arm of Calculus," *The College Mathematics Journal* 29, 5:376–386 (November 1998).

Section 13.3

13. (EX 23) Adapted from *Calculus Problems for a New Century*, Robert Fraga, ed., Mathematical Association of America, Washington, DC, 1993.

Section 13.4

49. (EX 58) Damien Gatinel, Thanh Hoang-Xuan, and Dimitri T. Azar, "Determination of Corneal Asphericity After Myopia Surgery with the Excimer Laser: A Mathematical Model," *Investigative Opthalmology and Visual Science* 42: 1736–1742 (2001).

Section 13.5

50. (EX 50, 53) Adapted from notes to the course "Dynamics and Vibrations" at Brown University (see http://www.engin.brown.edu/courses/en4/).

Section 14.8

51. (EX 34) Adapted from C. Henry Edwards, "Ladders, Moats, and Lagrange Multipliers," *Mathematica Journal* 4, Issue 1 (Winter 1994).

Section 15.3

52. (FIG. 11 COMPUTATION) The computation is based on Jeffrey Nunemacher, "The Largest Unit Ball in Any Euclidean Space," in *A Century of Calculus*, Part II, Mathematical Association of America, Washington, DC, 1992.

Section 15.4

53. (EX 68) Adapted from A. E. Richmond, *Calculus for Electronics*, McGraw-Hill, New York, 1958.

Section 15.5

33. (CONCEPTUAL INSIGHT) See R. Courant and F. John, *Introduction to Calculus and Analysis*, Springer-Verlag, New York, 1989, p. 534.

Section 16.2

54. Figure 8 was inspired by Tevian Dray and Corinne A. Manogue, "The Murder Mystery Method for Determining Whether a Vector Field Is Conservative," *The College Mathematics Journal*, May 2003.

Section 16.3

13. (EX 20) Adapted from *Calculus Problems for a New Century*, Robert Fraga, ed., Mathematical Association of America, Washington, DC, 1993.

Appendix D

33. [Proof of Theorem 7] A proof without this simplifying assumption can be found in R. Courant and F. John, *Introduction to Calculus and Analysis*, Vol. 1, Springer-Verlag, New York, 1989.

PHOTO CREDITS

PAGE 553, CO10
NASA/JPL/CALTECH/Photo Researchers, Inc.

PAGE 570
Mechanics Magazine London, 1824

PAGE 621, CO11
Alamy

PAGE 628
Renault Communication/All Rights Reserved

PAGE 664
Stockbyte

PAGE 673, CO12
Brand X Pictures

PAGE 702, FIG. 1
LBNL/Science Source/Photo Researchers, Inc.

PAGE 738, CO13
NASA

PAGE 761, FIG. 1
Robin Smith/Stone/Getty Images

PAGE 761, FIG. 2
Alfred Pasieka/Photo Researchers, Inc.

PAGE 771, FIG. 1
NASA

PAGE 775, FIG. 5
Bernd Kappelmeyer/zefa/Corbis

PAGE 784, HIST. PERSP. PHOTO
NASA, ESA, and the Hubble Heritage Team (STScl/AURA)-ESA/Hubble Collaboration

PAGE 790, CO14
Craig Aurness/Corbis

PAGE 791, FIG. 2
Krista Longnecker/Woods Hole Oceanographic Institution

PAGE 796, FIG. 15 (left)
Galen Rowell/Corbis

PAGE 796, FIG. 15 (right)
USGS/Maps A La Carte Inc.

PAGE 815, FIG. 5
SSPL/The Image Works

PAGE 863, FIG. 4
Bettman/Corbis

PAGE 873, CO15
blphoto/Alamy

PAGE 911, FIG. 1
Gary A. Glatzmaier (University of California, Santa Cruz)

PAGE 941, CO16
R. Palais

PAGE 941, FIG. 1
Bettmann/Corbis

PAGE 943
Bettmann/Corbis

PAGE 1005, CO17
Richard Berry/Alamy

PAGE 1009, FIG. 6 (left)
SSPL/The Image Works

PAGE 1039, HIST. PERSP. PHOTO
SSPL/The Image Works

INDEX

ELEMENTARY FUNCTIONS

Power Functions $f(x) = x^a$

$f(x) = x^n$, n a positive integer

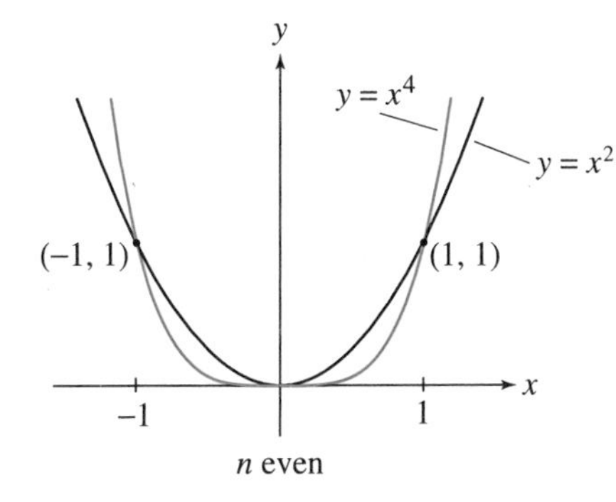

n even

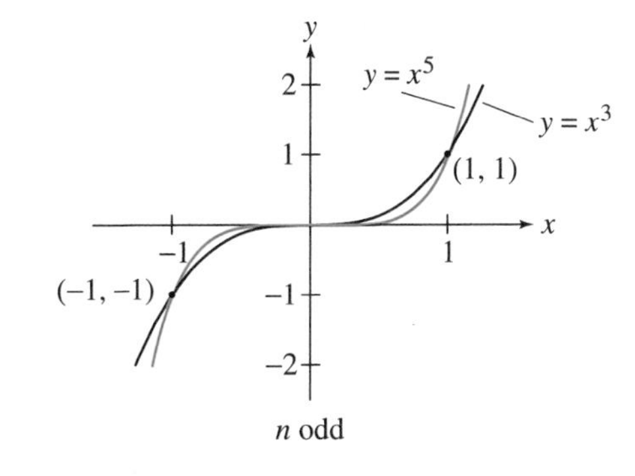

n odd

Asymptotic behavior of an even polynomial function

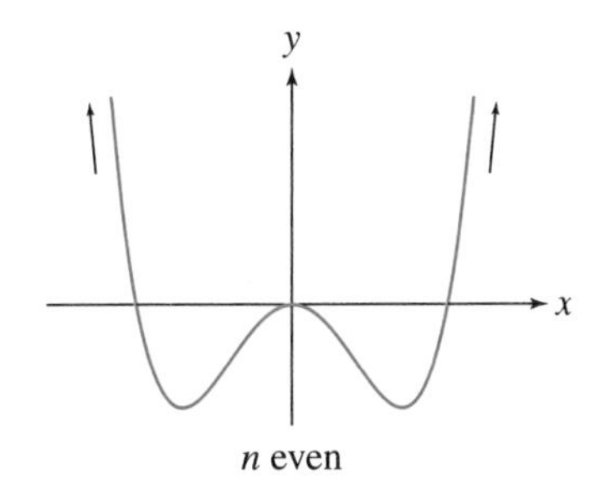

n even

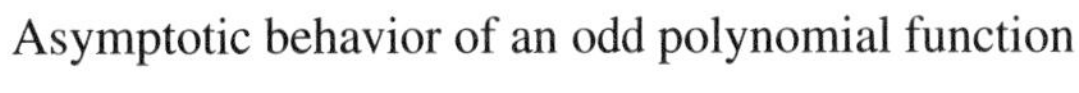
Asymptotic behavior of an odd polynomial function

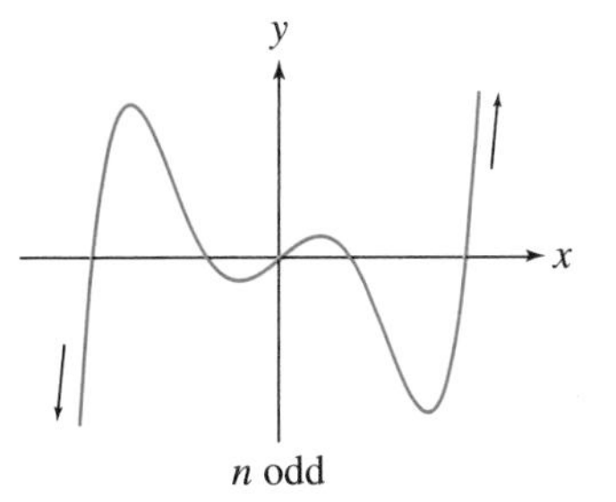

n odd

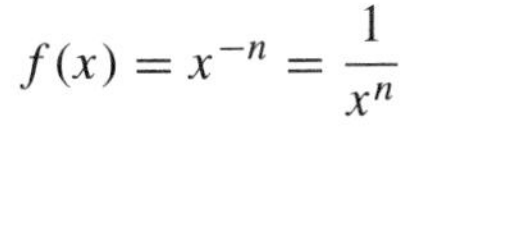
$f(x) = x^{-n} = \dfrac{1}{x^n}$

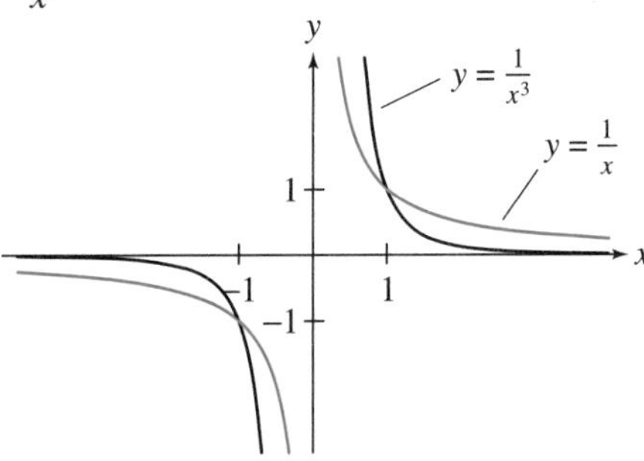

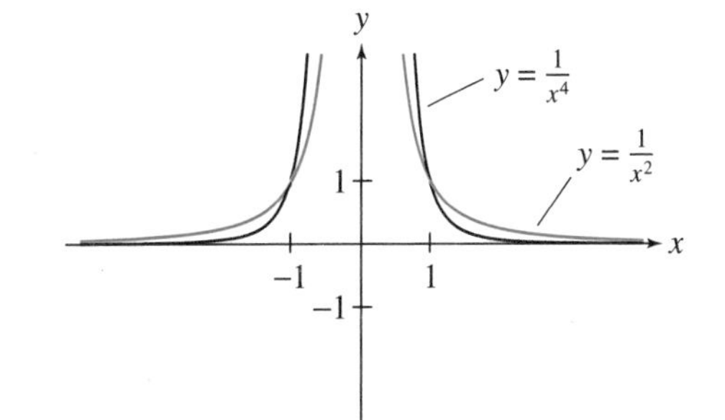

Inverse Trigonometric Functions

$\arcsin x = \sin^{-1} x = \theta$

$$\Leftrightarrow \quad \sin\theta = x, \quad -\frac{\pi}{2} \le \theta \le \frac{\pi}{2}$$

$\arccos x = \cos^{-1} x = \theta$

$$\Leftrightarrow \quad \cos\theta = x, \quad 0 \le \theta \le \pi$$

$\arctan x = \tan^{-1} x = \theta$

$$\Leftrightarrow \quad \tan\theta = x, \quad -\frac{\pi}{2} \le \theta \le \frac{\pi}{2}$$

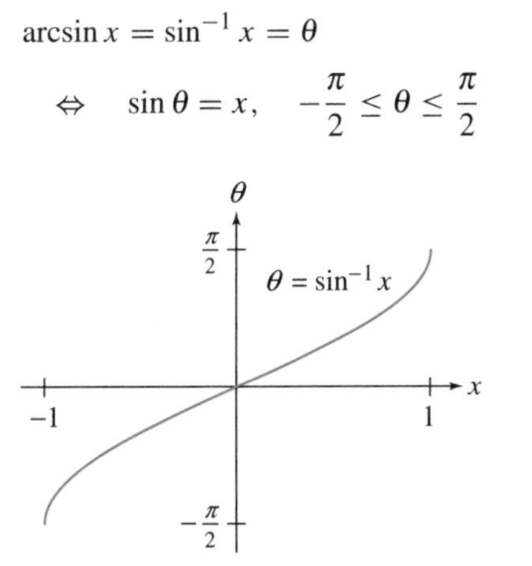

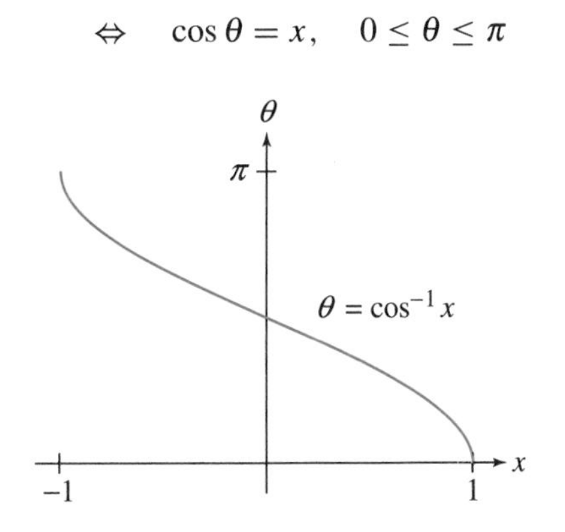

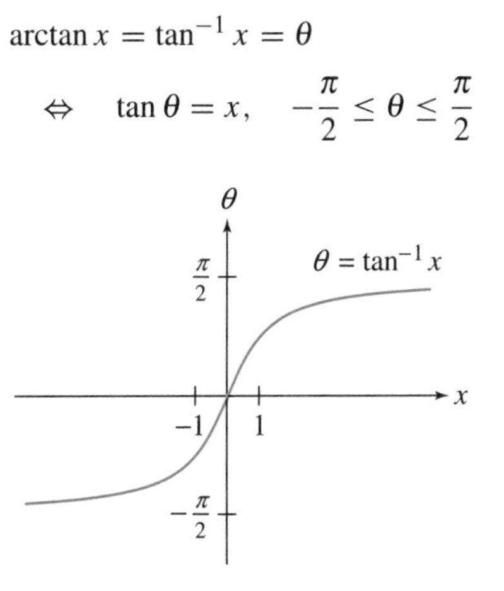

Exponential and Logarithmic Functions

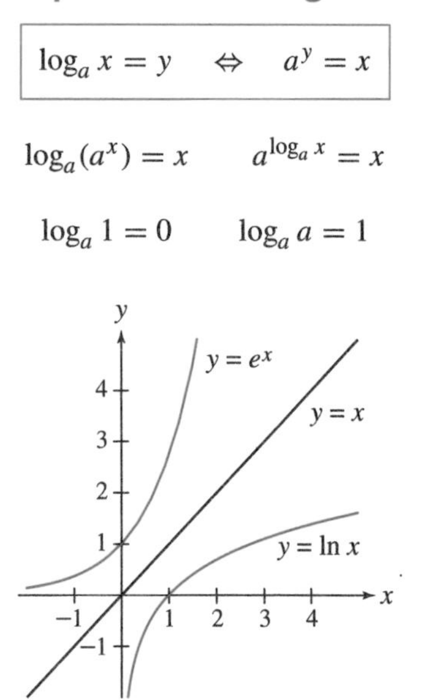

$$\log_a x = y \quad \Leftrightarrow \quad a^y = x$$

$$\log_a(a^x) = x \qquad a^{\log_a x} = x$$

$$\log_a 1 = 0 \qquad \log_a a = 1$$

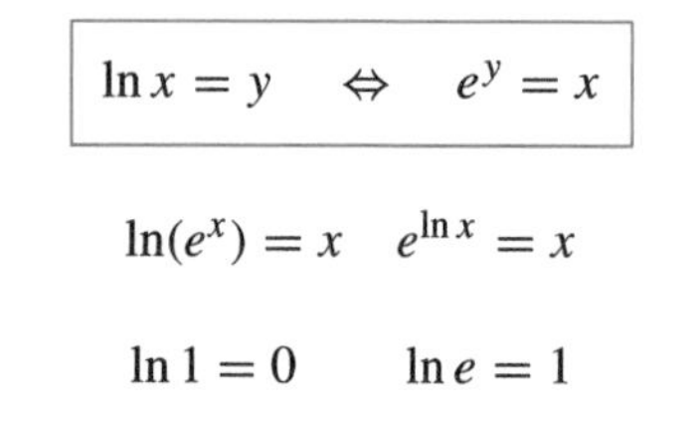

$$\ln x = y \quad \Leftrightarrow \quad e^y = x$$

$$\ln(e^x) = x \qquad e^{\ln x} = x$$

$$\ln 1 = 0 \qquad \ln e = 1$$

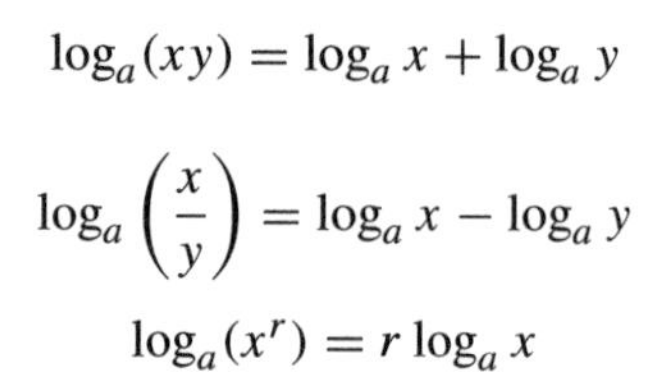

$$\log_a(xy) = \log_a x + \log_a y$$

$$\log_a\left(\frac{x}{y}\right) = \log_a x - \log_a y$$

$$\log_a(x^r) = r\log_a x$$

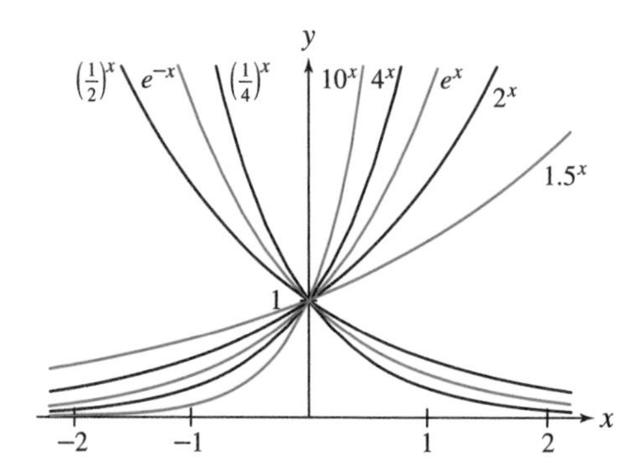

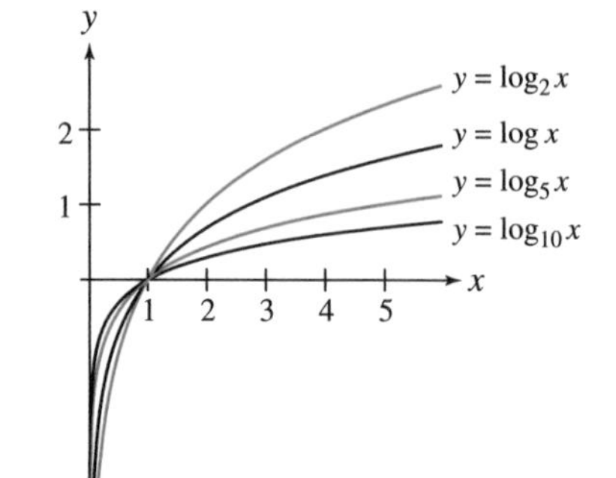

$$\lim_{x\to\infty} a^x = \infty, \quad a > 1$$

$$\lim_{x\to\infty} a^x = 0, \quad 0 < a < 1$$

$$\lim_{x\to-\infty} a^x = 0, \quad a > 1$$

$$\lim_{x\to-\infty} a^x = \infty, \quad 0 < a < 1$$

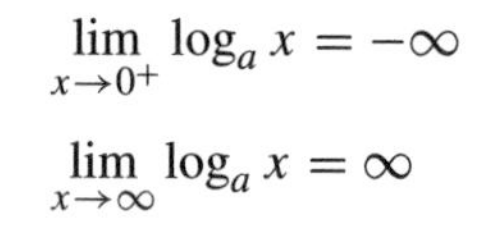

$$\lim_{x\to 0^+} \log_a x = -\infty$$

$$\lim_{x\to\infty} \log_a x = \infty$$

Hyperbolic Functions

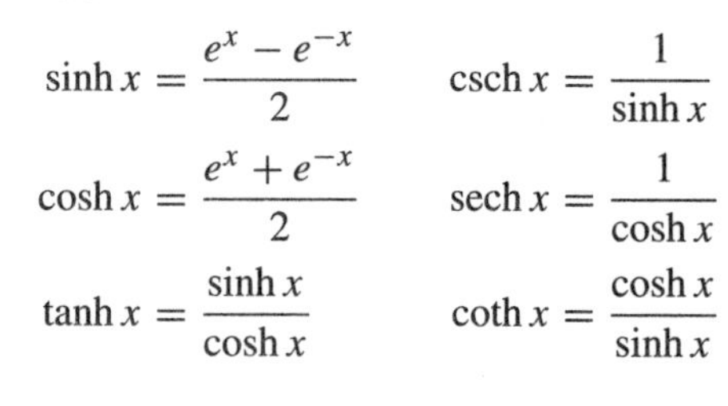

$$\sinh x = \frac{e^x - e^{-x}}{2} \qquad \operatorname{csch} x = \frac{1}{\sinh x}$$

$$\cosh x = \frac{e^x + e^{-x}}{2} \qquad \operatorname{sech} x = \frac{1}{\cosh x}$$

$$\tanh x = \frac{\sinh x}{\cosh x} \qquad \coth x = \frac{\cosh x}{\sinh x}$$

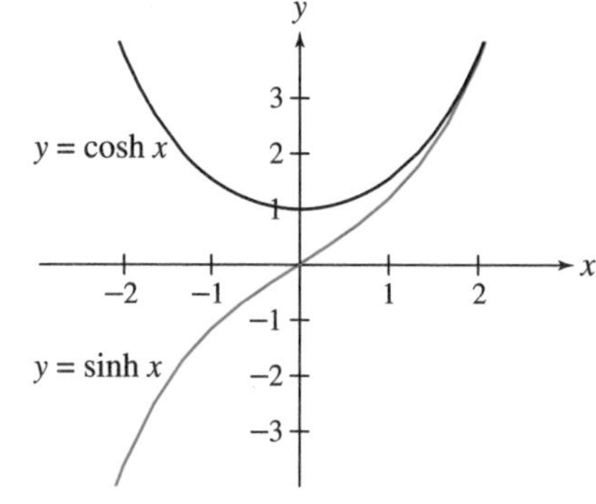

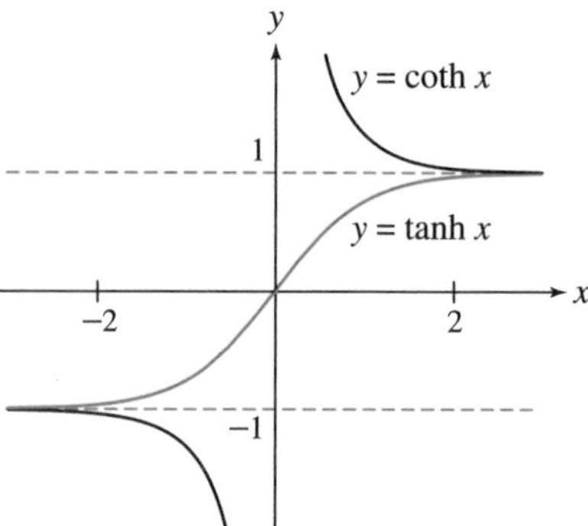

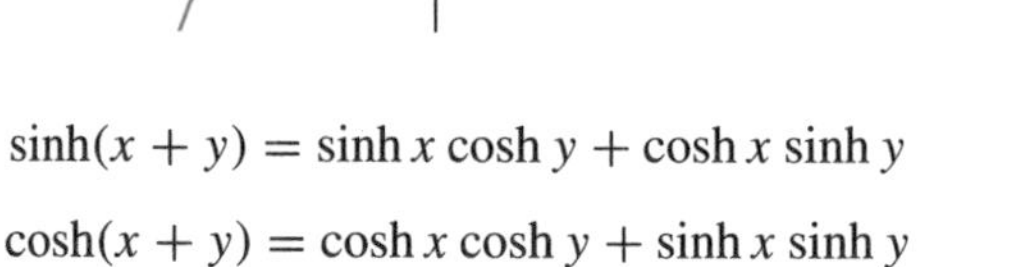

$$\sinh(x + y) = \sinh x\cosh y + \cosh x\sinh y$$

$$\cosh(x + y) = \cosh x\cosh y + \sinh x\sinh y$$

$$\sinh 2x = 2\sinh x\cosh x$$

$$\cosh 2x = \cosh^2 x + \sinh^2 x$$

Inverse Hyperbolic Functions

$$y = \sinh^{-1} x \quad \Leftrightarrow \quad \sinh y = x$$

$$y = \cosh^{-1} x \quad \Leftrightarrow \quad \cosh y = x \text{ and } y \geq 0$$

$$y = \tanh^{-1} x \quad \Leftrightarrow \quad \tanh y = x$$

$$\sinh^{-1} x = \ln\left(x + \sqrt{x^2 + 1}\right)$$

$$\cosh^{-1} x = \ln\left(x + \sqrt{x^2 - 1}\right) \quad x > 1$$

$$\tanh^{-1} x = \frac{1}{2}\ln\left(\frac{1 + x}{1 - x}\right) \quad -1 < x < 1$$

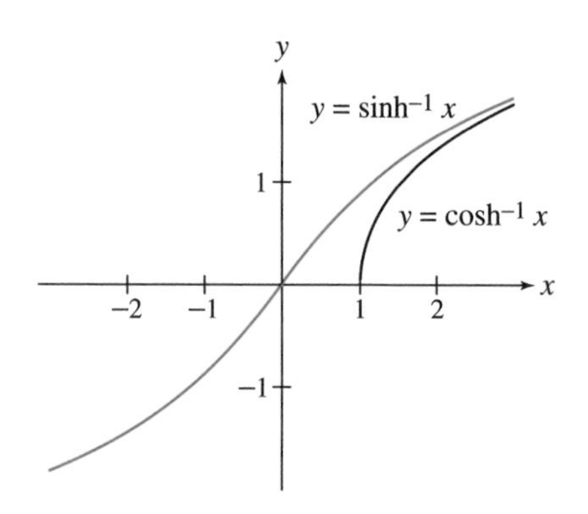

DIFFERENTIATION

Differentiation Rules

1. $\dfrac{d}{dx}(c) = 0$

2. $\dfrac{d}{dx}x = 1$

3. $\dfrac{d}{dx}(x^n) = nx^{n-1}$ (Power Rule)

4. $\dfrac{d}{dx}[cf(x)] = cf'(x)$

5. $\dfrac{d}{dx}[f(x) + g(x)] = f'(x) + g'(x)$

6. $\dfrac{d}{dx}[f(x)g(x)] = f(x)g'(x) + g(x)f'(x)$ (Product Rule)

7. $\dfrac{d}{dx}\left[\dfrac{f(x)}{g(x)}\right] = \dfrac{g(x)f'(x) - f(x)g'(x)}{[g(x)]^2}$ (Quotient Rule)

8. $\dfrac{d}{dx}f(g(x)) = f'(g(x))g'(x)$ (Chain Rule)

9. $\dfrac{d}{dx}f(x)^n = nf(x)^{n-1}f'(x)$ (General Power Rule)

10. $\dfrac{d}{dx}f(kx + b) = kf'(kx + b)$

11. $g'(x) = \dfrac{1}{f'(g(x))}$ where $g(x)$ is the inverse $f^{-1}(x)$

12. $\dfrac{d}{dx}\ln f(x) = \dfrac{f'(x)}{f(x)}$

Trigonometric Functions

13. $\dfrac{d}{dx}\sin x = \cos x$

14. $\dfrac{d}{dx}\cos x = -\sin x$

15. $\dfrac{d}{dx}\tan x = \sec^2 x$

16. $\dfrac{d}{dx}\csc x = -\csc x \cot x$

17. $\dfrac{d}{dx}\sec x = \sec x \tan x$

18. $\dfrac{d}{dx}\cot x = -\csc^2 x$

Inverse Trigonometric Functions

19. $\dfrac{d}{dx}(\sin^{-1} x) = \dfrac{1}{\sqrt{1 - x^2}}$

20. $\dfrac{d}{dx}(\cos^{-1} x) = -\dfrac{1}{\sqrt{1 - x^2}}$

21. $\dfrac{d}{dx}(\tan^{-1} x) = \dfrac{1}{1 + x^2}$

22. $\dfrac{d}{dx}(\csc^{-1} x) = -\dfrac{1}{x\sqrt{x^2 - 1}}$

23. $\dfrac{d}{dx}(\sec^{-1} x) = \dfrac{1}{x\sqrt{x^2 - 1}}$

24. $\dfrac{d}{dx}(\cot^{-1} x) = -\dfrac{1}{1 + x^2}$

Exponential and Logarithmic Functions

25. $\dfrac{d}{dx}(e^x) = e^x$

26. $\dfrac{d}{dx}(a^x) = (\ln a)a^x$

27. $\dfrac{d}{dx}\ln|x| = \dfrac{1}{x}$

28. $\dfrac{d}{dx}(\log_a x) = \dfrac{1}{(\ln a)x}$

Hyperbolic Functions

29. $\dfrac{d}{dx}(\sinh x) = \cosh x$

30. $\dfrac{d}{dx}(\cosh x) = \sinh x$

31. $\dfrac{d}{dx}(\tanh x) = \operatorname{sech}^2 x$

32. $\dfrac{d}{dx}(\operatorname{csch} x) = -\operatorname{csch} x \coth x$

33. $\dfrac{d}{dx}(\operatorname{sech} x) = -\operatorname{sech} x \tanh x$

34. $\dfrac{d}{dx}(\coth x) = -\operatorname{csch}^2 x$

Inverse Hyperbolic Functions

35. $\dfrac{d}{dx}(\sinh^{-1} x) = \dfrac{1}{\sqrt{1 + x^2}}$

36. $\dfrac{d}{dx}(\cosh^{-1} x) = \dfrac{1}{\sqrt{x^2 - 1}}$

37. $\dfrac{d}{dx}(\tanh^{-1} x) = \dfrac{1}{1 - x^2}$

38. $\dfrac{d}{dx}(\operatorname{csch}^{-1} x) = -\dfrac{1}{|x|\sqrt{x^2 + 1}}$

39. $\dfrac{d}{dx}(\operatorname{sech}^{-1} x) = -\dfrac{1}{x\sqrt{1 - x^2}}$

40. $\dfrac{d}{dx}(\coth^{-1} x) = \dfrac{1}{1 - x^2}$

INTEGRATION

Substitution

If an integrand has the form $f(u(x))u'(x)$, then rewrite the entire integral in terms of u and its differential $du = u'(x)\,dx$:

$$\int f(u(x))u'(x)\,dx = \int f(u)\,du$$

Integration by Parts Formula

$$\int u(x)v'(x)\,dx = u(x)v(x) - \int u'(x)v(x)\,dx$$

TABLE OF INTEGRALS

Basic Forms

1. $\displaystyle\int u^n\,du = \frac{u^{n+1}}{n+1} + C, \quad n \neq -1$

2. $\displaystyle\int \frac{du}{u} = \ln|u| + C$

3. $\displaystyle\int e^u\,du = e^u + C$

4. $\displaystyle\int a^u\,du = \frac{a^u}{\ln a} + C$

5. $\displaystyle\int \sin u\,du = -\cos u + C$

6. $\displaystyle\int \cos u\,du = \sin u + C$

7. $\displaystyle\int \sec^2 u\,du = \tan u + C$

8. $\displaystyle\int \csc^2 u\,du = -\cot u + C$

9. $\displaystyle\int \sec u \tan u\,du = \sec u + C$

10. $\displaystyle\int \csc u \cot u\,du = -\csc u + C$

11. $\displaystyle\int \tan u\,du = \ln|\sec u| + C$

12. $\displaystyle\int \cot u\,du = \ln|\sin u| + C$

13. $\displaystyle\int \sec u\,du = \ln|\sec u + \tan u| + C$

14. $\displaystyle\int \csc u\,du = \ln|\csc u - \cot u| + C$

15. $\displaystyle\int \frac{du}{\sqrt{a^2 - u^2}} = \sin^{-1}\frac{u}{a} + C$

16. $\displaystyle\int \frac{du}{a^2 + u^2} = \frac{1}{a}\tan^{-1}\frac{u}{a} + C$

Exponential and Logarithmic Forms

17. $\displaystyle\int ue^{au}\,du = \frac{1}{a^2}(au - 1)e^{au} + C$

18. $\displaystyle\int u^n e^{au}\,du = \frac{1}{a}u^n e^{au} - \frac{n}{a}\int u^{n-1}e^{au}\,du$

19. $\displaystyle\int e^{au}\sin bu\,du = \frac{e^{au}}{a^2 + b^2}(a\sin bu - b\cos bu) + C$

20. $\displaystyle\int e^{au}\cos bu\,du = \frac{e^{au}}{a^2 + b^2}(a\cos bu + b\sin bu) + C$

21. $\displaystyle\int \ln u\,du = u\ln u - u + C$

22. $\displaystyle\int u^n \ln u\,du = \frac{u^{n+1}}{(n+1)^2}[(n+1)\ln u - 1] + C$

23. $\displaystyle\int \frac{1}{u\ln u}\,du = \ln|\ln u| + C$

Hyperbolic Forms

24. $\displaystyle\int \sinh u\,du = \cosh u + C$

25. $\displaystyle\int \cosh u\,du = \sinh u + C$

26. $\displaystyle\int \tanh u\,du = \ln\cosh u + C$

27. $\displaystyle\int \coth u\,du = \ln|\sinh u| + C$

28. $\displaystyle\int \operatorname{sech} u\,du = \tan^{-1}|\sinh u| + C$

29. $\displaystyle\int \operatorname{csch} u\,du = \ln\left|\tanh\frac{1}{2}u\right| + C$

30. $\displaystyle\int \operatorname{sech}^2 u\,du = \tanh u + C$

31. $\displaystyle\int \operatorname{csch}^2 u\,du = -\coth u + C$

32. $\displaystyle\int \operatorname{sech} u \tanh u\,du = -\operatorname{sech} u + C$

33. $\displaystyle\int \operatorname{csch} u \coth u\,du = -\operatorname{csch} u + C$

Trigonometric Forms

34. $\displaystyle\int \sin^2 u\,du = \frac{1}{2}u - \frac{1}{4}\sin 2u + C$

35. $\displaystyle\int \cos^2 u\,du = \frac{1}{2}u + \frac{1}{4}\sin 2u + C$

36. $\displaystyle\int \tan^2 u\,du = \tan u - u + C$

37. $\displaystyle\int \cot^2 u\,du = -\cot u - u + C$

38. $\displaystyle\int \sin^3 u\,du = -\frac{1}{3}(2 + \sin^2 u)\cos u + C$

39. $\displaystyle\int \cos^3 u\,du = \frac{1}{3}(2 + \cos^2 u)\sin u + C$

40. $\displaystyle\int \tan^3 u\,du = \frac{1}{2}\tan^2 u + \ln|\cos u| + C$

41. $\displaystyle\int \cot^3 u\,du = -\frac{1}{2}\cot^2 u - \ln|\sin u| + C$

42. $\displaystyle\int \sec^3 u\,du = \frac{1}{2}\sec u \tan u + \frac{1}{2}\ln|\sec u + \tan u| + C$

43. $\displaystyle\int \csc^3 u\, du = -\frac{1}{2}\csc u \cot u + \frac{1}{2}\ln|\csc u - \cot u| + C$

44. $\displaystyle\int \sin^n u\, du = -\frac{1}{n}\sin^{n-1} u \cos u + \frac{n-1}{n}\int \sin^{n-2} u\, du$

45. $\displaystyle\int \cos^n u\, du = \frac{1}{2}\cos^{n-1} u \sin u + \frac{n-1}{n}\int \cos^{n-2} u\, du$

46. $\displaystyle\int \tan^n u\, du = \frac{1}{n-1}\tan^{n-1} u - \int \tan^{n-2} u\, du$

47. $\displaystyle\int \cot^n u\, du = \frac{-1}{n-1}\cot^{n-1} u - \int \cot^{n-2} u\, du$

48. $\displaystyle\int \sec^n u\, du = \frac{1}{n-1}\tan u \sec^{n-2} u + \frac{n-2}{n-1}\int \sec^{n-2} u\, du$

49. $\displaystyle\int \csc^n u\, du = \frac{-1}{n-1}\cot u \csc^{n-2} u + \frac{n-2}{n-1}\int \csc^{n-2} u\, du$

50. $\displaystyle\int \sin au \sin bu\, du = \frac{\sin(a-b)u}{2(a-b)} - \frac{\sin(a+b)u}{2(a+b)} + C$

51. $\displaystyle\int \cos au \cos bu\, du = \frac{\sin(a-b)u}{2(a-b)} + \frac{\sin(a+b)u}{2(a+b)} + C$

52. $\displaystyle\int \sin au \cos bu\, du = -\frac{\cos(a-b)u}{2(a-b)} - \frac{\cos(a+b)u}{2(a+b)} + C$

53. $\displaystyle\int u \sin u\, du = \sin u - u\cos u + C$

54. $\displaystyle\int u \cos u\, du = \cos u + u\sin u + C$

55. $\displaystyle\int u^n \sin u\, du = -u^n \cos u + n\int u^{n-1}\cos u\, du$

56. $\displaystyle\int u^n \cos u\, du = u^n \sin u - n\int u^{n-1}\sin u\, du$

57. $\displaystyle\int \sin^n u \cos^m u\, du$

$$= -\frac{\sin^{n-1} u \cos^{m+1} u}{n+m} + \frac{n-1}{n+m}\int \sin^{n-2} u \cos^m u\, du$$

$$= \frac{\sin^{n+1} u \cos^{m-1} u}{n+m} + \frac{m-1}{n+m}\int \sin^n u \cos^{m-2} u\, du$$

Inverse Trigonometric Forms

58. $\displaystyle\int \sin^{-1} u\, du = u\sin^{-1} u + \sqrt{1-u^2} + C$

59. $\displaystyle\int \cos^{-1} u\, du = u\cos^{-1} u - \sqrt{1-u^2} + C$

60. $\displaystyle\int \tan^{-1} u\, du = u\tan^{-1} u - \frac{1}{2}\ln(1+u^2) + C$

61. $\displaystyle\int u\sin^{-1} u\, du = \frac{2u^2-1}{4}\sin^{-1} u + \frac{u\sqrt{1-u^2}}{4} + C$

62. $\displaystyle\int u\cos^{-1} u\, du = \frac{2u^2-1}{4}\cos^{-1} u - \frac{u\sqrt{1-u^2}}{4} + C$

63. $\displaystyle\int u\tan^{-1} u\, du = \frac{u^2+1}{2}\tan^{-1} u - \frac{u}{2} + C$

64. $\displaystyle\int u^n \sin^{-1} u\, du = \frac{1}{n+1}\left[u^{n+1}\sin^{-1} u - \int \frac{u^{n+1}\, du}{\sqrt{1-u^2}}\right], \quad n \neq -1$

65. $\displaystyle\int u^n \cos^{-1} u\, du = \frac{1}{n+1}\left[u^{n+1}\cos^{-1} u + \int \frac{u^{n+1}\, du}{\sqrt{1-u^2}}\right], \quad n \neq -1$

66. $\displaystyle\int u^n \tan^{-1} u\, du = \frac{1}{n+1}\left[u^{n+1}\tan^{-1} u - \int \frac{u^{n+1}\, du}{1+u^2}\right], \quad n \neq -1$

Forms Involving $\sqrt{a^2 - u^2}$, $a > 0$

67. $\displaystyle\int \sqrt{a^2-u^2}\, du = \frac{u}{2}\sqrt{a^2-u^2} + \frac{a^2}{2}\sin^{-1}\frac{u}{a} + C$

68. $\displaystyle\int u^2\sqrt{a^2-u^2}\, du = \frac{u}{8}(2u^2-a^2)\sqrt{a^2-u^2} + \frac{a^4}{8}\sin^{-1}\frac{u}{a} + C$

69. $\displaystyle\int \frac{\sqrt{a^2-u^2}}{u}\, du = \sqrt{a^2-u^2} - a\ln\left|\frac{a+\sqrt{a^2-u^2}}{u}\right| + C$

70. $\displaystyle\int \frac{\sqrt{a^2-u^2}}{u^2}\, du = -\frac{1}{u}\sqrt{a^2-u^2} - \sin^{-1}\frac{u}{a} + C$

71. $\displaystyle\int \frac{u^2\, du}{\sqrt{a^2-u^2}} = -\frac{u}{2}\sqrt{a^2-u^2} + \frac{a^2}{2}\sin^{-1}\frac{u}{a} + C$

72. $\displaystyle\int \frac{du}{u\sqrt{a^2-u^2}} = -\frac{1}{a}\ln\left|\frac{a+\sqrt{a^2-u^2}}{u}\right| + C$

73. $\displaystyle\int \frac{du}{u^2\sqrt{a^2-u^2}} = -\frac{1}{a^2 u}\sqrt{a^2-u^2} + C$

74. $\displaystyle\int (a^2-u^2)^{3/2}\, du = -\frac{u}{8}(2u^2-5a^2)\sqrt{a^2-u^2} + \frac{3a^4}{8}\sin^{-1}\frac{u}{a} + C$

75. $\displaystyle\int \frac{du}{(a^2-u^2)^{3/2}} = \frac{u}{a^2\sqrt{a^2-u^2}} + C$

Forms Involving $\sqrt{u^2 - a^2}$, $a > 0$

76. $\displaystyle\int \sqrt{u^2-a^2}\, du = \frac{u}{2}\sqrt{u^2-a^2} - \frac{a^2}{2}\ln\left|u+\sqrt{u^2-a^2}\right| + C$

77. $\displaystyle\int u^2\sqrt{u^2-a^2}\, du$

$$= \frac{u}{8}(2u^2-a^2)\sqrt{u^2-a^2} - \frac{a^4}{8}\ln\left|u+\sqrt{u^2-a^2}\right| + C$$

78. $\displaystyle\int \frac{\sqrt{u^2-a^2}}{u}\, du = \sqrt{u^2-a^2} - a\cos^{-1}\frac{a}{|u|} + C$

79. $\displaystyle\int \frac{\sqrt{u^2-a^2}}{u^2}\, du = -\frac{\sqrt{u^2-a^2}}{u} + \ln\left|u+\sqrt{u^2-a^2}\right| + C$

80. $\displaystyle\int \frac{du}{\sqrt{u^2-a^2}} = \ln\left|u+\sqrt{u^2-a^2}\right| + C$

81. $\displaystyle\int \frac{u^2\, du}{\sqrt{u^2-a^2}} = \frac{u}{2}\sqrt{u^2-a^2} + \frac{a^2}{2}\ln\left|u+\sqrt{u^2-a^2}\right| + C$

82. $\displaystyle\int \frac{du}{u^2\sqrt{u^2-a^2}} = \frac{\sqrt{u^2-a^2}}{a^2 u} + C$

83. $\displaystyle\int \frac{du}{(u^2-a^2)^{3/2}} = -\frac{u}{a^2\sqrt{u^2-a^2}} + C$

Forms Involving $\sqrt{a^2 + u^2}$, $a > 0$

84. $\displaystyle\int \sqrt{a^2+u^2}\, du = \frac{u}{2}\sqrt{a^2+u^2} + \frac{a^2}{2}\ln\left(u+\sqrt{a^2+u^2}\right) + C$

85. $\displaystyle\int u^2\sqrt{a^2+u^2}\, du$

$$= \frac{u}{8}(a^2+2u^2)\sqrt{a^2+u^2} - \frac{a^4}{8}\ln\left(u+\sqrt{a^2+u^2}\right) + C$$

86. $\displaystyle\int \frac{\sqrt{a^2+u^2}}{u}\, du = \sqrt{a^2+u^2} - a\ln\left|\frac{a+\sqrt{a^2+u^2}}{u}\right| + C$

87. $\displaystyle\int \frac{\sqrt{a^2+u^2}}{u^2}\, du = -\frac{\sqrt{a^2+u^2}}{u} + \ln\left(u+\sqrt{a^2+u^2}\right) + C$

88. $\displaystyle\int \frac{du}{\sqrt{a^2+u^2}} = \ln\left(u+\sqrt{a^2+u^2}\right)+C$

89. $\displaystyle\int \frac{u^2\,du}{\sqrt{a^2+u^2}} = \frac{u}{2}\sqrt{a^2+u^2} - \frac{a^2}{2}\ln\left(u+\sqrt{a^2+u^2}\right)+C$

90. $\displaystyle\int \frac{du}{u\sqrt{a^2+u^2}} = -\frac{1}{a}\ln\left|\frac{\sqrt{a^2+u^2}+a}{u}\right|+C$

91. $\displaystyle\int \frac{du}{u^2\sqrt{a^2+u^2}} = -\frac{\sqrt{a^2+u^2}}{a^2u}+C$

92. $\displaystyle\int \frac{du}{(a^2+u^2)^{3/2}} = \frac{u}{a^2\sqrt{a^2+u^2}}+C$

Forms Involving $a+bu$

93. $\displaystyle\int \frac{u\,du}{a+bu} = \frac{1}{b^2}\left(a+bu-a\ln|a+bu|\right)+C$

94. $\displaystyle\int \frac{u^2\,du}{a+bu} = \frac{1}{2b^3}\left[(a+bu)^2-4a(a+bu)+2a^2\ln|a+bu|\right]+C$

95. $\displaystyle\int \frac{du}{u(a+bu)} = \frac{1}{a}\ln\left|\frac{u}{a+bu}\right|+C$

96. $\displaystyle\int \frac{du}{u^2(a+bu)} = -\frac{1}{au}+\frac{b}{a^2}\ln\left|\frac{a+bu}{u}\right|+C$

97. $\displaystyle\int \frac{u\,du}{(a+bu)^2} = \frac{a}{b^2(a+bu)}+\frac{1}{b^2}\ln|a+bu|+C$

98. $\displaystyle\int \frac{du}{u(a+bu)^2} = \frac{1}{a(a+bu)}-\frac{1}{a^2}\ln\left|\frac{a+bu}{u}\right|+C$

99. $\displaystyle\int \frac{u^2\,du}{(a+bu)^2} = \frac{1}{b^3}\left(a+bu-\frac{a^2}{a+bu}-2a\ln|a+bu|\right)+C$

100. $\displaystyle\int u\sqrt{a+bu}\,du = \frac{2}{15b^2}(3bu-2a)(a+bu)^{3/2}+C$

101. $\displaystyle\int u^n\sqrt{a+bu}\,du$
$\displaystyle= \frac{2}{b(2n+3)}\left[u^n(a+bu)^{3/2}-na\int u^{n-1}\sqrt{a+bu}\,du\right]$

102. $\displaystyle\int \frac{u\,du}{\sqrt{a+bu}} = \frac{2}{3b^2}(bu-2a)\sqrt{a+bu}+C$

103. $\displaystyle\int \frac{u^n\,du}{\sqrt{a+bu}} = \frac{2u^n\sqrt{a+bu}}{b(2n+1)} - \frac{2na}{b(2n+1)}\int \frac{u^{n-1}\,du}{\sqrt{a+bu}}$

104. $\displaystyle\int \frac{du}{u\sqrt{a+bu}} = \frac{1}{\sqrt{a}}\ln\left|\frac{\sqrt{a+bu}-\sqrt{a}}{\sqrt{a+bu}+\sqrt{a}}\right|+C, \quad \text{if } a>0$
$\displaystyle= \frac{2}{\sqrt{-a}}\tan^{-1}\sqrt{\frac{a+bu}{-a}}+C, \quad \text{if } a<0$

105. $\displaystyle\int \frac{du}{u^n\sqrt{a+bu}} = -\frac{\sqrt{a+bu}}{a(n-1)u^{n-1}} - \frac{b(2n-3)}{2a(n-1)}\int \frac{du}{u^{n-1}\sqrt{a+bu}}$

106. $\displaystyle\int \frac{\sqrt{a+bu}}{u}\,du = 2\sqrt{a+bu}+a\int \frac{du}{u\sqrt{a+bu}}$

107. $\displaystyle\int \frac{\sqrt{a+bu}}{u^2}\,du = -\frac{\sqrt{a+bu}}{u}+\frac{b}{2}\int \frac{du}{u\sqrt{a+bu}}$

Forms Involving $\sqrt{2au-u^2}$, $a>0$

108. $\displaystyle\int \sqrt{2au-u^2}\,du = \frac{u-a}{2}\sqrt{2au-u^2}+\frac{a^2}{2}\cos^{-1}\left(\frac{a-u}{a}\right)+C$

109. $\displaystyle\int u\sqrt{2au-u^2}\,du$
$\displaystyle= \frac{2u^2-au-3a^2}{6}\sqrt{2au-u^2}+\frac{a^3}{2}\cos^{-1}\left(\frac{a-u}{a}\right)+C$

110. $\displaystyle\int \frac{du}{\sqrt{2au-u^2}} = \cos^{-1}\left(\frac{a-u}{a}\right)+C$

111. $\displaystyle\int \frac{du}{u\sqrt{2au-u^2}} = -\frac{\sqrt{2au-u^2}}{au}+C$

ESSENTIAL THEOREMS

Intermediate Value Theorem

If $f(x)$ is continuous on a closed interval $[a,b]$, then for every value M between $f(a)$ and $f(b)$, there exists at least one value $c \in [a,b]$ such that $f(c)=M$.

Mean Value Theorem

If $f(x)$ is continuous on a closed interval $[a,b]$ and differentiable on (a,b), then there exists at least one value $c \in (a,b)$ such that

$$f'(c) = \frac{f(b)-f(a)}{b-a}$$

Extreme Values on a Closed Interval

If $f(x)$ is continuous on a closed interval $[a,b]$, then $f(x)$ attains both a minimum and a maximum value on $[a,b]$. Furthermore, if $c \in [a,b]$ and $f(c)$ is an extreme value (min or max), then c is either a critical point or one of the endpoints a or b.

The Fundamental Theorem of Calculus, Part I

Assume that $f(x)$ is continuous on $[a,b]$ and let $F(x)$ be an antiderivative of $f(x)$ on $[a,b]$. Then

$$\int_a^b f(x)\,dx = F(b)-F(a)$$

Fundamental Theorem of Calculus, Part II

Assume that $f(x)$ is a continuous function on $[a,b]$. Then the area function $A(x)=\int_a^x f(t)\,dt$ is an antiderivative of $f(x)$, that is,

$$A'(x)=f(x) \quad \text{or equivalently} \quad \frac{d}{dx}\int_a^x f(t)\,dt = f(x)$$

Furthermore, $A(x)$ satisfies the initial condition $A(a)=0$.